PERSONALIST BIOETHICS

Personalist Bioethics
Foundations and Applications

Elio Sgreccia

Translated by
John A. Di Camillo and Michael J. Miller

❧

With a foreword by
John M. Haas

The National Catholic Bioethics Center
Philadelphia

Personalist Bioethics is the English translation of the fourth edition of Cardinal Sgreccia's *Manuale di bioetica*, volume 1, *Fondamenti ed etica biomedica*, published in 2007 by Vita e Pensiero (Milan). Earlier Italian editions were published in 1988, 1994, and 1998.

Cover design: Heather Ermine
ISBN 978-0-935372-63-2

Unless otherwise noted, quotations from official Church documents are from the Vatican English translation, published online at www.vatican.va.

Library of Congress Cataloging-in-Publication Data

Sgreccia, Elio.
  [Manuale di bioetica. English]
  Personalist bioethics / Elio Sgreccia ; translated by John A. Di Camillo and Michael J. Miller ; with a foreword by John M. Haas.
       volumes cm
  Translation of: Manuale di bioetica.
  Includes bibliographical references and indexes.
  Summary: "Presents a metaphysical foundation for ethics grounded in non-relative, personalist values that can be communicated cross-culturally. Examines the philosophical bases of ethical criteria and applies them to issues in medical practice ranging from genetic engineering to euthanasia" -- Provided by publisher.
  ISBN 978-0-935372-63-2 (volume 1 : alk. paper) 1. Medical ethics--Philosophy. 2. Medical ethics--Italy. 3. Bioethics--Philosophy. 4. Bioethics--Italy. I. Title.
  R724.S45313 2012
  174.2--dc23
                            2012042304

CONTENTS

Chapter 3
Forms, Origins, and Meaning of Human Life

Chapter 4
The Human Person and His Body

Chapter 5
Bioethics and Its Principles

Chapter 6
Bioethics and Medicine

Chapter 7
Bioethics Committees

Part Two
Particular Bioethics

Chapter 8
Bioethics, Genetics, and Prenatal Diagnosis

Chapter 9
Bioethics, Sexuality, and Human Procreation

Chapter 10
Bioethics and Abortion

Chapter 12
Bioethics and Sterilization

Chapter 15
Bioethics, Euthanasia, and Death with Dignity

Chapter 16
Bioethics and Technology

Foreword

Elio Cardinal Sgreccia is a man of remarkable intellectual and managerial talents who placed himself and those talents at the service of the Church and his fellow human beings, most particularly through the discipline of bioethics. He is currently the president emeritus of the Pontifical Academy for Life. His interest and expertise in the discipline of bioethics grew out of his priestly ministry in the large Gemelli Polyclinic Hospital in Rome, where he served as chaplain and moral guide to health care professionals, clinicians, administrators, other chaplains, and patients. He served as an advisor in bioethics to Pope John Paul II and every Vatican dicastery dealing with that discipline, including the Congregation for the Doctrine of the Faith, the Pontifical Council for the Pastoral Care of Health Care Workers, and the Pontifical Council for the Family. He is a world-renowned expert in the field, with his publications appearing in many languages.

Cardinal Sgreccia was born on June 6, 1928, in the small town of Nidastore in the Marche region of Italy. He came from a humble family of farmers and was the youngest of six children. After his completion of primary school, his admission to the minor seminary in Fossombrone was delayed by the beginning of World War II in 1939. He assisted his father in the fields and attended a technical school in the meantime, eventually entering the seminary in Fano.

He was ordained on June 29, 1952, the Feast of SS. Peter and Paul. His first assignment was as a spiritual minister to youth members of Catholic Action, a lay organization founded in Italy in 1905 to promote Christian love, spiritual formation, service, and faith in local communities. He completed his studies in theology in 1963, and received a university degree in classical letters and philosophy from the historic University of Bologna, the first university in Europe (1088).

From 1954 to 1972, he served as vice rector, professor, and ultimately rector of Pius XI Regional Pontifical Seminary in Fano, in his native region of Marche, then from 1972 to 1973 as vicar general of the Diocese of Fossombrone, where he had requested to be transferred so he could serve alongside his bishop, the Most Reverend Vittorio Cecchi.

Cardinal Sgreccia then became a spiritual advisor at the Faculty of Medicine and Surgery at the Catholic University of the Sacred Heart in 1974, addressing the emerging need for moral guidance on medical issues, especially in the educational formation of health care professionals. He continued in that capacity until 1984, when he became an instructor of bioethics at the same university. He founded the university's bioethics center in 1985 and served as its director from then until 2006.

In 1992, he was also appointed director of the Institute of Bioethics, a role he fulfilled until 2000. Here his involvement in the discipline of medical and bioethics was most practical and direct, since the Agostino Gemelli Polyclinic where he served is the teaching hospital for the University of the Sacred Heart. The bioethics center is on the campus of the medical school, in a beautiful manorial villa surrounded by meadows which had once served as a convent. He was also the director of the Center for International Cooperation at the Sacred Heart University from 1998 to 2005.

In the 1980s, Cardinal Sgreccia served on the Ad Hoc Committee of Experts on Bioethics of the Council of Europe as an observer for the Holy See. He became a professor of bioethics at Sacred Heart University in 1990. He joined Italy's National Bioethics Committee (Comitato Nazionale per la Bioetica) in 1990 and contributed to its many important publications until 2006.

On November 5, 1992, he was named Titular Bishop of Zama Minor by Pope John Paul II, as well as secretary of the Pontifical Council for the Family, on which he served until 1996. He was officially consecrated as a bishop on January 6, 1993.

He devoted his full attention to his role as vice president of the Pontifical Academy for Life in 1996, a task that had been assigned to him in 1994. He served there alongside the French geneticist and Servant of God Jérôme Lejeune, the Academy's first president, and then as vice president to Lejeune's successor, Juan de Dios Vial Correa. Cardinal Sgreccia is now president emeritus of the Pontifical Academy for Life, having been named president on January 3, 2005 (succeeding Dr. Vial Correa), and having retired because of his age on June 17, 2008.

Cardinal Sgreccia used his managerial skills effectively as president of the Pontifical Academy for Life. Several months before their annual assembly, which is open to the public, he would gather all the speakers to present their papers to one another before the event. Controversial topics were debated later at the assembly, but the preparation ahead of time made it possible to avoid duplication and enhance the quality of the debates.

In 2001, Cardinal Sgreccia served as a member of the Committee for Guidelines on Genetic Counseling and Testing of the Italian Ministry of Health. In 2003, he founded the International Federation of Bioethics Centers and Institutes of Personalist Inspiration (Federazione Internazionale dei Centri ed Istituti di Bioetica di Ispirazione Personalista) and served as its president; he is currently honorary president. In 2004, he took on the role of president of the Ut Vitam Habeant Foundation and the Donum Vitae Association in the diocese of Rome and continues to serve in those capacities.

He was elevated to cardinal on November 20, 2010, by Pope Benedict XVI and installed on March 12, 2011. His official title is Cardinal-Deacon of Sant'Angelo in Pescheria. He is not eligible to vote in consistories because he is over the age of 80.

As a pioneer of bioethics on the international level and an inspiration to many aspiring Catholic bioethicists, he received an honorary doctorate from the Pontifical Athenaeum Regina Apostolorum in Rome on March 25, 2011. It was the first honorary degree granted by the Athenaeum, which is home to the first full-fledged bioethics department in the world. Its Faculty of Bioethics was founded in 2001 and awards undergraduate, licentiate, and doctoral degrees specifically in bioethics. Cardinal Sgreccia is currently a member of the Scientific Council of the Faculty of Bioethics at Regina Apostolorum.

His bioethics research and writing have focused on an ontologically grounded personalism. He is the director of the international journal *Medicina e Morale*

(*Medicine and Morality*), having first served as its editor (1974), then vice director and co-director. A highly regarded bioethics journal that was founded in 1951 by Agostino Gemelli, the founder of the Catholic University of the Sacred Heart, *Medicina e Morale* enjoys acclaim both in Italy and around the world.

Cardinal Sgreccia has published nearly four hundred works on bioethics in Italian and other languages. These include *The Human Embryo Before Implantation: Scientific Aspects and Bioethical Considerations* (2009); Italian monographs with A. Serra on the new genetics and human embryo creation (1991) and with A. G. Spagnolo on the ethics of clinical trials (1994); books with M. L. Di Pietro on assisted procreation and artificial fertilization in science, bioethics, and law (1999) and on the transmission of life in the teachings of Pope John Paul II (1989); and more recently a book on bioethics in daily life (2006).

Cardinal Sgreccia's monumental *Manuale di bioetica* has been published in two volumes and four editions, including various reprints, and translated wholly or in part into at least ten different languages. The National Catholic Bioethics Center is pleased to be able to make the first volume of this monumental work available in English, thus bringing Cardinal Sgreccia's scholarship to many more people throughout the world.

This English translation was made possible by the efforts of the Most Reverend Robert Morlino and by a generous contribution from a grateful alumnus of The University of Wisconsin Medical School through the Junck Family Charitable Gift Fund.

The National Catholic Bioethics Center also thanks Christopher J. Gergely, Esq., Colleen E. Gergely, Kenneth M. Kernen, MD, Christine A. Kernen, William Lawrence Mackle, Raymond T. Bauer, MD, Frances Bauer, Nicola M. Antakli, Virginia K. Antakli, Victor J. Zanolli Jr., and Patricia G. Zanolli, whose generosity made the publication of this volume possible.

I also want to express my profound gratitude for the extraordinary work done by the translators, John Di Camillo and Michael Miller, who worked with great skill to complete this massive project in as short a time as possible. The publications department of The National Catholic Bioethics Center also worked with great diligence to bring the new translation into print. This publication of Cardinal Sgreccia's seminal work required the cooperation of everyone at the Center and is an expression of our shared respect for the scholarship and industry of this great churchman.

JOHN M. HAAS, PhD, STL, KM
President of The National Catholic Bioethics Center

Translator's Note

Elio Cardinal Sgreccia's landmark publication on bioethics emerges from the Italian context; its scope, however, is international. A contribution of this magnitude, translated from its native language, warrants a word of introduction concerning its significance in the field of bioethics. I also lay out below a general roadmap for navigation, shed light on several linguistic matters, alert the reader to certain editorial decisions, and express my thanks and that of Mr. Michael Miller to those who were instrumental in the lengthy yet rewarding journey of its translation and preparation for publication.

Overview

Personalist Bioethics offers a consolidated approach to bioethics through *ontologically grounded personalism*, which gives due weight to natural reason and considers ethics as a science. It appeals to the natural moral law and realist or Aristotelian-Thomistic metaphysics in order to give substance to claims of moral obligation. Purely practical understandings of ethics, such as duty-based deontology or professional ethics strictly speaking, need not always recognize such objective grounding and can be readily manipulated by those with the power to determine the norms: the rules themselves become the ultimate reference for moral obligation, but are changeable at the whim of those with the power to establish or modify them. When the objective underpinnings of moral obligations are established through proper exploration of the reality of being, as expressed in the natural moral law, they can be defended against unjust manipulation.

Ontologically grounded personalism thus uncovers and defends the substantial unity of the human person as body and soul, or *corpore et anima unus* (Vatican Council II, Pastoral Constitution *Gaudium et spes*, n. 14). To touch the body is to touch the whole person, and vice versa. I do not merely have my body, but I am my body. At the same time, physical life is a fundamental but not an absolute good: there is a necessary hierarchy of values wherein the embodied spirit, the person as a whole—body, psyche, and spirit—takes priority over bodily life and health alone. The intrinsic, absolute, and unalienable value of the human person and his good are taken as the measure of ethical right and wrong. In light of these reasoned principles, the author shows great respect for the Magisterium of the Catholic Church, often citing documents from her rich tradition in concert with scientific studies and philosophical argumentation, while presenting a platform and an invitation for dialogue with the secular world in a context of intellectual honesty and integrity.

The focus on fundamental principles and their applications to broad topic areas, specific case examples aside, lends the work a universal and timeless character that transcends the scientific data. This edition of Cardinal Sgreccia's work takes into account the origins of bioethics as a discipline in the United States, the way that many nations in the African and Asian geographic regions now look to Europe or the United States on matters of ethics and law, and the reality of English as a practically "global" language, especially in medicine and government. At the same time, it provides a consolidated reference for person-centered bioethics not only in the United States but throughout the world. Its international dimension also provides an opportunity to analyze questions of public policy, health care, and professional ethics in the United States in light of an "outsider" perspective.

The book is divided into two major sections of sixteen chapters. Core concepts and principles can be found in chapters 4 through 6, while specific issues of interest can easily be identified by title in chapters 8 through 16. Part One constitutes the core of the work: a brief discussion of the history of bioethics, an exposition of the foundational principles of personalist bioethics and their philosophical background, considerations on the state of medicine, and a discussion of ethics committees. Part Two addresses major bioethical issues such as contraception, sterilization, abortion, reproductive technologies, genetic testing, cloning, eugenics, organ transplantation, clinical experimentation, euthanasia, physician-assisted suicide, palliative care, and technological development. The author approaches the topics with what he calls a "triangular method": visually, it is a two-dimensional pyramid with empirical science and philosophical anthropology as its two base points and ethics as its crown. It implies that ethical judgments must be grounded in sound science and the proper understanding of our essential constitution as human beings.

Terminology

A translation cannot do full justice to an author's work, since the original flavor of certain words, expressions, and styles can never be fully conveyed. Just as putting pen to paper may limit the author's expansive thoughts to a specific expression, so the translator's interpretation may limit the author's words to the translator's understanding of them. Further still, a translator may fully grasp the author's concept yet find no equivalent manner of rendering it in the new language or, perhaps more significantly, in the new culture. Yet by-and-large these limitations, common to all acts of translation and ultimately to communication itself, pertain to formal or stylistic matters more than to substance. Still, how is it possible to transmit an Italian love of the verbose to an English-speaking world that praises the concise? I have taken the utmost care in maintaining fidelity to the substance of the author's thoughts above all, while using prudential judgments as necessary to convey in English a style that is in some ways more direct yet respectful of the original.

In light of this, a few notes on terminology are in order. First, the English title *Personalist Bioethics: Foundations and Applications* departs from the Italian title, *Manuale di bioetica: fondamenti ed etica medica* (Manual of bioethics: foundations and medical ethics), in order to emphasize the relevance of its major contribution to the field, especially in an English-language arena more heavily populated with manifold approaches to bioethics. Next, the term "ontologically grounded personalism" was chosen over the more literal "ontologically founded personalism" (*personalismo ontologicamente fondato* in Italian) to emphasize its link with realism: it is a personalism anchored in the metaphysical truth and reality of being. On a similar though less significant point, the term "particular bioethics" is used to translate the Italian *bioetica speciale* (special bioethics), which describes the application of general principles to specific issues.

In various discussions of ethics, the terms "deontology" and "professional ethics" are both used to translate the term *deontologia*, which at its core transmits a single concept: a duty-based ethics having formal rules that are not necessarily established on the basis of an understanding of realist metaphysics. That is, rules or codes may be developed for purely practical ends without an appeal to the underlying natural moral law and without recognizing any grounding in being. Academic ethicists and philosophers will readily appreciate "deontology," yet practicing health care professionals may have never heard the term despite their familiarity with codes of medical professional ethics. In Italian, the same term is used for duty-based ethics in general and for specific codes of professional ethics, which may or may not appeal to the natural law.

The Italian adjective *creazionista*, literally translated as "creationist" and used to describe a particular understanding of the origin of the universe, is rendered as "creational" owing to potentially misleading political overtones associated with the English term "creationist." The latter brings to mind a literal interpretation of the Bible, according to which the world was created in seven days, in conflict with evolutionary theory. The concept conveyed by the author is instead any vision of a universe created *ex nihilo* (from nothing) by a Creator, neither ruling out evolutionary theory nor proposing a strict timeline on creation, as opposed to a "Creator-less" or "not created" universe.

Italian law, and much of the predominant Italian culture, uses the terminology *interruzione volontaria della gravidanza* (IVG) to talk about abortion. Literally it translates as "voluntary interruption of pregnancy," which is euphemistic in the use of an adjective—"interruption"—that does not connote a very real, definitive, or final termination. (Something "interrupted" can be expected to continue.) The closest English equivalent is "voluntary termination of pregnancy," which has been used throughout whenever IVG appears in the Italian.

Editorial Decisions

Turning to matters of editing, readers familiar with the Italian edition will notice a few changes of an aesthetic, organizational, or factual sort. At times, but without changing their meaning, certain paragraphs in the body text have been combined or separated for greater coherence, though sentences were never reordered or eliminated. The author's use of the plural "we" to express his personal opinion or thoughts was replaced with the singular "I," and phrases such as "a few years ago" or words such as "recently," when used extensively, were sometimes removed or modified. For the sake of accuracy some references to major international documents or to the "most recent" edition of works or events in the field of bioethics have been updated, with permission, insofar as they did not disrupt, confute, or otherwise change the author's purpose. Generally speaking, then, references to public facts or publications after 2006 (and earlier in rare cases) have been added and are not found in the Italian edition (2007).

Individual bibliographies for each chapter have been combined into a single bibliography at the end of the work; the majority of English language primary sources and official English translations of foreign language primary sources were researched and inserted in place of the original references to Italian translations of those same sources; citation format and style were adapted to house publishing rules based on *The Chicago Manual of Style*; and additional reference information was sought and added where necessary and feasible for completeness.

Appreciations

I wish to express my profound thanks to those who were instrumental to my work and that of Michael Miller on this volume: Dr. John Haas, president of The National Catholic Bioethics Center, who graciously offered me the opportunity to work on the project; Dr. Edward Furton, director of publications at the Center, whose editorial guidance, advice, and support at all stages of the project were essential; and Mr. Anthony Cunicelli, who introduced me to the Italian language and taught me with passion and clarity well before I knew where it might take me. I would like to especially thank Mr. Michael Miller, who had taken on the translation project before I learned of it, who accepted my assistance, and who, as a colleague, mentor, and friend, encouraged my work and generously shared with me the fruits of his extensive translation experience.

JOHN A. DI CAMILLO, BeL

PREFACE

The translation of the first volume of the *Manuale di bioetica* into English constitutes an important milestone for the work itself, which entered the arena in 1986 with a slightly different title: *Bioetica: Manuale per medici e teologi*—or, in English, *Bioethics: A Manual for Physicians and Theologians*—published by Vita e Pensiero. The work was published in 1988 in two volumes under the current Italian title, *Manuale di bioetica*. The first volume addressed the foundations of bioethics and the major issues in biomedicine, while the second volume focused in on the medico-social dimensions of bioethics. Three more Italian editions followed with the same title from 1994 to 2007.

The English version, *Personalist Bioethics*, which is now complete thanks to the efforts of The National Catholic Bioethics Center, is the translation of the first volume of the fourth Italian edition published by Vita e Pensiero in 2007.

Following the introductory section, this version offers its new audience several chapters of foundational content based on a Thomistic "ontological personalism," which is given new relevance through a look at more recent neo-Thomistic and neo-Scholastic philosophers such as Jacques Maritain, Etienne Gilson, Emmanuel Mounier, Gustavo Bontadini, and Sofia Vanni Rovighi, whose fundamental ideas I share regarding the centrality and dignity of the human person.

This basic philosophical approach, including its emphasis on an ontological foundation, came to me spontaneously in response to the cultural context in which bioethics arose in the United States at the Kennedy Institute of Ethics and how it was already understood in light of the first edition of the *Encyclopedia of Bioethics* (1978), edited by Warren Reich.

A metaphysical foundation allows the human person to maintain the possibility of grounding ethics on intangible values that are not subject to relativization yet are nonetheless communicable through intercultural dialogue.

With regard to the method by which the interdisciplinary character of the discipline is addressed, I felt the need to adopt a method I have called "triangular," because the scientific dimension represents the descriptive aspect of reality, the philosophical and anthropological dimension sheds a light of clarity on the summit of truth, and the science of ethics outlines the valuational dimension of moral acts, with which the legal dimension is also linked.

In the European context, this proposal has found kinship with classical Christian and neo-Thomistic thinking and has clearly distinguished itself from schools of liberal utilitarian or neo-Marxist thinking (Herbert Marcuse). There has been ongoing debate especially with secular and neo-utilitarian approaches to bioethics. Yet the number of Italian editions and the versions published in various languages (French, Spanish, Portuguese, Russian, Ukrainian, Romanian, Bulgarian, Arabic, Korean) offered me reassurance and led me to prepare subsequent editions.

I am particularly gratified by the regard of Dr. John Haas for the *Manuale di bioetica*. He proposed an English translation so that it could engage the subject in the country of origin of bioethics—the United States—as well as in the broader context of the English-speaking world.

My gratitude is therefore spontaneous and profound not only to Dr. Haas, who decided on this version and took on the responsibility of the project, but also to Michael Miller and John A. Di Camillo, Be.L., who translated and scrupulously and attentively edited the text with an eye to improvements. I hold great hope that this intercultural dialogue will be enriching for all readers.

ELIO CARDINAL SGRECCIA
Rome 2012

Part One

GENERAL BIOETHICS

Chapter 1

ORIGINS, DEVELOPMENT, AND DEFINITIONS OF BIOETHICS

The Rise of Bioethics as a "New" Discipline, and the Origins of the Term

About twenty-five years after the appearance of the term in the scientific literature, thanks to the oncologist Van Rensselaer Potter,[1] it became possible to trace the path of the intellectual movement referred to as "bioethics," which has had such rapid and great success. There are now clear historical and philosophical contours to its development.[2]

Since the publication of previous editions of this manual, the documents, studies, and publications relating to this new discipline have proliferated, making it possible to more precisely describe the various paths and directions it has taken.

Although it is necessary to review the great work that has been accomplished, this is not an easy task: the broadening of its speculative horizons and the increased number of authors, study centers, and scholarly works have led to increasingly varied proposals and sometimes antithetical philosophical foundations within this field of investigation. Anyone seeking to identify constants or sufficiently meaningful points of contact should therefore proceed with great caution.

It seems to me appropriate to provide a historical and cultural overview of this discipline at the outset, emphasizing several particularly prominent works and institutional developments, such as committees, centers, and academic programs, which illustrate its most significant developments and sufficiently survey its central questions.

Everyone acknowledges that bioethics, in the proper sense of the term, was born in the United States. While the merit does not belong exclusively to Potter, he coined the name and assigned a certain meaning to it. In introducing the term, he underscored that bioethics should constitute "a *new discipline*" that would combine

---

[1] See his 1970 article, "Bioethics: The Science of Survival," published in the journal *Perspectives in Biology and Medicine* 14, no. 1 (1982): 127–153. In the following year this article became the first chapter of his book *Bioethics: Bridge to the Future* (Englewood Cliffs, NJ: Prentice Hall, 1971).

[2] C. Viafora, ed., *Vent'anni di bioetica: Idee protagonisti istituzioni* (Padua: Gregorniana, 1990); D. Gracia, *Fundamentos de bioética* (Madrid: Eudema Universidad, 1989); A. R. Jonsen, *Birth of Bioethics* (Oxford: Oxford University Press, 1998).

biological knowledge with the study of human values[3]: "I have chosen," he wrote, "the root *bio* to represent biological knowledge, the science of systems of living things, and *ethics* to represent the knowledge of the system of human values."[4] Indeed, Potter pointed out that the entire ecosystem was put at risk by the split between two areas of knowledge: scientific knowledge and humanistic knowledge. The practice of making a neat distinction between ethical values, which are part of humanistic culture in the broad sense, and biological facts was, in Potter's view, at the root of an indiscriminate scientific and technological process that was endangering humanity and the very survival of life on earth. This was precisely the reason why he called bioethics the "science of survival." The instinct for survival was inadequate, so a new science had become necessary: bioethics.

Potter foresaw the urgent need for a new form of knowledge, not aimed solely at studying and explaining natural phenomena but also at discovering how to wisely use technological and scientific knowledge so as to foster the survival of the human species and improve the quality of life of future generations. The only possible way of preventing imminent catastrophe was to build a "bridge" between the two cultures, one scientific and the other humanistic and moral. Furthermore, according to Potter, bioethics should focus not only on man but also on the biosphere in its entirety, addressing every scientific intervention by human beings upon life in general.[5] In this sense the concept of bioethics was understood to have a broader meaning than traditional medical ethics. Potter defined bioethics as a "new wisdom": "A science of survival must be more than science alone, and therefore I propose the term bioethics to emphasize the two most important ingredients for the attainment of a new wisdom."[6]

Bioethics, as Potter conceived of it, therefore arises from a threatening situation and critical concerns about scientific and social progress, which in turn raise doubts about humanity's ability to survive. Paradoxically, this is the result of scientific progress itself.

The discoveries in the field of genetic engineering announced during the same time period and shortly thereafter brought the terrible potential for manufacturing biological weapons and altering the very constitution of life forms, species, and individuals. Consequently, the alarm that Potter sounded had great resonance and paved the way for a trend of alarmist thoughts and fears.

Alongside this original strand of bioethics, however, there is another branch to consider that has become even more prevalent than the one traceable to Potter. In fact, Warren Reich talks about a twofold origin of the term bioethics.[7] André Hellegers, a famous Dutch obstetrician, researcher in the field of demographics, and founder of the Kennedy Institute of Ethics, provided a strong impetus for this different approach during the same time period. Indeed, we must recognize the strong impetus pro-

---

[3] Potter, *Bioethics*, 1.

[4] Ibid.

[5] He would refer to this concept with the term *global bioethics*, which would also be the subject of a subsequent book: V. R. Potter, *Global Bioethics: Building on the Leopold Legacy* (East Lansing: Michigan University Press, 1988).

[6] Potter, *Bioethics*, 2.

[7] W. T. Reich, "The Word 'Bioethics': The Struggle over Its Earliest Meanings," *Kennedy Institute of Ethics Journal* 5, no. 1 (1995): 19–34.

vided during those same years. He considered bioethics a maieutic discipline, i.e., a science capable of discovering and compiling values through dialogue between representatives from the fields of medicine, philosophy, and ethics. Hence, according to Hellegers, the object of this new field of study is the ethical dimension implicit in clinical practice. It is certain that Hellegers was the first to introduce the term bioethics in a university setting—giving an academic structure to the discipline—and to apply it subsequently in other fields, such as the biomedical sciences, politics, and mass communications. His concept of bioethics would eventually become the prevalent one: most scholars would come to consider bioethics a specific discipline capable of synthesizing medical and ethical findings. Hellegers also deserves credit for having identified a specific methodology—the interdisciplinary method—for this new discipline, predicting that clinical bioethicists would become more expert than traditional moralists. With this in mind, the newer term *bioethics* was adopted in place of the term *medical morality* so as to distinguish it from the latter.

Potter's concept of bioethics was thereby eclipsed by the better-known bioethics of Hellegers. The developments that can be traced back to Potter have been modest but undoubtedly remain important, because the original vision of a global bioethics encompasses both the biosphere and mankind, as well as their reciprocal interactions over the short and long term. Furthermore, it was precisely this concept that would foster the birth of environmental bioethics over time.

For the sake of historical accuracy, we should note that the famous Hastings Center was founded by the philosopher Daniel Callahan and the psychiatrist Willard Gaylin in 1969, several years before Potter and Hellegers. Its purpose was to study and formulate norms, especially in the context of research and experimentation in the biomedical field, even though the term bioethics was not yet used. In fact, debates about the ethical issues involved in experimentation had already become heated in the United States before the discoveries in the field of genetics were announced; this was because of grievances and legal proceedings resulting from several highly publicized abuses in the field of human experimentation. In 1963, for example, tumoral cells were injected without consent into elderly patients during the course of an experiment at the Jewish Chronic Disease Hospital in Brooklyn. From 1965 to 1971, a series of studies was conducted on immunization against viral hepatitis at Willowbrook State Hospital of New York: several handicapped children who had been admitted to the hospital were inoculated with the virus. These experiments called to mind the savage experiments carried out in the concentration camps during the Nazi period.[8]

Major Bioethics Centers in the World

Paraphrasing Shakespeare in *A Midsummer Night's Dream* (V, I, 15–17), A. G. Spagnolo maintains that the contribution of theologians and philosophers "has given 'to airy nothing a local habitation and a name': the *habitation* is that of the centers for study which have spread bioethics rapidly by means of conferences and

---

[8] A. R. Jonsen, A. L. Jameson and A. Lynch, "Medical Ethics, History of North America in the Twentieth Century," in *Encyclopedia of Bioethics*, ed. W. T. Reich (New York: Free Press, 1978), 992–1001.

publications; the *name* is the one that sprang from the easily understood and almost spontaneous combination of the terms biology and ethics."[9]

Callahan and Gaylin had taken the initiative in assembling scientists, researchers, and philosophers to discuss these kinds of questions. Such reflections led, as we have noted, to the creation of an institute dedicated to the systematic study of bioethics: the Institute of Society, Ethics and the Life Sciences, headquartered in Hastings-on-Hudson, New York, which would soon become known as the Hastings Center. Its specific objective was to consider the ethical, social, and legal aspects of the medical sciences and health care practices. There was so much interest in these studies that little attention was initially paid to logistical and economic matters; at the outset, the center consisted of a room in Callahan's house and its activities were funded in part with moneys received from his mother.[10] By 1988, the center had an annual budget of 1.6 million dollars (which came partly from the government and partly from private funding) and numbered twenty-four members on the board of directors, thirty staff members, and around 130 grant recipients. The first director was the cofounder Callahan. The Hastings Center later moved its headquarters to Garrison, New York.

The center was designed as an independent, lay-run, nonprofit research institute; its principal activity (or social mission, so to speak) would be to educate the general public. The specific purposes of its activity were to address and attempt to solve ethical problems raised by advances in the biomedical sciences and by the medical profession itself; to educate the general public about the ethical relevance of many scientific discoveries; to contribute to the development of directives for many difficult moral questions in contemporary society, such as the handling of AIDS (acquired immunodeficiency syndrome), the discontinuation of treatment and removal of life support, the use of reproductive technologies and prenatal diagnosis, and the distribution of funding for health care.

In fact, the center would introduce far-reaching medical and medico-social themes to the bioethics debate, broadening its horizons beyond what Potter had anticipated and contributing to the development of educational projects and guidelines regarding many different topics in special bioethics.[11] The results of these studies are published in its official journal, the *Hastings Center Report*, and in numerous other monographs.

During the same years in which the Hastings Center was developing, Hellegers was involved in research on fetal physiology at Georgetown University in Washington, D.C. He had transferred to that university in order to inaugurate an interdisciplinary research program in bioethics. For this same purpose, Hellegers invited the Protestant moral theologian Paul Ramsey to teach several courses at Georgetown University's medical school in 1968 and 1969. The outcome of these morality courses was the

---

[9] A. G. Spagnolo, "Bioetica," in *Dizionario interdisciplinare di scienza e fede*, ed. G. Tanzella-Nitti and A. Strumia (Rome: Urbaniana University Press, 2002), 198.

[10] P. Quattrocchi, "La bioetica, storia di un progetto," in *Dalla bioetica ai comitati etici*, ed. C. G. Vella, P. Quattrocchi, and A. Bompiani (Milan: Ancora, 1988), 57–97.

[11] D. Callahan, ed., *The Hastings Center: A Short and Long 15 Years* (Hastings-on-Hudson, NY: Hastings Center, 1984).

publication of two volumes, *The Patient as Person* and *Fabricated Man*, both in 1970, which can be considered the books that launched bioethics in America.

During that same period, the Kennedy family decided to finance several research projects on the prevention of congenital mental handicaps. The additional ethical implications of this research prompted Hellegers to present his proposal to found an institute that would be concerned with both the physiology of reproduction and bioethics. The Joseph and Rose Kennedy Institute for the Study of Human Reproduction and Bioethics was thus founded in 1971; it was the first center to be formally called a bioethics institute. After Hellegers's death in 1979, the institute changed its name to the Kennedy Institute of Ethics and was officially annexed to Georgetown University. Edmund Pellegrino, MD, served as director at various times. The Center for Bioethics is headquartered within the Kennedy Institute and has its own director. It later developed collaborative activities with other centers at Georgetown University: the Division of Health and Humanities, within the Department of Community and Family Medicine; the Center for Population Research, which had been present at the university since 1964; the Asian Bioethics Program, which aims to evaluate the ethical implications of the impact of scientific and technological developments in the biomedical field in the countries of Asia; and the European Program in Professional Ethics, which has developed educational programs, first in Germany and then in other European countries.

The Center for Bioethics and the Kennedy Institute[12] are headquartered at Georgetown University, which was founded by the Jesuits in 1789 and, by its statutes, is open to students and researchers of all religious beliefs. The principal purpose of the institute is to conduct research by using an interdisciplinary method; its members hail from the natural sciences, the social sciences, and the humanities (predominantly philosophy and moral theology), and it encourages an interfaith, ecumenical approach.

The publications issued by the institute and the center are remarkable for their number and the range of subjects treated. One in particular deserves to be mentioned here: the *Encyclopedia of Bioethics* (1978), edited by Reich, which is unique in its class. After the second edition of the encyclopedia in 1995, a third edition was published in 2004 in five volumes totaling three thousand pages, with 464 articles written by 437 authors and arranged in alphabetical order. Every other month, the Kennedy Institute publishes *New Titles in Bioethics*, an updated bibliography of new books arranged by subject, and the *Scope Notes Series*, bibliographic monographs including articles about periodicals. Its official periodical is the *Kennedy Institute of Ethics Journal*.

Another important activity launched by the center is an online bibliographic information service, "Bioethicsline," which is supported by staff members of the National Library of Medicine in Bethesda, Maryland, and distributed throughout the United States and the world by means of the MEDLARS system. The library at Georgetown University, the National Reference Center for Bioethics Literature, has the most important English-language publications in its collection.

The U.S. "doctrine" of bioethics, as it could be called, emerged from the American cultural milieu, especially in the writings of Thomas Beauchamp and

---

[12] L. Walters, "The Center for Bioethics at the Kennedy Institute," *Georgetown Medical Bulletin* 37, no. 1 (1984): 6–8.

James Childress. Their famous work, *Principles of Biomedical Ethics*, outlines the theory of "principalism."[13] Another thinker who is counted among the pioneers of bioethics is Edmund Pellegrino, who served several terms as the director of the Center for Bioethics and as chairman of the President's Council on Bioethics. Together with David Thomasma, he formulated new ideas about the physician-patient relationship.[14]

After the first two centers, many others were founded in the United States, most of them affiliated with universities or hospitals.[15] Just a few of the ones that take unique approaches will be mentioned later. The Pope John XXIII Medical-Moral Research and Education Center, renamed The National Catholic Bioethics Center, is particularly worthy of note: it has published numerous monographs and is committed to fidelity to the Magisterium of the Catholic Church.

In Australia the activities of the Center for Human Bioethics at Monash University in Melbourne have been of note; it has been directed by Peter Singer, whose extreme secularism is well known. He has served as co-editor of *Bioethics*, the official journal of the International Association of Bioethics.[16] There are also two bioethics centers in Australia within the Catholic tradition: the Thomas More Center and St. Vincent's Clinic Foundation.

While the most important philosophical and moral systems had long since developed in the Old World and continued to inform social life for centuries, bioethics made a relatively late appearance in Europe. The delay might be attributed to the different structure of its health care and university systems with respect to the United States, the strong influence of professional ethics taught by forensic doctors, or its difficulty with organizing interdisciplinary work because of excessive academic specialization.[17]

During the 1975–1976 academic year, several seminars in various fields of bioethics were organized in Spain by the School of Theology at San Cugat del Valles

---

[13] T. L. Beauchamp and J. F. Childress, *Principles of Biomedical Ethics*, 4th ed. (New York: Oxford University Press, 1994); A. MacIntyre, *After Virtue: A Study in Moral Theory* (Notre Dame: University of Notre Dame Press, 1981).

[14] See especially the book by E. D. Pellegrino and D. C. Thomasma, *For the Patient's Good: The Restoration of Beneficence in Health Care* (New York: Oxford University Press, 1988). The same authors had previously published *A Philosophical Basis of Medical Practice: Toward a Philosophy and Ethics of the Healing Professions* (New York: Oxford University Press, 1981).

[15] The soaring number of centers for bioethics and medical ethics that have been founded in recent years throughout the world prevents us from mentioning them all here. For a detailed treatment of the situation in Europe, the reader is referred to the *Annuaire européenne de bioéthique* (Paris: Lavoiser, 1996). For worldwide information, see the complete list published yearly by the United Nations Educational, Scientific and Cultural Organization (UNESCO): *World Directory of Academic Research Groups in Science Ethics* (Paris: UNESCO, 1993).

[16] Among the principal works by P. Singer we recall: *Practical Ethics* (Cambridge: Cambridge University Press, 1979); *Animal Liberation: A New Ethics for Our Treatment of Animals* (New York: Review–Random House, 1975); *Rethinking Life and Death: The Collapse of Our Traditional Ethics* (New York: St. Martin's Press, 1994).

[17] See the article by D. C. Thomasma in *Medical Ethics in Europe*, special issue of *Theoretical Medicine* 3 (1988); A. Rogers and D. Durand de Bousingen, *Bioethics in Europe* (Strasbourg: Council of Europe Press, 1995).

in Barcelona; these seminars gave rise to the Borja Institute of Bioethics (Instituto Borja de Bioética), headed by a disciple and colleague of Hellegers, Francisco Abel, S.J., which was established as a private foundation in 1980. In addition to this center, which is the foremost in Spain for its interest and research in bioethics, we should note the efforts of Diego Gracia, director of the Department of Preventive Medicine, Public Health and History of Science (Departamento de Medicina Preventiva, Salud Publica e Historia de la Ciencia) at the medical school of the Complutense University of Madrid. He published an important work, *Fundamentos de bioética* (foundations of bioethics), which begins with a historical and philosophical examination of the evolution of ethical concepts in the biomedical field, from the Hippocratic school down to the present day, describing the foundations for ethical judgments in the biomedical field over the course of that development. The historical survey is very detailed, and the justification for the principles of doing good, doing no harm, autonomy, and justice is traced through the various stages of ethical thinking from antiquity to contemporary thought in the United States.

Gracia, an advocate of the personalist and phenomenological philosophy of his fellow countrymen L. Delgado and Xavier Zubiri, proposes a formal ethics of goods as the foundation for a universal ethical authority, while denying the possibility of a universal foundation at the level of the contents of ethical judgments. The author promised to write another work on clinical bioethics as a sequel to the *Fundamentos*. His is still among the most significant contributions in international circles.[18]

In Spain, furthermore, the Andalusian Society for Bioethical Research (Sociedad Andaluza de Investigación Bioética) was founded in 1993, and the publication of the periodical *Bioética y Ciencias de la Salud* (bioethics and health sciences) testifies to its activity. The activities of the Bioethics Investigation Group of Galicia (Grupo de Investigación Bioética de Galicia, GIB) have also been extensive.

In 1983, through the initiative of several professors from the Catholic University of Louvain, the Bioethics Center (Centre d'Études Bioéthiques) was created in Brussels; it is a nonprofit association affiliated with the University of Louvain.[19] There are other centers in France interested in bioethics: we note especially the National Institute for Health Care and Medical Research (Institut National de la Santé et de la Recherche Médicale, INSERM) and its affiliate, the Center for Documentation and Information in Ethics (Centre de Documentation et d'Information en Éthique, CDEI).

The first bioethics institute in the Netherlands, the Institute for Health Ethics (Instituut voor Gezondheidsethiek), was founded in Maastricht in 1985. In England the quarterly *Journal of Medical Ethics* has been published since 1975; it is produced by the Institute of Medical Ethics, headquartered in Edinburgh, which describes itself as an independent, nonpartisan organization. The journal is edited in collaboration with the British Medical Association.

The periodical *Ethics and Medicine* is published in London by the Centre for Bioethics and Public Policy, which is guided by Hippocratic and Christian values.

---

[18] Gracia, *Fundamentos de bioética*, 369–382.

[19] J. F. Malherbe, *Pour une éthique de la médecine* (Paris: Larousse, 1987). On bioethical problems, especially regarding demographics and public policy, see the works of the Louvain scholar M. Schooyans: *L'avortement: Enjeux politiques* (Longueuil, Quebec: La Préambule, 1990) and *La dérive totalitaire du libéralisme* (Paris: Ed. Universitaires, 1991).

We should also note the Linacre Centre for Health Care Ethics in London, founded in 1977, for its service to the Catholic community of Great Britain.

Within the European context, the German author Hans Jonas is worthy of mention. Having lived and worked in North America and Palestine, he made a substantial contribution to the bioethics debate with his book *The Imperative of Responsibility*, which is numbered among the major works in the field. The author's point of departure is similar to Potter's: he considers the increasing potential of technology and examines the threats to the survival of mankind that may result from it. Humanity has an obligation to survive—this imperative has the highest priority, according to the author—and therefore we need to develop a new ethics, what he calls the "ethics of the future," because it should be founded on the assessment of the effects of human interventions upon the biosphere that will fall upon future generations. The guiding criterion for biotechnological interventions should be the aversion of catastrophe.[20]

I should say a few words about the situation in Italy as well. The Center for Bioethics at the Catholic University of the Sacred Heart (Università Cattolica del Sacro Cuore, UCSC) was one of the first to be founded in Italy, in 1985. It is headquartered at the A. Gemelli School of Medicine and Surgery (Facoltà di Medicina e Chirurgia "A. Gemelli") in Rome. The organization is governed by a committee composed of the rector and the dean of the medical school (who are *de jure* members) and eighteen other members appointed by the rector, who are selected experts in medicine, biology, philosophy, law, ethics, and theology. A decision by the Faculty Council and the Academic Senate created—alongside the center—the Bioethics Institute (Istituto di Bioetica), which conducts its activities in the academic world, coordinating doctoral and postgraduate courses in addition to the entire curriculum for degrees in medicine and surgery and university certifications in the field of health care. The institute is headed by a full professor of bioethics and draws upon the work of researchers and grant recipients.

The center has run local training programs and has contributed to the creation of various associated centers in several Italian regions, through which it coordinates existing educational programs with residencies designed for health care personnel and any others who may be directly or indirectly interested in bioethical problems.

The official publication of the center is the bimonthly journal *Medicina e Morale* (medicine and morals), which publishes articles, notes, commentaries, and book reviews on various aspects of bioethics, professional ethics, and medical morals. The work of the center and the institute of the UCSC is reflected not only in the publication of the present manual but also in two series that, taken together, comprise dozens of volumes.[21]

---

[20] H. Jonas, *The Imperative of Responsibility: In Search of an Ethics for the Technological Age* (Chicago: University of Chicago Press, 1984), originally published as *Das Prinzip Verantwortung* (Frankfurt am Main: Insel, 1979). See also, by the same author, *Philosophical Essays: From Ancient Creed to Technological Man* (Chicago: University of Chicago Press, 1974).

[21] Here is a list of some of these works, all published by Vita e Pensiero in Milan: A. Serra and G. Neri, eds., *Nuova genetica: Uomo e società* (1986); E. Sgreccia, ed., *Il dono della vita* (1987); S. Mancuso and E. Sgreccia, eds., *Trattamento della sterilità coniugale* (1989); A. Bompiani and E. Sgreccia, eds., *Trapianti d'organo* (1989); M. L. Di Pietro and

The philosophical perspective characterizing the center and the institute is called ontologically grounded personalism. It follows in the Thomistic tradition and—on this foundation—continues to develop in harmony with Catholic thinking, yet without precluding or avoiding dialogue with other positions.

The Bioethics Center had already begun its work in Genoa, Italy, in 1984. It was characterized by its insistence on not restricting the focus to human life alone, but rather expanding bioethical reflection to include all living beings. This also meant addressing questions of environmental and animal ethics in depth.[22]

Another important approach to bioethics was taken by Saint Raphael Hospital Scientific Institute (Istituto Scientifico Ospedale San Raffaele) in Milan, which founded its Department of Medicine and Humanistic Sciences (Dipartimento di Medicina e Scienze Umane) in 1985. It also publishes a popular science magazine, *KOS*, and a journal about ethics in health care, *Sanare Infirmos* (heal the sick).[23]

In 1988, the Lanza Foundation in Padua started the Ethics and Medicine Project (Progetto Etica e Medicina), addressing a wide range of ethical problems posed by science and society, with a special focus on ethical themes in economics and trends in bioethics.[24] This center, adhering to the Catholic tradition, considers intercultural dialogue to be fundamental and is committed to promoting exchanges between the various philosophical perspectives in the field of bioethics.

Politeia, a center for research and training in politics and ethics located in Milan, with branches in other Italian cities as well, represents the secularist position (in the sense of being strongly critical of the Catholic perspective on bioethics). This center also has areas dedicated to economic ethics, environmental ethics, and bioethics. Its journal, *Notizie di Politeia* (Politea news), assembles different types of contributions from various sources, all of which, however, reflect the same analytical and utilitarian position and privilege the methodological individualism underpinning

---

E. Sgreccia, *La trasmissione della vita nell'insegnamento di Giovanni Paolo II* (1989); L. Antico and E. Sgreccia, eds., *Anzianità creativa* (1989); M. Petrini, *Accanto al morente: Prospettive etiche e pastorali* (1990); A. Serra, E. Sgreccia, and M.L. Di Pietro, *Nuova genetica ed embriopoiesi umana* (1990); E. Sgreccia, S. Burgalassi, and G. Fasanella, eds., *Anzianità e valori* (1991); E. Sgreccia and V. Mele, eds., *Ingegneria genetica e biotecnologie nel futuro dell'uomo* (1992); A.G. Spagnolo and E. Sgreccia, eds., *Lineamenti di etica della sperimentazione clinica* (1994); E. Sgreccia and A.G. Spagnolo, eds., *Etica e allocazione delle risorse nella sanità* (1996); A. López Trujillo and E. Sgreccia, eds., *Metodi naturali per la regolazione della fertilità: l'alternativa autentica* (1994); E. Sgreccia and V. Mele, *Rilevanza dei fattori etici e sociali nella prevenzione delle malattie professionali* (1994); M. Lombardi Ricci, *Fabbricare bambini? La questione dell'embrione tra nuova medicina e genetica* (1996).

[22] See, for example, S. Castignone, ed., *Etica ambientale: Atti della Giornata di etica ambientalista* (Naples: Guida, 1992).

[23] P. Cattorini, "Profilo della scuola di medicina e scienze umane: Educare ad un'intenzione antropologica," *Sanare Infirmos* 3 (1988): 19–23.

[24] Viafora, *Vent'anni di bioetica*; idem, *Fondamenti di bioetica* (Milan: Ambrosiana, 1989); idem, ed., *Centri di bioetica in Italia: Orientamenti a confronto* (Padua: Gregoriana, 1993).

all the research conducted by the group.[25] The department for bioethics would later give rise to the Bioethics Council (Consulta di Bioetica), which publishes the journal *Bioetica* (Bioethics).

Another reference point for the position just described is the Chair for Anthropology at the University of Florence (Università di Firenze), where Brunetto Chiarelli produced the journal *Problemi di bioetica* (bioethical issues) and founded an association, the Italian Bioethics Society (Società Italiana di Bioetica). Its philosophical viewpoint reflects the biologistic and evolutionist position.[26]

Other initiatives later enlivened the Italian scene, such as the Bioethics Center established at the International Institute for Human Rights (Istituto Internazionale per i Diritti dell'Uomo) in Trieste, and the Focus Group on Biotechnologies (Gruppo di attenzione sulle biotecnologie, GAB) in Milan, founded in 1988, which deals with biotechnologies and their ethical aspects. We should also mention the International Center for Studies on the Family (Centro Internazionale di Studi sulla Famiglia) in Milan, which conducts family-focused research from a Catholic perspective. In Sicily, the Sicilian Bioethics Institute (Istituto Siciliano di Bioetica) opened in 1991 at the Theological School of Sicily. This institute recently published the *Dizionario di bioetica* (dictionary of bioethics). In Messina, the Bioethics Workshop (Laboratorio di Bioetica) was founded in 1992 as an affiliate of the Pontifical Salesian University.

Academic instruction in bioethics was very quickly introduced in Italy at many pontifical universities, primarily from a theological perspective.[27] At the public universities, bioethics was introduced at first as an elective subject required by some schools of medicine and surgery, including the Catholic University of the Sacred Heart (Università Cattolica del Sacro Cuore) in Rome. It was later incorporated into clusters of subjects for candidates preparing to compete for positions as university professors and researchers, and then into the science and education curricula of universities to be taught in conjunction with the history of medicine, legal aspects of medicine, and moral philosophy.[28]

University courses in bioethics have undoubtedly helped to better define this discipline, and the same can be said for the establishment of bioethics committees, also referred to as ethics committees, which have offered strong incentives for bioethical reflection.

---

[25] Such an approach can be found in M. Mori, ed., *Questioni di bioetica* (Rome: Editori Riuniti, 1988); idem, *La bioetica: Questioni morali e politiche per il futuro dell'uomo* (Milan: Bibliotechne, 1991); U. Scarpelli, "La bioetica: Alla ricerca dei principi," *Biblioteca della libertà* 99 (1987): 7–32.

[26] B. Chiarelli, *Problemi di bioetica nella transizione fra il II e il III millennio* (Florence: Il Sedicesimo, 1990).

[27] Worth noting is the contribution in the theological field offered by D. Tettamanzi in the volume *Bioetica: Nuove frontiere per l'uomo*, 2nd ed. (Casale Monferrato: Piemme, 1996) and in many articles from the journals *Medicina e Morale* and *Anime e Corpi* (souls and bodies).

[28] I have described my experiences teaching bioethics at the medical school of the Catholic University of the Sacred Heart, in E. Sgreccia and A. G. Spagnolo, "L'insegnamento di bioetica nel Corso di laurea in Medicina e Chirurgia: L'esperienza nell'Università Cattolica del S. Cuore," *Medicina e Morale* 46, no. 4 (1996): 639–654. For a general treatment of the problems of teaching bioethics in medical school, see also P. Cattorini and V. Ghetti, eds., *La bioetica nelle facoltà di medicina* (Milan: F. Angeli, 1997).

From Medical Ethics to Bioethics

To give a full account of the present-day debates and the diverse positions in bioethics, we need to recall the historical course of ethical reflections in medicine before the term "bioethics" became widespread and underwent modifications. Some of the stages over the centuries have established criteria and defined philosophical categories and have often initiated and framed recurring discussions on specific themes.

Several stages are significant in this respect: Hippocratic medical ethics, medical morals in the theological tradition, the contribution of modern philosophy, and reflection on human rights in Europe, especially since the last world war. This survey cannot be as extensive and thorough as those of some authors;[29] it will instead focus on the historical data that are absolutely necessary for understanding the present cultural climate. The influence of principles and criteria from modern philosophies will be dealt with in a later chapter devoted to meta-ethics, models of bioethics, and the problem of the foundation for ethical thought.

At the origin of medical ethics in primitive societies, as well as in the more highly developed societies of antiquity, we always find three elements: the ethical requirements that the physician must observe, the moral significance of helping the sick person, and the decisions that the state had to make with regard to its citizens for the sake of public health. The Code of Hammurabi, dating from 1750 BC and influenced by earlier Sumerian precepts, already contained norms regulating medical practice and an initial levy of taxes for health care.[30] In tracing the development of Western ethical thinking in the medical field, Hippocrates (ca. 460–370 BC) and his oath cannot be overlooked.

It is known that there are problems of attribution and authenticity, not to mention difficulties of textual criticism, regarding the entire Hippocratic corpus, including the text of the Hippocratic Oath.[31] This body of writings was certainly the product of contributions from a tradition and not the work of a single thinker and teacher, yet the substantially Hippocratic origin of the ideas contained in the oath has withstood historical criticism.

The presentation by Gracia[32] seems to be thorough and well-documented in terms of the analysis performed on the text, its ethical and religious structure, and its historical and philosophical interpretation. According to Gracia's interpretation, the oath constitutes a distinctive expression of the culture of its time. It was pre-juridical in character, proper to a category of persons—i.e., physicians—who were considered above the law in a certain sense. The law was for those who practiced common trades as simple citizens, whereas the medical profession, like that of kings and priests, was supposed to be a "strong profession" governed by a "strong morality" that was expressed precisely in the religious sense of the oath.

---

[29] I refer especially to the extensive and well-documented survey by Gracia in *Fundamentos de bioética*.

[30] See D. von Engelhardt, "Storia dell'etica medica," in *Dizionario di bioetica*, ed. S. Leone and S. Privitera (Bologna: Dehoniane / Istituto Siciliano di Bioetica, 1994), 954–958.

[31] S. Spinsanti, ed., *Documenti di deontologia e etica medica* (Cinisello Balsamo: Paoline, 1985).

[32] Gracia, *Fundamentos de bioética*, 45–84.

The structure of the oath includes (a) an invocation of the deity as an introduction to characterize it as such; (b) a central part subdivided into two sections, one concerning the obligation to show respect for the teacher, to give free instruction to his sons, and to teach, in general, anyone subscribing to the oath, and the other devoted more specifically to therapy, requiring the physician to respect medical confidentiality and to rule out such actions as the administration of poison (even to someone who has requested it), "procured abortion," and any sexual abuse whatsoever of sick persons or members of the household; and (c) a conclusion invoking sanctions from the deity—in a rewarding sense (blessings) upon those who keep the oath and in a punitive sense (curses) upon those who violate it. According to this sort of historical analysis, therefore, we are not dealing with a timeless code as a written expression of a natural morality (as was thought until the eighteenth century), but rather with a reflection of the philosophy and culture of that time, which understood the medical profession in transcendent terms and invested it with a sacred character, considering it a physiological and charismatic priesthood. The conclusion following from this interpretation is that Hippocratic thought provided a philosophical and theological foundation for what is called today, with a negative connotation, medical paternalism.

Certainly the oath bases the morality of a medical act on the principle that was handed down through the centuries as the "principle of beneficence and nonmaleficence," that is, the good of the patient. The physician is always to act for the good of the sick person because this is his *ethos*, so what he prescribes has no need of further confirmation, not even from the patient.

Hence the text is neither a simple moral code protecting the interests of the caste of physicians, nor a sort of natural morality, but rather a morality founded on the sacred principle of the good of the patient, for whom the physician is the absolute and irrevocable caretaker and hence above the law and all suspicion. We cannot ignore, however, the way that the Hippocratic conception—particularly in relation to the subsequent development of ethical and philosophical thought in Socrates, Plato, and Aristotle—seeks to establish nonsubjective criteria for morality thus founded on objective truth: there is an awareness of the good in itself and of respect for the person, above and beyond one's own subjective desires.[33]

Everyone acknowledges that Hippocratic thought remained "canonical" for all of classical culture and throughout the medieval period. This almost universal influence of the Hippocratic Oath is evidenced by analogous formulas found in various cultures, such as the oath of Asaph ben Berechiah in sixth-century Syria, the physician's daily prayer of Moses Maimonides (1135–1204) in Egypt, and "A Physician's Ethical Duties" by Mohamad Hosin Aghili (1770) in Persia.

The appearance of the principle of autonomy in conjunction with the development of modern thought, the ethical liberalism of David Hume, Adam Smith, John Stuart Mill, and John Gregory, and the subsequent formulation of the rights of man and the rights of the citizen certainly represent a *medical antipaternalism*, as Gracia goes on to affirm, yet these new principles could not completely abolish the principle of beneficence as an element that validates and guarantees the autonomy of both patient and physician.

---

[33] G. Reale and D. Antiseri, *Il pensiero occidentale dalle origini ad oggi* (Brescia: La Scuola, 1983), 1:76–78.

The idea of justice propagated by contemporary social thought will also be unable to abolish this principle of beneficence, which I consider to be founded not on the ahistorical transcendence of the medical profession but rather on that idea of good and truth which is fundamental to the very existence of the other principles of autonomy and justice, as we will see in the chapter dealing with the foundation ethical judgments. Contemporary bioethics, especially the kind that harks back to the famous principles of beneficence and nonmaleficence, autonomy, and justice, therefore continues to draw on a historic tradition that comes from afar and runs through the entire development of Western thought.

The contribution of Christianity cannot be omitted in this historical survey: Christian theology, Christian health care practices, and Christian teachings as found especially in the Magisterium of the Catholic Church. Christianity did not limit itself to a favorable reception of Hippocratic ethics; rather, it introduced new concepts and values to it, just as it did with Platonic and Aristotelian thought, through both its teachings and its work of caring for the sick. These contributions can be found especially in the definitive establishment of the concept of the *human person*, in the new theological understanding of care for the sick and of the medical profession, and in the establishment and promotion of positive dialogue, especially in Catholic circles (after the initial difficulties in Galileo's day), between scientific reason and religious faith.

The value of the human person in Christianity is the result of overcoming the classic dualism between mind and body. Christian personalism considers not only the spiritual soul but rather the whole man, in his body-soul unity, as a creature of God. He is a steward who shares responsibility for the earth and for his life in the world, and he is held accountable to the Creator himself. Moreover, by virtue of the mystery of the Incarnation and Redemption, man—every human being, especially the neediest—is considered and valued as an expression of the Redeemer's presence. Caring for the sick will therefore be a criterion for the final, eschatological judgment, whereby what is done for the sick is judged according to the words, "you did it to me" (Mt 25:40).

This new vision of the world and of humanity in the personalist, creational, and redemptive sense prompted the Christian community to build hospitals embodying the parable of the Good Samaritan (see Lk 10:30–37) throughout the history of the Christianized world. For at least seventeen centuries, the Catholic Church and the Christian community would take charge of public health as a fraternal duty and as proof of the authenticity of their message.

Even after the French Revolution, with the introduction of the concept of a secular hospital and of the citizen's right to health care, the Christian communities sensed the right and the duty not to abandon care of the sick. They have continued not only by way of "supplementing" civil services in developing countries but also and above all as witnesses to the solidarity that Christ wants his followers to have with those who suffer.

The figure of the physician in the Christian sense, where this theological understanding has been cultivated, is not a priestly personage above the moral law; instead, he is called to be the servant (*diakonos*) of the suffering, representing the community of those who have the obligation to "take care" of their brothers and sisters. This theological understanding in fact calls him, if he is a believer, to model himself on the Good Samaritan, that is, on Christ himself who takes care of suffering humanity. Just as the Gospel commands us to see *Christus patiens* (the suffering Christ) in the sick patient, so too must we discern *Christus servus* (Christ the servant) in the physician.

In light of this theological insight, the Christian churches have developed a moral theology that proclaims the sacredness and inviolability of every human being and condemns abortion, infanticide, euthanasia, and mutilation. Within this moral theology, teachings on medical morality have become increasingly extensive, moving gradually from the medieval and renaissance treatises that addressed topics of medical ethics within their treatment of the virtue of justice or in their commentaries on the commandment "Thou shalt not kill," to the more recent works in which the foundation for ethical judgments on the physician's work is derived from divine revelation in addition to the precepts of Hippocrates, whose oath has always been recognized as an expression of ethical uprightness in both the Christian and Muslim worlds.

One continuation of this centuries-old interest is the constant magisterial teaching of the Catholic Church, shared by other Christian denominations, concerning the problems posed by medical science at the present time in particular. There was one important period in history in which great emphasis was placed on medical morals in Catholic circles: during the pontificate of Pope Pius XII. Anyone who browses through the talks and radio messages that Pius XII addressed to physicians—and these works merit a systematic rereading—will notice that they implicitly respond to two challenges: the Nazi crimes perpetrated in concentration camps and elsewhere, and the advancement of technological progress that, because of its manifold possibilities, could be used to oppress and suppress human life.

It is precisely at this historical crossroads that bioethics was born.[34] Yet the reflections of Catholic moral theology on the medical field have become continually enriched, even after the pontificate of Pius XII, in the writings of his successors. The pronouncements of the churches in general and of the Catholic Church in particular have been the object of careful consideration in international circles, too. Physicians cannot ignore them, whether because of their own possible religious affiliations, the religious beliefs of their patients, or the objective reasons on which moral norms and directives are based.

In addition to Pius XII's talks and radio messages addressed to physicians, one cannot overlook the documents of the Second Vatican Council, in particular the pastoral constitution *Gaudium et spes* (the part dealing with the concept of man and of the family), Paul VI's encyclical *Humanae vitae*, dated July 25, 1968; the declaration *Quaestio de abortu* on procured abortion by the Congregation for the Doctrine of the Faith, dated November 18, 1974; the declaration *Persona humana* on sexual ethics, dated December 29, 1975; and the letter to the U.S. bishops on sterilization in Catholic hospitals, *Quaecumque sterilizatio*, dated March 13, 1975. The many addresses by John Paul II concerning conjugal morality and his apostolic exhortation *Familiaris consortio*, dated November 22, 1981, should be mentioned, as well as the declaration on euthanasia *Iura et bona* by the Congregation for the Doctrine of the Faith, dated May 5, 1980, and the most compelling document in the field of bioethics, the instruction *Donum vitae* on respect for procreation and human life, dated February 22, 1987. Of fundamental importance among the encyclicals are *Veritatis splendor* (August 6, 1993) and *Evangelium vitae* (March 25, 1995), in

---

[34] E. Sgreccia, "La bioetica, fondamenti e contenuti," *Medicina e Morale* 34, no. 3 (1984): 285–306; idem, "La bioetica tra natura e persona," *La Famiglia* 108 (1985): 30–42.

which John Paul II explicitly first used the term *bioethics* and addressed current questions in that area.[35]

Other Christian churches and the other religious denominations have likewise offered guidelines to their own faithful and formulated proposals for the medical and political world; they will be mentioned in our discussion of particular issues. For now it will suffice to mention the guidelines concerning abortion and prenatal diagnosis titled *Manipulating Life: Ethical Issues in Genetic Engineering*, issued by the World Council of Churches (Geneva, 1982), and the Islamic Code of Medical Ethics, approved by the International Conference of Islamic Medicine held in Kuwait in January 1981.[36]

To complete this historical survey of the development of principles and criteria for conduct in the biomedical field, we must recall a very important legal and ethical contribution made in secular circles after and as a result of the Nuremberg trials (1945–1946). During these famous trials of Nazi criminals, the world learned about the crimes committed against prisoners and civilians, by order of the Nazi regime, with the collaboration of physicians. These crimes are well-known today from the transcripts of the trials,[37] and they stand as a horrific example of what can be done by an absolute power that heeds no morality or moral authority, with the collaboration of physicians who allow themselves to become instruments of that political power, claiming as their justification that they were "compelled."

Norms were developed along two lines as a result of that tragic moment, which according to some observers marks the birth of bioethics before the actual use of the term, because it involved the formulation of human rights and the approval and gradual updating of codes of medical ethics issued by international organizations such as the World Medical Association (WMA) and the Council for International Organizations of Medical Sciences (CIOMS). This legislation and these norms necessarily implied and required fundamental reflection upon the theory of and justification for medical interventions, and by the very nature of the subject this led to a systematic discipline: bioethics.

The first line of development involved extensive codification, beginning with the Universal Declaration of Human Rights, published by the United Nations on December 10, 1948, and the European Convention for the Protection of Human Rights

---

[35] The official editions of documents of the Catholic Magisterium are published by Libreria Editrice Vaticana. Documents by the popes appear in the following Italian anthologies: Pius XII, *Discorsi e radiomessaggi*, 20 vols. (Vatican City, 1948–1959); John XXIII, *Discorsi, messaggi; colloqui del S. Padre Giovanni XXIII*, 5 vols. (Vatican City, 1960–1964); Paul VI, *Insegnamenti di Paolo VI*, 16 vols. (Vatican City, 1965); John Paul II, *Insegnamenti di Giovanni Paolo II* (Vatican City, 1979). The most official publication for each document is, nevertheless, in the *Acta Apostolicae Sedis* (AAS). Another official publication of the Catholic Church is the newspaper *L'Osservatore Romano*. A collection of the addresses by Pius XII to physicians can be found in Italian in F. Angelini, ed., *Pio XII: Discorsi ai medici* (Rome: Orizzonte Medico, 1959).

[36] "The Islamic Code of Medical Ethics" (January 1981), in Spinsanti, *Documenti di deontologia*, 166–186 [Italian]. For an in-depth reflection on bioethics from the perspectives of various religions, see S. Spinsanti, *Bioetica e grandi religioni* (Milan: Paoline, 1987).

[37] See R.J. Lifton, *I medici nazisti: Lo sterminio sotto l'egida della medicine e la psicologia del genocidio* (Milan: Rizzoli, 1988).

and Fundamental Freedoms (dated November 4, 1950), which contain compelling statements on the protection of human life and physical integrity, together with a defense of and safeguards for other basic civil and political liberties. These led in turn to a whole series of declarations, agreements, recommendations, and charters. I wish to mention, by way of example, these recommendations made by the Council of Europe: no. 29/1978 on organ and tissue transplants, no. 79/1976 relating to the rights of the sick and dying, and nos. 1046/1986 and 1100/1989 on the utilization of human embryos and fetuses.

These and other documents will be cited during the discussion of individual issues because of their cultural authoritativeness and for the ethical values to which they refer. As to legal power, of course, the various documents have different normative weights. The recommendations in particular have full legal value once they are accepted by a member state, but they have a cultural and ethical impact even if they are not. Of notable ethical value is the Convention on Human Rights and Biomedicine adopted by the Committee of Ministers of the Council of Europe on November 19, 1996.

Major codes of medical ethics include the Nuremberg Code of 1947 and the Code of Medical Ethics published by the WMA in Geneva in 1948, containing the so-called Oath of Geneva, which was updated by the same association in London in 1949. Another famous document, again by the WMA, is the Declaration of Helsinki on biomedical experimentation and research, published in 1962, modified in Helsinki in 1964 and updated in Tokyo in 1975, in Venice in 1983, in Hong Kong in 1989, in Somerset West, South Africa, in 1996, and most recently in Seoul, South Korea, in 2008.

The Declaration of Sydney, issued by the WMA in 1968 and updated in Venice in 1983 and in Pilanesberg, South Africa, in 2006, addresses the issue of determining the moment of death and treatment in the final stage of terminal illness.

The International Conference of Medical Professional Associations, part of the European Economic Community (EEC), another authoritative organization in the medical field, published a document titled *Principles of European Medical Ethics* in Paris on January 6, 1987.[38] Its Italian affiliate, the National Federation of Medical and Dental Associations (Federazione Nazionale degli Ordini dei Medici Chirurghi e degli Odontoiatri, FNOMCeO), plans to update Italy's legal code of medical ethics.[39]

---

[38] An important collection of professional ethics codes can be found in Spinsanti, *Documenti di deontologia*. On the pertinence of human rights to medical activity, see M. Torrelli, *Le médècin et les droits de l'homme* (Paris: Berger-Levrault, 1983). The Council of Europe later published a volume on medicine and human rights (see bibliography).

[39] The stages in the historical development of the Italian *Codice di Deontologia Medica* (Code of professional medical ethics) are outlined in the book by F. Introna, M. Tantalo, and A. Colafigli, *Il Codice di Deontologia Medica correlato a leggi ed a documenti* (Napoli: Liviana Medicina, 1992). From this book we learn that a code of medical ethics was published by the Order of Physicians in Turin in 1912 because a national code did not exist. Following a referendum among Italian physicians in 1954, this code became the basis for the publication of the first Italian code, which was then modified in 1978 and again in 1989. A systematic treatment of professional ethics, together with the text of the 1995 code, can be found in FNOMCeO, ed., *Guida all'esercizio professionale per i medici chirurghi e gli odontoiatri* (Turin: Edizioni Medico Scientifiche, 1994), Part IV (revised in December 1995), 897.

The Problem of a Definition

The history of bioethics just sketched reveals a broad spectrum of problems, subjects, and criteria. Initially theoretical with a prevailing bioenvironmental interest (Potter, Jonas) and overtones of alarmism, calling into question the nineteenth-century concept of unidirectional and automatically beneficial progress, bioethics became enriched, thanks to the labors of various centers in the United States and Europe, by new ethical and philosophical reflections on old and new problems in medicine, demographics, and experimental research on human beings and animals. It then went on to emphasize the relation between human life and subhuman life and, finally, to address the contributions of classical medical ethics associated with various religious doctrines as well as the international concern with human rights.

Hence the problem arises of defining bioethics in the first place. Even today, it does not appear to have been solved. There are those who describe bioethics as an intellectual movement of ideas that change in both a historical and historicist manner; those who consider it a methodology for the interdisciplinary study of the biomedical and humanistic sciences; others who view bioethical reflections as a branch of moral philosophy; and still others who maintain that bioethics can be defined as an autonomous discipline distinct from medical professional ethics, medical law, human rights, and the older and better-known medical ethics, though it cannot avoid some crossover with those disciplines.[40]

From this geographical and cultural excursus describing bioethics and examining its multiple antecedents in medical ethics, religious morality, professional ethics, and human rights, we appear to be faced with a panorama that is certainly quite vast yet fairly well characterized. It should therefore be clear by now that the term bioethics is understood to include medical ethics, strictly speaking. It does not follow that bioethics is an addition to medical ethics, but quite the contrary: bioethics, as a form of ethics concerning interventions affecting life in general, is understood in a more comprehensive sense to include interventions affecting the life and health of mankind in addition to those affecting all other forms of life. Moreover, medical science increasingly avails itself of basic research in the field of biology as it advances, linking in a social sense with its environmental component.[41]

As noted in the introduction, Potter coined the term *bioethics* in 1971.[42] He had tentatively defined the new discipline as the combination of "biological knowledge and human values."[43] In other words, he considered bioethics a new type of wisdom that would have to show how to use scientific knowledge in order to safeguard the social good; bioethics therefore had to be the "science of survival."

---

[40] A. Bompiani, *Bioetica in Italia: Lineamenti and tendenze* (Bologna: Dehoniane, 1992).

[41] C. Iandolo, "Etica clinica e bioetica," *Giornale Italiano di Formazione Permanente del Medico* 15, no. 2 (1987): 88–103.

[42] Potter, *Bioethics*, 1. [Translator's note: An earlier usage of the term has been discovered and more widely examined since the time the author first published this work and since its most recent Italian edition in 2007. See H. M. Sass, "Fritz Jahr's 1927 Concept of Bioethics," *Kennedy Institute of Ethics Journal* 17, no. 4 (2007): 279–295.]

[43] Cited in Reich, "The Word 'Bioethics.'"

Reich gave two different definitions of bioethics in the successive editions of the *Encyclopedia of Bioethics*. In the 1978 edition he defined bioethics as "the systematic study of human conduct in the area of the life sciences and health care, insofar as this conduct is examined in the light of moral values and principles."[44] The scope of the life and health-related sciences therefore included a more general consideration of the biosphere and not just of medicine. The interventions could be those dealt with not only in the medical professions but also concerning populations, such as interventions directed to demographic and environmental problems. The specific nature of this systematic study was in reference to moral values and principles, and therefore to the definition of criteria, judgments, and limits regarding what is licit or illicit.

In the 1995 edition (reprinted in 2004), Reich offered a broader definition of bioethics, defining it as "the systematic study of the moral dimensions—including the moral vision, decisions, conduct, and policies, etc.—of the life sciences and health care, employing a variety of ethical methodologies in an interdisciplinary setting."[45] It is evident that this definition partially recovers Potter's original concept of global bioethics. Indeed, Reich himself explains that "the broad vision with which this new term was proposed, more than twenty years ago, has been confirmed. Unlike those who understand bioethics in a reductive sense—in practice, as medical ethics expanded enough to include the ethics of biomedical research—we have extended bioethics to include social, environmental, and global problems of health and of the life sciences. The scope of bioethics extends, therefore, beyond that of biomedical ethics."[46]

In the 1995 definition, therefore, the material object of bioethics was broadened to all its moral dimensions, including social behaviors and political decisions; in this sense the definition appears more complete. The formal object of bioethics was changed as well: it was no longer examined in the light of moral values and principles, but rather by means of "a variety of ethical methodologies." With this statement, Reich wished to eliminate a misunderstanding generated in the preceding years—namely, that its principles were only those promoted by Beauchamp and Childress—and ultimately sought to open the door to ethical pluralism.

In my opinion, this broadening of the subject was undoubtedly important, yet it conceals a real danger of ethical relativism that could interfere with the normative role of bioethics. While it may indeed be advisable to start by examining various viewpoints when confronting an ethical problem, it then becomes necessary—given the fact that bioethics has a practical purpose—to verify the validity of the arguments and of the criteria provided by each of the various positions when making decisions. Thus the validity of the choice should be argued rationally; this is the only way to avoid falling into ethical relativism, which would ultimately be the undoing of bioethics itself. In this sense, therefore, bioethics can be equated with neither professional medical ethics, nor medical law, nor simple philosophical speculation.

In order to explain this difference more precisely, a study group drew up the Erice Document during an international convention held in Erice, Sicily, in February 1991. It considered the object of bioethics and the relation between this discipline and medical ethics and law, followed by various debates between medical law experts

---

[44] Reich, *Encyclopedia of Bioethics* (1978), 1:xix.

[45] Ibid., rev. ed. (1995), 1:xxi.

[46] S. Spinsanti, "Incontro con Warren Reich," *L'Arco di Giano* 7 (1995): 219.

about the role of bioethics. In this document, which in large measure took the 1978 *Encyclopedia of Bioethics* as its point of departure, the competency of bioethics was acknowledged in the following four areas: (1) ethical problems in the health care professions; (2) ethical problems arising in the area of research on human beings, including research that is not directly therapeutic; (3) social problems connected with health care policies (national and international), medicine as an occupation, and policies of family planning and population control; and (4) problems related to human interventions affecting other life forms (plants, micro-organisms, and animals) and, generally, anything concerning the equilibrium of the ecosystem.

The document addresses the question of the relationship between medical law and professional ethics as follows:

> (1) Bioethics is an area of research which makes use of an interdisciplinary methodology and has as its object "the systematic examination of human behavior in the field of life sciences and health sciences, insofar as this behavior is examined in the light of moral values and principles," according to the widely accepted definition of the *Encyclopedia of Bioethics* (1978 edition). Its specific character results from the type of problems that it deals with, from the nature of the ethical questions and from the methodology used.

> Inasmuch as it is an application of ethics to the "biological kingdom"—which designates a much more extensive world than that of medicine—bioethics encompasses traditional medical ethics and more, including: (a) the ethical problems in all health care professions; (b) behavioral research, regardless of whether or not there are therapeutic applications; (c) the social problems associated with health care policies, medicine in the workplace, international health issues, and population control policies; (d) problems of animal and plant life in relation to human life.

> The *purpose* of bioethics consists of the rational analysis of the moral problems related to biomedicine and the study of their connections with the fields of law and the social sciences. This analysis involves the elaboration of ethical guidelines that are based on the values of the human person and on human rights, show respect for all religious faiths, and have a rational foundation and a methodology that is suitably scientific. Such ethical guidelines also have practical applications through the direction they impress not only upon personal conduct but also on the framing of legislation and the formulation of present and future professional codes of medical ethics.

> The *means* by which bioethics conducts its studies follow from its characteristic interdisciplinary methodology, which aims to thoroughly examine the nature of the biomedical fact (epistemological moment) using the latest findings, to discover the implications at the human level (anthropological moment), and to identify the ethical "solutions" and the rational justifications that support such solutions (pragmatic moment). . . .

> (2) Professional medical ethics is a discipline, the object of which is to study norms of professional conduct that are specific to the health care professions. This discipline includes three classes of norms:

>> (a) moral norms, which are the object of traditional medical ethics and are now included within bioethics, a "building," so to speak, for which "medical ethics paved the way"

21

(b) professional norms properly speaking, collected in codes and in the entire oral and written tradition of the medical profession

(c) the legal norms of each country

The *purpose* of medical deontology is the essential, in-depth study and updating of the norms and rules of conduct in the medical professions.

The *means* of studying these three areas are different:

(a) the study of moral norms and the work of interpreting and updating them is carried on in close collaboration with the findings of bioethics;

(b) updating deontological norms, properly speaking, involves constantly consulting the national and international codes of the medical professions;

(c) the legal norms concerning medical ethics are studied in light of the current law and proposed legislation of each country, for the purpose of finding ways to align these with the values that inform medical ethics. ...

(3) Medical law is by nature an interdisciplinary science, which applies a specific methodology to study the biological and medical contents of legal norms, for the purpose of approving the best interpretation, application and development of them; medical law also collaborates with the judicial system and with private citizens in finding solutions to cases that require investigations and evaluations of a biological and/or medical sort. ...

Forensic medicine is naturally connected with bioethics, by way of the teaching of medical professional ethics. Indeed, bioethics is a more eminently autonomous and broader discipline which, with its methodology and findings, contributes to the updating and epistemological justification of normative professional ethics, provides orientation for legislative efforts, and helps to situate interventions affecting human life within the broader context of the biosphere by discussing criteria and the limits of what is permissible.[47]

This definition, too, like the one from the *Encyclopedia of Bioethics*, deliberately does not specify what the values and moral principles are, given the plurality of philosophical positions which bioethics has the task of discussing and examining. The next chapter will provide a description of these various orientations, which are found, moreover, in the individual discussions of particular problems.

The definition of bioethics by Adriano Pessina is especially interesting: "Bioethics presents itself as the critical conscience of technological development, bringing into play all the intellectual resources that Western civilization has available."[48] Professor Pessina underscores the need to reflect on the technological process, so as to take back the meaning of Western civilization and the direction in which it is headed:

---

[47] Società Italiana di Medicina Legale e delle Assicurazioni, "Il Documento di Erice sui rapporti della bioetica e della deontologia medica con la medicina legale," 53rd Course on New Trends in Forensic Haematology and Genetics: Bioethical Problems (Erice, February 18–21, 1991), published in *Medicina e Morale* 41, no. 4 (1991): 561–567.

[48] A. Pessina, *Bioetica: L'uomo sperimentale* (Milan: Bruno Mondatori, 1999), 22.

That is to say, bioethics must present itself as *the conscience of technological civilization*. Hence the expression "critical conscience" indicates the level of moral clarification and evaluation of the specific practical and theoretical subject matter introduced by the technological sciences. ... From this perspective bioethics is configured as a philosophical activity, regardless of who is actually conducting the inquiry, because the questions (the formal object) encountered by the technological sciences (the material object) are by nature philosophical and concern the significance of how human identity is construed within the technological activity. ... The bioethical question concerns all of man and all men: technological science is more than a complex pragmatic operation; it is truly and properly a culture which claims to establish, by means of various instruments, the meaning and the purpose of life, and not only of human life.[49]

The bioethics proposed by analytic philosophy is characterized by its agnosticism, claiming that it is impossible to establish the truth or falsity of propositions that express value judgments. For example, Uberto Scarpelli maintains that bioethics can only make clarifications or point out the presuppositions of a given choice. He writes that

philosophical bioethics consists essentially of a meta-bioethics. ... It is an attempt to clarify concepts and bring presuppositions to light. ... In conclusion, it must be interdisciplinary, a river with many tributaries flowing into it. The disciplines that are involved vary in nature, and this poses particularly complex problems concerning the linguistic, logical, and methodological relations among them. At the center we find once more the fundamental and decisive questions in contemporary ethics: the *is-ought question*, or questions of the logical relations between descriptive propositions of fact and normative propositions that determine duties and assign values.[50]

Finally, let us mention so-called secular bioethics, which is "founded on reason and on the values of the conscience," supposedly in opposition to the Catholic variety, which is said to be founded on dogmas and faith.[51] Yet, as I will often repeat in this manual, the personalist, ontologically grounded position to which Catholics refer is not fideistic and does not dispense with a reasoned justification of values and norms; in fact, it is even accused of rationalism in some theological circles because of this.

As a final descriptive note we can say that the academic treatment of the subject of bioethics has produced a subdivision into three distinct areas: general bioethics, particular bioethics, and clinical bioethics.

- *General bioethics*, which is concerned with ethical foundations, is discourse about the original values and principles of medical ethics and the documentary sources of bioethics (international law, codes of professional

---

[49] Ibid., 41.

[50] U. Scarpelli, "Bioetica: Prospettive e principi fondamentali," in *Bioetica: Questione civile e problemi teorici sottesi* (Milan: Edizioni Glossa, 1998), 21.

[51] See the "Manifesto di bioetica laica," published on June 9, 1996, in *Il Sole 24 Ore* and signed by C. Flamini, A. Masserenti, M. Mori, and A. Petroni. See also the proceedings of the convention: "Quale base comune per la riflessione bioetica in Italia? Dibattito sul Manifesto di bioetica laica," *Notizie di Politeia* 12 (1996).

ethics, legislation). In practice, the fundamental and institutional part of general bioethics is truly and properly a moral philosophy.

- *Particular bioethics* analyzes major problems, which are always approached from a general perspective, both in the medical field and in the biological field: genetic engineering, abortion, euthanasia, clinical experimentation, etc. These are the major issues that make up the pillars of systematic bioethics and must obviously be resolved in light of the models and foundations that the ethical system assumes as a basis and justification for ethical judgment. Particular bioethics, therefore, can do nothing unless associated with the conclusions of general bioethics.

- *Clinical or decisional bioethics* concretely examines what values are at stake in medical practice and clinical cases; to put it differently, clinical or decisional bioethics assesses the right ways of determining a course of action without altering such values. The evaluation of the case will depend on whether or not an evaluating principle or criterion is selected.

In my opinion, one cannot separate clinical bioethics from general bioethics, although I recognize that concrete cases always or almost always present a plurality of aspects to be evaluated.

Bioethics, Anthropology, and Interdisciplinary Study

Based on the previous discussion, this new discipline cannot be understood as a simple comparison of the different opinions and the various ethical positions that exist; rather, it must propose standard values and effective decision-making approaches, providing objective answers based on rationally valid criteria.

The search for adequate answers demands an interdisciplinary approach to the problem, which is one of the unique characteristics of bioethics. The role played by the biomedical sciences and environmental science (ecology) is evident, but not everyone is aware that this area of study requires a standard philosophical anthropology, which is the framework within which an ethical value is assigned to bodily life, marital love and procreation, and suffering, sickness, and death, as well as to the relationships between freedom and responsibility, individual and society, and individual and nature. This complex interweaving of the experimental and humanistic sciences in search of a "wisdom of science," to use Potter's expression, also requires contributions from the philosophy of nature (to adequately establish the role, meaning, and value of the environment and ecosystems in bioethics), the philosophy of science, and law. Finally it is advisable that this interdisciplinary study be open to theology as a "horizon of meaning."[52] Although they are closely interrelated, each of these disciplines has its own epistemological status, independent of the others.

As far as anthropology is concerned, I will refer to the anthropological concept that, in my opinion, best does justice to the real and objective meaning of man and contributes to a respect for his intrinsic value: *ontologically grounded personalism*. This presents itself as an integral understanding of the human person, not subject

---

[52] F. D'Agostino, "La teologia del diritto positivo: Annuncio cristiano e verità del diritto," in *"Evangelium vitae" e diritto* (Vatican City: Libreria Editrice Vaticana, 1997), 121–131.

to reductive ideologies or biologistic thinking. Indeed, in order to solve the problems posed by scientific progress and the social organization of medicine and law, I believe it is first of all necessary to answer the question of the value of the human person, with his prerogatives and duties, so as to bar all possibility of exploitation.

The fundamental value of life, the transcendence of the human person, the integral concept of the human person (resulting from the synthesis of physical, psychological, and spiritual values), the relations of priority and complementarity between person and society, and a personalist understanding of marital love as a communion are valid points of reference for bioethics, as well as for human and social ethics in general. These values should enlighten those who attempt to solve the problems resulting from the advances in biomedical science—a science that seems inspired by an optimistic enthusiasm for progress while forgetting great challenges, such as fighting still-untamed diseases and stemming the evils that are typical of that same technological society, which are caused by environmental exploitation. Precisely for this reason, there is a need for a standard philosophical anthropology that takes into account the human person as a whole and the unique, twofold relations that tie the person to his existential conditions: the space in which he dwells and the time in which he lives and will live. In this perspective, then, one comes to understand the great importance of the category of responsibility to which Jonas refers in his previously cited book. The ontologically grounded personalist anthropology is very often criticized as an anthropology that can only be maintained by someone who allows for suprarational knowledge—by someone who admits the possibility of theology. As *Fides et ratio* reminds us, the importance of metaphysics and of the intelligibility of the faith should be emphasized:

> The word of God refers constantly to things which transcend human experience and even human thought; but this "mystery" could not be revealed, nor could theology render it in some way intelligible, were human knowledge limited strictly to the world of sense experience. Metaphysics thus plays an essential role of mediation in theological research. A theology without a metaphysical horizon could not move beyond an analysis of religious experience, nor would it allow the *intellectus fidei* to give a coherent account of the universal and transcendent value of revealed truth.[53]

In order to avoid equivocation, it seems necessary to establish a distinction between rational theology and revealed theology. *Rational theology*, traditionally called theodicy or the philosophy of God, is the science that studies, in the light of natural reason, what one can come to know about the Supreme Being through reason alone. *Revealed theology*, by contrast, has a material object (what it studies) and a formal object (the point of view that it adopts) different from those of rational theology; it is consequently a different science with a different epistemological status. Revealed theology studies the data of revelation in the light of reason as illuminated by faith. The material object partly coincides with that of rational theology, because it is the same God being studied, but is extended considerably to include everything that God has revealed to us about himself. Consequently, only those who have received the same faith can appropriately engage in theology.

---

[53] John Paul II, Encyclical Letter *Fides et ratio*, no. 83.

It is important to make clear that metaphysics and the rational philosophy of God have many points in common, because both of them manage to reach the ultimate foundation of reality: being. Having clarified this, it is necessary to add that the anthropology and ethics that I am proposing do not take reason illuminated by faith as their point of departure, since the resulting discourse would only be relevant to those who profess the same creed; rather, they take into account a whole range of rational philosophical findings, whether metaphysical or anthropological or ethical in nature. In my opinion, anyone who confuses ontology and ontologically grounded personalism with revealed theology shows that he has misunderstood the meaning of metaphysics itself and of theology.[54] Those who blockade themselves within the walls of an empiricist philosophy, which reduces man to his purely experiential aspects, exhibit intellectual prejudices against a large part of the philosophical tradition, from Plato to the present day, which considers man as body and as spirit. *Fides et ratio* mentions the important challenge "to move from *phenomenon* to *foundation*" (no. 83) and rediscover a way of thinking that is metaphysical in scope (see no. 81). John Paul II recalls that "reality and truth do transcend the factual and the empirical, and [I want] to vindicate the human being's capacity to know this transcendent and metaphysical dimension in a way that is true and certain, albeit imperfect and analogical" (no. 83).[55]

Every science manifests its own completeness within the parameters defined by the science itself. This does not prevent the sciences from being open to one another; in fact, interrelating the sciences—though each one preserves its distinct epistemological status—contributes to a richer understanding of the object of study. This occurs in much the same way that observing an object not only from the front, but also from the side, from within, and perhaps from above can lead to a thoroughly comprehensive understanding of the object, in keeping with an epistemological approach aimed at integration.[56]

Bioethics also has its own specificity that distinguishes it from the area of moral theology commonly referred to as "medical morals." This branch of moral theology, aimed at the formation of health care personnel, considers these interventions in the light of faith, and hence in the light of Christian Revelation, as specified by the Magisterium. Its purpose is to reflect on the content of the faith and on the application of divine law to human conduct; its applications are above all within the community of believers, though many of its conclusions happen to fully coincide with those of moral philosophy.

In my opinion, however, it would be inappropriate and not very useful for the faith itself to deny the legitimacy and necessity of rational and philosophical reflection on human life, and therefore on whether physicians and biologists are permitted to experiment on human subjects. Human life is first and foremost a natural value,

---

[54] See the article "Ragione" by A. Staglianò, in *Dizionario di scienza e fede*, Tanzella-Nitti and Strumia, 1167–1180, and the article "Metafisica" by A. Livi in the same volume, 939–957.

[55] See also the article by J. Ratzinger, "Fede e ragione," in the Italian edition of *L'Osservatore Romano* dated November 19, 1998, p. 8; J. Habermas and J. Ratzinger, *Ragione e fede in dialogo* (Venice: Marsilio, 2005).

[56] B. Lonergan, *Method in Theology* (London: Darton and Todd, 1972).

known rationally by all who make use of reason; the value of the human person is rendered even more precious by grace and by the gift of the Holy Spirit, but it remains an intangible value for everyone, believer or not. It is contrary to Church tradition to deny the value of reason and the legitimacy of rational ethics, also known as natural ethics.

In the debate over abortion, many people run the risk of supposing that it is a question of being a religious believer or not, yet human life is such for all people, and the obligation to respect it is a duty for man inasmuch as he is human, not just insofar as he is a believer. Supernatural reasons will reinforce this duty for a believer, but these reasons must not be used to excuse all people of good will and right reason from reflecting on the human facts in the light of reason.[57]

Over the centuries the Catholic Church herself has condemned any fideist position, which would deprive reason and intelligence of their weight and value, with the same vigor with which she has condemned any heresies in the realm of revealed truth. The Church has instead defended the principle of harmony between science and faith, between reason and Revelation; such a harmony is not always easy and immediate, however, whether because of the weakness of the human mind, ideological pressures, or the intrinsic difficulties of the problems.

This is a delicate yet essential point involving the relationship between man and God, natural and supernatural, and philosophy and theology. Reason and Revelation have the same author, who is God, and therefore merit equal respect and demand reciprocal support.[58] This encounter becomes all the more urgent and necessary the more one works within the field of the experimental sciences, which have this-worldly and corporeal realities as their object and avail themselves of a rational processes. It is

---

[57] "Respect for human life is not just a Christian obligation. Human reason is sufficient to impose it on the basis of the analysis of what a human person is and should be." Congregation for the Doctrine of the Faith (CDF), *Quaestio de abortu* [*Declaration on Procured Abortion*], November 18, 1974. With regard to the relationship between reason and faith, *Fides et ratio* is indispensable reading: it describes them as "two wings on which the human spirit rises to the contemplation of truth" (preamble). Also important is the study by R. Fisichella, "Da credente in difesa della ragione," in D. Antiseri, *Cristiano perché relativista, relativista perché cristiano: Per un razionalismo della contingenza* (Soveria Mannelli: Rubbettino, 2003), 133–153. See also the article "Fideismo" by P. Poupard in, *Dizionario di scienza e fede*, Tanzella-Nitti and Strumia.

[58] "If methodical investigation within every branch of learning is carried out in a genuinely scientific manner and in accord with moral norms, it never truly conflicts with faith, for earthly matters and the concerns of faith derive from the same God. ... We cannot but deplore certain habits of mind, which are sometimes found too among Christians, which do not sufficiently attend to the rightful independence of science and which, from the arguments and controversies they spark, lead many minds to conclude that faith and science are mutually opposed." Vatican Council II, Pastoral Constitution *Gaudium et spes*, December 7, 1965, no. 36. "This Sacred Synod, therefore, recalling the teaching of the first Vatican Council, declares that there are 'two orders of knowledge' which are distinct, namely faith and reason; and that the Church does not forbid that 'the human arts and disciplines use their own principles and their proper method, each in its own domain'; therefore 'acknowledging this just liberty,' this Sacred Synod affirms the legitimate autonomy of human culture and especially of the sciences." Ibid., no. 59.

ever more urgently needed in the wake of a the long period of silence from metaphysics, which abandoned human truth to the clutches of relativism and noncognitivism.

The dialogue between science and faith can take place only through the intermediary of reason, which is the common reference point for both. This led to the need for philosophical and moral reflection in the medical and biological fields as well.

On this topic, we must ask whether there can be a purely rational secular ethics, capable of prescinding from the affirmation of the existence of an Absolute, or whether, precisely in virtue of an ethics grounded rationally upon natural values, it might not be necessary to discover the existence of an Absolute within said values, especially in order to safeguard the value of the person. I share the position of those who affirm the legitimacy of a rational, "secular" ethics on the immediate phenomenological, epistemological, and ontological level; if, however, this discourse is impelled toward the ultimate and mediate metaphysical foundation, as is necessarily the case, then rationality itself must conclude in favor of the existence of the Absolute: God, the ultimate root of the transcendent value of the person and of ethical norms.[59]

This connection with rational ethics, which is based on metaphysics, reasonably affirms the existence of God by affirming the value of the person and sees in Christian revelation a suprarational—not irrational—knowledge, fostering dialogue between reason and Revelation, between science and faith.

In a cultural context characterized by a general "crisis of meaning," it is important to recover the "*sapiential dimension* as a search for the ultimate and overarching meaning of life" (*Fides et ratio*, no. 81). This wisdom-related dimension is indispensable because the immense growth of humanity's technological power demands a renewed consciousness of the ultimate values. If this technical know-how should become ordered to a merely utilitarian end, it could quickly prove to be inhuman and destructive.

In the field of scientific research, the prevailing positivist mentality ignores or even rules out all reference to a metaphysical or moral perspective. One consequence is that the person and the holistic understanding of his life have become unimportant. Some scientists, aware of the potential that is inherent in technological progress, seem to give in not only to the dynamics of the marketplace but also to the temptation of a godlike power over nature and over human beings themselves.

Alongside utilitarianism and rationalism there is also *nihilism*, which views research as an end in itself, with no hope or possibility of attaining the goal of truth: "In brief, there are signs of a widespread distrust of universal and absolute statements, especially among those who think that truth is born of consensus and not of a consonance between intellect and objective reality" (*Fides et ratio*, no. 56).

In this context, a personalist bioethics with an ontological foundation can contribute to the "sapiential dimension" that *Fides et ratio* evokes. Bioethics is therefore a discipline with a rational epistemological status that is open to theology, which is understood as a suprarational science, the ultimate authority, and the "horizon of

---

[59] See P. Valori, "Può esistere una morale laica?" *La Civiltà Cattolica* 3 (1984), 19–29. On the implicit affirmation of God in the ethical experience, see also the reflection by E. Levinas, *Etica ed infinito* (Rome: Città Nuova, 1984), and, in general, all the speculations of this Lithuanian philosopher, which differ, however, from the Thomistic approach to ethics on some important points.

meaning." Starting from a description of the scientific, biological, and medical data, bioethics rationally examines the liceity of man's interventions on man. This ethical reflection has its immediate point of reference in the human person and his transcendent value, and its ultimate point of reference in God, who is Absolute Value. It is both natural and necessary to investigate Christian revelation while pursuing this line of thought, and a comparison with current philosophical ideas will also be productive.

Summary Outline for Chapter 1

Origins of bioethics

Van Rensselaer Potter

Potter, an oncologist, introduces the term in two publications:
"Bioethics: The Science of Survival" (1970)
Bioethics: Bridge to the Future (1971)

A new discipline as the combination of "biological knowledge and human values."

"A science of survival must be more than science alone, and therefore I propose the term *bioethics* to emphasize the two most important ingredients for the attainment of a new wisdom."

Bioethics = a bridge between the two cultures
Bioethics = a look at the biosphere as a whole
Bioethics = biologically grounded wisdom (moral values should be identified in biology). Broader in scope than traditional medical ethics.

Hans Jonas

Philosopher who, somewhat like Potter, draws attention in his *Imperative of Responsibility* to the increased potential of new technologies and to the possible threats to the survival of mankind. The guiding principle for biotechnological interventions must be the prevention of catastrophes.

[bioecological current—"alarmism"]

André Hellegers

A young obstetrician, an expert in fetal physiology, who was asked in 1971 to direct the Kennedy Institute for the Study of Human Reproduction and Bioethics of Georgetown University.

Proposes a maieutic (Socratic) dimension of bioethics that understands values by way of the dialogue and contrast between medicine, philosophy and ethics (interdisciplinary method).

The object of bioethics = the ethical aspects implicit in clinical practice.

Bioethics = ethics applied to biomedicine (restricted in scope as compared to Potter). Maintained that the direct study of biological problems causes ethics to progress and that clinical bioethicists are more expert than pure moralists.

Dan Callahan

Catholic philosopher; as early as 1969, without using the term bioethics explicitly, he founds the Hastings Center with the psychiatrist Willard Gaylin.

Bioethics = intersection between ethics and life sciences.

[social humanism current]

Definitions and interpretations of bioethics

"The systematic study of human conduct in the area of the life sciences and health care, insofar as this conduct is examined in the light of moral values and principles." (W. T. Reich, *Encyclopedia of Bioethics*, 1978)

[bioethics of principles]

Potter's "global" concept is adopted in part:

"The systematic study of the moral dimensions—including the moral vision, decisions, conduct, and policies, etc.—of the life sciences and health care, employing a variety of ethical methodologies in an interdisciplinary setting." (Reich, *Encyclopedia of Bioethics*, 1995 and 2004)

[pluralistic bioethics that is also relativistic]

"This new discipline cannot be understood as a simple comparison of the different opinions and the various ethical positions that exist; rather, it must propose standard values and effective decision-making approaches, providing objective answers based on rationally valid criteria.

The search for adequate answers demands an interdisciplinary approach to the problem, which is one of the unique characteristics of bioethics."

"A discipline with a rational epistemological status that is open to theology, which is understood as a suprarational science, the ultimate authority, and the "horizon of meaning." Starting from a description of the scientific, biological, and medical data, bioethics rationally examines the liceity of man's interventions on man." (E. Sgreccia, *Personalist Bioethics*)

[ontological, realist, cognitive personalism]

Bioethics as the critical conscience of technological civilization

"From this perspective bioethics is configured as a philosophical activity, regardless of who is actually conducting the inquiry, because the questions (the formal object) encountered by the technological sciences (the material object) are by nature philosophical and concern the significance of how human identity is construed within the technological activity." (A. Pessina, *Bioetica: L'uomo sperimentale*)

[realism, cognitivism]

"Analytical" bioethics

Considers it impossible to establish the truth or falsity of value judgments. Bioethics can only clarify and point out the presuppositions of a specific choice.

"Philosophical bioethics consists essentially of a meta-bioethics. ... It is an attempt to clarify concepts and bring presuppositions to light. ... In conclusion, it must be interdisciplinary, a river with many tributaries flowing into it. The disciplines that are involved vary in nature, and this poses particularly complex problems concerning the linguistic, logical, and methodological relations among them. At the center we find once more the fundamental and decisive questions in contemporary ethics: the *is-ought question*, or questions of the logical relations between descriptive propositions of fact and normative propositions that determine duties and assign values." (U. Scarpelli, "Bioetica: Prospettive e principi fondamentali")

[noncognitivism, moral agnosticism]

"Secular" bioethics

"Founded on reason and on the values of the conscience."

Contrasted with Catholic bioethics, which is said to be founded on dogmas and faith. ("Manifesto di bioetica laica" [Manifesto of secular bioethics], published June 9, 1996, in *Il Sole 24 Ore* and signed by C. Flamini, A. Masserenti, M. Mori, and A. Petroni)

[The personalist, ontologically grounded approach that serves as a standard for Catholics is not fideistic and does not dispense with a rational justification of values and norms.]

Problems that contributed to the development of bioethics

Human experimentation
The discoveries of genetics
Organ transplantation
The beginning of life and human procreation
End-of-life issues

The first institutes of bioethics

1969: Founding of the Hasting Center (Hastings-on-Hudson, USA) by D. Callahan and W. Gaylin

1971: Founding of the Kennedy Institute of Ethics at Georgetown University (Washington, DC) by A. Hellegers (W. T. Reich, R. Veatch, R. McCormick)

Competencies of bioethics

The Erice Document—1991

Refers to the contents of the 1978 *Encyclopedia of Bioethics* and recognizes four areas in which bioethics has competence:

1. Ethical problems in the health care professions

2. Ethical problems arising in the field of human research, even if it is not directly therapeutic

3. Social problems connected with (national and international) public health policies, with health care jobs, and with family planning and population control policies

4. Problems related to interventions on the life of other living things (plants, microorganisms, and animals) and generally to anything having to do with the equilibrium of the ecosystem

Subdivision of bioethics

General bioethics: Ethical foundation, discussion of values and first principles, documentary sources of bioethics

Particular bioethics: Analyzes the major problems, always as part of a general approach, both in the medical field and in biology (genetic engineering, abortion, euthanasia, cloning, artificial fertilization, etc.)

Clinical bioethics: Application of ethical theories and accepted general principles to concrete clinical cases, seeking guidelines for action

Remote sources of bioethics

The Hippocratic writings and the Oath:
    "Do no harm" and paternalism
    Foundation of medicine on nonsubjective criteria

Christianity
    Foundation of the concept of person
    Theological significance of care (*Christus medicus et patiens*)

Recent sources of bioethics

The Nuremberg Code (1947): condemns all experimentation on human subjects without their consent

Alarm caused by several experiments in the USA:
- 1930s and 1940s, Tuskegee, Alabama: comparative study of a drug to treat syphilis versus a placebo on the colored population without informed consent
- 1962, Swedish Hospital, Seattle: a committee to decide on procedures for access to dialysis
- 1963, Jewish Chronic Disease Hospital in Brooklyn: tumor cells injected into elderly patients without their consent

- 1967–1971, Willowbrook State Hospital, New York: hepatitis virus in handi-capped children, with the parents' consent coerced

World Medical Association—Helsinki Declaration on clinical experimentation (1964–2008)

Documents of the Catholic Church on life and health

Pius XII

Address to a Symposium of the Italian Society of Anesthesiology (1957)

Paul VI

Humanae vitae (On the regulation of birth), Encyclical Letter 1968
Quaestio de abortu (On procured abortion), Declaration by the Congregation for the Doctrine of the Faith, 1974.

John Paul II

Iura et bona (On euthanasia), Declaration by the Congregation for the Doctrine of the Faith, 1980
Familiaris consortio, Apostolic Letter, 1981
Donum vitae, Instruction by the Congregation for the Doctrine of the Faith, 1987
Evangelium vitae, Encyclical Letter 1995

Theoretical presuppositions of ontologically grounded, personalist bioethics (having a "sapiential dimension")

The dignity of the human person
Unitotality of body and spirit and, from a faith perspective, image of God

Realism, cognitivism
The human person has the rational ability to know reality and the structure of values (denial of relativism and nihilism).

A metaphysical view of reality
The human intellect is able "to move *from phenomenon to foundation*" and to understand what ought to be from what is; it is necessary "to vindicate the human being's capacity to know this transcendent and metaphysical dimension in a way that is true and certain, albeit imperfect and analogical" (*Fides et ratio*, 83).

34

Chapter 2

EPISTEMOLOGICAL JUSTIFICATION, THE FOUNDATION OF BIOETHICAL JUDGMENTS, AND RESEARCH METHODOLOGY IN BIOETHICS

The Epistemological Justification of Bioethics

It is evident from the preceding chapter that bioethics does in fact exist. It is an attempt to reflect systematically on all of man's interventions that affect living beings, and it sets for itself the specific and arduous objective of identifying values and norms to guide human action and the interventions of science and technology on life itself and the biosphere.

Technological activity ought to be redefined so as to overcome the discrepancy between *purposeful meaning* and *technical meaning*, because there is no longer any corresponding intrinsic purposefulness to all the individual, technical meanings disseminated by technology. Bioethics ought to acquire the tools for reunifying what technology has divided.[1]

This is now the decisive question: Does this study have its own specific place in the panorama of the sciences? In other words, does it have its own reason for being that makes it indispensable? Does it have its own set of criteria on which to base its judgments? Does it have its own method of conducting research? Are we not dealing perhaps with a hodgepodge, a sort of "cocktail" prepared with ingredients from other sciences (biology, philosophy, medicine, ethics, etc.) without a precise identity or need for it?[2]

The question is clearly relevant and can be formulated more concretely into three distinct questions: first, the question of the role of and the epistemological justification for bioethics; second, the question of the foundation for its ethical judgments; and third, the question of its method.

Let us first address the inquiry concerning the epistemological justification for bioethics. Jean Bernard, the famous French hematologist and first president of the National Ethics Advisory Committee for Life and Health Sciences (Comité Consultatif National d'Ethique pour les Sciences de la Vie et de la Santé, CCNE) in France, has traced the history of the advances and discoveries in the field of biomedicine

---

[1] See F. D'Agostino, *Bioetica nella prospettiva della filosofia del diritto* (Turin: Giappichelli, 1997), 313–317.

[2] Objections of this sort have been made also in scholarly publications such as the *Rivista Italiana di Medicina Legale*. See, for example, G. Canepa, "Bioetica e deontologia medica: aspetti problematici e conflittuali," *Rivista italiana di medicina legale* 1 (1990): 3–6.

from 1930 to 1990. He talks about two major revolutions: the medical revolution and the biological revolution.[3]

The first revolution, after millennia of helplessness, came with the discovery of sulfa drugs (1932) and penicillin (1929), which conferred on mankind "the power to combat diseases which for a long time had been fatal, such as tuberculosis, syphilis, the major forms of sepsis, diseases of the endocrine glands and biochemical metabolic disorders."[4]

The second revolution came about two decades later. It started with the discovery of the genetic code and gave rise to so-called genomic medicine, which proceeded from the discovery of the laws that govern the formation of life.

The scientist goes on to remark that these discoveries revolutionized not only medicine but also our understanding of life and of man; consequently, they have shaken out of hibernation our reflections on the very destiny of humanity.

Bernard also recalls the great victories of medicine in those years: the development of methods for experimentation with new prescription drugs on healthy volunteers and sick patients; the transplantation of organs and tissues; the use of resuscitation techniques; the ongoing fight against tumors by means of chemotherapy, surgery, and radiation; and the deployment of vaccines on a large scale to combat major epidemics (typhus, smallpox, poliomyelitis, etc.), including the front against AIDS.

These discoveries have had an impact on medical ethics, stimulating it to develop in new and important areas. The more powerful and effective medicine becomes, the more rigorous and well-publicized the norms for protecting the individual must be.

As mentioned in the preceding chapter, the major international medical organizations have become the mouthpieces of these new norms, and international codes of ethics[5] and guidelines testify to them, including those developed in the Islamic world.[6]

This vein of therapeutic medicine has represented thus far the classical school and, I would say, the Hippocratic model, strengthened by the experimental method and protected by ethical codes and medical ethics in general.

With the advent of scientific discoveries in the field of genetics, and as a result of the application of these new findings to the areas of embryology and gynecology, which led to the advent of artificial procreation, medical science has entered a new phase. Its developments are not entirely predictable and often lack consistent ethical guidelines both for the conscience and in codes.

The initial appearance of genetic engineering, i.e., the possibility of transferring segments or portions of the genetic code from one cell to another (even between different species) by means of a twofold mechanism involving a *restriction endonuclease* and *recombinant DNA* (deoxyribonucleic acid), understandably created an alarming situation. The ability to engineer various life forms became a glimpse at the possibil-

---

[3] J. Bernard, *De la biologiè à l'éthique: Nouveaux pouvoirs de la science, nouveaux pouvoirs de l'homme* (Paris: Buchet Chastel, 1990).

[4] Ibid., 22.

[5] S. Spinsanti, ed., *Documenti di deontologia e etica medica* (Milan: Paoline, 1985), 37–43.

[6] Ibid., 166–188.

ity of creating a "biological bomb" that would be much less expensive than a nuclear bomb and more difficult to monitor. All these considerations also awakened the fear of potential alterations to the biosphere and the ecosystem through human agency.

And so there were calls for a new ethics so as to avoid what was described as a possible catastrophe for all mankind, an ethics of the whole biosphere that would be able to derive norms from within biological evolution itself. It was at this moment that bioethics arose; it was first mentioned by name in the work of Van Rensselear Potter.[7] Evidence of the same need became manifest in Germany through the afore-mentioned work by Hans Jonas, *The Imperative of Responsibility*.[8]

There was a fear of catastrophe and a perceived need for a moratorium and a universal system of scientific norms within the research community itself, to which papers presented at the Gordon (1973) and Asilomar (1975) conferences attest.[9] These lectures prompted the implementation of the first scientific-ethical oversight committees and the development of the first guidelines concerning experimentation on DNA. These guidelines were then adopted by various organizations throughout the world.[10]

Very soon, however, people began to look with greater realism and discernment at these new possibilities and technologies. As for genetic engineering as such, catastrophic fears aside, the possibility of performing gene therapy was quickly foreseen; it can now be readily performed on somatic cells. Its application to cells in the germ cell line, on the other hand, is very controversial because of the inherent risks of the procedure.[11]

Very soon after, industrial-type applications arose for the manufacture of new pharmaceuticals (insulin, interferon, etc.) and for increased agricultural and livestock production. The new discoveries in the genetic field have led to the more extensive use of prenatal and postnatal diagnostic techniques—a situation that is increasingly delicate and always fraught with ethical problems.

There has been talk for several years now about *genomic medicine* and *predictive medicine*, especially since the findings of the Human Genome Project and the development of *gene sequencing*, i.e., the reading of the human genetic code and hence of the structure of any individual's hereditary information and genetic patrimony. This achievement was a major initial conquest (and others continue to follow) in identifying the functions of individual genes or gene groups and of the millions of proteins that are encoded therein. The era of genomics gives way to the era of *proteomics* and *functional genomics*, which will enable researchers to pursue

---

[7] V. R. Potter, *Bioethics: Bridge to the Future* (Englewood Cliffs, NJ: Prentice Hall, 1971).

[8] H. Jonas, *The Imperative of Responsibility: In Search of an Ethics for the Technological Age* (Chicago: University of Chicago Press, 1984); originally published as *Das Prinzip Verantwortung* (Frankfurt am Main: Insel, 1979).

[9] R. Dulbecco, *Ingegneri della vita* (Milan: Sperling & Kupfer, 1988); E. Sgreccia, "Storia della bioetica e sua giustificazione epistemologic," in *La storia della medicina nella società e nella cultura contemporanea: Atti del convegno internazionale di studio all'Istituto di Studi Politici "S. Pio V"; Frascati, 21–30 giurno 1991* (Rome: Apes, 1992), 69–84.

[10] E. Sgreccia and V. Mele, eds., *Ingegneria genetica e biotecnologie nel futuro dell'uomo* (Milan: Vita e Pensiero, 1991).

[11] Concerning this argument, see the chapter on genetic engineering in this volume.

therapeutic goals more effectively but will also open the door to knowing the intimate secrets of the hereditary makeup of every person and family.

In the field of genetic engineering as such, possibilities for positive applications were also identified alongside the catastrophic, feared ones. The ethical problem therefore became important and apparent to researchers, politicians, and industrialists: finding a way to safeguard the applications of genetic engineering through which the genetic patrimony can be treated yet not altered, maintaining that equilibrium in the ecosystem, especially at the microbiological level, which is compatible with the health of present and future generations of mankind.[12] These concerns have been discussed by European legislative bodies and national bioethics committees.[13]

More serious fears have remained concerning the other major topic area, procreation, in which the frontiers advance ever farther and what is at stake is not only the lives of the artificially produced embryos but also the understanding of parenthood—of fatherhood and motherhood—and the purpose itself of human sexuality. The real possibility that these trends will end in selective eugenics, experimentation on human embryos, and the commodification of the human body and of human reproduction is a recognized fact that is dreaded by many.

It has been remarked, correctly, that by taking this route experimental science runs the risk of assuming the epistemological status of politics, which is to say that it would become "the art of the possible," the quest to do everything that is possible, no longer simply pursuing knowledge about reality.[14]

Thus it becomes indispensable to pose a preliminary question about the relation between science and ethics and to define what sort of contribution bioethics ought to provide to the biomedical sciences.

The Relationship between Biomedical Science and Bioethics

Robert Nozick's affirmation that "microscopes and telescopes reveal no ethical facts"[15] is certainly provocative, and Renato Dulbecco recalls that "for centuries scientists have kept themselves outside of the tragedies of history, defending the autonomy and neutrality of their role in society. With Baconian and Cartesian pride

---

[12] Sgreccia and Mele, *Ingegneria genetica e biotecnologie*, 131–166.

[13] See U.S. President's Commission for the Study of Ethical Problems in Medicine and Biomedical Research, *Summing Up: Final Report on Studies of Ethical and Legal Problems in Medicine and Biomedical and Behavioral Research* (Washington, DC: US Government Printing Office, 1983); European Parliament, "Resolution on the Ethical and Legal Problems of Genetic Engineering," doc. A0327-88, March 16, 1990, in *Official Journal of the European Communities* C 96 (1989): 165–171, available at http://codex.vr.se/texts/EP-genetic.html. For more extensive bibliography and documentation, see A. Serra, E. Sgreccia and M. L. Di Pietro, *Nuova genetica ed embriopoiesi umana* (Milan: Vita e Pensiero, 1990), 311–318.

[14] M. Serres, preface to J. Testart, *L'oeuf transparent* (Paris: Flammarion, 1986). F. D'Agostino considers it unproductive that bioethics should turn into a "rhetoric of impotence, imploring those with a capability (technology) not to use it." *Bioetica nella prospettiva della filosofia del diritto* (Turin: Giappichelli, 1997), 315.

[15] R. Nozick, *Philosophical Explanations* (Cambridge, MA: Belknap Press of Harvard University Press, 1981), 399.

they have rejected any claim to supervision or interference from any quarter whatsoever, whether government, Church, or civil authorities."[16]

In modern times, however, this mentality is not shared by the large majority of researchers themselves who work in the biomedical sciences, including Dulbecco; in fact, they are the first to pose the bioethical questions. What is at issue is the reason for bioethics and the precise scope of its epistemological foundation.

In this regard, it should be recalled that biology and medicine are experimental sciences because they follow a very precise method, the *experimental method*, which was proposed by Galileo Galilei and Sir Francis Bacon and continually perfected by scientists over the centuries.

This well-known method is based on a precise sequence of steps: the observation of phenomena, an interpretive hypothesis, experimental verification, and the evaluation of the experimental results. This methodological process has an inherent validity that allows for the organic and linear accumulation of findings: the experimenter who follows this sequence of steps can benefit from the results—whether positive or negative—obtained by the preceding experimenter and, in turn, make new contributions by using the same methodology. It is true, however, that science sometimes makes use of chance observations (such as Fleming's discovery of penicillin) or the repetition of natural phenomena, but what has allowed modern science to progress in the biomedical field has been above all the method based on the paradigm of methodologically defined experimentation.

The experimental method nevertheless has an inherent limit: it must perforce base itself on facts and data of a quantitative sort, which are capable of being observed, calculated, compared, etc. The experimental method is therefore *reductionist by definition*, and this fact must be taken into account.

Keeping this element in mind, let us ask ourselves: For what motive and by what necessity does the ethical question arise in the context of the biomedical sciences, which are experimental? Many have given what appears to be the most obvious answer: the need for bioethical reflection arises at the moment of application; hence it is supposed that experimental research per se is neutral, whereas applying the findings thereof would first require a bioethical assessment of the consequences and risks. This statement is true because no one can deny that, before proceeding to the application of a scientific discovery in the biomedical field such as recombinant DNA technology, it is necessary to address a series of bioethical questions concerning its purpose, consequences, risks, and so forth. Nevertheless, acknowledging the role of bioethics and its justification only at the moment of application would be insufficient and restrictive, as will be seen more clearly.

Other scholars also generally allow for a form of ethics as part of scientific research, but only in the sense of fidelity to the rules of research. Such an ethics would therefore play out in a scrupulous adherence to the methodology, exactitude in communicating the results, procedural transparency so the research could be checked and confirmed by the scientific world. This "intrinsic ethics" in research constitutes a valid professional ethics requirement for any type of science and, hence, also for bioethics, which pertains to biomedical research.[17]

---

[16] Dulbecco, *Ingegneri della vita*, 13–14.

[17] E. Sgreccia, "La risposta nella trascendenza," in *Scienza ed etica: Quali limiti?*, ed. J. Jacobelli (Rome: Laterza, 1990), 163–173; R.K. Merton, "Priorities in Scientific

But the ethics proper to scientific research in the field of the experimental sciences—and specifically bioethics in the field of the biomedical sciences—cannot be limited to such codes of procedural correctness. Indeed, one must learn to distinguish what is necessarily required for an action to be ethical from what is sufficient to judge it fully ethical. For example, professional ethics necessarily requires that a surgeon be quite capable of preparing for an intervention and precisely executing the technical steps of the operation, but this is not sufficient reason to say that his intervention is ethical in every respect and as a whole (e.g., in terms of the validity of the indications, obtainment of consent, and respect for the higher goods of the person). An organ transplant, for example, must satisfy many such ethical conditions in order to be performed.

Therefore, in addition to these two connections that undoubtedly exist between scientific research and ethics (one on the level of the application and one related to the researcher's professional adherence to the methodological procedures of the research itself), there are others that are no less important. First and foremost is the *researcher's intention*. The researcher, the research organizers, and the research sponsors are all human persons and can have intentions that are good, bad, or simply utilitarian. The coordination and foundation of the research are always planned and reveal or conceal a strategic purpose, which could be aimed at developing a treatment for an illness or increasing agricultural, industrial, or pharmacological production; it could also be aimed at manipulating or altering biological processes, as in the case of an experimental attempt to create chimeras (interspecies hybrids), altering a patient's genetic makeup.

This sort of *ethics in planning*, or lack thereof, besides being relevant in and of itself, has considerable implications for those who collaborate in a subordinate way. They have the right to know the purpose of the project and the right and duty of conscientious objection if they do not feel in good conscience that they can closely collaborate in a project they consider illicit. Neither scientific nor industrial confidentiality can take this right away from someone who is cooperating immediately in a project that is evil in itself or deliberately deviates from ethical norms. One can imagine many situations of this sort in the field of bioethics. One example is conducting research on an abortifacient drug, such as the RU-486 pill, which has unfortunately been put on the market.

Another connection between research and ethics is concerned with *experimental procedures*: this is the ethics—or more precisely, the bioethics—of biomedical experimentation, with all the problems related to experimentation on human subjects (consent, risks, experimentation on children, the mentally ill, the unconscious, fetuses, etc.) and even on animals.

Discovery," *American Sociological Review* 22 (1966): 235–259; P. Rossi, *La nascita della scienza moderna in Europa* (Bari: Laterza, 1997); C. Mitcham, "Philosophy of Technology," in *Encyclopedia of Bioethics*, ed. W. T. Reich (New York: Macmillan, 1995), 5:2477–2483; E. Agazzi, *Il bene, il male, la scienza* (Milan: Rusconi, 1992); E. Sgreccia, "Potenzialità e limiti del progresso scientifico e tecnologico," *Dolentium Hominum* 37, no. 1 (1998): 137–144; J. de D. Vial Correa and E. Sgreccia, eds., *Etica della ricerca biomedica: Per una visione cristiana; Atti della nona assemblea generale della Pontificia Accademia per la Vita; Città del Vaticano, 24–26 febbraio 2003* (Vatican City: Libreria Editrice Vaticana, 2004).

Indeed, it is not enough to have an ethics of ends; logical consistency also requires an ethics of means and methods. Even when the ends are good—for example, giving a child to a sterile couple—the procedures chosen can sometimes be illicit; they could be harmful to human life and dignity (e.g., the loss of "surplus" embryos). According to the well-known principle "*non sunt facienda mala ut veniant bona*," evil may not be done so that good may come of it.

Yet the more profound connection between research and ethics, in my opinion, which includes all of the preceding connections dealing with the operative aspect (ends, procedures, methods, and risks), is in a *holistic requirement*.

It was recalled that the experimental method by its very nature affords a reductive view of reality, inasmuch as it considers only the testable and quantitative aspect of it, whereas the deeper, more comprehensive aspects, including the ontological nature and axiological value of reality, elude the procedures of the experimental method. The ethical question arises from within the experimental methodology but seeks to transcend and interpret it coherently so as to embrace the complexity, profundity, and value of reality.[18]

If a scientist, for example, is conducting research on a human embryo, he cannot just observe the requirements and the aspects of procedural ethics governing methodological correctness or investigate the possible applications for his findings; rather, he must ask himself what the human embryo is, whether it is a human being, and whether or not it has the value of a human person.

It is the answer to this question that clarifies all other bioethical answers: when the entire depth of a reality has been assessed, then the ethical requirements concerning the ends, means, risks, and so forth can be understood. This aspect was grasped lucidly by Karl Jaspers, a scientist and philosopher, in his affirmation that experimental science is capable of knowing and perceiving neither the qualitative aspect of a reality nor its nature in terms of its deeper value; neither can science, by itself, through the experimental method, explain the ends of science and research, because all of this requires determining the ends of human activity and of human life itself.[19]

More recent epistemologists as well, such as Karl Popper and John Eccles, have underscored this limitation of experimental science in relation to its own methodological procedures, as well as with respect to its more global observation of reality.[20] Therefore the link between science and ethics—or better, between scientific research and ethics—is not an optional matter or a current fad; rather, it is a multifarious requirement springing from within scientific procedures themselves.

Of course, as alluded to earlier, while the question arises from within the research, the answer requires integrating the experimental aspect with reality as a

---

[18] J. Ladrière, *I rischi della razionalità* (Turin: SEI, 1978); E. Agazzi, *Il bene, il male e la scienza*; E. Sgreccia, "Il progresso scientifico-tecnologico di fronte all'etica," *Medicina e Morale* 33, no. 4 (1983): 335–342; A. Bompiani, *Bioetica in Italia: Lineamenti e tendenze* (Bologna: Dehoniane, 1992), 187–220; Sgreccia, "Potenzialità e limiti."

[19] K. Jaspers, *Der Arzt im technischen Zeitalter* (Munich: R. Piper, 1986); G. Reale and D. Antiseri, *Il pensiero occidentale dalle origini ad oggi* (Brescia: La Scuola, 1983), 3:457–462.

[20] Reale and Antiseri, *Il pensiero occidentale*, 3:707–779; D. Antiseri, *Trattato di metodologia delle scienze sociali* (Turin: UTET, 1996), esp. 220–237.

whole and therefore within the ontological and axiological perspective of the living being on which the research is performed. It therefore becomes necessary to develop criteria by which to judge; these cannot be defined exhaustively by scientific research itself but must be deduced from the final understanding and overall meaning of the reality under consideration.

To return to the previous example, if experimentation is conducted on an embryo, regardless of whether the purpose is therapeutic, one must ask first and foremost what the overall reality of the human embryo is (ontology) and what its value is (axiology). Once it has been convincingly concluded, for example, that we are dealing with a human being, with a human individual, then we must raise the question of the significance of experimentation on an individual human being and specify the researcher's obligations more precisely, as in the case of a child who is a minor.

In order to decide on the basis of these criteria, it is consequently necessary to clarify who man is, what his value is, and what his purpose is. And when speaking about man as such, about his origin and purpose, one is seeking that which is shared by every human being: dignity and transcendence.[21]

In conclusion, applying what has been said about the relation between science and ethics, and consequently about the relation between the biomedical sciences and bioethics, it can be stated that the justification for bioethics pertains not only to the moment when research becomes applied science, but also to the moment and method of the research itself. Bioethics encounters biomedical research with a holistic view.

If medicine is then considered from the perspective of health care and health care organizations, the ethical factor becomes even more relevant and the integration of scientific knowledge with ethical norms of conduct becomes obvious. The studies produced in recent years in the field of medical anthropology[22] have made quite evident the radical insufficiency of the unilaterally scientific approach to the very concepts of sickness, health, prevention, and so forth. The impact of the "personal," psychological, and spiritual factors throughout the health care field has become decisive not only in evaluating the well-being of the sick person but also in evaluating health care workers.

As far as the organization of health care is concerned, it is now well known that it presupposes health education and the collaboration of citizens. For those in authority, the organization of health care requires the concept of justice as its guiding criterion in allocating resources and determining what structures and services to offer. Economic ethics and health care ethics meet on the social level in one of the most important sectors of public spending in advanced democracies.[23]

---

[21] S. Vanni Rovighi, *Elementi di filosofia* (Brescia: La Scuola, 1963), 3:189–269.

[22] L. Delgado, *Antropologia medica* (Milan: Paoline, 1991); Jaspers, *Der Arzt*; J. Vedrinne, "Éthique et professions de santé," *Médecine et Hygiène* 11 (1984): 1171–1173; M. Vidal, "Etica de la actividad scientífico-técnica," *Moralia* 4 (1983): 419–443; L. Villa, *Medicina oggi: Aspetti di ordine scientifico, filosofico, etico-sociale* (Padua: Piccin, 1980); E. D. Pellegrino and D. C. Thomasma, *For the Patient's Good: The Restoration of Beneficence in Health Care* (New York: Oxford University Press, 1988).

[23] D. Gracia, *Fundamentos de bioética* (Madrid: Eudema Universidad, 1989), 199–311.

Bioethical Models and Questions of Meta-ethics

Bioethical models

Although the need for ethics in relation to the life sciences is almost universal, there are different ways of formulating ethical reference models and theories concerning the foundation of ethical judgments. The diversity is so great that some thinkers, such as Uberto Scarpelli, declare that in bioethics one should establish only formal rules based on the principle of tolerance of any ethical system or, alternatively, one could propose an additional principle of "the absence of considerable harm."[24]

In effect, anyone who surveys the panorama of bioethics will readily conclude that we are dealing here with a plurality—indeed, a pluralism—of systems of criteria that are difficult to reconcile with one another. This pluralism concerns both the anthropology chosen as a reference point and the various theories on the foundation of ethical judgments.

The pluralism and diversity of approaches to the foundation of bioethics are particularly evident in the English-language literature. For a long time the prevailing approach was so-called *principlism*, based on the application of the principles of beneficence, nonmaleficence, and justice; however, this approach has since come under heavy criticism from various camps. Other approaches have therefore been emerging gradually: virtue ethics, casuistic ethics, narrative ethics, interpretive or hermeneutic bioethics, care-taking ethics, and feminist bioethics.[25]

The discussion *about* bioethics, above and beyond the examination of individual problems *of* bioethics, has thus become a central concern so as to determine what values and principles can serve as the foundation for ethical judgments and the justification for the distinction between *licit* and *illicit*.[26]

It is necessary to add that, in such foundational discourse, it is not sufficient to elaborate conceptual paradigms based simply on a sort of pragmatic and flexible consensus that can be constantly adjusted, according to the circumstances, to solve extreme cases. One must instead insist on a true justification and, therefore, on the demonstration of the ultimate reason why a specific moral act should be considered right or wrong, licit or illicit, obligatory or prohibited.

In this context, the term *meta-ethics* is meant as precisely this type of foundational justification. It refers to the rational justification of the values, principles, and

---

[24] U. Scarpelli, "La bioetica: Alla ricerca dei principi," *Biblioteca della Libertà* 99 (1987): 7–32.

[25] The following works are pertinent: M. A. Grodin, ed., *Meta-Medical Ethics: The Philosophical Foundation of Bioethics* (Dordrecht: Kluwer, 1995); K. D. Clouser and B. Gert, "A Critique of Principlism," *Journal of Medicine and Philosophy* 15 (1990): 219–236; E. D. Pellegrino and D. C. Thomasma, *Virtues in Medical Practice* (New York, 1993); A. Jonsen, "Casuistry as Methodology and Clinical Ethics," *Theoretical Medicine* 12 (1991): 295–307; A. Carse, "The Voice of Care: Implications for Bioethical Education," *Journal of Medicine and Philosophy* 16 (1991): 5–28; S. Sherwin, *No Longer Patient: Feminist Ethics in Health Care* (Philadelphia: Temple University Press, 1992); V. Mele, *Bioetica al femminile* (Milan: Vita e Pensiero, 1998).

[26] Gracia, *Fundamentos de bioética*, 315–388; L. Palazzani and E. Sgreccia, "Il dibattito attuale sulla fondazione etica in bioetica," *Medicina e Morale* 42, no. 5 (1992): 847–870.

norms of bioethics. And meta-bioethics is built upon meta-ethics. "Meta-bioethics cannot limit itself to prescribing arbitrarily certain procedures, nor can it limit itself to elaborating a conceptual system in terms of practical requirements, although it should offer indications and guidelines in the 'strong' sense, striving to give a reasonable explanation for the prescriptive-axiological choice addressed to health care workers, to scientists, and also to mankind itself in relation to acts and interventions upon physical life."[27]

Neither can the proposal of "indifference" toward any reference system whatsoever be accepted under the pretext of tolerance, especially if we reflect on the human and social importance of bioethical issues. When it is a question of determining whether to eliminate or defend a fetus in the mother's womb, of legalizing the extermination of all unwanted preborn babies beneath a certain gestational age, of killing deformed newborns, of the "survival problems" of the human species, or of preserving the ecosystem for future generations, there is no escaping the moral duty—specifically the duty of ethical experts—to pursue valid guidelines grounded in reason, which can therefore be shared or at least debated responsibly.

The fact that there are several reference systems should not be a pretext for avoiding this search, but rather a challenge to theoretical and pedagogical efforts.

Cognitivism and noncognitivism: Hume's law

As a preliminary it is important to clarify a point of the debate that has become a sort of crossroads for all ethical discussions today. It is called Hume's law, and it is responsible for the basic alignment of ethicists and bioethicists into two opposing groups, noncognitivists and cognitivists. The law is derived from an observation found in David Hume's *Treatise of Human Nature*, and that observation was taken up by contemporary analytical philosophy starting with G. E. Moore, who defined it as the *naturalistic fallacy*.[28]

Hume's law states that there is "a great divide" between the two realms of natural facts and moral values: facts are knowable, can be described with verbs in the indicative mood, and can be proved scientifically, whereas values and moral norms are simply suppositions giving rise to prescriptive judgments that cannot be proved. It is therefore neither possible nor legitimate to pass or make an inference from being (where being is identified with observable facts) to what ought to be, from "is" to "ought," or from *Sein* (to be) to *Sollen* (to have to).

The noncognitivists maintain that values cannot be the object of knowledge or of statements that can be described in terms of "true" or "false." On the contrary, cognitivists seek a rational and "objective" foundation for values and moral norms. Justifying ethics (and hence bioethics) therefore means discussing first and foremost the possibility of overcoming this great divide or naturalistic fallacy.[29] Let us more closely examine Hume's thinking on this point.

---

[27] Palazzani and Sgreccia, "Il dibattito attuale," 849.

[28] Ibid.

[29] P. Zecchinato, *Giustificare la morale* (Trent: Verifiche, 1990). Concerning cognitivism and Hume's law, see G. Carcaterra, *Il problema della fallacia naturalistica: La derivazione del dover essere dall'essere* (Milan: Giuffré, 1969); U. Scarpelli, *Etica senza verità* (Bologna: Il Mulino, 1982); F. E. Oppenheim, "Non cognitivismo, razionalità e relativismo," *Rivista di*

Hume's law is the "golden rule" that declares the impossibility of deducing any duties from simple facts:

> In every system of morality, which I have hitherto met with, I have always remark'd, that the author proceeds for some time in the ordinary way of reasoning, and establishes the being of a God, or makes observations concerning human affairs; when of a sudden I am surpriz'd to find, that instead of the usual copulations of propositions, *is*, and *is not*, I meet with no proposition that is not connected with an *ought*, or an *ought not*. This change is imperceptible; but is, however, of the last [= utmost] consequence. For as this *ought*, or *ought not*, expresses some new relation or affirmation, 'tis necessary that it shou'd be observ'd and explain'd; and at the same time that a reason should be given, for what seems altogether inconceivable, that this new relation can be a deduction from others, which are entirely different from it.[30]

This law limits itself to stating that it is not legitimate to derive a norm (and hence an imperative, an *ought to be*) from a fact. In order to lead to noncognitivism, however, Hume's law necessarily presupposes another thesis, which historically was expressed first by empiricism and then by neoempiricism: only descriptive statements can be true or false. This is because the truth or falsity of a descriptive statement can be verified, whereas a normative statement expresses something that does not yet exist—it should come to be, hence there can be no verification whatsoever.

The theoretical supposition that legitimizes the great divide between "is" and "ought" is therefore the epistemological premise underlying Moritz Schlick's so-called *verification principle*, whereby a proposition is meaningful only when one can specify the method suited to verifying its significance. In this context of epistemological reflection today, this perspective has actually been abandoned in favor of a falsificationist perspective. The criterion of falsifiability is based on the

---

filosofia 78 (1987): 17–29; K. Baier, *The Rational and the Moral Order* (Chicago: Open Court, 1995); D. O. Brink, *Il realismo morale ed i fondamenti dell'etica*, ed. F. Castellani and A. Corradini (Milan: Vita e Pensiero, 2005); L. Ceri and S. F. Magni, eds., *Le Ragioni dell'Etica* (Pisa: Ets, 2004); F. D'Agostino, "Il diritto naturale e la fallacia naturalistica," in *Filosofia del diritto* (Turin: Giappichelli, 1996); W. Frankena, "The Naturalistic Fallacy," *Mind* 48 (1939): 464–467; R. Egidi, M. Dell'Utri, and M. De Caro, eds., *Normatività Fatti Valori* (Macerata: Quodlibet, 2003); W. Frankena, *Ethics* (Englewood Cliffs: Prentice Hall, 1973); S. Hampshire, "Fallacies in Moral Philosophy" (1949), reprinted in *Freedom of Mind* (Oxford: Clarendon Press, 1972); E. Lecaldano, *Hume e la nascita dell'etica contemporanea* (Rome-Bari: Laterza, 1991); E. Lecaldano, "'Grande Divisione,' 'legge di Hume' e ragionamento in morale," *Rivista di Filosofia* 67 (1976): 82; H. Putnam, *The Collapse of the Fact/Value Dichotomy and Other Essays* (Cambridge, MA: Harvard University Press, 2002); C. Stevenson, *Facts and Values* (New Haven, CT: Yale University Press, 1963); S. E. Toulmin, *An Examination of the Place of Reason in Ethics* (Cambridge: Cambridge University Press, 1950); G. H. von Wright, "Is and Ought," in *Man, Law and Modern Forms of Life*, ed. E. Bulygin, J. L. Gardies, and I. Niiniluoto (Dordrecht, 1985), 263–282; L. Wittgenstein, *Tractatus Logico-Philosophicus*, German text with new translation by D. F. Pears and B. F. McGuinness, 2nd ed. (London: Routledge & Kegan Paul, 1963); idem, "Conferenza sull'etica," in *Lezioni e conversazioni sull'etica, l'estetica, la psicologia e la credenza religiosa* (Milan: Adelphi, 1967).

[30] D. Hume, *A Treatise of Human Nature*, edited with analytical index by L. A. Selby-Bigge, 2nd ed. (Oxford: Clarendon Press, 1978), book III, part I, section I, p. 469.

logical asymmetry between confirming and disproving a theory, because refuting a hypothesis is logically conclusive, whereas confirming it is not: *"An infinite number of confirmations of a theory do not make that theory certain, whereas a single refutation (provided that there are no reasons to doubt the protocols and the supporting hypotheses) makes the theory logically false."*[31]

The passage from *is* to *ought* is illegitimate if being is understood as something static, analogous to mathematical being, and if one has a mechanistic and reductive understanding of the universe. Yet if we emerge from this bottleneck by means of a metaphysical perspective (see the encyclical *Fides et ratio* by John Paul II), we can grasp the *essences* and the *logos* of reality, that is, its *teleological orientation*: "We *find* the values of things … their nature. And we find an objective order of values in that the dignity of man is higher than that of a flower or of a worm."[32]

It is only by stopping at facts and fragments, without recovering the whole, that one can say, as Dario Antiseri does, that *"information does not produce imperatives. Therefore it is not logically possible to pass from what is to what ought to be. This, in brief, is Hume's law, the great divide between indicative statements and prescriptive statements, between facts and values. This law is a death knell to the natural law and tells us that values are not founded upon science; their foundation rests upon our choices in conscience."*[33] Antiseri explains that *"Hume's law is the logical basis for freedom of conscience. The highest values are subject to choices in conscience: they are neither 'proven' theorems nor 'self-evident' and 'self-grounding' axioms."*[34]

Hume's law was reformulated in the twentieth century by other authors, such as Henri Poincaré, who writes, "There can be no scientific morality, but neither can there be immoral science. And the reason for this is simple. There is a reason—how to put it?—that is purely grammatical. If the premises of a syllogism are both in the indicative, the conclusion will also be in the indicative. … The moral motor, that which can set into motion the whole apparatus of pistons and gears, cannot be anything but a sentiment."[35]

Other formulations of significance include those by Ludwig Wittgenstein (in his *Tractatus*), Rudolf Carnap, R. M. Hare, and Popper, who writes that "it is impossible to derive a sentence stating a norm or a decision, or, say, a proposal for a policy, from a sentence stating a fact; this is only another way of saying that it is impossible to derive norms or decisions or proposals from facts. … Ethics is not science."[36]

Moore's thinking on the naturalistic fallacy, often employed by noncognitivists, is different from Hume's law and opens the way to intuitionism: the good is something

---

[31] D. Antiseri, *Trattato di metodologia delle scienze sociali* (Turin: Utet, 1996), 63–64.

[32] J. Seifert, *The Philosophical Diseases of Medicine and Their Cure: Philosophy and Ethics of Medicine* (Dordrecht: Springer, 2004), 1:301, original emphasis.

[33] D. Antiseri, *Cristiano perché relativista, relativista perché cristiano: Per un razionalismo della contingenza* (Soveria Mannelli: Rubbettino, 2003), 62.

[34] Ibid., 64.

[35] H. Poincaré, *Dernières Pensées* (Paris: Flammarion, 1913), in *Poincaré*, ed. F. Severi (Florence: 1949), 59.

[36] K. Popper, *The Open Society and Its Enemies*, 5th ed. (London: Routledge & Kegan Paul, 1966), 1:64 ff.

analogous to the color yellow, and since the good is not a thing or a natural quality, there is no way of knowing what is good by way of argument.

> If I am asked "What is good?" my answer is that good is good, and that is the end of the matter. Or if I am asked "How is good to be defined?" my answer is that it cannot be defined. … My point is that "good" is a simple notion, just as "yellow" is a simple notion; that, just as you cannot, by any manner of means, explain to someone who does not already know it, what yellow is, so you cannot explain what good is. … You can give a definition of a horse, because a horse has many different properties and qualities, all of which you can enumerate. But when you have enumerated them all, when you have reduced a horse to his simplest terms, then you can no longer define those terms.[37]

"Good" and "value," not being things, are not definable and can only be intuited. Anyone trying to identify the good would fall precisely into the naturalistic fallacy.

> All these truths, however much they may differ, have this in common: that in them both the grammatical subject and the grammatical predicate stand for something which exists. Immensely the commonest type of truth, then, is one which asserts a relation between two existing things. Ethical truths are immediately felt not to conform to this type, and the naturalistic fallacy arises from the attempt to make out that in some roundabout way, they do conform to it. It is immediately obvious that when we see a thing to be good, its goodness is not a property which we can take up in our hands or separate from it even by the most delicate scientific instruments, and transfer to something else.[38]

Moore defines himself as a realist and is of the opinion that ethics is an objective discipline that allows us to determine univocally what human conduct can be defined as "good" or "bad," even though the goodness of a thing is inseparable from the thing itself and therefore indefinable. By virtue of intuitionism, this position can obviously drift toward subjectivism.

Noncognitivists make ample use of both Hume's law and critiques of the naturalistic fallacy,[39] but it cannot be said that *they alone* hinder the development of cognitive moral knowledge of an objective sort. The reduction of reality to mere factual data is what hinders some modern thinkers from perceiving a *teleological order* that legitimizes the passage *from what is to what ought to be*.

There is an *intermediate path* between the denial any relation whatsoever between fact and value and the flattening of moral values into empirical facts: *an Aristotelian-Thomistic finalism*, which seeks to break out of the schematic confines of the great divide. In order to fulfill our personal identity we must know what our *good* is, what our purpose or *end* is. This end ("ought to be") is identified by the

---

[37] G. E. Moore, *Principia Ethica* (1903; repr., Cambridge: Cambridge University Press, 1968), 6–7.

[38] Ibid., 124.

[39] Presently, however, the "great divide" is being overcome in the analytical field: H. Putnam, *The Collapse of the Fact/Value Dichotomy and Other Essays* (Cambridge, MA: Harvard University Press, 2002).

faculty of reason as the *foundation*, which is not simply at the origin of beings but is also their actual and necessary condition for existing or for continuing to exist.

Affirming the existence of an intelligent and transcendent Foundation of reality lends meaning to man's efforts, to his ability to project into the future, and even to his potential opposition to the individual duties his reason identifies. But metaphysical discourse is rejected nowadays because it lacks immediate exhortative value and persuasiveness: it addresses a need for justification that has now been plugged with "reasons of the heart" or "irrational faith."[40]

As will eventually be explained better, the whole problem lies in the attribution of the following meaning to the term *being*: knowable "factuality." If being is understood as merely empirical factuality, then Hume's law is certainly justified. For example, one of course cannot conclude that theft, homicide, and blasphemy are morally permissible from the fact that many men steal, kill, and blaspheme. If we want to show that these acts are illicit, we must invoke a criterion that is not a simple investigation of the facts.

Yet the idea of being underlying the facts can be understood in a way that is not simply empirical. In a deeper and more comprehensive sense, that is to say, in a *metaphysical* sense, being is understood as essence or nature or intrinsic purpose. What ought to be can thereby find a foundation in being—in that being which every conscious subject is called to fulfill.

Thus the term *men* can be understood in the empirical sense (in such a way that the expression refers to individuals who steal or do not steal, who kill or do not kill, and so forth), but one can also think in terms of the *essence* of man, the human *nature* proper to the rational person, or the *dignity of man*. At this point, one can and must find a rational foundation whereby a moral difference can be determined between someone who steals and someone who does not. Yet this apparently simple observation presupposes the demand for metaphysics: the necessity and capacity of our mind to go *beyond* the empirical facts and profoundly grasp the reason for being of things and the "truth" of behaviors, which is in their conformity to the dignity of the human person.

It is necessary to make this preliminary point before going on to describe the various models of bioethics, so as to understand the differences between them and to underscore the relevance of the possibility of a rational foundation for values. It would be pointless and disastrous to reason about bioethics without at least the hope of establishing it on a rational and solid basis, i.e., on the truth. As wearisome as the path may be that leads to a true foundation for moral action and values, it is worth the effort to set out on it. It is said that a society without values cannot survive; but if those values were mere opinions, what social bonds could they establish? "Ethics without truth" is an empty glass set before someone parched with thirst.

---

[40] See A. Pessina, "Linee per una fondazione filosofica del sapere morale," in *Identità e statuto dell'embrione umano*, ed. Pontifical Academy for Life (Vatican City: Libreria Editrice Vaticana, 1998), 210–236; G. Bontadini, *Saggio di una metafisica dell'esperienza* (Milan: Vita e Pensiero, 1995); A. Livi, "Metafisica," in *Dizionario interdisciplinare di scienza e fede*, ed. G. Tanzella-Nitti and A. Strumia (Rome: Urbaniana University Press, 2002), 939–957; V. Possenti, ed., *La questione della verità* (Rome: Armando, 2003); G. De Anna, *Realismo metafisico e rappresentazione mentale: Un'indagine tra Tommaso d'Aquino e Hilary Putnam* (Padua: Il Poligrafo, 2001).

All this notwithstanding, the difficulty of recognizing whether a course of actions adheres to the norm of what is good and true in concrete and sometimes complex situations is undeniable; yet this is precisely the task of practical reason, the *recta ratio agibilium*.[41] Jacques Maritain has correctly pointed out that knowledge of the moral norm is a type of "analogous" knowledge[42] compared with other forms of knowledge (e.g., mathematical or historical), but it is still knowledge and no less important or enlightening than the knowledge applied in other areas.

Descriptive ethics and the sociobiological model

An initial attempt to provide a foundation for ethical norms based on facts alone (in clear-cut opposition to Hume's law), which ends up relativizing values and norms with its proposal of a purely descriptive ethics, is the sociological and historicist approach.[43] According to this perspective, society evolves and produces and changes values and norms that are functional to its development. This is analogous to how living things have developed certain organs in their biological evolution with a view to the function and, ultimately, to the improvement of their own existence. Darwin's evolutionary theory is thus combined with the sociologism of Max Weber and the sociobiologism of H. J. Eysenck and E. O. Wilson.

Specialists in cultural anthropology as well as ecologists often take up similar positions. Translating it into simple language, they declare that societies evolve just as the universe and various forms of life on earth have evolved; as part of this biological and sociological evolution, moral values must also change. The evolutionary impulse, driven by "biological selfishness" or the instinct for self-preservation, takes on new forms through adaptation; law and morality are its cultural expression.

In the present evolutionary conditions, they say, man now has a new situation in the universe and in the biological world. A new system of values should therefore be devised because the previous one is no longer adapted to the configuration of the developing ecosystem. According to this reasoning, man's life is not essentially different from the other various life forms or from the universe itself with which he

---

[41] J. Maritain, *Neuf leçons sur les notions premières de la philosophie morale* (Paris: Téqui, 1951); Vanni Rovighi, *Elementi di filosofia*, vol. 3; Thomas Aquinas, *Summa Theologiae*, I-II, q. 55, a. 7, ad 3; A. Da Re, *L'etica tra felicità e dovere: L'attuale dibattito sulla filosofia pratica* (Bologna: Dehoniane, 1986).

[42] Maritain, *Neuf leçons*, 97–109.

[43] For the meaning of the term *descriptive ethics*, see the same author, S. Privitera, "Etica descrittiva," in *Nuovo dizionario di teologia morale*, ed. F. Compagnoni, G. Piana, and S. Privitera (Milan: Paoline, 1990), 354–358. For a presentation of sociological and bio-ecological concerns in ethics, see E. O. Wilson, *Sociobiology: The New Synthesis* (Cambridge, MA: President and Fellows of Harvard College, 1975); B. Voorzanger, "No Norms and No Nature: The Normal Relevance of Evolutionary Biology," *Biology and Philosophy* 2 (1987): 569–570; B. Chiarelli, "Storia naturale del concetto di etica e sue implicazioni per gli equilibri naturali attuali," *Federazione Medica* 37, no. 6 (1984): 542–546; idem, *Problemi di bioetica nella transizione fra il II e il III millennio* (Florence: Il Sedicesimo, 1990); F. Remotti, "La tolleranza verso i costumi," in *Teorie etiche contemporanee*, ed. C. A. Viano (Turin: Bollati Boringhieri, 1990), 165–185.

lives in symbiosis. In this view, ethics has the role and function of maintaining the evolutionary balance between mutation, adaptation, and the ecosystem.[44]

There is obviously a very close connection between nature and culture, and it is sometimes difficult to identify the boundary. For these thinkers, however, nature turns into culture and vice versa; culture is nothing more than a transcriptive development in the evolution of nature.

Accepting this model entails not only taking evolutionary theory for granted but also accepting the premise of *reductionism*, i.e., the reduction of man to a merely historical and natural element of the cosmos. This view therefore brings with it a relativization of all ethics and human values, submerging every living thing in the vast river of evolution. While man is indeed its highest point, that point is not definable as a permanent point of reference but is itself subject to active and passive mutations. In short, it is a Heraclitean ideology that acknowledges no stable unity and no universality of values; no norm is always valid for man in all times and places. In bioethics, this school of thought justifies eugenics for the purposes of selection and enhancement: Hitler's racism was itself rooted in this ideology.

If this ideology were true (for this is indeed an ideology), even the most atrocious crimes known to history, from those of Genghis Khan to those of Hitler, would be crimes only for us living in this time, posthumous crimes and not crimes against humanity. Any effort to define "human rights" would be meaningless and, at best, provisional.

In the perspective of this model, the mechanisms of adaptation and selection are necessary for the evolution and progress of the human species. Adaptation to the environment or ecosystem and selection for the qualities best suited to the progress of the species lead to the justification of both positive and negative eugenics. According to the followers of this theory, now that humanity has acquired the ability to scientifically control the mechanisms of evolution and biological selection through genetic engineering, attempts to improve and alter the genetic constitution of animal species—and even of man—through genetic engineering are justifiable. Several smaller currents can be identified within this major thought current, however. Some simply justify the existing values in societies while others, especially the sociobiological current, are also prone to justify innovative interventions on the biological patrimony of humanity. In any case, this thought current identifies with Vico's words *verum ipsum factum,* "truth itself is made," and hence *bonum ipsum factum*, "good itself is made."

It is obvious that while some components of a culture and its customs are subject to development, it is equally obvious that man remains man, different from all other living beings by nature and not by neurological complexity alone, and that good and evil are not interchangeable. Neither can the laws of being, science, and morality be false and true at the same time. Death, suffering, the thirst for truth, solidarity, and freedom are not cultural developments, but rather facts and values that accompany man in every historical age.

---

[44] E. Sgreccia, "Scienza, medicina, etica," in *Nuova genetica: Uomo e società*, ed. A. Serra and G. Neri (Milan: Vita e Pensiero, 1986), 7–11.

The subjectivist or liberal-radical model

Many thought currents coalesce today into moral subjectivism. Some examples include neo-Enlightenment thinking, ethical liberalism, nihilistic existentialism, neo-positivist scientism, emotionalism, and decisionism.[45] The main premise of all these currents is that morality cannot be founded on facts or on objective or transcendent values, but only on the autonomous choice of the acting subject. In other words, its point of departure is noncognitivism or the unknowability of values.

The principle of autonomy is thus understood in the strong sense. The sole foundation of moral action is autonomous choice, and the ethical and social dimension is wrapped up in the effort to liberalize society. The only limit is someone else's freedom (the freedom of someone who is capable of exercising it, obviously) or significant harm that should be avoided in the utilitarian sense.

Freedom is taken as the supreme and ultimate reference point: whatever is freely willed, accepted, and does not harm another person's freedom is licit. This was the message that sprang forth from the French Revolution with innovative energy. There is certainly an element of truth in this view, but not the entire truth about man nor even the entire truth about freedom. The demands of this current have been heard: liberalization of abortion; freedom to choose the sex of an unborn child and free choice for adults who desire a "sex change"; freedom to request extracorporeal fertilization, whether a woman is single, married, or widowed; freedom of experimentation and research; freedom to decide upon the time and manner of death by a living will; suicide as a sign and emphasis of freedom, and so forth.

In reality this freedom extends only halfway: it is freedom for some, usually for those who can command respect for their freedom and express it. For example, who defends the freedom of an unborn child? It is a *freedom from* ties and restraints and not a *freedom to* devise a plan for one's life and for society that is justified in terms of its purpose. In other words, it is freedom without responsibility.

During the 1960s, Herbert Marcuse had demanded three new freedoms so as to bring to completion the plans of the French and Russian revolutions, which, in his opinion, had focused only on civil liberties and freedom from want, respectively. According to Marcuse, the new frontiers of freedom would be freedom from work, because work enslaves human activity; freedom from family, because the family enslaves man's affections; and freedom from morality, because morality imposes ends on man's mind and ends restrict free choice itself. He speaks of free, polymorphic love in his book *Eros and Civilization*.[46] It is not difficult to understand that this freedom is a tragic game, even though he calls it a "celebration." It is *nihilism* because it assumes nothing before or within freedom.

---

[45] The subjectivist and decisionist perspective informs the thought of H. Kelsen and K. Popper, and in Italy this approach is found in the works of U. Scarpelli and D. Antiseri. The emotivism of A. J. Ayer and C. L. Stevenson can be traced back to the same intellectual system. The nihilistic existentialism of J. P. Sartre and the libertarianism of H. Marcuse eventually led to this current of thought. See Reale and Antiseri, *Storia del pensiero occidentale*, 3:508–779.

[46] H. Marcuse, *Eros and Civilization: A Philosophical Inquiry into Freud* (Boston: Beacon Press, 1966); V. Melchiorre, *Amore e matrimonio nel pensiero filosofico e teologico moderno* (Milan: Vita e Pensiero, 1976).

In reality, every free act presupposes the existing life of the human being who posits that act. Life comes before liberty, because someone who is not alive cannot be free. Freedom has content: it is always an act that aspires to something or affects someone, and it bears responsibility for its content. In sum, freedom presupposes being and existence for a life-project.[47] When freedom turns against life, it destroys itself and dries up its own roots; when it denies responsibility for a choice, it reduces itself to a blind force and risks making a joke of itself and, ultimately, destroying itself.

When we talk about responsibility, it is naturally the responsibility that springs from within freedom and is sustained by reason, which evaluates the means and ends for a freely pursued plan. It is not used, at least in this context, to mean responsibility with regard to civil laws and outside authority only. This can rightly be invoked for the sake of certain values and the common good, but it is neither the primary nor the greatest expression of responsibility.

Responsibility is first and foremost an inner reality. It confronts the faculty of reason and its reflections upon conscience and upon the ethical assessment of the values at stake. This responsibility persists even when civil law is silent and judges do not know and do not inquire. Sometimes this inner responsibility can even conflict with the civil law, when the latter happens to violate the fundamental and irrevocable values of the human person.

This is not the place to engage in all of the theoretical discourse about these themes down through the history of philosophy—which are among the most majestic and dramatic themes of human life—but it was necessary to note the existence of this ethical "model" affecting the mentality of our times. It is a model that influences culture, literature, news media, and especially morals.

Those who promote ethical subjectivism and decisionism meet with difficulties, however, when they face the need to propose a social norm, especially when confronting someone who, in deference to the principle of autonomy, does not accept any form of self-limitation. In order to avoid recourse to the "moderating" function in Hobbes's *Leviathan*, they propose the *principle of tolerance* or simply the criterion of doing no "significant harm" to others.[48] But this means renouncing any rational foundation for morality; in fact, especially in dealing with those who do not enjoy moral autonomy (embryos, fetuses, the dying), ethical liberalism has ended up sliding toward the legitimization of violence and of the principle that "might makes right."[49]

---

[47] See C. Vigna, ed., *La libertà del bene* (Milan: Vita e Pensiero, 1998); V. Possenti, *Essere e libertà* (Soveria Mannelli: Rubbettino, 2004); R. Spaemann, *Happiness and Benevolence* (Notre Dame, IN: Notre Dame University Press, 2000).

[48] Scarpelli, *La bioetica*.

[49] Criticism of subjectivist decisionism and of an exaggerated form of the tolerance principle can also be found even in so-called secular thought. For example, see M. Mori, "I limiti dell'etica senza verità," *Biblioteca della Libertà* 99 (1987): 67–76; S. Quinzio, "Perché la tolleranza non basta," *Biblioteca della Libertà* 99 (1987): 77–81. Concerning the "anti-liberal" results of ethical liberalism, see M. Schooyans, *La dérive totalitaire du libéralisme* (Paris: Editions Universitaires, 1991); V. Possenti, *Le società liberali al bivio: Lineamenti di filosofia della società* (Turin: Marietti, 1991); V. Possenti, "La cultura radicale," in *Filosofia e società: Studi sui progetti etico-politici contemporanei* (Milan: Massimo, 1983), 94–134.

The pragmatic-utilitarian model

The dead-end of noncognitivism and the inherent weakness of subjectivism on the social level have led to a recovery of intersubjectivity on the pragmatic level. Various formulas for *public ethics* have been developed in an effort to find a point of convergence that does not deny the individualistic foundation of the moral norm. This approach is widespread in English- and German-speaking countries and ends up being a sort of subjectivism of the majority.[50]

The common denominator among these various schools of thought is a denial of metaphysics and the resulting mistrust in reason's ability reach a universal truth and thereby declare a moral norm that is valid for everyone.

The basic principle is the calculation of the consequences of an action in terms of the relation between costs and benefits. It should be stated immediately that this relation is valid when it refers to one value and one person in a consistent and subordinate way, i.e., when it is not taken to be ultimate criterion, but rather as one assessment factor in reference to a human person and his or her values. Thus its application is valid when a surgeon or a physician needs to decide on a course of treatment—a decision that is rightly weighed on the basis of foreseeable harms (better described as "risks") and benefits to the life and health of the patient.

This principle cannot be used in an absolute and foundational manner, however, to "weigh" disparate goods against one another, such as when comparing financial costs with the value of a human life. Many formulas employed in the medical field and suggested as a means of making therapeutic decisions or allocating economic resources end up assuming a utilitarian character.

Traditional utilitarianism, traced back to Hume's empiricism, reduced the cost-benefit calculus to the individual subject's evaluation of what is pleasant or unpleasant. Neo-utilitarianism, inspired by Jeremy Bentham and John Stuart Mill, can be summarized in this threefold precept: maximize pleasure, minimize suffering, and broaden the sphere of personal freedoms for the greatest number of persons.[51] The concept of *quality of life*, used by some in opposition to the concept of the sacredness of life, has developed along these lines. Quality of life is assessed precisely in terms of the minimization of suffering and often of economic costs.

---

[50] Here is just some of the literature dealing with this subject: M. Mori, *Utilitarismo e morale razionale* (Milan: Giuffré, 1986); idem, "Bioetica: Una riflessione in corso," *L'informazione Bibliografica* 16, no. 3 (1990): 442–452; E. Lecaldano, "Il contributo di una filosofia 'laica,'" *Biblioteca della Libertà* 99 (1987): 57–66; idem, "Principi e basi razionali di un'etica non religiosa," in *Problemi di etica: Fondazione, norme e orientamenti*, ed. E. Berti (Padua: Gregoriana, 1990), 23–68; idem, "Etica e significato: un bilancio," in *Teorie etiche contemporanee*, Viano, 58–86.

[51] J. Bentham, *An Introduction to the Principles of Morals and Legislation* (1779; repr., London: Athon Press, 1970); J.J. Smart and B. Williams, eds., *Utilitarianism For and Against* (Cambridge: Cambridge University Press, 1973); J.S. Mill, *Utilitarianism* (1781; repr., New York: Bobbs-Merrills, 1957); J.C. Harsanji, "Rule Utilitarianism and Decision Theory," *Erkenntnis* 11 (1977): 25–53; R.M. Hare, *The Language of Morals* (London: Clarendon Press, 1952); R. Brandt, *Ethical Theory* (Englewood Cliffs, NJ: Prentice Hall, 1959); P. Singer, *Practical Ethics* (Cambridge, UK: Cambridge University Press, 1979); Palazzani and Sgreccia, "Il dibattito attuale."

Various quality-of-life formulas have been proposed, some inspired by "strong" utilitarianism and some by the "weaker" sort, to assess the efficacy and usefulness of treatments or even the expediency of allocating economic resources for the treatment of certain illnesses. Cost-benefit analysis (CBA), cost-effectiveness analysis (CEA), and quality-adjusted life-years (QALY) are formulas (especially the last-mentioned) that end up including, among the deciding factors for a therapeutic intervention and the allocation of resources in the health care field, the cost of various treatments, economic factors, and the patient's recovery of societal productiveness. These formulas, like many others invented for particular categories of patients—newborns with physical handicaps, adults with malignant tumors, etc.—try to compare disparate factors (health vs. productivity, therapy vs. availability of funds). They end up determining that treatment or care should be refused in the name of a unproductive expense or a quality of life based only on an evaluation of biological or economic factors.[52]

In order to temper this *act utilitarianism*, some have attempted to introduce some broader rules of beneficence, such as the concept of equity or of minimum assistance,[53] thus mitigating act utilitarianism with *rule utilitarianism*. The rules of "equity," "impartiality," "neutral observation," the "social extent of utility" or the "calculus of social happiness," and the "ethical minimum" cannot compensate for situation of relativism and the absence of a true foundation for the rules. The extreme difficulty of applying a calculus that can reconcile private and social interests on the empirical and pragmatic plane of happiness should also be emphasized.

Some authors who argue on these grounds—the search for happiness and quality of life—wind up reducing the category of "person" to that of a "sentient being," inasmuch as only the latter is capable of feeling pleasure and pain. The consequences of this redefinition are as follows:

> (a) ignoring the need to safeguard the interests of insensate individuals, i.e., those not endowed with sensitive faculties (such as embryos prior to nervous system development, individuals in a vegetative coma, etc.); (b) justifying the elimination of sentient individuals whose suffering is greater (or will become foreseeably greater) than their pleasure, or of individuals who cause others quantitatively more suffering than joy (the handicapped, deformed fetuses, the dying, etc.); and (c) justifying interventions that destroy human life, as long as suffering is avoided (legal abortion, even in the late stages of gestation, provided that the fetus is anesthetized).[54]

Thus, while on the one hand utilitarianism excludes some human beings from its consideration, on the other hand, paradoxically, it goes so far as to equate animals with human beings on the basis of their capacity to "feel"—to experience pleasure and pain.

One is thereby left with a utilitarian perspective that does not define *whose* utility should be pursued or on *what* basis; in other words, utilitarianism argues that

---

[52] See Seifert, *Philosophical Diseases*, 1:306–323.

[53] Gracia, *Fundamentos de bioética*, 276–281; G. Herranz, "Scienze biomediche e qualità della vita," *Vita e Pensiero* 6 (1986): 415–424; E. Sgreccia and A. G. Spagnolo, eds., *Etica e allocazione delle risorse in sanità* (Milan: Vita e Pensiero, 1996).

[54] Palazzani and Sgreccia, "Il dibattito attuale," 862.

human life is valued according to the presence or absence of suffering and according to the economic criteria of productive or unproductive expenditures.[55]

The contractualist model

Another trend in public ethics that resembles utilitarianism in certain respects (although there are differences) is *contractualism*. This approach is also founded on the criterion of intersubjective agreement reached by the ethical community, which is to say, by those who have the ability and power to decide. H. T. Engelhardt Jr.'s book *The Foundations of Bioethics* expresses this trend.[56] Engelhardt maintains that the social consensus of the "ethical community" justifies the undervaluation of those who are not yet part of the community (embryos, fetuses, and babies), whose rights therefore depend on adults, and who are ultimately not considered persons. In the same way, those who are no longer socially involved, such as sick people incapable of social interaction or the incurably demented, become undervalued to the status of "no longer persons." Ultimately the concept of human person ends up being a sociological concept.

The phenomenological model

The thought currents based on phenomenology and discourse ethics should also be part of this survey of intersubjective ethics. Phenomenological ethics, especially in the writings of Max Scheler and Nicolai Hartmann, exhibits an openness to ethical values that is defined as "intentional" and "intuitive." Ethical values, however, have their foundation on the emotional (the divine element in man, according to Scheler) and "religious" level. It therefore acknowledges the possibility of a foundation that means to be concrete and specific, yet on grounds that remain relative to and dependent upon emotional subjectivity; hence, universal validity cannot be claimed. The horizon remains social and difficult to formulate.

The theory of a *formal ethics of goods*, formulated by Diego Gracia, also fits in this spectrum of phenomenological views. It declares the formal and universal necessity of values inasmuch as the very knowledge of reality stimulates in the consciousness an awareness of realities as values, but this formal necessity is satisfied in acts of assessment and valuation that are subjective and dictated by circumstances. As a necessity, morality therefore has a foundation in the rational and universal sense; but as a concrete choice, it winds up being dictated by a subjective assessment.[57]

Even the attempt to overcome subjectivism in concrete choices by seeking a "procedural" agreement of a social sort through the adoption of shared norms, as in *egalitarianism*, or through the introduction of corrective concepts such as "the ideal observer," the "ethical minimum," or the "postulate of equiprobability," results in

---

[55] See P. Singer, *Rethinking Life and Death: The Collapse of Our Traditional Ethics* (New York: St. Martin's Press, 1994). The author looks forward to replacing the old commandments with new ones, such as "the value of human life varies," "bring only wanted children into the world," and "don't discriminate on the basis of species."

[56] H. T. Engelhardt, Jr., *The Foundations of Bioethics*, 2nd ed. (New York: Oxford University Press, 1986). A response to Engelhardt can be found in Seifert, *Philosophical Diseases*, 1:240–283.

[57] Gracia, *Fundamentos de bioética*, 363–382.

contrived procedures that fail to surmount the horizon of subjectivity and intersubjective conventions.[58]

The discourse theory proposed in the German-speaking cultural world by Karl-Otto Apel and Jürgen Habermas bases social consensus on communication. On the one hand it is supposed to overcome the "calculating reason" of utilitarianism, and on the other hand it should open up the possibility of agreement on the contents and beneficiaries of values.[59] It is necessary to recognize that some values are certainly implicit in communication itself, such as truthfulness, respect for others' opinions, and respect for freedom of opinion and expression, but these values are subject to and prepare the way for the foundation of a norm. The fundamental principle itself (*Grundnorm*) posited by this school of thought, according to which "the norms to be justified must be capable of obtaining a consensus about their foreseeable consequences for all interested parties," runs the risk of subordinating the validity of the norm to the consensus and of being unable to define who the interested parties are.

Principlist ethics

One approach that seems included within the scope of public ethics, in which the need for some moral principles is affirmed but with uncertain justification, is the so-called *principlism* formulated by Thomas Beauchamp and James Childress.[60]

The well-known principles of beneficence, nonmaleficence, autonomy, and justice, which are relevant when taken individually and certainly factor into the assessment of interventions in biomedicine and health care when taken together, in turn require a foundation. Indeed, what is good for a patient remains to be defined. (For example, if a boy is born with multiple serious deformities, is it good to care for him or to let him to die?) Furthermore, a hierarchy must be established among the principles themselves, especially between the principles of autonomy and beneficence. The former should be subordinate to the latter; otherwise the very autonomy of the acting subjects is not guaranteed, especially when the sick person is not in a position to exercise self-determination or when the autonomies of physician and patient are in conflict. In order to reconcile the principle of autonomy with the principle of beneficence, real common ground must be sought in pursuing the true good of the person.[61] This topic, limited here to a few brief remarks, will be addressed again in the chapter devoted to principles of bioethics.

---

[58] For the theory of the real observer, see R. M. Veatch, *Medical Ethics*, 2nd ed. (Boston: Jones and Bartlett, 1997); Hare, *The Language of Morals*; idem, *Freedom and Reason* (Oxford: Oxford University Press, 1963). The theory of the "postulate of equiprobability" is found in Harsanji, "Rule Utilitarianism." For the theory of egalitarianism, see J. Rawls, *A Theory of Justice* (Cambridge, MA: Belknap Press, 1971). A more extensive bibliography can be found in Palazzani and Sgreccia, "Il dibattito attuale."

[59] Here are just a few works: K.-O. Apel, *Comunità e comunicazione* (Turin: Rosenberg & Sellier, 1977); J. Habermas, *Theory and Practice* (Boston: Beacon Press, 1973); Gracia, *Fundamentos de bioética*, 558–591.

[60] T. L. Beauchamp and J. F. Childress, *Principles of Biomedical Ethics*, 4th ed. (New York: Oxford University Press, 1994).

[61] See the lucid analysis of the conflict between these principles in the essay by I. Carrasco de Paula, "L'etica dell'intervento medico: Il primato dell'interesse del paziente," in *L'assistenza al morente: Atti del congresso internazionale "Care for Dying Persons"; 15–18*

The discussion about a so-called *prima facie deontology*[62] is just as elusive. According to this theory there are no duties that are always valid in all circumstances but only duties that are valid prima facie, as a matter of principle, so to speak. Their concrete application admits exceptions and conflicts for which no uniform and certain solution can be given. I think that if we do not wish to declare concrete choices relative, under the guise of principled proclamations of formal value only, we must pay attention to the need and obligation of clarifying and resolving the conflicts by prioritizing and harmonizing the values at stake and eliminating the points of contention. Ethical science and the practice of ethical virtues draw meaning from this undertaking.

The personalist model

The model I consider capable of resolving the fundamental contradictions of the preceding models while providing an objective foundation for values and norms is called the personalist model.

There are at least three historical definitions of personalism, each with its own emphasis: relational personalism, hermeneutic personalism, and ontological personalism.[63] The relational and communicative definition primarily emphasizes the value of subjectivity and the relation between subjects, as seen in Apel and Habermas. The hermeneutic definition emphasizes the role of subjective consciousness in interpreting reality according to one's own prior understanding, as seen in Hans-Georg Gadamer. The ontological definition, without denying the relevance of relational subjectivity and of the conscience, emphasizes that there is an existence and an essence, a body–soul composite, at the foundation of subjectivity itself.

The person is understood as *ens subsistens ratione praeditum* (a subsistent being endowed with reason) or, as Boethius defines the person, *rationalis naturae individua substantia* (an individual substance of a rational nature). In man personhood consists in an individuality constituted by a body animated and structured by a spirit.[64]

The personalist tradition is rooted in man's reason itself and in the heart of his freedom. Man is a person because he is the only being in which life becomes capable of "reflection" upon itself, or of self-determination. He is the only living thing with the capacity to grasp and discover the meaning of things and to give meaning to his own expressions and conscious language. Reason, freedom, and

---

marzo 1992, ed. E. Sgreccia, A. G. Spagnolo, and M. L. Di Pietro (Vatican City: Libreria Editrice Vaticana, 2000), 333–342; A. G. Spagnolo, "I principi della bioetica nordamericana e la critica del 'principialismo,'" *Camillianum* 20 (1999): 225–246.

[62] D. Ross, *The Right and the Good* (Oxford: Oxford University Press, 1930).

[63] See C. Viafora, ed., *Vent'anni di bioetica: Idee protagonisti istituzioni* (Padua: Gregoriana, 1990), 45–48.

[64] For a more in-depth discussion of this view, the reader can consult Vanni Rovighi, *Elementi di filosofia*, vol. 3; J. Hervada, *Introduzione critica al diritto naturale* (Milan: Giuffré, 1990); S. Cotta, *Giustificazione e obbligatorietà delle norme* (Milan: Giuffré, 1981); Maritain, *Neuf leçons*; J. Seifert, *Essere e persona* (Milan: Vita e Pensiero, 1989); L. Palazzani, *Il concetto di persona tra bioetica e diritto* (Turin: Giappichelli, 1996); V. Possenti, "Noi che non sappiamo affatto che cosa sia la persona umana," *Filosofia Oggi* 27, no. 1 (2004): 3–28.

consciousness constitute, as Popper puts it, an "emerging creation"[65] that cannot be reduced to the flow of cosmic and evolutionary laws. This is so thanks to a spiritual soul that informs and gives life to his body and by which that body is contained and structured. His "I" cannot be reduced to a code or a number, to atoms, cells, or neurons. Jean-Pierre Changeux's "neuronal man" does not exhaust everything there is to man; on the contrary, a mind is required to structure his brain and a spiritual soul to structure, guide, and give life to his body. The ontological and axiological gap that distinguishes human persons from animals is incomparable to the distance that separates reptiles from plants and plants from rocks. In every man and woman, in every human person, the whole world is recapitulated and acquires meaning, yet the cosmos is at the same time crossed over and transcended. The meaning of the universe and the whole value of humankind are contained in every man and woman. The human person is a unity—a whole and not a part of a whole.

The point of reference for society itself is the human person: the person is the purpose and the source of society. Christian Revelation, with its truth about Creation (creation is also, within certain limits, a truth that can be attained by reason), Redemption and the communion of man with God, extends the horizons and values of this personalist vision toward the divine. For believers, every single human being is the image of God, a child of God, and a brother or sister of Christ. Yet even in terms of rational and secular reflection, the human person is the transcendent reference point—the end and not the means—for economics, law, and history itself. These philosophical premises should not be thought of as mere abstractions in the discussion of medical ethics or bioethics, because medicine and ethics are both at the service of man and must consider him in the fullness of his worth.

From the moment of conception until death, in every situation of suffering or health, the human person is the reference point and standard for distinguishing licit from illicit.

The personalism here discussed should not be confused with subjectivist individualism, which emphasizes almost exclusively the capacity for self-determination and choice as constitutive of the human person. This perspective is widespread in Protestant and existentialist contexts and is also influential in American theological circles. Classical personalism, which is Thomistic and realist, does not deny this existential component (the ability to choose, which forms the destiny and drama of the human person) but also affirms and prioritizes the objective and ontological constitution of the human person.[66] The person is first of all a spiritualized body or embodied spirit, valuable for who he is and not just for the choices he makes. In fact, the human person invests what he is—his existence and essence, his body and

---

[65] K. Popper and J. Eccles, *The Self and Its Brain* (New York: Springer International, 1977).

[66] From a different mold comes the ontological personalism of L. Pareyson, who interprets the person as a combination of relation to self and relation to others: man does not *have* but *is* relation with being; see L. Pareyson, *Esistenza e persona*, 7th ed. (Genoa: Il Melangolo, 2002). In an existentialist perspective, *being* is equated with *freedom* and hence the person becomes *choice*. See the objection by V. Possenti, *Essere e libertà* (Soveria Mannelli: Rubbettino, 2004). It should be noted that in Pareyson's thinking (as in the idealist perspective), the person is presented without his or her corporeal character.

spirit—in every choice. Every choice involves not just the exercise of choice or the faculty of choosing but also the *context* of the choice: the end, means, and values.

Realist personalism sees the human person as a *unity* or, as it is often called, a *unitotality* of body and spirit that constitutes his objective value. This objective value is placed in the charge of his subjectivity, and it cannot be otherwise for both his own person and the persons of others. The human person and his values cannot be melted down and liquefied into a series of choices without a source and without the value content they express.

In a personalist ethic the objective and subjective aspects of the human person refer to and implicate one another.[67] The ethical value of an act must be considered under the subjective aspect of intentionality but also in terms of its objective content and consequences. The natural moral law that urges every conscience to do good and avoid evil therefore becomes concrete in respect for the human person in all the fullness of his values and in his essence and ontological dignity. This is true in all areas of ethical conduct and also in bioethics.

If a surgeon were to make an involuntary careless mistake in a difficult and risky procedure that resulted in the death of the patient, he might not be subjectively accountable for it. The objective fact remains, however, that a human life was lost. This should prompt the surgeon to make a determined effort in the future not to repeat the mistake. The subjective evaluation prevails in the private, inner phase of judgment on the act, but the objective value prevails in the normative and deontological phase of judgment, and the subjective disposition should increasingly conform to it. Certitude requires searching more and more for the truth.

In my opinion, the personalist perspective can also include a trend among some thinkers in the Anglo-Saxon world who are trying to rehabilitate *virtue ethics*, which is understood to be in contrast with, or at least preferable to, principlist ethics.[68] I am convinced not only that certain acquired abilities are needed to put values into prac-

---

[67] See Seifert, *Philosophical Diseases*, vol. 1.

[68] L. Palazzani, "Bioetica dei principi e bioetica delle virtù: Il dibattito attuale negli Stati Uniti," *Medicina e Morale* 42, no. 1 (1992): 59–86; E. D. Pellegrino, "A Philosophical Source of Medical Morality," *Journal of Medicine and Philosophy* 4, no. 1 (1979): 1–7; idem, "The Caring Ethics: The Relationship of Physician to Patient," in *Caring, Curing, Coping: Nurse-Physician-Patient Relationships*, ed. A. H. Bishop and J. R. Scudder (Alabama: University of Alabama Press, 1985), 8–30; idem, "Philosophical Groundings for Treating Patient as a Person: A Commentary on Alasdair MacIntyre," in *Changing Values in Medicine*, ed. E. J. Cassell and M. Siegler (Frederick, MO: University Publications of America, 1985), 97–104; idem, "Professional Ethics: Moral Decline or Paradigm Shift?" *Religion and Intellectual Life* 4, no. 3 (1987): 21–45; idem, "Altruism vs. Self-Interest: Ethical Models for Medical Professions," *NYU Physicians* 45, no. 1 (1988): 41–43; idem, "Character, Virtue, and Self-Interest in the Ethics of the Professions," *Journal of Contemporary Health Law and Policy* 5 (1989): 53–73; idem, "Trust and Distrust in Professional Ethics," in *Ethics, Trust, and the Professions: Philosophical and Cultural Aspects*, ed. E. D. Pellegrino, R. M. Veatch, and J. P. Langan (Washington, DC: Georgetown University Press, 1991), 69–89; idem, "The Virtuous Physician and the Ethics of Medicine," in *Virtue and Medicine: Explorations in the Character of Medicine*, ed. E. E. Shelp (Boston: Reidel, 1985), 237–255; E. D. Pellegrino and D. C. Thomasma, *For the Patient's Good*. See also, by the same authors, *A Philosophical Basis of Medical Practice: Toward a Philosophy and Ethic of the Healing Professions* (New York: Oxford University Press, 1981).

tice during the application phase of an ethical judgment, but also that a sensitivity to the meaning and value of the human person springs from an area of the conscience inspired by virtue. Nonetheless, keeping in mind precisely the personalist model, the phase of clarifying and establishing values and norms must be integrated with the phase of applying them correctly and consistently. The role of the cardinal virtues in ethical action will be revisited later on, but one should not think of the two phases as separate; otherwise, the very concept of virtue or virtuous action could wind up unfounded.

Research Method in Bioethics

It is easy to see from our discussion of personalist bioethics thus far that the research and teaching method in bioethics does not fit squarely within either the inductive method (where norms are established through observation of biological or sociological facts) or the deductive method (where norms of conduct are immediately deduced from principles).[69] It seems necessary to propose a method that I describe as *triangular*, i.e., it is conducted through an analysis involving three connected points.

First of all, the biomedical facts should be set forth with scientifically ascertained substance and exactitude. (For example, human fertilization in vitro or the creation of recombinant DNA is possible.) This is point A of the triangle. One should then proceed from this point of analysis to a more in-depth study of anthropological significance and thereby determine what values are at stake with relation to the life, integrity, and dignity of the human person. This is point B, which is the more plainly philosophical point of synthesis. On the basis of this analysis, it is possible to determine which values to safeguard and which norms to apply to the action and agents on both the individual and social levels. Principles and norms of conduct must be in reference to this center, which is the human person-as-value and the hierarchically harmonized values within the person (life, health, personal responsibility, etc.).[70] The solution to these ethical problems must be sought in relation to the fundamental concepts and values of the human person. This is point C of our triangular method, which calls upon the philosophy of man as a whole.[71]

The justification for proposed solutions should be accompanied as much as possible by comparisons with solutions proposed by other thought currents. Confronting one's reference anthropology therefore occurs in a dynamic and ongoing way. Scientific discoveries and technological applications are continually opening up new possibilities and making new conquests, and these developments have continual repercussions on social development and the adequacy of society's laws.

---

[69] On the methodological problem, which obviously reflects the underlying anthropological view, see the extensive chapter by Gracia, *Fundamentos de bioética*, 395–503, where he restates the various solutions in the history of philosophy, and the following chapter (527–575) where he explains the problem of the "ethical minimum" or of "minimal ethics."

[70] A. Pessina, "Personalismo e ricerca in bioetica," *Medicina e Morale* 47, no. 3 (1997): 443–459; idem, "La relazione tra la ricerca biomedica, l'antropologia e l'etica filosofica: Appunti per una riflessione metodologica," in *Etica della ricerca biomedica*, Vial Correa and Sgreccia, 144–158.

[71] For a more complete examination of the tasks of moral philosophy in general, see Vanni Rovighi, *Elementi di filosofia*, 3:189–269.

Anthropology offers a criterion by which to distinguish between technological and scientific possibility and ethical permissibility. It also offers a criterion by which to discern what may be sanctioned by political majorities from what is licit and advantageous for the good of man.

Anthropology itself is clearly fostered and enriched by this triangular exchange (biology-anthropology-ethics), but it is also necessary for it to posit certain criteria and values that cannot be superseded or repressed insofar as they are the very reason behind the teleology of scientific progress and societal development.

The fundamental values of the person must be safeguarded not just morally but also legally: international courts and national constitutions make their decisions on the basis of these human values. The problem of the relation between ethics and law or between moral law and civil law arises here.[72]

The two trunks into which personalist bioethics is grafted are personalism and Aristotelian-Thomistic teleology, which is the metaphysical reference point. Personalist bioethics is a *dynamically thinking enterprise* because it deals with scientific discoveries that are revisited from time to time, but especially because it also takes into account the various ways in which contemporary man perceives his own identity and values in general. It has a critical and dialectical structure hinged upon a substantialist understanding of the human person.

The *truth* about man in his many dimensions is not just an *inheritance* from classical thought but also a theoretical *gain*. The bioethical method of research brings out the *analogous significance of truth* precisely because all disciplines deal with truth, according to their unique methods, and are able to understand and interact with one another along the journey of assessing what is at stake in human action.[73] Personalist bioethics is therefore a *concrete enterprise*, because it does not follow deductive reasoning alone; rather, it is a systematic interaction between various spheres of knowledge.[74]

Moral Law and Civil Law

The age-old philosophical debate on the relation between civil law and moral law is the essential element of the challenge facing Western democracies today. Saint Thomas pointed out that law cannot cover the entire sphere of morality. Moreover,

---

[72] *Ordine morale e ordine giuridico: Rapporto e distinzione tra diritto e morale* (Bologna: Dehoniane, 1985); P. Donati, "Il contesto sociale della bioetica: Il rapporto tra norme morali e norme di diritto positivo," in *Bioetica: Un'opzione per l'uomo; Atti del primo corso internazionale di bioetica; 15–16/29–30 aprile 1988* (Milan: Jaca Book, 1988). On the subject of the relation between bioethics and law, see F. D'Agostino, "Bioetica e diritto," *Medicina e Morale* 43, no. 4 (1993): 675–691; idem, *Bioetica* (Turin: Giappichelli, 1996); A. López Trujillo, G. Herranz, and E. Sgreccia, eds., *"Evangelium vitae" e diritto* (Vatican City: Libreria Editrice Vaticana, 1997); F. D'Agostino, *Bioetica nella prospettiva della filosofia del diritto* (Turin: Giappichelli, 1996); S. Viotti, "Il problema morale della legge civile," *Studia Moralia* 37 (1999): 321–356.

[73] See A. Pessina, "La relazione," in *Etica della ricerca biomedica*, Correa and Sgreccia, 144–158.

[74] See the Latin bioethics (referring to Roman law) mentioned by F. D'Agostino in *Parole di bioetica* (Turin: Giappichelli, 2004), 21–26, which declares that objectivity cannot be derived from abstract, formal principles, but can be acquired only by means of attentive observation in particular *cases*, with *prudence*.

law cannot be the foundation of morality; where possible, the former must recognize the requirements of the latter. One should therefore not wish for an "ethical state" presuming to be the source of what is right and wrong. Nevertheless certain fundamental values, which are necessary and indispensable to assuring the common good, must also be safeguarded by the law. When a law does not safeguard a good that is essential to society and the common good (as in the case of laws permitting abortion), that law is not law: it must be changed and can be grounds for *conscientious objection*.

The fundamental goods of the life of the individual (whether born or unborn), the family, and basic medical care are fundamental ethical requirements for safeguarding the common good. It is not an "ethical minimum" because this is no minimal matter; rather, it is the common good to be safeguarded for the benefit of all.

In today's pluralistic society, especially since the affirmation of bioethics and the problems it has raised in the field of law, there is an increasingly evident need to bring the axiological foundations of law back to light and to make inviolable human values sure and explicit so that, sanctioned by law, they may guide human conduct in the moral choices connected with the life and health sciences.

The split between the correlatives "truth" and "freedom" in the prevailing culture makes it increasingly difficult to defend human life effectively by means of state laws. Paradoxically, subtle forms of tyranny are emerging whereby a few people can dictate the destiny of many, many others.

For some time now, indeed, the laws on voluntary abortion in force in many Western countries have paradoxically transformed what was once considered a crime into a "right," thus legalizing the abuse of power by the strong against the lives of the weak and innocent.[75]

This muddled legal situation in which the value of human life rises and falls along a sliding scale,[76] like any other private interest, is nothing but the result of ethical relativism and legal positivism, which have transformed the theory of a distinction between law and morality into a theory of the radical separation of the two.

In the current legal debate, therefore, one often hears talk about the amorality of law.[77] This position ultimately leads to a system of law that ignores any criterion whatsoever of justice[78] or of the common good.

---

[75] John Paul II, Encyclical Letter *Evangelium vitae*, March 25, 1995, no. 11. G. Dalla Torre, "Le leggi contro la vita: Il loro significato politico-giuridico," in *"Evangelium vitae" e diritto*, Vial Correa and Sgreccia, 99–120; T. Styczen, "Le leggi contro la vita: Analisi etico-culturale," in *"Evangelium vitae" e diritto*, Vial Correa and Sgreccia, 813–227.

[76] The expression is from Ross, *The Right and the Good*; idem, *Foundations of Ethics* (Oxford: Oxford University Press, 1939).

[77] The expression is used in A. Levi, "Intorno ad un corollario del principio di socialità del diritto," in *Scritti minori di filosofia del diritto* (Padua: CEDAM, 1957), 1:3, in the author's critique of legal naturalism.

[78] H. Kelsen, *La dottrina pura del diritto* (Turin: Einaudi, 1966). He writes, "Justice is a requirement of morality, and the relation between morality and law includes the relation between justice and law." The author, however, starting from a simple distinction between law and morality, eventually comes to exclude morality from law entirely, an element of his teaching which is difficult to justify today: "Indeed, it should be kept in mind that, starting with Hobbes and continuing down to the doctrine of legal positivism, law has been considered the product of the will of the one who governs, whereas justice is construed either as an ethical ideal or as the mere product of positive law."

Many authors propose theories of an ethical minimum. They reserve a place for morality within certain limits: morality must recognize the absolute primacy of law and model itself upon universally accepted juridical principles sanctioned in the international declarations on human rights. Where this does not happen, morality is left to the imagination of individuals and deprived of its legitimacy at the collective level.[79]

Part of this doctrine actually seeks to affirm that law does not depend in any way upon the truth, but rather upon an act of normative will by whomever has legal power (legal positivism).[80]

The effects of such an approach have grave repercussions for legal and political systems. On the one hand, the law itself is actually emptied of its ethical content: rather than guiding the search for truth to promote the common good, it is reduced to a merely procedural mechanism for reaching a consensus. On the other hand, instead of being employed as a *means* or instrument for defending the rights of every individual at every stage and in every condition of life, the democratic system becomes the *end* that must be preserved so as to safeguard the interests of the majority.

In reality, as pointed out by authoritative proponents of the doctrine,[81] the modern concept of democracy should be marked not only by its formal procedures—as with the nineteenth-century model—but even more so by its emphasis on respect for individual rights and on safeguarding the dignity of the human person. This is what is described as the *rule of law*: the very power of government is checked by laws that are established to protect the individual and sanctioned by the principles enumerated in the constitution, which not even the ruling majority can alter.

Genuine democracy is in fact substantial, not merely formal,[82] and the value of democracy either

> stands or falls with the values which it embodies and promotes. Of course, values such as the dignity of every human person, respect for inviolable and inalienable human rights, and the adoption of the "common good" as the end and criterion regulating political life are certainly fundamental and not to be ignored. ... It is therefore urgently necessary, for the future of society and the development of a sound democracy, to rediscover those essential and innate human and moral values which flow from the very truth of the human being and express and safeguard the dignity of the person: values which no individual, no majority and no state can ever create, modify or destroy, but must only acknowledge, respect and promote.[83]

For some time now the legal world has perceived the urgent need to bring the ethical and axiological dimension of law back to light. It is a need to investigate not

---

[79] F. D'Agostino, "L'approccio morale al diritto," in *Scritti in onore di Angelo Falzea* (Milan: Giuffrè, 1991), 1:230.

[80] Following the well-known postulate of Hobbes, *auctoritas non veritas, facit legem* (authority, not truth, makes law).

[81] N. Blasquez, *Bioética fundamental* (Madrid: Biblioteca Autores Cristianos, 1996); *Ordine morale e ordine giuridico*; A. Caprioli and L. Vaccaro, eds., *Diritto morale e consenso sociale* (Brescia: Morcelliana, 1989).

[82] As D'Agostino puts it in his article, "Per un'ermeneutica della *Evangelium vitae*: Legge morale e legge civile," *Bioetica* 3 (1995): 406.

[83] John Paul II, *Evangelium vitae*, nos. 70–71.

so much its origins but rather its foundations, which can be accomplished through the recovery of those objective and universal values that support its norms and are founded on the ontological structure of man as a person.[84]

The task of constitutions is therefore to safeguard inalienable and fundamental values so as to guarantee an orderly civil life and the survival of society itself. This is called the constitutional minimum because it should be defined in the constitution of every state and is essential to the very life of a society.

Legislators are not expected to create, therefore, but rather to interpret the needs of man in society. They should seek not only consensus but also and more importantly the objective moral law "written in the human heart, [which] is the obligatory point of reference for civil law itself."[85] Law can only rediscover its intrinsic function on this basis and thereby steer clear of the dangers of ethical relativism, which too often throughout history has allowed the political authority to make abusive choices and has equated justice and freedom with arbitrary authoritarianism, especially when dealing with the weakest members of society. In the encyclical letter *Evangelium vitae*, the Magisterium of the Catholic Church therefore explicitly calls upon legislators to respect the "truth of the law," inviting them to say a courageous no to every form of violence and abuse against human life.

It is necessary to promote a genuine culture of human rights, with the awareness that they

> are incomprehensible without the presupposition that man, *qua* man, thanks simply to his membership in the species "man," is the subject of rights and that his being bears within itself values and norms that must be discovered— but not invented. Today we ought perhaps to amplify the doctrine of human rights with a doctrine of human obligations and of human limitations. This could help us to grasp anew the relevance of the question of whether there might exist a rationality of nature and, hence, a rational law for man and for his existence in the world.[86]

---

[84] S. Cotta, *Il diritto nell'esistenza: Linee di ontofenomenologia giuridica* (Milan: Giuffré, 1991), 194. As he puts it, "The most direct way of proving that a rule is obligatory is to demonstrate its conformity to a value."

[85] John Paul II, *Evangelium vitae*, no. 70. Interesting developments of this point are presented by J. Finnis, *Natural Law and Natural Rights* (Oxford: Clarendon Press, 1980), 314ff, and by J. Höffner, *La dottrina sociale cristiana* (Rome: Paoline, 1979), 57: "The content of most positive laws is not established by natural law, but merely obeys a universal requirement of the natural law, that is, the requirement of serving the common good. Such are most ordinances in civil, procedural, [and] penal ... law." See also J. Finnis, "Natural Law–Positive Law," in *"Evangelium vitae" e diritto*, Vial Correa and Sgreccia, 199–212.

[86] J. Ratzinger and J. Habermas, *Dialectics of Secularization: On Reason and Religion* (San Francisco: Ignatius Press, 2006), 71–72.

Secular Bioethics and Catholic Bioethics

Several centers and scholars have raised an issue—in many ways contrived—that sets so-called *secular bioethics* in opposition to *Catholic bioethics*.[87] The intent is to contrast a supposedly open and respectful view of everyone's choices—the secular position—with the Catholic view, which is characterized as closed, intolerant, and therefore unacceptable in a pluralistic and diverse society. Supposedly, secular bioethics is founded on reason and the values of conscience, whereas Catholic bioethics is founded on dogmas and faith, making them irreconcilable with one another.[88]

In reality, the question seems to be formulated improperly and addressed superficially. It should be clear from what has been said thus far that the ontologically grounded personalist approach, to which Catholics also subscribe, is far removed from fideistic positions. It does not ignore the rational justification of values and norms, but does just the opposite. Religious faith does not deaden appeals to reason but rather sharpens and reinforces them while fostering adherence to properly interpreted scientific data. Precisely through their respect for reality, which they maintain is created by God, Catholics take into account the scientific facts and derive from them elements for comparison with the principles of the faith, not vice versa.[89]

On the other side, there is an effort to propose an impoverished and distorted concept of the secular. It is equated with ethical relativism rather than with the affirmation of values common to all of humanity inasmuch as they spring from equal dignity and can be recognized by the light of reason alone through that ethical effort which is responsible for developing the doctrine of human rights. The opposition between Catholic bioethics and secular bioethics is therefore fictitious and misleading. The comparison should instead focus on their respective reference anthropologies and the problem of the foundation for ethical judgments without erecting a "dogmatic fence" around the dispassionate search for the truth.

The real difference in today's bioethics debate is between those who argue for an *ethics without truth* (to use the felicitous expression of Scarpelli) and those who maintain, to the contrary, that ethics (and hence bioethics) becomes an empty term unless rooted in truth. "To root ethics in truth is not easy. It involves humility, constant confrontation, and a willingness to recognize one's errors. It also involves the possibility of dialogue, for if we dialogue in the truth, which belongs to no one but is open to everyone, it is possible to avoid violence and abuses. But if we dialogue apart from the truth, then the temptation can become irresistible to substitute

---

[87] See the development of the debate in L. Palazzani, "Dall'etica 'laica' alla bioetica 'laica': Linee per un approfondimento filosofico-critico del dibattito italiano attuale," *Humanitas* 4 (1991): 413–446; A. Fiori, "Bioetica laica e bioetica cattolica," *Medicina e Morale* 46, no. 2 (1996): 203–207.

[88] A recent example of this interpretation is the "Manifesto di bioetica laica" (Manifesto of secular bioethics) signed by C. Flamigni, A. Massarenti, M. Mori, and A. Petroni. See the development of the extensive debate that followed in the monograph issue of *Notizie di Politeia* 41/42 (1996). See also the posthumous work by U. Scarpelli, *Bioetica laica*, ed. M. Mori (Milan: Baldini & Castoldi, 1998).

[89] Fiori, "Bioetica laica e bioetica cattolica," 203.

oneself, one's own authority, and one's own power for the hard but objective measure of truth itself."[90]

My intention is to promote an intellectual exchange on the basis of the reasons that support the personalist ethic here proposed. If I cite corroborating statements from magisterial documents, as I intend to do, it is because I find consonance and sometimes a prophetic intuition in them.

If I come to affirm that one finds the Creator and creation in the depths of the person, as the ultimate explanation for his existence and the ultimate reference for his dignity, it is also in light of demands that are not opposed to reason and are therefore rational. Thomas Aquinas's great confidence in the compatibility of the claims of reason with those of faith is the basis for my confidence that Catholics can dialogue with secular thinkers without feeling obliged to alter or reduce the claims of faith, which should be imposed upon no one but can be proposed with good reasons to anyone.

Moreover, there is no shortage of literature addressing bioethical problems from the starting point of revealed truth, in other words, from within a clearly theological framework. I think that major contributions both in the field of biomedical research and in the cause of humanizing medical care have come from the consideration of religious truth, in historical terms as well, especially in the Christian world and particularly in the Catholic Church.[91]

---

[90] F. D'Agostino, *Bioetica nella filosofia del diritto*, 312.

[91] L. Walters, "Religion and the Renaissance of Medical Ethics in USA, 1965–1975," in *Theology and Bioethics*, ed. E. E. Shelp (Boston: Reidel, 1985); D. Callahan, "Religion and the Secularization of Bioethics," *Hastings Center Report* 6–7, supplement (1990): 2–4; D. Tettamanzi, *Bioetica: Nuove frontiere per l'uomo* (Casale Monferrato: Piemme, 1990).

Summary Outline for Chapter 2

The experimental method

The experimental method consists of four phases:

- *Observation* of phenomena
- Formulation of a *hypothesis*
- Experimental *verification**
- *Evaluation* of results

---

*Currently the term *falsification* is used: the hypothesis is subjected to efforts at disproving it because *an infinite number of confirmations do not make a theory true, but one contradiction falsifies it.*

Reductionism

Although the experimental method makes it possible to accumulate extensive data, it is reductionist inasmuch as it evaluates only the quantitative aspects of reality:

"Microscopes and telescopes reveal no ethical facts." (R. Nozick)

From this perspective, what cannot be measured quantitatively does not exist or is not of interest.

Theoretical model of personalist bioethics

Personalist bioethics is *a dynamically thinking enterprise*, because

- it has to do with the *scientific discoveries* that are revisited from time to time, but especially because
- it also takes into account the various ways in which contemporary man perceives his own *identity* and *values* in general.

The *truth* about man in his many dimensions is not just an *inheritance* from classical thought but also a theoretical *gain*.

Method of personalist bioethics

This method has been presented by means of the image of *triangulation*:

- exposition of the biomedical reality
- more in-depth study of its anthropological significance
- identification of the values at stake

Dialectical aspect of personalist bioethics

Personalist bioethics does not follow the logic of deduction but rather the systematic interaction of various spheres of knowledge.

It has a *critical and dialectical structure*, which has its basis in a substantialist concept of the human person.

Purpose of personalist bioethics

The purpose of bioethics is not simply to

- *describe* the set of data involved in a more or less problematic relation, but to
- propose *prescriptive solutions.*

Cognitivist argumentation

The ethical and anthropological concept is *cognitivist*:

- It is possible to attain to some *truths* about man and about human acts which in principle *can be known* by all.
- From this follows *the analogous concept of truth*: precisely because all disciplines have something to do with truth, each according to the methods that characterize it, they are capable of understanding one another and of interacting as they go about evaluating what is at stake in human practices.

Trunks of personalist bioethics

The two trunks into which personalist bioethics is grafted:

- *personalism*
- Aristotelian-Thomistic *teleology* (*metaphysical reference*)

An ethics of ends is developed through the contributions of philosophical anthropology and metaphysics.

Ethical cognitivism, which is derived from this, aims to combine faithfulness and objectivity.

Metaphysical Necessity

In order for a human being to realize his own personal identity he has to know what his ontological identity is: what is his *good* and his *end.*

This good is identified by reason as the *Foundation*: a reality that is not simply at the origin of things but is also the actual necessary condition for their existence, or for keeping them in existence.

Affirming the existence of an intelligent and transcendent Foundation of reality delivers from meaninglessness man's duty, his own intentionality, and even his possible opposition to the purposefulness identified by reason.

Metaphysical discourse has no immediate exhortative value and has no persuasive capability: it corresponds to a demand for justification often substituted with "reasons of the heart" or "irrational faith" in the case of those who reject metaphysics.

Noncognitivism

Hume's law

This is the "golden rule" that declares the impossibility of deducing any duties from simple facts:

> "In every system of morality, which I have hitherto met with, I have always remark'd, that the author proceeds for some time in the ordinary way of reasoning, and establishes the being of a God, or makes observations concerning human affairs; when of a sudden I am surpriz'd to find, that instead of the usual copulations of propositions, *is*, and *is not*, I meet with no proposition that is not connected with an *ought*, or an *ought not*. This change is imperceptible; but is, however, of the last [utmost] consequence. For as this *ought*, or *ought not*, expresses some new relation or affirmation, 'tis necessary that it shou'd be observ'd and explain'd; and at the same time that a reason should be given, for what seems altogether inconceivable, that this new relation can be a deduction from others, which are entirely different from it." (Hume, *Treatise of Human Nature*, III, I, I)

Hume's law limits itself to stating that it is illegitimate to derive a norm (and hence an imperative, an *ought to be*) from a fact. Because Hume's law leads to noncognitivism, however, it must presuppose *another thesis* (historically expressed first by *empiricism* and then by neo-empiricism):

> Only descriptive statements can be true or false.

This is because the truth or falsity of a descriptive statement can be verified, whereas a normative statement expresses something that does not yet exist—it should come to be, hence there can be no verification whatsoever.

Therefore the theoretical presupposition that legitimizes the *great divide* between "is" and "ought" is the epistemological premise underlying the so-called *verification principle*, whereby a proposition would only have *meaning* when one could specify the method suited to verifying what is signified by that meaning. This perspective, within the context of epistemological reflection, has in fact been abandoned today in favor of the falsificationist perspective (trial and error).

The passage from "is" to "ought" is illegitimate if "being" is understood as something static, analogous to mathematical being, and if one has a mechanistic and reductive concept of the universe.

Moore's critique of the naturalistic fallacy

Moore's critique of the naturalistic fallacy opens the way to intuitionism:

> The good is something analogous to the color yellow, and since the good is not a thing or a natural quality, there is no way of knowing what is good by way of argument.

"If I am asked 'What is good?' my answer is that good is good, and that is the end of the matter. Or if I am asked 'How is good to be defined?' my answer is that it cannot be defined. ... My point is that 'good' is a simple notion, just as 'yellow' is a simple notion; that, just as you cannot, by any manner of means, explain to someone who does not already know it, what yellow is, so you cannot explain what good is. ... You can give a definition of a horse, because a horse has many different properties and qualities, all of which you can enumerate. But when you have enumerated them all, when you have reduced a horse to his simplest terms, then you can no longer define those terms.

"All these truths, however much they may differ, have this in common: that in them both the grammatical subject and the grammatical predicate stand for something which exists. Immensely the commonest type of truth, then, is one which asserts a relation between two existing things. Ethical truths are immediately felt not to conform to this type, and the naturalistic fallacy arises from the attempt to make out that in some roundabout way, they do conform to it. It is immediately obvious that when we see a thing to be good, its goodness is not a property which we can take up in our hands or separate from it even by the most delicate scientific instruments, and transfer to something else." (G. E. Moore, *Principia Ethica*, Cambridge 1903)

"Good" and "value," not being things, are not definable and can only be intuited. Anyone trying to identify the good would fall precisely into the naturalistic fallacy.

If the good can only be intuited, there is a danger of falling into subjectivism (implying a weak concept of reason that is incapable of a metaphysical view).

Clarifications

Ample use is made of both *Hume's law* and the *critique of the naturalistic fallacy* by noncognitivists, but it cannot be said that *they alone* hinder the construction of a system of cognitive moral knowledge of an objective sort.

The *rejection of metaphysical reason* and the reduction of reality to a mere factual datum are what hinders some modern thinkers from perceiving a *teleological order* which legitimates the passage *from* is *to* ought.

Between those who negate any relation whatsoever between fact and value and those who would like to place moral values on the same level as empirical facts there is an *intermediate path* which seeks to break out of the schematic confines of the *great divide*:

Aristotelian-Thomistic finalism

Models of ethics

The sociobiological model

This proposal is a purely descriptive ethics that relativizes values and norms.

The evolutionist theory of Darwin is combined with the sociology of Max Weber and with the sociobiologism of H. J. Heinsenk and E. O. Wilson:

- Society, as it evolves, produces and changes values and norms, which are instrumental in its development, just as living things in their biological evolution have developed certain organs with a view to the improvement of their own existence.

- The life of man is said to be essentially no different from the various life forms or from the universe with which he lives in symbiosis.

- Ethics then has the role and function of maintaining the evolutionary equilibrium.

- Nature is transformed into culture and vice versa.

- Man is reduced to a historicist and naturalistic element of the cosmos.

Critique: *Some components of a culture and its customs are subject to development, but man remains man, different from all other living beings by nature. Good and evil are not interchangeable; the laws of being, science, and morality cannot be false and true at the same time.*

The pragmatic-utilitarian model

The basic principle is the calculation of the consequences of an action in terms of the relation between costs and benefits; this relation is valid when it refers to one and the same value and to one and the same person.

Critique: *This principle cannot be used in an absolute and foundational manner to "weigh" disparate goods against one another, as when financial costs are compared with the value of a human life.*

The old utilitarianism going back to Hume's empiricism reduced the cost-benefit calculus to the single subject's evaluation of what is pleasant or unpleasant.

It is impossible to foresee all the consequences of an act so as to weigh them.

Neo-utilitarianism is inspired by Bentham and Mill and can be summarized in the threefold precept: maximize pleasure, minimize suffering, and broaden the sphere of personal freedoms for the greatest number of persons. And it is along these lines that the concept of "quality of life" has developed and is contrasted by some to the concept of "sacredness of life."

The quality of life is valued precisely in relation to the minimizing of suffering and often of economic costs.

Some have attempted to mitigate this "act utilitarianism" with "rule utilitarianism," by introducing some broader rules of beneficence, such as the concept of equity or the ethical minimum.

Critique: *It is not possible to establish rules of equity from an empirical and pragmatic perspective.*

Some authors (P. Singer) go so far as to reduce the category of "person" to that of "sentient being," in other words, one capable of experiencing pleasure or pain. Hence, *nonsentient human beings are no longer protected,* and, paradoxically, the authors end up *equating animals and human beings* on the basis of their capacity to "feel."

The subjectivist or liberal-radical model

Moral subjectivism is a common element in neo-Enlightenment thinking, ethical liberalism, nihilistic existentialism, neopositivist scientism, emotionalism, and decisionism.

- The main premise is noncognitivist: morality cannot be founded on facts or on objective or transcendent values, but only on the autonomous "choice" of the subject.
- The sole foundation of moral action is autonomous choice.
- The only limit is someone else's freedom.
- This is a freedom that extends halfway for some (for those capable of exercising it).
- It is "freedom from" ties and restraints and not "freedom to" devise a plan for one's life and for society that is justified in terms of its purpose; in other words, a freedom without responsibility.

Critique: *Every free act presupposes the life of the human being who posits that act; in other words, life comes before liberty, because someone who is not alive cannot be free. Freedom presupposes being and existence for a life-project. From this perspective it is difficult to impose a norm on someone who, in the name of autonomy, accepts no self-limitation.*

The abandonment of the rational foundation of morality tends toward the legitimization of violence, the law of the fittest ("might makes right").

The contractualist model

Contractualism is founded on the criterion of the intersubjective agreement reached by the ethical community, which is to say, by those who have the ability and power to decide (H. T. Engelhardt Jr.).

Critique: *Someone who cannot decide cannot belong to the ethical community and runs the risk of not being protected.*

Phenomenological model

Phenomenological ethics (M. Scheler, and N. Hartmann) exhibits an intentional and intuitive openness to values: ethical values have their foundation at the emotional and religious level.

Critique: *These grounds are still relativized, dependent upon emotional subjectivity.*

The theory of a "formal ethics of goods" formulated by D. Gracia also belongs in this spectrum; it declares the formal and universal necessity of values, inasmuch as the very knowledge of reality awakens in consciousness the awareness of values, but this formal requirement is realized in subjective acts of assessment.

Discourse ethics

The theory of discourse ethics, proposed by K. O. Apel and J. Habermas, bases social consensus on communication.

Critique: *There is the risk of subordinating the validity of the norm to consensus.*

Principlist ethics

This is the so-called theory of "principlism" formulated by Beauchamp and Childress, who start with four principles: beneficence, nonmaleficence, autonomy, justice.

Critique: *It remains to be defined what good or evil is for a patient, but above all it is necessary to establish a hierarchy among the principles themselves, especially between the principles of autonomy and beneficence. The possible common ground is the true good of the person.*

Virtue ethics

This is regarded as contrasting with principlist ethics because it emphasizes the importance of the *habit* of conscience inspired by the virtues.

Critique: *It becomes necessary to integrate the stage of clarifying and establishing values and norms with the stage of applying them (so that the virtuous act might not be deprived of a foundation).*

The personalist model

"Ontological" personalism emphasizes that at the foundation of subjectivity there is an existence and an essence constituted by *the body-soul composite*.

The person is *rationalis naturae individua substantia* [an individual substance of a rational nature] (Boethius), because in man personhood consists in an individuality constituted by a body animated and structured by a spirit.

Man is a person because he is the only being in which life becomes capable of "reflection" upon itself, or of *self-determination*. He is the only living thing with the capacity to grasp and *discover the meaning of things* and to give meaning to his own expressions and *conscious language*.

The spiritual soul that *informs and gives life to his bodily reality* causes the person to be irreducible to a code or a number, to cells or neurons.

The human person is a unity, *a whole and not a part of a whole*.

Christian Revelation broadens the horizons of this personalist vision: *every single human being is the image of God*, a child of God, a brother or sister of Christ.

Even from the secular perspective, the human person is always *an end and never a means*.

Ontological personalism is different from subjectivist individualism, which emphasizes almost exclusively the capacity for self-determination and choice. Personalism maintains that *the human person is valuable for who he is* (a body-soul unity) *and not only for the choices that he makes*; he is a source from which the choices proceed.

According to personalist ethics, the *ethical value of an act* will have to be considered under the subjective aspect of *intentionality*, but it will also have to be considered in terms of its *objective content* and its consequences.

The *natural moral law* that urges every conscience to do good and avoid evil therefore becomes concrete in *respect for the human person* in all the fullness of his values and in his essence and ontological dignity.

Moral law and civil law

Civil law ought to safeguard indispensable fundamental values, and when it is incapable of *guaranteeing the common good*, it should be changed.

The fundamental good is the life of the individual, whether born or not yet born.

The split between the correlatives "truth" and "freedom" makes it difficult to defend human life effectively by means of legislative statutes.

The law is emptied of its ethical content, and rather than guiding the search for truth so as to promote the common good, it is reduced to a mere procedural mechanism for seeking consensus.

The democratic system, instead of presenting itself as a means for defending the rights of every individual at every stage and in every condition of his life, becomes an end which is to be preserved so as to safeguard the *interests of the majority*.

What is needed is a substantial and not merely formal democracy that promotes the dignity of every human person and respect for his inviolable and inalienable rights.

The end and criterion of political life ought to be *the common good*.

The *legislator* is not expected to create but rather to *interpret* the needs of man in society, in search not only of consensus but more importantly of that *objective moral law* "written in the human heart, [which] is the obligatory point of reference for civil law itself" (*Evangelium vitae*, no. 70).

"Today we ought perhaps to amplify the *doctrine of human rights* with a doctrine of human obligations and of human limitations" (J. Ratzinger, 2005).

Secular bioethics and Catholic bioethics

The *opposition* between secular bioethics and Catholic bioethics is *fictitious* and misleading.

The intent is to contrast a *secular* view (supposedly *open* and respectful of everyone's choices) with the *Catholic* view (supposedly *closed* and intolerant).

Yet the *personalist* approach to which Catholics subscribe is not fideistic, nor does it ignore the *rational justification of values and norms*.

Religious faith does not deaden appeals to reason but rather sharpens them. Catholics, precisely through their respect for the reality that they maintain is created by God, take into account the scientific facts, and from these they derive elements to be compared with the principles of the faith.

Only *impoverished secular thought* is identical to ethical *relativism* and *is not* an affirmation of *values that are common to all men* (inasmuch as they spring from their equal dignity and can be recognized by the light of reason), which are the basis for the development of the doctrine of human rights.

Coming to affirm that one finds the *Creator* and creation in the depths of the person, as the *ultimate explanation* for his existence and the ultimate reference for his dignity, also means responding to demands that are rational or at least not opposed to reason.

One example of *the compatibility of the claims of reason with those of faith* is Thomas Aquinas, who maintained that faith should be imposed on no one but can be proposed with good reasons.

In the debate about bioethics the *real difference* is between those who argue for an *ethics without truth* and those who maintain, on the contrary, that unless it is *rooted in the truth*, ethics (and hence bioethics) becomes an empty term.

Chapter 3

FORMS, ORIGINS, AND MEANING
OF HUMAN LIFE

Life and Its Forms

To understand the topics addressed in this manual, a preliminary philosophical discussion of two questions is necessary: what is life in general and what is human life in particular? Our attention should shift from the perception and understanding of the cosmos as a whole toward a focus on life, then move from the wide variety of living things in the world to the consideration of the life of man. The spheres of reality implicated in this discussion are being, life, and man. It is then possible to examine the specific importance and meaning of these other notions: ethics (or the science of values) and ethics as applied to science and biomedical practice.

I cannot engage this twofold preliminary and philosophical discussion at length.[1] It will have to suffice, for the purposes of this book itself, simply to recall the fundamental concepts or notions. "Life is not an object to be investigated, but rather is the basis of all activity."[2] The highest form of life entails consciousness. "Quis non intelligit non habet perfecta vita": one who does not understand does not have perfect life.[3]

Consciousness and life cannot be put at odds as subject and object: "Whoever is not prepared to choose conscious life as the paradigm for interpreting the living in general must also deny the being-alive of the living and reduce it to an 'objective' structure of material being. What one does not realize is that that which is called

---

[1] On the philosophical problem of man, see J. Gevaert, *Il problema dell'uomo* (Turin: LDC, 1984); S. Vanni Rovighi, *L'antropologia filosofica di s. Tommaso d'Aquino* (Milan: Vita e Pensiero, 1965); *La filosofia dell'uomo: Atti del Congresso della Federazione Universitaria Cattolica Italiana* (Rome: 1961); J. Maritain, *Quatre essais sur l'esprit dans sa condition charnelle* (Paris: Aubier, 1965); G. Marcel, *L'homme problématique* (Paris: Aubier 1965); E. Mounier, *La personnalisme* (Paris: Seuil, 1950); R. Lucas Lucas, *L'uomo spirito incarnato: Compendio di filosofia dell'uomo* (Milan: Paoline, 1993); X. Zubiri, *Il problema dell'uomo: Antropologia filosofica* (Palermo: Augustinus, 1985); B. Mondin, *L'uomo: Chi è? Elementi di antropologia filosofica* (Milan: Massimo, 1989); A. Cavadi, N. Galantino, and E. Guarnieri, *Alla ricerca dell'uomo* (Palermo: Augustinus, 1988); N. Galantino, *Dire 'uomo' oggi: Nuove vie dell'antropologia filosofica* (Milan: Paoline, 1993).

[2] L. Melina, "Vita," in *Dizionario interdisciplinare di scienza e fede*, ed. G. Tanzella-Nitti and A. Strumia (Rome: Urbaniana University Press, 2002), 1521f.

[3] Thomas Aquinas, *Summa theologiae*, I, q. 18, a. 3.

'being' in reference to materiality can in its turn only be understood from the view-point of the living."[4] Man's conscious life is the place where it becomes apparent what life is; it is where his origin from God and his orientation to God emerge.

The question of life *is not a biological question*, but rather the *horizon* or *meaning* within which every other question is situated.

> Since the emergence of life is a prerequisite for the emergence of meaning, it is easy to understand how the first principle of bioethics arises: *defend life*, which is to say, defend that horizon which guarantees any possibility of meaning and consequently our possibility of existing as subjects open to meaning. It is a principle, it should be noted, and not yet a norm; thus an indication, not a prescription. But nevertheless, it is somehow an absolutely decisive indication: the wonder that we experience when we have the opportunity to observe and contemplate the birth of a new life is not merely the expression of some psychological disposition of ours, but rather something more profound and more personal: in every life that is born we perceive the manifestation of the same source of our own "I." And since it is possible to say "I" only in life and thanks to life, defending life is not only a bioethical principle, but also a fundamental ontological principle. It is that principle that biology, as a science, cannot investigate but can only presuppose, and which bioethics, as a philosophical discipline, is not called to presuppose but rather to propose resolutely, honestly, and courageously.[5]

Life and organic or biological life are not equivalent notions.[6] The concept of life is consequently analogous in and of itself: there are different levels of life, and the same term may denote even very different forms of life. "Life" is said in reference to animal life, human life, and superhuman life. Within man it designates corporeal life, psychological life, and spiritual life. In classical Greek there are three terms indicating various facets of the phenomenon of life: *zoé, bios*, and *psyché*.

Zoé refers to the *vitality* evidenced in all organic beings. It is the principle of life, the opposite of which is "nonlife" or "nonliving things" and not "death," since the things that die are individual, organic, living beings, and not the principle of life itself. For this reason spiritual language also uses the term *zoé* to designate eternal life (see Jn 12:25).

In contrast, *bios* is essentially *individual, plural*, and *mortal* and refers to the modalities or walks of life. Whereas *zoé* is the life by which we live (*qua vivimus*), *bios* is the life that we live (*quam vivimus*). *Bios* expresses the living thing in its empirical individuality, bound up with temporality. In reference to man, the Greek *bios* indicates metaphorically one's profession or trade and everything that qualifies a life in its fragile uniqueness. The various modes or walks of life, whether political, intellectual, contemplative, or otherwise, are designated in Greek by *bios* together with the pertinent adjective.

---

[4] R. Spaemann, *Happiness and Benevolence* (Notre Dame, IN: University of Notre Dame Press, 2000), 101.

[5] F. D'Agostino, "Vita," in *Parole di bioetica* (Turin: Giappichelli, 2004), 185.

[6] See V. Possenti, "Vita, natura e teleologia," in *Essere e libertà* (Soveria Mannelli, CZ: Rubbettino, 2004), 115–144; F. D'Agostino, "Bios/Zoe/Psyche," in *Parole di bioetica* (Turin: Giappichelli, 2004), 27–34.

Psyché refers to the vital breath, the soul, and hence life. The Romance languages have translated *psyché* with the word *anima*, because the soul animates the body, rendering it individual and distinct from other bodies. *Psyché* is much more than a merely animating principle of a physical sort because it allows the living being to recognize and affirm itself as an "I," opening the gates to the plane of meaning.

The first distinction that cleaves cosmic reality in two, in the qualitative and essential sense, is the distinction between living and nonliving beings. From a philosophical perspective, the peculiarity of living things is that they are capable of activity that starts from within the living subject and tends to perfect that same subject: *life is the capacity for immanent action*. One can set aside the investigation of the physical, chemical, and biochemical characteristics of living beings and examine the question from the philosophical perspective. The irreducible and qualitative leap in the phenomenon of "life" therefore consists in the real capacity of a living thing to be the cause and end of its own action: this is precisely what "immanent action" means.[7] On the first level of life, which is vegetative life, this immanent action is a threefold capacity: nutrition, growth, and reproduction. On the second level, sensitive life, sensory activity is added to the faculties of vegetative life and thereby the living thing has the ability to regulate its own activities. On the third level, the spiritual life that characterizes man, the capacity for intellectual consciousness and free will appear.

Contrary to the mechanistic view that sees only a difference of degree and complexity between living and nonliving beings, vitalism sees a qualitative and essential difference in living beings. This is not meant to deny that there are physical and chemical processes and exchanges within living beings, nor does it mean that a superior entity called a "soul" (whether vegetative, sensitive or intellectual) is situated on a higher, parallel plane alongside of these.

Vitalism—understood in its hylomorphic version—declares that the biochemical exchanges and processes in the living being are informed, pervaded, and guided by a new unifying principle by which *the whole regulates and determines the parts and their functions*. The living being therefore has its own specific and *substantial unity*. The living organism can be considered an enormous chemical laboratory in miniature where countless reactions take place, many of which are extremely complex and all of which tend toward the same end: the maintenance of the individual.[8] This unifying principle is the *soul* of the living thing.

---

[7] See Aristotle, *De anima*, II, 1, 403b, 16, and Thomas Aquinas, *De potentia*, q. 10, a. 1: "Actio immanens est tantum viventium."

[8] For a more in-depth discussion of the issues in this chapter, see the works V. Marcozzi, *La vita e l'uomo* (Milan: CEA, 1946); idem, *L'uomo nello spazio e nel tempo* (Milan: CEA, 1953); idem, *Le origini dell'uomo: L'evoluzione oggi* (Milan: Massimo, 1972); P. P. Grasse, *L'evoluzione del vivente* (Milan: Adelphi, 1979); G. Pastori, *Le leggi della ereditarietà biologica* (Brescia: La Scuola, 1958); Semaine des Intellectuels Catholiques, *Qu'est-ce que la vie* (Paris: P. Horay, 1958). For the philosophical dimension of the problems, see S. Vanni Rovighi, *Elementi di filosofia* (Brescia: La Scuola, 1963), 3:73–104; Gevaert, *Il problema dell'uomo*, 91–114; L. Lombardi Vallauri, "Le culture riduzionistiche nei confronti della vita," in *Il valore della vita: L'uomo di fronte al problema del dolore, della vecchiaia, dell'eutanasia; Atti del 54° corso di aggiornamento culturale dell'Università Cattolica; Roma, 2–7 settembre 1984*, A. Bausola et al. (Milan: Vita e Pensiero, 1985), 41–74; F. Facchini, *Il cammino dell'evoluzione umana* (Milan: Jaca Book, 1985); Lucas Lucas, *L'uomo spirito incarnato*; M. Artigas, *Le frontiere dell'evoluzionismo* (Milan: Ares, 1993).

The three so-called kingdoms of life are vegetative life, sensitive life, and intellective or rational life. From the philosophical perspective, the distinctions among them are based on two criteria: the living being's *autonomy* and *superiority* to the nonliving world. The levels of *autonomy* and *superiority* determine the three kingdoms.

In vegetative life, both the end of the immanent action and the form of that action are determined and not optional. An animal, through its cognitive and sensory life, chooses the form of its own action based on the cognitive form (e.g., feeding on grass or fleeing from a man, in the case of a rabbit). In addition to choosing the form of the activity and its execution thereof, man also chooses the end of his action; the purpose for which he acts is chosen through his free, intellective life. Since it is free, this choice entails ethics. Vegetative life, sensitive life, and intellective life therefore reveal not only differences of degree but also levels of superiority.

Life and Teleology

Today it is important to formulate a philosophy of life and organisms that is in keeping with the pace of biological discoveries[9] yet does not rule out *finality*, is not mechanistic, and therefore does not perceive living beings as "chemical machines."[10]

The panvitalism of antiquity has given way to the present situation in which life is considered extremely rare and improbable in the immensity of the universe. One might think that this is connected with the immense expansion of the known universe and the diminishing percentage of biosphere in the universe, but it is also important to note the modern tendency to reduce the organic to the inorganic.

The problem of the ancients was life and the organism; indeed, Plato considered the entire cosmos a living thing endowed with perfection and beauty. "When [the creator] was framing the universe, he put intelligence in soul and soul in body, that he might be the creator of a work which was by nature fairest and best. [In this way], using the language of probability, we may say that the world came into being—a living creature truly endowed with soul and intelligence by the providence of God" (*Timaeus*, 30b, Benjamin Jowett translation).

Although reality was a *res vitalis* (living thing) for the ancients, modern minds perceive a tension between *res extensa* (a thing having extension in space) and *res cogitans* (a thing that thinks). This dualism is completely unsuited to understanding the phenomenon of life, which cannot be reduced to either extension or thought. The modern vision has brought about the reduction of life to its mechanical aspect—to a series of efficient causes. This framework also rules out the concept of nature *"as the internal principle of movement and life."*[11]

Modern science has been built up in a long battle against Aristotelianism under the auspices of the exclusion of teleology. This has permitted the mathematical study of the vast realm of *res extensae*, leaving finality to the realm of the spirit.[12]

---

[9] See Possenti, "Vita, natura e teleologia," 115–144.

[10] See J. Monod, *Chance and Necessity: An Essay on the Natural Philosophy of Modern Biology*, trans. A. Wainhouse (New York: Knopf, 1971), 45.

[11] Possenti, "Vita, natura e teleologia," 118.

[12] See ibid., 126–132.

The idea of finality is linked to movements from what is less to what is more perfect, which can be grasped through the analysis of the intellect. The teleology that is intrinsic to the development of an organism is precisely one such movement from less to more. But in the mathematical analysis of the change, the very idea of "more or less perfect" no longer has any meaning. Generally speaking, the sciences that employ mathematics know how to express only the link between an efficient cause and its effect, excluding teleology from their outlook as a matter of principle.

It should instead be demonstrated that an end is neither just a concept projected onto things by man nor a concept that can only be employed within the sphere of consciousness; rather, there is finality throughout the world of living things, not just among those endowed with subjectivity and will.

In inanimate realities, the purpose or end is something external; it does not reside in the thing but rather in the mind of the planner. A knife is made to cut, and this teleology comes from the mind of the maker. It is not immanent to the knife, which is without ends of its own and has only those which the maker assigns to it, such that there is a separation between purpose and object. On the contrary, the purpose is immanent within animate realities.

Organisms that are not endowed with consciousness, or else those parts of the human body which function involuntarily, pose another problem. In both cases it appears difficult to deny an immanent teleology, intrinsic to the very operation of a plant, other organism, and/or single organ. "It should not be forgotten that finalism is not just external finalism, that is to say, a finalism of the whole that is directed toward something. It is also and primarily an internal finalism of the organism, even where there seem to be purposes without intentions. Final causes are to be found first of all within every being endowed with life, however small and insignificant it may be."[13] Hans Jonas also emphasizes that "there is no organism without teleology,"[14] and life cannot be reduced therefore to mathematically expressible phenomena.

It becomes impossible to guarantee the dignity of the human person in a cultural context where people fail to employ adequate concepts of life and living organisms: "If life is unthinkable, so is the person, for persons are living beings. The identity of the person is a function of the identity of a living thing. And where consciousness and matter are defined independently and opposed as incommensurables, we end up with quite different criteria for the identity of human beings and of persons."[15]

For and Against Evolutionary Theory

This question is still far from being resolved, and it is and will remain a scientific problem per se if considered in the terms in which it is usually posed, i.e., as an investigation of the immediate causes and the physical and material genesis of life. Yet this uncertainty does nothing to undermine the philosophical and metaphysical problem concerning the value of life.

Francesco Redi (seventeenth century) and Lazzaro Spallanzani (eighteenth century) proved that worms, insects, and infusorians do not spontaneously generate.

---

[13] Ibid., 129.

[14] H. Jonas, *Organismo e libertà*, ed. P. Becchi (Turin: Einaudi, 1999), 127.

[15] R. Spaemann, *Persons: The Difference between "Someone" and "Something,"* trans. O. O'Donovan (New York: Oxford University Press, 2006), 137.

In the nineteenth century, Pasteur proved that bacteria do not spontaneously generate either. If ultraviruses exist, and if they are living things, it cannot be determined at present whether they spontaneously generate. Yet even if spontaneous generation were ever proved, it would not undermine the position of those who maintain, by philosophical and metaphysical reasoning, that life constitutes a qualitative novelty of being and of cosmic reality. In other words, the theory of spontaneous generation does not contradict vitalism. Scientists will still have to explain *why* life was spontaneously organized as a typical and well-defined phenomenon from some inorganic matter through a complex interaction involving multiple reactions at a certain moment in time. And *why* here refers to the first and remote efficient cause as well as the final and ultimate cause of life, and specifically of human life.

The same holds true for the theory of transformism, or the evolution of species. As everyone knows, *transformism* or the *theory of evolution*, which declares that living species have descended from one another—the more complex from the less complex—through natural transformation or evolution, is the prevailing hypothesis today. This theory opposes a more ancient one called *fixism*, according to which the species that presently exist are the same ones that existed at the beginning of the world; it is summed up in the axiom attributed to Carl Linnaeus, an eighteenth-century Swedish botanist: "Tot numeramus species quot primum creavit Infinitum Ens."[16]

Over the centuries various versions of transformist or evolutionary theories have developed. Jean-Baptiste Lamarck provided an initial formulation in his work entitled *Philosophie Zoologique* (1809), basing his theory on two principles: transformations in species are induced by adaptation to the environment through the use or disuse of certain organs, and the transformations induced in individuals are transmitted by generation when shared by both sexes.

Charles Darwin, in his 1859 book *The Origin of Species*, bases the theory of evolution on the principle of the survival of the fittest with natural selection: competition occurs between the various individuals within a species, and changes are caused by the fact that the most fit individuals survive and multiply while the less fit disappear along with their characteristics.

Neo-Darwinism[17] distinguishes itself from the theories of Lamarck and Darwin, focusing its observation on the diversity encountered among the various individuals of the same species and denying the influence of the environment in the modification of hereditary traits. Causes for variation are sought in the so-called *germ plasm*.

The first to advance such arguments was August Weismann (1834–1914), who conducted the famous experiment of cutting off the tails of mice for several generations without ever obtaining tailless or short-tailed mice. Hereditary traits were therefore to be found in the reproduction of individuals and in their germ

[16] "We count the same number of species as the Infinite Being first created." Vanni Rovighi, *Elementi di filosofia*, 3:94; see E. Mayr, *Storia del pensiero biologico* (Turin: Bollati Boringhieri, 1990); A. La Vergata, *L'evoluzione biologica: da Linneo a Darwin* (Turin: Loescher, 1979); G. Sella and P. Cervella, *L'evoluzione biologica e la formazione delle specie* (Turin: SEI, 1987).

[17] G. Reale and D. Antiseri, *Il pensiero occidentale dalle origini ad oggi* (Brescia: La Scuola, 1983), 3:717; R. Morchio, "La biologia nel XX secolo," in *Storia delle scienze*, ed. E. Agazzi (Rome: Città Nuova, 1984), 2:367–385.

plasm. He presented his theory in *Essays upon Inheritance and Kindred Biological Problems* (1892).

Another neo-Darwinist was Hugo de Vries (1848–1935), who studied the ornamental plant *Oenothera lamarkiana*. He observed its normal variety and the appearance of individuals of a gigantic variety. He concluded that a certain percentage of abnormal individuals (1–2 percent) might arise in the transmission of hereditary traits. Thus he formulated the theory of evolution by leaps or "mutations": every species at some point undergoes abrupt mutations under certain yet-to-be-identified conditions, giving rise to individuals with new traits. This theory is still respected in the context of contemporary scientific opinion.[18]

The reasons for these abrupt variations were still unknown, however, and a new science would attempt to solve the riddle: genetics. Genetics was born with Gregor Mendel, who had discovered its governing laws in 1865 through his experiments on pea plants. By cross-pollinating peas of various heights he formulated the *law of segregation*; by cross-pollinating plants that differed in several traits he formulated the *law of independent assortment*. His experimentation led him to conclude that dominant and recessive traits are revealed at the moment of fertilization and, furthermore, that there must be discrete *elements* that separate and recombine according to the various possibilities determined by combinatorial calculations. At the time his discoveries did not receive the attention they deserved. It was the German scientist Walther Flemming (1843–1905), using chromatin dye, who discovered or rediscovered the *chromosomes* that are found in the cells of various species in a fixed number, joined in pairs; he also discovered that each germinal cell, or gamete, contains half of this chromosomal patrimony. He published his findings in 1882.

The Belgian scientist Edouard Van Beneden (1846–1910) described meiosis after studying the chromosomes of *Ascaris*; he also found that among the sex cells each one of the gametes had half of this patrimony. The American cytologist Walter Sutton (1877–1916) gave credit to Mendel in 1902 and confirmed the Mendelian laws in light of the discovery of chromosomes. In each pair of chromosomes in a new organism one chromosome comes from the father, transmitted by the spermatozoon, and the other comes from the mother, transmitted by the oocyte. In the meeting and mixing of chromosomes, each generation tends to bring to light recessive traits which are silenced by a dominant trait. These ever-new combinations therefore produce the variations that will then be utilized by natural selection.

Genetics took another step forward with the discovery of genes through the work of the American zoologist Thomas Hunt Morgan (1866–1945). He discovered that within the chromosomes—located in the nucleus of the cell—there are ordered particles, the size of a large protein molecule, which make up the hereditary patrimony of the individual. These genes are capable of reproducing while preserving their own individuality and independence from the other genes, are responsible for specific traits of the individual, and can recombine in every possible way. He conducted his experiments on the fruit fly, *Drosophila melanogaster*, which has four pairs of chromosomes and reproduces in twelve days.

Hermann Müller (1890–1967) made a sensational discovery in 1927: bombarding the gametes of animals and plants with x-rays resulted in mutations. Mutations

---

[18] G. F. Azzone, *Il senso della vita* (Bari: Laterza, 1994).

could be genetic, chromosomal, or genomic, depending on the elements involved in the mutation. The genomic variations implicate an extreme possibility: variations in the very number of chromosomes. This reinforced the theory of evolution on two key points: the possibility of spontaneous generation and the possibility of the evolution of life forms and species.

Additional hope for the spontaneous origin of life came from Stanley Miller's experiments, from 1953 to 1957, on the formation of complex organic compounds such as amino acids, which are the molecular building blocks of proteins, which in turn are the fundamental elements of protoplasm. These experiments were not considered sufficient proof of spontaneous generation and the question is therefore still open.

Meanwhile, with the discovery of deoxyribonucleic acid (DNA) that makes up genes, new light was shed on the mechanism by which hereditary traits are transmitted. The macromolecules of DNA are polymers made out of residues of phosphoric acid, a pentose sugar (deoxyribose), and the purine bases adenine, guanine, cytosine, and thymine. The double-helix structure of DNA was identified by Francis Crick and James Watson in 1953, in the wake of studies conducted some years before by Oswald Avery and Linus Pauling.

Another type of nucleic acid was later discovered, ribonucleic acid (RNA), which differs somewhat from DNA in its chemical composition. Its structure seems to be a single strand and its supposed function is to activate the mechanism for genetic transmission.

At this point molecular biology began to study the formation mechanisms of enzymes and of amino acid sequences in proteins, which are determined by the combinations of bases in DNA. In 1955, Severo Ochoa successfully synthesized RNA in vitro, and in 1956, Arthur Kornberg produced DNA in vitro. In 1961, François Jacob and Jacques Monod demonstrated the existence of messenger RNA synthesized according to the template of the DNA and appearing as a sort of ribbon containing the protein sequence, written in a code of triplet characters. With the findings of Marshall Nirenberg and Heinrich Matthaei in the 1960s, it became possible to define the various types of RNA (ribosomal RNA or rRNA, messenger RNA or mRNA, soluble or "transfer" RNA or tRNA), and more light was shed on the sequence

$$DNA \rightarrow RNA \rightarrow proteins$$
$$transcription \rightarrow translation$$

These discoveries relating to the genetic code led scientists to a mechanistic interpretation of all the phenomena involved in the origin of life, taking just one explanation as valid from the lowest forms of life, such as viruses and bacteria, to higher organisms, and on to man. Thus a strong incentive was given to the reductionist theory; furthermore, evolutionism also garnered significant support from the discovery of the mutations mentioned.[19]

The discovery of the "second genetic code," published in the journal *Nature* in May 1988 by two researchers from the Massachusetts Institute of Technology in Boston, Paul Schimmel and Ya-Ming Hou, could be interpreted as support. Special-

---

[19] See Pius XII, Encyclical Letter *Humani generis*, August 12, 1950; see John Paul II, "Message to Participants in the Plenary Assembly of the Pontifical Academy of Sciences," October 22, 1996, in *Orizzonte Medico* 5 (1996): 4–5.

ists know that this second genetic code clarifies an important phase in the process of intracellular protein synthesis, allowing scientists to decipher the ways in which tRNA transmits instructions that specify which amino acids should be bound and then transported to the site where the synthesis occurs. This logical system is used by all life forms and, according to the same researchers, the more progress is made in deciphering this code, the greater success there will be in ascertaining the mechanisms of the evolution of species.[20]

But at this point we need to note the progressive reduction of the intrinsic significance of the various forms of life to a simple measurement. This explanation is understood in an increasingly elementary way with regard to the mechanisms involved, while the finalism that pervades the enormous range and wealth of the life forms themselves is bracketed off. This leads to a sort of epistemological divergence between biology and anthropology. Yet the further down the mechanistic explanations reduce reality, the more necessary the philosophical and finalistic perspective becomes in order to interpret the whole vital reality.

These evolutionary theories that arose in the scientific field were fostered along those lines by more general philosophical ideas that found their support in idealism and dialectical materialism, intellectual currents that are based on the dialectic concept of reality, and also in some spiritualistic currents of thought, such as the philosophy of Henri Bergson.

We should recall the more recent perspectives of Pierre Teilhard de Chardin, SJ (1881–1955), and Karl Popper (1902–1994). The former offers a creationist and Christocentric vision of evolution which maintains that, from creation on, the evolutionary plan of the cosmos moves toward the appearance of man (hominization), while the evolutionary history of humanity is incorporated into the divine plan of redemption because it is finalistically centered upon and projects toward Christ, the Omega Point of humanity. Rev. Teilhard also relied on arguments from paleontology regarding the hominization phase.

Popper formulates the hypothesis of *emerging* or *creative* evolution. He explains his theory as follows: it starts from the assumption that "matter turns out to be highly packed energy, transformable into other forms of energy; and therefore something of the nature of a process, since it can be converted into other processes such as light and, of course, motion and heat." Thus he goes so far as to declare: "There appear to be at the very least the following stages in the evolution of the universe, some of which produce things with properties that are altogether unpredictable or emergent: (1) The production of the heavier elements (including isotopes), of liquids and the emergence of crystals. (2) The emergence of life. (3) The emergence of sentience. (4) The emergence of the consciousness of self (together with human language) and of death (or even of the human cerebral cortex). (5) The emergence of human language,

---

[20] For an overview of the various authors who have recently dealt with the problem of evolutionism in a philosophical light as well, see H. Rolston, ed., *Biology, Ethics and Origin of Life* (Boston: Jones & Bartlett, 1995). This volume is an anthology of papers that various authors, all of them eminent scientists and philosophers of science, presented during the conference held at State University of Colorado in 1991, which was also entitled "Biology, Ethics and Origins of Life."

and of theories of the ego and of death. (6) The emergence of such products of the human mind as explanatory myths, or scientific theories, or works of art."[21]

Popper accepts to some extent the ideas of another recent evolutionist, Jacques Monod (1910–1976), a Nobel prize winner in medicine, who in his work *Le hasard et la nécessité* (*Chance and Necessity*) declares the unpredictability of the appearance of life on earth and the unpredictability of the various species, particularly the human species: we were unpredictable before our appearance.[22]

Monod explains evolution as an event due to chance and to the defects in the mechanism of invariance.[23] But since it is related to chance, evolution is not "natural"; it is an exception. Necessity, invariance, and stability are the rule, so much so that teleonomy is traced back to invariance: "The essential teleonomic project ... [consists] in the transmission from generation to generation of the invariance content characteristic of the species."[24]

Vittorio Possenti incisively notes that Monod, though vehemently hostile toward finality, is forced to admit it: *finality is the transmission of invariance.*[25] Monod abolishes "nature"—as activity and non-mechanistic construction from within—and omits the relationship between agent, action, and end, thereby reducing life to nonlife.

Evolution therefore emerges as something like a pathology of what is normal and pre-existent:

> In Monod's scheme invariance is affirmed so as to allow for stability, as much as it is denied so as to allow for the mutations which are precisely the opposite of invariance. Invariance seems to be a revolving door that at one moment blocks the mutation and at the next accepts it and establishes it within itself. Invariance is denied in order to make room for the mutation and then suddenly affirmed in order to make room for the transmission of the mutation. If we wished to translate the scientific language of Monod into the philosophical terminology of hylomorphism, we would have to say that in Monod the formal cause is invariance—or the DNA that preserves the invariance of the species.[26]

In Monod's framework, man emerges by chance; hence it is opposed to theories of natural theology in which the universe is created by a wise intelligence and man is the image of God.

---

[21] K. Popper and J. Eccles, *The Self and Its Brain* (New York: Springer Verlag, 1977), 7, 16.

[22] See Monod, *Chance and Necessity*, 42–44. "The biosphere does not contain a predictable class of objects or of events but constitutes a particular occurrence, compatible indeed with first principles, but not *deducible* from those principles and therefore essentially unpredictable. ... The theory would anticipate the existence, the properties, the interrelations of certain *classes* of objects or events, but would obviously not be able to foresee the existence or the distinctive characteristics of any *particular* object or event" (p. 43).

[23] The reason is that "it stems from the very *imperfections* of the conservative mechanism" (Monod, *Chance and Necessity*, 116), i.e., genetic invariance.

[24] Monod, *Chance and Necessity*, 14.

[25] See Possenti, "Vita, natura e teleologia," 143.

[26] Ibid., 142–143.

In addition to their philosophical underpinnings, evolutionary theories rely on various forms of argumentation for support. They are listed and described as follows:

- Paleontology confirms that the plethora of species did not appear contemporaneously. Less complex living things appeared first, then more complex, and finally—among the mammals—human beings appeared. Present-day species exhibit differences with respect to those that have disappeared, even though they are of the same order and degree; the differences between the species that disappeared in one period and others that disappeared in later periods are minimal, so as to indicate slow transformations; paleontologists have found fossils (e.g., *Archaeopteryx*) belonging to intermediate species and individual fossils that seem to have played the role of "links" between one species and another.

- Comparisons between various geographical environments show that various species developed in different ways. Such a comparative study was done on about one hundred species of marine fauna of the Pacific and Atlantic oceans east and west of the isthmus of Panama, species that did not exist before the Miocene period: these species exhibit paired (i.e., eastern and western) varieties, which suggests that there was only one species before the Miocene period and that local groups evolved differently in different environments.

- The morphology of the various species reveals something like a single scheme of vital organization at the anatomical and physiological level that has been progressively modulated. One macroscopic example often cited is analogous organs (a human hand has a structure that is fundamentally analogous to the front paw of an ape, etc.).

- Arguments are also drawn from embryology. While certain differences between the organs of individuals of different species appear major and irreducible when compared in adult individuals, such differences are minimal in the embryos of the respective species. For example, one can find a rudimentary form of fish gills in mammalian embryos.

This fact has led some researchers to the hypothesis of biogenetics: *ontogeny repeats phylogeny*, that is, the formation of the individual in his embryonic state is said to repeat the formation of the species.[27] Arguments in favor of evolutionary theory can be found in the field of genetics, as mentioned, especially thanks to the discovery of mutations that can be produced by means of particular types of radiation and the vast combinatorial potential ensconced within the laws of genetic transmission.

It should immediately be noted that these arguments have met with significant scientific critiques that are just as weighty as the conjectural proofs adduced. It is not possible here to cite them analytically or in their entirety.

Serious paleontologists have clarified in recent studies that orders, classes, and varieties of the same species have existed together in the same period and that, at times, certain progenitor forms existed only in the minds of the authors who

---

[27] E. Haeckel, "Zellseelen und Seelenzellen," *Deutsche Rundschau* 16 (July–September 1978): 40–59.

formulated the theories.[28] The arguments derived from geographical distribution seem valid within a very limited range of variation.

The arguments derived from the morphology and physiology of organs are countered by a more precise and complete observation not limited to the external observation of single pieces or features of organs. If anything, the unity of the organizational plan of life can prove the existence of a hierarchical continuity in living nature, but not the derivation of one species from another.[29] The most formidable objections to evolutionary theory, however, come precisely from the geneticists, who point to the determination of the genetic code or the fixed number of chromosomes for each species.[30] Those who advance the hypothesis of a sudden, unforeseen genetic mutation must support it by identifying the precise physical or environmental conditions that could cause it.

As a scientific hypothesis, evolution nonetheless remains an issue open to discussion and scientific research. This does not allow the theory to be presented on the philosophical level as though it excluded the problems of first cause, value, and final end; if anything, evolution may more strongly emphasize the philosophical problem of the "why" in this process and the value of its culminating point, which is the appearance of man in the universe.

It may be useful to borrow a page from St. Thomas Aquinas, who discusses the ascending perfection of forms (and substances) in the process of generation and the related substantial transformations, but also clears the path to a philosophy of the evolution of life: "Therefore man is the end of all generating."[31] Aquinas's position seems to suggest that the evolution of life has terminated with man. Jacques Maritain interprets it from this perspective in a study in which he proposes the hypothesis that evolution was completed with the appearance of the human species.[32] Even if evolution has finished with man, this does not imply that every other form of evolution concerning inferior levels of life has finished.

Against Reductionism

In the debate for and against evolutionism, it is necessary to avoid two forms of philosophical and epistemological reductionism: (1) the reductionism of those who contrast evolutionary theory not to "fixism"—which is a legitimate contrast that can be proposed on the scientific level—but rather to "creationism," thus turning a

---

[28] See G. Pastori, "Il centenario dell'opera di C. Darwin: L'origine della specie per selezione naturale," *Pedagogia e vita* 21 (1959–1960): 24–40 and 99–110.

[29] S. Vanni Rovighi, *Elementi di filosofia*, 3:98; Lucas Lucas, *L'uomo spirito incarnato*, 57–64; S. Muratore, *L'evoluzione cosmologica e il problema di Dio* (Rome: AVE, 1993), 8–11; Artigas, *Le frontiere dell'evoluzionismo*, 159–205.

[30] For a critique of evolutionism, see G. Sermonti and R. Fondi, *Dopo Darwin: Critica all'evoluzionismo* (Milan: Rusconi, 1980); V. Marcozzi, "'Sorella scimmia' e controversie evoluzionistiche," *La Civiltà Cattolica* 1 (1985): 134–145; idem, *Però l'uomo è diverso* (Milan: Rusconi, 1981); L. Palazzani, "La natura nel dibattito bioetica," in *La tecnica, la vita, i dilemmi dell'azione*, ed. V. Possenti (Milan: Mondadori, 1998), 204–226.

[31] Thomas Aquinas, *Summa contra Gentiles*, 1, III, c. 22.

[32] See J. Maritain, "Vers une idée tomiste de l'évolution," in *Approches sans entraves* (Paris: Fayard, 1973), 105–162.

scientific theory on the origin of living things into a philosophy of becoming in the materialistic and deterministic sense, in which the first cause is negated along with the ontological difference among the various forms of life, including human life; and (2) the related reductionism of explaining what is complex by what is more elementary, reducing anthropology, sociology, and psychology to biology, biology to chemistry, chemistry to physics, and so forth, until one arrives at the "unknown" which consists of sub-elementary particles, thus annihilating the values proper to each level.

This is an opportune moment to quote the observations of Popper, which are beyond suspicion on this point. He breaks down the stages of cosmic evolution into the following arrangement:

World 3 (the products of the human mind)
6. Works of art and of science (including technology)
5. Human language: theories of self and of death

World 2 (the world of subjective experiences)
4. Consciousness of self and of death
3. Sentience (animal consciousness)

World 1 (the world of physical objects)
2. Living organisms
1. The heavier elements, liquids and crystals
0. Hydrogen and helium

Reductionism tries to explain what happens at higher levels by means of what happens at lower levels; what happens in the whole is explained by means of what happens in its parts, in accordance with the principle of "upward causation." Popper says, "This reductionist idea is interesting and important; and whenever we can explain entities and events on a higher level by means of those of a lower level, we can speak of a great scientific success and can say that we have added much to our understanding of the higher level. As a research programme, reductionism is not only important, but is part of the programme of science, whose aim is to explain and to understand." [33]

The same author declares, however, that there is also a "downward causation," as he calls it, through which the whole, as a structure, influences the individual parts. He mentions several examples: stars, in which the mass exerts a terrible gravitational pressure on the elementary particles in the central zone, so that some atomic nuclei merge and form the nuclei of heavier elements. Furthermore, "an animal may survive the death of many of its cells and the removal of an organ, such as a leg (with the consequent death of the cells constituting the organ), but the death of the animal leads, in time, to the death of its constitutive parts, cells included," and, he concludes, "I believe that these examples make the existence of downward causation obvious; and they make the complete success of any reductionist programme at least problematic." [34]

While the reduction of anthropology to biology, of biology to chemistry, and of chemistry to physics can be a useful program for specialized research, it therefore cannot be a framework for a comprehensive interpretation of reality.

---

[33] Popper and Eccles, *The Self and Its Brain*, 18.

[34] Ibid., 20.

This downward reduction, used by the mechanistic interpretation of life often present in evolutionary theories, is even less valid when discussing the level of higher life: the human level. In this regard Popper himself, though a professed agnostic, admits the insufficiency of the natural sciences in interpreting human reality as a whole. He writes:

> Two things, Kant says near the conclusion of his *Critique of Practical Reason*, fill his mind with an always new and increasing admiration and respect: the starry heavens above him, and the moral law within him. The first of these two things symbolizes for him the problem of our knowledge about the physical universe, and the problem of our place in this universe. The second pertains to the invisible self, to the human personality. ... The first annihilates the importance of man, considered as a part of the physical universe. The second raises immeasurably his value as an intelligent and responsible being.
>
> I think that Kant is essentially right. As Josef Popper–Lynkeus once put it, every time a man dies, a whole universe is destroyed. ... Human beings are irreplaceable; and in being irreplaceable they are clearly very different from machines. They are capable of enjoying life, and they are capable of suffering, and of facing death consciously. They are selves; they are ends in themselves, as Kant said.[35]

Recently, especially in the field of the philosophy of biology, much use has been made of the *paradigm of complexity*.[36] At the basis of this paradigm, though not always recognized explicitly, is Aristotle's intuition that "the whole is greater than the sum of its parts." Hence the life of a living thing is not simply the result of a composition of organs capable of living; rather, it requires organization and a "form" (*entelécheia*).[37]

The following elements characterize the notion of *complex*: order, coherence, unity, structure, and organization. The philosopher Edgar Morin expressed the notion of complexity in terms of three principles: the *dialogical principle*, the *principle of recursivity*, and the *hologrammatic principle*.[38]

According to the *dialogical principle*, duality is present within unity, making it possible to associate contrasting data that actually prove to be reciprocally constitutive. One example of this is DNA in relation to amino acids: a stable hereditary "memory" together at the same time with the chemical action of a protein that breaks down and is reconstituted.

The *principle of recursivity* recognizes that there is not just a linear relation between cause and effect, but also a recursive relation; in other words, products

---

[35] Ibid., 3.

[36] A *paradigm* is a model for a vision of reality and of reasoning. See G. Del Re, "Complessità," in *Dizionario di scienza e fede*, ed. G. Tanzella-Nitti and A. Strumia (Rome: Urbaniana University Press, 2002), 259–265.

[37] See M. T. Russo, *Corpo, salute, cura: Linee di antropologia biomedica* (Soveria Mannelli, CZ: Rubbettino, 2004), 58–60.

[38] See E. Morin, *Introduzione al pensiero complesso*, 3rd ed. (Milan: Sperling & Kupfer, 1995).

and effects often become causes and producers of that which produces them. One example is the cellular membrane in living organisms: it is simultaneously the product of cellular metabolism but also an integral part of the metabolic components themselves.[39]

The *hologrammatic principle* states that the whole is present in the part, inasmuch as the one is inconceivable without the other. Cellular biology confirms this, since every cell contains in its totality the genetic information of the entire organism.

The Philosophical Problem within the Biological Problem

The range of biological causes within which the various life forms came into being requires a metaphysical explanation, whether one subscribes to fixism or is inclined toward evolutionism, whether one considers upward causation or downward causation. *Biological reductionism*, even if it should prove useful in attaining the programmatic goals of scientific research, would be unacceptable as a comprehensive explanation of the origin of life and, in particular, of the origin of man.

Within and beyond the mechanistic or combinatorial forms of causality, two metaphysical steps must be postulated that are indispensable to logic as a whole and to an explanation of reality, of living reality, and of man in particular. The first is the *principle of creation*; the second is the *principle of human spirituality*.

The causality that develops its activity in contingent worldly reality requires an intelligent cause to explain the transition from nonbeing to being, from nonexistence to the existence of all reality within the world, which appears precisely as contingent, i.e., not able to exhaust in itself the fullness of being or existence.

The first and subsistent cause must possess within itself the explanation and fullness of being; it must be capable of imparting existence that is distinct from its own (that contingent existence which is not comparable to its own subsistent existence), and it must be definable as capable of purposeful, organizing intelligence.

The principle of causality or sufficient reason, which is grounded in reality and not just a product of the human mind, is ultimately based on the first principles of identity and non-contradiction: being cannot originate from nonbeing; what is more cannot be explained by what is less. This key philosophy, already present in classical thought, was corroborated and clarified by Christian revelation, yet is still a truth of reason and of metaphysical reasoning.[40] It is the truth affirmed in the concept of *creation*.

The fact that life bloomed from cosmic reality through a complex combination of chemical elements or spontaneous generation or the fact that the various forms of life developed through evolution does not eliminate the First Cause: a creative, provident, intelligent, and ordering causality. If anything, the evolutionary hypothesis—should it ever be proved—would be an even more impressive testimony to the profundity of the creative Intelligence. The unpredictability of the countless possible combinations of DNA is called "chance," and the de facto stabilization of the genetic code of individual species is called "necessity"; this terminology demonstrates the inability to foresee and define the elective combinations, but it does not eliminate the fact that

---

[39] See H. Maturana and F. Varala, *L'albero della conoscenza* (Milan: Garzanti, 1987).

[40] See Vanni Rovighi, *Elementi di filosofia*, vol. 2, with bibliography.

such possibilities had to have the concrete potential of being realized in a substrate of reality requiring a causal explanation—unless one wants to explain existence with nothingness, which would be offering no explanation at all.[41]

It should also be added that the various forms of life (vegetal, animal, human), though they exhibit ties not only of relation but also of dependence, would not and do not lose their specificity in their various degrees of autonomy. However one may explain the mechanisms by which a plant, an animal, or man originated, their respective levels of life remain distinct and unique to each and, if anything, reveal a well-ordered hierarchy leading up to an ever richer and more varied fullness of life. The fact that a common material substrate and discernible elementary mechanisms exist within the various levels of life should not prevent us from grasping the specificity of their substantial *forms.*

It is not my intention to impose on biologists or physicians a particular religious viewpoint or reflections that would require lengthier treatment and that have occupied and tormented the minds of so many scientists, philosophers, and artists. Rather, I simply intend to maintain the distinction between experimental scientific discourse, which examines how life comes about and is propagated, and philosophical discourse (which is no less valid for the purposes of knowing truth) about the why, the end, and the value of life itself.

The other metaphysical step concerns the value of human life, the human person, his spirituality (which cannot be derived from matter), and the unity of spirituality and corporality in man.[42] The next chapter will survey this topic, however, because it deserves a more thorough, point-by-point treatment and not just a passing reference.

It was necessary meanwhile to explain this distinction between the viewpoint of the experimental sciences and that of philosophy. The former aim to provide an explanation for the origin of life, but by "explanation" is meant (and it is impossible for it not to mean) a description of *how* the data and the facts are connected experimentally. The latter seeks the *meaning* of the same facts, individually and as a whole, i.e., the causal "why," the first Cause, the final Reason why, and the ultimate End. The philosophical point of view is fundamental for an interpretation of meaning and therefore of value. It is essential to address these difficulties and make these distinctions in order to engage in ethical discourse.

Having made these clarifications, it also seems appropriate to me to cite embryologist G. Goglia's observation regarding the reductionist and determinist theory, with particular reference to Monod:

> Based on these achievements, today we are able to explain with a certain approximation the chemical dynamic that leads to the performance of specific cellular activities. The mechanisms that lead to the specialization of the various cell families in a multi-celled organism are sufficiently clear to us as well. There are questions, however, to which these notions are unable to respond. They are questions relating to the general development of the

---

[41] On the scientific theories of Monod, see the purely scientific observations made by G. Goglia, "Jacques Monod," *L'Osservatore Romano*, August 28, 1976. See also G. Blandino, "L'argomentazione casualistica di Jacques Monod," *La Civiltà Cattolica* 2 (1978): 557–565; idem, "Caso e finalità," *La Civiltà Cattolica* 2 (1977): 366–368.

[42] See Artigas, *Le frontiere dell'evoluzionismo.*

embryo and the completion of the morphological and functional structure in prodigiously complex organs such as the eye and the brain. In these cases the understanding of the genes in molecular biology, which Monod headlines in elaborating his theory, offers only marginal assistance. It is possible, of course, to concede that even the general architecture of the body is codified in the genes, and it can readily be admitted that the development of the embryonic forms is the immediate or mediate expression of a coordinated series of information packets encoded in the genome. But for precisely this reason the idea that the random mutation of a gene could promote an evolutionary phenomenon (such the transformation of a fin into a paw or of a paw into a wing) is unacceptable.

Actually the transition from a fin to a paw and from a paw to a wing does not imply the chemical transformation of one or more proteins (the proteins in a fin are, in fact, almost the same as those in a paw or a wing, and the muscles, skin, and bones of the fin, the paw, and the wing are structurally identical), so much as a phenomenon of morphological and morphogenetic reprogramming, in the context of which the mutations of individual genes are not strictly necessary or else assume an entirely marginal importance.

Indeed, the genetic determination of the cellular structures must remain almost unvaried and at most might involve a marginal variation affecting the specific proteins of the cellular membrane, which serve to determine the immunological character of the new species.[43]

In other words, the paw, the fin, and the wing involve essentially the same genetic equipment, varying only in what is necessary to give species-related specificity to the cells. What has to change profoundly in the three cases is the *enteléchia* and the program that regulates its actualization in time. "In order to go from the frantic rhythm of jazz to the majestic strains of the 'Hymn to Joy' there is no need to change the musical notes: it is enough to arrange them differently."[44]

The debate between evolution and creation entered a new phase between 2005 and 2006 in particular, largely owing to several authoritative interventions by the Catholic Church and the subsequent development, especially among Catholics in the United States, of efforts to introduce the teaching of *intelligent design* in the classroom. Intelligent design involves the acceptance of an "evolutionary creation," by which evolution is justified in its origin as part of the intelligent design of the Creator.

In his address at the conclusion of the International Symposium on Christian Faith and the Theory of Evolution on April 26, 1985, Pope John Paul II affirmed that "a properly understood faith in creation and a properly understood teaching of evolution are not in conflict: in fact, evolution presupposes creation and creation, in light of evolution, presents itself as an event stretched out over time—like an 'ongoing *creatio*'—in which God becomes visible to the eyes of the believer as the Creator of heaven and earth."

The *Catechism of the Catholic Church* adds that "creation ... did not spring forth complete from the hands of the Creator" (no. 302). In this sense, God created

---

[43] Goglia, "Jacques Monod."

[44] Ibid.

the world "'in a state of journeying' towards its ultimate perfection. In God's plan this process of becoming involves the appearance of certain beings and the disappearance of others, the existence of the more perfect alongside the less perfect, both constructive and destructive forces of nature" (no. 310). In his message to the Pontifical Academy of Sciences in October 1996, John Paul II had already recognized the scientific nature of the evolutionary understanding of the universe while at the same time rejecting those theories with materialistic underpinnings.

Another authoritative document published by the International Theological Commission presumes the validity of evolution and insists on the essential theological problem: the dependence of creation on the Creator.[45]

The intelligent design proposal for schools in the United States sought to emphasize this underlying dependence while giving credit to the theory of evolution. The way it was presented, however, moved into teaching certain ideas of a scientific nature in themselves, such as the way in which this design was achieved (the synergic action of genetic mutations and environmental changes) that are not accepted by most scientists. The Supreme Court of Pennsylvania, before which the proposal was brought, therefore defined it as unscientific.

It should be concluded that this openness of the Catholic Church, understood as openness to the possibility in principle of compatibility between creation and evolution, must be kept free of explanations regarding how evolution specifically happened. This is a matter that must be left to the scientific domain and supported by scientific evidence, as anthropologist Fiorenzo Facchini maintains in his article of January 16–17, 2006, in *L'Osservatore Romano*. On the other hand, it must be admitted that since evolution is a theory, however widely accepted, it is not exempt from scientific critiques against it.[46] Such critiques can always bring to light scientific arguments, which their authors believe to be solid, and support them with scientific reasoning. This is not off limits even to Catholics, especially after the openness shown in the documents cited above.

Anthropocentric Ethics and Anti-anthropocentric Ethics

To complete this discussion, I think it will be helpful to consider further the role of man within the natural world. I have maintained the thesis that man must be assigned a central role within nature because he is ontologically different from the rest of natural realities. This position has been recovered within the discipline of physics with the formulation of the *anthropic principle*, which is directed to the

---

[45] International Theological Commission, "Communione e Servizio" [Communion and service], *La Civiltà Cattolica* 4 (2004): 254–288.

[46] Goglia, "Jacques Monod." For the more recent phase of the debate on creationism and evolutionism, see G. Tanzella Nitti, "Creazione ed evoluzione: Chi ha rimescolato le carte?" *Documentazione Internazionale di Scienza e Fede*, September 2005, http://www.disf.org/Editoriali/Editoriale0509.asp; F. Facchini, "Evoluzione e Creazione," *L'Osservatore Romano* (January 15–16, 2006), 4; G. De Rosa, "L'evoluzione dei viventi: Il fatto e i meccanismi," *La Civiltà Cattolica* 3 (2006): 232–241; idem, "Caso o finalismo nella evoluzione dei viventi?" *La Civiltà Cattolica* 2 (2006): 483–492. For the scientific literature against evolution, see D. Raffard De Brienne, *Per finirla con l'evoluzionismo: Dichiarazioni su un mito inconsistente* (Rome: Minotauro, 2003); G. Sermonti and R. Fondi, *Dopo Darwin: Critica all'evoluzionismo* (Milan: Rusconi, 1980).

scientific explanation of the cosmos, signaling a distinct trend reversal regarding the image of the universe that empirical science—starting with Copernicus—has given to us: a picture in which human beings had lost their central and privileged position.

In 1974, Brandon Carter formulated the so-called *anthropic principle*.[47] Especially in its *strong form*, i.e., as a concept in which the whole cosmos is characterized by a strict correlation with the phenomenon of life, it reconsiders man not only as a part of nature but also as the sole being capable of grasping the intrinsic intelligibility of the universe. Indeed, as Saturnino Muratore puts it,

> only an anthropology that emphasizes the ontological girth and the metaphysical salience of the mind is able to explain the direction of cosmic evolution toward intelligent life. Inasmuch as it is an essentially ontological reality—constitutively capable of *logos*—the mind reveals simultaneously the closure and incompleteness of a cosmos characterized by a constitutive intelligence yet lacking any illumination of the spirit, unable to utter itself with the language of being and in relation to being. Hence cosmic evolution completely unravels at the *ontic* level until the emergence of the *mind*, which alone can grasp and bring to light the intrinsic intelligibility of natural processes, for it is constitutively all-inclusive intentionality, openness to the absolute, the foreknowledge of being.[48]

Man evidently comes to assume a unique position in a construct of this sort, inasmuch as he is placed in a central, preeminent role with respect to all the other components of the natural world, whether animate or inanimate.[49] Nevertheless this idea is not shared universally. Indeed, within the realm of philosophical reflection, particularly in the Anglo-Saxon world, numerous anti-anthropocentric theories have been developed and have spread especially through attempts to resolve the serious problems related to environmental pollution.[50] An anti-anthropocentric approach starts precisely with the negation of man's centrality and proposes to resolve these issues through a rehashing of ethics: rather than being built only around human beings, it should also take nature into account. Man's strict dependence upon the surrounding environment is thereby emphasized, and it therefore becomes impossible to ignore that the fact that his own interests are tightly interlinked with those of the physical world, which are thus endowed with a degree of moral consideration not traditionally extended to those sectors. It is consequently denied that man has his own spirituality—not derivable from matter—and that there is in him a profound

---

[47] B. Carter, "Large Number Coincidences and the Anthropic Principle in Cosmology," in *Confrontation of Cosmological Theories with Observational Data*, ed. M.S. Longair (Boston: Reidel, 1974). With reference to the formulation of the anthropic principle, whether in the strong form or in the weak form, the reader can consult Muratore, *L'evoluzione cosmologica*.

[48] Muratore, *L'evoluzione cosmologica*, 204.

[49] See M. Zatti, "Biologia antropica," in *Il principio antropico: Condizioni per l'esistenza dell'uomo nell'universo*, ed. B. Giacomini (Ferrara: Istituto Gramsci, 1991); C. Porro, "'I cieli narrano la gloria di Dio': Note su cosmologia e teologia," *La rivista del clero italiano* 6 (1996): 453–463.

[50] See M.B. Fisso and E. Sgreccia, "Etica dell'ambiente I," *Medicina e Morale* 46, no. 6 (1996): 1057–1082; idem, "Etica dell'ambiente II," *Medicina e Morale* 47, no. 1 (1997): 57–74.

unity between the *res extensa* and the *res cogitans*. He is thus reduced to a mere component of the natural world endowed with the same moral relevance as any other component.

A change of this kind has been wrought by forms of so-called environmental-ist ethics, including holism, eco-feminism, animal rights ethics, and biocentrism. Of particular note is Aldo Leopold's famous theory, dubbed *land ethics*,[51] which is a radically eco-centric and holistic concept of morality according to which the so-called *biotic community*, understood globally, has value in and of itself.

It should be noted that, while it is important to admit that natural components have moral relevance regardless of their usefulness to man, surpassing anthropo-centrism cannot not run to the extreme of conferring rights on all natural entities. "The replacement of anthropocentrism with a more dynamic model of the relation between man and environment is certainly supported by the praiseworthy intention of saving the environment and preserving biological diversity, but it must not end in a sterile extremism. The proposed abandonment of anthropocentrism is supposed to safeguard the intrinsic value of all beings in the biosphere. Unfortunately this desire tends to be expanded into a total prohibition against man's use of the beings that are inferior to him, whereas it belongs to the natural order of things that each species in the biosphere should live on other species."[52] The real problem is determining the value to be attributed to natural objects and whether this value should be considered intrinsic or simply instrumental.

Most philosophers of the anti-anthropocentric mindset, such as J. Baird Callicot and Tom Regan,[53] admit that the most serious challenge for environmental ethics is developing a theory that can adequately explain how nonhuman entities can also be bearers of intrinsic value—hence deserving respect—in and of themselves, independently of any eventual human recognition.

The biologistic perspective presumes to interpret a complex and multilayered reality such as a human being by reducing him to simpler, lower realities.[54] The most classic form of this reductionism is *radical Darwinism*, which declares that there is an evolutionary continuity between man and the rest of the natural world of living beings. "After Darwin, we can no longer think of ourselves as occupying a special place in creation—instead, we must realize that we are products of the same evolutionary forces, working blindly and without purpose, that shaped the rest of the animal kingdom."[55]

---

[51] See A. Leopold, *A Sand County Almanac and Sketches Here and There* (New York: Oxford University Press, 1949). For general observations on all anti-anthropocentric theories, see P. C. List, *Radical Environmentalism: Philosophy and Tactics* (Belmont, CA: Wadsworth, 1993), 15–133; S. Bartolomei, *Etica e ambiente* (Milan: Guerini, 1989), 35–136; idem, *Etica e natura* (Bari: Laterza, 1995).

[52] B. Przewozny, O. Todisco, and F. Targonski, *Etica ambientale* (Rome: Miscellanea Francescana, 1991), 130.

[53] See T. Regan, "The Nature and Possibility of an Environmental Ethics," *Environmental Ethics* 3 (1981): 19–34; idem, *The Case for Animal Rights* (Berkeley: University of California Press, 1983).

[54] See Russo, *Corpo, salute, cura*, 53–56.

[55] J. Rachels, *Created from Animals: The Moral Implications of Darwinism* (New York: Oxford University Press, 1990), 1.

Another form of *extreme naturalism* is *radical* (anti-anthropocentric) *ecologism*, which affirms a substantial continuity and therefore equality between human and nonhuman life forms. Nature is said to have value in itself and human beings do not have a central role. In some cases, if human beings are reduced to a simple, natural datum, nature actually ends up becoming anthropomorphized, i.e., considered to be a living totality with a kind of soul of its own. These sorts of expressions are used both in *deep ecology*[56] and in the mentalist biology of the *ecology of mind*,[57] which start from radicalized Darwinist premises and lead to pantheistic results. The *ecology of mind*[58] considers the human mind a system, a function of natural complexity. A thinking man has neither a soul nor a brain and is nothing but a particular function of the natural organism in a certain environment; the human subject ends up dissolving into a totality to which he must simply conform and adapt himself.

From this perspective, life is understood as a global phenomenon with no internal differentiation. According to James Lovelock, the earth is a single system or even a living organism (the *Gaia hypothesis*) with a self-regulatory capacity for confronting aggression from external agents. Indeed, the life of Gaia—the earth—will continue because a new equilibrium will be constantly reestablished, leading to the extinction of any species that damages the environment. From this viewpoint, what matters—that is to say, the criterion for ethical action—is determined by the life of Gaia, not by the value of the individual species.[59]

Some versions of this biocentric concept consider all beings endowed with sentience to be moral subjects with rights. Not only does this thereby place men and animals on the same level, but it also excludes some human beings—"nonpersons"—from having rights.[60]

The attribution of intrinsic value to nonhuman entities has led to an extension of the boundaries of the moral community beyond the unique category of human beings. This broadening can be considered essentially correct so long as it is interpreted as the need to establish *moral duties* for man not only toward other human beings but also toward natural entities. Conversely, matters become remarkably problematic and even unacceptable from both the philosophical and scientific standpoints with the affirmation that all natural entities possess the same moral value.[61] This road clearly leads to absurd results because this value winds up being defined in such a way that the moral agent—the human being—is not only neutralized but also eliminated. It is quite difficult to imagine that anyone else could protect nature if man lost his importance and prominence within the natural world.

---

[56] See J. Ballesteros, "La costruzione dell'immagine attuale dell'uomo," in *Immagini dell'uomo: Percorsi antropologici nella filosofia moderna*, ed. I. Yarza (Rome: Armando, 1997).

[57] See G. Bateson, *Steps to an Ecology of Mind* (San Francisco: Chandler Publishing Co., 1972).

[58] See A. Llano, *La nuova sensibilità* (Milan: Ares, 1995).

[59] See J. Lovelock, *The Ages of Gaia: A Biography of Our Living Earth* (New York: Norton, 1988).

[60] See P. Singer, *Practical Ethics*, 2nd ed. (New York: Cambridge University Press, 1993); idem, *Animal Liberation*, 2nd ed. (New York: New York Review of Books, 1990).

[61] See C. V. French, "Against Biospherical Egalitarianism," *Environmental Ethics* 12 (1995): 41–57.

Hence, after this necessarily brief examination, I cannot refrain from critiquing the initial assumption of these anti-anthropocentric notions. It is futile and harmful to try to establish a new ethics, distinct from traditional morality, in order to correctly establish the relation between man and nature; rather, it is much more convenient to expand traditional ethics in light of the new scientific and cultural achievements, without trying to ignore the human role. Man has a preeminent role in the world, grounded in his profound ontological diversity with respect to the rest of creation. His spirituality is not derivable from matter and sets him at a higher level; however, this supremacy does not excuse him from respecting nature but rather obliges him to do so.

Man's harmony with nature cannot be achieved by equating him to other beings, but rather by first of all changing his manner of thinking and acting with regard to all nonhuman entities. The extremism of certain eco-philosophies must therefore be rejected and an *anthropocentric, creational framework*—essentially a philosophical conception according to which man has a central and determining role with respect to the rest of nature—must be resolutely affirmed. Personalist reflection on the subject maintains that

> at the present cultural moment, philosophical thought about the person is called to take up the ecological and environmental challenge—and therefore responsibility for animals and the ecosystem—in both the synchronic [present] and diachronic [future] senses. This entails developing a sort of "personalism of the biosphere" that would broaden and increase that responsibility in a "plenary" sense, recognizing that our obligation to respect life includes all living beings (animals, plants, and the earth).
>
> Such a personalism, then, without attenuating its realistic, ontological foundation, is open to all the new challenges resulting from scientific and technological progress. The human person is the author of society and the steward of the biosphere, the artisan of the ecological and social environment, by which he is in turn conditioned and stimulated. It is therefore necessary to establish an *ethics of responsibility* understood not only as individual responsibility but also responsibility toward future generations.[62]

In short, I therefore believe the preferable approach is anthropocentric, understood to mean a moderate anthropocentrism that reserves for man the role of steward, thereby charging him with a serious responsibility. This is also the intellectual framework of Catholic environmental ethics: it is classified with the anthropocentric theories but has some peculiarities and a profoundly different approach with respect to all the others.

It cannot be forgotten, naturally, that the true cause of every environmental problem is a skewed understanding of the relation between man and nature brought about by man's desire to substitute himself for God the Creator, thereby disregarding the idea of limitation. Hence it becomes most opportune to reaffirm the Christian understanding of this topic: man is the only creature who can participate in the lordship of the Creator; nevertheless, while firmly acknowledging his superiority

---

[62] E. Sgreccia, "La persona umana," in *Bioetica*, ed. C. Romano and G. Grassani (Turin: UTET, 1995), 194–195.

with respect to the rest of the world, he remains and must consequently be considered only the steward of creation. In conclusion, let us recall the words on this subject by John Paul II in his encyclical letter *Evangelium vitae*:

> Man has a specific responsibility towards *the environment in which he lives*, toward the creation which God has put at the service of his personal dignity, of his life, not only for the present but also for future generations. ... In fact, "the dominion granted to man by the Creator is not an absolute power, nor can one speak of a freedom to 'use and misuse,' or to dispose of things as one pleases. The limitation imposed from the beginning by the Creator himself and expressed symbolically by the prohibition not to 'eat of the fruit of the tree' (Gen 2:16–17) shows clearly enough that, when it comes to the natural world, we are subject not only to biological laws but also to moral ones, which cannot be violated with impunity."[63]

This approach can be shared both by those who work within a philosophical construct open to transcendence, as is the case with Christians, and by those who reject this openness, provided that they agree to assume the same responsibility.[64]

---

[63] John Paul II, Encyclical Letter *Evangelium vitae*, March 25, 1995, no. 42, emphasis added; internal citation from John Paul II, Encyclical Letter *Sollicitudo rei socialis*, December 30, 1987, no. 34.

[64] See H. Jonas, *The Imperative of Responsibility: In Search of an Ethics for the Technological Age* (Chicago: Chicago University Press, 1984).

Summary Outline for Chapter 3

The concept of life

Life is a spontaneous motion that tends toward an end.

Its central characteristic is that it functions as a principle of *self-movement* and as a principle of *internal change*: life consists in the fact that a certain substance moves itself, in such a way that its action proceeds from itself and remains in itself, that it is the principle and end of its action.

A subject endowed with life appears as an individual being that has its own center within it and possesses an essential boundary between inside and outside.

A living being appears, however, as an "open system" within which is established a complex equilibrium of fluxes, a system endowed with individuality and capable of exchanges with the environment.

The phenomenology of life manifests features of continuity with respect to the inferior order of physical and chemical phenomena, but it also has dimensions indicating a qualitative leap. Related to this is the classic debate between the "mechanists," who try to trace back the properties of life exclusively to phenomena of chemical and physical exchanges, and the "vitalists," who emphasize instead the positive irreducibility of vital phenomena to that level of explanation.

Life *is not an object to be investigated; rather, it is the basis of all activity*, the horizon within which we situate all problems, and it is *fraught with meaning.*

The highest form of life entails human consciousness. Consciousness and life cannot be contrasted as subject and object; consciousness is the most perfect level of living.

Modern mechanism

There is plainly a modern tendency to reduce what is organic to inorganic factors. This puts biology in a dialectical situation: on the one hand it is a marginal science, since the biosphere represents a very small part of the universe; on the other hand it is a central science, since it deal with what matters most: life, including human life.

The modern problem is the Cartesian dualism between *res extensa* and *res cogitans*. This dualism is completely unsuited to an understanding of life, which cannot be reduced to either extension or thought. Hence the problematic concept of life has been reduced to a mechanical element, to a series of efficient (and not final) causes.

Concomitantly modern thought has abolished the concept of *nature* as *the internal principle of movement and of life.*

Vitalism in antiquity

The problem of the ancients was life and the organism, and the whole cosmos was understood as a living thing endowed with perfection and beauty:

"When [the creator] was framing the universe, he put intelligence in soul and soul in body, that he might be the creator of a work which was by nature fairest and best. [In this way], using the language of probability, we may say that the world came into being—a living creature truly endowed with soul and intelligence by the providence of God" (Plato, *Timaeus*, 30b).

Therefore matter was not interpreted by them merely as a *res extensa* (thing having extension in space) but rather as a *res vitalis* (living thing).

Finalism

Modern science, which was built up in a long battle against Aristotelianism and teleology, allows scientists to study mathematically the vast zone of *res extensae*, leaving out the zone of the spirit and consciousness, where finality alone governs.

The idea of finality is connected with those movements that go from what is less to what is more perfect, movements that can be grasped by the analysis of the intellect. The teleology that is intrinsic to the development of an organism is a typical example of such a movement from less to more.

The end is not just a concept projected by man onto things, nor a concept that can be employed only within the sphere of consciousness; in fact, there are finalities throughout the world of living things, not only in those that are endowed with subjectivity and will.

In inanimate things, the purpose or end is something external; it does not reside in the thing but rather in the mind of the planner. A knife is made in order to cut, and this teleology comes from the mind of the maker; it is not immanent to the knife. On the contrary, in "animate objects" the purpose is immanent.

Laws and analyses of a mathematical type exclude finalism. It should be emphasized that finality appears or disappears depending on whether one adopts a philosophical-ontological approach or a scientific and mathematical one. The latter adopts a method that is completely de-ontologized, which can be useful at its own level—that of purely scientific "explanations"—although it generally involves the exclusion of purpose. The scientific method considers neither the intrinsic connection between potency and act nor the concept of action/agent but only the efficient cause (science seems to exclude finality).

The evolution of life

Creation and becoming

Creation and becoming are not antithetical.

Christian theology asserts that *creatio non est mutatio*: there is an insurmountable difference between creation, the total and absolute bringing of everything into being (*creatio ex nihilo*) and change or transformation (*mutatio*), which presupposes the existence of that which is subject to becoming and transformation.

Divine causality operates differently in creation and becoming: in creation He is not only the first cause, but the *unique and absolute* cause of all being; in the area of becoming He still operates as the first cause (transforming, directing, elevating, etc.) but not as the sole cause, since finite secondary causes come into play which are also endowed with the capacity for action and causality.

Natural philosophy is the sphere of *mutatio* of any order, type and degree. This means that during evolution, in relation to the insurmountable difference between creation and change in something that has been created, there is no longer any form of *creatio ex nihilo*, but rather processes of accidental and substantial change.

Reductionism

According to reductionism, a complex system is nothing but the sum of its parts, and hence one can account for the system by "reducing" one's consideration of it to a consideration of the individual constituents.

It must be admitted that reductionism has allowed science to identify better its proper object and to understand more profoundly the fundamental parts of a reality, such as the human body, and the laws that govern its functioning.

Yet the reductionist perspective is being abandoned, because it proves inadequate if one intends to understand a phenomenon in its totality. For example, it is impossible to understand the function of a gene without situating it within the wider context of the cell's dynamics; these in turn are studied as part of intercellular dynamics, which can be understood only by taking into account the role that groups of cells have in an organ, the role of which, in turn, is considered as part of the whole organism.

Theories of evolution

Two possible meanings can be assigned to the concept of evolution, each leading to its respective theory:

Some understand evolution as a teaching that proposes to explain scientifically all spiritual and material phenomena by tracing them back to the laws of physics and biology (reductionists). In this view the variance that gives rise to evolution is random and therefore unforeseeable. Evolution appears as a simple unfolding of what is already given, and therefore creative evolution never ends and we do not know where it will lead us. This position does not anticipate that evolution could bring about substantial transformations.

For others, evolution means instead not just change, but also directional change or "change toward." In this view, "man is the end of all generating" (Thomas Aquinas); the process of generation entails, through these substantial transformations, the ascending perfection of life forms.

Reductionism (anti-anthropocentric theories)

The most classic form of this biologistic anthropological reductionism is *radical Darwinism*: a form of *naturalism* that asserts an evolutionary continuity between man and the rest of the natural world. The Darwinist James Rachels, in his book *Created from Animals* (1990), writes: "After Darwin, we can no longer think of ourselves as occupying a special place in creation—instead, we must realize that we are products of the same evolutionary forces, working blindly and without purpose, that shaped the rest of the animal kingdom."

Radical ecologism (Deep Ecology) also presents an underlying extreme naturalism, declaring that there is substantial continuity and therefore equality between human and nonhuman life forms. Nature is said to have value in itself and human beings have no central role in it; man is thought to be simply biological, *homo naturalis*, on a par with other living things.

The paradigm of complexity

The notion of complexity is central to reflection in the field of biology, and it has been transformed into a *paradigm*, a model for a vision of reality and for reasoning about it.

The notion of a "complex" is characterized by order, coherence, unity, structure, and organization.

The philosopher E. Morin has expressed the notion of complexity in terms of three principles, detecting them in reality and making out of them a paradigm for scientific reasoning: the *dialogical principle*, the *principle of recursivity*, and the *hologrammatic principle*.

According to the *dialogical principle*, duality is present within unity. One example of this is DNA in relation to amino acids: in DNA there is a stable hereditary "memory" and at the same time the chemical action of a protein that breaks down and is reconstituted.

The *principle of recursivity* recognizes that there is not just a linear relation between cause and effect, but also a recursive relation; in other words, products and effects often become the cause and producers of that which produces them. One example is the cellular membrane in living organisms: this should be considered simultaneously as the product of cellular metabolism and also as an integral part of the metabolic components themselves.

Based on the *hologrammatic principle*, we note that the whole is present in the part, inasmuch as the one is inconceivable without the other. An example: every cell contains in its totality the genetic information of the entire organism.

Chapter 4

The Human Person and His Body

Humanizing Medicine

The phenomenon of life, in its unfolding in various forms throughout the world, has its summit in the life of man: even for biologists and naturalists, man constitutes the richest, most independent, and most active form of life, at the highest level of the kingdom of living things and the peak of the natural history of the universe.

Biology studies the life of man along with other forms of subhuman life, yet it cannot fail to notice the peculiarities of human life, especially in its applied, diagnostic, or therapeutic phase: man differs from animals and primates in terms of something more than just morphology or number of chromosomes.[1]

As for medicine, its central goal is service to man and his health. Even though physicians deal most immediately with the human body, it is impossible for any physician to prescind from the freedom and responsibility of the individual, the totality of the person who is the patient, and the human community and social environment as a whole.

Today there is talk of the "humanization of medicine," but distinct or, if you will, complementary concepts are found within this term. Some use it to underscore the intersubjective relationship between the patient and health care personnel in the face of technological encroachment or hospital standardization. For others it means the introduction of the humanities, particularly psychology, into medical school curricula. Yet the most profound meaning of this appeal, which in a way sums up the preceding aspects, consists in recognizing the personal dignity of every human subject, beginning, as will be shown, at the moment of his conception and continuing until the moment of death. This recognition is also accompanied by an awareness of human spirituality and immortality. Robert Spaemann recalls that "there can, and must, be one criterion for personality, and one only; that is biological membership of the human race. The beginning and end of personal existence cannot be taken apart from the beginning and end of human life. If someone exists, that someone has existed since the individual human organism existed, and will continue to exist for as long as the organism continues to live. What it is to be a person is to live a

---

[1] Personal existence is the most perfect form of existence: "Persona significat id quod est perfectissimum in tota natura, scilicet subsistens in rationali natura." Thomas Aquinas, *Summa theologiae*, I, q. 29, a. 3.

human life."[2] Despite the highly acclaimed "anthropological revolution" thought to characterize modern culture, these three meanings are not so clear in medical practice as they appear to be in stating the principle.

Yet in theoretical discourse even the concept of person does not always have a uniform meaning, especially when set in relation to corporality and when examining the unitive link between body and person in greater detail. This is why it will be necessary to reflect in particular upon this reality, which is so central to the universe and to life, to society and especially to medicine, which is the main topic of interest here.

The Human Person and His Centrality

No one, whether scientist or philosopher, evolutionist or fixist, materialist or idealist, denies that man represents a summit in the life of the universe and in the kingdom of the various forms of life.

The natural sciences (paleontology, biology, etc.), the human sciences (psychology, sociology, etc.), philosophy, and many religions of diverse origin and creed converge in support of this centrality. The human organism itself is a summary and representation of cosmic reality; man's consciousness is able to consciously reprocess the realities of the universe, and man's work involves the capacity to exercise dominion over the realities that surround him.

The crucial point is defining the person in his true constitution, going beyond the awareness of it that any individual may (or may not) have and beyond whatever expressive abilities any single person acquires in the process of maturing.

Let it be made immediately clear that the perspective presented in this preliminary discussion is not a psychological one. In psychology the terms "personality" and "persona" are often equated to the concepts of temperament and character, and their specific meanings, classifications, and dynamics are understood in different ways by various authors and schools of thought. As early as 1937, Gordon Allport had reviewed fifty different descriptions of personality including ethnological, theological, juridical, sociological, biosocial, and psychological connotations of the term.[3]

This does not mean that these various approaches to the reality of the person are irrelevant. In a particular way, psychology grasps the relational aspects of personality, whether displayed or hidden, and this knowledge is helpful for the appropriate practice of medicine in relation to the various categories and situations of patients. But the purpose of pointing this out at this time is to ensure a focus on the metaphysical perspective: the objective value of the dignity of the human person and his ontological structure.

In this light, the first aspect to be highlighted is the spiritual, intellectual, and moral character of the human person: a person is a unity of spirit and body. Of course, the meaning of spirituality, soul, or spirit must be clarified along with how this spiritual life is united to corporality. Seeking out this profound understanding of the person should not appear idle or irrelevant to medicine. Indeed, many people

---

[2] R. Spaemann, *Persons: The Difference between "Someone" and "Something,"* trans. O. O'Donovan (New York: Oxford University Press, 2006), 247.

[3] G. W. Allport, *Personality: A Psychological Interpretation* (New York: Holt, 1937); G. Grasso, "Personalità," in *Dizionario enciclopedico di pedagogia* (Turin: SAIE, 1959), 3:680–682.

today point out that as the number of academic disciplines studying man and the number of specialized fields within the human sciences increase, the crisis of man's loss of identity worsens. Suffice it to recall what Max Scheler wrote: "In the history of more than ten thousand years, ours is the first epoch in which man has become universally and radically problematic to himself: man does not know who he is and even realizes that he no longer knows."[4] Joseph Gevaert notes that "within this context of a loss of identity, of uncertainty and perplexity regarding the image of man, critical and systematic philosophical reflection on the being and meaning of man is one of the most urgent tasks of our time. Attempts to develop a new philosophical anthropology are therefore typical of many thinkers."[5]

The first stage of this reflection, at least insofar as medicine and human biology are concerned, involves overcoming materialism and biological monism in anthropology. This can happen through the rediscovery and reaffirmation of the spirituality of the human soul. In other words, the first problem is the "essence" of man. It is true that existentialism, like actualism and spiritualism, has singled out the existential aspect and used it as a basis for accentuating man's creativity, freedom, and risk. For the existentialist school, the most human aspect of man is found in his *ex-sistere*, in his ability to detach himself from the determinism of the world and *to be*, in his unrepeatable uniqueness, through consciousness and free will. Thus existentialist morality is also characterized by the "optional" and "dramatic" nature of choice, whether in the currents heading toward spiritualism as with Gabriel Marcel, or in those that sink back into nihilism as with Jean-Paul Sartre and Albert Camus, or even in the currents more firmly anchored to ontologism as with Martin Heidegger and Karl Jaspers.[6] This emphasis on existence is supposed to entail an emergence from the schematic limitations of "essences" and of essentialist morality.

In reality there is no such thing as existence or being except in something definite and real, and that reality appears as a *synolos*, or "whole": the simultaneous realization of being in an essence (the *synolos* of *essentia et esse* in Thomistic terms). Yet while *being* or existence marks the passage from the realm of sheer possibility to reality and therefore has a more primary function at the level of "realization," *essence* is the first thing into which the philosopher must inquire at the level of the subsequent "value assessment" and definition of the particular being (addressing the questions of what it is and what it is worth).

Is human existence, observed in man as he exists and is realized concretely, corporality and spirituality or mere corporality? This is the first question that the philosopher must answer, especially the philosopher of biology and medicine. This is the *magna quaestio*, as Saint Augustine defined it,[7] which becomes critical in the face of death—an experience that touches every human being, necessarily coming to the daily attention of physicians and health care as a whole and against which

---

[4] M. Scheler, *Philosophische Weltanschauung* (Bonn: F. Cohen, 1929), 62; see G. Marcel, *Problematic Man*, trans. B. Thompson (New York: Herder and Herder, 1967), 52–53.

[5] J. Gevaert, *Il problema dell'uomo* (Turin: LDC, 1984), 8; J.Y. Jolif, *Comprendre l'homme: Introduction à une anthropologie philosophique* (Paris: Cerf, 1967), 19–20.

[6] See G. Reale and D. Antiseri, *Il pensiero occidentale dalle origini ad oggi* (Brescia: La Scuola, 1983), 3:1.

[7] Augustine, *Confessiones*, chap. 4: "Factus eram ipse mihi magna quaestio."

physicians continue to fight. Outbreaks of violence in society and trends in favor of suicide and euthanasia pose the serious problem of the meaning of death for man, whether it means just the end of suffering or whether it might still open up the hope of life. In the famous words of Camus, "There is but one truly serious philosophical problem, and that is suicide. Judging whether life is or is not worth living amounts to answering the fundamental question of philosophy."[8]

The problem of *essence* and *spirituality* remains radically in the forefront even in those philosophies which have emphasized man's social aspect, his "being in dialogue with others," as in the Christian existentialism or neo-Socratic philosophy of Marcel and the socialist vision of humanity according to which the individual is merely a social being, because the question continues to resurface. Not only does it ask what becomes of us after death, but also who we are—who each one of us is—in relation to others and who those others are for me. Simply saying "I-Thou-we" is not enough to calm the human mind. Even a purely relational personalism limiting itself to a definition of the "I" and the "Thou" in virtue of the interpersonal relationship, without the metaphysical requirement of a definition of essence in relation to existence and of existence in relation to a concrete and real essence, would produce a fleeting definition of person.[9]

Defining the essence of the human person in general and defining it specifically as corporality and spirituality united together does not imply an "objectification" or "reification" of man, depriving the person of his character as a "subject." It simply means defining or exploring the real depths that are hidden in the depths of the "I" and of the "Thou" in a social relationship.

Let us take a moment to simply recall the arguments pertaining to the proof that the self is spiritual, from the classical proofs taken up and corroborated by neo-Thomist and personalist philosophy to the more recent and complementary proofs.[10] The classical proof of the spirituality of the soul, and hence of the self, is based on the principle of proportionality between cause and effect, that is, between the activities of man and the principle from which they proceed. As in the case of animals, there are activities in man of a biological and corporeal sort that can be

---

[8] A. Camus, "The Myth of Sisyphus," in *The Myth of Sisyphus and Other Essays*, trans. J. O'Brien (New York: Vintage Books, 1991), 3.

[9] See Martin Buber's contributions to dialogical personalist thought: *I and Thou* (New York: Scribner, 1970), *Werke* (Munich: Kösel-Verlag, 1962); A. Babolin, *Essere e alterità in M. Buber* (Padua: Gregoriana, 1965).

[10] An exhaustive philosophical justification for the substantial union of soul and body and for the spirituality of the rational soul cannot be offered in this context. To this end, which is beyond the scope of this manual, the reader can consult an extensive bibliography on the subject, which includes the following works: Aquinas, *Summa theologiae*, I, qq. 75 and 76; idem, *Questiones disputatae: De spiritualibus creaturis*, articles 2 and 9; idem, *De Anima*, art. 2 and 3. See the development of the argument in S. Vanni Rovighi, *Elementi di filosofia* (Brescia: La Scuola, 1963), 3:157–183; F. Locatelli, "Alcune note sulla dimostrazione dell'immortalità dell'anima di S. Tommaso," *Rivista di filosofia neoscolastica* 33 (1941): 413–418; A. Coccio, "Il problema dell'immortalità dell'anima nella *Summa theologica* di S. Tommaso d'Aquino," *Rivista di filosofia neoscolastica* 38 (1946): 298–306; J. Maritain, *The Range of Reason* (New York: C. Scribner's Sons, 1952), originally published as *Raison et raisons: Essais détachés* (Paris: Egloff, 1947); V. Possenti, "Noi che non sappiamo affatto che cosa sia la persona umana," *Filosofia oggi* 27, no. 1 (2004): 3–28.

explained by the vegetative and sensorial faculties; yet the same subject, the same "self," also manifests activities of an immaterial sort, i.e., such that even if they are evoked by the sensorial faculties, they find expression at a higher, immaterial level. Examples include the understanding of universal ideas, the ability to reflect, and free will (and hence love in the spiritual and altruistic sense). These activities cannot be explained except by a principle, a source of energy of a higher order, an immaterial source—not identified with matter—which is therefore spiritual.

On the other hand this source of higher energy reveals itself by emanating from the same subject that carries out the sensory activities. These same sensory activities are perceptible and can enter into the spiritual self-consciousness of the human being. In the spirit there is also a consciousness, coexisting with the intellect, that unifies and reflects corporality: "Idem ipse homo est qui percipit se intelligere et sentire; sentire autem non est sine corpora" (it is the selfsame man who perceives that he understands and senses; sensation, moreover, does not occur without the body).[11]

> We must conclude that the essence or substance of man is single, but that this single substance itself is a compound, the components of which are the body and the spiritual intellect: or rather matter, of which the body is made, and the spiritual principle, one of the powers of which is the intellect. Matter—in the Aristotelian sense of prime matter, or of that root potentiality which is the common stuff of all corporeal substance—matter, substantially united with the spiritual principle of the intellect, is ontologically molded, shaped from within and in the innermost depths of being, by this spiritual principle as by a substantial and vital impulse, in order to constitute that body of ours. In this sense, Saint Thomas, after Aristotle, says that the intellect is the form, the substantial form of the human body.
>
> That is the Scholastic notion of the human soul. The human soul, which is the root principle of the intellectual power, is the first principle of life of the human body, and the substantial form, the *entelechy*, or [*sic*] that body. And the human soul is not only a substantial form or entelechy, as are the souls of plants and animals according to the biological philosophy of Aristotle; the human soul is also a spirit, a spiritual substance able to exist apart from matter, since the human soul is the root principle of a spiritual power, the act of which is intrinsically independent of matter. The human soul is both a soul and a spirit, and it is its very substantiality, subsistence and existence, which are communicated to the whole human substance, in order to make human substance be what it is, and to make it subsist and exist.[12]

I will return in a moment to the topic of the spirit-body union to quote the words of Jacques Maritain once more. Here I wish to add that more recent demonstrations of the soul's spirituality have underscored and founded it more strikingly and existentially on the *dialogical nature* of the self: the ability to relate with another, of an "I" to relate with a "Thou," is what more thoroughly grounds the relation.

Gevaert says, "Anthropologies inspired by existentialism, and especially those inspired by personalism, seek to characterize man's being in various ways, resorting to a more complete formulation that is fundamentally open to religious mystery. Man

---

[11] Aquinas, *Summa theologiae*, I, q. 76, a. 1.
[12] Maritain, *Range of Reason*, 57.

is seen as an *ego* existing together with others in the world in order to be fulfilled."[13] Heidegger, Marcel, Martin Buber, Emmanuel Lévinas, and others have powerfully described the personality and spirituality of man as "being with others," "being in the world," "being for others," or the "I-Thou dialogue"; these formulas attempt to express the existential and active aspect of spiritual personality in the modern sense. As mentioned earlier, this perspective presupposes and does not replace the substantialist discourse about the spiritual essence of the spirit, which is the source of the rational and volitional faculties.[14]

In order to avoid possible misunderstandings, a further note concerning the terms *soul, psyche*, and *spirit* is important. In the philosophical sense, *soul* is understood as the principle of vital operations and can be used to denote the principle of vegetal, sensorial, and rational life functions. These correspond to the vegetative, sensitive, and rational souls, respectively, though the terms *soul* and *ensoulment* more commonly refer to the rational and spiritual soul. Maritain, however, rightly and more appropriately applies the term *spirit* (in French, *esprit*) to the rational soul.

In experimental and human psychology, the term *psyche* is also used to designate certain manifestations that more often reflect the sensorial and emotional aspect of animal and human life (whether plants have sensibility is debated and doubted). On the other hand, the term *psyche* or *psychic* is sometimes used to indicate the spiritual and higher aspect of the spirit, but from a functional perspective: the realm of higher psychic activity, mind, psychology, intelligence, and so forth.

It is important to avoid thinking of man as constitutionally made up of three principles—soma, psyche, and spirit—because psyche denotes the life of both the soma and the spirit in functional terms. This is especially true in recent times as psychology focuses on the study of the emotional and sensorial dimension that is rooted in both the soma and the spiritual self.

The human being still has only two constitutive principles, each of a profoundly different origin and nature: the soma and the spirit (or spiritual soul). The spirituality of the soul, once recognized, entails two other pertinent conclusions: if the soul is spiritual, then it cannot be derived from the body and, furthermore, it is immortal.

Furthermore, by the principle of the metaphysical impossibility of deriving spirit from matter, it should be concluded that in man—a being composed of a soul and a body—the soul has a different and spiritual principle of origin. Hence the thesis of the direct creation of the individual soul by God: the same Creator who brought life into existence as a general phenomenon, rich in variety and hierarchical levels, is also the Author of the individual soul of every human being. "Creando infunditur et infundendo creatur," as Saint Thomas Aquinas declares:[15] by creation it is infused

---

[13] Gevaert, *Il problema dell'uomo*, 7; J. F. Malherbe, "Médecine, anthropologie et éthique," *Médecine de l'homme* 156/157 (1985): 5–12.

[14] See P. Nepi, *Il valore persona: Linee di un personalismo morale* (Rome: Editrice Universitaria di Roma/La Goliardica, 1993); A. Rigobello, ed., *La persona e le sue immagini* (Vatican City: Urbaniana University Press, 1999); idem, ed., *L'altro, l'estraneo, la persona* (Vatican City: Urbaniana University Press, 2000); E. Baccarini, *La persona e i suoi volti: Etica e antropologia*, 2nd ed. (Rome: Anicia, 2003).

[15] Aquinas, *Summa theologiae*, I, q. 90, art. 2.

and by infusion it is created. The manner and moment of this union of the soul with the body in the unity of the person will be explained more precisely further on.

The immortality of the soul results from its immateriality. Some people consider this conclusion insufficiently sound, saying that immortality has more to do with religious belief than rational proof. It is true that the immortality of the soul and its spiritual origin imply and presuppose the concept of creation and the existence of the Creator, but this conclusion is not just within the province of faith. It also stands to reason, though the Christian faith illuminates and corroborates it. Once the spirituality of the soul and its creation by God are proved, one can only conclude that it is immortal. This is because, being immaterial, the soul cannot undergo corruption in itself; it could only lose its proper existence by the will of the Creator, who "non subtrahit rebus id quod est proprium naturis earum" (does not deprive things of that which is proper to their natures).[16] This is not to deny the relevance of truths of faith and the broadened horizon that follows for all of anthropology. The Christian faith, based on Revelation, proposes an understanding of man as a subject created "in the image and likeness of God," who becomes the recipient of divine life through the Redemption carried out by Christ and is called to eternal life, understood as a participation in divine life. But this should not lead to underestimating the importance of arguments derived from reason, from ethics, and from the natural and universal desire of humanity for what is eternal.[17]

The imaginative or conceptual transference of spatio-temporal notions to life outside of time should also be avoided: reason can readily affirm the fact of immortality, but this immortality should not therefore be imagined as a chronological prolongation of temporal life.[18]

The Body and Its Values

It is impossible to draw ethical lines in the biomedical field—to engage in bioethics—without first clarifying the value inherent in human corporality and, therefore, the body-spirit relation in the unity of the person.

In the fifth century BC, the school of Kos had already noted in Hippocrates' *The Places in Man* that "the nature of the body is the principle of medical discourse." Every physician knows intuitively that in approaching the body of the patient he is really approaching the person and that patient's body is not properly speaking the *object* of the medical or surgical intervention but rather a *subject*. But this body-person relationship must be more profoundly investigated especially in reference to questions concerning the beginning of embryonic life as well as the topics of health, illness, and death.

---

[16] Aquinas, *Summa contra gentiles*, ch. 55.

[17] See P. Schoonenberg. "Je crois à la vie éternelle," *Concilium* 5 (1969): 91ff; M. Blondel, "Le problème de l'immortalité de l'âme," *Supplément de la vie spirituelle* 61 (1939): 1–15; J. Maritain, *Bergsonian Philosophy and Thomism*, trans. M. L. Andison (New York: Philosophical Library, 1955), 84–116; Gevaert, *Il problema dell'uomo*, 270–282. Among recent authors, Maritain and Blondel argue that the immortality of the soul is demonstrable, whereas Jaspers and Scheler refer this truth to religious faith.

[18] J. T. Ramsey, *Freedom and Immortality* (London: SCM, 1971), 91–148; Gevaert, *Il problema dell'uomo*, 273.

This discussion will necessarily be philosophical and ethical in character. It will recall, or at least mention, contemporary ideological thinking regarding the "culture" or "counter-culture" of the body, just as it will make brief reference to biblical and theological reflections on corporality.

Indeed, remaining on philosophical terrain alone, there are three different concepts of corporality, each of which entails a different anthropology and a correspondingly different ethics: the dualist concept, the monist concept, and the personalist concept.

The dualist or intellectualist concept

This concept has distant origins in Greek thought, which is cosmocentric: reality is centered on the *kósmos*, the ordered world, in which matter, the fluctuating, blind, and indeterminate element, the seat of irrationality and fate, comes to be ruled and to a certain extent organized by divine ideas of a superior and opposite nature. Reality is dualistic and tragic in itself, and man is an "instance" of this tension[19] between the material world and the ideal, divine world. Anthropological dualism specifically affirms that there is a conflict between soul and body. Plato (427–347 BC), a primary proponent of this dualistic concept,[20] maintains that soul and body are accidentally united. Since the soul is the eternal and divine element, the body proves to be the chief obstacle to knowledge of ideas; the ideal for man is therefore to detach himself from bodily reality and estrange himself from the world. This attitude in the Platonic concept influences more than just his theory of knowledge and his general vision of life: since Plato situates morality in the context of an organicistic concept of the absolute state, he even goes so far as to justify physician-assisted euthanasia of seriously ill adults (*Republic*, 460b). This dualistic understanding is mitigated in the thinking of Aristotle (384–322 BC) but does not disappear completely.

Aristotle understands the union of soul and body through the *substantial* relation of *form* and *matter*, *act* and *potency*. The soul is the substantial form of the body for Aristotle, meaning that the body is human in all its parts inasmuch as it is

---

[19] C. Squarise, "Corpo," in *Dizionario enciclopedico di teologia morale*, ed. L. Rossi and A. Valsecchi (Rome: Paoline, 1981), 149–166; G. Giannini, *Il problema antropologico* (Rome: PUL, 1965); G. Mazzantini, *Storia del pensiero antico* (Turin: Bottega d'Erasmo, 1965); N. A. Luyten, "Das Leib–Seele Problem in philosophischer Licht," reprinted in *Ordo Rerum* (Freiburg: Universitätsverlag Freiburg, 1969), 285–287; G. Reale, *Corpo anima e salute: Il concetto di uomo da Omero a Platone* (Milan: Vita e Pensiero / Raffaello Cortina Editore, 1999).

[20] In Greek philosophical anthropology, the body is one of the ontological principles that make up the human being; it is the principle of materiality, the locus of the unintelligibility that must be subdued by the soul, which is in contrast to the seat of intelligence. The body (*soma*) is the tomb (*sema*) of the soul, as Plato declares in *Cratylus* (400a–c), and hence the soul must escape from its bodily prison (*Phaedo*, 65a, 66d–67a). The soul gives life and movement to the body; it is what organizes and unifies the body. The soul is shown to be the seat of intelligence, memory, sensation, and the spiritual faculties that demonstrate the superiority of man over the material and natural world. Therefore the soul is indissolubly connected with ideas (*Phaedo*, 77a); it governs, oversees, and gives life to the body (*Republic*, I, 353d), into which it has been cast down from the celestial regions, as the famous myth of the charioteer explains (*Phaedrus*, 246a).

informed by the soul; the soul actualizes the body and makes it a human body. The strong influence of dualism nevertheless remains inasmuch as the body is matter of foreign origin, opposed to the spirit, for Aristotle as well. The two co-principles of man do not entail a single origin, since matter is eternal and opposed to God. The soul unifies the body and is its *entelécheia*, conferring life upon it, but it does so as the principle of unification of vital functions and is therefore not identified with the body. Furthermore, as the active principle for understanding truth (*nous poieticós*), the soul remains foreign to the body and is identified instead with the divine.[21]

Moreover the works of Aristotle, prior to their interpretation in a Thomistic light, do not clearly affirm the immortality and afterlife of the soul of every single human being: it is the divine intellect that is eternal and eternally active within the personal soul. The Aristotelian vision of the body is organicistic and the human organism is what attracts the philosopher's scientific interest.

In an attempt at unification what loses weight is the personal soul and what gains prominence is the body: its form and its unity. This would lead Aristotle to a great interest in the organicistic school of medicine.[22] The unitary approach and the concepts that justify it (act and potency, matter and form) would be employed again later in the Thomistic and personalistic concept, with better safeguards for the spiritual and substantial weight of the personal soul.

Yet Aristotelianism is unable to conceive of morality as an all-embracing morality of man. Aristotle, given his organicistic view of man, also justifies the elimination of deformed newborns (*Politics*, VII, 1335b) on the basis of their lack of physical perfection. An anthropological regression is also found in the post-Aristotelian schools of thought, whether neo-Platonic or stoic: in the monistic overestimation of the universal reason (stoicism) of the One, from which everything emanates by way of progressive steps of degradation down to matter (neo-Platonism), the body remains an obstacle to which the divine is essentially foreign and the spirit is accidentally united.

The exaltation of suicide as a reasonable act of freedom, frequently carried out by the philosophers of those schools, should not be surprising. Nonetheless, the voices reaffirming the inviolable worth of human life in medicine (Hippocrates and Galen) or its precious value in the context of civic work (Cicero) testify to the persistence of an ethics of respect for individual human beings that rejects a purely instrumental consideration of the body.

The dualistic concept influenced certain sectors of early Christian literature, especially thinkers of the Platonic school and particularly with regard to their explanations for the present and active sin in man's temporal life. Nevertheless, they did not go so far as to deny that creation makes man, in his totality, *created* by God and therefore one in both his origin and destiny.

In order to find another period of pronounced anthropological dualism, one must look ahead to the triumph of rationalism with René Descartes,[23] Nicolas Malebranche,

---

[21] See Aristotle, *De anima*, B, 1, 412a 20; *Nicomachean Ethics*, K 7.

[22] A. Roselli, "La medicina e le scienze della vita," in *Storia delle scienze*, ed. Agazzi (Rome: Città Nuova, 1984), 1:105–107.

[23] See R. Descartes, *Discourse on Method and Meditations on First Philosophy*, trans. D. A. Cress (Indianapolis: Hackett Publishing, 1981), 6.

and Gottfried Leibniz[24] in the Age of Reason. For Descartes the body is united concretely and physically to the soul by means of the epiphysis or pineal gland, but the two realities differ in essence and value: the body is a machine and the study of it pertains to the science of mechanics and natural science; the spirit is consciousness and is what gives worth to man. The body acquires a distinctly instrumental significance and Descartes perceives no operational conflict in man, just as there is no conflict between an operator and his machine.

Knowing the laws of nature is the task of reason, and science—as its method was defined by Galileo according to the criteria of observation and experiment—aims at knowledge in order to control. The mechanistic worldview, as an interpretation of physical and corporeal reality, would lead medicine to embrace the experimental method of observation and to make considerable progress in understanding the structure and functioning of the human body. According to Cartesian thinking, the human body has no need of the soul (in the sense of a vital principle) in order to be understood: the body is physical and mechanical. The human spirit is not required to explain the functioning of the body, but rather to explain self-consciousness, reason, and rational interpretation of the world.

Malebranche accentuates this dualism, declaring that the spirit does not even control the body directly in the manner of an instrument or a machine. The separation between them is such that Malebranche uses God to explain the harmonization of bodily life with spiritual life in cognitive processes and practical activity (occasionalism). Leibniz resorts to the concept of a pre-established harmony between the spiritual order and the physical order in man so as to eliminate the idea of God's continuous intervention, while emphasizing the same structural dualism.

This extreme dualism, which now belongs to the past, is exactly what would foster materialistic monism, which views so-called psychic or spiritual phenomena as a reflection of bodily organization.[25]

The monistic concept

Setting aside the ancient notions of Epicureanism, the materialistic and monistic interpretation since the time of Karl Marx—particularly the neo-Marxism of Jean-Paul Sartre and Herbert Marcuse—has offered a reductionistic and political view of the body.

Classical Marxism subordinated the body to the species and to society; neo-Marxism orients it toward a second, more individualistic revolution. The body exhausts the totality of the human being ("J'existe mon corps," Sartre declares) and his experiences: both what man is within himself and what man attains in experience with others are corporality and bodily experience. Marcuse, whose thinking is not always systematic, generally affirms that the body is the locus as well as the means of liberation: to take back the body means to liberate the human person from the organization of employment that depends on and belongs to bourgeois and industrial society, from extrinsic morality, and from the institutionalization of marriage; it means making the body a place for pleasure, recreation, and the expression of all

---

[24] See "Monadology," in *Monadology and Other Philosophical Essays*, G. W. Leibniz, trans. P. Schrecker and A. M. Schrecker (New York: Macmillan, 1965), § 81.

[25] Gevaert, *Il problema dell'uomo*, 53–64.

that it can be. These ideas are more explicit in the work *Eros and Civilization* (1955). The body must be liberated from money, the *logos* of the law, and social constrictions; once liberated it will be the starting point for a new society.[26] The first feminist movement, associated with Simone de Beauvoir and her famous work, *The Second Sex*, was grafted into this revolutionary political trunk. It had and still has great influence in terms of demands for sexual freedom and campaigns for contraception, abortion, voluntary sterilization, and so forth.[27]

Some contemporary psychological trends such as behaviorism and psychoanalysis, while helping to overcome the organicistic and mechanistic concept of the body, look no further than the limited horizon of the materialistic and monistic concept of man, yet they propose to investigate the mystery of the human psyche and of the "body as it is experienced," whether through behavioral studies or the analysis of subconscious dynamics and social pressures. These schools have surely made some remarkable contributions to their field and offer keys of interpretation for the reality of man in terms of his subjectivity and his pathology, but they do not emerge from the monistic and time-bound view of the human being except by way of a certain meta-psychological correction and integration.

Obviously these remarks constitute a simple sketch of complex cultural movements that include other names and variations on these basic ideas. I have noted only what is absolutely necessary for understanding the importance of these themes in the ongoing discussion within ethical science, medicine, and the social management of health care issues.

A more recent affirmation of monistic reductionism can be found in the scientific and biological thought of Jacques Monod, mentioned earlier in reference to his book, *Le hasard et la nécessité* (1970). Not only are human beings reducible to their biology, according to this author, but biology itself is nothing more than physics. François Jacob's *La logique du vivant* (1970) and Jean-Pierre Changeux's *L'homme neuronal* (1983), in which the author declares that the mind is identical to the brain, present positions along these same lines. I will return to this biological reductionism in a later chapter.

The personalist concept of man and of corporality

At various times over the course of the history of theological and philosophical thought, many contributions have been made to this concept, which is of fundamental importance for any ethics of the body and all ethical problems in the areas of sexuality and medicine.

First of all, it is universally recognized that Christianity introduced into the history of Western thought the very notion of *person*, understood to mean a subsistent, conscious, free, and responsible being. The teachings of Revelation concerning creation and the understanding of God as a personal Being who transcends worldly reality; the concept of man as "created in the image and likeness of God," that is to say, capable of entering into dialogue with God and called to stewardship over

---

[26] R. Nebuloni, "Crisi dell'eros e crisi della civiltà nel pensiero di H. Marcuse," in *Amore e matrimonio nel pensiero filosofico e teologico moderno*, ed. V. Melchiorre (Milan: Vita e Pensiero, 1976), 319–344; S. Spinsanti, *Il corpo nella cultura contemporanea* (Brescia: Queriniana, 1983), with an extensive bibliography.

[27] S. Cremaschi, "Il concetto di eros in *Le deuxième sexe* di Simone de Beauvoir," in *Amore e matrimonio*, Melchiorre, 296–313.

creation in God's name and by His authority; the concept of sin and Redemption; the teachings concerning the gift of divine life granted to the human person who thus becomes joined to the very life of God; and the mysterious fact of the Passion and Resurrection of Christ that guarantees the bodily resurrection of humanity all represent a vision that has established the following principle in the heart of Catholic theology: *caro cardo salutis* (flesh is the hinge of salvation). The very body–soul dichotomy inherent in man becomes powerfully welded into unity, as it were, to such an extent that biblical language uses the corresponding terms *basar* (flesh; in Greek, *sarx*) and *ruah* (spirit; in Greek, *pneuma*) to designate not so much the two ontological components of man, but rather his two opposing dispositions or propensities: the idolatrous or worldly propensity and the propensity of filial obedience to God.[28]

Leaving aside the contributions of the Church Fathers of various schools (influenced by Platonism or Aristotelianism), it is important to underscore Aquinas's systematization of the problem of the soul-body relation, which would become indispensable for the entire personalist school as well as for neo-Thomism and neo-scholasticism. In addition to Christian thought, which is presupposed, Aquinas proceeds by using the interpretive lenses of Aristotelian metaphysics: matter and form, essence and existence.[29]

First of all, Aquinas states that the soul is united to the body *substantially* and not accidentally, as a co-principle of the person, since the soul is the substantial form of the body. This implies that the body is human because it is animated by a spiritual soul: it is what it is because it receives the ontological structure of humanity from the spiritual principle, the same principle by which human beings know and are free. This same principle—the soul—is the body's substantial form, created for the twofold and connatural capacity of being in itself and of animating its body. According to Aquinas, only this explanation successfully accounts for the unity of human activity, which is always both physical and spiritual, and rules out the aporias of dualism. The soul is therefore the substantial form of the body and of the whole human individual. The principle that *operari sequitur esse* (activity is proportional to being) gives an account of this argument: man, even when he acts with his body, acts in a human and spiritual form. The substantial form is what allows a body to carry out its activity. Indeed, something must *be* in order to act and it must *be in a definite way* in order to act in a definite way—it must have a definite nature. The principle by which a body has a definite nature—and hence a definite activity—is the substantial form.[30] If the body were alongside the soul and the soul alongside the

---

[28] Squarise, "Corpo," 159–164; W. Mork, *The Biblical Meaning of Man* (Milwaukee: Bruce, 1967); Congresso dei Biblisti e Moralisti dell'Italia Meridionale, ed., *Antropologia biblica e morale: Atti del 1° congresso dei biblisti e moralisti dell'Italia meridionale; Castellammare, 1–2 giugno 1971* (Naples: Dehoniane, 1972); F. Baumgärtel, R. Meyer, and E. Schweiser, "Sarx," in *Grande lessico del nuovo testamento*, ed. G. Kittel and G. Friedrich (Brescia: Paideia, 1965–1992), 11:1265–1398; H. Kleinknecht, "Pneuma," in *Grande lessico*, Kittel and Friedrich, 10:767–849.

[29] See A. Ghisalberti, "Anima e corpo in Tommaso d'Aquino," *Rivista di Filosofia Neoscolastica* 97 (2005): 282–296.

[30] Vanni Rovighi, *Elementi di filosofia*, 3:164–166. Aquinas sets forth his famous arguments in *Summa contra gentiles*, chaps. 56–67 (also in *De anima*, art. 2 and 3, in the *Summa theologiae*, I, q. 76, art. 1 and 2, and in the *Quaestiones disputatae: De spiritualibus*

body, there would be no explanation for the unity of activity or, as one might say with more modern terminology, for the unified source of energy and information in the individual.

Saying that the soul is the substantial form of the body also means it is the only substantial form, because a plurality of forms would entail a plurality of beings and sources of activity. Therefore the soul, with its energy and unifying force, also activates and informs the faculties—they are different *faculties*, not beings—proper to vegetative and sensitive life. This has great ethical significance in matters concerning the unity of life in man: man remains man even when he does not yet or can no longer exercise his mental faculties for accidental reasons.

As will be seen more clearly in the chapter on abortion, human embryos are certainly not in a position to display typically human activities, and not even fetuses and neonates are able to express themselves through their mental faculties; nevertheless, it cannot be denied that the real capacity to activate these higher activities is established at the moment of fertilization. It is not even necessary to wait for the formation of the primitive streak or the first nucleus of the nervous system because the embryo, in its reality, possesses the active capacity to realize both the organ— the brain—and the function. This real capacity is rooted in the very essence of the individual human being whose corporality is informed and structured by the spirit that vivifies it. This is why one cannot make an ontological distinction—a real distinction—between a human individual and a human person at any stage of biological development from fertilization onward or at any stage of intellectual maturity.

While the notion of person does not necessarily indicate the species *Homo sapiens*, neither does it indicate any other species: no other species is capable of recognizing the right to life that belongs to a person. A person is often defined as a self-conscious, rational being capable of moral activity and endowed with autonomy, yet this overlooks the *particular* fact that human beings are not *only* this; indeed, another characteristic is that the human being is a corporeal being and a being in a state of becoming.

The duties of persons are to be exercised even toward those who are persons but do not live as such, i.e., who do not exercise their personal faculties. The fact that there are persons who are currently able and obliged to decide for others who cannot do so is because other persons have allowed them to reach the stage in which they can fully (yet only temporally) exercise this power. "Reciprocity is constitutive not only of morality but also of the very possibility of personal existence. ... This reciprocity is the source of all bioethical discussion and the point to which all such discussion must be traced." [31]

---

creaturis, art. 2 and 9). Here are two important citations from Aquinas on the matter: "The human soul is a form that does not depend on the body in its being" (*Summa contra gentiles,* II, 79), and "The body and the soul are not two substances existing in act, but from them results only one substance in act" (*Summa contra gentiles,* II, 69). See S. Vanni Rovighi, *L'antropologia filosofica di s. Tommaso d'Aquino* (Milan: Vita e Pensiero, 1965); see also Gevaert, *Il problema dell'uomo,* 47–58; F. Nuyens, *L'évolution de la psychologie d'Aristote* (Louvain: Inst. supérieur de philosophie, 1948).

[31] A. Pessina, *Bioetica: L'uomo sperimentale* (Milan: Bruno Mondadori Editore, 1999), 88.

Limiting respect for life and the right to life only to those stages in which moral life is exercised means forgetting that morality is exercised in intermittent cycles and circumscribed in time. The ethical problems encountered in the field of health care often concern precisely those individuals who do not fit the mythical description of self-conscious, autonomous, rational, and free persons. The human being is structurally a person: this condition does not depend on his will but rather on his origin.

The question of human corporality must be addressed in a radical, profound manner. Why is it that deformed human bodies, or bodies visible only under the microscope (as in the case of embryos), or inert bodies without clear signs of consciousness, *are* human beings? Why *are* they persons? What is the basis for the anthropological equivalence that unites the healthy and the sick, the deformed and the nondeformed?

The humble but decisive argument for determining who is or is not a human being is "to look to the origin: a human being is someone who is born of other human beings ... man is always someone who is born of other men; this is the prerequisite for proceeding to any further and more in-depth definition of man."[32]

Anyone who is generated by other human persons, whether directly or by means of their genetic patrimony, must be considered a human person. Hence the human body is central. There is no way of protecting one's self, one's dignity and integrity, unless the concrete corporality of others is also protected and respected.

The second pivotal idea in Thomistic philosophy, even though it is not fully explained, is based on the two principles of every real being—namely, essence and existence (the Thomistic terms are *essentia* and *esse* in Latin). The composite essence of man (soul and body) passes from the potential and hypothetical state to the real state—just as it happens with every reality—following an existential act that concretely realizes the potentiality. Now the existential act that realizes the human body is the very same existential act that realizes the soul: it is one existential act, not two; this is implied by the unity of the form. The existential act proper to the form is what actualizes the matter, which in this case is the body. Since the existential act of the soul proceeds (and can only proceed) from the Creator, the same act vivifies and actualizes the body, and this occurs (and can only occur) simultaneously with the secondary causes—that is, at the moment of procreation.

Despite these principles, given the imperfect knowledge about human embryology in his day, Aquinas was led to postulate two stages in the formation of the human being in utero: one preceding and the other following the infusion of the soul. This presumed discontinuity in receiving the "human form" does not invalidate his principles, which now lead personalists to a different conclusion based on the discoveries of embryology, nor does it allow us to interpret Aquinas as favoring abortion prior to bodily ensoulment.[33]

The conclusion to what has been said here can be summed up in the words of Marcel: "What is unique about my body is that it does not and cannot exist by itself."[34] It can therefore be said that "I am my body," so long as this is not understood in the

---

[32] Pessina, *Bioetica*, 91.

[33] V. Fagone, "Il problema dell'inizio della vita del soggetto umano," in *Aborto: Riflessioni di studiosi cattolici*, ed. A. Fiori and E. Sgreccia (Milan: 1975), 149–179.

[34] G. Marcel, *Du refus à l'invocation* (Paris: Gallimard, 1940), 30; idem, *Being and Having*, trans. K. Farrer (Westminster: Dacre Press, 1949), 154–157; idem, *Metaphysical Journal*, trans. B. Wall (Chicago: H. Regnery, 1952), 241–248.

exhaustive sense to mean that I am *only* my body; it can also be said that "I have a body," so long as this does not mean it is merely an object. In this regard, Maritain declares, "Each element of the human body is human, and exists as such, by virtue of the immaterial existence of the human soul. Our body, our hands, our eyes exist by virtue of the existence of our soul."[35]

The work done by Josef Seifert in his book *Essere e persona* (Being and person)[36] appears to be of great importance. It is an attempt to demonstrate the immense value of the connection between the subject, the person, and classical metaphysics. This connection brings to light the metaphysics of the person, whether finite or absolute, and the ability of a redeveloped phenomenology to contribute to a personalist metaphysics.

The phenomenological approach to corporality has made notable contributions to a comprehensive and thorough understanding thereof, but it is open to criticism on many counts unless it is supplemented and integrated with an ontological vision of corporality. Existence is unintelligible except in terms of an essence, and man's essence is defined by the substantial union of the spiritual soul with the body, which owes its existential form to the spiritual soul itself. Indeed, these more recent contributions to the concept of man leave the ontological discussion open; they are situated on the relational level—that is, they consider the value of the body in its mediation with the world, society, and history.

The phenomenological concept of corporality,[37] which arose as an attempt to overcome both the over-intellectual and materialistic views, experienced a development in its consideration of the body. Edmund Husserl[38] introduced the original distinction between *Körper*, understood as the bodily organism, an object of study, and *Leib*, the experience or consciousness of one's own body. He posits the constant awareness of one's own corporality, a certain "co-presence" of the body in every other perception, an interior experience of one's own body that is different from the perception of other bodies or *Körper*.

Maurice Merleau-Ponty, making an argument analogous to Heidegger's and achieving a synthesis between the progress made in psychological experiments and Husserl's approach, overcomes the subjectivity of corporality by leading it toward the idea of a concrete being-in-the-world. From this point of view, bodily perception becomes a description of the relation between consciousness and world as well as of its manifestations or expressions. The body not only sets up the relation with the world but also confers ever-new meanings upon the world by continually transcending its own experiences and the previous meanings. One's own body is in the world as the heart is in the organism: it continually keeps the visible spectacle alive, entirely

---

[35] Maritain, *Range of Reason*, 57.

[36] J. Seifert, *Essere e persona: Verso una fondazione fenomenologica di una metafisica classica e personalistica* (Milan: Vita e Pensiero, 1989), cited in chapter 2 of this work.

[37] For a thorough analysis of this perspective, see V. Melchiorre, *Il corpo* (Brescia: La Scuola, 1984), 187–230; A. Ales Bello, "L'analisi della corporeità nella fenomenologia," in *Corporeità e pensiero: Atti dell'VIII Convegno Studium; Roma, 21–23 ottobre 1999* (Rome: Studium, 2000).

[38] See E. Husserl, *Ideas: A General Introduction to Phenomenology*, trans. W. R. Boyce Gibson (New York: Macmillan, 1931), II, II, 39; idem, *Cartesian Meditations: An Introduction to Phenomenology*, trans. D. Cairns (The Hague: M. Nijhoff, 1977), § 44, p. 92–99.

animating and nourishing it, forming a system with it. The body is our general means of having a world[39]; it is "of the Being-seen."[40] According to Merleau-Ponty, the world of experience is the realm constructed in the body and by the body. This assumes that the transcendence of purpose and meaning has an explanation beyond the body itself; in my view, in the spirit. According to Heidegger, the *Da-sein* (being-there) proper to human existence—as spatio-temporal existence—marks what is proper to man and is a property of his being embodied in a body.[41]

The thinking of Marcel emphasizes the function of the body in terms of its social mediation. If human existence is human inasmuch as it is a "being-with" others, being open to others, this is possible through corporality and its language: the body is a "presence" set before others; it is a memorial synthesis of the past, present, and future set before others and before society. It requires that human beings recognize one another as persons and entails the possibility of communion.[42]

Marcel brings to light the ambiguity of the body in the expressions "I *have* a body" and "I *am* a body,"[43] both of which are valid yet neither of which can be used in the absolute sense. To say "I *have* a body" is correct but necessarily requires an explanation, because the way in which I possess my body is not the same way in which I possess what is not my body. I do not have or possess a body as I have or possess other objects in that I cannot distance myself from my body and, in the strict sense, I cannot *make use* of it without the consequences affecting my entire person. In the same way, the expression "I *am* a body" should be accompanied by the clarification that "I am not *just* a body," but also something more.

For this reason, philosophical anthropology prefers to use the term *corporality* rather than *body*: it better expresses the corporeal–spiritual unity of the person. In contrast to the term *body*, which refers to a part of the person, evoking again the classic dichotomy between body and soul, *corporality* has greater breadth: it designates human subjectivity in its corporeal condition, which constitutes personal identity.

Scheler, the chief exponent of the phenomenological school, has emphasized the value of the body as "expression" and hence as culture, civilization, and ability to transform the world and matter by means of technology. He sees the human person's very capacity for fulfillment in the cultural mediation of the body.[44] The body, moreover, is capacity for language precisely because it is expressiveness, epiphany, and phenomenology of the self; this is true not only in spoken, written, or artistic

---

[39] M. Merleau-Ponty, *The Structure of Behavior* (Boston: Beacon Press, 1963), 185–224; idem, *Phenomenology of Perception* (New York: Humanities Press, 1962); Reale and Antiseri, *Il pensiero occidentale*, 3:467–469; C. Bruaire, *Filosofia del corpo* (Cinisello Balsamo: Paoline, 1975); Melchiorre, *Il corpo.*

[40] M. Merleau-Ponty, *The Visible and the Invisible: Followed by Working Notes*, ed. C. Lefort, trans. A. Lingis (Evanston, IL: Northwestern University, 1968), 250.

[41] Reale and Antiseri, *Il pensiero occidentale*, 3:445–453.

[42] See G. Marcel, *Homo Viator: Introduction to a Metaphysic of Hope*, trans. E. Craufurd and P. Seaton (South Bend, IN: St. Augustine's Press, 2010), 23–62.

[43] G. Marcel, *Metaphysical Journal*, 241–248. Translator's note : More precisely, Marcel compares the expressions "I ... make use of my body," which is entailed by the concept of having a body, and "I am my body."

[44] M. Scheler, *The Nature of Sympathy*, trans. P. Heath (London: Routledge & Paul, 1954), 238–264.

language, but in all of the person's gesturing[45]: from smiling to weeping, from gazes to facial expressions. Finally the body is the principle and sign of individual differentiation: man or woman, this individual or another.

The debate about the body–soul relation has focused recently in on the brain–mind relation. Here, too, a unified concept emerges that entails simultaneously interdependence and transcendence of the mind with respect to the cerebral organ.[46] The more recent expression *mind–body relation* is a manifestation of what has described as an *epistemological reduction*: the term *mind* is not used in the same sense as the traditional understanding of the term *soul*. It is a reference to the term *mind* as used by the empiricist philosopher David Hume in his *Treatise on Human Nature*, designating the whole of man's conscious states and/or functions. The mind-body relation has thus come to indicate the relations between fundamentally psychological and neurophysiological functions, not the relations between the subjects of those functions (soul–body). This is a new, reductionistic approach that excludes the metaphysical dimension.

The philosophers Karl Popper and John Eccles, who remain moderately dualistic in their theory of the "interaction" between mind and body, consider the existence of thought to be an enigma for the biological and cosmological sciences.

The body is normally overlooked in silence, as if transparent. In pain, fatigue, and illness, however, the body becomes opaque, heavy, strange, and menacing. Health has been defined as "life when the organs are quiet" or "unawareness of one's own body."[47] Being well means not realizing that one has a body.

In the body one experiences a dialectic between activity and passivity, a dialectic between *praxis* and *pathos*: undergoing the body that one governs.[48]

In conclusion, the personalistic and human meanings of the body can be summed up with these expressions and qualifiers: spatiotemporal embodiment, individual differentiation, expression and culture, relation with the world and with society, instrumentality and principle of technology. Recall that technology is nothing other than the extension and reinforcement of the body's muscular system (cars), senses (audiovisual devices), and brain (computer processing).

Yet the body is also a *limit*—a sign of limitation in space and time. This limitation, which is highlighted in particular by existentialism and personalism, carries with it the concepts of suffering, illness, and death.

This reflection on the relational, phenomenological, and historical values of corporality derives all of its substance and richness from the ontological fact of the relation of the body with the spirit, and therefore with the person, who is existential activity endowed with thought, freedom, and self-determination. Every medical procedure and every intervention upon corporality cannot fail to keep in mind this rich endowment and this tie: it is the act of one person upon another person through bodily mediation.

---

[45] J. Mouroux, *Sens chrétien de l'homme* (Paris: Aubier, 1945), 43–74; Gevaert, *Il problema dell'uomo*, 70–80.

[46] See the major dialogue between Popper and Eccles in *The Self and Its Brain*.

[47] G. Calguilhem, *Il normale and il patologico* (Turin: Einaudi, 1998), 65.

[48] P. Ricoeur, *Oneself as Another*, trans. K. Blamey (Chicago: University of Chicago Press, 1992), 152–168.

A final reflection should accompany this conclusion on the values of corporality: corporeal values are in harmony with one another and hierarchically arranged. This is a corollary to the unity of the body which, though composed of parts, becomes one living organism and is animated by the soul. The unity of many living parts both requires and structures the hierarchy of the parts and their intercommunication—that is, their harmony. One can only speak of a living organism on this condition. This is what moralists call *unitotality*.

Important consequences for morality and medical ethics follow from this fact. *Life* is the first good that presents itself as essential to the living organism. Whatever takes away life destroys the organism as such, and this is the greatest privation that can be inflicted upon the person. Only the spiritual good that transcends the person, the moral good, can entail the risk of a voluntary loss of life. A person's physical life can be placed in danger and indirectly sacrificed (the direct lethal action is the work of others) only for the sake of moral goods that pertain to the totality of the person or in order to save other persons; such is the case of martyrdom or the defense of one's loved ones or one's fellow human beings from an unjust aggressor. Even in these cases, however, there is always a responsibility that rests with those who created the situation and caused the loss of life.

Next in importance after life itself is the *integrity* of life, which we are permitted to take away only if its removal is required to safeguard physical life in its totality or for a higher moral good.

The goods of interpersonal life, such as emotional or social goods, must be subordinated to the two preceding goods: life and its integrity. A surgical procedure justifies recovery in the hospital and separation from family life. There is no social reason, on the other hand, that could justify self-mutilation or sterilization, much less the direct suppression of a human life.

It must also be kept in mind, however, that the values of the person are in harmony with one another: when one is violated, the others suffer as a result. For example, the total deprivation of social relations becomes burdensome for the whole person. This is why any violation of a value, even a temporary one temporarily, demands an objective justification in terms of a higher good.[49]

The Transcendence of the Person

"To all those who still wish to talk about man, about his reign or his liberation, to all those who still ask themselves questions about what man is in his essence, to all those who wish to take him as their starting-point in their attempts to reach the truth … to all these warped and twisted forms of reflection we can answer only with a philosophical laugh."[50] This remark tells us that the reduction of man to nothing, which elicits a heavy-hearted philosophical laugh, is a characteristic not only of the cruelty of some totalitarian regimes—those built upon concentration camps and gulags and those built upon a profit-based, hedonistic society—but has also pervaded

---

[49] E. Sgreccia, "Valori morali per la salute dell'uomo," *Rassegna di teologia* 5 (1979): 390–396; M. V. Nodari, ed., *Uomo e salute* (Vicenza: Edizioni del Rezzara, 1979).

[50] M. Foucault, *The Order of Things: An Archaeology of the Human Sciences* (New York: Vintage, 1973), 342–343.

contemporary thought, which to a large extent proves to be nihilistic and shut off from transcendence.

In the face of this crisis, personalism vindicates the transcendence and inviolable dignity of the human person, who ontologically sums up all the values of the cosmos and is the center of society and history.

The term *transcendent* is not used here in the absolute sense reserved for the Creator, who calls the world into being while remaining infinitely distinct and different from it. The world's being is causally distinct from the Creator's and ontologically dependent on His creative act. In this respect, even though the human person is created in His image and likeness, as Christian Revelation confirms, God infinitely transcends him as well. The transcendence of the human person is to be understood in reference to subhuman realities and social and political interests.

The person is transcendent with respect to subhuman realities from the ontological and axiological perspectives. The person, capable of self-consciousness and self-determination, surpasses the material world in terms of novelty and ontological level and value; it is the world that acquires meaning in the human person, who represents the purpose of the universe.

> Whenever we say that a man is a person, we mean that he is more than a mere parcel of matter, more than an individual element in nature, such as is an atom, a blade of grass, a fly or an elephant. … Man is an animal and an individual, but unlike other animals or individuals. Man is an individual who holds himself in hand by his intelligence and his will. He exists not merely physically; there is in him a richer and nobler existence; he has a spiritual superexistence through knowledge and through love. He is thus in some fashion a whole, not merely a part; he is a universe unto himself, a microcosm in which the whole great universe can be encompassed through knowledge; and through love he can give himself freely to beings who are, as it were, other selves to him. For this relationship no equivalent is to be found in the physical world. All this means, in philosophical terms, that in the flesh and bones of man there lives a soul which is a spirit and which has a greater value than the whole physical universe. However dependent it may be on the slightest accidents of matter, the human person exists by virtue of the existence of its soul, which dominates time and death. It is the spirit which is the root of personality.

> The notion of personality thus involves that of totality and independence; no matter how poor and crushed a person may be, as such he is a whole, and as a person, subsists in an independent manner.[51]

The definition of person offered by Boethius, *rationalis naturae individual substantia* (an individual substance of a rational nature), has appeared too essentialist and static for the contemporary existentialist mindset. It should be noted, however, that the concept of substance must be understood in a dynamic sense: it is the center of activity, movement, and tension, especially in living beings and in living beings endowed with reason. When medieval thinkers spoke about substance, they did

---

[51] J. Maritain, *The Rights of Man and Natural Law* (New York: C. Scribner's Sons, 1943), 2–4.

not intend "a static inert substratum; [rather] it is the first root of a thing's activities and, while remaining the same as to its substantial being, it ceaselessly acts and changes—through its accidents, which are an expansion of itself into another, non-substantial, dimension of being."[52] Boethius emphasizes substantial metaphysical identity because the person is original and irreducible to other things or to the cosmos. The focus on substance brings to light the person's character as an existing subject (*substratum*) and not just as simple activity.

Similar definitions are provided by Richard of Saint Victor (*rationalis naturae individua existentia*: an individual existence of a rational nature) and Aquinas (*individuum subsistens in rationali natura*: a subsisting individual in a rational nature), reflecting the tension between the universal reference to the species (*rationalis naturae*) and the individual character (*individua substantia*). We all belong to humanity, but each one in his or her own way. The fundamental notion of *individual* appears in all the formulas cited; it does not mean that the individual is indivisible, but rather that he is undivided—in other words, endowed with unity. Individuality does not imply indivisibility as if the individual were *a-tomon*; rather, it implies existence as one and the same, divided from others. In other words, the individual is undivided in himself and divided from all others.[53]

At its core, the philosophy of the person is linked to the philosophy of substance, toward which a portion of philosophical thought has harbored profound suspicion for centuries, preferring to replace it with the philosophy of function.

In *Categories*, Aristotle defines first substance as "that which is neither in a subject nor predicated of a subject" (3a 8s).[54] Substance is existence-in-itself, self-sufficient and independent from other subjects in its existence. The act of being is the *first and radical* act of the individual substance: all other (secondary) acts of the person are rooted in this act and derive their life from it.

The person is an *in se existens* (being existing in itself) and a *per se existens* (being existing for itself). The first expression indicates that it does not exist in anything else or as a mode of anything else; the second means that it does not exist for the sake of something else but for its own sake (*propter se, non propter aliud*). The person is set as an end, not a means. In addition to these characteristics of *inseitas* (*in se*) and *perseitas* (*per se*), the metaphysical tradition employs the concept of *aseitas* (*a se*, from itself) as a prerogative of the divine person alone. It does not depend on another, whereas *esse ab alio* (being from another) is the mark of a contingent being, the sign of its radical inability to be the principle of its own existence.

While the person is said to be a set of successive selves and successive states devoid of a common substratum in empiricist theories (John Locke, David Hume, Derek Parfit), the same is not true of the substance-based approach. Furthermore, the latter safeguards the notion that the person is more than just his actions.

"Becoming a person" in the sense of possessing one's own fundamental ontological status is not a *process* but rather an *event* or instantaneous act whereby one is established in personhood once and for all. On the other hand, *personality*—in

---

[52] Maritain, *Range of Reason*, 36.

[53] See Possenti, "Noi che non sappiamo," 1–12.

[54] *Metaphysics* 1017 b, 10–25.

the psychological sense—is acquired gradually through the accomplishment of personal (secondary) acts.

Contemporary thinking places much emphasis on *relationality*, and some ask whether the person could be considered as relationality only (and not substantiality) and whether, consequently, a human being no longer able to relate to others can be considered a person. "Rarely is it understood that rationality necessarily implies relationality: the life of the spirit is openness and it is relational per se." [55]

This ontological greatness and worth of the human person also becomes evident when viewed in relation to society. The person cannot be considered just a part of society, just as society cannot be considered a "living organism" (the organicistic conception). Society emerges from the heart and center of the person, who opens himself with all his being toward other persons; yet while he is the originator of society, the person does not invest himself entirely in the social, temporal, or political realm. The dissolution of the person in social and collective enterprises has been and still is the most serious catastrophe for humanity.

This is not the place to explain how immanentism, whether in its idealist or materialist form, after having dissolved the person away in historical determinism, has justified the most atrocious forms of absolutism and provided the theoretical foundations for the physical and legalized suppression of those who do not comply with the government or party or sector of society in power at any given time. Ultimately, the physical suppression of so many persons at the hands of others should not be surprising since modern intellectual history has erased the concept of person from individual consciences. As already mentioned at the beginning of this chapter, the practice of medicine does not address mere bodies or machines, but rather persons in all of their majesty and moral greatness. The social organization of health care services cannot allow a state of affairs in which assistance is guaranteed—even free of charge—to some persons, while death is dealt out—perhaps under the guise of assistance—to others (abortion, euthanasia, killing by starvation, etc.).

The whole world and the meaning of the universe are summed up in every person, and social organization and even legal order find their justification in the human person. The very notion of the *common good* should not be understood as the statistical average of goods belonging to or appropriate for individuals in a quantitative understanding of social transactions; rather, it should be understood as the good that is realized in all the individuals and components of society in an adequate and just manner. "The common good of society is neither a mere collection of private goods, nor the good proper to the whole, which (as in the case of the species with regard to its individual members, or the hive with regard to the bees) draws the parts to itself alone, and sacrifices these parts to itself. It is the good human life of the multitude, of a multitude of persons, the good life of totalities at once carnal and spiritual, although they more often happen to live by the flesh than by the spirit." [56]

It was already noted at the beginning of this chapter how this primacy of the person in the universe and in society comes to assume a new and even richer dimension within the discipline of theological anthropology. The person is able to touch the divine by virtue of the gift of divine and supernatural life given to man

---

[55] Possenti, "Noi che non sappiamo," 17.

[56] Maritain, *Rights of Man*, 8.

gratuitously through the Incarnation and Redemption of Christ and by virtue of the Christian hope in the Resurrection.

A treatise on bioethics should insist above all on what is rationally valid for every human being, believer or nonbeliever, yet I cannot silence this vision that is open to everyone and shared by such a large percentage of people of faith. To this end I gladly quote a passage from Vatican Council II's Pastoral Constitution *Gaudium et spes*, which speaks about the dignity of man: "Believers and unbelievers agree almost unanimously that all things on earth should be ordained to men as to their center and summit."[57]

The Person, Health, and Illness

The importance of what we have discussed thus far becomes evident in light of the concepts of health and illness. Health and illness do not concern the physical body alone, nor can the two terms be defined in a purely organicistic sense.

Everyone is familiar with the definition of health offered by the World Health Organization (WHO) in 1946: "Health is a state of complete physical, mental and social well-being and not merely the absence of disease and infirmity."[58] This definition already surpasses the limits of an organicistic understanding of health because it includes mental state and social condition. It nevertheless appears to leave room for improvement and warrants further reflection. First of all, it is quite difficult to conceive of health as a static fact or perfect measure. Health can be understood more precisely as a *dynamic equilibrium* among the different organs and different functions of the organism's unity within the soma, between the soma and the psyche at the individual level, and between the individual and the environment. Another point to be added to the definition is precisely the concept of environment: it should be understood as not only social but also ecological precisely because of the unity of mutual exchange between the human organism and its biophysical environment.

A distinction must be made between *perfect health* and *relative health*. The first results from the concurrence of a subjective sense of well-being with physiological and behavioral normalcy; it is a simple, ideal concept that can only be approached and never fully attained. The second, as Pedro Laín Entralgo declares, consists in the "physical ability to carry out one's life plans with minimal trouble, with minimal injury and, if possible, with a certain well-being and enjoyment."[59]

---

[57] Vatican Council II, Pastoral Constitution *Gaudium et spes*, no. 12.

[58] World Health Organization, *Constitution of the World Health Organization*, July 22, 1946, http://www.who.int/governance/eb/who_constitution_en.pdf. The World Health Organization confirmed its definition of health on the occasion of its International Conference in Ottawa (1986). It explains that "to reach a state of complete physical, mental and social well-being, an individual or group must be able to identify and to realize aspirations, to satisfy needs, and to change or cope with the environment." World Health Organization, *Ottawa Charter for Health Promotion*, November 21, 1986, WHO/HPR/HEP/95.1, http://www.who.int/hpr/NPH/docs/ottawa_charter_hp.pdf.

[59] P. Laín Entralgo, *Antropología medica para clínicos* (Barcelona: Salvat, 1985), 199, quoted in M. T. Russo, *Corpo, salute, cura: Linee di antropologia biomedica* (Soveria Mannelli, CZ: Rubbettino, 2004), 186.

Man's health is therefore not equivalent to his perfection, though it is a condition favoring its attainment. Richard Siebeck's point on this matter is helpful:

> The concept of health is not complete without the question: health to what end? The bottom line is that we do not live in order to be healthy, but we are and want to be healthy in order to live and act. Health is not a good that is entrusted to us just for certain functions and abilities. Health is not an ultimate end, but rather is determined and limited by the meaning of life itself. And the meaning of life is availability, self-gift and sacrifice. The meaning of health lies beyond health; it is found in that plan for a good life that each person proposes to carry out, for which health is a necessary but not indispensable condition. Not only does illness not necessarily defeat a plan for a good life, but conversely, one can say that without that plan health itself has no value. For this reason Plato declared that virtue comes first and then health and wisdom follow, just as a god is followed by his cortege. Indeed, health is not a property of the organism but of the whole person; it will therefore be subordinate to whatever constitutes the genuine good for the person.[60]

The equation "health = salvation" can lead to despair when one is faced with the prospect of chronic or fatal illness; on the other hand, the awareness that health is a *penultimate* or relative good can be fostered by the very discovery that, even in a situation of obvious limitation such as illness, one is *capable*—that is, still has the possibility—of willing and loving.

Health is a balance that concerns the whole person; "it is the rhythm of life, a permanent process in which equilibrium reestablishes itself."[61] Yet the greatest completion of the definition should be considered precisely at the ethical level because there is also an ethical dimension of health that is rooted in man's spirit and freedom: many illnesses result from misguided ethical choices (drugs, alcoholism, AIDS, violence, a lack of goods necessary for health). A person responsibly cares for the balance of health by treating it as a good of the person himself. Even when the origin of a condition is independent of the person's ethos and responsibility, prevention, therapy, and rehabilitation involve the subject's will and freedom as well as the responsibility of the community—and responsibility entails the ethical dimension.

Furthermore, the "manner" in which the patient confronts his illness or the citizen manages his health is influenced by the person's framework of ethical and religious values. This is why a physician has not finished his task with the patient after having offered physical treatments; he must also offer the patient human assistance of a moral nature, as mentioned even in codes of professional ethics.

In this light, the presence of chaplains and other *pastoral care professionals* in hospitals is justified not only in the name of religious freedom but also as a part

---

[60] R. Siebeck, *Medicina en movimento: Interpretaciones clínicas para médicos* (Barcelona: Editorial Científico-Médica, 1953), 514, originally published in German as *Medizin in Bewegung: Klinische Erkenntnisse ärtzliche Ausgabe* (Stuttgart: Thieme, 1949), quoted in Russo, *Corpo, salute, cura*, 186 [Italian].

[61] H.-G. Gadamer, *The Enigma of Health: The Art of Healing in a Scientific Age*, trans. J. Gaiger and R. Walker (Stanford: Stanford University Press, 1996), 114. For Gadamer, however, the essence of health consists in being hidden, and there are no criteria for examining it.

of health care: their presence, always respectful of individual freedom, affects the spirit of sick persons.

Finally, health education aimed at maintaining the balance of health and preventing illness is based entirely on an ethical premise of responsibility—both of the community that teaches and educates and of the citizen who endeavors to use those means to maintain the good of health for himself and for others.[62]

Health has four dimensions that intersect and interpenetrate one another: (1) biological, (2) psychological and mental, (3) social and environmental, and (4) ethical. There are just as many corresponding dimensions that cause or shape illness. Each dimension warrants investigation and has its own history: biological diseases and their causes have been studied from the very beginning; psychological and mental illnesses (psychiatry and psychopathology) became the subject of attention later on; and the study of social and environmental components is more recent but appears to warrant increasing consideration, to the point where people are speaking with ever greater alarm about environmental damage as a threat to health inasmuch as it causes a large number of diseases and alters the essential conditions for life. Perhaps contemporary humanism will keep developing this dimension of man's responsibility for balance in the cosmos in proportion to man's increasing capacity for technological control of the environment.

What has been said thus far does not mean that the "origin" of a particular illness is always to be attributed to the simultaneous presence of these various components; rather, it means that the general, complex balance of health involves them all and that, in any case, prevention, treatment, rehabilitation, care for the environment, health education, and the ability to respond and collaborate in a time of illness all entail an ethical dimension requiring great commitment at both the personal and social levels. I want to affirm, above all, that the very concepts of health and illness should be related to the whole person.

I would like to conclude this chapter with an observation by Romano Guardini:

> Human existence is structured upwards and downwards in degrees, through various ranks. The initial development of one degree, however, sometimes depends on the fact that the degree above, which is independent with respect to the others and out of their reach, comes to meet them on its own initiative.

> Thus for example physical health is certainly not identical to upright behavior; nevertheless, the ultimate safeguard for bodily health is keeping the moral sphere, which is above the biological sphere, well ordered. But this, if considered from the perspective of the lower sphere, is grace.[63]

---

[62] From the constitution of the World Health Organization; B. Häring, *Free and Faithful in Christ*, vol. 3 (New York: Crossroad, 1981); G. Burani, *Il passaggio dalla assistenza sanitaria alla tutela della salute* (Brezzo di Bedero: Salcom, 1985); E. Sgreccia, "Uomo e salute," *Anime e Corpi* 91 (1980): 419–444; idem, "Salute e salvezza cristiana nel contesto dell'educazione sanitaria," *Medicina e morale* 32, no. 3 (1982): 284–302; S. Spinsanti, *L'etica cristiana della malattia* (Rome: Paoline, 1971); B. Häring, *Healing and Revealing: Wounded Healers Sharing Christ's Mission* (Slough: St. Paul, 1984), originally published as *Vom Glauben, der gesund macht: Ermutigung der heilenden Berufe* (Freiburg: Herder, 1984); M. D. Grmek, *Le malattie all'alba della civiltà occidentale* (Bologna: Il Mulino, 1985).

[63] R. Guardini, *Fede, religione, esperienza* (Brescia: Morcelliana, 1984), 163.

Health and Illness: A Brief Historical Overview

The classical period[64]

Illness or sickness presents itself as a privation—the negation of the positive element of health. A sick person not only lacks something and suffers that lack (a plight that man shares with the animals), but also suffers because of that lack, because he is aware of it, perceives it as a limit, and seeks a cure for it. Illness is therefore a complex notion that goes beyond the notion of *broken*, which refers to machines, and beyond the notion of *alteration* or *defect*, which refers to any biological substance. The classical languages identified various aspects of illness: deficiency (*asthéneia* in Greek or *infirmitas* in Latin), objective harm (*nósos* or *morbus*), and subjective harm (*páthos, aegrolatio, dolentia*).

In Greek civilization, health was the proper balance of the organism's forces, the suitable measure and proportion: "neither more nor less than necessary, nor insufficiently tempered"[65]; Plato says, "To produce health is to establish the elements in a body in the natural relation of dominating and being dominated by one another, while to cause disease is to bring it about that one rules or is ruled by the other contrary to nature."[66]

These concepts of proportion and symmetry are closely connected with the concept of "measure" or, rather, "right measure," which is the key concept of Platonic metaphysics.

"More" and "less," "excess" and "defect" can be determined in terms of two different relationships and hence situated on two different planes. Proceeding according an *arithmetic* criterion, they can be reciprocally commensurate; using a more complex *ontological* and *axiological* method, they can be measured in terms of the right measure.

The "measure" to which Plato refers to does not have a quantitative but rather a qualitative, axiological character implying a relation of value. Reality is structured in terms of this measure and human beings use it to distinguish what is good and what is evil, what has value and what has no value, what is suitable, what is obligatory, and what is proportionate.[67]

Health is the right proportion, natural harmony, or intrinsic agreement of the organism with itself and with what is outside of it. The proportionate measure for the body should be determined in relation to the soul and to the man as a whole.[68]

For Plato, in light of the unity of the whole, a part of the body can only be cured in terms of the whole body and the body cannot be cured without the soul (that is, the entirety of the individual): "Then he who best blends gymnastics with music and applies them most suitably to the soul is the man whom we should most rightly

---

[64] See G. Reale, *Corpo, anima e salute: Il concetto di uomo da Omero a Platone* (Milan: Vita e Pensiero, Raffaello Cortina Editore, 1999).

[65] Hippocrates, *Antica medicina*, 5.

[66] Plato, *Republic*, IV, 444d.

[67] See Plato, *Politics*, 284e.

[68] See Plato, *Phaedrus*, 270b–d.

pronounce to be the most perfect and harmonious musician, far rather than the one who brings the strings into unison with one another." [69]

Curing the soul is bound up with virtue, the essence of which is structurally connected with the Good—that is, the measure of all things. Virtue is therefore mediation between excess and defect; in other words, it is the right measure between too much and too little. Virtue is the health of the soul: "To produce health is to establish the elements in a body in the natural relation of dominating and being dominated by one another, while to cause disease is to bring it about that one rules or is ruled by the other contrary to nature. ... Virtue, then, as it seems, would be a kind of health and beauty and good condition of the soul, and vice would be disease, ugliness, and weakness." [70]

Here, then, is Plato's therapeutic proposal: "Therefore the mathematician or anyone else whose thoughts are much absorbed in some intellectual pursuit, must allow his body also to have due exercise, and practice gymnastics, and he who is careful to fashion the body should in turn impart to the soul its proper motions and should cultivate the arts and all philosophy if he would deserve to be called truly fair and truly good." [71]

The thought expressed in the famous Latin maxim, *mens sana in corpore sano* (a healthy mind in a healthy body) was already formulated by Plato. According to him, however, having a healthy body without a healthy mind is in no way possible: "I, for my part, do not believe that a sound body by its excellence makes the soul good, but on the contrary that a good soul by its virtue renders the body the best that is possible." [72]

The modern and postmodern period [73]

The birth of clinics in the modern period involved a radical change of perspective, focusing attention solely on the evidence of the facts (the positivist paradigm). In analyzing illness, the clinical approach rejected any consideration whatsoever of a biographical or, more generally, humanistic sort, regarding medicine as a truly and strictly *natural* science.

In this perspective, typical of the nineteenth century, illness was always considered a quantitative alteration with respect to a norm (a functional deficit), able to be traced back in every case to a physiological dysfunction.

The structural and functionalist view of the early twentieth century emphasized the social system in which each person has a definite role and a specific function; illness alters it, threatening the stability of the system. In this perspective, medicine is a means of social control for rectifying such deviations from the norm and illness is again interpreted according to the biological paradigm.

The radical nominalist model, on the other hand, completely overturns the positivist model. It considers illness a merely arbitrary construction, the product of

---

[69] Plato, *Republic*, III, 411e.

[70] Ibid., IV, 444c–e.

[71] Plato, *Timaeus*, 88c.

[72] Plato, *Republic*, III, 403d.

[73] See Russo, *Corpo, salute, cura*.

an observation that creates its object, whereby the boundary between normal and pathological, between a healthy person and a sick person, is entirely dependent upon social or cultural criteria.

The phenomenological paradigm considers the health-illness concept pair to be a human experience whose meaning is sought out by the subject. Although the objective factors of the pathological condition are not denied, emphasis is placed on the subject's interpretation of his own situation, on the change in his relationship with his own body image, on the alteration of his ability to communicate, and on the different perception of space and time.

The most recent relational paradigm for health adopts a biopsychosocial inter-pretation of illness rather than an exclusively biomedical one, which is considered reductive. Health is considered a comprehensive vital condition of the subject in relation to his surroundings. The boundary between normal and pathological is therefore said to be based on objective parameters, but also upon how the subject understands himself in reference to his own system of relations. Medicine is no longer to be understood only as *cure* but above all as *care*—in other words, not as simple prescription and therapy, but as a true and proper relation of care in which the patient and the caregiver form a therapeutic alliance.

The postmodern paradigm is founded on a *fluid* notion of health, which oscil-lates between *being well* and *well-being*. Well-being implies the utopian purpose of attaining normality based on a norm that can be revised at any moment. *Well-being* is more than *being well*: it is understood to include psychological balance, emotional satisfaction, and the ability to have fulfilled relationships. *Health* has expanded into *fitness*, with constantly increasing performance standards—products for everyone and for no one.

Health refers to a norm that, albeit with a certain flexibility, is established on the basis of constant, measurable parameters. "To be in good health" means to be in a position, given one's own psychophysical situation, to perform one's social and professional tasks adequately. However liable to change its dynamic balance may be, the notion of good health always corresponds to some objective criteria that can be precisely described from the outside.

"Being fit," on the other hand, does not have those characteristics. *Fitness* is a fluid notion, a sort of "fuzzy health," as it has been aptly described, in the sense that it is neither precise nor measurable from the outside. It is a notion fraught with subjectivity that designates a potential state of being. By its very nature, it is projected toward the future; in other words, it is always a work in progress and, by definition, never fully realizable. "Being fit" means having a flexible, adaptable body, ready to experience sensations never felt before and impossible to define in advance.

Fitness means being ready to confront what is unusual, nonstandard, extraor-dinary, and above all new and surprising. Health sticks with "adhering to the norm," whereas physical fitness is the ability to break the rules and leave behind any standard already attained; it feeds a constant sense of inadequacy. *Fitness* is demanded and at the same time threatened, so it becomes once more a source of growing anxiety.

The notion of *quality of life* is a sort of code word invented by postmodern sensibility. It refers not to the absence of illness but to a situation of well-being. *Well-being* is more than the "silence of the organs"; it is also psychological balance, emotional satisfaction, and the ability to have fulfilled relationships. The strictly biomedical model of medicine ignored these important aspects; however, in the name of the quality of life, the postmodern model risks casting doubt on whether the life of a handicapped or chronically ill person is worth the trouble of living.

The human body: commercialized and de-formed[74]

Framing the philosophical issues related to the commodification of the human body requires a preliminary clarification of an anthropological and ethical nature. The question of whether the human body can be bought and sold as a commodity can only be answered after establishing what the body is and what value is attributed to corporality.[75]

Indeed, "market" discussions in relation to the human body presuppose its classification as "thing," "object," or "property." The market—purchase and sale transactions—pertains to goods or objects that are owned. If the body is not something that man "owns," but rather the being and embodiment of the human person, it cannot be considered a commodity. If the body is not an object but rather a *subject*, then it has a dignity making it infinitely superior to *things*. There is an immeasurable, ontological, and qualitative gap between a human body and a thing.

These philosophical principles form the basis for the negative assessment of the commodification of the body. Inasmuch as it is endowed with intrinsic value, the body is priceless. On the basis of this assertion, a subject may not treat his own body as a contractual object and third parties may not use it as an object of negotiation.

The instinctively emotional reaction of repugnance to the idea of considering the human body as a mere object of exchange is an indication on the experiential level of the ontological and moral richness that common sense attributes to corporality. The attribution of an ontological and qualitative meaning to corporality rules out the application of economic and monetary criteria.

If the value of the human body is such that it can have no price, and therefore cannot be commercialized, then the only possibility of exchange is by way of *gift*—the kind of gift that is made in the context of selfless donation, solidarity, and altruism. Even then, when it is a question of a living body, this is only possible on certain conditions, such as informed consent and other safeguards for the essential integrity of the donor, as required in transplantation ethics.

A fundamental principle thus emerges: respect for the dignity of the body, which entails the non-commercialization of the body. This is true first of all in reference to the living human body (in its totality or in its individual elements—whether organs, tissues, or cells—as part of the whole) but also in reference to a body that is no longer living, as a "memorial" of the lived experience of the value of corporality.

Certainly there are different degrees of seriousness depending on whether it involves an individual (as in the case of trafficking children or living embryos), organs, or tissues. Yet it seems that the respect due to human dignity forbids these latter forms of trafficking as well.

---

[74] See F. D'Agostino, ed., *Il corpo de-formato: Nuovi percorsi dell'identità personale* (Milan: Giuffré, 2002).

[75] See E. Sgreccia, "Corpo e persona," in *Questioni di bioetica*, ed. S. Rodotà (Rome-Bari: Laterza, 1993), 113–122.

A historical reconstruction has revealed various forms of body trafficking:[76] slavery, prostitution, and the exploitation of women and children for labor. Yet more recent forms have been taking shape, such as those associated with reproductive technologies (surrogate motherhood, sperm banks, and others), trafficking in children for prostitution, and trafficking in organs for transplantation. It has been rightly noted that the market has a unique power to create incentives in medicine.

Slavery usually comes to mind when we talk about individuals being purchased or sold; however, attention should also be focused on phenomena that could be linked or likened to that sort of market. In the past, slaves were considered by many to be non-persons without personal dignity—a description that could now be applied to embryos that are frozen and used for experimentation. According to personalist anthropology, however, the human fetus and embryo are valued as persons.

Trafficking human organs—a topic most often dealt with in the literature addressing the commercialization of the body[77] and in the opinions of ethics committees, especially the one in France—is plainly illicit. Although it seems a less serious matter at first glance, it is illicit not only because of the potential abuses but also because of the relation between the organ and the person: it appears that only donation under specific conditions can be licit.

The abuses entailed in the commercialization of human organs can lead to full-fledged crimes against persons—a sad lesson taught by news stories about the abduction and mutilation of children and the implantation of organs taken from patients with cancerous tumors.

I would not even make an essential distinction between organs and tissues taken from living persons. Even the parts of a cadaver—though the cadaver is not a person—preserve their symbolic dignity as *res sacra* (sacred matter) and a memorial of the living person.

Certainly what happened in England at the time of the Anatomy Act is incomprehensible: from 1809 to 1813 thousands of cadavers were desecrated by "body snatchers" who sold the pieces to laboratories, and at the same time forms of slavery were still practiced in the colonies.

The rejection of commercialization should not be based only on the fear of its harmful consequences. Here I refer to the positions of several English-speaking authors such as Pranlal Manga[78] and Leon Kass.[79] The position of the French National Ethics Advisory Committee for the Life and Health Sciences (CCNE), expressed in its *Avis sur la non-commercialisation du corps humain* (Opinion on the non-commercialization of the human body),[80] seems more suitable. In this document, which is part of a research project on the general topic of the relation between

---

[76] G. Berlinguer, *Il corpo come merce o come valore* (Bari: Laterza, 1993).

[77] P. Manga, "A Commercial Market for Organs? Why Not," *Bioethics* 1, no. 4 (1987): 321–338; L. R. Kass, "Organs for Sale? Propriety, Property and the Price of Progress," *The Public Interest* 107 (1992): 65–86; S. Nespor, R. Santosuosso, and R. Satolli, *Vita, morte e miracoli* (Milan: Feltrinelli, 1992), 149–165.

[78] See Manga, "Commercial Market."

[79] See Kass, "Organs for Sale?"

[80] French National Ethics Advisory Committee for the Life and Health Sciences, *Avis sur la non-commercialisation du corps humain*, December 13, 1990.

bioethics and money, the committee declares the commercialization of the human body illicit by citing the body's status as a subject; therefore, it cannot be objectified ("the body is not a thing," as the document repeats several times). On this basis, the CCNE declares that "neither the human body nor a part of the human body can be bought or sold." It therefore forbids "trafficking by individuals in fetuses, embryos, gametes, tissues, or cells."

The CCNE makes a distinction between *harvesting* organs, tissues, etc., and *transplanting* them. While it is not permitted to apply economic criteria in the buying and selling of the human body or its parts, it is nonetheless considered permissible to offer remuneration to health care personnel for their work in procurement, analysis, and surgical transplantation.[81] The compensation, however, is for their work and does not constitute payment for the organ.

Interesting points for ethical reflection on this subject can be found in the CCNE's *Avis sur la transfusion sanguigne au regard de la non commercialisation du corps humain* (Opinion on blood transfusion in relation to the non-commercialization of the human body), dated December 2, 1991, which proposes extending the ethical criteria used for blood transfusions to the general topic of the commercialization of the human body and its parts. These are the proposed criteria: (1) the gratuitousness of the donation, (2) the non-lucrative character of consequent operations, (3) respect for the donor, and (4) respect for the interest of the patient. These criteria can also be found in the recommendation concerning blood transfusions (no. R 90/9 dated March 29, 1990): "For both ethical and clinical reasons, blood donation must be voluntary and without remuneration."

The prohibition against commercializing the human body and its parts was made explicit also by the more recent French law on bioethics and by the European Convention on Human Rights and Biomedicine.[82]

Since dissolving the subject, contemporary culture has concentrated its attention on the body in an absolute manner.[83] In constructing one's self-image, the primary duty is to attend to the body. This becomes both cause for satisfaction and a secular form of asceticism insofar as it involves renunciation and sacrifices.

In the postmodern world, the new shameful sin is devoting little attention to one's body. The body has gone from being a friend, the messenger of one's identity to others, to being an enemy: liable at any time to betray us and evade our control, it must be placed under constant surveillance.

With the decline of the notion of a natural teleology of the body, attending to one's body is no longer aimed at just fostering nature, but rather at modifying or

---

[81] "That which constitutes the object of the price is not a part of the human body, but the work provided and the substance which results." French Ethics Committee, *Avis sur la non-commercialisation*.

[82] Council of Europe, Convention for the Protection of Human Rights and Dignity of the Human Being with Regard to the Application of Biology and Medicine: Convention on Human Rights and Biomedicine, April 4, 1997, http://conventions.coe.int/Treaty/en/Treaties/html/164.htm.

[83] See A. Komersaroff, *Troubled Body: Critical Perspectives on Postmodernism, Medical Ethics and the Body* (Durham: Duke University Press, 1995).

even editing it. "The body is no longer a given but rather a task to be accomplished, subjected to construction or even deconstruction." [84]

There is a tendency to project a plan onto the body, as though it were a machine; yet it is not a machine, and "every de-formation that one may wish to impose on it will encounter *resistance* in the body itself, making the accomplishment of the planner's intentions uncertain and jeopardizing the identity he intended to bring into play." [85]

---

[84] Russo, *Corpo, salute, cura*, 19.
[85] D'Agostino, *Il corpo de-formato*, 11.

Summary Outline for Chapter 4

Person = substance

Ontology of the person

An understanding of the person comes about through metaphysical investigation, which addresses the human being in his entirety. The ontological approach to the person seeks a substantial and not merely functional definition of him, although it does not underestimate the *signa personae*, all those elements or indications that can signal the presence thereof.

"Classic" definitions of person

In the well-known Aristotelian definition, which is that "man is a rational animal" (*zoon logon echon*), man is defined by means of the proximate genus and the specific difference for the purpose of determining the essence or nature of man, in a way that perhaps does not account fully for his originality, which makes him more than just another element in the cosmos.

After the advent of Christianity we encounter in Boethius the first decisive definition of a person: *rationalis naturae individua substantia* (an individual substance of a rational nature). Boethius emphasizes the substantial metaphysical identity, because the person is original and irreducible to something else, to the cosmos. The reference to substance brings to light the person's character as an existing subject (substrate) and not just mere activity.

Similar definitions are given by Richard of Saint Victor (*rationalis naturae individua existentia*: an individual existence of a rational nature) and Aquinas (*individuum subsistens in rationali natura*: a subsisting individual in a rational nature), which reflect the tension between the universal reference to the species (*rationalis naturae*) and the individual character (*individua substantia*): we all belong to humanity, but each one in his or her own way. In all the formulas cited the fundamental notion of individual appears, which does not mean that the individual is indivisible, but rather that he is undivided—in other words, endowed with unity. Individuality does not imply indivisibility as though the individual were *a-tomon*; it implies, rather, existence as one and the same, divided from others (in other words, the individual is undivided in himself and divided from all others).

Substantiality and relationality

Contemporary thinking asks whether the person can be considered as relationality alone (and not substantiality) and whether, consequently, a human being no longer capable of relating to others can be considered a person.

"Rarely is it understood that rationality necessarily implies relationality: the life of the spirit is openness and it is relational per se" (V. Possenti).

Person and substance

Substance is not a function

The philosophy of the person is connected at its root with the philosophy of substance, toward which one vein of philosophical thought has harbored profound suspicion for centuries, preferring to replace it with a philosophy of *function*.

In *Categories*, Aristotle defines prime *substance* as "that which is neither in a subject nor predicated of a subject" (3a 8s; see also *Metaphysics* 1017b 10–25). Substance is existence-in-itself, self-sufficient and independent from other subjects in its existence. The act of being is the *first and radical* act of the individual substance: all other (secondary) acts of the person are rooted in this act and derive their life from it.

Existing in itself and by itself and not in something else is the most radical property of substantial existence, and personal existence is the most perfect form of it: "*Persona significat id quod est perfectissimum in tota natura, scilicet subsistens in rationali natura*" (Aquinas, *Summa theologiae*, I, q. 29, a. 3).

The person is an *in se existens* (being existing in itself) and a *per se existens* (being existing for itself). The first expression indicates that it does not exist in anything else or as a mode of anything else; the second means that it does not exist for the sake of something else but for its own sake (*propter se, non propter aliud*). *The person is set as an end, not a means.* In addition to these characteristics of *inseitas* (*in se*) and *perseitas* (*per se*), the metaphysical tradition employs the concept of *aseitas* (*a se*, from itself) as a prerogative of the divine person alone. It does not depend on another, whereas *esse ab alio* (being from another) is the mark of a contingent being, the sign of its radical inability to be the principle of its own existence.

The person is more than his own acts

While the person is said to be a set of successive selves and successive states devoid of a common substratum in empiricist theories (Locke, Hume, Parfit), the same is not true of the substance-based approach. Furthermore, the latter safeguards the notion that *the person is more than his actions.*

Becoming a person, in the sense of possessing one's own radical ontological status, is not a *process* but rather an *event* or instantaneous act whereby one is established in personhood once and for all; on the other hand, personality—in the psychological sense—is acquired gradually through the accomplishment of personal (secondary) acts.

Person = unitotality of body and soul

The spiritual essence of the person

Interiority of the person

Inasmuch as he is a substantial subject of a spiritual nature, endowed with intellect, freedom, and love, *the person lives in openness to the totality of being.* This openness to the All includes the person's openness to himself, to others, to God, and to the universe.

The *person* appears above all as *a center of dynamic unification t*hat proceeds from within, as a unity that lasts over time, beyond all the changes and beyond the psychological fluxes of a multiplicity of sensations and beyond the temporal and spatial scattering of the self. The person is therefore capable of turning *toward himself* through self-reflection.

Interiority is an exclusive property of the person as a subject capable of turning toward himself, and it is precisely his interiority that makes the individual irreducible to the genus or to the totality (Kierkegaard).

The person is a totality

The person is a *whole*, a concrete *totality*. He is not a part, he is not a component of a larger totality that would become foundational with regard to him. The human being as an individual is one among many, individuated by that portion of matter that he makes his own, and therefore he has the status of something particular.

Man as a person is not part of a fragment, but is a totality, and as such is something that is *unique* and *unrepeatable*.

Corporality is not unessential

The human individual is corporeal and must not consider *corporality as an unessential addition* (as the idealistic approach does).

The hylomorphic concept of the person

The soul is the form of the body, the subsisting form

To say that *the soul is the substantial form* means to attribute it the ability to make the individual belong to a particular species of living things, the one marked by an organic body that is suited to express rational life.

The *Thomistic teaching about the oneness of the substantial form* made a decisive contribution to solving the problem of the intrinsic unity of the subject, which seemed to be compromised in Platonic anthropology. The problem still remained, however, of "justifying" the teaching about the immortality of the soul.

Aquinas arrived at the statement that *the human soul has the privilege of surviving the dissolution of the body*: besides being the substantial form, the soul *is the subsisting form* because it has an autonomous being, as is evident from the fact that it *performs operations independently of the body*. These operations can be identified in the consciousness that the soul has of all bodies, in its *knowledge of what is universal, and in self-consciousness.*

The thesis of the *"subsistence" of the human soul,* combined with the thesis of *the intellective soul as the unique substantial form of the body*, is decisive in the development of anthropology, since it explains *the value of the soul not just in terms of its function as "mind,"* as a principle of understanding, or as a function appointed to render an account of properly human intellectual activities. Aquinas speaks about *the soul in its metaphysical function as well,* as an ontological principle, based on a *strong isomorphism between the operation of knowing and the being of the subject who knows: operari sequitur esse.* Operation depends on being; in other words, the operative capacity of a subject is conditioned by the ontological nature of the subject that operates, namely, by the formal perfections that allow it to enact those operations.

Corresponding to this different level of "immaterial" operations there must be a different level of "immaterial" being, by virtue of the isomorphic axiom.

To summarize: *understanding and will express two faculties of the soul, whereby it is not formally bound up with the body.* The dissolution of the body is not necessarily followed by the dissolution of the soul, because the latter proves to be subsistent.

Person = human being

"Person" and *Homo sapiens*

According to Peter Singer, not all members of the species *Homo sapiens* are persons, and not all persons are members of the species *Homo sapiens*.

If the notion of person does not imply the species Homo sapiens, *then it does not imply any other species either: no other species is capable of recognizing the right to life that is proper to a person.*

The person is defined as a self-conscious, rational entity capable of moral activity and endowed with autonomy (i.e., free will), but this overlooks the *particular fact* that human beings are not just that, and that an additional characteristic of human beings is being corporeal, being in a state of becoming.

The duties of persons are performed toward those who are persons but do not live as such, in other words, who do not exercise their personal faculties. *The bioethical question arises precisely because there are cases in which some human persons (e.g., human beings in the embryonic, fetal or infantile state, those who are chronically or mentally ill, the terminally ill, persons in a persistent vegetative state) cannot live as persons without the help of other persons.*

The fact that there are persons who are currently able and obliged to decide for others who cannot do so is because other persons have allowed them to reach the stage in which they can fully (yet only temporally) exercise this power. *"Reciprocity is constitutive not only of morality but also of the very possibility of personal existence. ... This reciprocity is the source of all bioethical discussion and point to which all such discussion must be traced"* (A. Pessina).

Limiting respect for life and the right to life only to those stages in which moral life is exercised means forgetting that morality is exercised in intermittent cycles and circumscribed in time.

The ethical problems encountered in the field of health care often concern precisely those individuals who do not fit the mythical description of self-conscious, autonomous, rational, and free persons.

The human being is structurally a person: this condition does not depend on his will but rather on his origin.

From the human body to the human person

The question of human corporality must be addressed in a radical, profound manner. Why is it that deformed human bodies, or bodies visible only under the microscope (as in the case of embryos), or inert bodies without clear signs of consciousness, *are* human beings? Why *are* they persons? What is the basis for the anthropological equivalence that unites the healthy and the sick, the deformed and the nondeformed?

The humble but decisive argument for determining who is or is not a human being is *"to look to the origin: a human being is someone who is born of other human beings. ... Man is always someone who is born of other men; this is the prerequisite for proceeding to any further and more in-depth definition of man"* (A. Pessina).

Anyone who is generated by other human persons, whether directly or by means of their genetic patrimony, must be considered a human person. Hence the human body is central. There is no way of protecting one's self, one's dignity and integrity, unless the concrete corporality of others is also protected and respected.

"There can, and must, be one criterion for personality, and one only; that is biological membership of the human race. The beginning and end of personal existence cannot be taken apart from the beginning and end of human life. If someone exists, that someone has existed since the individual human organism existed, and will continue to exist for as long as the organism continues to live. What it is to be a person is to live a human life" (R. Spaemann).

Body = epiphany of the person

From the body to corporality

The personal character of corporality

The present context brings to light a *dissociation between the exteriority of the body and the interiority of the person*, a break in the unity of the human being, which is a synthesis of a corporeal element and a spiritual one. Indeed, our culture often confronts us with images of the body that are inadequate, because they do not sufficiently express the personal character of corporality.

The body necessarily enters into the definition of the self, since it is impossible to perceive oneself without it. And yet the self is not identified with the body but rather goes beyond it. The enigma of man is, therefore, precisely the enigma of the relation between the body and that *something else* which the body is not and which philosophical tradition has called spirit or soul. *The person is corporeal but transcends the body, not in the sense of being able to do without it, but in the sense of having the capacity to perform operations that are not exclusively corporeal.* The peculiar nature of the person is therefore certified by experience, which shows us the human being as that subject who is capable of sensations and, at the same time, also capable of conceiving universal ideas. This peculiarity is, however, guaranteed only if one affirms that there is a principle that allows the body to operate in a certain way yet is not corporeal. This ambivalence is shown in the expressions "I have a body" and "I am a body" (G. Marcel), which are both valid, though neither one can be used in an absolute sense.

Saying "I *have* a body" is quite correct but requires a necessary clarification, because the modality of this possession of my body is not on a par with the manner in which I possess that which is not my body. I do not have, I do not possess a body in the same way as I have or possess other things, in the sense that *I cannot distance myself from my body and, in the strict sense, cannot* make use *of it without consequences that have repercussions for the whole person.* In the same way, the expression "I *am* my body" must be accompanied by the clarification that "I am not just a body," but also something else.

For this reason, philosophical anthropology prefers to use the term *corporality* rather than *body*: *it better expresses the corporeal-spiritual unity of the person.* In contrast to the term *body*, which refers to a part of the person, evoking again the classic dichotomy between body and soul, *corporality* has greater breadth: it designates human subjectivity in its corporeal condition, which constitutes personal identity.

Man is not an animal organism, but corporality

We owe the phenomenological school a more attentive and profound reflection on the significance of corporality, which has allowed us to partially overcome the traditional anthropological dualism between soul and body, which has given rise at different times to disembodied spiritualism or physicist materialism. The distinction introduced by Edmund Husserl between *Körper*, the body understood in a purely material sense, and *Leib*, the living body, the faculty of beings endowed with psycho-physical characteristics, has brought to light the essential participation of the Leib in the functions of consciousness and in the relation that a human being has with the world (see Husserl, *Cartesian Meditations*, §44).

Man is not a simple animal organism to which a consciousness endowed with complex mechanisms has been added, accounting for his emergence from among other living things. He is *corporeal intelligence, "incarnate spirit"* (R. Lucas Lucas): this means that the very dynamics of the human body manifest a complexity that corresponds to the intimate unity of the person. As opposed to the concept of *organism*, which is proper to animals, man is characterized by *corporality*.

The body as friend or foe

The postmodern body

Since dissolving the subject, *postmodernity* has concentrated its attention on the body in an absolute manner. In constructing one's self-image, the primary duty is to attend to the body. This becomes both cause for satisfaction and a secular form of asceticism insofar as it involves renunciation and sacrifices.

In the postmodern world, the new shameful sin is devoting little attention to one's body. *Instead of being a friend or the messenger of one's own identity to others, it becomes an enemy to keep under surveillance, because it is liable to betray us and elude our control at any time.*

With the decline of the notion of a natural teleology of the body, attending to one's body is no longer aimed at just fostering nature, but rather at modifying or even editing it. "The body is no longer a given but rather a task to be accomplished, subjected to construction or even deconstruction" (M. Russo).

The mind-body relation today

The expression *mind-body relation* is not understood as a simple variation on the ancient debate about the *soul-body relation*, in which the soul was considered *enteléchia*, the principle of a series of functions such as thinking and willing.

With the more recent expression *mind-body relation*, the problem has undergone what has been described as an *epistemological reduction*: indeed, in common usage the term "mind" *is not equivalent to* "soul" in the traditional sense of the term. This goes back to the connotation of "mind" in the *Treatise on Human Nature,* by the empiricist philosopher David Hume, where it designates the totality of man's conscious states and functions.

Today the mind-body relation means the relation between functions that are fundamentally psychological and neurophysiological, *not the relation between the subjects of these functions (soul-body).* This is a new, reductionistic approach that excludes the metaphysical dimension.

But there is yet another reduction: the oversimplification of understanding the mind-body relation as a *mind-brain* relation, arbitrarily excluding those aspects of human corporality that have remarkable importance in the expression of the psychological and intellectual faculties, e.g., the function of the hands.

The body as epiphany

The body undoubtedly has an epiphanic character, that is, it reveals or manifests; it is "of the Being-seen" (Merleau-Ponty). The body reveals the person to others; at the same time, however, it veils or conceals it. Through the body we show ourselves to others and are perceived by others; at the same time, in some way the body hides us from others: in this sense we talk about the ambivalence of the human body.

The body as limit

The body is normally overlooked in silence, as if transparent. In pain, fatigue, and illness, however, the body becomes opaque, heavy, strange, and menacing.

Health has been defined as "life when the organs are quiet" or as *"unawareness of one's own body"* (G. Calguilhem). Being well means not realizing that one has a body. In the body one experiences a dialectic between activity and passivity, between *praxis* and *pathos*: undergoing the body that one governs (see P. Ricoeur).

Body-soul: a historical review

The body

Plato

In Greek anthropology, the body is one of the ontological principles that constitute the human being; it is the principle of materiality, the site of irrationality, and must be tamed by the soul, which in contrast is the seat of the intellect. The body (*soma*) is the tomb (*sema*) of the soul, as Plato declares in *Cratylus* (400a-c), and hence the soul must escape from the prison of the body (*Phaedo*, 65a, 66d-67a).

Aristotle

For Aristotle, too, the body is the seat of the passions that disturb the purity of contemplative reflection. The body is the animal bond of the human being, which alone can affect the rational soul negatively (*Nicomachean Ethics*, K 7).

The body is "substrate, that is, material" (*De Anima*, B, 1, 412a 20), in other words potentiality in the spheres of being and acting, the disposition to receive a form. For this reason the body is no longer the tomb of the soul, but rather its real instrument. In Aristotle, too, the body acquires philosophical dignity only negatively, as the place where the soul's faculties are exercised.

The soul

Plato

The soul gives life and movement to the body and is what organizes and gives unity to the body. The soul is identified as the seat of the intellect, the memory, emotions, the spiritual faculties that show the superiority of man to the material and natural world.

Therefore the soul is indissolubly linked to ideas (*Phaedo*, 77a); it governs, supervises, and gives life to the body (*Republic*, I, 353d), into which it was cast down from the celestial regions, as the famous myth of the charioteer explains (*Phaedrus*, 246a).

Aristotle

According to Aristotle, "the soul is the first entelechy [substantial form] of a natural body that has life in potency" (*De anima*, B, 1, 412 a 27). Like everything that exists in nature, the human being, too, is a *"synolos" or whole, composed of a body* that, as an organized body, is potentiality in the spheres of being and acting, *and of a soul* that becomes the organizing principle and intrinsic finality of that corporeal substrate. But the soul is tied to the body and shares its destiny of corruption, being incapable of surviving the physical death of the body.

Thomas Aquinas

The requirements that motivate the reflection of Aquinas are to safeguard the unity of the human being, which is compromised by Platonic dualism, without excluding the immortality of the soul, as in Aristotle.

"The human soul is a form that does not depend on the body in its being" (*Summa contra gentiles*, II, 79).

The soul is therefore *form*; it gives the body its reality, actualizes the materiality thereof, to which it is naturally related. As for its *being*, however, it does not depend on the body; it is an autonomously existing substance, as intellectual knowledge testifies. The soul is substantial form and at the same time full substance and a reality that subsists alone as the form of a body.

"The body and the soul are not two substances existing in act, but from them results one substance in act," i.e., the person (*Summa contra gentiles*, II, 69). This does not rule out the immortality of the soul.

This unitary insight also prevents us from considering the body an instrument; the body, in fact, is the necessary condition for the historicity and the very personality of a human being.

The Thomistic solution results in a metaphysical foundation for the individuality of the human being: *soul and body are the constitutive ontological principles of a single, individual human reality.*

Descartes

The body has an autonomy of its own, is a *res extensa*, an extended substance without consciousness but perfectly organized "like a clock" (*Meditations on First Philosophy*, 6).

Leibniz

The body is presented now as being endowed with laws of intrinsic development, as Leibniz thinks: bodies act as though, absurdly, souls do not exist; souls act as though they influence one another (see *Monadology*, §81). The soul-body connection proves to be problematic (*occasionalism*) and can be guaranteed by God through his harmonious plan.

Descartes (and occasionalism)

For Descartes soul and body are two substances having different natures: the soul is consciousness, it is *res cogitans*, a substance without extension, whose sole function is thought (*Meditations on First Philosophy*, 6).

Health and illness in the classical era

Balance of forces in the organism

Illness or sickness presents itself as a privation—the negation of the positive element of health. A sick person not only lacks something and *suffers that lack* (a plight that man shares with the animals), but also *suffers because of that lack*, because he is aware of it, perceives it as a limit, and seeks a cure for it. Illness is therefore a complex notion that goes beyond the notion of *broken*, which refers to machines, and beyond the notion of *alteration* or *defect*, which refers to any biological substance. The classical languages identified various aspects of illness: deficiency (*asthéneia* in Greek or *infirmitas* in Latin), objective harm (*nósos* or *morbus*), and subjective harm (*páthos, aegrolatio, dolentia*).

In Greek civilization, health was *the proper balance of the organism's forces*, the *suitable measure and proportion*, "neither more nor less than necessary, nor insufficiently tempered" (Hippocrates, *Antica medicina*, 5).

"To produce health is to establish the elements in a body in the natural relation of dominating and being dominated by one another, while to cause disease is to bring it about that one rules or is ruled by the other contrary to nature" (Plato, *Republic*, IV, 444d).

These concepts of proportion and symmetry are closely connected with the concept of "measure," or rather of "right measure," which is *the key concept of Platonic metaphysics*.

"More" and "less," "excess" and "defect" can be determined *in terms of two different relationships* and hence situated on two different planes. Proceeding according an *arithmetic* criterion, they can be reciprocally commensurate; using a more complex *ontological* and *axiological* method, they can be measured in terms of the *right measure*.

The "measure" to which Plato refers to does not have a *quantitative* but rather a *qualitative, axiological* character implying a relation of *value*. Reality is structured in terms of this measure and human beings use it to distinguish what is *good* and what is *evil*, what has *value* and what has *no value*, what is *suitable*, what is *obligatory*, and what is *proportionate* (Plato, *Politics*, 284e).

Harmony with the whole

Health is the *right proportion, natural harmony,* or *intrinsic agreement of the organism with itself and with what is outside of it.* The proportionate measure for the body should be determined in relation to the soul and to the man as a whole (Plato, *Phaedrus*, 270b-d).

For Plato, in light of the unity of the whole, a part of the body can only be cured in terms of the whole body and the body cannot be cured without the soul (that is, the entirety of the individual): "Then he who best blends gymnastics with music and applies them most suitably to the soul is the man whom we should most rightly pronounce to be the most perfect and harmonious musician, far rather than the one who brings the strings into unison with one another" (Plato, *Republic*, III, 411e).

Curing the soul is bound up with virtue, the essence of which is structurally connected with the Good—that is, the measure of all things. Virtue is therefore mediation between excess and defect; in other words, it is the right measure between too much and too little. Virtue is the health of the soul: "To produce health is to establish the elements in a body in the natural relation of dominating and being dominated by one another, while to cause disease is to bring it about that one rules or is ruled by the other contrary to nature. ... Virtue, then, as it seems, would be a kind of health and beauty and good condition of the soul, and vice would be disease, ugliness, and weakness" (Plato, *Republic*, IV, 444c-e).

Here, then, is Plato's therapeutic proposal: *"Therefore the mathematician or anyone else whose thoughts are much absorbed in some intellectual pursuit, must allow his body also to have due exercise, and practice gymnastics, and he who is careful to fashion the body should in turn impart to the soul its proper motions and should cultivate the arts and all philosophy if he would deserve to be called truly fair and truly good"* (Plato, *Timaeus*, 88c).

The thought expressed in the famous Latin maxim, *mens sana in corpore sano* (a healthy mind in a healthy body) was already formulated by Plato. According to him, however, having a healthy body without a healthy mind is in no way possible: "I, for my part, do not believe that a sound body by its excellence makes the soul good, but on the contrary that a good soul by its virtue renders the body the best that is possible" (Plato, *Republic*, III, 403d).

*Health and illness in the modern
and postmodern period*

Modern paradigms

Positivist paradigm

The *birth of clinics* in the modern period involved a radical change of perspective, focusing attention solely on the evidence of the facts (the positivist paradigm). In analyzing illness, the clinical approach rejected any consideration whatsoever of a biographical or, more generally, humanistic sort, regarding medicine as a truly and strictly *natural* science.

In this perspective, typical of the nineteenth century, illness was always considered a quantitative alteration with respect to a norm (a functional deficit), able to be traced back in every case to a physiological dysfunction.

Structural-functionalist paradigm

The structural and functionalist view of the early twentieth century emphasized the social system in which each person has a definite role and a specific function; illness alters it, threatening the stability of the system. In this perspective, medicine is a means of social control for rectifying such deviations from the norm and illness is again interpreted according to the biological paradigm.

Nominalist or radical paradigm

The radical nominalist model, on the other hand, completely overturns the positivist model. It considers illness a merely arbitrary construction, the product of an observation that creates its object, whereby the boundary between normal and pathological, between a healthy person and a sick person, is entirely dependent upon social or cultural criteria.

Phenomenological paradigm

The phenomenological paradigm considers the health-illness concept pair to be a human experience whose meaning is sought out by the subject. Although the objective factors of the pathological condition are not denied, emphasis is placed on the subject's interpretation of his own situation, on the change in his relationship with his own body image, on the alteration of his ability to communicate, and on the different perception of space and time.

Biopsychosocial (relational) paradigm

The most recent relational paradigm for health adopts a biopsychosocial interpretation of illness rather than an exclusively biomedical one, which is considered reductive. Health is considered a comprehensive vital condition of the subject in relation to his surroundings. The boundary between normal and pathological is therefore said to be based on objective parameters, but also upon how the subject understands himself in reference to his own system of relations. Medicine is no longer to be understood only as *cure* but above all as *care*—in other words, not as simple prescription and therapy, but as a true and proper relation of care in which the patient and the caregiver form a therapeutic alliance.

Postmodern paradigm: A kind of health for everyone and for no one

The postmodern paradigm is founded on *a fluid notion of health*, which oscillates between *being well* and *well-being*. Well-being implies the utopian purpose of attaining normality based on *a norm that can be revised at any moment. Well-being* is more than *being well:* it is understood to include psychological balance, emotional satisfaction, and the ability to have fulfilled relationships. *Health* has expanded into *fitness*, with constantly increasing performance standards—products for everyone and for no one.

"Health" refers to a norm that, albeit with a certain flexibility, is established on the basis of constant, measurable parameters. "To be in good health" means to be in a position, given one's own psychophysical situation, to perform one's social and professional tasks adequately. However liable to change its dynamic balance may be, the notion of *"good health" always corresponds to some objective criteria* that can be precisely described from the outside.

"Being fit," on the other hand, does not have those characteristics *Fitness is a fluid* notion, a sort of "fuzzy health," as it has been aptly described, in the sense that it *is neither precise nor measurable from outside*. It is *a notion fraught with subjectivity* that designates a potential state of being. By its very nature, it is projected toward the future; in other words, it is always a work-*in-progress* and, by definition, never fully realizable. *"Being fit"* means having a flexible, adaptable body, ready to experience sensations never felt before and impossible to define in advance.

Fitness means being ready to confront what is unusual, nonstandard, extraordinary, and above all new and surprising. *Health sticks with "adhering to the norm," whereas physical fitness is the ability to break the rules and leave behind any standard already attained; it feeds a constant sense of inadequacy. Fitness* is demanded and at the same time threatened, so it becomes once more a source of growing anxiety (M. Russo).

The notion of *quality of life* is a code word invented by postmodern sensibility. It refers not to the absence of illness but to a situation of well-being. *Well-being* is more than the "silence of the organs"; it is also psychological balance, emotional satisfaction, and the ability to have fulfilled relationships. The strictly biomedical model of medicine ignored these important aspects; however, in the name of the quality of life, the postmodern model risks casting doubt on whether the life of a handicapped or chronically ill person is worth the trouble of living.

Definition of health

Definition of the WHO

Health: State of complete well-being?

The definition of health offered by the World Health Organization in 1946, that *"health is a state of complete physical, mental and social well-being and not merely the absence of disease and infirmity,"* should be considered utopian. What does it mean "to be in good health"? One cannot ignore the fact that the concept of health is closely connected to cultural and social elements: the *health* of a young person is different from that of an elderly one, just as the health of a European is different from that of an African.

In the postmodern context, *being well* becomes a synonym of *well-being*, and therefore it tends to treat as the absence of health all sorts of transitory situations such as discomfort, stress, and dissatisfaction. But this mentality creates a further sickness: the suffering of being normal human beings and therefore mere mortals, subject to aging and decline.

For this very reason attempts have been made to correct the WHO definition by contrasting the notion of health as a *state* with health as a *dynamic process* or as an *equilibrium* and contrasting the notion of *complete well-being* with that of *relative ability.*

In this perspective, a situation that reduces one's ability to work, such as maternity, is not an illness, because it does not annul the capacity for a broader life project; whereas a condition of general malnutrition that makes a person incapable of performing his own duties cannot be tolerated as normal.

It is not possible to understand the notion of health without combining objective, subjective, and sociocultural criteria.

The notion of health has a complexity that eludes reduction to only one of these component aspects (objective, subjective, sociocultural): it is an integral notion that, furthermore, can be found only to a relative degree at the concrete level of a particular life.

Ottawa Charter (1986)

The WHO repeated and refined its definition of health at its International Congress in Ottawa (1986).

This definition explains that "to reach a state of complete physical, mental and social well-being, an individual or group must be able to identify and to realize aspirations, to satisfy needs, and to change or cope with the environment."

Personalist definition: Equilibrium with respect to the whole person

A distinction must be made between *perfect health* and *relative health*:

> The first results from the concurrence of a subjective sense of well-being with physiological and behavioral normalcy; it is a simple, ideal concept that can only be approached and never fully attained.

> The second consists, as Laín declares, consists in the "physical ability to carry out one's life plans with minimal trouble, with minimal injury and, if possible, with a certain well-being and enjoyment."

The clarification by Richard Siebeck in this regard is helpful: "The concept of health is not complete without the question: *health to what end?* The bottom line is that we do not live in order to be healthy, but we are and want to be healthy in order to live and act. Health is not a good that is entrusted to us just for certain functions and abilities. *Health is not an ultimate end*, but rather is determined and limited by the meaning of life itself. And the meaning of life is availability, self-gift, and sacrifice. *The meaning of health lies beyond health*; it is found in that plan for a good life that each person proposes to carry out, for which health is a necessary but not indispensable condition. Not only does illness not necessarily defeat a plan for a good life, but conversely, one can say that without that plan health itself has no value. For this reason Plato declared that virtue comes first and then health and wisdom follow, just as a god is followed by his cortege. Indeed, *health is not a property of the organism but of the whole person*; it will therefore be subordinate to whatever constitutes the genuine good for the person."

Health is a balance that concerns the whole person; "it is the rhythm of life, a permanent process in which equilibrium reestablishes itself" (Gadamer).

The equation "health = salvation" can lead to despair when one is faced with the prospect of chronic or fatal illness; on the other hand, the awareness that health is *a penultimate or relative good* can be fostered by the very discovery that, even in a situation of obvious limitation such as illness, one is *capable*—that is, still has the possibility—of willing and loving.

According to the personalist vision, the concepts of health and illness must refer to the whole person.

Hence there are four dimensions of health that overlap one another: (1) biological, (2) psychological and mental, (3) social and environmental, and (4) ethical.

Chapter 5

BIOETHICS AND
ITS PRINCIPLES

Ethics as Life and as a Science

Within the study of anthropology, distinctions must first be made between *ethical life*, *ethos*, and *ethics*.[1] Ethical life means man's intrinsic tendency or propensity to implement good or "values"; ethos means effective behavior that can be sociologically identified within a given culture, which man has achieved or attempted in the pursuit of certain values; ethics, or moral philosophy, is the science of what man *ought* to do, of the values that he *ought* to live out.[2]

The following explanatory definitions can therefore be employed: *ethics* is the science of human behavior in relation to moral values, principles, and norms; *descriptive ethics* (ethos) is the examination of customs and behaviors reflecting the moral values, principles, and norms of a given population or of several populations, either in general or relative to a specific act (marriage, abortion, theft, etc.); *normative ethics* is the discipline that studies values, principles, and norms of conduct in relation to what is right or wrong (good or evil) and seeks out their foundations and justification.

There is a distinction within normative ethics between *general ethics*, which deals with fundamentals, values, principles, and norms, and *particular ethics*, which deals with the application of those principles, norms, and values in specific areas such as economics (economic ethics), politics (political ethics), or working professions (professional ethics). As seen in the first chapter, bioethics deals with the area of life and health sciences. Bioethics is a form of particular ethics but has also developed a fundamental branch corresponding to general ethics. This branch is called *metabioethics* because it investigates the ultimate and metaphysical foundations of bioethics itself.

Due to the need for topic boundaries, it is not here possible to discuss the sociology of the human ethos among various peoples at various levels of civilization or

---

[1] There is also talk of ethology, the science that studies animal behavior and compares it with human behavior, often with the aim of pointing out and emphasizing their similarities.

[2] The Thomistic definition of the object of ethics is stated in these words: "Subiectum moralis philosophiae est operatio humana ordinata in finem, vel etiam, homo prout est voluntarie agens propter finem." Thomas Aquinas, *Sententia libri ethicorum Aristotelis* (Casale Monferrato: Marietti, 1949), book I, lectio 1, 3; S. Vanni Rovighi, *Elementi di filosofia* (Brescia: La Scuola, 1963), 3:189–191.

in relation to individual human actions or values, such as suicide, euthanasia, and marriage. References will be made to the individual issues only as necessary for understanding value judgments within the context of ethical science.

It is necessary, however, to say a few words about ethical life when it is understood as the primary source of the human person's propensity toward good and toward the values that embody it. Reconnecting to the topic of life, then, ethicality is not considered an external qualification of human actions but an irrepressible need of human life, present in all of man's actions. Yet discussing ethical life also means treating ethics as a science.

There is an unquenchable urge in man that could be defined as the "insatiability" of the human being for fulfillment by "being more." This tendency toward the fulfillment of one's own being is called the will and entails variously nuanced factors: desire, aspiration, joy, effort, action and duty, movement toward completeness and encounters with others, achievement, and life planning.[3]

The will is the naturally operative faculty of the human spirit that expresses this tension of action. Just as the intellect opens itself to reality and to others through a necessary and ceaseless urge for knowledge, so the will opens itself to the good present in the reality of things and in persons. Man is drawn to intellectual knowledge, as soon as his psychological development allows it, and is able to connect concepts and knowledge into an interpretative ordering of reality; likewise, he tends to express himself in action through his will by establishing ends and ways for accomplishing them. This is in order to express his own being and be gratified by the experience, whose summit holds the content of the word *good*. That content can be accepted, overlooked, or denied, but is always present.

That which has always fascinated many thinkers, mystics, men of action, and great achievers is also the daily and immediate experience of every human being, of every person: the Good, sought in contingent and insufficient realities through actions that only partially realize the good and leave a remnant of unfulfillment.

Given the limits on the discussion, just a few points will be mentioned that should be kept in mind regarding the human will as the source of ethical life. First of all, it must be noted that there is a constant disproportion between life and will, between action and the ultimate end, which is the Good plain and simple. Every human action only partially expresses its source, even when the will is intense and deeply felt. In the same way that the will does not express the entire personal being of man, but only his operative tendency, so single actions do not exhaust the human will or attain its total end. The will thus transcends action and there is a disproportion between individual actions and the ultimate end of life, which transcends particular ends.

This fact, which cannot be discussed here at length,[4] simply and conclusively means that being or life comes before the will and before action. It also means that man's personal being is definitively open to a good that individual goals can never attain.

---

[3] A thoughtful description of ethical and "political" life is carried out philosophically in passionate and almost poetic language by G. Capograssi, *Introduzione alla vita etica* (Rome: Studium, 1976).

[4] To explore this topic further, see the following works that closely share our approach: Vanni Rovighi, *Elementi di filosofia*, 3:189–243; J. Maritain, *Introduction to the Basic Problems of Moral Philosophy* (Albany, NY: Magi Books, 1990), originally published as *Neuf leçons sur les notions premières de la philosophie morale* (Paris: Téqui, 1951); J. Gevaert, *Il*

Life as a source is richer than the river of its actions, and the goal, the ultimate end, always transcends the individual step taken, because particular ends do not realize the ultimate end, which is the Good plain and simple, the fullness of happiness. This means that though a person may deprive himself of a particular good, he can neither be deprived nor deprive himself of the ultimate Good or of the tendency to pursue his own good.

It should also be noted that the will, though necessitated by, open to, and drawn toward the Good, remains free as to particular goods. Precisely by virtue of this insatiable openness of the human being, the will cannot shut off its willing: even in a state of non-action, even when yearning to receive (receiving is also activity), or even when forcibly restraining its attraction, the will aspires to a good and remains open to action. Even suicide is an action, however tragic and harmful. At the same time, precisely because it is open to the boundless Good and never exhausts or fully experiences the whole in any particular end, the will remains free from entrapment in particular objectives and remains self-determining in every instance.

Freedom is the profound experience of every human being as a faculty of self-possession reflected in one's own acts. This is because the will does not invest itself totally in any act and knows that it can want or not want. Even in the case of external physical coercion or of a sudden and strong internal urge, a human being with awareness knows that the act originates from within and remains in the possession of its author. Just as life is a principle that proceeds from within, so freedom is self-determination; both thought and freedom represent the highest expressions and the summit of human dignity. Yet thought and freedom, the hallmarks of the human spirit, presuppose the life that they express and from which they draw their strength and take their origin. Life is expressed through freedom but is presupposed by freedom itself. This fact has great significance in bioethics.

A further observation is necessary and already implicit in what has been said thus far. The will originates in the depths of human life, of the human being, in a different manner than the intellect. The intellect lives to know; the will expresses its vitality in action. The intellect collects partial truths but aspires to the whole and entire Truth; the will expresses itself in particular actions in order to attain particular goods, but aspires to the full Good. Despite this distinction, which could lead to thinking in terms of a separation between thought and action, they nonetheless maintain a very lively, reciprocal relationship in the unity of the same spirit—and also in the unity of action. The good as the goal of activity is actually discovered and proposed by the intelligence, whether intuitively or through reasoned reflection. Therefore the *true* good can never be separated from the truth. The first synthesis that qualifies an action as moral is determined by this coincidence of truth and good: an action is morally sound and constructive for a human being only when the will fulfills what is truly good through action.

This intervention of the intelligence upon the will or in guiding action is a form of estimative knowledge ordered to the good, which is a degree of knowledge

---

problema dell'uomo (Turin: Leumann LDC, 1978), 147–190; J. de Finance, "L'ontologia della persona e della libertà in Maritain," in *Jacques Maritain oggi: Atti del Convegno, Milano, 20–23 ottobre 1982*, ed. V. Possenti (Milan: Vita e Pensiero, 1983); A. Bausola, *Natura e progetto dell'uomo* (Milan: Vita e Pensiero, 1977); P. Ricoeur, *Philosophie de la volonté*, vol. 1, *Le volontaire et l'involontaire* (Paris: Aubier, 1949), 37–81.

distinct from theoretical or speculative contemplation.[5] The light that the intellect projects onto the ends and means of the action and onto their mutual congruency is a judicious knowledge that resides in the estimative and prudential intelligence; it is a capacity developed through effort, mental rigor, and intellectual honesty. In the footsteps of Aristotle, Saint Thomas Aquinas called it the virtue of *prudence*, describing it as the *auriga virtutum* (charioteer of the virtues).[6]

The will nonetheless remains free even when faced with this estimative and imperative information provided by the light of the intellect, precisely because every particular end does not completely realize the total Good *ex toto et ex omni parte* (wholly and in all parts), and the estimative intelligence can fail to accurately perceive the connection of a particular end *hic et nunc* (here and now) with the complete end and Supreme Good. Moreover, insofar as the urges of the body and the cultural influences of society weigh upon the person, the intellect can lose sight of the truth of the act and the rectitude of its relation to the Good to be accomplished. Maintaining a clear and correct view of the good in individual actions requires a profound unity of the human person, an internal harmony, and a freedom from the promptings of ego, self-interest, and current ideologies. These endowments are "ethical": they are found in mature personalities with inner clarity and an awareness of their surroundings. Other virtues that must therefore accompany this activity of evaluation and the ethicality of actions were also treated by Aquinas: *fortitude* in pursuing the good in the face of difficulty and opposition, *justice* in considering the goods to be equitably evaluated in relation to those who fall within the sphere of action, and *temperance* that masters the instinctive nature of individual emotions and interests and social and environmental pressures.

Through its doctrines on original sin, which subtly and instinctively inclines the individual toward egoism, and on the gift of redemptive grace, which reestablishes the internal harmony of the person, Christianity enlightens and profoundly sustains the "toil of being," and of being upright, experienced by every human person.

It is essential to add that, in this complex relationship between will and intellect in man's corporeal and historical embodiment, there is a reciprocal influence of the will on the intellect, or rather of the whole man on the intellect. The influence can be positive, making the truth more attractive and binding when it has been embodied— albeit with great difficulty—in a good action; on the other hand, it can become an aggravating factor when the burden of negative experiences manages to obscure or distort the truth. If this happens, the evil becomes all-encompassing within the person: it is no longer just a question of acting against a clearly perceived truth, but also of calling something "true" when it is not—that is, falsity is added to the evil.

At this point it seems necessary to pause and emphasize the concept of responsibility, underscoring in particular the value of the *responsibility of scientists and physicians*.

There is no free act that does not involve *responsibility* (in the sense of moral responsibility, because sometimes the concept of legal responsibility may also enter

---

[5] This is what Maritain points out in reflecting on the "degrees of knowledge." Moral knowledge is distinct but not separate from theoretical or scientific knowledge.

[6] Vanni Rovighi, *Elementi di filosofia*, 3:223–230; A. Bausola, *Libertà e responsabilità* (Milan: Vita e Pensiero, 1980).

into play). The term responsibility includes the idea of an assessment of the goods (*rem ponderare*) in question with regard to the free decision, as well as having to answer (*respondere*) to the conscience.

The *conscience* represents the awareness of the ethical value of a given action. It is the place and the moment in which the appropriateness or inappropriateness of the act becomes apparent within the framework of moral reference proper to and present in the subject. It is not a question of sentiment or emotion, although such intrasubjective reactions can be concomitant; rather, it is a judgment by the reason regarding the idea of good and evil as applied to the specific act performed by the judging subject or by others.

The conscience can be defined as the aptitude or act of knowledge and discernment aimed at the assessment of moral actions. The conscience's object of judgment is therefore human action, which is evaluated in reference to moral values, principles, and norms.

The conscience is the inner and proximate tribunal of the moral act. The truer and more comprehensive its judgment is while assessing, the more objective and valid the moral judgment will be; the more its judgment is clouded, distorted, or deprived of necessary information, the more the moral judgment could be false or erroneous.

One can speak of a *true* or *erroneous* conscience in terms of this objective agreement between the subjective ethical judgment and the objective conformity of the act to the good end.

In terms of the degree of certainty with which the conscience perceives the moral value of an act, one can speak of a *certain*, *doubtful*, or *perplexed* conscience. Man can err in good faith with an erroneous but certain conscience. In this case the obligation remains to do everything possible—in proportion to the realities in question and in relation to one's own possibilities—so that a *certain conscience is also true*, i.e., so that the subjective judgment corresponds to objective fact. Moral formation is obligatory in order to achieve this; the greater the goods in question, the more compelling the obligation.

There is also a distinction between an *upright* conscience and a *non-upright* or *bad* conscience, depending on the coherence between the dictates of the conscience and the conscious behavior. One acts in bad conscience when one knows that an action is not right yet nonetheless carries it out, perhaps even defending it as good.

In conclusion, there is a moral obligation not only to act according to conscience but also to form a true, upright, and certain conscience.

It was mentioned earlier that it is impossible to separate a free and conscious human act from ethical judgment—and therefore from responsibility—because every free act has content: it is acting for something and on something or toward someone, and the act can either be suitable or unsuitable—yet never objectively indifferent or irrelevant—to the person who carries it out and to the beings on the receiving end. Whether we are aware or not, we are responsible for what we do. Awareness or a lack thereof can make the subject more or less morally or legally accountable for the actions, but the objective burden and responsibility remain. One can kill a person accidentally—not deliberately and therefore unaccountably—but the reality is that the person is objectively dead, and the void created by that death remains objectively assessable and burdensome. However, as will be noted later on, greater or lesser subjective responsibility for the action does not annul the objective morality of the act in itself, which remains harmful regardless of the subject's awareness or intention.

At any rate, everyone is obliged to act with great care so that good is done and evil is avoided and so that the margin of error is reduced as much as possible in proportion to the good in question. The idea of freedom without responsibility is an existential joke—a *ludus*. It decapitates and dishonors reason and freedom at the same time.

Objective Morality and Subjective Morality

Ethicists and, more importantly, educators must always keep two methodological requirements in mind: (1) distinguishing the *objective value* of an action from the *subjective phase* in which that action is devised and decided internally by the subject; and (2) emphasizing the moral obligation to align the subjective judgment, or inner compass, with the objective value of the action itself. The *singleness of the morality of the act* is guaranteed in virtue of this obligation to align the subjective aspect with the objective aspect.

The formulation of a judgment on the objective value of a given action is the result of a knowledge process that can have various degrees of certainty and various modes of understanding. It may first manifest as an immediate and preconscious mental reaction, almost "connaturally," as Jacques Maritain puts it,[7] then becoming conscious and reflective knowledge; sometimes it may remain marked by doubt. Various factors contribute to the formulation of this judgment on the objective value of the action, and they can have a greater or lesser impact on the gradual process of knowing that leads the subject to form that judgment. In fact, this presupposes first and foremost a comparison with a norm or law, which makes clear the objectivity of the values and their hierarchy. This law can be the *natural moral law* inherent in the human conscience, which is innately knowable at the fundamental level of awareness and can be articulated later on by reason; this is why it can also be defined as a *rational* norm. The natural moral law can therefore be described as an set of general moral principles that man's natural reason spontaneously discovers on the basis of his own mode of being or own nature, and this law is universal and immutable like human nature itself.

Personalist philosophy, which owes much to the Christian tradition, understands the natural law as a reflection of the *eternal law*, which is the very order of reality as it is in the mind of the Creator, who by creating and ordering reality gave it its worth and ordered its values.

A subject may or may not be aware of the Creator's existence (which is accessible to reason in theory but is sometimes ignored in practice), but in any case he can successfully become aware in his own conscience of the natural or rational norm—in accordance with the gradual process mentioned—and manage to formulate his judgment on the currently proposed course of action. Thus, a physician can have the immediate awareness of being unable to comply with a suffering and desperate patient's request for euthanasia, whether by an immediate inner reaction of the conscience or, more consciously, by thoughtful reference to a rational norm that affirms the sanctity of human life.

This fact can be recognized as a reflection of the Creator's eternal law, as it truly is, whereby man was created but not granted arbitrary control over life and death, whether his own or that of others. The same physician may, on the other hand,

---

[7] J. Maritain, *Nove lezioni sulla legge naturale* (Milan: Jaca Book, 1985).

be perplexed and therefore obligated to make a more thorough rational and cultural study of the treatments he may be required to administer in relation to the patient's pain and the conditions of the illness.

One factor that can significantly contribute to the formulation and illumination of this judgment is *religious faith*—in other words, the revealed law proper to revealed religions such as Judaism, Islam, and Christianity. Religious faith offers itself as a light that allows reason to observe the human act in a broader perspective of meaning, thereby helping the conscience to formulate judgments regarding the objective value of a particular action.

This religious faith may be held by the patient, who is therefore obliged to follow it, or by the physician, who is likewise obliged to observe it if he wishes to adhere. It could also be shared by both, in which case it affords immediate and facilitating help. In any case, remaining within the parameters of the physician-patient relationship, the physician can neither ignore the patient's religious beliefs nor legitimately act against his own religious conscience.

Although religious norms cannot theoretically conflict with the natural law or with rational ethics, some conflicts may nonetheless arise in practice (for example, patient refusal of certain treatments in the name of a particular religious belief). They must be resolved through deeper reflection and on the basis of ethical principles that will be explained at the end of this chapter. A rational and "secular" morality, with its own foundations and judgment criteria, is therefore conceivable. But in cases where the person—patient, physician, or both—has a religious faith, it cannot be ignored in order to be reasonably respectful of others' consciences.

The *positive law*, issued by the legitimately constituted authority charged with the common good, can be another point of comparison in the formulation of an objective judgment. That authority could be either national or international, as in the case of "human rights,"[8] and it could also be religious. The national or supranational authority is charged with interpreting the requirements for the common good, which must be understood as the good of the community pursued through the good of individual persons. A religious authority, in contrast, based on its own framework of religious or faith values, issues instructions that interpret the revealed norm and the faith, as seen especially with regard to the instructions of the Catholic Church. Although the Church automatically addresses its own believers directly by its teaching authority, it may also propose values, directions, and moral guidance to all people willing to consider them.

This source of information can facilitate and clarify the ethical judgment, but there may still be a conflict between civil laws and religious norms or between the positive laws of the state and the individual conscience of the person, which is internally guided by the framework of natural or rational values.[9] It is necessary to

---

[8] As mentioned in the first chapter, the declarations, conventions, and recommendations of the various supranational and international bodies (UN, WHO, EEC, Council of Europe), as well as the professional ethics codes of medical associations, while distinct from the science of ethics and far from perfect with respect to ethical values, are very helpful for the purpose of recognizing basic values.

[9] On this topic of the relationship between ethics, norms, and conscience, see the following moral treatises: F. Böckle, *Morale fondamentale* (Brescia: Queriniana, 1979); J. de Finance, *Etica generale* (Cassano Murge: Tipografia Meridionale, 1982); T. Goffi and

refer to general values and principles in order to overcome these potential conflicts, and a complex and in-depth study is often required.

The path to formulating an objective judgment, though it may often seem immediate, always demands commitment, rigor, and rational development, and can lead to doubts and sometimes dramatic uncertainties. Ignorance can sometimes interfere with this process, leading to *error*. Both ignorance and error can alter the interpretation of the norm and, consequently, the hierarchy of values in question. They can also affect the historical and concrete configuration of the action. One can err, for example, in judging how to act when safeguarding a value such as the life of an unborn child conflicts with defending the mother's life during a high-risk birth—an issue that will be addressed again later. One could also be mistaken about the existence of an important circumstance, such as a serious heart disease in the mother, whose death results from inadequate assistance during the delivery. Indeed, this possibility of error must be duly taken into account in order to assess the responsibility of the acting subject in both ethical and legal terms. It must nonetheless be remembered that a responsible subject must make every effort to avoid error in proportion to the type of responsibility exercised.

From an ethical point of view, error—and the ignorance that causes it—can be *inculpable* and *invincible* if the subject has not neglected any reflection and knowledge that was required of him and therefore finds it impossible to judge differently; otherwise, when this effort has been omitted due to *inexperience*, *imprudence*, or *negligence*, the error or ignorance can be *culpable "in causa"* (in the cause). Think of the responsibility that could result from a lack of ongoing study and education in the case of an architect, a judge, or, in the field of bioethics, a physician.

Another dimension in the assessment of the moral act must be discussed at this point: the *subjective phase*, in which the action is devised and decided within the inner life of the subject. Subjective consciousness is not a computer, but a vital and lived act of *awareness*, *freedom*, and *responsibility*, which touches upon the very mystery of the person.

Freedom, which fulfills being when it means freely choosing according to truth, can mean choosing to go against its own being and that of others, depriving the person of values and beneficial experiences and causing harmful inclinations. Faced with the choice between an arduous good or an easier and more appealing deprivation of good—that is, an evil—playing on human self-interest, freedom can slacken its tension toward good and surrender to evil. The experience of every human being confirms the real influence of this inclination to evil, which Christianity has defined as the result of original sin. The same Christianity proposes the remedy and healing of grace, which is God's aid offered by Christ—an aid that allows the gradual and painstaking restoration of the capacity for good through communion with God.

While faith and conscious access to redemption are a gift and a hope for believers, the experience of fragility and the toil of freedom are shared by all people. In reality, freedom is incomplete if it is understood only as freedom from external

---

G. Piana, eds., *Corso di morale*, vol. 1, *Vita nuova in Cristo: Morale fondamentale e generale* (Brescia: Queriniana, 1983); A. Günthor, *Chiamata e risposta*, 3 vols. (Rome: Paoline, 1982); B. Häring, *Free and Faithful in Christ: Moral Theology for Clergy and Laity*, 3 vols. (New York: Crossroad, 1978–81); M. Vidal, *L'atteggiamento morale*, 2 vols. (Assisi: Cittadella, 1976).

constrictions, whether political, economic, or of some other sort. It becomes complete when it is a freedom "freed" from inner obstacles, which pit the self against itself—the body against the spirit—in the depths of the divided human being. It is neither logical nor foreseeable, moreover, that someone who has not "freed" his own freedom from egoism could be an authentic promoter of others' freedom.[10]

In assessing man's subjective responsibility for his actions, this *frailty of being* and the harmful possibilities of freedom must be kept in mind. Good and evil are possible, though never equivalent or of equal value, due to this mystery of freedom and the possibility of making choices that are not in accordance with truth.

But there is more: this already fragile freedom may be weighed down by *limitations* that can reduce the capacity for self-determination and choice. Such limitations can diminish and sometimes eliminate responsibility. One of these limitations has already been mentioned: ignorance, which can cause judgment errors. The limitation is indirect in this case because it acts primarily on the intellectual assessment, but it nonetheless reduces the scope of freedom.

Those who have written on these topics speak of so-called circumstances that moralists, educators, pastors, and even judges (to a certain degree) must keep in mind in order to more adequately assess a subject's responsibility. A crime such as homicide committed with the aim of extortion is one thing, while homicide committed by an insane person or a driver whose brakes malfunctioned are another.

It is only possible here to briefly touch on the long list of circumstances and categories of limitation about which psychologists, moralists, and legal scholars have written at length and which in their totality, in certain situations, can render an ethical judgment on subjective responsibility entirely problematic and sometimes remove all responsibility.

Awareness of these limitations and qualifications is especially necessary for physicians and medical professionals in general because their profession compels them to maintain a decision-making relationship with the patient, whose consent may be free and informed or heavily vitiated.

Limitations to freedom can arise from the personality and psychology of the patient in relation to age, level of education, character, or psychological and mental health. Traditional moralists described this circumstance with the interrogative pronoun *quis* (who).

Some circumstances arise from the very nature of the accomplished action (*quid* [what]) being judged: its difficulty, complexity, novelty, or the fear it instills in the subject. It is one thing for a mother to accept a pregnancy that appears to be normal and another to accept it when the fetus appears sick or when the pregnancy involves a risk to the mother's life; in such a case, accepting the child can become an arduous and sometimes heroic duty, though it remains obligatory.

The circumstances of place and cultural setting (*ubi* [where]) also have a certain weight. For example, certain forms of killing may be a matter of honor in certain countries.

---

[10] On the concept of freedom and original sin, the reader may consult J. Mouroux, *Sens chrétien de l'homme* (Paris: Aubier, 1945); P. Valori, *Il libero arbitrio: Dio, l'uomo, la libertà* (Milan: Rusconi, 1987); G. Piana, "Libertà," in *Dizionario enciclopedico di teologia morale* (Rome: Paoline, 1981), 562–574; M. Flick and Z. Alszeghy, *Il peccato originale* (Brescia: Queriniana, 1972).

Great weight is attributed to intentionality or motivation (*cur* [why]), which is like the soul of the action: it often happens that actions devised with a good end lead to real harm and hence objective evil. Euthanasia, for example, is very often depicted today as an "act of mercy." Some authors have tried to encapsulate all of morality in the intention (*intentionalism*), but the action obviously remains what it is. Nonetheless, the intention with which it is carried out may partially or totally exempt the agent from subjective responsibility, as in the case of a parent who provides an addicted child with a dose of heroin in order to calm a withdrawal crisis, which then leads to the child's death. If the intention—which is certainly important when assessing subjective responsibility—were enough to constitute goodness, then the world would be entirely good.

Still other circumstances have been drawn from experience and the literature of traditional moralists, such as the circumstances referring to manner (*quomodo* [how]) and time (*quando* [when]). It is not here necessary to describe example situations that can easily be intuited, especially since the list of them can always be expanded but never determined *a priori*.

It must be clearly affirmed, in any case, that the *moral value of subjective responsibilities does not annul the objective moral value of the act*. The moral obligation therefore remains to act so as to make subjectivity align with objectivity, and not vice versa, when determining the value of the action in itself and establishing norms for behavior. To put it simply, a person may die in an automobile accident without the subjective responsibility of the driver who causes the collision (though this is not always the case), yet the loss of that life remains an objectively grave evil; therefore, this creates an obligation for drivers to take all necessary precautions to avoid this physical, moral, and social harm. The same could be said of an unsuccessful surgical operation if the death of the patient could have been avoided.

Truth establishes the foundation of the good: *verum et bonum sunt idem* (the true and the good are the same). This also means that the first good act that can and must be performed is seeking the objective truth and the objective foundation of the good.

Ethical theories that prescind from the objective truth of the person and attempt to ground ethics *solely* on freedom, utility, or the advancement of the species or of science, ultimately end up sharing the lowest common denominator of relativism— that is, they occlude truth and objective good, which are the reference point for everyone regardless of the subjective interests of a few or of many. Criteriologies based on so-called *situation ethics* or devised by analytical philosophy are equally relativistic and likewise end up uncoupling ethics from objective truth.[11]

Based on these premises, it is necessary to reiterate at this point the responsibilities incumbent upon scientists and physicians. In a positive and meritorious sense, they are responsible for whatever good may be done to prevent illnesses, disasters, and epidemics; in a negative and harmful sense, they are responsible for whatever

---

[11] Many thought currents flow into the vast stream of reflection understood by the term *situation ethics*, some of which make reference to questions about the validity of the moral law raised by so-called *dialectical theology*. For further study of this aspect of twentieth-century theology, see H. Brouillard and K. Barth, *Genèse et évolution de la théologie dialectique* (Paris: Montagnier, 1957); J. Moltmann, *The Beginnings of Dialectical Theology*, ed. J.M. Robinson (Richmond, VA: John Knox Press, 1968).

evil they may cause, such as the loss of human lives, permanent disabilities or health deficiencies, and so forth.

Many times physicians will be called upon to reconcile truth and life or the life of the individual and the good of the community; however, the ability to accomplish this reconciliation must be grounded in a firm ethical character and rigorous reflection.

The responsibility of the physician—who acts on other persons, on the very life of others—is therefore one of the most striking and recognized forms of responsibility. For this reason, the medical or health care class has distinguished itself throughout the centuries for its authoritativeness, strong conscience, and meticulousness.

In whatever time or place it is situated, therefore, the medical act has an ethical dimension in addition to its technical and scientific one. Indeed, just as it is impossible to separate a concept from the faculty that forms it, so it will never be possible to separate medical responsibility from the medical act that entails it. It remains true for all professions that "man is defined before all else by his responsibility to his brothers and at the court of history,"[12] but this is uniquely true for the physician, who must also learn to make others responsible—the patient, the community, and the culture itself—without abdicating his own responsibilities.

Human Freedom

According to the Thomistic tradition,[13] freedom proceeds from the intelligence and the will: a free choice is an act of the will informed by the intelligence. For Aquinas, freedom is not a faculty distinct from reason and will but an extension of both: it unites and specifies them in order to create a concrete action, just as conclusions are derived from principles.

Francesco Botturi[14] distinguishes three meanings of freedom: freedom as self-determination, freedom as self-realization, and freedom as a relationship.

Freedom as self-determination

According to this perspective, acts of the subject are measured only in reference to themselves; this presupposes perfect autonomy. The flaw with this idea, however, is that the human subject not only has the power to choose, but also the need to do so: man cannot just choose for the sake of choosing; rather, he must choose in order to obtain—if possible—certain goods, certain realities that he needs and are advantageous to him. In this understanding, freedom is identical to undirected choice, and

---

[12] Vatican Council II, Pastoral Constitution *Gaudium et spes*, December 7, 1965, no. 55; L. Villa, *Medicina oggi: Aspetti di ordine scientifico filosofico, etico-sociale* (Padua: Piccin, 1980); J. Vedrinne, "Éthiques et professions de santé," *Médecine et Hygiène* 11 (1984): 1171–1177.

[13] S. Pinckaers, "Human Freedom according to St. Thomas Aquinas," in *The Sources of Christian Ethics*, trans. M. T. Noble (Washington, DC: Catholic University of America, 1995), 379–399, esp. 381; C. Vigna, *La libertà del bene* (Milan: Vita e Pensiero, 1998).

[14] F. Botturi, "Libertà e formazione morali," in *Alla ricerca delle parole perdute: La famiglia e il problema educativo*, ed. G. Borgonovo (Casale Monferrato: Piemme, 2000), 36–53; idem, "Formazione della coscienza morale: un problema di liberta," in *Per una libertà responsabile*, ed. G. L. Brena and R. Presilla (Padua: Ed. Messaggero, 2000), 73–95; G. L. Brena, ed., *La libertà in questione* (Padua: Edizioni Messaggero, 2002).

according to some advocates of the philosophy of freedom (Luigi Pareyson), being is in itself freedom.[15]

Freedom as self-realization

Freedom means a *journey toward a/the fulfillment of the agent*, and therefore has the connotation of *liberation* from the poverty and slavery of imperfection, lack of fulfillment, and so forth.

Freedom means *adherence to the good*, fulfillment, and *liberation* of the subject. On this level, there is a significant *paradox*: the *linking of necessity and freedom*. One is drawn toward the good, the object of free choice, by a certain need for it; the attainment of this necessary good in turn frees us.

Freedom as a relationship [16]

Freedom is also a *relationship with others, and more precisely with another freedom*. Not only must freedom take another freedom into account, but, more profoundly, freedom is a constitutive requirement for another's free existence.

Man exists (ex-sistere) in virtue of the recognition he receives. He needs recognition not in order to be a person, but in order to *exist as a person*—that is, in order to fully activate his emotional and intellectual capacities and reach a strong and stable sense of his own identity. Recognition thus becomes the relationship through which the subject identifies and affirms himself.

Integral definition of freedom

An *integral definition of freedom* could be "*dependent self-mastery* and *deliberate fidelity.*"[17] It combines an idea of autonomy wherein one accepts one's dependence on others and on the good and an idea of choice that is fulfilled in fidelity to the freedom of others together with what attains the good of the subject.

The need for moral formation

It only makes sense to speak of moral formation if freedom is not understood as pure initiative and exclusive autonomy. In fact, the common element in the different meanings of freedom is the *coexistence* of *activity* and *passivity* within it, and freedom is an active ability only if it recognizes that it does not come from itself.

Moral formation is not so much training for a certain *conduct* as formation of a moral conscience, understood as an education in the truth of freedom and the desire for it: "A proper moral formation is therefore directed not toward a morality of law but toward a morality of fulfillment or, according to R. Spaemann's expression, of a 'successful life.'"[18]

---

[15] L. Pareyson, *Ontologia della libertà: Il male e la sofferenza* (Turin: Einaudi, 1995). See also J. L. Nancy, *The Experience of Freedom* (Stanford, CA: Stanford University Press, 1993), which talks about the "spontaneity" of the occurrence of being.

[16] C. Vigna, "Etica del desiderio umano," in *Introduzione all'etica* (Milan: Vita e Pensiero, 2001), 130–138.

[17] F. Botturi, "Libertà e formazione morali," in *Alla ricerca delle parole perdute: La famiglia e il problema educativo*, ed. G. Borgonovo (Casale Monferrato: Piemme, 2000), 47.

[18] Ibid., 51.

Norms, Values, and Natural Law

Contemporary moral philosophy is returning to a conception of ethical life as something more than a set of norms, laws, and ends: it is a call to attain certain values that lead to the fulfillment of the human person. Morality was viewed for several centuries from a primarily formal and normative perspective that obscured the content and purpose of an upright moral life. This ultimately led to the rejection of norms and morality altogether. We are currently witnessing a recovery of the integral meaning of morality with an emphasis not so much on the normative or coercive aspect, but rather on the attainment of the person's end: *eudaimonia* or happiness.[19]

It is necessary, however, to specify the meaning of ethical values, norms, and the moral law, and above all to identify their foundation, since the contours of these concepts are often rather unclear.[20] It must be recognized, nonetheless, that such language is by now prevalent and, if properly understood, can express the direction of free and responsible activity in a more personalistic, dynamic, and proactive way. An ethical value can be understood as "anything that allows us to give meaning to human life."[21] What gives meaning to human life obviously depends on one's understanding of human life. The close relationship between anthropology and ethics immediately emerges: differences will arise in values, norms, and ethics in general on the basis of one's understanding of man. I do not intend to present a treatise on anthropology from which to deduce an ethics suitable to the hypothetical concept of man, but rather to present the fundamental coordinates around the human person that allow for an understanding of the terms "value" and "moral norm."

Historically speaking, the concept of value was established in philosophical language by way of a transposition of economic and financial language as part of a reaction against positivism. Positivism recognizes only facts;[22] phenomenological thought (Edmund Husserl, Max Scheler) affirms the importance in man's life of what manifests itself as striving, aspiration, and duty in addition to what exists in fact alone. From this perspective, values understood in a philosophical sense are anything that elicits esteem, admiration, a sense of perfection, and, as Paul Ricoeur would say, values appear at the crossroads of our infinite desire to be and the finite conditions of its realization. Moreover, there are not only facts but also values throughout all of human history; there are not only economic values based on utility and convention, but also cultural, spiritual, religious, and moral values.

---

[19] This topic and the return to a moral philosophy understood in terms of the fulfillment of the person is treated exhaustively in Pinckaers, *Sources of Christian Ethics*.

[20] For further study of the concepts of fundamental ethics, consult Maritain, *Nove lezioni*; D. Von Hildebrand, *Ethics* (Chicago: Franciscan Herald, 1953); A. Rodriguez Luño, *Etica* (Florence: Le Monnier, 1992).

[21] Gevaert, *Il problema dell'uomo*, 149; A. Lalande, *Vocabulaire technique et critique de la philosophie* (Paris: Presses Universitaires de France, 1968), 1182–1186; P. Valori, "Valore morale," in *Nuovo dizionario di teologia morale*, ed. F. Compagnoni, G. Piana, and S. Privitera, 1416–1427; idem, *L'esperienza morale* (Brescia: Morcelliana, 1985); P. Ricoeur, *The Conflict of Interpretations: Essay in Hermeneutics* (Evanston: Northwestern University Press, 1974).

[22] The dichotomy between facts and values has also been debated in the field of analytical philosophy. Of great interest in this regard is H. Putnam, *The Collapse of the Fact/Value Dichotomy and Other Essays* (Cambridge, MA: Harvard, 2002).

In particular, moral value is specific to human activities and to moral experience and denotes the quality or perfection of an action or behavior insofar as it conforms to the good or to the dignity of the human person. Love of neighbor, respect for life, generosity, a spirit of sacrifice, and so forth are moral values. Values exist as an inherent appeal within the very reality of the human person, as an ideal that continually attracts the personal subject—an *ought to be* toward which to direct one's being. As such, they are the presupposition and fundamental condition for the existence of moral discourse, in terms of both personal action and scientific reflection.[23]

Values must first be known, however, if they are to be attained. In order to know them, one must profoundly contemplate that ideal realm of values which is before and above man, which preexists man. Indeed, values are neither created nor invented but only discovered, known, recognized, accepted, or rejected.

They do not change over time, inasmuch as their essence transcends time and history. What changes, if anything, is man's awareness of the value, his way of relating to it and placing it in his own hierarchy.

A moral value therefore has its own specificity, which keeps it from being confused with the inventiveness and culture of the subject or even with religious values (faith in God, spirit of prayer). Nonetheless, since a definite concept of God logically implies a coherent vision of man and of the world, religious values reflect upon moral values, but not necessarily in a sufficient or consistent way. It is possible to find individuals filled with religious sensibility yet incapable of expressing values in consistent moral conduct.

The most important clarification to be made regarding values is in the problem of their foundation—that is, whether they have a purely subjective origin and justification (whereby the subject and experiences that go beyond organic life simply transcend material reality) or whether they correspond to reality and have an objective ontological foundation. Subjectivism conflicts with ontologism in this area, too, and subjectivist personalism with ontologically grounded personalism. A subjective ethics will result in norms and values that change with the subject and over time. On the other hand, an objective ethics founded on the objective reality and meaning of the human person will result in objective norms and values, regardless of the subject or time period.

If there were not a structure of reality or essence within a value serving as its foundation, then it would cease to be a value and would simply be an illusion. If solidarity among people did not correspond to a definition of man and, therefore, to a structural and natural requirement of the person, it would have no reason to be or to obligate. In the final analysis, a value presupposes certain realities and has an objective foundation: things are values, persons are values, and God is the highest value. Moreover, the greater the ontological richness and perfection of the reality to which the value refers, the more lofty the value is.

Values also have a subjective resonance: the subject recognizes in them goodness, a good, a constructive factor, and the meaning to impress upon life. Aquinas spoke of *ens* (being) and *bonum* (good); contemporary thinkers speak about reality (things, persons, institutions, or expressive forms such as art, technology, and

---

[23] S. Privitera, "Casistica," "Deontologia/teleologia," "Etica normativa," and "Valori," in *Dizionario di bioetica*, ed. S. Leone and S. Privitera (Bologna: Dehoniane, 1994).

religion) and values. What contemporary philosophy brings to light is the personalist element: values have meaning for man and do not exist without man. This is also evident in the vocabulary change that has occurred in reference to man's ultimate end: antiquity spoke of *eudaimonia*, the Middle Ages spoke of the *beatific vision*, and contemporary language speaks of *self-realization*. Yet these perspectives are not mutually incompatible so far and nothing prevents us from adopting the new terminology, given the change in cultural sensibilities. The delicate point here is the *foundation of values*.

The personalist ethical model admits and posits the existence of values, but sees them as founded on metaphysical realities: a value cannot be such without real content and without the inherent capacity of persons to perceive and assess. The reality becomes a value to the person, who intuits and assesses, when it takes on the quality of "goodness" or of correspondence to the being and life of the person.

It is necessary to specify, moreover, that it is one thing to affirm that values intuitively present themselves as such and are therefore attractive and morally appealing to the person, while it is another to affirm that moral experience is an absolute given that cannot be further justified. It is one thing to proceed methodologically from the historical importance of moral actions rich in values or from the ascertainment of the existence of moral values with universal validity, which are almost inherent to human living, and it is another to affirm that these values are incapable of being rationally grounded or critically justified. Moreover, the work of the intellect and rational reflection, more or less systematically developed and more or less intuitively present, cannot be excluded—and in fact must be included—in the foundation of ethical values.

That values can be fraught with affective repercussions and spontaneous emotional implications can be readily confirmed, but that in itself does not conflict with the need for verification on the real and rational level. Moreover, social life and the influence of the social environment can deform values, emphasizing some and obscuring others, even to the detriment of true objectivity.

A further clarification is necessary concerning the *natural law*, which is a pivotal concept at the heart of extensive debates and theoretical discussions—often fraught with problems—that cannot be examined here in great detail.

Clarity is important regarding this concept, because over the centuries it has been used with different meanings, sometimes to legitimize the arbitrary will of the rich and powerful, and sometimes to proclaim the inferiority and relativity of all political power as part of an awareness of a higher justice and a universal idea of man. The natural law has been considered a reflection of the eternal law of God and, in contrast, as a dogmatic secular state of reason to be served as an *institutio* either of the prince or the erudite, both repositories of the goddess of reason. It has been identified with the biological law, considered bereft of any substance outside the realm of positive law, and has served both to reinforce the power of the mighty and defend the claims of human rights. Suffice it to state here that, from a personalist point of view, the natural law is not understood merely as physical or biological law or individual spontaneity, though biological facts can and must enter into the assessment of human facts; neither is it a set of legal and moral precepts, which if anything could be considered specifications of the natural law.

The expression *natural law* or, more precisely, *natural moral law*, designates a fact rather than a theory: the fact that man by his nature is a moral being, and that human reason is, in and of itself, practical and moral reason. The moral law springs forth from human nature, finding in the latter a supporting structure without which it would be an external, extrinsic, repressive, and unbearable authority—indeed, an unintelligible one. The natural law is therefore "the light of our intellect through which moral realities are accessible to man."[24] It is the natural light of the intellect. It could not be otherwise, given that it is a question of moral rules, because no moral requirement can govern the will without passing first through the reason.

The natural light of practical reason can attain some kinds of knowledge clearly and immediately: the first principles (*fac bonum, vita malum*: do good, avoid evil) and the virtues considered generally. Furthermore, through reasoned reflection on moral experience, it arrives at other truths with a necessary connection to the first principles or the virtues.

The natural law is man's participation in the *divine law*, and as such it can be traced back to and explained only in terms of a creationist perspective. Therefore, through their reason illuminated by the supernatural light of Revelation, believers seek the methods best suited to realizing in their actions the fullness of being human.

All of this is verified on the religious level through comparison with the Word of God, which has created man and reveals him to himself, and it can be formulated in the twofold precept of love of God, the Highest Good, and of one's neighbors, who are created in his image. For this reason, from the Middle Ages until the Enlightenment, there was no hesitation in identifying the natural law with this twofold precept of charity.

It is also true that the first deductions of the natural law are intuitively ascertainable from this fundamental requirement. With the help of sociology and ethnology, some scholars recognize them in respect for others as persons and in the prohibition of homicide, incest, and so forth, but reason must always reflect—sometimes imperfectly and with difficulty and sometimes vacillating—and learn to find the specification of the natural law in particular situations, striving to purify social behaviors of false "natural" justifications. The necessary mediation of reason has been emphasized because, once again, there is no ethics without objective supporting reasons and without rational refinements to one's own behavior.

Christian experience and the Magisterium of the Catholic Church, with its sense of realism concerning man, who is subject to egoism and sin, bring to bear the light of Revelation as something not merely useful but actually necessary for a full understanding of human good in its totality, even in situations where the content of this moral law is per se reasonable and accessible to reason.

In other words, the natural law appears as a profound need of the whole human being for the full realization of his or her own life in harmony with the lives of others and the full realization of values—even when they appear arduous and laden with suffering.[25]

---

[24] Rodriguez Luño, *Etica*, 213.

[25] Thomas Aquinas, *Summa theologiae*, I-II, q. 18–19; II-II, q. 10 a. 3; I-II, q. 72; J. Maritain, *The Rights of Man and Natural Law* (New York: C. Scribner's Sons, 1943); idem, *Introduction to the Basic Problems*; V. Possenti, "La vita preconscia dello spirito nella filosofia della persona di J. Maritain," in *Jacques Maritain oggi*, Possenti, 228–242; S. Mosso, "Il ruolo della connaturalità affettiva nella conoscenza morale secondo J. Maritain," in *Jacques Maritain oggi*, Possenti, 525–546; F. Viola, "La conoscenza della legge naturale

The fact that the natural law is part of a "preconscious" intuition and is partly connatural knowledge, subsequently explicated through reflection, is not an exemption from having to verify and explain its applications. In the international arena and in the documents of the Magisterium of the Catholic Church, several obvious and indispensable applications of this natural law have been formulated into so-called *human rights*, especially after the founding of the United Nations. Yet while this constitutes a step of enormous ethical importance, the interpretative definitions (for instance, the case of the rights of the unborn) are often neither easy nor universally recognized.[26]

The Natural Moral Law as the "Way to the End"[27]

According to the Thomistic tradition, the natural moral law expresses the energy and finality that pervades all creatures. It is the law of their activity; "it is the way for attaining the end."[28]

The end is the *good*, which "is not a thing, but is that *relation* which exists between something and the human condition: a relation that allows man to realize certain aspects of his humanity in some way. ... Good and evil name those actions which are or are not conformed to the human condition, or more precisely, to the structure of man. It is therefore clear that one must begin to understand *who* man is in order to understand what is good for man or not."[29] Hence the natural moral law draws it specific characteristics from human nature, of which it is the expression, as compared to the natural laws of other creatures. Since man is a reasonable, self-conscious being, "the natural moral law is *known* and not merely lived."[30]

Man is free; therefore, the natural moral law is a duty and not a necessity (*Sollen* and not *Müssen*). It is an obligation that can be unmet, a commitment that can be betrayed: "The fact that the moral law is *valid* does not depend on me but on the human nature that God gave me—on the *lex aeterna* (eternal law) in which man participates. Whether the moral law *is actuated* depends on me. If man does not implement the moral law, then he is lost and ruined, for he deprives himself of the attainment of the end, which is perfection; yet he continues to exist, whereas a plant that does not follow the laws of its nature will die."[31]

---

nel pensiero di J. Maritain," in *Jacques Maritain oggi*, Possenti, 560–582; J. F. Malherbe, "Médecine, anthropologie et éthique, " *Médecine de l'Homme* 156/157 (1985): 5–12.

[26] Regarding the Church's Magisterium on human rights, see John XXIII, Encyclical Letter *Pacem in terris*, April 11, 1963, and Vatican Council II in its pastoral constitution *Gaudium et spes*. As already noted, the Council of Europe has compiled the indications of the international pronouncements on human rights, as applied to the medical field, into a single manual.

[27] See Vanni Rovighi, *Elementi di filosofia*, 3:214–233.

[28] Ibid., 3:214.

[29] A. Pessina, "Operatori sanitari come agenti morali," in *Etica e giustizia in sanità*, ed. A. G. Spagnolo et al. (Milan: McGraw-Hill, 2004), 22.

[30] Vanni Rovighi, *Elementi di filosofia*, 3:217.

[31] Ibid., 3:218.

The Essential Unity of Nature and Reason

Martin Rhonheimer criticizes the dualistic fallacy that

> obfuscates the fact that reason as well, as a cognitive faculty, and thus the very subjectivity of the moral agent, is a part of what we call the "nature of man." It is precisely the intellectual acts, of which reason is the discursive part, which open the human subject to an understanding of the human good according to the truth of his or her "being a person" (which is a corporeal-spiritual unity), and this is a good that reveals itself to be a "good of the person" only in the face of intellectual cognition. This good, therefore, is not a simple *object* which is said to be face to face with the knowing subject as a simple "natural given fact." It is also a part of the same knowing subject because in the cognitive acts it is manifested, and in a certain sense constituted, through its intelligibility.[32]

The natural law is part of and expresses the natural order. "However, this natural order," Rhonheimer reaffirms, "is not an entity that man as a knowing and acting agent finds himself, so to speak, in front of. It is a natural order of which the same natural cognitive acts—the natural acts of practical reason—form a part. Thus one discovers a reason that is also specifically nature (a kind of *ratio ut natura*). It is for this reason that the natural law can really be called 'inside man' and that one can say that it is 'engraved in his soul.'"[33]

Again according to Rhonheimer, counterposing *objective* "nature" ("natural order") and *subjective* "reason" ("moral knowledge") favors a "physicist" interpretation of the natural law, identifying it with merely natural structures to which a morally normative character is immediately attributed.[34]

Man is capable of *shaping* nature, but by *participating in it*. Adriano Pessina points out that the *nature-culture* pairing can delimit the anthropological situation because it is able to describe the peculiar nature of the human person. He maintains that experimental science has transformed the *spontaneous* way of understanding *nature*, which "seems to have lost both its intrinsic *intelligibility* and its essential *teleological character* in favor of human subjectivity. Largely through the Kantian critique, the conviction spread that it is the cognitive activity of man which *provides* meaning (sense and direction) to *nature*. In this perspective, it is no longer a question of extracting truth from reality but of attributing truthfulness to reality."[35] If the dualism betwen nature and reason prevails, the subject becomes the source of intelligibility rather than its conscious recipient.

---

[32] M. Rhonheimer, "Natural Moral Law: Moral Knowledge and Conscience; The Cognitive Structure of the Natural Law and the Truth of Subjectivity," in *The Nature and Dignity of the Human Person as the Foundation of the Right to Life: The Challenges of the Contemporary Cultural Context; Proceedings of the Eighth Assembly of the Pontifical Academy for Life: Vatican City, 25–27 February 2002*, ed. J. de D. Vial Correa and E. Sgreccia (Vatican City: Libreria Editrice Vaticana, 2003), 126.

[33] Rhonheimer, "Natural Moral Law," 139.

[34] Ibid., 125–126.

[35] A. Pessina, "Il contesto dello sviluppo della biomedicina," in *Etica e giustizia in sanità*, Spagnolo et al., 10–11.

Knowledge of the Natural Moral Law[36]

Both speculative and practical knowledge start from immediately evident universal principles. The supreme principle in the speculative realm is the principle of identity, from which the principle of non-contradiction immediately follows. Similarly, the supreme principle in practical knowledge is the principle of finality: every being has a definite end for its activity and a perfection to attain: *bonum est faciendum, malum vitandum* (do good, avoid evil). But how can the more specific precepts of the natural moral law be deduced from this principle?

Aquinas and traditional ethics in general respond that we know the moral law *with our reason*, by reasoning (we are individual substances of a rational nature): "Man must bear witness to his spirituality even in the most commonplace actions. One should not then believe that man must preserve himself just like any other being, reproduce like any animal, and then additionally speculate about truth."[37] The attainment of spiritual values does not occur only in contemplation; rather, it is the form of every authentically human value. In general, the law is the way to the end and the various duties and precepts are nothing but ways of actuating human values.

Since knowledge of the moral law is obtained through reasoning, it can lead to error. Knowledge of more general precepts is immediately evident, yet the application of these principles to specific questions is subject to error; this is even truer in applying them to concrete cases.

Medieval scholars called the knowledge of first principles of the moral order *synderesis*, whereas the application of those principles to a concrete case is called moral *conscience*.[38]

Conscience is the most precise knowledge—applied to the concrete situation—of the finality of our nature, which reveals itself to us in the more general principles of the natural law. It is therefore wrong to think of the moral law and conscience as two criteria of morality standing side by side; rather, it is the same criterion—ordering to the end—that is manifested first generally and then more precisely.[39]

Conscience is not the subjective criterion of morality set in opposition to the moral law, which is the objective criterion. From this perspective, the conscience would be a sentiment, a mysterious voice or, as Robert Spaemann says, an "oracle."

Rather, "the moral law (*lex naturalis*) is human finality as manifested to the faculty of reason, and conscience is the result of rational reflection seeking how to actuate that finality in concrete situations. Conscience is also objective because it is the knowledge of what a man's action should be, and the moral law is also subjective, where 'subjective' is understood (quite poorly) to mean that which is known. The statement that conscience is a subjective criterion and law is an objective criterion

---

[36] See Rhonheimer, "Natural Moral Law," 123–159.

[37] Vanni Rovighi, *Elementi di filosofia*, 3:211–212.

[38] See S. Pinckaers, "Human Freedom," 384–385.

[39] R. Spaemann, *Basic Moral Concepts*, trans. T. J. Armstrong (New York: Routledge, 1989), esp. 57–67; idem, *Happiness and Benevolence* (Notre Dame, IN: University of Notre Dame, 2000); idem, *Persons: The Difference between "Someone" and "Something,"* trans. O. O'Donovan (New York: Oxford University Press, 2006), esp. 164–179.

can have only one acceptable meaning: conscience, being a more specific form of knowledge, is more readily subject to error than a more general form of knowledge." [40]

The Natural Law Is "Living" [41]

The natural law is living because it is the root of all moral norms, indicating what is good and to be pursued. According to John Finnis,

> Aquinas's theory of natural law, in its essentials, is living because it reso-
> lutely recognizes and affirms that what Aquinas calls interchangeably the
> first principles of natural law and the first principles of practical reason
> are principles picking out and directing us towards the primary forms of
> human good—primary goods such as life itself, knowledge, friendship,
> marriage, and practical reasonableness (the *bonum rationis*) itself. These
> principles are truths. They are, as he frequently says, indemonstrable, that
> is, self-evident, *per se nota*—not of course data-less intuitions, but insights
> into the experience of desires (e.g. curiosity) and knowledge of possibilities
> (e.g. that questions can be answered and that answers can hang together as
> knowledge of truth and reality). As isolated insights their directiveness is
> not yet moral, but it becomes moral when, as directed by the principle that
> practical reasonableness itself is a good to be pursued and done, one consid-
> ers the question what to choose in light of the *combined* directiveness of all
> the principles as they bear both on one's own individual good and the good
> of everyone else. [42]

Finnis points out the need for a consistent understanding of the connection between practical principles and the *ultimus finis* (final end) from the position of the acting subject, recovering a virtue ethic. He says that the encyclical letter *Veritatis splendor* teaches not only that "the principles and leading moral norms of the natural law direct us in relation to '*the good* of the person ... by protecting his *goods*,'" [43] but also that "it is therefore necessary to place oneself *in the perspective of the acting person*. ... By the object of a given moral act, then, one cannot mean a process or an event of the merely physical order, to be assessed on the basis of its ability to bring about a given state of affairs in the outside world. Rather, that object is the proximate end of a deliberate decision which determines the act of willing on the part of the acting person." [44]

For Finnis, every living theory of natural law must be a virtue ethic, since virtues are nothing other than dispositions for making and carrying out just and rea-sonable decisions while avoiding wrong and unreasonable ones. Virtues "correspond to the intransitive aspect of one's choice and action, its significance as not only transitively shaping the world outside one's will but also shaping one's character,

---

[40] Vanni Rovighi, *Elementi di filosofia*, 3:224.

[41] J. Finnis, "Nature and Natural Law in Contemporary Philosophical and Theological Debates: Some Observations," in *Nature and Dignity*, Vial Correa and Sgreccia, 81–109.

[42] Ibid., 93.

[43] Ibid., 101, citing John Paul II, Encyclical Letter *Vertitatis splendor*, August 6, 1993, no. 13.

[44] John Paul II, *Vertitatis splendor*, no. 78.

one's dispositions and thus one's very self, because choices last *in* one's will unless and until reversed by another contrary act of will."[45] Hence virtue ethics is not in conflict with natural law ethics.

Teleological Ethics and Deontological Ethics

What was stated in the previous section on moral values and the natural moral law is the explanatory premise for understanding some of the attitudes that dominate modern ethics, which is characterized by a rational, *duty*-based morality, disconnected from the context of its meaning—that is, disconnected from a purpose in life, from a true understanding of man, and from a community that makes this truth clear and convincing.[46]

The alleged irreconcilability of teleological ethics, whose main pillar is the concept of value and the realization of an end, and deontological ethics, which is founded primarily on the realization of the norm as such, therefore remains to be clarified. In my opinion, one does not exclude or replace the other; rather, they must be integrated in an understanding of the complementariness of values and norms. In other words, a norm and a human value need one another: the norm is aimed at the realization of a human value and the human value needs the norm in order to become concrete in individual and social action. Essentially, it must be concluded that deontological (duty-based) ethics and teleological ethics are to be integrated and not seen as two independent modes of moral life. Let us now briefly examine these two ways of understanding morality.[47]

Teleological ethics considers man's supreme task to be the realization of values and of his plan for himself. This is expressed as a tension toward a "not yet" that awaits fulfillment. This tension is certainly not extraneous to life, which is always to be understood as a calling to fulfillment of humanity, but it is unclear how it could prescind from a definition of man, of the human person, and of the values consistent with this vision—that is, without a definition of the nature and truth of man. The realization of such a plan must clearly be gauged by respect for the ontological truth regarding man himself; otherwise it would be reduced to an exercise in inventive and domineering Titanism. Finally, every realization of the plan should translate into concrete acts and objective behaviors represented by norms and deontology.

Classical ethical theories, such as those of Aristotle and Aquinas, were widely considered by philosophical historiography to be teleological ethics inasmuch as they dealt primarily with the supreme good or happiness of man. The meaning of happiness emerges from the perspective of human action seen from within the acting subject and therefore from within his intrinsic dynamic of volition. Indeed, it is said that they are ethics developed from the first-person perspective that necessarily pay significant attention to the desire for the complete human good. First-person ethics presupposes

---

[45] Finnis, "Nature and Natural Law," 101.

[46] L. Melina, *Morale: Tra crisi e rinnovamento* (Milan: Ares, 1993), 26ff.

[47] For an interesting essay on deontological and teleological ethics, see R. Spaemann, "La responsabilità personale e il suo fondamento," in *Etica teleologica o etica deontologica? Un dibattito al centro della teologia morale odierna* (Rome: 1983).

that the truth about what is good for man actually exists and is attainable.[48] Therefore, this type of ethics is incompatible with relativism and epistemological and ethical skepticism. Consequentialism and proportionalism, which emerge within the context of contemporary ethics but have classical roots, have been erroneously and improperly considered teleological ethics on a par with the classical theories; in reality, they are diametrically opposed to classical teleological ethics.[49]

On the foundational level, the teleology of consequentialism and proportionalism entails that the greatest good or happiness is the criterion for establishing right and wrong actions: the good or the end is defined first, and then right or wrong actions. What is right is nothing other than the maximization of the good (John Stuart Mill's utilitarian ethics). On the concrete level of moral judgment, consequentialism reasons in this way: acts and rules are always and primarily to be assessed on the basis of their consequences for the optimization of reality. In this way a new ethics emerges, organized as a normative science for the production of a good state of affairs. This way of grounding ethics has nothing to do with practical teleological ethics.[50]

Deontological ethics is any understanding of ethics that involves certain categorical duties and prohibitions taking unconditional precedence over other moral concerns and considerations about finality and function. In addition to Immanuel Kant, contemporary liberals who uphold the primacy of the just over the good, such as John Rawls, are deontologists.[51] Modern and contemporary ethics abandons the problem of man's ultimate good and focuses on the definition of right or wrong actions and the identification and foundation of norms. For this reason, it is considered an ethics developed from a third-person perspective, which loses sight of the dynamic of intentionality proper to moral action as such. The underlying approach is the following: Harry has performed action *x*—is this action licit or illicit, obligatory or morally prohibited? In this way, ethics becomes an ethics of acts and norms rather than of man's good and his end.

Deontology is a form of justification by which the first principles are derived without presupposing any ultimate human aim or end, nor any set idea about man's ultimate end: "the right is prior to the good not only in that its claims take precedence, but also in that its principles are independently derived."[52] The classical model of deontological ethics along these lines is Kantian ethics.

All of these considerations allow us to conclude that in teleological ethics, as in Christian morality, the foundation of what is right or wrong is not independent of the ultimate good, just as the understanding of the good is not independent of

---

[48] A. Rodriguez Luño, "*Veritatis splendor* un anno dopo: Appunti per un bilancio (II)," *Acta Philosophica* 1, no. 5 (1996): 66. This article offers an exhaustive and in-depth analysis of the teleological and deontological ethical theories mentioned in *Veritatis splendor*.

[49] For further study of this issue, see ibid. The author distinguishes between practical teleology (Aquinas, Maritain, Dietrich von Hildebrand) and normative teleology (Mill) or consequentialism and proportionalism.

[50] Many observations could be made in this regard, so we again refer to the article by Rodriguez Luño, "*Veritatis splendor* un anno dopo," 61–69.

[51] J. Rawls, *A Theory of Justice* (Cambridge, MA: Belknap Press, 1999).

[52] M. Sandel, *Liberalism and the Limits of Justice* (New York: Cambridge University Press, 1982), 2.

the right. The fact that teleological ethics attributes absolute value to certain ethical requirements does not mean that it is deontological. In the teleological perspective, the ultimate end is the center of moral life. The end is not a good that can be "maximized" by just actions, nor is it possible to deduce which actions are just based on the idea of the end. Ethical virtues are principles of the practical reason and the foundation of ethical norms. In this sense, teleological ethics is simultaneously deontological, but with an anthropological and ontological foundation; this means that norms spring forth from man's good and his end.

In conclusion, these two forms of reasoning are not necessarily mutually exclusive in the context of a complete and balanced ethical discussion: on the contrary, they should complete one another in keeping with a comprehensive vision of human action, which is always directed toward the realization of an end. This must happen through means that are destined to become realized in concrete acts, which are in turn represented by values and deontological norms.

It is therefore necessary, particularly in bioethics, to make "concerted use"[53] of deontological and teleological judgments in order to avoid the risks associated with the extremes of moralistic rigorism on the one hand, which winds up becoming pharisaical by rendering norms absolute, and of opportunism on the other, which is preoccupied with drawing the greatest possible good out of the consequences of the action without adequately justifying the moral understanding of the "good" that is the basis for acting.

Perspectives in Contemporary Ethics

V. Possenti[54] highlights several recurring themes in contemporary ethics: the isolation of ethics both from metaphysics and religion; the growth of procedural and abstract elements to the detriment of the problem of good and virtue; a focus on problems of public ethics while disregarding both individual ethics and the problem of the relationship with the Absolute; the inability to overcome the central postulate that ends and values are irrational (Hume's law); the eclipsing of the good in connection with the crisis of final causes among modern thinkers and the associated emphasis by the sciences on efficient causality; and the separation of deontology and teleology. It is necessary to recall that the deontological approach to norms cannot be primary—ethics cannot be made to revolve exclusively around the idea of norms and laws as self-sufficient; they must instead be based on a teleological ethics of the good and of the good life. The concept of a norm is never primary in the development of moral science;[55] of the predicates "good" and "obligatory," the former is more primary.

Above all, the separation of the doctrine of freedom from the doctrine of the good makes it futile to justify freedom of choice itself, since it is possible to deduce freedom of choice only if the will of the subject is necessarily attracted to the infinite Good: it consequently remains free with respect to the various finite goods—or rather, not necessarily attracted to them. According to an only apparent

---

[53] C. Viafora, *Fondamenti di bioetica* (Milan: Ambrosiana, 1989).

[54] V. Possenti, "Prospettive sull'etica," in *Essere e libertà* (Soveria Mannelli, CZ: Rubbettino, 2004), 207–246.

[55] F. Botturi, ed., *Le ragioni dell'etica: Natura del bene e problema fondativo* (Milan: Vita e Pensiero, 2005).

paradox, freedom of choice with respect to finite goods is rigorously deduced from its necessary attraction to the infinite Good (see Aquinas, *Summa theologiae*, I-II, q. 0, a. 2). Contemporary philosophy of freedom is tempted to tack all of reality, including God, onto freedom—this unknown and obscure abyss. The inversion of the relationship between freedom and the Good leads to vertigo, positing the abyss of an irrational and blind freedom as the First Principle (*Abgrund* as *Urgrund*).[56]

Anthropological silence

There is a tendency to develop ethics *without reference to man*, in a way that is disconnected from anthropology, from the idea of man as he is and how he would have to be if he were to realize his essence and attain his end—that is, ethics without an idea of the nature and end of man. It looks to the subject, ignoring the person. There is a difference between the notion of subject and the notion of person: a subject, having an essence, exercises existence and action; person implies a certain degree of ontological perfection that manifests spirituality, knowledge, and freedom. A person is a microcosm that acts by assigning ends to itself. Contemporary ethics knows the speaking, dialoguing, communicating subject, but not the existing person with his or her individual substance, freedom, spirituality, and totality. It is necessary to combine ethics and anthropology and to recognize that ethics goes beyond the foundation of norms. Without a connection between the doctrine of man and the doctrine of norms, the latter seem unintelligible. No procedural practice can compensate for what lacks on the level of the real perception of the good[57] and of man's nature.

Distortion of the notion of "transcendental"

Modern philosophy, setting the subjective perspective in opposition to the objective, has lost sight of the correct notion of transcendental, which is an attribute predicated of everything that exists. Transcendentals are universal modes of being in relation to the intending soul-mind-conscience; the transcendental quality of the subject-person-soul in general is always included in the transcendental of being. According to medieval theory, the transcendentals of being (one, something, true, good, beautiful) are such in relation to the person, who is the archetype of being.

Modern philosophy proceeded to detach the soul-person from being and from others, turning it into a formal "self." We have lost the way to wholeness and to the possibility of knowing the universal modes of being, and the isomorphism between thought and reality has become unhinged.[58] But once the limitless intentional openness of the person is eliminated, it is difficult to establish any degree of openness to others, whose otherness will be constantly in jeopardy of being understood reductively.[59]

---

[56] L. Pareyson, *Ontologia della libertà: Il male e la sofferenza* (Turin: Einaudi, 1995).

[57] V. Soloiev recalls that "the proper subject of moral philosophy is the idea of the *good*; the purpose of this philosophical investigation is to bring to light everything that reason conceives—under the influence of experience—concerning that idea, thereby providing an answer to the question—which is essential to us—of what the object and meaning of our life should be." *La justification du bien* (Geneva: Ed. Skatkine, 1997), 2, cited in Possenti, "Prospettive sull'etica," 210.

[58] G. De Anna, *Realismo metafisico e rappresentazione mentale: Un'indagine tra Tommaso d'Aquino e Hilary Putnam* (Padua: Il poligrafo, 2001).

[59] See Possenti, "Prospettive sull'etica," 235–236.

Discourse Ethics (Habermas) [60]

Presuppositions of discourse ethics

Discourse ethics arises from the "disenchantment" with the world manifested by Weber, the plurality of conflicting moral views of the world, and the adoption of the post-religious and post-metaphysical perspective. The work of reason directed toward understanding in the field of moral discourse is seen as a "last resort" worth pursuing for lack of anything better, though real results require the necessary and full-bodied support of the positive law in civil life and may be modest.

Characteristics of discourse ethics

Through theoretical and communicative means, discourse ethics intends to reformulate the Kantian doctrine on the problem of the foundation of norms. In the Kantian treatise, the central element to be explained is the prescriptive character of norms and commands, or rather their validity: norms that can be universalized according to the formula of the categorical imperative are valid.

According to Habermas's account, discourse ethics (1) emphatically distinguishes between deontological questions of justice and questions of living a good life, assuming a priority of justice over good that can only be constituted at the level of metaphysical thought; (2) defines practical reason as essentially procedural and formal; (3) reformulates the concept of theoretical truth and practical validity terms of intersubjective universal agreement: "only such legal norms are morally justified that are arrived at in an agreement by all concerned," [61] and the norm requiring the mutual recognition of the participants in the discussion entails the norm of recognizing all people as subjects; (4) discourse ethics, directed to norms and their justification in terms of public procedures and validity, entirely neglects the will, the virtues, and the teaching thereof. It provides a restricted, procedural, "weak" understanding of morality and of philosophical ethics, withholding any "substantial" contributions. It emphasizes formality and the proceduralist approach, which are far from substantial moral intuitions. The foundation of material norms is left up to "practical discourse."

Procedural reason

Comprehensive worldviews have dissolved in pluralistic societies and the morality of conscience does not provide a sufficient basis for natural law, which previously had a religious or metaphysical foundation. The necessary basis for legitimacy comes from procedural reason and from the democratic procedure by which laws and ethical

---

[60] J. Habermas, *Moral Consciousness and Communicative Action*, trans. C. Lenhardt and S. Weber Nicholsen (Cambridge, MA: MIT Press, 1990), originally published as *Moralbewusstsein und kommunikatives Handeln* (Frankfurt am Main: Suhrkamp, 1983); idem, *The Theory of Communicative Action* (Boston: Beacon Press, 1984–1987); idem, *The Inclusion of the Other: Studies in Political Theory*, ed. C. Cronin and P. De Greiff (Cambridge, MA: MIT Press, 1998); idem, *Between Facts and Norms: Contributions to a Discourse Theory of Law and Democracy* (Cambridge, MA: MIT Press, 1996); V. Possenti, "Metafisica, problema della verità, pragmatica trascendentale," in *Annuario di filosofia 2000* (Milan: Mondadori, 2000); idem, "Postazione," in J. Maritain, *La filosofia morale* (Brescia: Morcelliana, 1999); Possenti, "Prospettive sull'etica," 219–230.

[61] Habermas, *Moral Consciousness*, 72.

systems are generated. In this manner, the democratic principle tends to have its own roots, independent of moral principles. "A significant impoverishment of the idea of the person is another part of post-metaphysical apriorism. The term 'person' itself disappears and is replaced with 'subject,' which is then defined through linguistic acts and intersubjective agreement."[62] *Procedure* and *consensus* become the two criteria for legitimacy. The post-metaphysical assumption implies that no moral order preexists intersubjective decisions and that a just legal order must be ideologically neutral.

Intersubjectivity takes the place of universality

Since the supreme rule of ethical reason is intersubjective agreement on respect for what everyone wants, the form or procedure prevails over the content—that is, the latter adapts to the former. There is a *separation between values and norms (with their corresponding obligations)*, and it is impossible to establish norms on the basis of values. Given this separation, the guiding criterion for the norm is no longer a value but an intersubjective agreement reached procedurally.

Proceduralism does not assist with the examination of substantial issues, such as human relationships and the question of moral responsibility.

Some Corollaries and Principles of Personalist Bioethics

Following this introduction to the meaning and content of personalist ethics, it is now possible to enumerate and explain some of the principles and directives of bioethics—in other words, those regarding man's interventions on human life in the biomedical field. The impact of these principles will become evident when examining their application to specific questions in the second part of this volume, but it seems useful at this point to provide a presentation and justification of the principles.

The justification of these principles follows from what has been presented in the previous chapters. It will therefore suffice to recall the reasons set forth and the motivations adduced, whereas the applications to the various cases in biomedicine, or to the primary ethical issues connected with biomedical practice, will be better explored in the later chapters.

The principle of defense of physical life[63]

It has already been discussed how bodily, physical life is not something extrinsic to the person; rather, it is the *fundamental* value of the person. I say "fundamental" because it must be understood that bodily life does not exhaust all of the richness of the person, who is also and most importantly spiritual, therefore transcending temporality and the body itself. The body is nonetheless coessential to the person; it is the first embodiment, the *unique foundation* in which and by means of which the person is realized and enters into time and space, expresses and manifests himself, and develops and expresses other values, including freedom, sociality, and his own plan for the future.

---

[62] Possenti, "Prospettive sull'etica," 226.

[63] I use the expression "physical life," which is by now widespread, recognizing that it is reductive and poorly expresses the holistic understanding of the person as a unity of body and soul. By physical life I mean organic life, the unique and unified foundation for the whole development of the person.

Only the total and spiritual good of the person is above and beyond this fundamental value. The sacrifice of bodily life could be required only if that spiritual and moral good were unattainable except through the sacrifice of life. In such a case, since it is a spiritual and moral good, it could never be imposed by other people—it must develop as a free gift. Martyrs legitimately give their lives only when there is no other way of actuating the moral good of the person and of the society. In that case, however, those who are responsible for the situation demanding martyrdom are also responsible for the loss of that life. Finally, the martyr is not the one who actually takes a life, strictly speaking; he is only prompted to expose himself to a justified risk by fidelity to a higher good. I alluded earlier to this case, which seems at first glance to contradict the moral precept concerning the inviolability of human life.

The importance of this principle therefore emerges in regard to the assessment of various ways of suppressing human life: homicide, suicide, abortion, euthanasia, genocide, wars of conquest, and so forth. Some of these emergent cases pertaining to the field of biomedicine will be analyzed in later chapters, along with all of their implications, but up to this point it was necessary to demonstrate how respect for life, as well as its defense and promotion, constitute man's first ethical imperative toward himself and others. Perhaps it is necessary to emphasize that this is not just a question of respect, but also of active defense and promotion. International charters of rights dealing with human rights place life and its inviolability at the forefront.[64]

There is hardly need to recall how inconceivable it is to contemplate the idea of the direct and deliberate taking of someone's life (which is different from the case presented above concerning the voluntary exposure of oneself to the risk of losing one's life for the total moral good of the person and of the community) in order to promote the lives of others or to improve the political and social conditions of others. This is because the person is valued as a whole individual and not just as a part of the society.

The Catholic Church has gathered rich teachings on the fundamental value of human life from throughout its tradition, further enriched by revealed truth and expressed in official documents,[65] which constitute its theological anthropology and cannot be adequately examined here. It is also important to emphasize, however, that the ethical obligation to respect, defend, and promote life has rational and universal validity.

The value of life at lower levels, such as in the plant and animal kingdoms, could also be mentioned at this point; it is a topic to which the environmental movements

---

[64] The Universal Declaration of Human Rights, http://www.un.org/en/documents/udhr, approved and promulgated by the United Nations General Assembly on December 10, 1948, affirms in article 3 that "everyone has the right to life, liberty and security of person." The Convention for the Protection of Human Rights and Fundamental Freedoms for the European region (http://conventions.coe.int/treaty/en/treaties/html/005.htm), promulgated in Rome on November 4, 1950, affirms in article 2 that "everyone's right to life shall be protected by law."

[65] The Congregation for the Doctrine of Faith expresses itself in the following manner in its declaration on procured abortion, *Quaestio de abortu* (November 18, 1974): "The first right of the human person is his life. He has other goods and some are more precious, but this one is fundamental—the condition of all the others. ... It is not recognition by another that constitutes this right. This right is antecedent to its recognition; it demands recognition and it is strictly unjust to refuse it" (no. 11).

are very sensitive. This type of life also has its value. Balance among the various forms of life in the cosmos is linked to the health and survival of man, and we have a consequent a duty to preserve this balance. Nevertheless, it cannot be forgotten that man is at an ontologically superior level, transcending lower life forms, and therefore plants and animals can and should—by their connatural biological connection—be utilized by man. Such utilization, however, does not mean plundering, violence for violence's sake, devastation, or the endangerment of global equilibrium. Neither must the defense of the life of plants and animals be overemphasized, as sometimes happens, to the point of demanding a higher position for them than the one that society assigns to human life, or to the point of banning the use of animal life for experimentation and the advancement of science.

The topic of the defense of human *health* comes into play within the context of promoting human life. The concepts related to the definition of health have already been addressed. What must now be added to this presentation of general principles can be summarized in two statements, consistent with what was said a little while ago: (1) the right to life precedes the so-called right to health,[66] and (2) there is a moral obligation to defend and promote the health of all human beings in proportion to their need.

The first affirmation seems obvious: one can only talk about health in reference to a living person, and health is a quality of a person who is alive. However, some people understand the matter differently and distort it, such as when the life of another is endangered or taken for the sake of someone's health. This is the case with the legalization of so-called therapeutic abortions, supposedly justified by medical and medico-social factors, wherein the relationship between life and health is distorted by those who maintain that life need only be accepted if it has a sufficient "quality of life." It should not be forgotten that the Western world emphasizes health understood in a hedonistic sense, whereby health is considered the supreme good of temporal well-being, to the point that societies find themselves overburdened by ever-increasing health care costs.

Certain risks are evident in this way of thinking. Above all there is a risk that economic well-being, with its excesses, will lead to new health threats (illnesses of well-being, such as drugs, alcoholism, prescription drug abuse, sexual and eating disorders, and so forth). Another risk is that the excessive spending to treat these health threats will absorb economic resources that ought to be devoted to preventing and curing major diseases of biological origin: excessively protecting the health of some leads to neglecting the health of millions of people in underdeveloped countries and even marginalizing the most defenseless, such as the handicapped, the elderly, victims of incurable illnesses, or fetuses rejected for fear that they will threaten the well-being of the "well-to-do."

Here, then, is the other requirement: health, a value subordinate to life and resulting from it, must be promoted for all according to each one's need. It is not a question of a *right to health*, which no nation can guarantee, but rather a *right to essential means and care* for the defense and promotion of health. In this light, the affirmation in the Universal Declaration on Human Rights is acceptable: "Everyone has the right to a standard of living adequate for the health and well-being of himself and of his family, including food, clothing, housing and medical care and

---

[66] E. Sgreccia, "La posizione della Chiesa di fronte alla vita e alla salute nell'attuale contesto socioculturale," *Camillianum* 13ns (2005): 9–31.

necessary social services, and the right to security in the event of unemployment, sickness, disability, widowhood, old age or other lack of livelihood in circumstances beyond his control."[67] The constitution of the World Health Organization (1946) also confirms the criterion of equality among all peoples in protecting the health of individual citizens: "The enjoyment of the highest attainable standard of health is one of the fundamental rights of every human being without distinction of race, religion, political belief, economic or social condition. The health of all peoples is fundamental to the attainment of peace and security."[68]

Although they address an ethical and existential issue of primary importance, what the international charters do not say is that it is also necessary, together with the right to the promotion of health, to teach individuals how to accept inevitable suffering and death within a personalist and transcendent understanding of man. Thanks to many specific studies, it is now known that not accepting pain, sacrifice, and death when they appear on the existential horizon—in spite of everything and because of the inherent limitations of human life—in turn leads to disturbances of the whole person[69] and within the cultural world itself, especially in countries with a high level of economic development. But this topic will be discussed later; for now, it suffices just to mention these implications of the principle of defending life.

The defense and promotion of life has its limit in death, which is part of life, and the promotion of health has its limit in illness, which is to be treated and cured and in any case actively addressed even when it is incurable.

The principle of freedom and responsibility

Freedom and responsibility have already been considered as the source of the ethical act. Now let us consider some implications of this in the field of bioethics.

The first point of note in applications to bioethics is the following: the right to the defense of life precedes the right to freedom. In other words, freedom must mean taking responsibility for one's own life first and foremost, as well as the lives of others. This affirmation is justified by the fact that one must be alive in order to be free; therefore, life is the necessary condition—for everyone—for exercising freedom.

As obvious as it may seem, this affirmation poses many problems today in the field of medical ethics. Take the so-called right to euthanasia, for example: one does not have the right, in the name of freedom of choice, to decide to take a life. Other applications arise in matters of obligatory care for the mentally ill or the refusal of treatment for religious reasons.

More generally, this principle confirms a patient's moral obligation to cooperate with the ordinary treatments necessary for protecting his or her own life and health and that of others (through what is known as the *therapeutic alliance* between physician and patient). In certain cases, such as when a patient refuses a treatment essential to life and survival when the physician has deemed the treatment necessary in good

---

[67] Universal Declaration of Human Rights, December 1948, art. 25. See also the Declaration of Alma-Ata on September 12, 1978, and the Charter on Health Care Development in the African Region, published in Maputo on September 24, 1979.

[68] World Health Organization, *Constitution of the World Health Organization*, July 22, 1946, http://www.who.int/governance/eb/who_constitution_en.pdf.

[69] E. Becker, *The Denial of Death* (New York: Free Press, 1973).

conscience, this principle must govern the procedure for obligatory care. Another typical case involves parents who refuse to feed a physically handicapped newborn, practicing so-called *neonatal euthanasia*, in what amounts to an evident abuse of the parents' freedom with respect to the life of the newborn child.

On the other end, the principle of the freedom and responsibility of the patient—circumscribed by the principle of the defense of physical life, which is a value superior to freedom that calls first on responsibility—in turn limits the freedom and responsibility of the physician, who cannot transform care into compulsion in all other cases where life is not in question. This is the challenge of patient consent. There is an implicit consent the moment the patient entrusts himself to the physician or health care facility so that the physician can do what is necessary for care and recovery of health. This consent, however, does not dispense the physician from the duty of informing the patient about the progress of the treatment and asking additional and explicit consent every time there may be unforeseen developments, such as a treatment involving risk or harm, an experimental treatment when other options have proved ineffective, or the testing of a new drug. It is always necessary to recall that life and health are entrusted first and foremost to the responsibility of the patient and that the physician has no other rights over the patient greater than the patient's own. If the physician considers the expectations or wishes of the patient to be ethically unacceptable, he can—and sometimes must—responsibly part ways with the patient while inviting him to reflect and seek out other hospitals or physicians. The patient's conscience cannot be violated by the physician, nor can the physician's be forced by the patient: both are responsible for life and health as both personal and social goods. Many of the ramifications of this principle will be clarified when applied to specific cases.[70]

The principle of totality, or the therapeutic principle

This is one of the fundamental and characteristic principles of medical ethics. It is based on the fact that human corporality is a *unitary whole* resulting from distinct parts, organically and hierarchically unified by a single, personal existence.

The principle of the inviolability of life, which has been shown to be primary and fundamental, is not disproved but in fact applied when it becomes necessary to intervene in a harmful manner on part of the body in order to save the whole and the very life of the subject. This principle ultimately upholds all of the legitimacy and obligatoriness of medical and surgical treatment. The surgeon who removes an appendix is morally justified, and also obligated, to the extent that this removal is necessary to protect the body. For this reason, this principle is also referred to as the *therapeutic principle*.

The same principle can have more important applications when there is a question of tumor removal, high-risk operations, or surgery that could involve significant impairment, as in the case of therapeutic sterilization resulting from the excision of a uterine tumor. In such cases, the concomitant damage resulting from a direct intervention that is aimed at another good end may be considered ethically

---

[70] R. Limat et al., "Du soin à la contrainte: Quelques interrogations éthiques vécues par l'infirmier(e) dans la pratique des soins," *Médecine et Hygiène* 42 (1984): 1177–1182; G. Bonjean et al., "Le refus de soins: La dimension éthique du problème," *Médecine et Hygiène* 42 (1984): 1184–1190.

acceptable in accordance with the *principle of double effect*. This principle, though its formulation appears simple, sometimes presents delicate moral questions that will be understood best in the context of particular scenarios.

In order to apply the therapeutic principle, certain conditions must first be met: (1) the intervention in question is on an afflicted part, or a part that is the direct cause of the harm, in order to save the healthy organism; (2) there is no other way or means of correcting the condition; (3) there is a good and proportionally high chance of success; (4) the patient or legally responsible decision maker has provided consent. In these cases, it is understood that what is in question is not so much life as *physical integrity*, which is also a very important good inherent in corporality; it is therefore a personal value that can only be put at risk or harmed for the benefit of the higher good to which it is linked.[71]

The therapeutic principle has unique applications not only in the general cases of surgical operations, but also in more specific cases such as therapeutic sterilization, organ transplantation, and gene therapy.

Some ethicists, including theologians, have tried to extend the concept of totality beyond the physical body and corporality, thereby including the psychological dimension and the subjective, psychosocial well-being of the person in the purview of the therapy, or even including the whole set of final results in the concept of totality while disregarding the means and methods of intervention. These remain heated topics in medical ethics debates. Examples include questions of contraceptive sterilization, in vitro fertilization, and so-called therapeutic abortion: in these cases, one cannot legitimately apply the concept of totality understood organically and in its proper sense. (The individual arguments will be examined in the particular ethics section of the manual.)

As indicated above, some interpret this principle in an organicistic sense: a part of the organism can be harmed only if it is helpful to the same organism, physically understood. Others offer a broad interpretation, understanding totality as psychological or psychosocial well-being without regard for the physical organism and its harmonious recomposition together with the spiritual good. Finally, in what seems to me the best interpretation, others understand totality as the physical, spiritual, and moral totality of the person—that is, a personalist totality in which the good of the physical organism is also to be respected.

The body is therefore not to be taken in an exclusive sense (ignoring the rest), but in an assertive and holistic sense—that is, considering the bodily good in the whole context of the spiritual and moral good of the person. These clarifications will be helpful especially in discussing the questions of contraception and sterilization.

It is necessary at this point to state that accepting concepts of psychological and psychosocial motivation in order to justify physical mutilations, without also considering the good of the physical organism and without referring to the total, moral, and spiritual good of the person, would mean abandoning objective criteria and allowing the arbitrary manipulation of corporality. Through psychosocial reasoning, one can justify even violent actions against the person such as forced sterilization, euthanasia, and abortion.

---

[71] Häring, *Free and Faithful in Christ*, 3:92–93; M. Zalba, "Totalità (principio di)," in *Dizionario enciclopedico di teologia morale*, 1141–1149; idem, "La portata del principio di totalità nella dottrina di Pio XI e Pio XII e la sua applicazione nei casi di violazioni sessuali," *Rassegna di Teologia* 9 (1968): 225–237.

Finally, this principle of totality or therapeutic principle is linked to another practical norm that can be defined as the *proportionality of treatment*. This norm states that a treatment must be evaluated within the totality of the person in order to apply or continue it; moreover, there must be a certain proportion between the risks and damages it entails and the benefits it secures. Pursuing disproportionate treatments—to deceive the patient while giving the impression of being efficient, to fulfill the patient's or relatives' request to "do everything possible" without foreseeable results, or to surreptitiously test certain treatments with no benefit to the patient and without his knowledge—can express therapeutic aggressiveness or stubbornness. This will be clarified in discussing aid for the dying or incurably ill.

The principle of totality or therapeutic principle is also linked to the above-mentioned principle of double effect, which is a criterion for justification in cases involving an action with a so-called double effect—one positive and one negative, from the ethical perspective. This situation will be mentioned in upcoming pages.

The principle of sociality and subsidiarity

This principle was first developed within moral theology (as was the preceding principle of totality), but it is also widely shared by international directives and used in the development of health care assistance programs. Indeed, there is more and more discussion about the socialization of medicine.

The ethical *principle of sociality* must first be distinguished from the organizational and political dimension of socialization. The principle of sociality commits each individual person to self-realization through participation in achieving the good of their neighbors. In promoting life and health, this implies that all citizens work toward respecting their own lives and the lives of others as a good—not only a personal one but also a social one—and that they engage the community to promote the life and health of all, promoting the common good by promoting the good of each individual.

The person is essentially open to society, and sociality is an intrinsic characteristic of personhood. In the case of life and health, which are primary goods of the person, the same situation actually proves that the life and health of each individual also depend on the aid of others. In order to recognize the importance of this ethical principle, suffice it to consider the health care situation with respect to pollution and contagious epidemics or to observe the array of services making up the health care industry, in which it is possible to recover health insofar as there is collaboration at many levels between professions, specialties, and legislative regulations.

It will be seen how the principle of sociality can justify organ and tissue donations (which also involve a certain mutilation of the donor), stimulate volunteering, and lead—as has happened almost throughout the world—to the rise of health care initiatives (hospitals, care facilities, leprosariums, etc.) just through a sense of fraternal service rendered by the healthy to the sick. In terms of social justice, however, the principle obligates the community to guarantee everyone the means of accessing necessary care, even at the cost of sacrifices for the well-to-do.

At this point, however, the principle of sociality melds with the *principle of subsidiarity.*[72] According to this principle, the community on the one hand must help

---

[72] The principle (or doctrine) of subsidiarity was proposed in the encyclical *Quadragesimo anno* (March 15, 1931) by Pius XI, and was taken up again by Vatican Council

more where the need is greater (caring more for those in greater need of care and spending more on those who are more ill), and on the other hand must not supplant or replace the free initiatives of individuals and groups; rather, it should guarantee that they can function.

These principles are so obvious that it seems entirely superfluous to cite them in a manual of bioethics. Yet it should be noted that the global health care situation does not merit a "satisfactory" judgment—just think of how few means of care and how few health care structures exist in developing countries, where the need is greater. Furthermore, even in industrialized countries, voices emerge every once in a while claiming that the state cannot cope with the vast costs of health care, and that it should manage expenditures by adopting the economic principle of costs and benefits. This leads to attempts at directing the greater portion of health care spending not to the most seriously ill (who may have no chance of recovery), but to citizens who can still be "productive." If anything, given the increasing costs of health care, greater sacrifices should be asked of those who can best sustain them; however, precisely in the name of subsidiarity, health care must never be denied to the most seriously ill.

Indeed, the idea of so-called social euthanasia is creeping in along this path, motivated by the dramatic and unfortunate choice of societies to the detriment of the incurably ill, severely handicapped, and mentally ill.[73] In this manner, sociality turns nonsensical and its true meaning is perverted.

The socialization of medicine as a health care policy—different from liberal and collectivist medicine—is instead a model for a health care system whose ideal objective is to provide free means of health care and assistance to everyone equally, simultaneously promoting respect for the freedom of citizens and their active participation. This is one way of applying the principle of sociality that puts the burden of organizing health care services on the democratic state, while leaving citizens free to choose their physicians and health care institutions. The intention is certainly considered a good one, though it is not free of the dangers of bureaucratic centralism, excessive costs, leveling of health care standards, and politicization of the administrative bodies of the health care system. But a more thorough discussion will be necessary regarding the situation of medicine as a science, as care, and as a health care system.

Principles of North American Bioethics

In specifically bioethical literature, especially from the English-speaking world, it is easy to find references to other principles that should guide physicians in their relations with patients and in general in every action or choice in the biomedical field.

It was already mentioned, in the chapter concerning the ethical foundations of bioethics, that within the context of the debate in the Americas there is a widespread

---

II in the pastoral constitution *Gaudium et spes*, nos. 31, 63, 65, and above all in the encyclicals by John XXIII, *Mater et magistra* (March 15, 1961), nos. 40–44, and *Pacem in terris*, no. 74. See Häring, *Free and Faithful in Christ*, 3:277–278.

[73] A. Franchini, "Le grandi scoperte della medicina," in *Storia delle scienze*, ed. E. Agazzi (Rome: Città Nuova, 1984), 2:388; P. Rentchnick, "Euthanasie: Evolution du concept d'"euthanasie' au cours de ces cinquante dernières années," *Médecine et Hygiène* (February 29, 1984): 653–666.

use of so-called *principlism*, which was devised by T. L. Beauchamp and J. F. Childress in their well-known book, *Principles of Biomedical Ethics* (sixth edition, 2009). The authors developed a sort of ethical model geared toward those who work in the health care field, aiming to provide a practical and conceptual reference point to guide them in concrete situations—a moral alphabet by which to justify their decisions. This model, to which the greater part of English-language bioethical literature has referred for more than twenty years, entails the formulation of the principles of autonomy, beneficence/nonmaleficence, and justice, which are supposed to be independent of any foundational ethical theory, serving as a common language for international and pluralistic bioethics. This is in spite of the fact that they have already been interpreted in light of two theories: utilitarianism and prima facie deontology.

A closer examination of the context in which the "principlist" system began reveals that it was initially proposed within the limited context of experimentation on human subjects.[74] In fact, between 1974 and 1978 the National Commission for the Protection of Human Subjects of Biomedical and Behavioral Research was given the specific task, by mandate of the United States Congress, of identifying some general ethical principles, which were then disseminated through its final document, *The Belmont Report*.[75] The reason for these principles was clearly expressed in the document: to overcome the ethical conflicts that particular norms pose for experimenters at the time of application.

Thus the following principles were proposed: (1) the *principle of respect for persons* involved in the experimentation, which implies treating them as autonomous subjects—where autonomy is understood as the capacity to act consciously and without constraint—and protecting them when their autonomy is limited or even absent, the immediate corollary of which is the need to obtain informed consent from the subject or from his or her legal representative for medical treatments; (2) the *principle of beneficence* in experimental procedures, which means doing no harm, minimizing risks, maximizing advantages, and assessing the risk-benefit relationship before any experimentation begins; and (3) the *principle of justice* in allotting the burdens and risks of experimentation.

The Belmont Commission proposed these principles without any intention of basing them on an ethical theory; in reality, however, they were expressly derived from deontological norms.

An attempt to ground these principles was instead proposed by Beauchamp and Childress in their above-mentioned book, which was first published in 1979. The authors' intention was twofold: on the one hand, to extend the model—based

---

[74] For a historical reconstruction and the considerations that follow, see A. G. Spagnolo, "Bioetica (Fondamenti)," in *Dizionario di teologia pastorale sanitaria* (Turin: Camillianum, 1997); idem, "Principios de la bioética norteamericana y critica del principlismo," *Bioética y ciencias de la salud* 3 (1998): 102–110; idem, "I principi della bioetica Nord-Americana, e la critica del 'Principlismo,'" *Camillianum* 20 (1999): 225–246.

[75] U.S. National Commission for the Protection of Human Subjects of Biomedical and Behavioral Research, *The Belmont Report: Ethical Principles and Guidelines for the Protection of Human Subjects of Research* (Washington, DC: US Government Printing Office, 1979), http://ohsr.od.nih.gov/guidelines/belmont.html. Comments on the Belmont Report and on the meaning of the principles were made in C. Viafora, "I principi della bioetica," *Bioetica e Cultura* 3 (1993): 9–37.

on the principles for the field of experimental research—to the entire biomedical field; on the other hand, to propose the use of these principles within various and even conflicting ethical theories.

The basic points of principlism can be summarized in this way: (1) there are no intrinsic norms in medical practice that can guide decisions; (2) there are four fundamental principles (beneficence, nonmaleficence, autonomy, and justice) that should guide medical actions; and (3) these principles must be applied to specific situations in order to formulate particular moral judgments.

Essentially, in order to make a decision concerning an action to be carried out in a particular case, the *ultimate practical judgment* results from the application of certain *practical rules*, which are generalizations about what should or should not be done in a given context and for a specific aim. Such rules follow in turn from *general principles*, which are ultimately justified by *ethical theories* that serve to guide decisions, especially cases of conflict between two or more principles. This means that radically opposing theories on the methodological or theoretical level can nonetheless come together and reach an overlapping consensus on identical principles and rules and therefore on the actions to recommend.

Its methodological simplicity was one of the main reasons for the success of the principlist paradigm: it offers even non-experts the possibility of a framework on the basis of which they can engage the various ethical problems encountered in medical practice.

These four principles are also able to cover many facets of biomedicine. The principle of respect for autonomy is considered foundational with respect to all the other principles, grounding considerations about informed consent, truthfulness to the patient, and the refusal of treatment. It also includes such great semantic variety (autonomous decision, right to freedom, privacy, individual choice) that there has been talk of a real and distinct *theory of autonomy*.[76] However, this principle is only a *prima facie principle* that can be limited by the principles of beneficence and justice: if the individual's autonomous decision were to threaten public health or entail a significant economic cost that he could not sustain, shifting it to the state, it would be justifiable to limit the individual's autonomy. The same can also be said for the principles of beneficence/nonmaleficence and justice: the former can be limited when avoiding harm and doing good are connected to social obligations of distributive justice; the latter, which is related to the classical formulas of *suum cuique tribuere* (to each his due) and *alterum non laedere* (do not harm another), is even considered by some authors to have no identity of its own, because it sometimes draws on the principle of beneficence and other times on the principle of autonomy.[77]

Now let us see how the authors address the question of conflict between principles and, more fundamentally, between the often contradictory theories that justify them.

Thomas Beauchamp and James Childress propose incorporating the principles into a *composite ethical theory* that allows each basic principle to have a certain weight without however giving any of them priority (any objective hierarchy of principles is

---

[76] J.F. Childress, "The Place of Autonomy in Bioethics," *Hastings Center Report* 20 (1990): 12–17.

[77] H.T. Engelhardt Jr., *The Foundations of Bioethics*, 2nd ed. (New York: Oxford University Press, 1996).

thereby excluded). Which principle takes precedence in the event of a conflict will depend on the particular context, which always has unique characteristics.

The risk of falling into situational ethics is evidently very high; additionally, there is an explicit element of intuitionism in this "balancing" of goods, to use William David Ross's metaphor.[78] According to Ross, the weight of the principles in a situation of conflict rises and falls as though on a scale. In this perspective, Ross distinguishes *prima facie duties*, which are binding in all circumstances unless they are in conflict with equal duties or duties that are more important in the concrete situation, from *actual duties*, which are to be fulfilled in the concrete situation and are determined by balancing the different weights of the implied prima facie duties in that situation.

Beauchamp and Childress take from Ross this distinction between the principles, which in fact derives from a deontological, anti-utilitarian theory: for Ross, an action should be chosen based on whether it fulfills a duty that is judged better than others in the concrete circumstances (intuitionism) and therefore becomes obligatory.

In the event of a conflict between principles, Beauchamp and Childress talk about balancing them by assessing the *consequences* linked to the decisions that would be inspired by one or another of the principles (*rule utilitarianism*). Any reference to objective criteria for assessing potential decisions is excluded; a subjective criterion is therefore proposed when only a single person is involved, or a collective consensus when more people are involved.

Thus, with the "contemporary reference to a theory of deontological character (prima facie duties) and teleological character (rule utilitarianism), Beauchamp and Childress were convinced that they had designed a suitable method for solving many ethical problems in the biomedical field."[79]

It is important to note, however, that the two authors recognize—in the 2001 edition their book—the particular importance of various theories besides the two mentioned here, such as those founded on character, virtue, experience, and solidarity, thereby signaling what has been described as "the beginning of the end of principlism," even in North American bioethics.[80] Furthermore, while this ethical paradigm has been remarkably widespread and particularly so in the Anglo-American regions, there has been no shortage of critical observations in the literature.

There are undoubtedly elements of validity in all three of these principles, and upon close inspection there could be a certain correspondence with the principles enumerated in personalist bioethics (the therapeutic principle, the principle of freedom and responsibility, and the principle of subsidiarity, respectively). But while the latter seem coherently interconnected by a grounded, personalist anthropology, which ultimately refers to the integral good of the person as derived from an analysis of the connatural characteristics of his essence, the same cannot be said for the former, which fail to clarify what should be understood by the good of the person or the autonomy of the individual, for example. In fact, this lack of reference to a unified

[78] W. D. Ross, *The Foundations of Ethics: The Gifford Lectures, delivered in the University of Aberdeen 1935–6* (Oxford: Clarendon Press, 1968), esp. 84–86, 108–112.

[79] Viafora, "I principi della bioetica."

[80] E. J. Emanuel, "The Beginning of the End of Principlism," *Hastings Center Report* 25, no. 4 (1995): 37–38.

186

theoretical framework means that different conclusions will be reached depending on which principle is emphasized over the others (relativism).

Indeed, the physician-patient relationship and the physician-society relationship cannot be limited to a single, horizontal dimension that exhausts the relationship itself: the ultimate reference for all of them (physician, patient, and society) must be situated outside of them, transcending them. In this way, only a reference to an integral, objective good can avoid the major risk of falling into extreme relativism. Such principles are therefore deceptive from both the theoretical and the practical perspectives inasmuch as they are disconnected from a unified ethical theory, thereby lacking a systematic correlation and often coming into conflict with one another. Their alleged universality[81] is therefore not justified, since they only allow for the identification of a *collection of concrete cases*, or rather a casuistry of ethics, which automatically limits the universal applicability of the decisions.

Many authors have looked to this excessive simplification of method as a sort of "magic formula."[82] Regarding an individual principle such as autonomy, it is paradoxically nonexistent if it is not reciprocal. In reality, there is a radical reciprocity of the human being whereby autonomy, correctly understood, is connected to heteronomy, which is its dialectic opposite consisting in the responsible exercise of freedom. Furthermore, the principle of beneficence contains no indication of what the "good" of man is, and the principle of justice does not state what the "due" is, to whom it is due, and why.

In reference to subjectivity in resolving conflicts between principles in concrete situations, it is also necessary to acknowledge the objective element alongside the subjective one: moral autonomy requires the unity of subject and object. Furthermore, it has already been shown how the recognition of moral objectivity is a prerequisite for the full and universal justification of moral principles.

Other critical points that have emerged in the literature about principlism will be briefly mentioned here. Some authors have pointed out the incomplete nature of these principles, which do not explain moral experience exhaustively. The infinite richness of the moral life eludes schematic and rigid enclosure within these principles. The principlist paradigm, focusing its attention on the application of its principles in practice, risks losing sight of moral experience.

Furthermore, the formulation of principles tends to foster a passive attitude of obedience rather than active, morally committed behavior. It is not enough to apply some principles in the situation; rather, it is necessary that the agent understand the moral significance inherent in the act he is performing.

In conclusion, the formulation of principles without an ontological and anthropological foundation renders them sterile and ambiguous. A systematization and hierarchy is necessary in order to harmonize and unify their meaning. This can only

---

[81] Various particularly telling examples have been adduced to illustrate the limits of the principle-based paradigm; for comparison, see R. Gillon, *Philosophical Medical Ethics* (Chichester: John Wiley, 1986).

[82] In one of its monographic issues several years ago, the important *Journal of Medicine and Philosophy* presented some interesting critiques of North American principlism. See in particular K. D. Clouser and B. Gert, "A Critique of Principialism," *Journal of Medicine and Philosophy* 15, no. 2 (1990): 219–236, and the recent book *Bioethics: A Return to Fundamentals*, by B. Gert, C. M. Culver, and K. D. Clouser (New York: Oxford University Press, 1997).

happen by reformulating them and defining them within a unified ethical theory that has the human person as its ultimate criterion, from which several corollaries follow: respect for physical life and essential integrity, respect for freedom in connection to the responsibility of the person, therapeutic justification for medical interventions, and an understanding of the common good not as the good of the majority but as the sum of the good of individual persons. Potential conflicts between these principles are in fact only apparent and are resolved by their *harmonization* within the ethical theory that inspires them. The reference to the person in his or her entirety actually helps to identify a hierarchy among the principles and therefore to harmonize them when they appear to be in conflict.

The validity and meaning of these principles would reemerge by reinterpreting them in a hierarchical, ontologically grounded arrangement. One would proceed from the principle of beneficence to the principle of autonomy, and finally to the principle of justice (in the event of a conflict in the application of each preceding principle).

The *principle of beneficence* would be at the summit, as the ultimate reference. It corresponds to the primary aim of medicine in a naturalistic perspective, which is to promote the good of the patient and of society and to avoid harm. It is of course something more than the Hippocratic *primum non nocere* (first, do no harm), which is also called the principle of nonmaleficence, insofar as it implies not only refraining from doing harm but also and above all the imperative to actively do good and even to prevent harm. The term *beneficence* is more appropriate than *benevolence*, insofar as it emphasizes the need to effectively *do* good and not merely to *want* or *desire* it.

The *principle of autonomy* refers to the respect due to fundamental human rights, including self-determination. This principle is inspired by the maxim "do not do to others what you would not have them do to you," and is therefore at the basis of a morality inspired by mutual respect. The therapeutic alliance between physician and patient and consent for diagnostic and therapeutic treatments are founded on this principle; it is an integral part of beneficence and at the service of beneficence. Obviously this principle may not be applicable to psychiatric patients—for example, in situations of dementia or acute psychosis—or to those who are unable to express consent (patients in comas, minors, etc.). In such cases, then, one would have to appeal to the principle of beneficence or to the third principle: justice.

The *principle of justice* refers to the duty of equal treatment and equitable distribution of funds for health care, research, and so forth at the state level. While it certainly does not mean treating everyone in the same way, because there are different clinical and social situations, it should nonetheless entail adherence to certain objective facts, such as the value of life and respect for proportionality in interventions.

Returning to the principlist paradigm in light of the priority on the fundamental good of the human person, attention is now given to experiential moral reality and there is agent intentionality (the action is not a merely extrinsic application of principles, but a behavior directed toward the realization of one's own good and the good of others). The principles provide general guidelines for behavior, but the ethical value of the good of the person as the ultimate end to be attained is what confers final meaning on the action.[83]

---

[83] For another critique of principlism in addition to the sources already cited, see E. D. Pellegrino and D. C. Thomasma, *For the Patient's Good: The Restoration of Beneficence in Health Care* (New York: Oxford University Press, 1988); L. Palazzani, "Bioetica dei

Situations of Conflict and the Principles for Resolution

Ethical experience has nothing to do with mathematical formulas. Moral experience has to do with historical situations and subjectivity: conflicts in judgment and perplexities regarding the proper course of action arise in even the clearest consciences.

It is debated whether this possibility of conflict is due to value conflicts or assessment difficulties. Catholic theology denies the possibility of truly irresolvable conflicts as a matter of principle, because that would mean admitting contradiction in God himself, who is at the same time the author of reality and of the moral law. The conflict is therefore due to limitation, imperfection, and influences that affect the conscience making the assessment. Protestant theology sees certain situations of irresolvable conflict as a symptom of humanity's sinful state. The discussion has a theoretical tone, however, because in reality conflicts of conscience and even serious perplexities arise in the sphere of medical decisions as well.[84]

The theological tradition, more specifically regarding moral questions, has developed and discussed some secondary principles that can shed light on some of these situations of conflict: the principle of the lesser evil and the principle of double effect. It will be useful here to say a few words recalling their proper meanings.

The lesser evil (the possible good)

Situations of conflict involving a choice between two evils—including acts of omission—cannot be generalized because we are not obligated to fulfill all duties simultaneously: in order to keep from omitting one, we would necessarily omit another one that could never again be recovered. Situations of conflict are fortunately rare, but they do exist. It is therefore important to find a principle of priority or hierarchy by which to clarify such situations.

We are first of all aided by a distinction that allows for an initial determination of precedence and hierarchy: the distinction between physical and moral evil. Since *moral evil* compromises the higher, spiritual good, and ultimately—more or less consciously—our relationship with God, when there is a conflict in a serious decision between a physical or material harm and a moral one, there is no doubt that the material good or goods must be sacrificed. Material goods are not just economic but also social (e.g., harmony with others or a job). When faced with the alternative of committing moral evil, physical life itself is objectively deemed a justified sacrifice (martyrdom). This is not equivalent to suicide because the blame falls on those who create the situation of conflict.[85]

When there is a question of two moral evils, the obligation is to refuse both, because evil can never be the object of a choice. This is also true even when a greater evil is brought about as a consequence of refusing the one that appears to be a lesser evil. An often-used example is ordering someone to commit a robbery or tamper with documents on threat of sexual violence or the death of other people. With all of the attenuating circumstances that must be considered from the subjective viewpoint, objectively the robbery should not be committed because it is an evil. If refusal to

---

principi e bioetica delle virtù: Il dibattito attuale negli Stati Uniti," *Medicina e Morale* 42, no. 1 (1992): 59–85.

[84] A. Günthor, *Chiamata e risposta* (Rome: Paoline, 1982), 1:437.

[85] Aquinas, *Summa theologiae*, I, q. 48, a. 6.

commit the robbery leads to a form of revenge entailing a more serious moral evil, then this would be imputable solely to the person deciding to commit the evil.

A simpler and more common example is that of a family physician or gynecologist who is faced with the dilemma of a patient requesting a prescription for contraceptives as an alternative to the future prospect of an abortion (a greater evil with respect to the contraception). The potential abortion, if it occurred, would not be imputable to the physician, particularly if the physician had informed the patient that both are wrong and that there are ways of avoiding both situations.

Then there is the possibility of being forced to choose between (and therefore undergo) one of two physical evils, one lesser and one greater. The guideline is clearly that one can and should normally prefer the lesser physical evil, whether it pertains to others or to oneself. There may be cases, however, in which a subject can legitimately choose a greater physical evil in view of a reasonable and proportionate motive of a higher order. For example, a person with a tumor might refuse painkillers, thereby enduring greater physical suffering, because he wants to remain clear-minded so as to speak with relatives or to have the opportunity to lend religious meaning to the suffering. This does not, however, deny the legitimacy of using painkillers in general.[86]

The principle of double effect (indirect voluntary)

Just as pharmacological treatment often involves side effects in connection with the directly intended and primary therapeutic effect, so in moral experience a good and sometimes even necessary action is bound up with foreseeable negative consequences.

Our everyday activity itself, which each of us carries out faithfully and diligently out of a sense of duty, can sometimes have negative effects on our health. It would be necessary to leave the world or condemn oneself to inactivity in order to avoid every possible negative repercussion. The treatises of moralists over the centuries have clarified certain principles in order to overcome doubts in situations involving an action with a double effect, one positive and the other negative, in an effort to avoid paralyzing perplexity on the one hand and Machiavellian unscrupulousness on the other.[87]

Here is a summary of the appropriate guidelines in these situations. Assuming that the complex action has no alternatives without the evil effects, it is licit to perform the action (or deliberately omit it) under the following conditions:

1. The act in itself, regardless of the consequent evil, is good or at least indifferent.

2. The intention of the agent is to achieve the good effect only.

3. The good effect is not achieved by means of the evil effect.

---

[86] Günthor, *Chiamata e risposta*, 1:435–440; E. Quarello, "Male fisico e male morale nei conflitti di coscienza," *Salesianum* 34 (1972): 295–318.

[87] Günthor, *Chiamata e risposta*, 1:530–534; R. Frattallone, "Persona e atto umano," in *Nuovo dizionario di teologia morale*, 936–952; S. Privitera, "Principii morali tradizionali," in *Nuovo dizionario di teologia morale*, 987–996; S. Leone, "'Divinum est sedare dolorem': Risorse mediche e implicanze etiche dell'analgesia," *Camillianum* 11ns (2004): 275–312; B. Lo and G. Rubenfeld, "Palliative Sedation in Dying Patients," *Journal of the American Medical Association* 14 (2005): 1810–1816.

4. There is a proportionately grave reason for allowing the evil effect to occur.[88]

As should be apparent, these guidelines or norms are based on the premise that evil can never be the object of a direct choice and a good end cannot be pursued through evil actions. This is the framework of a moral theory that assesses actions first and foremost based on the direct object of the act. It does not fit, therefore, with the theory of intentionalism or the idea of success at all costs.

For example, it is an action with a double effect if a physician operates on a patient with a tumor in the procreative organs, indirectly causing sterility (therapeutic sterilization). Another example is alleviating the acute pain of a patient with bone cancer, which could require the administration of morphine, leading to the negative effects of tolerance (requiring ever higher doses to relieve the pain) and a potential shortening of the subject's life and physical resistance.

It is obvious, without need for further explanation, that the aim should always be to avoid the negative effect—wherever possible—without seriously harming the person, thereby overcoming any link to the negative effect.

This discussion, addressed primarily to researchers and professionals in the biomedical sciences, will not address some of the notions found more specifically in moral theology such as epicheia and the *fundamental option*.[89]

---

[88] For an analysis of the conditions making an action with a double effect licit, see Günthor, *Chiamata e risposta*, 1:530–534. The topic will be addressed again here in chapter 10, "Bioethics and Abortion."

[89] *Epicheia* is a primarily legal concept based on the reasonable assumption that a specific case, never before encountered, would have been decided in a certain manner if the legislators had considered it at the time the law was written. This assumption is based on the idea of inquiring into the law. The *fundamental option* is a concept in more recent moral theology according to which the important thing to grasp or assess in the moral life is the underlying consistency of a believer's orientation toward God. Only the lack of this orientation or its rejection would be a situation of genuine "sin." This orientation could therefore coexist with actions that are not always consistent with such an opinion—because they are performed out of weakness or because of a doubtful conscience—without implying a rejection of God. According to this view, in the context of the fundamental option the conscience enjoys a certain interpretive freedom with regard to concrete situations and is not entangled by rigid norms. The danger in this approach is the risk of falling into a way of thinking in which values are independent of norms and teleology has nothing to do with deontology. In reality, the concrete action is the confirmation of the fundamental option and the norm should protect and guarantee it. Of course, one should not deny the pedagogical efficacy of the fundamental option in terms of direction for one's personality and an overall perspective on life. K. Demmer, "Opzione fondamentale," in *Nuovo dizionario di teologia morale*, 854–861.

Summary Outline for Chapter 5

Ethics of the good life

Two paths for normative ethics

There are two ways to develop a normative ethics:

by moral intuitionism

by deducing ethics from metaphysics.

Moral intuitionism consists of admitting that the moral norm is an immediate given that is not further justifiable; it opens the door to relativism.

Finalistic concept of reality

The eternal truth of Platonism is that every entity has an idea to realize by its activity, because reality is rational. But, as Aristotle maintains, in order to guide the activity of the entity, the idea must be immanent, the *form* of the thing itself; otherwise the world of things becomes a shadow, an empty appearance.

Further progress is made by Christianity, in which the forms of things, the rationality immanent in things is the reflection of ideas, of a rational plan of the world existing in the divine mind.

In order to have a basis for ethics, in other words, to know what man ought to be, it is not enough to know what ends are in fact proposed, but one must know what the end of his nature is.

The *end* is the *value*, not a thing, but the *quality* by which something is good, considered in its nature or essence and not in its actual being-there.

In order to recognize values, it is necessary to think of reality as being rational, created by an Intelligence.

The natural moral law

The natural moral law as the "way to the end":

According to Thomistic tradition, the natural moral law expresses the dynamism and purposefulness that pervades all creatures; it is the law of their activity, it is the way leading to the end.

The end is the good, *which "is not a thing, but is that* relation *which exists between something and the human condition*: a relation that allows man to realize certain aspects of his humanity in some way" (A. Pessina).

Hence the natural moral law draws its specific characteristics from human nature, of which it is the expression, as compared to the natural laws of other creatures. Since man is a reasonable, self-conscious being, the natural moral law is *known* and not merely lived. Man is free; therefore, the natural moral law is a duty and not a necessity (*Sollen* and not *Müssen*).

Essential unity of "nature" and "reason"

The natural law is part of and expresses the natural order. "However, this natural order is not an entity that man as a knowing and acting agent finds himself, so to speak, in front of. It is a natural order of which the same natural cognitive acts—the natural acts of practical reason—form a part. Thus one discovers a reason that is also specifically nature (a kind of *ratio ut natura*). It is for this reason that the natural law can really be called 'inside man' and that one can say that it is 'engraved in his soul'" (M. Rhonheimer).

Knowledge of the natural moral law

We know the moral law *with our reason*, by reasoning (we are individual substances of a rational nature). Since knowledge of the moral law is obtained through reasoning, it can lead to error. Knowledge of more general precepts is immediately evident, yet the application of these principles to specific questions is subject to error; this is even truer in applying them to concrete cases. Medieval scholars called the knowledge of first principles of the moral order *synderesis*, whereas the application of those principles to a concrete case is called moral *conscience*. Conscience is the most precise knowledge—applied to the concrete situation—of the finality of our nature, which reveals itself to us in the more general principles of the natural law. It is therefore wrong to think of the moral law and conscience as two criteria of morality standing side by side; rather, it is the same criterion—ordering to the end—that is manifested first generally and then more precisely.

Two metaphysical presuppositions for normative ethics

The metaphysical doctrines that are at the foundation of ethics are

the finalistic conception of reality

the affirmation of human freedom

Human freedom

According to the Thomistic tradition, *freedom proceeds from the intelligence and the will*: a free choice is an act of the will "informed" by the intelligence.

F. Botturi distinguished three meanings of freedom:

1. *Freedom as self-determination*

   According to this perspective, acts of the subject are measured only in reference to themselves; this presupposes perfect autonomy.

   The *flaw* with this idea, however, is that the human subject not only has the power to choose, but also the *need* to do so: man cannot just choose for the sake of choosing; rather, he must choose in order obtain—if possible—certain goods, certain realities that he needs and are advantageous to him.

2. *Freedom as self-realization*

   Freedom means a *journey toward the fulfillment of the agent*, and therefore has the connotation of *liberation* from the poverty and slavery of imperfection, lack of fulfillment, and so forth.

Freedom means *adherence to the good*, fulfillment, and *liberation* of the subject. On this level, there is a significant *paradox*: the *linking of necessity and freedom*. One is drawn toward the good, the object of free choice, by a certain need for it; the attainment of this necessary good in turn frees us.

3. *Freedom as a relationship*

Freedom is also a *relationship with others, and more precisely with another freedom*. Not only must freedom take another freedom into account, but, more profoundly, freedom is a constitutive requirement for another's free existence.

Man exists (ex-sistere) in virtue of the recognition he receives. He needs recognition not in order to be a person, but in order to *exist as a person*—that is, in order to fully activate his emotional and intellectual capacities and reach a strong and stable sense of his own identity. Recognition thus becomes the relationship through which the subject identifies and affirms himself.

An *integral definition of freedom* could be "*dependent self-mastery* and *deliberate fidelity*" (F. Botturi). It combines an idea of autonomy wherein one accepts one's dependence on others and on the good, and an idea of choice that is fulfilled in fidelity to the freedom of others together with what attains the good of the subject.

The need for moral formation

It only makes sense to speak of "moral formation" if freedom is not understood as pure initiative and exclusive autonomy. In fact, the common element in the different meanings of freedom is the *coexistence* of *activity* and *passivity* within it, and freedom is an active ability only if it recognizes that it does not come from itself.

Moral formation is not so much training for a certain *conduct* as formation of a moral conscience, understood as an education in the truth of freedom and the desire for it.

"A proper moral formation is therefore directed not toward a morality of law but toward a morality of fulfillment or, according to R. Spaemann's expression, of a 'successful life'" (F. Botturi).

Contemporary ethics "without substance"

Recurring aspects

The isolation of ethics both from metaphysics and from religion.

The growth of the procedural and abstract element to the detriment of the problem of good and virtue.

A focus on problems of public ethics while disregarding those involving the ethics of the individual subject and the problem of the relationship with the Absolute.

The inability to overcome the central postulate that *ends and values are irrational* (Hume's law).

The separation of freedom from the good with an emphasis on the former: in the moral act the freedom with which it is performed is considered more important than the perfection of the act itself.

The eclipsing of the good in connection with *the crisis of final causes* among modern thinkers and the associated emphasis by the sciences on efficient causality.

The separation of deontology and teleology. It is necessary to recall that the deontological perspective of norms cannot be primary—ethics cannot be made to revolve exclusively around the idea of norms and laws which, instead, must be based on a teleological ethics of the good and of the good life. In the development of moral science, the concept of a norm is never primary; of the predicates "good" and "obligatory," the former is more primary.

Above all, the separation of the doctrine of freedom from the doctrine of the good makes it futile to justify freedom of choice itself, since it is possible to deduce freedom of choice only if the will of the subject is necessarily attracted to the infinite Good: it consequently remains free with respect to the various finite goods—or rather, not necessarily attracted to them. According to an only apparent paradox, freedom of choice with respect to finite goods is rigorously deduced from its necessary attraction to the infinite Good (see Aquinas, *Summa theologiae* I-II, q.10, a.2). Contemporary philosophy of freedom (Pareyson) is tempted to tack all of reality, including God, onto freedom—this unknown and obscure abyss. *The inversion of the relationship between freedom and the Good leads to vertigo, positing the abyss of an irrational and blind freedom as the First Principle (Abgrund as Urgrund).*

Anthropological silence

There is a tendency to develop ethics *without reference to man,* in a way that is disconnected from anthropology, from the idea of man as he is and how he would have to be if he were to realize his essence and attain his end—that is, ethics without an idea of the nature and end of man. It looks to the subject, ignoring the person. There is a difference between the notion of subject and the notion of person: a subject, having an essence, exercises existence and action; person implies a certain degree of ontological perfection that manifests spirituality, knowledge, and freedom. *A person is a microcosm that acts by assigning ends to itself.* Contemporary ethics knows the speaking, dialoguing, communicating subject, but not the existing person with his or her individual substance, freedom, spirituality, and totality. It is necessary to combine ethics and anthropology and to recognize that ethics goes beyond the foundation of norms. Without a connection between the doctrine of man and the doctrine of norms, the latter seem unintelligible. No procedural practice can compensate for what lacks on the level of the real perception of the good and of man's nature.

Distortion of the notion of "transcendental"

Modern philosophy, setting the subjective perspective in opposition to the objective, has lost sight of the correct notion of transcendental, which is an attribute predicated of everything that exists. Transcendentals are universal modes of being in relation to the intending soul-mind-conscience; the transcendental quality of the subject-person-soul in general is always included in the transcendental of being. Modern philosophy proceeded to detach the soul-person from being and from others, turning it into a formal "self." But once the limitless intentional openness of the person is eliminated, it is difficult to establish any degree of openness to others, whose otherness will be constantly in jeopardy of being understood reductively (Possenti).

Discourse ethics (Habermas)

Presuppositions of discourse ethics

Discourse ethics arises from the "disenchantment" with the world manifested by Weber, the plurality of conflicting moral views of the world, and the adoption of the post-religious and post-metaphysical perspective. The work of reason directed toward understanding in the field of moral discourse is seen as a "last resort" worth pursuing for lack of anything better, though real results require the necessary and full-bodied support of the positive law in civil life and may be modest.

Characteristics of discourse ethics

Through theoretical and communicative means, discourse ethics intends to reformulate the Kantian doctrine on the problem of the foundation of norms. In the Kantian treatise, the central element to be explained is the prescriptive character of norms and commands, or rather their validity: norms that can be universalized according to the formula of the categorical imperative are valid.

According to Habermas's account, discourse ethics (1) emphatically distinguishes between deontological questions of justice and questions of living a good life, assuming a priority of justice over good that can only be constituted at the level of metaphysical thought; (2) defines practical reason as essentially procedural and formal; (3) reformulates the concept of theoretical truth and practical validity in terms of intersubjective universal agreement: "only such legal norms are morally justified that are arrived at in an agreement by all concerned" (Habermas), and the norm requiring the mutual recognition of the participants in the discussion entails the norm of recognizing all people as subjects; (4) discourse ethics, directed to norms and their justification in terms of public procedures and validity, entirely neglects the will, the virtues, and the teaching thereof. It provides a restricted, procedural, "weak" understanding of morality and of philosophical ethics, withholding any "substantial" contributions. It emphasizes formality and the proceduralist approach, which are far from substantial moral intuitions. The foundation of material norms is left up to "practical discourse."

Procedural reason

Comprehensive worldviews have dissolved in pluralistic societies and the morality of conscience does not provide a sufficient basis for natural law, which previously had a religious or metaphysical foundation. The necessary basis for legitimacy comes from procedural reason and from the democratic procedure by which laws and ethical systems are generated. In this manner, the democratic principle tends to have its own roots, independent of moral principles. "A significant impoverishment of the idea of the person is another part of post-metaphysical apriorism. The term 'person' itself disappears and is replaced with 'subject,' which is then defined through linguistic acts and intersubjective agreement" (V. Possenti). *Procedure and consensus* become the two criteria for legitimacy. The post-metaphysical assumption implies that no moral order preexists intersubjective decisions and that a just legal order must be ideologically neutral.

Intersubjectivity takes the place of universality

Since the supreme rule of ethical reason is intersubjective agreement on respect for what everyone wants, the form or procedure prevails over the content—that is, the latter adapts to the former. There is a *separation between values and norms* (*with their corresponding obligations*), and it is impossible to establish norms on the basis of values. Given this separation, the guiding criterion for the norm is no longer a value but an intersubjective agreement reached procedurally.

Proceduralism does not assist with the examination of substantial issues, such as human relationships and the question of moral responsibility.

Principles of bioethics

Principles of personalist bioethics

The principle of defense of physical life

Man's bodily, physical life is not something extrinsic to the person; rather, it is the *fundamental* value of the person. The body is coessential to the person, it is the *unique foundation* in which and by means of which the person is realized and enters into time and space, expresses and manifests himself, and develops and expresses other values, including freedom, sociality, and his own plan for the future. Only the total and spiritual good of the person is above and beyond this fundamental value. The sacrifice of bodily life could be required only if that spiritual and moral good were unattainable except through the sacrifice of life. In such a case, since it is a spiritual and moral good, it could never be imposed by other people—it must develop as a free gift. The right to life precedes the so-called right to health; on the other hand, there is a moral obligation to defend and promote the health of all human beings in proportion to their need.

The principle of freedom and responsibility

The right to the defense of life precedes the right to freedom. In other words, freedom must mean taking responsibility for one's own life first and foremost, as well as the lives of others. This affirmation is justified by the fact that one must be alive in order to be free; therefore, life is the necessary condition—for everyone—for exercising freedom.

The principle of totality or therapeutic principle

This is one of the fundamental and characteristic principles of medical ethics. It is based on the fact that human corporality is a *unitary whole* resulting from distinct parts, organically and hierarchically unified by a single, personal existence.

In order to apply the therapeutic principle, certain conditions must first be met: (1) the intervention in question is on an afflicted part, or a part that is the direct cause of the harm, in order to save the healthy organism; (2) there is no other way or means of correcting the condition; (3) there is a good and proportionally high chance of success; (4) the patient or legally responsible decision maker has provided consent. In these cases, it is understood that what is in question is not so much life as *physical integrity*, which is also a very important good inherent in corporality; it is therefore a personal value that can only be put at risk or harmed for the benefit of the higher good to which it is linked. The body is therefore not to be taken in an exclusive sense (ignoring the rest), but in an assertive and holistic sense—that is, considering the bodily good in the whole context of the spiritual and moral good of the person.

The principle of sociality and subsidiarity

This principle implies that all citizens work toward respecting their own lives and the lives of others as a good—not only a personal one but also a social one—and that they engage the community to promote the life and health all, promoting the common good by promoting the good of each individual. In terms of social justice, however, the principle obligates the community to guarantee everyone the means of accessing necessary care, even at the cost of sacrifices for the well-to-do. The principle of sociality melds with the *principle of subsidiarity*, whereby the community must help more where the need is greater.

Principles of North American bioethics ("principlism")

The basic points of "principlism" can be summarized in this way: (1) there are no intrinsic norms in medical practice that can guide decisions; (2) there are *four fundamental principles (beneficence, non-maleficence, autonomy, and justice)* that should guide medical actions; (3) such principles must be applied to specific situations in order to formulate particular moral judgments.

Its methodological simplicity was one of the main reasons for the success of the principlist paradigm: it offers even non-experts the possibility of a framework on the basis of which they can engage the various ethical problems encountered in medical practice.

The principle of respect for autonomy is considered foundational with respect to all the other principles, grounding considerations about informed consent, truthfulness to the patient, and the refusal of treatment. This principle, however, can be limited by the principles of beneficence and justice.

There could be a certain correspondence with the principles enumerated in personal-ist bioethics (the therapeutic principle, the principle of freedom and responsibility, and the principle of subsidiarity, respectively). It should be noted that *the personalist principles are interconnected by a fundamental personalist anthropology*, which ultimately refers to the integral good of the person as derived from an analysis of the characteristics connatural to his essence, whereas *the North American principles lack a coherent framework* and offer no clarification as to what should be understood, for example, by "the good of the person" or "individual autonomy."

The "principlist" paradigm, focusing its attention on the application of its principles in practice, runs the risk of forgetting moral experience.

Furthermore, the formulation of principles tends to foster a passive attitude of obedience, rather than active, morally committed behavior. It is not enough to apply some principles in the situation, but rather it is necessary that the agent understand the moral significance inherent in the act he is performing.

The formulation of principles without an ontological and anthropological foundation renders them sterile and ambiguous. A systematization and hierarchy is necessary in order to harmonize and unify their meaning. This can only happen by reformulating them and defining them within a unified ethical theory that has the human person as its ultimate criterion, from which several corollaries follow: respect for physical life and essential integrity, respect for freedom in connection to the responsibility of the person, therapeutic justification for medical interventions, and an understanding of the common good not as the good of the majority but as the sum of the good of individual persons. The validity and meaning of these principles would reemerge by reinterpreting them in a hierarchical, ontologically grounded arrangement. One would proceed from the principle of beneficence to the principle of autonomy, and finally to the principle of justice (in the event of a conflict in the application of each preceding principle).

Situations of conflict and principles for resolution

The lesser evil (the possible good)

Moral experience has to do with historical situations and with subjectivity, therefore conflicting judgments arise. Above all *it is necessary to distinguish between physical evil and moral evil.*

Since *moral evil* compromises the higher, spiritual good, and ultimately—more or less consciously—our relationship with God, when there is a conflict in a serious decision between a physical or material harm and a moral one, there is no doubt that *the material good or goods must be sacrificed.*

When there is a question of *two moral evils, the obligation is to refuse both, because evil can never be the object of a choice.* This is also true even when a greater evil is brought about as a consequence of refusing the one that appears to be a lesser evil.

The example most frequently given is that of a family physician or gynecologist who is faced with the dilemma of a patient requesting a prescription for contraceptives as an alternative to the future prospect of an abortion (a greater evil with respect to the contraception). The possible abortion, if it occurred, would not be imputable to the physician, particularly if the physician had informed the patient that both are wrong and that there are ways of avoiding both situations.

Then there is the possibility of being forced to choose between (and therefore undergo) one of two physical evils, one lesser and one greater. The guideline is clearly that one can and should normally prefer the lesser physical evil, whether it pertains to others or to oneself. There may be cases, however, in which a subject can legitimately choose a greater physical evil in view of a reasonable and proportionate motive of a higher order. For example, a person with a tumor might refuse painkillers, thereby enduring greater physical suffering, because he wants to remain clear-minded so as to speak with relatives or to have the opportunity to lend religious meaning to the suffering.

The principle of double effect (indirect voluntary)

In moral experience it often happens that a good and sometimes even necessary action is bound up with foreseeable negative consequences.

Assuming that the action has no alternatives without the evil effects, it is licit to perform the action (or deliberately omit it), even when this choice involves a harmful effect, under the following conditions:

1. The act in itself, regardless of the consequent evil, is good or at least indifferent,

2. The intention of the agent is to achieve the good effect only,

3. The good effect is not achieved by means of the evil effect, and

4. There is a proportionately grave reason for allowing the evil effect to occur.

The presupposition of these guidelines is that an evil can never be made the object of a direct choice and that a good end cannot be reached through evil actions.

For example, it is an action with a double effect if the doctor operates on a patient with a tumor in organs connected with procreation and indirectly causes sterility (therapeutic sterilization).

Another example is alleviating the acute pain of a patient with bone cancer, which could require the administration of morphine, leading to the harmful effects of tolerance (demand for ever higher doses to relieve the pain) and a possible shortening of the subject's life and physical resistance.

Chapter 6

BIOETHICS AND MEDICINE

The Complexity of Medicine and Ethical Convergence

The term *complexity* is used today in the experimental sciences, including physics, to indicate certain factors and relationships of interdependence within which a particular datum or, more precisely, a particular event is situated.[1]

In the contemporary historical period, medicine includes various areas of interest: scientific research, to which the future physician's education is linked; the development of technology to support it; the management of social services; and the provision of health care itself, personally administered by physicians and their coworkers (nurses and technicians).

The discussion becomes far-reaching and fascinating if the historical development of all these aspects is examined from a diachronic perspective—that is, in terms of changes throughout the history of medicine—because it situates the stages of development and progress within the cultural conceptions and social developments characterizing the historical development of the West. It would also be interesting to examine a realm that has remained untouched by this sort of historical development and is today the subject of research under the name *alternative medicine*.[2]

From the bioethical perspective, however, it is possible and necessary to find a unifying thread or point of interest among the various sectors where the ethical question joins them together, even though it may initially emerge from one or another of the contributors mentioned: science, technological development, service management, and the therapeutic or health care relationship between physician and patient.

---

[1] P. Quattrocchi, *Etica, scienza, complessità* (Milan: F. Angeli, 1984); E. Morin, "La via della complessità," in *La sfida della complessità*, ed. G. Bocchi and M. Ceruti (Milan: Feltrinelli, 1985).

[2] For a description of the various periods in the history of medicine, particularly in terms of scientific aspects, see the sections concerning medicine in E. Agazzi, ed., *Storia delle scienze* (Rome: Città Nuova, 1984). See also P. Lain Entralgo, *Historia Universal de la Medicina*, 8 vols. (Barcelona: Salvat, 1972); G. Montalenti, "Storia della biologia e della medicina," in *Storia delle scienze*, ed. N. Abbagnano (Turin: UTET, 1962), 3:1; L. Premuda, *Storia della medicina* (Padua: CEDAM, 1960); idem, *Metodo e conoscenza da Ippocrate ai nostri giorni* (Padua: CEDAM, 1971); L. Stroppiana, *Storia della medicina tra arte e scienza* (Rome: Edizioni dell'Ateneo, 1982); W. F. Bynum and R. Porter, eds., *Companion Encyclopedia of the History of Medicine*, 2 vols. (London: Routledge, 1994).

It seems that this point of intersection on which the various ethical issues converge is the professional care phase—that is, physician-patient relationship. In effect, science, the management of technology and health care, and professional education have always had as their final goal and ultimate purpose the offering of support—in the person of the professional physician and his or her coworkers—to sick persons, whether by preventing or curing illness, rehabilitating patients, or accompanying the dying.

It is equally relevant, however, to emphasize the various perspectives of medicine and the complex origins of ethical questions. This is precisely because physicians, in the course of their practice, recapitulate the scientific, educational process yet find themselves in the midst of the psychological, social, and managerial factors that influence health care.[3]

I must repeat that I consider it very important to grasp the general outline of the development and direction of medicine within its "complexity," without entering into the particular historical stages that led to the emergence of specializations, which is beyond the scope of this manual. I will therefore refrain from tracing historical developments, as illuminating as they might be, since it would limit a more detailed examination of the present situation.

Medicine as Science and Ethical Demands

The development of medical science, particularly in recent years, has been marked by the progressive and accelerated growth of research sectors—with their respective methods of investigation—to the point that it would be more accurate to talk about *medical sciences* in the plural rather than simply *medicine.*

The progress of specialization has been especially rapid since the 1950s, both in terms of new branches of study and in terms of success and achievements. It should be added that some of the new areas of specialization, such as genetics, psychiatry, radiology, nuclear medicine, immunology, and so on, involve approaches and interpretative parameters that are neither always nor immediately connected with the traditional disciplines of anatomy, pathological anatomy, physiology, and medical pathology.

This well-known reality of the increasing subdivision and super-specialization of medical knowledge is a source of epistemological, educational, and ethical problems. The first thing that begins to evaporate is the overall perspective or holistic understanding of the patient and his or her personal history. Yet practicing medicine in a human way is not possible unless it is personalized. Occasional remarks in medical language highlighting the name of the "case" rather than acquaintance with the individual ("the gall bladder in bed nine") are a symptom of the fragmentation of the sick person and of medical knowledge.

A rethinking of teaching methods is incumbent upon those who teach the individual disciplines: perhaps addressing a topic from different scientific perspectives could help restore unity. The heart could be studied from the perspectives of anatomy, physiology, pathology, and so on, bringing the findings of the various specialized disciplines together on a single topic. This would be insufficient, however,

---

[3] P. Cattorini, "Terapia e parola: Il rapporto medico-paziente come nucleo essenziale della prassi medica," *Medicina e Morale* 35, no. 4 (1985): 781–799.

unless the interconnections with the whole of the organism and the person of the patient—where the various systems interact and merge into one—were kept in mind by the professors of the individual specialties and therefore by their pupils, who from their partial knowledge would have to compose an understanding of the whole. The greater the specialization and descent into particulars, therefore, the more urgent is the ascending path back toward unity. The didactic tools for synthesis are just as important as the ones for research and detailed analysis.

A pathological event such as a heart attack can and must be explained mechanically, just as a broken pump is explained, but it must also be connected with neuropsychological conditions and stresses to which the patient may be subjected in his environment. This means that data on triglyceride, cholesterol, and other levels are important, but the patient's dietary regimen and other conditions cannot be underestimated, nor can the individual's constitution and habits. An understanding of the whole is therefore required in order to practice medicine along personalistic and ethically acceptable lines.

Super-specialization also causes another difficulty: assigning to each patient *one* physician who can maintain the diagnostic and therapeutic dialogue and assume the resulting responsibilities. For every sick person there is a string of specialists, from the diagnostician to the physician who prescribes a treatment or performs an operation.

In this situation, which nonetheless has many advantages, it is also difficult to make the individual aware that he or she is in fact the subject and not the object of diagnosis and care. A patient finds it difficult to see himself as the principal agent, aware of and responsible for his condition (many tests will remain incomprehensible to him); he is therefore more inclined to seek the help of a professional, to whom he entrusts his life and health.

In the extensive battery of chemical and radiological analyses and specialized consultations that occur in general hospitals, it is easy for patients to abdicate their own responsibility. The work of the primary care physician therefore becomes even more necessary in order to remedy this shattering of the patient's inner integrity. The primary care physician or family physician interprets or helps interpret the specific findings, develops a summary, and through dialogue—which requires even more effort—restores the patient's awareness of his own condition and his ability to make the decisions that no one else can make for him.

Super-specialization definitely has the advantage of increasing the amount of data but requires greater effort to synthesize it all into an overview of the object of the research (the illness) while maintaining the patient's awareness of unity and the duality of the physician-patient relationship. These factors have not only epistemological but also ethical importance, since ethics assumes a truth in the comprehensive sense ("this is the situation") and requires the possibility of agreement between physician and patient.[4]

Yet medicine as science reveals another more serious situation, also fraught with both benefits and risks, which is essentially implicit in what has been said thus far. The issue, which was already touched upon in the preceding chapters, is *scientific reductionism*. It will be addressed now at greater length.

---

[4] L. Villa, *Medicina oggi: Aspetti di ordine scientifico, filosofico, etico-sociale* (Padua: Piccin, 1980).

Reductionism can be either a scientific method or an ideology.[5] As a scientific method, it is the procedure whereby complex facts and phenomena are explained through simpler and, if possible, more elementary components. The ideal model for this method is applied in physics, where all natural phenomena are explained by invoking fundamental entities (atoms, electrons, nuclei, elementary particles, etc.) that are governed by rigorous mathematical laws and relationships.

In biology, this same method aims to explain every phenomenon pertaining to the life sciences through purely chemical and physical—that is, molecular— mechanisms. The application of reductionism to biology has led to tremendous and striking advances, especially after the discovery of the molecular structure of deoxyribonucleic acid (DNA). This discovery, which shed light on the genetic code, allowed for the occurrence of a radical reduction: all of the functions of all the types of cells in living beings could be explained with the genetic information found in the structure of DNA.

There are three central areas of interest in the progress made over the past thirty years or so: molecular genetics, the explanation of evolution through molecular theory, and the explanation of the nervous systems of animals and human beings. There are names and works that have left their mark on these areas of interest: Jacques Monod with his *Chance and Necessity*, François Jacob with *The Logic of Life*, and Jean-Pierre Changeux with *Neuronal Man*.[6] Changeux, a disciple of Monod, famously announced the end of a need for the spirit in the conclusion of the fifth chapter of his book: "At the theoretical level there is no longer any reason not to describe human behavior in terms of neuronal activity. It is high time for Neuronal Man to appear on the scene," for indeed, "what good is it to talk about a spirit? ... 'Man no longer has a need for the "Spirit"; it is enough for him to be Neuronal Man.'"[7] The explanation of the genetic code, the explanation of the origin of life and its evolution, and the explanation of thought are the major steps taken by current biological reductionism.

Man thus ends up being "explained" as a mechanical aggregate of parts: he is a human machine. Given this idea of man, it should not be surprising that modern science is currently attempting to produce life and achieve conception—perhaps even pregnancy—in the laboratory, just as a machine is produced in a factory.

There is no denying that the application of this method has led to great advances in biology and genetics, and even more are expected in the field of gene therapy, in

---

[5] B. Lamotte, "Le réductionisme: Méthode ou idéologie?" *Lumière et vie* 172 (1985): 5–19; H. Jonas, "Technique, morale et génie genétique," *Communio* 9, no. 6 (1984): 45–65; L. Ruiz de la Peña, "Anthropologie et tentation biologist," *Communio* 9, no. 6 (1984): 66–80; E. Sgreccia, "Il riduzionismo biologico in medicina," *Medicina e Morale* 35, no. 1 (1985): 3–9.

[6] J. Monod, *Chance and Necessity: An Essay on the Natural Philosophy of Modern Biology*, trans. A. Wainhouse (New York: Knopf, 1971), originally published as *Le hasard et la nécessité: Essai sur la philosophie naturelle de la biologie moderne* (Paris: Seuil, 1970); F. Jacob, *The Logic of Life: A History of Heredity*, trans. B. E. Spillman (Princeton: Princeton University Press, 1993), originally published as *La logique du vivant: Une histoire de l'hérédité* (Paris: Gallimard, 1970); J. P. Changeux, *Neuronal Man: The Biology of Mind*, trans. L. Garey (Princeton: Princeton University Press, 1997), originally published as *L'Homme neuronal* (Paris: Fayard, 1983).

[7] Lamotte, "Le réductionisme," 9, citing Changeux, *Neuronal Man*, 169.

the treatment of sterility, and perhaps in the discovery of the biological expression mechanisms for widespread, incurable diseases such as cancer.

But is the reduction of life to biophysical mechanisms a true explanation? Where does the leap from description to explanation occur? If man is nothing other than a neuronal machine, what "human" concept can we have of medicine, human suffering, and death? What ethical boundary might there be between one human being and another, between a physician and a patient? It is not surprising that Monod and Jacob were the first two signers of the 1974 Manifesto on Euthanasia.

The main critique of these positions comes from Karl Popper, who summarized his thinking about the "incompleteness of all scientific knowledge" according to the principles of verisimilitude and falsifiability in the following statements: (1) scientists must be reductionists, and (2) scientists must be reductionists only in their method, but there are no arguments in favor of philosophical reductionism.[8]

Popper bases his position on the fact that every scientific theory leaves unresolved margins or residues that are not integrated into the explanatory system. These are precisely what has allowed the reformulation of theories, replacing them with others and stimulating scientific progress in explaining the universe, which must always be considered an open question because of the inherent incompleteness of every science.[9]

Regardless of this incompleteness, which cannot help but astonish scientists, it must be added that the *description* of the elementary mechanisms at the origin of complex realities is not exhaustive; rather, it demands an *explanation* in terms of *meaning*, which is to say in terms of sufficient reason, primary efficient cause, and ultimate final cause. Without trivializing such a demanding topic, I must insist on one point: a house that has been constructed cannot be "explained" only by the bricks and the pattern in which they are laid one on top of the other. The problem is no longer merely scientific and descriptive; it thus becomes philosophical. This distinction holds and is all the more valid for the presumed identification of brain and thought, where the operative capacities for thought cannot be explained by neuronal function—as stated in the preceding chapters—without being related to the immaterial and spiritual character of their source, i.e., the mind or spirit.

This excursion into reductionism was necessary in order to understand and observe how the reductionist temptation is present in medicine not only in the scientific and preliminary research phase but also in the applied phase of health care. It happens any time the physician-patient relationship is viewed reductively, eliminating the "spirit," communication, dialogue, and the multidimensional rather than merely biologistic understanding of illness, pain, death, and treatment. There is no doubt that each of us projects his own idea of being and of human beings in general onto his actions. Thinking of life and man as machinery sets the stage for treating man like a machine.

---

[8] K. Popper's works on this subject include *The Logic of Scientific Discovery* (London: Hutchinson, 1968); *Objective Knowledge: An Evolutionary Approach* (New York: Oxford University Press, 1979); and *The Open Universe: An Argument for Indeterminism* (Totowa: Rowman and Littlefield, 1982).

[9] J. Ladrière, *L'articulation du sens* (Paris: Aubier-Montaigne, 1970).

The Temptation of Technology

There is no need to explain how much the progress of medicine owes, especially since the time of Galileo, to the technology developed in the diagnostic, experimental, therapeutic, medical and surgical fields. Some specialties would be unthinkable without the support of technology: think of microbiology, genetics, radiology, nuclear medicine and biochemistry. And a new technological era is now dawning in which technology is no longer just the enhancement of physical strength, whether manual or sensorial, but also intellectual enhancement through the applications of computer science. Revolutionary assistance can be expected in the field of scientific research and in health care services, but this resource also involves some risks and therefore needs to be supplemented and corrected.

There are those who think that the use of technology, by reducing diagnosis time and making it surer and more incisive, allow a physician to have more time for the relationship with the patient. Experience does not bear out such uncritical optimism, and the reason for it has been identified.[10] There is certainly no need to fall into a reactionary refusal reminiscent of the Luddites, nor to forget how much medical progress owes to the use of technology, but it is necessary to keep in mind what has often occurred in the cultural development of peoples: the invention of a tool not only modifies the conditions in which work is performed, but also ends up "inducing" a culture in response, giving rise to a different mentality and culture.

Think of what happened in the history of civilization with the discovery of tools for working the soil and subsoil or with the invention of machines. What we call agricultural and industrialized or urban civilization began with these technological methods, which changed earlier values and ways of life. Technological means ultimately express the relationship between man and nature, and changing the means establishes the conditions for changing the relationship. Technology entails a cyclical law: man creates the technology but the technology changes man. Returning to the field of medicine, the use of diagnostic equipment—which by its nature involves the characteristics of repeatability, standardization, classification, and data storage and retrieval—certainly implies a two-part epistemological operation: the compartmentalization of the diagnosis and the depersonalization of the disease. The disease has a history, an environment in which it emerged, and a subject in whom it lives and manifests itself not only through "data" but also through signs. In other words, diagnostic equipment can compromise the relational, holistic, and personal understanding of diagnosis.

---

[10] S. J. Reiser, *Medicine and the Reign of Technology* (London: Cambridge University Press, 1978). See the review by P. Cattorini in *Medicina e Morale* 35, no. 1 (1985): 235–237; M. Timio, *La storia tecnologica del guarire* (Rome: Borla, 1990); A. Pessina, *Bioetica: L'uomo sperimentale* (Milan: Bruno Mondadori, 1999); idem, "La relazione tra la ricerca biomedica, l'antropologia e l'etica filosofica: Appunti per una riflessione metodologica," in *Etica della ricerca biomedica: Per una visione cristiana* (Vatican City: Libreria Editrice Vaticana, 2004), 144–158; H. Jonas, *Technik, Medizin und Ethik: Zur Praxis des Prinzips Verantwortung* (Frankfurt am Main: Suhrkamp, 1985); J. Hersch, "Nuovi poteri dell'uomo, senso della vita e salute," *Aut aut* 318 (2003): 180–187; J. Habermas, *The Future of Human Nature* (Cambridge, UK: Polity, 2003).

Furthermore, technology can change the way a physician examines his patient. Being led to believe that he knows everything even before having spoken with or listened to the patient, the physician's manner may become more visual, objective, and distant. The sick person may be told, "Send me the test results!" In this way, the use of technology reduces rather than increases the time that physicians have for listening and dialogue. In other words, technology can create a mentality and a culture, and therefore a sort of unique reductionism in medicine.

Other ethical risks have been identified in this area. The first is a greater temptation to view everything in terms of *can* and *do*: technology allows for so-called *therapeutic stubbornness* and engineered attempts to manipulate human corporality (e.g., the proposal to replace organs that express personal identity, like the head and the gonads, through transplantation).

Finally, the link between the economic question and economic criteria for evaluating treatments cannot be hidden. On the one hand, manufacturing companies tend to continually update and perfect their equipment, causing earlier models to become obsolete in the market with the aim of making a profit; public health care spending increases exponentially where health care is socialized; and unsustainable costs lead to the total abandonment of certain categories of patients who are too costly to treat and care for. This collision of economic forces with medicine will be addressed again shortly.

I have to agree with Stanley Reiser's conclusion on this topic: physicians "must regard them all [technologies] with detachment, as mere tools, to be chosen as necessary for a particular task."[11] I would like to add that the more the use of technology increases, along with the accompanying mentality, the more the interpersonal relationship between physician and patient should intensify and become meaningful. This is not only because the equipment can sometimes make mistakes, but also because it does not tell us about the nature of the illness, much less about the human depth of the sick person.

Impact on Society

This issue has not yet been clearly defined in the scholarly literature. Journalistic reactions and the tensions felt particularly in societies at the forefront of innovation are expressed more clearly. The problem involves the ideologies present in the society and reflected in the laws, the management models provided by various nations and their impact on the freedom of physicians and citizens, and the established economic dynamics in health care spending and cost-benefit calculations. All of these aspects converge on the ethical and deontological side of the physician, who is the hinge connecting society and the patient, the state's interests and the patient's interests.

It can already be said that physicians are torn between two loyalties: their loyalty to the patient, to whom they offer their services and with whom they are bound by a contract having moral value and legal weight; and their loyalty to organized society, which is a society of goods and services that invests them with responsibilities. They are safeguarded by and subject to the law, which holds them responsible and even determines their compensation in many countries; the state therefore considers them

---

[11] S.J. Reiser, *Medicine and the Reign of Technology*, 231.

its functionaries and representatives. As the rifts between individual conscience and organized society grow, so does the physician's loyalty conflict. This brings with it the risk of losing his professional identity.

To the best of my knowledge, not a single sociological study has been done in Italy on what the feeling is in regard to this conflict in the professional conscience of physicians or on the reactions caused by changes in this area. The discussion will therefore be limited to examining the facts and problems in the most verifiable and perhaps most obvious way, which is already meaningful enough.

The ideologizing of medicine and laws is one disturbing factor that affects the role of physicians. *Ideology* can be defined as a plan or program that aims to acquire a certain power while ignoring any assessment of the objective and total good of persons. Ethics and ideology are antithetical terms. The reference point for ideology is the will for power and for the efficacy of a plan, whereas the reference point for ethics is the person in his total good and objective truth. Suffice it to cite the words of Friedrich Engels: "Ideology is a process accomplished by the so-called thinker consciously, it is true, but with a false consciousness." [12] According to this understanding the components of ideology are rooted in the intellect, which moves "either with an ineffective consciousness or a consciousness that is false because of its wrong motivation, or one that obfuscates the real motives, or one that is false because it places the primacy on pure thought to the detriment of the human and social being." [13] According to Friedrich Nietzsche, ideology is based on the fact that "the criterion of truth resides in the enhancement of the feeling of power." [14]

There are still other definitions of ideology, but the two factors that seem to characterize every definition are (1) the will to carry out a plan and (2) a plan that ignores or circumvents the question of truth. Ideologies—whether Marxist, bourgeois, nihilist, or otherwise—thrive in cultural and political debates within the society, and political parties are often—though not only—their vehicles and mediators with regard to the positive law of the state, which is where ideological abuses can enter.

According to the Thomistic definition, the positive law is *ordinatio rationis ad bonum commune promulgata* (an ordinance of reason promulgated for the common good).[15] Hence the law presumes the primacy of reason, and therefore of truth, and has the common good as its aim. Yet there were ideological laws during the Nazi period and there are other such laws even in regions with parliamentary democracies. Legislation making abortion legal is ideological because it betrays the truth about the human status of the unborn life and its aim is not to defend that life. Hence there has also been talk about *ideological medicine* as opposed to Hippocratic medicine, referring precisely to the exploitation of the medical profession as it is forced to carry out ideological aims, whether legalized or not, that are present in the society.[16]

---

[12] F. Engels to F. Mehring: London, July 14, 1893, in *Marx Engels: Selected Correspondence* (Moscow: Foreign Languages Publishing House, 1953), 541.

[13] I. Mancini, *Teologia, ideologia, utopia* (Brescia: Queriniana, 1974), 286.

[14] F. Nietzsche, *The Will to Power* (New York: Vintage, 1967), aphorism 534, p. 290.

[15] S. Vanni Rovighi, *Elementi di filosofia* (Brescia: La Scuola, 1963), 3:235–246.

[16] A. Fiori, "Medicina ippocratica, medicina ideologica, obiezione di coscienza," *Medicina e Morale* 25, nos. 1 and 2 (1977): 167.

Faced with the confirmed abuses of ideology in the law, and consequently in the medical profession, physicians are compelled to defend their consciences by conscientious objection. This is a physician's right and duty out of fidelity to his relationship with service to man as such and to human life considered as a value in itself, which is even superior to the patient's free will. The fact that a physician must resort to this form of self-defense denotes the presence of a conflict between persons and legal society, subjecting the physician to pressures that are often more than just psychological.

This is why deontological codes were formulated. They are sets of behavioral norms that guarantee citizens the impartial and nonmanipulable conscientiousness of physicians, protecting them from the pressures of special interest groups or political ideologies.

The reality is that even these deontological codes, since they must function within a pluralistic society and are in turn formulated by persons—the physicians themselves—who are also affected by ideological influences and various cultural trends, often contain clauses or formulations that do not guarantee the unexceptionable observance of the common good and the defense of human life always and in every case. This means there is still a gap and an open dialectic between ethical values and written codes of medical deontology. Codes of medical deontology certainly are and remain the palisades of the medical profession's autonomy from ideological and social pressures, expressing the right of the medical class to prioritize the service of patients. Despite its frequent mention in these normative codes, *ethics* nonetheless remains a more comprehensive perspective, less restricted by legislative interference, with the twofold function of justified reasoning and critical judgment. This is precisely because ethics has a direct reference to human values.[17]

The encounter between the medical profession and society also occurs at another level: health care service management. There are three major models of health care management in the world: the liberal model, the collectivist model, and the socialized medicine model.

The liberal model is based on freedom for private initiatives in the system of services (hospitals, ambulatory care centers, etc.), freedom for the patient and/or family to choose a physician, free determination of compensation, and government oversight to ensure legality and self-assessment. Switzerland and many countries in the Americas use this model.

In the collectivist model, typical of Eastern European and communist countries, health care—like education—is managed by the state, which organizes and manages services, and assigns physicians in accordance with local needs. Physicians are functionaries of the state. Citizens receive services free of charge but cannot choose their physician or hospital.

The socialized model, active currently in England and Italy, is based on the principles of public management of services, free and equal services for all, planned regional management of services, respect for contracted or recognized private initiatives, and freedom to choose a physician.

---

[17] See the Italian *Codice di deontologia medica* (code of professional medical ethics) of the National Federation of Medical and Dental Associations (Federazione Nazionale degli Ordini dei Medici Chirurghi e degli Odontoiatri, FNOMCeO), approved on December 16, 2006, http://portale.fnomceo.it/PortaleFnomceo/showVoceMenu.2puntOT?id=5.

It is not within the scope of this manual to delve into an analysis of the law that established the National Health Care Service (law no. 833 of December 22, 1978) in Italy,[18] nor to examine its actual application or shortcomings. As a quick summary it can be said that, based on this law, funding comes at the tail end of the list; in other words, the amount of money to be allotted for health care spending is determined after the fact, meaning that spending is based on fulfillment of the need or demand.

The need to ensure greater efficiency led legislators to profoundly rethink the system. This culminated in legislative decree no. 502 of 1992: Reorganization of Regulations in Health Care. It was later implemented with legislative decree no. 517 of 1993: Modifications to Legislative Decree no. 502. The latter established an annual ceiling for resources devoted to health care, favored competition between service providers, and promoted accountability and incentives among the many workers involved. The *unità sanitari locali* (USL), or local health care units, were transformed into businesses, and private structures were accredited to compete with the public sector. The cost-needs dynamic was thereby inverted: available finances determine the needs met.

Further modifications were made by the so-called *ter* reform with legislative decree no. 229 of 1999: Rationing the National Health Care Service. The right to health care was reaffirmed, and the process of regionalization and transformation into companies continued. The role of local structures was expanded, and *livelli essenziali di assistenza* (LEA), or basic care levels, were defined along with guidelines for monitoring the appropriateness of services and for accrediting health care structures.

The LEAs and their associated monitoring systems were further defined by the decree of the president of the Council of Ministers (DPCM) on November 11, 2001, defining basic care levels, and ministerial decree of December 12, 2001, regarding a system of guarantees for health care monitoring. A reductive interpretation of the basic care levels seems to prevail: they are considered a justification for reducing the allotment of health care services. Consequently, there is an increasingly widespread distinction between medical care entailing technological services—therefore warranting public funding—and personal services left up to families and volunteers.

The doorway to federalism in health care was opened through the promulgation of legislative decree no. 56 of 2000 on fiscal federalism and law no. 3 of 2002: Reform of Title V of the Constitution. This has allowed for the passage from a *welfare state* to a *welfare community*, which gives value to the development of autonomy.[19]

It should be noted that these laws are an attempt to reconcile the principles and values of individual freedom with the social nature of health care services. In this light, the pertinent points of the discussion that pose ethical questions are discussed below.

The collectivist model eliminates the freedom of both physicians and citizens. Only the bodily and functionalistic aspects of health care are considered. It is not the citizens who handle their own health or illness with the help of physicians, but rather the state that manages the bodies of citizens, ensuring their functionality, through physician functionaries and their services.

---

[18] R. Ziglioli, ed., *Riforma sanitaria e comunità cristiana* (Brezzo di Bedero: Salcom, 1979).

[19] See D. Sacchini, "Sistemi-sanità: Una lettura etica 'integrata,'" in *Etica e giustizia in sanità*, A. G. Spagnolo et al. (Milan: McGraw-Hill, 2004), 79–119.

In the liberal model, sociality and the principle of equal services are not always guaranteed. Those in greater need of care may have fewer opportunities for treatment. Hospitals and health care service organizations in general could turn into businesses supplying costly services. Physicians could be taken in by the dynamics of private profit and, if they lack a certain moral stature or ethical standards, they could become instruments of the highest bidders even for ethically unacceptable aims (sterilization, abortion, euthanasia on demand, etc.).

There are shortcomings with ethical repercussions even in the socialized model: physicians come to depend primarily on the public structure (though private practice is not precluded under certain conditions) and become the expression of the state bureaucracy. Health care services can also become bureaucratized and sometimes politicized.

Since management is inherently political, the danger of politicizing health care services and the resulting subjugation of health care to political parties are currently major problems for health care management in Italy. These problems can be corrected and resisted only by a strong ethical consciousness, shared by physicians and citizens, or by changing the law itself.

Society also affects the practice of medicine and influences the medical profession in another way: funding. The conflict between ethics and health care finances is intensifying. Health care costs continue to increase as the ideology of well-being turns citizens' health into the summit of well-being itself. All nations that currently manage the financing of health care are experiencing great difficulties with sustaining this increase in public spending. Health care is accused of causing a crisis in government finances, so remedies are devised. In the name of the so-called *cost-benefit principle*, some propose eliminating "unproductive" health care expenses, which may affect care for the most seriously or terminally ill.

> The acquisition of sophisticated means of treatment capable of preventing people from dying of illnesses that were fatal or incurable until recently involves a cost that prevents their widespread use, meaning that recovery and life have become so highly priced that society can no longer allow itself the luxury of sustaining them. It is also for this reason that one could dismally observe that the more medicine progresses, the more difficult it is to care for the sick. The inevitable conflict between society and the individual leads to the tragic moment of having to decide which patients should be left to die. The serious matter of the physician's social and individual responsibility then presents itself. I am convinced that the concept of individual responsibility cannot be replaced by the rather ideological concept of social responsibility; in doing so, the physician would be simultaneously a physician for persons and a physician for the society.[20]

There has been no lack of support for so-called "social" euthanasia based on the scarcity of economic resources. In this regard, society must be treated more than the

---

[20] A. Franchini, "Le grandi scoperte della medicina nel XX secolo," in *Storia delle scienze*, ed. E. Agazzi (Rome: Città Nuova, 1984), 3:388. A chapter in the second volume of this work is devoted specifically to topics pertaining to the relationship between economy, society, and health care management. See E. Sgreccia, *Manuale di bioetica*, 3rd ed. (Milan: Vita e Pensiero, 2002), 2:559–596.

211

ill person: by correcting the economic criterion of costs and benefits and replacing it with the proportionality of treatment criterion, it can adopt policies modeled on the duty to help those with greatest need for care. This is another topic that needs to be address more specifically later on.

In conclusion, it is apparent that the problem of the physician's dual loyalty—to the patient and to society—must be resolved. This means that the society itself must be understood to serve the person, including the neediest of persons. Physicians, who express this social service to the human person, can therefore become educators of society by preserving their fidelity to the person of the patient.

The Environmental Component

Environmental issues pertaining to the life and health of human beings were included in the description of the areas of bioethics provided in the first chapter of this book. In fact, no one can deny that medicine is strongly linked to the study of the social and environmental conditions of the physical world, including climatic conditions, whether or not the atmosphere is polluted, the presence of pathogenic chemical or biological factors in the surrounding world, the equilibria of animal species and their health, and the viral and microbiological agents present in certain living conditions.

It is common knowledge that politics and decisions on environmental matters have paradoxically led some people to thinking it is necessary to block demographic population growth in order to avoid atmospheric pollution. The idea is that the polluters are humans, and especially children. In this perspective, something that was always considered an issue of family ethics (number of children) turns into an issue of environmental policy. All this is mentioned to give some examples of how bioethics and environmental ethics overlap.

Since the environment is continually influenced by technological and industrial activities and the economic decisions of human beings, an environmental ethics relative to the equilibrium of the biosphere and the ecosystem is gaining adherents. Moreover, the first bioethicists—such as Van Rensselaer Potter and Hans Jonas— were well aware[21] of the environmental component of bioethics.

In my opinion, there are two tiers of ethical issues in this area. The first tier is of a general philosophical nature and consists in defining the underlying anthropology. The second tier entails identifying the particular bioethical problems involving the responsibility of society and of medicine in the environmental realm.

The first level of environmental ethics, focusing on the anthropological premise, is developing reflections on anthropocentrism and biocentrism. Indeed, it is necessary to determine first of all whether the equilibrium of the biosphere should be achieved by making man—along with the promotion of his living conditions and health—the central point of reference (anthropocentrism), or whether man should instead be considered just one among many animal species with the global biosphere as the central point of reference. Those who take the latter position ultimately end up affirming that environmental ethics must be an ethics in itself and not a part or a chapter of bioethics.

---

[21] A chapter in the second volume of this work is devoted to bioethics in the working world, which is certainly a significant part of environmental bioethics. See E. Sgreccia, *Manuale di bioetica*, 3rd ed. (Milan: Vita e Pensiero, 2002), 2:335–444.

In order to be consistent with my approach, I must declare myself in favor of anthropocentrism and the primacy of man. This does not mean that man should be a despotic ruler of the biosphere; rather, he must be the responsible steward of creation and of life for the good of man and of future generations, with the greatest respect for the biosphere compatible with that good.

Aims, Limits, and Risks of Medicine [22]

Respect for reality [23]

In medicine, as in every other field of learning, knowledge also depends on *respect for reality*, which is divided into at least three specific dimensions.

First of all, respect for reality implies respect for *empirical facts*: if the data contradict the expectations (for example, diagnostic expectations) of the researchers or if the calculations do not confirm them, then the initial working hypotheses must be abandoned or radically reformulated in the name of respect for reality.

Second, respect for reality implies respect for the multifaceted nature of various linguistic usages. In other words, it must be recognized that no dimension of language is absolute and respect for reality implies knowing how to recognize the intrinsic limits of every means of expression. *Medical paternalism* also involves the physician's attempt to force the patient to conform to medical language and abandon other linguistic dimensions through which the patient could best express an authentic sense of his or her suffering.

Third and finally, respect for reality implies respect for the relational dynamics through which reality is perceived. The modern formalization of knowledge seems to disregard the dimension of relationality as cognitively irrelevant, yet the criterion of the verifiability and falsifiability of theories entails testing and checking not only by the scientific community, but also more generally by all persons.

The disposition that is opposite to respect for reality is *ideology*, which is not only *false knowledge* but also *false consciousness*. Ideology falsifies reality because it does not respect it, and this lack of respect for what is real perpetuates the falsifications.

When knowledge truly respects reality it elicits gratitude:

> Taken in itself, gratitude has a revealing value: regardless of any subjective intention, those who are able to nurture authentic gratitude provide an authentic ontological testimony, revealing the intrinsic unity of truth and good within reality. Those who feel in their heart the duty to thank the physician for what he has done for them are grateful for both his skilled knowledge

---

[22] See the opinion published by the Italian National Bioethics Committee titled *Scopi, limiti e rischi della medicina* [*Aims, Limits and Risks of Medicine*], December 14, 2001 (Rome: Presidenza del Consiglio dei Ministri, Dipartimento per l'Informazione e l'Editoria, 2001), http://www.governo.it/bioetica/pdf/51.pdf, English abstract available at http://www.governo.it/bioetica/eng/opinions/abstracts/AIMS_RISKS_AND_LIMITS.pdf; Hastings Center, "Gli scopi della medicina: Nuove priorità," *Notizie di Politeia* 45 (1997); D. Callahan, *False Hopes: Why America's Quest for Perfect Health is a Recipe for Failure* (New York: Simon and Schuster, 1998); I. Cavicchi, *Ripensare la medicina: Restauri, reinterpretazioni, aggiornamenti* (Turin: Bollati Boringhieri, 2005).

[23] See F. D'Agostino, *Parole di bioetica* (Turin: Giappichelli, 2004), 69–76.

[*scienza*] and his conscientiousness [*coscienza*], and thank the scientist and the man respectively. Physicians who receive such thanks—and perceive its authenticity—will note that they are being praised not only for their learning, but also for the personal way they were able to apply it in that particular case. Under this aspect as well, then, ethics seems to enter necessarily into the logic of scientific knowledge not as an external or supplemental dimension but as an internal requirement.[24]

Medicine as relational knowledge[25]

According to Francesco D'Agostino, there is an irresolvable paradox and an inherent contradiction in medicine: "wanting to remove a contingent being like man from the intrinsic logic of his contingency."[26] Medicine was established as part of humanity's implicit and universal rebellion against evil. Yet at the same time illness and death are a constitutive part of human nature, and those who fight them know from the beginning that they are ultimately doomed to failure. The paradox of medicine is no different from the paradox of philosophy itself, "which is irresistibly drawn to taking categories that cannot be made dialectical and rendering them dialectical: contemplating the eternal on the basis of the contingent, being itself on the basis of existing beings, the unfathomable on the sole basis of the very limited strength of the fathomable."[27]

It is important not to lose an awareness of medicine's specific contribution to human self-understanding, because medicine studies man in a dimension that is *constitutive*, not accidental: illness. In this way, medicine effectively balances philosophy's ever-present temptation to think abstractly of man as *disembodied*. In this perspective, medicine fulfills a function that is not only practical but also epistemological. Precisely when it is underestimated, this function causes a sort of confusion about the very reasons for this discipline. In this way, medicine has contributed to the identification of a particular paradigm of man:

> Man is a being who acquires awareness of the finite structure of his own being only through confrontation with illness and death, and who only through reference to medicine acquires the further awareness that this finitude is not symbolic, mystical, or—worse yet—magical. Illness and death are not the sign of an impenetrable destiny or of his oppression by occult and malicious transcendent forces, but rather the sign of his own ontological createdness. Though it cannot be suppressed, this createdness can nonetheless be understood and confronted by reason (when formulated as medical knowledge) thanks to the specific dimension of relationality uniting physician and patient.[28]

---

[24] D'Agostino, *Parole di Bioetica*, 75–76.

[25] See F. D'Agostino, *Bioetica* (Turin: Giappichelli, 1997), 15–46.

[26] Ibid., 18; see also I. Cavicchi, *L'uomo inguaribile: Il significato della medicina* (Rome: Editori Riuniti, 1998).

[27] D'Agostino, *Bioetica*, 19; G. Cosmacini and C. Crisciani, *Medicina e filosofia: Nella tradizione dell'Occidente* (Milan: Episteme, 1998).

[28] D'Agostino, *Bioetica*, 21–22.

Acting on human subjects, medicine is inextricably *also* friendship and justice, and the patient is not a mere object of the physician's art but always and in all cases a subject in a therapeutic relationship. The well-known distinction drawn by Plato between physicians of free men and physicians of slaves is also enlightening in this regard.[29] Physicians of slaves, who do not speak with the patient, who hurriedly move around the city with the sole desire of maximizing their profit, and who are like tyrants toward their patients by forcing them to take unfamiliar or unknown medications, *are not authentic physicians*. Physicians of free men, in contrast, attempt to understand the cause of the illness before treating, ask patients and their friends about their customary lifestyle, and do not prescribe any medications before convincing patients of the appropriateness of the prescription.

Physicians of free men demonstrate a dialectic capacity in accordance with medicine, which is *relational knowledge* aimed at restoring health, where health is not definable in strictly biological terms. In fact, according to Georges Canguilhem, health is to be understood as a set of circumstances that enables one to live in confidence and security.[30] "'Confidence' and 'security' are 'hot' or polemical categories that are extremely difficult to combine appropriately in a technologically shaped and therefore 'cold' culture like our contemporary one, which aims to pedantically replace the naturalness of experiences (and therefore their relative unpredictability) with an absolute expectation that necessarily follows from the artificial reconstruction of experiences themselves."[31]

Medicine as the ability to restore[32]

In contemporary Western society, the boundaries of medicine extend beyond the simple dimension of care and are difficult to identify. The danger is that the only recognized limits are dictated by economics and the ethical questions are reduced to the problem of resource allocation alone.[33] Given a definition of health (a condition of perfect well-being) vague enough to cover all aspects of the human condition, it is difficult to establish the nature of medicine.

For a long time, the health care relationship was thought of and conducted as an interpersonal relationship only partly mediated by the physician's instruments. Today a great deal of treatment occurs through instruments that reduce personal interaction to a minimum. The achievement of *scientific* results is becoming increasingly important, but these are only knowledge results that do not always or necessarily correspond to the achievement of patient well-being. "Obviously this is knowledge that either potentially or actually has proximate or remote therapeutic consequences, but the centrality of the patient as a *concrete individual* (this patient here and now)

---

[29] Plato, *Laws*, IV. X., 720a ff.

[30] See G. Canguilhem, *The Normal and the Pathological* (New York: Zone Books, 2007).

[31] D'Agostino, *Bioetica*, 31.

[32] See H.G. Gadamer, *Enigma of Health: The Art of Healing in a Scientific Age* (Stanford: Stanford University Press, 1996); A. Pessina, "Il contesto culturale dello sviluppo della biomedicina," in *Etica e giustizia in sanità*, Spagnolo et al., 3–20.

[33] See the opinion published by the Italian National Bioethics Committee titled *Scopi, limiti e rischi della medicina*; Hastings Center, "Gli scopi della medicina."

inevitably fades in all of this, replaced by a more or less abstract universal (*the patient*, if not *the pathology* or *humanity*)."[34]

Medicine is practical knowledge that can be defined with the notion of *téchne*, or art, meaning the capacity to make or produce a work on the basis of a form of knowledge. But "[in this field] there is no 'work' produced by art, and no 'artificial' product. Here we cannot speak of a material which is already given in the last analysis by nature, and from which something new emerges by being brought into an artificially conceived form. On the contrary, it belongs to the essence of the art of healing that its ability to produce is an ability to re-produce and re-establish something. This signifies a special modification of what 'art' means, and one which is unique to the knowledge and practice of the physician. One can indeed say that physicians 'produce' health by means of their art, but this is not a very precise way of speaking."[35]

The value and meaning of the physician's art are not contingent on his plan, but on *submission* to the proper dynamic of nature. Restoring health does not mean "creating" something new, but allowing the organism to regain its lost balance. Though illness is just as *natural* as health, in the sense that it appears without man's desire for it, it is considered in terms of a *deprivation* of something that should be there; hence it is fought.

It is necessary to address the question of *nature* and the question of the *human condition* in order to define health and illness. In contrast to the *natural environment*, the human condition is characterized by both *naturalness* (its being "given") and the ability to plan for and integrate the ways of *nature* (that is, its being *free*).[36] The affirmation ultimately arises from this that man is *beyond* nature, insofar as he is able to *think of* and *shape* nature even as he participates in the processes of nature itself. "Modern science facilitating technological application does not understand itself as something that steps in to occupy the open domain yielded by nature itself or as something that must integrate itself into the entire process of nature. ... As a science this knowledge allows us to calculate and to control natural processes to such an extent that it finally becomes capable of *replacing* the natural by the artificial."[37]

There is a problem—neither medical nor scientific—regarding the understanding of *nature*. If nature is only what is known through the instruments of science, then it is difficult to understand why it should be normative. Otherwise, if the *nature* or *reality* of which man is a part maintains its intelligibility (the condition of possibility for the artificial), then the philosophical question about *limits* and *ends*, which man must voluntarily set for his actions, makes sense.[38]

The therapeutic relationship[39]

The therapeutic relationship is always difficult because it is an asymmetrical relationship governed by need and illness. Illness always has two sides: the third-

---

[34] Pessina, "Il contesto culturale," 7.

[35] Gadamer, *Enigma of Health*, 32.

[36] See Pessina, "Il contesto culturale," 10–11.

[37] Gadamer, *Enigma of Health*, 39.

[38] See Pessina, "Il contesto culturale," 16.

[39] See A. Pessina, "Operatori sanitari come agenti morali," in *Etica e giustizia in sanità*, Spagnolo et al., 21–37; G. Cosmacini and R. Satolli, *Lettera a un medico sulla*

person side corresponding to what the physician can diagnose in various ways, which can be represented objectively and impersonally; and the lived experience of the sick person, because illness entails a new way of existing and understanding one's condition.

Today this therapeutic relationship is often marked by impersonal practices, procedures, or medicines that reduce the time and space for a relationship to a bare minimum. In reality, those who turn to hospital structures today enter into a process in which it makes little sense to use the concept of *taking care*, understood as concern for others, attending to them as persons or sharing their expectations.

The contemporary culture has often decried the limits of medical paternalism in the name of patient autonomy. Responding to the potential abuses of medical paternalism by appealing to the principle of autonomy alone does not serve to reestablish equilibrium in the physician-patient relationship; instead it seems to create a condemnation to conflict in which two autonomies and two perspectives on illness battle one another. Proceeding along purely contractual lines is undesirable as well: the existential weight of illness—its ethical and anthropological meaning—leads to the impossibility of immediately translating the therapeutic relationship into the terms of a "contract" between specialist and patient.

> In fact, it is not only physical pain, dysfunction and disability that emerge in illness, but the experience of suffering as a question about the meaning of existence also enters in. Sickness exists only as an abstraction: what truly exists is a sick person, and relational values that are not exhausted by treating the organ or the dysfunction enter into play in the therapeutic process. Without wishing to overload physicians with demands and duties that go beyond the limits of their profession, the fact remains that in treating the human body, care is being given to a human being, a human person, whether we like it or not. Indeed, human corporality marks the continuity of each one's personal life, and any violence done to the human body is always a violence done to the human person as well. It is always necessary for a medical act carried out on another's body to occur in the context of a morally qualified relationship in order to prevent it from becoming an act of violence.[40]

Moreover, the concept of respect for patient autonomy cannot be reduced to respect for the *exercise of autonomy* by others. Recognizing autonomy means recognizing the value of the subject of autonomy, that is, the human being and person. It is often forgotten that the *conditions for exercising autonomy*, which are limited in time,

---

cura degli uomini (Rome-Bari: Laterza, 2003); S. Manghi, *Il medico, il paziente e l'altro: Un'indagine sull'interazione comunicativa nelle pratiche mediche* (Milan: F. Angeli, 2005); V. Masini, *Medicina Narrativa: Comunicazione empatica ed interazione dinamica nella relazione medico-paziente* (Milan: F. Angeli, 2005); M. Gensabella Furnari, *Tra autonomia e responsabilità: Percorsi di bioetica* (Soveria Mannelli: Rubbettino, 2000); R. Tatarelli, E. De Pisa, and P. Girardi, *Curare con il paziente: Metodologia del rapporto medico–paziente* (Milan: F. Angeli, 1998); G. Tuveri, ed., *Saper ascoltare, saper comunicare: Come prendersi cura della persona con tumore* (Rome: Il Pensiero Scientifico Editore, 2005).

[40] Pessina, "Operatori sanitari," 28–29; see also P. Bellavite, P. Musso, and R. Ortolani, eds., *Il dolore e la medicina: Alla ricerca di senso e di cure* (Florence: Società Editrice Fiorentina, 2005).

refer back to the *ontological conditions for autonomy*: that is, the fact that we are human beings and persons.

There are certain *moral postulates* in the art of medicine that cannot be eliminated without damaging medical practice itself: the *dignity* of the human being, *equality* among human beings, and *respect* for persons and for the will of others. These postulates, which are linked to the philosophical justification underlying the art of medicine, express the essential prerequisites for every interpersonal relationship that can be described as morally good. They can be expressed in various ways in different historical periods, but they remain an essential component of intersubjective relationality and are of interest to the art of medicine precisely because it is an irreducibly relational activity.

The relationship between physician and patient, in the changed cultural conditions of today, is mediated by a legal instrument known as *informed consent*. In principle, informed consent is not reducible to a simple informative paper that the patient (or patient's representative) has to sign: it constitutes the reference point for the physician-patient relationship, because interpersonal respect, concern for others, recognition of professionalism, and attention to dialogue must be practiced precisely with regard to what must be done here and now from the clinical point of view. However, informed consent is often used as a form of legal protection—there is an encounter between two autonomies, two wills, two rights, but no longer on the basis of trust. It has been observed that this model does not work because the therapeutic relationship can never guarantee its own results. For as much as medicine is capable of controlling, replacing, and supplementing biological and physiological processes, the possibility of failure through no fault always remains inscribed in the therapeutic relationship. Awareness of the condition of finitude and vulnerability that characterizes the human condition is important. This awareness is at the root of the relationship of trust and cannot be replaced by any contract.[41]

Impossible medicine and false hopes [42]

Medicine has become "impossible" in the current context because it sets goals that are not (economically) sustainable. This is the thesis of Daniel Callahan, who maintains that it is not just a question of making health care institutions more efficient and better managed, but also of reconsidering the fundamental values of modern medicine and especially the idea of limitless medical progress.

The meaning and aims of medicine should be thought of in a different way that can reconcile greater equity with economic compatibility. According to Callahan, it

---

[41] William F. May makes it quite clear that there are two very different fundamental socio-psychological models that can govern the physician-patient relationship: one based on the idea of *contract*, the other on the idea of *covenant*; one cold and formal, the other warm and involving. See "Code or Covenant, or Philanthropy and Contract," in *Ethics in Medicine: Historical Perspectives and Contemporary Concerns*, ed. S. J. Reiser, A. J. Dyck, W. J. Curran (Cambridge, MA: MIT Press, 1977), cited in D'Agostino, *Bioetica*, 40.

[42] See D. Callahan, *False Hopes: Why America's Quest for Perfect Health is a Recipe for Failure* (New York: Simon & Schuster, 1998); idem, *What Kind of Life: The Limits of Medical Progress* (New York: Simon and Schuster, 1990); A. G. Spagnolo, "La relazione medico-paziente nella sanità aziendalizzata," in *Etica e giustizia in sanità*, Spagnolo et al., 155–183.

is necessary to adopt a more measured approach to medical possibilities due to the economic impossibility of paying for everything that medicine can make possible.

The Physician-Patient Relationship

The crux of the ethical issues in medicine thus becomes ever clearer: the physician–patient relationship understood in terms of the physician's subordinate fidelity to the absolute value of the human person and the constant valuation and reevaluation of this relationship. This is what a personalist approach to medicine requires.

This topic will now be examined in greater depth so as to specify the nature of the ethical relationship and the demands and salient contexts in which it finds expression, and so as to define the fundamental characteristics of the medical act and of the patient's consent to it. Reference to the premises set forth in the preceding paragraph remains decisive in describing the existence and quality of the moral relationship between physician and patient.

The nature of the physician-patient relationship

Life and (secondarily) health are goods entrusted to the person: each person has the right and duty to watch over these goods responsibly; in other words, patients are responsible for their life and health but do not have arbitrary moral control over them. Instead, they have the duty to guard their life and to promote their health.

Physicians are professionals who are called, freely chosen, and accepted by patients (or at least summoned by the family or offered by the society) to help prevent or treat illness or to rehabilitate the subject's strength and skills, and they take on the role of providers of a qualified service (qualified service providers).

Patients, therefore, are always the primary agents (in the event of impairment, the duty to act passes to relatives or legal representatives) in managing their own health. Medical acts and medical examinations therefore become a synergic relationship.

In order to recover from or prevent harm to their autonomy, sick persons (or others acting for them) who have become aware of their state of health and its limits and recognize that they are not competent to address the illness that threatens them and diminishes their autonomy, take the initiative of turning to someone else, a physician, who is capable of helping them because of his training and experience in the medical profession. The patient remains the principal agent in managing his health.[43] The physician who agrees to help the sick person is also an agent, but not in the sense of someone who acts upon an object; rather in the sense of one who works together with a primary subject or for a determined aim. The fact that in practice many patients are passive and many physicians act like the only agent does not make it the proper structure of the relationship.

This does not mean the patient can request any action of the physician or that the patient can usurp the physician's rightful competencies. The conscience and competence of the physician remain intact and both parties are still obliged to act ethically. If the patient—the primary agent and person responsible but surely not the master of his own life or of the physician's conscience—demands illicit services

---

[43] J. F. Malherbe, "Médecine, anthropologie et éthique," *Medecine de l'Homme* 156–157, nos. 3–6 (1985): 11.

(e.g., euthanasia), the physician can and must refuse to provide them, just as he must refuse when in conscience and on the basis of his medical competence he does not feel he should proceed with treatments he considers unsuitable or harmful. It is therefore a question of a pact between two people, one of whom is the person primarily responsible for the initiative and the other of whom is more competent in the means of resolving the problem; it is a pact or contract that can be rescinded if one of the two parties no longer considers the other capable of sustaining the therapeutic action.

The *Charter for Health Care Workers* specifies the particular nature on which the interpersonal relationship in health care activity is founded: it is "'a meeting between trust and conscience.' The 'trust' of one who is ill and suffering and hence in need, who entrusts himself to the 'conscience' of another who can help him in his need and who comes to his assistance to care for him and cure him."[44]

Jean-François Malherbe applied Aristotle's categories of action to medical acts. The four causes proper to every action of change are: (1) the material cause, (2) the formal cause, (2) the efficient cause, and (4) the final cause.[45] In the case of a medical act, the material cause is the patient who presents himself to the physician; the material cause is therefore heteronomous—as opposed to autonomous—with respect to the physician. The formal cause is the parameter of normality to which the medical act leads back, and as difficult as this parameter is to determine, it is also heteronomous with respect to the physician. The physician's autonomy is manifested most at the level of efficient cause, though it is never absolute or solitary with respect to either the patient—who is called to cooperate—or the other health care workers (pharmacologists, analysts, specialists, nurses, etc.). The final cause is recovery from or prevention of illness, which is an aim that is obligatory for both patient and physician.

Having analyzed the medical act in its dynamic components and in its network of collaboration, the primary and general responsibility of the patient and the sectorial and specialized responsibility of the physician are evident. The physician's responsibility is subsidiary; however, he also has his own realm of autonomous responsibility.

The bioethical discussion of the medical act requires further reflection, especially concerning the range of values the physician should contemplate in his relationship with the patient. In light of the fitting clarification by Paolo Cattorini,[46] it must be said that the therapeutic encounter naturally progresses toward ever higher and ever richer levels. The first level is predominantly objective and bodily: the physician observes the affected organ or physiological dysfunction that impedes normal organic life; at this first level, the physician "narrows" his objective and "reduces" his attention to the particular object. That attention is then brought from this first level to the somatic whole within which the particular area of illness is inscribed: this is a holistic attention that allows an understanding of the particular within the organic whole. A diachronic and historical review of the subject's medical history then follows: the whole organism and the particular fact of illness have a history in which the etiology and pathogenesis of the illness are found. In the next step,

---

[44] Pontifical Council for Pastoral Assistance to Health Care Workers, *Charter for Health Care Workers* (Boston: Pauline Books, 1995), no. 2.

[45] Malherbe, *Médecine, anthropologie et éthique*, 10–11.

[46] Cattorini, "Terapia e parola. "

the physician turns to his science, gathering the facets of his knowledge and using them to evaluate the symptoms and objective data. The physician's mind processes this data with a view to making a judgment. If time allows, this judgment should be communicated to the patient, precisely because the patient is not an object but a subject—the primary subject of the process of sustaining life and health. It is during this phase that the physician's perspective is raised and extended to the point of understanding the sick person's psychology: the illness is one thing and the way in which that illness is experienced is another, in accordance with emotional, psychological, and spiritual states that can sometimes be unconscious and sometimes verbalized. This leads to the interpersonal relationship in which two free existences communicate. The importance of dialogue between physician and patient thus emerges: dialogue has an informative value, a therapeutic value, and a decisional value. It is a dialogue that sets two consciences before a good that transcends both: human life and the person with his or her values. For Christian believers this is also the place of the encounter with Christ, who is God made man: "Truly I tell you, just as you did it to one of the least of these who are members of my family, you did it to me " (Mt 25:40). In this sense the dialogue becomes broader, reaching from the sphere of "health" into that of "salvation" without losing anything of the scientific objectivity and human significance of the therapeutic act.

Concerning the aims of the dialogue between physician and patient, and without attempting to enter into a discussion of precepts or methodological details,[47] it can simply be stated that this dialogue must have the following aims in order to be ethically complete: (1) informative, (2) therapeutic, and (3) decisional. Each of these aims could be discussed at length, but they will be addressed later in the context of specific situations.

Suffice it to recall that informative dialogue must be understood in a dual and interpersonal sense and involves the duty of confidentiality as well as the difficult task of informing patients about their true health conditions, including the worsening or incurability of the illness. These are all ethically important aspects that cause the dialogue phase to be fraught with values and responsibilities. The technical as well as psychological preparation required for this dialogue must not be neglected: dialogue techniques are useful aids for good preparation. Ethical maturity is required of the physician so that, in the face of realities such as illness, death, and potential failure, no psychological dynamics will be set in motion in the physician that might activate defense mechanisms, such as flight or aggressiveness, which often harm and leave a mark on both patient and physician. Physician training in psychology is therefore an inherent requirement for practicing the profession, though this does not eliminate the need for the intervention of a psychologist or psychiatrist for patients in the case of specific psychotherapeutic needs.

The dialogue is therapeutic inasmuch as it is part of a therapeutic process; however, this requirement is intended to emphasize the therapeutic effectiveness of dialogue—when it involves both listening and speaking—in dealing with the

---

[47] C. Iandolo, *L'approccio umano al malato* (Rome: Armando, 1979); idem, *Parlare col malato* (Rome: Armando, 1983).

psychological component present in every illness and particularly in certain forms of neurosis.[48]

The significance of dialogue becomes increasingly ethical as it enters into the decisional phase, which must always be conscious of patients' primary right to manage their own health. The issue of informed, implied, or presumed consent—according to the given situation—arises in this context. These situations will be examined in the chapters on particular topics in bioethics dealing with specific therapeutic issues (transplants, experimentation, etc.). The need for consent, which will be addressed shortly, and respect for what are defined as patients' rights must be kept in mind at every stage of treatment.

The foundation of the medical act

The relationship between physician and patient has always been very complex since both diagnosis and treatment depend heavily on subjective factors. Nonetheless, from the origins of medicine to the beginning of the twentieth century, this relationship only raised real problems by way of exception—that is, only when physicians did not respect the rules or when, for one reason or another, patients lost respect for and confidence in those who cared for them.

At the beginning of the twentieth century, however, the situation was reversed: the conflictual character of the physician-patient relationship became the norm to such an extent that the physician–patient relationship was described as "troubled."[49]

It would be interesting to study the physician-patient relationship further from a historical point of view, but to this end the reader is referred to the works of experts.[50] The aim here is to consider this relationship from the ethical point of view, which is to say from the perspective of the ethical values and principles justifying the conduct of two moral subjects: the physician and the patient. It should not go unnoticed that in Hippocratic medical ethics the physician-patient relationship was based on the beneficence model:[51] the physician's fundamental duty was to relieve the patient of illness, suffering, and injustice by resolving to bring about the good of the patient. In the name of this good, the objective was often pursued authoritatively by the physician, whose primary responsibility was to make all decisions in the patient's best interest. It was very easy to fall into what has been called—with a negative connotation—*medical paternalism*. In this model, for example, the information given to the patient was formulated by the physician so as to convince and potentially even force the patient to "do what is best for him," even if this came at the cost of sacrifices that the patient may not have wished to take on at the time. In certain circumstances, furthermore, the physician could justify withholding certain

---

[48] V. Frankl, *The Doctor and the Soul: From Psychotherapy to Logotherapy* (New York: Knopf, 1965); B. Häring, *Free and Faithful in Christ* (New York: Crossroad, 1981), 3:56–69.

[49] E. Shorter, *Bedside Manners: The Troubled History of Doctors and Patients* (New York: Simon & Schuster, 1985).

[50] In addition to the work mentioned in the preceding note, see P. Lain Entralgo, *Il medico e il paziente* (Milan: Mondadori, 1969).

[51] T.L. Beauchamp, "The Promise of the Beneficence Model for Medical Ethics," *Journal of Contemporary Health Law and Policy* 6 (1990): 145–155.

information, manipulating the truth, and intervening without full consent in the name of the patient's best interest.

As Diego Gracia aptly recalls, the figure of the physician and his decisions had a sacred and therefore unquestionable character in antiquity.[52] The Hippocratic Oath does not talk about consent; however, some passages in the Hippocratic corpus recommend that the physician give instructions calmly, accurately, and completely. This was certainly not to receive authorization to prescribe, but rather to have the patients accept the instructions themselves. This teaching remained unchanged in medieval and modern medicine up through the first half of the twentieth century and the majority of deontological codes took inspiration from it.

The modern era, by contrast, is characterized by the affirmation of the principle of autonomy, which has been gradually incorporated into the deontological codes. In fact, medical ethics and health care policies were involved in a profound transformation immediately after the mid-twentieth century. Some medical, legal, and juridical events, such as the trend toward affirming certain civil rights (the right to health care and information, the right to not undergo experimentation without consent, etc.) that were promoted by social movements and opinion groups, constituted a new turning point in medical ethics, which was itself marked by the appearance of the new term *bioethics*, expanding the areas within its scope to include the whole realm of *bios* (life). The historical prevalence of the duty of beneficence then began to be challenged by a patient-centered ethics emphasizing patient autonomy. In the autonomy model, a medical act is judged posivitely not so much because it brings about the good of the patient, but rather because it results from the patient's free decision. Information is of no value here other than to allow the patient to make a decision, even if that decision is unreasonable with respect to the clinical condition.

It should be mentioned that Thomas Percival's fundamental work, *Medical Ethics* (1803), already began talking about the duty to inform the patient's friends of the prognosis, with the understanding that the patient should be informed only "if absolutely necessary."[53] Later, the principles of the American Medical Association (*Principles of Medical Ethics*, 1912) required that physicians facilitate the patient's understanding of the benefits that could derive from their professional capacity. The council of the same association continued in this direction later on (1969), making it the physician's duty to obtain the patient's consent any time new drugs were used. In the meantime the Nuremberg Code (1949) and the Declaration of Helsinki (1964) were also promulgated, requiring informed consent for experimentation. In the United States, a law was passed in 1957 making informed consent obligatory.

The medical act thereby comes to have this additional foundation: patient consent. The physician may intervene with the patient's informed consent and never against it. But can it be said that informed consent constitutes the foundation and sole basis of the ethical justification of medical acts? While it is true that a medical act ordinarily requires patient consent, it is also true that such consent has limits. This is because some patients do not have the capacity to consent, patients have a

---

[52] D. Gracia, *Fundamentos de bioética* (Madrid: Eudema Universidad, 1989).

[53] T. Percival, *Medical Ethics; or, A Code of Institutes and Precepts Adapted to the Professional Conduct of Physicians and Surgeons* (Manchester: S. Russell, 1803), chap. 2, §3.

duty to safeguard their own life and health, and society has the duty—which also binds physicians—to preserve the life and health of its citizens.

On the topic of society, it must also be said that the legal element is the third fundamental element of the medical act, albeit *ab extrinseco*. The medical profession is legally recognized and qualified and cannot be legitimately practiced without this recognition, even if the act is intentionally aimed at the good of the patient and has the patient's consent.

Furthermore, society and its competent authorities regard life and health as a social good. Defending this good is a duty that follows from, among other things, its direct and valuable consideration enshrined in the society's constitution and laws.

The ethical foundation for a medical act must therefore consider all three of these components or reasons: (1) the good of the patient as the objective and ultimate component, (2) the consent of both patient and physician as the subjective component, and (3) legal recognition as the social component and guarantee of its ethically required legitimacy.

The discussion about legal foundations and the various associated theories is a different though complementary one, for which the reader is referred to more specialized publications.[54]

Models of the physician-patient relationship

The two contrasting models of the physician-patient relationship emphasized in ethical reflection over the past twenty years have stigmatized and perhaps oversimplified that moment of unique encounter and interaction between two individuals, physician and patient, both with rights to claim and duties to fulfill. More recently, some authors[55] have maintained that the rigid opposition of the two models could be better articulated in four models, taking certain characteristics into account: the various objectives assigned to the physician-patient relationship, the physician's obligations to the patient, the role of the patient's values, and, finally, the significance attributed to the concept of patient autonomy.

The first model they describe is the *paternalistic* model, which is also called *parental* or *priestly*. Establishing the relationship with the physician constitutes the assurance that patients will receive whatever interventions best promote their health and well-being. It is up to the physician to identify the diagnostic and therapeutic interventions best suited to the aim, and at the appropriate time the physician selectively presents this information to the patient with the intention of obtaining consent. In extreme circumstances, the physician authoritatively informs the patient when the interventions will begin. This model assumes that there are objective means of determining what is best for the patient and therefore that the physician can make decisions with minimal patient participation. In practice, the physician acts as the patient's legal guardian and his decisions prevail over the patient's autonomy. It is

---

[54] On this subject, see G. Iadecola, *Consenso del paziente e trattamento medico-chirurgico* (Padua: Liviana, 1989).

[55] E.J. Emanuel and L.L. Emanuel, "Four Models of the Physician-Patient Relationship," *Journal of the American Medical Association* 267, no. 16 (1992):2221–2226; A.G. Spagnolo, "La relazione medico-paziente," 121–154.

apparent, however, that such a model can only be valid in emergency situations. In all other normal situations of diagnostic and therapeutic activity, it absolutely cannot be taken as the ideal model.

The second model described by these authors is the *informative* model, also called the *scientific* or *engineering* or *consumer* model. Interaction enables the physician to provide the patient with all of the information regarding the diagnosis, potential treatments, and the risks of each; afterward, with the necessary patient consent, the physician carries out the specific interventions requested. This model assumes that the facts are strictly distinct from values in determining the diagnostic and therapeutic measures to be taken. Assuming that the patient's values are well defined, only the facts are missing; it is therefore the physician's duty to provide the patient with all the information that will allow the patient to decide on one treatment rather than another. There is no room for the physician's values or what the physician thinks of the patient's values. As a technical expert, the physician has the duty to provide the most complete information possible, to maintain and update his knowledge and skills, and to consult other experts when necessary. The patient's autonomy therefore exercises tight control over the physician's decision-making process.

With the highly technological dimensions of today's professional practice, patients are increasingly led to expectations that focus on the technical quality of the intervention, which then further accentuates the *contractual* aspect of the patient's relationship with the physician as the provider of qualified services. All of this is often to the detriment of conditions that once favored a therapeutic relationship based on trust. In this model, the medical act concludes with the simple execution—suitable precisely on the technical and scientific level—of the patient's expressed requests. The only limitation is an act that conflicts with the conscience of the individual health care worker. When the patient approaches the physician with suspicion regarding respect for the contract, he runs the risk of demanding or receiving information from the physician not in order to decide on his own response to the illness, but rather to stipulate a safe and assured contract. Likewise, the physician runs the risk of providing redundant information so as to protect himself from the potential legal consequences of his interventions.

There are clear limitations in this model as well. First of all, the fact that physicians limit themselves to implementing the patient's decision-making authority not only rules out the possibility of attempting to thoroughly understand the values that motivate the patient's request for a particular intervention, but above all it prevents them from asking how the illness might conflict with those values. Furthermore, fear of imposing their own convictions prevents physicians from offering any advice to the patient; yet in practice, patients often ask for such advice. Once again the duty to remain strictly within the confines of their specialty and competence makes a relationship with physicians impersonal and dehumanizing. Finally, the very concept of autonomy in this informative model does not hold up on the philosophical level. In fact, such a concept would presuppose the existence of immutable values and knowledge in the individual; however, it is well known that reflection allows man to review and change his desires and preferences, and reflection is based on a decisional process in which the input of other persons can be of great assistance.

The third model described is the *interpretive* model. Here the role of the physician-patient relationship is to help patients reflect on values and give meaning to their decisions so that, in addition to providing information on the risks and benefits of individual interventions, physicians can help patients articulate their values and choose the interventions that best respect specific values. Patients therefore do not have predetermined or fixed values, and the meaning of their autonomy is a self-understanding that becomes increasingly clearer through the assistance of the physician-consultant. This model in itself is very interesting, but here again certain limitations must be noted: most significantly, the meager preparation of physicians in the field of counseling prevents them from effectively fulfilling this role. Furthermore, if physicians lead too strongly, some maintain that they run the risk of influencing patients, almost persuading them, which thereby limits their ability to judge for themselves.

Finally there is the fourth model, the *deliberative* model, in which the physician is supposed to act as a teacher or friend toward the patient, informing the patient of the clinical aspects and values entailed in each particular intervention. It is a truly moral journey on which the physician and patient should embark in order to reach a decision. The physician has a very active role in this dialogue, indicating what the patient should do and which methodology to use in order to reach the decision. Patient autonomy means being able to proceed with this moral self-realization, which is attained after having examined the various conflicting values and their involvement in the treatment to be applied. The major criticism of this model is that the physician does not have a privileged knowledge of the priorities of the different values. Furthermore, in the context of a pluralistic society in which incommensurable values emerge, it would be unacceptable for a physician to offer the patient a methodology for resolving value conflicts based on the physician's own hierarchy of priorities. Not to mention that patients go to a physician to receive help with an illness, not to engage in a moral debate or to reconsider their own values. In conclusion, this model—like the interpretive model—can unintentionally hide the danger of paternalism.

Despite the limits of this last model, the authors who devised it maintain—and it seems to me reasonably so—that it should be privileged over the others; they advance several arguments to justify the claim. First of all, the concept of a methodologically guided reflection on the decisions to be made is the ideal expression of patient autonomy, since an autonomy that presumes to call itself such based on arbitrary choice or the ability to influence a physician's decision cannot be considered complete and correct. Then there is our society's ideal image of a physician: someone who is not limited to providing technical information but who can advise, helping to make a decision by discussing the different values involved. As far as the risk of paternalism is concerned, it seems remote in practice: a physician's proper attempt to persuade does not entail imposition. Indeed, this is precisely where the role of patient autonomy comes into play: the final decision rests with the patient. It should also be said that the physician's personal values are important to the patient, since this is often the very reason why the patient has chosen that physician. The physician's proposal of certain values can even prove advantageous for treatment, as in the case of illnesses linked to certain lifestyle choices: AIDS, alcoholism, etc. Finally, the fear of potentially poor physician training for fulfilling this role in the decisional model can be overcome by appropriately structured university medical programs, which are currently dominated by the informative model that privileges compartmentalized instruction, lacking a holistic understanding of the patient.

The meaning of the patient's good

Many of the models just described reveal profound conflict, as seen, which contemporary bioethics often seems to resolve by favoring patient self-determination and autonomy. In reality, not all scholars agree on resolving the conflict by privileging autonomy as an absolute or at least prioritized value. An apparently autonomous choice by the patient that the physician believes to be in conflict with the patient's best interest should not constitute a "principle of abstention"; rather, it should lead the physician to consider the appropriateness of intervening as much as possible in order to restore the good of the patient, allowing the patient to recover his full autonomy. The person should therefore constitute the true foundation of the duty to promote the good of others while respecting their autonomy.

For this reason, Edmund Pellegrino and David Thomasma[56] broadened the concept of beneficence by founding it on trust between physician and patient (*beneficence-in-trust*): the common objective for both is therefore to act in the best interest of the other. In this sense, the patient's best interest is also achieved by acting so as to restore his ability to regain the autonomy that is threatened in some way by the illness. Dialogue—communication between physician and patient—thus becomes indispensable in order to create trust between them; it is therefore necessary for the patient to be able to express his expectations in relation to the illness. In this relationship, the patient can ask the physician to act in the interest of the patient's good and to inform him only when there are important or risky interventions to consider. On the other hand, when the patient fears that the relationship with the physician does not have or no longer has that element of trust that should characterize it, he can clarify in a written document what value he attributes to the medical efforts regarding his pathology. Indeed, that value cannot be assumed to be the same for all patients, and it has been rightly affirmed that "one cannot impose on anyone the obligation to have recourse to a technique which is already in use but which carries a risk or is burdensome. Such a refusal is not the equivalent of suicide; on the contrary, it should be considered as an acceptance of the human condition, or a wish to avoid the application of a medical procedure disproportionate to the results that can be expected, or a desire not to impose excessive expense on the family or the community."[57]

The patient's involvement in managing his own illness and the personalization (where possible) of treatment plans and health care protocols are therefore all objectives that should be pursued according to an ethics that looks to the dignity of the person, promotes the humanization of medicine, and strives to replace the paternalistic model with the model of beneficence based on trust.

A crucial knot nonetheless remains to be untangled in the decisional phase of the diagnostic and therapeutic intervention: differing interpretations of the good or the best interest of the patient. Acting for the good of the patient is the most ancient and universally recognized principle of medical ethics, but the various ethical theo-

---

[56] E. D. Pellegrino and D. C. Thomasma, *For the Patient's Good: The Restoration of Beneficence in Health Care* (New York: Oxford University Press, 1988).

[57] Congregation for the Doctrine of Faith, Declaration *Iura et bona*, May 5, 1980, part IV.

ries and persons involved in the decision (patient, family, state) can interpret this promise of good in different ways.

According to the analysis by Pellegrino and Thomasma,[58] there is certainly a biomedical good to consider that includes all of the effects of clinical interventions on the natural course of the illness. It is a good directly connected to the technical competence of the physician, who is morally bound to provide it since it is the instrumental good requested by the patient. However, two ethical errors can arise if the good of the patient is equated with the biomedical good alone. The first is making the patient victim to a medical imperative: if a certain procedure implies a physiological or therapeutic advantage, then it *must* be adopted. In this way, all of the patient's values and the ethical issues that could result from the intervention are ignored, reducing the ethicality of the intervention to merely technical correctness. The second error is related to the medical indication of an intervention based on the physician's judgment of the tolerability of the quality of life. It is clear, therefore, that the concept of the patient's good must also include the patient's perception of the good—that is, the patient's idea of his own good.

In order to be good in the fullest sense of the term, the decision must also conform to what the patient considers valid in relation to the circumstances and alternatives contingent upon his illness. When the patient is capable of expressing it, no one better than he can determine what his best interest is; and when a patient cannot express it, those who decide in his stead must adhere as faithfully as possible to what the patient would have chosen for his own good had he been able to do so. There is still another component of the good: the one that allows a patient to exercise the ability to reason in order to make decisions. Physicians bring about the patients' good when they act so as to free them from all obstacles that prevent them from making free decisions. All three of these aspects of the particular good of the patient are clearly related to the idea of a supreme good that constitutes the standard by which patients regulate their decisions. This type of good has an ontological nature and hence has objective content in some way. In making clinical decisions in situations of conflict, this ultimate idea of good—which is the least negotiable and often the least explicit—becomes the element permeating the decision.

The thinking of H. Tristram Engelhardt Jr. is in outright opposition to the idea of a metaphysically grounded good: he maintains that it is necessary to decide together what the good is when making a decision to do good.[59] For him, since profound metaphysical truth does not exist, the good is not a principle with solid content at all. Science is just a game that can be played together once all of the rules are accepted. Respect for the patient's good only means respecting an agreement, not the person in terms of any sort of substantial content. It is a respect devoid of content leading only to the affirmation that nothing can be decided unless all others have agreed. The cardinal virtue in the decision is therefore tolerance or pleasantness toward others, devoid of any predetermined content. It is clear how these two opposing views of the patient's best interest in decision-making lead to an unbridgeable gap, since the first remains in the realm of a rationalistic objectivism and the other is open to a clear and dangerous relativistic subjectivism.

---

[58] Pellegrino and Thomasma, *For the patient's good*, 78 ff.

[59] H. T. Engelhardt Jr., "La bioetica nell'era post-moderna," interview edited by C. Botti, *Notizie di Politeia* 7, no. 24 (1991): 9–16.

A problem emerges in the drafting of certain documents, so-called *living wills*, which record in advance the patient's wishes concerning treatment for his illness, especially during its final stages. As will be seen in the specific chapter on the topic, what raises perplexities in practice is not *whether* a patient can intervene in the decision-making process regarding the illness, but rather *what* can legitimately be the object of that expression of wishes and what the best ways of expressing it are.

Extent and quality of information and consent

The various types of consent will be discussed at this point: implied (or tacit), explicit, personal, represented, and presumed (*ex silentio*). Suffice it to say that at the moment when the physician–patient relationship is established, the patient gives implied consent for the physician to do whatever is within his competence for the patient's good. This is what happens on admission to the hospital: the diagnostic tests and subsequent treatments are implicitly requested and authorized. Nevertheless, this is not always sufficient either in ethical or deontological terms.

An important element to consider in obtaining patient consent is the extent of the information that must accompany and precede the request for informed consent. It is generally agreed that this information must be complete with respect to the treatment, effects, alternatives, associated risks, and potential complications. Completeness of information is the norm for diagnosis; however, when this involves informing a patient of a particularly grave situation, prudence is required in relation to the psychological consequences that the news could provoke. Prudence concerns the manner of informing the patient more than anything else; the more inauspicious the prognosis, the greater the prudence that must be used to communicate it.

In one of its documents, the Italian National Bioethics Committee (NBC) provides important directives that seek to reconcile two principles: patient self-determination and protection of the physician's diagnostic and therapeutic autonomy.[60] In particular, the guidelines of the NBC can be summarized as follows:

1. In the case of serious illnesses or complex diagnoses, the relationship must not be fleeting or momentary.

2. The physician should have some psychological training so as to be able to understand the patient's personality.

3. Information significant enough to cause serious concerns or imply inauspicious prognoses must be given with prudence.

4. Information given about the therapeutic procedure must allow for a substantial and objective understanding of the situation, avoiding overly detailed facts or data that could lead to confusion in terms of basic comprehension.

5. Recommendations by relatives aimed at hiding the truth are not binding for physicians. The truth must be prudently stated in order to allow the patient to make his own decisions in matters that are important to him and

---

[60] Italian National Bioethics Committee, *Information and Consent to Medical Treatment* (Rome: Presidenza del Consiglio dei Ministri, Dipartimento per l'informazione e l'editoria, 1992), http://www.governo.it/bioetica/eng/pdf/Informazione_e_consenso _all%27atto_medico_2.pdf.

to others. Parents or legal representatives are not generally substitutes for the patient.

6. The attending physician at a public health care institution, or the physician coordinating diagnosis and treatment in any case, has the responsibility of informing the patient.

7. When the type of consent becomes particularly relevant, it should normally be written.

8. The written form of consent is particularly important when a representative's consent is necessary for someone who is incapable of providing consent (a minor or an incompetent individual).

Consent, whether written or verbal, does not entirely free the physician of the professional responsibilities that precede or extend beyond the request for consent.

Let us conclude by considering some extreme cases. The preceding indications include all of the problems that arise in the normal physician-patient relationship, both when the patient is capable of consenting and when the patient is incompetent but can be represented. There are some unique situations, however, in the case of urgency and imminent danger to the life of a patient who cannot give consent at that moment and in the case of a deliberate refusal of treatment that could result in serious danger to the patient's life.

If there is an imminent and serious danger to the life of the patient, the physician is authorized to intervene without consent in the case of minors, mentally handicapped individuals with impaired reasoning, or patients in a state of unconsciousness. If time allows, the physician nonetheless has the duty to request the legal guardian's consent in the case of minors; if it is refused (e.g., Jehovah's Witnesses refusing a blood transfusion for their child in imminent danger of death), the physician must appeal to the judicial authority.[61]

In the case of an unimpaired, adult patient who refuses medical treatments, the physician cannot consent to acts of euthanasia (interruption of efficacious and proportionate treatments or the performance of active euthanasia) because he cannot act against the life and good of the patient. However, the physician can request a consultation and attempt to make the patient aware of both his duty to seek and accept appropriate care and the consequences of refusal. If the patient persists, the physician cannot force him but must request to be released from his own responsibilities, alerting the proper health care authorities and (where required by law) legal authorities, because they have the duty to defend the lives of citizens.

The mandatory treatment of mentally ill patients is a separate issue. This scenario has a precise legal configuration and the procedures are regulated by law. In principle, however, the ethical approach is always to seek the mentally ill person's assent, which would rule out any physically coercive treatment even in this case.[62]

---

[61] See G. Perico, "Testimoni di Geova e trasfusioni di sangue," *Aggiornamenti sociali* 5 (1986): 323–336; S. Finfer et al., "Managing Patients Who Refuse Blood Transfusions: An Ethical Dilemma," *British Medical Journal* 308 (1994): 1423–1426.

[62] On this topic see G. Iadecola, "I trattamenti sanitari obbligatori," in *Potestà di curare e consenso del paziente* (Padua: CEDAM, 1998), 71–75; L. Eusebi, "Sul mancato

The Call for Virtue in the Physician-Patient Relationship

When faced with the potential conflict between what physicians think should be done—in light of their skilled knowledge (*scienza*) and in good conscience (*coscienza*)—and what patients might request in the name of autonomy, a physician-patient relationship based on an ethics of rights and duties alone may prove problematic. The specificity that this theoretical basis seems to promise actually becomes illusory in attempts to reach an agreement about the right and good thing to do in given circumstances. So it is the concept of virtue, though it has become worn down, that is still the essential reality at the core of moral transactions and therefore in the physician-patient relationship.

The call for virtue and for ethical and spiritual qualities in physicians and health care workers in general is a topic of debate mentioned in the philosophical discussion on the paradigms and principles of bioethics. This debate has developed especially in the United States[63] and has led to a reassessment of the ethical virtues of physicians, returning to the original philosophy of medicine that would lead physicians (and health care workers in general) to act virtuously—in other words, so as to attain the values intrinsic to the action, realizing their specific end.

In this way, the physician's ethical profile as such can be concisely summarized in the list of previously mentioned cardinal virtues proper to every responsible human being, synthesizing nearly everything that has been explained in this chapter. This concept derives from what the Second Vatican Council proposes to all lay persons active in temporal affairs and civic duties in the pastoral constitution *Gaudium et spes*; these demands also apply to the medical profession.[64] They can be easily enumerated as follows: (1) scientific and professional competence, (2) awareness of values, (3) consistent behavior, and (4) collaboration. It will not take many words to explain these requirements, which are evident from a theoretical perspective.

Competency marks the primary ethical demand regarding the specific scope of the profession, which includes the complexity and unity of medical knowledge mentioned at the beginning of the chapter. Since this competency is becoming increasingly specialized and compartmentalized, physicians will need to make an effort to integrate, stay up-to-date, and continue their education, while at the same time knowing how to benefit from the specialized competencies of others.

The awareness of values is related to the underlying anthropology that the physician has assimilated. As evidenced, there is always a temptation to think of human beings reductively. Yet everyone is aware that the values at stake in this profession are those of the human person and, within the person, the values of life and health in the broad and harmonious meaning here presented. It is clear that as a health care professional's awareness of values—which can be enlightened in a perspective of faith—becomes richer, his or her conscience will become more attentive and sensitive.

---

consenso al trattamento terapeutico: Profili giuridico-penali," in *Bioetica in medicina*, ed. A. Bompiani (Rome: CIC Edizioni Internazionali, 1996), 221–228.

[63] L. Palazzani, "Bioetica dei principi e bioetica delle virtù: Il dibattito attuale negli Stati Uniti." *Medicina e Morale* 42, no. 1 (1992): 59–85.

[64] Vatican Council II, Pastoral Constitution *Gaudium et spes*, December 7, 1965, nos. 41–43. See also E. Sgreccia, "Per l'esercizio cristiano della medicina," *Medicina e Morale* 29, no. 2 (1979): 161–190.

Yet ethics is not a purely speculative science, and even bioethics becomes active when it enters the medical field: ethical life unfolds and values are attained in the moment of action. The important thing to note is that this active character, when implemented with coherence between scientific competence and awareness of values, first of all makes the action in itself ethical; at the same time, it contributes to the enrichment of the personal being of the health care professional, the patient, and the community. If it is true that a society is determined by its values, and that a profession like medicine is fraught with responsibilities because of the values it necessarily effects, it is also true that those values must be actuated and embodied and not merely asserted.

Collaboration, which was emphasized in talking about the physician–patient relationship, must be extended to the whole group of persons involved in the course of a single therapeutic process: nursing personnel, relatives, various specialists, etc. Only this collaboration can prevent the dispersive fragmentation of the medical act, which would thereby lose its efficacy as well as its human meaning.

Needless to say, ethical refinement—like personal enrichment—is tested and sealed in professional practice. In the words of Karl Jaspers, who can hardly be accused of religious bias, "There is perhaps a certain analogy between the callings of the minister and the physician. In both cases, the practical aspect of the work takes precedence over theoretical knowledge, which is only an auxiliary. The future of the physician's art is not determined in the laboratories devoted to medical research, nor is the future of Biblical faith decided by academic theology. ... Today the crucial practical elements of the physician's art and of Biblical religion are passed over in silence, while medical research and theological speculations are loudly publicized. Thus, a kind of acoustic illusion is produced, which misleads as to the true state of affairs."[65]

---

[65] K. Jaspers and R. Bultmann, *Myth and Christianity: An Inquiry into the Possibility of Religion without Myth* (New York: Noonday Press, 1958), 35–36; see F. F. Casson, "Dignità della professione medica," *Federazione Medica* 37, no. 10 (1984): 936–941.

Summary Outline for Chapter 6

Models of health care management

There are *three major models* of health care management in the world: the liberal model, the collectivist model, and the socialized medicine model.

- The *liberal model* is based on freedom for private initiatives in the system of services (hospitals, ambulatory care centers, etc.), freedom for the patient and/or family to choose a physician, free determination of compensation, and government oversight to ensure legality and self-assessment. Switzerland and many countries in the Americas use this model.

- In the *collectivist model*, typical of Eastern European and communist countries, health care—like education—is managed by the state, which organizes and manages services, and assigns physicians in accordance with local needs. Physicians are functionaries of the state. Citizens receive services free of charge but cannot choose their physician or hospital.

- The *"socialized" model*, active currently in England and Italy, is based on the principles of public management of services, free and equal services for all, planned regional management of services, respect for contracted or recognized private initiatives, and freedom to choose a physician.

Critiques

- The *collectivist model* eliminates the freedom of both physicians and citizens. Only the bodily and functionalistic aspects of health care are considered. It is not the citizens who handle their own health or illness with the help of physicians, but rather the state that manages the bodies of citizens, ensuring their functionality, through physician functionaries and their services.

- In the *liberal model*, sociality and the principle of equal services are not always guaranteed. Those in greater need of care may have fewer opportunities for treatment. Hospitals and health care service organizations in general could turn into businesses supplying costly services. Physicians could be taken in by the dynamics of private profit and, if they lack a certain moral stature or ethical standards, they could become instruments of the highest bidders even for ethically unacceptable aims (sterilization, abortion, euthanasia on demand, etc.).

- Even within the *socialized model* one can find shortcomings with ethical repercussions: physicians come to depend primarily on the public structure (though private practice is not precluded under certain conditions) and become the expression of the state bureaucracy. Health care services can also become bureaucratized and sometimes politicized.

Nature of the physician-patient relationship

Life and (secondarily) health are goods entrusted to the person: each person has the right and duty to watch over these goods responsibly; in other words, patients are responsible for their life and health but do not have arbitrary moral control over them. Instead, they have the duty to guard their life and to promote their health.

- Patients, therefore, are always the primary agents (in the event of impairment, the duty to act passes to relatives or legal representatives) in managing their own health. Medical acts and medical examinations therefore become a synergic relationship.

- The patient remains the principal agent in managing his health. The physician who agrees to help the sick person is also an agent, but not in the sense of someone who acts upon an object; rather, in the sense of one who works together with a primary subject or for a determined aim. The fact that in practice many patients are passive and many physicians act like the only agent does not make it the proper structure of the relationship.

The *Charter for Health Care Workers* (1995) specifies the particular nature on which *the interpersonal relationship in health care activity* is founded: it is "'*a meeting between trust and conscience.*' The 'trust' of one who is ill and suffering and hence in need, who entrusts himself to the 'conscience' of another who can help him in his need and who comes to his assistance to care for him and cure him."

The importance of dialogue between physician and patient:

Dialogue has an *informative* value, a *therapeutic* value, and a *decisional* value.

This dialogue sets two consciences before a good that transcends both: human life and the person with his or her values.

Informative dialogue must be understood in a dual and interpersonal sense and involves the duty of confidentiality as well as the difficult task of informing patients about their true health conditions, including the worsening or incurability of the illness. The significance of dialogue becomes increasingly ethical as it enters into the decisional phase, which must always be conscious of patients' primary right to manage their own health. The issue of consent arises in this context.

Models of the physician-patient relationship

- The first model is the *paternalistic* model, which is also called *parental* or *priestly*. This model assumes that there are objective means of determining what is best for the patient and therefore that the physician can make decisions with minimal patient participation. In practice, the physician acts as the patient's legal guardian and his decisions prevail over the patient's autonomy. It is apparent, however, that such a model can only be valid in emergency situations. In all other normal situations of diagnostic and therapeutic activity, it absolutely cannot be taken as the ideal model.

- The second model described by these authors is the *informative* model, also called the *scientific, engineering*, or *consumer* model: interaction enables the physician to provide the patient with all of the information regarding the diagnosis and potential treatments, and the risks of each; afterward, with the necessary patient consent, the physician carries out the specific interventions requested. In this model, it is assumed the facts are strictly *distinct* from values in determining the *diagnostic and therapeutic measures* to be taken. Assuming that the patient's values are well defined, only the facts are missing; it is therefore the physician's duty to provide the patient with all the information that will allow the patient to decide on one treatment rather than another. There is no room for the physician's values or what the physician thinks of the patient's values. The patient's autonomy therefore exercises tight control over the physician's decision-making process. There are clear limitations in this model as well. First of all, the fact that physicians limit themselves to implementing the patient's decision-making authority not only rules out the possibility of attempting to thoroughly understand the values that motivate the patient's request for a particular intervention, but above all it prevents them from asking how the illness might conflict with those values. Furthermore, fear of imposing their own convictions prevents physicians from offering any advice to the patient; yet in practice, patients often ask for such advice.

- The third model described is the *interpretive* model. Here the role of the physician-patient relationship is to help patients reflect on values and give meaning to their decisions so that, in addition to providing information on the risks and benefits of individual interventions, physicians can help patients articulate their values and choose the interventions that best respect specific values. Patients therefore do not have predetermined or fixed values, and the meaning of their autonomy is a self-understanding that becomes increasingly clearer through the assistance of the physician-consultant. This model in itself is very interesting, but here again certain limitations must be noted: most significantly, the meager preparation of physicians in the field of counseling prevents them from effectively fulfilling this role. Furthermore, if physicians lead too strongly, some maintain that they run the risk of influencing patients, almost persuading them, which thereby limits patients' ability to judge for themselves.

- Finally there is the fourth model, the *deliberative* model, in which the physician is supposed to act as a teacher or friend toward the patient, informing the patient of the clinical aspects and values entailed in each particular intervention. It is a truly moral journey on which the physician and patient should embark in order to reach a decision. The physician has a very active role in this dialogue, indicating what the patient should do and which methodology to use in order to reach the decision. The major criticism of this model is that the physician does not have a privileged knowledge of the priorities of the different values. This last model should be privileged over the others. First of all, the concept of a methodologically guided reflection on the decisions to be made is the ideal expression of patient autonomy, since an autonomy that presumes to call itself such based on arbitrary choice or the ability to influence a physician's decision cannot be considered complete and correct. Then there is our society's ideal image of a physician: someone who is not limited to providing technical information but who can advise, helping to make a decision by discussing the different values involved.

The meaning of the patient's good

Pellegrino and Thomasma broadened the concept of *beneficence* by founding it on *trust* between physician and patient (*beneficence-in-trust*): the objective is to act in the best interest of one another. In this sense, the patient's best interest is also achieved by acting so as to restore his or her ability to regain the autonomy that is in some way threatened by the illness. According to the analysis done by Pellegrino and Thomasma, the following points should be considered:

- A *biomedical good* that includes all of the effects of clinical interventions on the natural course of the illness. This could lead, however, to the error of making the patient the victim of a medical imperative: if a certain procedure implies a physiological or therapeutic advantage, then it *must* be adopted. In this way, all of the patient's values and the ethical issues that could result from the intervention are ignored, reducing the ethicality of the intervention to mere technical correctness.

- *The patient's perception of the good*, the idea that the patient has of his or her own good.

- The good understood as the possibility of *exercising the ability to reason in order to make decisions.*

All three of these aspects of the particular good of the patient are clearly related to the idea of a supreme good that constitutes the standard according to which patients regulate their decisions. This type of good has an ontological nature and hence in some way has objective content. In making clinical decisions in situations of conflict, this ultimate idea of good—which is the least negotiable and often the least explicit—becomes the element permeating the decision.

Aims, limits, and risks of medicine

Respect for reality

In medicine, as in every other field of learning, knowledge also depends on *respect for reality*, which is divided into at least three specific dimensions.

- First of all, respect for reality implies respect for *empirical facts*: if the data contradict the expectations (for example, diagnostic expectations) of the researchers or if the calculations do not confirm them, then the initial working hypotheses must be abandoned or radically reformulated in the name of respect for reality.

- Second, respect for reality implies respect for the multifaceted nature of various linguistic usages. In other words, it must be recognized that no dimension of language is absolute and respect for reality implies knowing how to recognize the intrinsic limits of every means of expression. *Medical paternalism* also involves the physician's attempt to force the patient to conform to medical language and abandon other linguistic dimensions through which the patient could best express an authentic sense of his or her suffering.

- Third and finally, respect for reality implies respect for the relational dynamics through which reality is perceived. The modern formalization of knowledge seems to disregard the dimension of relationality as cognitively irrelevant, yet the criterion of the verifiability and falsifiability of theories entails testing and checking not only by the scientific community, but also more generally by all persons.

The disposition that is opposite to respect for reality is *ideology*, which is not only *false knowledge* but also *false consciousness*. Ideology falsifies reality because it does not respect it, and this lack of respect for what is real perpetuates the falsifications. When knowledge truly respects reality it elicits gratitude: "gratitude ... reveal[s] the intrinsic unity of truth and good within reality. Those who feel in their heart the duty to thank the physician for what he has done for them are grateful for both his skilled knowledge [*scienza*] and his conscientiousness [*coscienza*], and thank the scientist and the man respectively" (F. D'Agostino).

Medicine as the ability to restore

Medicine is practical knowledge that can be defined with the notion of *téchne*, or art, meaning the capacity to make or produce a work on the basis of a form of knowledge. But "the essence of the art of healing [is] ... its ability ... to re-produce and re-establish something. This signifies a special modification of what 'art' means, and one which is unique to the knowledge and practice of the physician" (Gadamer).

The value and meaning of the physician's art are not contingent on his plan, but on *submission* to the proper dynamic of nature. Restoring health does not mean "creating" something new, but allowing the organism to regain its lost balance. Though illness is just as *natural* as health, in the sense that it appears without man's desire for it, it is considered in terms of a *deprivation* of something that should be there; hence it is fought. It is necessary to address the question of *nature* and the question of the *human condition* in order to define health and illness. In contrast to the *natural environment*, the human condition is characterized by both *naturalness* (its being "given") and the ability to plan for and integrate the ways of *nature* (that is, its being *free*).

Impossible medicine and false hopes

Medicine has become "impossible" in the current context because it sets goals that are not (economically) sustainable. This is the thesis of Callahan, who maintains that it is not just a question of making health care institutions more efficient and better managed, but also of reconsidering the fundamental values of modern medicine and especially the idea of limitless medical progress. The meaning and aims of medicine should be thought of in a different way that can reconcile greater equity with economic compatibility.

Medicine as relational knowledge

There is an irresolvable paradox and an inherent contradiction in medicine: "wanting to remove a contingent being like man from the intrinsic logic of his contingency" (F. D'Agostino). Medicine was established as part of humanity's implicit and universal rebellion against evil. Yet at the same time illness and death are a constitutive part of human nature, and those who fight them know from the beginning that they are ultimately doomed to failure.

It is important not to lose an awareness of medicine's specific contribution to human self-understanding, because medicine studies man in a dimension that is *constitutive*, not accidental: illness.

While acting on human subjects, medicine is inextricably *also* friendship and justice, and the patient is not a mere object of the physician's art but always and in all cases a subject in a therapeutic relationship.

The therapeutic relationship

The therapeutic relationship is always difficult because it is an asymmetrical relationship governed by need and illness. Illness always has two sides: the third-person side corresponding to what the physician can diagnose in various ways, which can be represented objectively and impersonally; and the lived experience of the sick person, because illness entails a new way of existing and understanding one's condition.

Today this therapeutic relationship is often marked by impersonal practices, procedures, or medicines that reduce the time and space for a relationship to a bare minimum.

The contemporary culture has often decried the limits of medical paternalism in the name of patient autonomy. Responding to the potential abuses of medical paternalism by appealing to the principle of autonomy alone does not serve to reestablish equilibrium in the physician-patient relationship; instead it seems to create a condemnation to conflict in which two autonomies and two perspectives on illness battle one another.

Moreover, the concept of respect for patient autonomy cannot be reduced to respect for the *exercise of autonomy* by others. Recognizing autonomy means recognizing the value of the subject of autonomy, that is, the human being and person. It is often forgotten that the *conditions for exercising autonomy*, which are limited in time, refer back to the *ontological conditions for autonomy*: that is, the fact that we are human beings and persons.

Informed consent

The relationship between physician and patient, in the changed cultural conditions of today, is mediated by a legal instrument known as *informed consent*. In principle, informed consent is not reducible to a simple informative paper that the patient (or patient's representative) has to sign: it constitutes the reference point for the physician-patient relationship, because interpersonal respect, concern for others, recognition of professionalism, and attention to dialogue must be practiced precisely with regard to what must be done here and now from the clinical point of view. However, informed consent is often used as a form of legal protection—there is an encounter between two autonomies, two wills, and two rights, but no longer on the basis of trust. It has been observed that this model does not work, because the therapeutic relationship can never guarantee its own results. For as much as medicine is capable of controlling, replacing, and supplementing biological and physiological processes, the possibility of failure through no fault always remains inscribed in the therapeutic relationship. Awareness of the condition of finitude and vulnerability that characterizes the human condition is important.

Chapter 7

BIOETHICS COMMITTEES

Why Bioethics Committees?

Any treatment of the topic of bioethics committees, even one as concise as this intends to be, must take into consideration several key points concerning their proper scope and characteristics. In particular, it is important to mention the preconditions that might allow and call for the establishment of bioethics committees; this includes whatever serious reasons there may be for instituting them. The parameters for formulating ethical judgments once these committees have been established will likewise be examined, as will the topics or areas on which value judgments can be applied in reference to situations and issues in medical research and health care. Finally, the characteristics necessary for the proper function of bioethics committees will be examined.

This discussion is presented on the basis of the abundant literature available on the subject,[1] reflections from national and international meetings, and some of the

---

[1] P. Allen and W. E. Waters, "Attitudes to Research Ethics Committees," *Journal of Medical Ethics* 9 (1983): 61–65; N. Fost and R. E. Cranford, "Hospital Ethics Committees: Administrative Aspects," *Journal of the American Medical Association* 253 (1985): 2687–2692; G. G. Griener and J. L. Storch, "Hospital Ethics Committees: Problems in Evaluation," *HEC Forum* 4, no. 1 (1992): 5–18; B. Hosford, *Bioethics Committees* (Rockville: Aspen, 1986); F. A. Isambert, "De la bioéthique aux comités d'éthique," *Études* 358 (1983): 671–683; Judicial Council of the American Medical Association (Chicago), "Guidelines for Ethics Committees in Health Care Institutions," *Journal of the American Medical Association* 253 (1985): 2698–2699; M. J. Kelly and D. G. McCarthy, eds., *Ethics Committees: A Challenge for Catholic Health Care* (St. Louis, MO: Pope John Center, 1984); L. W. Osborne, "Research on Human Subjects: Australian Ethics Committees Take Tentative Steps," *Journal of Medical Ethics* 9 (1983): 66–68; Royal College of Physicians, *Guidelines on the Practice of Ethics Committees in Medical Research Involving Human Subjects*, 2nd ed. (London: RCP, 1990); F. Rosner, "Hospital Medical Ethics Committees: A Review of Their Development," *Journal of the American Medical Association* 253 (1985): 2693–2697; E. Sgreccia, "Etica, ma su quale fondamento?" *Orizzonte Medico* 52, no. 1 (1987): 1–2; E. Sgreccia, "Il comitato etico tra assistenza e ricerca," *Orizzonte Medico* 52, no. 4 (1987): 2–3; E. Sgreccia, "L'etica: Presupposto di affidabilità dell'ospedale," *Sanare Infirmos* 1 (1987): 12–16; A. G. Spagnolo, "I comitati etici negli ospedali: Sintesi e considerazioni a margine di un recente simposio," *Medicina e Morale* 36, no. 3 (1986): 566–583; A. G. Spagnolo and E. Sgreccia, "Comitati e Commissioni di bioetica in Italia e nel mondo," *Vita e Pensiero* 12 (1989): 802–818; idem, "I comitati di bioetica: Sviluppo storico, presupposti e tipologie," *Vita e Pensiero* 78 (1989):

instructions found in the regulations already approved at the national and regional European levels as well as in the publications of bioethics committees themselves.

I prefer to use the term *bioethics committee*, rather than ethics committee or ethical committee, because medicine has lately come to place significant emphasis on its biological foundation. This has implicated the expansion of traditional medical ethics to include bioethical reflections. I therefore believe that these committees must take into account the new demands of biological and medical issues, with a consequent broadening of the foundations of the biomedical sciences and a need for the interdisciplinary methodology that bioethics entails. This interdisciplinary approach, however, should not be cause for forgetting the existence of a common reference to the ethical values of the person.

Even the international literature on the topic of these committees is heading in this direction: the English term *bioethics committee* is increasingly used to denote them.[2] On the one hand, this is precisely because of the need to assign uniform nomenclature to these bodies which, having been instituted to carry out specific tasks, used to have very different names. On the other hand, the phrase is increasingly used because of the emphasis it places on the need for the "engines" of these committees to be the various centers for bioethical research and reflection: privileged places in which the education and training of committee members is also conducted.[3]

It is difficult to provide a single, concise definition of a bioethics committee that takes into account all the functions that such a body can have in theory and in practice. In general, bioethics committees should be considered *groups in which different members of the various sectors of activity connected with human life and health can meet in a pluralistic context with an interdisciplinary methodology*, whether in connection with health care facilities, clinical research institutes, or laboratories dedicated exclusively to experimentation.[4] In this context, sufficiently

---

500–514; R.M. Veatch, "Hospital Ethics Committees: Is There a Role?" *Hastings Center Report* 7 (1977): 22–25; L. Walters, "Bioethics Commissions: International Perspectives," *Journal of Medicine and Philosophy* 14, no. 4 (1989): 363–462; R.J. Levine, "Research Ethics Committees," in *Encyclopedia of Bioethics*, ed. W.T. Reich, rev. ed. (New York: Macmillan, 1995), 4:2266–2270.

[2] See, for example, the following publications: Hosford, *Bioethics Committees*; B.J. Edwards and A.M. Haddad, "Establishing a Nursing Bioethics Committee," *Journal of Nursing Administration* 18, no. 3 (1988): 30–33; S. Theoret, "The Role of the Bioethics Committee Dealing with HIV Infection," *Canadian Journal of Nursing Research* 84, no. 7 (1988): 41–47.

[3] Contributions supporting this perspective can be found in C.G. Vella, P. Quattrocchi, and A. Bompiani, *Dalla bioetica ai comitati etici* (Milan: Ancora, 1988). Various denominations are encountered in the literature: biomedical ethics committee, medical ethics committee, ethics committee or ethics advisory committee, hospital ethics committee, institutional ethics committee or institutional review board (IRB), ethics forum or bioethics study group, patient care review committee, prognosis committee, critical care committee, terminal care committee, brain death committee, child protective committee, infant care review committee or infant care advisory committee, optimum care committee, nursing bioethics committee, nursing home bioethics committee, unit for human values in medicine, human rights committee, institutional animal care and use committee, and so on.

[4] This is the way P. Cattorini expresses it in "I Comitati d'etica negli ospedali," *Aggiornamenti Sociali* 6 (1986): 415–429. Limited to the specific function of reviewing

qualified members will be called upon to confront the various ethical issues that gradually surface, attempting to find an effective solution as consistent as possible with the underlying values and principles that the committee itself has explained and declared in its constitution.[5]

Necessary Preconditions for the Establishment of Bioethics Committees

By preconditions I mean the de facto historical and cultural conditions in which medicine is practiced that give rise to the need for both multidisciplinary reflection and unitary judgment regarding an individual, concrete event, or particular situation to be addressed, with the aim of determining the best course of action to promote the patient's good.

The same thing that sometimes occurs in calamitous political or military emergencies also happens in medicine: in order to determine what decision to make when faced with a plurality of factors in the concrete situation, all the individuals capable of contributing to the decision are convened in a council. Most times this common effort to clarify the situation provides solid support for innovative strategies and programs being tested for the first time, even outside of extreme or emergency situations.

Bioethics committees first arose in response to highly bewildering and dramatic situations (the case of Karen Ann Quinlan or similar situations). They are now proposed as support structures for making decisions about experimentation protocols or situations that may appear novel or otherwise ethically unclear. In other words, a transition has taken place from an emergency response structure to a support structure serving as a constant reference point.

Although the 1976 decision of the Supreme Court of the State of New Jersey is usually referenced as the historical date of the first formal establishment of a bioethics committee, it should be noted that in 1971 the Canadian Catholic Bishops had already advanced a proposal in their *Medico-Moral Guide* to establish ethics committees in every Catholic hospital with certain fundamental tasks, including education and training, primarily in order to uniformly apply the *Ethical and Religious Directives for Catholic Health Care Facilities* issued by the United States Conference of Catholic Bishops (formerly the National Conference of Catholic Bishops) that same year.[6]

---

clinical experimentation protocols, an independent ethics committee is defined in the Guideline for Good Clinical Practice of the European Medicines Agency as follows: "An independent body ... constituted of medical professionals and non-medical members, whose responsibility it is to ensure the protection of the rights, safety and well-being of human subjects involved in a trial and to provide public assurance of that protection." European Medicines Agency, *Note for Guidance on Good Clinical Practice* (London: EMEA, 2006), article 1.27, p. 7, http://www.emea.europa.eu/docs/en_GB/document_library/Scientific_guideline/2009/09/WC500002874.pdf.

[5] Rosner, "Hospital Medical Ethics Committees."

[6] Those directives were revised in 1975, 1994, and 2009. See United States Conference of Catholic Bishops, *Ethical and Religious Directives for Catholic Health Care Services*, 5th ed. (Washington, DC: USCCB, 2009), http://www.usccb.org/issues-and-action/human-life-and-dignity/health-care/upload/Ethical-Religious-Directives-Catholic-Health-Care-Services-fifth-edition-2009.pdf). Concerning bioethics committees, directive 37 establishes

Some significant statistics confirm this initial stimulation and expansion of bioethics committees in Catholic hospitals: a report by the US President's Commission in 1983 showed that only 1 percent of public hospitals had a bioethics committee, whereas a study by the Catholic Health Association in the same year showed the presence of bioethics committees in 36 percent of all Catholic hospitals.[7]

One of the first bioethics committees was formally established by the decision of the Supreme Court of New Jersey on March 31, 1976; from that moment on, numerous questions have emerged concerning the function these committees should have.[8]

Let us briefly recall how the New Jersey judges' decision to establish a bioethics committee arose in the case of Karen Ann Quinlan. This young lady had been in a coma for a year due to severe neurological trauma and had been rejected by numerous hospitals and private clinics because her condition was considered irreversible. She was finally accepted at a clinic, the Morris View Nursing Home, and was kept alive in a state of total unconsciousness by sophisticated equipment. The question being asked by public opinion, in the now distant year of 1976, was whether survival at all costs for a person in such a condition was licit, or whether it would be better to let nature take its course.

In establishing the committee, the supreme court assigned its members the task of assessing the reasonable chances of Karen Ann Quinlan coming out of her comatose state, with the specific aim of approving or rejecting the decision to disconnect her once and for all from the equipment keeping her alive. It immediately became apparent that the committee had not been asked whether it approved or disapproved of the decision to suspend treatment, but rather to issue a plain and simple clinical judgment regarding prognosis. The committee requested by the judges was therefore formed at the clinic where the young woman was receiving care. It was composed of two priests, the health care facility director, a social worker, a physician, and a legal consultant. Doubts immediately surfaced, however, regarding the competence of this committee: if the assigned task was indeed strictly prognostic, then why was there only one physician among the members, who was neither a specialist in neurology nor directly involved in the woman's care? Hence, from the moment of their first emergence in the United States, there were questions regarding the composition and role of bioethics committees.

In addition to establishing a bioethics committee for this contingent and dramatic reason, these committees were subsequently proposed for ordinary situations as well, such as drafting protocols for experimentation on patients or for situations that may arise in health care and in biomedical research in general. In other words, bioethics committees changed from having an emergency role to functioning as supports and constant reference points in everyday practice.

---

that "an ethics committee or some alternate form of ethical consultation should be available to assist by advising on particular ethical situations, by offering educational opportunities, and by reviewing and recommending policies. To these ends, there should be appropriate standards for medical ethical consultation within a particular diocese that will respect the diocesan bishop's pastoral responsibility as well as assist members of ethics committees to be familiar with Catholic medical ethics and, in particular, these Directives."

[7] Kelly and McCarthy, *Ethics Committees*.

[8] Veatch, "Hospital Ethics Committees."

The perspective of Catholic ethics also identifies the need to not abandon health care workers under the weight of unbearable responsibilities in increasingly complex and problematic clinical cases. This is why the *Charter for Health Care Workers* issued by the Pontifical Council for Pastoral Assistance to Health Care Workers also calls attention to the role of bioethics committees in facilitating and monitoring the decisions of health care workers.[9] For a bioethics committee, which is a support structure and constant reference point for the health care organization and its employees and acts as a kind of conscience that monitors the organization in which it is established, referencing the moral principles of the institute itself—which are also the basis for verifying adherence to its mission—is not a limitation. The task is anything but easy, as the National Office for Pastoral Assistance in Health Care of the Italian Bishops' Conference emphasized, since it is not always possible to "confront [bioethical challenges] with the necessary courage and effective decision making that flows from it." For this reason, Catholic health care institutions are especially called to witness through courage, conscious determination, and ethical attention to respect for fundamental human values, but also to offer "intelligent and creative solutions to the new questions posed by technology" and to "decisively overcome the reductive understanding of Catholic thought as being in rigid opposition to secular thought. It is instead necessary to rediscover the great wealth that results from a calm exchange, identifying and applying points that are agreed upon while simultaneously establishing the basis for a common journey toward the potential convergence of values."[10]

This shift from an emergency body to a daily consultation tool resulted from a whole series of preconditions that continue to foster their gradual spreading into every country. They are actually the same preconditions that led to the establishment of bioethics as a discipline.

If the history of the emergence of bioethics committees is closely scrutinized in an effort to grasp the underlying causes, there seem to be four reasons behind the development and gradual spread of bioethics committees. The first reason is epistemological, the second ethical-philosophical, the third political, and the fourth legal: (1) there is a need to recover a lost or at least obfuscated anthropological unity in medical practice; (2) there is a need for clarification regarding the emergence of various ethical reference models; (3) there is a need to defend the physician's professional autonomy over against the dangers of the bureaucratization or politicization of medicine; and (4) patients need some sort of arbitration to safeguard their rights in times of illness that will make legal support rapid and simple without needing to turn to courts and the penal code.

---

[9] "In these commissions, medical competence and evaluation is confronted and integrated with that of other presences at the patient's side, so as to safeguard the latter's dignity and medical responsibility itself." Pontifical Council for Pastoral Assistance to Health Care Workers, *Charter for Health Care Workers* (Boston: Pauline, 1995), no. 8.

[10] Ufficio Nazionale per la Pastorale della Sanità [National Office for Pastoral Assistance in Health Care], "Le Istituzioni sanitarie cattoliche in Italia: Identità e ruolo," July 7, 2000, http://www.chiesacattolica.it/pls/cci_new_v3/cciv4_doc.edit_documento?id_pagina=9068&p_id=5324.

Reintegrating the Anthropological Unity of
Medical Knowledge and Medical Practice

It is well known that medicine is no longer a science but a group of sciences; in other words, medicine is a complex science. To use an antiquated distinction that is hardly applicable anymore and certainly not applicable in medicine, it is both an experimental science and a human science. The demand for specialization has introduced hundreds of disciplines into the education curriculum for medical students, and many of these specialties operate autonomously in large hospitals. The organization itself of the medical curriculum tends to dissociate the various aspects (morphological, physiological, pathological, etc.) of what is a unified human reality. Students in particular experience the great difficulty of reintegrating them.

Yet the great teachers of medicine can overcome this: the more they study a specialized branch in depth, the more they perceive within it the need to invoke the whole and a sort of unity that transcends the naturalistic plane.

I would like to confirm this need for a unitary knowledge of man with the words of a well-known medical pathologist, Franz Büchner, which are found in the first German edition of his volume on general pathology:

> More than all the particular questions concerning medicine, there are three that should keep lively the restless and impatient mind which distinguishes and ennobles physicians: the question of the essence of the living being, the question of the essence of illness, and the question of the essence of man. The first connects him to all other creatures, the second awakens in him the awareness of guarding life that is in danger, and the third sets before him the enigma of the most mysterious being that lives under the sun, man himself. ... We therefore begin our book on general pathology with a discussion of these three questions. ... Human pathology certainly cannot be contained within the confines of biology, if we basically define the human *bios* as a unity of body and spirit and we are well aware that man's psyche is continually and immediately involved in the concrete reality of the human body, and therefore in the state of illness also, just as in the state of health. ... In human pathology, naturalistic medicine therefore needs to be tempered by anthropological medicine. And with that, we find ourselves faced with the question about the essence of man.[11]

The conviction that individual medical specialties—not unlike pathologies—need to find anthropological unity precisely in virtue of their unique perspectives sits at the basis of any effort to overcome the fragmentation of medical knowledge, which becomes increasingly less meaningful if separated from its comprehensive significance. This is analogous to the way a fragment of an epigraph is only meaningful to the extent that one is able to reconstruct the whole text, and with it the historical event that was engraved in the stone and its meaning in the history of time.

---

[11] E. Mascitelli, "Per una lettura antropologica della medicina," in *Saggi di medicina e scienze umane*, by E. Mascitelli et al. (Milan: Istituto Scientifico Ospedale S. Raffaele, 1984), 20–22, citing F. Büchner, *General Pathology*, vol. 1 (Wiesbaden: Office of Military Government for Germany Field Information Agencies Technical, 1948).

Overcoming complexity, which does not entail eliminating individual specialties but rather understanding them in their anthropological unity, also requires overhauling the body-mind dualism or the tripartite division of individuality into soma, psyche, and self, as if they were three floors of a building, one on top of the other. It also requires a critique of the tendency to reduce all of human reality to the biological world. The strong, unitary stamp that the human self imprints upon the whole of his physical, psychological, and spiritual energies and the close connection between man and the ecological and social environment must be reexamined as a whole in a reinterpretation of what it means to be human that simultaneously preserves the person's transcendence, his phenomenological and existential unity, and his place within cosmic and worldly reality.

This discussion about restoring a unitary anthropology, which has been threatened by Cartesian dualism, biologistic reductionism, and the complexity produced by specialization, should not appear to be an abstract demand foreign to the concrete, everyday practice of medicine or to the specific topic of bioethics committees. If we were to stop, for example, to think about how to stem the AIDS crisis or how to regulate the problem of organ removal and transplantation, or if we wanted to address the issues related to reproductive technologies, we would find ourselves faced with numerous clinical aspects and we would have to consider juridical, moral, medico-legal, and other factors requiring consultation with experts in various fields. In the case of reproductive technologies, for example, we would be dealing with a topic that involves the most all-encompassing issues affecting the biological, psychological, and spiritual dimensions of the human person and his responsibilities toward his own life and the lives of others, as well as his understanding of married and conjugal life and the responsibilities of procreation.

It is no longer possible not to pose the overarching questions about man and his life whenever there is a nonroutine or nonstandard intervention on his body, his health, and his illness, although even routine interventions could be reexamined in light of a more complete medical anthropology with sufficient interest in exploring the matter. Health and illness themselves are also concepts with horizons for which the definition of medicine as an experimental science is necessary but insufficient.

In setting up this anthropological study, which intends to overcome the dangers of a dualistic or reductionistic medicine splintered into many different specialties, it is also necessary to establish a valid methodology that should be taken into consideration in all bioethical reflection, and therefore within bioethics committees as well.

I have defined this method as *triangular* (figure 1, *next page*).[12] First of all, it provides for consideration of the *scientific and experimental data* in its objectivity, the study of which can require various medical competencies. Then, in order to acquire its unitary and comprehensive meaning, the data from this investigation must be compared with *anthropology*—that is to say, with the understanding of the human person in his richness and uniqueness. For example, the significance of experimenting on animals is one thing, while the significance of experimenting on human beings is another; furthermore, in this second case, which is of particular

---

[12] See chapter 2 of this work; see also E. Sgreccia, "Problemi dell'insegnamento della bioetica," *Giornale Italiano per la Formazione Permanente del Medico* 15, no. 2 (1987): 104–117.

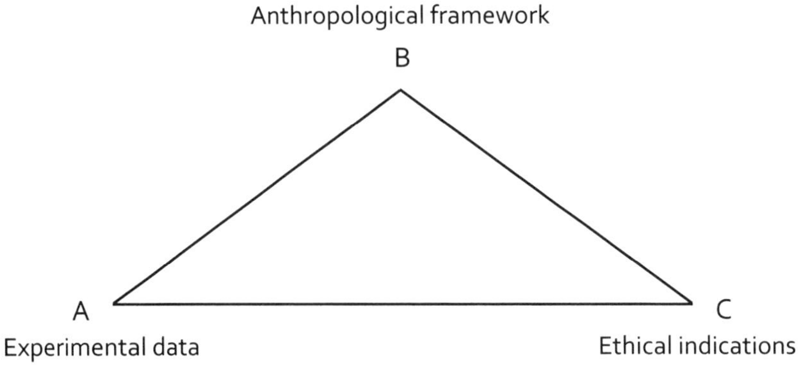

Anthropological framework

Figure 7-1. Formation of judgments in bioethics.

concern, experimenting on a conscious patient or a healthy volunteer is one thing, whereas experimenting on a child, an embryo, or a mentally disabled person takes on an entirely different meaning. From this high point of the triangle, the anthropological point, it is then necessary to think in terms of the third point: *ethical indications for action.*

The Problem of Standard Values and Criteria

A previous chapter of this manual illustrated the de facto existence of a plurality of ethical models and guiding criteria in bioethics today: the liberal-radical approach of individualistic nature, ethical utilitarianism of social and positivist nature, sociobiologism, so-called secular morality, and religious morality (or moralities).[13] I believe that the best ethical model to adopt, in conformity with the human rights conventions and declarations affirmed over the past fifty years, starting with the Universal Declaration of Human Rights in 1948, is the personalist one. Unified Europe, with its political and judicial bodies such as the Council of Europe, the European Court of Human Rights in Strasburg, and the European Parliament, is founded on the human rights expressed in the Convention for the Protection of Human Rights and Fundamental Freedoms of 1950 (the Pact of Rome).

In its most valid and well-founded formulations, the personalist model recognizes the duty to respect the person from the first moment of conception, requires the participation of the patient—as a person and in the first person—in making ethical decisions, and sees physical and bodily life as the fundamental value in which the other values of the person can be grounded and through which they can be expressed. This is the only model I consider wholly human and capable of serving as a foundation and inspiration for anthropological medicine.

In principle, this model can be applied not only in religious hospitals but also in public ones. It must nonetheless be noted that some laws, such as law no. 194/1978 making abortion legal in Italy, in fact permit certain practices in public hospitals that are very from removed from this ethical vision. Indeed, the personalist standard,

---

[13] See chapter 2.

248

which considers corporality a constitutive component of the person himself as the embodiment, language, and manifestation of the entire self, entails defending man's physical life—as a fundamental value—from the first moment of conception until natural death.

This same vision of human existence gives medicine a statute inspired by the therapeutic principle, whereby interventions are justified if there is a therapeutic efficacy that benefits the subject on whom they are carried out. The principle of freedom and responsibility always enters into this same personalist perspective: the patient can never be considered an object but is always a participating subject responsible for his decisions. The principle of sociality and subsidiarity should ensure that the health care system is guided by the duty to give greater assistance to those in greater need and less assistance to those who can count on their own strength or resources.

Personalism entails a set of guidelines for medical conduct, forming a bulwark against all potential forms of political utilitarianism or practical exploitation of patients and of medicine itself.

However, in cases where the personalist ethical model cannot be fully integrated into a bioethics committee and its constitution, the problem arises of what the meeting point should be for the differing ethical views present within the committee. It will be necessary to return shortly to this problem of conflicts among the various ethical positions and conflicts between ethics in general and professional or legal norms.

At this point it should be noted that, depending on the underlying philosophy, two different types of bioethics committees can take shape. In one, the participants share a set of ethical criteria, and the committee proceeds to a multidisciplinary assessment of the facts, determining the relative duties from this assessment. In the other, the "pluralistic" type of committee, there are two fronts or two separate levels of work: assessing the facts from the different perspectives and seeking a common standard by which value judgments can be made. This entails ethical study of the individual facts and an ethical search for the general criteria that will serve as a starting point and basis for comparison. The solution to these concurrent problems is not simple; if unresolved, the work of a bioethics committee that tries to remain ideologically and culturally pluralistic can be rendered unreliable and uncertain.

A proposed solution, which does not intend to be a mere compromise, will be examined later on. This search for criteria and common parameters is certainly one of the reasons for—and difficulties with—the existence of bioethics committees.

Overcoming the Politicization of Medicine: Patients' Rights and Physicians' Rights

A third precondition favoring the establishment of bioethics committees is the increasingly evident need to safeguard the decisional autonomy and ethical-deontological responsibility in the physician-patient relationship against the dangers of the politicization and bureaucratization of medicine. A move toward a sort of negative synergism has begun, in fact: on the one hand government is increasingly managing health care, and therefore indirectly managing the health of citizens; on the other hand, medical professionals risk losing their ethical and professional autonomy by accepting the identification of ethics with law. Obscuring the independence of the judicial branch would certainly be harmful to a society, and it would surely be a serious step backward for a judge to take orders from the executive branch or a political party; the loss of autonomy of conscience in the physician-patient relationship in favor of a medical practice bureaucratically directed from without, to the detriment of the responsibilities of the true and primary protagonists of health and illness, is equally to be feared.

Society, which organizes services, guarantees their means, monitors their quality, and therefore has its own responsibility as defined by law, cannot replace the decisions of persons, who are autonomous in accordance with sanctioned and recognized constitutional values.

The citizen is the person primarily responsible for his own life and health: no decision about his life can be made without his consent (implied or expressed) or the consent of his legitimate representative. The citizen himself is not the arbiter of his own life, but is responsible for this objective and transcendent good that he has received, which constitutes the fundamental good of the person and the society. The physician is a *qualified service provider*: as such, he is responsible to the citizen—whether healthy or sick—within the limits of his professional competence, so long as the contract-alliance remains valid in his conscience in light of the transcendent duty to protect the life and person of the patient.

This ideal research model implies that bioethics committees should be subsidiary or facilitating tools for exercising medical responsibility with regard to the patient. The interdisciplinary analysis of the complex experimental science data in light of the other human sciences specifies a set of pertinent values on the basis of which the physician can make a decision and assume his responsibilities toward the patient.

If this approach based on differing roles and the ethical-deontological autonomy of the various parties involved in the medical act (health care workers, patient, society) is lacking, the bioethics committee can easily sink to the level of a bureaucratic structure of compromise—both internally among the various contingent positions and in relation to the changeable cultural climate predominant in the society. In other words, a bioethics committee would turn into a sort of observatory from which to study the social tolerability of certain specific behaviors in the biomedical field.

In the situation I have proposed, which includes protecting both patients' rights and physicians' rights and establishing the ethical and deontological duties of each, bioethics committees would have to be structured as bodies that protect those respective autonomies and responsibilities. Bioethics committees should protect the rights of the patient against potential incursions by the physician and the rights of the physician against potential situational pressures, helping determine the duties of both.

Viewed in this way, bioethics committees can become a factor promoting humanization and the acceptance of responsibility in medical care, thus helping to improve the quality of the care itself. In the field of research they can protect the goods of both the person and society.

Judging from the experiences of these bodies at the national and international levels, the tasks of bioethics committees pertain to both monitoring experimentation and protecting patients' rights. In the first case, abuses may occur due to economic incentives, profit-making considerations by pharmaceutical companies, or the experimenter's utilitarian quest for success at all costs. In the second case, the cause of potential violations of patients' rights could be linked to many different instances of shirked responsibility that could be ascribed to administrators, physicians, the patients' relatives, the patients themselves, or even endemic shortcomings of the health care facility or organization.

In this same perspective, there is a new foreseeable danger for health facilities coming from health care economics, especially in systems of socialized medicine where the state plans and finances health care.[14] Funding providers have an interest in avoiding costly treatments, but such treatments may be necessary for the patient and included in the right to treatment. This is not only an allusion to the dangers of social euthanasia through withholding treatment from categories of incurable patients receiving costly care, but also of less conspicuous administrative measures that omit one drug or another, one form or another of intensive care. In protecting patients' rights, a bioethics committee could tie together patient needs and the needs of society, preventing planning and funding from following strictly budgetary logic rather than the *principle of subsidiarity*, whereby the more seriously ill deserve more care.

Parameters for the Formulation of Ethical Judgments

It is necessary to provide a procedure within bioethics committees for formulating a judgment on a situation. It should into account all of the parameters contributing to the determination of the responsibilities of physicians and researchers regarding patients or healthy citizens.

After having gathered, processed, and examined the scientific data in their complexity, the civil and criminal laws of the state should be considered first, since both physicians and other citizens are obliged to respect and observe them. This comes first precisely because bioethics committees have competence in areas not regulated by law or where the law must be interpreted.

This obligation does not conflict per se with conscientious objection, because the latter is provided for by the law. There may be a need for civil disobedience with the aim of changing the law in the case of an objectively unjust law; nonetheless, the law creates moral obligations. In addition to the law, the parameter of medical deontology or so-called professional ethics, as codified at the national and international levels, must be taken into consideration. This parameter does not take into account all the ethical aspects of the health care problem because it considers the question primarily from the perspective of health care workers; nevertheless, it has a fundamentally ethical inspiration and provides indications that are meant to be binding in the consciences of professionals.

Some international organizations of physicians, such as the World Medical Association and the World Federation of Catholic Medical Associations, continually issue updated versions of their codes of ethics. Some nursing associations do likewise.

Beyond this, the ensemble of what are defined as human rights should be considered as an overarching framework. They have been declared and elucidated in international charters and conventions starting with the UN's Universal Declaration of Human Rights on December 10, 1948 and the Convention for the Protection of

---

[14] E. Sgreccia, "Economia e salute: Considerazioni etiche," *Medicina e Morale* 36, no. 3 (1986): 31–46; C. Lucioni, "Economia e salute," *Medicina e Morale* 36, no. 4 (1986): 777–786; A. Bompiani, "Medico, servizio sanitario, economia," *Medicina e Morale* 35, no. 4 (1985): 691–716; Fost and Cranford, "Hospital Ethics Committees"; see E. Sgreccia, *Manuale di bioetica: Aspetti medico-sociali*, 3rd ed. (Milan: Vita e Pensiero, 2002), 2:491–524; A. G. Spagnolo et al., *Etica e giustizia in sanità* (Milan: McGraw-Hill, 2004).

Human Rights and Fundamental Freedoms (Pact of Rome) on November 4, 1950,[15] up through all of the conventions, declarations, charters, and recommendations (of varying legal weight and binding force) instructing states with increasing frequency on how to deal with ethical problems in biomedicine as well: suffice it to recall recommendations 1046 and 1100 of the Council of Europe, the resolutions of the European Parliament in March 1989, and the European Convention on Bioethics.

Instructions concerning "religious morals," issued by the official and competent bodies of various religions (Muslim, Jewish, Christian), constitute an additional and higher level. For a religiously affiliated hospital, the instructions from its own religious authority (for example, the instructions of the Magisterium of the Catholic Church for a Catholic hospital) clearly become the last word in terms of guidance. This operative freedom of the entity must be recognized and protected in democracies in the name of the principle of religious freedom.

These instructions are still valuable and necessary for public hospitals, above all because they should be duly kept in mind for patients and physicians identifying with a particular religious morality (all citizens should be able to choose a public facility, and those seeking and sharing the approach a particular type of health care practice should be able to choose the religious facility), but also because valuable, timeless guidelines of rational morality are found in these moral teachings. Ignoring the words of Pope Pius XII or—in more recent documents—Pope John Paul II regarding ethical problems in medicine would not just be a gap in general cultural knowledge but a serious lacuna for anyone in the field.

As far as this comparison of the various parameters is concerned (law, deontology, human rights, religious morals), there is an ongoing experience that could serve as a model: the composition of a European manual titled *Le médecin face aux droits de l'homme* (The physician and human rights). This recently published work[16] compiles documents regarding the practice of medicine (international laws, deontological codes, human rights, religious morals) and compares them in individual cases, which are thereby examined in the light of those same parameters.

I think that taking this path could lead to certain difficulties arising from the context of ethical pluralism in which we find ourselves. From an ethical point of view, therefore, it seems absolutely necessary to me that every hospital bioethics committee declare its own ethical identity—and therefore the ethical parameters to which it will strive to adhere—in its constitution. This could facilitate the work of the committee itself, and constitutes an act of honesty toward the citizens who will use the hospital's services. The citizens' doubts can turn into a sort of "philosophy of suspicion" regarding treatment of the terminally ill, organ transplants, or gynecological care. Clarity and transparency, not just in politics or financial management, is once again the criterion for ethicality.

---

[15] On this topic, see M. Torrelli, *Le médicin et les droits de l'homme* (Paris: Berger-Lévrault, 1983).

[16] Institut International d'Études des Droits de l'Homme and Council of Europe, *Le médecin face aux droits de l'homme* (Padua: CEDAM, 1990). In 1996, the European Scientific Co-operation Network "Medicine and Human Rights" produced the volume *La santé face aux droits de l'homme, à l'éthique et aux morales* (Strasbourg: Éditions du Conseil de l'Europe), which presents 120 practical cases and examines them in terms of human rights, ethics, and religious morals.

In the case of a religious hospital, as already indicated, the last word is ultimately pronounced in light of the religious morality by which the hospital is guided. This does not mean that the judgment is quick and simple, because the complexity often arises from the concrete situation, the plurality of values in question, and the constant need for correct mediation between the principles of moral theology and the concrete reality.

Those wishing to know the position of Catholic morality concerning the specific issue at hand can find significant help in the official documents of the Magisterium of the Catholic Church, such as the document on reproductive technologies.

In the context of ethical pluralism and public hospitals, on the other hand, it will be necessary to work harder—comparatively researching all parameters—in order to defend and promote the human person as such. This comparative research should not be conducted, however, with the criterion of a sociologically shared *minimum*, but with the intention of discovering the *maximum* respect owed to the human person—individual patient, healthy citizen, or physician—and his conscience in light of all the parameters set forth above. If it is to be called an "ethical minimum," then, it must be understood in a strong and anthropological sense: that is, it should include the defense of every person and of the family, since they are the fundamental basis of society.[17]

Every parameter in this pursuit can contribute something to the assessment, but I think that deontological codes and human rights should be acceptable points of reference for all—regardless of the philosophy one follows—for the purpose of practically defining a valid and universal medical anthropology and medical ethics. Religious morality can also make its own meaningful contribution to this universal truth about man.

Finally, I would like to briefly mention a problem that is becoming a topic of reflection for legal experts regarding bioethics committees: What are the potential legal responsibilities of the bioethics committee as a whole and of its individual members with regard to an issued ethical opinion that may cause harm to persons or things? Considering that the opinions of a bioethics committee are nonbinding, national and international legal thinking currently seems to be leaning toward the exclusion of legal responsibility. If this holds true, the introduction of liability disclaimers for bioethics committees—as suggested by some—would appear unnecessary.[18]

Functions and Characteristics of Bioethics Committees

An examination of the guidelines for bioethics committees and the practices established in some hospitals and scientific research institutes, particularly in English-speaking countries, reveals that there are various types of bioethics committees or various goals entrusted to them by their constitutions.

Above all, an educational goal is found in all bioethics committees regarding issues of the humanization of medicine, patients' rights, and new questions in medi-

---

[17] On this point, see part III of the Instruction *Donum vitae* by the Congregation for the Doctrine of Faith (CDF), February 22, 1987.

[18] On this topic see M. Zanchetti, "La responsabilità giuridica del Comitato di etica ospedaliero," in *Una verità in dialogo*, ed. P. Cattorini (Milan: Istituto Scientifico Ospedale San Raffaele–Europa Scienza Umane, 1994), 78–94.

cal ethics. In this respect, bioethics committees seem to share some characteristics with so-called tribunals for patients' rights. In Italy, an old proposed bill aiming to set forth and define patients' rights by law also provided for a sort of ethical commission to serve as an educational body that would also defend these rights, which by now are identified in more or less identical lists.

This is an aim that certainly allows for educational activity in every hospital, whether public or private, and it appears to be urgently needed given the disorganization and dehumanization of care in many facilities today.

It will be important to provide for practical ways of reporting eventual hospital inadequacies to the bioethics committee and ways of making its opinions effective. I believe that these opinions should subsequently be brought to the health care administration, which has the legal authority to promote the observation of policies and rules (e.g., the modification of meal times).

The *educational* task of a hospital bioethics committee could focus on promoting debates, conferences, and courses on ethical issues. These initiatives would promote ethical awareness.

Another foreseeable task for hospital bioethics committees is the analysis of *protocols for research and clinical experimentation*. There are bioethics committees specifically limited to this purpose at certain university polyclinics, scientific institutes for rehabilitation and treatment, hospitals, and medical practices (at times they may have other names, such as "science and technology committee" or "ethical committee for clinical experimentation"). Other bioethics committees, in accordance with their constitutions, intend to carry out this task along with other functions in the structure with which they are affiliated. The review of protocols for clinical experimentation is the only legally required function of a bioethics committee in Italy and in other nations as well.[19] This has led to the progressive growth and spread

---

[19] With the issuing in Italy on July 15, 1997, of a ministerial decree accepting the European Union's Guidelines for Good Clinical Practice for Trials on Pharmaceutical Products (CPMP/ICH/135/1995), it became obligatory for research protocols to be scrutinized by an "independent ethics committee." It is also stipulated that "no subject should be admitted to a trial before the IRB/IEC issues its written approval/favourable opinion of the trial" (art. 3.3.6). This normative device and later ministerial decrees and circulars—in particular the ministerial decrees on March 18 and March 19, 1998—have further specified and defined the organizational procedures, characteristics, tasks, modes of operation, and worthiness of accreditation of ethics committees for clinical trials. Legislative decree no. 211 on June 24, 2003, which accepted European directive 2001/20/EC, further defined the operative areas and duties of ethics committees in the assessment of clinical trials, requiring strict time limits for issuing an opinion. This therefore requires greater organizational effort by bioethics committees in order to punctually comply with assessment requests. Additional responsibility was also attributed to bioethics committees in clinical trial review by the ministerial decree on December 17, 2004, which established specific regulations for the practice of so-called nonprofit trials: within the sector of National Health Care Service activities, these norms should give a particular boost to trial protocols that do not have commercial or profit aims but are aimed at improving clinical practice. In this case, bioethics committees are called to identify and recognize not only the value and validity but also the particular importance of these protocols for improving clinical practice, to verify that the ownership of the scientific data is held by a noncommercial sponsor and to ensure that there are no conflicts of interest or shared interests with the industry.

of committees devoted primarily or exclusively to this task. Bioethics committees receive accreditation for this task from the competent offices of the Italian Ministry of Health, and it is of essential importance due to the countless manifest and hidden interests at play in clinical research, the complexity of the required competencies, and the new frontiers opening up every day for pharmacological, medical, and surgical experimentation. Suffice it to think of experiments with AIDS vaccines or the technical possibilities of experimentation on embryos and fetuses, children, and the mentally or terminally ill. Experimentation is necessary and should enjoy the support and protection of society, but by the same token it should be directed toward the good of both the individual person and society; in other words, it should prevent the individual from being considered an object or a means with respect to society.[20]

Finally, the third task of a hospital bioethics committee is to be an *advisory* body for specific cases of care. These cases may arise in such a way that the physician wishes to request the opinion of the bioethics committee. Sometimes the law itself will provide for such opinion requests. In this regard, it is interesting to note that the decree on May 12, 2006, which redefined the institutional and organizational criteria for Italian ethics committees, indicates—unlike previous ministerial pronouncements—the possibility of "also fulfilling an advisory function with regard to ethical questions related to scientific and health care activities, with the aim of protecting and promoting the values of the human person. Furthermore, ethics committees may propose educational initiatives for health care workers regarding issues in the field of bioethics" (art. 1, §3).

I deliberately use the term *opinion* because I maintain that a bioethics committee cannot *decide* in place of those who have the real responsibility (patients and physicians), even if the law requires such an opinion prior to the decision. Bioethics committees can never take the place of someone's conscience; if anything, they can lend an extra hand to those who are looking to make reasoned and informed decisions.

---

[20] As already mentioned, there are numerous international guidelines regarding the function of bioethics committees, and since the 1990s—with the Italian ministerial decree on April 27, 1992—this function of reviewing clinical trial protocols has been the object of constant and specific legislative activity in Italy. In May 1996, the steering committee of the International Conference on Harmonization (ICH) issued a new version of the Guideline for Good Clinical Practice, which was called "tripartite" because it was developed by experts from the three primary geographic regions concerned with controls for clinical experimentation: North America, Japan, and Europe. In fact, these guidelines currently constitute the standard for all three geographic areas. The Italian Ministry of Health accepted them with a decree (July 15, 1997), and they have therefore become binding both for research trials and the functioning of bioethics committees. See the text of the decree in *Medicina e Morale* 47, no. 6 (1997): 1162–1215. The specific reference point remains the Declaration of Helsinki, which is the true international guide on ethics and deontology for clinical researchers. This document, drafted by the World Medical Association, which was last amended in Seoul in 2008, explicitly states that "the well-being of the individual research subject must take precedence over all other interests (no. 6) and that "no national or international ethical, legal or regulatory requirement should reduce or eliminate any of the protections for research subjects set forth in this Declaration" (no. 10). World Medical Association, "Declaration of Helsinki: Ethical Principles for Medical Research Involving Human Subjects," June 1964, amended October 2008, http://www.wma.net/en/30publications/10policies/b3/17c.pdf.

These three tasks can be combined in a single bioethics committee, or some bioethics committees may be limited to some of these objectives (e.g., the assessment of research protocols only).

The ideal characteristics of a hospital bioethics committee are now well known: competence, independence, impartiality, and subsidiarity (that is, of an advisory nature). It is important for these characteristics to become a reality through the choice of personnel and adherence to the operating procedures outlined in its constitution and policies.

Regarding its composition, the Good Clinical Practice guidelines of the European Economic Community (EEC) state, without further clarification, that "medical professionals and non-medical members" must be present on bioethics committees. The compositional aspect has since been widely examined in the various guidelines issued and updated, and by now there are some fixed points that every bioethics committee should take into account. The multidisciplinary character is indispensable and must be a minimum requirement: this is because professional experts have greater difficulty transcending their fields and evaluating the full range of problems posed by a given situation. Furthermore, a multidisciplinary committee will be more credible in the eyes of public opinion, feeling that it is somehow represented in this way. In addition to physician members knowledgeable in clinical research, the following groups should always be represented: (1) legal experts (lawyers, judges, and law professors); (2) representatives of the human sciences; (3) philosophers and ethical experts; (4) teachers, who bring the aspirations of the younger generations and can transmit messages to them in turn; and (5) journalists, who are the interpreters of and for public opinion.

In the case of uniquely challenging issues, an ad hoc convocation of experts in the field is appropriate. Often questioned or questionable groups include representatives for administrators of the hospital or research institute (due to the danger of losing independence and impartiality) and patient representatives (due to the danger of representing oneself and not the patient).[21]

All of the work of a bioethics committee presupposes some reference points and ethical study to back it. Bioethics centers or centers for medicine and the human sciences, which collect data, furnish guidelines, research solutions, and culturally prepare the members of the bioethics committee, can fulfill this role. Without such a resource, the committee would lack the sound support of evidence and research.

The Ideal Characteristics of Bioethics Committees

Reflection on the various points considered so far leads to the discovery of some ideal characteristics that should be pursued. It also reveals ways in which bioethics committees can become corrupted that must be avoided at all costs in order for these structures to continue working in accordance with the original spirit of their institution.

First of all it seems that a fundamental prerogative of these committees, acknowledged by many, should be to have an advisory and not a decision-making

---

[21] For an in-depth analysis of the establishment and operating procedures of bioethics committees, see A. G. Spagnolo, "La protezione dei soggetti di sperimentazione: ruolo e procedure operative dei Comitati di Etica," in A. G. Spagnolo and E. Sgreccia, eds., *Lineamenti di etica della sperimentazione clinica* (Milan: Vita e Pensiero, 1994), 113–140.

role. Their place is to provide support for the decision-making process of the individual. In other words, the bioethics committee offers its competence to facilitate ethical decisions without replacing the individual physician, researcher, or patient. While the law may require seeking the opinion of a bioethics committee, the opinion formulated by the committee should never be compulsory: its advisory role implies that the final responsibility belongs to the person who requested the opinion.

Another important characteristic is consistency with the ethical parameters set forth in the committee's constitution. From an ethical standpoint, it seems obligatory and absolutely necessary that every bioethics committee make clear its ethical identity and consequently the ethical parameters by which it will strive to abide. This can facilitate the work of the committee itself and constitutes an act of honesty toward those who request an opinion.

It also seems that impartiality and independent thinking and deliberation should be guaranteed. It is therefore essential that bioethics committees have no influential ties to the administration of the structure in which they work or to the pharmaceutical industries that propose drug trials. Furthermore, the members that make up the committees should have no direct or indirect connection, for example, to the researchers who submit their experimentation protocol for an ethical opinion.

Bioethics committees also entail many hidden dangers, which tend to weaken the political interest in establishing them by law.[22] I believe these dangers are actually

---

[22] In one of its documents on the topic of establishing local bioethics committees, the Italian National Bioethics Committee (NBC) addressed the controversial question of creating a complex "network" of bioethics committees by law. The NBC observed that an argument favoring legislative measures for the obligatory establishment of such committees is that the rationality of the system would improve and public opinion would attribute greater authoritativeness to the opinions. The aim of avoiding unequal treatment for health care service recipients on the basis of whether or not their local structure has a bioethics committee is even more persuasive. But the most serious and substantiated objections of the NBC appear to be the following: establishment by law "would allow for [their] establishment out of mere obedience to a legal precept and not out of effective sensitivity to the issues of bioethics; 'institutional' demands on the formal level would multiply, but without guaranteeing substantial commitment to them; the danger of corruption and practical compromises would become more real." Even if the law encouraged and promoted the spontaneous organization of bioethics committees in a way that would extol their autonomy and independence, its purpose—the NBC goes on to say—could even be to define parameters that the constitution of the committee must incorporate and respect when responding to mandatory requests for consultation and opinion. On the other hand, providing for obligatory opinions does not necessarily translate into the need to establish committees by law, and this is why the ministerial decree cited does not say much more about the characteristics of bioethics committees. See Italian National Bioethics Committee, *Comitati ctici* [*Ethics committees*], February 27, 1992 (Rome: Presidenza del Consiglio Dipartimento per l'Informazione e l'Editoria, 1992), http://www.governo.it/bioetica/pdf/8.pdf, English abstract at http://www.governo.it/bioetica/eng/opinions/abstracts/abstract_the_ethics_committees.pdf). This opinion was proposed again later (April 1997) with some modifications taking into account the developments in the bioethics committee situation in Italy. The opinion *Orientamenti per i comitati etici* [*Guidelines for ethics committees in Italy*], July 13, 2001 (Rome: Presidenza del Consiglio Dipartimento per l'Informazione e l'Editoria, 2001), http://www.governo.it/bioetica/pdf/47.pdf, English abstract at http://www.governo.it/bioetica/eng/pdf/abstract_guidelines_for_ethics_committees.pdf) draws

due to certain corruptions of their tasks and mode of operation rather than to the nature or origins of the committees. For example, it can be seen how these committees lead to a dispensation from moral and civic responsibility for the health care workers making the decisions; this tendency is particularly evident in the United States. They can also arrogate roles that are not theirs, finding themselves in conflict with the professional associations already responsible for holding their members to a professional code of ethics through specifically designated commissions. The politicization of bioethics committees—in the sense of drawing party lines and in the sense of "factiousness" among researchers—could lead to the creation of new sorts of unions and new headquarters for the division of power.

Furthermore, the race to institute countless peripheral committees for the use of the structures that institute them should be avoided.[23] As far as the assessment of clinical trials is concerned, the ministerial decree on April 27, 1992 (now abrogated), provided that "wherever instituted in Italy, ethics committees must always be in keeping with the guidelines for good clinical practice ... [and] must be based in facilities of proven credibility for health care or scientific research." The ministerial decree on March 19, 1998, specified that clinical trials on patients cannot be carried out in private health care facilities, except in cases where multiple centers are involved in the study, including at least one public structure; the private structure is accredited and deemed qualified by the competent local health care agency (Azienda Sanitaria Locale, ASL) for that area; and the study has received prior approval from the competent ASL ethics committee or, when this is lacking, from the relevant public ethics committee.

Finally, it is important to underscore that a bioethics committee can be established and managed and can function in the best way possible through the help of Standard Operating Procedures (SOPs). This need is demonstrated precisely where there is a lack of regulatory uniformity in operating procedures, as in Italy; this will be addressed later on.[24] The functions, responsibilities, and operating procedures of bioethics committees, above all in relation to their advisory function regarding the protocols of clinical trials, are therefore moving toward uniformity in accordance

---

attention to the risk of "pharmaceutical intoxication," or the risks tied to supervising drug research protocols, along with the possibility of separating ethics committees for research from those for medical practice.

[23] The National Ethics Advisory Committee in France addressed the spontaneous and uncoordinated emergence of local bioethics committees, often arising only as a result of the needs and interests of the people involved, in a specific report. It concluded with several recommendations for facilitating and coordinating the emergence of such local committees. See French National Ethics Advisory Committee for Life and Health Sciences, *Rapport et Recommandations sur les Comités d'Éthiques Locaux*, November 7, 1988.

[24] Leigh and Barron Consulting and Christie Associates for NHSTD on behalf of the UK Department of Health, *Standard for Local Research Ethics Committees: A Framework for Ethical Review* (London: NHS Training Division, 1994); C. Bendall, *Standard Operating Procedures for Local Research Ethics Committees: Comments and Examples* (London: McKenna, 1994); Food and Drug Administration, *Information Sheets for Institutional Review Boards and Clinical Investigators* (Rockville, MD: FDA Office of the Associate Commissioner for Health Affairs, 1995).

with the guidelines outlined in the 1996 harmonized, tripartite (Europe, Japan, North America) version of the guidelines for Good Clinical Practice (see note 19).

It therefore seems, in returning to an earlier premise, that all of the functions and activities requested of bioethics committees cannot prescind from a direct or indirect link to certain centers of study where especially the fundamental ethical and philosophical issues are addressed. Such research must then become concrete through the teaching of medical ethics in the medical studies curriculum. Indeed, a systematic presentation of the fundamental criteria and standard anthropologies is necessary for those preparing to work in the field of biomedicine. As mentioned in the introduction, this seems to be the task of bioethics as an academic discipline. It should be an intellectual movement founded on a group of specialists—an invisible school—that starts from basic principles and shows how they are to be applied in certain relevant and current problems selected from among the most well-known and controversial issues in medical practice. Without this fundamental study, bioethics committees would become tools with no content, probably linked only to practical circumstances, which would develop dissimilar solutions derived extemporaneously from the values of their individual members.

The International Situation

Since the formal institution of the first bioethics committees, such as the one set up by the Supreme Court of New Jersey in March of 1976, numerous bioethics committees have continued to spring up in various countries at the national, regional, and local levels.

On the basis of the documentation available to the bioethics center and the available literature, an attempt will be made here to outline the situation of these committees in countries with experience in the field. Points of interest include their distribution at various levels, their composition, and the opinions they have already formulated concerning some particular problems in bioethics. These opinions undoubtedly constitute an important reference point, as will be seen later on, especially when coming from national bioethics committees. The Italian situation will be examined in greater detail, including the legal and operative frameworks in which the committees come to be situated.

It is relevant to note that the various ad hoc commissions formed to address particular issues should also be considered within the broad topic of bioethics committees.[25] While bioethics committees are generally understood as permanent structures that from time to time address the ethical questions raised by clinical practice, experimentation protocols, or the application of biomedical technology to man in general, commissions for the most part consist of a group of experts designated by a given authority for a limited time—although there are numerous and important exceptions, including the permanent United States Presidential Commission for the Study of Bioethical Issues—and charged with evaluating a certain issue and expressing a reasoned ethical opinion. Various commissions of this sort were instituted much earlier than 1976 yet should plainly be considered true bioethics committees before

---

[25] See Walters, "Bioethics Commissions," previously cited in note 1. A thorough survey of the international situation of bioethics commissions outside the United States is presented in D. Wikler, "Bioethics Commissions Abroad," *HEC Forum* 6, no. 4 (1994): 290–304.

the widespread use of the term. Ad hoc ethics commissions continue to be instituted even today in many countries, as will be seen in the discussion to follow,[26] but first I would like to make a few comments on the theoretically possible institutional levels that have actually come into existence in various countries.

The *first level* is the *central, national (federal)*, or even *international* level. It entails the duty of confronting broad issues involving the population in general (e.g., genetic manipulation, reproductive technologies, embryo protection, national health care policies, and so forth). In other words, it can serve as a technical reference for government action in developing framework legislation. This level includes the National Ethics Advisory Committee in France, the National Bioethics Committee in Italy, and the President's Commission for the Study of Bioethical Issues in the United States. Moreover, one of its specific tasks should be to issue recommendations and guidelines, eventually strengthening the ethical and deontological line at lower levels. Finally, it could have the task of accreditation and procedural review regarding second- and third-level bioethics committees.

The *second level* is the *institutional, academic*, or *associational* level. It includes individual research institutes, whether financed by the central government or not, university communities, associations of physicians or nurses, or regional administration structures. The specific task of these committees is primarily connected with research and clinical trials, strictly deontological and professional problems, and the issuing of guidelines for the protection of patients' rights in central and local hospitals.

The *third level*, finally, is the *local hospital* or *health care agency* level, with more specific functions relating to clinical cases and the education and training of health care workers.

Let us now briefly examine bioethics committees at the international level, keeping in mind that almost every country has established such a committee by now and it would be impossible to provide an exhaustive list. The following are examples of bioethics committees that were among the first to be established and that regularly publish reports on various topics.

United States of America

As mentioned before, the United States is the place where bioethics committees first sprang up and from which they spread to the rest of the world. An examination of the American situation is therefore particularly informative.

In reference to the three-level subdivision previously described, let us begin by considering central or first-level bioethics committees. The most important of these is currently the President's Commission for the Study of Bioethical Issues, which was instituted in 2009 by President Barack Obama to replace the President's Council on Bioethics, instituted by President George W. Bush in 2001—and renewed in 2003, 2005, and 2007—which was the "descendant" of the President's Commission for the Study of Ethical Problems in Medicine and Biomedical and Behavioral Research

---

[26] Obviously, this analysis cannot be an exhaustive survey of the entire international dimension of bioethics committees; nonetheless, it seems revealing enough and sufficient for grasping the pertinent issues.

(1978–1983).[27] Already in 1995, President Bill Clinton had established the National Bioethics Advisory Commission to replace the latter commission. The advisory role and tasks of these national committees instituted in the United States during subsequent presidencies have remained essentially the same as those of the first commission.

Taking into account the urgency of numerous ethical problems in the realm of biomedicine, in November 1978 the Congress of the United States authorized the creation of a permanent presidential commission to study and offer guidelines on these problems. In June 1979, President Jimmy Carter nominated the eleven members of the commission: three members chosen from among distinguished scientists in the area of experimental research; three members chosen from among eminent clinicians; and five members chosen from among particularly distinguished individuals in the fields of ethics, theology, law, natural sciences, social sciences, health care administration, government, and public administration.

The commission officially began working in January 1980, and these specific tasks were identified: (1) help clarify the problems, providing specific instructions especially for problems of particular relevance at the decision-making level; (2) suggest guidelines for public policy at various levels and by various means, not necessarily to be turned into legislation; (3) offer a guide for those who are the direct decision makers, yet without dictating particular choices regarding the different degrees of morality.

From the time it began working, the commission published various reports on numerous problems (determination of brain death, how to treat patients in a persistent vegetative state, genetic engineering and its applications on man, relationships between economics and health, etc.). From the economic point of view, the commission could count on a budget of $5 million per year. An educational and training program was organized around it, developed and run by the Institute of Society, Ethics and Life Sciences, better known as the Hastings Center, which was established in 1969 as one of the first bioethics centers in the world.

At the second and third levels, still in the United States, there are two different types of bioethics committees: (1) hospital ethics committees, which are instituted within individual hospitals, and (2) institutional review boards (IRBs), which are established in research institutes primarily to make assessments and express ethical opinions on human experimentation protocols.

The following specific tasks are assigned to hospital ethics committees: assessment of decisions regarding the treatment of terminally ill or incompetent patients; assessment of all medical decisions with ethical implications; consultation for patients, relatives, physicians, and any other member of the hospital staff regarding the delicate and unique problems affecting the diagnostic, therapeutic, or rehabilitation phases; issuance of guidelines for diagnostic techniques, particular treatments, or economic or administrative problems involving the distribution of funds; and organization of educational and training programs for all hospital members, including physicians,

[27] United States President's Commission for the Study of Ethical Problems in Medicine and Biomedical and Behavioral Research, *Final Report on Studies of the Ethical and Legal Problems in Medicine and Biomedical and Behavioral Research* (Washington, DC: US Government Printing Office, 1983); B. A. Brody, "The President's Commission: The Need to Be More Philosophical," *Journal of Medicine and Philosophy* 14, no. 4 (1989): 369–383.

nurses, and administrators. Regarding the composition of these bioethics committees, a 1983 survey by the President's Commission revealed that physicians were the most-represented category of individuals on bioethics committees, constituting approximately 60 percent of the total members. Some committees were composed exclusively of physicians (committees "of equals"), but for the most part at least one member was religious and two or three members were representatives of other professions. Administrative personnel and nurses were also frequently represented.

In January 1994, President Clinton also established the Advisory Committee on Human Radiation Experiments in response to increasing reports of unethical conduct by federal or government-affiliated institutions regarding the exposure of human beings to radiation energy during the Cold War. In 1995, the Advisory Committee produced a voluminous final report.[28]

On a smaller scale, the Protestant American Health Care Association recently developed guidelines defining the role of the professional figure of the hospital chaplain on bioethics committees. In reality, even though chaplains can have ethical training, their role is different from that of moralists: they identify and pinpoint spiritual perspectives as essential components of the process of bioethical reflection.[29]

The role of IRBs is more clearly defined and codified. Their composition, operating procedures, and responsibilities have been regulated by the Food and Drug Administration (FDA; the central organization that deals with all substances introduced into the human body: food additives, drugs, biological products, etc.) since the 1970s. In July 1981 the regulations became federal law, describing in detail the characteristics that members should have, what their responsibilities are in scrutinizing experimentation protocols, what authority they have with regard to experimenters, and finally what criteria must be met in order to approve a study.[30]

Australia

In the 1970s the Australian government began a rigorous analysis of the ethical problems involved in medicine and the biological sciences.[31] The first government body charged with examining these issues was the Australian Law Reform Commission, which developed a report on human tissue transplants in 1977.

---

[28] United States Advisory Committee on Human Radiation Experiments, *Final Report* (Washington, DC: US Government Printing Office, 1995). The report is divided into four parts: (1) history, (2) case studies, (3) current projects, and (4) final recommendations. The committee also has an internet website: http://www.seas.gwu.edu/.

[29] For further analysis of these guidelines see E. Tripaldi, *Il cappellano ospedaliero nei Comitati di Bioetica* (Rome: Centro Studi S. Giovanni di Dio, 1995); W. T. Reich, ed., *Encyclopedia of Bioethics*, rev. ed. (New York: Macmillan, 1995).

[30] M. Sherman and J. D. van Vleet, "The History of Institutional Review Boards," *Regulatory Affairs* 3 (1991): 615–628. A survey of the characteristics, policies, and procedures of IRBs in category I American universities is provided in G. J. Hayes, S. C. Hayes, and T. Dykstra, "A Survey of University Institutional Review Boards: Characteristics, Policies, and Procedures," *IRB* 17, no. 3 (1995): 1–6.

[31] P. Kasimba and P. Singer, "Australian Commissions and Committees on Issues in Bioethics," *Journal of Medicine and Philosophy* 14, no. 4 (1989): 403–424; Wikler, "Bioethics Commissions Abroad."

The report was rich with ethical indications despite the fact that its aim was of an exclusively legal nature (all the members of the commission were legal experts).

After this first official institution, numerous other committees were formed at both the state and federal levels; in fact, Australia has more commissions—in proportion to the population—for problems concerning reproductive technologies than any other nation. This can be explained in part by the fact that Australia's system of federal government confers considerable autonomy on the states in matters of biomedical research. Nonetheless, a large part of the research is conducted through the National Health and Medical Research Council (NH&MRC), and it is precisely within this council that the Medical Research Ethics Committee was instituted in 1983 with the aim of forming a permanent advisory body analogous to the President's Commission in the United States. The specific tasks of this committee include assessing individual problems that arise in experimentation on human beings, expressing the principles that should be observed in particular fields of research from time to time by means of recommendations, facilitating the work of local bioethics committees by providing guidelines and reviewing their decisions as appropriate (the relationship with local bioethics committee appears to be very important since in Australia, as in other countries, public research funds are assigned to individual institutes only with approval from the local bioethics committee or, for some fields of research, from the national bioethics committee), and responding to specific inquiries formulated by the government or the Department of Health and Ageing.

Later, in 1988, the departments of health and social affairs of the individual states and the federal government established a joint advisory body, the National Bioethics Consultative Committee (NBCC). It was nonetheless short-lived, disbanding after just three years. In place of the two committees just mentioned, a new Australian Health Ethics Committee (AHEC) was established in 1991 with the task of not only guiding social and educational policies but also coordinating the individual, local bioethics committees.

Finally, individual states established their own commissions on specific topics. The work of several important bioethics committees provided direction on the topic of in vitro fertilization and embryo transfer (IVF-ET) and in general on all reproductive technologies. The Waller Committee, instituted in the state of Victoria in 1982 and composed of nine members (a law professor, Louis Waller, as president; two theologians, one Catholic and one Protestant; a teacher; two professors of medicine, one of whom was an obstetrician–gynecologist; a general practitioner; a lawyer with expertise in family law; and a social worker), drew up an Interim Report in 1982 that considered so-called *simple case IVF* (also called homologous IVF, where a husband and wife are the gamete donors) to be ethically acceptable.[32] It later published the Report on the Disposition of Embryos Produced by In Vitro Fertilization in 1984 on the topic of embryo freezing, evincing strong disagreement among committee members concerning the need to respect the intrinsic value of the embryo. The Queensland Bioethics Advisory Committee, in the state of Queensland, provided observations on the use of embryos in non-therapeutic experimentation. The New South Wales Law Reform Commission was established in 1988 and issued various reports on IVF and surrogate motherhood.

---

[32] As a side note, even this scenario was considered ethically unacceptable by the Congregation for the Doctrine of the Faith. See CDF, *Donum vitae*, part II, no. 5.

Japan

In Japan, primarily due to the ethical tradition of Confucianism, medical paternalism was never questioned until recent years. For millennia, medical practice was considered an art of *jin* (meaning love and benevolence in the teachings of Confucius), reflecting the benevolent action of the physician on whom the patient depended entirely.[33]

Naturally, this situation gradually changed in relation to the affirmation of new sociocultural values promoted by the younger generations. References to the Universal Declaration of Human Rights and the Declaration of Helsinki began to enter into the fields of health care and scientific research as well. The first bioethics committee was established in 1982 at the School of Medicine of the University of Tokushima to address the concerns raised by the first in vitro fertilization experiments. By July 1987, there were bioethics committees at 41 out of 42 state universities and 2 out of 29 private universities.

Given their ability to debate, assess, and approve or disapprove of experimentation protocols, bioethics committees fostered a very positive image of scientists and physicians in public opinion. It should be pointed out, nevertheless, that the lack of a common standard of judgment for the various committees is cause for numerous contradictions. Finally, it is interesting to note some peculiarities of these Japanese committees as identified by a recent survey: their members are professionals only, they are dominated by the presence of men only, and they follow a rigid procedure that defends the experimenter's power in conducting his research, definitively perpetuating medical profession's ever-present paternalism that—so it seemed—should have disappeared.

The Japanese government also instituted an Ad Hoc Committee on Brain Death and Organ Transplantation as an advisory body for the prime minister. A report on the topic was produced in 1992 with many differences of opinion among the members, and it was never made public.[34]

In the year 2000, the Bioethics Committee Council for Science and Technology (BCCST), established at the Science and Technology Policy Bureau, published *Fundamental Principles of Research on the Human Genome*. Since 2001, the same national committee has published numerous guidelines for various areas of research: the human genome, embryonic stem cells, gene therapy, epidemiology, and clinical research. These guidelines emphasize above all the role of informed consent and of ethical committees for experimentation. This intense guideline publication activity shows the Japanese government's considerable and growing attention to bioethics.

---

[33] R. Kimura, "Ethics Committees for 'High Tech' Innovation in Japan," *Journal of Medicine and Philosophy* 14, no. 4 (1989): 457–464.

[34] Wikler, "Bioethics Commissions Abroad." It is common knowledge that organ transplants are taboo in Japan, particularly for organs such as the heart.

Europe

Since the emergence of the first bioethics issues in the 1970s, the Council of Europe has produced numerous recommendations on specific topics[35] through its Parliamentary Assembly and Committee of Ministers, carrying out the functions of a true bioethics committee.[36]

One important step for bioethical reflection was the Parliamentary Assembly's recommendation 934 (1982), which expressed four essential recommendations for the individual member states and the Committee of Ministers on the basis of their competencies: (1) ensure the protection of human rights in the field of genetics, especially the right to have a genetic pattern free of any sort of manipulation and the right to confidentiality concerning the genetic information about individuals contained in databases; (2) compile a list of serious illnesses that can be treated with the consent of a legally responsible individual; (3) prepare a European agreement defining the licit applications of genetic engineering on human beings, envisioning a European registry of all genetic research conducted; and finally, (4) examine the possibility of patenting genetically modified microorganisms. As a result of this recommendation and by the mandate of the Committee of Ministers, the Ad Hoc Committee of Experts on Ethical and Legal Problems relating to Human Genetics (Comité ad hoc d'experts sur les problèmes éthiques et juridiques de la génétique humaine, CAHGE) was instituted in 1983 for the following purpose: "in order to study the questions raised by genetic manipulation techniques, particularly in the light of recommendation 934 (1982) by the Council of Europe, with a view to defining a common policy among member states and possibly developing appropriate legal instruments." Representatives from all the member states were chosen to be part of this committee from among four groups of specialists: biologists, physicians, lawyers, and ethicists. Furthermore, there are representatives from other countries, in the capacity of observers, that are not part of the Council of Europe, such as Australia, Canada, the United States, Japan, and the Holy See.

The CAHGE has examined the problems of protecting patient information, prenatal diagnosis, and therapeutic applications, as well as reproductive technologies and experimentation on embryos and fetuses.

Following the broadening of its scope and the results of the European Ministerial Conference on Human Rights held in Vienna in March 1985, which in resolution 3 expressed the need to provide the public with clear information—up until then kept reserved—and to promote international action in the field of biomedicine, the CAHGE developed into a new committee toward the end of 1985: the Ad Hoc Committee of Experts on Progress in the Biomedical Sciences (Comité ad hoc d'experts sur le progrès des sciences biomedicales, CAHBI). It was given the task of coordinating the activity of the Council of Europe in this field. Specifically, the objective of this

---

[35] Taken from A. G. Spagnolo, "Il progetto di 'Convenzione' Europea sulla bioetica," *Vita e Pensiero* no. 4 (1995): 249–268.

[36] For a reconstruction of the development of bioethics at the European level, see P. Riis, "Medical Ethics in the European Community," *Journal of Medical Ethics* 19 (1993): 7–12; A. Rogers and D. Durand de Bousingen, *Bioethics in Europe* (Strasbourg: Council of Europe Press, 1995).

multidisciplinary body was to help fill some political and legal gaps forming in the wake of the rapid development of biomedical science (experimentation on human beings and embryos, prenatal diagnosis, the use of biotechnologies, etc.). The CAHBI was aware of the difficulties of reaching a consensus among the member states on these extremely delicate topics, and its activity was therefore aimed at exploring every means of promoting constructive dialogue between member states and avoiding points on which no one would budge.

Taking up the mantle of the CAHGE, the CAHBI promoted important recommendation proposals that later developed into recommendation 1046 (1986) and recommendation 1100 (1989), both by the Parliamentary Assembly and pertaining to the use of human embryos and fetuses. In these two important documents, the Council of Europe sought to affirm the values and fundamental principles that should guide every set of regulations in bioethics, while also noting the limits that should be respected at all costs.[37]

But the issuing of recommendations alone was not considered sufficient to achieve the established objective of harmonization, so in a second phase of the development of bioethics in Europe it was considered necessary to proceed by means of a convention to be signed by all member states. So, during the course of the Seventeenth Conference of European Ministers of Justice in Istanbul in 1990, it was recommended (resolution no. 3) that the Committee of Ministers provide the CAHBI with instructions to identify some bioethical questions to be considered of primary importance and "to examine the possibility of preparing a framework convention, open to non-member States, setting out some common standards for the protection of the human person in the context of the development of the biomedical sciences."[38]

During the course of the forty-third ordinary session in June 1991, through recommendation 1160 (1991) on the preparation of a bioethics convention, the Parliamentary Assembly invited the Committee of Ministers to make a concrete proposal for such a framework convention, which was to consist of "a main text with general principles and additional protocols on specific aspects." It was further specified that "the convention should provide a flexible formula with regard to its form, but must not constitute the lowest common denominator as to its content. It must include human rights aspects and take into account the previous work of the Council of Europe."[39] It was therefore recommended that the CAHBI be authorized and encouraged to proceed with the preparation of this proposed convention by also listening to repre-

---

[37] The three recommendations of the Council of Europe cited above—no. 934 of 1982, no. 1046 of 1986, and no. 1100 of 1989, which concern the issues of genetic manipulation, the use of embryos and embryonic material, and experimentation on human embryos in utero and outside the uterus, respectively—were also published in the journal *Medicina e Morale*: the first in 34, no. 1 (1984): 93–96 (French); the second in 36, no. 4 (1986): 902–906 (Italian); and the third in 39, no. 2 (1989): 397–403 (English) and 404–412 (French).

[38] Seventeenth Conference of European Ministers of Justice, resolution no. 3, "Bioethics," June 5–7, 1990, cited in R. Andorno, "The Oviedo Convention: A European Legal Framework at the Intersection of Human Rights and Health Law," *Journal of International Biotechnology Law* 2 (2005): 133–143.

[39] Council of Europe Parliamentary Assembly, recommendation 1160, "Preparation of a Convention on Bioethics," June 28, 1991, http://assembly.coe.int/Main.asp?link=/Documents/AdoptedText/ta91/EREC1160.htm.

sentatives from Third World countries, scientific organizations, in particular those of the European community, and governmental and nongovernmental organizations specializing in the field. Finally, the recommendation required the Committee of Ministers to send in the finalized convention proposal for a formal opinion from the Parliamentary Assembly before its definitive approval.[40]

In 1992, the CAHBI turned into the Steering Committee on Bioethics (SCB).[41] It had the specific task of drafting the convention proposal, which was distributed to the public for debate prior to the opinion of the Parliamentary Assembly. In fact, with the authorization of the Committee of Ministers, the Committee on Legal Affairs of the Assembly of the Council of Europe made the SCB draft of the convention proposal public in July 1994,[42] allowing the SCB itself and individual national governments to proceed with the necessary consultations and take the expressed opinions into account during the preparation of the final text.[43] On November 19, 1996, after five years of work, the Committee of Ministers of the Council of Europe finally passed the Convention for the Protection of Human Rights and Dignity of the Human Being with Regard to the Application of Biology and Medicine: Convention on Human Rights and Biomedicine.[44]

Remaining at the European level, the European Forum for Good Clinical Practice (EFGCP) merits attention. It is a nonprofit organization constituted in accordance with Belgian law that published a document in June 1995 titled *Guidelines and Recommendations for European Ethics Committees*.[45] The aim of this document "is to establish a greater degree of scientific efficacy and procedural responsibility in the practices of Ethics Committees (ECs) in Europe. The document is intended

---

[40] The formal opinion referred to here was approved on February 2, 1995, during the course of the sixth session of the assembly (opinion no. 184).

[41] In French, the other official language of the Council of Europe, the designated name for this committee is the Comité Directeur de Bioéthique (CDBI).

[42] Document DIR/JUR (94) 2. This document consists of a convention proposal (thirty-two articles) and a report proposal explaining the same convention proposal.

[43] On the meaning of the Council of Europe's proposed convention on bioethics, see A. G. Spagnolo, "Il progetto di 'convenzione' europea sulla bioetica," *Vita e Pensiero* no. 4 (1995): 249–268.

[44] Document DIR/JUR 96 (14) of the Directorate of Legal Affairs of the Council of Europe. Prior to its approval by the Committee of Ministers, the text had already been approved by the SCB in June 1996, as well as by the Council of Europe's Parliamentary Assembly—with some amendments—in September (with two opinions, nos. 180 and 198). Even after approval by the Committee of Ministers, the assembly can nonetheless reserve the right to make further amendments to the text. For an analysis of the document, which is acceptable on some points and open to criticism on others, see A. Bompiani, "Una valutazione della 'Convenzione sui diritti dell'uomo e la biomedicina' del Consiglio d'Europa," and E. Sgreccia, "La Convenzione sui diritti dell'uomo e la biomedicine," both published in *Medicina e Morale* 47, no. 1 (1997) on pp. 37–55 and 9–13, respectively. The text of the convention was published in French with an Italian translation in the same issue of the journal, on pp. 128–149. The updated version is available in English at http://conventions .coe.int/Treaty/en/Treaties/html/164.htm.

[45] The text of the guidelines, in both the original English and an Italian translation, can be found in *Medicina e Morale* 45, no. 5 (1995): 1064–1085.

as a basis upon which bioethics committees can develop their own specific written procedures for their functions within biomedical research."[46]

To conclude, there is also a European Ethical Review Committee at the European level.[47] It is a private body, however, instituted in 1977 with the specific aim of reviewing drug trial protocols presented by the major multinational pharmaceutical companies, from which the committee claims to be independent. It is made up of thirty-one members from nine European countries (Belgium, France, Germany, United Kingdom, Holland, Italy, Norway, Sweden, and Switzerland). Its standard ethical and deontological principles are the ones found in the Declaration of Helsinki, while its composition and procedural methodology are in keeping with US government policies regarding experimentation on human subjects.[48]

The individual states that are part of Europe, however, each have various national and local bioethics committees. These will now be mentioned by way of example, ultimately concluding with the situation in Italy.

Belgium. The diversity of experiences, mentalities, and philosophical traditions distinguishing the French region from the Flemish region is also a characteristic of Belgian bioethics committees.[49] The Foundation for Medical and Scientific Research (FMSR), part of the most important organization in the country for subsidizing research, created a medical ethics commission in 1976 to provide opinions on particular bioethical problems (genetic manipulation, human experimentation, organ transplants) and to encourage and organize the establishment of ethics committees in medical schools and university hospitals. Two local bioethics committees emerged in the two Catholic universities in the area of Leuven: the Flemish-speaking Katholieke Universiteit Leuven (KU Leuven) and the French-speaking Université Catholique de Louvain (UCL). The ethics committee at the medical school of the Flemish KU Leuven was immediately entrusted with several tasks: first of all, to express opinions on broad ethical issues with concrete relevance to scientific and educational activity, then to review experimentation protocols and analyze clinical cases of particular ethical complexity, and finally to provide the institutions with updates on topics of bioethics.

Conflicting positions regarding the functions to be adopted arose within the committee at the medical school of the French UCL. At first, the intention was for the committee to provide an ethical assessment before conducting research projects, which would help physicians and ensure their protection in dealing with public

---

[46] European Forum for Good Clinical Practice, *Guidelines and Recommendations for European Ethics Committees*, rev. ed. (Brussels: EFGCP, 1997), 2, http://www.efgcp.be/Downloads/EFGCP%20Guidelines%20and%20Recommendations%20for%20EEC.pdf.

[47] J. M. Faccini, P. N. Bennett, and J. L. Reid, "European Ethical Review Committee: The Experience of an International Ethics Committee Reviewing Protocols for Drug Trials," *British Medical Journal* 289 (1984): 1052–1054. This committee usually meets once a month. The location for the meetings is often London, but other continental European cities are also chosen from time to time. It should be pointed out that although it considers itself an ethics committee, it does not include an ethical expert or a philosopher among its members!

[48] "Final Regulations Amending Basic HHS Policy for the Protection of Human Research Subjects," *Federal Register* 46, no. 16 (January 26, 1981): 8366–8391.

[49] M.-L. Delfosse, "I Comitati di Etica in Belgio," in *I Comitati di etica in ospedale*, ed. S. Spinsanti (Cinisello Balsamo, Italy: Paoline, 1988), 101–110.

opinion while at the same time safeguarding patients' rights. It later sought to draw up general rules as a necessary preamble to the assessment of research projects.

Ethics committees and commissions are also present in the secular hospitals of the country, whether affiliated with universities or not. These committees are as pluralistic and interdisciplinary as possible in terms of their composition, and their primary reference point was the Medical Ethics Commission of the Foundation for Medical and Scientific Research, which had in turn incorporated the deontological principles spelled out by the National Council of Medical Associations.

In 1993, through a cooperation agreement between the federal states and the three Belgian linguistic communities, the Belgian Advisory Committee on Bioethics (Comité Consultatif de Bioéthique de Belgique, CCBB) was created with the task of providing institutions with opinions on issues inherent to their research and its applications in the fields of biomedicine and health care, and of informing the public and the authorities about those issues.

At the local level, ethics committees are currently the most important, practical tool for biomedical ethics and perform three functions: review of experimentation protocols, formulation of opinions on the ethical aspects of clinical practice, and ethical consultation.[50]

Denmark. In 1979, a central committee was established—the Central Research Ethics Committee (CREC)—as the apex of an entire system of institutional bioethics committees assessing the individual protocols for experimentation on human beings, along with seven local committees to review experimentation. This system was established in compliance with the Declaration of Helsinki regarding experimentation on human beings.[51] The CREC was originally made up of a group of professionals working on a volunteer basis, albeit in the government sector, but as of 1992 its constitution and authority are well defined. This committee coordinates the local bioethics committees and intervenes where there are difficulties in making decisions.[52]

The Danish parliament issued provisions for the formation of another national bioethics committee with law no. 353 on June 3, 1987: the Danish Council of Ethics for Health Care Services and Biomedical Research Conducted on Human Subjects. This committee, which began its activity in January 1988, has the task of working with health care authorities and individual institutional ethics committees.

It is significant that the Danish government already stated the committee's standard ethical principles in the legislative act establishing the committee. Thus the first article declares that experiments must be conducted according to the principles of the Declaration of Helsinki and that the council "shall carry out its work on the assumption that human life begins at the moment of fertilization."[53]

---

[50] T. Meulembergs, J. Vermylen, and P. T. Schotsmans, "The Current State of Clinical Ethics and Heathcare Ethics Committees in Belgium," *Journal of Medical Ethics* 31 (2005): 318–21.

[51] U. H. Petersen, "The Danish Committee System," in *Funzione e funzionamento dei comitati etici*, ed. G. Gerin (Padua: CEDAM, 1991), 131–141.

[52] Wikler, "Bioethics Commissions Abroad."

[53] Danish Folketing, law no. 353, "Establishment of an Ethical Council and the Regulation of Certain Forms of Biomedical Experiments," June 3, 1987, § 1 (1), http://etiskraad.synkron.com/sw297.asp.

The council is made up of seventeen members nominated by the Minister of the Interior and Health, ensuring that there is only one more member of one sex than the other. In particular, articles 4 through 7 require the council to study and provide recommendations on the status and protection of the human embryo, the possibilities of gene therapy, the use of techniques for prenatal diagnosis, and cryopreservation, respectively.

Moreover, together with local bioethics committees, the council must provide information on the issues of experimentation on human beings and the most salient ethical questions regarding health care services in relation to the use of new diagnostic and therapeutic methods. Finally, articles 11 and 12 respectively contain a list of prohibited experiments that the the council must obviously take into account and the associated penalties for violating them.

Every year this council publishes a volume composed of two parts: (1) a report on the activity carried out during the year and (2) in-depth studies and recommendations on particular topics. Over the years, numerous reports have been published. Examples topics include the protection of human gametes, fertilized oocytes, embryos, and fetuses (1990); euthanasia (1996); ethics and fetal diagnosis (1990); ethics and mapping the human genome (1992); screenings (2000); anonymity and selection in sperm donation (2003); and the beginning of human life and the moral status of the human embryo (2004).[54]

France. The National Ethics Advisory Committee for Life and Health Sciences (Comité Consultatif National d'Éthique pour les Sciences de la Vie et de la Santé, CCNE) has existed for many years now in France, instituted through a decree by President of the Republic F. Mitterand on February 23, 1983.

The decree, consisting of fifteen articles, charges the committee (article 1) with the task of giving opinions on moral problems raised by research in the fields of biology, medicine, and health care—problems regarding both individual human subjects and specific social groups or society in general.

The committee is composed of thirty-two members in addition to the president, who is directly appointed by the President of the Republic. The latter also directly nominates members representing the major currents of philosophical and spiritual thought.

Fourteen members are chosen from among the most distinguished individuals with competence and interest in the ethical field and in areas pertaining to the ministries of education, research, industry, work, health care, justice, family, and communication. Another fourteen members are representatives from the research sector and the major national research institutes and universities.

To date, the French National Ethics Advisory Committee has published over eighty-nine reports on subjects of great importance, classified by topic (ethics committees, epidemiology and prevention, end of life, genetics, neurosciences, transplants, reproductive technologies and embryos, experimentation on human beings, AIDS, sports, drug addiction, etc.).[55] The first president was Jean Bernard.

---

[54] The annual volumes can be requested directly from the Danish Ethics Committee: Det Etiske Råd, Ravnsborggade 2–4, DK-2200 Copenaghen N (fax: +4535375755), at http://www.etis kraad.dk.

[55] The complete list of all documents was published in *Les cahiers du Comité Consultatif National d'Éthique pour les Sciences de la Vie et de la Santé*, the official publication of the

United Kingdom. In this nation, the initiative for a national bioethics committee came from the British Medical Association. Since 1849, the association autonomously availed itself of a medical ethics committee that later became the National Ethical Research Committee. Following deliberations on January 4, 1984, and January 8, 1986, this committee issued some norms on the composition of local committees and of the national committee itself.

One of the primary functions assigned to this bioethics committee was the review of all biomedical clinical research (in a footnote in the approval document, it was specified that all projects involving the use of clinical techniques or treatments and the personal information of patients are included in the context of this research). In this field, therefore, the committee was a reference for quality control and procedural assessment as well as for preliminary opinions.

Another committee of great importance in the area of research on embryos was the Warnock Committee, instituted by the government of the United Kingdom to address problems regarding in vitro fertilization.[56] This is also addressed in another chapter of this work, where the reader can find a more detailed discussion of the content.[57]

There are currently two types of local ethics committees in the United Kingdom: ethics committees (ECs) for clinical practice and ethical consultation at hospitals and research ethics committees (RECs) for the evaluation of clinical experimentation. The national (or first-level) ethics committees reflect that distinction: the Central Office for Research Ethics Committees (COREC), which is part of the National Patient Safety Agency, has the task of coordinating the research ethics committees and backing them in implementing standards and training members; the Nuffield Council on Bioethics, which is an independent body instituted in 1991 by the Nuffield Foundation, the Medical Research Council, and the Wellcome Trust, has the task of identifying and assessing bioethical issues in biology and medicine and issuing reports (the most recent ones concern the patentability of DNA, research on stem cells, pharmacogenetics, and experimentation on animals) addressed to both legislators and the general public. The members of the Nuffield Council are experts in medicine, law, philosophy, scientific research, and theology.[58]

Portugal. Here there is a National Council on Ethics and Medical Deontology linked to the Federation of Medical Associations. It is composed of a president and six members, two each from the most important Portuguese universities. The National Council is closely connected with the Portuguese Bioethics Center: they are both recognized by the government with constitutions approved by public act.

---

committee, whose headquarters is located at 71, rue Saint-Dominique, 75007 Paris. The work is also available online at http://www.ccne-ethique.fr.

[56] M. Warnock, "A National Ethics Committee," *British Medical Journal* 297 (1988): 1626–1627.

[57] See the chapter "Bioethics and Abortion" in this volume; for an overall perspective on the situation of bioethics committees in the United Kingdom, see A. V. Campbell, "Committees and Commissions in the United Kingdom," *Journal of Medicine and Philosophy* 14, no. 4 (1989): 385–401; J. Metters, "Regulations on Bioethics in the United Kingdom," in *Funzione e funzionamento*, Gerin, 107–116.

[58] The reports of the Nuffield Council on Bioethics and the COREC are available at http://www.nuffieldbioethics.org and http://www.corec.org.uk, respectively.

In 1990, the national government instituted the National Council of Ethics for the Life Sciences (Conselho Nacional de Ética para as Ciências da Vida, CNECV), in response to a parliamentary initiative. The president is nominated by the prime minister; the other twenty members, who hold their posts for five years, are nominated by other related ministries and by scientific and professional organizations.[59]

Then there are bioethics committees in various central hospitals and university clinics. For example, a local bioethics committee was formed at the Hospital of the University of Coimbra called the Medical Ethics Committee. It is charged with the task of keeping watch within the hospital over the proper observance of the ethical and deontological norms that guide medical practice. Its specific reference principles are the ones contained in the Declarations of the World Medical Assembly and in the deontological code of the Portuguese Medical Association. Members by right include the clinical director of the hospital, a head physician with clinical research experience, an assistant with clinical research experience, the professor of pharmacology for the medical school, and the professor of medical deontology for the medical school. From time to time, the collaboration of legal experts, moralists, or any other persons considered competent on particular problems may be requested for purposes of consultation.

The committee, which meets periodically, draws up a report within thirty days on all experiments regarding diagnostic and therapeutic procedures not yet in use. Additionally, the committee responds with its own report whenever there is an explicit inquiry from the health care management of the hospital.

Spain. The Spanish Bioethics Committee (Comité de Bioética de España, CBE) was established by law no. 14 on July 3, 2007.[60] Prior to this, there had been ad hoc advisory commissions for discussing particular problems that were later dissolved. One of these was the commission provided for by law no. 35 in 1988 on reproductive technologies, which affirmed in article 21 that a national committee should be instituted by a royal decree. This committee had not yet been established as of 1993, but an advisory committee on assisted reproductive technologies was instituted by decree in 1992 in Catalonia, where there is an autonomous government.[61] At the central level, the Central Committee on Deontology is an advisory body dependent on the General Council of the Spanish Medical Association. It does not have executive powers and all of its decisions and recommendations become effective only after having been approved by the executive council or general assembly.

The Central Committee is composed of eight members elected from the general assembly of the General Council. Its specific function is the development, updating, and interpretation of the Spanish code of medical deontology. Another one of its functions is to organize the national convention of deontology committees instituted within each local medical association.

---

[59] Wikler, "Bioethics Commissions Abroad"; R. Bandeira, "Hospital Ethics Committees in Portugal," *HEC Forum* 3, no. 6 (1991): 347–348; J. Biscaia and W. Osswald, "Bioethics in Portugal: 1991–1993," in B. A. Lusting, ed., *Bioethics Yearbook: Regional Developments in Bioethics: 1991–1993* (Dordrecht: Kluwer, 1995), 4:285–289.

[60] Spanish Cortes Generales, law no. 14, "Biomedical Research," July 3, 2007.

[61] F. Abel et al., "Bioethics in Spain: 1991–1993," in *Bioethics Yearbook*, 4:269–283.

The ethical evaluation of cases of unprofessional conduct and the issuing of recommendations or directives regarding deontological conduct are instead the responsibility of the General Council.[62]

Regarding second- and third-level committees, the law requires that hospitals—public or private—have two types of bioethics committees: one for clinical drug trials and one for evaluating the legal requirements for abortion. The manner of formation, the composition, and the functions of these committees were indicated in article 64 of law no. 25/1990 and later in Royal Decree 561/1993.

The ethics committee of San Juan de Dios Hospital in Barcelona is an example of a hospital bioethics committee. This committee is composed of four groups of people: (1) members by right (the superior of the community, an ethics expert, a representative for the health pastoral care program, and the medical director, who is the highest representative of health care services in the hospital), (2) ordinary members (two physicians, a professional nurse, and a representative from outside the health care sector), (3) additional members (convened from time to time according to the needs of the case or the protocols under scrutiny), and (4) a secretary, who should not belong to any of the other groups mentioned.

The committee has three specific functions: making decisions on strictly ethical issues, advising on questions in which ethical issues might be involved, and promoting educational and informational initiatives on ethical questions and issues. Decisions can be made unanimously or by a relative majority of the members, but the vote of the ethics expert weighs the most in cases of decision-making consultation.[63]

Switzerland. Here there is a Central Medical Ethics Committee (Commission Centrale d'Éthique Médicale) that was established by the Swiss Academy of Medical Sciences (Académie Suisse des Sciences Médicales, ASSM) in 1979. Its basis is in the previous *Directives pour la recherche expérimentale sur l'homme* (Guidelines for experimental research on human beings) by the same ASSM (1970, updated in 1981) and in the first revised edition of the Declaration of Helsinki (1975) by the World Medical Association. There is no federal law regulating human experimentation, so the indications given by the ASSM are taken into very serious consideration by the courts in legal proceedings on harm resulting from experimentation. The ASSM's indications are therefore much more than merely "paralegal" recommendations.

The Central Committee has various well-defined tasks: responding to medical ethics inquiries posed by the Swiss Confederation, cantons, international organizations, Switzerland's Federation of Physicians, or individual researchers and private citizens; putting the reasoned opinion into writing; maintaining close contact with the ethics committees of university institutes and hospitals; carrying out coordination activities; promoting information exchange and constantly updating the list of local bioethics committees; functioning as an ethics committee for hospitals and research

---

[62] G. Herranz, "Il Comitato Centrale di deontologia spagnolo," in *I Comitati di etica in ospedale*, 141–148. For the situation of ethics committees in Spain, see J. Egozcue, "The Ethical Committees in Spain," in *Funzione e funzionamento*, Gerin, 145–151.

[63] The internal regulations of the ethics committee of San Juan de Dios Hospital are published in *Labor Hospitalaria* 209, no. 3 (1988): 216–217. Two important documents by this committee merit particular attention: one regarding care for children affected by spina bifida cystica (myelomeningocele), in *Labor Hospitalaria* 210, no. 4 (1988): 301–303; and the other on the criteria for brain death in children, in *Labor Hospitalaria* 212, no. 2 (1989): 148–151.

institutes that do not have their own committee; and issuing recommendations and guidelines after approval by the ASSM Senate.[64]

Italy

Several years later than in other countries (particularly the United States), some regional and local bioethics committees began to spring up in Italy as well, albeit sporadically and with difficulty. Only later was the National Bioethics Committee (NBC) instituted by a decree of the president of the council of ministers.

Proceeding chronologically, some Italian regions—in the context of the constitutionally afforded autonomy in managing health care—had already instituted advisory or mediatory commissions by 1975, primarily in the area of patients' rights but also for clinical drug trials.[65] Initially, as mentioned, these bioethics committees experienced difficulties getting started—difficulties that were likely due more to fear of their functions being corrupted than of their nature or origins. This is why it will

---

[64] For a more detailed look at the situation of bioethics committees in Switzerland, see J.-M. Thévoz, "Research and Hospital Ethics Committees in Switzerland," *HEC Forum* 4, no. 1 (1992): 41–47

[65] In a study by the Italian Catholic Physicians Association (Associazione dei Medici Cattolici Italiani, AMCI) financed by the National Research Council (Consiglio Nazionale delle Ricerche, CNR), in which the Bioethics Center of the Catholic University of the Sacred Heart participated, the role of bioethics committees in the National Health Care Service (Servizio Sanitario Nazionale, SSN) was examined and a comparative analysis with foreign institutions was conducted. The work containing the contributions to this study was published under the title *I Comitati di Bioetica: Storia, analisi, proposte* (Rome: Orizzonte Medico, 1990). More recently, a study by the Center for Social Investment Studies (Centro Studi Investimenti Sociali, CENSIS) more systematically revealed the number and distribution of bioethics committees in Italy, presenting its results in CENSIS Forum on Biomedical Research, *La ricerca biomedica in Italia*, vol. 2, *Industria farmaceutica ed università verso l'integrazione europea* (Milan: F. Angeli, 1993). A summary of these results is presented in C. Vaccaro, "I Comitati di Etica in Italia," in *Una Verità in Dialogo: Storia, metodologia e pareri di un comitato di etica*, ed. P. Cattorini (Milan: Istituto Scientifico Ospedale San Raffaele–Europa Scienza Umane, 1994), 100–111. In short, the CENSIS study revealed that a good 33 percent of Italian bioethics committees rapidly emerged around the time of the issuing of the cited ministerial decree on April 27, 1992 (see note 20). The same survey showed extreme variety in the understanding, composition, tasks, functions, and operating procedures characterizing these committees. This is to be attributed especially to the improvisation and lack of coordination that have characterized the emergence of a large number of these bodies. Furthermore, in many cases these bioethics committees for experimentation arose more as means of administrative monitoring than as true bioethics committees; they were composed primarily of technical experts and lacked, for the most part, the "lay" members who ensure the interdisciplinary character of bioethics committees. Indeed, according to the CENSIS study, only 45 percent of the committees had an ethics expert, 21 percent had a psychologist, and 35 percent had a theologian or religious member. This was precisely because the activity of those committees, as mentioned earlier, was the administrative monitoring of pharmacological trials in 91 percent of cases, and therefore did not require the presence of a philosopher, ethicist, or legal expert. Another negative characteristic that surfaced from the CENSIS study was the complete lack of contact between the various bioethics committees, even though they operated just a few miles from one another: there was no form of cooperation, no shared discussion about approaches, and no attempts to standardize operating procedures, judgment criteria, etc.

be important to identify and avoid such corruptions, as indicated earlier, in order to restore the true identity of bioethics committees and thereby promote the activation and function of an ever greater number of them in Italy. The ministerial decree on April 27, 1992, contributed to this end, as did the ministerial decree on July 15, 1997, which accepted the EEC Directive on Guidelines for Good Clinical Practice and subjected the conducting of experiments to approval by the bioethics committee.

The National Bioethics Committee (NBC) was instituted by decree of the prime minister on March 28, 1990, with the following aims:

> To develop an overall summary of the programs, objectives, and results of research and experimentation in the field of life sciences and human health, making use of the ability to access the necessary information at existing and active centers at the national level and in conjunction with similar committees instituted in other countries and other international organizations working in the field;
>
> To formulate opinions and indicate solutions, including with the aim of preparing legislation, to address ethical and legal problems that may arise with the advancement of research and the appearance of new potential applications of clinical interest, having respect for the protection of the fundamental rights and dignity of human beings and for other values expressed by the constitution and by international documents to which Italy adheres;
>
> To propose solutions for oversight aimed at protecting both human safety and the environment in the production of biological material as well as protecting patients who are being treated with genetically engineered products or subjected to gene therapy from potential risks;
>
> To promote the drafting of codes of conduct for workers in the various sectors concerned and to foster properly informed public opinions.[66]

In recent years, the NBC has produced various documents on issues of bioethics, the first of which examined gene therapy, followed by publications on biotechnology, bioethics training, organ transplants, the definition of death, informed consent, and other topics.[67]

Some ministerial commissions for advisory purposes had been established before the NBC; they addressed far-reaching issues such as genetic engineering and assisted reproductive technologies. One example is the Santosuosso Commission, which was active from October 31, 1984, to November 22, 1985, and provided a report and a draft of a bill on assisted reproductive technologies. Another is the Polli

---

[66] Decree of the President of the Council of Ministers, March 28, 1990, article 1, http:// unipd-centrodirittiumani.it/public/docs/28_1990.pdf.

[67] A summary of all the opinions issued in the first two years of activity is contained in Italian National Bioethics Committee, *Rapporto al Presidente del Consiglio sui primi due anni di attività del Comitato nazionale per la Bioetica* (Rome: Presidenza del Consiglio dei Ministri Dipartimento per l'Informazione e l'Editoria: 1992). The documents of the NBC can be directly requested from the main office of the NBC (via della Mercede, 96-00187 Roma), found on the NBC website (http://www.governo.it/bioetica/), or read in the journal *Medicina e Morale*, which publishes all of them.

Commission, which was active from September 13, 1985, to February 9, 1987, and drew up a report on the topics of genetic engineering and prenatal genetic diagnosis.

It has been said that outside the realm of reviewing clinical experimentation, there is no law in Italy regulating the institution of local bioethics committees. Faced with a national legislative void, some bills were presented but never enacted over the course of the subsequent legislatures. Just two of them are listed here because they are more directly connected to the establishment of ethics committees in hospitals: (1) bill no. 236, presented July 21, 1987, by Bompiani and others, titled Protection of Patients' Rights with Particular Regard to the Situation of Hospital Stay; and (2) the Framework Act on the Rights of Sick Citizens (Chamber of Deputies bill no. 4181) proposed by Aniasi and others and presented on November 14, 1986. Article 14 of the first bill provided for the establishment of so-called deontological ethics committees. The committees were designed to advise the health care director of the recovery and treatment facility, who would be responsible for enacting the law once approved. It is interesting to note here that the advisory role and the scope of any bioethics committee intervention would have regarded "everything that lies beyond the scope of administrative, civil, and criminal laws and regulations that govern health care service." In other words, the competence of bioethics committees would pertain to something "new," not otherwise codified in laws or regulations. This is a very appropriate clarification in order to avoid the "corruptions" mentioned previously. In particular, the opinions requested of these committees would primarily regard the topic of biomedical experimentation. However, they would also be assigned the task of promoting seminars and continuing education courses on topics of deontology and medical ethics directed to personnel working in health care facilities.

The proposal by Aniasi and others, on the other hand, talked about the institution of "mediation committees" (article 11) that would intervene in proceedings initiated by the ombudsman, who would receive and prepare complaints regarding the violation of the patient's rights. The Democratic Federation Movement (Movimento Federativo Democratico, MFD), which had already fought for the institution of a tribunal for patients' rights in 1978, was actually the real proponent of this bill. And the entire proposal for the framework law was based precisely on the years of experience with these tribunals. Ultimately this bill remained entirely within the legal arena, identifying new "powers" that were different from the traditional ones—even in terms of disciplinary aspects—and therefore turned out to be very far, in my opinion, from the original approach and reasons for bioethics committees.

In the absence of a national set of norms which, according to many people, would be inappropriate in that it would restrict bioethics committees, there have instead been second- and third-level entities already operating for some time with particular functions different from those of the national committee.

In the area of constitutionally afforded autonomy in health care management, there are laws at the level of some Italian regions that have established advisory commissions. These commissions work just as much in the area of the use of new drugs and diagnostic techniques in the hospital setting—in practice in the field of clinical trials—as in the area of protecting the rights of patients or, more generally, of those who use the services of health care structures.[68]

---

[68] See the previously cited volume edited by the AMCI, and the articles "I comitati di bioetica" and "Comitati e Commissioni di bioetica" by Spagnolo and Sgreccia.

University committees, research institutes, hospitals, local health care agencies, and scientific and professional associations are more or less autonomous and independent from the provisions of regional laws and have operating procedures that more closely resemble the original features of bioethics committees. The CENSIS study cited earlier[69] presented a very accurate picture of the reality of Italian bioethics committees, counting 114 of them, though with the awareness—declared by the study itself—that this is neither a precise nor definitive number because it only refers to the committees that responded to the questionnaire they were sent. It is nonetheless a starting point for developing potential proposals and for identifying new approaches to the activation and coordination of local bioethics committees. Some of the characteristics of the committees surveyed in the study are indicated in the tables presented at the end of the chapter.

As far as experimentation reviewing is concerned, the national situation has been evolving over the years[70] by way of a lively debate between bioethics committees throughout the nation and the Ministry of Health, and with the issuing of numerous ministerial decrees and circulars that, in addition to accepting the guidelines of the European Union,[71] have gradually sought to improve and render uniform the standards and quality of evaluations by bioethics committees. The National Monitoring Centre on Clinical Drug Trials (Osservatorio Nazionale sulla Sperimentazione Clinica dei Medicinali) was instituted in 1998, which is a computerized system for collecting data on clinical drug trials conducted on national territory.

All ethics committees accredited by the Ministry of Health and public and private sponsors are connected by a network, and through the information in this network they can learn about the characteristics of clinical research in Italy: the centers conducting experiments, approved research, and the areas of research that are currently significant.[72] Before these decrees were issued, several initiatives had been given the green light in attempting to fix shortcomings stemming from a lack of standard ethical review procedures for experimentation protocols, including in the composition and functioning of bioethics committees. Starting in 1986 there were many successive conventions and seminars on the topic of bioethics committees,

---

[69] CENSIS Forum on Biomedical Research, *La ricerca biomedica in Italia*.

[70] In addition to the activities of civil bioethics committees in Italy, which will be discussed shortly, the institution of a bioethics committee by decree of the minister of defense on March 13, 1996, is on record at the General Directorate of Military Health (Direzione Generale della Sanità Militare, DIFESAN). Its purpose is to review biomedical experimentation protocols in the area of military health care as well as the diagnostic and therapeutic protocols followed in health care activity. This bioethics committee has already published an opinion on the ethics of obligatory health care procedures for Italian military personnel, which is published in *Medicina e Morale* 47, no. 2 (1997): 363–373. See also A. G. Spagnolo, "Necessità, opportunità, utilità della istituzione di un Comitato Etico presso la Direzione Generale della sanità militare," *Giornale di Medicina Militare* 146, no. 3 (1996): 300–305.

[71] See A. G. Spagnolo, A. A. Bignamini, and A. de Franciscis, "I Comitati di Etica fra linee-guida dell'Unione Europea e decreti ministerial," *Medicina e Morale* 47, no. 6 (1997): 1059–1098.

[72] The normative documents and the list of ethics committees in Italy can be accessed on the site of the National Monitoring Centre on Clinical Drug Trials: http://oss-sper-clin .sanita.it.

but all of them were of a theoretical nature and never addressed the subject of defining standard operating procedures. The Italian Society for Bioethics and Ethics Committees (Società Italiana di Bioetica e dei Comitati Etici, SIBCE) was formed in July 1989, and one of the goals in its constitution was to "promote the development of widespread sensitivity to ethical issues in scientific and biomedical research." [73] The SIBCE organized two conventions for the purpose of promoting standard operating procedures for Italian bioethics committees: one in 1993, aimed at promoting development and debate within bioethics committees in general, and the other in 1997, geared to training and accrediting individual members of bioethics committees and proposing some concrete operating procedures for them, such as the previously cited guidelines offered by the European Forum for Good Clinical Practice (EFGCP).

In 1993, the bioethics committee of the National Association for Drug Control (Associazione Nazionale per la Farmacovigilanza) was one of the first in Italy to publish its standard operating procedures (SOPs).[74] They are a formal requirement for bioethics committees in accordance with the specifications of the GCP guidelines. In reality, they have tremendous binding power that guarantees uniformity of criteria and impartiality of assessment, making them a sure practical reference both for the members of bioethics committees and for the people and groups requesting opinions. Furthermore, SOPs can be gradually and dynamically modified over time in terms of the composition and competence of bioethics committee members as well as of the particular assessment procedures for the protocols under examination.

Since 1994, the Lanza Foundation of Padua has organized training courses for members of bioethics committees. However, they are designed for committees dealing with biomedical practice and not specifically with the ethical review of clinical trial protocols.

In June 1995, the Italian Society of Pharmacology (Società Italiana di Farmacologia) organized a convention at the Greater Institute of Health (Istituto Superiore di Sanità, ISS) in which the Ministry of Health developed regulation proposals for the establishment and function of bioethics committees that review clinical trials.

In the meantime, a spontaneous association was forming among many ethics committees. This "network" took shape under the name of the National Federation of Ethics Committees (Federazione Nazionale dei Comitati di Etica, FNaCE).[75]

---

[73] Società Italiana di Bioetica e dei Comitati Etici, "Statuto," July 1989, http://www.sibce.org/statuto.html.

[74] V. Berté and A. A. Bignamini, eds., *Comitati di etica e farmacovigilanza* (Bologna: Health, 1994).

[75] In particular, the FNaCE proposes—in accordance with its constitution—to help define the tasks and minimum organizational and procedural standards of bioethics committees; to promote and coordinate the activities of federated bioethics committees with the aim of guaranteeing bioethical debate; and to promote the exchange of information and working documents between federated bioethics committees. In its constitution, the FNaCE also allows bioethics committees on animal research, which defend the quality of life and death of animals by acting as third parties between the sponsors/experimenters and the passive subjects, to be members of the federation. Furthermore, technical and scientific commissions that overlap in function, aim, and composition are also considered bioethics committees. Other tasks undertaken by the federation include identifying and updating the criteria for monitoring the function of federated bioethics committees and preserving decorum without overriding the independence of the committees. The federation proposes

Moreover the Agency for Regional Healthcare Services (Agenzia per i Servizi Sanitari Regionali)[76] developed its *Linee guida per i comitati di etica* (Guidelines for ethics committees) in 1996 with the primary objective of ensuring a high degree of scientific efficacy and procedural correctness for the activities of bioethics committees in Italy. In the same year a CENSIS commission on bioethics committees and drug trials drew up an ad hoc document that, among other things, underscored the function of the NBC as a tool for the education and guidance of local bioethics committees.

For its part, the NBC perceived the need—in light of such a rapidly changing situation—to return to the topic of bioethics committees with a new opinion after its first document published in 1992 (see note 22).[77]

---

to be a reference point for the bioethics committees and a guarantor of the decorum and seriousness of federated bioethics committees within institutions and in public opinion. The regulations of the FNaCE state that federated bioethics committees should represent health care or scientific facilities of proven reliability that are in accord with the goals of the federation and accredited where appropriate according to specific norms. They cannot therefore be constituted as private associations having the sole aim of providing opinions on matters of biomedical ethics. The federation further holds that the articulation of the various levels of territorial competence of the bioethics committees is a delicate organizational juncture for bioethics committee function; it therefore reserves the right to offer the competent bodies a proposal—through a specific document—for the accreditation of local bioethics committees and the determination of the role of regional commissions. The FNaCE further reaffirms that each bioethics committee must work on the basis of clear and publicly available operating procedures, in accordance with the provisions of the guidelines for Good Clinical Practice, and explicitly promotes the *Guidelines and Recommendations for European Ethics Committees* published by the European Forum for Good Clinical Practice. One of the aspects addressed by the FNaCE is the auditing of bioethics committees. Bioethics committees belonging to the federation in fact commit themselves to accepting in loco visits from FNaCE delegates with the aim of ensuring the compliance of bioethics committee activities with the declared standard operating procedures and the spirit of the federation (on the topic of audits, see P. Gobel, "FDA Audits of Institutional Review Boards," *Applied Clinical Trials* [Oct. 1995]: 54–59). In conclusion, the FNaCE organized a convention to present the institution in April 1997. Its headquarters is with the ethics committee of the G. D'Annunzio School of Medicine and Surgery of Chieti (via dei Vestini, snc - 66013) and has both a web page (http://www.unich.it/fnace) and an e-mail address (fnace@unich.it).

[76] The agency, established by legislative decree no. 266 on June 3, 1993, aimed to ensure—in the guidelines cited above, which are the product of the work of an ad hoc group of experts in the various disciplines involved in the matter—a high degree of scientific efficacy and procedural correctness in the activities of ethics committees in Italy. The document was the basis on which the committees could develop their own specific procedures for the performance of their respective functions. This means was intended to promote the adoption of procedures—valid throughout the entire national territory—that would ensure a standard of quality, transparency, and efficacy of the activities conducted by bioethics committees. These guidelines considered not only the bioethics committees conducting ethical reviews of biomedical research but also those conducting ethical reviews of health care.

[77] See Italian National Bioethics Committee, *I Comitati Etici in Italia: Problematiche recenti* [*Ethics Committees in Italy: Recent Issues*], April 18, 1997 (Rome: Presidenza del Consiglio dei Ministri Dipartimento per l'Informazione e l'Editoria: 1997), http://www.governo.it/bioetica/testi/180497.html, English abstract at http://www.governo.it/bioetica/eng/opinions/abstracts/abstract_ETHICS_COMMITTEES.pdf. This document again underscores that in addition to their educational role, bioethics committees have the functions of monitoring

The institution of ethics committees in Italy has been almost exclusively linked to the ethical assessment of clinical drug trials, in compliance with European norms on the matter. In the wake of so-called pharmaceutical intoxication, there has been a dangerous tendency to conceptually separate clinical trials from health care practice, envisioning two different *ethical bodies* (with different names) to assess the associated problems.

> Instead, those areas absolutely cannot be separated inasmuch as the common denominator of both is the human person, who is both the subject of experimentation and the patient. It is therefore dangerous to introduce different names for these bodies on the basis of different functions, since they are nonetheless required to operate with the same methodology proper to bioethics. For this reason, we have preferred to use the now widespread term *ethics committee* even for different functions, in order to underscore the same methodological meaning, regardless of the specific tasks that might be required of them. Since today these bodies are the expression of the bioethical sensibilities of the research or health care organization in which they are established, it is more fitting to talk about *institutional ethics committees*.[78]

The so-called San Macuto Charter, proposed at the conclusion of a convention organized by the Politeia Center for Studies (Centro Studi Politeia) and the NBC, enters into this debate: it affirms that along with the role of evaluating clinical trials, which is by now well-regulated by law, the advisory role of institutional ethics committees in clinical activity should also be given normative value. To this end we propose the following:

> Local health care agencies, hospital administrations, and university institutes for rehabilitation and care, both public and private, must be responsible for establishing an ethics committee for ethical consultation on clinical activities.

> In addition to this function, the committee will also be responsible for advancing the culture of bioethics, training in bioethics for health care personnel, and ethical consultation for administrative bodies regarding what policies to enact.

> Ethics committees will work so that the top management levels of health care structures not only take into account the technical, professional, and economic aspects of services, but also include among their priorities the ethical training of health care workers, the health needs of the social context, and the ethical implications implicit in the various organizational decisions.

---

scientific research—ensuring that it is conducted according to rules of certified competence—and monitoring health care in order to ensure that care is humanized and the dignity of citizens is safeguarded. These functions are increasingly found all together within a single bioethics committee. Finally, it reaffirms the following: (1) the opinions of bioethics committees must not be binding, even when the opinion is mandatory; (2) the bioethics committee must remain independent of potential sponsors and promoters; and (3) hierarchical relationships between different bioethics committees must be ruled out.

[78] A.G. Spagnolo, "Un nuovo ruolo per il comitato etico istituzionale," in *Etica e giustizia in sanità*, Spagnolo et al., 218.

Ethics committees should have a good degree of independence from the health care structures and should be geared toward interdisciplinarity and ethical pluralism.

The independence of the ethics committee will be safeguarded through the following institutional measures: (a) the presence of no fewer external members than internal members; (b) the entrustment of the presidency to an external member; and (c) the incompatibility of "top management" positions in health care structures with the duty of committee membership.

The structures in which ethics committees work must make available to the committee the resources necessary for it to function: (a) personnel qualified to carry out secretarial functions; (b) sufficient financial resources for the committee's activities; and (c) adequate office space.

The appointment of ethics committee members should occur on the basis of documented competencies not only in their specific professions, but also in the field of bioethics. The importance of defining the general organizational framework for this type of ethics committee seems evident, and it is to be hoped that such a definition follows serious public debate involving the institutions and is not left up to the initiatives of individuals or to personal good will. There is already a document on this subject by the NBC, the 2001 *Orientamenti per i Comitati Etici in Italia* (*Guidelines for Ethics Committees in Italy*), but the enactment of our proposal must involve all concerned parties: the NBC, the various committees already operating on national territory, the regional commissions for bioethics, the cultural associations in the sector, health care workers, and anyone else who is interested in the matter. By now the time has come for ethics committees to fully and concretely achieve the mandates entrusted to them by law and by society, not just in the area of experimentation but also in health care—the time to be bodies that protect the safety, rights, and dignity of persons who use health care facilities and instruments for spreading the culture of bioethics.[79]

I consider it appropriate to envision an "integrated" model of ethics committees "with a well-identified core that functions as the institution's 'critical' conscience. Through subcommittees and ad hoc supplemental members it will be possible to fulfill the various tasks of evaluating experimentation protocols, developing guidelines for practice, constantly rethinking the mission of the institution, organizing ethical courses and training for personnel and, finally, advising in clinical cases, possibly in conjunction with the Clinical Ethics Service (Servizio di Etica Clinica)."[80]

Through a publicly discussed preparatory document, the National Bioethics Committee initially considered entrusting the various tasks to distinct organs so that each one could be optimally addressed. Nonetheless, such an idea presented numerous conceptual and practical difficulties because research and practice are tightly connected: the subjects to whom they apply are the same and the same organization carries them out. Therefore, after intense debate, the NBC underscored

---

[79] Palazzo San Macuto (Rome), March 21, 2003. Organizers: Mariella Immacolato (coordinator of the local ethics committee of Local Healthcare Agency Unit 1 in Massa), Maurizio Mori (University of Turin), and Søren Holm (University of Manchester).

[80] Spagnolo, "Un nuovo ruolo," 221.

in its conclusive opinion (*Guidelines for Ethics Committees in Italy*, 2001) that, even in light of the wealth of local experience, making a distinction between two organs whose tasks, composition, and structure are different is not necessary.

Conclusions and Outlook

These new public institutions, bioethics committees, merit particular attention as a whole: their very existence indicates the importance that bioethics has acquired, together with the issues it investigates, in the fields of both research and health care and even in the lives and cultures of populations. Bioethics committees will increasingly constitute a sort of "hinge" between political and civil society and the world of scientific research. In fact, in democratic societies, every major decision should be the product of confrontation and synergy between scientists, legislators, and the consent of the people. This dialogue will not be easy, and legislators will often be tempted to follow public opinion even against the public's own good in order to curry favor and build consensus. Especially in these cases, bioethics committees can serve as a critical voice and a developing conscience, provided they do not allow themselves to become politicized or manipulated by special interests, remaining anchored instead to principles that safeguard the values of the human person.

The spread of ethics committees can be interpreted as a sign of a growing need to reflect on the ethical dimension of medicine.

In recent years, there has been a change in the way decisions are approached in the medical field: they were once made by at most two persons, the physician and the patient, with a clear emphasis on the decision-making power of the physician; the situation has been turned upside-down in more recent years, with many voices intervening before a decision is made and more figures involved in the care process.

Increasingly attentive to research and development in biomedical technology, medicine finds itself having to deal with patients, the world of pharmaceutical industries, and society as a whole.

Certainly helpful are the rules of democracy, which provide for participation in decision making and therefore the sharing of responsibilities to guarantee full respect for the subjects involved, but they are not the only point of reference that can be adopted for the activity of ethics committees.

Ethics committees could be described as observatories: observation and evaluation posts watching over biomedicine and its current developments. But they could also be thought of as laboratories of biomedicine, places of reflection on the changes in medicine today, where it is possible to rethink the characteristics of medicine in its three different dimensions of research and experimentation, clinical work and care, and health in general—especially public health.

They are no longer characterized by the idea of "monitoring" medical acts, but are becoming privileged places for reflection on medicine—an occasion to rethink the ways in which it is defined today.[81]

I think it important to conclude that the work of ethics committees is made possible by certain shared values:[82] respect for human life, individuality, and

---

[81] See L. Galvagni, *Bioetica e comitati etici* (Bologna: EDB, 2005).

[82] See Italian National Bioethics Committee, *Bioetica e formazione nel sistema sanitario* [*Bioethics and professional training in the health care system*], September 7,

autonomy; responsibility; and an anthropologically integral understanding of the corporality of man and the sacredness and quality of human life, not necessarily understood within the definite framework of religious faith.

As Francesco D'Agostino affirms, there are some fundamentally Hippocratic understandings in biomedicine that enormously reduce the relevance of ethical pluralism. Physicians still feel that they are essentially Hippocratic, and therefore the first principle is still *do no harm*.

> Health and illness have bioethical meaning only from the Hippocratic perspective, which is to say that there is something we call health, there is something that we call illness, and the ethic of the physician is to promote health and combat illness. If we abolish the epistemological autonomy of the concept of health and the concept of illness, as many authors do … if the epistemological autonomy of these concepts is abolished … bioethics entirely evaporates, disappears, has no more reason for being, or turns into something else through the social legitimization of manipulative practices. The concept of manipulation in turn makes sense only in a Hippocratic perspective; outside of a Hippocratic perspective, it is unclear what manipulation means.[83]

UNESCO and Its Bioethics Committee

It would be remiss to omit mentioning the intent of the United Nations Educational Scientific and Cultural Organization (UNESCO), which is the specialized agency of the United Nations for culture and dialogue between cultures, to establish itself as a global point of reference in bioethics.

The starting point was UNESCO's publication of the Universal Declaration on the Human Genome and Human Rights in November 1997. Article 1 states that "the human genome underlies the fundamental unity of all members of the human family, as well as the recognition of their inherent dignity and diversity. In a symbolic sense, it is the heritage of humanity."[84]

This definition and other points in the declaration have incited great contention because they ground human dignity in the genome, expressing a biologistic understanding of man, when the opposite is actually true: it is the spiritual dignity of man that privileges the human genome; hence, having a defective genome does not diminish the dignity of an individual. Moreover, the declaration does not clarify who is entitled to the rights proclaimed, such as nondiscrimination. There was no interest in defining whether the bearers of these rights are human beings at the moment of fertilization or whether genetic discrimination can be readily perpetrated—and it currently is—at this level.

1991 (Rome: Presidenza del Consiglio dei Ministri Dipartimento per l'Informazione e l'Editoria: 1991), http://www.governo.it/bioetica/testi/070991.html, English abstract at http://www.governo.it/bioetica/eng/opinions/abstracts/abstract_bioethics_and _professional.pdf.

[83] "Intervista a F. D'Agostino," in *Bioetica e comitati etici*, Galvagni, 177.

[84] UNESCO, "Universal Declaration on the Human Genome and Human Rights," November 11, 1997, http://unesdoc.unesco.org/images/0012/001229/122990eo.pdf.

Yet on this problematic basis of the greatly trumpeted declaration, two bio-ethics committees have been formed within UNESCO's bioethics program: (1) the International Bioethics Committee (IBC), made up of thirty-six independent experts, and (2) the Intergovernmental Bioethics Committee (IGBC), made up of representatives from the thirty-six member nations. These committees are charged with submitting recommendations and advice to the director-general. The purpose of the bioethics program, as far as it is discernible, is also to guide the national committees and produce a sort of common code of bioethics.

Concerns persist in principle about a political organization being the basis for conducting ethics and about seeking to promote a sort of global ethics, fostering uniformity and generalized relativism rather than dialogue.

Summary Outline for Chapter 7

Ethics committees

Identity and functions of ethics committees

The spread of ethics committees can be interpreted as a sign of a growing need to reflect on the ethical dimension of medicine. Various *functions* are assigned to ethics committees:

- Evaluating research protocols
- Analyzing complex clinical cases
- Developing guidelines on individual questions
- Conducting training programs

There is still debate over whether these various functions should be entrusted to *distinct committees* or whether they should be recognized as different activities belonging to a *single organization.*

Attempts to classify ethics committees can be boiled down to three main types:

- *Committees that review research protocols*, which have a standard model in the so-called institutional review boards (IRBs) in the United States
- *Clinical ethics committees*, which deal with single cases that arise in clinical practice (comparable to hospital ethics committees)
- Ad hoc commissions and *regional and national committees*, which address "cutting-edge" topics in bioethics and publicize them in training programs

Legislative questions

In Italy, following the European guidelines for regulation and experimentation and the so-called guidelines for good clinical practice, norms have been issued concerning the institution of ethics committees:

> *As far as the research and experimentation sector is concerned, the opinion of an ethics committee is obligatory and binding today* inasmuch as it is decisive for the purposes of obtaining authorization for the research.

> *With regard to ethical questions related to clinical practice and the insurance and organizational dimensions thereof, no specific, unified set of norms exists yet in Italy* requiring the institution of places in which to discuss them, even though committees that deal with such questions are present today in increasing numbers in various institutional health care settings.

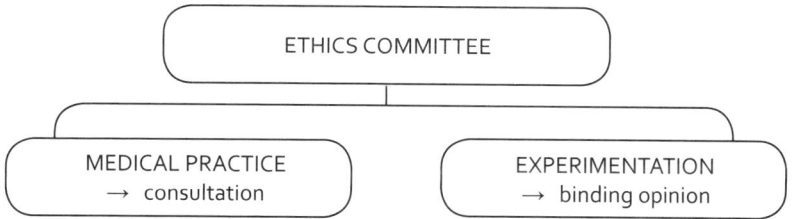

Ethics committees for confrontation and reflection on medicine

In recent years, there has been a change in the way decisions are approached in the medical field: they were once made by at most two persons, the physician and the patient, with a clear emphasis on the decision-making power of the physician; the situation has been turned upside-down in more recent years, with many voices intervening before a decision is made and more figures involved in the care process.

Increasingly attentive to research and development in biomedical technology, *medicine* finds itself having to deal with

- Patients
- The world of pharmaceutical industries
- Society as a whole

Certainly helpful are the rules of democracy, which provide for participation in decision making and therefore the sharing of responsibilities in order to guarantee full respect for the subjects involved, but they are not the only point of reference that can be adopted for the activity of ethics committees.

Ethics committees could be described as observatories: observation and evaluation posts watching over biomedicine and its current developments. But they could also be thought of as laboratories of biomedicine, places of reflection on the changes in medicine today, where it is possible to rethink the characteristics of medicine in its three different dimensions of research and experimentation, clinical work and care, and health in general—especially public health.

They are no longer characterized by the idea of "monitoring" medical acts, but are becoming privileged places for reflection on medicine—an occasion to rethink the ways in which it is defined today.

Definition of ethics committee

"An independent body, an expression of a health care institution or a center for scientific research, made up of physicians and nonphysicians, whose responsibility is to ensure the protection of the safety, well-being and rights of persons and to provide opinions and to create opportunities for training in the ethical aspects of medical practice and research in the biomedical sciences, thus providing a public safeguard, and consulting with the related professional organizations on aspects pertaining to the professional code of ethics." (Italian National Federation of Ethics Committees, FNaCE)

Historical and epistemological reasons
for the rise of ethics committees

- Abuses in experimentation
- Applications of biomedical technologies
- Discrepancies between practice, ethics and legislation.

Birth of ethics committees in connection with research review

In 1938, in the United States, the Federal Food, Drug and Cosmetic Act allowed physicians to experiment with new drugs without an ethical review of the protocol.

In 1962, this legislation was emended as a result of the tragic harm caused to fetuses by experimentation with thalidomide on pregnant women.

As early as 1953, a federal document ("Group Consideration of Clinical Research Procedures Deviating from Accepted Medical Practice or Involving Unusual Hazard") requested ethical review of research by an independent commission. Such guidelines, however, were applicable only to research conducted in public clinical centers affiliated with the National Institutes of Health.

In 1966, the Surgeon General of the United States issued a regulation concerning the responsibility of institutions with regard to research activity conducted within them.

In accepting public funds for research, research institutes also agreed to share public responsibility for their use.

All research institutes had to guarantee that proposals for research involving human subjects would be subjected to an independent ethical review.

But this did not put an end to the Tuskegee Syphilis Study (1932–1972, in which many colored persons were not cured with available antibiotics).

As a public reaction to the Tuskegee Syphilis Study, President Nixon signed the National Research Act, which in section 474 called for the formation of institutional review boards (IRBs) in all settings where research was being conducted with public funding.

The same law also instituted the National Commission for the Protection of Human Subjects of Biomedical and Behavioral Research, with the responsibility of providing ethical guidelines for research (particularly on vulnerable subjects), and providing regulations for the operation of the IRBs (*The Belmont Report*).

In the United Kingdom, the establishment of Research Ethics Committees (RECs) was promoted by professional societies and not by decree on the part of the authorities because they thought that this was a matter of professional ethics.

Within the context of this non-juridical system, the Royal College of Physicians (RCOP) recommended as early as 1967 that all clinical experiments be approved in advance by a group of physicians including experts in clinical experimentation.

In the following years the RCOP published guidelines to standardize objectives, structures and procedures so as to improve the functioning of the RECs.

Birth of the ethics committee connected to medical practice

In 1962, at Swedish Hospital in Seattle, a commission was formed, made up for the most part of nonphysician members, to establish criteria by which to grant patients access to dialysis, a procedure that was still experimental and costly. *Life* magazine (November 1962) headlined the story: "They Decide Who Lives, Who Dies: Medical Miracle Puts Moral Burden on Small Committee."

In 1971, the Canadian Bishops issued their *Medical-Moral Guide* instituting commissions with the task of education and applying the *Directives for Catholic Health Care Facilities* that had been issued by the bishops of the United States.

In 1976, the sentence of the Supreme Court of New Jersey was issued in the case of Karen Ann Quinlan, which instituted an ethics committee to evaluate the prognosis. (This highlights the problem of the composition and the purposes of such committees.)

United States Conference of Catholic Bishops, *Ethical and Religious Directives for Catholic Health Care Services* (2009):

> "An ethics committee or some alternate form of ethical consultation should be available to assist by advising on particular ethical situations, by offering educational opportunities, and by reviewing and recommending policies. To these ends, there should be appropriate standards for medical ethical consultation within a particular diocese that will respect the diocesan bishop's pastoral responsibility as well as assist members of ethics committees to be familiar with Catholic medical ethics and, in particular, these Directives." (no. 37)

Pontifical Council for Pastoral Assistance to Health Care Workers, *Charter for Health Care Workers* (1995):

> "Health care workers, especially doctors, cannot be left to their own devices and burdened with unbearable responsibilities when faced with ever more complex and problematic clinical cases arising from biotechnical possibilities—many of which are at an experimental stage—open to modern medicine, and from the socio-medical import of certain questions.

> "To facilitate choices and to keep a check on them, the setting up of 'ethical committees' in the principal medical centers should be encouraged. In these commissions, medical competence and evaluation is confronted and integrated with that of other presences at the patient's side, so as to safeguard the latter's dignity and medical responsibility itself." (no. 8)

Documents

Nuremberg Military Tribunals, *Nuremberg Code* (Nuremberg, 1946).

World Medical Association, *Declaration of Helsinki: Ethical Principles for Medical Research Involving Human Subjects* (Helsinki, 1964).

National Commission for the Protection of Human Subjects of Biomedical and Behavioral Research, *The Belmont Report: Ethical Principles and Guidelines for the Protection of Human Subjects of Research* (Washington, DC, 1979).

President's Commission for the Study of Ethical Problems in Medicine and Biomedical and Behavioral Research, *Summing Up: Final Report on*

Studies of the Ethical and Legal Problems in Medicine and Biomedical and Behavioral Research (Washington, DC, March 1983).

President's Commission for the Study of Ethical Problems in Medicine and Biomedical and Behavioral Research, *Decisions to Forego Life-Sustaining Treatment* (Washington, DC, 1983).

Warnock Commission, *Report of the Committee of Inquiry into Human Fertilisation and Embryology* (London, 1984).

Arbeitsgruppe des Bundesministers für Forschung und Technologie und des Bundesministers des Justiz [Task force of the National Ministry for Research and Technology and the National Ministry of Justice], *In-vitro Fertilisation, Genomanalyse und Gentherapie: Gericht der Gemeinsamen* (Munich, 1985).

Council of Europe, *Recommendation R (90) 3 of the Committee of Ministers to Member States concerning Medical Research on Human Beings* (1990).

Council for International Organizations of Medical Sciences (CIOMS), *International Ethical Guidelines for Biomedical Research Involving Human Subjects* (Geneva, 1993).

European Forum for Good Clinical Practice (EFGCP), *Guidelines and Recommendations for European Ethics Committees* (Louvain, 1995); rev. ed. (Brussels, 1997).

Council of Europe, *Convention for the Protection of Human Rights and Dignity of the Human Being with Regard to the Application of Biology and Medicine: Convention on Human Rights and Biomedicine* (Oviedo, April 4, 1997).

UNESCO, *Universal Declaration on the Human Genome and Human Rights* (November 11, 1997).

Council of Europe, *Additional Protocol to the Convention on Human Rights and Biomedicine, on Biomedical Research* (Strasbourg, August 31, 2001).

Erice Declaration of the Ethical Principles of Pharmacogenetic Research (Erice [Trapani], March 2001).

Nationaler Ethikrat, *Stellungnahme zum Import menschlicher embryonaler Stammzellen* [Position paper on the importation of human embryonic stem cells] (Berlin, 2001).

UNESCO, *International Declaration on Human Genetic Data* (October 16, 2003).

Italian National Bioethics Committee—DPCM March 28, 1990

Responsibilities:

- To develop an overall summary of the programs, objectives, and results of research and experimentation in the field of life sciences and human health

- To formulate opinions and indicate solutions, including with the aim of preparing legislation . . . having respect for the protection of the fundamental rights and dignity of human beings and for other values expressed by the constitution and by international documents to which Italy adheres

- To propose solutions for oversight aimed at protecting both human safety and the environment in the production of biological material as well as protecting patients who are being treated with genetically engineered products or subjected to gene therapy from potential risks

- To promote the drafting of codes of conduct for workers in the various sectors concerned

- To foster properly informed public opinions

Assorted documents (opinions and motions; see http://www.governo.it/bioetica)

> *Bioethics and Professional Training in the Healthcare System*, September 7, 1991.

> *The Ethics Committees*, February 27, 1992.

> *Ethics Committees in Italy: Recent Issues*, April 18, 1997.

> *Cloning*, October 17, 1997.

> *Bioethical Guidelines for Genetic Testing*, November 19, 1999.

> *I comitati etici in Italia: orientamenti per la discussione (Ethics committees in Italy: Guidelines for discussion)*, April 2000.

> *Psychiatry and Mental Health*, November 24, 2000.

> *Bioethical Guidelines for Equal Access to Healthcare*, May 25, 2001.

> *Bioetica interculturale (Intercultural bioethics)*, June 22, 2001.

> *Guidelines for Ethics Committees in Italy*, July 13, 2001.

> *Aims, Risks and Limits of Medicine*, December 14, 2001.

Ethics committees in Italy

Risks of separate ethics committees

"Pharmaceutical intoxication"

In Italy, the institution of ethics committees has been almost exclusively connected with the duty of ethical evaluation of clinical pharmacological experiments, following European norms on that subject.

Separate ethics committees?

One consequence of "pharmaceutical intoxication" has been a dangerous tendency to conceptually separate clinical trials from health care practice, envisioning two different *ethical bodies* (with different names) to assess the associated problems. "Instead, those areas absolutely cannot be separated inasmuch as the common denominator of both is the human person, who is both the subject of experimentation and the patient. It is therefore dangerous to introduce different names for these bodies on the basis of different functions, since they are nonetheless required to operate with the same methodology proper to bioethics. For this reason, we have preferred to use the now widespread term *ethics committee* even for different functions, in order to underscore the same methodological meaning, regardless of the specific tasks that might be required of them. Since today these bodies are the expression of the bioethical sensibilities of the research or health care organization in which they are established, it is more fitting to talk about *institutional ethics committees*." (A.G. Spagnolo)

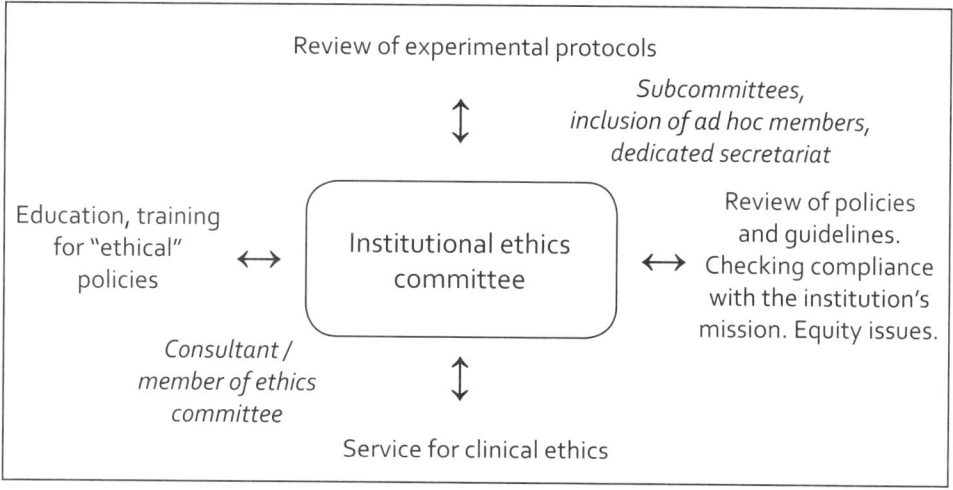

Figure 7-2. Proposal of an integrated function model for an institutional ethics committee. SOURCE: A.G. Spagnolo, 2004.

The "integrated" model

As A.G. Spagnolo says, "it makes sense to envision an 'integrated' institutional ethics committee with a well-identified core that functions as the institution's 'critical' conscience. Through subcommittees and ad hoc supplemental members it will be possible to fulfill the various tasks of evaluating experimentation protocols, developing guidelines for practice, constantly rethinking the mission of the institution, organizing ethical courses and training for personnel, and finally advising in clinical cases, possibly in conjunction with the Clinical Ethics Service."

San Macuto Charter

The so-called San Macuto Charter, proposed at the conclusion of a convention organized by the Politeia Center for Studies (Centro Studi Politeia) and the NBC, enters into this debate: it affirms that along with the role of evaluating clinical trials, which is by now well-regulated by law, the advisory role of institutional ethics committees in clinical activity should also be given normative value.

"To this end we propose the following:

> Local health care agencies, hospital administrations, and university institutes for rehabilitation and care, both public and private, must be responsible for establishing an ethics committee for ethical consultation on clinical activities.

> In addition to this function, the committee will also be responsible for advancing the culture of bioethics, training in bioethics for health care personnel, and ethical consultation for administrative bodies regarding what policies to enact.

> Ethics committees will work so that the top management levels of health care structures not only take into account the technical, professional, and economic aspects of services, but also include among their priorities the ethical training of health care workers, the health needs of the social context, and the ethical implications implicit in the various organizational decisions.

"Ethics committees should have a good degree of independence from the health care structures and should be geared toward interdisciplinarity and ethical pluralism.

"The independence of the ethics committee will be safeguarded through the following institutional measures: (a) the presence of no fewer external members than internal members; (b) the entrustment of the presidency to an external member; and (c) the incompatibility of 'top management' positions in health care structures with the duty of committee membership.

"The structures in which ethics committees work must make available to the committee the resources necessary for it to function: (a) personnel qualified to carry out secretarial functions; (b) sufficient financial resources for the committee's activities; and (c) adequate office space.

"The appointment of ethics committee members should occur on the basis of documented competencies not only in their specific professions, but also in the field of bioethics. The importance of defining the general organizational framework for this type of ethics committee seems evident, and it is to be hoped that such a definition follows serious public debate involving the institutions and is not left up to the initiatives of individuals or to personal good will. There is already a document on this subject by the NBC, the 2001 *Orientamenti per i Comitati Etici in Italia* (*Guidelines for Ethics Committees in Italy*), but the enactment of our proposal must involve all concerned parties: the NBC, the various committees already operating on national territory, the regional commissions for bioethics, the cultural associations in the sector, health care workers, and anyone else who is interested in the matter. By now the time has come for ethics committees to fully and concretely achieve the mandates entrusted to them by law and by society, not just in the area of experimentation but also in health care—the time to be bodies that protect the safety, rights, and dignity of persons who use health care facilities and instruments for spreading the culture of bioethics."

Functions of ethics committees:
Opinion of the Italian National Bioethics Committee

Through a publicly discussed preparatory document, the National Bioethics Committee initially considered entrusting the various tasks to distinct organs so that each one could be optimally addressed.

Nonetheless, such an idea presented numerous conceptual and practical difficulties because research and practice are tightly connected: the subjects to whom they apply are the same and the same organization carries them out. Therefore, after intense debate, the NBC underscored in its conclusive opinion (*Guidelines for Ethics Committees in Italy*, 2001) that, even in light of the wealth of local experience, making a distinction between two organs whose tasks, composition, and structure are different is not necessary.

Bioethics and Professional Training in the Healthcare System
(NBC, September 7, 1991)

Shared values:

Respect for human life, individuality, and autonomy

Responsibility

Anthropologically integral understanding of human corporality

Sacredness and quality of life, not necessarily understood in terms of a specific religious faith

There are some fundamentally Hippocratic understandings in biomedicine that enormously reduce the relevance of ethical pluralism. Physicians still feel that they are essentially Hippocratic, and therefore the first principle is still *do no harm*.

"Health and illness have bioethical meaning only from the Hippocratic perspective, which is to say that there is something we call health, there is something that we call illness, and the ethic of the physician is to promote health and combat illness. If we abolish the epistemological autonomy of the concept of health and the concept of illness, as many authors do . . . if the epistemological autonomy of these concepts is abolished . . . bioethics entirely evaporates, disappears, has no more reason for being, or turns into something else through the social legitimization of manipulative practices. The concept of manipulation in turn makes sense only in a Hippocratic perspective; outside of a Hippocratic perspective, it is unclear what manipulation means." (F. D'Agostino)

Part Two

PARTICULAR BIOETHICS

Chapter 8

Bioethics, Genetics, and Prenatal Diagnosis

Genetic Engineering and Manipulation

Relevance and delimitation of the topic

As the earlier presentation of its historical origins goes to show, bioethics received its initial impetus precisely from the discoveries of genetic engineering. Even today, the perception in public opinion and widely distributed publications is that "genetic manipulation" and "genetic engineering" are surrounded by uncertainty, and, in any event, call for extreme vigilance. Uncertainty and alarm grow where neither the meanings of the terms nor the actual possibilities for application are well known.

A first terminological ambiguity found even in many publications is the de facto inclusion of other interventions on life—especially on nascent life—within the general meaning of genetic manipulation. Artificial reproduction is an example of such an intervention: strictly speaking it does not directly involve intervention on the genetic code, but rather on gametes and embryos for the purpose of achieving procreation or conducting experiments. It is nonetheless true that one can exploit artificial reproduction, and especially in vitro fertilization, in order to perform interventions on the genetic code of the embryo or of the gametes as well. In my opinion, however, these two major issues—genetic engineering and artificial reproduction—should be kept distinct from one another. The very term *genetic manipulation* is highly generic and can mean any sort of intervention (manipulate = to treat or operate with or as if with the hands) on the genetic patrimony, whereas *genetic engineering* more strictly means "the set of techniques aiming to transfer certain genetic information into the cellular structure of a living being that otherwise would not have been present."[1]

It is also necessary to report a variety of attitudes among scholars. There are those who are optimistic about the great prospects opening up in the field of gene therapy: this attitude is more common among those who work in the field, such as molecular biologists and geneticists, who sometimes want no legal obstacles or restraints in exploring these prospects and do not care what methods they employ, as in the case of experimentation on embryos, which some consider necessary in order to achieve the goal.

---

[1] E. Sgreccia and V. Mele, "Gli aspetti etici dell'ingegneria genetica," in *Ingegneria genetica e biotecnologie nel futuro dell'uomo* (Milan: Vita e Pensiero, 1992), 131–166.

There are those who are more concerned about potential errors in genetic engineering and manipulation, which could change the genetic constitution of humanity once initiated, causing a more dangerous revolution than any political or military revolution. This attitude is more commonly found among legal scholars and moralists and is clearly the main source of pressure for precise regulations that will be binding for researchers—whose activities are not in public view—and reassuring for public opinion.

Nevertheless, we are no doubt confronting a crucial moment in the history of science and the history of humanity. Indeed, the current period in the history of human civilization is considered momentous because of the exceptional capacity for change that it bears and can unleash; upon deeper reflection, it also seems a "paradoxical" moment.[2]

On the one hand there is talk of man's *autopoiesis* or self-creation: man has acquired the ability to elicit the emergence of human life in the laboratory and even to alter the genetic make-up of his own species. There has been talk of a new stage of Darwinism: phylogenesis, or the evolution of species by which the various forms of life in the universe proliferate, would no longer occur through mutations in the genetic code caused by environmental influences; man, who is at the apex of this evolution—or, more specifically scientists—would manipulate the mysterious genetic code, robbing it of its mystery, and would cause, can cause, desired mutations along predetermined lines.

On the other hand, this same phase in our history presents a different and opposing possibility: the self-destruction of humanity. In 1982, the Pontifical Academy of Sciences developed and presented to the heads of state its Declaration on Prevention of Nuclear War, declaring among other things that there are roughly fifty thousand explosive nuclear devices accumulated in arsenals around the world, some of which are thousands of times more destructive than those unleashed on Hiroshima, and furthermore that the explosive potential of these devices is equivalent to three tons of trinitrotoluene (TNT) for each inhabitant of the globe, which is more than sufficient to destroy humanity. It has been written that at the end of the first millennium the European people spent sleepless nights thinking that God would bring about the end of the world; at the end of the second millennium there is a fear that human beings will destroy the world.[3]

I have taken atomic power as a point of comparison so as to reason by analogy: just as atomic energy can also be used for peaceful purposes, so genetic engineering can be directed to therapeutic aims, treating illnesses that have been otherwise indomitable and resistant to medical science. The problem is whether to take the peaceful and constructive approach or the manipulative and destructive approach; in sum, the outcome depends on ethics and the ethical decisions that will be made.

Never has ethics been so important in medicine, biology, and society. As a result of these scientific discoveries, morality—specifically morality dealing with

---

[2] L. Lombardi Vallauri, "Manipolazioni genetiche e diritto," *Rivista di Diritto Civile* (Jan.–Feb. 1985): 1–23.

[3] Pontifical Academy of Sciences, "Declaration on Prevention of Nuclear War," September 23–24, 1982, 13–15, http://www.vatican.va/roman_curia/pontifical_academies/acdscien/2010/documenta_3_4_7_11_pas.pdf.

matters of life—has become everyone's concern, a problem of primary importance in society, and a problem for society at the global level.

This importance is even more evident and well documented in briefly recalling the rapid progress of the biological sciences, the possibilities they hold for the future of humanity, and the ethical problems they pose for the responsibilities of scientists and others in the civil community. Due to its importance and breadth, this topic is being treated in its own right even though its primary purpose is related to procreation and infertility treatment.

It should be added that this topic receives the attention not only of moralists but also of national and international bodies, which have developed certain guidelines. The first formulation took its cue from the conclusions reached in February 1975 by the Asilomar Conference on Recombinant DNA,[4] which was made up of experts on in vitro genetic modification from all over the world. It provided an initial classification of risks (four containment levels), listed experiments that should be banned, and suggested the need for oversight and regulatory measures while underscoring the positive potential for human progress.

The most significant regulations are listed and discussed as follows.

1. The *Guidelines for Research involving Recombinant DNA Molecules*[5] were first published in July 1977 by the National Institutes of Health (NIH) of the United States. In this work, experiments are classified according to several charts; prohibited experiments are listed, responsibilities are assigned, and so forth; and funding is granted to those who observe these provisions.

These guidelines were later updated (1984) to further specify the area of research application and to outline very specific tasks and procedures for the Recombinant Advisory Committee (RAC) and the Federal Register. They have since undergone many more revisions.

2. In 1985, the same NIH issued some "Points to Consider in the Design and Submission of Human Somatic Cell Gene Therapy Protocols."[6] The applicability of the points is limited to research conducted at institutions already receiving NIH funding (according to the previous guidelines) for research on recombinant DNA, including the research of the NIH itself. All of the requirements that the experimental research must meet are listed there (specific objectives, preclinical studies, detailed clinical procedures, researcher qualifications, patient selection, informed consent, and confidentiality), as well as all of the documentation that must be produced and the duty to immediately communicate the results obtained.

---

[4] S. Krimsky, *Genetic Alchemy: A Social History of the rDNA Controversy* (Cambridge, MA: MIT Press, 1982). For a review of the development of the first national and international ethics committees that specifically dealt with genetics, see A. G. Spagnolo, "Comitati di etica per la genetica," in *Ingegneria genetica*, Sgreccia and Mele, 411–451.

[5] Most recent version: NIH Department of Health and Human Services, *Guidelines for Research Involving Recombinant DNA Molecules*, May 2011, http://oba.od.nih.gov/oba/rac/Guidelines/NIH_Guidelines.pdf.

[6] National Institutes of Health, "Points to Consider in the Design and Submission of Human Somatic Cell Gene Therapy Protocols," *Recombinant DNA Technical Bulletin* 8, no. 4 (1985): 181–186.

3. The Williams Report, published in the United Kingdom in 1976,[7] even offers a classification of laboratory experiments according to risk level. It established the Genetic Manipulation Advisory Committee (GMAC) as an obligatory reference point and defined its mandate, requiring explicit approval for protocols and experiments exposed to the highest risks (classes III and IV). The Williams Report provides for the self-disciplining of researchers through the development of a code of conduct. It reserves oversight for the government through laboratory inspections and provides for an appropriate system of supervision in agreement with the scientists.

4. The report by the Commission of Enquiry of the Parliament of the Federal Republic of Germany (West Germany) on the prospects and risks of genetic engineering (1987) analyzed the areas and applications of genetic engineering thoroughly and at length.[8]

In the wake of these first guidelines, others began to emerge in other countries: Australia, Canada, France, the Netherlands, and the United Soviet Socialist Republic.

The Regulation of Genetic Engineering Act (Gesetz zur Regelung von Fragen der Genetechnik) of the Federal Republic of Germany, dated June 20, 1990,[9] and its successive modifications in 1993 and 1994,[10] are judged to be of particular importance, partly because of certain restrictive measures.

5. The French law no. 94–653 on respect for the human body, dated July 29, 1994, should also be mentioned. It specifically deals with topics of genetics in Title II, article 5.[11]

6. Two ministerial commissions were instituted in Italy as well. One was established at the Greater Institute of Health (Istituto Superiore di Sanità, ISS) in 1977 to examine problems related to research through in vitro genetic manipulation; its concluding report offered several proposals for national regulations. The other commission, on which I served, was established at the Ministry of Health (Ministero della Sanità)[12] and concluded its work in February 1987; it dealt with ethical, scientific, and health care problems relative to genetic engineering. The Italian National Bioethics Committee (NBC) has published several important documents on the matter:[13]

---

[7] Working Party on the Practice of Genetic Manipulation, *Report of the Working Party on the Practice of Genetic Manipulation* (London: Her Majesty's Stationary Office, 1976).

[8] German Bundestag, *Report of the Commission of Enquiry on Prospects and Risks of Genetic Engineering* (Munich: Bundestag, 1987), 151b–152b.

[9] Taken from F. Luzi, ed., *L'ingegneria genetica nella Germania Federale: Raccomandazione del Bundestag, Legge, Regolamenti* (Rome: Senato della Repubblica Italiana, Servizio Studi, Ufficio Ricerche nel Settore Sociale, 1992).

[10] Bundesgesetzblatt (Official Gazette) 1993, part I, p. 2066, and Bundesgesetzblatt 1994, part I, p. 1416.

[11] Published in Italian in *Politica del diritto* 26, no. 2 (1995).

[12] Italian Ministry of Health, "Relazione della Commissione di Studio per l'ingegneria genetica," in *Documenti conoscitivi sulla riproduzione umana assistita, embriologia ed ingegneria genetica*, ed. F. Luzi (Rome: Senato della Repubblica Italiana, Servizio Studi, Ufficio Ricerche nel Settore Sociale, 1990), 187–325.

[13] The English titles provided here are taken from the official website of the NBC, www.governo.it/bioetica. The documents are published in Italian by the Presidenza del Consiglio dei Ministri Dipartimento per l'Informazione e l'Editoria in Rome. (The NBC has its main offices at Via Veneto 56.) All the documents are also published in Italian under "Documentazione" in the journal *Medicina e Morale* (Medicine and morals) and are made available online at

Document on the Safety of Biotechnology (May 28, 1991), *Gene Therapy* (February 15, 1991), *Prenatal Diagnosis* (July 18, 1992), *Report on the Patentability of Living Organisms* (November 19, 1993), *Human Genome Project* (March 18, 1994), *Cloning as a Bioethical Problem* (March 21, 1997),[14] *Bioethical Guidelines for Genetic Testing* (November 19, 1999), *NBC Statement on the Patentability of Human Embryonic Cells* (February 25, 2000), *The Therapeutic Use of Stem Cells* (October 27, 2000), *Ethical and Juridical Considerations on the Use of Biotechnologies* (November 30, 2001), *Opinion of the NBC on the Draft Protocol on Human Genetics* (March 6, 2002), and *From Pharmacogenetics to Pharmacogenomics* (April 21, 2006).

7. As far as the institutions of the European community are concerned, Recommendation 934 (1982) of the Parliamentary Assembly of the Council of Europe[15] is of particular importance. This recommendation, together with the complementary institutions of the Assembly of European Communities, constitutes a basic reference document for the development of guidelines. Its various points within the context of the individual problems will therefore be examined in the following pages.

It seems appropriate to note immediately some general points from the document that are also important on the ethical level. The document recognizes that genetic engineering techniques can be of great help in the industrial and agricultural fields and can even contribute to resolving serious therapeutic difficulties.

The recommendation provides for the solemn recognition—through additional clarification within the European Convention on Human Rights, among other ways—of the right of individual persons to have their genetic heritage protected against manipulation: "the right to inherit a genetic pattern which has not been artificially changed"[16] and "the right to a genetic inheritance which has not been artificially interfered with."[17]

Therapeutic treatment for the cure of genetic disorders is legitimate when it is recognized as valid and compatible with respect for human rights and meets the conditions legally required for other therapeutic procedures.[18] These rights are based on articles two and three of the European Convention on Human Rights, which talk about the person's right to life and physical integrity.[19] The conclusion is therefore obvious that the right to life is intended for and extended to the moment of fertilization, since both life and the possibility of genetic manipulation begin at that moment.

It therefore seems that these affirmations implicitly and truly recognize the right and duty to protect embryonic life as well. This fact is also clear on the ethical level: it would be absurd to speak about the defense of genetic inheritance without

---

http://www.governo.it/bioetica/pareri.html. Titles, abstracts, and some complete texts can be accessed online in English at http://www.governo.it/bioetica/eng/opinions.html.

[14] This document in particular was published as *La clonazione come problema bioetico* in *Medicina e Morale* 47, no. 2 (1997): 360–362.

[15] Council of Europe, Recommendation 934, "On Genetic Engineering," January 26, 1982, http://assembly.coe.int/Main.asp?link=/Documents/AdoptedText/ta82/EREC934.htm.

[16] Ibid., article 4(a).

[17] Ibid., article 7(b).

[18] Ibid., articles 4(c) and 7(b), (c).

[19] Ibid., article 4(a).

recognizing the right to defense of the subject who carries that inheritance from the moment it is constituted in an individual sense.

The freedom of science and of scientific research, defined as a "basic value of our societies and a condition of their adaptability to the changing world environment," must be combined with responsibilities and the duties to protect the health and safety of workers and of the public in general and not to contaminate the cradle of life.[20]

The recommendation calls for an assessment of the real risks and for the regulation of in vitro experimentation on DNA, with regard to both the safety of the laboratories and the type of microorganisms that can be used in experimentation; furthermore, this risk must be periodically reassessed.[21] The constitution of the Ad Hoc Experts Committee on the Ethical and Legal Problems of Human Genetics (Comité ad hoc d'Experts sur les problèmes éthiques et juridiques de la génétique humaine, CAHGE) at the Council of Europe took its inspiration from this recommendation. For the history of this committee, which studies bioethical issues at the level of the Council of Europe, see the chapter on ethics committees.

8. Recommendation 82/472 of the European Economic Community (EEC) established safety standards for laboratories and proposed that the so-called right to difference—the prohibition against modifying human individuals through genetic engineering—be included in the European Convention.[22]

In the context of the already-mentioned Convention on Human Rights and Biomedicine (also called the Convention on Bioethics),[23] the Council of Europe also addressed the ethical problems emerging in the field of genetics (all of chapter 4 is dedicated to the human genome). Later, on November 6, 1997, the Committee of Ministers approved an additional protocol (among the additional protocols provided for by the convention itself) that prohibits human cloning.[24] The same Committee of Ministers had previously issued a specific recommendation on diagnosis and genetic screening for health care purposes.[25]

On March 16, 1989, the European Parliament published its Resolution on the Ethical and Legal Problems of Genetic Engineering,[26] which contains ideas that can

---

[20] Ibid., article 3(iii).

[21] Ibid., article 6(a), (b).

[22] European Economic Community, recommendation 82/472, "The Registration of Work Involving Recombinant Deoxyribonucleic Acid (DNA)," June 30, 1982, http://eur-law.eu/EN/82–472-EEC-Council-Recommendation-30-June-1982,96347,d.

[23] Council of Europe, Convention for the Protection of the Rights and Dignity of the Human Being with Regard to the Application of Biology and Medicine: Convention on Human Rights and Biomedicine, April 4, 1997, http://conventions.coe.int/Treaty/en/Treaties/Html/164.htm. The document was approved by the Committee of Ministers of the Council of Europe on November 19, 1996, and signed in Oviedo, Spain, on April 4, 1997.

[24] Council of Europe, Additional Protocol to the Convention on Human Rights and Biomedicine on the Prohibition of Cloning Human Beings, January 12, 1998, http://conventions.coe.int/treaty/en/treaties/html/168.htm.

[25] Council of Europe, "Recommendation R (92) 3 of the Committee of Ministers to Member States on Genetic Testing and Screening for Health Care Purposes," *International Journal of Bioethics* 3, no. 4 (1992): 255–257.

[26] European Parliament, Resolution A2–327/88 on the Ethical and Legal Problems of Genetic Engineering, March 16, 1989, http://www.codex.uu.se/texts/EP-genetic.html.

be widely agreed upon, followed later by its first resolution on cloning on March 12, 1997[27] and a new resolution on cloning dated January 15, 1998.[28]

Various other directives concerning engineered microorganisms, biotechnological inventions, the protection of the human genome, and genetic testing were also issued by the authorities of the European Community, including the following:

- EEC Council directive 90/219 of April 23, 1990, on the Contained Use of Genetically Modified Micro-Organisms.[29]

- EEC Council directive 90/220 of April 23, 1990, on the Deliberate Release into the Environment of Genetically Modified Organisms.[30] Other documents—including Church documents—concerning some particular issues will be cited from time to time.

- European Parliament and Council directive 98/44/EC of July 6, 1998, on the Legal Protection of Biotechnological Inventions,[31] which clarifies the distinction between what is patentable and what is not. This document reaffirms that the human body in its various phases of formation and development cannot be considered a patentable invention; techniques of human cloning or procedures for modifying the genetic identity of human germ cells are not patentable either and are expressly prohibited. In order to protect biotechnological inventions, member states must ensure that national patent laws adhere to the provisions of the directive. The ethical aspects of the directive were entrusted to an independent ethics committee charged with providing advice to the European Commission on these topics.

- The opinion on the Protection of the Human Genome by the Council of Europe, approved by the Parliamentary Assembly of the Council of Europe on March 19, 2001.[32]

- The European Commission published a document titled *Twenty-five Recommendations on the Ethical, Legal and Social Implications of*

---

[27] European Parliament, document B4–209/97, "Resolution on Cloning," March 12, 1997, http://wfrt.org/humanrts/instree/cloning1.html.

[28] European Parliament, document OJ (C 34) 164, "Resolution on Human Cloning," January 15, 1998, http://wfrt.org/humanrts/instree/cloning2.html.

[29] European Economic Community, council directive 90/219, "Contained Use of Genetically Modified Micro-Organisms," April 23, 1990, http://eur-lex.europa.eu/LexUriServ/LexUriServ.do?uri=CELEX:31990L0219:EN:HTML.

[30] European Economic Community, council directive 90/220, "Deliberate Release into the Environment of Genetically Modified Organisms," April 23, 1990, http://eur-lex.europa.eu/LexUriServ/LexUriServ.do?uri=CELEX:31990L0220:EN:HTML.

[31] European Parliament and Council, directive 98/44/EC, "Legal Protection of Biotechnological Inventions," July 6, 1998, http://eur-lex.europa.eu/LexUriServ/LexUriServ.do?uri=CELEX:31998L0044:EN:HTML.

[32] Council of Europe, Doc. 9002, "Protection of the Human Genome by the Council of Europe," March 19, 2001, http://assembly.coe.int/main.asp?Link=/documents/workingdocs/doc01/edoc9002.htm.

Genetic Testing,[33] which discusses the quality of available genetic testing, accuracy of results, conditions for access to testing and treatment (particularly for persons with rare disorders), proper use of samples and data, informed consent, respect for privacy, genetic counseling, and the risk of discrimination.

- The World Health Organization ultimately made a pronouncement as well on the topic of cloning on March 18, 1997.[34]

- The Universal Declaration on the Human Genome and Human Rights was issued by UNESCO on November 11, 1997.[35] It is an important international instrument with many valuable components such as the rejection of any genetic reductionism (art. 2[b] and art. 3), the affirmation of the primacy of respect for the human person over research (art. 10), the rejection of discrimination (art. 6), the confidentiality of private information (art. 7), the promotion of independent ethical committees (art. 16), and the commitment of states to promoting bioethics education and discussion that is also open to the various religions (arts. 20 and 21). However, it also has several debatable and questionable aspects: it does not define who is entitled to the rights it proclaims, it fails to affirm that those rights belong to all human beings from the moment a unique genetic patrimony is constituted, and it makes no reference to embryos or fetuses.[36]

Some important stages in the development of genetic engineering[37]

The basic facts of biology and genetics, such as the description of the cell and notions about chromosomes, genes, and DNA chains (which were touched on in the chapter

---

[33] E. McNally and A. Cambon-Thomsen, *Twenty-five Recommendations on the Ethical, Legal and Social Implications of Genetic Testing* (Luxembourg: Office for Official Publications of the European Communities, 2004), available online at http://ec.europa.eu/research/conferences/2004/genetic/pdf/recommendations_en.pdf.

[34] World Health Organization, WHA Resolution 50.37, "Cloning in Human Reproduction," May 14, 1997.

[35] United Nations Educational, Scientific and Cultural Organization (UNESCO), Universal Declaration on the Human Genome and Human Rights, November 11, 1997, http://unesdoc.unesco.org/images/0011/001102/110220e.pdf#page=47.

[36] The Holy See developed a document that contains observations on this declaration: Informal Working Party of Bioethics of the Section for Relations with States of the Vatican Secretariat of State, "Observations on the Universal Declaration on the Human Genome and Human Rights," November 11, 1997, http://www.vatican.va/roman_curia/pontifical_academies/acdlife/documents/rc_pa_acdlife_doc_08111998_genoma_en.html. See also the Italian editorial "L'UNESCO sul genoma umano: Un segnale di forte significato bioetico," *Medicina e Morale* 48, no. 1 (1998): 9–14. The French text of the document is provided in the same issue: 158–166.

[37] For greater historical, technical, and scientific detail see A. Serra, *A cento anni dalla morte del fondatore della genetica: Nuove conquiste, nuove responsabilità*, in *Nuova genetica, uomo e società*, ed. A. Serra and G. Neri (Milan: Vita e Pensiero, 1986), 13–17; Idem, "La 'nuova genetica': Attualità, prospettive, problemi," in *Medicina e genetica verso il futuro*, ed. S. Cotta (L'Aquila: Japadre, 1986), 5–23; E. Brovedani, "L'ingegneria genetica: Aspetti scientifico-tecnici," *Aggiornamenti Sociali* 7, no. 8 (1986): 517–534; M. Milani Comparetti, "Introduzione alla nuova

on life and its origins), will not be addressed here. It would be a repetition of widely known information more suitably described in textbooks on biology, embryology, and human genetics. The scope here is limited to what is strictly necessary for understanding the ethical aspects of the problem, starting with a summary of the historical stages in the development of genetics.

The first important stage—the most important in the field of genetics since the discoveries of Gregor Mendel (between 1856 and 1865)—was in 1956, with the rediscovery of human chromosomes as the basic structures carrying genetic material. Mendel called them *elements*, and in 1910, Thomas Hunt Morgan studied their chemical composition.

The first cell fusion came to pass in 1965: the fusion of human and mouse cells with the transmission of genes (*assignment*) to the human chromosomes. That was precisely when Rollin Hotchkiss first introduced the term *genetic engineering* with the meaning indicated at the beginning of this section.[38]

The use of prenatal diagnosis techniques in the field of genetics began in 1967. As everyone knows, these allow for the early identification of the genetic conditions of the fetus either by revealing the fetus's chromosomal makeup or through biochemical tests.

The discovery of restriction endonuclease, an enzyme suitable for acting as a "scalpel" to split DNA at definite points made up of specific, short sequences of bases, was announced in 1969. Since then, over four hundred different restriction enzymes have been identified that are capable of recognizing a wide variety of sequences.

The synthesis of the first artificial gene was announced in 1970. In 1971, recombinant DNA was created for the first time: a portion of DNA could be incorporated into a bacterium serving as a vector (e.g., *Escherichia coli*), and that portion could be transferred and combined with the genetic information of a host cell with the possibility of recombining and replicating into a new genetic structure.

In 1981, the first mice were born by means of *cloning*. This term indicates the process whereby a group of genetically identical subjects (clones), all derived from a single organism, can be obtained. Several other attempts at cloning animals of various species were made later, in the 1990s, and some were successful. Such attempts were also repeatedly carried out in the case of human cloning.[39] The increasing speed with which science passes through its developmental stages, continuously conquering new frontiers and increasing the technological potential for manipulating life, is worthy of note. Hybridization, transfer, cloning, and selection are terms that evoke the possibility of genetic interventions, many of which have already been implemented in the plant and animal kingdoms and are also conceivable for humans.[40]

---

genetica," in *Ingegneria genetica*, Sgreccia and Mele, 3–23; F. Vogel and A. G. Motulsky, *Human Genetics: Problems and Approaches* (Berlin: Springer, 1986).

[38] R. D. Hotchkiss, "Portents for a Genetic Engineering," *Journal of Heredity* 56, no. 5 (September 1965): 197–202.

[39] The chapter on human fertilization techniques (chapter 11) amply addresses the scientific and technical aspects of the topic of cloning.

[40] For the subject matter addressed in this section, see G. Sansone, "Recenti progressi in genetica medica," *Medicina e Morale* 32, no. 1 (1982): 14–24; A. De Flora, "Strategie per la correzione di malattie genetiche," *Medicina e Morale* 32, no. 1 (1982): 25–35; H. Jonas,

All of the successes that have been achieved should not be automatically viewed in a negative light because they also create the potential for positive interventions. First and foremost, much significant progress has been made in acquiring knowledge about the genetic information contained in individual chromosomes: specifically, immunoglobulin production mechanisms and the structures of the X and Y chromosomes are now better known. Technological and industrial applications have been found for the production of very important polypeptide molecules: human insulin, interferon, somatostatin, somatotropin, vaccines against influenza and hepatitis A and B, and many others.

Another recent application arises from the possibility of identifying pathogenic genes in healthy subjects through DNA probes or through the study of so-called dimensional polymorphisms. Gene therapy is therefore a prospect for researchers. As early as 1982, one researcher in this area wrote that "it is possible to isolate normal genes, to put them into mammalian cells at any stage, including the fertilized embryo, and to obtain gene function in a random way and at a low level. However, it is not yet possible to ensure that the 'therapeutic' newly inserted genes function under normal control in the animal, in time, space or quantity."[41]

Yet the possibilities for gene therapy that have opened up in the field of genetics should not lead to underestimating the ethical implications in other areas of application, including the implications of in vitro fertilization.

Before examining the ethical issues more closely, here is an outline of genetic engineering technologies arranged according to the objectives they pursue:[42]

a. *Mapping.* This was the first phase prior to the Human Genome Project (1973–1988). It aims to identify the position on the chromosomes of genes whose products or effects are already known. The strategies employed range from cellular hybridization to the use of specific probes capable of revealing the position of pieces of DNA corresponding to a given gene. The most important results in this area involve the localization of coding genes, including the ones for disorders such as Duchenne muscular dystrophy and cystic fibrosis.

b. *Isolation.* This objective was achieved through the use of a set of biological "scalpels," or so-called restriction enzymes (endonucleases) capable of "cutting" the

---

"Technique, morale et génie génétique," *Communio* 9, no. 6 (1984): 46–65; J.L. Ruiz de la Peña, "Anthropologie et tentation biologiste," *Communio* 9, no. 6 (1984): 66–79; P. Caspar, "Les fondements de l'individualité biologique," *Communio* 9, no. 6 (1984): 80–90; A. Serra, "Interrogativi etici dell'ingegneria genetica," *Medicina e Morale* 34, no. 3 (1984): 306–321; J. Harris, *Wonderwoman and Superman: The Ethics of Human Biotechnology* (New York: Oxford University Press, 1992); J.R. Nelson, *On the New Frontiers of Genetics and Religion* (Grand Rapids, MI: William B. Edermans, 1994); R. Pollack, *Signs of Life: The Language and Meanings of DNA* (Boston: Houghton Mifflin, 1994); C. Munthe, *The Moral Roots of Prenatal Diagnosis* (Göteborg: Centre for Research Ethics Graphic Systems, 1996); R. Dulbecco, *La mappa della vita* (Milan: Sperling and Kupfer, 2001).

[41] B. Williamson, "Gene Therapy," *Nature* 298 (1982): 418.

[42] A. Serra, "La rivoluzione genomica: Conquiste, attese, rischi," *La Civiltà Cattolica* 2 (2001): 439–453; A. Bompiani, "Genomica funzionale e proteomica: Recenti svilppi delal ricerca nelle malattie poligeniche e considerazioni etiche," *Medicina e Morale* 53, no. 5 (2003): 797–840.

DNA chain at very precise points, thereby isolating the genes included in the base sequence between the two cuts.

c. *Cloning.* This is a sort of "biological replication" of individual genes in order to have a large quantity available for analysis or various other uses. Cloning is possible by inserting the gene of interest into the genome of microorganisms through so-called vectors (plasmids, bacteriophages, viruses) in the technique of recombinant DNA. When the microorganisms replicate, the genes inserted into their genome are also replicated.

d. *Sequencing.* This was the second phase (1989–2001), which involved a powerful and very extensive international, scientific, and industrial organization (the Human Genome Organization, HUGO). It culminated with the development of a major research program called the Human Genome Project, whose main goal was precisely to "sequence"—that is, to determine the molecular structure of the genes by exactly identifying the sequence of bases composing them. This offered the possibility of better understanding their mechanisms of function and modification.

After the initial announcements in 2000, first by Francis Collins, director of the National Human Genome Research Institute of the NIH, and shortly afterward by Craig Venter of the private company Celera Genomics, the results of the two research groups were published in 2001 in the two prestigious journals *Nature* and *Science*.[43] This work provided a first draft of the human genome that would lead to the Human Genome Database, which contains the complete text of the human genome edited by the public consortium under Collins' direction.

e. *Annotation and function of genes* is the third phase, which is still in progress. The genome must be annotated, i.e., the individual genes must be assigned to the various chromosomes. This phase has already provided important information such as the genetic origins of hundreds of diseases, especially tumors.

Thanks to this intense and coordinated effort, the large Genome Database (GDB)[44] listed 12,084 annotated genes as of the year 2001, 10,341 of which were definitively mapped, meaning localized to well-defined points on certain chromosomes. It is estimated that there are tens of thousands of additional putative genes that the research will have to continue investigating.

Some important information has come out of this vast and ongoing effort: every gene has coding segments (exons) and noncoding segments (introns); there can be various specific alterations (mutations) in the code of a given gene, leading to pathologies of varying severity.[45] This is the case for Duchenne and Becker muscular dystrophies: the former is much more serious than the latter, and they are due to

---

[43] International Human Genome Sequencing Consortium, "Initial Sequencing and Analysis of the Human Genome," *Nature* 409 (February 15, 2001): 860–941; J. C. Venter et al., "The Sequence of the Human Genome," *Science* 291 (February 16, 2001): 1304–1351.

[44] Human Genome Database (GDB), "Count of Mapped Genes by Chromosome," accessed Apr. 8, 2001. [The GDB was a key database in the Human Genome Project, and the "Count of Mapped Genes," formerly available online, was updated continually as the project progressed. The project was completed in 2003 and the database shut down in 2008.]

[45] P. F. R. Little, "Structure and Function of the Human Genome," *Genome Research* 15 (2005): 1759–1766.

two different mutations of the same gene coding for dystrophin,[46] which is a protein necessary for normal muscle cell activity.

f. Research is now in the *postgenomics* phase, which provides a glimpse of even more challenging and remarkable prospects.[47] The identification and mapping of all the genes, the study of their specific functions, and the analysis of their possible pathogenic mutations constitute the backbone of research in this area. But other, broader prospects for research and clinical applications are beginning to open up. Specifically, there are three highly interesting and important areas: (1) *Proteomics* should enable the identification of the thousands of proteins produced by the human organism and an understanding of their distribution and function in the complexity of cells and tissues. (2) *Functional genomics*, thanks to the automation and computerization of procedures, allows the simultaneous screening of vast numbers of genes and networks of DNA fragments through certain tools for bioinformatics—microarrays and genetic chips—in order to store the complete genetic structure in a tiny space. It contributes to an understanding of not only the functions of individual genes but also their interactions and integral functions, which are responsible for controlling development from the zygotic stage onward and for the physiological processes of cells, tissues, and organs. This in turn contributes to an understanding of the causes of many pathologies, especially the more complex ones, which are called polygenic or polyfactorial.[48] (3) Lastly, *pharmacogenetics and pharmacogenomics* have the tasks of identifying how individual responses to a drug vary according to heredity and personal and family genetic characteristics, and of developing new, personalized, and more effective drug therapies with lower risks of side effects, through an understanding of the genes involved in metabolizing drugs.[49] The development of these new genetic applications will probably result in a substantial change in medical practice as well, with new diagnostic criteria and especially with the possibility of administering safer and more effective forms of drug therapy while managing to predict their tolerability and effectiveness in subjects based on the presence or absence of particular genetic polymorphisms.

A first risk to be feared in the pharmacogenetics sector, which is in full experimental development and therefore requires the application of scientific and ethical monitoring for all biomedical research, stems from a warped understanding that could increase the drive toward genetic determinism or radical genetic reductionism, which is already widespread, with the affirmation that every expression of phenotype and even of human behavior is associated with (and determined by) a particular genetic structure. The negative consequences would include fatalism and

---

[46] E. P. Hoffman, R. H. Brown, and L. M. Kunkel, "Dystrophin: The Protein Product of the Duchenne Muscular Dystrophy Locus," *Cell* 51 (1987): 919–928.

[47] A. E. Guttmacher and F. S. Collins, "Realizing the Promise of Genomics in Biomedical Research," *Journal of the American Medical Association* 294, no. 11 (2005): 1399–1402.

[48] Bompiani, "Genomica funzionale"; Serra, "La rivoluzione genomica."

[49] See A. D. Roses, "Pharmacogenetics and the Practice of Medicine," *Nature* 40 (2000): 857–865; W. E. Evans and M. V. Relling, "Pharmacogenomics: Translating Functional Genomics into Rational Therapeutics," *Science* 286 (1999): 487–491; M. Ingelman-Sundberg, "Pharmacogenetics: An Opportunity for a Safer and More Efficient Pharmacotherapy," *Journal of Internal Medicine* 250 (2001): 186–200.

a loss of responsibility at the ethical and social levels, and the pursuit of erroneous objectives at the level of research policies.[50]

It is also necessary to consider the risk that small segments of the population with pharmacogenetically unfavorable genetic characteristics (such as nonresponders and/or those who are intolerant) might be ignored by pharmacogenetic research—which is largely sponsored and commercial—and turn into "pharmacologically orphaned" groups.[51] The previously cited UNESCO declaration of 1997 underscores this particular responsibility for researchers—but also for those who deal with research policies, whether in the public or private sector—to respect genetically vulnerable individuals, families, and populations and promote solidarity toward them (art. 13 and art. 17).

The possibility of archiving biological samples and associated data (in so-called *biobanks*), from which important clinical indications can be derived in correlation with the development of genetic findings, whether in diagnosis or treatment, for individuals or entire populations with similar genetic characteristics, leads to other prospects and expectations. The maintenance and use of such archives, however, demands the implementation of strict measures to safeguard the confidentiality of individuals and their right to know the purposes of the research and the ways in which their samples and personal information will be used.[52]

The topic of informed consent regarding the current and future use of samples and the possibility—however remote—of discovering secondary or undesired information (such as predispositions for disorders, being a nonresponder for a drug, or ruling out paternity) also takes on particular and specific relevance in the area of pharmacogenetics and biobanks.[53]

---

[50] Italian National Bioethics Committee, *From Pharmacogenetics to Pharmacogenomics* (Rome: Dipartimento per l'Informazione e l'Editoria, 2006), available at http://www.governo.it/bioetica/eng/opinions/Pharmacogenetics.pdf.

[51] A. G. Spagnolo and R. Minacori, "Farmacogenetica e farmacogenomica: Aspettative e questioni etiche," *Medicina e Morale* 52, no. 5 (2002): 819–866; J. Bell, "The New Genetics in Clinical Practice," *British Medical Journal* 316 (1998): 618–620; D. Housman and F. Ledley, "Why Pharmacogenomics? Why Now?" *Nature Biotechnology* 16 (1998): 492–493; A. Persidis, "The Business of Pharmacogenomics," *Nature Biotechnology* 16 (1998): 209–210.

[52] J. A. Robertson, "Consent and Privacy in Pharmacogenetic Testing," *Nature Genetics* 28 (2001): 207–209; HUGO Ethics Committee, "Statement on DNA Sampling: Control and Access," February 1998, http://www.hugo-international.org/img/dna_1998.pdf; National Bioethics Advisory Commission, *Research Involving Human Biological Materials: Ethical Issues and Policy Guidance*, vol. 1 (Rockville, MD: US Government Printing Office, 1999).

[53] See Italian National Bioethics Committee, *From Pharmacogenetics to Pharmacogenomics*; A. G. Spagnolo, M. Cicerone, R. Minacori, "Biobanche: Aspetti clinici della conservazione del materiale biologico umano," *Iustitia* 1 (2006): 63–78; Spagnolo and Minacori, "Farmacogenetica e farmacogenomica"; G. J. Annas, "Reforming Informed Consent to Genetic Research," *Journal of the American Medical Association* 286, no. 18 (2001): 2326–2328.

Levels and aims of intervention:
Prospects and risks for the new postgenomic medicine

In order to offer ethical guidelines, it is necessary to discuss the various levels of gene intervention and the various goals that medicine attempts to pursue in applying the discoveries about the human genome.

As with any progress in scientific research, the development of genomics has created high hopes in the medical field. Several areas of interest in the medical field will therefore be considered in order to understand their significant value but also the risks of potential abuse that have begun to emerge.

LEVELS OF INTERVENTION. These levels must be kept firmly in mind and distinct from one another. An intervention aimed at bringing about an alteration in the genome could occur at the somatic cell level, the germ cell level, or at the embryonic level during the early stages of development. The ethical significance of the intervention itself consequently changes for each of these levels.

Somatic cell gene therapy, an application to which thousands of researchers and many pharmaceutical companies have turned their attention, aims to develop new therapeutic procedures for genetic disorders. Its implementation was already proposed in 1984;[54] at the time, it was already thought that normal genes could be transferred (by *transfection*) into affected cells and function well there.

An intervention aimed at altering some degeneration or defect in *somatic cells* could be theorized, for example. Let us suppose that researchers succeed in correcting the genetic defects of microcytic anemia in *hematopoietic* cells in such a way that normal cells would be produced, replacing the defective ones as they replicate. This would be a very important success, and the potential loss of living cells would neither constitute harm to the individual nor pose ethical problems.[55] Such a procedure should be classified as therapeutic, as will be seen more clearly. Unfortunately, however, the initial optimism has not met with significant results in the clinical setting, though a courageous effort accompanied by prudent warnings has continued.[56] In the year 2000, the American Society of Human Genetics (ASHG) published a statement on gene therapy in which it stressed the need for an attitude marked by prudence and gradualness with the aim of preventing abuses: "Gene

---

[54] W. F. Anderson, "Prospects of Human Gene Therapy," *Science* 226 (1984): 401–409.

[55] The first authorization to practice gene therapy was provided in the United States on September 4, 1990, by the Food and Drug Administration (FDA) in order to attempt a cell transplant into a 4-year-old girl with the correct gene to fix an adenosine deaminase (ADA) enzyme deficiency. Additional authorization has always been given by the FDA in the United States for tumor necrosis factor (TNF) gene transplants. In Italy, a gene transplant for ADA deficiency was done for the first time in 1992 at Saint Raphael Hospital Scientific Institute (Istituto Scientifico Ospedale S. Raffaele) in Milan. About three years later, both patients who underwent the gene therapy showed normalization in their immunological reports and functional restoration of both cellular and humoral immunity. These results indicate the complete success of the procedure. See C. Bordignon et al., "Gene Therapy in Peripheral Blood Lymphocytes and Bone Marrow for ADA-Immunodeficient Patients," *Science* 270, no. 5235 (1995): 470–475.

[56] G. Ross et al., "Gene Therapy in the United States: A Five-Year Status Report," *Human Gene Therapy* 7 (1996): 1781–1790.

therapy holds much promise. However, this promise will only be achieved through continued rigorous research on the most fundamental mechanisms underlying gene delivery and gene expression in animals. Clinical trials should be undertaken only after solid evidence of both safety and efficacy has been obtained in an appropriate animal model."[57] The prospect of somatic cell gene therapy nonetheless remains one of the most fascinating and promising[58] in all genomic research. The most significant advances in this field have been made through experimentation with in utero gene therapy,[59] mostly limited to animal studies at this stage.

Another application method for genomic research is *germ-line gene therapy*. *Gene transfection* would have to be carried out in oocytes, in the zygote itself, or in the very early embryonic stages. Although attempts at germ-line gene therapy in laboratory animals have produced positive results, the prospect of preventively correcting erroneous genetic information in human beings at a very early developmental stage appears remote. The procedure for modifying *germ cell lines* is still precluded in reality because of the impossibility—at least at the current time—of guiding the insertion of the correct gene.

There is strong resistance to beginning this approach with human beings as well, even among researchers themselves, because of the serious risks it could involve. In a commentary on the subject, Paul Billings wrote, "Responsible scientists should resist the application of powerful genetic tools that will tamper with the human germline. ... It would be a tragedy if once again the Siren of reductionism blinded well-meaning interventionists, allowing the use of inadequately refined genetic methods to modify highly evolved and interactive human systems."[60]

Given the current ineffectiveness of gene therapy on germ-line cells, some documents from other countries propose the need for experimentation on germ cells, and therefore on the embryos derived from the manipulated cells, with the aim of improving the technique. Nontherapeutic experimentation on human embryos remains seriously wrong from the perspective of personalist bioethics, regardless of the end pursued (such as altering physiological and genetic inheritance or even improving the future possibilities of science).[61]

---

[57] Board of Directors of the American Society of Human Genetics, "ASHG Statement on Gene Therapy: April 2000," *American Journal of Human Genetics* 67 (2000): 272–273.

[58] See T. Gura, "After a Setback, Gene Therapy Progresses ... Gingerly," *Science* 291 (2001): 1692–1697.

[59] See E. D. Zanjani and W. F. Anderson, "Prospects for In Utero Human Gene Therapy," *Science* 283 (1999): 2084–2088; A. L. Caplan and J. M. Wilson, "The Ethical Challenges of In Utero Gene Therapy," *Nature Genetics* 24 (2000): 107; S. N. Waddington et al., "In Utero Gene Therapy: Current Challenges and Perspectives," *Molecular Therapy* 11, no. 5 (2005): 661–676; C. Coutelle et al., "Gene Therapy Progress and Prospects: Fetal Gene Therapy—First Proofs of Concept—Some Adverse Effects," *Gene Therapy* 12, no. 22 (2005): 1601–1607.

[60] P. R. Billings, "In Utero Gene Therapy: The Case Against," *Nature Medicine* 5 (1999): 255.

[61] This position also comes out in the NBC document on gene therapy. See for comparison the commentaries on the document in V. Mele, "La geneterapia ed il parere del Comitato Nazionale per la Bioetica," *Vita e Pensiero* 5 (1991): 362–367. For a comparison with documents on the topic by other ethics committees, see A. G. Spagnolo and M. L. Di Pietro, "Terapia genica: Il documento 15.2.91 del Comitato nazionale per la Bioetica ed un'analisi

Finally, the notion of a medical intervention *on the human embryo* becomes an even more delicate matter because there is a high risk of compromising the embryo's life or biological future in a genetic sense. Such a procedure poses even more ethical problems when it is deliberately planned for experimental purposes.

This topic has been addressed in several important speeches by Pope John Paul II, such as his discourse at the conclusion of the 35th General Assembly of the World Medical Association on October 29, 1983,[62] his address to the participants in the Conference on Experiments in Biology organized by the Pontifical Academy of Sciences on October 23, 1982, and his address to the participants in the first International Medical Conference of the pro-life movement on the topic of prenatal diagnosis and the surgical treatment of congenital malformations.[63] But above all the instruction *Donum vitae* (February 22, 1987) by the Congregation for the Doctrine of Faith, on respect for human life and the dignity of procreation, contains precise ethical guidelines regarding experiments and all other manipulative interventions on embryos.[64] This possibility will be further explained after the discussion of in vitro fertilization, which is the primary method for obtaining human embryos on which to carry out any type of experimentation.

AIMS. As can be easily imagined, the *aims* of use for genetic advances are the main determinant of the morality of many interventions. The aims of intervention in the field of genetics can be classified into the following categories (see also the summary table at the end of the chapter): (1) diagnostic, (2) therapeutic, (3) production, (4) modification, and (5) experimentation (by destruction of the embryo).

The uses of genetic engineering for *diagnostic* aims are manifold and the range of new applications has increased in recent years. Aside from prenatal genetic diagnosis, which will be addressed separately and at length, adult applications are being developed to (a) verify the presence of disorders of a presumably genetic origin at the premarital and preconception stages so as to assess the risk of healthy individuals being potential carriers and having unhealthy offspring, (b) verify biological paternity in civil proceedings, (c) identify criminals in criminal investigations, and (d) conduct screenings. In fact, screenings are used not only for pregnant women within a certain at-risk population, but also for other adults—in the field of workplace medicine, for example—in the form of generalized aptitude tests for a certain working environment. These diagnostic methods and applications will have to be discussed shortly from the ethical perspective. Particular attention should be given to the project of sequencing the entire *human genome*.

---

comparativa con le esperienze di altri Comitati etici nazionali ed internazionali," *Il Diritto di Famiglia e delle Persone* 21, no. 2 (1992): 323–363.

[62] John Paul II, Address to Members of the World Medical Association, October 29, 1983, in *Insegnamenti di Giovanni Paolo II* (Vatican City: Libreria Editrice Vaticana), 6.2:917–923 [Italian].

[63] John Paul II, Address to Participants in the First International Medical Conference of the Pro-Life Movement on Prenatal Diagnosis and the Surgical Treatment of Congenital Malformations, December 3, 1982, in *Insegnamenti di Giovanni Paolo II*, 5.3:889–898 and 1509–1513 [Italian].

[64] Congregation for the Doctrine of Faith, Instruction *Donum vitae*, February 22, 1987, part I.

Therapeutic aims in the field of genetics, when they are feasible, must be understood as therapeutic in reference to the subject on whom the procedure is carried out and certainly not with a view to sacrificing one person in order to help another. As mentioned earlier in talking about levels of intervention, gene therapy can be planned on somatic cells (the lymphatic system, blood cells, bone marrow cells),germ line cells, or even early-stage embryos. As will be shortly seen, the ethical judgment may vary considerably.

The aim of *production* through genetic engineering is already developing, particularly in the field of pharmacology, with the production of hormones such as human insulin; interferon; vaccines against bacteria, viruses, and parasites; clotting factor VIII, which is lacking in people with hemophilia A; and other ongoing developments. The technique employed requires the use of recombinant DNA. In the laboratory synthesis of human insulin, for example, a section of DNA from the beta cells in the human pancreatic islets of Langerhans is extracted by cutting the DNA at precise points with those "biological scalpels"—the restriction enzymes (restriction endonucleases)—so that the bases on the ends of that segment of deoxyribonucleic acid can connect in the desired order with the DNA bases of the recipient cell. In the case of insulin, the target is the plasmid of the recipient cell *Escherichia coli*. The *E. coli* DNA is opened up at the right point by the same "enzymatic scalpel" of the restriction endonuclease so the bases can reconstruct the complete sequence with the inserted segment. *E. coli*, replicating in culture, then produces human insulin that is purified and marketed.

Understandably the question is primarily technical, but it also has ethical implications: What are the effects of the product (which still seem positive in this case)? What type of bacterium is being used? It is not dangerous in this case but could be so in the case of other bacteria and viruses.

The area of production also includes many biotechnologies used in the food and agriculture sector, where vaccines, bacteria, and viruses are actually used both to improve agricultural food products and to protect crop production. Some ethical problems requiring scrutiny appear in this field as well, particularly in relation to environmental protection.

The prospect of *modification* (where modification means a change that is not therapeutic but elective and selective) can be envisioned for use in creating modified species or classes of engineered individuals among human beings or animals and plants. Regarding the first scenario, everyone agrees that the modification of human beings should be ruled out regardless of real technical possibility. The second scenario is currently in practice and poses the problem of the potential influence of biotechnologies on the ecosystem or on ecosystems. In this context, an attenuated form of modification called species *enhancement* is also debated. For some, genetic enhancement should not be ruled out in principle, even for human beings.[65]

Ethical approaches and ethical judgment criteria

It is important to pause and reflect on the *fundamental criterion* adopted in the field of bioethics in order to affirm that some procedures are permissible, such as gene therapy, whereas other procedures are illicit, such as genetic manipulation

---

[65] M. Cuyás, "Problematica etica della manipolazione genetica," *Rassegna di Teologia* 5 (1987): 471–497.

for eugenic aims. On the basis of what principles should this boundary between licit and illicit be established? The scientist working at the level of technical and scientifically grounded possibilities can pose the question, but cannot ask molecular biology for the criterion distinguishing licit from illicit; otherwise, everything that is technically and scientifically possible would be declared licit ipso facto. The latter statement is supported by those who hold that man is entirely the master of himself and his destiny (in this case, only the scientist happens to be the fortunate master) and that, just as certain combinations were made possible at the beginning of life by chance—as affirmed by some thinkers already mentioned—and thus made us human beings what we are, so today certain possibilities open new outcomes to the human species and new changes in its biological destiny. In this case there would be no morality, unless morality were understood as a radical efficientism considering whatever works to be moral. Philosophical historicism would thus be transplanted into biology.[66] In this perspective, the Hiroshima bomb would be just as moral as the invention of the polio vaccine.

Others might think it is up to national or international positive law to set acceptable limits for the majority of humanity at a given moment in history. For example, it could be said that research with the aim of gene therapy is permissible today, but research with eugenic and racist aims in search of the *neanthropos* (new man) is not. Tomorrow, when things are more certain and safer and the risks and benefits of a procedure have been weighed, the community could decide on other laws. This position is encountered in the context of many scientific or legal and scientific committees, such as the previously mentioned Warnock Committee.

This is clearly an empiricist and pragmatic position that seeks to adapt current laws to new situations.[67] It is important to add that in the context of this school of thought, utilitarianism makes its presence felt also by justifying any method that could lead to a result. For example, some are unwilling to accept the criterion of respect for embryos when the destruction of embryos is considered necessary for achieving the goal of advancing scientific research in prestigious and promising areas that hold great hope for the good of humanity.

It is clear that the national and international committees dealing with this subject are indeed affected at least to some extent by this approach, which is positivistic in the legal realm and pragmatistic in the philosophical sense. It is not difficult to imagine that the limits placed on scientific research are actually based more on majority vote than on objective and universally valid criteria. Even the appeal to human rights has trouble making headway and hardly constitutes a criterion understood by everyone in the same way. It must nonetheless be said that the problem is shifted but not resolved in this positivist and empiricist perspective as well: On what basis should lawmakers and committee majorities or minorities rest their opinions?

If, as it seems, the criterion is usefulness at this historical moment, then it must be said that utilitarianism is not a good morality, especially when applied to subjects who cannot decide (embryos) or at the expense of future generations that will have

---

[66] This position is supported by radical adherents to the ideas expressed by Jacques Monod in *Chance and Necessity.*

[67] See the chapter on ethical reference models.

to live with decisions made by others. In sum, it is necessary to find an objective reference criterion not just for morality but also for the law.[68]

According to some, referencing theology would be inconclusive because the sources of Revelation do not offer detailed guidance on this subject, but only shed light on the sacred value of human life in general. In addition, this type of theological guidance would only have value for believers.[69] Therefore, while the intervention of the Church's Magisterium in this area of knowledge in which the future of man is at stake is completely legitimate, as well as normatively and theologically valid,[70] rational reflection remains not only legitimate but dutiful. Some theologians, basing their arguments on revealed truths of a general nature such as man as the image of God, the value of life, and so forth, affirm that theological facts allow the deduction that man, precisely because he is the image of God, can have dominion not only over the things in the world for his own good, but also over his own physical reality for his own good.[71] Naturally, though, the question remains: Wherein lies the good of man in a specific case of genetic manipulation or genetic engineering?

I also consider the guidance proposed by legal scholars and philosophers of law concerning the basic criterion for protecting the life of the embryo—namely, anything that does not seriously endanger the life of the embryo is licit—to be insufficient. If firmly upheld, this point is certainly an essential and fundamental one; however, it must be admitted that not all legislation upholds it resolutely. While this objective may be the last line of defense in law, it cannot suffice for morality, because the fullness

---

[68] These questions are posed by the whole of personalist legal thought and by moralists as well. See Lombardi Vallauri, "Manipolazioni genetiche"; C. Cirotto and S. Privitera, *La sfida dell'ingegneria genetica: Tra scienza e morale* (Assisi: Cittadella, 1985); E. Agazzi et al., *Manipolazioni genetiche e diritto: Atti del XXXV Convegno di studio; Roma, 7–9 dicembre 1984*, Quaderni di Iustitia 34 (Rome: Giuffrè, 1986).

[69] K. Rahner, "The Problem of Genetic Manipulation," in *Theological Investigations* (New York: Seabury Press, 1972), 9:225–252, esp. 225–227; A. Dumas, "Fondements bibliques d'une bioétique," *Le supplément* 142 (1982): 353. Their thinking is presented in Cirotto and Privitera, *La sfida*, 146.

[70] The instruction *Donum vitae*, which primarily addresses artificial procreation and deals with genetic engineering only in some passages, constitutes the conclusion of a whole line of magisterial teachings developed through bishops' documents relating to the topic of IVF (see note 73 in chapter 11, "Bioethics and Human Fertilization Technologies), in addition to the addresses by John Paul II cited in this chapter. I do not agree with what Privitera affirms in *La sfida dell'ingegneria genetica* on the question of the value of these magisterial documents: he agrees with J. Fuchs, who holds that since these are questions involving scientific, medical, and biological data, "the judgments it formulates on this matter do not have the same weight that is attributed to the truth it proclaims; they can be true, but they can also be false" and are valid "not for the authority they bear, but for the arguments they offer" (Privitera, *La sfida*, 167–168). I think it must be kept in mind that if these documents intend to be normative—and the author holds that they are—then they have the weight of truth, authoritatively taught by virtue of the charism of the Magisterium, to the extent that are not making pronouncements on scientific data but on ethical norms and moral values, which is the area in which the Magisterium is fully competent. It should instead be recalled that the pronouncements of the Magisterium do not exempt moralists from explaining and illustrating the reasons behind them.

[71] See S. Spinsanti, "Vita fisica," in *Corso di morale*, ed. T. Goffi and G. Piana (Brescia: Queriniana, 1983), 2:127–267, especially 223.

of man is not encapsulated in physical survival. There are various needs and values on the spiritual level that must be jointly protected in procreation.[72]

So what should the ethical or bioethical criterion for discernment be, then? We need to refer back to what was explained in the chapter on the principles of bioethics, and specifically to the unity of person and body and the therapeutic principle or principle of totality.

The body, first and foremost its genetic program or genome, is essentially united to the spirit so as to constitute the existential and essential unity of the person. This means that any intervention on corporality—therefore including on the genome—is an intervention on the whole person and can be justified only if there is a therapeutic reason, since any other intervention on the person, who is an intangible good, is arbitrary or ideological or involves the domination of one human being over another.

This does not mean relying on a biologistic understanding of the natural law, but rather on a rational deduction based on the essence of man understood as a body-soul unity. The body does not exhaust the reality of humanness but is an essential component thereof, and it is up to human reason to preserve man in his integrity and fullness.

This leads to reaffirming the recognition of the createdness of man and the dominion of God over the human person, but I maintain that the ethical criterion should also be based on the personalist vision of corporality, of which the genetic code is the fundamental pillar.[73]

The nonmanipulability of causal origin

Robert Spaemann passes a negative judgment on the *genetic engineering of human beings*. Those who support it appeal to the fact that man, in his current state, is the result of a natural evolutionary history and that deliberate human activity, in the final analysis, is no less worthy a cause than the blind mutations resulting from cosmic radiation. Why should the *homme de l'homme* (man-made man) be worse than the *homme de la nature* (natural man)?

> We do not have any criteria to distinguish a noncontingent part of ourselves, called person or subjectivity, from a contingent part which would then be disponible for arbitrary reconstruction. In the service of which purposes shall we carry out this reconstruction? For we would transform the purpose by the reconstruction. Such a transformation of human nature, for example for the purpose of making the human more fit for an interplanetary sojourn, means to degrade the future human to a mere means for the satisfaction of the purposes of the present manipulators, for the satisfaction of their creative fantasy or their idea of what human happiness consists in, for example, The dignity of the human is inseparably connected with its natural spontaneity. Its nature is

---

[72] Lombardi Vallauri, "Manipolazioni genetiche."

[73] Rahner, "Genetic Manipulation," 225–252. See the address by John Paul II on October 29, 1983, upon which the instruction draws and which establishes the following conditions for the permissibility of genetic engineering: that the life of the embryo is protected, that only strictly therapeutic interventions are involved, that fundamental human rights are respected including the right to genetic identity, and that emotional and spiritual as well as biological and bodily values are protected in the act of procreation.

contingent, certainly. But any consciously planned reconstruction of human nature would not avoid this contingency; rather it would make it unbearable.[74]

However, there are those like the German philosopher Peter Sloterdijk who, launching harsh attacks on Heideggerian humanism, propose "rules for the human park"[75]—a "park" that resembles an animal farm or even a zoo. Man has never freed himself of his animal component, though he has sought to "domesticate" it, and Heidegger's humanism focused on arts and letters like classical humanism. This form of humanism, in Sloterdijk's opinion, is definitively defunct.

Referring repeatedly to Friedrich Nietzsche's *Zarathustra*, Sloterdijk identifies the solution in the new prospects opened up by genetic engineering: he speaks openly about the "production of human beings" and "anthropotechnology." He maintains that human beings will look inept if they allow a higher power to work in their stead, whether it be God or chance. He hopes to move from the fatalism of natural birth to optional birth and prenatal selection. Sloterdijk also makes reference to Karl Rahner, "when he asserted that 'modern self-made man' makes use of a freedom of 'categorical self-manipulation,' which emerged thanks to the Christian liberation from frightful natural necessity."[76] He cites the opinion of Rahner, a Jesuit, according to whom wanting and having the duty to shape oneself in a self-manipulative way is part of the ethos of speech-endowed man; at this point, the future of "mankind's self-manipulation … has already begun."[77] But in "The Problem of Genetic Manipulation" (1968), Rahner revises his "openist" position, expounded in "The Experiment with Man: Theological Observations on Man's Self-Manipulation" (1966), and writes that "'we do not *want* to manipulate man genetically'; but this will is meaningful in spite of the fact that it neither claims nor is obliged to be exhaustively analysable by theoretical reason; measured against the opposite will, this will is more deeply meaningful and more genuinely human."[78] He does not propose any reasoning or argumentation, but only a decision on the basis of a sort of "*moral* instinct of faith"[79]

---

[74] R. Spaemann, *Happiness and Benevolence* (Notre Dame: University of Notre Dame Press, 2000), 171.

[75] P. Sloterdijk, *Non siamo stati ancora salvati*: *Saggi dopo Heidegger* (Milan: Bompiani, 2004), 239–266, originally published in German as *Nicht gerettet: Versuche nach Heidegger* (Frankfurt: Suhrkamp, 2001); A. Musio, "La casualità dell'uomo e la rinascita del pensiero aristotelico in margine ad alcune voci della filosofia pratica tedesca," *Rivista di Filosofia Neoscolastica* 1 (2005): 105–130; see also G. M. Del Re, "Le tesi del filosofo Sloterdijk fanno rabbrividire i tedeschi," *Avvenire*, September 30, 1999, available at http://www.swif.uniba.it/lei/rassegna/990930b.htm; A. Galli, "Nietzsche, profeta dell'eugenetica," *Avvenire*, September 21, 2005; G. Coccolini, "Peter Sloterdijk e la genetica," *Rivista di Teologia Morale* 148, no. 4 (2005): 461–472.

[76] P. Sloterdijk, "La domesticazione dell'essere," in *Non siamo stati ancora salvati*, 177.

[77] K. Rahner, "The Experiment with Man: Theological Observations on Man's Self-Manipulation," in *Theological Investigations*, trans. G. Harrison (New York: Seabury Press, 1972), 9:224.

[78] Rahner, "Genetic Manipulation," 251.

[79] Ibid., 238; see also 243: "All the 'reasons' which are intended to form the basis for rejecting genetic manipulation are to be understood, at the very outset, as only so many references to the moral faith-instinct."

that weakens the rational value of the acceptance of the necessarily heteronymous determination that characterizes man's being.

Positions opposing Sloterdijk's theses have come from all over—Jürgen Habermas's is a prominent one—because the idea of a "stockbreeder" man and a "bred" man clearly violates the principle of equal dignity. In fact, Sloterdijk's reasoning that the humanistic project should be eliminated through the stockbreeding of human beings is already contradicted by the fact that geneticists would have to agree on the objectives, in other words, on what type of human being is truly desirable.

Genetics has certainly disproved those who thought they could turn morality into something private: now that nature offers ever less resistance, it is man himself who must set limits.

The World Transhumanist Association, founded at Oxford University in 1998 by Nick Bostrom and David Pearce, intends to bring the pairing of Nietzsche and eugenics back into favor. As early as 1909, the eminent popularizer of Nietzsche in the British world, Maximilian Mügge, wrote in the *Eugenics Review* that Nietzsche had the honor of having founded the religion of eugenics.

In opposition to this, Francis Fukuyama draws attention to the dangerous consequences of the "biotechnology revolution" in his book, *Our Posthuman Future* (2002). He maintains that eugenic and biogenetic research can truly undermine the progressive ideals of free democracy through a radical alteration of human nature. The position of Jürgen Habermas is exemplary:[80] in a postmetaphysical perspective, he emphasizes the risks of so-called liberal genetics. According to Habermas,

> As soon as adults treat the desirable genetic traits of their descendants as a product they can shape according to a design of their own liking, they are exercising a kind of control over their genetically manipulated offspring that intervenes in the somatic bases of another person's spontaneous relation-to-self and ethical freedom. This kind of intervention should only be exercised over things, not persons. For this reason, later generations can demand an account from the programmers of their genome; they can hold these producers responsible for what they, the offspring, consider the unwanted consequences of the organic starting point of their life histories. This new structure of attribution results from obliterating the boundary between persons and things. ... A previously unheard-of interpersonal relationship arises when a person makes an irreversible decision about the natural traits of another person. This new types of relationship offends our moral sensibility because it constitutes a foreign body in the legally institutionalized relations of recognition in modern societies. When one person makes an irreversible decision that deeply intervenes in another's organic disposition, the fundamental symmetry of responsibility that exists among free and equal persons is restricted. ... The adult would remain blindly dependent on the nonrevisable decision of another person, without any opportunity to establish the symmetrical responsibility required if one is to enter into a retroactive ethical self-reflection as a process among *peers*.[81]

---

[80] J. Habermas, *The Future of Human Nature* (Malden: Blackwell, 2003), originally published as *Die Zukunft der menschlichen Natur: Auf dem Weg zu einer liberalen Eugenik?* (Frankfurt am Main: Suhrkamp, 2001).

[81] Ibid., 13–14.

This "premeditated quality control" manipulates human life, which is generated with reservations on the basis of the preferences and values harbored by third parties.

Habermas had always distinguished the *just*, which leads to democratic agreement, from the *good*,[82] which is of subjective and not public relevance. But in *The Future of Human Nature*, to the "ethics of discourse" he adds an "ethics of the species": an ethics of the *non-instrumentalization of the causal origin of the human species*. "Knowledge of one's own genome being programmed might prove to be disruptive, I suspect, for our assumption that we exist as a body or, so to speak, 'are' our body, and thus may give rise to a novel, curiously asymmetrical type of relationship between persons."[83]

Taking his cue from Aristotle, Habermas confirms that what the classical practices of care, treatment, and breeding have in common is *respect* for an intrinsic self-regulatory dynamic of nature. All protective, therapeutic, and selective procedures should be oriented toward this if they are not to fail in their objectives. For Hans Jonas as well, "In the case of organisms, one activity enters into another activity: the biological technique collaborates with the proper activity of an active 'material'—with the biological system functioning according to its nature—into which a new determining element must be incorporated ... The form of the technical act is an intervention, not a construction."[84]

After having been conquered by technology, according to Jonas, nature will ultimately re-encompass man, who previously set himself against her as the "master of technology." The interventions of genetic engineering constitute a "self-imprisonment": an enslavement of the living to the dead.

In order for the plan of so-called liberal genetics to be compatible with the foundations of political liberalism, the eugenic interventions must not compromise either the genetically modified person's potential to lead an autonomous life or the conditions for equal relations with other persons.

The proponents of liberal genetics liken the *genetic* modification of heritable traits to the *pedagogic* modification of attitudes and expectations. They intend to demonstrate that there are no significant differences between genetics and education from the moral perspective.

Those who propose eugenic liberty are reassured by the fact that genetic inclinations always interact unpredictably with the environment, without ever automatically translating into phenotypic qualities. Therefore, genetic programming would not in itself signify an illicit alteration of the future life plans of the programmed person. But this argument only holds water if a very debatable parallelism is postulated based on the elimination of the difference between what is naturally "grown" and what is technically "produced," between subjective and objective.

But in this way the parents decide on the basis of their own preferences, without presuming any consent, as they might arbitrarily do with anything at their disposal.

---

[82] See ibid., 32–33.

[83] Ibid., 42.

[84] H. Jonas, *Tecnica, medicina ed etica: Prassi del principio responsabilità* (Turin: Einaudi, 1997), 125. Originally published as *Technik, Medizin und Ethik: Zur Praxis des Prinzips Verantwortung* (Frankfurt am Main: Suhrkamp, 1985).

By putting natural fate and educational socialization on the same level and parallel to one another, liberal geneticists oversimplify the matter.

According to Habermas, in order for the person to be able to feel that he is entirely one with his body, the body must be experienced as spontaneous. He cites Hannah Arendt,[85] who introduced the concept of "natality" in the framework of her action theory, starting from the observation that the birth of a child does not simply represent another life story, but rather a *new* life story. "We can achieve continuity in the vicissitudes of a life history only because we may refer, for establishing the difference between what *we* are and what happens *to us*, to a bodily existence which is itself the continuation of a natural fate going back beyond the socialization process. The fact that this natural fate, this past before our past, so to speak, is not at our human disposal seem to be essential for our awareness of freedom—but is it also essential for the capacity, as such, of being oneself?"[86]

Genetic interventions to "improve" or "enhance" compromise ethical freedom inasmuch as they fasten the concerned party to the intentions of third parties (intentions that remain irreversible even if rejected) and prevent him from understanding himself as the inseparable author of his own life. "The programmed person, being no longer certain about the contingency of the natural roots of her life history, may feel the lack of mental precondition for coping with the moral expectation to take, even if only in retrospect, the *sole* responsibility for her own life."[87]

In summary, postmetaphysical Habermasian discursive reason agrees with Jonas and the Aristotelian description of the world-of-life in reaffirming all prohibitions on the manipulation of human life.

Habermas proves that law and morality based on reciprocal recognition of equality are lost through the attack on causal origin; but perhaps this is not enough to prevent the realization of the "human park" hoped for by Sloterdijk. In response to Habermas one might object that not everyone necessarily wants a society of equals, that man's autonomy is total independence and therefore he should strive to become independent even of the human condition, which entails its causal origin.

As Jonas warned, even if "God is dead," man must continue to be understood and respected as being in his image and likeness.[88]

Finally, it is important to recognize that only a teleological view of reality based on creation can protect causal origin, which in this case is not seen as blind chance but as Providence. Dependence on an equal can enslave us, whereas dependence on God—due to the absolute distance between Creator and creature—is a source of freedom.

Only a philosophy of transcendence can justify respect for the dignity of the person, even when the person is dependent: "Inadequacy, understood as a need, is in fact a limit, and a limit defers to something more, but a more that limits my being

---

[85] H. Arendt, *The Human Condition* (Garden City, NY: Doubleday, 1959), esp. 158.

[86] Habermas, *Future of Human Nature*, 60.

[87] Ibid., 81–82.

[88] H. Jonas, *The Imperative of Responsibility: In Search of an Ethics for the Technological Age* (Chicago: University of Chicago Press, 1984), originally published as *Das Prinzip Verantwortung: Versuch einer Ethik für die technologische Zivilisation* (Frankfurt am Main: Insel-Verlag, 1979).

only insofar as it fulfills it; that is, it defers to the principle that is the foundation of my being,"[89] because the relationship with God is the only instance of a relationship with a higher power that does not suppress the independence of the person, but instead acknowledges the person in his autonomy and respects his freedom.

It seems that postmetaphysical bioethics has perceived that so-called bioethical problems are often more accurately anthropological problems, but it does not consider it feasible to take an ontological approach to anthropology. It prefers an ethical route—meaningful, but insufficient for preventing science from being considered the only form of knowledge and human nature from being subjected to a continual process of redefinition. For Habermas it is sufficient to emphasize ethics of species in which the principles of equality and reciprocity are essential.

Perhaps the reluctance to engage in ontological discussion about the person arises from prejudicially considering it a religious and nonrational question, but this postmetaphysical preconception runs the risk of confining anthropological reflection to the sciences and attributing only ethics to philosophy. We need to renew our knowledge of man's nature and place in the universe:[90] it is in itself a rational, ontological, nonreligious question, yet religion can still offer its perspective of meaning.

Ethical principles

Ethical principles that flow out of this approach can be summarized as follows.

1. *Protecting the life and genetic identity of every human individual.* Any intervention involving the destruction of the physical individuality of a human subject, even when directly desired in order to obtain a benefit for others, constitutes an offense against the fundamental value of the human person because it deprives the human subject of the fundamental value on which all others rest: the value of bodily life. Even the concept of the quality of life cannot be placed above life itself, since quality is an additional attribute of life. The assumption of the quality of life criterion in order to discriminate against and destroy certain lives constitutes an offense against the principle of equality and equal dignity.

Therefore the genetic inheritance of the human individual should also be considered untouchable except in the case of the therapeutic principle, which will be discussed later. The possibility that some human beings—even scientists—might alter the genetic constitution of other human beings would constitute the domination of man over man and, furthermore, the exploitation of some human beings by others, with irreversible consequences. For example, if it were possible to alter a genetic code in order to produce a sex change in early stage embryos through microinjections of appropriate DNA segments, as it seems to be, this would be an abuse and an offense against the identity of the subject and a manipulation[91] at the hands of

---

[89] L. Pareyson, *Esistenza e persona* (Genoa: Il Melangolo, 2002), 173.

[90] See V. Possenti, *Essere e libertà* (Soveria Mannelli, CZ: Rubbettino, 2004), 115–144 and 359–368.

[91] E. Brovedani, "Verso la terapia genica umana: Prospettive e implicazioni etiche," *Aggiornamenti Sociali* 9, no. 10 (1988): 591–610.

human subjects. For this reason, both recommendation 934 of the Council of Europe in 1982 and the 1989 resolution of the European Parliament explicitly prohibit it.[92]

2. *The therapeutic principle.* The validity of the therapeutic principle based on the concept of totality, which is proper to the living organism and the human person, is also recognized in the case of genetics. It is licit to carry out even an invasive procedure for the benefit of the living subject in order to correct a defect or eliminate an otherwise incurable condition. As with every therapy, gene therapy has its foundation and justification here. The applications of this principle will be examined shortly.

3. *Protecting the ecosystem and the environment.* The justification for this principle is twofold: first of all because the environment, which is made up of a set of individual ecosystems that constitute the global ecosystem, is necessary for the life and health of man; second, in the creational understanding of the universe, the created world is indeed ordered to the good of man, who is its center and steward, yet serving man's welfare is not its only reason for being: it is still a good that has its reason for being in God.[93] If it is true that being comes prior to having in the human person, it is also true that the existence of other living beings is not exhausted in being an instrument.

In the debate between anthropocentric environmentalists and biologistic environmentalists, I therefore propose a theocentric and creationary personalism in which man is the steward, in addition to the beneficiary, of the created world and in particular of other living creatures. All of this has a significant impact on the biotechnology sector.

4. *The ontological and axiological difference between man and other living beings.* While recognizing the bond of intimate and vital exchange between living beings and man, it is nonetheless impossible to overlook the real and profound difference in man by virtue of his capacity for reflective knowledge, freedom, and responsibility—in short, his being endowed with a spirit.

Within the realm of Catholic thought, let us recall a phrase from Vatican Council II[94] that affirms the ordering of creatures to man, whereas man is created as an end in himself. This affirmation is also supported by reason itself, which is a distinctive component of man. This fact prevents the use of the same criterion for interventions on man and on other living beings, such as the criterion of *feeling pain*,[95] because the

---

[92] European Parliament, Resolution A2–327/88, on the Ethical and Legal Problems of Genetic Engineering (see note 26), holds that "even [if] a recombination of genes only partly alters the genotype [*sic*], the identity of the individual is falsified which is both irresponsible and unjustifiable because a very individual legal asset is involved" (§ 30). Council of Europe recommendation 934 on genetic engineering affirms that "the rights to life and to human dignity protected by Articles 2 and 3 of the European Convention on Human Rights imply the right to inherit a genetic pattern which has not been artificially changed" (art. 4[a]).

[93] These concepts are presented in John Paul II's message for World Peace Day in 1990, "Peace with God the Creator, Peace with All of Creation," January 1, 1990. The German Bishops' Conference expressed itself on the same topic with a declaration titled "The Future of Creation and the Future of Humanity," published in *Medicina e Morale* 40, no. 2 (1990): 366–380 [Italian].

[94] Vatican Council II, Pastoral Constitution *Gaudium et spes*, December 7, 1965, no. 12.

[95] P. Singer, *Practical Ethics*, 2nd ed. (New York: Cambridge University Press, 1993); idem, *Animal Liberation: A New Ethics for Our Treatment of Animals*, 2nd ed. (New York: Random House, 1990).

way that animals feel pain differs fundamentally from the way that man feels pain: animals suffer and man knows that he suffers and seeks meaning in the suffering. Moreover, genetic interventions can be seriously harmful even when carried out during phases of life in which there is no pain.

5. *The competence of the community.* The search for solutions to the problem of interventions on the genetic patrimony of human beings and other living beings as well cannot be entrusted only to certain experts, whether scientists or politicians: it is a question that in certain ways regards humanity as a whole. The future of humanity often demands the responsible participation of the community. This is why the principle of freedom of science and research should be recognized but also combined with the fact that populations need information and share in responsibility. Think for example of the issues of genetic screening, the manipulation of the environment, and the sequencing of the human genome.

In accordance with the principle of self-regulation, it will be necessary to find forms of association and shared responsibility among scientists themselves in the interests of the common good and in collaboration with public opinion. The experience of bioethics committees and self-imposed regulations, bolstered by the law where necessary, is strengthening this principle in various countries.

Suffice it to mention once more the problem in developing countries, which should also be heard and included in this community. Indeed, these countries could benefit from biotechnological progress on the one hand, yet be forced to live in greater economic dependence on the developed countries holding the biotechnological power on the other. It is certain that the development of biotechnologies has already launched a new era of relations between the North and South of the world.

Specific ethical norms

The consequences of these criteria and general principles must now be analyzed concisely. This will be done by generally following the outline of potential applications.

First and foremost is the ethically justified need to ensure the *safety* of biotechnologies, both in relation to laboratories and in relation to the environment. This was the first alarm sounded regarding genetic engineering: the possibility of destroying the biosphere through a biological bomb. The first norms and guidelines for the safety of the environment and the security of laboratories were formulated in light of this. Indeed, microorganisms can be engineered in laboratories in such a way that they may constitute a greater, lesser, or equal danger to those who work there; in the event of a release of those microorganisms into the environment, even inadvertently, entire populations could be harmed. An additional safety concern is the risk of a deliberate release of engineered microorganisms into the environment.

The above-cited Italian NBC document on the science of biotechnology summarizes the directives and ethical demands that I believe can be agreed upon. A definition of the term and concept of biotechnology is first of all provided: "any technique that uses living organisms (or parts of organisms) to make or modify products, to improve plants or animals, or to develop micro-organisms for specific uses."[96] Other more technical and targeted definitions have been provided by the

---

[96] The definition is taken from United States Congress Office of Technology Assessment (OTA), *Commercial Biotechnology: An International Analysis* (Washington,

Organisation for Economic Co-operation and Development (OECD) and the European Federation of Biotechnology (EFB).[97] The development of biotechnologies involves the implementation of genetic engineering techniques, in vitro cell cultures, and the production of monoclonal antibodies. Areas in which these technologies are applied include the chemicals and pharmaceuticals industry and the agricultural foods industry. In 1990, there were approximately 1,500 biotechnology businesses in the United States and 134 in Italy.[98]

The general ethical guidelines in this matter ultimately take concrete form in (1) the defense of human life and health and (2) the protection of the environment. These two principles, which must be taken together in assessing biotechnologies, require that researchers keep in mind not so much the technique per se as the product and the dangerousness of the consequences.

For the risk of biotechnologies in *closed environments*, the Italian NBC recommends that two factors be taken into account: (1) the genetic, physiological, and ecological characteristics and potential pathogenicity of the parent organism; and (2) the nature of the modifications or the characteristics of the recombinant DNA with which it was transformed.

To this end, starting with the United States NIH guidelines and continuing with the regulations of other individual health care authorities in various nations, classifications of microorganisms according to their risk level have been developed. Risk level is also studied with respect to oncogenes. Appropriate safety measures are established based on risk level. In Italy, the guidelines are provided by the Ministry of Health.[99]

Next, as far as the *release of genetically modified organisms* is concerned, various types can be distinguished: plants (as early as 1989 more than thirty genetically modified species were being reviewed); higher animals for the enhancement of individuals and products; and microorganisms, namely bacteria and viruses.

For plants, "the application of research is aimed at the modification of several characteristics that are fundamental for agriculture, such as the ability to form

---

DC: US Government Printing Office, 1984), http://www.fas.org/ota/reports/8407.pdf.

[97] The definitions of biotechnology formulated by the OECD and the EFB are cited in Italian National Bioethics Committee, *Documento sulla sicurezza delle biotecnologie* [*Document on biotechnology safety*], May 28, 1991 (Rome: Presidenza del Consiglio dei Ministri Dipartimento per l'Informazione e l'Editoria, 1991), 9, available at http://www.governo.it/bioetica/pdf/4.pdf, English abstract available at http://www.governo.it/bioetica/eng/opinions/abstracts/abstract_document_on_biotechnlogy.pdf: According to the OECD, biotechnology is "the application of science and technology to living organisms, as well as parts, products and models thereof, to alter living or non-living materials for the production of knowledge, goods and services" (http://www.oecd.org/document/42/0,3746,en_2649_34537_1933994_1_1_1_1,00.html, accessed November 28, 2011) while for the EFB it is "the integrated use of biochemistry, microbiology, and engineering in order to develop technological applications from the properties of microorganisms, cell cultures, and other biological agents" (cited in Italian in National Bioethics Committee, *Documento sulla sicurezza*).

[98] C. Spalla, *Il progresso delle biotecnologie in Italia* (Milan: Federchimica-Assobiotech, 1990); Italian National Bioethics Committee, *Documento sulla sicurezza*, 10.

[99] See the Polli Document on genetic engineering in which said guidelines are mentioned (in Italian Ministry of Health, "Relazione della Commissione").

atmospheric nitrogen, the enhancement of nutritional properties, and resistance to viruses, parasites, insects, herbicides, and difficult environmental conditions."[100] As far as microorganisms are concerned, research is conducted on bacteria and viruses not just for potential uses in the food sector, but also for the biodegradation of waste.

In relation to the release of genetically modified microorganisms into the environment, *the first ethical condition is risk assessment*. It has been determined that the introduction of such modified microorganisms into the environment ends up having negative effects 10 to 20 percent of the time,[101] where "negative effects" means a significant disturbance of the ecological balance, the disappearance of some wild species or a strongly invasive behavior. The negative effects are linked to three factors: (1) a lack of natural enemies, (2) harmfulness to other organisms, and (3) the direct destruction of ecosystems.

Therefore, before proceeding with the release of modified organisms, the competent bodies (such as the OECD, the directives of EEC commissions, etc.) and petitioners are required to answer five questions:

1. Is the modified organism capable of survival?

2. Is it capable of reproducing?

3. Is it capable of forming a population and spreading?

4. Is the organism capable of transmitting its artificially conferred traits to other organisms?

5. Are the factors listed above (survival, reproduction, establishment, and propagation of the organism) capable of producing undesired effects on human beings and the environment?[102]

The identification of risk factors should be assessed on a case-by-case basis and requires a sophisticated monitoring ability and appropriate equipment,[103] since it

---

[100] Italian National Bioethics Committee, *Documento sulla sicurezza*, 13.

[101] Ibid., 14; R.I. Sailer, "Our Immigrant Fauna," *ESA Bulletin* 24 (1978): 3; D. Simberloff, "Community Effects of Introduced Species," in *Bioethics Crises in Ecological and Evolutionary Time*, ed. M.H. Nitecki (Chicago: Academic Press, 1981); USDA Agricultural Research Service, *Plant Introduction Service* (Washington, DC: US Government Printing Office, 1986).

[102] C. Mantegazzini, *The Environmental Risks from Biotechnology* (London: Frances Printer, 1986); V. Capuano, *Rilascio ambientale di organismi geneticamente modificati: la valutazione del rischio* (Rome: ENEA, 1991).

[103] H. Halvorson, ed., *Engineered Organisms in the Environment: Scientific Issues* (Washington, DC: American Society for Microbiology, 1985); J. Fiksel and V.T. Covello, eds., *Biotechnology Risk Assessment* (Oxford: Pergamon Press, 1986); Mantegazzini, *Environmental Risks*; M. Kvistgaard and A.M. Olsen, *Biotechnology and Environment: A Literature Review of Environmental Applications and Impacts of Biotechnology* (Lyngby: Technical University of Denmark, 1986); OECD, *Recombinant DNA Safety Considerations: Safety Considerations for Industrial, Agricultural and Environmental Applications of Organisms Derived by Recombinant DNA Techniques* (Paris: OECD 1986); OTA, *New Developments in Biotechnology: Field Testing Engineered Organisms; Genetic and Ecological Issues* (Washington, DC: US Government Printing Office, 1988), available at http://www.fas.org/ota/reports/8816.pdf; P. Regal, "The Adaptive Potential of Genetically Engineered Organisms in Nature," *Trends in Biotechnology* 6 (1988): 36–38; M. Sussmann

is necessary—as an ethical requirement—that there be no environmental or health risks. Another requirement is that it must entail *real utility for the good of the current society without harm to future societies*. Experimentalism as an end in itself should not be encouraged.

It is likewise necessary that the *public be sufficiently informed* in order to dispel false fears or to provide information about the real advantages.

Furthermore, provisions should be made for the *preservation of species on the verge of extinction* and the *protection of biodiversity* through the preparation of seed banks and the improvement of artificial reproduction techniques.

Finally, it is necessary to consider the macroethical problems involved in *making biotechnologies available in developing countries* in order to solve their food problems without creating further adverse dependence on the countries that produce biotechnologies.

Even though the eagerly awaited beneficial effects in this field of biotechnology have perhaps fallen short of expectations, it is certain that the sector will experience solid growth in the future with major implications for both agriculture and zootechnics as well as for the food industry.

Problems with postnatal genetic diagnosis and screening

There are many ethical issues connected with genetic diagnosis: up until a few years ago, these issues were centered primarily on the issue of prenatal diagnosis. This is undoubtedly an area that is especially fraught with ethical problems, and for this reason I wish to devote a specific treatment to the subject. Nevertheless, there is an increasingly prolific debate today over the applications of genetic diagnosis to adults as well. I wish to mention here some of the areas in which ethical problems are encountered in applying genetic diagnosis to adults.

The issues can be grouped into the following applied areas of postnatal genetic diagnosis: (a) genetic diagnosis for the purpose of identifying the genetic cause of a clear pathological condition with evident clinical symptoms; (b) premarital and preconception genetic diagnosis—also including screenings, whether en masse or for at-risk populations—in order to identify carriers of genetic disorders; (c) genetic testing of employees for sensitivities to the chemical agents present in the workplace environment; (d) the human genome project and its ethical implications; (e) forensic applications; and (f) genetic testing requested by insurance companies.[104]

POSTNATAL GENETIC TESTING TO VERIFY A DIAGNOSIS OF ILLNESS. Let us suppose, for example, that a doubt arises as to whether a 6- or 7-year-old child, who is gradually exhibiting a loss of leg muscle strength, might have an illness of genetic origin such

---

et al., eds., *The Release of Genetically Engineered Micro-Organisms* (New York: Academic Press, 1988); J. Defaye et al., eds., *Risk Management in Biotechnology* (Paris: Technique et Documentation, 1990); OECD, *Good Developmental Practices for Small Scale Field Research* (Paris: OECD, 1990); Capuano, *Rilascio ambientale*.

[104] The literature on these individual topics has been growing and parliaments have begun to provide indications on them. Reserving specific references for the treatments of individual topics, at present I will mention the work by W. Kennet, *Parliaments and Screening* (Paris: Libbey, 1995), which collects the proceedings of two European conferences that addressed the ethical and social issues of both genetic and AIDS screening, which are areas that present similar problems.

as Duchenne muscular dystrophy. Only a genetic test can provide certainty within a few days. The ethical problem concerns the communication of the truth to the interested party and his relatives. The problem is more serious when the test is done for other reasons—by chance or deliberately—and a pathology is discovered that is not yet clinically evident but will manifest itself some time later in adulthood, such as cystic fibrosis or Huntington's disease. In these complex cases, it becomes necessary to link the *right to know* with the *right not to know*. The right not to know, and therefore the obligation not to communicate, is certainly valid for patients who are minors; the right to know could be valid for relatives who request it. The advisability of knowing the answer and not revealing it to the child until he reaches the age of majority should nonetheless be assessed in relation to the psychological capacity of the parents. When an adult patient asks to be informed of the result of the diagnosis, it is necessary to examine—on a case-by-case basis—the potential effects of giving information about an unpromising diagnosis of future onset.[105]

PREMARITAL AND PRECONCEPTION DIAGNOSIS. This is advisable for an individual who could be a carrier of the gene responsible for a recessive genetic disorder. If he or she were matched with another carrier, they would run the risk that one out of four (25 percent) of their children would be afflicted with the disorder in question.

These genetic tests are permissible and recommended in populations at risk of transmitting the gene responsible for thalassemia, for example. In some geographic areas, broad tests of this type accompanied by the necessary counseling seem to have led to significant results in primary prevention, causing a drop in incidence.[106]

This system merits the name *prevention* and can be a true alternative to selective abortion, which some improperly pass off as prevention.[107] It must nonetheless be added that these tests cannot be required through general screenings or laws imposed on those about to marry. The fact that they are permissible does not make them morally obligatory. One might emphasize the sense of responsibility from the ethical perspective, but these tests cannot be made mandatory.[108]

If then, for example, a couple planning to marry were to accept this risk, they should certainly take into serious consideration the consequences of a marriage with a very high risk of transmitting the genetic disorder, but they could not be compelled—either legally or morally—not to marry. The possibility thus remains for them to marry all the same, aware however that the voluntary abortion of a fetus

---

[105] E. Sgreccia and V. Mele, "La diagnosi genetica postnatale," in *Ingegneria genetica*, Sgreccia and Mele, 251–277; R. Chadwich, M. Levitt, and D. Shickle, *The Right to Know and the Right Not to Know* (Chippenham: Antony Rowe Ltd., 1997).

[106] For example, see the thalassemia prevention program, which works through the identification of carriers, run by the Latium region of Italy. The data is published in I. Bianco et al., "Updated Results of the Thalassemia Prevention Programme Carried Out in Latium," *Journal of Medical Genetics* 26 (1989): 667 ff.

[107] A.G. Spagnolo et al., "Significato della diagnosi prenatale nella prevenzione delle malattie congenite: Aspetti etico-sociali," in *Atti del LXVII Congresso della Società Italiana di Ostetricia e Ginecologia* (Brescia: Clas International, 1990), 39–40.

[108] Italian National Bioethics Committee, *Diagnosi prenatali* [*Prenatal Diagnoses*], July 18,1992 (Rome: Dipartimento per l'Informazione e l'Editoria, 1992), 37–46, available at http://www.governo.it/bioetica/pdf/10.pdf, English abstract available at http://www.governo.it/bioetica/eng/pdf/abstract_prenatal_diagnoses.pdf.

affected by the disorder can never be considered permissible, especially when there has been a conscious acceptance of the risk.

GENETIC SCREENING OF EMPLOYEES. This type of test, which is occurring with increasing frequency as an application of so-called predictive medicine, is also fraught with ethical problems.

Three types of employee screening can be identified, with different purposes: (a) identification of genetic predispositions causing hypersensitivity to particular substances present in the workplace; (b) ascertainment of genetic predispositions to illnesses unrelated to the job that could become manifest in the future (e.g., arteriosclerosis); and (c) identification of a genetic disorder unrelated to the job with future phenotypic expression (e.g., Huntington's disease).

Screenings to identify a genetic predisposition (a) causing hypersensitivity to particular substances present in the workplace provide the data for ecogenetics. This branch of genetics studies the genetically determined reactions of the human organism to environmental factors of a physical, chemical, biological, and social nature and is based on the premise that individual biochemical characteristics affect the organism's reaction to an external stimulus.[109] Examples of conditions of genetic origin and development that are responsible for weakening the organism's defense mechanisms against harmful substances present in some work environments include alpha-1-antitrypsin deficiency and glucose-6-phosphate dehydrogenase (G6P-DH) deficiency. G6P-DH deficiency seems to predispose employees who are exposed to naphthalene to hemolytic crises.[110] In the homozygotic form, alpha-1-antitrypsin deficiency results in severe emphysema; the heterozygotic form appears to be a predisposing factor that results in pathology through interaction with specific environmental agents (such as pulmonary irritants).

The essential ethical problem in these cases is the possibility that the employer might launch a genetic investigation program in order to "prevent" higher costs deriving from compensation for eventual harm to employees. Such a plan, based solely on the cost-benefit ratio and considering no other value than the economic advantage of the industry, would not be ethically acceptable.

The ethical opinion is instead essentially different, and therefore positive, if the understanding and the object of the assessment refer to the risks and benefits to the health of the employee and the genetic tests are carried out essentially to prevent the onset of a pathology.[111] In this case, the end being pursued is ethically acceptable in principle, always assuming that the association between the genetic status and the occupational pathology has been proven certain, and also considering the recurrence

---

[109] V. Mele, G. Girlando, and E. Sgreccia, "La diagnosi genetica sui lavoratori: recenti acquisizioni scientifiche, problematiche etiche ed etico-giuridiche," *Medicina e Morale* 40, no. 2 (1990): 301–329. The Nuffield Council on Bioethics also expressed its opinion on the various ethical aspects of genetic screening, including the screening of employees, with its report *Genetic Screening: Ethical Issues* (London: Nuffield Council on Bioethics, 1993). Its conclusions and recommendations were published in *Bulletin of Medical Ethics* (December 1993): 8–11.

[110] G.S. Omenn, "Predictive Identification of Hypersusceptible Individuals," *Journal of Occupational Medicine* 24, no. 5 (1982): 369–374.

[111] International Commission on Occupational Health (ICOH), "International Code of Ethics for Occupational Health Professionals," *Bulletin of Medical Ethics* 82 (1992): 9.

of the illness (in the case of extremely rare pathologies, the predictive value of the test is minimal), the severity of the illness, and its forms of clinical expression (for illnesses that are not severe and can be controlled at early clinical stages, there may be no ethical reasons to perform the diagnosis inasmuch as there is little chance of significant harm to the health of the employee).[112]

Continuing with this second scenario, once the permissibility of the diagnostic act has been determined on those conditions at the medical and scientific level, other conditions are required in order for the test to be free of any job discrimination and therefore completely permissible. These conditions as a whole are directed toward safeguarding respect for the health, autonomy, and self-determination of employees and the criteria of impartiality in hiring. These conditions are extensively emphasized in national and international European documents.

Several ethical criteria are taken into consideration in the international documents: the employee's *informed consent* is referenced first and foremost. Numerous groups and various documents attribute decisive importance to consent. European Parliament resolution A2–327/88 on genetic engineering reaffirms the right of concerned employees to be fully informed and consulted prior to proceeding with such exams, and the right to refuse to be subjected to genetic analyses at any time without stating a reason and without negative consequences.[113] The German parliament's report recommends that professional societies establish legally binding regulations for the use of tests and their consequences. Such regulations should ensure the following conditions: the employee concerned must be informed, prior to beginning the test, about the planned genetic analysis and the significance that the analysis could have for him; the content of the interview and the employee's consent to undergoing genetic analysis must be put in writing and signed by the employee in question and by the physician who informed him; and willingness to undergo a genetic analysis must be considered invalid in the event that the employee in question was not previously informed of the pertinent circumstances for the purpose of making the decision.[114]

Moreover it seems unlikely that the employee would refuse to undergo the test after having been accurately informed of the extent of the risk and the severity of the condition and particularly if alternate work opportunities were offered to him. In this regard, in fact, I concur with the opinion stated by the United States Congress's Office for Technology Assessment (OTA), namely that the employee's informed consent is the only possible way to reconcile the *principle of beneficence*, which would suggest the exclusion of employees who are genetically susceptible to harmful environments, with the *principle of autonomy*.[115] With reference to this consent, it is my opinion that a duty nonetheless remains to accept the consequences of an eventual positive result of the diagnostic examination: that is, not being hired for the job after having received the appropriate information and having agreed to undergo the exam.

---

[112] Omenn, "Predictive Identification."

[113] European Parliament, "Resolution on Genetic Engineering."

[114] German Bundestag, *Report on Genetic Engineering.*

[115] OTA, *Genetic Monitoring and Screening in the Workplace* (Washington, DC: US Government Printing Office, 1990), 103.

On the subject of the non-compulsory nature of genetic tests even on employees, Recommendation R (92) 3 of the Council of Europe[116] affirms the principle that hiring and the opportunity to continue working cannot be made dependent upon undergoing genetic tests or screenings. The recommendation then affirms that exceptions to this principle must be justified by reasons involving the direct protection of the person in question or of third parties and must be directly correlated to the specific conditions of employment activities. Finally, it recommends that the tests may be conducted in a compulsory fashion for the protection of individuals or of the group only where explicitly allowed by law.

The *employer's right to know* is considered next, and it is requested that this right be limited by appropriate normative provisions. The American College of Occupational Medicine's Code of Ethical Conduct, for example, states that the information given to the employer should provide only certain strictly necessary elements.[117]

Finally, regarding the *exclusion of employees* who are found to be genetically susceptible, the European Parliament's resolution notes that under no circumstances can such a decision constitute an alternative to improving the workplace environment.[118] The document by the German parliament, on the other hand, demands that the eventual exclusion of employees with those predispositions be permitted only if it is impossible to sufficiently improve working conditions for the employee in question.[119]

Employee screenings to identify a genetic predisposition or predetermination for an illness unrelated to the job that could or will emerge in the future (b, c) seem unacceptable in principle from both the ethical and legal perspectives. This opinion against it, which reflects the lack of proximity of the pathology's occurrence and the fact that it is not job-related, is clearly expressed in both the resolution by the European Parliament and the document by the German parliament.

Specifically, the resolution requires that genetic tests on employees be carried out in relation to their current state of health. The report by the Federal Republic of Germany recommends excluding genetic analyses and other tests that diagnose the future pathological situations and illnesses of employees. This is to be applied even if the future pathology might affect the necessary abilities of the employee for a specific job. Only the employee's current state of health can be a reason for permissible tests.[120]

From the legal perspective, furthermore, the attempt to take probable future illnesses into account at the time of hiring would violate the principle requiring balance between the interests of the two parties in the context of an employment contract. According to this principle, the employer must also take on the risk of an

---

[116] Council of Europe, Recommendation R (92) 3, "On Genetic Testing and Screening," principle 6.

[117] See American College of Occupational Medicine, "Code of Ethical Conduct for Physicians Providing Occupational Medical Services," *Journal of Occupational Medicine* 18, no. 8 (1976): 1. This position was reaffirmed by the board of directors of the American College of Occupational Medicine in 1988.

[118] European Parliament, "Resolution on Genetic Engineering."

[119] German Bundestag, *Report on Genetic Engineering.*

[120] Ibid.

eventual future illness of the employee, whereas with genetic selection that risk would, in practice, fall solely on the employee.[121]

An ethical exception to this prohibition is provided, in my opinion, in the case of a potential risk to the life of third parties caused by the clinical expression of the predetermined future genetic illness in specific employment situations (c). In this case, genetic testing prior to hiring should indeed be conducted as compulsory, on penalty of not being hired. Huntington's disease provides an example in this regard: it is a genetic pathology that does not have a phenotypic expression for many years, and is characterized by moments of mental confusion and memory loss upon initial clinical onset. It is evident that at the time of clinical onset for an individual working in some particular professions—an air traffic controller, for example—the disease puts the lives of many persons at serious risk. In this case, the principle of individual autonomy (no obligation to undergo genetic testing) must be subordinated to the principle of the defense of physical life and the common good.[122]

The Human Genome Project and related ethical issues. Knowledge of the human genome has continually expanded since the 1950s,[123] particularly in relation to the identification of pathogenic genes. In recent years, the Human Genome Project—an ambitious international project for sequencing the entire human genome—was conducted and concluded with the aim of identifying both normal and pathological genes. Italy participated in this project as well. A special commission of experts from the seven most industrialized countries examined the ethical and social implications of this project during a study week in Rome in 1988.[124] This author participated in the study group on behalf of the Italian government.

The benefits expected from this project, which was brought to completion with the collaboration of approximately three thousand scientists, relate to the possibility of identifying the genes responsible for hereditary illnesses and then being able to proceed with gene therapy; the creation of an international archive of all the nitrogen bases that compose and code the human genome; and the typing of certain individuals through the use of DNA polymorphisms, primarily for criminological use or to determine paternity or else to understand the predispositions to illness in a given workplace.

Yet certain ethical risks and problems connected with the possibilities of new findings in genetics become apparent alongside these desired benefits. The first problem that has been pointed out arises from the possibility of a broad application of prenatal diagnosis with eugenic aims. Precisely because abnormal genes and individuals who are carriers can be identified on a larger scale in the prenatal phase,

---

[121] Ibid.

[122] Sgreccia and Mele, "La diagnosi genetica postnatale."

[123] V. L. Pascali and E. D'Aloja, "Il progetto genoma e le conoscenze sui geni normali e patologici dell'uomo: Problemi etici e deontologici," *Medicina e Morale* 42, no. 2 (1992): 219–232; J. Maddox, "The Case for the Human Genome," *Nature* 352 (1991): 11–14; C. Anderson and P. Aldhous, "Human Genome Project: Still Room for HUGO?" *Nature* 355 (1992): 4–5.

[124] International Conference on Bioethics, *International Conference on Bioethics: The Human Genome Sequencing; Ethical Issues; Roma, 10–15 April 1988* (Brescia: Clas International, 1989): 286–292. A summary and recommendations can be found in Italian in *Medicina e Morale* 38, no. 2 (1988): 308–315.

the eugenic drive could find a broader area of application; prenatal diagnosis, instead of being aimed at the best treatment for the fetus or a greater acceptance of the fetus or at least serving as a call to responsibility on the part of the woman and the couple, might therefore be distorted for other objectives.

Furthermore, if the prenatal genetic exam were to reveal that the subject is expected to suffer from a genetic disorder with a late onset, such as Huntington's disease, then the ethical problems would multiply: in addition to the possibility of abortion—which in this case would probably not even be legal in accordance with Italy's law no. 194/78, since there would not yet be any "malformation" or "anomaly" present in the fetus, and which is always illicit from the moral perspective in any case—there is also the complex problem of information. This information cannot be given to an individual who is still healthy unless it is requested and the person is legally an adult; in the case of an individual who is not yet an adult, the information can instead be given to the parents or legal guardians or siblings, assessing the situation case by case.[125]

Another problem that has been identified is the building of *databases* with the information of individuals who have undergone the tests. These databases should be accessible only for scientific purposes or by court order. The information must be kept inaccessible to private individuals, industries, and insurance companies.[126]

The third type of risk is that genetic monitoring may enable discrimination in hiring against all those individuals who might be genetically sensitive to certain chemical agents in the workplace. This problem was already addressed a shortly before.

Italy's NBC also devoted a document to this topic[127] in which the following recommendations were formulated:

- In order to develop suitable ethical reflection on responsibility at the employee level, it is recommended that those who coordinate such research in Italy promote, as is already done at the international level, parallel investigations into their ethical, social, and legal impact.

- Information and public debate are another important antidote against distortions of scientific reality that can generate unfounded fears, but also against potential improper uses of the knowledge acquired during the course of the Human Genome Project. It is therefore recommended that the competent authorities promote information, debate, participation, public and professional education on the basic concepts of genetics, and

---

[125] V. Mele, "Diritti e doveri della medicina predittiva," *Medicina e Morale* 41, no. 3 (1991): 71–90; Sgreccia and Mele, *Ingegneria genetica.*

[126] C. McGourty, "Profiles Bank on the Way," *Nature* 339, no. 6223 (1989): 327; Pascali and D'Aloja, "Il progetto genoma," 220; C. Anderson and P. Aldhous, "Secrecy and the Bottom Line," *Nature* 354 (1991): 96.

[127] Italian National Bioethics Committee, *Progetto Genoma Umano* [*Human Genome Project*] (Rome: Presidenza del Consiglio dei Ministri Dipartimento per l'Informazione e l'Editoria, 1994), available at http://www.governo.it/bioetica/pdf/15.pdf, English abstract available at http://www.governo.it/bioetica/eng/pdf/abstract_Human_genome_2.pdf.

heightened sensitivity to the reality of the Human Genome Project and the ethical, social, and legal issues connected with it.[128]

Again on the subject of the Human Genome Project, UNESCO also drew up its Universal Declaration on the Human Genome and Human Rights.[129]

GENETIC TESTS IN FORENSIC MEDICINE. This is a recent application of genetic analysis employed since the year 1985, which has developed in two directions: (1) confirmation or denial of paternity, which was previously done only through blood groups; (2) identification of a criminal culprit based on traces of DNA left behind by the suspect at the scene of the crime.

The basic principle is that DNA allows not only the identification of tissues (e.g., blood and sperm) and the species to which they belong (e.g., human or nonhuman animal), but even—under certain specified conditions and with precise and more modern methods—unique individual information (a *genetic fingerprint*).[130] The techniques have undergone developments precisely in terms of the possibility of providing individual information with sufficient certainty based on even modest evidence.

The first phase in the development of these techniques allowed only the generic identification of tissue type and species; when the material was abundant and collected shortly after its production, however, individualization was possible through the use of multilocus probes and the technique of Southern blot analysis. Some famous cases were actually solved using this method together with other traditional evidence.[131]

The second phase involved the introduction of molecular single locus probes (SLPs). This method, while more specific than the first, revealed some limits especially in terms of the quantity and freshness of the material needed for analysis. It was nonetheless used up until the 1990s and served to solve important legal cases. In the wake of numerous criticisms about the dependability of the technique, the OTA drew up a report[132] on its reliability, controls, and standards in July 1990.

The third phase in the development of these methods was marked by the discovery of the polymerase chain reaction (PCR).[133] Its use became even more widespread, allowing the analysis of polymorphisms with a very high identification capacity. It

---

[128] Italian National Bioethics Committee, *Progetto genoma*, 8.

[129] UNESCO, Universal Declaration on the Human Genome and Human Rights. The French text of the Declaration Universelle sur le Genome Humain et les Droits de l'Homme can be found in *Medicina e Morale* 48, no. 1 (1998): 158–166, or online at http://unesdoc .unesco.org/images/0011/001102/110220f.pdf#page=47.

[130] This part draws from A. Fiori, "Genetic Tests in Legal Medicine and Criminology," in *Human Genome: Human Person and the Society of the Future*, ed. J. de D. Vial Correa and E. Sgreccia (Vatican City: Libreria Editrice Vaticana, 1999).

[131] See, for example, the Pitchfork case: P. Gill and D. J. Warret, "Exclusion of a Man Charged with Murder by DNA Fingerprinting," *Forensic Science International* 35 (1987): 145–148.

[132] OTA, *Genetic Witness: Forensic Use of DNA Tests* (Washington, DC: US Government Printing Office, 1990), available at http://www.fas.org/ota/reports/9021.pdf.

[133] See H. A. Erlich and N. Arnheim, "Genetic Analysis Using the Polymerase Chain Reaction," *Annual Review of Genetics* 26 (1992): 479–506; G. Umani Ronchi and C. Vecchiotti, *Il laboratorio medico-legale* (Rome: Lombardo, 1994).

is now possible to identify not only gender but also the individual person of interest based on tiny samples such as a hair or traces of blood or sperm. These techniques nonetheless require high levels of experience and accurate monitoring as well, not just to avoid potential contamination of the material under examination but also to ensure that procedures are applied carefully.

The ethical demands in this field can be summarized as follows:

a. Paternity tests must be done at public institutions accredited for experience and authoritativeness and can only be conducted when requested by a court or an advisory institution for legitimate purposes provided for by law or to promote family unity. Various countries in Europe exclude private laboratories and do not accept requests from private individuals.[134] The ethical reason is easy to understand: it is a question of respecting the rights of children, privacy, and family unity, which can be threatened in certain cases by the results of such tests.

b. Regarding criminological use, the tests must be performed by personnel trained to use the appropriate instruments and should be considered permissible as a means for ascertaining the truth about the perpetrator of a crime and avoiding the conviction of innocent persons. The test must also refer only to the genetic aspects pertinent to the judicial inquiry and not to other biological information.

c. Criminological records and associated databases must be protected as secret. The reason is based on the right to privacy, which acquires unique importance in the genetic realm because of the fact that DNA also provides information about lineage.

GENETIC TESTS FOR INSURANCE PURPOSES. In the ethical and legal realm, the principle of protection of individual self-determination proves to be of the utmost importance. In other words, the potential economic advantages for insurance companies resulting from prior knowledge of future pathologies, or of the risk that those pathologies might occur in the insured person, should not be considered more important than the right of insured persons to not have prior knowledge of said pathologies. The requirement of a genetic test also seems to contradict the "essential" meaning of the insurance system: the manifestation of a predisposition at some undetermined future point is a risk that should be covered by health insurance, whose task should be precisely to take on risks and not to avoid them.[135] The Nuffield Council of Bioethics also addressed the topic of genetic tests for insurance purposes,[136] recommending that insurance companies regulate themselves by not requiring any genetic tests as a prerequisite for obtaining a policy and, furthermore, that the government clarify the legitimacy of using genetic information for this purpose.[137]

---

[134] The Bundestag Recommendation of the Federal Republic of Germany affirms in its *Report on Genetic Engineering* that genetic analyses may only be authorized on judicial order; the order must specify not only the objective of the analyses but also the method of genetic analysis (see art. 11 §1). It further provides that they may only be conducted on persons who are soundly suspected with evidence and not merely under suspicion.

[135] German Bundestag, *Report on Genetic Engineering*.

[136] Nuffield Council on Bioethics, *Genetic Screening*.

[137] For further study of the problems related to genetic diagnosis, for insurance purposes as well, see L. B. Andrews et al., eds., *Assessing Genetic Risks: Implications for Health and Social Policy* (Washington, DC: National Academy Press, 1994); A. G. Spagnolo, "Predictive and Presymptomatic Genetic Testing: Service or Sentence?" in *Human Genome: Human Person*, Vial Correa and Sgreccia.

Gene therapy

The Italian NBC defines gene therapy in the following terms: "the introduction a gene, which is to say a DNA fragment, into human bodies or cells with the effect of preventing and/or curing a pathological condition."[138]

It has already been mentioned that the greatest hopes for genetic engineering applications center on gene therapy. The first authorizations came from the Food and Drug Administration (FDA) of the United States of America on September 14, 1990 for the transplantation of adenosine deaminase (ADA) enzyme-producing cells into a girl with a compromised immune system. The lymphocytes were infected in vitro with a weakened retrovirus containing the normal ADA gene for replication. The therapy was carried out successfully: though she was not entirely cured, her T-cell count rose to normal levels within six months and she was able to live a fairly normal childhood with continued low doses of a synthetic form of the ADA enzyme.[139] Another approval was given, again by the United States, for the introduction of the gene for Tumor Necrosis Factor (TNF) into a particular category of lymphocytes for the treatment of melanoma. In Italy, a somatic therapy procedure was carried out in 1992 by Saint Raphael Hospital and Scientific Institute in Milan.

The fundamental distinction in the field of gene therapy is between gene therapy on *somatic cells*, such as lymphocytes or bone marrow, with the anticipated effect of restoring normal function to defective cells in the patient, and gene therapy on *germ-line cells* (on gametes or early-stage embryos), where the anticipated modification would affect the genome of a subject who will be conceived or has already been conceived and that subject's descendents as well.

The fundamental ethical principle for evaluating these procedures is the nonmanipulability of a subject's genetic inheritance, which is in turn founded on respect for the physical integrity of the person. This principle is compatible with—and actually necessitates, in my opinion—the patient's right to preserve or recover the integrity and effectiveness of his genetic endowment in accordance with the therapeutic principle. Based on this twofold principle, therefore, both somatic and germ line gene therapy are generally admissible when the aim and objective are therapeutic.

Germ line gene therapy is prohibited for two different reasons: (1) because current methods pose incalculable risks yet do not allow the achievement of therapeutic results (this reason is connected with the present state of scientific knowledge); and (2) when research goes beyond the therapeutic aim and seeks to modify the genetic makeup, the impermissibility is absolute and does not depend on the state of knowledge because it constitutes an alteration contrary to the principle of respect for biological life and identity and equality among persons.

---

[138] Italian National Bioethics Committee, *Terapia genica* [*Gene Therapy*], February 15, 1991 (Rome: Presidenza del Consiglio dei Ministri Dipartimento per l'Informazione e l'Editoria, 1991), 7, available at http://www.governo.it/bioetica/pdf/1.pdfm English abstract available at http://www.governo.it/bioetica/eng/opinions/abstracts/abstract_gene_therapy.pdf; W. M. Moore O'Connor, "Gene Therapy: Clinical Aspects," in *Human Genome: Human Person*, Vial Correa and Sgreccia.

[139] R. Naam, "More Than Human," *New York Times*, July 3, 2005, http://www.nytimes.com/2005/07/03/books/chapters/0703–1st-naam.html?pagewanted=print.

All of this does not authorize us to consider gene therapy, even somatic gene therapy, to be free of difficulties and conditions. The ethical conditions for somatic gene therapy are ultimately linked to the fact that the treatments are experimental and there may be associated risks such as harmful induced mutations, proliferation disorders, rejection, and the potential spreading of viral vectors. The NBC stipulates two sets of conditions:

1. Conditions regarding the *choice of a treatable disorder*

- The illnesses must be due to a defect in a single recessive structural gene, which can therefore be treated completely with a single healthy gene.

- The illness must be serious, with low life expectancy.

- There must be no alternative treatment for the illness today.

  Examples of such disorders include ADA deficiency, Lesch-Nyhan syndrome, cystic fibrosis, and others.

2. Conditions for *monitoring the techniques*

- There is a duty to notify the authorities in charge at the national and regional levels about the protocols used for the techniques employed and for the availability of facilities.

- Facilities must be assessed.

- The severity of the illness, lack of alternatives, and negligibility of side effects must be documented.

- A probable, positive outcome must be expected.

By now the European regulations,[140] the Italian NBC, and the opinions of authoritative committees such as the study group of the top seven industrialized countries [141] are in agreement on these points.

It is obvious that patient consent must also be obtained for these treatments because of the risks and the experimental stage of the treatment. Since the procedure is therapeutic, this consent may be provided by representatives in the case of minors or those who are otherwise unable to give consent.

As is apparent in many countries, the unique nature of this type of treatment makes it desirable for a central authority to verify and monitor protocols and facilities.

The Catholic Bishops' Joint Committee in Great Britain also made a pronouncement on the ethical aspects of genetic intervention on human beings with a report that was the fruit of several years of work.[142]

---

[140] European Parliament, "Resolution on Genetic Engineering."

[141] International Conference on Bioethics, *Human Genome Sequencing.* The conclusions of the study group of the top seven industrialized countries were published in *Medicina e Morale* 38, no. 2 (1988): 308–315.

[142] Catholic Bishops' Joint Committee on Bioethical Issues, *Genetic Intervention on Human Subjects* (London: Linacre Centre, 1996). On the general topic of the ethical aspects of gene therapy, see also L. Walters and J.G. Palmer, *The Ethics of Human Gene Therapy* (New York: Oxford University Press, 1997).

Genetic engineering for the modification and enhancement of human beings

The use of genetic engineering in the realm of plant and animal life, aimed not just at enhancing but also altering the genetic heritage of living beings in order to enhance or improve product quality or even to create new pharmacological products, has already been addressed. It was concluded that there are risks (deadly epidemics, damage to the ecosystem, etc.), but they are manageable and must be kept in check.

In fact, these present and future risks are the criterion for determining what is licit or illicit in the plant and animal realm and for the environment. The simple argument that transformations also occur in nature by means of environmental mechanisms is insufficient in itself to constitute a norm, because biological function alone (natural law understood in a biologistic sense) does not constitute a moral norm for man, who is not merely a biological being. Moreover, not everything that happens in nature by natural mechanisms (e.g., contracting an illness) is good in itself.[143]

Even the ethical and economic problem resulting from productivity in the field of biotechnology is no different, particularly at the international level, from the problem of justice affecting all technological progress and economic relations between developed and developing countries. In fact, some maintain not only that the eventual inequalities of economic means and resources can be controlled, but also that biotechnologies could open up new possibilities for developing countries. The stage of ethical and cultural reflection on genetic engineering that began with serious fears of potential catastrophes is therefore quietly coming to an end, making way for calmer and more balanced considerations.

The problem is couched in different terms when it is a question of engineering to modify or enhance/improve human beings. Everything stated thus far on gene therapy, which in my opinion does not fall under the concept of modification, still holds true. In fact, therapeutic genetic engineering aims to restore normal integrity to the subject, considered as a member of the human species. There is therefore a very precise distinction between gene therapy and engineering interventions for the purpose of modification. John Paul II also pointed this out in a speech to the members of the World Medical Association on October 29, 1983: "A strictly therapeutic intervention whose explicit objective is the healing of various maladies such as those stemming from deficiencies of chromosomes will, in principle, be considered desirable, provided it is directed to the true promotion of the personal well-being of man and does not infringe on his integrity or worsen his conditions of life. Such an intervention, indeed, would fall within the logic of the Christian moral tradition."[144]

I also maintain that the prohibition in Recommendation 934 (1982) of the Parliamentary Assembly of the Council of Europe, where it requires that member states protect citizens' rights to have "a genetic inheritance which has not been artificially

---

[143] Cuyás, "Problematica etica"; R. Spaemann, "Genetic Manipulation of Human Nature in the Context of Human Personality," in *Human Genome: Human Person*, Vial Correa and Sgreccia, 340; H. Watt, "Human Gene Therapy: Ethical Aspects," in *Human Genome: Human Person*, Vial Correa and Sgreccia, 255–270.

[144] John Paul II, Address to the World Medical Association.

interfered with," [145] does not refer to therapeutic interventions but to those involving modification.

It can therefore be asserted that any intervention not aimed at restoring genetic integrity must be either a modification or an enhancement/improvement. Conceptually, modification implies the production of a new characteristic, whereas improvement or enhancement would involve the strengthening of an already existing characteristic. Some of the examples cited include improvement of genes responsible for either growth in height or prolonging of age.

With the purpose of finding a line of demarcation within genetic engineering between modification and enhancement/improvement in the sense of "genetic recombination as an enhancing or eugenic therapy," [146] Manuel Cuyás describes four scenarios of genetic engineering. The different degrees depend on the differences of purpose: (1) correcting a deficiency that puts the subject in a condition of inferiority with respect to the statistical average, (2) improving the subject to be above the statistical average in one or more qualities, (3) giving descendents superiority with respect to the norm by maximizing certain qualities (height, strength, intelligence, etc.), and (4) endowing subjects with qualities that are foreign to the human species either in themselves or because of their intensity.

The author rightly associates the first case with a form of therapy; however, I do not think that all the other types of intervention can automatically be called "enhancements." It is instead a matter of eugenics. The second case presents some ethical difficulties for the author because the subject cannot be asked for consent, even though the induced enhancement could affect the future of that subject himself; therefore, in this area, the author only considers the case of possible experiments with consenting adults.

Nonetheless, if there were such a nontherapeutic possibility of enhancing some physical qualities beyond the normal average, then in my opinion one would have to ask, in addition to requiring the absence of risks and the subject's consent to undergo such an augmentation, whether this election/selection of physical or intellectual qualities offends the principle of equality and crosses over into arbitrary dominion over one's own body. The author himself even questions the very classification of qualities and defects, which often obeys cultural fads or manipulative fervor.

The argument that human beings have always pursued this enhancement and selection of physical qualities through the choice of a marriage partner is not quite fitting: in such cases, it is a matter of setting the conditions for improvement in place and certainly not of directly determining the degree of a physical quality through intervention on the biological structure. This type of hypothetical intervention seems more serious than what happens, for example, in the sphere of cosmetic medicine. The fact that a person enjoys a certain power over his own life and that the concept of health is not limited exclusively to the absence of illness but extends to "complete well-being" in a physical, psychological, and social sense—assuming but not

---

[145] Council of Europe, Recommendation 934, 7(b).

[146] Cuyás, "Problematica etica," 487–488; J. Seifert, "Respect for the Nature and Responsibility of the Person in Acquiring Knowledge about the Human Genome and in the Application of Human Biotechnology," in *Human Genome: Human Person*, Vial Correa and Sgreccia, 351–395.

conceding that this completeness of well-being exists—does not solve the problem of the limits within which this objective must be contained.

The words of Pope Pius XII praising the efforts of science to "promote what is good and eliminate what is harmful" do not exempt scientists from the duty to respect those "moral barriers that no human power has the right to break down" in human beings.[147]

Although some might consider it licit under certain conditions, I am very skeptical about the enhancement of progeny in the sense of augmenting physical and intellectual qualities beyond therapeutic aims and above the norm. Indeed, requiring the possibility of extending the presumed benefit to the majority of persons or working to prevent the potential implementation of racist strategies by those in political power will be to no avail.

I believe that we must ask ourselves, even more than in the previous case, about the potential exploitation of the human person through his genome. Even more evidently, I do not see how this enhancement of the qualities of the progeny (aside from cases of therapy) can be achieved at the genetic level without intervening on the germ line and without doing so through in vitro fertilization.

In this case, all of the ethical and deontological positions, including the Italian NBC document, prohibit such attempts not only in a conditional sense—because this type of intervention is currently and technically uncertain and fraught with enormous risks even when the intention is therapeutic—but also in an absolute sense, because of the limit imposed by respect for man's biological identity. And as far as IVF-ET is concerned, all the ethical problems of the case have to be considered.

Aside from ideas about hybridizing multiple human genomes, which nearly everyone condemns, I believe this third type of descendant engineering leads to the creation of a mindset of domination over the unborn child and an increasing presumption to have "designer babies," which is well outside the logic of respect for others and for the gift of the life of the unborn child.

Conferring qualities that are foreign to the human species upon human individuals through genetic engineering is essentially a scenario of modification. It must be rejected on ethical grounds whenever it involves the acquisition of animal qualities through hybridizations (but there are some who do not rule out this possibility if the qualities do not conflict with those of human beings, such as the eyesight of an eagle).

Aside from what was said above about the means and methods for reaching this objective, I believe that some objections also remain at the level of the principle of equality between human beings. John Paul II's warning in this regard is quite pertinent: he urges that genetic interventions be directed so as to avoid "manipulations that tend to modify genetic inheritance and to create different groups of men at the risk of causing new instances of marginalization in society."[148] The same line of reasoning is expressed in the European Parliament resolution in March 1989[149] and the Italian NBC document addressing intervention on germ lines.[150]

---

[147] Pius XII, Address to the First International Symposium on Genetic Medicine, September 7, 1953, in *Acta Apostolicae Sedis* 45, no. 12 (1953): 596–607 [French].

[148] John Paul II, Address to the World Medical Association.

[149] European Parliament, "Resolution on Genetic Engineering."

[150] Italian National Bioethics Committee, *Terapia genica*.

Patenting the products of biotechnology

In all legal systems, an industrial patent takes the form of a procedure designed to recognize the inventor's intellectual ownership of the resulting invention and, at the same time, to guarantee him remuneration. The ultimate aim of this recognition is to create incentives for industrial development. In order to have a legal basis for patenting, there are three conditions: the product must be new, imply inventive activity, and have an industrial application.[151] In this respect, a patent reveals precise commercial value and confirms ownership, though it is limited by law to a set number of years.

From the 1930s to the 1980s, the extent of patent rights was delimited by the distinction between living and nonliving: a patent could be requested and granted only for inventions involving nonliving matter. It is also true that until recently the progress of biological technologies was so minimal that patenting probably never even came to mind. There is information indicating that a vegetable gardener in Nice was denied a patent in 1921 for a special variety of vegetable (*ocillet*).[152] Article 53 of the Munich Agreement on the Grant of European Patents, which was signed on October 5, 1973, prohibits the patenting of animal breeds and vegetable varieties but not of products obtained through microbiological processes.[153]

Legal developments since then have moved toward allowing inventions that involve living beings to be patented as intellectual property. Two separate paths were taken by Europe and the United States.

In Europe, the *certificat d'obtention végétale* (COV) system was initiated, guaranteeing the certificate holder the exclusive right to produce, market, and import vegetal reproduction and multiplication material for the newly produced variety for a period of 20–25 years. This system, often applied to grains, had the feature of not preventing the protected product from being used for later research and inventions.

In 1963, the Strasburg Convention on the Unification of Certain Points of Substantive Law on Patents for Invention nonetheless affirmed the patentability of products obtained through the use of microorganisms. Not until 1972 did the problem arise of whether to grant patents for the invention and creation of the microorganisms themselves through genetic engineering.

In the United States, on the other hand, they were already proceeding with the recognition of the patentability of ornamental plants (Plant Patent Act), and in 1970, other measures of recognition were conceded for patenting plant species.[154]

---

[151] E. Brovedani, "Il brevetto di organismi viventi ottenuti con l'ingegneria genetica: Aspetti scientifici, giuridici ed etici," *Aggiornamenti Sociali* 39, no. 4 (1988): 245–259, which provides an extensive bibliography; G. Ancora, *Biotecnologie animali e vegetali: Nuove frontiere e nuove responsabilità* (Vatican City: Libreria Editrice Vaticana, 1999), especially V. Buonomo, "Brevetti e brevettabilità delle biotecnologie: Alcune considerazioni sugli aspetti etici e giuridici," 101–148.

[152] M.D. Chevallier, *Rapport sur les applications des biotecnologies à l'agricolture et à l'industrie agroalimentaire* (Paris: Office Parlementaire des choix scientifiques et technologiques, Assemblée Nationale Francaise, Sénet, 1990–1991), 1:48–59.

[153] Chevallier, *Rapport sur les applications*, 50.

[154] Ibid., 51–52.

The turning point in the granting of patents occurred when patents were applied to biotechnologies, namely, to new living products obtained by inventive activity through the application of genetic engineering techniques.

The innovative event occurred in the United States in 1972, when a researcher at General Electric Company, Ananda Chakrabarty, produced a bacterial strain of the genus *Pseudomonas* in the laboratory that was capable of breaking down various components of petroleum hydrocarbons, offering the potential to provide an effective means of combating pollution.

The company requested patent protection for the invention from the Patent and Trademark Office in Washington, DC. The patent office granted the patent on the technique used but not on the product (the bacterial strain), basing its decision on the boundary between living and nonliving.[155]

The company lawyers petitioned a higher government body, the Board of Patent Appeals and Interference, which accepted the lawyers' petition and held that the right to patent was founded. Ultimately, the United States Federal Supreme Court intervened with a decision on June 16, 1980, in which it sanctioned the patentability of genetically manipulated microorganisms. The innovative aspect was overcoming the boundary between living and nonliving and developing the new criterion defined as "whether living or not"—that is, whether existing or not existing in nature. A later decision on April 3, 1987 extended patentability to higher, multicellular organisms (the actual case involved giant oysters). And hence patentability is accepted to this extent in the United States.

On the European scene, the European Commission presented a draft directive on the legal protection of biotechnology inventions in 1988. In March 1995, the European Parliament rejected that proposal, leading the commission to present another proposal—not much different from the previous one—that was approved on May 12, 1998, after a heated debate.

The debate on this topic in Europe also began with the principle—which in itself is always valid—that patents indicating ownership of human life must be ruled out, and therefore the patentability of human beings must be ruled out.

The discussion clarified that a patent should not refer to the *object* of the experiment, but to the *technique* as such; for this reason, many have stated that it is unclear why the patenting of a technique—such as a diagnostic technique—that might make it possible to identify an illness should be prohibited, given that patenting provides incentives for research.

On the other hand, the potential applications for patenting diagnosis or treatment techniques in the context of the human genome, after news of requests of this kind in the USA, led to the categorical exclusion of all patents presupposing research on embryos or germ cell lines, which could presage the possibility of dominion over heredity.[156]

---

[155] Brovedani, "Il brevetto di organismi viventi," 251; Chevallier, *Rapport sur les applications*, 52–53; OTA, *Patenting Life*.

[156] European Parliament and Council, Directive 98/44/EC. For a summary of the ethical debate at the European level, see A. G. Spagnolo, *Bioetica nella ricerca e nella prassi medica* (Turin: Camilliane, 1997), 307–312.

Aside from this fundamental norm, the ethical requirements for patenting other living organisms have yet to be determined. Some of the problems posed by this type of patenting are of a more specifically philosophical and metaphysical nature: the risks of promoting a materialistic understanding of life, of increasing man's control over other living beings, or of violating the integrity of species. Strictly speaking, such risks are connected to the ideological and cultural context in which the patenting of new living species is proposed and conducted. Other issues are of a more pointedly environmental or social nature. In reference to the environmental context, the need arises for in-depth study of the potential impact of releasing genetically modified microorganisms into the environment. This scientifically and ethically important aspect is nonetheless independent of patenting; rather, it refers to the release of said organisms into the environment, whether they are patented or not.[157]

The socio-ethical issues, on the other hand, are more numerous. "Industrial confidentiality" runs the risk of depriving the international scientific community of factors essential to its growth, such as debate, comparison, and critiquing of new scientific achievements.[158] The possibility of patenting in some countries and not in others could escalate the already existing situation of international economic competitiveness, potentially widening the economic gap between industrialized countries and developing countries. The scant economic resources of universities could contribute to diverting many researchers toward private industries in view of patent protection. In the long term, this could make the ideological manipulation of new discoveries possible, causing research to lose its proper characteristics of freedom and independence.

The Italian NBC also expressed its opinion on this subject with its report on patenting living organisms on November 19, 1993.[159] Biotechnologies can be applied to microorganisms, plants, or multicellular animals, and the report addresses the topic of patentability accordingly, in a nuanced way. The NBC specifically maintains the following:

> *In relation to microorganisms*: (a) it is appropriate to promote a set of regulations at the community level that rigorously defines the legal notion of them; (b) it is appropriate to promote an in-depth analysis of the reasons that could warrant the exclusion or limitation of patentability for certain categories of them; (c) it is important to promote the establishment of genetic preservation banks in our country as well, which would be valuable in addressing the risk of impoverished biodiversity caused by the expediency of the predominant or exclusive commercial use of the new, patented genes.

> *In relation to plants*: (a) it is appropriate to reconsider the national regulations for implementing the Paris Convention for the Protection of Industrial Property in light of the most recent changes to it; (b) it is appropriate to keep

---

[157] OTA, *Patenting Life*.

[158] Brovedani, "Il brevetto di organismi viventi."

[159] Italian National Bioethics Committee, *Rapporto sulla brevettabilità degli organismi viventi* [*Report on the patentability of living organisms*], November 19, 1993 (Rome: Presidenza del Consiglio dei Ministri Dipartimento per l'Informazione e l'Editoria, 1993), available at http://www.governo.it/bioetica/pdf/12.pdf, English abstract available at http://www.governo.it/bioetica/eng/pdf/Report_patentability_2.pdf.

the protection of plant innovation, under whatever name, from taking on overly rigid forms borrowed from those proper to the patent protection of inventions; (c) at the same time it is necessary to prevent the regulation of intermediary rights from penalizing only the bearers of the exclusive rights to the new plant varieties; (d) it is urgent that Italy not only ratify the Rio de Janeiro Convention on Biological Diversity, as it already has, but also hasten the ratification, acceptance, approval, or adherence by other countries to the same convention, in order to allow, through its entering into force, the initiation of coordinated national policies aimed at pursuing certain global objectives proposed therein (e.g., avoiding the drastic impoverishment of cultivated plant varieties, preventing the destruction of millenarian balances among plant species in shared environments, promoting a more civil exchange of germplasm with Third World countries, etc.).

In relation to multicellular animals: (a) it is appropriate that national authorities support the approaches of the "modified proposal" of the community directive on the protection of biotechnological inventions at the part in which it calls for the exclusion from patentability of "processes for modifying the genetic identity of animals which are likely to cause them suffering without any substantial medical benefit to man or animal"; (b) it is appropriate that national authorities take it upon themselves to preventively coordinate the cited text from the modified proposal with that of EEC Directive 609/86 (which went into force in Italy with Legislative Decree No. 116 on January 27, 1992) in which the acceptability of experimentation on multicellular animals is subordinated to the condition that it be directed toward particular socially useful aims, which the directive itself names; (c) it is important for national authorities to encourage the approval of legislative measures aimed at avoiding the inappropriate extension of patent protection to biotechnological inventions, and that in particular they: (c.1) provide that the protection of patented animals should not extend past the second generation after the one on which the genetic modification was carried out; (c.2) provide that the inventor's right over the genetically modified animals should extend only to the explicitly claimed functions of the invention and only to the animals in which the modification is explicitly expressed and not merely incorporated; (c.3) provide that the inventor's right over the genetically modified animals should not extend to animals belonging to varieties other than the one to which the genetically modified animal belongs; (c.4) provide for the automatic granting of a concession of rights to those who, with a royalty payment to the patent holder, request the nonexclusive use of the invention incorporated into the animal in relation to a function other than the one already claimed.[160]

In conclusion, it can be said that the patenting of human beings as such is to be ruled out, whereas patenting other living beings can be considered permissible provided that suitable legal and legislative measures are enacted in order to ensure a high level of freedom for scientific research, economic cooperation at the international level with a particular focus on developing countries, and environmental balance. The measures capable of addressing potential ideological corruptions (e.g., a materialistic understanding of life) are certainly of a cultural and philosophical nature.

---

[160] Italian National Bioethics Committee, *Rapporto sulla brevettabilità*, 9–11.

Prenatal Diagnosis

I am addressing this topic together with the topic of genetics because many of the diagnostic procedures during the prenatal period aim to investigate the genetic makeup of the embryo/fetus. Furthermore, I wish to devote special consideration to this topic because it involves not only genetic testing but also a practice that is very frequently, though not necessarily, linked to it, namely, the abortion of malformed fetuses. In reality, given the current state of knowledge and medical resources, the diagnostic phase is rarely able to be connected with the therapeutic phase regarding chromosomal or genetic malformations and defects. For this reason, only the following alternatives remain open in nearly all cases once the defect has been verified: either accept the unborn child with the defect or resort to an induced abortion. Because of this, there are many who maintain, as we will see, that either genetic testing is pointless or else it opens the door to abortion; they base a de facto moral obligation to reject such tests on this reasoning.

Prenatal genetic tests are performed during a fixed period of embryonic and fetal development to ascertain whether the unborn child is affected by malformations or defects that could influence his or her life in the future. Trisomy 21, which is responsible for Down syndrome, is the most frequent case, but the spectrum of diagnosable disorders is continually expanding.

The recent practice of IVF-ET and reproductive technologies in general have made it possible to diagnose a fertilized embryo in vitro before transferring it into the uterus (preimplantation diagnosis). It can also be done with a naturally conceived early-stage embryo by removing it through washing out the uterus and later reimplanting it after genetic testing.[161] In reality, the removal of a single cell from even an eight- or sixteen-cell embryo is sufficient to perform a chromosomal or genetic test through appropriate genetic probes.[162] The motivation put forth for this

---

[161] J. E. Buster et al., "Biologic and Morphologic Development of Donated Human Ova Recovered by Non-surgical Uterine Lavage," *American Journal of Obstetrics and Gynecology* 135 (1985): 211–217; R. Penketh and A. McLaren, "Prospects for Prenatal Diagnosis during Preimplantation Human Development," *Bailliere's Clinical Obstetrics and Gynaecology* 1 (1987): 747–764; C. Bacchus and W. Buselmaier, "Blastomere Karyotyping and Transfer of Chromosomally Selected Embryos Implications for the Production of Specific Animal Models and Human Prenatal Diagnosis," *Human Genetics* 80 (1988): 333–336; H. Li et al., "Amplification and Analysis of DNA Sequences in Single Human Sperm and Diploid Cells," *Nature* 335 (1988): 414–417; M. Adinolfi, C. Camporese, and T. Carr, "Gene Amplification to Detect Fetal Nucleated Cells in Pregnant Women," *Lancet* 2 (1989): 328–329; K. A. Hodgkinson et al., "Adult Polycystic Kidney Disease: Knowledge, Experience, and Attitudes to Prenatal Diagnosis," *Journal of Medical Genetics* 27 (1990): 552–558; Italian National Bioethics Committee, *Diagnosi prenatali*, 16–19.

[162] C. Julien et al., "Rapid Prenatal Diagnosis of Down's Syndrome with In Situ Hybridization of Fluorescent DNA Probes,"*Lancet* 2 (1986): 863–864; D. Pinkel, T. Straume, and J. W. Gray, "Cytogenetic Analysis Using Quantitative, High Sensitivity, Fluorescence Hybridization," *Proceedings of the Natural Academy of Sciences of the USA* 83 (1986): 2934–2938; S. H. Embury, S. J. Scharf, and R. K. Saiki, "Rapid Prenatal Diagnosis of Sickle Cell Anaemia by a New Method of DNA Analysis," *New England Journal of Medicine* 316 (1987): 646–651; R. Kozma and M. Adinolfi, "In Situ Hybridization and the Detection of Biotinylated DNA Probes," *Molecular Biology and Medicine* 4 (1987): 357–364; R. G. Cotton, N. Rodrigues, and R. D. Campbell, "Reactivity of Cytosine and Thymine in Single-Base-Pair

type of diagnosis is to select embryos unaffected by genetic disorders or to test the chromosomal sex; testing chromosomal sex can in turn be motivated by the search for potential genetic disorders linked to chromosomal sex or by the desire to choose a preferred sex.

Preimplantation genetic diagnosis can be performed using invasive and noninvasive methods. Noninvasive methods culture the blastocysts in an artificial medium. Data are collected from the composition of the culture medium, allowing the analysis of the embryo's metabolic processes without altering its physical integrity.

Invasive methods, by contrast, make use of an extracted portion of the early-stage embryo's tissue—by so-called *embryo splitting*. Embryo splitting involves dividing a 4- or 8-cell embryo in two: one part is tested and the other is cryopreserved and eventually implanted, whereas the part used for diagnosis is then discarded. As is known, twinning is possible during this stage by virtue of the totipotency of the cells; therefore even the implantation of a "half" of the embryo can develop into a whole individual.

There are basically two ethical justifications advanced by those who practice preimplantation genetic diagnosis: when the aim is to discover potential defects or abnormalities, the intention is to "prevent" the birth of an unhealthy subject without resorting to a postimplantation abortion; the other justification is that of responding to a couple's desire to indicate a preferred sex. In addition to these stated pseudo-justifications, there is also a tenaciously pursued interest in experimentation during the preimplantation phase.

Regardless of the judgment on IVF-ET, which presents numerous difficulties as we will see, neither the reality nor the gravity of abortion is changed when embryo selection is performed at an early stage. Interrupting a life at an earlier or later stage does not substantially alter the real effect and the ethical assessment except in the minds of those who, following the so-called gradualist doctrine, hold that the early-stage embryo is not yet an individual with the value of a human person.[163]

Mismatches with Hydroxylamine and Osmium Tetroxide and Its Application to the Study of Mutations," *Proceedings of the Natural Academy of Sciences of the USA* 85 (1988): 4397–4401; T. Cremer et al., "Detection of Chromosome Aberrations in Metaphase and Interphase Tumor Cells by In Situ Hybridization Using Chromosome-Specific Library Probes," *Human Genetics* 80 (1988): 235–246; A. Dilella, W. M. Huang, and S. L. C. Woo, "Screening for Phenylketonuria Mutations by DNA Amplification with the Polymerase Chain Reaction," *Lancet* 2 (1988): 479–499; P. Lichter et al., "Delineation of Individual Human Chromosomes in Metaphase and Interphase Cells by In Situ Suppression Hybridisation Using Recombinant DNA Libraries," *Human Genetics* 80 (1988): 224–234; B. Trask et al., "Fluorescence In Situ Hybridization to Interphase Cell Nuclei in Suspension Allows Flow Cytometric Analysis of Chromosome Content and Microscopic Analysis of Nuclear Organisation," *Human Genetics* 78 (1988): 251–254; A. H. Handyside et al., "Biopsy of Human Pre-Implantation Embryos and Sexing by DNA Amplification," *Lancet* 1 (1989): 347–349.

[163] Singer, *Practical Ethics*; M. Warnock, *A Question of Life: The Warnock Report on Human Fertilization and Embryology* (Oxford: Blackwell, 1985); H. T. Engelhardt, Jr., *The Foundations of Bioethics*, 2nd ed. (New York: Oxford University Press, 1996); M. Mori, "Il feto ha diritto alla vita? Un'analisi filosofica dei vari argomenti in materia con particolare riguardo a quello di potenzialità," in *Il meritevole di tutela*, ed. L. Lombardi Vallauri (Milan: Giuffrè, 1990), 735–840; Idem, "Per un'analisi dei problemi morali relativi agli interventi che comportano la morte degli embrioni umani," in *Quale statuto per l'embrione umano: Problemi*

Even for postimplantation diagnosis on the fetus, the fundamental ethical problem—albeit not the only one—is selective abortion: it is well known that, given the current legal status of elective abortion, a test result that is positive for the presence of a disorder of genetic origin or a severe, difficult-to-treat malformation can very often, though not necessarily, lead to a *selective abortion*.

Some laws, such as the Italian abortion law, explicitly provide for the selective abortion of fetuses under the legal scenario of a "therapeutic" abortion, inasmuch as the existence of a malformation or disorder in the fetus could induce a "pathological" psychological state in the mother. But this pretext has the hallmarks of a legal contrivance designed to slip selective, eugenic abortion in under the constitutionally acceptable and culturally less odious heading of treatment. But the essence of the reality remains what it is: selection is in fact practiced, and in the vast majority of cases, malformed fetuses are rejected because of their malformation.

It may also happen that a social organization offering "prevention" services is created, which deliberately and systematically intends to promote such selection (by screening) in order to avoid the socioeconomic burden that such individuals could cause. The ethical problem therefore involves families, the professionals who perform the diagnosis and find themselves involved in a given procedure, and the organization providing the services.[164]

In order to better clarify these complex and acute ethical problems, I wish to proceed with a brief discussion of the following aspects: the history of genetic diagnosis and medical indications for genetic testing, its methods and procedures, and post-analysis results and outcomes, all from the ethical perspective.

Before specifically examining the individual aspects, for the sake of completeness I wish to mention that prenatal diagnosis tests are also performed together with other techniques in discovering pathologies that do not necessarily have a genetic cause but can be identified somatically.

Ultrasonography, for example, allows the detection of somatic malformations which, for the most part, can be addressed by suitable treatments at birth or even during the intrauterine phase when it is necessary and possible to perform the intervention earlier. In any case, ultrasonography is also used in conjunction with genetic diagnosis during the phase involving study of the fetus and extraction of fetal cells or tissue.

---

e prospettive; Convegno Internazionale, Milano, gennaio 1991 (Milan: Bibliotechne, 1992), 75–91; P. Singer and K. Dawson, "Individuals, Humans and Persons: The Issue of Moral Status," in *Embryo Experimentation*, ed. P. Singer et al. (London: Cambridge University Press, 1990); S. Maffettone, "Proposte per uno statuto morale e giuridico dell'embrione," in *Quale statuto*, Mori, 96–107.

[164] For a complete treatment of this complex subject, see the issue of the journal *Medicina e Morale* 34, no. 4 (1984), which is entirely devoted to the topic; in particular, see the following contributions: A. Serra, "La diagnosi prenatale di malattie genetiche," 433–448; C. Caffarra, "Aspetti etici della diagnosi prenatale," 449–457; L. Leuzzi, "Indicazioni etiche per la diagnostica prenatale," 458–463. See also A. Serra et al., "La diagnosi prenatale di malattie genetiche: Esperienze, prospettive e problemi," *Il progresso medico* 37, no. 15 (1981): 1–18. The Italian National Bioethics Committee published the extensive document *Diagnosi prenatali*, which has already been cited.

Based on the use of ultrasounds, ultrasonography per se does not present veri-fied risks, although prudence and economic ethics dictate that normally—that is, if there are no special reasons—it is not advisable to undergo more than three during gestation.

History and medical indications

The practice of prenatal genetic diagnosis (PGD) began around the years 1968–1969, once the techniques were perfected for culturing fetal cells suspended in amniotic fluid for the study of chromosomes and for biochemical investigations relative to the study of metabolic disorders.

Once these culturing and testing techniques became available, it was not dif-ficult to fine-tune amniocentesis techniques for withdrawing the amniotic fluid where those cells are suspended. During those same years, other techniques for the somatic study of the fetus were perfected with sonography and fetoscopy.[165]

When performing a prenatal genetic diagnosis, particularly if the use of invasive techniques is expected, which involve a certain element of risk, as we will see, it is deontologically and ethically required that there be well-founded medical reasons, which in technical terms are called *indications*. The presence or absence of such indications is ascertained through *genetic consultation*, which precedes the interven-tion and examines how well-founded and necessary it would be. Proceeding with a test involving amniocentesis (or fetoscopy or placentocentesis, as the case may be) without prior genetic consultation, solely on the patient's request, would in any case signal a lack of ethical and deontological responsibility.[166]

At the end of the International Conference on prenatal diagnosis held at Val David in Quebec in 1979, a sort of deontological code was drawn up to protect women's freedom to pursue prenatal diagnosis or not and to define the indications mentioned above.[167] These indications, supported by objective reasons, should be mentioned briefly.

The first indication, which is sufficient in itself, is if the *mother is older than 36 years of age*. The reason is that "the risk of discovering a chromosomal abnormal-ity associated with serious pathological conditions in a fetus at 16 to 20 weeks of gestation increases with the age of the mother, rising from 0.9 percent in mothers at 35 to 36 years of age to 8.2 percent in mothers at 45 years of age. Considering a risk greater than 1 percent to be significant, a medical indication for prenatal cytogenetic diagnosis can be offered and the procedure can be practiced on mothers having reached 36 years of age."[168]

The second indication is the presence of *a child with trisomy 21*, or Down syndrome, who has already been born in the family. In this case, there may be a risk that another individual affected by the same syndrome, or by trisomy 18 or other chromosomal abnormalities, will be born. The total risk of potential abnormalities

---

[165] The historical and technical information in this and the previous paragraph is taken from Serra et al., "La diagnosi prenatale: Esperienze."

[166] A. Serra, "La consulenza genetica prima della diagnosi prenatale: Un obbligo deontologico," *Medicina e Morale* 47, no. 5 (1997): 903–921.

[167] Serra, "La diagnosi prenatale," 447.

[168] Serra et al., "La diagnosi prenatale: Esperienze," 2.

in these cases for a family that already has one individual with Down syndrome is 1.4 percent. When other close relatives outside the immediate family are affected, the statistics do not confirm the presence of this risk.

A third indication is *a balanced structural chromosomal abnormality in one of the parents*. This occurs when one of the parents exhibits a chromosomal abnormality due to translocation (a spontaneous breaking and subsequent recombination of chromosomal fragments): if the translocation is balanced, no phenotypic or functional malformation will be evident, but there will be a risk of descendents having not only identical balanced translocations (40 to 70 percent risk), but also of having severe, uncompensated, or unbalanced abnormalities with consequent pathologies in the unborn child. This risk varies depending on whether the mother or the father is the carrier: if it is the mother, the risk is higher and can reach up to 12 to 16 percent. The percentage of risk is also relative to the type of translocation. In any case, this type of situation constitutes a valid indication for prenatal genetic diagnosis.

A fourth indication is the presence of an already born *child affected by a metabolic disorder*, or the case in which one of the parents is a carrier (heterozygous) of the gene responsible for that disorder. In this case as well the risk of recurrence is high (about 25 percent).

A fifth indication is the presence of *a child already affected by severe hemoglobinopathy*, particularly sickle-cell disease and alpha- or beta-thalassemia, or parents who are heterozygous carriers of the gene for the condition. The risk of recurrence in these cases according to 1978 worldwide statistics is 23.7 percent.

A sixth indication is the presence of *a child already affected by an X-linked condition*, or if the mother is a heterozygous carrier of the gene in question. Examples of conditions of this type that can be diagnosed in utero include Fabry's disease (deficiency of the alpha-galactosidase A enzyme), Hunter syndrome (deficiency of the iduronase-2-sulphatase enzyme) and Lesh-Nyhan syndrome (deficiency of the hypoxanthine-guanine phosphoribosyltransferase enzyme). The risk of recurrence is about 46 percent.

A seventh indication is the presence of *a child affected by a serious neural tube closure defect*. This malformation usually involves anencephaly or spina bifida. The rate of recurrence in these cases ranges from 3.4 to 5.5 percent according to the most reliable and well-known statistics.

These are the official, recognized indications, but there could be other situations that might be taken into consideration, such as prolonged exposure to ionized radiation or chemical mutagens; suspicion or certainty of a maternal infection such as toxoplasmosis, rubella, or cytomegalovirus; the presence of Rh-isoimmunization; or the risk of recurrence of genetically transmitted diseases such as cystic fibrosis (or mucoviscidosis) and beta-thalassemia.

In Italy, the National Committee for Biosafety and Biotechnologies of the Presidency of the Council of Ministers formulated several guidelines for genetic tests in which two major categories of indications are distinguished for prenatal diagnosis: "(1) the presence of a foreseeable a priori reproductive risk (advanced maternal age, parents who are heterozygous carriers of structural chromosomal abnormalities, parents who are carriers of genetic mutations); (2) the presence of a fetal risk that has become evident during the course of gestation (defects discovered by ultrasound, infectious diseases arising during pregnancy, positive results from biochemical tests for chromosomal abnormalities, family history of genetic pathologies)." More precisely, the factors indicating cytogenetic tests for chromosomal abnormalities are "advanced maternal age (35 years or older); parents with a previous child affected by a chromosomal pathology; a parent with a structural rearrangement not associated with phenotypic effects; a parent with sexual chromosome aneuploidy compatible with fertility; malformation abnormalities evidenced by ultrasound; probability of 1/250 or higher that the fetus is affected by Down syndrome (or some other aneuploidies) on the basis of biochemical parameters in maternal blood analysis or ultrasounds, performed with specific regional programs in centers identified by the individual regions and subjected to constant quality control."[169]

The presence of these indications is verified during the genetic consultation interview, which therefore becomes ethically required and necessary before proceeding to an actual genetic diagnosis. Genetic consultation therefore has the object of verifying the presence of an indication and furthermore of providing all the information about the risks, problems, and limits of the requested tests so that the woman or the couple becomes aware of all the factors involved in the request itself.

In the experience related by Professor Angelo Serra concerning the activity of the Cytogenetic Service of the School of Medicine at the Catholic University of the Sacred Heart in Rome from 1977 to 1980,[170] for example, the interviews concluded with the signing of a form wherein the requestor certified that she had been informed about the fundamental elements and risks involved in the test.

From the perspective of the overall ethical assessment of the practice of PGD, it is important to point out how in the same experience just cited it turns out that 28.5 percent of the people who had requested testing refused that testing after consultation. This is due to the fact that requests are often prompted by superficial or inadequate information about the need for testing, or else to the ethical convictions of the woman, who is unwilling to consider under any circumstances the idea of an eventual abortion as the next step after testing; moreover, this allows her to live out her pregnancy free of anxiety.

Methods and technical procedures

The techniques currently used for prenatal diagnosis in general[171] can be distinguished according to two fundamental criteria: (1) the *type of diagnostic information*

---

[169] See appendix C of the Ministerial Decree of September 10, 1998, published in Serie Generale, *Gazzetta Ufficiale della Repubblica Italiana* 245 (October 20, 1998). Also available at http://www.salute.gov.it/imgs/C_17_normativa_1653_allegato.pdf.

[170] Serra et al., "La diagnosi prenatale: Esperienze," 7.

[171] For an exhaustive description of the techniques and issues relative to the various methods, from the perspective of later fetal treatments as well, see G. Noia, A. Caruso, and S. Mancuso, *Le terapie fetali invasive* (Rome: SEU, 1998); A. Serra and G. Bellanova, "Accertamento prenatale di rischio di patologia cromosomica fetale: Aspetti scientifici, etici e deontologici," *Medicina e Morale* 47, no. 1 (1997): 15–35; C. Giorlandino, *Trattato di diagnosi prenatale e terapia fetale* (Rome: CIC Edizioni Internazionali, 1997); A. Pachì, R. Paesano, and F. Torcia, "Diagnosi prenatale: Clinica e tecniche diagnostiche," in *La clinica ostetrica e ginecologica*, ed. G. B. Candiani, V. Danesino, and A. Gastaldi (Milan: Masson, 1996), 1:133–145; M. E. Dalton and A. H. DeCherney, "Prenatal Diagnosis," *New England Journal of Medicine* 328, no. 2 (1993): 114–120; G. Loverro, L. Selvaggi, and F. M. Boscia, "Procedure di indagine prenatale: Significato diagnostico e pericolosità," *Medicina e Morale*

sought, whether morphofunctional, cytogenetic, biochemical and metabolic, or clinical and disease-related; (2) the *invasiveness* of the diagnostic procedure, meaning the need to physically enter into the bodily area in question (uterine cavity, placental tissues, body of the fetus) in order to explore, extract, and sample biological material for subsequent analysis. Closely connected to the concept of invasiveness is the concept of *risk*, which is always present (though to various degrees, for objective and subjective reasons), foreseeable, and quantifiable, and is therefore one of the most important concepts from an ethical standpoint.

Using the parameter of invasiveness and therefore of risk, all of the methods and techniques of prenatal diagnosis can be classified as noninvasive, minimally invasive, or invasive techniques.

NONINVASIVE TECHNIQUES. This group essentially includes *ultrasonography*, which makes use of ultrasounds, and it has now reached remarkable and amazing diagnostic levels thanks to sophisticated equipment making it possible to obtain optimal image resolution. This technique, as its classification suggests, is neither traumatic nor apparently harmful, although it is recommended that it be performed only when strictly necessary, generally no more than three tests during the course of a healthy pregnancy.

The first ultrasound is generally performed during the first trimester, when it is possible to assess the vitality of the embryo, its effective gestational age, and the presence of a multiple pregnancy, as well as to rule out problems preventing proper implantation and placental development. The second ultrasound is performed most often between the twentieth and twenty-fourth weeks of gestation; it allows for an analytical examination of the anatomical and morphofunctional structure of the fetus and is able to detect up to 95 percent of the structural and somatic malformations that may be present. The third ultrasound, in contrast, is performed after the thirtieth week of gestation and allows the assessment of fetal growth, proper functioning of the placenta (fetal fluximetry), and potential umbilical cord problems that might affect the completion of the delivery.

The three-test limit is often exceeded when there are medical reasons to do so. The technique, as explained above, does not reveal genetic or chromosomal abnormalities but proceeds by morphofunctional study of the embryo/fetus, highlighting external or structural malformations for which, wherever possible, early treatment or even an intrauterine surgical operation can be arranged in the neonatal phase.

Additionally, ultrasonography is a supportive technique for the majority of invasive techniques that will be described later; these are performed for both diagnostic and therapeutic purposes and are therefore described as ultrasound-assisted and ultrasound-guided. The ultrasound is also performed before any invasive procedure in order to determine gestational age, evaluate the vitality of the embryo and the site of placental attachment, and identify the best way of accessing the fetus as well as potential fetal abnormalities.[172]

34, no. 4 (1984): 464–487; F. Mantegazza, "Tecnologie per la diagnosi prenatale," *Medicina e Morale* 32, no. 1 (1982): 72–75.

[172] American College of Obstetricians and Gynecologists (ACOG), *Ultrasound in Pregnancy*, Technical Bulletin 116 (Washington, DC: ACOG, 1988), 1–3; Dalton and DeCherney, "Prenatal Diagnosis"; S. Campbell and J. M. Pearce, "Ultrasound Visualization

Because its purpose is primarily therapeutic, ultrasonography is less frequently used in connection with abortion; however, there have been attempts to identify certain early ultrasound markers for use essentially for screening with a view to additional invasive tests and ultimately, therefore, to the termination of pregnancy.[173]

I include the so-called *triple test* (tri-test) among the noninvasive techniques even though it is not a diagnostic method, properly speaking, but rather a predictive and therefore probabilistic test for fetal chromosomal pathologies, especially Down syndrome. It is a test carried out between the fifteenth and eighteenth weeks of gestation and is based on the concentrations of three biochemical markers present in maternal serum: AFP (alpha-fetoprotein), uE3 (unconjugated estriol) and hCG (human chorionic gonadotropin).[174] The test provides statistical information, namely the values for the three biochemical markers, which is used in combination with maternal age to estimate the woman's risk of having a fetus affected by Down syndrome. The risk is used as a screening variable, and a threshold value has been set (equivalent to 1:250 according to most sources)[175] in order to identify those pregnant women who should proceed with amniocentesis. Women with a risk level above the threshold value are considered positively screened and are sent to undergo amniocentesis, while those whose risk level is below threshold are considered negatively screened and thereby discouraged from having the invasive prenatal diagnosis performed.

In the manner in which it is performed, then, the *triple test* is practically devoid of any risk to the mother and the fetus; nonetheless, as alluded to earlier, the type of information it provides is partial and has merely statistical value. The use of such a test therefore raises a series of very important ethical questions.[176] First, it raises the question of the range of prenatal diagnosis: until now, only women who presented precise indications (age, family history, etc.) would proceed with amniocentesis, whereas any pregnant woman can undergo the triple test to learn her risk of giving birth to a child with Down syndrome and eventually proceed to amniocentesis. In theory this should lead to a reduction in the number of invasive tests, but since the number of women who are tested is continually increasing, in the end the number

---

of Congenital Malformations," *British Medical Bulletin* 39 (1983): 322–333; B. G. Ewigman et al., "Effect of Prenatal Ultrasound Screening on Perinatal Outcome," *New England Journal of Medicine* 329 (1993): 821–827.

[173] M. Wittle, "Ultrasonographic 'Soft Markers' of Fetal Chromosomal Defects," *British Medical Journal* 314 (1997): 918; P. Taipale et al., "Increased Nuchal Translucency as a Marker for Fetal Chromosomal Defects," *New England Journal of Medicine* 337 (1997): 1654–1658.

[174] N. J. Wald et al., "Maternal Serum Screening for Down's Syndrome in Early Pregnancy," *British Medical Journal* 297 (1988): 883–887; B. Nørgaard-Pederson et al., "Maternal Serum Markers in Screening for Down Syndrome," *Clinical Genetics* 37 (1990): 35–43; J. E. Haddow et al., "Prenatal Screening for Down's Syndrome with Use of Maternal Serum Markers," *New England Journal of Medicine* 327 (1992): 588–593; J. E. Haddow et al., "Reducing the Need for Amniocentesis in Women 35 Years of Age or Older with Serum Markers for Screening," *New England Journal of Medicine* 330 (1994): 1114–1118; Serra and Bellanova, "Accertamento prenatale."

[175] N. J. Wald and A. Kennard, "Screening biochimico prenatale per la sindrome di Down," *The Ligand Quarterly: Edizione Italiana* 12, no. 4 (1993): 519–523.

[176] Serra and Bellanova, "Accertamento prenatale."

of amniocenteses is rising—along with the corresponding risk of abortion due to the technique itself or to the woman's decision in the event of an undesired result.

Another consideration is the spreading of eugenic alarm: practically every pregnant woman is gradually put on the alert, even when there are no indications of age or family history, because the potential that there might be a general risk of trisomy 21 in an ever smaller but always possible percentage of cases is not ruled out. This alarm reaches the point that any doubt (test values too close to the threshold value, though negative, cause the pregnant woman anxiety and distress!) becomes a risk, and any risk becomes a reason to request an invasive prenatal diagnosis or a termination of the pregnancy anyway. In the facilities where the tests are performed, this is aggravated by a general lack of proper counseling that would explain to the couple the relative nature and actual value of the information provided by this test, noting the high percentage of false positives and false negatives. In this regard, it is important to state that in many cases the triple test indicates the presence of risk factors that do not then correspond to an actual pathology, or "will not reliably detect other chromosomal abnormalities that occur more frequently with advancing maternal age."[177] In practice, the ethical admissibility of the triple test could only be considered if it were used together with valid counseling to limit recourse to amniocentesis.

MINIMALLY INVASIVE TECHNIQUES. These are methods based on the prenatal analysis of trophoblast cells; such methods are strictly limited to applied research and still need clinical confirmation.[178] During the first phase of gestation, these cells are spontaneously shed in the cervical canal and can be sampled through four types of techniques.[179] The techniques are applied between the sixth and the thirteenth weeks of gestation. This intuition of Landrum Shettles in 1971 was revived and significantly advanced once it became possible to obtain and analyze fetal DNA through molecular techniques such as PCR[180] (polymerase chain reaction) or FISH[181] (fluorescent in situ hybridization). The safety and reliability of these procedures seem very high[182] but

---

[177] M. L. MacDonald, R. M. Wagner, and R. N. Slotnik, "Sensitivity and Specificity of Screening for Down Syndrome with Alpha-Fetoprotein, HCG, Unconjugated Estriol, and Maternal Age," *Obstetrics and Gynecology* 77 (1991): 68.

[178] A. Massari et al., "Non-Invasive Early Prenatal Molecular Diagnosis Using Retrieved Transcervical Trophoblast Cells," *Human Genetics* 97, no. 2 (1996): 150–155.

[179] C. Rodeck et al., "Methods for the Transcervical Collection of Fetal Cells during the First Trimester of Pregnancy," *Prenatal Diagnosis* 15 (1995): 933–942.

[180] L. B. Shettles, "Use of the Y Chromosome in Prenatal Sex Determination," *Nature* 230 (1971): 52; M. D. Griffith et al., "Detection of Fetal DNA in Transcervical Swabs from First Trimester Pregnancies by Gene Amplification: A New Route to Prenatal Diagnosis?" *British Journal of Obstetrics and Gynaecology* 99 (1992): 508–511.

[181] M. Adinolfi et al., "Detection of Trisomy 18 and Y-Derived Sequences in Fetal Nucleated Cells Obtained by Transcervical Flushing," *Lancet* 342 (1993): 403–404; M. Adinolfi, P. Soothill, and C. Rodeck, "A Simple Alternative to Amniocentesis?" *Prenatal Diagnosis* 14 (1994): 231–233; B. Pertl et al., "Detection of Fetal Cells in Endocervical Samples," *Annals of the New York Academy of Sciences* 731 (1994): 186–192.

[182] M. Adinolfi et al., "Molecular Evidence of Fetal Derived Chromosome 21 Markers (STRs) in Transcervical Samples," *Prenatal Diagnosis* 15 (1995): 35–39; Pertl et al., "Detection of Fetal Cells," 186–192; B. Tutshek et al., "Isolation of Fetal Cells from Transcervical Samples

still need verification in developing pregnancies, since they were tested in studies involving women who decided to terminate their pregnancies in order to then check, after the abortion, the accuracy of the diagnosis that had been done on the cells. These techniques essentially involve considerable ethical reflections, from scientific validity to the way in which the experimental studies are carried out, and from the methods for comparing results with those obtained from invasive procedures, such as chorionic villus sampling or amniocentesis, to the need to seek genetic counseling. In fact, despite an apparent lack of risk, a precise and ethically acceptable indication is always necessary in order to perform any diagnostic study.[183]

INVASIVE TECHNIQUES. *Fetoscopy* involves inserting a fiber-optic fetoscope into the uterus, whether to observe the somatic structure when the ultrasound is insufficient, to make sampling fetal blood possible by inserting a needle into a vessel of the chorion or of the umbilical cord, or to sample fetal tissue for the purpose of running tests for genetic disorders. It is performed during the period between the eighteenth and twentieth weeks.

Fetoscopy has also been used to perform intrauterine treatments (digitalins, diuretics, etc.). It involves a high risk of miscarriage and an increased likelihood of premature birth as well as of the potential isoimmunization of an Rh-negative mother. The incidence of abortion varies based on the degree of specialization of the centers that practice it. The lowest percentages are above 2 percent when done within the first five weeks. The increase in premature births is 8 percent. Others calculate an average risk of 4 to 6 percent.[184] With the qualitative refinement of ultrasound images, however, fetoscopy is being gradually abandoned for all of its indicated uses, especially fetal blood sampling,[185] which is now done by cordocentesis with the aid of ultrasound localization of the umbilical cord.

Placentocentesis allows fetal blood to be obtained from the chorion through the insertion of a needle into the placenta so as to run genetic tests. Since the fetal blood obtained is often contaminated by maternal blood, it is repeated multiple times. It involves a high risk of miscarriage (7 to 10 percent), which is why it has fallen into disuse and can be replaced by amniocentesis, chorionic villus sampling, or cordocentesis.

Embryoscopy involves the direct viewing of the embryo and its morphology during the first trimester of pregnancy through the use of a fiber-optic instrument (a small, flexible endoscope that was initially 2–3 mm in diameter but has been reduced

---

by Micromanipulation: Molecular Confirmation of Their Fetal Origin and Diagnosis of Fetal Aneuploidy," *Prenatal Diagnosis* 15 (1995): 951–960.

[183] Spagnolo, *Bioetica nella ricerca*, 182–184.

[184] Loverro, Selvaggi, and Boscia, "Procedure d'indagine prenatale," 473; C. Rodeck, "Fetoscopia e prelievo di sangue fetale," in *La diagnosi prenatale dei difetti congeniti: Atti del 5° Corso Nazionale di Aggiornamento in Medicina Prenatale* (Palermo: 1981), ed. G. B. Candiani et al., 273–280; Serra, "La diagnosi prenatale," 441.

[185] Pachì, Paesano, and Torcia, "Diagnosi prenatale," in *La clinica ostetrica e ginecologica*, 1:142–143; E. A. Reece et al., "Embryoscopy: A Closer Look at First-Trimester Diagnosis and Treatment," *American Journal of Obstetrics and Gynecology* 166 (1992): 775–780.

to 0.45 mm in diameter).[186] It is a second-level diagnostic technique: that is, it is applied only after previous tests (e.g., ultrasound) have raised a suspicion that cannot be otherwise confirmed. Moreover it makes it possible to perform fetal biopsies, to draw fetal blood for diagnostic and therapeutic purposes, and to carry out aspiration and electrocoagulation procedures.[187] Given the relatively recent introduction of the procedure, the risk for miscarriage cannot yet be determined precisely. In the early 1990s, a miscarriage rate of 6 to 10 percent was reported;[188] more recently, authors have reported values around 2 to 2.5 percent.[189]

In addition to its intrinsic risks, the ethical assessment of this technique is also dependent on the way it is used. If one reads the literature on the subject, it is apparent that most authors understand this test to be aimed at the termination of pregnancy following the discovery of some pathology affecting the embryo/fetus that would have been identified by ultrasound only at a later point. Therefore, as Yves Dumez wrote on the topic in one of his articles, the precociousness with which a precise diagnosis can be made via embryoscopy as early as the first trimester offers patients the "potential for undergoing repeated terminations of pregnancy."[190] One could argue that this method would be ethically acceptable only if it were put to therapeutic use, with genuine respect for the aim that prenatal diagnosis should have in the first place.[191]

Chorionic villus sampling (CVS) allows for earlier genetic testing (generally as early as the eighth week of pregnancy, more recently even as early as the sixth week) in order to promote a supposedly greater "acceptability" of the eventual choice to terminate the pregnancy. The technique involves removing small quantities of chorion (extraembryonic tissue derived from the trophoblast that provides nourishment to the embryo) transcervically or transabdominally with the aid of an ultrasound, in order

---

[186] M. T. Cullen et al., "Embrioscopy: Description and Utility of a New Technique," *American Journal of Obstetrics and Gynecology* 162 (1990): 82–86; R. A. Quintero, K. S. Puder, and D. B. Cotton, "Embrioscopy and Fetoscopy," *Obstetrics and Gynecology Clinics of North America* 20, no. 3 (1993): 563–581; C. Giorlandino et al., "L'embrioscopia: Nuova tecnica di diagnosi prenatale invasiva," *Ultrasonica* 4 (1994): 103–106.

[187] R. A. Quintero et al., "Percutaneous Fetal Cystoscopy and Endoscopic Fulguration of Posterior Urethral Valves," *American Journal of Obstetrics and Gynecology* 172 (1995): 206–209; A. R. Morgani, "Diagnosi prenatale invasiva: 'Stato dell'arte' dell'embrioscopia," *Ultrasonica* 3 (1995): 59–61.

[188] Y. Dumez et al., "Meckel-Gruber Syndrome: Prenatal Diagnosis at 10 Menstrual Weeks using Embryoscopy," *Prenatal Diagnosis* 14, no. 2 (1994): 141–144; M. Dommergues et al., "Prenatal Diagnosis of Cleft Lip at 11 Menstrual Weeks Using Embryoscopy in the Van der Woude Syndrome," *Prenatal Diagnosis* 15 (1995): 378–38.

[189] E. A. Reece et al., "First-Trimester Needle Embryofetoscopy and Prenatal Diagnosis," *Fetal Diagnosis and Therapy* 12 (1997): 136–139.

[190] Dumez et al., "Meckel-Gruber Syndrome," 142.

[191] A. R. Morgani and V. Mele, "L'embrioscopia: Stato dell'arte sul piano scientifico e valutazione sul piano etico-deontologico," in *Le radici della bioetica*, ed. E. Sgreccia, V. Mele, and D. Sacchini (Milan: Vita e Pensiero, 1998) 2:71–77.

to proceed with subsequent disease tests and genetic tests [192] on the basis of the fact that the trophoblast and fetus originate from the same tissue.

The chorionic villus biopsy involves an even higher risk for miscarriage, which at the initial applications of the technique was estimated to be from 4 to 10 percent.[193] Smaller percentages as low as 3.2 percent have been reported more recently.[194] This lower percentage of miscarriages caused by CVS is partly due to the fact that a distinction is made between the total miscarriage rate, which also takes into account the rate of theoretical natural miscarriages during that period of pregnancy, and a miscarriage rate specifically linked to the procedure (calculated by subtracting the rate of theoretical natural miscarriages). But the captiousness of this calculation is evident: it leads to an ethics deduced from the statistical average, which is used in an attempt to justify abortions that actually occur after the procedure and that no one could ever say with certainty would have occurred anyway.

Even with a lower percentage, the procedure remains nonetheless very problematic from the ethical perspective, despite the fact that its use is continually spreading because of the previously mentioned fact that it allows testing in a very early phase of the fetus's life.

Furthermore, the possibility of causing malformation abnormalities in the fetus through CVS procedures performed too early in the pregnancy (around the fifty-fifth to sixtieth day of gestation) [195] has been reported in recent years.

Genetic tests following chorionic villus sampling involve biochemical studies (enzyme analysis), traditional cytogenetic studies, and biomolecular tests for the investigation of the DNA through marked genetic probes, FISH, and PCR. It must be stated that a certain percentage of error (0.3 to 0.5 percent) is possible even with these sophisticated tests.

---

[192] W. Foulon et al., "Detection of Congenital Toxoplasmosis by Chorionic Villus Sampling and Early Amniocentesis," *American Journal of Obstetrics and Gynecology* 103 (1990): 1511–1513; C. Grose, O. Itani, and C. P. Weiner, "Prenatal Diagnosis of Fetal Infection: Advance from Amniocentesis to Cordocentesis: Congenital Toxoplasmosis, Rubella, Cytomegalovirus, Varicella Virus, Parvovirus and HIV," *Pediatric Infectious Disease Journal* 8 (1989): 459–468.

[193] Medical Research Council Working Party on the Evaluation of Chorion Villus Sampling, "MRC European Trial of Chorion Villus Sampling," *Lancet* 337 (1991): 1491–1499. L. G. Jackson, R. A. Warner, and M. A. Barr, "Safety of Chorionic Villus Biopsy," *Lancet* 1 (1986): 674; W. A. Hogge, S. A. Schonberg, and M. S. Golbus, "Prenatal Diagnosis by Chorionic Villus Sampling: Lessons of the First 600 Cases," *Prenatal Diagnosis* 5 (1985): 393–400; G. Simoni et al., "First Trimester Fetal Karyotyping: One Thousand Diagnoses," *Human Genetics* 72 (1986): 203–209. In this report, the results regarding induced abortions are provided by the Centro di Milano (8. 3 percent) and the Centro di Genova (10. 2 percent).

[194] A. Fortuny et al., "Chorionic Villus Sampling by Biopsy Forceps: Results of 1580 Procedures from a Single Centre," *Prenatal Diagnosis* 15 (1995): 541–550.

[195] H. V. Firth et al., "Severe Limb Abnormalities after Chorion Villus Sampling at 56–66 Days' Gestation," *Lancet* 337 (1991): 762–763; H. V. Firth et al., "Analysis of Limb Reduction Defects in Babies Exposed to Chorionic Villus Sampling," *Lancet* 343 (1994): 1069–1071; P. Mastroiacovo et al., "Transverse Limb Reduction Defects after Chorion Villus Sampling: A Retrospective Cohort Study," *Prenatal Diagnosis* 13 (1993): 1051–1056.

Already from the description of the procedures, it is clear that some of them—in particular chorionic villus sampling but also fetoscopy and placentocentesis (no longer in use)—involve a significant risk level and raise an ethical question as to the legitimacy of their use until experience acquired from animal studies indicates an acceptable risk level. In reality, the refinement of these techniques is occurring through their use on human beings themselves, and this experimental procedure is ethically wrong.

Amniocentesis is the method that has been in use for the longest amount of time, and this is the procedure usually referenced when talking specifically about genetic diagnosis. Regarding the time frame for its use, amniocentesis can be performed very early (between the eleventh and fourteenth weeks of gestation), early (between the fifteenth and eighteenth weeks) or late (after the twenty-fifth week).[196]

Traditional amniocentesis is performed early and involves the withdrawal of 15–20 mL of amniotic fluid, which contains fetal cells that have been shed. The withdrawal of amniotic fluid is performed by ultrasound-guided or ultrasound-assisted needle aspiration; this allows the fetus to be seen and the fluid to be withdrawn with the minimum risk. The intervention is performed on an out-patient basis after the administration of betamimetic drugs to the woman, which relax the uterus at the time of fluid extraction and reduce contractions to a minimum.

The risk for a miscarriage induced by the procedure is from 0.5 to 1.5 percent[197] and is therefore defined as acceptable. For very early amniocentesis, however, the risk of fetal loss is greater.[198] Sometimes it is necessary to repeat the fluid sampling procedure, and if so it must be done fifteen days later. These results clearly vary if the individual performing the withdrawal lacks sufficient experience.

The second phase of the test involves the cytogenetic analysis of the cells in the amniotic fluid belonging to the fetus. These cells are derived from epithelial tissue or from the gastrointestinal or urogenital tracts of the fetus. The cells must be separated by centrifuge and treated in a culture, which takes thirteen to fifteen days, for the purpose of making the chromosomes of those cells recognizable and identifiable.

At this point it can be determined whether the sampling has succeeded or failed and needs to be repeated. If the culture was successful, observation of the chromosomes can thereby indicate the presence of abnormalities. For a certain number of cases, however, this test is replaced or combined with the biochemical genetic test because the defect cannot be sought in the form of the chromosomes. Biochemical and genetic tests are necessary to find certain defects in the fetal central nervous system or congenital metabolic disorders.

The biochemical test reveals the presence in the amniotic fluid of cells that manifest the damage and defects indicated above, in accordance with determined and corroborated laboratory experience that has been recognized internationally.

---

[196] For further analysis of the procedure, see M. R. Harrison, M. S. Golbus, and R. A. Filly, *The Unborn Patient: Prenatal Diagnosis and Treatment*, 2nd ed. (Philadelphia: Saunders, 1991); C. Giorlandino, D. Cannone, and A. Vizzone, "L'amniocentesi," in *Trattato di diagnosi prenatale*, Giorlandino, 1–31; Noia, Caruso, and Mancuso, *Le terapie fetali invasive*, 151 ff.

[197] A. Tabor et al., "Randomized Controlled Trial of Genetic Amniocentesis in 4604 Low-Risk Women," *Lancet* 327, no. 8493 (June 7, 1986): 1287–1292.

[198] V. Choo, "Early Amniocentesis," *Lancet* 338, no. 8769 (Sept. 21, 1991): 750–751.

Thanks to the techniques of molecular biology, it is also currently possible to perform diagnoses—in addition to traditional studies— of many genetic conditions (e.g., cystic fibrosis) and infectious diseases (e.g., rubella and toxoplasmosis) that might be feared during gestation.

Late amniocentesis is performed for the most part in the case of maternal-fetal immunization or in order to assess the pulmonary development of the fetus in the event of a preterm birth.

Cordocentesis (or funiculocentesis) is an ultrasound-guided needle insertion into the umbilical cord, preferably into the vein,[199] for sampling fetal blood. It is therefore an invasive technique. The blood is usually drawn around the eighteenth week of gestation. It allows the doctors to acquire diagnostic information important to the fetus' well-being and in certain cases it can be a means of conveying intravascular treatments to the fetus.

The reported rate of fetal loss oscillates between 0.5 and 1.9 percent.[200] Complications associated with the procedure, though rare, are as follows: fetal bradycardia, transitory tachycardia, and umbilical cord hemorrhage. Indications for cordocentesis include Rh isoimmunization; monitoring fetal status during the course of infections, caused for example by cytomegalovirus, toxoplasma, and parvovirus B19; hemocoagulatory pathologies (alloimmune and autoimmune thrombocytopenia, if there was previous maternal splenectomy); delayed intrauterine growth; malformations (to karyotype the fetus rapidly); anemia and fetal dyscrasia; and the ascertainment of the metabolic status of the fetus.[201] The assessment of the risk associated with cordocentesis depends significantly on the indication for which the test is performed,[202] and its use, in any case, should be limited to the acquisition of information that is really necessary for modifying obstetric therapeutic action where possible.

Results and outcomes after genetic testing

It is important to understand just what is discovered by the genetic tests performed following an indication identified during consultation. In other words, what is the percentage of reassuring (negative) results with respect to inauspicious (positive) ones? But it is also a matter of knowing what happens, statistically speaking, in cases where the test result is positive and indicates the presence of a malformation

---

[199] F. Daffos, M. Capella-Pavlowsky, and F. Forestier, "A New Procedure for Fetal Blood Sampling In Utero: Preliminary Results of Fifty-Three Cases," *American Journal of Obstetrics and Gynecology* 146 (1983): 985–987; C. P. Weiner, "Cordocentesis," *Obstetrics and Gynecology Clinics of North America* 15 (1988): 283–301.

[200] K. H. Nicolaides, "Cordocentesis," *Clinical Obstetrics and Gynecology* 31 (1988): 123–135; F. Daffos, M. Capella-Pavlowsky, and F. Forestier, "Fetal Blood Sampling During Pregnancy with Use of a Needle Guided by Ultrasound: A Study of 606 Consecutive Cases," *American Journal of Obstetrics and Gynecology* 153 (1985): 655–660; M. J. Whittle, "Cordocentesis," *British Journal of Obstetrics and Gynaecology* 96 (1989): 262–264; G. Noia et al., "La cordocentesi: Indicazioni, utilità e rischi," *Medicina e Morale* 41, no. 4 (1991): 625–640.

[201] Noia et al., "La cordocentesi"; L. Mobili and A. R. Morgani, "Ruolo della cordocentesi in medicina prenatale," *Ultrasonica* 2 (1995): 37–39; C. Giorlandino, "La funicolocentesi," in *Trattato di diagnosi prenatale*, Giorlandino, 49–63.

[202] D. J. Maxwell et al., "Fetal Blood Sampling and Pregnancy Loss in Relation to Indication," *British Journal of Obstetrics and Gynaecology* 98 (1991): 892–897.

or serious condition. All of this is relevant for the purposes of an ethical judgment. It goes without saying that, in the case of genetic tests, we are dealing mainly with conditions that are incurable as of now. By contrast, in the case of merely somatic malformations identifiable through other procedures such as ultrasounds—e.g., cardiac or renal malformations or the presence of hernias—it is possible to intervene with neonatal surgery, and the de facto link to abortion is rarer.[203]

Nonetheless, some malformations due to genetic defects—specifically spina bifida and hydrocephaly, which are usually associated with one another—can be surgically treated in the first days of life with good results not just for survival but also for quality of life.[204] This means that in this last case the theoretically available paths after diagnosis do not consist solely in the possibility of abortion or, on the other hand, the pure and simple acceptance of the birth of a malformed child (or an individual affected by a genetic disorder), but there is also a therapeutic possibility. This is of great significance in terms of the ethical judgment on whether the genetic test can be used by those who do not accept in conscience the possibility of the abortion outcome. It is expected that, with the progress of science, the possibility of following up genetic diagnosis with treatment will become increasingly frequent.

Ethical guidelines for prenatal genetic diagnosis

Among the various forms of prenatal diagnosis, prenatal genetic diagnosis is the one most fraught with ethical problems, and it has already been shown why: genetic disorders offer few prospects of recovery as of yet (malformations due to spina bifida were mentioned), and therefore an inauspicious diagnosis leads to selective abortion so frequently that it could almost be thought of as a consequence.

That being the case, the ethical problem arises not only with regard to the family but also with reference to the specialists who collaborate in the diagnosis: indeed, it is a matter of understanding whether or not this sort of diagnosis becomes a form of cooperation with abortion. This is an acute problem strongly experienced by those specialists who, whether believers or nonbelievers, are against procured abortion and specifically against selective abortion. Sometimes the connection with an abortion outcome, in the case of an unfavorable diagnosis, is not only determined by the family's choice but also carried out by programs set up by local health care authorities that disguise their real intent under the pretext of the "prevention" of genetic disorders.

Before presenting the ethical principles in this area, in light of the recognition of the fetus's real dignity as a human subject, even when the fetus is the genetic carrier of a malformation or a disorder, I wish to list the various policy positions expressed in the literature and put into practice in different countries.

1. First of all there is the position of those who take it for granted, on the basis of cost-benefit reasoning, that these subjects should not be allowed to live because they are a burden on their families and on society. They also affirm that it is not worthwhile to perform prenatal genetic diagnosis in accordance with that same

---

[203] A. Calisti, "Diagnosi prenatale e possibilità terapeutiche chirurgiche," *Medicina e Morale* 34, no. 4 (1984): 493–497; Noia, Caruso, and Mancuso, *Le terapie fetali invasive*.

[204] C. Di Rocco, "Problemi decisionali relativi al trattamento chirurgico delle malformazioni congenite del Sistema Nervoso Centrale," *Medicina e Morale* 34, no. 4 (1984): 488–492.

principle of the assessment of costs (and risks) and benefits. They say that prenatal genetic diagnosis is costly and furthermore risky for the mother and for healthy fetuses in a given percentage of the cases; therefore it is more economical and less risky to bring the pregnancy to term and then, in the event of a severe malformation, to perform *neonatal euthanasia*, that is, to refrain from feeding the newborn with the consent of the parents.[205] The thinker behind this theory, Henry David Aiken, clearly expresses the underlying reasons:

> Claims regarding the right to biological survival are entirely contingent upon the ability of the individual in question to make, with the help of others, a human life for himself. This means that in circumstances where there exists the possibility of anything approaching a truly human life, the right to biological or physical survival loses its own *raison d'etre* and, hence, that the merciful termination of life, in the biophysical sense, is acceptable or perhaps even obligatory. … The rights of parents to accept the onus of caring for a radically defective child which has no capacity for enjoying a human life may be acknowledged. But then this is owing not to the child's right but to the human rights of its parents. And when the care of such a child seriously endangers the well-being of others, including in particular the members of its own family, parental rights … must give way to other, more exigent claims.[206]

It is very rare to read such clear statements filled with inhuman cruelty. In the United States in 1983, the case of Jane Doe, who was born with spina bifida and hydrocephaly and left to die without nourishment, made this theory very timely and controversial.

2. The second position, which is common in the utilitarian mentality that is widespread everywhere and not just in English-speaking countries, can be summarized in position taken by Neil MacIntyre, the geneticist who first studied the possibility of in vitro culturing of cells suspended in amniotic fluid: in order to prevent the birth of a terribly defective child and the emotional and economic destruction of the family, abortion is the better of the two unfortunate choices.[207] Although the position implicitly recognizes the fetus's right to life, the latter is subordinated to the consideration of the subject's physical quality of life and to the family's economic and emotional stability.

3. The position of the Protestant churches, expressed by the World Council of Churches in 1981, is very close to the previous position: the right to life is subordinated to the quality of life; wherever the latter is seriously deficient, the right to request an abortion belongs to the parents (right of the parents). This position was mitigated two years later by no longer appealing to the right of the parents, but primarily to freedom of conscience and the cost-benefit comparison based on the definition of the fetus as a potential person.[208]

---

[205] See H. D. Aiken, "Life and the Right to Life," in *Ethical Issues in Human Genetics*, ed. B. Hilton (New York: Plenum, 1973), 173–183; see also A. Serra, "Problemi etici della diagnosi prenatale," *Medicina e Morale* 32, no. 1 (1982): 52–61.

[206] Aiken, "Life and the Right to Life," 180.

[207] M. N. MacIntyre, "Professional Responsibility in Prenatal Genetic Evaluation," *Birth Defects Original Article Series* 8, no. 4 (1972): 35.

[208] C. Birch and P. Abrecht, *Genetics and the Quality of Life* (Oxford:Pergamon Press, 1975).

4. It is also important to note the position of several pro-life movements, particularly in the United States, where abortion following prenatal diagnosis has been more widespread: the pure and simple rejection of genetic diagnosis was suggested as a reaction to this, because it was said that genetic diagnosis was either useless or else directed to the selection of fetuses. This position became attenuated after the publication of the instruction *Donum vitae* by the Congregation for the Doctrine of the Faith, which considers prenatal diagnosis acceptable under certain conditions, as will be seen in the next point.

5. The fifth position is open to prenatal diagnosis, but with certain conditions, and is based on full consideration for the fetus as a person and on the geneticist's rightful freedom of conscience, even with respect to the eventuality—inadmissible a priori—of the woman and the family having recourse to abortion. Since this is the position I consider most in keeping with resolving the problem, even though it is the most difficult to apply, I wish to devote a more extensive discussion to it.[209]

This position ascribes certainty to the biologically and rationally demonstrable fact that the embryo or fetus is a human subject and enjoys both the full human dignity and the full right to life that must be acknowledged in every human being. The fact that the growth and development of the subject are not yet fully complete can allow talk of potential development but not of potential personhood or potential humanization: in the embryo, as in the fetus, an individual *already exists* who, if allowed to live, completes the development proper to the human person. Regardless of any philosophical reflections on the moment of ensoulment and on the various theories of personhood, it is beyond a doubt that this biological reality, embryo or fetus, constitutes that individual and fundamental value of individual life without which there can be no human subject. The subject is biologically and ontologically defined in this reality and therefore constitutes the primary and indispensable foundation for the realization of all the other values and rights of the person. These concepts were already made clear by the Catholic Church in the 1974 declaration *Quaestio de abortu* on procured abortion by the Congregation for the Doctrine of the Faith; moreover, the position of the Catholic Church leaves no uncertainty about the condemnation of voluntary abortion, which is made evident through a whole series of documents and pronouncements and has a rational foundation.[210]

At the objective level, the fact that the fetus may have malformations or be a carrier of even a serious illness does not diminish—and in fact aggravates—the offense against human life and dignity. All of the international documents relative to the recognition of the rights of handicapped persons affirm the full dignity of human subjects with handicaps as equal to the dignity of those without handicaps and, if anything, they also establish the need to provide greater assistance to those who are less physically autonomous. The selection of fetuses represents a mindset and a practice of domination by those who are physically able over those who are less so, and it shares in the seriousness of racism despite its roots in hedonism.

It has been fairly noted that affirming a *right* to quality of life could lead to the eventual legal situation of someone considering himself to be in a condition of

---

[209] This position was also expounded in the previously cited *Medicina e Morale* 34, no. 4 (1984), which provides interdisciplinary contributions.

[210] See G. Caprile, *Non uccidere: Il Magistero della Chiesa sull'aborto* (Rome: La Civiltà Cattolica, 1973).

wrongful life pressing charges in court against those (parents or health care workers) who did not intervene to end his existence, with a corresponding request of compensation for damages.[211]

Precise guidelines on the acceptability of prenatal diagnosis are found in the Magisterium of the Catholic Church,[212] especially in two documents. The instruction *Donum vitae* by the Congregation for the Doctrine of Faith on February 22, 1987, affirms that "such diagnosis is permissible, with the consent of the parents after they have been adequately informed, if the methods employed safeguard the life and integrity of the embryo and the mother, without subjecting them to disproportionate risks. But this diagnosis is gravely opposed to the moral law when it is done with the thought of possibly inducing an abortion depending upon the results: a diagnosis which shows the existence of a malformation or a hereditary illness must not be the equivalent of a death-sentence."[213] In the encyclical letter *Evangelium vitae* on March 25, 1995, John Paul II expressed the following warning:

> Special attention must be given to evaluating the morality of prenatal diagnostic techniques which enable the early detection of possible anomalies in the unborn child. In view of the complexity of these techniques, an accurate and systematic moral judgment is necessary. When they do not involve disproportionate risks for the child and the mother, and are meant to make possible early therapy or even to favour a serene and informed acceptance of the child not yet born, these techniques are morally licit. But since the possibilities of prenatal therapy are today still limited, it not infrequently happens that these techniques are used with a eugenic intention which accepts selective abortion in order to prevent the birth of children affected by various types of anomalies. Such an attitude is shameful and utterly reprehensible, since it presumes to measure the value of a human life only within the parameters of "normality" and physical well-being, thus opening the way to legitimizing infanticide and euthanasia as well.[214]

It follows from all of these premises that it is not only the woman or couple that commits a wrongful act by proceeding with genetic testing with the intention of having an abortion in the event of an unfavorable result: those associated with the decision who consent to cooperate with it also act, objectively speaking, in a gravely wrongful way.

Therefore the geneticist who is called to provide his professional services would commit a wrongful act if he *were certain* beforehand that this service of his would become a preparatory act for an abortion. The case of cooperation with an

---

[211] W. H. Shaw, "To Be or Not To Be? That Is the Question," *American Journal of Human Genetics* 36 (1984): 1–9, referenced in Serra, "La diagnosi prenatale," 445.

[212] Congregation for the Doctrine of Faith, *Donum vitae*, part I, no. 2; Catholic Bishops' Joint Committee on Bioethical Issues, *Antenatal Tests: What You Should Know* (London: Incorporated Catholic Truth Society, 1989), 50–56; *Catechism of the Catholic Church*, nos. 2274–2275; John Paul II, Encyclical Letter *Evangelium vitae*, March 25, 1995, no. 63; Pontifical Council for Pastoral Assistance to Health Care Workers, *Charter for Health Care Workers* (Boston: Pauline, 1995), nos. 59–61.

[213] Congregation for the Doctrine of the Faith (CDF), *Donum vitae*, part I, no. 2

[214] John Paul II, *Evangelium vitae*, no. 63

evil action comes up again here: it is not only wrong to commit an evil action, but it is also wrong to cooperate in one.

As is known, cooperation with evil can be *formal* or *material*. The former happens when there is participation not only at the level of actions but also at the level of intentions; the latter happens when the connection occurs through the actions, but the cooperator may either be unaware of that connection or in any event does not seek it, want it, or cause it. Material cooperation in turn can be *direct* or *indirect* (others say proximate or remote): in the former case, the action of the cooperator is in operative unity with the action of the primary agent; in the latter case, there is a gap between the action of the primary agent and that of the cooperating agent such that the primary agent's activity can have multiple aims and not a single and inevitable outcome. In the case of indirect material cooperation, the action of the primary agent could take several directions and is not limited to the single one in accordance with his own decision, to which the cooperator's action is neither necessarily nor deliberately connected. In itself, the cooperation of those who perform the diagnosis can be included in this latter situation.

Nevertheless the concrete cases of prenatal genetic diagnosis can involve various situations in relation to the subsequent acceptance of life, which I would like to illustrate and summarize as follows:

a. There can be the case of someone sharing the willingness to abort of the woman who requests the diagnosis. In the event of an unfavorable result, the woman reveals the decisive and explicit will to abort and, if the specialist performing the diagnosis shares this attitude, it is formal cooperation: the action is wrong and wrongly performed by both. Even though the action of abortion in fact depends on the woman's decision and not on the diagnosis, the shared intention joins the two agents in a single action.

b. There can be the case—and it is the most frequent one—where the woman's conduct after the prenatal diagnosis cannot be predicted with certainty and it is impossible to ask the woman ahead of time what her future conduct in that regard will be. Above all, the result of the diagnosis itself is unknown when beginning the diagnosis, and in fact about 94 percent of diagnoses provide a reassuring or negative result; furthermore, the woman does not have a definitive position when she feels anxiety and even the most firm and negative affirmations should not be considered unalterable. Finally, from the objective viewpoint, the diagnostic action is not directly connected with the abortion outcome: the woman, particularly if assisted, could always accept the unborn child even if the child is deformed, and the abortion outcome does not necessarily follow from the diagnosis. Asking the woman what her attitude would be in the event of an unfavorable result does not prove to be a practicable course of action because it could lead to a hypocritical position and, moreover, it is difficult to predict a state of mind when the real conditions are lacking.

c. Another situation is where, regardless of the intentions and good will of the specialist, there is already a prearranged and preestablished directive in the program enacted by the health care structure, which the woman has accepted de facto, whereby termination of pregnancy is prescribed in the event of an unfavorable result of the diagnosis. In this case of screenings followed by planned termination of pregnancy, the cooperation provided by the specialist is direct cooperation with the preestablished program; it is therefore the equivalent of a wrongful act on the part of someone who is bound to respect the life of the unborn child.

The condemnation in this regard found in the cited instruction *Donum vitae* is firm: "In conclusion, any directive or programme of the civil and health authorities or of scientific organizations which in any way were to favour a link between prenatal diagnosis and abortion, or which were to go as far as directly to induce expectant mothers to submit to prenatal diagnosis planned for the purpose of eliminating foetuses which are affected by malformations or which are carriers of hereditary illness, is to be condemned as a violation of the unborn child's right to life and as an abuse of the prior rights and duties of the spouses."[215]

As stated, the second case mentioned above is the most common case: many women, about 20 percent, refuse the diagnosis after a simple consultation, recognizing that there are no reasons for apprehension; many others, in fact most (over 90 percent), are relieved after the diagnosis since it gives a high percentage of reassuring results. The percentage of women who receive inauspicious results can find assistance from the genetic consultation team in order to confront the decision responsibly.[216] In order for the specialist to feel he is complying with his duty and with respect for life in cases where the woman has not yet made a decision either for or against an abortion, but also in the cases where the woman appears to be leaning toward abortion, he must inform the woman herself not just of the truth about the diagnostic report, but also of the *complete truth about the life of the fetus*: that the fetus is a human being who can be helped, also pointing out the means of support, social assistance, and structures available for welcoming the life of the unborn child upon birth.[217]

An eventual negative decision by the woman—that is, for abortion—should be of concern to the geneticist who respects the truth and life of the unborn child, just as the surgeon is concerned about the possible failure of one of his surgical interventions through the death of the patient after an operation in which he had done everything in his power to save the patient. In this case, full responsibility for the negative decision would fall on the woman or the couple that decided in favor of termination and death after having requested to know the health status of the unborn child, after learning the scientifically confirmed truth, and after having been offered the help necessary to accept that child dutifully. And this responsibility is as objectively grave as taking the life of an innocent person.

For scientists, it is a question of facilitating the *meeting of truth and life*: it is a role that cannot be carried out with aseptic neutrality, even when the freedom of others must be respected. This is because, as explained in the chapter on the principles of bioethics, dutiful respect for the freedom of others cannot entail the abdication of one's own freedom or responsibility. Scientific truth and the values of human life, freedom, responsibility, the patient's conscience, and the physician's conscience are

---

[215] CDF, *Donum vitae*, part I, no. 2

[216] See the contributions by Caffarra and Leuzzi in *Medicina e Morale* 34, no. 4 (1984) as well as in the editorial written by Sgreccia in the same issue.

[217] For further study on the topic of prenatal genetic counseling as a unique relationship of assistance, see Serra, "La consulenza genetica"; V. Mele, A. R. Morgani, and C. Giorlandino, "Etica 'pratica' e counseling genetico," in *Trattato di diagnosi prenatale*, Giorlandino, 171–183.

the most fundamental terms of the physician's ethical conduct and are all called into play in this complex and delicate question. That is why I have defined it as almost emblematic in the field of bioethics.

In finishing its work on February 9, 1987, the ministerial commission on ethical problems in genetics established at the Italian Ministry of Health by Minister Degan's decree on September 13, 1985, offered the following general principles for guidance:

The Commission,

- *Given* that, in principle and in the ethics shared by civilized countries and the Italian legal system, the right of a conceived human individual to life and health is recognized;

- *Given* that "healing" or "preventing" by destroying disadvantaged lives is a defeat not just for humanity, but for scientific progress, both in terms of organized treatment and psychological support;

- *Considering* that in the majority of cases prenatal diagnosis can reassure the mother and the couple, and even in the event of unfavorable results always leaves open the possibility of therapeutic and supportive interventions;

- *Expresses* a positive opinion, in principle, regarding the promotion of research and services related to prenatal diagnosis with therapeutic aims;

- *Requests* that diagnosis be mandatorily accompanied, in accordance with articles 1, 2(d) and 5 of law no. 194 (May 22, 1978), by exhaustive notification by health care personnel of the existing public, private, and volunteer forms of assistance for disadvantaged children and their families, in the context of a general attitude, on the part of the medical profession and the civil community, of responsible and supportive acceptance of human life that has begun;

- *Requests*, additionally, that in this context premarital and preconception diagnosis be encouraged, and more generally that every appropriate initiative intended to form a civil and technical culture favorable to the recognition of the fundamental value of life be taken and developed;

- *Maintains* that the principles delineated above must be confirmed by the legislature and not left up to ethical committees deciding on a case-by-case basis, since they are principles that are independent of the development, however rapid it may be, of diagnostic and therapeutic procedures.[218]

And in its document published on July 18, 1992, the Italian NBC maintains a certain line of neutrality: it notes the different positions on the matter of preimplantation diagnosis and in the area of potential links to abortion. The earlier controversy between pro-abortion and anti-abortion proponents is further complicated today by the divergence between supporters of the gradualist theory and the personalist position on the subject of the identity and status of the embryo and fetus,[219] which is a point that will have to be examined more thoroughly in the chapter on abortion.

---

[218] Italian Ministry of Health, "Relazione della Commissione," in *Documenti conoscitivi*, Luzi.

[219] Italian National Bioethics Committee, *Diagnosi prenatali*, 41–44.

Genetics

Beginnings

G. Mendel (1856–1863) studied the laws governing the transmission of phenotypical characteristics.

In 1902, a 26-year-old American researcher, W.S. Sutton, who was studying the behavior of chromosomes in the processes of mitosis and meiosis, aware of Mendel's hypothesis, postulated that the "distinct particulate units" described by Mendel must be found in the chromosomes, each of which, therefore, must carry more than one of such "character units"; and in 1909 W. Johansen gave to these the name of "genes."

By 1913, it was no longer possible to doubt the material existence of the "gene," which had to carry a specific bit of "information."

There were four phases in the research aimed at *analyzing genetic material at the molecular level*:

1. The *first line of research* led to the hypothesis, which was proved definitively between 1940–1952, that the *nature* of the substance that carries "genetic information"—and hence is an essential constitutive part of the gene—is chemically deoxyribonucleic acid (DNA).

2. The *second line of research* contributed to knowledge of the particular chemical *structure* of DNA, which then opened the way to an analysis of the gene. E. Chargaf declared that it was made up of four nitrogenous molecules (Adenine, Guanine, Cytosine, Thymine); in 1953 D. Watson and F.H. Crick developed a hypothetical model of its *structure*, the double helix.

3. The *third line of research*, guided in particular by H.G. Khorana and M. Nirenberg, was aimed at *deciphering the genetic code*, that is, the language in which that information is written, and explaining how the body *decodes* it.

4. The *fourth line of research*, the *artificial synthesis* of individual genes and the analysis *of their activity*, was to be further evidence for what had been proved to date. After H.G. Khorana and his twelve collaborators synthesized the DNA sequence that coded for alanine tRNA, in 1976 T.H. Maugh synthesized a perfectly functioning little gene that coded for tyrosine tRNA. The artificially produced gene, incorporated into a bacteriophage that had undergone a mutation at that same gene, functioned perfectly, correcting the error that was preventing its proliferation. It was irrefutable evidence of the structure and function of the gene.

The Human Genome Project

As early as the 1970s, the need was felt to be able to read the entire original "text" of the DNA, in other words, the sequence, one after the other in the individual chromosomes, of the almost three billion molecules of the four types—adenine (A), thymine (T), cytosine (C), guanine (G)—that make up the code's alphabet, to count the genes and to define their code, and finally to analyze and understand their activity. In 1973, the era of *genetic engineering* began.

From 1973 to 1989, this focused on *genetic mapping*, that is, determining the exact position of individual genes on the various chromosomes. The results of these fifteen years are presented in the eleven volumes of the International Workshops on Human Gene Mapping.

This work of mapping genes was intensified in 1989, when the Human Genome Project (HGP) was begun as a powerful, widespread organization that facilitated international cooperation and allowed for input from major industrial groups.

The project achieved its ambitious goal: *the sequencing of the human genome* (International Congress "Beyond the Genome 2000," San Francisco, California).

In the design and development of this undertaking, which was carried on competitively by two prestigious institutions, the National Human Genome Research Institute (a public institution) and Celera Genomics (a private company), the oncologist and Nobel prize-winner *Renato Dulbecco* initially played a decisive role. This led to a complete reading of human DNA, in other words, to knowledge of the exact sequence, in terms of individual chromosomes, of the almost three billion molecules of the four types (adenine, thymine, cytosine, and guanine) that make up the alphabet (*"genetic code"*).

Then the phase of recognizing and completely mapping the thirty or forty thousand genes that are "written" in this alphabet began.

Annotation and function of genes is the third phase, which is still in progress. The genome must be "annotated," that is to say, the individual genes must be assigned to the various chromosomes. This phase has already provided significant information such as the genetic causes of hundreds of diseases, especially tumors.

Research is now in the *postgenomics* phase, which provides a glimpse of even more difficult and remarkable prospects.

The identification and mapping of all the genes, the study of their specific functions and the analysis of their possible pathogenic mutations constitute the backbone of research in this area. But other, broader prospects for research and clinical application are beginning to open up. Specifically, the three most interesting and important areas would be: (1) *proteomics*, (2) *functional genomics*, and (3) *pharmacogenetics and pharmacogenomics* (see A. Serra in www.portaledibioetica.it).

Ethical problems connected with genetics

Essential problems

Limiting ourselves to the field of human studies, the essential problems that require particular ethical reflection can be grouped into three categories:

- Problems connected with *research*, as such
- Problems connected with *diagnostic applications*
- Problems connected with *therapeutic applications*

366

The ethical perspective can be traced back to four principles.

1. An ever broader and deeper knowledge of the structure and the functions of the human genome is *an invaluable good.*

2. *Scientific study must be subject to behavioral criteria based on ethical reflection in the most rigorous sense of the word.* The very *structure of science* requires observation, analysis, hypothesis, and experimentation; in other words, it implies the contribution of the researcher. Therefore, since every act of the person must be a *responsible act,* it cannot be independent of a *norm of action,* which must proceed from *within* man himself and not merely be imposed from outside.

3. *Technology,* in other words the application of the knowledge acquired from science, is a great *power* of man, which derives from his *right* to use the results of scientific research and his *duty* to apply them to the advantage of man himself and of society. Technology is an inestimable good. But the use of power, too, even more than research activity, requires a norm rooted *within* the individual who exercises it.

4. The human person with his *nature* and the *dignity* that follows from it, in other words, the integral reality that is the human subject, *dictates the norm of his action by his own intrinsic being.*

Risks for the embryo

There is a risk that the embryo will be considered an object with no ethical significance. Some maintain that it is necessary to experiment on the human embryo for the sake of the good that may result from it for mankind, and they go so far as to consider this a new requirement and a duty for science and medicine; the reasons that they offer are comprehensible only within a scientific and technological mindset which denies that the embryo has any value or *human* significance.

The eugenic *selective destruction* of embryos is declared a *good.*

This growing sense of capacity and power to rack up conquests has weakened the *sense of the boundaries* between what is *good* and what is *evil*; a *sense of responsibility* has been lost. The return to a just equilibrium requires a return to full, true knowledge of man, and this implies moving to a higher level of analysis that scientists or technicians cannot disregard without compromising their own great responsibilities toward society.

John Paul II referred to this vision in his address to the participants in the plenary meeting of the Pontifical Academy of Sciences on October 28, 1994, when he declared, "Nevertheless, we must not allow ourselves to be beguiled by the myth of progress, as though the possibility of conducting research or of applying a technique would immediately qualify them as morally good. The moral goodness of all progress is measured by its genuine benefit to man, considered in relation to his twofold corporal and spiritual dimension. ... Since man is at issue, the problems go beyond the area of science, which cannot take into account the transcendence of the subject nor lay down moral norms deriving from the subject's central place and primordial dignity in the universe."

Prenatal diagnosis

We can classify prenatal diagnosis (PD) techniques into two groups: invasive and noninvasive.

- A *noninvasive* technique allows for the analysis of the fetus "from the outside," without risks of altering or harming the mother or the unborn child. The noninvasive methods of PD most often used today are *the sonogram, the tri-test*, the evaluation of *nuchal translucency,* and in some cases the *bi-test.*

- An *invasive* prenatal diagnosis technique involves penetration of the uterine cavity.

The invasive PD methods most often used are *amniocentesis* (removal of a sample of the amniotic fluid), *chorionic villus sampling*, and *sampling of fetal blood*. These are not techniques of analysis but procedures for obtaining sample materials of fetal origin: cells, fluids, or biological tissues. The sample is then analyzed in the laboratory using biochemical, cytogenetic, or molecular techniques, depending on the case.

Noninvasive techniques

- An *ultrasound examination* is based on the ability of tissues to reflect particular sound waves called ultrasounds. This makes it possible to image the inside of the uterus and the fetus. Sonograms provide much information about the course of a pregnancy and the condition of the fetus.

- One examination recently developed is a sonogram for evaluating *nuchal translucency*, which shows whether there is an accumulation of liquid in the posterior part of the nape of the fetus. That condition can indeed be associated with a genetic problem.

- The *tri-test* is done preferably between the 15th and the 17th weeks of pregnancy; it is based on the concentrations of three biochemical markers present in maternal serum: AFP (alpha-fetoprotein), uE3 (unconjugated estriol) and hCG (human chorionic gonadotropin). The stage of pregnancy must be determined exactly, since the concentration of the hormones varies over the course of pregnancy. The combined evaluation of these three substances and of other parameters (maternal age, weight, smoking, etc.) can identify those women with an increased risk of giving birth to a fetus afflicted with Down syndrome, neural tube defects, or other chromosomal anomalies.

- The *bi-test* is preferable to the tri-test, although in this case, too, there is some risk.

Cytogenetic testing techniques

These are diagnostic tests that make it possible to determine the number and characteristics of an individual's chromosomes. For prenatal diagnosis, the chromosomes that are analyzed come from the fetal cells obtained from the amniotic fluid or chorionic villi.

Invasive sampling techniques

- *Amniocentesis* consists in the sampling of the amniotic fluid (the fluid surrounding the fetus inside the uterus) by means of a thin needle. Usually the needle is introduced through the abdomen; the whole operation is guided by ultrasound. In the amniotic fluid there are some fetal cells (called amniocytes), which are extracted and used for cytogenetic and/or molecular analyses.

Analysis of the amniotic fluid can yield important indications concerning the presence of genetic diseases. One protein in the amniotic fluid that is often measured is alpha-fetoprotein (AFP), which is produced by the fetus.

When the AFP level is elevated, this can indicate the possibility of fetal malformations, such as defects of the neural tube (spina bifida, anencephaly, or meningocele), defects of the abdominal wall, etc.

As in all invasive procedures, amniocentesis involves a certain percentage of risk of spontaneous abortion.

• *Chorionic villus sampling* is a procedure that consists of removing a minuscule piece of tissue from the placenta (or more precisely, from the chorion, the part of the placenta that "belongs" to the fetus).

This removal takes place transcervically (i.e., by way of the uterine cervix) or transabdominally (as in amniocentesis), depending on the position of the placenta. The fetal cells obtained by the sampling are then used for cytogenetic and/or molecular investigations.

The removal of chorionic villi involves a higher risk of abortion.

• *Cordocentesis* (funiculocentesis) involves the sampling of fetal blood from the umbilical cord. It is performed in order to diagnose some hereditary blood diseases, or to verify the state of the fetus's health in a case where the mother has contracted a serious infectious disease during pregnancy.

Molecular testing techniques

These are methods that study the DNA directly. They make it possible to diagnose known genetic alterations that are otherwise invisible to a cytogenetic examination. Most of the alterations, indeed, are so slight as to cause no visible modification in the structure of the chromosomes.

Genetic tests

Types of tests

We can classify genetic tests into four types:

1. *Diagnostic tests* serve to confirm the presence of a pathology (today they are quite common at the prenatal stage). The cognitive value and medical usefulness of a diagnostic test are well known: no one can ignore, however, the fact that they are also used as selective instruments.

2. *Presymptomatic tests* serve to identify the risk of pathology, in the absence of symptoms, in persons who belong to families in which a specific disease has already been manifested.

3. *Tests to identify healthy carriers* serve, through mass screening systems, to identify subjects who could be carriers of hereditary diseases (e.g., Mediterranean thalassemia or cystic fibrosis in Caucasians).

4. *Predictive tests* serve, in theory, to identify possible exposures to very common diseases that usually cannot be correlated with the presence of a single anomalous genetic factor. But *these tests have no therapeutic function per se*: knowing whether a disease arises, and how, does not always mean knowing how to cure or treat it as well.

Problems

Unlike diagnostic tests, all the other types indicate "a risk of disease," and hence the first effect of predictive tests is to transform onerously the lives of the persons who are placed in the "at risk" category.

These are persons who are not sick but, to the extent that they believe in the predictive power of science, will experience their years of health with the anxiety of losing it.

As a matter of principle, a citizen has both the right to forego genetic tests and the right to confidentiality, but how can these rights be exercised if a culture is created in which genetic tests are viewed as a sort of duty toward oneself and others?

All the resources employed in so-called *predictive medicine* are resources diverted from the treatment and care of real illnesses.

Diagnostic tests can be an aid to medical practice, but in order to intensify these lines of research it is not necessary to invent a predictive practice that runs the risk of becoming a means of social discrimination.

> "There is no need for a predictive genetic test to know that sooner or later someone is going to get sick and that sooner or later he is going to die; it suffices to be aware of our human condition as finite persons." (A. Pessina)

The nonmanipulability of causal origin

The German philosopher Peter Sloterdijk proposes "rules for the human park," which resembles an animal farm, or even a zoo.

Positions opposing Sloterdijk's theses have come from all over—Jürgen Habermas's is a prominent one—because the idea of a "stockbreeder" man and a "bred" man clearly violates the principle of equal dignity. In fact, Sloterdijk's reasoning that the humanistic project should be eliminated through the stockbreeding of human beings is already contradicted by the fact that geneticists would have to agree on the objectives, in other words, on what type of human being is truly desirable.

According to Habermas, "premeditated quality control" manipulates human life, which is generated with reservations on the basis of the preferences and values harbored by third parties. He demonstrates that interference with causal origin ignores the morality and rights that are based on mutual recognition as equals. From this it necessarily follows that only a teleological view of reality based on creation can protect causal origin, which in this case is not seen as blind chance but as Providence. Dependence on an equal can enslave us, whereas dependence on God—due to the absolute distance between Creator and creature—is a source of freedom.

It seems that postmetaphysical bioethics has perceived that so-called bioethical problems are often more accurately anthropological problems, but it does not consider it feasible to take an ontological approach to anthropology. It prefers an ethical route—meaningful, but insufficient for preventing science from being considered the only form of knowledge and human nature from being subjected to a continual process of redefinition.

Chapter 9

Bioethics, Sexuality, and Human Procreation

Scope and Current Relevance of the Topic

I must necessarily and deliberately remain within the perspective of bioethics, which is moral reflection strictly connected with biology and medicine, and therefore it will not be possible to devote space in this chapter to other perspectives that can be taken on the topic of sexuality. I am therefore obliged to refer the reader to more specialized studies for the biblical and theological perspective,[1] which will be mentioned only briefly even though it is of great interest even from the ethical and cultural standpoint. Similarly, I cannot devote time to treating the subject from the ethnological, historical, cultural, sociological, or even psychological standpoints—all areas in which we have witnessed the burgeoning of a large number of highly interesting research efforts, studies, and other contributions.

The purpose of this chapter is only to reconnect the topic of sexuality to the vision of corporality and of the person presented in the general section of this volume, and thus to establish the premises for the ethical analysis of the questions of responsible procreation, abortion, surgical sex-change operations, the treatment of sexual disturbances, sterilization, and reproductive technologies. This is therefore a delimited perspective, but one which pertains to the objective and perennial structure

---

[1] For basic information in this field, the reader is referred to the documents of the Magisterium that will be indicated later in this chapter. Here are just a few selections from the very extensive bibliography on the matter: M. Flick and Z. Alszeghy, *Fondamenti di un'antropologia teologica* (Florence: Libreria Editrice Fiorentina, 1970); B. Häring, "Sessualità," in *Dizionario enciclopedico di teologia morale* (Rome: Paoline, 1981), 993–1006; F. Salvoni, *Sesso e amore nella Bibbia* (Genoa: Lanterna, 1970); J. M. Aubert, *Sexualité, amour et mariage* (Paris: Beauchesne, 1970); J. Mouroux, *Sens chrétien de l'homme* (Paris: Aubier, 1945); D. Tettamanzi, "La sessualità umana: Prospettive antropologiche, etiche e pedagogiche," *Medicina e Morale* 2 (1984): 129–154; Centro Italiano Femminile (CIF), ed., *Uomo-Donna: Progetto di vita* (Rome: CIF / UECI, 1985); P. Barberi and D. Tettamanzi, eds., *Matrimonio e famiglia nel Magistero della chiesa* (Milan: Massimo, 1986); M.L. Di Pietro and E. Sgreccia, *La trasmissione della vita nell'insegnamento di Giovanni Paolo II* (Milan: Vita e Pensiero, 1989); S. Palumbieri, *Antropologia e sessualità: Presupposti per un educazione permanente* (Turin: SEI, 1996); A. Scola, *Il mistero nuziale*, vol. 1, *Uomo-Donna* (Rome: PUL / Mursia, 1998); A. Scola, *Il mistero nuziale*, vol. 2, *Matrimonio e famiglia* (Rome: PUL / Mursia, 2000); C. Caffarra, *Non è bene che l'uomo sia solo*, vol. 1, *Creati per amare* (Siena: Cantagalli, 2006).

of human life and is consequently valid regardless of cultural changes and the influence of ideologies. Important aspects regarding the ethics of premarital and marital sexual conduct are also left out of our analysis; however, moralists have already discussed and debated these questions at length.[2]

In order to historically contextualize these reflections, with their necessary limits, there are two cultural components important for understanding social attitudes and changes that cannot be overlooked. The first and more widespread component is hedonistic permissivism, of bourgeois origin and coextensive with the industrial and postindustrial culture, which can be summarized in the philosophy of "sexuality as consumption": sexuality without risks and without regrets.[3] It is the most effective vehicle for the contraceptive and pro-abortion mentality. This is the sociological context in which several factors of cultural change have been observed in the past fifty years, since the end of World War II, coinciding with the abovementioned phenomenon of industrialization. These cultural factors have had the positive effect of bringing discussion on the topic of sexuality to the anthropological level and overcoming the idea of sexuality understood as genitality, but on many social and cultural levels they have also done much to justify and increase the detachment of sexual conduct from any ethical norms (sexual liberalization) or even to regard this liberalization as the basic condition for total human liberation.

Let us report some of the phases of the move in this direction. The contributions of Sigmund Freud[4] were fundamental in affirming sexuality as a dimension of the whole person and in identifying the importance of sexuality in the individual's process of maturation and socialization. For Freud, however, this means not that the person expresses himself in the dimension of sexuality, but that sexuality expresses itself and, by its dynamics linked to the depths of the subconscious, structures personality. Normal and pathological manifestations of personality are "determined" by these dynamics, among which the Oedipus complex is primary. Even the most genius and spiritual manifestations of personality, including culture and art, are the products and constructs of these dynamics and of so-called defense mechanisms (sublimation, repression, flight, aggression, etc.). The very image of the family comes under suspicion as a web of repressions, instincts, and deeply rooted tensions.

This view overturned the classic understanding of sexuality in its relation to the person and above all engendered a new line of conduct. It would nonetheless be inaccurate to charge Freud with teaching libertinism and licentiousness in the area of sexuality. He explained the mechanisms that inhibit instinctual forces and proposed

[2] M. Vidal, *Morale dell'amore e della sessualità* (Assisi: Cittadella, 1973); B. Häring, *Free and Faithful in Christ* (New York: Crossroad, 1978–81), 2:493–571; G. Piana, "Orientamenti di etica sessuale," in *Corso di morale*, ed. T. Goffi and G. Piana (Brescia: Queriniana, 1983), 2:271–366; G. Rossi, "Il rapporto uomo-donna," in *Trattato di etica teologica*, ed. L. Lorenzetti (Bologna: Dehoniane, 1981), 2:399–458; A. Günthör, *Chiamata e risposta* (Rome: Paoline, 1984), 3:620–717; B. M. Ashley, *Theologies of the Body: Humanist and Christian* (Braintree, MA: Pope John XXIII Center, 1985).

[3] H. Schelsky, *Soziologie der Sexualität* (Hamburg: Rowohlt, 1955), referring to Häring, "Sessualità," 999; W. Barclay, *Ethics in a Permissive Society*, Glasgow-London 1971.

[4] See S. Freud, *The Standard Edition of the Complete Works of Sigmund Freud*, ed. J. Strachey, 24 vols (London: Hogarth, 1953–1974); A. Milano, "Freud critico della religione per l'autonomia della morale," *Aspernas* 3, no. 4 (1981): 397–425.

the mechanism of sublimation, but the fact remains that a pansexualistic and deter-ministic understanding of the person took its cue from Freud: sex is everything, sex takes orders from no one, and the psychological disturbances and sufferings of the person are due to nothing other than sexual repression. All traditional education is presented as one big repression, and the culture of duties is depicted as the result of a collective neurosis. Textbooks of psychology, pedagogy, art history, and religion, along with journals and films, have brought the world this revolution bereft of any transcendent vision, weighed down with a bitter "suspicion" and disquieting deter-minism at the expense of all manifestations of ascetic and religious life.

The existentialist philosophy of Jean-Paul Sartre and similar tendencies, which are present in literature and films, have contributed much to the nihilistic under-standing of morality and the exaltation of sexual experiences as a free expression and privileged form—if not the only form—of communication.

The explosive and justificatory effect produced by the publication of the so-called Kinsey Reports cannot be ignored.[5] The primary author, a zoologist, presents the results of a vast survey on sexual behavior—even in its most abnormal and aberrant forms—conducted in American society with statistics and percentages, concluding that sexual behavior is nothing more than a relatively simple mechanism that causes an erotic reaction when there is adequate physical and psychological stimulation: the "goodness or the badness, the morality, the immorality, the unmo-rality, or the amorality, the guilt or the innocence of the acts or the persons he has interviewed are no concern of his."[6] In the Weberian understanding of culture, the changing of social structures that occurs with the passage from a traditional agrar-ian society to an industrial society is inevitably accompanied by a changing of the cultural framework and therefore of behavior in terms of family life, premarital and extramarital sexuality, and conjugal life itself. A presupposition of sociological determinism lies at its root and emerges as the conclusion of this description: what happens to be is equated with what should be, that is, sociological reality is equated with morality, which can be expressed by the already-mentioned $is = ought$ formula, and has the potential for a "terrific impact."[7]

Another incentive on the path toward the liberalization of sex and its detachment from procreative responsibilities was the spreading and liberalization of contraceptives and in particular of the "Pinkus pill." Other than its effect of encouraging hedonism and challenging traditional conjugal sexual morality, this phenomenon—which burst on the scene right during the years of controversy—also had the effect of providing a weapon, as will be explained, for the politicization of plans for birth control.

---

[5] A. C. Kinsey, W. B. Pomeroy, and C. E. Martin, *Sexual Behavior in the Human Male* (Philadelphia: Saunders, 1948); A. C. Kinsey, W. B. Pomeroy, and C. E. Martin, *Sexual Behavior in the Human Female* (Philadelphia: Saunders, 1953). See the summary of these developmental stages of the culture regarding sexuality in L. Ciccone, "Etica della sessualità," in *Corso di Bioetica*, ed. E. Sgreccia (Milan: Angeli, 1986), 69–94, with bibliography; V. Melchiorre, ed., *Amore e matrimonio nel pensiero filosofico e teologico moderno* (Milan: Vita e Pensiero, 1976); Schelsky, *Soziologie*.

[6] D. P. Geddes, "New Light on Sexual Knowledge," in *About the Kinsey Report: Observations by 11 experts on "Sexual Behavior in the Human Male,"* ed. D. P. Geddes and E. Curie (New York: New American Library, 1948), 6; Ciccone, "Etica," 73.

[7] A. Kardiner, *Sex and Morality* (Indianapolis, NY: Bobbs-Merrill, 1954), 80, see 73–83.

Another innovatory contributing factor, still taken into consideration by some schools of thought although with less brutal forms and methods than the original ones, cannot be left out: the works in the area of sexological therapy by the American sexologists William Masters and Virginia Johnson, whose theories were then taken up by Helen Singer Kaplan and Robin Skynner.[8] Drawing on the precepts of positivistic behaviorism, the work by these two offered a study of the neurophysiological reactions in men and in women during sexual intercourse, "reducing" the dynamics of the unitive gesture to a complex network of recordable reactions. Measurements of these reactions were provided using scientific instruments, filmed recordings of them were distributed, and viewing them in the laboratory was suggested as a method of "treatment."

Some schools of thinking in Europe still rely on this positivist view in their approach to treating sexual dysfunction. Although they have abandoned certain methods, such as substitute partners, they nonetheless emphasize exercises and "prescriptions" that involve a heavy dose of biological reductionism, ignoring the inner emotional and spiritual component of the sexual and conjugal act and assuming a division between the therapeutic technique and the ethical values of the person.

Finally we must take into consideration the other ideology that has fomented the cultural climate in the same direction: the neo-Marxist ideology that was more limited to the educated and ideologically active elite, especially during the years of cultural upheaval and protest. Already for Marx, the family should be linked to productivity. The classic Russian ideologues strongly insisted on the need for family unity and integrity, but always as an element of the collective and in relation to productivity. Even children's games have to be understood as preparation for productive activity and training for work. A tragic indicator of this functional quality of the family in relation to productivity is precisely the fact that the first Marxist society would consider it necessary for women to enter the working world and for abortion to be legal when pregnancy interfered with the working woman's fulfillment.

For the neo-Marxists, the sexual revolution—perceived as the second phase of the social freedom revolution—must be the major factor in establishing a new society. In the new society, man should be freed not only from dependence on factory work but also from the erotic and emotional dependence that comes about in marriage and the dependence of the spirit that occurs in moral life. The primary representative of this ideology is Herbert Marcuse, who even ends up theorizing the liberation of sexuality from heterosexuality and talking about "polymorphism" and hence "free choice of sex."[9]

The rejection of the sexuality-conjugality-family nexus breaks the bond between love and life within the family and makes the reality of procreation entirely nonessential. The ideologized and radical vanguard of the feminist movement, starting with Simone de Beauvoir, theorized the right to abortion and contraception

---

[8] W. H. Masters and V. E. Johnson, *Human Sexual Response* (Boston: Little, Brown, & Co., 1966); Masters and Johnson, *Human Sexual Inadequacy* (Boston: Little, Brown, & Co., 1970); H. S. Kaplan, *The New Sex Therapy: Active Treatment of Sexual Dysfunctions* (New York: Brunner / Mazel, 1974); B. Abraham and R. Perto, *Psicoanalisi e terapie sessuologiche* (Milan: Feltrinelli, 1979); R. Skynner, *One Flesh, Separate Persons* (London: Constable, 1976).

[9] H. Marcuse, *Eros and Civilization: A Philosophical Inquiry into Freud* (Boston: Beacon Press, 1966); Melchiorre, *Amore e matrimonio*, 458 ff.

as a woman's right, adopting a neo-Marxist concept of women's liberation and demanding a sociopolitical role for women as an alternative to the domestic and family role.

The absolutizing of freedom and the affirmation of freedom without responsibility and purpose, which are extreme but "logical" consequences of the ideologies of Marcuse and de Beauvoir, have been among the predominant features of recent international conferences of the United Nations, particularly the one in Cairo in 1994 and the one in Beijing in 1995.

On these occasions, in fact, the theme of freedom "from" all conditions was proposed again by several feminist organizations supporting sexual permissiveness and abortion. In these same conferences, freedom "from" was therefore also the starting point for the ideological use of the expressions "women's rights," "sexual rights," and "reproductive rights," which were included among human rights in the final document in Beijing and extended to mean the possibility "to have control over and decide freely and responsibly on matters related to their sexuality, including sexual and reproductive health, free of coercion, discrimination and violence."[10] This does not eliminate the possibility that the so-called right to abortion might be recognized anyway, though covertly, in the name of such sexual or reproductive rights.

During the preparatory phase of the Beijing Conference, in the wake of this radical and liberal paradigm, the proponents of so-called ethics of social behavior—which equates historical and social reality with morality—even attempted to eliminate the biological difference between the two sexes—male and female—and to reduce it to a mere question of culture through the frequent use of the term *gender* instead of sex.

In other words, starting from the premise that women's lack of fulfillment is caused by their "difference" from men and that this difference does not depend on nature but is only the result of socially produced male and female roles, gender ideology holds that these roles may be changed.[11] Femininity, masculinity, heterosexuality, and maternity are therefore not "states of nature" but rather "artificial roles"; hence they are neither final nor determined.

Consequently, it has been maintained that each individual should feel free not just to choose his way of living, but also his own nature and sexual orientation: male, female, transsexual, homosexual male, and homosexual female. It is therefore a reductive and individualistic interpretation of sexuality, which is considered an exclusive product of the culture and is completely uprooted from corporality.[12] But even the body itself loses consistency in this process of deconstruction that is part and parcel of the more recent gender theories. Suffice it to think of the interpretation of Judith Butler, who opposes the division that associates sex with matter and gender

---

[10] United Nations, *Report of the Fourth World Conference on Women: Beijing, September 4–15, 1995* (New York: United Nations, 1996), no. 96, p. 36, available at http://www.un.org/womenwatch/daw/beijing/pdf/Beijing%20full%20report%20E.pdf.

[11] M. L. Di Pietro, "Identità sessuale e genere," *Anthopotes* 20, no. 1 (2004): 83–105.

[12] For further study see P. J. Elliot, "La prospettiva etica nella Conferenza ONU di Pechino sulla donna," *Medicina e Morale* 6 (1995): 1175–1182; M. L. Di Pietro, "Temi bioetici nelle Conferenze Internazionali delle Nazioni Unite," *Vita e Pensiero* 3 (1996): 175–188; Pontifical Academy of Sciences, *Population and Resources: A Report* (Vatican City: Pontificia Academia Scientiarum, 1994).

with the cultural element and proposes applying the same critique of gender to sex.[13] Consequently, if gender is a social and cultural construct imprinted upon the surface of the "sex-matter," then matter is also a construct, and if sex is matter, then sex too is a construct. Once bodies are deconstructed, they can change and transform in order to give rise to artificial "postmodern bodies," open to mixing and manipulation until the "cyborg myth" is achieved: "A cyborg is a cybernetic organism, a hybrid of machine and organism, a creature of social reality as well as a creature of fiction." [14]

Procreation, repressed by hedonism and devalued by Marcusean ideology, has also been made the object of control and planning in the context of global demographic policies. This is the second historical and cultural factor or component that must be taken into account.

Taking the theories of Thomas Robert Malthus [15] as its reference point, neo-Malthusianism went further and provided a theoretical basis for the antinatalist policies currently enacted in the world. The theories expounded by neo-Malthusians on the socioeconomic level are no longer even consistent with those of Malthus: the lack of energy sources or worsening pollution, which are the most recent motives for supporting birth control, are not inevitably connected to population increase and are not valid reasons for a worldwide birth control policy. However, despite the uncertainty of the demographic data (for example, the Population Reference Bureau predicted a global population of about 10 billion people by 2046,[16] whereas others predicted about 11 billion as early as 2025 [17]), some have predicted scenarios of extreme poverty—at the limits of survival—for all the inhabitants of the planet.[18] Various authors have declared Malthusian and neo-Malthusian theories unfounded, affirming that the available resources of the Earth are sufficient for everyone both now and in the future.[19]

---

[13] J. Butler, *Bodies That Matter: On the Discursive Limits of "Sex"* (New York: Routledge, 1993).

[14] D. Haraway, *Simians, Cyborgs, and Women: The Reinvention of Nature* (New York: Routledge, 1991), 149.

[15] T. R. Malthus was the first to propose the need for birth control because of the presumed lack of resources on the planet. His theory of geometric population growth and arithmetic resource growth, however, very quickly proved unfounded. Nonetheless, he proposed nothing other than marrying at a later age, continence, and coitus interruptus as solutions for population containment. See T. R. Malthus, *Essay on the Principal of Population* (London: J. Johnson, 1798), republished in electronic format in 1998 by Electronic Scholarly Publishing Project and available at http://www.esp.org/books/malthus/population/malthus.pdf.

[16] Population Reference Bureau, *World Population Data Sheet 1991* (Washington, DC: PRB, 1991).

[17] United Nations Department of Social and Economic Affairs Population Division, *World Population Prospects 1990* (New York: United Nations, 1991).

[18] See K. M. Leisinger and K. Schmitt, *All Our People* (Washington, DC: Island Press, 1994); M. L. Di Pietro and A. G. Spagnolo, "Sovrappopolazione come sottosviluppo? Alcune riflessioni per capire la posizione cattolica alla Conferenza de Il Cairo," *Bioetica* 2 (1995): 216–228.

[19] See "Population: Battle of the Bulge," *The Economist*, September 3, 1994, 19–21. See also the following reviews: L. Cantoni, *Il problema della popolazione mondiale e le politiche demografiche: Aspetti etici* (Piacenza: Cristianità, 1994); L. Cantoni, "Popolazioni e risorse: Alcune pietre d'inciampo," *Agricoltura* 269, no. 270 (1995): 2–19; R. Cascioli, *Il complotto demografico* (Casale Monferrato: Piemme, 1996).

The real reasons for these catastrophic predictions lie with the prevailing fear that the baby boom in developing nations might threaten the well-being and political welfare of the major economic powers. And neo-Malthusian politics, which are "aggressive" with respect to the Malthusian proposal, take shape concretely through the imposition of birth control—particularly on developing societies but also on the whole world—as a condition for economic assistance in development.

As early as 1968, Robert McNamara, then-president of the World Bank, declared, "As a developmental body we must give priority to problems of demographic growth and ask that governments wishing to invoke our aid do likewise, adopting policies that are able to stabilize the rate of demographic growth."[20]

Data supporting these theories emerged, for example, during the international symposium on world demographic policies organized by the Federation of Catholic Universities:[21] the population of Latin America doubled from 1950 to 1975 and the growth rate was the highest in the world; the rate of population increase during the same period ranged from a minimum of 35 percent to a maximum (in Venezuela) of 115 percent. The reduction in the rate of increase during recent years is due to private family planning organizations and subsequent programs adopted by the health departments in certain nations like Brazil. BEMFAM, a nongovernmental organization for "family welfare," operates in Brazil and carries out active propaganda for contraception.

In 1970, the United States government approved the Family Planning Services and Population Research Act, earmarking enormous sums of money for propaganda promoting contraceptive methods and characterizing the campaign as a fight against poverty. Despite the opposition of the Catholic world, the Agency for International Development (AID) promotes active support for birth control campaigns. The United States government, although its policies change and it proclaims the right of spouses to decide freely on the number of children to have, is nonetheless still the world leader in birth control campaigns.[22]

In this context, the pill has become a political weapon and a means of economic control. Confirmation of United States policy came at the third United Nations International Conference on Population and Development in Cairo, during which the Mexico City Policy—supported by the Reagan administration during the previous International Conference on Population (1984)—wavered. In fact, the autonomous right of families to plan births had been recognized in Mexico City, and it had been affirmed that "abortion would in no case be promoted as a means of family

---

[20] Cited in E. De Lagrange, M. M. De Lagrange, and R. Bel, *Il complotto contro la vita* (Milan: Ares, 1987), 114–115, originally published as *Un complot contre la vie: L'avortement* (Paris: Société de Production Littéraire, 1979). See also M. S. Teitelbaum and J. M. Winter, *The Fear of Population Decline* (Orlando: Academic Press, 1985).

[21] A. Fonseca, "Politiche demografiche e contraccezione oggi nel mondo," *La Civiltà Cattolica* 4 (1983): 134–148.

[22] On this international establishment of birth control policies, see M. Schooyans, *Aborto e politica* (Vatican City: Libreria Editrice Vaticana, 1991); M. Schooyans, *L'évangile face au désordere mondial* (Paris: Fayard, 1997), 225–255; E. Tremblay, *L'affaire Rockfeller: L'Europe occidentale en danger* (Reuil-Malmaison: UPN, 1979); De Lagrange, De Lagrange, and Bel, *Un complot contre la vie.*

planning."[23] One can understand better today, therefore, the value of the position of the Magisterium of the Catholic Church, particularly Pope Paul VI's declaration in *Humanae vitae* that life is inseparable from conjugal love, placing both of them outside man's dominion.[24] In his concern about similar policies, Pope John Paul II also denounced the catastrophic, unrealistic tone used by those who deal with demographic issues in *Evangelium vitae*.[25] He also pointed out the gravity of the policies enacted, which lead to "the destruction of so many human lives still to be born or in their final stage" and harm the dignity of spouses.[26]

In China the population exceeds one billion, and from 1975 to 1980 China had a birth rate of 21 percent; the government set a zero growth limit for the year 2000 through the legalization and imposition of abortion for a second child as well as through sterilization, social pressure, postponing marriage, the free distribution of contraceptives, and the entrance of women into the workforce.

The population of India (700 million inhabitants) doubles every 30 years, despite the desperate efforts of the government. It reached the point of requiring sterilization after a second child, but the law failed due to the fierce opposition of the people. The policies are now more often carried out through a cultural approach. The proposal of natural methods has met with great acceptance and considerable success in India.

Africa is the place with the highest birth rate and the highest infantile death rate—the total population grows 3 percent annually—along with the highest rate of poverty.

Europe, by contrast, has been experiencing a marked demographic decline over the past few decades. In Italy the trend is close to zero population growth, with sharp differences between the North and the South. Similar situations are found in France, Germany, and Northern Europe. In Eastern Europe, where Russia was the first to introduce legalized abortion in 1920, there is now an effort to limit abortions and government encouragement of demographic growth; nonetheless, there is still a decline in birth rates and population growth in these countries. It is increasingly common on the contemporary scene for many different voices to sound the alarm about the phenomenon of demographic decline, to the point that today people are talking about a second demographic revolution and pronatalist policies have been proposed and enacted in some countries.[27]

The summary of these considerations is that a uniform policy capable of guiding these phenomena has not been found anywhere in the world and, more importantly, it is not anticipated that government policies will be able to reconcile complex and diametrically opposed variables: the poorest countries are the wealthiest in terms

---

[23] Di Pietro and Spagnolo, "Sovrappopolazione"; Cantoni, *Il problema della popolazione*, 87–102.

[24] For considerations on the dominion over the process of procreation, see E. Sgreccia, "Divieto morale e profezia," in *Il dono della vita* (Milan: Vita e Pensiero, 1987), 205–211.

[25] John Paul II, Encyclical Letter *Evangelium vitae*, March 25, 1995, no. 16.

[26] Ibid., no. 4. See also *Commento interdisciplinare alla "Evangelium vitae,"* ed. E. Sgreccia and R. Lucas Lucas (Vatican City: Libreria Editrice Vaticana, 1997).

[27] See Pontifical Council for the Family (PCF), *Ethical and Pastoral Dimensions of Population Trends* (Vatican City: Libreria Editrice Vaticana, 1994), no. 8. See also Teitelbaum and Winter, *Fear of Population Decline*.

of increasing numbers of births and demographic growth; the affirmation of the freedom of spouses is not supported with matching policies, which instead tend to impose conditions and choices made by the government.[28]

Global demographic policies are torn between these two slogans: contraception fights poverty and economic development is the best form of contraception.

Behind this problem, which touches the lives and fates of many peoples, I believe there is an ethical call. In other words, the problem can only have an ethical solution, through education toward responsible procreation: not biologistic spontaneity, not freedom without ethical norms, not coercive legislation, but rather the responsible and conscientious conduct of love and life in marriage.[29]

Medicine has gradually been compromised by ideologies and antinatalist policies and ultimately by the laws, at the risk of being reduced to an obsequious tool either for the prevailing policies or the current mentality: contraception, abortion, and sterilization require the intervention of medicine and of physicians.

It is important to note the escalation that has ensued in the implementation of birth control policies: it began with propaganda and the distribution of contraceptives, moved to the second phase of abortion, then reached the point of voluntary sterilization.[30] This gradual progression was not by chance: the introduction of abortion requires a *contraceptive* cultural context. In other words, it is necessary to first create the mentality regarding births as something bad that should be avoided—the *anti-life mentality*; once this mentality is established, abortion becomes the way out for unplanned pregnancies; finally, sterilization is presented as both the safer and less traumatic option for avoiding pregnancies and ultimately as an ideologically liberating event as well. These cultural changes have required propaganda campaigns, particularly where the cultural traditions were or are more resistant. The search for the most effective and easiest means of social dissemination has led to stimulating pharmacological research for the production of contraceptives with fewer health risks and abortifacient drugs that can shift the problem of abortion into the home.[31]

In the first years of the new century, however, an initial review of international policies seemed to be occurring in the United States with regard to the choice of methods to be promoted and in particular to containing the spread of abortion, but it is too early to determine the effectiveness and duration of this new approach.

---

[28] PCF, *Ethical and Pastoral Dimensions*; M.L. Di Pietro, "Bioetica e demografia," in *A sua immagine e somiglianza?*, ed. A. Mazzoni (Rome: Città Nuova, 1997): 323–344.

[29] See John Paul II, "All Demographic Policies Must Respect Man," in *Insegnamenti di Giovanni Paolo II* (Vatican City: Libreria Editrice Vaticana, 1984), 7.1:1625–1636 [Italian]; PCF, *Ethical and Pastoral Dimensions*; John Paul II, *Evangelium vitae*, nos. 16 and 91.

[30] See the decision by Italy's Supreme Court of Cassation on June 18, 1987, in which it was pronounced that voluntary sterilization even for hedonistic aims does not constitute a crime. This will be discussed at greater length in chapter 12.

[31] Bioethics Center of the Catholic University of the Sacred Heart, "Sulla cosiddetta 'contraccezione d'emergenza,'" *Medicina e Morale* 3 (1997): 582–589; M.L. Di Pietro and R. Minacori, "'Contraccezione d'emergenza': Problema medico, etico, giuridico," *Vita e Pensiero* 5 (1997): 353-361; M.L. Di Pietro and M. Casini, "Il mifepristone," *Medicina e Morale* 52, no. 6 (2002): 1047–1079; M.L. Di Pietro et al., "Norlevo e obiezione di coscienza," *Medicina e Morale* 53, no. 3 (2003): 411–455.

Developments in Catholic Morality on the Topic of Sexuality

Before providing an overview in the field of bioethics and embarking on a discussion of the specific issues, I also think it useful to mention the contributions offered by Catholic thought in the area of sexual ethics: without this component, it would be difficult to understand the debate as a whole, and it is necessary to recognize that many natural and rational foundations for sexual ethics find their most extensive development and fullest respect within the context of Catholic thought. There has been doctrinal development and continuity in this area of thought and reflection. The doctrinal development already began with the teachings of Pope Pius XII, but it is present above all in the Magisterium of Vatican Council II, particularly in the pastoral constitution *Gaudium et spes* [32] and in the teachings of Popes Paul VI, John Paul II, and Benedict XVI. Documents of great breadth and impact include the encyclical *Humanae vitae* by Paul VI; the apostolic exhortation *Familiaris consortio*, the apostolic letter *Mulieris dignitatem*, the *Letter to Families*, and the encyclical letter *Evangelium vitae* by John Paul II, as well as his "theology of the body," speeches commenting on the first chapters of the Book of Genesis, which drew the attention of theologians and philosophers due to the significant development they afforded to the philosophy and theology of corporality; and finally the recent encyclical *Deus caritas est* by Benedict XVI. [33]

There have been numerous documents by the conferences of bishops that have echoed or developed the doctrinal teachings of the popes, particularly on the topics of abortion, contraception, and sterilization. [34] Official documents by various organs of the Holy See are also important, such as those by the Congregation for the Doctrine of the Faith, the Congregation for Catholic Education and the Pontifical Council for the Family. [35]

In addition to its specific topic, the instruction *Donum vitae* on questions of artificial fertilization also entails a development of the vision of corporality and sexuality as inalienable elements of spousal love and procreation.

Those with enough patience and intellectual honesty to read these documents carefully will find enrichment not just from the theological and biblical perspectives, but also from the philosophical, anthropological, and ethical perspectives on the

---

[32] Vatican Council II, Pastoral Constitution *Gaudium et spes*, December 7, 1965, nos. 47–52.

[33] Paul VI, Encyclical Letter *Humanae vitae*, July 25, 1968. John Paul II, Apostolic Exhortation *Familiaris consortio*, November 22, 1981; Apostolic Letter *Mulieris dignitatem*, August 15, 1988; *Letter to Families*, February 2, 1994; and Encyclical Letter *Evangelium vitae*, March 25, 1995. John Paul II, *The Theology of the Body: Human Love in the Divine Plan* (Boston: Pauline Books and Media, 1997); this is a series of 129 Wednesday papal audience addresses given from September 5, 1979, to November 28, 1984. Benedict XVI, Encyclical Letter *Deus caritas est*, December 25, 2005.

[34] See Barberi and Tettamanzi, *Matrimonio e famiglia*.

[35] Congregation for the Doctrine of the Faith (CDF), Declaration *Persona humana*, December 29, 1975; CDF, Declaration *Quaestio de abortu*, November 18, 1974; CDF, *Homosexualitatis problema*, October 1, 1986; CDF, *Donum vitae*; Congregation for Catholic Education, *Educational Guidance in Human Love: Outlines for Sex Education*, November 1, 1983; PCF, *Ethical and Pastoral Dimensions*; PCF, *Truth and Meaning of Human Sexuality*, August 8, 1995.

subject of sexuality, along with a precise understanding of the terms of the ongoing cultural debate.

Sexuality is recognized as a dimension of the whole person according to a personalist vision that has always been vibrant within the Church yet has been better elucidated in recent documents. The contributions of the human sciences such as psychology and pedagogy to a healthy balance and a suitable sexual education are recognized; the complementary role and equal dignity of women are acknowledged (the conciliar documents describe the aspiration to recognize the equal dignity of women as a "sign of the times"); the role of responsibility that the spouses have in regulating births is recognized (responsible parenthood), as is the role of the family in the society and in the Church. Conjugal love is understood as the path to holiness for the spouses and the basis of harmony for human persons and for the family. Any form of sexophobia vanishes with the full recognition of the dignity of the body and its coessential meaning in spousal love and procreation. As noted, this latter development is found especially in the instruction *Donum vitae*, above and beyond the relevant topics specifically addressed therein, which are among the most serious ones of the present historical moment, and it later finds full confirmation in the *Letter to Families* and the encyclical letter *Evangelium vitae*.

It is important to recognize that even in Catholic circles there have been disagreements and statements by individual theologians who hold positions that differ from official Church teaching and clearly favor "subjectivist" approaches. This has been the case with the thinking of the major American schools and particularly of Charles Curran and John McNeill, who use plain and simple subjectivism to justify many abnormal forms of sexual conduct such as masturbation, homosexual acts, and premarital sexual relations.[36]

Although it has been enriched culturally and theologically as mentioned, Catholic magisterial teaching has followed a continuous line of development with regard to several fundamental points: the bond between sexuality and legitimate marriage, in the sense that the exercise of the sexual faculty finds meaning and uprightness, human fulfillment and justification, only in the context of a legitimate marriage; the connection between the unitive and procreative dimensions in the exercise of the conjugal act, in the sense that the conjugal act must always remain open to procreation and procreation must be anchored to the conjugal act, and that this must be the case in every individual conjugal act and not just in the totality of conjugal life; the reaffirmation of the legitimacy of the spouses' procreative responsibility to be pursued by mutual agreement and with means that are "not artificial"; and consequently the condemnation of contraception, abortion, contraceptive sterilization, and in vitro fertilization. In this broad spectrum of momentous topics and issues, the special guidance of Catholic morality—often contested initially—has nonetheless been recognized later by many

---

[36] CDF, "Letter to Father Charles Curran," July 25, 1986; CDF, *Homosexualitatis problema*. For the writings to which these documents refer, see J.J. McNeill, *The Church and the Homosexual* (Kansas City, MO: Sheed, Andrews, and McMeel, 1976); C. E. Curran, *Critical Concerns in Moral Theology* (Notre Dame: University of Notre Dame, 1984). For information on the specific questions of moral judgments on homosexuality in reference to Rev. McNeill's positions, see G. Perico, *Problemi di etica sanitaria* (Milan: Ancora 1992), 408; CDF, "La cura pastorale delle persone omosessuali in un recente documento della S. Sede," *Aggiornamenti Sociali* 1 (1987): 79–87.

for the educational weight of its positions, the authoritativeness of its pronouncements, and the accuracy of its predictions about the inadequacy of certain paths often undertaken because of the push of trends and protest movements.

Anthropological Presuppositions Regarding Sexuality and Procreation

The first truth that must be mentioned concerns the relationship between the person and sex. This relationship implies and mirrors the relationship between the person and the body.

The differentiation and complementarity of the sexes is discerned first of all in corporality. The body, within an essentially uniform structure, exhibits a series of differentiating factors that are imprinted upon the entire basic personality: there are chromosomal factors (the presence of X or Y in the last chromosome pair); neuroendocrine factors linked to the latter and marked by major differences in the gonads: internal for women (ovaries) and external for men (testicles); differentiation continues with the ducts, which are also distinct (Wolffian ducts in men, Müllerian ducts in women); finally, there are phenotypic sex characteristics or primary and secondary sex characteristics. Cases of abnormalities such as intersexuality, homosexuality, and transsexuality will be discussed in volume two of this text. Corporality as a whole, in its morphology, in the voice, in movement, and in sensorial and perceptive traits, is nonetheless marked by sexual differentiation[37] in its fundamentally identical and homogeneous structure.

Being sexed is therefore an *original fact* about men and women, since personal experience cannot avoid passing through masculinity or femininity from its very beginning, which is fertilization. Being sexed, furthermore, acquires a unique *originality* in men and women in that they are male or female in a different dimension and on a different level than animals: the femininity and masculinity of the human person, precisely because it is expressed *in* and *by* the body, carries the richness and vitality of the entire being—of the spirit first and foremost—and is a reflection of the image of God.[38]

It is quite simple to conclude that corporality exists only when it is sexually differentiated into a male body or female body, and it is just as intuitively obvious that this differentiation is not limited to certain accessory characteristics, but profoundly marks all of one's corporality over time. This does not mean, however, that corporality is only sexuality: the body also has other functions and dimensions.

It must also be concluded from this premise that sexuality marks the entire person as well: it is the spirit and the "personal self" that is man or woman and not just the body; this is precisely because it is the spirit (personal self) that animates,

---

[37] L. De Marinis, A. Barbarino, and A. Serra, "Biologia della differenziazione sessuale," *Medicina e Morale* 34, no. 2 (1984), 155–165; G. Boiardi, "Sessualità maschile e femminile tra natura e cultura," *Medicina e Morale* 33, no. 1 (1983): 12–24; A. Serra, "Le componenti biologiche della sessualità umana," in *Uomo-Donna*, ed. CIF, 103–136.

[38] See John Paul II, "Original Unity of Man and Woman," November 7, 1979, in *Theology of the Body*, 42–45. See also C. Caffarra, introduction to *Uomo e donna lo creò: Catechesi sull'amore umano*, by John Paul II (Rome: Città Nuova, 1985), 5–24; A. Scola, *Identidad y differencia* (Madrid: Encuentro, 1989).

informs, and enlivens corporality. Therefore the person does not merely *have* a given sex, but *is* a man or a woman. One's whole personal vocation in the world can be harmoniously fulfilled only by accepting and appreciating this definite mode of being.

Human sexuality therefore cannot be reduced to a thing or an object; it is a *structural conformation* of the person, an expressive structure rather than a function. As a fundamental component of the person, sexuality demands respect and acceptance.

Manipulating sexuality in order to turn it into its opposite conformation is therefore equivalent to manipulating the genome with the aim of modifying it. The body is to be received, the body is what it is; the same must be said of sexuality. We can apply to sex what Karl Rahner has said on the subject of in vitro fertilization by appealing to the principle of "transcendental deduction": man must freely accept his nature just as it is, including the fact that it is already predetermined. Indeed, he is not a being called into existence through his own initiative. Consequently the acceptance of this necessary determination, made by a hand outside of his own existence, is and perennially remains an essential task incumbent upon the free moral life of man. All of this, referring back to transcendental deduction as well, appears necessarily connected to man's very nature.[39] Those who see life in the light of the truth of creation know that this transcendental deduction is nothing other than the gift of personal and spiritual being made by the Creator Who gives life and actualization to this body, with this sex, in this existence marked by sexual differentiation, without in any way diminishing the role of secondary causes and the processes of procreation and fertilization, but acting within them just as He acts within every potentiality that becomes actualized.

But we must go one step further. What was said about corporality in general must also be said about sexuality: sexuality does not exhaust all of the richness of corporality, nor does it exhaust the entirety of the person; therefore, it does not fulfill all of the person's values. One's spirit and self transcend one's sex in terms of fullness of life and wealth of values: one's person is more than one's body, and one's body is more than one's sex.

On this subject, let us recall what John Paul II said in the previously cited papal audience on November 7, 1979: "Bodiliness and sexuality are not simply identical. Although in its normal constitution, the human body carries within itself the signs of sex and is by its nature male or female, the fact that man is a 'body' belongs more deeply to the structure of the personal subject than the fact that in his somatic constitution he is also male or female."[40]

No one can live in this world except with a body and no one can live except as a man or a woman, but one's personal being is grander than both one's body and one's sex. This self-evident truth implies that, in the hierarchy of personal goods, sex inheres in the person's totality but does not exhaust him in his fullness. The whole good of the person, with all of his spiritual and transcendent richness, is foremost; physical life is the fundamental good through which the person expresses himself in time, and sexuality comes directly into play in physical life. Therefore, while it is true that no one can refuse to be either a man or a woman, it is not equally necessary or possible for sex to express every aspect of life or for every person to be required

---

[39] K. Rahner, "The Problem of Genetic Manipulation," in *Theological Investigations* (New York: Seabury Press, 1972), 9:243–245.

[40] John Paul II, *Man and Woman He Created Them: A Theology of the Body*, trans. and ed. M. Waldstein (Boston: Pauline Books and Media, 2006), 157.

to express the full extent of his sexual capacities. Just as the act does not exhaust the faculty (an act of thought does not exhaust the faculty of thought) and the faculty does not express the whole person (man does not consist of thought alone), it must be stated by way of analogy that performing individual acts of the sexual life does not express the whole of sexual life and sexual life does not actualize the whole person.[41]

As a fundamental component of the human person, sexuality nonetheless influences even the way in which the person reveals himself and relates with others: "If the person is an 'I' open to a 'you,' a 'being in relation,' then sexuality has an essential relational dimension. It is the *sign* and the *locus* of openness, encounter, dialogue, communication, and unity of persons with one another."[42]

Understood in this way, sexuality becomes the need to come out of one's solitude, to communicate with others, to find oneself in others: men and women perceive sexual difference and feel attracted and oriented toward one another. And each sex wishes to discover the most enigmatic mysteries of the other. Instead there is always a difference, an unbridgeable abyss between man and woman:

> As a human being, man is always in communion with his counterimage, woman, and yet never reaches her. The converse is true of woman. If we take this man/woman relationship as a paradigm, it also means that the human "I" is always searching for the "thou," and actually finds it ("This at last …"), without ever being able to take possession of it in its otherness. Not only because the freedom of the "thou" cannot be mastered by the "I" using any superior transcendental grasp—since, in its proper context, all human freedom only opens up to absolute, divine freedom—but also because this impossibility is "enfleshed" in the diverse and complementary constitution of the sexes.[43]

Male-female sexual duality therefore takes on a different meaning: "The sexual distinction, which appears as a determination of human being, is diversity, but in equality of nature and dignity. The human person, through his or her intimate nature, requires a relation to another, implying a reciprocity of love. The sexes are complementary: similar and dissimilar at the same time; not identical, yet equal in personal dignity; they are alike so that they may understand each other, diverse so as to be reciprocally complementary."[44]

It must also be added that sexuality is not equivalent to the exercise of genitality: in order to clarify this difference, it is necessary to distinguish between a "sexed relationship" and a "genital sexual relationship."[45] The sexed relationship comes about through sexuality understood in a general sense, without recourse to genitality: this is what happens in each and every phase of life. It involves the common interrelation of persons of different sexes marked by esteem, respect, friendship, and affection without the involvement of the sexed body at the level of physical genitality.

---

[41] John Paul II, "Original Unity," 42–45.

[42] D. Tettamanzi, "L'etica sessuale," in *Sessualità da ripensare*, ed. C. Dastoli (Milan: Vita e Pensiero, 1990), 28.

[43] H. U. von Balthasar, *Theo-Drama: Theological Dramatic Theory* (San Francisco: Ignatius Press, 1990), 2:366.

[44] Congregation for Catholic Education, *Educational Guidance in Human Love: Outlines for Sex Education*, November 1, 1983, no. 25.

[45] Tettamanzi, "L'etica sessuale," 29–30.

The unique characteristic of the genital sexual relationship is, in contrast, the totality of the components of the person, which give life to openness, encounter, dialogue, communion, and unity: it is a personal, total, and mutual self-giving.

It follows from what has been said that the practice of genital activity, which is expressed through the genital organs, is not the only way of expressing oneself as a man or woman, and it is not necessary in a deterministic sense for an individual subject to practice genital activity in order to consider himself fulfilled as a person.

It therefore remains true that being a woman does not necessarily mean being a mother or even a spouse, and likewise being a man does not necessarily mean being a husband or father. Well before the current feminist movements, Saint Thomas already affirmed that marriage is a natural right but a secondary one: the primary natural right is the right to life; marriage is naturally necessary for the human race but not for the individual human being. This does not mean that marriage and parenthood are in opposition to being man or woman. If this difference is not grasped, then the choice of virginity made by a Catholic religious or priest would be incomprehensible: the fact of having chosen to live one's sexuality without genital activity certainly does not make a person any less a man or any less a woman.

Summarizing this first reflection, it must be said that sexuality is inscribed in human beings. It is therefore rooted in human nature and marks the entire human being, but at the same time it does not exhaust everything about him. This is far from the thinking of de Beauvoir and those who reduce sexuality to a cultural matter, yet also far from the pansexualism of certain psychological schools of thought.[46]

Sexuality, like the body, belongs to that being and having—"I am and I have"—in which the person is present but with respect to which the person remains transcendent. The transcendence of the personal being with respect to his or her sex does not devalue sex, but enriches it; likewise thought, which need not be expressed in verbal form, does not devalue speech but enriches it.

The fact that sexuality is neither necessarily nor always expressed in the exercise of genitality does not mean that once genitality is exercised it does not involve the full commitment of sexuality and of the whole person.[47] To offer another example, I am neither obligated nor required to place my signature on a contract, nor do I live only to make real estate transactions, but once I decide to sign the contract, my signature commits my whole person not just in the moment that I trace the few letters that make up my name, but also in everything that follows from it. I am worth more than my signature and that is what makes the written marks so precious. I am not obligated to use it on a contract, but if I use it then I am committing myself.

Sexuality as Reciprocal Recognition

Phenomenological reflection has explored the relationality of the person and the dynamics of reciprocal recognition[48] that clarify existential experience, which

[46] S. De Beauvoir, *The Second Sex* (New York: Knopf, 1952); M. Mead, *Sex and Temperament in Three Primitive Societies* (New York: Morrow, 1963); Melchiorre, *Amore e matrimonio.*

[47] A. Valeriani, *Il nostro corpo come comunicazione* (Brescia: La Scuola, 1964).

[48] See F. Botturi and C. Vigna, eds., "Affetti e legami," *Annuario di etica* 1 (2004); C. Vigna, "Sostanza e relazione: una aporetica della persona," in *L'idea di persona*, ed.

remains distinct from the metaphysical plane while participating in it. Its language is different from Aristotelian-Thomistic language but does not deny its structure because there is no dichotomy between substance and relation: saying that a *person is an individual substance of a rational nature* also means affirming his relationality, because the logos (reason; from *legein*) relates, binds (*lega*), and brings the multitude back to unity.

Common sense connects happiness above all with the encounter between man and woman and, in particular, to exclusive loving relationships. In addition, it holds that when one is happy there is a contact with the realm of the absolute because time and space disappear (the manifold character and development of things and awareness). Our desire finally feels it has obtained what it had always more or less mysteriously awaited. The situation of happiness is always linked to an interpersonal relationship, a "relationship of recognition."

The human person, precisely because he is an immensely layered and complex psychophysical whole, has empirical qualities (ascertainable by the senses) and transcendental qualities (surpassing space and time).

All of modern thought has explored this transcendental aspect: man has gradually been revealed to himself as openness to wholeness and infinity. The person is seen as a situated yet transcendentally open existence. This person does not live in solitude: he is in a constitutive condition of relation and, from the perspective of generation, the awareness of his own existence (self-consciousness) arises after and through a relation with another consciousness: the child opens his eyes to see not himself but his mother, who cares for him and calls him to life, and only later is he able to say "I."

Twentieth-century philosophy focused its attention on the relationship with the other, which is revealed in the *face* and even more specifically in the *look* of the eyes. Without a look that welcomes the person into the world, no consciousness could awaken. Just as the body is nourished by food, so the consciousness is nourished in its transcendental sphere by the self-giving of the consciousness of others.

This twofold (empirical and transcendental) existential situation also occurs in sexual relationships—that is, a physical relationship and a union of recognition. One "recognizes" the other as unique and incapable of being reified: a person open to the Absolute. For this reason, the person's *object of desire* must be an *absolute object*, that is, a person. The person can be understood only within an absolute horizon, but *understanding a consciousness as an absolute horizon means allowing oneself to be understood.*

A sort of crossing over occurs in sexual relationships worthy of persons: while one person recognizes the other, the latter does likewise with the former. The person feels fulfilled when recognized in his or her entirety and individuality. This condition of absolute exclusivity is only fulfilled in loving relationships: "Only you forever." It is necessary to go beyond the historical horizon in every relationship of recognition, and for this reason a "metaphysical protection" is necessary: the God-Person willing to be in relation with man. The relationship between spouses implies a willingness for dependence and self-abandonment but not a symbiosis, which instead amalgamates.

V. Melchiorre (Milan: Vita e Pensiero, 1996), 175–203; C. Vigna, ed., *Introduzione all'etica* (Milan: Vita e Pensiero, 2001); G. Borgonovo, "Essere e amore: Per un approccio personalista alla bioetica," *Annales Teologici* 17 (2003): 43–76.

Nothing helps us understand our *ontological dependence*, our original state of passivity, better than our affections. The word *affection* is derived from the Latin *afficio*: he who is *affected* is *struck* and reacts with emotions (from *ex-moveo*). Contemporary thought understands affectivity as all-encompassing and unbridled, and so *serial* experiences of affectivity multiply: so-called histories. This is because affections are perceived as disconnected from the logos (reason) and therefore ungovernable. Instead, *the world of affectivity is within the logos, although it is not the logos*, precisely because the logos can free affectivity and, as the Western tradition has recognized, can distance itself from it in order to direct and "govern" it.

Being "struck" or "affected" is the fundamental ontological structure of a finite and corporeal rational subject, and for this reason the world of affections is ostensibly experienced as our most proper world, or the world that we are. The original character of bonds—in addition to affections—is felt, albeit secondarily, along with the need to eliminate every dichotomy. Affections without bonds are a source of painful grieving because they fade quickly; bonds without affections are a source of conflict because they easily fall prey to mistrust and suspicion. Only the synergy of bonds and affections makes possible the tenderness of the good life, lived in the strength of stability. The family is its privileged bosom for this reason.

Not all bonds are positive. Positive bonds allow us to offer to the other what we would want for ourselves (the golden rule[49])—namely, recognition—which is possible only within the open mindset of reciprocity and self-giving[50] that is not a calculated exchange. In fact, the golden rule requires putting oneself in another's shoes: "the true sense of the rule is that the place of another is the true point of view for equitable judgment when we attempt it."[51]

Aside from the sorts of reductionism that tend to see in sexuality only those aspects linked to pleasure and reproduction,[52] it is absolutely necessary to frame the question of *sexuality* as a question of *truth*,[53] because sexuality is not an element that qualifies the human subject by its addition, but rather the way in which that unique form of ontological subjectivity—human subjectivity—manifests itself. Human sexuality must be thought of in reference to the entirety of the person, aimed not just at the *reproduction of the species* but first of all at the very *production of the*

---

[49] See J. Wattles, *The Golden Rule* (New York: Oxford University Press, 1996).

[50] See also D. Tettamanzi, *Nuova bioetica cristiana* (Casale Monferrato: Piemme, 2000), 43: "The category of the 'gift' becomes essential and decisive in order to define the logos, the truth, and the reality of human life, and consequently in order to understand the ethos, the history, and the task of human life. Precisely along this path, which appears so distant and abstract with respect to the concrete and complex issues of bioethics, these same issues can be cast in a unique and rather fruitful and stimulating light."

[51] G. Leibniz, *New Essays on Human Understanding* (New York: Cambridge University Press, 1996), book I, chapter II, § 4, p. 89.

[52] A. Shopenhauer sees sexuality as a cruel deception that makes the individual the laughingstock of the species. "Metafisica dell'amore sessuale," in *Supplementi al "Mondo come volontà e rappresentazione,"* vol. 2 (Rome-Bari: Laterza, 1986).

[53] F. D'Agostino, *Bioetica* (Turin: Giappichelli, 1997), 127–139; P. Ventura, *Problemi attuali della filosofia del diritto* (Turin: Giappichelli, 1991), 61 ff (cited in D'Agostino, *Bioetica*); F. D'Agostino, *Una filosofia della famiglia* (Milan: Giuffrè, 1999).

self. Unlike other biological functions, human sexuality needs the *other* to express itself completely; for this reason, unlike animal sexuality,

> *it needs laws*, it needs rules, it needs to affirm itself in its constitutive duality. This is proven by the fact that when this duality is attenuated or even eliminated one ends up in a realm of indeterminacy regarding not only sexuality but subjectivity itself. The psychological sciences apply various names to it, but they all lead back to and are rooted in a relational deficiency that is expressed socially and relationally in *rifts* within the self, which always ask (desperately at times) to be mended; rifts that can translate into the search for a *mimesis* of unresolved duality (as in the case of homosexuality) or else into pretensions of *dominion* over the other (the extreme case being sadism); that is, into pretensions that structurally contradict the logic of coexistence and therefore the logic of law.[54]

Relationality, understood as balance in difference, is therefore the criterion for the genuine possibility of reciprocal recognition.

In conclusion, a sexuality worthy of the human person allows the recognition of the person's dignity, respects the person's constitutive relationality and the ontological determinations (*inclinationes*) of our being as a unitotality of body and spirit (affections, will, intellect), and respects the *law* inscribed in our *nature.* This is possible if we do not stop short at the empirical layer but rather enter the transcendental one, which allows for communication and rules. This level is constitutive of the person, who contains that universal which is the basis for a possible encounter between an *I* and a *you*—between two personal individuals.

Physical Sex and Psychological Sex

It was said that sex is inscribed in the body, yet with and in the body it marks the entire personal being, which nonetheless remains transcendent through its spirituality. The role of psychological activity in the human person was already examined and it was recalled that the psyche, according to the personalist understanding, neither belongs totally to the corporeal organism—though it has its sensory, nervous, and emotive roots here—nor totally to the spirit, which consciously exercises vigilance over this complex psychological activity and receives stimuli and influences from it. This concept of the psyche should therefore be understood in a *hylomorphic* sense as the product of the two human co-principles: the physical and the spiritual or, as some psychologists call it, the metapsychological. Therefore, the spirit cannot even be reduced to psychological activity, nor can man be considered the product of three ontological co-principles. Man is ontologically spirit and body united; psychological activity is inscribed and develops within this vital unity, that is, within the entire biological and conscious person.[55]

Sexuality therefore has an emotional drive and a psychological world of its own as well. It was mentioned that, from the vantage point of this psychological activity

---

[54] D'Agostino, *Bioetica*, 134.

[55] G. Cesari and M.L. Di Pietro, *L'educazione della sessualità* (Brescia: La Scuola, 1996), 17–38.

and subconscious psychological activity, some psychologists have sought to interpret all human and social behavior in an all-encompassing vision.[56]

Although this all-encompassing view is exaggerated, nonetheless it cannot be denied that psychological life enriches, influences, and dramatizes sexual life as well. Conflicts and tensions, passions and eros meet within the psychological life of persons and, though they certainly do not constitute all of psychological life, they profoundly affect it. The complementarity of the sexes and their reciprocal attraction comes alive in psychological activity just as in corporality.

It is not within our scope here to recall the phases of the psychosexual development of the person or to identify the distinguishing traits of the sexual psyche: the authors are in more agreement in the field of evolutionary psychology than in the field of differential psychology.[57] It seems clear, however, that psychological differences cannot be eliminated or attributed entirely to sociocultural influences. It is not culture alone that establishes differing male and female psychologies. The culture can certainly influence the emphasis on certain roles and lead to certain prejudices and false sensibilities (active man / passive woman, strong man / weak woman, etc.), but not all of the psychology proceeds from culture: the psyche is rooted in the soma as well as in the spirit, which permeates it with its vitality.[58]

It was necessary to recall these elementary ideas in order to note the possibility that conflicts and difficulties in harmonizing bodily and psychological sexuality can occur, also because the influence of the cultural environment weighs heavily on the psychological dimension. This may then lead not only to developmental difficulties but also to real and deeply rooted abnormalities in psychological activity in which the physical sex is not accepted and an appeal is made to the psychological sex in order to change the physical.[59]

Aside from the usual problematic nature of harmony within the human being, a situation that the Catholic Church explains with the doctrine of original sin, these cases involve genuine disturbances that will be addressed later. For now it can be said that the solution cannot be sought in the pure and simple suppression of one of the two elements (the physical or the psychological), assuming it were possible; rather, the solution lies in harmonizing both of them as much as possible through the spirit, education, and the aid of psychiatric science where necessary.

Freedom and Responsibility with regard to Sex

Within these complex psychophysical and cultural dynamics and beyond them lies the spirituality of man: his freedom and responsibility. Sexuality cannot be deprived of this spiritual vitality which enriches it, harmonizes it in terms of its elements and dynamics, and expresses it in an interpersonal relationship and in the overall process of personal growth. To ignore this dimension and the stream of spiritual

---

[56] Freud, *Complete Works*.

[57] R. Simon, "Amore e sessualità, matrimonio e famiglia," in *L'ateismo contemporaneo*, ed. Pontifica Università Salesiana (Turin: SEI, 1969), vol. 2; A. Jannière, *Anthropologie sexuelle* (Paris: Aubier-Montaigne, 1964).

[58] J. Gevaert, *Il problema dell'uomo* (Turin: LDC, 1984), 80–85.

[59] F. Castagnet, *Sexe de l'âme et sexe du corps* (Paris: Le Centurion, 1981).

vitality in the study of sexuality would be to reduce it to a psychophysical mechanism or a cultural construct or, in other words, to apply yet another form of reductionism.

It is important to recall that there can be no freedom without responsibility in this area just as in other areas of human activity: freedom cannot responsibly ignore the good that it directs in this case (sex and sexual activity), nor can it ignore all of the personal richness that sex brings with it, the personal life that it directly involves, and the potential impact on other persons and on the family that may be formed through it. Sex, though always accompanied by the flutter of spontaneity, is never merely a game and cannot disregard the compelling richness of spirituality.

All of sexual life should therefore be accompanied by responsibility, and all the more so when sexuality is committed to the context of conjugal love and the procreation of other persons. Responsibility also means accepting sexuality for what it is and for what it involves in its meanings and in its consequences.

I wished to recall the personalist foundations of sexuality without which it would be impossible to understand the subsequent steps of conjugality and procreativeness.

Conjugality and Procreativeness

Human sexuality, as already mentioned, has a complementary structure and presents itself as the whole being's capacity of openness to conjugality. This statement, however obvious, is of the utmost importance and should therefore be carefully weighed.

The structure of the body, in which the dimension of the whole person is read, indicates that differentiated and complementary sexuality is directed toward hetero-sexual union: from the chromosomes to the endocrine components to the anatomical and functional structure of the internal ducts and the phenotypic sex, everything indicates that the man-woman personal being, within a fundamental identity of body and spirit, also exhibits a complementarity that is simultaneously a sign of a *lack* that requires completion and a *gift* that offers completion.

This complementarity is fully actualized—always by a strong tendency though not deterministically (in the sense that it is not necessary to exercise it)—in conjugality, that is to say in the physical, psychological, and spiritual union with the opposite sex. When this complementarity is actualized in conjugality by free choice, the union involves the *totality* of the person and not just a part. When a man and woman unite, if the act is human and complete, it involves the body, the heart, and the spirit; if one of these dimensions is missing, then it is a humanly incomplete and objectively false union, because the body has no meaning if not as the expression of the totality of the person. Conjugality therefore implies totality, a unity that expresses a reciprocal and total gift of persons. Since it involves a gift of the person, it is fully human when it is total in terms of content and total in terms of a stable bond: the person cannot be broken down into separable parts in either an ontological or a chronological sense.

Conjugality therefore implies the permanence and stability of the unitive bond. This is not just out of the need to provide a stable educational environment for potential offspring but above all because it involves a gift of self, of one's whole subjectivity. It is still true that sexuality does not exhaust all of the person's being, but it is still true also that sexuality signs, implicates, and involves the totality of the person. The body-person relationship can be recalled in order to understand the sex-person subject.

392

Let us return to the comparison with the signature that I freely affix to a contract. I am not obligated to do so, I do not live only for this contract, but if I affix my signature it involves not just the hand that traces the lines but the whole person and my freedom and responsibility, and it is not valid for just one moment but for the entire time indicated in the terms of the contract.

So it is with the existential language of conjugality: I am not deterministically obligated, I only have the faculty to engage in sexual acts, but if I set this union into action then it implies everything that it means and supposes and expresses the total and stable gift of the person. Once again, ethicality for man means the realization of the whole in the harmony and hierarchy of values.

We are often led to think that morality is an expression of constraint. In reality it does require the mastery of personal drives so that they are not wasted in anarchy, but morality is the realization of the fullness of being in the awareness of action.

In a handshake with a friend I do not just make physical and mechanical movements, but I express a feeling and recognize the dignity of my friend and welcome him; if this were not the case, it would be a false and nonhuman gesture.[60]

Several consequences at the bioethical level follow from this all-inclusive—or better yet, all-expressive—structure of conjugality in which sexuality tends to become realized.

First of all it makes no sense to exercise sexuality in an autoerotic (masturbation), self-centered, or provisional way, which is to say before or outside of conjugality, precisely because the gift of the body is a sign of the stable and committed gift of the person. For Christian believers, the sacrament of marriage enriches this totality with the gift of participation in divine love and the grace of consecration. But apart from faith and the sacrament and the demands of the supernatural plane, it can be said that the very structure of conjugal sexuality renders the exercise of sexuality outside of conjugality and human fullness senseless and unfulfilling.[61]

This indivisible unity of the gift of self also justifies the unity of marriage and condemns every form of polygamy, polyandry, divorce, and even occasional extramarital relations, as indicated before.

As far as masturbation is concerned, I will not enter here into questions of pathological states or particular conditions of neuropathology, which can subjectively reduce culpability; instead I will simply refer to the objective facts.[62] As an objective fact, it is an egocentric behavior contrary to the proper meaning of sexuality and is therefore illicit. The delicate question of homosexuality, at a certain stage of its practice, can also present more as an illness to be treated through psychotherapy—and results thus far have been encouraging—than as a deliberate vice. From an objective viewpoint and without making judgments on subjective culpability, which are not

---

[60] For a reflection on the philosophy of human acts and the conjugal act in particular, see E. Sgreccia, "Il riduzionismo biologico in medicina," *Medicina e Morale* 35, no. 1 (1985): 3–9.

[61] Häring, "Sessualità"; B. Schlegelberger, *Rapporto sessuale prima e fuori del matrimonio* (Rome: Paoline, 1974); B. Häring, *Rapporti sessuali prematrimoniali e morale* (Rome: Paoline, 1973).

[62] See CDF, *Persona humana*; L. Rossi, "Masturbazione," in *Dizionario enciclopedico*, 614–625; *Elementi di medicina e psicologia pastorale* (Brezzo di Bedero: Salcom, 1969), 205 ff; A. Nalesso, *L'autoerotismo nell'adolescente* (Turin: Marietti, 1970).

always possible and not always attributable to the subject in question, homosexuality should be considered an abnormality to be prevented, healed, and corrected—so long as it is possible and within the limits of possibility—because sexuality has an objectively heterosexual orientation toward conjugality and only attains its fullness in this configuration.[63] Homosexuality, as well as other sexual abnormalities or disturbances, will be addressed in the second volume of this work.

The Meanings of Conjugality and the Conjugal Act

The other consequences flowing from conjugality—that is, from the stable, full, and conscious encounter and union of two persons of different sexes—are the following: the *unitive dimension* and the *procreative dimension*, which are connected with the sexual act, and the *family dimension* and the *social dimension*, which proceed from the covenant of union and interpersonal choice.

These are topics of great cultural interest in the present day, particularly in their impact on debates about questions of responsible procreation, so-called sexual freedom, cohabitation, in vitro fertilization, and so on.

The sexual act is suited to the very constitution of sexuality: the philosophy of being has repercussions on the philosophy of doing; therefore the physical act of sexual union simultaneously involves the capacity to signify and actuate the union of the two sexes (the unitive logos) and the objective aptitude for procreation (the procreative logos). If the individual acts do not always result in actual procreation or psychological unity because of some desired or undesired impediment, this does not detract from the finality of the gesture: sexual union can be achieved only with the sexual act; procreation is likewise inscribed in the very constitution and finality of that same conjugal act.

On the ethical level, a sexual act will therefore be comprehensible and fully and truly human when, in and of itself, by its objective and intentional reality, it simultaneously aims to express the total union (physical, psychological, and spiritual) of the two persons, man and woman, and at the same time remains objectively open to procreation.[64]

The repercussions of this principle of totality on the question of responsible procreation will be seen in the following pages. For now, suffice it to say that a sexual act between a man and woman is not ethically right when it is merely a biological act without unitive, affective, and spiritual love or, vice versa, when it aims at unity but interferes with its biological and procreative completeness. In other words, *love* and *life*—values that are simultaneously expressed and inherent in sexual activity—cannot be separated.

---

[63] See CDF, *Persona humana*; see also B. Häring, "Omosessualità," in *Dizionario enciclopedico*, 682–688; A. Massone, *Cause e terapie nell'omosessualità* (Brezzo di Bedero: Salcom, 1970); B. Häring, *Medical Ethics*, (Notre Dame: Fides, 1973); G. Cesari, "Natura e interpretazione dei disorientamenti sessuali: L'omosessualità," in *Interrogativi per la bioetica*, ed. M. L. Di Pietro and E. Sgreccia (Brescia: La Scuola, 1998), 78–96.

[64] A. Cappella, ed., *Liberi e responsabili collaboratori del Dio Creatore* (Rome: Università Cattolica del Sacro Cuore, 1980); Häring, *Free and Faithful*, 2:520–530 and 3:12; C. Caffarra, "La trasmissione della vita nella *Familiaris consortio*," *Medicina e Morale* 33, no. 4 (1983): 391–399.

Another feature of conjugality that follows from this is the *family dimension*. By family I mean the stable union of the two persons, sanctified by marriage, and the willingness to welcome children. It is common knowledge that certain sociological theories taking a historicist approach have declared the family, as a stable institution, to be merely cultural and historical and therefore unconnected to the nature of conjugal sexuality. The reasons in favor of the unity and stability of the marital bond proceed from the fact that the reciprocal giving of the man and woman, if it is human, involves the totality of the individual persons: totality understood in the immediate sense (physical, psychological, and spiritual) as well as in the broader sense.

The person cannot give himself or herself on loan or temporarily in sexuality, because the two persons come to form an interpersonal union: "The two of them become one flesh" (Gen 2:24), as Sacred Scripture says. Yet reason, too, is able to understand this truth. This unity and indissolubility is also demanded by the procreative dimension: the potential birth of children in accordance with the exercise of sexuality requires, due to the need for unity in upbringing, the continuity of the original home. Educational development, even at the subconscious level, and the processes by which the child's personality is formed demand the stability of the original parental home. Therefore divorce, "free love," and cohabitation are contrary not only to the Christian faith and the sacrament of marriage but also to the harmonious character and intrinsic dynamics of male and female sexuality.

The last characteristic feature is *sociality*. The union of a man and a woman in marriage and in the family is already a typical form of sociality; at the same time, it is the basic unit of society and the source of the first socialization of its members. This also means that the society as a whole, even as an authoritative community, is affected by and involved with the family, precisely because the family is its origin and perennial spring.

This implies that society should be informed about the constitution of the family and should recognize it as such in some way. Legal and public recognition should not be seen as a constraining shackle but as an act of commitment and appreciation. Society recognizes the needs and fundamental rights of such a bond and commits itself to defend and promote them. The family's need for food, work, housing, and social services are neither gifts nor impositions: they are duties of the social body toward the family. For the couple, on the other hand, there is an obligation to make public the constitutive will to form a family and to give reason to trust the grounds for this intention.

Therefore, cohabitation—"marriage without papers"!—does not respect or allow this public guarantee and does not recognize the social importance of marriage.

Having clarified these dimensions of sexuality inscribed in the comprehensive and enriching concept of the human person, the individual problems arising within conjugal life and the exercise of sexuality can be better understood.

Ethics of Responsible Procreation and Contraception

It is first of all necessary to recall a general principle regarding the morality of every human decision: for an action to be right the end must be right, which is to say consistent with the good of the human person, and it is likewise necessary that the means be right, which is to say that it too must be in conformity with the whole good of the persons and consistent with the end. The same principle must therefore be applied in the context of human procreation.

Human procreation is the act that is most greatly invested with ethical significance: it is one of the couple's most important choices, one of the primary ends of marriage, and it results in the birth of a new human person.

Procreation, which is inscribed in the finality of sexuality and the complementarity of the sexes, is therefore licit and cannot be excluded by those who have made the decision for conjugal life. Voluntarily excluding fertility from a union that is directed toward fertility means contradicting the finality of the conjugal act. It is nevertheless incorrect to talk about a "right to a child": the right deriving from marriage is the right to be able to carry out acts that are fertile in themselves; actual fertility can depend on other causes. In any event the "right to a child" is an improper expression because no one has the "right" to have a person as though that person were an object.

The procreative act takes on an even greater meaning for Christian believers insofar as it involves a special intervention of God the Creator: "At the beginning of every human life there is a creative act of God: no man comes into existence by chance; he is always the end of God's creative love. It follows from this essential truth of faith and reason that the capacity for procreation, inscribed in human sexuality, is in its deepest truth a cooperation with the creative power of God. It also follows that man and woman are not the arbiters of this same capacity, nor are they its masters, since they are called in it and through it to be participants in God's creative decision." [65] "In affirming that the spouses, as parents, cooperate with God the Creator in conceiving and giving birth to a new human being, we are not speaking merely with reference to the laws of biology. Instead, we wish to emphasize that *God himself is present in human fatherhood and motherhood* quite differently than he is present in all other instances of begetting 'on earth.' Indeed, God alone is the source of that 'image and likeness' which is proper to the human being, as it was received at Creation. Begetting is the continuation of Creation." [66]

There is therefore a twofold movement in the procreative act: a reaching out of the human power of generation within the creative power of God, made accessible since the creation of Adam and Eve; corresponding to and preceding this human outreach is an initial condescension of God, Who from the beginning makes His creation dependent on a process placed in the hands of the creature itself. "What does man reproduce? ... We should speak of human generative power, in its natural operation, extending into the divine creative power, which opens up and makes itself available in the creation of man. From God's point of view, this constitutes a 'humiliation' of God. ... Thus he hands on his creatorship, making it dependent on events initiated at the will of creatures. The real depth of this mystery only emerges when the child is seen no longer as '*res patris*' but as a personality in direct relationship with God." [67]

There is therefore a consignment of uncreated freedom (God) to created freedom (man); the spouses can never forget this reality or the great task with which they have been charged: "Therefore, when the spouses remove the potential creative capacity from the exercise of their conjugal sexuality through contraception, they attribute to

---

[65] John Paul II, Address to Priests Participating in the Study Seminar on Responsible Procreation, September 17, 1983, in *Insegnamenti di Giovanni Paolo II*, 6.2:561–564 [Italian].

[66] John Paul II, *Letter to Families*, no. 9.

[67] Balthasar, *Theo-Drama*, 371.

themselves a power that belongs to God: the power to have the *last word* in deciding over the coming into existence of a human person. They attribute to themselves the title, not of cooperators in the creative power of God, but of the *ultimate* keepers of the source of human life." [68] Such arrogance implies contempt not only for God but also for human life itself. The ethics of procreation should therefore be based upon principles described in the following sections.

The principle of responsibility

The principle of responsibility implies that the decisions of whether or not to procreate and how many children to have is up to the man and woman as a couple, legitimately constituted in marriage for the founding of a family: this decision cannot be coerced by the state since it is a personal right and a couple's right.

Any legislation establishing a maximum or minimum number of children—or, worse still, imposing sterilization, or in any event applying sanctions or penalties based on the number of children—is unjust legislation. For the common good, the state can educate or encourage with valid reasons, but it cannot establish direct or indirect injunctions.

Seen from the perspective of the individual spouses and of the couple, this responsibility must be adjusted to the values that are involved and the conditions in which they can be fulfilled. The primary responsibility is that of knowledge, and therefore of tending toward the *truth* of the act of procreation:

> Only true knowledge of a person makes it possible to commit one's freedom to him or her. Love consists of a commitment which limits one's freedom— it is a giving of the self, and to give oneself means just that: to limit one's freedom on behalf of another. Limitation of one's freedom might seem to be something negative and unpleasant, but love makes it a positive, joyful and creative thing. *Freedom exists for the sake of love.* If freedom is not used, is not taken advantage of by love it becomes a negative thing and gives human beings a feeling of emptiness and unfulfilment. Love commits freedom and imbues it with that to which the will is naturally attracted—goodness. The will aspires to the good, and freedom belongs to the will, hence freedom exists for the sake of love, because it is by way of love that human beings share most fully in the good.[69]

What is the truth about the act of procreation? As already stated, the *total truth* of this act is that it should express its entire objective reality—psychological, corporeal, and procreative—regardless of whether or not procreation actually occurs in each gesture of spousal love: "In the conjugal act, husband and wife are called to confirm in a responsible way *the mutual gift* of self which they have made to each other in the marriage covenant. The logic of the *total gift of self to the other* involves a potential openness to procreation: in this way the marriage is called to even greater fulfillment as a family. Certainly the mutual gift of husband and wife does not have

---

[68] John Paul II, Address on Responsible Procreation, September 17, 1983.

[69] K. Wojtyla, *Love and Responsibility* (New York: Farrar, Straus, Giroux, 1981), 135.

the begetting of children as its only end, but is in itself a mutual communion of love and of life." [70]

A responsible attitude, insofar as it is oriented by and to the truth and is the result of a free choice, is present when the integrity of the conjugal act is respected, whether it is performed with a view to becoming pregnant or during the times that would allow the couple to space their pregnancies. *It is irresponsible*, on the other hand, *to manipulate the conjugal act in its total and personal objectivity* so that it expresses only the psychological and affective dimension and not the procreative one, or so that it simply constitutes a biophysical event and not an affective and spiritual union as well.

The conjugal act entails these two dimensions, which are intrinsically interdependent on one another and not just secondarily linked. In this case, *manipulating* means dividing and separating the aspect of *love* from the aspect of *life* in an individual act, reducing this personal and hence total act of love, and not recognizing the truth of a reality and a precept inscribed in man's very nature: "The Church, nevertheless, in urging men to the observance of the precepts of the natural law, which it interprets by its constant doctrine, teaches that each and every marital act must of necessity retain its intrinsic relationship to the procreation of human life." [71]

A responsibility follows from the recognition of this truth: a responsibility toward oneself, one's spouse, the child, and the Creator. The responsibility toward oneself and one's spouse involves the reciprocal need to recognize each other as persons who must be respected and not exploited. It follows, for example, that a husband who saddles his wife with the burden of birth control, imposing on her the choices of contraception or even abortion, does not recognize her full dignity.

Both spouses have a responsibility with respect to the sexual act in itself and their fertility in general. "In relation to *biological* processes, responsible parenthood means knowledge and *respect* for their functions: in the power of giving life, the intellect discovers biological laws that are part of the human person. When dealing with the psychological dimension of the tendencies of the instincts and passions, responsible parenthood means the necessary mastery that the reason and will must exercise over them." [72]

According to this view, the couple is challenged and guided to establish an understanding consisting of mutual knowledge of vital rhythms and biological needs, shared and harmonious decisions, as well as renunciations decided upon and pursued together, without imposition or exploitation.

In the context of the objective dimension of responsibility, the evaluation of the good of the children who have already been born or who may potentially be born also comes into play. Being responsible parents, especially for Christian believers, does not strictly mean having few children; rather, it means evaluating objectively and according to "a conscience dutifully conformed to the divine law itself, and ... submissive toward the Church's teaching office, which authentically interprets that

---

[70] John Paul II, *Letter to Families*, no. 12.

[71] Paul VI, *Humanae vitae*, no. 11.

[72] John Paul II, "Responsible Parenthood in the Light of *Humanae Vitae*," August 1, 1984, in *Insegnamenti di Giovanni Paolo II*, 7.2:144–151 [Italian].

law in the light of the Gospel"[73] regarding whether "to have more children ... [or] for serious reasons and with due respect to moral precepts, decide not to have additional children for either a certain or an indefinite period of time."[74]

Along with the right to procreate, the right "not to procreate" is also recognized in the sense that the spouses under certain conditions have the right to engage in a predictably infertile conjugal act or abstain from fertile conjugal acts and can even in certain cases entirely exclude fertile acts for reasons independent of the will (risks to life or health). In these cases, the right must be expressed in a way that requires legitimate ways of not engaging in fertile acts: for example, abortion in the name of the right not to procreate would be unjustifiable.[75]

It should be noted that the couple's subjective propensity to fecundity is effectively influenced by love for life and trust in society, so those who do not love their own lives and those who fear living also express this fear by limiting births. This is also a sign of the need for the harmony and connection of moral values; it is important that a society seeking to encourage life should know how to create a "culture" that favors life and its growth.

Subjective and objective responsibility are prompted and unified by the *transcendental* dimension. If a spouse who is a Christian believer recognizes the value of the other as person and creature, he or she cannot but fully accept being a partner of the Creator Himself, participating in His fruitful and unitive love. The person recognizes in him- or herself the gift of a transcendent love and a procreative responsibility and that the life of the child is a gift from the Creator even before being the fruit of conjugal love: the procreative design thus comes to be perceived as a "transcendental deduction" and not as a merely temporal function, and even less as a this-worldly and manipulative operation. And it is for this reason that believers sense that the conjugal act is not just any gesture and that procreation is not simply reproduction and therefore cannot be either counterfeited or contradicted in its structure.

The principle of the truth of love

This principle concerns the content of human love and therefore the method of pursuing it. Fecundity or procreation should be the expression of a *true* love between two persons, objectively and completely true.[76]

In order to comprehend thoroughly this integral truth of conjugal love, it is first of all necessary to recall that the conjugal act has a particular ethical and ontological dignity that distinguishes it from other human activities (playful, poetic,

---

[73] Vatican Council II, *Gaudium et spes*, no. 50.

[74] Paul VI, *Humanae vitae*, no. 10.

[75] On the concept of responsible procreation and the reference criteria for this responsibility, see Paul VI, *Humanae vitae*, nos. 10 and 13.

[76] Caffarra, "La trasmissione della vita"; J.M. Finnis, "Personal Integrity, Sexual Morality and Responsible Parenthood," *Anthropos* 1 (1985): 43–45; G. Grisez, *Contraception and the Natural Law* (Milwaukee: Bruce, 1964); H. Lio, *"Humanae vitae" e coscienza: L'insegnamento del card. Wojtyla teologo e papa* (Vatican City: Libreria Editrice Vaticana, 1980); see *Procreazione responsabile: Quale realtà per la famiglia oggi? Atti del II Congresso Internazionale per la Famiglia d'Africa e d'Europa; Roma 12–15 marzo 1988*, ed. Centro Studi e Ricerche sulla Regolazione Naturale della Fertilità (Rome: Facoltà di Medicina e Chirurgia dell'Università Cattolica del S. Cuore, 1989).

or productive): it is an act that in itself is open to initiating and giving life not to an object, but to a person; it is also an act that involves the totality of the two persons (physical, affective, and spiritual dimensions) in the form of a union which is therefore all-encompassing. For the Christian believer—but also for human reason—this act involves a special intervention by God the Creator; indeed, as already stated, we talk about "procreation" by the spouses.

As already recalled, it is up to the spouses whether or not to perform the conjugal act and it is also their responsibility to do so deliberately either during the times that would allow for an actual pregnancy or not. It is not however a moral option for the spouses to manipulate the act itself in its total and personal objectivity in such a way that it expresses only the psychological and affective dimension and not the procreative one, or such that it expresses merely a biological event and not the affective and spiritual union of a higher order as well.

The moral connection between procreation and the conjugal act was also reiterated in the instruction *Donum vitae* by the Congregation for the Doctrine of Faith, aligning the contraceptive mentality with the mentality at the root of artificial fertilization:

> Contraception deliberately deprives the conjugal act of its openness to pro-creation and in this way brings about a voluntary dissociation of the ends of marriage. Homologous artificial fertilization, in seeking a procreation which is not the fruit of a specific act of conjugal union, objectively effects an analogous separation between the goods and the meanings of marriage ... The conjugal act by which the couple mutually express their self-gift at the same time expresses openness to the gift of life. It is an act that is inseparably corporal and spiritual. It is in their bodies and through their bodies that the spouses consummate their marriage and are able to become father and mother. In order to respect the language of their bodies and their natural generosity, the conjugal union must take place with respect for its openness to procreation; and the procreation of a person must be the fruit and the result of married love.[77]

From this it can be understood why the so-called natural methods of fertility regulation are thought to respect the totality and uniqueness of the conjugal act, whereas contraceptives effectively reduce that totality and uniqueness and are therefore to be considered illicit and not completely human in the full sense of the word. This must be said regardless of their danger to physical and psychological health.

The contraceptive effect can have two modes of action: (1) blocking the release of the ovum from the ovary (estrogen-progestin pill) or (2) preventing the encounter of the ovum and a spermatozoon during sexual intercourse (condom, diaphragm, spermicide, etc.). Also to be counted among methods of contraception is the practice of coitus interruptus, which entails the interruption of sexual intercourse immediately before ejaculation so that the spermatozoa are not deposited in the vagina.

Sterilization is also erroneously included among contraceptive methods. However, it must be clarified—and this is why the subject is addressed in a separate chapter—that while contraceptive methods render an individual sexual act or series of acts infertile, sterilization definitively alters the bodily integrity of the man or the woman, even though a certain reversibility may be possible.

---

[77] CDF, *Donum vitae*, part II, nos. 4(a)–4(b).

Contraceptives need to be distinguished from interceptives and contragestives: the former act by preventing the implantation of the embryo in the early stages of development (intrauterine device [IUD] or coil, the "morning-after pill," progestin administered by intramuscular injection or subcutaneous implantation), and the latter act by causing the detachment of an embryo that has already implanted in the uterus (RU-486, prostaglandins, anti-chorionic gonadotropin vaccine).[78]

1. *Barrier contraceptives.* Barrier contraceptives therefore include the condom and the diaphragm. The *condom*, named after the English physician who first suggested using it for contraceptive purposes, is a rubber sheath that fully covers the penis during sexual intercourse. In this manner, it prevents seminal fluid from being deposited in the vagina. The condom's *Pearl index*,[79] which depends on manufacturing quality, usage capacity, and coupling with spermicidal chemicals, is 7 to 10 percent.

In comparison with several "advantages" (low cost, availability without a prescription, partial protection against contracting sexually transmitted diseases), the use of a condom exposes the man and the couple to unquestionable disadvantages: the interruption of the sexual relationship which seems to lose spontaneity, the appearance of vaginal irritations in the woman or allergies to latex or to the spermicides in the man, and the creation of a psychological barrier between the spouses that can even alter the entire harmony of the marriage.

A *diaphragm* consists of a rubber disc that is soft in the central zone, rigid but flexible around the edges, and once inserted in the vagina it also carries out the role of a barrier by impeding the encounter between the ovum and the spermatozoa. In addition to the diaphragm described above, there are also *cervical caps* on the market made of rubber or hard plastic or metal, which fit over the cervix and perfectly adhere to it. There is also a so-called *female condom* consisting of a sort of polyurethane hood applied to the female perineal area to cover the vulva, impeding contact with spermatozoa. Aside from the question of the woman's psychological acceptance of these means, the possibility should be kept in mind that the spermicides and the rubber itself out of which the diaphragm is made may cause vaginal irritation and facilitate the occurrence of urinary tract infections. Its Pearl index is 14 to 15 percent.

---

[78] E. Diczfalusy and M. Bygdeman, eds., *Fertility Regulation Today and Tomorrow* (New York: Serono Symposia, 1987); G. Pescetto et al., *Manuale di Ginecologia e Ostetricia*, vol. 1 (Rome: SEU, 1989); J. G. Raymond, R. Klein, and L. Dumble, *RU-486: Misconceptions, Myths and Morals* (Cambridge, MA: Institute on Women and Technology, 1991); D. T. Baird and A. F. Glasier, "Hormonal Contraception," *New England Journal of Medicine* 328, no. 21 (1993): 1543–1549; R. Peyron et al., "Early Termination of Pregnancy with Mifepristone (RU-486) and the Orally Active Prostaglandin Misoprostol," *New England Journal of Medicine* 328, no. 21 (1993): 1509–1513; M. L. Di Pietro and R. Minacori, "Sull'abortività della pillola estroprogestinica e di altri 'contraccettivi,'" *Medicina e Morale* 5 (1996): 863–900; M. L. Di Pietro and R. Minacori, "'Contraccezione d'emergenza': Problema medico, etico e giuridico," *Vita e Pensiero* 5 (1997): 353–361; Bioethics Center of the Catholic University of the Sacred Heart, "Sulla cosiddetta 'contraccezione d'emergenza,'" *Medicina e Morale* 3 (1997): 582–589.

[79] The Pearl index is a formula used to determine the "effectiveness" of a contraceptive and is calculated based on the number of pregnancies that occur per 100 women using a given method of contraception for one year (12 months); the lower the Pearl index, the more effective the contraceptive is at preventing pregnancy. Pearl index = (number of pregnancies / number of months of contraception use) × 12 months × 100.

2. *Hormonal contraceptives: The estrogen-progestin pill.* The estrogen-progestin pill, as the name suggests, is made up of two synthetic hormones, an estrogen and a progestin, and acts as a "contraceptive" by (a) inhibiting the hypothalamic-pituitary system and consequently the release of the ovum and hormones; (b) altering the normal sequence of uterine endometrial changes such that if ovulation and conception were to occur it would be impossible for the embryo to implant in the uterus; (c) modifying the motility of the Fallopian tubes and impeding the passage of spermatozoa to the ovum and, in the case of a conception, impeding the descent of the embryo into the uterus; and (d) altering the composition of the cervical mucus so that it becomes "inhospitable" to spermatozoa and impedes their ascent toward the cervical canal. Effects (b) and (c), given that they occur after conception, are to be considered *abortifacient* and not contraceptive. In the light of recent studies on the subject, it is estimated that the possibility of an abortion occurring with the use of the estrogen-progestin pill, after its contraceptive action has failed, is roughly equal to one abortion for every ten years of use.[80] Relating this statistic to more than one woman, there would be a probability of one abortion for every ten women that use the estrogen-progestin pill for twelve months. This rate may appear low, but the gravity of the occurrence should not be weighed in statistical terms.

Regular use of the estrogen-progestin pill exposes the woman to the risks of deep vein thrombosis; pulmonary embolism and cerebral thrombosis (from three to six times higher than those who do not use it); cardiovascular incidents including heart attack; arterial pathologies (atherosclerosis) and coronary cardiopathies; somatic malformations in the child if it was accidentally used during pregnancy; higher exposure—though this is not certain—to breast tumors, particularly if the pill was used at an early age and for many years; development of a malignant tumor in the cervix, which is associated not so much with ingesting the pill but rather with the promiscuous sexual behavior that results from reduced concern about becoming pregnant; and complete blockage of hypothalamic function even after suspension of use.

It is also possible for those who take the estrogen-progestin pill to have the following *side effects*, some of which disappear after several months of use while others persist for the entire duration: nausea and vomiting, swelling and tenderness of the breasts, swelling and weight gain due to retention of liquids, headache, depression, loss of sexual desire, vaginal dryness causing painful intercourse, and bleeding between periods. With regard to psychological problems, it should be emphasized that the estrogen-progestin pill can cause negative effects on the woman's behavior when treatment stops, even if done gradually; create forms of misunderstanding and psychological barriers between her and her partner when she begins taking it, particularly if the partner is unaware; or lead to a fear of parents finding out if the woman taking it is an adolescent still living in a family that might not agree with her choices. The Pearl index for the estrogen-progestin pill is less than 1 percent.

3. *Spermicides, sponges, and vaginal douching. Spermicides* are chemical substances capable of immobilizing spermatozoa deposited in the vagina, before they reach the cervical canal, and they also act as a mechanical barrier against

---

[80] Di Pietro and Minacori, "Sull'abortività." See also B. Bayle, "L'activité antinidatoire des contraceptifs oraux," *Contraception Fertilité Stérilité* 22, no. 6 (1994): 391–395; W. Rella, *Die Wirkunsweise oraler Kontrazeptiva und die Bedeutung ihres noidationshemmenden Effekts* (Vienna: IMABE Studien, 1994).

spermatozoa. They have little contraceptive effectivenes; in fact, considering that ejaculation immediately propels the sperm to nearby the cervix, no spermicide in the vagina—however potent—can prevent a large number of sperm from reaching the cervical canal within a few seconds, rendering its effects futile. For this reason, spermicides are usually used together with other barrier methods.

Sponges are made of polyurethane soaked with a potent spermicide (non-oxynol-9) and are inserted deep into the vagina near the cervix. They act mechanically, with little contraceptive effectiveness, to impede the access of spermatozoa to the cervical canal, immobilizing them with spermicide and absorbing the entire ejaculate.

Vaginal douching is used with the aim of removing spermatozoa that have been deposited in the vagina, flushing it out with water and other liquids that may or may not contain spermicides; strange liquids such as vinegar, Coca-Cola, and others are often utilized. This is an entirely ineffective practice from the contraceptive point of view, since the sperm are already able to reach the cervical mucus within 90 seconds from the time of ejaculation and can no longer be removed.

4. *Interceptives.* Among interceptive methods there is a distinction between hormonal interceptives (mini-pill, morning-after pill, progestin depot administrations or subcutaneous implantation) and mechanical interceptives (IUD or coil).

The *mini-pill* contains only progestins in small quantities and is taken daily. The mini-pill has several mechanisms of action: it (a) blocks ovulation in 30 to 40 percent of cases; (b) renders the cervical mucus impenetrable to spermatozoa; and (c) alters the uterine endometrial structure, making it unfavorable for the implantation of an embryo that may have been conceived. It has been calculated that a woman using the mini-pill could experience one abortion in every five years of use.[81] The mini-pill is not often used because it causes very irritating side effects, including irregular bleeding and, in the event of conception, a higher incidence of tubal pregnancies. The mini-pill has a Pearl index of 1 to 6 percent.

Depot administration of estroprogestins is performed through non-oral alternative modes, which is to say in the form of "depository" preparations that are injected intramuscularly or contained in capsules implanted under the skin, which then gradually release their active ingredient day after day. Since the progestin quantity released on a daily basis is very low, there is not a total inhibition of ovulation and other "defense" mechanisms therefore come into play: the alteration of the permeability of the cervical mucus against spermatozoa (a *contraceptive mechanism*) and the alteration of the endometrial mucosa, which becomes inhospitable to any embryo that may be conceived (an *abortifacient mechanism*). The primary disadvantage to these preparations is the occurrence of repeated intermenstrual bleeding caused by the steady emission of small quantities of progestins into the blood circulation, while the contraindications and risks are otherwise similar to those described for the estrogen-progestin pill. The Pearl index for the mini-pill is 0 to 2 percent.

The *morning-after pill* is a product whose mechanism of action has nothing to do with the contraceptive estrogen-progestin pill. Do not let the word "pill" or its chemical composition (estrogen, progestins, or both), which is similar to that of the contraceptive pill, deceive you. The morning-after pill did not exist as such on the market and this name generally referred to taking massive doses of estrogens

---

[81] Di Pietro and Minacori, "Sull'abortività."

and estroprogestins within 72 hours after sexual relations thought to have caused a conception; this dose was repeated after 12 hours from the first. Today there is an emergency contraception product available on the market whose active ingredient is levonorgestrel, a synthetic progestin, and it is administered according to the modality described above. Once they reach the bloodstream, high doses of orally administered hormones can cause—if taken prior to ovulation—the prevention of ovulation in a high percentage of cases. But if ovulation occurs or the drug is administered during the postovulatory phase, there is a total disruption of the delicate hormonal balance usually responsible for preparing the uterine mucosa to receive the embryo that may be conceived. So it happens that if conception has occurred, the embryo is unable to implant in the endometrium, which has been significantly altered by the emergency contraceptive.

An *intrauterine device (IUD)* or *coil* is a device made of plastic or another material (silver, for example) in various styles (*coils* or *intrauterine devices*) which is gently inserted through the cervix into the uterine cavity.

There is more than one mechanism of action for the IUD: induction of a foreign body reaction that causes chronic inflammation of the endometrial mucosa, rendering it inhospitable to an eventual embryo ready for implantation (since the IUD does not interfere with ovulation, it is always possible that conception may occur: *abortifacient action*); alteration of the composition of the cervical mucus impeding the ascent of spermatozoa toward the cervical canal and uterine cavity (this effect occurs when the IUD is treated with copper or a progestogen); in addition there may be the characteristic effects of progesterone if the IUD is treated with this hormone.

The risks and consequences linked to the use of intrauterine devices are manifold. Examples include syncope reactions or perforations of the uterus during insertion; increased risk of contracting a pelvic inflammatory disorder; higher incidence of tubal pregnancy both during and after the period of IUD use. Its Pearl index is 1 to 3 percent.

5. *Contragestives.* RU-486, also known as the "month-after pill," is composed of a substance—mifepristone—that acts as an antiprogestogen. Progesterone is the pregnancy hormone and acts at the level of the uterine endometrium, binding to specific receptors. RU-486 irreversibly binds to those progesterone receptors, interrupting their action. The lack of progesterone activity causes the death and detachment of the embryo from the uterine lining and later expulsion. RU-486 does not cause abortions in 100 percent of the cases, and this seems due to the fact that the drug is not able to reach sufficient levels in the blood to completely counteract the progesterone in the circulation: in fact, the contragestive effect occurs at a higher frequency in the early phases of pregnancy (within 49 days of conception), when the progesterone levels are still low.

For this reason, the administration of RU-486 is accompanied by *prostaglandins*, which are also used alone as abortifacients and, among other things, act by inhibiting the action of the corpus luteum.

The *anti-hCG* vaccine consists of a fragment of the beta chain of chorionic gonadotropin transferred by a vector or carrier. Its administration induces the formation of antibodies against the subunits of the hormone. It consequently hinders the action of chorionic gonadotropin, which is secreted by the trophoblast at the beginning of its implantation in the uterine wall and is intended to maintain the production of progesterone by the corpus luteum. Because of the interference in the action of chorionic gonadotropin and the drop in progesterone levels, the possibility of embryo implantation and development is compromised. In fact, the corpus luteum tends to regress, with consequent shedding of the endometrium and embryo loss.

The terms *fertility awareness methods* and *natural methods* or *methods of natural fertility regulation* refer to different sets of methods that allow a woman to become aware of the fertile and infertile phases of her menstrual cycle so as to be able to regulate sexual intercourse according to whether she and her husband wish to achieve or temporarily or definitively avoid a pregnancy. These include the basal temperature method, the ovulation or Billings method (from the surname of the spouses who discovered it: John and Evelyn Billings), and symptothermal methods.

Each fertility awareness method is based on the gathering of certain signs and symptoms, called *fertility indexes*, which allow the woman to know whether ovulation is about to happen or has already happened. The fertility indexes usually referenced are (a) *basal body temperature*, which normally presents a diphasic course in connection with the hormonal changes of the menstrual cycle and provides the possibility of identifying an effective ovulation after a stabilized rise in temperature with respect to the previous phase; and (b) the *cervical mucus* produced by specialized cells in the cervical canal, which is not just a fertility index but also a *factor* in fertility. In fact, the woman should be considered infertile if there is no cervical mucus. The quantity and characteristics of the cervical mucus undergo significant change during the menstrual cycle in response to the varying levels of ovarian hormones in the bloodstream: the woman is made capable of assessing these variations in the cervical mucus, which are objective (changes in appearance, consistency, and clarity) or subjective (sensations of dryness, moisture, or wetness).

The following should be added to the abovementioned fertility indexes: changes in the cervix and, although they are less reliably correlated with the moment of ovulation, bleeding between menstrual periods, abdominal pain halfway through the cycle, breast tenderness, weight gain, and changes in desire, mood, and appetite.

The so-called rhythm method or *Knaus-Ogino method*, named after the two physicians who developed it, is based on a mathematical probability calculation for the time of ovulation. It has been rendered obsolete by methods implementing objective gathering of fertility indexes.

Although initially they were not acknowledged and too often underestimated, fertility awareness methods—in particular the Billings method—have experienced a revival, particularly after the World Health Organization subjected them to testing in various research centers and published the results concerning their effectiveness in pregnancy spacing as well: fertility awareness methods are now being proposed as a preferred strategy for health care reasons, out of bioenvironmental sentiment, and because they are more respectful of the couple's freedom and responsibility.[82]

---

[82] The World Health Organization (WHO) study on natural methods and the results are presented and discussed in A. Cappella, V. Navarretta, and E. Giacchi, "Il metodo della ovulazione Billings: Dati e valutazioni dello studio multicentrico della Organizzazione Mondiale della Sanità," *Medicina e Morale* 4 (1982): 371–387; J. Brown et al., "Correlation Between the Mucus Symptoms and the Hormonal Markers of Fertility throughout Reproductive Life," in *The Ovulation Method: The Achievement or Avoidance of Pregnancy by a Technique which Is Reliable and Universally Acceptable*, ed. J.J. Billings (Melbourne: Advocate, 1983), 104 ff; S. Mancuso and P. F. van Look, *La regolazione naturale della fertilità*

At the same time, confidence in hormonal contraception has been declining for medical and health care reasons, as pointed out earlier.

The field of natural fertility regulation is looking to the future, which is to say to the possibility that the woman will be able to rapidly and independently measure levels of indicator hormones, estrogens and progesterone, to determine when ovulation occurs. To this end, a device called an *ovarian monitor* has been developed for measuring the amount of ovarian hormones present in urine during the menstrual cycle. The method, perfected by Professor James Brown, allows the detection and measurement of ovarian hormone catabolite levels in the urine: estrone glucuronidean (an estrogen derivative) for determining the beginning of the fertile phase, and pregnanediol glucuronide (a progestogen derivative) for determining the end of the fertile phase. The ovarian monitor designed by Brown is able to identify the ovulation event within a twelve-hour period, and shows without a doubt the moment that ovulation occurred. This diagnostic tool, which can be used both by couples seeking to space their pregnancies and by couples actively trying to achieve a pregnancy, does not replace the natural fertility regulation method but confirms and completes it. Furthermore, it offers verification in unusual situations that may arise during unclear cycles (coinciding with stress, illness, travel, lactation, the premenopausal period, or the period after use of the contraceptive pill).[83]

Recognizing her fertile period allows the woman to identify the days in which there is a greater probability of conception and to have fertility-focused sexual intercourse. There are still few studies available on this matter in the scientific literature. A study of the symptothermal method involving 346 women in Germany reported that 38 percent conceived during the first cycle of use, 68 percent during the third cycle, and 92 percent during the sixth cycle. A study of the Creighton model—a standardized version of the Billings method—involving 50 women in the United States reported that 76 percent conceived during the first cycle, 90 percent in the third, and 98 percent in the sixth.

Natural methods can also have an important role in the diagnosis and prevention of some cervical and vaginal pathologies (infections or phlogistic and dysplastic pathologies of the cervix) or endocrine pathologies (ovulatory disorders, especially a short or insufficient luteal phase) that are often causes of infertility. The results of the use of natural methods by couples with low fertility, specifically with regard to the effectiveness of their guidance for focused intercourse and of their support in the diagnostic and therapeutic process, remains to be assessed by specific studies.

---

oggi (Milan: Vita e Pensiero, 1992); A. López Trujillo and E. Sgreccia, ed., *Metodi naturali per la regolazione della fertilità: L'alternativa autentica* (Milan: Vita e Pensiero, 1994); A. W. Cappella et al., "Metodi naturali e cultura della vita: Valutazione di un'esperienza di insegnamento," *Medicina e Morale* 4 (1996): 669–682.

[83] A. W. Cappella and A. Saporosi, "Impiego di moderne tecnologie nella regolazione naturale della fertilità: Il monitor ovarico di Brown," *Medicina e Morale* 47, no. 6 (1997): 1119–1128. For predicting the possibility of conception through the use of natural methods, see G. Gnoth et al., "Time to Pregnancy: Results of the German Prospective Study and Impact on the Management of Infertility," *Human Reproduction* 18, no. 9 (2003): 1959–1966; T. W. Hilgers et al., "Cumulate Pregnancy Rates in Patients with Apparently Normal Fertility: Focused Intercourse," *Journal of Reproductive Medicine* 37 (1992): 864–866.

These methods, which can be useful for predicting ovulation in women and cause no harm to the woman herself, are considered licit in themselves. Obviously the purpose in using them must also be morally irreproachable: not self-centered, but for a human and responsible management of procreation.

In order to better understand this point, which is difficult for the modern technology- and efficiency-oriented mentality to accept, let us examine several objections that are usually raised against this point, about which the Catholic Church above all has provided specific instructions.[84]

The following objection is raised first: If the result is the same, then what difference is there between those who responsibly seek to avoid a pregnancy by using a contraceptive technique and those who find themselves in the same ethical situation but use a "natural method"?

I respond that this is obviously a mentality that looks primarily to the results (efficientism) and hardly at all to the person; however, it must also be considered that the morality of an act is not derived solely from the end, which in this case is assumed to be good and justified, and must also be understood in terms of the means or methods used. Now, in choosing natural methods, the decision is respectful of the expressive and ontological totality of the person. In the use of contraceptive means the person is not expressed in his totality; rather, he is divided in his act with the aim of accepting one part and rejecting another: accepting the dimension of erotic satisfaction and blocking the dimension of procreation. Whatever the means and at whatever level of the body it acts, the disorder becomes ethical and ontological and not merely physical and biological.

Another objection repeatedly raised by various authors references the principle of "totality," which they apply in an extended sense. It is said that those who use contraceptive methods in order to space pregnancies with reason do not intend to reject the procreative dimension as a whole, but only in that specific period and for valid reasons; hence a partial value or moment of procreation is sacrificed in the name of the principle of totality.

The response to this is not difficult and is bipartite. First of all, the principle of totality in this case is arbitrarily extended to the overall situation of the couple; however, it is only valid—in accordance with the therapeutic principle—within a single, identical, physical organism. If the principle were extrapolated it could even wind up justifying sterilization, the killing of a given individual for the good of the group, and so forth. Furthermore, it must also be said that if a moral precept is objectively valid for the whole of life, it is also valid for individual acts: if stealing is wrong as a general act, it is because stealing is wrong as a particular act.

Another objection often raised appeals to the so-called lesser evil or greater good. There are two "greater" evils taken into consideration: (1) the danger of a rift in the conjugal bond when on the one hand there is a need to express conjugal

---

[84] Here are just a few works from the plethora of literature available: D. Tettamanzi, *"Humanae vitae": Commento all'Enciclica sulla regolazione delle nascite* (Milan: Ancora, 1968); L. Ciccone, *"Humanae vitae": Analisi e orientamenti pastorali* (Rome 1970); *Il matrimonio dopo l'"Humanae vitae,"* ed. T. Goffi (Bologna: Dehoniane, 1969); B. Häring, *Paternità responsabile* (Rome: Libreria Editrice Salesiana, 1970); J. M. Finnis, "Natural Law in *Humanae Vitae*," *Law Quarterly Review* 84 (1968): 467–471; J. M. Finnis, "Personal Integrity."

affection yet the result of using natural methods is difficult or uncertain on the other hand; and (2) abortion. It is then said that contraception is preferable to abortion.

The response to this series of objections has already been given, both generally and with regard to the two individual hypotheses of greater evils. On the general level, it must be noted that the principle of the lesser evil, which some prefer to call the "greater good" (since it is never ethical to choose evil), requires first of all that there be a true moral dilemma, that is, a situation with no third possibility. Now in this case the use of natural methods constitutes precisely that third possibility which allows both the expression of affection and the regulation of births, provided there is a reason to do so. The greater or lesser certainty should be overcome, if necessary, through a commitment to learning and verification.

Furthermore, still on the general level, the abovementioned principle does not apply to *moral* evil, but only to the assessment of a greater or lesser *physical* harm in those cases where the harm is nonetheless justified. For example, when it is necessary to surgically intervene because of a tumor (therapeutic reason), I am required to go ahead with the lesser evil, that is, to cause the least harm possible to the body. But one cannot justify, for example, a smaller lie in order to avoid a bigger lie, or a lesser offence against a person in order to avoid a more serious one.

In any case, the subjective situation must be distinguished from the objective reality: there could be subjective situations in which, due to a lack of information or agreement between the spouses, certain acts—unsuitable in themselves—become "subjectively" experienced as inevitable in order to keep the peace. Objectively, however, the matter remains what it is and must be overcome first and foremost through information and education, which may require a gradual journey, as in the case of someone climbing a mountain, together with the undesired eventuality of taking some backward steps. The law of pedagogic graduality does not, however, admit the affirmation of a "graduality of the law" understood in terms of its objectivity.

Regarding abortion, the second evil feared, it is well known—in addition to everything just stated—that a bad habit is never the solution to prevent a worse situation. It is well known, in fact, that the mental habit of setting oneself against procreation and considering it a negative event is the best way to foster abortion. In reality the various "planning" campaigns have always proceeded gradually in this direction, which has proven psychologically and socially effective: first contraception, then abortion, then sterilization. In his encyclical letter *Evangelium vitae*, John Paul II writes the following on this subject:

> It may be that many people use contraception with a view to excluding the subsequent temptation of abortion. But the negative values inherent in the "contraceptive mentality"—which is very different from responsible parenthood, lived in respect for the full truth of the conjugal act—are such that they in fact strengthen this temptation when an unwanted life is conceived. Indeed, the pro-abortion culture is especially strong precisely where the Church's teaching on contraception is rejected. Certainly, from the moral point of view contraception and abortion are specifically different evils: the former contradicts the full truth of the sexual act as the proper expression of conjugal love, while the latter destroys the life of a human being; the former is opposed to the virtue of chastity in marriage, the latter is opposed to the virtue of justice and directly violates the divine commandment "You shall not kill."

But despite their differences of nature and moral gravity, contraception and abortion are often closely connected, as fruits of the same tree. It is true that in many cases contraception and even abortion are practised under the pressure of real-life difficulties, which nonetheless can never exonerate from striving to observe God's law fully. Still, in very many other instances such practices are rooted in a hedonistic mentality unwilling to accept responsibility in matters of sexuality, and they imply a self-centered concept of freedom, which regards procreation as an obstacle to personal fulfilment. The life which could result from a sexual encounter thus becomes an enemy to be avoided at all costs, and abortion becomes the only possible decisive response to failed contraception.

The close connection which exists, in mentality, between the practice of contraception and that of abortion is becoming increasingly obvious. It is being demonstrated in an alarming way by the development of chemical products, intrauterine devices and vaccines which, distributed with the same ease as contraceptives, really act as abortifacients in the very early stages of the development of the life of the new human being.[85]

The real task ahead of us is therefore to offer methods at the scientific level that allow the personal and integral exercise of conjugal love united with procreative responsibility, and to help couples understand the rhythms and the richness of their sexuality and procreative capacity at the educational level, allowing for a personalist, ethically upright, and healthy use of conjugal sexual life. Moreover, the couple's free and responsible self-management of conjugal sexuality, harmonious in its affections and purposes, constitutes a major value and a source of richness that should be pursued.

The principle of sincerity

The principle of sincerity refers to the motivations of the spouses with respect to the real situation of conjugal responsibility, whereas the principle of truth refers, as stated, to the reality of the sexual act and of sexual expression.

By this I mean that the reasons for accepting or spacing pregnancies must be sincerely and ethically grounded even in the use of ethically licit methods. In other words, if someone wanted to pursue self-centered aims, he could also do so even with natural methods by using them with a "contra-ceptive" mentality. A subjectively and intentionally unsuitable attitude in a specific scenario means that the use of natural methods is not justified always and in every case.

However, this should not lead to forgetting or ignoring the ethical and objective difference between the use of natural methods and the use of artificial means of contraception. This is in keeping with what was stated about the need, in the context of ethics, for both the end and the means or methods to be objectively licit.

I therefore think that the cooperation of the physician with a couple asking for contraceptives should be regulated as follows:

- First of all, it should be remembered that all interceptives can act like abortifacients because they block the implantation of the already-fertilized

---

[85] John Paul II, *Evangelium vitae*, no. 13.

ovum, which is an embryo in the initial phases of development,[86] while contragestives always cause its detachment from the uterine lining.

- It is not objectively licit to prescribe estroprogestins for contraceptive purposes for the reasons explained above. Furthermore, it seems that this is not properly a part of the physician's duties even from a deontological perspective, since his role is to treat or prevent illnesses:[87] a contraceptive does not prevent or treat an illness except in the broad and equivocal sense of impeding a pregnancy that might potentially cause an illness. It is instead the duty of the physician to inform the patient of the health risks and harms caused by contraceptives and, more generally, to inform the patient about the possibility of using methods of natural fertility regulation, making sure that the information is up-to-date.

- The physician can and must prescribe a drug that is also a contraceptive only when its primary aim is therapeutic, which is to say necessary for treating disorders (e.g., polycystic ovary syndrome). If a period of pharmacologically induced infertility derives from this estroprogestin-based therapy as a side effect of the treatment, it does not constitute an illicit act (*principle of double effect*).[88]

---

[86] In relation to the interceptive mechanism of action, the issue of regulations on access to this form of "contraception" is discussed in the context of Italy's Law no. 194/78, which regulates abortion requests (see the chapter on abortion).

[87] In Italy, the case of a physician who refused to prescribe a contraceptive pill that was urgently requested by a patient brought the issue before the courts. The judges initially dismissed the case based on the fact that, given a lack of tests and verifications, the physician was justified in her unwillingness to write the prescription. Later, when the decision to dismiss was contested, the physician had to stand trial before the Court of Milan. During the course of the trial, however, she was nonetheless absolved—by a panel of all female judges—because the contested action "does not exist" as a crime: when "presented with a patient asking for a prescription for a drug ... that cannot be freely self-administered because the legislators have considered it dangerous and prohibited its use without medical supervision ... a healthcare professional who refuses to prescribe it, regardless of the reason provided, does not commit a legally wrongful refusal" (Decision of the Court of Milan, June 25, 1997). In the decision, on the other hand, there is no reference to the possibility of conscientious objection to prescribing the contraceptive pill, which was one of the arguments that the defense had also advanced, maintaining that the conscientious objection exercised by the physician—even if the actual reason the physician had not prescribed the contraceptive had been revealed—should not be considered a crime. A dissenting opinion on the possibility of invoking conscientious objection in prescribing the pill was expressed by P. Benciolini, "Obiezione di coscienza alla pillola?" *Bollettino dell'Ordine dei Medici di Padova* 1 (1994): 8–11. In the case of abortifacient drugs as well as those with only contraceptive actions, ultimately the issue of conscientious objection should also be considered by the pharmacist who sells these products. See J. Melgar Riol, "Objeción de conciencia y farmacia," *Cuadernos de Bioética* 2 (1993): 37–47; J. Lopéz Guzmán, *Objeción de conciencia farmacéutica* (Barcelona: Ediciones Internacionales Universitarias, 1997); M. L. Di Pietro et al., *Obiezione di coscienza in sanità: Nuove problematiche per l'etica e il diritto* (Siena: Cantagalli, 2005); J. Noríega, *El destino del eros: Perspectivas de moral sexual* (Madrid: Palabra, 2005); M. L. Di Pietro, *Sexualidad y procreación humana* (Buenos Aires: Educa, 2005).

[88] In *Humanae vitae*, no. 15, Paul VI says the following: "The Church does not consider at all illicit the use of those therapeutic means necessary to cure bodily diseases, even if a

It is obvious that the principle of sincerity in conjugal life does not just apply to procreative intentionality but embraces the full range of interpersonal relationships, and above all it is the reflection of a conscience illuminated by the values that enrich and define marriage and the family.

These principles clearly imply and refer to one another, since responsibility presumes truth and requires sincerity.

No discussion will be provided here on the sacramental dimension of conjugal love, which is a deeply theological subject that broadens and enriches the vision of sexuality, placing it in communion with Life, Trinitarian Love, the mystery of the Incarnation, and the mystery of Christ's union with His Church. This aspect is fundamental in a theological approach to bioethics and certainly not in conflict with the anthropological, philosophical, and ethical aspects here explored. Given the design of this book, which is focused on highlighting the elements of bioethics through the eyes of medical science and moral philosophy, this aspect has been mentioned here simply and briefly.[89]

---

foreseeable impediment to procreation should result therefrom—provided such impediment is not directly intended for any motive whatsoever."

[89] For further discussion, see John Paul II, *Man and Woman*; T. Goffi and G. Piana, eds., *Corso di morale* (Brescia: Queriniana, 1984), 3:261 ff; D. Tettamanzi, *Il procreare umano: Verità e responsabilità* (Casale Monferrato: Piemme, 1985); D. Tettamanzi, *Alle sorgenti della vita* (Casale Monferrato: Piemme, 1993); E. Sgreccia, "Procreazione responsabile e metodi di regolazione naturale della fertilità: Aspetti teologici," in *Metodi naturali*, ed. Lopéz Trujillo and Sgreccia, 63–72.

Summary Outline for Chapter 9

Sexuality: The personalist vision

The sexual inclination of the human person is connected with love and cannot be reduced to a biological, psychological, or physiological impulse. Sexuality is a structural conformation of the person and not just a function, even though it does not exhaust the fullness of the human person. It is the sign and locus of one's openness and self-gift to another. Human love always develops through voluntary acts performed at the level of the person where he or she is able to make a free gift of self.

It is necessary to recall the distinction between a sexed relation and a genital sexual relationship: the former is a relation between persons of different sex without any genital or physical involvement; the latter involves all the components of the person in reciprocal personal donation. A genital sexual relationship is not necessary for a person's fulfillment. The genital sexual relationship involves the totality of the person and is full of meaning under these conditions: heterosexuality, unity and exclusivity, indissolubility, mutual edification, openness to the gift of life, and openness to social recognition.

Phenomenology of sexual life

A phenomenology of the sexual inclination and a correct interpretation of its significance are the necessary prerequisite for the metaphysical analysis of love which, in its truth, affirms the value of the person and the communion of persons.

The behaviorist and purely biophysiological models prove to be incomplete understandings because human sexuality includes meanings that the personalist point of view stresses by highlighting the moral perspective.

Evident in the sexual impulse is the human need for an existential complement, the sign of an ontological lack. Whereas in all other animal species the sexual impulse acts as a strong, irrational instinct, in human beings it has the natural tendency to be transformed into love.

In order that its own natural finality might be preserved, it is therefore necessary (paradoxically) that the sexual impulse be closely connected to love: only in this way do the playful components of sexual life, too, which are associated with the impulse, rise to the level of the dignity of the person. Enjoying sexual pleasure without treating the person as an object of pleasure is the crux of the sexual moral problem (see K. Wojtyla, *Love and Responsibility*)

> "The logical consequence is therefore a rejection of both the rigorist notion and the permissive, libidinous notion of sexuality, which are one-sided and disharmonious. Indeed, despite their apparent antithesis, both assume it is possible to separate pleasure (the subjective element), the biological end of sexuality (the objective element), and love (the specifically human element). In contrast, the fundamental principle of ethical personalism safeguards love as something owed to the person from an objective point of view and refers us to man's essential ontological structure for the qualitative truth regarding love itself" (G. Borgonovo, "Essere e amore").

412

Metaphysics of love

Whereas concupiscence can be summed up as "I desire you because you are a good for me," in a love of benevolence the subject opens himself completely to the Thou: "I desire you as a good," "I desire your good," "I desire what is good for you."

Only benevolence, therefore, can constitute a solid basis for love between persons, which in its essence is described as *spousal*, as a *gift*.

Spousal love is self-gift and self-fulfillment, because the mutual belonging of persons cannot be guaranteed by a calculated balance of self-interests but only by a free choice of mutual donation. In the donation each one's desire for the absolute is satisfied, and in the reciprocal recognition each one knows the other because he or she allows himself to know and to be known; hence this recognition is a gift—not of an object, but rather a gift of self, an absolute gift. This donation is safeguarded by a "metaphysical protection": the common belonging to the root of Being.

The human person cannot be reduced in any way to an object. This is proved by the sense of shame that prevents the inversion of the hierarchical order of values: appreciation for the sexual values of the body, indeed, is a result of appreciation for the value of the person, and not vice versa.

> "It is not a question of lowering the person to the status of an object of sexual pleasure, but rather of recognizing his or her dignity, which is structurally his or her due (personalist norm), while respecting the finalistic order in which the sexual inclination is inscribed (law of nature). Sexual modesty, therefore, is not a flight from love; it is rather the appropriate means of entering into it" (G. Borgonovo, *Essere e amore*).

The unitive and procreative meaning of conjugal love: physical, psychological, and spiritual dimensions (chastity and modesty)

Conjugal love entails the reciprocal gift of the spouses, *a personal, total gift*: the spouses give themselves not as gifts that they *have* but rather as the persons that they *are*. The human person is indeed a unique, undivided, and indivisible whole of body, mind, and spirit. A gift that is not total is immediately evident in the separation of the two indivisible meanings implied in conjugal love (the unitive and procreative meanings).

The process of human generation—the moment in which the spouses are "cooperating with the love of God the Creator" (*Gaudium et spes*, no. 50; see also *CCC* 2367 ff.)— is articulated in several successive phases; separating them in fact alters the truth of human love in its specific existential meanings. *Humanae vitae* teaches that union and procreation are the two inseparable meanings of the marital act: from the experience of the gift of self in conjugal love that is *open* to the transmission of life we recognize that the human being is not the master of life but the steward of the plan inscribed by God in the nature of his person. By safeguarding both of these essential aspects, the unitive and the procreative, the conjugal act preserves integrally the sense of mutual, true love and the ordering of that love to the exalted vocation of the man and woman to parenthood. Periodic continence and methods of birth regulation based on self-observation and recourse to infertile periods respect the bodies of the spouses and foster their growth in authentic freedom.

The relation of the *body* with the *person* is central. Standing against all forms of anthropological dualism, which are disproved by experience, this relation is the observation point from which to determine the morality or immorality of behaviors connected with the world of sexuality and procreation.

In order to understand the pertinence of ethical judgments, we need to go back again and again to the basic anthropological structure that is their foundation. The criterion is the nature of the human person, which can be recognized for what it is. In contrast, today's culture tends to emphasize freedom and to misunderstand the intellect, maintaining that the intellect cannot grasp the meaning of reality but can only organize the world in terms of subjective ends. It is necessary to stress that the intellect is capable of adhering to being and that man's freedom is defined in terms of the message of being, and hence the first object of the intellect is the acceptance or rejection of that message.

Instrumental reason alone is no longer able to grasp the wonder of being, of love, and of the gift.

The sexual revolution

The sexual revolution has led to idolatry of the body, behind which the person has disappeared, to the trivialization of sexual life, which is understood as devoid of meanings or purposes, detached from procreation and from mutual self-giving. In this mindset, corporality and sexuality are not original, constitutive conditions but rather cultural constructs; hence, even sexual identity can be chosen.

The ideologies of the sexual revolution

- According to *libertinism*, sexuality is one of the fundamental instincts that govern life; consequently it must be exempt from moral control. Hence there is one sexuality oriented to pleasure and another sexuality oriented to procreation.

- The *post-Freudian* interpretation sees sexuality as the fundamental driving force of the person and of all his expressions (pansexualism). It is therefore thought necessary to satisfy every impulse; neuroses in the personality are said to be the result of sexual repression.

- According to sexual liberation theorists (*W. Reich and H. Marcuse*), sexuality must be lived without setting purposes and conditions: eros without ethos, partying without rules (monogamous marriage is thought to be repressive).

- According to the *feminism of Simone de Beauvoir*, a woman must be made capable of planning her own life autonomously, and hence of freeing herself from the trappings of motherhood (scission of sexual activity from procreation).

- For the Kinsey Report (the first major investigation of sexual behaviors in the United States), the behavior of the majority becomes the moral norm.

- According to *behaviorism* (Masters and Johnson), sexual activity should be interpreted from the perspective of stimulus/response, and sexual dysfunctions can be corrected by exercise.

These are positions that no longer interpret freedom as responsibility and as the product of intelligence and will, of truth and good, but rather as absolute arbitrariness, in other words, *freedom from* and not *freedom for*. In this perspective the Good is no longer the foundation and the end, but only a subjectivist option.

Fertility regulation

It is important to distinguish *natural methods of fertility regulation* from *contraceptives*, which are deemed illicit because they are reductive and do not respect the totality of the person.

Kinds of contraceptives

1. *Barrier contraceptives (condom and diaphragm)*

   The *condom* is a rubber sheath that fully covers the penis during sexual intercourse. In this manner, it prevents seminal fluid from being deposited in the vagina.

   The *diaphragm* consists of a rubber disc that is soft in the central zone, rigid but flexible around the edges, and once inserted in the vagina it also carries out the role of a barrier by impeding the encounter between the ovum and the spermatozoa.

2. *Hormonal contraceptives*

   The estrogen-progestin pill, a combination of two synthetic hormones, an estrogen and a progestin, acts as a "contraceptive" by (a) inhibiting the hypothalamic-pituitary system and consequently the release of the ovum and hormones; (b) altering the normal sequence of uterine endometrial changes such that if ovulation and conception were to occur it would be impossible for the embryo to implant in the uterus; (c) modifying the motility of the Fallopian tubes and impeding the passage of spermatozoa to the ovum and, in the case of a conception, impeding the descent of the embryo into the uterus; and (d) altering the composition of the cervical mucus so that it becomes "inhospitable" to spermatozoa and impedes their ascent toward the cervical canal.

3. *Spermicides, sponges and vaginal douching*

   Spermicides are chemical substances capable of immobilizing spermatozoa deposited in the vagina, before they reach the cervical canal, and they also act as a mechanical barrier against spermatozoa.

Sponges are made of polyurethane soaked with a potent spermicide (nonoxynol-9) and are inserted deep into the vagina near the cervix.

Vaginal douching is used with the aim of removing spermatozoa that have been deposited in the vagina, flushing it out with water and other liquids that may or may not contain spermicides.

4. *Interceptives*

There are two kinds: hormonal interceptives (mini-pill, morning-after pill, progestin depot administrations or subcutaneous implantation) and mechanical interceptives (IUD or coil).

The *mini-pill* contains only progestins in small quantities and is taken daily. The mini-pill has several mechanisms of action: it (a) blocks ovulation in 30 to 40 percent of cases; (b) renders the cervical mucus impenetrable to spermatozoa; and (c) alters the uterine endometrial structure, making it unfavorable for the implantation of an embryo that may have been conceived.

Depot administration of estroprogestins is performed through non-oral alternative modes, which is to say in the form of "depository" preparations that are injected intramuscularly or contained in capsules implanted under the skin, which then gradually release their active ingredient day after day.

Since the progestin quantity released on a daily basis is very low, there is not a total inhibition of ovulation and other "defense" mechanisms therefore come into play: the alteration of the permeability of the cervical mucus against spermatozoa (a *contraceptive mechanism*) and the alteration of the endometrial mucosa, which becomes inhospitable to any embryo that may be conceived (an *abortifacient mechanism*).

The *morning-after pill* is in the true and proper sense an abortifacient: it consists of taking massive doses of estrogens and estroprogestins within 72 hours after sexual relations thought to have caused a conception.

An *intrauterine device (IUD)* or *coil* is a device made of plastic or another material (silver, for example) in various styles (*coils* or *intrauterine devices*) which is gently inserted through the cervix into the uterine cavity.

The mechanism by which the IUD acts is the induction of a foreign body reaction, which causes chronic inflammation of the endometrial mucosa, rendering it inhospitable to an eventual embryo ready for implantation (*abortifacient action*).

5. *Contragestives*

RU-486, also known as the "month-after pill," is composed of a substance—mifepristone—that acts as an antiprogestogen. Progesterone is the pregnancy hormone and acts at the level of the uterine endometrium, binding to specific receptors. RU-486 irreversibly binds to those progesterone receptors, interrupting their action. The lack of progesterone activity causes the death and detachment of the embryo from the uterine lining and later expulsion.

The *anti-hCG vaccine*: its administration induces the formation of antibodies and consequently hinders the possibility of embryo implantation and development.

416

Methods of natural fertility regulation
(fertility awareness methods)

Methods of natural fertility regulation or *fertility awareness methods* are a set of methods that allow the woman to become aware of the fertile and infertile phases of her menstrual cycle.

The fertility indexes usually referenced are (a) *basal body temperature*, which normally presents a diphasic course in connection with the hormonal changes of the menstrual cycle and provides the possibility of identifying an effective ovulation after a stabilized rise in temperature with respect to the previous phase; and (b) the *cervical mucus* produced by specialized cells in the cervical canal, which is not just a fertility index but also a *factor* in fertility. In fact, the woman should be considered infertile if there is no cervical mucus. The quantity and characteristics of the cervical mucus undergo significant change during the menstrual cycle in response to the varying levels of ovarian hormones in the bloodstream: the woman is made capable of assessing these variations in the cervical mucus, which are objective (changes in appearance, consistency, and clarity) or subjective (sensations of dryness, moisture, or wetness).

Finally, the *ovarian monitor* designed by Brown is able to identify the ovulation event and shows without a doubt the moment at which ovulation occurred.

The fertility awareness methods are respectful of the expressive and ontological totality of the person, whereas with the use of contraceptive means the person in his totality is not expressed, but is divided in his act with the aim of accepting one part (the unitive dimension) and rejecting another (the procreative dimension).

417

Chapter 10

BIOETHICS AND
ABORTION

The Bioethical Vantage Point

The tragic subject of procured abortion can be addressed from various angles. It can be discussed from the historical perspective to see whether it has been practiced and by which peoples in antiquity, the Middle Ages, and more recent times; how it was judged by the moral norms of those times; and what the reasons were for its spread or abolition.[1]

The topic can be approached in sociological terms so as to examine the current incidence of the phenomenon, both clandestine and legal, and also to determine what social and economic conditions foster it.[2]

---

[1] This aspect was addressed in E. Nardi, *Aborto procurato nel mondo greco-romano* (Milan: Giuffrè, 1971); idem, "L'eredità del mondo antico," in *L'aborto: Riflessioni di studiosi cattolici*, ed. A. Fiori and E. Sgreccia (Milan: Vita e Pensiero, 1975), 23–47; E. Sgreccia, "L'insegnamento dei Padri della Chiesa," in *L'aborto*, Fiori and Sgreccia, 49–68; G. Caprile, *Non uccidere: Il Magistero della Chiesa sull'aborto* (Rome: La Civiltà Cattolica, 1974); E. Sgreccia and M.L. Di Pietro, "L'interruzione volontaria di gravidanza nel pensiero cattolico," *Affari sociali* 4 (1989): 75–92.

[2] On clandestine abortion, see B. Colombo, "Sulla diffusione degli aborti illegali in Italia," *Medicina e Morale* 26, nos. 1/2 (1976): 17–78. See also E. Quintavalla and D.E. Raimondi, eds., *Aborto, perché?* (Milan: F. Angeli, 1989); E. Spaziante, "L'aborto provocato: Dimensioni planetarie del fenomeno; Aspetti epidemiologici, demografici e considerazioni di bioetica sociale," *Medicina e Morale* 46, no. 6 (1996): 1083–1134. For the purpose of monitoring the progress of the phenomenon, the Italian law on voluntary termination of pregnancy (VTOP) directs the Minister of Health to present a yearly report on the application of that same law. The report summarizing the final statistics for 2003 and the general figures for 2004 was published in October 2005. *Increasing in 2004:* In 2004 around 137,000 abortions were reported and, although there was a decrease of 41.8 percent since 1982 (the year in which 234,000 cases were reported, the highest number ever), the number of abortions performed in Italy increased by 4,500 in comparison to the final figure for 2003 (approx. 132,000). In particular, the greatest contribution to the increase came from the central (+6 percent) and northern regions (+4.8 percent), whereas in the south and on the islands a slight reduction was noted (-0.1 percent). The trend can be evaluated more precisely by using the abortion rate, which represents the number of abortions per 1,000 women of childbearing age (15–49 years); this rate went from 9.6 per thousand in 2003 to 9.9 per thousand in 2004: an increase of 2.6 percent. *The figures for 2003:* From 1983 to 2003, the

Another vantage point from which the matter can be considered is that of civil and criminal law. There are many works on this subject even for Italy,[3] including commentaries both on law no. 194/78 and more general concepts of criminal law.

Then there is the theological, moral, and canonical point of view, which was amply addressed in the past, especially within the sphere of the Catholic Church, both when abortion was a hidden deed as well as more recently, as a result of the legalization and liberalization that have occurred in various nations.[4]

---

abortion rates decreased in all age groups, with less marked reductions for women under 20 years of age (-12.5 percent); furthermore, a slight increase has been observed since 1995 for the age groups of 20–24 and 25–29 years. The distribution according to level of education follows a trend already discovered in the preceding years, with a great majority of the women who obtained an abortion holding middle-school diplomas (46.4 percent) or high-school diplomas (40.4 percent). As for occupational status, 48.9 percent of the women were employed, while 27.1 percent were homemakers, and 10.1 percent students. An important figure concerns foreign women who had abortions in Italy: the number went from around 9,000 in 1995, the year in which systematic efforts began to collect information about citizenship, to 29,000 in 2002, for an overall increase of 226.3 percent; the number reached approximately 32,000 in 2003. Thus over the years the number of procedures performed on women of foreign citizenship has been growing, and in 2003 these procedures represented 25.9 percent of the total number of abortions, whereas in 1998, for example, it was 10.1 percent. It has been estimated that the abortion rate among foreign women has been three times greater than that of Italian women, and this negatively affects the picture of elective abortion in Italy. Contrary to expectations, clandestine abortion has not been eradicated; estimates place the number of cases per year at around 20,000.

[3] F. Stella, "L'aborto come illecito penale," *Medicina e Morale* 24, no. 2 (1974): 243–249; F. Introna, "Legalizzazione dell'aborto: riflessioni medico-legali," in *L'aborto*, Fiori e Sgreccia, 297–312. See also L. Eusebi, "Tutela giuridica dell'embrione ed esigenze irrisolte di prevenzione dell'aborto," in *Ingegneria genetica e biotecnologie nel futuro dell'uomo*, ed. E. Sgreccia and V. Mele (Milan: Vita e Pensiero, 1992), 329–358; M. Zanchetti, *La legge sull'interruzione della gravidanza: Commentario sistematico alla legge 22 maggio 1978 n. 194* (Padua: CEDAM, 1992); F. D'Agostino, *Bioetica* (Turin: Giappichelli, 1996); A. Tarantino, ed., *Per una dichiarazione dei diritti del nascituro* (Milan: Giuffrè, 1996); V. De Paolis, "La protezione penale del diritto alla vita," in *Commento interdisciplinare alla "Evangelium vitae,"* ed. E. Sgreccia and R. Lucas Lucas (Vatican City: Libreria Editrice Vaticana, 1997), 501–520; L. Eusebi, "Corresponsabilità verso le scelte giuridiche della società pluralista e criteri di intervento sulla c.d. norma imperfetta," in *"Evangelium vitae" e diritto*, ed. A. López Trujllo, J. Herranz, and E. Sgreccia (Vatican City: Libreria Editrice Vaticana, 1997), 389–406.

[4] D. Tettamanzi, *Comunità cristiana e aborto* (Alba: Paoline, 1975); B. Häring, *Free and Faithful in Christ*, 3 vols. (New York: Crossroad, 1978–81); C. Caffarra, "Il problema morale dell'aborto," in *L'aborto*, Fiori and Sgreccia, 313–320. See Congregation for the Doctrine of the Faith, Declaration *Quaestio de abortu*, November 18, 1974; idem, Instruction *Donum vitae*, February 22, 1987; L. Ciccone, "L'aborto," in *Non uccidere: Questioni di morale della vita fisica* (Milan: Ares, 1984), 144–254; M. Schooyans, *L'avortement enjeux politiques* (Longueil: Préambule, 1990); E. De Lagrange, M. M. De Lagrange, and R. Bel, *Un complot contre la vie: L'avortement* (Paris: Société de Production Littéraire, 1979); A. Rodriguez Luño, "La valutazione teologico-morale dell'aborto," in *Commento interdisciplinare*, Sgreccia and Lucas Lucas, 419–434; L. Ciccone, "I problemi etici dell'aborto nell'enciclica *Evangelium vitae*," in *"Evangelium vitae" e bioetica: Un approccio interdisciplinare*, ed. E. Sgreccia and D. Sacchini (Milan: Vita e Pensiero, 1996), 59–76.

It cannot be overlooked that the phenomenon of abortion, whether clandestine or not, is the object of study in terms of psychology as well: psychology is interested in the motivations that can push a choice for life or, alternatively, to the choice of death.[5] It is also psychology that can evaluate the psychological repercussions in the woman after a voluntary abortion.[6]

One could also examine a cultural aspect of the topic, because no doubt the authorization or legalization of abortion is seen by some cultures or subcultures as a civilized choice, while for others it is one of the most alarming signs of the culture of death.

In this volume, I intend to consider the bioethical aspect. The discussion will therefore be based on the findings of biology and genetics so as to evaluate them in the light of rational ethics, stopping short of the evaluations of the Church's Magisterium and of moral theology, though they are also important. This perspective of bioethics should serve as a meeting point for believers and nonbelievers and should define the line of professional ethics for the physician.[7]

This vantage point is therefore limited yet fundamental: it involves inquiring whether or not the human embryo is an individualized human life from the first moment of fertilization and, if so, whether there are circumstances in which it becomes lawful or ethically acceptable to end or voluntarily terminate pregnancy. The repercussions at the level of behavior and with regard to the law and current customs depend on this answer, which comes from biological science and ethics jointly.

The behavior to be examined is above all that of the physician and his direct collaborators, while admitting that the problem arises first within families.

I wish to make yet another clarification: this discussion will work within the realm of objective values and objective truths, that is to say, without getting into or passing judgment on subjective responsibilities. Indeed, every ethical choice can be considered in its objective reality or in its subjective dimension: the subject may be in extenuating circumstances, coerced, or ignorant of what is involved, and in any case should be helped even if he has committed an immoral act, yet the objective reality remains what it is and help, if has a direction, must be given morally in the direction of objective truth, which in this case means not repeating the error committed and preventing it whenever possible and to the extent required by the responsibilities of each person.

---

[5] L. Ancona, "Prospettive psicologiche dell'aborto," in *L'aborto*, Fiori and Sgreccia, 235–263. See also L. Zichella and L. Cersosimo, eds., *Aspetti medici psico-sociali e legali dell'interruzione volontaria di gravidanza* (Rome: Meditalia, 1986).

[6] M. T. Mannion, ed., *Post-Abortion Aftermath* (Kansas City, MO: Sheed and Ward, 1994); S. Gindro et al., eds., *Aborto volontario: Le conseguenze psichiche* (Rome: CIC Internazionali, 1996).

[7] I recommend the previously mentioned volume Fiori and Sgreccia, *L'aborto*, especially the articles A. Serra, "Il neoconcepito alla luce degli attuali sviluppi della genetica umana," 115–148 (also published in *Medicina e Morale* 34, no. 3 [1974]: 333–366); V. Fagone, "Il problema dell'inizio della vita del soggetto umano," 149–180 (also published in *Medicina e Morale* 34, no. 2 [1974]: 212–242); and A. Fiori, "Strage di Stato degli innocenti," 9–22. See also A. Serra, E. Sgreccia, and M. L. Di Pietro, *Nuova genetica e embriopoiesi umana* (Milan: Vita e Pensiero, 1990).

From this tried and true, rational perspective, then, the following topics will be addressed: the biological reality of the new conceptus; the objective value of human life from the moment of conception, in itself and in relation to contingent situations (illness of the mother, illness of the unborn child, combating clandestine abortion with legalized abortion); and the behavior of the physician with regard to a request for voluntary termination of pregnancy and with regard to the law (conscientious objection). These are four aspects I propose to develop within the framework of bioethics.

The New Conceptus in Light of Genetics and Human Biology

Today human conception is no longer a natural mystery hidden behind impenetrable walls, obscured by shadows of doubt resulting from imprecise observations or wrapped in veils of beguiling syllogisms or deceptive sophisms. Today this reality cannot be mystified at the pleasure of those discussing it. As with any conquest of knowledge, much still remains to be understood and researched in order to arrive at an ever more accurate understanding of this reality, yet the observations accrued thus far are already sufficient to clarify the aspects that are of interest to us here.[8]

The first incontestable fact brought to light by genetics is this: at the moment of fertilization, that is, the penetration of the spermatozoon into the egg cell, the two gametes of the parents form *a new biological entity*, the zygote, which bears within it a new *individualized design program*—a new individual life.

It is commonly observed that the first event in the formation of an individual human being is the fusion of the two highly specialized cells, the oocyte and the spermatozoon, through the process of fertilization. This is a highly complex process in which two extraordinary and teleologically programmed cells, which constitute two systems that are self-sufficient yet ordered to one another, interact with one another, thus giving rise to a new system. In an initial stage of the encounter—with the assistance of species-specific receptors found in the pellucid zone that surrounds the egg cell and by binding proteins present on the external membrane of the sperm cells and by proteolytic and glycolytic enzymes released by unique structures called *acrosomes* in the head of the sperm cells—the head of a spermatozoon penetrates into the cytoplasm of the oocyte.

As soon as this has occurred, a chain of activities begins that clearly indicates there are no longer two systems operating independently of one another, but rather a "new system" has been formed, which begins to operate as a "unity" that is more precisely termed a "zygote" or "one-celled embryo."[9]

---

[8] Serra, "Il neoconcepito," in *L'aborto*, Fiori and Sgreccia, 115.

[9] A. Serra, "Dalle nuove frontiere della biologia e della medicina nuovi interrogativi alla filosofia, al diritto, e alla teologia," in *Nuova genetica*, Serra, Sgreccia, and Di Pietro, 69–70; A. Serra, "Per uno statuto integrato dell'embrione umano: Alcuni dati della genetica e dell'embriologia," in *Nascita e morte dell'uomo*, ed. S. Biolo (Genoa: Marietti, 1993), 55–105, A. Serra, "Pari dignità all'embrione umano nell'enciclica *Evangelium vitae*," in *"Evangelium vitae" e bioetica*, Sgreccia and Sacchini, 147–173; A. Serra, "Lo stato biologico

The two respective gamete cells have in themselves a well defined patrimony, the genetic program, gathered around the twenty-three pairs of chromosomes: each of the gamete cells has one half of the genetic material compared with the somatic cells of the parent organism along with *genetic information* that is qualitatively different from that of the somatic cells of the paternal or maternal organism. These two gametes differ one from one another and from the somatic cells of the parents, yet they complement one another: once they are united they activate a new *design program* by which the new conceptus is defined and individualized.[10] There is not the slightest doubt about the novelty of this design program that results from the fusion of the twenty-three pairs of chromosomes, and to deny it would mean denying the sure findings of science:

> The forty-six chromosomes of the zygote are indeed a qualitatively new combination of instructions, called a *genotype* in technical terms; this new combination is capable of imprinting a new pattern of structure and activity on the cell that possesses it. For a clinical proof of its novelty, suffice it to mention the results of combining the various blood types: although the two parents have type A and type B respectively, and these types are definitively determined by genetic instructions, the new conceptus can have type O blood (if the parents are both heterozygous).[11]

To sum up and reaffirm what has been said thus far, I here recall the contents of the document "Identità e statuo dell'embrione umano" (Identity and Status of the Human Embryo) by the Bioethics Center of the Catholic University of the Sacred Heart (Rome), which discusses this subject as follows:

> The first level of facts comes from the study of the zygote and its formation. From these facts it is clear that during the fertilization process, as soon as the egg and the sperm—two teleologically programmed cellular systems— interact with one another, a new system immediately begins that has two fundamental characteristics:
>
> 1. The *new system* is not the simple sum of the two subsystems but a combined system, which as a consequence of the loss of the individuality and autonomy of the two subsystems begins to operate—given all the necessary conditions—as a new, intrinsically determined unity, toward attaining its specific and final form. Hence the classic and still current term *one-celled embryo.*
>
> 2. The biological center or coordinating structure of this new unity is the *new genome* with which the one-celled embryo is endowed; in other words, those complex molecules—visibly recognizable at the cytogenetic level in the chromosomes—that contain and conserve as if by memory a well-defined project design, with the essential and permanent "information" for the gradual

---

dell'embrione umano: Quando inizia l'"essere umano'?" in *Commentario interdisciplinare*, Sgreccia and Lucas Lucas, 573–597.

[10] A. Serra, "Embrione umano, scienza e medicina: In margine al recente documento vaticano," *La Civiltà Cattolica* 2 (1987): 247–261; idem, "Quando comincia un essere umano," in *Il dono della vita*, ed. E. Sgreccia (Milan: Vita e Pensiero, 1987), 99–105.

[11] Serra, "Il neoconcepito," 117.

and independent realization of that project. This genome is what identifies the one-celled embryo as biologically human and specifies its individuality. This genome is what confers on the embryo enormous morphogenetic potential that the embryo itself will gradually actuate throughout its entire development by means of constant interaction with its environment, whether cellular or extracellular, from which it receives signals and materials.[12]

The fact that should be most noted here is that this new program *is neither inert nor "executed"* by means of maternal physiological organs using the program in the way an architect uses a blueprint as a passive pattern; rather, it is a *new project* that builds itself and is its own principal moving force. Although the informational systems of maternal origin that had brought the egg cell to maturity remain active for some time, the zygote's control systems are nevertheless set in motion from the first moment of fertilization, assuming complete control of it even before implantation: from the formation of the blastomeres by replication and duplication through the formation of the blastocyst and on to implantation, the genetic information intrinsic to this new entity is as the pilot or architect of the construction project.

Let us think of the trajectory of this self-building dynamic from DNA to RNA to proteins in simple terms: the DNA can be imagined as a speaking telecaster, RNA as a transcribing teleprinter, and the proteins as the printed result of the transmission.

As early as the two-cell stage—therefore before implantation—the so-called *rDNA* and *tDNA cistrons* are activated, which direct or dictate, so to speak, their messages into specific molecules of ribonucleic acid: rRNA is contained in the ribosomes, which are minute endocellular organs that form the nuclear machinery for synthesizing proteins, and tRNA is absolutely necessary in the same mechanism of protein synthesis and has the task of transporting the amino acids.

The recognition of the specific amino acids is said to occur by means of a "second genetic code" that was mentioned in the chapter on the origin of life. Once these molecules, which serve as synthesizing mechanisms, are ready in a certain sufficient number of cells, the part of the genome called mRNA is activated: these molecules transport the instructions for the synthesis of proteins and specific proteinacious molecules necessary for the formation of the blastocyst.

This process occurs between the second and third days after fertilization, continues over the following days at a dizzying speed and, by means of signals that are transmitted from cell to cell, the cells gradually and selectively arrange themselves in the appropriate regions of the embryo. The various cells, still commanded by the instructions of the genetic program, come to assume distinct forms and functions with the precise characteristics required.

This process of development has three unique biological properties that are summed up as follows in the aforementioned document by the bioethics center:

---

[12] Bioethics Center of the Catholic University of the Sacred Heart, "Identità e statuto dell'embrione umano," supplement, *Medicina e Morale* 39, no. 4 (1989): 665–666; Pontifical Academy for Life, "Comunicato finale dell'Assemblea Plenaria su 'Identità e Statuto dell'embrione umano,'" *L'Osservatore Romano*, February 21, 1997, 4. For the debate on the status of the human embryo, see also Italian National Bioethics Committee, *Identity and Status of the Human Embryo*, June 22, 1996 (Rome: Presidenza del Consiglio dei Ministri Dipartimento per l'Informazione e l'Editoria, 1996), http://www.governo.it/bioetica/eng/pdf/human_embryo_19960622.pdf.

(1) *Coordination.* During this whole process, from the formation of the zygote on, there is a succession of molecular and cellular activities guided by the information contained in the genome and controlled by signals originating from interactions that unceasingly multiply at every level, within the embryo itself and between it and its environment. Precisely this guidance and control result in the coordinated expression of thousands of structural genes, a process that involves and confers a strict unity on the organism, which is developing in space and time. (2) *Continuity.* The "new life cycle" that begins at fertilization proceeds—if the required conditions are satisfied—without interruption. The individual events, for example cellular replication, cellular differentiation, the differentiation of tissues, and the formation of organs, obviously appear to be successive. But the process per se of the formation of the organism is continuous. It is always the same individual that is acquiring its definitive form. If this process were to be interrupted at any moment, the death of the individual would occur. (3) *Gradualness.* It is a law intrinsic to the formation process of a multicellular organism that it should attain its final form by passing from simpler forms to increasingly complex forms. This law of gradualness in acquiring the final form implies that the embryo, from the one-celled stage on, permanently keeps its own identity and individuality throughout the entire process.

This developmental process therefore involves the simultaneous actuation of a quantitative expression and a qualitative expression, with a succession of phases and a determination of successive places that always correspond to a definite program.

The life cycle of the new living being proceeds moment by moment, with predetermined rhythms, as it constructs itself by establishing its own direction, differentiated structures, and quality of growth according to the design inscribed in the genome from the moment of fertilization (histogenesis, regionalization, morphogenesis). Studies have been conducted on animals (rabbits) and on aborted fetuses, allowing scientists to track the percentage of development by the mRNA genes in various zones (brain, liver) to discover how the genome proceeds with the construction of different organs at various speeds through mechanisms of replication or repression. In research done on aborted and malformed fetuses, they have traced down precise segments of DNA and of missing or aberrant chromosomes responsible for the malformations themselves.

In conclusion,

> the development of organization is so rapid that at the end of the eighth week—unless deviations occur owing to a more or less significant alteration of the "program" itself or attributable to forces resulting from an anomalous interaction between the genome, the cytoplasm, and the environment—organization is complete and the embryo possesses, albeit in miniature, all the characteristic structures of a human being (including well-defined sex) as they are recognized at the end of pregnancy.[13]

The following characteristics have therefore been highlighted: programmed differentiation, finality, continuity without qualitative interruption, and the self-production and self-direction of the program within the genome and the embryo. The autonomy here mentioned is not to be understood in an absolute sense, however, because there

---

[13] Serra, "Il neoconcepito," 123.

is no complete independence even after birth: even adults depend on the living environment that surrounds us (atmosphere, food, etc.). Suffice it to recall that when the blastocyst arrives in the uterus its development stops if the uterus has not been prepared by estrogens to accept its implantation. This is a matter of an *extrinsic* dependence, however, analogous to that of an adult with regard to his environment: the maternal environment provides nutrition and oxygenation and removes the products of metabolism that can be toxic. The quality, impetus, and direction of development, however, do not depend on directive maternal organs but rather on the autogenetic composition of the embryo itself.

To say that the embryo is a part of the mother's body is therefore an error or an anti-scientific mystification. The biological evidence for this fact is now found in a more striking way through experiments involving *in vitro fertilization*, which demonstrate precisely that fertilization can be actuated by the union of the two gamete cells and that from that moment, in a suitable environment, the embryo develops by way of self-constructing mechanisms. Another proof had already been discovered in animals when some blastomeres were able to develop into advanced-stage embryos after being transferred into male organs (liver, kidney, spleen, brain). This means that the development of the embryo depends on the mother only in an *extrinsic* way.

The fact that extrinsic or pathological events can be observed in these initial phases whereby two homozygotic twins are derived from one fertilized cell, or two fertilized eggs combine into one (hybridization), or interruptions or mini-abortions even occur, does not change the nature and the intrinsic quality of the genome and of its individualized dynamic.

The Human Character of the Embryo

Since biological development is uninterrupted and occurs without any intrinsic qualitative mutation and without the need for any further causative intervention, it must be said that the new entity constitutes a new human individual who from the moment of conception continues his life cycle, or more precisely, his vital trajectory. The self-development of the embryo occurs in such a way that the successive phase does not eliminate the preceding one but rather absorbs and develops it according to an individualized and ordered biological law.

Even when the human form is not yet recognizable, there are hundreds of thousands of muscle cells already making a primitive heart beat and tens of millions of nerve cells assembling into circuits and arranging themselves to form the nervous system of a specific person.

The distinction between a human being and a "humanized being"—the one distinguished from the other by the appearance of the human form—collapses in this light, as does the other objection hinging on the analogy between *ontogeny* and *phylogeny*.

This theory, based on evolutionism, claims that the formation of the individual recapitulates and condenses the history of the evolution of the life forms in the world; human beings appeared last in this history and so humanization is preceded by vegetative or animal life forms. Yet at no instant in the development of a human embryo is there a biological dynamic of a vegetative type or of an undifferentiated nature belonging to a different species. The whole that will appear at the end (if "end" is understood to mean birth or adult life) is already causatively and genetically present at the beginning and in an individual sense as well.

426

Also ruled out, therefore, is the notion of *thresholds* or *levels* representing moments in which continuity is broken (gradualist theory), after which there would supposedly be a change of program. This intermediate state does not exist from a genetic perspective.

Moreover, the uncertainty of legal scholars as to whether or not the concept of *person* can be applied to the initial stages becomes a pointless distraction when we consider that it matters little how one wishes to define it legally: *this* embryo is nevertheless already the same individual-in-development who will be defined as a person. One can very well respond—together with Tertullian, who among other things was a lawyer—that "he who will be a man already is one."

It will be possible to explain this same point more precisely later on in connection with the problem of simultaneous or successive ensoulment, which is a question of philosophical and speculative nature that turns out to be irrelevant and contradictory when confronted with the reality of ontogeny.

Some difficulties are nevertheless adduced by the scientific world concern the beginning of an individual human life. Indeed, there are those who maintain that an individual human life does not begin until the recombination of the two genetic patrimonies, one of paternal origin and the other of maternal origin, in the new genome of the embryo (theory of karyogamy), which is said to take place at around twenty-one or twenty-two hours after fertilization.[14] Others go so far as to claim that if a one-celled embryo is characterized as an individual human being, then this assessment should be extended to the single gametes as well.[15]

According to the Warnock Report, on the other hand, human embryos may be used for experimental purposes up to the fourteenth day after conception, clearly implying that the human status of the embryo is not acknowledged or else it is subordinated to adult life until that time.[16] The fourteen-day period was first proposed in 1979 by the Ethics Advisory Board of the Department of Health, Education, and Welfare (DHEW) in the United States, which justified it by the fact that implantation is complete on the fourteenth day.[17]

In 1984, the Waller Committee in Australia repeated "not more than fourteen days," because after that stage the *primitive streak* forms and differentiation can be observed in the embryo.[18]

---

[14] K. Dawson, "Fertilization and Moral Status: A Scientific Perspective," in *Embryo Experimentation*, ed. P. Singer et al. (Cambridge: Cambridge University, 1990), 43–52.

[15] M. Mori, *Aborto e morale* (Milan: Il Saggiatore, 1996).

[16] Department of Health and Social Security Committee of Inquiry into Human Fertilisation and Embryology (Warnock Committee), *Report of the Committee of Inquiry into Human Fertilisation and Embryology* (London: Her Majesty's Stationery Office, 1984), http://www.hfea.gov.uk/docs/Warnock_Report_of_the_Committee_of_Inquiry_into_Human_Fertilisation_and_Embryology_1984.pdf.

[17] DHEW Ethics Advisory Board, *HEW Support of Research involving Human In Vitro Fertilization and Embryo Transfer* (Washington, DC: DHEW, 1979).

[18] Committee to Consider the Social, Ethical and Legal Issues Arising from In Vitro Fertilization, *Report on the Disposition of Embryos Produced by In Vitro Fertilization* (Melbourne: Government Printer, 1984).

The Warnock Committee, or more accurately one part of that study group, therefore confirmed this date as the time at which the primitive streak forms and the individual development of the embryo simultaneously begins.[19]

Anne McLaren, a member of the Warnock Committee, declared the following in her article: "The point at which I began as a total whole individual human being was at the primitive streak stage, the formation of the embryo."[20] The appearance of the *primitive streak* is said to indicate that the cells destined to make up the embryo proper have by now been differentiated from the cells that will instead form the placental and protective tissues.[21]

Embryonic development up to the fourteenth day would therefore be, according to McLaren, a period of preparation in which all the protective and nutritive systems develop that are required to sustain the future embryo; the embryo itself can only begin to develop as an individual entity when the support systems have been established.[22]

McLaren's claims are repeated by Clifford Grobstein, who asserts, "The human pre-embryo has a special set of characteristics that distinguish it biologically from the ovum that precedes it and the embryo that succeeds it. It is genetically but not developmentally (or in any other scientific sense) an individual."[23] They are also repeated by Norman Ford, who declares, "The appearance of one primitive streak signals that only one embryo proper and human individual has been formed and begun to exist. Prior to this stage it would be pointless to speak about the presence of a true human being in an ontological sense."[24]

As can be noticed, these authors introduce the term "pre-embryo" to designate the period of prenatal human life between the moment of fertilization and the appearance of the primitive streak, since only in this phase is there said to be "a spatially defined entity that can develop directly into a foetus and thence into a baby."[25]

The Warnock Report also declared the introduction of the term "pre-embryo" was the result of a decision ("we agreed"). The report affirms that "the timing of the different stages of development [of the embryo] is critical, once the process has begun, there is no particular part of the developmental process that is more important

---

[19] Warnock Committee, *Report*, chap. 11, §22, p. 66.

[20] A. McLaren, "Prelude to Embryogenesis," in *Human Embryo Research: Yes or No?*, ed. G. Bock and M. O'Connor (London: Tavistock, 1986), 5–23.

[21] M. Zatti, "La prospettiva del biologo: Statuto biologico dell'embrione," in *Procreazione artificiale ed interventi nella genetica umana: Atti del convegno di Verona; 2–3-4 e 25 ottobre 1986*, ed. Ordine degli Avvocati e Procuratori di Verona (Padua: CEDAM, 1987), 181.

[22] McLaren, "Prelude," 5–23.

[23] C. Grobstein, "Biological Characteristics of the Preembryo," *Annals of the New York Academy of Sciences* 541 (1988): 347–348.

[24] N. M. Ford, *When Did I Begin? Conception of the Human Individual in History, Philosophy and Science* (New York: Cambridge, 1989), 172, emphasis removed.

[25] McLaren, "Prelude." Elsewhere in the article, McLaren describes a "pre-embryo" (or the embryo from fertilization until the primitive-streak stage) as "a different entity, which includes and gives rise to the 'embryo' that grows into a foetus and neonate but is in no way coextensive with it" (14).

than another; all are part of a continuous process, and unless each stage takes place normally, at the correct time, and in the correct sequence, further development will cease. Thus biologically there is no one single identifiable stage in the development of the embryo beyond which the in vitro embryo should not be kept alive."[26] It then continues, "However, *we agreed* that this was an area in which some precise *decision* must be taken, in order to allay public anxiety."[27]

The term *pro-embryo* is also used alongside pre-embryo. It is an old term, introduced in botany as early as 1847, that refers to the alternation of generations in lower plant species. In the contemporary sense, however, *pro-embryo* is used in much the same way as pre-embryo for the purpose of introducing a discriminatory criterion.[28]

The fourteenth day—the seventh day according to some authors[29]—is also said to be the limit beyond which the phenomena of *twinning* and *hybridization*[30] are no longer possible. As for *monozygotic twinning*, confirmed laboratory experiments on animal zygotes (mice, rabbits) have proved that in the earliest days the cells in the DNA (rDNA, tDNA)- and RNA (rRNA, tRNA)-building phase still retain the character of totipotent cells: if, through the weakness of the zygote's membrane, a division accidentally occurs in the earliest phases (at the sixty-four-cell stage it seems that totipotency no longer exists), the portions resulting from the division give rise to as many developmental programs and as many individuals.[31]

In the case of *hybridization*, there is a possibility that two fertilized cells may fuse at the earliest stages of development, giving rise to a unique individual.

According to some authors, individual human life begins at the moment of implantation.

The fertilized ovum starts to implant in the uterine wall between the sixth and seventh days, and the process is completed on the ninth day; by the fourteenth day the endometrium has covered the implanted embryo.[32] For these authors, individual human life therefore only begins on the sixth day after conception. Based on this theory, the blastocyst passes from the state of *totipotency* to *unipotency* at the moment of implantation, developing only as a human being and only into that human being from that moment on.[33]

For still other authors, finally, the formation of the nervous system and the beginning of brain life are considered fundamental in embryonic development. "Human life can be seen as a continuous spectrum between the onset of brain life in utero (eighth week of gestation), until the occurrence of brain death. At any point human

---

[26] Warnock Committee, *Report*, chap. 11, §19, pp. 64–65.

[27] Ibid., 65, emphasis added.

[28] J. D. Biggers, "Arbitrary Partitions of Prenatal Life," *Human Reproduction* 5 (1990): 1–6. See also A. Sutton, "Ten Years after the Warnock Report: Is the Human Neo-Conceptus a Person?" *Medicina e Morale* 44, no. 3 (1994): 475–490.

[29] See the data reported in Serra, "Per uno statuto integrato."

[30] Zatti, "La prospettiva del biologo," 185–186.

[31] W. Ruff, "Individualität und Personalität in embryonalen Werden: Die Frage nach dem Zeitpunkt der Geistbeseelung," *Theologie und Philosophie* 45 (1970): 25–49.

[32] Zatti, "La prospettiva del biologo," 182.

[33] J. F. Malherbe, "L'embryon est-il une personne humaine?" *Lumière et Vie* 172 (1985): 19–31.

tissue or organ systems may be present, but without the presence of a functional human brain, these do not constitute a 'human being,' at least in a medical sense."[34]

Joseph Donceel also states, "I do not know when the human soul is infused into the body, but I, for one, am certain that there is no human soul, hence no human person, the first few weeks of pregnancy. ... The least we may ask before admitting the presence of a human soul is the availability of these organs: the senses, the nervous system, the brain, and especially the cortex. Since these organs are not ready during early pregnancy, I feel certain that there is no human person until several weeks have elapsed."[35]

In connection with this, studies of the development of the nervous system of the embryo for the purpose of identifying the moment at which the soul is infused should be noted.[36] What the authors try to prove is that brain activity would in fact allow the passage of embryonic life from the "cellular level" to the "holistic level": according to their theory, "holistic functioning ... means the beginning of life of a new individual,"[37] and so "a new human life begins when the newly built body organs and systems begin to function as a whole,"[38] which is only with the onset of neural functioning.[39]

In response to these objections, and in particular with regard to the *primitive streak*, the words of Angelo Serra should be kept in mind: "This represents nothing other than the point of arrival in a sequentially ordered process with unbroken continuity that began at the moment in which the zygote was formed. The cell or cells from which the cell layers making up the embryonic primitive streak will originate are already present during the period of preparation for the nutritive and protective systems. It does not appear somehow by chance as if by external causes and separate from the whole process that begins unfolding at the moment of syngamy: it is a product of that process."[40]

As far as the use of the term "pre-embryo" is concerned, let us again read what Serra writes:

It is no doubt permissible—and perhaps sometimes advantageous and useful from a terminological point of view, which always has active value—to introduce new symbols in order to emphasize new aspects.

Therefore the term *preembryo* proposed by McLaren and others could be used to indicate the precocious phase of the embryo, that is, the period from

---

[34] J.M. Goldenring, "The Brain-Life Theory: Towards a Consistent Biological Definition of Humanness," *Journal of Medical Ethics* 11, no. 4 (1985): 198.

[35] J.F. Donceel, "Immediate Animation and Delayed Hominization," *Theological Studies* 31 (1970): 101.

[36] See M.L. Di Pietro, "La fecondazione extracorporea," *Medicina e Morale* 34, no. 4 (1984): 570–571.

[37] M.C. Shea, "Embryonic Life and Human Life," *Journal of Medical Ethics* 11, no. 4 (1985): 206.

[38] Ibid., 209.

[39] "For the whole to function as one organism there needs to be sufficient brain activity to regulate and coordinate the workings of the various body organs and systems." Ibid., 206.

[40] Serra, "Embrione umano," 255–256.

the formation of the zygote to the moment in which the embryonic primitive streak appears. But it would be erroneous if one tried to imply by this distinction that the two phases, from the zygote to the primitive streak and from then on, are two discontinuous processes with no correlation between them and that the two structures—the one before the primitive streak and the one after—belong to two different subjects, or that the first is an aggregate without a subject.[41]

In the case of monozygotic twinning, the eventual division of the embryo does not contradict what has been said about its continuity of development, but rather proves it. Indeed, the moment of division requires the intervention of a cause that interferes with the design; that is to say, it does not happen by virtue of the mechanism of development, but contrary to it. Furthermore, the outcome is still consistent with the development here described, and this self-constructive and determinate development is repeated in each of the divided portions. The nature of these portions of the zygote that come to behave like as many zygotes is still a human design (not a plant or animal).

> Let us suppose that by some mechanism or other, spontaneous or induced, one part should become detached from this already-determined developmental system: one cell if the system consists of two, or a few or a relatively large number of cells if the detachment occurs in the more advanced stages, and that among these cells one or more do not undergo a sudden reduction of their totipotency, thus remaining capable of resuming or continuing the development that has begun. If the development occurs, then logically one can only declare that another system "had its origin" from the first system; the new one can be similar to the first or different as well, as already proved by various cases with monozygotic twins of differing chromosomal makeup.

> It cannot be affirmed, however, that the first system became or included another system. ... The first still remains the first system, modified, if you will, as an adult would be whose limbs were amputated (the analogy here is exclusively for the purpose of illustration); and the second system will be the one that originates from the first, even if the order of temporal succession could not be determined. Moreover, whereas the second begins its own ontological existence at the moment in which it separates, the first continues its development without losing its own biological and ontological identity.[42]

Finally, there is a limit within which even the accidental and external cause loses its efficacy: so great is the guiding determination of the design that twinning is only possible in human beings up until the end of the second week at the most.

A similar difficulty is alleged by some authors based on the phenomenon of hybridization, as already mentioned; that is, based on the possibility that two fertilized cells, in the earliest hours after fertilization, before implantation or during the implantation phase, if they are synchronized in phase and in proximity to one

---

[41] Serra, "Dalle nuove frontiere," 82.

[42] A. Serra, "Quando è iniziata la mia vita?" *La Civiltà Cattolica* 3348 (1989): 582. On this topic, see also P. Caspar, "Individuazione genetica e gemellarità: L'obiezione dei gemelli omozigoti," *Medicina e Morale* 44, no. 3 (1994): 453–467.

another, can combine into a single zygote that behaves like a single totipotent cell and leads to the development of a single individual. Given this fact, the authors in question try to draw the conclusion that one cannot speak of complete individuation or the occurrence of ensoulment before implantation. The issue would have great importance for embryos produced in vitro (and not implanted), whom some would like to define not as "persons with potential" but rather as only "potential persons."[43]

But in this case, too, there is no denying that the circumstance does not contradict but rather confirms the existence of a well-defined program in each of the two fertilized cells—a program that would develop independently in each if it were not disturbed by an external cause. External causes can also interrupt embryonic development, but we cannot conclude from this that the embryo is incapable of autonomous development if allowed to live in normal conditions.

The phenomenon of hybridization, on the other hand, "is not known well enough to be able to serve as the basis for a deeper reflection; nevertheless ... it must be said that the life of one individual unfortunately comes to an end. Perhaps we should define this as an accident and the recombination as the premature death of one twin,"[44] but that does not authorize us to postpone the beginning of the individual life of the survivor.

As far as implantation is concerned, the embryo certainly cannot continue to live without it just as a child could not survive after birth without food. But it is not implantation that causes the embryo to be an embryo, just as it is not the mother's milk that causes the baby to be a baby. As a result of this, it cannot be concluded that individuation is not yet complete.

Finally, the statement that the organs and systems already present in the embryo cannot constitute a human being without a functioning human brain does have meaning and validity when the human subject, having arrived at the end of his life cycle, ceases to live: brain death is then considered a definitive sign of the conclusion of this cycle.[45]

The situation of the developing embryo is different:

We are dealing (in this case) not with the terminal phase of a dynamic life process in which the individual begins to disintegrate, but instead with a single dynamic process unifying all the parts that are gradually being prepared. This is the developing human subject who, by the ontogenetic law, requires a gradual differentiation and hence the gradual formation of the cerebral structures as well. This gradualness does not cause qualitative leaps but only richer expression of the potential already inscribed in the zygote.[46]

In conclusion, the new conceptus has its own well-defined biological reality: it is a developing but totally human individual that constructs its own form autonomously, moment by moment, and without any discontinuity, while

---

[43] Malherbe, "L'embryon," 19–31; J. Mahoney, *Ethics and Belief* (London: Sheed and Ward, 1984).

[44] R. Di Menna, "Umanizzazione ed animazione del concepito umano," in *Scienza ed origine della vita* (Rome: Orizzonte Medico, 1980), 55.

[45] Serra, "Embrione umano," 256.

[46] Ibid.

implementing—by means of intrinsic activity—a design that is programmed in its genome.[47]

At this point I would like to consider one further point that touches on an even vaster set of problems: the question of the origin of life. It is undeniable from the scientific perspective that in the formation of the new conceptus, from the first moment of fertilization until birth and throughout the process of subsequent growth and development, there is a certain determinism at work that is oriented toward a finalism of design. The determinism that reveals itself in the earliest phases is directed toward a very precise and purposeful design: determinism and finalism encounter and affect one another.

How is it possible, then, to talk about "chance and necessity" as the explanation for living phenomena in general and human life in particular? Chance is said to be present when there are many open possibilities, and necessity cannot spring from chance, because that would be a contradiction in terms. We see that there is no chance here; nothing is left up to multiple eventualities.[48] Is it not more logical to think of the creational explanation—which is the only plausible one—whereby the design is the reflection of a mind that orders and causes the biological processes?

But this problem should be discussed more fully on a general level that takes into account not only ontogeny but also phylogeny. I have mentioned it here, however, because the horizon of philosophical reflection proper to bioethics cannot stop at the description of the phenomenon in terms of its mechanisms. It has to asks its own questions, which have the same form of expression—the whys: (1) the causal why (i.e., by what producing cause or agent?) and the finalistic why (i.e., for what end or purpose?).

These questions go beyond the boundaries of purely physical facts and are therefore metaphysical questions. They are born from within the physical facts and are not random or escapable.

In order to summarize the discussion on the status and development of the new conceptus from a genetic perspective, I will have to use a comparison that I owe in a way to the essay by Serra cited above. Imagine that we have to build a house. We need an architect who makes the blueprint, a contractor who oversees the project, workers who carry it out, and materials with which to do so. In the zygote these various functions—planning, managing, construction, and building materials—are found and activated internally. The zygote is the planner, the contractor, the builder, and the manufacturer of the building materials. Moreover, just as the house already reveals the plan on which it is based from the first appearance of the foundation, so too the zygote, once it becomes an embryo, manifests the entire structure of the individual: the mother furnishes only the work environment and the necessary supplies to manufacture the building materials.

The essential difference is only this: the architect, the contractor, and the workers build an object outside of their own persons, whereas the embryo builds itself.

This fact and this process examined thus far amount to a proven scientific finding that no one can deny and that must be taken as a "datum" and not an opinion.

---

[47] Serra, "Il neoconcepito," 130.

[48] J. Monod, *Chance and Necessity: An Essay on the Natural Philosophy of Modern Biology* (New York: Knopf, 1971).

Slogans such as "the embryo is part of the mother's body," "procured abortion is just like any other operation," "the woman has full rights over her uterus," and so on, are an offense against the competency of science even more than against morality.

The Ontological and Ethical Value of the New Conceptus

What has been carefully explained so far leads to the conclusion that the embryo is a developing human individual and therefore deserves the respect due to every human being. One could say that the problem already has its ethical solution and that the question as to the liceity of procured abortion already has its answer. Indeed, without diminishing the importance of the findings of genetics and embryology, rational reflection of the philosophical and ethical sort is equally necessary for many reasons.[49]

This field will be explored by examining a series of philosophical and sociological questions that are not only hypothetical but have actually been posed in recent years by experts on the problem and in the context of political debates, which are always fraught with presumptions and emotions.

The law as a preventive measure

First of all, here is an objection—used as a diversion, I would say—that is invoked as an alibi: *the law as a preventive measure*. It is said that abortion should not be discussed "moralistically" or "in the abstract," but rather in social and concrete terms: what is in question is not the reality of abortion in itself, which everyone considers or ought to consider a negative one or else something to be left up to the moral and religious conscience of each individual; what is in question here is instead the need to find a law to regulate the de facto practice of clandestine abortion.

Clandestine abortion, the argument goes, is an economic exploitation of a woman in a difficult situation, it is dangerous to the woman's health and life, and there is no way to fight it other than the legal regulation of abortion. Various "preventive" solutions should then be devised, such as *legalization*, *regulation*, and *liberalization*, but a law is required: a law that aims to end the tragedy of clandestine abortion—they say—would be deemed a positive and morally good law. Some have gone so far as to declare, in the midst of the polemics, that any opposition to such a law would be an ideological imposition shielding the interests of various profiteers (e.g., the doctors performing clandestine abortions) or else an effort to achieve partisan consensus on matters of personal and individual conscience. The battle to pass the law has therefore become a battle for freedom of conscience, civil rights, and the vindication of individual freedom over and against a medieval state. Slogans of this sort have piled up and contributed to the lack of calm, objective thinking. Other more cautious commentators use the expression "lesser evil." Legalization would be the lesser evil compared with the spread of clandestine abortions.

---

[49] On this topic, see also L. Palazzani, *Il concetto di persona tra bioetica e diritto* (Turin: Giappichelli, 1996); A. Pessina, "Bioetica e antropologia: Il problema dello statuto ontologico dell'embrione umano," *Vita e Pensiero* 6 (1996): 402–424; V. Possenti, "La bioetica alla ricerca dei principi: La persona," *Medicina e Morale* 42, no. 6 (1992): 1075–1096.

Let us seek to respond rationally. Let us disregard for the moment the fact that this sort of reasoning has often disguised a theoretical presumption of the following sort: the fetus or embryo is a part of the woman's body ("It's my uterus!") and it is therefore the woman's right to decide whether she should accept its development or terminate it. Considered in this light, the radical position is much clearer and more sincere, albeit illicit. In the meantime it should also be noted that as far as the situation in Italy is concerned, the whole argument has even distorted the statistical data itself in two respects: the estimated number of clandestine abortions in the period preceding the law and an objective appraisal of the law's ability to eliminate clandestine abortion.[50]

At the European level, the statistics also show that legalizing abortion does not eliminate illegal abortion but sometimes increases it. The reason or series of reasons for this is understandable: the clandestine character depends neither exclusively nor primarily on fear of punishment by the state, but rather on reasons of familial and social secrecy that the law cannot safeguard against: think of conceptions resulting from adultery or conceptions by unmarried or very young women. Moreover, once the law admits that a person can eliminate a fetus in plain daylight, there is no longer any comprehensible reason why the same thing cannot be accomplished in the secrecy of a doctor's office or a house, given that the question of moral value has been set aside.

Finally, the law imposes certain formalities and restrictions that do not always coincide with the immediate interests of the woman or couple (time limits, registration, etc.). Yet even admitting, for the sake of argument, that the law would reduce the number of clandestine abortions by transferring them to legal facilities, this would not render the procedure licit because the law does not make something morally licit; the law may suppose the morality of an act, if anything, but cannot determine it.

Thus when a law approves morally illicit conduct, the law itself becomes a negative law and distorts ethical judgment, especially in a matter of such important value as human life. Consider the example of theft or violence. No one would approve reasoning of this sort: because theft in Italy is increasing (financial fraud, burglary, extortion, kidnapping, etc.), it should be legalized and then theft will disappear. Theft would not cease to be theft if it were approved by the law, nor would violence cease to be violence; it would lead to the breakdown of the social order. Legal positivism betrays its limitations and contradictions here.

Moral decline increases with laxity and legal permissiveness, especially when this permissiveness occurs at the expense of a fundamental value such as human life. What legal system would still be appreciated and respected if it did not serve to defend the lives of those who cannot defend themselves? The very function of the law and therefore of the state is undermined when an illicit act of this type is deemed

---

[50] Groundbreaking statistical research on clandestine abortions in Italy before the abortion law can be found in B. Colombo, "Sulla diffusione degli aborti illegali in Italia," *Medicina e Morale* 26, nos. 1/2 (1976): 17–78. Colombo estimated the figure to be between 200,000 and 250,000 at the most, as opposed to the figures given in parliament (1 million or as many as 2 million). As for the persistence of clandestine abortions after the law (keeping in mind that it is difficult to determine an exact number), some estimates from 1991 place it at 60,000, as compared to 100,000 in 1983—a 40 percent reduction.

legal: the teaching role of the law ceases along with the reason for the existence of law in the first place.

It is one thing for an act to occur against the law and quite another for it to occur with the law's approval: it is no longer a question of numbers but of the ethical character and social function of the law.

In any case, given what has been said, the moral problem of determining whether or not abortion is licit, objectively and per se, would still exist in any legal situation whatsoever since the law frequently approves of abortion. Yet this conundrum remains and weighs even more heavily upon the conscience: it will always be a serious decision for the acting subject whether to take advantage of the permissive law or else, notwithstanding those accommodations, to remain faithful to moral duty. Furthermore I would say that, in addition to the ethical problem of every individual who is in a difficult situation, there is the ethical and social problem of determining whether or not the law itself is just and whether or not the physician must obey it, as will be seen more clearly in reference to conscientious objection.

And if we then think about how the life of the unborn child is not a private but a social good, just like the life of any human being, then the moral problem spreads on the basis and by the force of a law that is shirking its principle duties, namely, to defend and promote the life of every human being, especially of those who are defenseless. Even the Italian law concerning voluntary termination of pregnancy, law no. 194/78, had to recognize this fact by declaring in article one that the purpose of the law itself is to defend nascent human life, even though it goes on to extensively contradict that assumption in subsequent articles.

The moment of ensoulment

The notion of "mediate" or successive ensoulment was introduced by some of the Church Fathers to combat so-called *traducianism*, which was Tertullian's theory explaining the transmission of original sin by supposing that not only the body but also the soul was transmitted by the parents.[51] In order to refute that theological hypothesis, other Fathers and eventually Saint Thomas Aquinas proposed the theory of successive ensoulment.

The latter theory affirms that the soul, though destined to substantial union with the body, has an ontologically different origin and is created directly by God. Moreover the Thomist hypothesis supposes that a certain "form" or organization of the body is necessary for the infusion of the soul, since the soul is the form of the body. Aquinas supposed that there was a vegetable and animal soul at the beginning of the fertilization process and only later a rational soul. The chronological problem thus took on an ontological relevance.[52]

The time of infusion of the rational soul was reckoned at the thirtieth or fortieth day by analogy with the biblical prescriptions concerning the purification of a woman after she gives birth. It should immediately be added that not all the Church

---

[51] Tertullian, *De anima*, chap. XIX, PL II, 682; see also Fagone, "Il problema dell'inizio."

[52] Thomas Aquinas, *Quaestiones disputatae: De potentia* (Casale Monferrato: Marietti, 1953), III, a. IX, ad 12, 13, 15; idem, *Quaestiones disputatae: De spiritualibus creaturis* (Rome: Gregorianum, 1946), ad 13.

Fathers were of the same mind, especially the Greek Fathers and in particular Saint Gregory of Nazianzus, who had taught ensoulment from the first instant along with followers such as Saint Maximus.[53] It has instead been the constant position of the Church, both in teaching and in canonical provisions (which even sanctioned severe penalties), that abortion is illicit in every instance. Abortion has always been a sin, indeed a crime, at whatever stage of development it is committed; the doubt, if any, concerned the classification of the crime—whether it should be defined as *plain and simple homicide* or merely as a specific crime against life[54]—and the varying severity of canonical penalties has been linked to this question, but not to the judgment of the act as morally illicit.

It should be remembered, moreover, that Aquinas himself declared with respect to the person of Christ that the Incarnation of the Word was simultaneous with the moment of his conception.[55]

Indeed, aside from the advances of genetics and embryology that now have definitively refuted all hypotheses of discontinuous succession in embryonic development, it needs to be made clear that this question does not diminish the illicit character of the act of abortion on the ontological and moral level. Without meaning to trivialize the matter, it does not work for someone who steals a donkey tied by a rope to excuse himself by saying he was really just carrying a rope in his hand and was unaware of the donkey tied to the other end. What is being concealed is that fundamental value required by the definition of human life. What is being taken away is the essential element of an individual's life, and the one who does it knows full well how to describe his own intention: in a procured abortion, the act in itself and the intention unite to prevent the birth of a developing human life.

In the final analysis, the chronological factor does not affect the ontological value and hence not even the ethical value; or at least it does not affect them in a substantial and objective manner.

To confirm all of this, let us cite the words of the instruction *Donum vitae*: "Certainly no experimental datum can be in itself sufficient to bring us to the recognition of a spiritual soul; nevertheless, the conclusions of science regarding the human embryo provide a valuable indication for discerning by the use of reason a personal presence at the moment of this first appearance of a human life: How could a human individual not be a human person? The Magisterium has not expressly committed itself to an affirmation of a philosophical nature, but it constantly reaffirms the moral condemnation of any kind of procured abortion. This teaching has not been changed and is unchangeable."[56]

---

[53] Maximus the Confessor, *De variis difficilibus locis Sanctorum Dyonisii et Gregorii seu ambiguorum liber*, PG XCI, 1335 a.

[54] In addition to the quoted article by Fagone in the volume Fiori and Sgreccia, *L'aborto*, see also Sgreccia, "L'insegnamento dei Padri"; B. Honings, "L'aborto e il momento della ominizzazione," in *Il diritto alla vita*, ed. G. Concetti (Rome: Logos, 1981), 23–41; Caprile, *Non uccidere*.

[55] On this topic, see M. Pangallo, "Actus essendi tomistico e spiritualità dell'anima," *Medicina e Morale* 36, no. 2 (1986): 407–414; S. J. Heaney, "Aquinas and the Presence of the Human Soul in the Early Embryo," *The Thomist* 56, no. 1 (1992): 19–48.

[56] Congregation for the Doctrine of the Faith, *Donum vitae*, part I, no. 1.

Self-conscious awareness

The tendency to diminish the biological status of the embryo so as not to consider it an individual human being except on the basis of some arbitrarily determined features is linked to the attempt to deny the embryo's human personhood.

The response to the question of ensoulment given by the Congregation for the Doctrine of the Faith in the instruction *Donum vitae* has already been seen. Let us now try to analyze the other explanations of why some human beings should be persons and others not and why the same individual would be a person when he is recognized as such by others yet not from the moment his existence begins as a unique and unrepeatable individual.

According to some authors, the new conceptus does not yet have a human reality or dignity. It is said to be only a potentially human reality or even a mere possibility of humanity that is not yet endowed, in any event, with a developed consciousness.[57]

For others, the decisive fact to be emphasized is that this "potential" human being is not even endowed with self-consciousness, making it incapable of free and intentional activity and subordinating its existence to the ability to express thought and use language. Something that does not have self-conscious awareness, they say, cannot perceive that its development is suppressed and therefore cannot suffer either if it is eliminated: in order to suffer from it, that entity would have to consider deliberately the value of life, comparing it with nonlife. But this ability to reflect on oneself, to compare something with its antithesis, to take a dialectical position on the value of life through negating the value of its opposite, cannot belong either to the embryo or to the fetus; the former does not have consciousness, and the second has consciousness but not self-consciousness. Why worry, then, if it is destroyed?[58]

One of the major proponents of this theory is H. Tristram Engelhardt Jr. Since he does not find in the embryo or fetus those characteristics which, according to him, are proper to a person, i.e., self-consciousness, rationality, and a moral sense, he maintains that "not all humans are persons. ... Fetuses, infants, the profoundly mentally retarded, and the hopelessly comatose provide examples of nonpersons. Such entities are members of the human species. They do not in and of themselves have standing in the moral community. ... They are not prime participants in the moral endeavor. Only persons have that status."[59]

And Peter Singer also maintains that a human person possesses self-consciousness, self-control, a sense of the past, a sense of the future, the ability to relate with others, respect for others, communication ability, and curiosity. Therefore, if the value of life depends on being persons, it is necessary to

> reject the doctrine that places the lives of members of our species above the lives of members of other species. Some members of other species are persons: some members of our own species are not. No objective assessment can support the view that it is always worse to kill members of our species who are not persons

---

[57] See the opinions reported in A. Bausola, introduction to *Il dono della vita*, Sgreccia, 45–49. See also M. Tooley, "Abortion and Infanticide," *Philosophy and Public Affairs* 2, no. 1 (1972): 37–65.

[58] See Tooley, "Abortion and Infanticide," 62–65.

[59] H.T. Engelhardt Jr., *The Foundations of Bioethics*, 2nd ed. (New York: Oxford University, 1996), 107.

than members of other species who are. On the contrary, as we have seen there are strong arguments for thinking that to take the lives of persons is, in itself, more serious than taking the lives of non-persons. So it seems that killing, say, a chimpanzee is worse than the killing of a human being who, because of a congenital intellectual disability, is not and never can be a person.[60]

Essential behavior and relation

Behaviorist philosophical trends, sociology, and *phenomenology* converge on one point: the negation of metaphysics, or at least the affirmation of its irrelevance. Starting from this assumption, they declare that the sole criterion for recognizing personhood or human individuality is the examination of behavior: a human subject is present if one can observe human behavior. Since it is not possible to note human behavior in the case of the fetus, at least up to a certain stage, what is left as the "objective" criterion is the mother's attitude. The way that one can prove the presence of a new human subject, they say, is by the mother's attitude. In other words, this new acting subject exists if there is acceptance by the mother. Sociologists talk about a social relation: personhood exists if the social relation exists.[61] Here the social relation would be the consent expressed by the mother or by the parents.

Clearly there are two observations to be made here. One, in the scientific perspective, comes from psychoanalysis, in particular of the Freudian variety: the fetus, too, is in a social co-relation with the mother, storing in its depths experiences, sensations, and refusals, in such a way as to remain marked by them well into adulthood. It is not true, therefore, that there is no evidence for an intersubjective relationship between the fetus and the mother.

The second observation is from the psychophysical perspective: from the very first days the embryo enters into a dialogue with the maternal organism in its own way, blocking the production of hormones, sending messages to her pituitary gland and hypothalamus, to the ovaries and to the implantation site of the fertilized egg, gradually calling for a whole set of modifications in the maternal organism, which is forced to "recognize" the presence of that individual, quite apart from any conscious consent.

The philosophical reason, however, is more important: being in relation is not what constitutes the reality of the acting subject, but rather the reality of the acting subject is what makes an interpersonal relationship possible. There could not be an intersubjective relationship if it were not for the reality of the acting subjects.

Returning to the Thomistic principle *operari sequitur esse* (functioning follows being), it could be translated with these equivalent words: a things needs to exist in order to relate. The concept of relation as a prerequisite for being is an absurdity, just as the attempt to eliminate all metaphysical principles from reality was absurd. Moreover what was said before holds true: it would be morally illicit to eliminate a being that is on the way to becoming a person and entering into relation if allowed to live and grow.

---

[60] P. Singer, *Practical Ethics*, 2nd ed. (New York: Cambridge University Press, 1993), 117. See also M. F. Goodman, *What Is a Person?* (Clifton, NJ: Humana Press, 1988); J. Feinberg, *The Problem of Abortion* (Belmont, CA: Wadsworth, 1984).

[61] J. Raes, "A propos de l'avortement," *La Revue Nouvelle* (1971): 90; S. Muret, "La réflexion chrétienne et l'avortement," *La Revue Nouvelle* (Jan. 1973).

Recognizing the human image

The perception of personhood, some psychologists say, depends on the recognition of a human image; but when there is only a desire for or a denial of that image, there is no recognition of personhood. When parents imagine the fetus in the form of a little baby they are conjuring a fictitious image. Recognition begins when a name is given: "There is no name for what remains anonymous until a human decision has already made someone out of it."[62]

The problem is shifted and not resolved through this clever approach: the real issue is knowing whether or not the parents must regard the embryo as human life, not making human life depend on their recognition of it.

But hiding behind this reasoning is the philosophical presupposition of relativism: the object is established by knowledge of it; this is to deny the objective value of cognition. In this phenomenological and subjectivist approach, how would the parents be able to recognize the embryo as a human subject—assuming they welcomed it—if it were not one already? Ultimately this is a reformulation of the preceding thesis. In light of these theories it must be said that while it is true that every human being enters into relation, it must be remembered that being is what constitutes relation and not vice versa. If the students and the teacher in a classroom enter into a reciprocal relation, it is because they exist, whereas it is unthinkable that their existence should depend on the scholastic relation! This in no way diminishes the fact that every person lives and grows while immersed in relations that provide nourishment for enrichment and development.

Let us go one step further: the social relation has not only a pedagogical and cultural value but is inscribed in the ontology of the person as something coessential to the person. This is what Martin Heidegger calls "*Mitsein*" (being-with), Karl Jaspers calls "communication," and Gabriel Marcel, the Christian existentialist, calls "communion"; they mean to say that personal existence is what requires and establishes social openness, and not the contrary (i.e., that relation makes persons exist objectively).

The ontological and objective supposition, which is the condition for psychological reality, comes back again and again. It is on this terrain that the distinction between "human being" and "humanized being" has also been set up—a distinction designed to make room for the liceity of abortion:

> For our part, we think it necessary to distinguish between human life and humanized life. ... The relation of recognition reveals, if not establishes, the fully human character of the being in gestation. In other words, just as a human being does not exist without a body, so it is not humanized without this relation to others.[63]

Virgilio Fagone rightly notes that this assertion is true if the relation has "revelatory" meaning; it would no longer be true if it were supposed to mean that the

---

[62] L. Beirnaert, "L'avortement est-il un infanticide?" *Études* (Nov. 1970): 520; Fagone, "Il problema dell'inizio," 161.

[63] B. Ribes, "Pour une réforme de la législation française relative à l'avortement," *Études* (Jan. 1973): 69. See also Fagone, "Il problema dell'inizio," 164–165.

relation has "constitutive" meaning, which the proponents of this distinction seem to want to insinuate by their thesis.

In the same way the statement that "humanity is what humanizes," meaning that the mother's attitude of acceptance is what confers human value on the embryo (and therefore that when this acceptance is lacking, the voluntary termination of pregnancy would become "socially justifiable"), is contradictory and false. If the embryo is a human being even before "humanization," it is not clear how one can justify destroying it; if it is not human, then let them tell us what it is and how it can be humanized by a simple psychological attitude of the mother. Here again it is necessary to repeat that ontology conditions psychology and not vice versa.

Another captious distinction of similar inspiration is the one introduced together with the concept of "quality of life": the right to life is supposedly bound up with the "quality of life," that is, with the possibility of normal development. It is a distinction of the "selective" sort intended to justify destroying an embryo that is affected by abnormalities leading to foreseeable defects or handicaps, or one that for extrinsic reasons would be impaired in its human development (economic concerns within the family, psychological anxiety of the couple, etc.).

This even introduces a racist principle that, if also applied to newborns or adults, could lead to the worst possible consequences. Additionally, it introduces an ontological and moral relativization of human life, which would no longer have transcendent value in itself: its value would depend on greater or lesser physical well-being or even on its assessment by others.

The intention to procreate and contraceptive abortion

This thesis is aimed at legitimizing elective abortion when conception has occurred against or without the will of the spouses (or of one of them). The proponents of this thesis reason as follows: it is permissible and legitimate for the spouses to freely decide about the birth of children; it is likewise lawful and obligatory that they express their love in the conjugal act even during the times when they do not intend to give life to a new child, and there are various contraceptive methods (among which the authors in question see only differences of efficacy and not of moral value) to prevent conception; sometimes these methods fail and an unwanted and unintended conception occurs; absent the will to procreate, the unwanted pregnancy may be voluntarily terminated.[64]

This actually brings us back to the link already examined: the subordination of the right to life to the will of the mother. The right not to procreate, which is a real right within certain limits, can legitimately demand an equal right—i.e., a right on the same plane of values—to employ those measures that are technologically and morally suited to not procreating, but it cannot go so far as to demand the destruction of a life that has already been conceived. Otherwise this would end up legitimizing the destruction of an unwanted newborn as well.

Then there is a fact to keep in mind: subjective intentions cannot fail to take into account the objective structure of one's own acts. If I punch a friend in the face and his teeth fall out, I cannot get out of it and exempt myself from culpability and

---

[64] B. Quelquejeu, "La volonté de procréer: Réflexion philosophique," *Lumière et vie* (Aug.–Oct. 1972): 64.

from making reparations by saying that it was not my intention for his teeth to fall out, because punching him entailed the objective possibility of causing the harm inflicted, regardless of my intention.

Thus, when a couple performs a conjugal act the two know that the procreative structure is present in the very act being posited, and therefore subjective intentions or preventive measures cannot nullify the objective structure of the conjugal act.

But the underlying and more serious flaw in the reasoning described above is found in a philosophical assumption that draws on the historicist philosophy found in various trends of cultural anthropology.

These trends of thinking tend to confuse the problem of the origin of a fact with the question of its intrinsic value. I will explain what I mean with a simple example: if by chance I find a precious stone while traveling in the desert, I will not say that it is a worthless thing just because I found it by chance in an arid, sandy place; whatever the manner and place in which it was found, I will value the precious stone for what it is worth in itself, for what intrinsically belongs to it. The new conceptus should be considered according to its human and individual value, whatever the manner in which it has been conceived. There are children conceived outside of marriage, as a result of a rash adventure or a betrayal, just as there are others who are conceived within a marriage but in ways and at times not planned or foreseen by the spouses. Yet the new conceptus should be valued in him- or herself and not subordinated to the will of others.

This recurring theme of trying to make the value of a human life dependent on the will of others, even if they are the parents, is extremely dangerous, and if it were extended to society as a whole it would become downright monstrous. The ontological value of an individual cannot depend on factors external to him or her.

Philosophical conclusion: the biological and the human

All that is left at this point is to draw the conclusion: What value should be attributed to the embryo–fetus, which by a continuous process of internally and autonomously programmed development arrives at the moment of birth as a new human individual, a human person? The rigorous logic of realist and personalist philosophy leads to just one conclusion: the embryo has the value proper to the human person. Stated in a negative construction: elective abortion is a crime against personal human life, or more precisely, against the human person; in factual reality it is homicide, even though it may not be perceived as such subjectively or psychologically and the law may not recognize this external description of the facts. There are two reasons, which were already explained: one of a biological and the other of a philosophical nature.

1. *From the perspective of biology*, considering the programmed, continuous, and intrinsically autonomous development of the embryo-fetus from the corporeal point of view, it follows that there is no substantial difference but only a developmental difference between the first moment of conception and the moment of birth.

2. *From the perspective of philosophy* or of values, it must be concluded that the entire value of the individual human person is ontologically present from the moment of conception for two reasons: (a) *the connection between body and soul* is essential and not accidental—the body is the transcription, manifestation, and instrument of the person and not simply its gar-

ment or accessory—so the person *is* a corporeal person, an incarnate "I" and not just an entity that *has* a body;[65] and (b) *personhood* in the human race is identical with the existential *act* that realizes the human nature which is made up of soul and body, psyche and soma. The existential act operates at the very same moment in which the new being is *actuated*. The fact that the manifestation of this ontological and existential reality occurs gradually and continues throughout life does not justify thinking that what happens *next* is not rooted in and caused by what is *already*: between what is *already completed* and what is *not yet* developed there is the trajectory of gestation and life, but there is no qualitative—or more precisely, onto-logical—leap: it is the same existential act that nourishes development, it is the truly present and operative "I", even though there may as yet be no self-consciousness or social recognition.[66] The developmental unity and ontological unity of the human being in formation lead to the same conclu-sion: this is an individual human life in a state of development.

Hence the human embryo, even though it happens to be in a particular phase of its existence in which the human form as commonly envisioned has not yet been expressed, is not pure potentiality but rather a living and individualized substance. From the moment of fertilization the embryo is capable of guiding to maturity a corporality that serves to express, as though in a historical earthly epiphany, the incomparable greatness of the human spirit. Indeed, the human embryo is a being in which the principle of development and change is, as in all living substances, internal to the substance itself. Therefore the statement that the embryo is potentially a human being is ambiguous and misleading; the embryo is potentially a child or an adult or an old man, but it is not potentially a human individual: it is actively that already.

The inherent substantial unity of the fertilized egg cell reveals substantial continuity in its development precisely because the principle of development and change is internal to the substance itself. One cannot therefore postulate different and successive existences of the same living embryo, and this is fully in keeping with the experimental data and embryology: as it develops, the same subject maintains in each successive phase its ontological unity with the preceding phase without any break in continuity.

If this is true, it must be logically and reasonably concluded that there is ontological identity throughout the course of development of that unique individual who, once born, is recognized by everyone to possess the character and dignity of a human person. And, even though from the psychological and social point of view a personality matures through a long journey of relational and cultural exchanges

---

[65] On this subject Pope John Paul II had occasion to say, "Each human person, in his absolutely unique singularity, is constituted not only by his spirit, but by his body as well. Thus, in the body and through the body, one touches the person himself in his concrete reality. To respect the dignity of man, consequently, amounts to safeguarding this identity of the man 'corpore et anima unus,' as Vatican Council II says." John Paul II, Address to Members of the World Medical Association, October 29, 1983, in *Insegnamenti di Giovanni Paolo II* (Vatican City: Libreria Editrice Vaticana, 1983), 6.2:917–923 [Italian].

[66] J. Gevaert, "L'esistenza corporea dell'uomo," in *Il problema dell'uomo* (Turin: LDC, 1984). On this topic, see also R. Lucas Lucas, *L'uomo, spirito incarnato* (Cinisello Balsamo: Paoline, 1993); Possenti, "La bioetica alla ricerca."

with the environment, his existence should be posited from the moment in which his biological individuality comes to be: "How could a human individual not be a human person?"[67] Moreover, even if we do not perceive in the embryo all of the characteristics we consider typical of a person, we must nevertheless keep in mind that the embryo per se is destined to become that person. And since the end is not just the conclusion of a path or of a development, as in this case, but is also what guides and determines that development, it follows that this human zygote, destined to become that person, already is that person from the first moment of its appearance.

The Instruction *Donum vitae,* on respect for human life and the dignity of pro-creation, did not intend to make explicit pronouncements on particular philosophical theories concerning the definition of the term *person,* nor did it wish to take up or settle the ancient question concerning the moment at which the soul is infused, nor did it try to deal specifically with the various objections to the anthropological status of the human embryo, which it nevertheless says that it acknowledges. Instead, while deliberately remaining above the fray, it declares that the human embryo should be respected as a person from the moment of conception:

> Certainly no experimental datum can be in itself sufficient to bring us to the recognition of a spiritual soul; nevertheless, the conclusions of science regarding the human embryo provide a valuable indication for discerning by the use of reason a personal presence at the moment of this first appearance of a human life … The Magisterium has not expressly committed itself to an affirmation of a philosophical nature, but it constantly reaffirms the moral condemnation of any kind of procured abortion. This teaching has not been changed and is unchangeable. … The human being is to be respected and treated as a person from the moment of conception; and therefore from that same moment his rights as a person must be recognized, among which in the first place is the inviolable right of every innocent human being to life.[68]

And this is also reaffirmed by Pope John Paul II, who states the following in his encyclical *Evangelium vitae*: "The result of human procreation, from the first moment of its existence, must be guaranteed that unconditional respect which is morally due to the human being in his or her totality and unity as body and spirit. … Human life is sacred and inviolable at every moment of existence, including the initial phase which precedes birth."[69]

On the moral plane this means that the voluntary termination of pregnancy, even though it acts upon a biological entity that is still in formation, *afficit personam* (affects the person) because of the unity of the human person. Just as the person is shot when a wound is inflicted to the chest or to the head by a firearm, so too the action committed against the embryo that is not yet visible or fully developed is a crime against the person.

---

[67] Congregation for the Doctrine of the Faith, *Donum vitae,* part I, no. 1.

[68] Ibid.

[69] John Paul II, Encyclical Letter *Evangelium vitae,* March 25, 1995, nos. 60–61. See also *Catechism of the Catholic Church,* nos. 2270–2275.

Ethical conclusion

From the ethical perspective it is possible to do without the preceding philosophical discussions to a certain extent: it is sufficient to verify that the fertilized egg has an *intrinsic connection* and shares an *intrinsic destination* with the developing personal being in order to rule out any act of destruction or alteration of its integrity.

Moreover even if there were a *dubium facti* (doubt about a fact) about the aforesaid connection, one would be obliged in conscience to refrain from any sort of destructive act, just as one must refrain from shooting a rifle in the forest if there is any doubt that it could hit a person instead of the prey.[70]

> Doubt as to the personal identity of the product of conception is sufficient to oblige us morally to decide on the safer behavior, which consequently avoids any and all danger or risk to the human person. Indeed, morality requires that one refrain not only from an act that is *certainly* wrong, but also from an act that could *probably* be wrong.

> In reality, to act while in doubt as to whether or not the product of conception is a human person means running the risk of killing a human being, which constitutes a moral disorder in and of itself.[71]

And John Paul II also declares, "Furthermore, what is at stake is so important that, from the standpoint of moral obligation, the mere probability that a human person is involved would suffice to justify an absolutely clear prohibition of any intervention aimed at killing a human embryo."[72]

This is why John Paul II, in number 57 of the same encyclical letter, confirms that "*the direct and voluntary killing of an innocent human being is always gravely immoral.* ... Before the moral norm which prohibits the direct taking of the life of an innocent human being *there are no privileges or exceptions for anyone*."[73] The teaching in question—as Pope Paul VI had already explained—is "unchanged and unchangeable."[74] With regard to the behavior called for when dealing with the human embryo, the following can be read in the document by the Italian National Bioethics Committee that was approved on June 27, 1996:

> [The committee] has unanimously come to recognise the moral duty to treat the human embryo, since fertilisation, according to criteria of respect and protection that must be adopted towards human beings who are recognised as persons, and this regardless of whether we attribute to the embryo from the beginning, and with certainty, the characteristic of person in its technically philosophical sense, or whether this characteristic can be attributed only with a high level of plausibility, or we prefer to use the technical concept of person and refer only to their belonging to the human species that cannot be

---

[70] See also Congregation for the Doctrine of the Faith, Declaration *Quaestio de abortu*, no. 13.

[71] Bioethics Center of the Catholic University of the Sacred Heart, "Identità e statuto."

[72] John Paul II, *Evangelium vitae*, no. 60.

[73] Ibid., no. 62, emphasis added.

[74] Ibid.

contested to the embryo from the first instants and does not change during his/her subsequent development.[75]

Within the committee, however, different positions emerged regarding the protection of *preimplantation embryos*.

Abortion from the Viewpoint of Legal Ethics

Francesco D'Agostino appropriately reflects on the ethicality of abortion from the legal perspective, i.e., as a *specific social practice*. Abortion is a *social practice* inasmuch as it is a *relational* practice with *public relevance*: "Relationality will have a public character whenever the social consciousness perceives that the possibility of violence, fraud, or more generally the oppression of the weak by the strong, can make headway in and through that relationality. Only in these cases will the social practice ask the law to safeguard its identity so it can remain what it is."[76] It is clear that the decision to abort involves not only the woman but also the unborn child she carries in her womb, not to mention the father.

The law must intervene in the matter of abortion because in general it has the duty to intervene every time the relational equilibrium is disturbed, every time that the concerns of the weak are threatened by the pretensions of the strong. Hence the law cannot remain neutral with regard to a social practice: "It is not neutral when it is rightful law, that is, when it defends the concerns of the weak against the abuses of the strong; it is most certainly not neutral when it is wrongful law, that is, when through hypocrisy or at best through weakness it sanctions or at least tolerates the abuses of the strong against the concerns of the weak."[77] The pretension of the Italian abortion law (law no. 1978/194) to handle abortion with a neutral approach, leaving any decision to the mother, therefore goes against the very nature of law.

Before that legislation was passed, the Italian penal code prohibited procured abortion, classifying it as a *crime against the integrity and health of the lineage*. It considered the *passive subject* of the crime—the victim of the criminal act—to be not an individual subject (the unborn child) but rather a collective subject (the *stirpe*, or lineage). The life of the unborn child was safeguarded indirectly by virtue of the protection that the juridical ordinance accorded to the lineage.

---

[75] Italian National Bioethics Committee, *Identity and Status*, §10, p. 20. As for the debate on the legal status of the human embryo, see Center for Bioethics and Human Rights of the University of Lecce, *I diritti del nascituro e la procreazione artificiale* (Vatican City: Libreria Editrice Vaticana, 1995); Pontifical Council for the Interpretation of Legislative Texts, "I diritti del nascituro secondo la legislazione canonica e il Magistero della Chiesa," *Medicina e Morale* 45, no. 3 (1995): 532–542; A. Tarantino, "Sul fondamento dei diritti del nascituro: alcune considerazioni bioetico-giuridiche," *Medicina e Morale* 45, no. 5 (1995): 951–984; F. Mastropaolo, "Lo statuto dell'embrione," in *A sua immagine e somiglianza?* ed. A. Mazzoni (Rome: Città Nuova, 1997), 142–179; J. de D. Vial Correa and E. Sgreccia, eds., *Identity and Status of the Human Embryo: Proceedings of the Third Assembly of the Pontifical Academy for Life; Vatican City, February 14–16, 1997*, 2nd ed. (Vatican City: Libreria Editrice Vaticana, 1999).

[76] D'Agostino, *Bioetica*, 275.

[77] Ibid., 277.

The Italian law currently in force does speak about the unborn child but the mother is the one who has the right—but not the duty—to bring him or her into the world. It should instead be highlighted that the unborn child has the right to be born: "The right to be born is truly the first of all rights, not only in the chronological sense—as it might appear at first—but above all in the *axiological* sense: that is, the right to be born includes within itself and serves as the foundation for all the constitutive rights of the person, the right to be respected in one's own identity, the right not to be exploited for any reason, the right to be considered the bearer of a specific dignity that cannot be reduced to that of any other human being whatsoever—in sum, the right to be recognized as *person*."[78]

The *ethical* problem of abortion is often reduced to a merely *psychological* problem, as though it were a *self-referential* practice concerning the woman alone and not a *relational* dynamic that the woman brings about by killing her own child: "Hence the misleading insistence ... on prenatal life as a mere *possibility*: you do not *kill* a possibility, you simply renounce it. In abortion, however, there is only one way of renouncing this 'possibility,' and that is by taking a life."[79]

As moral subjects were are called to take *otherness* seriously, even when the other does not have the ability to defend his interests. We are called to remain clear-sighted about the truth of things. Ethical reflection on abortion is complex because it is objectively a matter of *killing* a human life that is still in the womb; it is therefore necessary to avoid resolving the problem too hastily, that is, by equating its phenomenology with that of homicide. We cannot fail to look at the question from the psychological and social point of view *as well*, but abortion is *first and foremost* an ethical problem, and from the legal perspective it is the denial of relationality. D'Agostino sees in the practice of abortion "an *archetypal* emergence of that profound desire to deny relationality, in which the human intolerance of one's own creatureliness (and that of others) is manifested."[80]

"Therapeutic Abortion": The Conflict between the Life of the Mother and the Life of the Conceptus

It is necessary to immediately offer some clarifications and distinctions regarding this particular topic. First of all it must be said that the adjective *therapeutic* is inaccurate: in fact, it is not a matter of therapy except in a broad and improper sense. The conditions for the therapeutic principle have already been mentioned. One of them is that the medical or surgical intervention be aimed directly at treating or removing the sick part of the body; it is not a matter of treating an active illness in the case in question, but rather of killing the fetus to avoid aggravating the mother's health or risking her life. The movement is not that of a therapeutic action on the pathology in order to attain health, but rather an action upon what is healthy (the fetus, who could also be healthy) so as to prevent an illness or the danger of death. It

---

[78] Ibid., 286.

[79] D'Agostino, "Aborto," in *Parole di bioetica* (Turin: Giappichelli, 2004), 6.

[80] D'Agostino, *Bioetica*, 275.

would therefore be better to talk about the termination of pregnancy *in the presence of a danger to the life or health of the mother.*[81]

A further clarification is necessary in order to understand a so-called indirect abortion, which could instead be classified as a properly therapeutic treatment. This is the case in which a tumor on the uterus is removed, indirectly entailing the death of the fetus. Today this distinction is no longer used and the problem is not even posed in ethical terms, but the expression *indirect abortion* is sometimes used to designate the "therapeutic" abortion that I am about to discuss, which is quite a different matter from both the ethical and medical perspectives.

The acceptation and extension intended to be given to the expression *therapeutic abortion* are thus remarkably important not only in the medical literature but also in more recent legislation. Following the framework offered by Adriano Bompiani,[82] the following distinctions can be made:

1. Therapeutic abortion is proposed as the sole means of *saving the life of the mother*, because the continuation of pregnancy would cause the death of the mother with scientific *certainty*. This scenario can entail one of two situations: (a) continuation of the pregnancy would involve the certain death *of both mother and child*; (b) continuation of the pregnancy would certainly involve the death of the mother, but with *the hope of saving the child.*

2. Therapeutic abortion is proposed to *safeguard the health of the mother.* This scenario is also interpreted in practice over a spectrum of different situations: (a) continuation of pregnancy could pose a lethal risk to the life of the mother rather than some damage to her health; (b) continuation of pregnancy could involve a *permanent worsening of the health* of the mother (here the boundaries are already becoming blurred and prognoses are difficult and almost impossible, especially if health is understood to refer not only to the body but also to the psychological or personal dimension); (c) there is simply a generic *concern about health*, understood as a "state of complete physical, psychological and emotional wellbeing"; or (d) the *psychological repercussions* are considered impacts on health, whether they result from a foreseeable deterioration in financial circumstances, from the fact that the pregnancy is unwanted, from fear, or from the prognosis that the fetus has serious defects or is deformed.

These are therefore scenarios in which cases of elective abortion for eugenic reasons (fetal deformity or pathology), contraceptive motives (unwanted child), and socio-economic motives are categorized as "therapeutic abortions." It should be acknowledged that this gradual extension beyond cases in which the procedure is medically indicated has often been driven by political reasons in an effort to bring the entire casuistry of the demand for contraception and "liberalization" under the heading of "therapeutic" and under the approach of abortion "regulation."

The text of Italy's law no. 194/78, article 4, reads as follows:

---

[81] Häring, *Free and Faithful*, 3:33–34.

[82] A. Bompiani, "Indicazioni dell'aborto 'terapeutico': Stato attuale del problema," in *L'aborto*, Fiori and Sgreccia, 191–215.

> For voluntary termination of pregnancy during the first ninety (90) days, a woman claiming circumstances that would make continuation of the pregnancy, childbirth, or motherhood a serious danger to physical or mental health in terms of her state of health, her socioeconomic or family circumstances, the circumstances in which conception occurred, or predictions of abnormalities or malformations in the conceptus, shall apply to a public counseling center (established under article 2[a] of law no. 405 on July 29, 1975) or to a social health care agency authorized by the region or to a trusted physician.[83]

It is easy to see the breadth of these so-called "indications," which embrace every sort of socioeconomic motive, and the very wording of "serious danger" does not allow for either a medical or legal interpretation.

In effect this is a transition from therapeutic abortion to abortion understood as a means of birth control, which is a motive the law itself excludes in article 1 but reintroduces in practice in article 4. The history of the application of the law has confirmed this surreptitious and broad interpretation. As far as abortion after the first ninety days is concerned, let us look at what article 6 of the Italian law states: "Voluntary termination of pregnancy after the first ninety (90) days can be carried out: (a) when pregnancy or birth involves a serious danger to the woman's life; (b) when pathological processes have been ascertained, including those related to significant abnormalities or malformations of the fetus, that pose a serious danger to the physical or psychological health of the woman."

The "indications" for therapeutic abortion

Limiting ourselves to this part of the topic, it is necessary to provide additional clarifications concerning the substance and relevance of the medical indications:

1. First of all, there are in fact biological conditions that complicate pregnancy or in which pregnancy worsens health; however, they are increasingly manageable and open to amelioration with adequate treatment. Advances in medicine and medical treatment continue to reduce the dangers to the life and health of the mother. It is clear that termination is not justified in these cases, even from the perspective of professional medical ethics alone.

2. There are health circumstances that are usually taken into consideration for elective abortion in which the termination of pregnancy would nevertheless be even more harmful to the woman's health than continuing the pregnancy or in which termination would not afford substantial help. Here, too, the lack of a medical justification for termination is clear.

3. Finally, there are conditions in which worsening health is real but can be dealt with by *methods other than termination* (periodic dialysis in pregnant women with serious renal problems, heart surgery in women with heart defects, etc.). There is no need for lengthy explanations to understand that the real therapy in these cases, the one that directly eliminates the condition without harming the life of the fetus, is the only licit one.

---

[83] Italian Republic, law no. 194, On the Social Protection of Motherhood and the Voluntary Termination of Pregnancy, May 22, 1978, published in *Gazzetta Ufficiale della Repubblica Italiana* 140, May 2, 1978, part 1, 3642–3646 [Italian].

Leaving out of the discussion the so-called socioeconomic indications, which cannot compare—however real they may be—to the life of the unborn child, it is above all necessary to rigorously review, in the context of medicine and deontology, the series of so-called health indications for voluntary termination of pregnancy on which its practice in health care is founded. In light of advances in science and medical care, many of these "indications" have lost their compelling force. Tuberculosis, heart diseases, vascular diseases, disorders of the hematopoietic system (some forms of anemia), renal diseases, liver and pancreatic diseases, gastrointestinal diseases, pregnancy-related chorea, myasthenia gravis, and tumors (except those of the genital system): all these conditions are listed "indications" justifying elective abortion. But a careful study of each one of them, in light of what has been said above, leads to the conclusion that the medical basis of these "indications" has eroded and there has been a progressive and sweeping reduction of the number of cases in which there is no alternative therapy and a real risk to the life or health of the mother remains.

The conclusions of Bompiani can be cited verbatim: "Understood as an act capable of removing the female patient from the danger of imminent death and as an irreplaceable therapeutic intervention to achieve that purpose, therapeutic abortion has really lost much ground and finds no logical place among modern health care criteria; on the contrary, it has proved to be more harmful than helpful in many critical circumstances precisely because of the condition of maternal 'decompensation.'"[84]

Ethical assessment of "therapeutic abortion"

It must be acknowledged or at least presumed that cases may arise in which pregnancy can constitute an aggravating circumstance:

- *Socioeconomic conditions*, with repercussions on the patient's psychological health

- *Conditions of physical health* that are liable to worsen permanently

- *A state of real and serious danger* to the life of the mother to the point where, in the worst case, a choice would have to be made between the life of the mother and the loss of both mother and child.

Indications of an ethical and rational nature will have to be guided by the following lines of thought and behavior, in keeping with a personalist vision of man:

1. It is necessary to start from this fundamental fact and ethical principle: the human person is the highest value in the world and transcends every other temporal good and every economic consideration.[85] Arguments invoking

---

[84] Bompiani, "Indicazioni," 214.

[85] Starting from this premise, the Catholic Church has always defended the inviolability of human life. Pius XII declared that "there is no man, no human authority, no science, no 'medical indication' or eugenic, social, economic, or moral reason that could produce or give a valid legal claim to a direct, deliberate control over an innocent human life, which is to say, a control aimed at its destruction, whether as an end or as a means to another end that is perhaps in no way illicit in itself." Pius XII, Address to the Participants in the Conference of the Italian Catholic Union of Obstetricians, October 29, 1951, in *Discorsi e Radiomessaggi* (Vatican City: Tipografia Poliglotta Vaticana, 1940–1969), 13:211–221 [Italian]. Pius XII also intervened on the subject of abortion on other occasions: Address to

economic reasons should therefore be taken into consideration by public authorities and by the community in the sense that it is necessary to fit the economy to the person and not to sacrifice the person to the economy. This is all the more true when it is recalled that the life of each individual is not just an inalienable personal good but also a social good belonging to all: society therefore has the obligation to defend and promote it.

2. Even the so-called social reason (number of children, educational plans, etc.) cannot trump the value of human life, not even the life of a single person. Ontologically and axiologically, the person precedes society, because society takes its origin from human persons and finds its reason for being in the act of perfecting and helping the growth of individual persons. Society is therefore of persons and for persons. Consequently even the so-called principle of balancing values is inconsistent from the ethical perspective when applying it to a social justification of abortion. There is no balancing but rather harmony and subordination of social values with regard to the human person. In addition to the philosophy of medicine, the philosophy of law is also at stake here. It is not possible to compare a single person and society as a whole, because the value of the person is not numerical or quantitative but rather ontological and qualitative. Therefore anyone who legitimizes the direct killing of an innocent person strikes at the foundational value of all society and of every person.[86]

3. Although it does not represent the totality of the values of the human person, physical life is the *primary and indispensable foundation* of all the other personal values. Therefore the destruction of the physical life of the unborn child by abortion, even by "therapeutic" abortion, jeopardizes all temporal values that are necessarily based upon physical life.

---

the Saint Luke Medical-Biological Union, November 12, 1944, in *Discorsi e Radiomessaggi*, 6:181–196 [Italian]; Address at the Seventh International Surgery Conference, May 21, 1948, in *Discorsi e Radiomessaggi*, 10:96–100 [Italian]; Address at the Conference of the Family Front and the Federation of Large Family Associations, November 27, 1951, in *Discorsi e Radiomessaggi*, 13:411–418 [Italian]. Subsequent teachings of the Magisterium that should be recalled include John XXIII, Encyclical Letter *Mater et magistra*, May 15, 1961; Paul VI, Encyclical Letter *Humanae vitae*, July 25, 1968; idem, Address to the Participants in the Twenty-Third National Conference of the Italian Catholic Jurists Union, December 9, 1972, in *Insegnamenti di Paolo VI* (Vatican City: Tipografia Poliglotta Vaticana, 1963–1979), 10:1260–1264 [Italian]; John Paul II, Apostolic Exhortation *Familiaris consortio*, November 22, 1981; idem, Address to Participants in the International Medical Conference Promoted by the Italian Movement for Life, October 12, 1985, in *Insegnamenti di Giovanni Paolo II* (Vatican City: Libreria Editrice Vaticana, 1979–2000), 8.2:933–936 [Italian]; idem, *Letter to Families*, February 2, 1994; idem, Encyclical Letter *Evangelium vitae*, March 25, 1995. For an analysis of papal teaching on the subject of abortion, see M. L. Di Pietro, "La Lettera Enciclica *Evangelium vitae* e l'aborto procurato: Nuovi elementi di riflessione nella continuità di un insegnamento," *Vita e Pensiero* 10 (1995): 653–676.

[86] I will not examine here the question of the death penalty or the sacrifice of individuals for the defense of the community, nor cases in which the killing is not direct or the person killed is not innocent; however, these cases should also be ethically reexamined for a different interpretation with respect to the historical paradigm.

4. The therapeutic principle is invoked here erroneously and improperly, as already noted, not only because alternatives to the elimination of the fetus are very often not examined, but also because the therapeutic end is attained indirectly by way of the destruction of a supreme good: life. Therefore the comparison between the health of the mother and the life of the fetus is unbalanced and subverted; in any case, the life of the unborn child cannot be made instrumental to the mother's health (a secondary good in comparison to life). It should also be kept in mind that maternity per se, like any other task in life, entails risks to one's own health.

5. The ethical obligations of society, science, and individuals require working to prevent situations of risk and health loss in pregnant women, through legitimate and licit means, in order to guarantee the best hospital and technological care for women giving birth—that is, to direct health care policies toward support for life and not toward its simple destruction. Science is for the sake of life and society is for the sake of the human person: this is the fundamental ethical commitment.

The dramatic cases

Given all of this, it is necessary to admit—even with the availability of more rigorously scientific assessments and even for ethically informed and upright consciences—the existence of cases (more limited in scope than those provided for in abortion laws) in which the conflict between the life of the unborn child and the survival of the mother becomes a human drama for the parents, the medical professionals, and the health care staff.

These dramatic cases must now be addressed while making sure to pay due attention to their many facets: the subjective drama; the personal and professional involvement of the physician, surgeon, or gynecologist; and the objective vision of the values in question and the ethical path to be followed.

Theoretically, and perhaps not just theoretically, there can be two gradations in the event of a conflict between the life of the mother and the life of the fetus: (a) continuation of pregnancy not only causes the death of the mother but does not even save the baby, while procured abortion effectively saves the mother; (b) continuation of pregnancy involves the death of the mother, while there is hope of saving the baby. Ethical opinions concerning the first case will be cited first because it is more complex and serious and automatically sheds light on the second case.

In reference to the first case, some moralists, and even some Catholic moralists, have sought arguments to justify abortion for the purpose of saving the mother, but always for the case in which the alternative would be the loss of both mother and child. Let us examine these arguments individually:

1. *Conflict of duties.* The physician has the duty to preserve the life of the mother and the duty to assist the birth of the child; since he is not able to carry out both, and since the conflict is in the facts and not in the will of the persons involved, the most accessible duty is chosen.[87] It should be

---

[87] E. Pousset, "Étre humain déjà," *Études* (Nov. 1970): 512–513. See also Tettamanzi, *Comunità cristiana*, 298–299.

noted that the choice here does not proceed by way of giving priority to the mother's treatment, involuntarily causing the death of the fetus; rather, it is a decision to kill by means of a direct act of destruction on the living fetus. The author himself who proposes this thesis says that it can occur only "with fear and trembling."[88]

2. *The subordinate classification of an already doomed fetus.* A fetus that is already doomed to die on its own cannot be called a human life in the full sense: in these cases abortion can be considered a hastening of death to save the mother's life.[89] It is not difficult to identify an objection to this thesis: the fact that the baby is doomed to die on its own is not sufficient reason to kill it, because natural death cannot be compared with direct killing; otherwise this principle would wind up justifying any act of euthanasia.

3. *Global assessment.* The at-risk mother-child health care question is considered as a global problem, as is the action of the responsible physician. Since full and complete success is not possible in the global situation, the possible good is sought.[90] Furthermore, this directive is based on the commitment to protect life, which is better safeguarded by saving the mother's life.

This solution also lends itself to criticism for being empiricist because we are not actually dealing with a single global reality if there are two lives. Furthermore, the commitment to defend life (the mother's) does not authorize the disproportionate and abnormal means of destroying the fetus; nor can the intention of the agent (*finis operantis*) ignore the objective reality of the act itself (*finis operis*).

4. *Indirect abortion.* The well-known principle of double effect applies to actions with two effects, one good and one bad. In these cases it is licit to posit the action with a view to achieving the good effect, even if a negative, undesired effect results indirectly from it. This position cites the admissibility and lawfulness of so-called indirect abortion in the case of the removal of a tumor as support.[91] But the scenario is different in the case being considered here: the direct action is the destruction of the fetus while the indirect effect is saving the mother's life. The old precept *non sunt facienda mala ut veniant bona* (evil must not be done to bring about good) should be applied and considered here so that both the end and the means are licit.

---

[88] Pousset, "Être humain déjà," 512–513.

[89] R. Troisfontaines, "Faut-il légaliser l'avortement?" *Nouvelle Revue de Théologie* 103 (1971): 500.

[90] G. Davanzo, "L'aborto nella problematica etico-cristiana," *Anime e Corpi* 38 (1971): 550–551, presents this hypothesis as an open, speculative question.

[91] A. Günthor lists several conditions for the liceity of an action (or an omission) that also produces a bad effect: "(1) the action must be good in itself, or at least morally indifferent; (2) besides the bad effect there is also a good effect, and the will directly intends the good effect without counting on the bad one either as a means or as an end; (3) the good effect is not obtained through the bad effect; the latter at most springs from the action in parallel to the good effect; (4) the allowance of the bad effect is justified by a proportionate reason." A. Günthor, *Chiamata e Risposta* (Rome: Paoline, 1982), 1:531.

5. Finally the other argument: *the norm is not absolute.* "Thou shalt not kill" is not an absolute norm because there have always been justified exceptions such as legitimate defense against an unjust aggressor, sacrifice for the good of one's neighbor, and the death penalty.[92] This argument, which is hardly new, does not apply to the case at hand either: here it is a question of an innocent life and not of an unjust aggressor or of a culprit who deliberately violates the law even though he knows that the crime is punishable by death. As for someone who sacrifices himself for the good of his neighbor, he does so seriously for a higher motive, and strictly speaking he does not kill himself—others unjustly kill him.

Conclusion

Our conclusion, in keeping with the personalist position and with the norms of objective ethics, is as follows:

- It is the physician's duty to preserve life, both of the mother and of the child, and to apply all therapeutic measures to save both. Direct killing, which is neither a medical nor an ethical act, is not included among these measures. Human life can and does in fact cease for many reasons, but innocent life cannot be directly destroyed for any reason, since it is a transcendent value, and it cannot be directly sacrificed by others, not even in order to save someone. By allowing exceptions to this principle and introducing value judgments such as "worthless life," "subordinate value," or "a life that is not fully human," one opens the way to euthanasia and any other discriminatory procedure.

- The simplest case, at first glance, is when the death of the mother is foreseen and the continuation of the pregnancy is linked with the hope of saving the child. One cannot choose the life of the mother with a direct action of destroying the life of the child, because no human being has the right to make life-or-death decisions about another person. A cesarean section, which is a normal intervention, can certainly be attempted in this case when there is hope of saving the child in a woman already near the end of life; however, if it is possible to wait until the moment of clinical death with the use of a respirator, then one must wait for the natural death of the mother.

It is also possible that there might be a need to keep a pregnant woman artificially "alive" after brain death for the purpose of bringing the fetus to a stage of development at which life outside the womb would be possible.[93]

The ruling by the Italian Constitutional Court on this question, decision no. 27 on February 18, 1975, established the priority of the life and health of the mother, opening the door to a discriminatory process and a progressive undermining of the defense of human life that has ended up making the life of the fetus relative to the will of others in every case: the first relativizing step was favoring maternal life,

---

[92] J. M. Pidier, "La Chiesa e l'aborto," *Il Regno attualità* 2 (1973): 16–17.

[93] On this subject, see A. G. Spagnolo, *Bioetica nella ricerca e nella prassi medica* (Turin: Camilliane, 1997), 357–359.

then the same criterion was used for safeguarding maternal health, then for the sake of psychological health, and then for social reasons.

The involvement of the medical profession is such that medicine itself is no longer related exclusively to the defense of life but also to directly lethal actions. The need for conscientious objection arises in the face of this ethical and moral upheaval.

Eugenic Abortion

This expression is no longer used for two reasons: because it evokes Nazi ideology, implying comparisons that are sought to be avoided at all costs, and because this type of "indication" for abortion is in fact repackaged as "therapeutic" abortion inasmuch as the presence of a malformed or defective fetus constitutes a threat to the psychological health and social balance of the family.

Actually the case remains what it is: one proceeds with abortion, often with legal authorization, to prevent the birth of sickly or malformed or handicapped individuals, allegedly so as to prevent those individuals from embarking on an inhuman life, but above all so as to avoid the burden of sacrifice on the families and on society. This differs from Nazi ideology in its purpose: for Nazism the purpose was the purification of the race, whereas for contemporary culture the motivation has a socioeconomic and even hedonistic character.

Prenatal diagnosis techniques have sometimes been used and so-called preventive policies have been implemented in health care plans for this reason.

From the ethical perspective the presence of a deformity or a handicap in no way diminishes the ontological reality of the unborn child; indeed, the presence of an impairment—such as a disease—in a human subject calls for protection and assistance all the more, in the name of solidarity.

Since this problem was already addressed under the heading of prenatal diagnosis, I do not intend to dwell on it again at length. It should simply be added that from the ethical point of view there is an obligation in these cases and generally to promote research and support programs so as to prevent the causes of such deformities and to support families with suitable means whenever such an individual is born, whose condition unquestionably imposes human and economic burdens that are sometimes difficult.

A society qualifies as such by its ability to help the weak and the sick and not by its arrogance in causing their premature death.

Abortion Law and Conscientious Objection

Abortion is a traumatic experience not only for the woman and for the family but also in the context of law and legislation. In fact, a legal system that tries to authorize or liberalize abortion is compelled to deny and traumatize itself, abandoning its commitment to be equal for all in terms of what it permits medical professionals to do.

The topic being addressed here concerns not only bioethics but also the philosophy of law, constitutional law, criminal law, and legal medicine.[94] Legal medicine is the discipline most concerned with this matter.

---

[94] A. Fiori and E. Sgreccia, eds., *Obiezione di coscienza e aborto* (Milan: Vita e Pensiero, 1978).

What I would like to clarify in this brief discussion can be summed up in the following aspects: how the conflict between moral conscience and human law arises and is justified; whether there is an ethical obligation for conscientious objection on the part of the physician and of those who are called on to collaborate in the abortion procedure when it is requested; and what the extent and complications of conscientious objection are in this matter.

Conscience and its demands of freedom and truth

It is first of all necessary to reach an agreement on the precise meaning and inner dynamic of the conscience: What do we mean when we say "act according to your conscience"? And what conscience are we talking about?

We must distinguish *psychological conscience* from *moral conscience*; the first is the awareness of the human act as it is being performed, and it is the indispensable prerequisite for the second. Moral conscience is the awareness of the *moral value* of that act. This moral judgment is in turn twofold: it includes a previous evaluation of the act and a subsequent evaluation upon its completion. The two agree when the previous judgment is taken as a norm and followed; the conflict arises whenever freedom does not follow the dictates of conscience.

But the more important problem lies in answering this question: Why are we supposed to "follow our conscience" in the first place and never to act "against our conscience," thus binding human freedom itself to the conscience? What it is that binds conscience from within?

Conscience is the rational judgment, which may be more or less systematic or intuitive, of the value of a given act. This moral value, on the other hand, is based on ontological truth: in other words, objective truth binds reason and reason binds the conscience. In the case under discussion, the judgment as to the lawfulness of abortion, objective truth is represented by the value of man, which is the greatest of all temporal values, transcending even the temporal realm itself. The judgment of reason, if it is *upright* and sincere, and moreover if it is *certain* (i.e., without elements of subjective doubt), creates the ethical demand of "thou shalt not kill" an innocent human life.

It should be noted, incidentally, that this adherence of reason to the truth and this subsequent adherence of the will to reason seem to suggest something constraining; in reality this adherence is liberating, because it makes human freedom dependent only on one's own conscience and frees it from all risk of external manipulation or of social or ideological conformism:

> Consequently *in the practical judgment of conscience*, which imposes on the person the obligation to perform a given act, *the link between freedom and truth is made manifest*. Precisely for this reason conscience expresses itself in acts of "judgment" which reflect the truth about the good, and not in arbitrary "decisions." The maturity and responsibility of these judgments—and, when all is said and done, of the individual who is their subject—are not measured by the liberation of the conscience from objective truth, in favor of an alleged autonomy in personal decisions, but, on the contrary, by an insistent search for truth and by allowing oneself to be guided by that truth in one's actions.[95]

---

[95] John Paul II, Encyclical Letter *Veritatis splendor*, August 6, 1993, no. 61.

Some moralists note that man transcends himself in a judgment of conscience, or more accurately, he perceives a value that transcends and commits him: ultimately this judgment implies that there is a guarantee of absoluteness within reality that coincides with the absolute.

Certainly affirming the Absolute, God, who is rationally postulated even in theoretical terms, brings with it a clearer guarantee, especially if it is illuminated by a revealed religion.

It is to this purpose that we should recall the first "conscientious objection" declared by the Apostles before the Sanhedrin: "We must obey God rather than any human authority " (Acts 5:29).

"When confronted, therefore, with a law that puts itself in direct opposition to the good of the person, which even denies the person and itself by destroying the right to life, the Christian cannot help but oppose it with his civil but firm refusal, mindful of the words of the Apostle Peter before the Sanhedrin: 'We must obey God rather than any human authority.'"[96] Nevertheless, I do not think that the eyes of faith are necessary per se in order to perceive the unacceptability of certain situations that are detrimental to the dignity of man. There have always been secular consciences that, as a protest against oppressive laws or against situations of violence, have raised the protest of their conscience in defense of freedom, even paying for it with their lives. And besides, even in the view of a secular and rational conscience, the killing of innocents and the attempt by the law to use the medical profession for that end are more than sufficient reasons to make conscientious objection not only possible but also mandatory.[97]

The normative character of the law and the bond of conscience

"Human law is the determination and expression of the legitimate authority of some requirements of the common good of a given society at a given historical moment."[98]

This definition implies that the law, too, should be founded on reason and that it should seek the common good: Thomistic philosophy defined the law as *ordinatio rationis* (the ordering of reason). The common good is to be pursued in the ways that are legally and constitutionally provided for in particular societies. In a democratic society this pursuit is accomplished through consultation of the constitutional organs and with respect for the plurality of ways of thinking and religious inspirations, and therefore with respect for freedom of conscience and religion (principle of religious freedom). Nevertheless the idea of the *common good* should be understood not in the sense of the good of the majority (that would be dictatorship), but rather as the pursuit of the *conditions* by which every person can experience personal fulfillment

---

[96] John Paul II, Address to Midwives Participating in the Conference on Defense of Life and Family, January 26, 1980, in *Insegnamenti di Giovanni Paolo II*, 3.1:191–195 [Italian].

[97] J. De Finance, "La coscienza e la legge," in *Obiezione di coscienza*, Fiori and Sgreccia, 19–37.

[98] Ibid., 26. On this topic, see alo R. Garcia De Haro, "La legge morale e le norme civili," in *Persona, verità e morale: Atti del congresso internazionale di teologia morale; Roma, 7–12 aprile 1986*, ed. A. Ansaldo (Rome: Città Nuova, 1986), 361 ff.

in his own life. Realizing one's own potential in life and perfecting oneself morally remains the task of every human person.

Therefore the law neither creates ethics nor imposes its own morality (the ethical state of Hegelian memory), but must be respectful and capable of creating the conditions for the fulfillment of persons. In defining the common good, the law will often have to ask for sacrifices even in the exercise of freedom by individuals, within certain limits, and will also have to permit certain things that some would not consider good in themselves so as to avoid worse evils. The law is not identical with ethics; it cannot always prevent every evil and every abuse in the exercise of personal freedoms, but it must create the objective conditions for the ethical conduct of everyone and for the fulfillment of individual persons.

In his encyclical *Evangelium vitae*, John Paul II explained how the only possible basis for a civil law is the common good and respect for fundamental values of an ethical order when he stated that "values such as the dignity of every human person, respect for inviolable and inalienable human rights, and the adoption of the 'common good' as the end and criterion regulating political life are certainly fundamental and not to be ignored." He also said that "the basis of these values cannot be provisional and changeable 'majority' opinions, but only the acknowledgment of an objective moral law which, as the 'natural law' written in the human heart, is the obligatory point of reference for civil law itself."[99] Hence "laws which authorize and promote abortion and euthanasia are therefore radically opposed not only to the good of the individual but also to the common good; as such they are completely lacking in authentic juridical validity."[100]

Now there is no doubt that the following two objective conditions must be included among the essential and objective conditions that the law must safeguard for the good of persons and for the common good (guarantees of constitutionality and legitimacy):

- The law must defend the lives of all, especially of the innocent and the most defenseless. If the law does not create this condition—the right to life—it is no longer law and becomes unjust: it must be combated with all legitimate means by everyone for the sake of those who cannot defend themselves.

- The law cannot force anyone to take the lives of other persons, except as a legitimate defense against an unjust aggressor (and the embryo/fetus cannot be considered an aggressor); much less can it require the physician to lend his own assistance in order to kill; the physician by profession is not called to do that.[101]

The argument of the so-called lesser evil, an expression that is inherently ambiguous, cannot be applied in these cases because there is no greater evil than to take a life.[102]

---

[99] John Paul II, *Evangelium vitae*, no. 70

[100] Ibid., no. 72.

[101] See the entire third part of the instruction *Donum vitae* by the Congregation for the Doctrine of the Faith, titled "Moral and Civil Law."

[102] C. Caffarra, "Aborto e obiezione di coscienza," *Medicina e Morale* 27, nos. 1/2 (1977): 101–109; E. Sgreccia, "La obiezione di coscienza e le implicazioni nella prassi

Conscientious objection and elective abortion

From the foregoing review of the relation between conscience and law it can be concluded that, of all the various forms of protest (against military service, for example), conscientious objection to abortion on the part of physicians is most unequivocally *lawful and obligatory*. *As a human being*, the physician cannot perform or directly collaborate in an act of taking the life of a human individual, even in the formative stage; *as a physician* he is called by his profession and professional code of ethics to care for and sustain life, and the autonomy of his profession should be respected accordingly.[103]

"Disregard for the right to life, precisely because it leads to the killing of the person whom society exists to serve, is what most directly conflicts with the possibility of achieving the common good. Consequently, a civil law authorizing abortion or euthanasia ceases by that very fact to be a true, morally binding civil law. ... There is no obligation in conscience to obey such laws; instead there is a *grave and clear obligation to oppose them by conscientious objection*."[104] Such conscientious objection must be considered a duty on the part of those—physicians, health care workers, or hospital, clinic, or nursing home administrators—who might have the occasion to take part in the administrative procedures or in the very performance of abortion. The same should be said about pharmacists since—as will be seen—the majority of "contraceptives" distributed in pharmacies have abortifacient mechanisms of action.[105]

Moreover, John Paul II introduced a further novel consideration in the encyclical *Evangelium vitae* by extending this right and duty of *noncooperation with evil acts* to elected officials in the legislative sphere as well,[106] in other words, to the way that legislators should act with regard to laws that do not respect the right to life of

---

assistenziale e nei consultori familiari," *Anime e Corpi* 77 (1978): 295–315; B. Honings, "Doveri e responsabilità degli operatori sanitari alla luce dell'enciclica *Evangelium vitae*," in *"Evangelium vitae" e bioetica*, Sgreccia and Sacchini, 125–146.

[103] It should be emphasized, however, that this duty always to care for and sustain human life is not taken into consideration at all in the 2006 Italian Code of Medical Ethics of the National Federation of Medical and Dental Associations (Federazione Nazionale degli Ordini dei Medici Chirurghi e Odontoiatri, FNOMCeO), which, while it condemns clandestine abortion (at least this can be read between the lines), accepts recourse to elective abortion according to the provisions of the law no. 194/78. Article 43 reads as follows: "Termination of pregnancy, excepting the cases provided for by law, constitutes a very serious infraction of the code of professional ethics, especially if it is also performed for reasons of financial gain. Conscientious objection by the physician is expressed in the context and within the limits of the law in force and does not exempt the physician from the obligations and duties inherent in the relationship of care with the woman."

[104] John Paul II, *Evangelium vitae*, nos. 72–73, emphasis added. The *Charter for Health Care Workers* of the Pontifical Council for Pastoral Assistance to Health Care Workers repeats this statement, maintaining that physicians and nurses are obligated to invoke conscientious objection (see *Charter*, no. 143).

[105] J. L. Guzmán, *Objeción de conçiencia farmacéutica* (Barcelona: Ediciones Internacionales Universitarias, 1997).

[106] John Paul II, *Evangelium vitae*, nos. 73–74.

the unborn child. This addresses a situation that is more and more common today, because many countries are calling into question permissive laws that already exist[107] and also because disputable proposed laws are often encountered in a legislative session. The central question is whether one may vote for a law that is an improvement yet still permissive: "In the case like the one just mentioned," replies John Paul II in paragraph 73 of *Evangelium vitae*, "when it is not possible to overturn or completely abrogate a pro-abortion law, an elected official, whose absolute personal opposition to procured abortion was well known, could licitly support proposals aimed at *limiting the harm* done by such a law and at lessening its negative consequences at the level of general opinion and public morality. This does not in fact represent an illicit cooperation with an unjust law, but rather a legitimate and proper attempt to limit its evil aspects."

To return to the physician's duty: it is necessary to recall that a political authority cannot compel a physician to perform a surgical intervention that he considers unnecessary or harmful; much less can the law compel him to destroy a life.

Italian law allows abortion at the woman's request. The physician is requested to intervene in two phases: a "verification" or "procedural" phase and an "operational" phase. If the legal requirements are satisfied, the woman alone—not even her husband necessarily—is given the authority to request a termination of pregnancy.

> When the physician at the consulting room or at a social service or health care facility, or a trusted physician, verifies the existence of conditions making the intervention urgent, he may immediately issue the woman a certificate attesting to the urgency. With this certificate the woman herself can visit one of the facilities authorized to practice termination of pregnancy.

> If there is no proof that the case is urgent, at the end of the examination the physician at the counseling center or the social service or health care facility, or the trusted physician, in response to the woman's request to terminate the pregnancy on the basis of circumstances enumerated in article 4, should issue her a copy of a document, signed by the woman as well, attesting to her state of pregnancy and her request, and ask her to postpone her decision for seven (7) days. When the seven (7) days have passed, the woman can present herself at one of the authorized facilities with the document issued as described in the present clause and undergo a termination of pregnancy.[108]

Evidently the law acts as though the unborn child were a part of the woman's body, and even something anomalous. The woman's "right," though not explicitly formulated, is specifically made to prevail even over the acknowledged competence of the physician.

It was evident that conscientious objection would be claimed as a right and a safeguard for the physician's professional conscience. Moreover, this right is recognized more or less extensively, with greater or lesser restrictions, by most European

---

[107] On this subject, compare Eusebi, "Tutela giuridica"; M. L. Di Pietro and M. B. Fisso, "La tutela dell'embrione umano in Germania: Dalla legge del 1990 alla Sentenza della Corte Costituzionale del 28 maggio 1993," *Vita e Pensiero* 4 (1994): 269–283; C. Casini, "Prospettive di riforma dell'attuale legislazione sull'aborto: Il dibattito italiano ed europeo," supplement, *La Speranza* (Jan. 1995); Eusebi, "Corresponsabilità."

[108] Law no. 194/78, art. 5, §§ 3–4.

abortion laws as well.[109] The conscientious objection formula of the Italian law is found in article 9 of law no. 194/78.

> Health care and auxiliary personnel are not bound to participate in the administrative procedures described in articles 5 and 7 or in interventions to terminate pregnancy when they raise a conscientious objection by way of a prior declaration. The declaration of the objector must be communicated to the provincial physician,[110] or to the health care facility director in the case of hospital physicians or physicians practicing in nursing homes, within a month of when this law goes into effect, or the objector's date of licensing, or the objector's date of hiring by an entity that is required to provide termination of pregnancy services, or the date of entering into an agreement with insurance companies that involves the performance of such services.

> The objection can always be revoked or proposed even outside of the limits specified in the preceding paragraph, but in that case it goes into effect one month after it is presented to the provincial physician.

> Conscientious objection exempts health care personnel who carry out auxiliary activities from performing procedures and activities directly pertaining to abortion but not from the care that precedes and follows abortions.

> Conscientious objection cannot be invoked by health care and auxiliary personnel when, given particular circumstances, their intervention is indispensable to save the life of a woman in imminent danger.[111]

An exemption similar to conscientious objection is granted to health care institutions run by religious bodies unless they ask (article 8) to perform such termination procedures.

In addition to its overall character as an abortion-permitting law, by which it therefore fails to effectively defend the unborn child's right to life, a commentary on this law would have to include many ethical remarks that can only be briefly mentioned here.

In reality, the range of "indications" prior to ninety days is so broad, as seen earlier, that in practice it includes medical, socioeconomic, therapeutic, and eugenic indications by which a contraceptive and liberalized use of abortion takes shape. This has, in fact, come about, notwithstanding declarations to the contrary.

Articles 6 and 7, which provide for termination after ninety days, consider the following conditions: (a) serious danger to the life of the woman; (b) serious danger

---

[109] F. Stella, "La situazione legislativa in merito alla obiezione sanitaria in Europa," *Medicina e Morale* 2 (1985): 281–301; on this topic, see V. M. Caferra, *Diritti della persona e stato sociale* (Bologna: Zanichelli, 1987), 96–99; R. Venditti, *Le ragioni dell'obiezione di coscienza* (Turin: Gruppo Abele, 1986), 48–52; P. Benciolini, "Deontologia e obiezione di coscienza," in *Bioetica in medicina*, ed. A. Bompiani (Rome: CIC Edizioni Internazionali, 1996), 366–382.

[110] With the reform of the National Health Care Service (Servizio Sanitario Nazionale, SSN), the role of provincial physician has been replaced by the Head of the Public Hygiene Sector (or the equivalent authority according to regional norms) of the appropriate Local Health Care Unit (Unità Sanitaria Locale, USL).

[111] Law no. 194/78, art. 9.

to the physical and psychological health of the woman, including the case of *fetal malformation.*

The father figure is undervalued in assigning decisional authority; the physician listens to him only "when appropriate and requested by her" (article 5). And decision no. 389 of the Italian Constitutional Court on March 31, 1988, confirmed the legitimacy of this exclusion of the father from any chance of intervening in the woman's decision to abort. *Consultation* with the physician and *counseling* in family counseling centers is ultimately reduced to a functional formality.

As far as conscientious objection is concerned, it is important to note some restrictions of a punitive character for those who invoke it. At the time it was said that such restrictions were established to discourage clandestine abortions and the exploitation of conscientious objection for financial profit. Certainly a physician performing clandestine abortions in private while protected by conscientious objection in public facilities would commit a doubly illicit immoral act. But that could be avoided in another way: with the penalties discussed in articles 19 and 20, which provide for serious sanctions for those who perform abortions clandestinely.

Yet the following restrictions should also be noted: article 9 implicitly excludes an objecting physician from the *procedural* phase, since the physician who examines the woman *is bound* to issue the signed declaration even on the basis of the woman's will alone, even when the physician is of the contrary opinion; a physician who thus participates in the procedural step would lapse from his conscientious objection and, by law, would become ipso facto a cooperator in the abortion. This was done deliberately to prevent objectors from having counseling contact with a woman who is inclined to abort.[112]

---

[112] It is well known that there is no unequivocal interpretation as to the activities from which the conscientious objector *must* and *can* refrain: hence the possibility of *tacitly* revoking the objection in the case of participation or interventions to terminate pregnancy, as provided for by law no. 194/1978, apart from the state of necessity. Based on a second interpretation, the activities from which the conscientious objector *must* and *can* refrain do not coincide, because otherwise one would fall into too strict an interpretation of the law: "Indeed, there are activities in which the conscientious objector could decide to participate in conscience, even if they are only incidentally aimed at abortion, for the sole purpose of limiting terminations of pregnancy (consider, for example, the phase of certifying the conditions in the woman and the subsequent possibility of refusing to issue the document when the prerequisites are not verified). Yet there seems to be no justification in connecting participation in these particular activities—which are subjectively aimed at limiting abortions—with the negative effect of revocation. Indeed, revocation is a *sanction* or penalty and, that being so, it seems right to interpret the assumption restrictively: the *reason* for revocation is to make sure that there is a way of assessing the seriousness of conscientious objection and of punishing health care workers who conveniently raise objections so as to perform abortive interventions anyway clandestinely" (Zanchetti, *La legge sull'interruzione*, 253). The main point of debate, therefore, is in particular whether or not it is possible for the physician objector to have access to the consultation with the woman provided for in article 5 of the law, a consultation that is followed by the issuance of the document signed by the woman and countersigned by the physician. In fact, quite apart from the opinions of various legal and medical scholars (see P. Benciolini and A. Aprile, *L'interruzione volontaria della gravidanza: Compiti, problemi, responsabilità* [Padua: CEDAM, 1990]), physician objectors are excluded from the consultation and even from "raising" the issue of abortion

As to whether cooperation is formal or material, proximate or remote, the guidelines presented in the chapter on prenatal diagnosis hold true here as well. It is not just performing the intervention that is morally illicit, but also any *formal* (i.e., intentional) cooperation: whether it is expressed by the health care workers or approved by the parents, relatives, or partner of the woman, who naturally is still the person most responsible, although she is sometimes the person most subject to social pressure.[113]

Any direct cooperation is also illicit, even if it is *material* (non-intentional): this occurs when the act of cooperation is expressed in activities that have no other purpose than to prepare or accompany the abortion procedure (assistant surgeons, medical and gynecological assistants present during the termination, surgical nurses, anesthetists).

Other forms of material cooperation not directly connected with the event and with the abortion procedure are illicit in principle and the subject may express his refusal when he is aware that his service—though not evil in itself—will be utilized for evil, unless there are proportionate reasons that compel or recommend the service.[114]

In particular, there ought to be room for conscientious objection by the tutelary judge appointed by law for female minors and mentally incompetent females (articles 12 and 13). Unfortunately the Italian Constitutional Court[115] has repeatedly ruled that a tutelary judge is not entitled to conscientious objection in the case of an abortion request by a minor, reaffirming that this right is allowed only to health care personnel and their assistants. It is well known in Italy that the constitutionality

---

at all (see L. Eusebi, "Il dibattito nella bioetica: Obiezione di coscienza all'interruzione della gravidanza," *Medicina e Morale* 41, no. 6 [1991]: 1068–1070), even if such consultation does not appear to be incompatible with conscientious objection *insofar as it does not imply revocation of one's conscientious objection*. Naturally if one recognizes that the physician objector may participate in the consultation, one cannot, however, oblige him to issue the document, which would constitute direct material cooperation in abortion, inasmuch as it is an act necessary and sufficient to grant the woman access to that procedure. Since it seems difficult, on the practical level, to introduce into public administrative structures physician objectors who might offer consultations without issuing certifications, the proposal has already been made to sever the phase of consultation from the issuance of the document—a solution that would "overcome any problem whatsoever of the involvement of physician objectors in the consultation strictly speaking and furthermore would constitute a tangible manifestation of a real preventive orientation of the statutory norm" (Eusebi, "Il dibattito," 1069; see also the proposed modifications to law no. 194/1978 in Eusebi, "Tutela giuridica").

[113] The *Catechism of the Catholic Church* addresses this question as follows: "Formal cooperation in an abortion constitutes a grave offense. The Church attaches the canonical penalty of excommunication to this crime against human life. 'A person who procures a completed abortion incurs excommunication *latae sententiae*,' 'by the very commission of the offense,' and subject to the conditions provided by Canon Law." But it also reaffirms that "the Church does not thereby intend to restrict the scope of mercy. Rather, she makes clear the gravity of the crime committed, the irreparable harm done to the innocent who is put to death, as well as to the parents and the whole of society" (*CCC*, no. 2272).

[114] E. Sgreccia, "Aborto e responsabilità dei credenti," in *Il diritto alla vita*, ed. Concetti, 101–128.

[115] Italian Constitutional Court, decision no. 196, May 26, 1987.

of articles 9 and 12 of law no. 194/78 was first questioned during legal proceedings initiated on behalf of a female minor by a tutelary judge from Naples in 1984, who invoked articles 2, 3, 19, and 21 of the Italian Constitution. The judge from Naples had declared that when, due of a judge's profound and deeply rooted conviction against abortion, there is an "irresolvable conflict between his own conscience and the obligations resulting from his official duties," a tutelary judge should be allowed to raise a conscientious objection. The Italian Constitutional Court, however, citing article 5 of law no. 194/78, ruled that once the procedural verifications have satisfied the technical criteria required for voluntary termination of pregnancy, the judge can no longer distance himself from them; it therefore declared the claim of unconstitutionality to be unfounded.

In 1993, the Italian Constitutional Court reaffirmed this position,[116] rejecting an appeal by a lower court judge of Cuneo. The latter, maintaining that any distinction between prenatal and postnatal life is arbitrary, had questioned the constitutionality of law no. 194/78 because it does not guarantee the right to life of the conceptus and thus affords the mother an essentially limitless power of self-determination. He had particularly reaffirmed the need to provide for conscientious objection by tutelary judges with respect to the authorization of pregnancy termination for female minors.

Without addressing the question at length, the Constitutional Court limited itself to declaring the appeal inadmissible and reaffirming that provisions for conscientious objection are not necessary inasmuch as the judge still has the power to deny authorization (for example, when he doubts the minor's capacity to understand and decide freely).

Conscientious objection is regulated in various ways in other countries and, in general, can be expressed on a case-by-case basis without all the punitive restrictions of the Italian law.

"Hidden" Forms of Abortion

A significant and troubling phenomenon is developing at the scientific and social level based on the use of confusing and misleading terminology. There is a tendency to mask the term *abortion* under other names so as make it less conspicuous.

"In order to facilitate the spread of abortion," writes John Paul II in no. 13 of the encyclical *Evangelium vitae*, "enormous sums of money have been invested and continue to be invested in the production of pharmaceutical products which make it possible to kill the fetus in the mother's womb without recourse to medical assistance. On this point, scientific research itself seems to be almost exclusively preoccupied with developing products which are ever more simple and effective in suppressing life and which at the same time are capable of removing abortion from any kind of control of social responsibility." John Paul II continues by saying that we need "to have the courage to look the truth in the eye and to *call things by their proper name*, without yielding to convenient compromises or to the temptation of self-deception."[117]

---

[116] Italian Constitutional Court, decision, June 24, 1993.

[117] John Paul II, *Evangelium vitae*, no. 58, emphasis added.

The deceptive trend is evident especially in popularized books designed for broad dissemination and aimed at influencing public opinion,[118] but the same trend is even found in scientific journals.[119]

In practice, we are dealing with several methods of birth control that are incorrectly described as contraceptive because they do not prevent the union of the gametes, i.e. fertilization, which is what the term *contraceptive* would lead one to think. Their mechanism of action involves preventing the already fertilized egg cell from implanting in the uterus. Those who advertise these techniques are very careful not to call them *abortifacient* (for many people the term abortion still entails something dramatic); they are instead defined as *interceptive* if they intercept the zygote, thus preventing it from implanting, or *contragestive* (from *contragestion* and by analogy with *contraception*) if they prevent the continuation of pregnancy once the embryo has already implanted in the uterus.[120]

Procured abortion is not, however, the only manifestation of anti-life sentiment; the widespread and deeply rooted contraceptive mentality, which drives researchers to find ever "safer" techniques to avoid conception, can also be considered a danger to human life.

And even if it is undeniable—we read again in *Evangelium vitae*—that contraception is not equivalent to abortion from a moral standpoint,[121] it cannot be overlooked that both abortion and contraception are *fruits of the same tree*, which is "rooted in a hedonistic mentality unwilling to accept responsibility in matters of sexuality, and they imply a self-centered concept of freedom, which regards procreation as an obstacle to personal fulfillment. The life which could result from a sexual encounter thus becomes an enemy to be avoided at all costs, and abortion becomes

---

[118] See the study by A. G. Spagnolo, "Aborto e nuova sessualità: Situazione oggi sul piano sociale e politico," *Anime e Corpi* 129 (1987): 33–48. In a popular volume by E. Lauricella, *La riproduzione della specie umana* (Rome: Riuniti, 1986), for example, sexuality, birth control, and artificial reproduction are explained to the public in simplistic and often imprecise terms (the scientific studies on the Billings method and the positive statistics provided by the World Health Organization are completely ignored), especially through falsification of terminology, causing serious confusion for the inexperienced reader.

[119] E. E. Baulieu, "Contragestation by Antiprogestin: A New Approach to Human Fertility Control," in *Abortion: Medical Progress and Social Implications*, Ciba Foundation Symposium 115, ed. R. Porter and M. O'Connor (London: Pitman, 1985), 192–210.

[120] Here are the most common "contraceptives" today that act by interception or contragestion and are therefore really abortifacient: the pill with a low dose of progesterone (the so-called mini-pill); trimestral injections and subcutaneous implants based on progestins or estro-progestins; intrauterine devices (IUDs); the vaccine against human chorionic gonadotropin (hCG vaccine); the so-called morning-after pill; prostaglandins and luteolitic drugs (including RU-486). The use of these methods does not require hospitalization of the woman in most cases. See the preceding chapter for a systematic discussion of the subject. See also M. L. Di Pietro and R. Minacori, "Sull'abortività della pillola estro-progestinica e degli altri contraccettivi," *Medicina e Morale* 46, no. 5 (1996): 863–900.

[121] "Certainly, from the moral point of view contraception and abortion are *specifically different* evils: the former contradicts the full truth of the sexual act as the proper expression of conjugal love, while the latter destroys the life of a human being; the former is opposed to the virtue of chastity in marriage, the latter is opposed to the virtue of justice and directly violates the divine commandment 'You shall not kill.'" John Paul II, *Evangelium vitae*, no. 13.

the only possible decisive response to failed contraception."[122] Furthermore, one of the signs of the close connection between abortion and contraception is the finding that greater awareness and understanding of contraceptives nevertheless does not prevent abortion. Indeed, it can be ascertained that some women or couples use contraceptives with the intention of avoiding the temptation to abortion in the future, but it is also true that the distorted values inherent in the contraceptive mentality will increase that very temptation when an unwanted life is conceived.

"Indeed, the pro-abortion culture is especially strong precisely where the Church's teaching on contraception is rejected,"[123] and campaigns to legalize abortion have been led by the same persons who have long been involved in the spread of contraception.

In addition to the ethical problem raised by the fact that interceptives and contragestives cause the killing of a human individual, which was already dealt with in the preceding paragraphs, there are also legal implications. According to Italian law no. 194/78, only a physician is technically qualified to perform an abortion; the introduction and use of abortifacients effectively ends this exclusive control and an abortion can be obtained easily, safely, and privately by ingesting a pill at ever earlier stages of pregnancy.

That being the case, all the legislative provisions concerning the procedures to be followed by a woman intending to abort will no longer hold. This obviously includes verification of the legal conditions to obtain an abortion, the actual fulfillment of those conditions, and the doctor's position with regard to conscientious objection. Termination of pregnancy then becomes a private matter between the pregnant woman and the physician, who most times would limit himself to prescribing the drug and then intervening as needed in the event of serious complications.

The question thus arises: How should the matter be regulated and what guidelines should be given to a health care worker who is unwilling to prescribe, administer, or dispense these abortifacients, possibly raising a conscientious objection? Is there an obligation to dispense such products or can a health care worker refuse to do so? Law no. 194/78 made no provisions for the case of so-called emergency contraception: it is possible, however, to draw some inferences from an analysis of the purpose of the law itself.

In the name of safeguarding human life from its beginning, the law in question does not liberalize voluntary abortion, at least in theory, yet provides that it shall not be punishable under certain circumstances (see articles 4 and 6). This nonpunishment must therefore be considered an exception and is related to conditions and procedures established by the law in light of a general prohibition against actions that are harmful to unborn human life. In fact, the law provides that certain procedures must be observed in order to allow access to abortion: consultation, confirmation of the state of pregnancy, determination of the stage of pregnancy, a waiting period in which to reconsider, etc. Outside of these administrative procedures, abortion becomes a criminal, illegal act and is therefore not obligatory.

Applying what was stated above to the case of "emergency contraception," it is easy to show how the minimal procedures required by law no. 194/78 are disregarded

---

[122] Ibid.

[123] Ibid.

at the time the product is prescribed, administered, or dispensed. For example, there is no confirmation of the state of pregnancy because the usual level of β-hCG (the first hormonal sign of pregnancy) in the mother's blood does not produce positive results until the seventh or eighth day after fertilization, yet abortifacient products—particularly those that act by intercepting the zygote—must be used within and not beyond the first six days after fertilization in order to be effective.

Consequently a health care worker who is convinced of the abortifacient effects of these products and is unwilling to prescribe, administer, or dispense them, can act accordingly and no one can oblige him to act differently, since that would be to oblige him to perform an illegal act. This seems to be an aspect that should be kept in mind particularly by those, for example, who think that a physician working at a public institution, or in a contract agreement with it, should be obliged to perform this professional service.

Starting from these premises, a pharmacist—a professional not taken into account by law no. 194/78—can also refuse to dispense these products, which could make him jointly responsible for an illegal abortion, as provided in article 19 of the same law.

It is clear, however, that widespread recourse to "emergency contraception" provides further occasion to reconsider—even in legal terms—the question of abortion and the need to protect unborn human life, which is increasingly exposed to the dangers of manipulation and destruction, even in circumstances not foreseen by law no. 194/78.[124]

---

[124] M. L. Di Pietro and R. Minacori, "'Contraccezione d'emergenza': Problema medico, etico e giuridico," *Vita e Pensiero* 5 (1997): 353–361. Some jurists and scholars of legal medicine have pointed out that since these means are administered prior to the possibility of diagnosing pregnancy—by the usual tests—there would be no way of invoking the application of the procedures provided for by law no. 194/78 requiring the preliminary confirmation of pregnancy (this is therefore a contingency the legislators had not taken into consideration and that would need to be regulated). As for the possibility of refusing to provide professional services involving the prescription and/or administration of such substances for reasons of conscience, on the other hand, this is legitimate—according to the same scholars—not only from the ethical but also from the legal point of view. Indeed, once one been scientifically convinced of the abortifacient action of the aforementioned means, the subsequent decision not to prescribe, administer, or dispense them could not be characterized as a failure to do one's professional duty. In other words, no norm compels outright the performance of an abortion, since even in the case of abortion within the first ninety days definite procedures have been established for a meeting between the woman and the social or health care worker, along with a waiting period for a possible change of mind. And in order to refuse services connected with emergency contraception it is not even necessary to be a conscientious objector, inasmuch as it can very well happen that nonobjecting physicians may nevertheless notice that employing those means would be against the dictates of their own conscience, even though they might be in favor of performing an abortion in conformity with the procedures provided for by the law itself. See Benciolini and Aprile, *L'interruzione volontaria*, 40–42. In conclusion, the spread of the practice of "emergency contraception" is yet another reason to review the entire text of law no. 194/1978 and to initiate a real campaign to prevent abortion, which is always a serious trauma for the woman. Moreover it is additional evidence that the boundary lines between contraception and abortion have become further eroded since both are manifestations of one and the same anti-life mentality.

The Prevention of Spontaneous Abortion

In conclusion I will also mention the ethical aspects connected with spontaneous abortion, which is an event that occurs with some frequency: around 15 percent of all clinically recorded pregnancies end in spontaneous abortion.[125]

It is difficult, however, to determine the number of spontaneous abortions that occur in early, preclinical stages. Based on some studies it appears that at least 60 percent of conceived zygotes do not continue their development and are lost.[126]

Spontaneous abortion or miscarriage can be caused by various factors: chromosomes, endocrine glands, infectious diseases, metabolic disorders, malformations of the genital tract, immunological factors, or psychological reasons. In most cases, the miscarriage is an isolated episode and it is often impossible to identify its etiology.

In 0.3 to 0.5 percent of cases, the event can occur three or more times consecutively, which is clinically referred to as recurrent spontaneous abortion (RSA). Primary RSA occurs in a couple of which the female partner has not had a full-term or premature birth with living offspring, as distinguished from secondary RSA, in which the woman has first had one or more children and then the repeated miscarriages.

The causes of RSA, which are not always readily identifiable, can include: (1) environmental factors (living or work environment); (2) chromosomal factors; (3) hormonal factors; (4) malformations, fibromas or synechias of the uterus, endometritis; (5) infections (caused by *Ureaplasma urealyticum, Mycoplasma hominis,* or *Chlamydia trachomatis*); (6) autoimmune factors; and/or (7) psychological causes (the woman can, for example, wish for a child at the conscious level but at the same time reject it subconsciously).

Once the cause has been identified, it is possible to intervene either by eliminating it or providing supportive measures to counteract it. In fact, as a result of specific treatment, it has been possible to observe the birth of a higher percentage of living fetuses than in groups of patients who were not treated.

At this point it becomes legitimate to ask whether or not there is a moral obligation to avoid spontaneous abortion. Those who do not acknowledge such an obligation claim that if nature resorts to abortion to maintain genetic stability by discarding deformed embryos, there is no reason why one should need to intervene to prevent it.[127] Now while it is true—as already stated—that the errors of nature should not be repeated, it is also true that the reality of diseases occurring in nature does not mean that a fetus exposed to RSA should not be cared for like someone who is already born: this is why every woman and every couple, as soon as there is some thought that a pregnancy may have begun, must take all of the practical precautions to reduce the risk of spontaneous abortion.

---

[125] For a more in-depth discussion of the subject, see B. Cinque et al., "Aborto ripetuto spontaneo: Aspetti scientifici e obbligazioni morali," *Medicina e Morale* 42, no. 5 (1992): 889–910.

[126] D.K. Edmonds et al., "Early Embryonic Mortality in Women," *Fertility and Sterility* 38 (1982): 447.

[127] C.J. Robert and C.R. Lowe, "Where Have All the Conceptions Gone?" *Lancet* 305 (1975), 498–499.

When the risk factors are environmental, the obligation to intervene applies not just to the individual but to society as a whole. Once a duty to intervene is recognized, however, it needs to be clear that it should be carried out using proportionate measures so as to avoid therapeutic stubbornness. Indeed, RSA can sometimes arise as the consequence of a pathological inability to continue the pregnancy, whereby insisting upon heroic therapies could mean therapeutic stubbornness with regard to the fetus, whose death in such a case would be postponed by only a few hours or days.

Summary Outline for Chapter 10

Abortion

Definition of Techniques

Procured abortion (or "voluntary termination of pregnancy," VTOP) terminates a pregnancy directly and intentionally by destroying the life of the unborn child.

Procured abortion is justified by asserting that until a certain stage of development the embryo is not yet a human person, or that his rights are subordinate to the will of the mother, or else that his life and death are indifferent inasmuch as he does not yet have his own interests or decision-making autonomy.

Procured abortion is carried out by means of the techniques of *suction* and *curettage*. It can also be induced by *contragestive chemicals* that are capable of causing the detachment, death, and elimination of an embryo that has already implanted.

Other methods of abortion include chemical products or devices that have an *interceptive or antinidatory* effect; in other words, they alter the physiology of the movement of the already formed embryo in the Fallopian tube and cause its death by preventing its implantation in the uterus.

Antinidatory substances, acting before implantation in the uterus, are not considered abortifacients by those who hold that pregnancy begins only with implantation (i.e., around 6 to 7 days after fertilization; complete around 14 days). The WHO, which provided its official definition of pregnancy in 1985, determined that it begins at implantation. However, this in no way changes what is pursued through the use of antinidatory methods: the death of the already existing unborn child, even if not yet implanted in the uterus.

"Therapeutic" abortion

Some talk about "therapeutic abortion" when voluntary termination of pregnancy is carried out with the intention of safeguarding the life or health of the mother.

"Danger to the life or health of the mother," however, can include a wide range of situations having different degrees of seriousness; this opens the door to the possibility of extending the justification also to situations that have little to do with maternal health. In fact, there is the risk of talking about "therapeutic abortion" in cases of eugenic abortion connected with the malformation or illness of the fetus, or simply because the child is not wanted, or because of socioeconomic difficulties.

Therapeutic abortion is almost always carried out after diagnostic tests on the health of the unborn child. The life of the unborn child is increasingly put at risk for the purpose of "knowing," and the slightest anomaly or suspicion about the qualities of the child elicits from the parents a reaction of rejection.

Abortion can be *direct* when it is directly intended, as an end or a means. This is different from *indirect abortion*, which occurs as a consequence of an action. It is brought about neither as an end nor as a means, but is foreseen and permitted inasmuch as it is inevitably connected to what is intended and caused directly. It is not an abortive action aimed directly at the fetus but rather an action necessary to preserve the life of the mother which is aimed at some part of her body. It is an action that causes abortion as a collateral effect, which therefore is not the direct object of the will or of the specific action.

We cannot talk about true therapy in the case of a direct abortion, even if it is called therapeutic, since it does not treat an actual illness but rather causes the destruction of the fetus in order to avoid burdening the health or endangering the life of the mother.

Arguments against procured abortion

The life of every human being starts at conception

With the fusion of the gametes, a new human cell begins to operate as a unity. From that moment on there is a unique, well-identified being that is intrinsically oriented toward a defined pattern of development. This new entity, through continual and gradual coordinated development, maintains its own biological and genetic identity over time, without interruptions in continuity, until the moment of its death. It is a new being with its own identity conferred on it by a unique, unifying substantial principle. The essential qualitative leap occurs at fertilization, in the transition from two substances (the gametes) to a single substance (the zygote). The latter, in its biological development, exhibits a substantial continuity because the principle of change and growth is intrinsic to the substance itself.

All human beings have an equal right to live

Every human individual, once conceived, has the right to be respected in his physical integrity, which for him is the necessary condition for life and growth.

Respect for the biological life of every human being (regardless of his attributes or stage of development) is the prerequisite for any other type of respect that should be given to him.

Since all human beings have the same nature in common, the same fundamental rights must be granted to all, first and foremost the right not to be killed (right to life).

The perspective of the law

The legal perspective sees abortion as a *specific social practice*, inasmuch as it is a *relational* practice with *public relevance*. It is obvious that the decision to abort involves not only the woman but also the unborn child that she has in her womb, not to mention the father.

The law cannot be neutral; it must intervene on the issue of abortion because in general it has the duty to intervene whenever the relational equilibrium is disturbed, whenever the concerns of the weak are threatened by the claims of the powerful.

The pretension of the Italian abortion law (law no. 1978/194) to handle abortion with a neutral approach, leaving any decision to the mother, therefore goes against the very nature of law.

Before that legislation was passed, the Italian penal code prohibited procured abortion, classifying it as a *crime against the integrity and health of the lineage.*

It considered the *passive subject* of the crime—the victim of the criminal act—to be not an individual subject (the unborn child) but rather a collective subject (the *stirpe*, or lineage). The life of the unborn child was safeguarded indirectly by virtue of the protection that the juridical ordinance accorded to the lineage.

The Italian law currently in force does speak about the unborn child but the mother is the one who has the right—but not the duty—to bring him or her into the world. It should instead be highlighted that the unborn child has the right to be born: "The right to be born is truly the first of all rights, not only in the chronological sense—as it might appear at first—but above all in the *axiological* sense: that is, the right to be born includes within itself and serves as the foundation for all the constitutive rights of the person, the right to be respected in one's own identity, the right not to be exploited for any reason, the right to be considered the bearer of a specific dignity that cannot be reduced to that of any other human being whatsoever—in sum, the right to be recognized as *person*." (F. D'Agostino)

The human embryo is a (fragile) human person

The human being has vegetative life in common with plants and sensitive life in common with animals. There is a qualitative difference, however, between the life of plants or animals and the life of a human being. This difference results from man's faculty of the intellect, which is also proof of the spirituality of his soul. The word "person" means, according to Thomas Aquinas, what is noblest in all the universe, and it indicates the dignity of the individual having a rational nature (*Summa Theologiae* I, q. 29, a. 3), which, as such, is free to choose its final end and has full responsibility for its actions.

Peter Singer dissolves the bond that unites human being to person, declaring that there can be nonhuman "persons" and human beings deprived of the dignity of personhood. What gives personal dignity to a living thing, for Singer, is its capacity for "self-consciousness" and "desire." Singer's error is to replace the fundamental distinction between human life and nonhuman life with a distinction between human life and person. Such positions can be countered with scientific and philosophical arguments.

From the scientific perspective, the embryo is a human being, because the fertilized ovum contains a new life program, defined and inscribed in a stable fashion in the genome of the zygote.

From a philosophical perspective, the embryo is a person because its vital principl is the same as that of an adult human being. This vital principle is not of a biological nature but of a spiritual nature. It is an immaterial principle, which in the Western tradition has been called the soul.

Biological examination shows that the embryo is not merely a clump of cells but rather is an individual of the human species, identified by its own genetic heritage, which contains a unique, unrepeatable life plan.

The humanity of the embryo can be demonstrated from the biological perspective, since from conception it starts a development that will lead to the adult individual without any breaks in continuity.

Philosophical reflection, however, deduces that within the individual of the human species there is a spiritual vital principle that constitutes it as an "individual substance of a rational nature," in other words, as a person.

The person exceeds his or her own acts, and this is the basis for the difference between being a person (the first act) and personality (in the psychological sense). Personality depends on the progressive acquisition of qualities which belong to the person inasmuch as they flow from his essence, but which do not necessarily accompany him from the beginning of the person's existence. Becoming a person is not a process but an instantaneous event or act, by which the entity is establish as being a person once for all, whereas personality is something that is acquired in a gradual process, through the performance of personal (second) acts. The act of being is the first and fundamental act of the individual substance: all other (second) acts of the person are rooted in it and derive their life from it.

In order to be a human being, in order to have the dignity of a person, there is no need to already have in act the capacities of knowing, willing, and loving that are typical of an adult; it is enough to have them in potency. Every human being is a person in act, even if he or she has the faculties of intellect and will in potency.

To be a human person means to have a human nature, which develops gradually; the embryo must not be considered as a thing, an object that may be produced and manipulated, but rather as a *subject*, as a personal substance that "stands beneath" and hence "sustains" the organs and faculties that are his components and the relations that concern him. Part of the nature of the human person is being a unitotality of body and spirit, being constitutionally fragile and in need of care.

The embryo is a person, even if he is not capable of thinking and willing, precisely because a human being is a person even though he is unconscious and depends on the care of the human family. Self-consciousness is certainly an indispensable premise for the exercise of human freedom, but it is not the constitutive characteristic of human nature; a human being is such simply because he or she shares in human nature, in other words, because he or she is a unity of body and spirit, constitutionally vulnerable and tending toward self-realization.

Therefore the embryo is a person even before developing a cerebral cortex; the brain is the organ of consciousness, but it does not produce thought and will; it is simply the organ indispensable for the exercise of these two faculties, which nevertheless remain in potency in the soul; for this reason it is not licit to destroy an embryo on the grounds that it lacks a brain or that its brain does not function.

The human person is neither thought nor self-consciousness; on the contrary, both of these presuppose the existence of life, of the unitotality of body and spirit. Where there is no life there can be no self-consciousness, but where there is no self-consciousness there can still be human life.

A human being remains a person even in the case where the exercise of his intellectual or volitional faculties is suspended, as in the case of sleep, or irremediably impaired, or not yet manifested.

It must be remembered that the human person (who is not divine) is also corporeal, fragile, and limited. The life of the human person is not bound up with the actuation of his faculties. The dignity of the human person does not depend on a greater or lesser "quality" of life. Human life has its own value, prior to all specific qualities; its dignity is bound up with being a "person" (a substance of rational nature) and not with its "personality" (realization of that rational nature).

Chapter 11

BIOETHICS AND HUMAN
FERTILIZATION TECHNOLOGIES

Definition of the Ethical Problem

Fertilization means and entails bringing into existence a new living being, a new individual: when applied to a human being, fertilization is synonymous with procreation. Now a type of biomedical and technological intervention into procreation cannot be assessed by the same standard as any other physiological and technological act—renal function, for instance, can be artificially accomplished outside the body by means of hemodialysis if it cannot take place normally because of a pathology, and this does not cause ethical problems per se.

Human fertilization or procreation is a personal act of the couple and results in a human individual: this involves the responsibility of the spouses and the very structure of marital life, as well as the destiny of the person who is called into being.

For believers, the process of human fertilization—which should more accurately be called *procreation*—involves a direct act of the Creator, creating the spiritual soul and thereby bringing to life a new person who can be raised up through the Mystery of the Incarnation to become a part of the Mystical Body of Christ.

The topic is within the sphere of medicine for one reason: treatment of male or female infertility. The ethical problem requiring clarification is the following: to what point does the medical act, the intervention of the physician or the biologist, as the case may be, have the character of a therapeutic act, and when does it become a substitutive or manipulative act? Treating means removing obstacles and assisting natural processes; it does not mean replacing the responsibility of persons, of the couple in this case, in that which pertains properly, exclusively, and inalienably to them.

Acknowledging the possibilities revealed by science and technology, we must therefore ask ourselves to what extent in utero or in vitro artificial fertilization is included within the legitimate scope of activity of biologists or physicians. This entails a central reference to the human values (the unborn person, the nature of marriage) that are at stake. The ethical solution will also shed light on the social and legal problem raised by this sort of biomedical intervention. Bioethics is fully and legitimately called upon to intervene. It is obvious that the moral responsibilities of the couple requesting the procedure are also involved in this reflection.

It should be added that the human values implicated in the problem under examination have a much more far-reaching cultural importance than it might seem at first glance. At stake are the relations of harmony and balance between *love* and *life* in marriage, between freedom and responsibility in the medical profession, between nature and person within human life, and between technology and morality in medicine and biology. These are problems that pervade the whole cultural crisis of our times and characterize the difficult encounter between *Homo sapiens* and *Homo faber*. We are not therefore dealing with a simple set of medical "cases"; rather, our ethical conclusions must be situated within a broader ethical and cultural dimension.[1]

We find ourselves at a crossroads of man's scientific and technological applications, where it is necessary now more than ever to make the distinction between what is technically possible—and even helpful—and what is morally licit. Moreover, we are dealing with a key point in the behavior of a couple and of a physician, in which morality must be based on criteria of ethical objectivity and not merely on good intentions.

It is known that the morality of the intention is not enough to define an act as ethically good: a good intention in an act that is not right in and of itself can excuse someone from subjective guilt or legal punishability, but it does not safeguard the integral good of the person. A good act that is omitted (and still more an evil act that is committed) will leave the world worse for it, whatever the intention the subject may have had in making that choice.[2] The intention of treating infertility does not justify any method and any process whatsoever in order to achieve conception.

---

[1] In an address given December 3, 1982, Pope John Paul II reaffirmed what he had said to UNESCO on June 2, 1980: "One of the most serious dangers to which this present era is exposed is in fact the divorce between science and morality, between the possibilities offered by a technology that is aimed at ever more astonishing goals and the ethical norms that arise from an increasingly ignored nature. It is necessary that all responsible persons agree in reaffirming the priority of ethics over technology, the primacy of the person over things, the superiority of the spirit over matter. Only on this condition will scientific progress, so many aspects of which fill us with enthusiasm, not turn into a sort of Moloch that devours its unwitting followers." Address to Participants in the Conference of the Movement for Life, in *Insegnamenti di Giovanni Paolo II* (Vatican City: Libreria Editrice Vaticana, 1982), 5.3:1513 [Italian]. Moreover, the instruction *Donum vitae* by the Congregation for the Doctrine of the Faith (CDF) (February 22, 1987) reads, "Science and technology are valuable resources for man when placed at his service and when they promote his integral development for the benefit of all; but they cannot of themselves show the meaning of existence and of human progress. Being ordered to man, who initiates and develops them, they draw from the person and his moral values the indication of their purpose and the awareness of their limits. It would on the one hand be illusory to claim that scientific research and its applications are morally neutral; on the other hand one cannot derive criteria for guidance from mere technical efficiency, from research's possible usefulness to some at the expense of others, or, worse still, from prevailing ideologies. Thus science and technology require, for their own intrinsic meaning, an unconditional respect for the fundamental criteria of the moral law: that is to say, they must be at the service of the human person, of his inalienable rights and his true and integral good according to the design and will of God" (introduction, no. 2). On this subject, see also E. Sgreccia, "Autonomia e responsabilità della scienza," in *Lineamenti di etica della sperimentazione clinica*, ed. A. G. Spagnolo and E. Sgreccia (Milan: Vita e Pensiero, 1994), 39–49.

[2] With specific regard to human procreation, Vatican Council II declares, "When there is question of harmonizing conjugal love with the responsible transmission of life, the moral aspects of any procedure does not depend solely on sincere intentions or on an evaluation of motives, but must be determined by objective standards. These, based on the nature of

Some Distinctions

According to the instruction *Donum vitae* by the Congregation for the Doctrine of the Faith,[3] an initial distinction can be made between *homologous* artificial fertilization and *heterologous* artificial fertilization. In addition to being technically necessary, the distinction is also ethically significant.

Homologous artificial fertilization is the term for techniques aimed at achieving a human conception with gametes from two spouses who are united in marriage, or—according to many laws—with gametes from a couple requesting the procedure. It can be carried out by two sorts of methods: homologous *in vitro fertilization and embryo transfer* (homologous IVF-ET), in which the meeting of the gametes occurs in vitro (*extracorporeal fertilization*), and homologous *artificial insemination* (artificial insemination by husband, AIH), which involves the transfer of the husband's previously collected sperm into the female genital tract (*intracorporeal fertilization*). Variations on IVF-ET include so-called micromanipulations of the gametes (partial zonal dissection, PZD; subzonal injection, SUZI; and intracytoplasmic sperm injection, ICSI).

Heterologous artificial fertilization, by contrast, is the term for techniques aimed at achieving a conception with gametes coming from at least one donor other than the spouses who are united in marriage (or the couple requesting the procedure). It can be done via the intracorporeal method (heterologous artificial insemination with sperm previously collected from a donor other than the husband: artificial insemination by donor, AID) or by the extracorporeal method (heterologous in vitro fertilization and embryo transfer, or heterologous IVF-ET).

An additional possibility has developed with a corresponding distinction in the case of in vitro artificial fertilization: the transfer of the embryo can be done directly into the "maternal" uterus by use of a *surrogate mother* or "rental mother." The possibilities become more complicated when turning to a *sperm bank*, in which donated sperm are frozen in liquid nitrogen, or to donated egg cells, or when proceeding to freeze not only gametes but also embryos. In the last-mentioned case there have been further instances of legal complications, such as the death of the parents who "commissioned" the procedure—for whom the child was intended—after the embryo had already been implanted in the surrogate mother. All this forms the basis for complex legal cases concerning paternity, inheritance rights, the use of frozen and "abandoned" embryos, and so on. The various ethical and legal implications will be addressed in the pages to follow.

Other techniques for artificial fertilization involving an intracorporeal union of the gametes have since been developed, including *gamete intrafallopian transfer* (GIFT), which entails the simultaneous but separate transfer of male and female gametes into the fallopian tube; *low tubal ovum transfer* (LTOT), or the transfer of the egg cell to the proximal part of the fallopian tube; and *gamete intrauterine transfer* (GIUT).

---

the human person and his acts, preserve the full sense of mutual self-giving and human procreation in the context of true love." Pastoral Council *Gaudium et spes* (December 7, 1965), no. 51.

[3] CDF, *Donum vitae*, part II, "Interventions upon Human Procreation."

In addition to the techniques for artificial procreation just mentioned, there are other ways "of coming into the world," as listed in Table 1. These technologies will therefore be considered and assessed as well.

Artificial Insemination

Historical notes and statistics

After studies and unsuccessful experiments by Marcello Malpighi and Francesco Bibiena on silkworms in 1600, after not entirely verified successes by Ludwig Jacobi and Baron Weltheim de Barbke in 1725 on salmon and trout eggs, and after studies by the Swedish researcher Carl Alexander Clerck on spiders, the first to succeed at artificial insemination was Lazzaro Spallanzani in 1782 on a female dog. In 1785, Michel Thouret managed to cure the infertility of his own wife by means of an intravaginal injection of seminal fluid. Dr. Girault later had several successes over thirty years of attempts; and the gynecologist James Marion Sims had only one

---

Table 1. *Possible methods of artificial fertilization*

Artificial insemination (AI)

Direct intrafollicular insemination (DIFI)

Embryo intratubal transfer (EITT)

Gamete intrafallopian transfer (GIFT)

Gamete intraperitoneal transfer (GIPT)

Gamete intrauterine transfer (GUIT)

Intracervical insemination (ICI)

Intracytoplasmic sperm injection (ICSI)

Intraperitoneal insemination (IPI)

Intratubal insemination (ITI)

Intrauterine insemination (IUI)

In vitro fertilization and embryo transfer (IVF-ET)

Lower tubal ovum transfer (LTOT)

Preembryo tubal transfer (PRETT)

Pronuclear-stage [tubal] transfer (PROST)

[Insemination by] partial zona dissection (PZD)

Subzonal insemination (SUZI)

Transcervical gamete intrafallopian transfer (TC-GIFT)

Vaginal intratubal insemination (VITI)

Zygote intrafallopian transfer (ZIFT)

SOURCE: Adapted from J. Testart, "Le chaparnaum des fécondations artificielles," *Le Monde*, January 3, 1990.

success out of fifty-five attempts at direct injection into the uterus. William Pancoast performed the first heterologous insemination in 1884, at which point the method took off in medical practice and literature.

It seems certain that the percentage of successes is considerably higher with heterologous semen than with homologous semen. Donor sperm thus appears to offer a greater advantage: surmounting the obstacle of male infertility. Giarola (1976) speaks of 895 pregnancies from 1,351 inseminations with donated sperm and 265 from 1,324 with the husband's sperm; Gardelli (1981) reports 972 pregnancies from 1,605 inseminations with donated sperm and 389 pregnancies from 3,050 with the husband's sperm. It would need to be verified on a case-by-case basis whether infertility was due to defects in the female genital apparatus, the low fertility of the husband's sperm, or both.[4]

The practice of artificial insemination experienced a slower development in Italy than in other countries, especially the United States. It seems that there were more than a half million children in the world conceived through unknown donors as early as 1981.[5]

In that same year the numbers in Europe were one-tenth of those in the United States, and in Italy around 300 heterologous artificial inseminations had been performed. Obviously, these figures and statistics are approximate, difficult to calculate, and in constant flux. Artificial insemination was mentioned for the first time in Italy at the twenty-second Congress of the Italian Gynecology and Obstetrics Society in 1923 in a presentation by P. Alfieri, and then in 1935 at the thirty-fifth congress of the same society in a presentation by G. Tesauro.

During the two-year period from 1933 to 1935, Berretti performed eight cases of artificial insemination in Turin without recording the results. In Bari, G. Traina gained renown for the activities of the Center for the Study and Treatment of Infertility (Centro per lo Studio e la Terapia della Sterilità) at the obstetrics and gynecology clinic of the university there: artificial insemination was performed there on sixteen infertile couples from 1936 to 1942, resulting in three pregnancies (18 percent); the procedure in each case was homologous insemination.[6]

While researchers abroad were working on insemination with frozen semen (sperm banks), Agostino Gemelli, O.F.M., joined the debate in 1949 with the second edition of his study *La fecondazione artificiale* (Artificial fertilization),[7] which had a great influence not only on the ethical positions of moralists but also of gynecologists.

---

[4] For these historical and statistical details, see M. Di Ianni, "Fecondazione artificiale," in *Dizionario enciclopedico di teologia morale*, ed. L. Rossi and A. Valsecchi (Rome: Paoline, 1981), 401–403.

[5] L. Leuzzi, "Il dibattito sulla inseminazione artificiale nella riflessione medico-morale in Italia nell'ultimo decennio," *Medicina e Morale* 32, no. 4 (1982): 343–370, with an extensive bibliography. See also M. S. Frankel, "Artificial Insemination," in *Encyclopedia of Bioethics*, ed. W. T. Reich (New York: Free Press, 1978), 1444–1448; W. Wadlington, "Artificial Insemination," in *Encyclopedia of Bioethics*, ed. W. T. Reich, rev. ed. (New York: Macmillan, 1995), 2216–2221.

[6] F. Clauser, P. Bailo, and P. Alfieri, *Problemi medico-morali* (Bergamo: Edizioni della Rotonda, 1958); see the bibliography contained in Leuzzi, "Il dibattito."

[7] A. Gemelli, *La fecondazione artificiale*, 2nd ed. (Milan: Vita e Pensiero, 1949).

At a conference on the topic of artificial insemination held in Salice Terme on June 28, 1959, the Italian Society of Social Medicine took the following position:

> Intramarital artificial insemination is licit when it is the sole means of correcting a malformation that is congenital or caused by illness; hence it can be compared to the corrective intervention of the surgeon on what nature has produced. Artificial insemination with heterologous sperm is illicit, dangerous, and the precursor of serious biological and moral harm to humanity; it should therefore be decisively rejected and banned from the legal standpoint as well as from the biological and ethical standpoint. [8]

For a time these conclusions influenced private medical and gynecological practice in Italy, which lacked any pertinent legislative norms whatsoever at that point. In 1977, at the Fourth Continuing Education Course on Conjugal Infertility in Florence, Traina presented the results of 241 conjugal infertility treatments with heterologous artificial insemination. At the fifth course (again in Florence) in 1978, he announced the first results with frozen sperm and the opening of a sperm bank in Bari. After the one in Bari, other banks were started contemporaneously in Rome, Palermo, and Verona. Now they are present in many cities.

Medical indications, techniques, and results

MEDICAL INDICATIONS. Homologous artificial insemination is "indicated" for some forms of female infertility (vaginismus, malformations and inflammatory pathologies of the vagina, cervix, or uterus), male infertility (impotentia coeundi, whether of psychological or biological origin, retrograde ejaculation, hypospadias, oligoasthenospermia), or for some couples (production by the woman of antibodies against the husband's sperm).[9] Sex selection is another indication for homologous

---

[8] V. Traina, "Inseminazione artificiale eterologa," in *L'inseminazione artificiale umana: Atti del II Seminario Internazionale; Bari, 12–15 maggio 1980*, ed. R. Schoysmann, S. Bettocchi, and F. M. Boscia (Palermo: Cofese, 1981), 133–150. See Leuzzi, "Il dibattito," 348.

[9] Concerning infertility and artificial insemination techniques, see M. Diamond, C. Christianson, and J. F. Daniell, "Pregnancy following Use of the Cervical Cup for Home Artificial Insemination utilizing Homologous Semen," *Fertility and Sterility* 4 (1983): 480 ff; V. Traina, C. Mancini, and G. Miniello, "L'inseminazione artificiale: Timing e tecnica," in *Atti del XIII Congresso Nazionale della Società Italiana per lo Studio della Fertilità e la Sterilità (SIFES)* (Salsomaggiore Terme: 1986); M. Garcea, "La procreazione assistita," in *Trattamento della sterilità coniugale*, ed. S. Mancuso and E. Sgreccia (Milan: Vita e Pensiero, 1988), 145–159; Ethics Committee of American Fertility Society, "Ethical Considerations of the New Reproductive Technologies," supplement 1, *Fertility and Sterility* 46, no. 3 (1986): 345–385; A. Rodriguez Luño and R. López Mondéjar, *La fecondazione in vitro: Aspetti medici e morali* (Rome: Città Nuova, 1986), 40–42; A. Bompiani, "Problemi biologici e clinici dell'ingegneria genetica," in *Bambini in provetta*, ed. G. Concetti (Rome: Logos, 1986), 63–73; R. Curson and J. Parsons, "Disappointing Results with Direct Intraperitoneal Insemination," *Lancet* 329.8524 (January 10, 1987): 112; A. Isidori, "L'inseminazione artificiale omologa e eterologa nella sterilità maschile: Aspetti medici e psicologici," *Medicina e Morale* 43, no. 1 (1993): 75–96; P. Vandekerckhove et al., "Infertility Treatment: From Cookery to Science; The Epidemiology of Randomised Controlled Trials," *British Journal of Obstetrics and Gynaecology* 100 (1993): 1005–1036; M. G. R. Hull, "Efficacia dei trattamenti per l'infertilità: Scelta e analisi comparativa," *Giornale Italiano di Ostetricia e Ginecologia*

artificial insemination when it is for the purpose of avoiding the transmission of particular genetic illnesses connected with the sex of the offspring.

Before resorting to the use of artificial insemination, however, the possibility of interventions aimed at removing the causes of infertility must be assessed. Artificial insemination would have to be the last resort, and even then it does not always prove ethically acceptable.

In the case of artificial insemination by donor (moral questions aside), the proposed medical indications obviously concern the man and are related especially to aspermia, asthenospermia, or oligospermia, or to the prevention of sexually transmitted diseases (e.g., AIDS) or genetic disorders.

It should be noted that eugenic criteria are constantly mentioned among the expected qualities of the donor in the case of donated sperm. One often finds requests such as these: "He should be the father of good-looking, healthy children," or "We prefer a donor who resembles the father in his physical, racial, intellectual, and moral qualities."[10] There will be an opportunity to analyze this phenomenon further on.

Obviously, there are also contraindications to artificial insemination: tubal blockage, the presence of endocrine disorders in the woman, active transmissible diseases, or the advanced age of the woman.

TECHNIQUES. The male semen to be used in artificial insemination can be collected immediately before it is transferred to the female genital tract, thus using "fresh" sperm, or it can be collected well in advance, preserved by freezing (*cryopreservation*), then unfrozen shortly before it is transferred into the female genital tract.

The semen can be obtained by different methods: together with sexual intercourse, following sexual intercourse, or without sexual intercourse:

a. Together with sexual intercourse:

After coitus interruptus, by collecting the semen immediately in a sterile vial

By using a condom

By using a perforated condom that allows a portion of the semen to be collected

b. After marital intercourse:

By retrieving the sperm deposited in the vagina

(1995): 567–576; Italian National Bioethics Committee, *La fecondazione assistita* [*assisted fertilisation*] (Rome: Presidenza del Consiglio dei Ministri Dipartimenti per l'Informazione e l'Editoria, 1995), 11–46, http://www.governo.it/bioetica/testi/feconfazione170295.pdf, English abstract at http://www.governo.it/bioetica/eng/opinions/abstract_assisted percent-20fertilisation.pdf; K. Sterzik, "Assistierte Reproduktion," *Sexualmedizin* 5 (1996): 148–151. A complete review of the different fertilization techniques can be found in A. Bompiani, *Le tecniche di fecondazione assistita: Una rassegna critica* (Milan: Vita e Pensiero, 2006).

[10] R. Cardente, "La fecondazione artificiale," *Iustitia* 3 (1950): 21; A. Simili, *La fecondazione artificiale umana* (Turin: Minerva Medica, 1961), 47; J. de D. Vial Correa and E. Sgreccia, eds., *The Dignity of Human Procreation and Reproductive Technologies: Anthropological and Ethical Aspects; Proceedings of the Tenth Assembly of the Pontifical Academy for Life; Vatican City, 20–22 February 2004* (Vatican City: Libreria Editrice Vaticana, 2005).

By collecting the sperm remaining in the male urethra

In the case of retrograde ejaculation, by collecting the semen inside the seminal vesicle together with urine, which is preventively treated with an antacid base and then separated out

c. Separate from marital intercourse:

By means of masturbation

By obtaining the sperm in the urethra after involuntary nocturnal emission

By electroejaculation

By squeezing the prostate and the seminal vesicles

By puncturing the epididymis (microsurgical epididymal sperm aspiration, MESA) or the vas deferens

By means of testicular biopsy

It should be noted that the male gametes obtained in a biopsy of the testicle have not completed their proper maturation.

Once it has been obtained, the semen is transferred to the female genital tract at the moment in the menstrual cycle closest to ovulation (whether spontaneous or induced through the administration of hormones, in accordance with the different protocols).[11]

The sperm can be deposited in various parts of the female genital tract depending on the type of obstacle that is to be overcome: in the vagina in the case of impotentia coeundi, for example (intravaginal insemination, IVI), at the intracervical level for blockage of the external orifice of the uterus (intrauterine insemination, IUI), in the fallopian tube (intrafallopian insemination, IFI) for serious oligoasthenospermia, at the intraperitoneal level (direct intraperitoneal insemination, DIPI) where the sperm enter the tube from the other side, or at multiple sites simultaneously.

In the case of a lawfully wedded couple, when the procedure uses the husband's semen as obtained together with or immediately after the conjugal act, some moralists claim that it should be referred to as *homologous artificial insemination "improperly speaking."* This is to distinguish it from homologous artificial insemination "properly speaking," which involves obtaining the semen outside of sexual relations.[12]

RESULTS. The success of homologous artificial insemination varies depending on the technique used and the medical indications. Better results are obtained in couples where the male partner has normal sperm production. On average, pregnancy is achieved in 25 percent of the cases treated. According to some authors the percentages are lower: no more than 15 to 16 percent.

The percentage of success of heterologous artificial insemination in terms of babies born per cycle of therapy varies from 10 percent with frozen semen to 20 percent with fresh semen; this discrepancy seems to be attributable to the fact

---

[11] Z. Serebrovska and M. L. Di Pietro, "La sindrome da iperstimolazione ovarica: Tra clinica e etica," *Medicina e Morale* 56, no. 2 (2006): 327–347.

[12] Leuzzi, "Il dibattito," 350–353.

that the potency of frozen semen is reduced by ultrastructural damage caused by the freezing process.[13]

Ethical assessment of intracorporeal artificial insemination

The problems that arise in the ethical assessment of artificial insemination are related to the *goal* pursued through intervention on the life and relationship of the couple, but they also concern the *methods* used and the techniques applied. With regard to the goal in the case of heterologous artificial insemination, the eugenic aim—even in the absence of infertility in the husband—should be kept in mind in addition to the "therapeutic" aim.

It is obvious that first and foremost the Catholic Church, through its Magisterium and theological reflection, has taken a position on this matter. This means that many of those testimonies will need to be referenced, such as the above-cited instruction *Donum vitae* and later the encyclical *Evangelium vitae*, which offer rational and anthropological reasoning as well.[14]

---

[13] N. Garcea, "Tecniche di procreazione assistita," *Medicina e Morale* 43, no. 1 (1993): 59–66; M. S. Mills et al., "A Prospective Controlled Study of In Vitro Fertilization, Gamete Intra-Fallopian Transfer and Intrauterine Insemination Combined with Superovulation," *Human Reproduction* 7, no. 4 (1992): 490–494; K. Zikopoulos et al., "Homologous Intrauterine Insemination Has No Advantage over Timed Natural Intercourse When Used in Combination with Ovulation Induction for the Treatment of Unexplained Infertility," *Human Reproduction* 8, no. 4 (1993): 563–567.

[14] Pius XII, Address to Participants in the Fourth International Congress of Catholic Physicians, September 29, 1949, in *Discorsi e Radiomessaggi* (Vatican City: Tipografia Poliglotta Vaticana, 1940–1969), 11:221–225 [Italian]; idem, Address to Participants in the Conference of the Italian Catholic Union of Obstetricians, October 29, 1951, in *Discorsi e Radiomessaggi*, 13:211–221 [Italian]; idem, Address to Participants in the Second World Congress on Fertility and Sterility, May 19, 1956, in *Discorsi e Radiomessaggi*, 18:211–221 [Italian]; idem, Address to Participants in the Seventh International Congress of the Italian Hematology Society, September 12, 1958, in *Discorsi e Radiomessaggi*, 20:341–352 [Italian]; John XXIII, Encyclical Letter *Mater et magistra*, May 15, 1961; Vatican Council II, *Gaudium et spes*, no. 51; Paul VI, Encyclical Letter *Humanae Vitae*, July 25, 1968, no. 12; CDF, Declaration *Persona humana*, December 29, 1975; idem, *Donum vitae*. Here is what Paul VI declares in paragraph 12 of *Humanae vitae*, which serves as a summary: "This particular doctrine, often expounded by the magisterium of the Church, is based on the inseparable connection, established by God, which man on his own initiative may not break, between the unitive significance and the procreative significance which are both inherent to the marriage act. The reason is that the fundamental nature of the marriage act, while uniting husband and wife in the closest intimacy, also renders them capable of generating new life—and this as a result of laws written into the actual nature of man and of woman. And if each of these essential qualities, the unitive and the procreative, is preserved, the use of marriage fully retains its sense of true mutual love and its ordination to the supreme responsibility of parenthood to which man is called. We believe that our contemporaries are particularly capable of seeing that this teaching is in harmony with human reason." See also John Paul II, Encyclical Letter *Evangelium vitae*, March 25, 1995, and the related commentaries on *Evangelium vitae*: E. Sgreccia and D. Sacchini, eds., *"Evangelium vitae" e bioetica: Un approccio interdisciplinare* (Milan: Vita e Pensiero, 1996); E. Sgreccia and R. Lucas Lucas, eds., *Commento interdisciplinare alla "Evangelium vitae"* (Vatican City: Libreria

In accordance with the nature and structure of human procreation within the family, a moral investigation of the problem is framed differently depending on whether the artificial insemination is homologous or heterologous.

Let us here recall what was set forth in an earlier chapter on the general issue of procreation. Procreation is not a merely biological event for man; it is a "personal" act: this means that, in order to be human, procreation must be an act that freely and responsibly involves the totality of the individual persons of the spouses in an exclusive manner; procreation is an essential, exclusive and personal duty of spouses, who are called to participate in it by the total gift of their own personal being: body, heart, and soul. The biological component comes to be inscribed in the totality of the person and in the psychological and spiritual component and vice versa. To detach the biological component of procreation from the emotional and spiritual component is to produce an unnatural division within the person and within the sexual act that expresses the conjugal gift. It is tantamount to separating life from love.

There is no person if there is no body, mind, and spirit; there is no human and personal exercise of sexuality that does not involve body, mind (heart), and spirit; furthermore, there is no sexual act that is morally honest and upright except within the context of marriage and in the reciprocal, exclusive, and total self-giving of the persons.

> Thus, fertilization is licitly sought when it is the result of a "conjugal act which is per se suitable for the generation of children to which marriage is ordered by its nature and by which the spouses become one flesh." But from the moral point of view procreation is deprived of its proper perfection when it is not desired as the fruit of the conjugal act, that is to say of the specific act of the spouses' union. [15]

As mentioned in earlier chapters in this regard, it should be recalled that three levels of activity can find expression in the human person: (1) the biological level, which pertains to the physiological and "involuntary" functions of life (digestion, metabolism, etc.); (2) the level of productivity, which proceeds from the person and has things as its object; and (3) the specifically personal level, which involves the revelation of the person and the relation of the soul and of the "self" by means of corporeal signs and corporeal language, as in all human relations.

The procreative act cannot be a purely biological act, like the mixing of biochemical elements, nor a production activity proper to the manufacture of objects; rather, in order to be at the personalistic level of responsible sexuality and interpersonal reciprocity, it must be accomplished through the gift of the persons—a gift that transcends and transfigures the biological fact in a spiritual dimension that is incommensurate with a production technique or a mere combination of gametes.

Morality therefore entails unity and totality. This premise must be maintained and taken into consideration not only with regard to the problem of birth control but

---

Editrice Vaticana, 1997). See especially the essay therein by A. G. Spagnolo, "Tecniche di fecondazione artificiale e inizio della vita umana," 599–615.

[15] CDF, *Donum vitae*, part II, no. 4, emphasis removed. See also John Paul II, *Evangelium vitae*, part III; C. Caffarra, *Non è bene che l'uomo sia solo*, vol. 1, *Creati per amare* (Siena: Cantagalli, 2006).

also in relation to the present topic of artificial insemination. In each case, one of the points solidifying and unifying the development of humanity—the bridge between *nature* and *person*—is called into question.

What was noted in the chapter on procreation in general should be repeated here: morality—and especially Catholic morality—should not be understood as the repression of man's vitality but, on the contrary, as the total and unified expression of the components of human life.

Moral aspects of homologous artificial insemination

There are generally no contraindications for homologous artificial insemination, and it presents no moral difficulties so long as we are talking about therapeutic and holistic assistance to ensure that the conjugal act, which is complete in itself with all of its components (physical, psychological, and spiritual), can have a procreative result.

This practice does not pose ethical problems even for the Magisterium of the Catholic Church, provided that the techniques applied (in particular for obtaining the semen) are moral in themselves. The semen can also be licitly treated to improve capacitation once it has been obtained.

The moral assessment of homologous artificial insemination nonetheless depends on whether it is true and proper artificial insemination or simply a form of *assistance* to the conjugal act. In the latter case, which is artificial insemination "improperly speaking," technical *assistance* is provided so that the semen ejaculated within the context of and at the same time as the conjugal act can be united to the egg cell and thus bring about fertilization.

Based on the earlier teachings of Pope Pius XII and of Pope Paul VI in *Humanae vitae*, the Catholic Magisterium later drew a line that must not be crossed in the instruction *Donum vitae*: the protection of the physical and spiritual unity of the conjugal act. In this light, the intervention of a gynecologist is declared licit provided that it *assists* the efficacy of that act and its procreative completion but does not *substitute* for it. As already mentioned, the repeated interventions of Pius XII on this matter, which cannot be extensively reproduced here, emphasized that

> in its natural structure, the conjugal act is a personal action, a simultaneous and immediate cooperation on the part of the husband and wife, which by the very nature of the agents and the proper nature of the act is the expression of the mutual gift that, according to the words of Scripture, brings about union "in one flesh." This is much more than the union of two gametes, which can also be carried out artificially, that is, without the natural action of the spouses. The conjugal act, ordered and intended by nature, is an act of personal cooperation the right to which is exchanged by the spouses in contracting marriage.[16]

Yet on several occasions the same Pontiff repeated, in almost the same words, "By that, however, we do not necessarily intend to exclude the use of certain artificial means aimed either at *facilitating the conjugal act* or at causing the *natural act as normally performed* to achieve its proper end."[17] The Catholic position on artificial insemination was repeated and clarified in the instruction *Donum vitae*.

---

[16] Pius XII, Address to Obstetricians.

[17] Pius XII, Address on Fertility and Sterility, emphasis added.

The intention of the spouses to have a child, though quite legitimate, is not sufficient to declare a reproductive procedure licit; the means and methods must be licit as well. Conception is therefore licit when—as already mentioned—it is the conclusion of "a conjugal act in itself apt for the generation of offspring. To this act marriage is by its nature ordered and by it the spouses become one flesh."[18]

An act of procreation is deprived of its perfection if it is not the fruit of the union, both physical and spiritual, of the spouses: "The procreation of a new person, whereby the man and the woman collaborate with the power of the Creator, must be the fruit and the sign of the mutual self-giving of the spouses, of their love and of their fidelity. The fidelity of the spouses in the unity of marriage involves reciprocal respect of their right to become a father and a mother only through each other."[19]

It follows that homologous artificial insemination, which involves dissociating the *union* of the spouses from *procreation*, "cannot be admitted except for those cases in which the technical means is not a substitute for the conjugal act but serves to facilitate and to help so that the act attains its natural purpose."[20] And, as already noted, the only form of intervention that can be considered an "aid" or assistance to the conjugal act is artificial insemination *improperly speaking*.

At this point, what the Catholic Magisterium means to say when it demands respect for the natural structure of the procreative act should be explained. Indeed, those who defend artificial procreation as licit, at least in the homologous form, accuse the Catholic Magisterium of relying on a fixist and biologistic understanding of the natural law, adding that there is nothing inhuman about achieving procreation by means of technology since human technology is a product of man. They maintain, ultimately, that the Creator has given man the ability to master not only external nature but also his own human nature, as happens in many medical procedures and technologies.

I would simply like to offer several points for reflection on this topic, since it would require an extensive treatment starting with a sufficiently thorough discussion of what is meant by "natural law."

a. The "nature" that the Catholic Magisterium mentions in its documents and for which it demands respect is not automatically biological nature alone (*bios*), but nature in the metaphysical sense: the structural peculiarity of the human person thanks to which the human person is what he is, an individual in whom spirit and body are united in such a way that the spirit manifests itself in the body and the spirit informs, structures, and vivifies corporality. Hence nature is what constitutes the human person in his unitotal essence.

b. The law that the Catholic Magisterium mentions in its documents is the natural *moral* law, not a physical and biological law. The biological law is always obeyed, even in artificial procreation, because if it were not observed in combining the male and female gametes, there would be no fertilization; what is not being observed is the natural *moral law*, which obliges us to consider man as a person in his totality and the procreative sexual act as an expression of the spirit and of personal love in

---

[18] *Code of Canon Law*, can. 1061.

[19] CDF, *Donum vitae*, part II, no. 1; see also *Catechism of the Catholic Church*, nos. 2375–2379, which fully relates the information already contained in *Donum vitae*.

[20] CDF, *Donum vitae*, part II, no. 6, emphasis removed.

the giving of one's corporality. What is being condemned is not technology per se, nor the application of technology to the human body, but rather the fact that this type of application introduces a dualism by separating the biological-fecundatory dimension from the spiritual dimension of the spousal "self."

The criteria listed in the instruction *Donum vitae* are therefore intended to rule out the legitimacy of a purely subjective morality (*intention-based morality*).

In the case of two spouses living in a legitimate and complete conjugal union and desiring to have a child, one crucial point to be ascertained and clarified was whether the procurement of semen outside the individual marital act, *bypassing* an obstacle and operating at a moment distinct from the complete performance of the conjugal act, effects a *real separation* of the physical element of procreation from its psychological and spiritual component in the single conjugal act or whether it should be thought that the separation pertains not to the physical dimension of the act but simply to a particular moment or effect of that dimension, namely, insemination.

The instruction *Donum vitae* noted that, when prescinding from the individual acts performed within conjugal life, the totality of conjugal life is not sufficient to ensure the dignity that is a constitutive element of human procreation.[21]

Indeed, citing *Humanae vitae* no. 14, the instruction emphasizes that the connection between the unitive and the procreative dimensions in the conjugal act must be realized in the individual acts and not just over the whole course of conjugal life, considered globally; the document thereby intends to correct this "principle of totality" that has been extensively proposed by some theologians. Moreover, it is obvious that if a behavior is morally unacceptable, it is also and above all unacceptable in the individual acts that express it.

The case of impotence requires special consideration, and it should be examined not only from a medical standpoint but also from an ethical and canonical standpoint. When impotence is "psychological" and hence reversible, the marriage remains valid, and medical and psychological assistance should be offered to the couple as therapeutic support;[22] this is also a case in which there may be a possibility of artificial insemination. When impotence is total, however, it posits another precedent reality: the invalidity of the marriage in question. It is licit to assist the fertility of the conjugal act only if the marriage is valid and contracted legitimately.

The permissibility of the various means
and methods of obtaining semen

It is common knowledge that an act is licit when both the ends and the means are licit. What remains to be examined from the ethical viewpoint are the means of obtaining semen, which were listed above from a technical point of view. The debate centers primarily on masturbation, considered in this case not as an egotistical act of sexual self-satisfaction but rather as a way of obtaining semen that is richer and more potent than the semen obtained, for example, by puncturing the epididymis or squeez-

---

[21] CDF, *Donum vitae*, part II, no. 5.

[22] On this subject, see C. Zaggia, ed., *Progresso biomedico e diritto matrimoniale canonico* (Padua: Veneta, 1992); M.L. Di Pietro and S.M. Correale, "Valutazione delle terapie medico-chirurgiche e protesiche dell'impotentia coeundi nell'uomo ai fini della validità del matrimonio canonico," *Apollinaris* 66 (1993): 273–314.

ing the seminal vesicles. Let us examine in detail the techniques for obtaining semen from a moral perspective. Obtaining semen by an act between spouses that is *contra naturam* (contrary to nature), such as coitus interruptus, is clearly an illicit action from the moral perspective. On the other hand, mechanical means of collecting semen are not morally objectionable in themselves provided that the conjugal act is respected in its integrity: for example, the use of a perforated condom is not contraceptive in this case since it is still possible for the semen to pass into the female genital tract.

As far as masturbation is concerned, it is common knowledge that this is an intrinsically immoral act in itself; indeed, the structure of sexual life orients the voluntary activity of sexual acts toward conjugal life, and the declarations of the Magisterium repeatedly condemn masturbation as an "intrinsically" disordered act, that is, outside of the moral order. Particular examples include the documents by Pius XII and the declaration *Persona humana* on sexual ethics by the Congregation for the Doctrine of the Faith (December 29, 1975).

These texts are taken up in the instruction *Donum vitae*, which reads, "Masturbation, through which the sperm is normally obtained, is another sign of this dissociation: even when it is done for the purpose of procreation, the act remains deprived of its unitive meaning: 'It lacks the sexual relationship called for by the moral order, namely, the relationship which realizes "the full sense of mutual self-giving and human procreation in the context of true love."'"[23]

Men who are supposed to procure semen through masturbation often refuse to do so for psychological, moral, or religious reasons, even if it is requested for diagnostic purposes. That being the case, the document by the Italian National Bioethics Committee (NBC), *Problemi della raccolta e trattamento del liquido seminale umano per finalità diagnostiche* (Problems in gathering and treating human seminal fluid for diagnostic purposes), demands that when a patient must provide a semen sample, "full respect for the patient's cultural or religious convictions and personal dignity should be strictly safeguarded in the physician-patient relationship," guaranteeing "complete information about any scientifically valid alternative methods, for which he must nonetheless provide proper consent."[24]

In reference to obtaining semen for diagnostic and therapeutic purposes, as in the aforementioned document, there are permissible alternative ways of obtaining the semen (e.g., the use of clinical vibrators that stimulate semen emission without constituting a masturbatory act).[25] The same cannot be said for artificial reproduction techniques, because here it is important to safeguard the unity of the conjugal act, which would not be respected regardless of the method for obtaining semen.

---

[23] CDF, *Donum vitae*, part II, no. 6, which cites the CDF's Declaration *Persona humana*, no. 9, which in turn cites Vatican Council II, *Gaudium et spes*, no. 51. See also *Catechism of the Catholic Church*, no. 2352.

[24] Italian National Bioethics Committee, *Problemi della raccolta e trattamento del liquido seminale umano per finalità diagnostiche* [*Problems of the Collection and Handling of Human Seminal Liquid for Diagnostic Purposes*], May 5, 1991 (Rome: Presidenza del Consiglio dei Ministri Dipartimento per l'Informazione e l'Editoria, 1991), 9, http://www.governo.it/bioetica/pdf/3.pdf, English abstract at http://www.governo.it/bioetica/eng/opinions/abstracts/abstract_problems_of_the_collection.pdf.

[25] A. G. Spagnolo et al., "Valutazione scientifica ed etica di un metodo per il prelievo diagnostico del liquido seminale," *Medicina e Morale* 43, no. 6 (1993): 1189–1202.

It should be added, furthermore, that the technique of masturbation for obtaining semen is often used only because it is thought to be more expeditious: it could be avoided in many or most cases by using methods closely connected with the conjugal act and therefore less objectionable, although they require more effort.

Moreover, by collecting semen from within the vagina or by means of a perforated condom, one can also avoid falling into artificial insemination properly speaking and remain within homologous artificial insemination improperly speaking. Furthermore, masturbation can itself be a problem in certain stressful conditions, for persons with particular temperaments, or because of an instinctive and delicate modesty that is nothing to be ashamed of.

In the use of these methods for obtaining semen as well, the dignity of persons should be preferred to the expeditious ease of the method. I repeat, however, that the real problem in this whole issue is safeguarding the union of love and life, of the unitive and procreative dimensions in the conjugal act and the reciprocal self-gift of the persons as spouses.

Ethical assessment of heterologous artificial insemination

There is no doubt that obtaining semen from donors is *illicit*. The reasons pertain to the unity of the couple and the indissolubility of marriage. Legal scholars have debated whether heterologous artificial insemination is tantamount to the crime of adultery as foreseen by the Italian criminal code in article 559, which has been weakened by two rulings of the Constitutional Court: no. 126 on February 19, 1968, and no. 147 on December 3, 1969. Yet from the ethical standpoint, a personal separation occurs in this case between those who are living in matrimony and those who accomplish procreation: it is said to result in a new type of sui generis, multi-parent family that is no longer monogamous. Thus, the exclusive "right [of the spouses] to become a father and a mother only through each other" is violated.

This means that even if the spouses agree to heterologous fertilization, their spousal and parental unity is still changed as a result of the donation of gametes: "Heterologous artificial fertilization is contrary to the unity of marriage, to the dignity of the spouses, to the vocation proper to parents, and to the child's right to be conceived and brought into the world in marriage and from marriage. Furthermore, it offends the common vocation of the spouses who are called to fatherhood and motherhood: it objectively deprives conjugal fruitfulness of its unity and integrity; it brings about and manifests a rupture between genetic parenthood, gestational parenthood and responsibility for upbringing." [26]

The consequences should also be considered in relation to the child, who will have to learn about and accept this anomalous situation; from a psychological viewpoint, the child will have to navigate the difficulties of "identification" with the father.

Specific legal formulas have been introduced into the established laws of some sovereign nations for the attribution of paternity, donor confidentiality, and "commerce" regulations on semen.[27] Doubts remain, however, as to the legitimacy of an ordinance that shrouds the true paternity of the child in secrecy: every citizen

---

[26] CDF, *Donum vitae*, part II, no. 5, emphasis removed.

[27] See M. L. Di Pietro, "Analisi comparata delle leggi e degli orientamenti normativi in materia di fecondazione artificiale," *Medicina e Morale* 43, no. 1 (1993): 231–282; and

has the right to know who his parents are, and this right is explicitly recognized in many nations.

From a psychological perspective, it should also be noted that there is a possibility of subconscious conflicts within the spouses, especially in the "putative" father, who knows that he is the father only in a legal and emotional sense but not in the biological sense; by that very fact, he also comes to find that he is not in the same position as the mother with respect to the child. This can also be reflected in the relationship between the two spouses, who are not equally parents with respect to the child.[28]

Eugenic tendencies can also creep in along this path of semen donation and sperm banks through the acquisition of "selected semen." Aside from the obvious case of a sperm bank in California that was reserved for Nobel Prize winners, the information that must be provided at the moment when the semen is obtained is increasingly consonant with biological selectivity. It is necessary to clarify, however, that the demand for semen selected on the basis of genetic disorders is more theoretical than practical. Indeed, as many people know, it is currently possible to diagnose genetically transmitted diseases—not necessarily caused by dominant genes (and hence transmitted by healthy carriers as well)—for a vast number of loci on the human genome, and also to test for predispositions to diseases of varying degrees of seriousness, including tumors, which can manifest when paternal and maternal chromosomes with similar characteristics meet. One consequence of this, on the scientific level, is that the preliminary health examination of the donor for purposes of artificial reproduction can no longer be confined to common clinical assessments of an apparent state of health, but must also be extended to genetic tests.[29] Yet in order for this investigation to have any real efficacy, the genetic tests must clearly be performed on the recipient as well and, in the case of multiple inseminations of the same semen in different women, on each of the recipients. The practical result of these examinations will then be the rejection of heterologous fertilization matches that involve risk and a search for others by examining the semen of several donors. If there are multiple candidates for receiving the semen of a particular donor—up to five, according to the laws in some countries—the combinations would multiply, and one could be offered a selection of various donors for each case. This would plainly be entering into the vast, unexplored territory of eugenics, which is ethically unacceptable.

Consequently, the very progress made in genetic research—mostly in the period following the passage of laws in many countries allowing heterologous fertilization—leads me to think that, given present knowledge of the ever-widening circle of consequences, this practice has negative medical and ethical implications that are serious and insuperable enough to make its legalization unacceptable in countries that have not yet done so, and may lead to its eventual abandonment in other countries that allowed it too hastily. Even compiling a profile of the donor's psychological health may involve practical and ethical problems, because it is a matter of determining

---

M. L. Di Pietro and E. Sgreccia, *Procreazione assistita e fecondazione artificiale tra scienza, bioetica e diritto* (Brescia: La Scuola, 1998).

[28] See Isidori, "L'inseminazione artificiale omologa"; P. Marrama et al., eds., *L'inseminazione della discordia* (Milan: F. Angeli, 1987).

[29] A. Fiori and L. C. Caprioli, "La fecondazione artificiale eterologa: Un altro ritorno al matrilineare?" *Medicina e Morale* 44, no. 2 (1994): 213–230; A. Fiori, "Contro la fecondazione eterologa," *Medicina e Morale* 47, no. 2 (1997): 241–266.

what psychological characteristics make for an ideal donor: Does it mean the absence of psychotic symptoms only, that is, of serious psychiatric illnesses (which are not always easy to diagnose), or also the absence of other psychological disorders along the whole gamut of possibilities?

Another ethical and legal complication with heterologous artificial insemination is that when a so-called sperm bank is established, or at the time a "withdrawal" is made, it can happen that a single semen sample is used—as already mentioned—for different insemination procedures, giving rise to children related by blood on the father's side. The danger is that in future generations it may no longer be possible to know one's paternity—not legal but genetic paternity—and marriages could therefore occur between blood relatives, with easily predictable hereditary consequences.

The governing authorities in many countries have begun to study precautions in this matter and to issue regulations to avoid unfavorable consequences in the use of sperm banks, with regard to health and the prevention of transmissible diseases and also with regard to imprudent economic speculation.

The instruction *Donum vitae* condemns recourse to heterologous artificial insemination, which is "contrary to the unity of marriage, to the dignity of the spouses, to the vocation proper to parents, and to the child's right to be conceived and brought into the world in marriage and from marriage."[30] The child, fruit of the mutual self-giving of the spouses and a living image of their love, "has the right to be conceived, carried in the womb, brought into the world and brought up within marriage: it is through the secure and recognized relationship to his own parents that the child can discover his own identity and achieve his own proper human development."[31]

Gamete intrafallopian transfer (GIFT) and other techniques of intracorporeal fertilization

GIFT is a technique of intracorporeal artificial fertilization that involves the simultaneous *but separate* transfer of the male and female gametes into the fallopian tube. It is medically indicated for some forms of female infertility (pelvic endometriosis, idiopathic infertility) or male infertility (oligoasthenospermia) that cannot be treated by artificial insemination, provided that the woman has at least one pervious (i.e., not blocked) tube. Proponents of this technique claim that of cases of infertility due to a known or undiagnosed pathology that is not always surmountable through artificial insemination, around 40 percent can be cured by means of this technique.

There are three phases in the GIFT procedure: (1) induction of ovulation and egg cell procurement by means of laparoscopy or aspiration guided by ultrasound imagery, (2) procurement and preparation of sperm, and (3) transfer of the gametes into the fallopian tube by means of a small catheter in which the egg cells and the spermatozoa are separated by an air bubble. In this way, fertilization does not occur until the spermatozoa and the egg cells are freed to come in contact with each other inside the fallopian tube. The success rates reported by the authors who first carried out the GIFT procedure averaged around 20 to 27 percent in terms of completed

---

[30] CDF, *Donum vitae*, part II, no. 2, emphasis removed.
[31] Ibid., no. 1.

pregnancies out of the total number of gamete transfers performed. The low rate is supposedly due to a high incidence of miscarriage.[32]

Attempts have been made in recent years to perform GIFT by a transcervical transuterine route, aspirating the oocytes by puncturing the posterior vaginal fornix under local anesthesia, with guidance by ultrasound imagery, and then introducing them into the fallopian tube with a suitable catheter through the uterus. The results seem promising but often do not agree between different groups of researchers or between one time and the next within the same group.[33]

Another reason for using this technique would be to avoid ethical misgivings with regard to IVF in both Catholic circles and other sectors of the medical and legal world. Indeed, some ethicists assert that GIFT could be used by a lawfully wedded couple, reducing the manipulation of the gametes to a minimum and respecting the conjugal act. In this case, a few oocytes would need to be obtained and then introduced into the fallopian tube after a short interval of time together with the spermatozoa obtained during or immediately after a conjugal act.

According to some ethicists, GIFT could constitute a form of *assistance* and not a *substitute* for the conjugal act by observing these criteria, since the gametes remain outside the body for brief periods of time and fertilization is intracorporeal without any embryo manipulation. Indeed, the moral conscience "does not necessarily proscribe the use of certain artificial means destined solely either to the facilitating of the natural act or to ensuring that the natural act normally performed achieves its proper end."[34]

Not all moralists agree with this interpretation, either because they do not consider GIFT to be an *aid* to the conjugal act or because they take no clear stand as to the moral status to be assigned to this technique.[35] According to others, by

---

[32] R. H. Asch et al., "Gamete Intra-Fallopian Transfer (GIFT): A New Treatment for Infertility," *International Journal of Fertility* 30 (1985): 41–45; idem, "Preliminary Experiences with Gamete Intrafallopian Transfer (GIFT)," *Fertility and Sterility* 45 (1986): 366–371; J. X. Wang et al., "Gamete Intra-Fallopian Transfer: Outcome following the Elective or Non-elective Replacement of Two, Three or Four Oocytes," *Human Reproduction* 8 (1993): 1231; J. P. Balmaceda, A. Manzur, and R. H. Asch, "Gamete Intrafallopian Transfer," in *Infertility: Evaluation and Treatment*, ed. W. R. Keye et al. (Philadelphia: Saunders, 1995), 772–779. On the loss of embryos with GIFT, see M. L. Di Pietro, A. G. Spagnolo, and E. Sgreccia, "Meta-analisi dei dati scientifici sulla GIFT: Un contributo alla riflessione etica," *Medicina e Morale* 40, no. 1 (1990): 13–40.

[33] N. Garcea, S. Campo, and P. L. Cannella, "Current Therapeutic Possibilities of GIFT," *Acta Europea Fertilitatis* 1 (1988): 315 ff; N. Garcea, S. Campo, R. D'Argenio, "Is GIFT a Possibility for Catholics in the Assisted Reproduction Field?," in *From Basics to Clinics*, ed. G. L. Capitanio et al. (New York: Raven Press, 1989), no. 63, p. 313; D. Meirow and J. G. Schenker, "Appraisal of Gamete Intrafallopian Transfer," *European Journal of Obstetrics and Gynecology and Reproductive Biology* 58, no. 1 (1995): 59–65, D. T. Kenny, "In Vitro Fertilization and Gamete Intrafallopian Transfer: An Integrative Analysis of Research 1987–1992," *British Journal of Obstetrics and Gynecology* 102 (1995): 317–325.

[34] Pius XII, Address to Catholic Physicians, cited in CDF, *Donum vitae*, part II, no. 6.

[35] On this subject, see J. Seifert, "Substitution of the Conjugal Act or Assistance to It? IVF, GIFT and Some Other Medical Interventions: Philosophical Reflections on the Vatican Declaration *Donum vitae*," *Anthropotes* 2 (1988): 273–286; J. W. Carlson, "*Donum Vitae* on Homologous Interventions: Is IVF-ET a Less Acceptable Gift than GIFT?" *Journal of Medical Philosophy* 5 (1989): 523–540; N. Tonti-Filippini, "*Donum Vitae* and Gamete Intra

contrast, this technique could satsify the sensibilities of believers and those who have qualms about entrusting the moment of fertilization and the manipulation of the human embryo to laboratory technicians.

I have deliberately used the conditional mood of the verbs in the preceding paragraphs: in fact, the GIFT technique causes—as already noted—the loss of embryos at a higher rate. Suffice it to think that only two-thirds of these clinically initiated pregnancies reach full term and that there is no information whatsoever regarding the "fate" of the gametes introduced into the fallopian tube—whether or not fertilization has taken place and hence whether abortions have taken place before implantation—until the first signs of clinical pregnancy appear with a significant rise in the levels of human chorionic gonadotropin (hCG).[36]

The instruction *Donum vitae* makes no pronouncement with regard to GIFT, either implicitly or explicitly, because of doubts and misgivings about the method due to its novelty at the time. In fact, in practice it would be easy not to follow the guidelines that the instruction provides for classification as simple assistance to procreation, and so the procedure could easily turn into a substitution for the conjugal act.

Other intracorporeal fertilization techniques that were refined over the course of time but are no longer used include lower tubal ovum transfer (LTOT) and gamete intrauterine transfer (GIUT). The first is a technique based on the transfer of oocytes into the fallopian tubes or the uterus.[37] The oocytes are obtained from the woman by laparoscopy following stimulation and are subsequently transferred into the highest part of the uterus by a transcervical route or to the proximal third of the fallopian tube by a transabdominal route. The couple is then invited to have sexual intercourse; if fertilization takes place, normal implantation occurs. LTOT is medically indicated in the following cases: bilateral tubal occlusions, bilateral tubal aplasia, uterine endometriosis, or idiopathic female infertility. The rate of pregnancies achieved is very low (around 15 percent).

GIUT consists in the simultaneous but separate transfer of the egg cell and the spermatozoa into the uterus after the gametes have stayed in a test tube for around thirty minutes. The success rate for GIUT is 10 percent. It is medically indicated especially for idiopathic infertility.

While LTOT seems to be a method of *assisting* the conjugal act, inasmuch as there is no separation between the unitive and procreative dimensions of the conjugal act when sperm is procured in connection with the conjugal act, GIUT does not seem to respect that unity inasmuch as the technical procedure would require semen procurement solely through masturbation. It is therefore impossible to consider GIUT as *assistance* to procreation, since it has the essential elements of a *substitution* for the conjugal act. Moreover, both LTOT and GIUT become "occasions" for increased embryo loss: indeed, only a few clinically initiated pregnancies reach full term.

Fallopian Transfer," in *"Humanae vitae": 20 anni dopo; Atti del II congresso internazionale di teologia morale; Roma, 9–12 novembre 1988*, ed. A. Ansaldo (Milan: Ares, 1989), 791–802; J. F. Doerfler, "Is GIFT Compatible with the Teaching of *Donum Vitae?" Linacre Quarterly* 64, no. 1 (1997): 16–29.

[36] Di Pietro, Spagnolo, and Sgreccia, "Meta-analisi."

[37] "Future Offers Hope for Infertile Women," *Journal of the American Medical Association* 245, no. 6 (1981): 565.

In Vitro Fertilization
and Embryo Transfer

Importance and current relevance of the topic

On the pages of Italian and international newspapers, as well as in specialized publications, in vitro fertilization and embryo transfer (IVF-ET) is a popular subject. The papers exalt the successes of biological science, and they worry about the serious consequences, some of which have already happened. It is like a race to conquer space—not the interstellar kind, but the mysterious space of life.

Yet, emotions aside, the practice is fraught with consequences at all levels—scientific, ethical, social, and genetic. There is talk of the autopoiesis of man and technological power, and the ectopia of human generation has been accomplished (from the uterus to the test tube and from the couple to experimental biologists). There is danger of a further reification of life when used as "human experimental material." (This expression is used in the literature of specialists; it is identical to the terminology used by the Nazis in the extermination camps.) A legal and social transition is taking place, from the existence of one father and one mother to a "parent cooperative" and the manufacturers of the embryo.

From the ethical point of view, it should be noted that the *escalation* of experimental success has been accompanied by a *de-escalation* of consideration for the anthropological value of the human embryo.

Indeed, with abortion laws the life of the embryo and the fetus has come to be considered less important than the life or even just the health of the mother, through a sort of predominance of the adult—and often of the adult's contingent interests—over the newly conceived child.

A further step was taken down the slippery slope with IVF: because of technical requirements that will be seen later, more embryos are usually produced than could possibly be needed for eventual transfer to the uterus, thereby resulting in a "surplus" of frozen embryos that are destined for "experimentation" and destruction. Moreover, it is claimed that such experimentation is necessary for scientific progress and particularly for the study of DNA; this necessity is heightened through the hope of thereby correcting altered genomes and developing treatments for chromosomal abnormalities and genetic disorders.

John Paul II writes the following:

> Apart from the fact that they are morally unacceptable, since they separate procreation from the fully human context of the conjugal act, these techniques have a high rate of failure: not just failure in relation to fertilization but with regard to the subsequent development of the embryo, which is exposed to the risk of death, generally within a very short space of time. Furthermore, the number of embryos produced is often greater than that needed for implantation in the woman's womb, and these so-called "spare embryos" are then destroyed or used for research which, under the pretext of scientific or medical progress, in fact reduces human life to the level of simple "biological material" to be freely disposed of.[38]

---

[38] John Paul II, *Evangelium vitae*, no. 14.

Some nations have therefore legalized experimentation on human embryos, from which the killing and destruction of the embryos themselves follows.[39] This is tantamount to equating experimentation on the embryo with experimentation on animals and amounts to a plain and simple denial of the human character of the embryo.

Once this juncture has been reached, one runs the risk of no longer seeing any problem with the utilization of live embryos, even those conceived in utero (an embryo is still an embryo, no matter how it is conceived), in order to obtain fresh living tissues for therapeutic or nontherapeutic purposes or even for industrial or commercial purposes. Thus the chain reaction starts with manufactured man and ends up with reified man: a human being whose uniqueness and personal value have been destroyed.

The trajectory "construct in order to destroy" is not new; it was applied in the development of atomic energy. This juxtaposition is not meant to be rhetorical or forced: both practices—the use of atomic weapons in warfare and the destruction of human fetuses and embryos—are born of one and the same philosophy, in which truth is subordinated to utility, being is subjugated to doing, and the value of the person is subjugated to the power of others.

By pointing out these negative aspects, I do not mean to imply that IVF-ET was not proposed with the intention of remedying the infertility of couples, which is an eminently human and praiseworthy purpose. I wish to emphasize that technology has a dynamic of its own, beyond the intentions, and has in fact opened up further possibilities for devaluing the human embryo.

Historical notes on IVF-ET

Experiments on animals to achieve IVF began after methods of intracorporeal insemination had been developed and after the *New England Journal of Medicine* had hypothesized IVF in 1937.[40] The experiments began with the study of culture media for embryos, which were formed in the uterus, then transferred to the culture medium, and finally reimplanted in the uterus. The experiments were first performed on rats and then on rabbits.

The first to report success was Min Chueh Chang in 1959, who achieved the birth of rabbits by this method. Other successes followed with other species of experimental animals.[41] The one who began experiments in human biology (1963) and first published results (1965) was Robert Edwards.[42] This time the experiments concerned the maturation in vitro of human oocytes that were obtained from the ovaries at various stages of maturation. Edwards concluded these experiments with

---

[39] For a detailed analysis of the problem, see Di Pietro, "Analisi comparata"; M. Vega, J. Vega, and P. Martinez Baza, "Comentarios a la legislación española sobre Reproducción Asistida," *Cuadernos de Bioética*, 21, no. 5 (1995): 57–64.

[40] "Conception in a Waterglass," *New England Journal of Medicine* 217 (1937): 678.

[41] M. C. Chang, "Fertilization of Rabbit Ova In Vitro," *Nature* 184 (1959): 466–467; D. G. Whittingham, "In-vitro Fertilization, Embryo Transfer and Storage," *British Medical Bulletin* 35 (1979): 105–111.

[42] R. G. Edwards, *A Matter of Life: The Story of the Medical Breakthrough* (London: Hutchinson, 1980); idem, "Maturation In Vitro of Human Ovarian Oocytes," *Lancet* 2 (1965): 926–929.

this comment, "Perhaps the greatest challenges of the present work lie, however, in the prospect of obtaining fertilised human eggs. ... Fertilisation of mammalian eggs in vitro is not easy. ... Nevertheless, we may shortly be able to obtain numbers of human embryos in the process of cleaving. ... The plentiful supply of oocytes from one ovary could ultimately allow us to grow human embryos in vitro, and even control some of the genetic disorders of man." [43]

In 1969, Edwards offered conclusive data on the possibility of fertilizing human oocytes in vitro.[44] Of fifty-six oocytes that were obtained and preserved in follicular fluid or in a simple culture medium, then inseminated with suspensions of spermatozoa, eighteen were fertilized. In 1970 and 1971, Edwards and his collaborators again achieved success with growing embryos in vitro up to the blastocyst stage (110 to 116 cells). The goal was still *embryo transfer* to a suitably prepared woman for the purpose of achieving a natural birth.

In 1973, another team of researchers (Alex Lopata, Alan Trounson, and Carl Wood) at Monash University in Australia attempted the first human embryo transfer at the Queen Victoria Centre in Melbourne: the embryo did not mature to a suitable stage for transfer. The first success was achieved in England by the Edwards team in July 1978, when Louise Brown was born at Oldham General Hospital in Manchester after an embryo transfer procedure performed by Edwards and Patrick Steptoe. Candice Reed was born shortly afterward in Australia. In Melbourne in 1980, 13 pregnancies were recorded among 103 patients who had been treated; one of them had twins.

The first success in Italy was achieved in Naples in 1983 by Dr. Abbate with the help of Australian biologists. At present it is difficult to give exact figures on how many babies have been born since 1978 as a result of IVF, since not all countries have pertinent records at the national level, but there are thought to be tens of thousands of cases.[45]

---

[43] R. G. Edwards, "Maturation In Vitro of Human Ovarian Oocytes," *Lancet* 2 (1965): 929.

[44] R. G. Edwards, B. D. Bavister, and P. C Steptoe, "Early Stages of Fertilization In Vitro of Human Oocytes Matured In Vitro," *Nature* 221 (1969): 632–635. See also R. G. Edwards: *Life Before Birth: Reflections on the Embryo Debate* (New York: Basic Books, 1989). For a brief historical overview, see L. Leuzzi, "Riflessione etico-morale sulla fecondazione in vitro," *Ospedale Miulli* 3 (1986): 1–166; A. L. Bonnicksen, "In Vitro Fertilization and Embryo Transfer," in *Encyclopedia of Bioethics*, Reich, 2221–2225; H. W. Jones, "History of In Vitro Fertilization," in *Infertility*, Keye et al., 736–744.

[45] D. M. Saunders, M. Mathews, and P. A. L. Lancaster, "The Australian Register: Current Research and Future Role: A Preliminary Report," in *In Vitro Fertilization and Other Assisted Reproduction*, ed. H. W. Jones and C. Schraedere (New York: New York Academy of Sciences, 1988), 7–21; Medical Research International and American Fertility Society Special Interest Group, "In Vitro Fertilization/Embryo Transfer in the United States: 1990 Results from the National IVF-ET Registry," *Fertility and Sterility* 57 (1992): 15–24; Interim Licensing Authority (ILA) for Human In Vitro Fertilisation and Embryology, *Fifth Report of the Interim Licensing Authority for Human In Vitro Fertilisation and Embryology* (London: ILA, 1990); "French In Vitro Fertilization Registry: 1994 Report," *Contraception, Fertilité, Sexualité* 23, nos. 7–8 (1995): 490–493; "Assisted Reproductive Technology in the United States and Canada: 1994 Results Generated from the American Society for Reproductive Medicine/Society for Assisted Reproductive Technology Registry," *Fertility and Sterility*

Medical indications, methods, and success rates

In an effort to highlight ethical ramifications, further technical details will be set aside; only the fundamental data necessary for understanding the moral issue will be mentioned. The term "medical indications" presupposes a reference to therapeutic, pharmacological, or surgical acts; in the present case, as will be seen, the procedure goes beyond therapeutic action per se and does not stop there. Nevertheless, since there is generally an intentional aim of remedying the infertility of a couple in the case of IVF-ET, it is common to speak of medical indications for IVF-ET.

Proponents of IVF-ET make much of the fact that many cases of male and female infertility are otherwise insurmountable; this proves true especially in cases of infertility of tubal origin. Indeed, artificial insemination is no longer sufficient or applicable in such cases, because the ovum does not pass through the fallopian tube and therefore cannot be reached by the semen, even when introduced artificially. To achieve fertilization in these cases, the ovum must be obtained and fertilized outside the body and then placed back into the uterus.

In addition to cases of tubal and ovarian pathology, IVF-ET is considered clinically indicated for the treatment of idiopathic infertility, immunologically based infertility, and infertility associated with endometriosis; in men, oligospermia and oligoasthenospermia are considered valid indications. It can be hoped that an alternative therapy will become practicable with the progress of medical and surgical techniques, making it pointless to resort to IVF-ET, which presents complex ethical and legal problems.

To head in this direction, the development of tubal microsurgery techniques should be considered and promoted. These employ a microscope-operator and technical instruments designed to minimize trauma to delicate tissues, specifically the fallopian tube and the ovary, allowing the surgeon to restore the anatomical integrity of these organs.[46]

Once physical examinations and male and female fertility tests have shown that IVF-ET is indicated, practitioners notify the couple as to the practicability of

---

66, no. 5 (1996): 697–705; "Assisted Reproductive Technology in the United States: 2000 Results Generated from the American Society for Reproductive Medicine/Society for Assisted Reproductive Medicine/Society for Assisted Reproductive Technology Registry," *Fertility and Sterility* 81 (2004): 1207–1220; A. N. Andersen et al., "Assisted Reproductive Technology in Europe, 2001: Results Generated from European Registers by ESHRE," *Human Reproduction* 20, no. 5 (2005): 1158–1176; M. Van den Bergh et al., "Ten Years of Swiss National IVF Register FIVNAT-CH, "Are We Making Progress?" *Reproductive BioMedicine Online* 5, no. 11 (2005): 632–640; J. Gumby, S. Daya, and IVF Directors Group of the Canadian Fertility and Andrology Society, "Assisted Reproductive Technology (ART) in Canada: 2001 Results from the Canadian ART Register," *Fertility and Sterility* 84, no. 3 (2005): 590–599.

[46] To ensure greater efficacy before intervening with tubal microsurgery, it is also good to assess the integrity of the mucus in the fallopian tubes; if compromised, it could impair the results of the surgical intervention. A technique called salpingoscopy is used for this purpose, allowing a view of the inside of the fallopian tubes. On this subject, see R. Marana, G. F. Catalano, and L. Muzii, "Trattamento chirurgico della sterilità di origine tubarica," *Medicina e Morale* 43, no. 1 (1993): 67–74; R. Marana et al., "Prognostic Role of Laparoscopic Salpingoscopy of the Only Remaining Tube after Controlateral Ectopic Pregnancy," *Fertility and Sterility* 63, no. 2 (1995): 303–306.

IVF-ET with homologous semen or donated sperm. The woman must be no older than 36 years of age. Women who have already become mothers obviously have better prospects, generally speaking.

At this point it is important to briefly and summarily say a few words about the methods and techniques for procuring the ovum and semen and their preliminary treatment.[47]

The first step is to ascertain in general whether the couple is able to handle the IVF-ET program (pregnancy, birth, and the subsequent growth of the baby); if there are emotional problems, psychological treatment is recommended.

The tests performed on the woman are normally as follows. First, a laparoscopy is performed to determine whether the ovaries can be accessed for harvesting oocytes; if necessary, adhesiolysis is carried out to free the ovaries. Then the curve for the woman's basal temperature is charted over three cycles to ascertain the regularity of ovulation. Whenever anovularity or hyperprolactinemia is encountered (the latter can be determined from the results of blood tests), suitable treatment must begin before the patient is accepted into the IVF-ET program. There is no guarantee that an egg cell suitable for fertilization can be obtained from an infertile woman; if not, the program provides for the use of a donor ovum so long as the other intrauterine conditions prove compatible with subsequent implantation.

It is important to keep all of this in mind in order to grasp the complexity of the IVF-ET procedure and how homologous IVF-ET can easily turn into the heterologous form.

Preliminary assessments of the husband pertain first to his seminal fluid, which is examined in appropriate tests at least three times in different laboratories, the last of which is the one where the IVF-ET procedure will take place. Meanwhile, a reserve of semen is frozen in case there are difficulties in producing suitable semen at the time of IVF-ET. If the semen proves unusable, the next step involves the possibility of a donor, who must undergo the same preliminary assessments as for artificial insemination.

In some centers, besides the above-mentioned examinations performed separately on the wife and the husband, additional combined examinations are carried out through a postcoital test, and the decision is made to proceed with IVF-ET when it is

---

[47] For the technical aspects of IVF-ET, see Rodriguez Luño and López Mondejar, *La fecondazione in vitro*, 9–51; Bompiani, "Problemi biologici," 74–90; A. Serra, "Il concepimento umano in vitro: Aspetti e problemi biologici," in *Fecondazione artificiale embryotransfer: Problemi biologici, clinici, giuridici, etici*, ed. G. F. Zuanazzi (Verona: Cortina International, 1986), 11–24; N. Pasetto, "Aspetti biologici e clinici della fecondazione in vitro," in *Fecondazione artificiale embryotransfer*, 38–48; E. Cittadini, "Aspetti biomedici ed etici della fertilizzazione in vitro," in *La fertilizzazione in vitro: Atti del convegno della Societas Ethica; Palermo, 2–6 settembre 1985*, ed. S. Privitera (Palermo: EDI OFTES, 1986), 7–34; A. Bompiani and N. Garcea, "La fecondazione in vitro: Passato, presente, futuro," *Medicina e Morale* 36, no. 1 (1986): 79–102; G. T. Kovacs et al., "In Vitro Fertilization and Embryo-Transfer," *Medical Journal of Australia* 144 (1986): 682 ff; E. Tosti and B. Dale, *Fecondazione in vitro* (Naples: Edizioni Scientifiche Italiane, 1995); Italian National Bioethics Committee, *La fecondazione assistita*; O. K. Davis and Z. Rosenwaks, "In Vitro Fertilization," in *Infertility*, Keye et al., 759–771; E. S. Surrey and J. F. Kerin, "Extended Techniques in Assisted Reproductive Technologies," in *Infertility*, Keye et al., 788–797.

determined that infertility is due to an immunological barrier between the oocytes and spermatozoa despite normal genital tracts.

While a sperm donor is the only possible donor in artificial insemination, a donated ovum may be used as well in IVF-ET if the woman cannot produce her own egg cells. IVF-ET may even make possible, or necessary, the donation of an embryo who is completely unrelated to both spouses.

Once the preliminary assessments of the woman and man have been completed, the treatment of the patients begins. Since spontaneous ovulation brings about the release of a single egg cell, it is preferable to have several oocytes available at the same time so as to increase the success rates of IVF-ET or GIFT (and so as not to subject the woman to repeated procedures to obtain them); as we will see, this fact has ethical significance. The procedure is then to induce ovulation pharmacologically by administering clomiphene citrate, human menopausal gonadotropin (hMG), or human chorionic gonadotropin (hCG).

The ova are obtained by means of laparoscopy; this must be preceded, however, by an ultrasound examination. In the case of clomiphene, the drug is administered from the fifth to the ninth day of the ovarian cycle; the ultrasound exams are conducted on the tenth, eleventh, and twelfth days; hCG is administered on the thirteenth day. Meanwhile tests are conducted on the cervical and vaginal secretions to determine whether there are any infections. If so, everything is postponed to the following cycle.

The woman is usually hospitalized when a sonogram shows a follicular diameter of 1.7 cm. From then on, urine samples must be taken every three hours and analyzed quickly to gauge the levels of urinary estrogen and to monitor ovulation. The laparoscopy procedure to obtain the ova takes place 35 to 36 hours after the administration of hCG, or after the spontaneous peak of luteinizing hormone (LH). Variations in the cervical mucus are also used to confirm the laboratory results.

Timing is important. To obtain a mature egg, the particular moment of ovulation must be determined: all the procedures—laparoscopy, ultrasound examinations, blood tests, urinalysis, correlation with the cervical mucus—must be carried out with minute precision. Laparoscopy requires a high degree of dexterity and experience. The egg cell can also be obtained by the transcervical route under the guidance of ultrasound imagery.

Once the follicle has been punctured and the oocyte has been aspirated, the oocyte is placed in a culture medium, the composition of which has obviously been studied and adjusted to resemble the environment of the fallopian tube.

As the techniques become more refined, the trend is toward faster surgical procedures, which are thought to be less stressful for the woman.

The collection of the semen—unless it is frozen sperm from a bank—is carried out a few hours before fertilization. The method is generally masturbation (the ethical assessment of which has already been done), as is usual when the procedure is homologous IVF.

The ovum must remain in the culture medium for five or six hours prior to being fertilized. The culture medium for the ovum remains the same after fertilization, which occurs following the transfer of a large quantity of spermatozoa. Edwards and Trounson have described the phases of the development of an embryo in vitro. Pronuclei are visible twelve to twenty-four hours after fertilization, and the following times are listed for cellular division: two blastomeres at twenty-five to twenty-six hours; four at twenty-six to thirty-six hours; and eight at thirty-six to fifty-six hours.

The transfer procedure requires embryos that have developed to the four-, eight-, or even sixteen-cell stage; it is thought that with further development the embryo would exceed the dimensions for tubal development and might suffer damage.

The transfer takes place by the transcervical and transuterine route under anesthesia (the technique is described in the literature for specialists) by means of catheters or coaxial tubes. After twenty-four hours of rest in the hospital, if deemed necessary, and after four to five days at home, the woman undergoes blood tests beginning around the tenth or twelfth day to determine whether pregnancy has in fact been achieved, and monitoring begins after the second week. The laboratory results will indicate any eventual treatments that may be needed to support the pregnancy.

As for success rates, it is necessary to distinguish the rates achieved in harvesting a mature oocyte (95 percent), the fertilization procedure (90 percent), initiation of development (58.8 percent), initiation of pregnancy by implantation (17.1 percent), and bringing the pregnancy to term (6.7 percent). In light of this data, the total rate of embryo loss is 93 to 94 percent.[48] It is difficult to estimate total embryo loss, however, since most studies calculate the rate of success in relation to the number of embryos transferred and not the number fertilized. According to the latter criterion, the success rate of IVF-ET is somewhere between 14 and 20 percent.[49]

Extracorporeal artificial fertilization techniques also include micromanipulations, that is, the insertion of the spermatozoon into the egg cell or the act of facilitating its penetration. Recorded micromanipulation techniques include PZD (partial zonal dissection), or the opening of the pellucid zone with a microneedle followed by contact with spermatozoa in a culture medium; SUZI (subzonal injection), which injects spermatozoa into the perivitelline region; and ICSI (intracytoplasmic sperm injection), which injects a spermatozoon into the cytoplasm of the egg cell.[50] The success rates for these methods, in terms of births, range from 22 percent with SUZI and ICSI to 45 percent with ICSI alone.[51] The only technique currently used is ICSI.

---

[48] M. Seppala, "The World Collaborative Report on In Vitro Fertilization and Embryo Replacement: Current State of the Art in January 1984," *Annals of the New York Academy of Sciences* 442 (1985): 558–563; D. S. Guizick, C. Wilkes, and H. W. Jones, "Cumulative Pregnancy Rates for In Vitro Fertilization," *Fertility and Sterility* 46 (1986): 63 ff; H. C. Liu, H. W. Jones, and Z. Rosenwaks, "The Efficacy of Human Reproduction after In Vitro Fertilization and Embryo-Transfer," *Fertility and Sterility* 49 (1988): 649–653; M. G. R. Hull, "Infertility Treatment: Relative Effectiveness of Conventional and Assisted Conception Methods," *Human Reproduction* 7 (1992): 785–796; Medical Research International and American Fertility Society, "In Vitro Fertilization/Embryo Transfer."

[49] See "French In Vitro Fertilization Registry"; "Assisted Reproductive Technology in the United States and Canada: 1994 Results"; "Assisted Reproductive Technology in the United States: 2000 Results."

[50] Tosti and Dale, *Fecondazione in vitro*, 73–83; J. Cohen, "Micromanipulation of Human Embryos, Zygotes, and Embryos," in *Infertility*, Keye et al., 841–858.

[51] See A. C. Van Steirteghen et al., "Higher Success Rate by Intracytoplasmic Sperm Injection Than by Subzonal Insemination: Report of a Second Series of 300 Consecutive Treatment Cycles," *Human Reproduction* 8, no. 7 (1993): 1055 ff; G. B. Candiani, V. Danesino, and A. Gastaldi, *La clinica ostetrica e ginecologica* (Milan: Masson, 1996) 2:1222–1223.

Complications of IVF-ET

Before addressing the strictly ethical problems, it is necessary to make a note about the uncertainty in safeguarding the life and health of the woman who undergoes artificial fertilization and the conceptus resulting from it. Indeed, the clinical complications of IVF for both the women and the embryos are hardly negligible.[52]

A first set of complications is related to the use of drugs to stimulate ovulation, with the following possible consequences: ovarian hyperstimulation syndrome, increased incidence of fetal malformations, and asynchrony in endometrial and embryonic development. These complications should be taken into account both for judgments in terms of medical deontology and for broader ethical considerations.

OVARIAN HYPERSTIMULATION SYNDROME. Ovarian hyperstimulation syndrome is a serious complication that can occur following ovarian stimulation; it usually occurs after treatment with hMG or clomiphene citrate.[53] It is characterized by ovarian growth, accumulation of fluids in the interstitial spaces, electrolytic imbalances, hypovolemia, and oliguria. The etiology, its clinical signs, and the various therapeutic interventions are a highly specialized area of discussion; here I wish to draw attention to the fact that along with its mild and moderate forms, which require only close observation and symptomatic treatment, there are severe forms accompanied by serious complications, including ascites, hydrothorax, shock, increased risk of thromboembolism, stroke, and adult respiratory distress syndrome.[54]

POSSIBLE INCREASE IN THE INCIDENCE OF FETAL MALFORMATIONS. It is known that 60 to 80 percent of spontaneous abortions in the first trimester of pregnancy are caused by chromosomal alterations either in the male or female gametes or within the embryo. While a high percentage of chromosomal abnormalities are caused during gametogenesis in oocytes (trisomy 21, for example, in which the extra chromosome

---

[52] See the following reviews: J. G. Schenker and Y. Ezra, "Complications of Assisted Reproductive Techniques," *Fertility and Sterility* 61, no. 3 (1994): 411–422. M. L. Di Pietro and R. Minacori, "Qual è il rischio delle tecniche di fecondazione artificiale?," *Medicina e Morale* 48, no. 3 (1998): 465–497.

[53] M. Germond, M. C. Gaillard, A. Senn, "Syndrome d'hyperstimulation ovarienne," *Archives of Gynecology and Obstetrics* 246 (198): S53–S64; Y. Fujita et al., "Methods for Monitoring Follicle Maturation Decrease: The Ovarian Hyperstimulation Syndrome during Gonadotropin Treatment," *European Journal of Obstetrics and Gynecology and Reproductive Biology* 32 (1989): 223; A. F. Haney, "Controlled Ovarian Hyperstimulation and Intrauterine Insemination," in *Infertility*, Keye et al., 745–758; A. P. Ferraretti, L. Gianaroli, C. Magli, "Sindrome da iperstimolazione: Prevenzione e trattamento," in *Attualità sulla procreazione medico assistita*, ed. C. Ragni and W. Vegetti (Rome: CIC Edizioni Internazionali, 1997), 47–61.

[54] For a review, see Serebrovska and Di Pietro, "La sindrome da iperstimolazione ovarica"; H. I. Demey et al., "Acute Oligo-Anuria during Ovarian Hyperstimulation Syndrome," *Acta Obstetricia et Gynecologica Scandinavica* 66 (1986): 741; S. Mashiach et al., "Adnexal Torsion of Hyperstimulated Ovaries in Pregnancies after Gonadotropin Therapy," *Fertility and Sterility* 53, no. 1 (1990): 76–80; M. G. Wagner and P. A. St. Claire, "Are In Vitro Fertilization and Embryo Transfer of Benefit to All?," *The Lancet* 2 (1989): 1027–1030; Società Italiana per lo Studio della Fertilità e Sterilità, *Libro bianco sulla riproduzione assistita* (Palermo: SIFES, 1991), 80–85.

is often of maternal origin), chromosomal abnormalities occur in spermatozoa at a rate of only 5 to 10 percent.[55]

While it has been found that there are more chromosomal alterations in the infertile population than in the fertile population, the increased incidence of chromosomal alterations in oocytes obtained by means of induced ovulation has led scientists to hypothesize that there is a correlation between these abnormalities and the dosages and administration methods of hMG or clomiphene citrate.[56] The rate of chromosomal abnormalities found in oocytes procured by induced ovulation and then fertilized during IVF procedures varies in different reports. After stimulation with follicle-stimulating hormone or clomiphene, multiple alterations causing an unbalanced chromosomal complement (aneuploidy) occur at rates from 4 to between 21 and 57 percent, and single alterations causing an unbalanced chromosomal complement (hypoploidy) occur at rates up to 28 percent, with a total incidence of altered karyotype around 50 percent.[57] Similar data have been obtained through studies conducted on guinea pigs and mice.[58] These data are interpreted by many as one among a series of causes contributing to the low success rate of IVF.

It is also possible that an increase in the incidence of genetic abnormalities results from the use of micromanipulation techniques, especially ICSI: there has been heated debate on this subject.[59]

---

[55] J. F. Mattei et al., "Origin of the Extra Chromosome in Trisomy 21," *Human Genetics* 46 (1979): 107–110; B. Brandiff et al., "Detection of Chromosome Abnormalities in Human Sperm," *Progress in Clinical and Biological Research* 209B (1986): 469–476.

[56] A. Bongso et al., "Chromosome Anomalies in Human Oocytes Failing to Fertilize after Insemination In Vitro," *Human Reproduction* 3, no. 5 (1988): 645–649; L. Hens et al., "Chromosome Aberration in 500 Couples Referred for In Vitro Fertilization or Related Fertility Treatment," *Human Reproduction* 3, no. 4 (1988): 451–457; H. Wramsby et al., "Chromosome Analysis of Human Oocytes Recovered from Preovulatory Follicles in Stimulated Cycles," *New England Journal of Medicine* 316 (1987): 121–124; J. Santalo et al., "The Genetic Risks of In Vitro Fertilization Techniques: The Use of an Animal Model," *Journal of Assisted Reproduction and Genetics* 9 (1992): 462–474.

[57] R. H. Martin et al., "Chromosomal Analysis of Unfertilised Human Oocytes," *Journal of Reproduction and Fertility* 78 (1986): 673–678; M. Plachot et al., "Chromosome Investigation in Early Life. I. Human Oocytes Recovered in an IVF Programme," *Human Reproduction* 1, no. 8 (1986): 547–551; Wramsby et al., "Chromosome Analysis."

[58] R. H. Martin, "Comparison of Chromosomal Abnormalities in Hamster Egg and Human Sperm Pronuclei," *Biology of Reproduction* 31, no. 4 (1984): 819–825; Santalo et al., "Genetic Risks."

[59] J. Testart, "Procréation médicalement assistée: L'éthique et la loi," *Études* 3816 (1994): 599–610; idem, "La fécondation de mieux en mieux assistée," *Les Cahiers du CCNE* 3 (1995): 24–25; A. Kahn, "Essais sur l'homme et essais d'hommes," *Le Cahiers du CCNE* 3 (1995): 25–26; for a commentary on these essays, see A. G. Spagnolo, "Procreazione sempre più 'assistita' con l'uso degli spermatidi," *Medicina e Morale* 45, no. 5 (1995): 1107–1111. See M. Bonduelle et al., "Comparative Follow-Up Study of 130 Children Born after Intracytoplasmic Sperm Injection and 130 Children Born after In-Vitro Fertilization," *Human Reproduction* 10 (1995): 3327–3331; J. W. Persson, G. B. Peters, and D. M. Saunders, "Genetic Consequences of ICSI," *Human Reproduction* 11 (1996): 921–924; R. H. Martin, "The Risk of Chromosomal Abnormalities following ICSI," *Human Reproduction* 11 (1996): 924–925; T.-H. Bui and H. Wramsby, "Micromanipulative Assisted Fertilization: Still Clinical Research," *Human Reproduction* 11 (1996): 925–926; A. A. Baschat, E. Schwinger, and K. Diedrich, "Assisted

ASYNCHRONY OF ENDOMETRIAL AND EMBRYONIC DEVELOPMENT. In order for the embryo to successfully implant in the uterus, the crucial factor is the synchronous development of the embryo and the inner mucosa of the uterus (endometrium).[60] Endometrial development is induced by estrogens that stimulate the proliferative activity of the supporting tissue and glandular tissue and cause an increase in blood flow. The progesterone produced by the corpus luteum during the postovulatory phase blocks the proliferative activity and induces glandular secretion, thus preparing the uterine wall for the implantation of the blastocyst.

It is known that high doses of estrogen have an abortive effect by preventing embryo implantation. This abortive effect of estrogens, according to the literature, seems to be due on the one hand to their counteracting the function of the corpus luteum (*luteolytic action*), thus reducing the production of progesterone, and on the other hand to the cessation of endometrial maturation caused by a disruption of cellular metabolic equilibrium. During the ovulation-inducing phase, the drugs administered increase endogenous estrogen levels to three to six times the naturally occurring levels; while these levels are subsequently neutralized by progesterone, they could be the cause of a lower rate of pregnancies due to the death of the embryo in the pre- or post-implantation phase.[61] It has been discovered, however, that this asynchrony between the embryo and the endometrium would not occur if clomiphene citrate were used, alone or together with hMG, to induce follicular development.[62]

Reproductive Techniques: Are We Avoiding the Genetic Issue?" *Human Reproduction* 11 (1996): 926–928; B. Rosenbusch, E. Strehler, and K. Sterzik, "Micro-Assisted Fertilization and Sperm Chromosome Abnormalities," *Human Reproduction* 11 (1996): 928–930; A. C. Chandley and T.B. Hargreave, "Genetic Anomaly and ICSI," *Human Reproduction* 11 (1996): 930–932.

[60] A. A. Gidley-Baird et al., "Failure of Implantation in Human In Vitro Fertilization and Embryo-Transfer Patients: The Effects of Altered Progesterone-Estrogen Ratios in Humans and Mice," *Fertility and Sterility* 45 (1986): 69–74; G. T. Fossum, A. Davidson, and R. J. Paulson, "Ovarian Hyperstimulation Inhibits Embryo Implantation in the Mouse," *Journal of In Vitro Fertilization and Embryo Transfer* 6 (1989): 7–10.

[61] S. Smith, L. Scott, and S. Hosid, "Combined Intrauterine Triplet and Ectopic Pregnancy following Pronuclear Embryo Transfer in a Patient with Elevated Serum Progesterone during Ovulation Induction," *Journal of Assisted Reproduction and Genetics* 10, no. 7 (1993): 478–480; Y. Yaron, J. B. Lessing, and M. R. Peyser, "Expectant Management of Ectopic Pregnancy in the Presence of Ovarian Hyperstimulation Syndrome," *Acta Obstetricia et Gynecologica Scandinavica* 74 (1995): 80–81.

[62] P. Sundstrom, O. Nilsson, and P. Liedholm, "Scanning Electron Microscopy of Human Preimplantation Endometrium in Normal and Clomiphene Citrate–Human Chorionic Gonadotropin Stimulated Cycles," *Fertility and Sterility* 40 (1983): 642–647. Concerning the possibility that asynchrony in endometrial and embryonic development may cause an increased incidence of ectopic pregnancies, see A. Lewin et al., "Second-Trimester Heterotopic Pregnancy after In Vitro Fertilization and Embryo Transfer: A Case Report and Review of the Literature," *International Journal of Fertility* 36, no. 4 (1991): 227–230; J.C. Chang, T. T. Sun, and Y. C. Lin, "Simultaneous Ectopic Pregnancy with Intra-Uterine Gestation after In Vitro Fertilization and Embryo Transfer," *European Journal of Obstetrics and Gynecology and Reproductive Biology* 44, no. 2 (1992): 157–160; J.C. Peek and F. M. Graham, "Ectopic Pregnancy in a Non-patent Fallopian Tube following Transfer of Embryos to the Contralateral Tube," *Human Reproduction* 7, no. 1 (1992): 136–137; Y. F. Chan and P.C. Ho, "Bilateral Tubal Pregnancies after Pronuclear Stage Embryo Tubal Transfer to One Tube," *Australian and New Zealand Journal of Obstetrics and Gynaecology* 33, no. 3 (1993): 315–316; Y. Zalel

Indeed, the anti-estrogenic action of clomiphene citrate would block the effects of higher estrogen levels on the endometrium.

Given what has been said thus far about ovarian stimulation therapies, I draw the conclusion—from the perspective of professional ethics and ethics in general (setting aside for the moment ethical judgments about particular artificial fertilization techniques)—that they should be used only in connection with a real need (pathologies causing ovarian insufficiency) and not primarily to increase the success rates of artificial fertilization techniques. The proportionality between risks and benefits must always be calculated.

The complications of artificial fertilization techniques also include so-called at-risk pregnancies, where the risk affects both the mother and the child.

ECTOPIC PREGNANCIES. Statistics show that the rate of ectopic pregnancies with the use of GIFT varies from 2.5 to 8.3 percent, with a median of around 5.5 percent; the figures are double the naturally occurring rate of 2.8 percent. Equal if not higher rates also occur with IVF-ET, so they are undoubtedly a consequence of the artificial fertilization techniques.[63]

Given the difficulty with which ectopic or extrauterine pregnancies reach full term, these pregnancies must be added to the count of lost embryos, inasmuch as they actually terminate spontaneously or are terminated in order to avoid serious complications (maternal hemorrhages that can be fatal).[64]

MULTIPLE PREGNANCIES. Another complication of artificial fertilization techniques is multiple pregnancy. While the incidences of twins and triplets in nature are 1:80 (1.25 percent) and $1:80^2$ (0.01 percent), respectively, the multiple pregnancy rate after GIFT is, according to the statistics examined, somewhere between 7.1 and 23 percent, and between 4.3 and 22 percent after IVF-ET.[65]

---

et al., "Heterotopic Pregnancy: An Usual Case Report following In Vitro Fertilization and Embryo Transfer," *Journal of Assisted Reproduction and Genetics* 10, no. 2 (1993): 169–171.

[63] Australian in Vitro Fertilization Collaborative Group, "High Incidence of Preterm Birth and Early Losses in Pregnancy after In Vitro Fertilization," *British Medical Journal* 291 (1985): 1160; A. Nazari et al., "Embryo Transfer Technique as a Cause of Ectopic Pregnancy in In Vitro Fertilization," *Fertility and Sterility* 60, no. 5 (1993): 919–921; E. Pyrgiotis et al., "Ectopic Pregnancies after In Vitro Fertilization and Embryo Transfer," *Journal of Assisted Reproduction and Genetics* 11, no. 2 (1994): 79–84; A. Aanesen and F. Flam, "Bilateral Tubal Pregnancy following In Vitro Fertilization and Transfer of Two Embryos," *European Journal of Obstetrics and Gynecology and Reproductive Biology* 64 (1996): 235–236.

[64] On the management of ectopic pregnancies, see A. G. Spagnolo and M. L. Di Pietro, "Bioetica clinica: Quale decisione per l'embrione in una gravidanza tubarica?" *Medicina e Morale* 45, no. 2 (1995): 285–310.

[65] G. M. Stirrat, *Aids in Obstetrics and Gynaecology* (London: Churchill Livingstone, 1987), 83; Società Italiana per lo Studio della Fertilità e Sterilità, *Libro bianco*, 86–89; N. Bollen et al., "The Incidence of Multiple Pregnancy after In Vitro Fertilization and Embryo Transfer, Gamete, or Zygote Intrafallopian Transfer," *Fertility and Sterility* 55 (1991): 314–318; P. Doyle, "The Outcome of Multiple Pregnancy," supplement, *Human Reproduction* 11, no. 4 (1996): 110–120; D. R. Meldrum, "Assisted Reproductive Technology: Choice of Patient, Program, and Procedure," in *Infertility*, Keye et al., 879–885.

The possibility of pregnancies with twins, triplets, or more after GIFT or IVF-ET depends on the number of oocytes (GIFT) or embryos (IVF-ET) transferred into the fallopian tubes (GIFT) or uterus (IVF-ET). In the case of GIFT, there may be a higher number of implanted embryos than foreseen (that is, more than the number of oocytes placed into the tubes); this happens either because of the cellular division of an embryo in the initial phases of development to produce a twin or because not all the oocytes induced to maturity were aspirated.[66] Let us examine the disposition of the woman and the physicians after a diagnosis of multiple pregnancy following recourse to artificial fertilization techniques.

Some women, aware of the complications (anemia, preeclampsia, hydroamnios, premature birth, etc.), choose to carry a multiple pregnancy term. In other cases, since the physicians consider it risky to continue a multiple pregnancy, a reduction procedure is performed, that is, a so-called selective abortion or fetal reduction.[67] This procedure is said to give the embryos that remain alive a greater chance of survival and growth and the mother a less risky gestation period, but it overlooks the fact that these aims are pursued at the cost of the lives of the destroyed embryos. Sometimes, the abortion outcome occurs spontaneously.

The possibility of inducing a multiple pregnancy is a possibility of great ethical significance. Indeed, it is impossible to justify the destruction of lives that were sought with an insistence that bordered on stubbornness. It seems, moreover, that artificial fertilization techniques are associated with increased maternal and neonatal morbidity.[68] This is why it is increasingly recommended that not more than one or two embryos be transferred into a woman's genital tract, since this ultimately results in a rate of pregnancy similar to the rate achieved by transferring more embryos.

The possibility of the aforementioned complications poses dilemmas from an ethical point of view. Nevertheless, the ethical problem arises also and above all from other parameters that principally concern the nature and fate of the human embryo in general and the nature and structure of human procreation in relation to marriage.

---

[66] F. R. Batzer et al., "Multiple Pregnancies with Gamete Intrafallopian Transfer (GIFT): Complications of a New Technique," *Journal of In Vitro Fertilization and Embryo Transfer* 5, no. 1 (1988): 35–37.

[67] K. Still et al., "Early Third Trimester Selective Feticide of a Compromising Twin," *Fetal Therapy* 4 (1989): 83–87; R. L. Berkowitz, "From Twin to Singleton," *British Medical Journal* 313 (1996): 373–374; A. Radestad, T. H. Bui, and K. G. Nygren, "Multifetal Pregnancy Reduction in Sweden: Utilization Rate and Pregnancy Outcome (1986–1992)," *Acta Obstetricia et Gynecologica Scandinavica* 73, no. 5 (1996): 403–406. On this subject see also Bioethics Center of the Catholic University of the Sacred Heart, "Sulla cosiddetta 'riduzione' embrionale," *Medicina e Morale* 47, no. 2 (1997): 374–381; J. Aznar, "Multifetal Pregnancy Reduction," *Acta Obstetricia et Gynecologica Scandinavica* 75, no. 1 (1996): 90–91.

[68] P. Rufat et al., "Task Force Report on the Outcome of Pregnancies and Children Conceived by In Vitro Fertilization (France: 1987–1989)," *Fertility and Sterility* 61, no. 2 (1994): 324–330; J. X. Wang et al., "The Obstetrics Outcome of Singleton Pregnancies following In Vitro Fertilization/Gamete Intra-Fallopian Transfer," *Human Reproduction* 9, no. 1 (1994): 141–146; C. P. Tallo et al., "Maternal and Neonatal Morbidity Associated with In Vitro Fertilization," *Journal of Pediatrics* 127, no. 5 (1995): 794–800; Doyle, "Outcome of Multiple Pregnancy."

Ethical problems concerning IVF-ET

As mentioned at the beginning of this chapter, IVF-ET is at the frontier of medical ethics, where science and technology on the one side are in a difficult confrontation with ethics on the other. This is not a reference just to the positions of Catholic morality; the problem is debated in the field of secular ethics as well.

In the United States, an appropriate ethical and scientific commission was established for the purpose of examining this problem in depth along with other interventions in the medical and biological fields; in England, the government's Warnock Committee worked on such problems.[69] The National Bioethics Committee of Italy issued a document on assisted reproduction,[70] and an international ad hoc committee of the Council of Europe published a document called the Convention for the Protection of Human Rights and Dignity of the Human Being with regard to the Application of Biology and Medicine: Convention on Human Rights and Bio-medicine, which also addresses this subject both in general terms and with a view to developing more extensive protocols.[71] Committees and commissions for the study of these issues have been created in every member state.[72]

The Catholic Magisterium has made official pronouncements on the problems concerning IVF-ET in the instruction *Donum vitae* and the encyclical *Evangelium vitae*, prompted also by the publication of several documents on the subject by various bishops' conferences: a first and then a second document by the Catholic Bishops' Joint Committee on Bioethical Issues of the England and Wales; a document by the

---

[69] Of the documents by the U.S. President's Commission, see *Splicing Life: A Report on the Social and Ethical issues of Genetic Engineering with Human Beings* (Washington, DC: President's Commission for the Study of Ethical Problems in Medicine and Biomedical and Behavioral Research, 1982), http://bioethics.georgetown.edu/documents/pcemr/splicinglife .pdf. See also the U.K. Department of Health and Social Security Committee of Inquiry into Human Fertilisation and Embryology, *Report of Inquiry into Human Fertilisation and Embryology* (London: Her Majesty's Stationery Office, 1984), http://www.hfea.gov.uk/ docs/Warnock_Report_of_the_Committee_of_Inquiry_into_Human_Fertilisation_and _Embryology_1984.pdf.

[70] Italian National Bioethics Committee, *La fecondazione assistita.* See also idem, *Parere sulle tecniche di procreazione assistita: Sintesi e conclusioni* [*Opinion on medically assisted procreation techniques: synthesis and conclusions*], June 17, 1994 (Rome: Presidenza del Consiglio dei Ministri Dipartimento per l'Informazione e l'Editoria, 1994), http:// www.governo.it/bioetica/pdf/16.pdf, English abstract at http://www.governo.it/bioetica/ eng/opinions/NBC_opinion_on_medically_assisted_procreation_techniques_55.pdf; for a commentary on the latter document by the NBC, see M.L. DiPietro, "Fecondazione artificiale: Sul parere del Comitato Nazionale per la Bioetica," *La rivista del clero italiano* 12 (1994): 837–853.

[71] Council of Europe, Convention for the Protection of Human Rights and Dignity of the Human Being with Regard to the Application of Biology and Medicine: Convention on Human Rights and Biomedicine, April 4, 1997, http://conventions.coe.int/Treaty/en/ Treaties/html/164.htm.

[72] On this subject, see A.G. Spagnolo, "Comitati di bioetica in tema di procreazione artificiale," *Medicina e Morale* 43, no. 1 (1993): 183–204; A. Serra, E. Sgreccia, and M.L. DiPietro, *Nuova genetica ed embriologia umana* (Milan: Vita e Pensiero, 1990), 247–270.

Bishops of Victoria (Australia) and another by the Bishops' Conference of France,[73] to cite just a few.

There is no medical journal or cultural periodical that has not dealt with the subject by now; the coverage has been so extensive in weeklies and dailies that it is no longer possible to count the number of articles. It is even more difficult to evaluate the circumspection and scientific accuracy of the popular coverage.[74]

The main ethical issues will be summarized by referring to the more significant cases.

---

[73] In particular, the document of the Catholic Bishops' Joint Committee on Bioethical Issues titled *In Vitro Fertilization: Morality and Social Policy*, March 2, 1983, was addressed to the Warnock Committee before the *Warnock Report* was published. As a critical response to the report itself, a subsequent document was addressed to the committee, titled *Comments on the Warnock Report on Human Fertilisation and Embryology* (London: Catholic Media Office, 1984), dated July 27, 1984. Moreover, the Ontario Conference of Catholic Bishops (Canada) published a document titled *Directives on Family Life* (Toronto: 1983). There is also a letter by the Bishops' Conference of the State of Victoria (Australia) addressed to the Secretariat of the IVF Department in Melbourne and, finally, a document by the French Bishops' Conference titled *Life and Death on Demand,* issued in November 1984.

[74] To limit ourselves to just a few of the many contributions on the subject, here are several that appeared in *Medicina e Morale* 33, no. 4 (1983), which was dedicated to the topic: D. Tettamanzi, "Problemi etici sulla fertilizzazione in vitro e sull'embryotransfer," 342–364; L. Leuzzi, "Deontologia medica e fecondazione in vitro," 365–390; Caffarra, "La trasmissione della vita nella *Familiaris consortio*," 391–400. See also *Atti del Convegno Nazionale su Famiglia e fecondazione artificiale*, organized by the Bioethics Center of the Catholic University of the Sacred Heart in Rome and by the magazine *La Famiglia* and published by Editrice La Scuola in Brescia, as reported in *Medicina e Morale* 43, no. 1 (1993). See also L. Mastroianni, "In Vitro Fertilization," in *Encyclopedia of Bioethics*, Reich, 1448–1451; H. Wattiaux, "Insémination artificielle, fécondation in vitro et transplantation embryonnaire: Repères éthiques," *Esprit et vie* 24 (1983): 354–364; Häring, *Free and Faithful*, 3:23–27; S. Spinsanti, "L'inseminazione artificiale: Scelte deontologiche e interrogativi etici," *Res Medicae* (March–April 1970): 134–138; P. Verspieren, "L'aventure de la fécondation in vitro," *Études* (November 1982): 479–491; R. A. McCormick, "Notes on Moral Theology," *Theological Studies* 40 (1979): 107 ff. Finally, for the sake of thorough documentation and bibliography, see also D. Tettamanzi, *Bambini fabbricati* (Casale Monferrato: Piemme, 1985); F. Giunchedi, "Considerazioni morali sulla fecondazione artificiale," *La Civiltà Cattolica* 1 (1984): 223–241; G. Tagliapietra, "Le banche del seme: il caso italiano: È necessaria subito una rigida regolamentazione," *Prospettive nel mondo* (Jan.–Feb. 1984): 95–98; P. Ramsey, *Parenthood and the Future of Man by Artificial Donor Insemination: Fabricated Man* (New Haven: Yale University Press, 1970), 104–160; E. P. Flynn, *Human Fertilization: A Catholic Perspective* (Lanham, MD: University Press America, 1984); G. Berlinguer, ed., *La procreazione assistita: Aspetti scientifici, etici e giuridici; Atti della terza Giornata Europea di Bioetica; Milano, sabato 26 novembre 1988* (Turin: LDC, 1989); E. D. Pellegrino, J. C. Harvey, and J. P. Langan, eds., *Gift of Life* (Washington, DC: Georgetown University Press, 1990); V. Mattioli, *Laboratorio umano* (Palermo: Augustinus, 1990); L. V. Gutierrez, J. V. Gutierrez, and P. M. Baza, *Reproducion asistida en la comunidad europea: Legislación y aspectos bioeticos* (Valladolid: Universidad de Valladolid, 1993); E. Sgreccia, "Fecondazione artificiale: Problemi etici," *Medicina e Morale* 43, no. 1 (1993): 183–204; M. Lombardi Ricci, *Fabbricare bambini? La questione dell'embrione tra nuova medicina e genetica* (Milan: Vita e Pensiero, 1996); I. Carrasco De Paula, "Dal dono al Vangelo della vita: Per una lettura teologica dell'*Evangelium vitae*," in *"Evangelium vitae" e bioetica*, 113–124; Spagnolo, "Tecniche di fecondazione."

EXTRACORPOREAL HOMOLOGOUS ARTIFICIAL FERTILIZATION: LOSS OF EMBRYOS AND OF CONJUGAL UNITY. The topic of homologous extracorporeal artificial fertilization is the most heavily debated, especially among Catholic moralists.

First, let us clarify a matter that is not of terminology alone: in IVF-ET, it is proper to speak of "artificial fertilization" rather than "artificial insemination." The reason is that what is artificial in artificial insemination is only the introduction—or at most the procurement and introduction—of the semen, while the moment of fertilization remains natural as to both cause and location. In IVF, however, fertilization (the union of the gametes) is controlled and achieved in vitro in an artificial way.

Proceeding to the moral evaluation, it is necessary to recall two fundamental ethical requirements of human procreation about which all Catholic moralists are in agreement, and these requirements themselves are derived from rational morality: the first is that *the life of the embryo should be safeguarded*; the second is that procreation should be the result *of the union and personal relation of legitimate spouses*.

Concerning this point, we must refer back to what was written in the chapters on the value of the human embryo (and hence the illicitness of abortion) and on the subject of sexuality and procreation. There are serious difficulties in safeguarding the life of the human embryo in IVF-ET, at least in the methods currently employed. Indeed, the procedure generally calls for the fertilization of several embryos because the rate of implantation and continued pregnancy is very low (one or two out of ten); therefore, in order to guarantee success, the technicians make sure to have several embryos available so as to repeat the attempt in case of failure. The "spare" embryos pose an ethical and legal problem: they could be eliminated or used in experiments or for the production of cosmetics or for transfer to another woman (in which case, obviously, it would no longer be homologous fertilization).[75]

From the standpoint of morality, and not just Catholic morality, the planned elimination or destruction of embryos is a procured loss of a human being, just like a voluntary termination of pregnancy. Some laboratories have been developing a process for freezing the embryo[76] so as to avoid the frequent asynchrony between the ovarian cycle and the menstrual cycle of women who have been hyperstimulated. In this case, too, the reserves and the "surplus" are clearly foreseen. As an excuse or moral justification for this destruction of spare embryos, some advocates point to the fact that many "mini" abortions occur in natural fertilization, either immediately

---

[75] A. Dyson and J. Harris, *Experiments on Embryos* (London: Routledge, 1990); G.R. Dunstan and M.J. Seller, eds., *The Status of the Human Embryo* (London: King's Fund, 1988); A.G. Spagnolo and E. Sgreccia, "Il feto umano come donatore di tessuti e di organi," *Medicina e Morale* 38, no. 6 (1988): 843–875; A. Serra, "La sperimentazione sull'embrione umano: Una nuova esigenza della scienza e della medicina?" *Medicina e Morale* 43, no. 1 (1993): 97–116; L. Walters, "Fetal Research," in *Encyclopedia of Bioethics*, Reich, 857–863; Bioethics Center of the Catholic University of the Sacred Heart, "Contro la sperimentazione sugli embrioni umani," *Medicina e Morale* 46, no. 4 (1996): 802–809; E. Sgreccia, "Interventi su embrioni e feti umani," in *Commento interdisciplinare*, Sgreccia and Lucas Lucas, 617–635.

[76] Z. Serebrovska, M.L. Di Pietro, and A. Bompiani, "Fecondazione artificiale e crioconservazione degli embrioni," *Medicina e Morale* 56, no. 1 (2006): 13–39.

after fertilization and before implantation or immediately after implantation, because of anomalies or incompatibility.[77]

It is said that if nature itself practices selection and ensures that among various embryos only those with greater vitality are implanted and grow, then it is also licit to make several attempts in the laboratory in search of the best results. The physician would be doing nothing but "mimicking" what happens in nature.

It is not difficult to see the inconsistency and speciousness of the argument. The inconsistency lies in the fact that nature is invoked to justify the loss of embryos, while it is said that one should not have a "biologistic" concept of nature when it comes to justifying artificial fertilization as a whole. But above all a distinction must be made in this context between *natural death* and *death inflicted at the hands of human beings*. If the reasoning cited above were valid, then it could be applied to other cases as well. For example, since many people die accidentally in traffic accidents, is it no longer a crime to willfully cause death by means of a staged traffic accident? Or again, since many people die a natural death, is it no longer immoral to bring about a death deliberately?

Another argument proposed to gloss over this problem is the following: the loss of embryos is a temporary phenomenon; once the techniques have improved, this risk will be reduced to the normal rate of risk that exists for every therapeutic act. Here, too, we are dealing with an argument steeped in efficientism: a technique that is responsible for many deaths is used in the meantime as a transitional experiment for developing a better technique.

Who would accept this criterion, for example, in pharmacological experimentation? For the purpose of refining and perfecting the dosage of a drug, it would certainly not be permissible to perform experiments on a man "in the meantime" if those experiments involved an 80 percent risk of death. In any case it must at least be admitted that IVF-ET, as it stands now, involves an unjustifiable loss of human embryos, in which human reason—and not just the Catholic faith—identifies the structure and value of human beings.

Responses to the other objections—that we are dealing with a potential human being who is not yet capable of social relations, and such—were provided in the chapter on procured abortion.[78]

---

[77] A response to this argument has already been given in the chapter on abortion, but for further clarification, see B. Cinque et al., "Aborto ripetuto spontaneo: Aspetti scientifici e obbligazioni morali," *Medicina e Morale* 42, no. 5 (1992): 889–910.

[78] The instruction reads, "Such deliberate destruction of human beings or their utilization for different purposes to the detriment of their integrity and life is contrary to the doctrine on procured abortion already recalled. The connection between IVF and the voluntary destruction of human embryos occurs too often. This is significant: through these procedures, with apparently contrary purposes, life and death are subjected to the decision of man, who thus sets himself up as the giver of life and death by decree. This dynamic of violence and domination may remain unnoticed by those very individuals who, in wishing to utilize this procedure, become subject to it themselves. The facts recorded and the cold logic which links them must be taken into consideration for a moral judgment on IVF and ET (in vitro fertilization and embryo transfer): the abortion-mentality which has made this procedure possible thus leads, whether one wants it or not, to man's domination over the life and death of his fellow human beings and can lead to a system of radical eugenics." CDF, *Donum vitae*, part II.

Yet there is another reason of an ethical sort that should be considered and that has already been noted with regard to artificial insemination. In vitro fertilization, even when it is homologous, separates the unitive and affective dimension of the conjugal act from the procreative and physical dimension. The conjugal life is constituted in such a way and the phenomenology (or "language," as it is often called) of human sexuality is structured in such a way that the sexual act simultaneously unites the spouses (physically and emotionally, i.e., as "persons") and opens them to the possibility of procreation. Separating the unitive component from the procreative component is tantamount to breaking the unity of love and life in the conjugal act.

In connection with this, the instruction *Donum vitae* says,

> The conjugal act by which the couple mutually express their self-gift at the same time expresses openness to the gift of life. It is an act that is inseparably corporal and spiritual. It is in their bodies and through their bodies that the spouses consummate their marriage and are able to become father and mother. In order to respect the language of their bodies and their natural generosity, the conjugal union must take place with respect for its openness to procreation; and the procreation of a person must be the fruit and the result of married love. The origin of the human being thus follows from a procreation that is "linked to the union, not only biological but also spiritual, of the parents, made one by the bond of marriage." Fertilization achieved outside the bodies of the couple remains by this very fact deprived of the meanings and the values which are expressed in the language of the body and in the union of human persons.[79]

Hence an act of procreation without bodily expression deprives that act not of the biological element (which is supplied technologically by the transfer of the gametes), but of the interpersonal communion that can be expressed completely through the body in its fullness and unity. The characteristic of conjugal love is its totality and the fullness of the gift of the two persons. Replacing the bodily gesture with technology constitutes a *reduction* of the conjugal act to the level of a technical act.

In a purely technical act, the object constructed remains ontologically dissimilar to the parent subject and the constructing subject has ownership and dominion over it. In acts of communion and communication, such as the conjugal act, the subject *expresses himself* to another subject, whose equality he respects and with whom free expression is allowed. Among these expressive acts or "languages of the body," the conjugal act is characterized by *fullness* and *totality*: reducing procreation to a technological fact means establishing a relation of dominion of the "subject/producer" over the "object/product," which impoverishes and degrades the procreative act in both the theological and anthropological senses.

It seems to me that a further, deeper consideration must be added: in the case of in vitro "fertilization," a causality that is extra and foreign to the couple is introduced. In homologous artificial insemination improperly speaking, on the other hand, the gynecologist performs a subsidiary and complementary act prior to fertilization, which remains within the will and the union of the couple. In IVF, the moment of fertilization—which culminates a process of several connected steps in the process of constituting new human being—is brought about by an operator external to the couple. This is true to such an extent that a mix-up of semen or an error or a

---

[79] Ibid., part II, no. 4.

risk could not be blamed on nature, and legal action could be brought against the biologists.[80] From the perspective of proximate cause and immediate responsibility, extracorporeal fertilization becomes extramarital fertilization, albeit partially: the components that remain marital are the genetic patrimony, the will to commission the procedure, and maternal gestation. In the case of IVF, the terms "generation" and "fertilization" become improper terms with respect to the parents: the one who actually accomplishes the *decisive phase of procreation* is an outsider, or a team of outsiders, not the couple. For this reason, which should be added to the preceding one yet is sufficient in itself, one can understand why the act of conjugal love is considered in the teachings of the Catholic Church to be the only place worthy of human procreation. Duty requires notifying the reader that, notwithstanding the position of the instruction *Donum vitae*, some Catholic moralists are still "possibilists" with regard to the so-called *simple case*, that is, the procedure of homologous IVF-ET devoid of any compromise with the practice of abortion by embryo destruction.[81]

Having made this proviso—which in fact cannot be put into practice yet—they maintain that the unity between the unitive and the procreative dimensions of homologous IVF would be assured in the same way as in homologous artificial insemination improperly speaking: unity would be provided by the "context of marital love" from which the request comes and in which the conceptus takes his place.

As already noted, however, the instruction reaffirmed that the totality of conjugal life does not suffice to assure the dignity that belongs to human procreation. The technique therefore remains morally unacceptable even in simple case IVF-ET because it deprives human procreation of its proper and connatural dignity. The instruction emphasizes that the simple case is certainly not aggravated by all the ethical negatives encountered in extramarital procreation,[82] yet homologous IVF would still be morally inadmissible even if everything were done to avoid the death of the human embryo. As stated in the encyclical *Evangelium vitae*, "Apart from the fact that [the various techniques of artificial reproduction] are morally unacceptable, since they separate procreation from the fully human context of the conjugal act, these techniques have a high rate of failure: not just failure in relation to fertilization but with regard to the subsequent development of the embryo, which is exposed to the risk of death, generally within a very short space of time."[83]

Catholic moralists are the exception, of course, since many others care less about ethical demands than about efficacy and results; not only do the latter have no ethical worries or hesitations, but they do not even stop at homologous IVF-ET.

---

[80] To prove that such ideas are not merely hypothetical, suffice it to cite the case of a woman in the United States who, after asking to be inseminated with semen deposited in a sperm bank by her dying husband, was surprised to give birth to a mulatto baby (although both spouses were white). The woman sued the sperm bank and the physician who inseminated her for damages. This episode was reported in *La Repubblica*, March 10, 1990.

[81] See, for example, J. L. Brugues, *La fécondation artificielle au crible de l'éthique chrétienne* (Paris: Fayard, 1989).

[82] CDF, *Donum vitae*, II, no. 5.

[83] John Paul II, *Evangelium vitae*, no. 14; see *Catechism of the Catholic Church*, nos. 2376–2377.

From a legal standpoint, homologous IVF-ET is permitted even by the strictest laws. It is known, however, that the legal perspective might not coincide with the moral perspective and that positive human law is many times in conflict with morality.[84]

Continuing with the legal theme, it must be added that one cannot appeal to a "right to generate" that supposedly belongs to the couple. In reality, between the spouses there is only a right to have marital relations that are open to generation; if that were not so, then every infertile marriage would be illegal or invalid. Nevertheless, as recalled in the discussion on the procreative dimension, one cannot speak of a "right to a child."

It should also be added, as several commentators have noted, that this issue points out one of the most striking contradictions in contemporary culture: an *anti-life mentality* is being propagated through abortion and contraception on the one hand, while on the other hand the right to a child is claimed at all costs and by all means through IVF-ET. In both cases, one has to wonder whether the spouses might not be considering the child an "accessory" and an "object" rather than a "subject" with his own value who should be desired and accepted for his own sake, following the route charted by the design of life and love that passes through the family.

The path of adoption, which is thought to be impractical nowadays, would be much more accessible if children were always welcomed responsibly and if there were also the possibility of "legal pre-adoption" in the case of mothers wishing to abort or who would not be able to keep the child after birth.

Alarms have also been sounding for some time now in secular circles about technologized human reproduction. Nicola Abbagnano, for example, has written, "In this field, as in many others, science and technology can and should correct and assist the natural processes but must not replace them with artificial means annulling the effects that only nature can ensure. The whole world of nature needs to be defended today against the massive contaminations and distortions caused by the abuse of technology. Man is an integral part of this world, and his life, starting at birth, is the most precious thing to be defended against all manipulation that diminishes its dignity."[85]

---

[84] Some countries have passed laws regulating artificial reproduction techniques (Sweden, Spain, Germany, United Kingdom, Austria, France, Norway, Australia, United States, etc.); others have developed recommendations and proposals (Italy, Israel, Bulgaria, etc.). On this topic, see G.B. Ascone and L. Rossi Carleo, *La procreazione artificiale: Prospettive di una regolamentazione legislativa del nostro Paese* (Naples: Edizioni Scientifiche Italiane, 1986); E. Sgreccia and M.L. Di Pietro, "Manipolazioni genetiche e procreazione artificiale: Orientamenti giuridici e considerazioni etiche," *Il Diritto di Famiglia e delle Persone* 3, no. 4 (1987): 1351–1447; Di Pietro, "Analisi comparata." See also R.A. Charo, "Reproductive Technologies: Legal and Regulatory Issues," in *Encyclopedia of Bioethics*, Reich, 2241–2248; Di Pietro and Sgreccia, *Procreazione assistita e fecondazione.* To illustrate the contrast between ethics and positive law, suffice it to observe that while recourse to techniques that substitute for the conjugal act is considered unethical and even simple-case IVF-ET is rejected from the perspective of personalist ethics, not one of the laws currently in force accepts these ideas. Only the German law requires a gradual approach to these techniques: one should have recourse to IVF—with the fertilization of only three embryos, all of which are transferred to the uterus—only if the other techniques have proved ineffective in overcoming the couple's infertility.

[85] N. Abbagnano, "Addio cicogna, ora avremo il neonato ordinato per telefono," *Corriere della Sera*, October 15, 1984. Abbagnano refers to heterologous IVF in his

Other articles appearing in the literature discuss dilemmas in connection with criteria for health care resource distribution. While these are indeed morally important concerns, they are not a priority from the personalist perspective.[86]

Franz Böckle, a Catholic theologian, has written, "We have evidently reached a point in which we can do more than we are allowed to do, and that is why we are not allowed to do everything that we can do."[87]

EXTRACORPOREAL HETEROLOGOUS ARTIFICIAL FERTILIZATION. This type of fertilization is "indicated" to overcome more complex pathologies than those for which homologous IVF-ET is considered. Indeed, a couple's infertility may be due to the inability to produce gametes (spermatozoa on the man's part and/or egg cells on the woman's part); the intervention of a donor is requested in this situation.

A *donation of egg cells* is made by a relative (for example, a sister) or friend or by an anonymous woman undergoing another IVF-ET. The main indication for donation of egg cells is female anovularity.[88]

One of the biggest difficulties in donating oocytes is preserving them; for this reason, oocytes are currently procured only when it is possible to fertilize them. However, studies have shown that it is possible to freeze egg cells, and technicians have successfully frozen oocytes that were later used to achieve pregnancies.[89] Yet the results have not been encouraging: damage to the genetic material of the oocyte caused by freezing was sufficient to discourage further research. (Many embryos subsequently obtained by the fertilization of those oocytes had genetic abnormalities and were therefore destroyed.)

Researchers have traced the causes of this genetic damage back to the fact that at the time of ovulation, the mature oocyte is in a state, *metaphase*, in which all its chromosomes are "spread out" to form the spindle of the second meiotic division. Under these conditions, the low temperatures and the substances used to freeze the cells can easily cause the destruction of the proteins to which the chromosomes are connected, causing them to disperse and resulting in serious genetic alterations.

To avoid these problems with the freezing of mature oocytes, research has been conducted into the possibility of freezing oocytes at an immature stage, *prophase I*,

---

reflections, but what he says is valid—beyond what the philosopher intended—for the homologous variety as well.

[86] G. Berlinguer, *Questioni di vita: Etica scienza e salute* (Turin: Einaudi, 1991); G. Haan and R. Van Steen, "Costs in Relation to Effects of In Vitro Fertilization," *Human Reproduction* 7, no. 7 (1992): 982–986; P. J. Neumann et al., "The Cost of a Successful Delivery with In Vitro Fertilization," *New England Journal of Medicine* 331 (1994): 239–243.

[87] F. Böckle, "Le pouvoir de l'homme sur l'homme," in *L'homme manipulé*, ed. C. Robert (Strasbourg: Cerdic, 1974), 185. The words pronounced by John Paul II according to press reports upon hearing news of Miss Brown's birth—namely, that he shared the joy of the parents—in no way signify his approval of the matter itself or of IVF. The incident is reported in J.-M. Moretti, "L'insémination artificielle," *Études* 351, no. 6 (1979): 619–629.

[88] C.L. Schmidt-Sarosi, "In Vitro Fertilization with Donor Oocytes," in *Infertility*, Keye et al., 780–787.

[89] C. Chen, "Pregnancy after Human Oocyte Cryopreservation," *Lancet* 1 (1986): 884–886; A. Dalla Serra, "Il congelamento di gameti ed embrioni," in *Attualità sulla procreazione medico assistita*, ed. C. Ragni and W. Vegetti (Rome: CIC Edizioni Internazionali, 1997), 99–106.

in which the chromosomes are less vulnerable, proceeding then (after unfreezing) to their maturation in vitro and their fertilization.[90] Researchers have sought, moreover, to improve the technique of cryopreservation through the use of cryoprotectants and the method of *vitrification*.[91] There seems to be less damage to the chromosomes of the oocytes with this method, as well as the possibility of regular IVF and subsequent development of the embryo.

The donation of spermatozoa is technically simpler, since male semen can be frozen for a long time without any problem. Semen from a donor is used in the case of aspermia or oligospermia in the man or, in connection with a pathology in the woman justifying the need for extracorporeal fertilization (e.g., blockage of the fallopian tubes), to avoid genetic disorders.

It is necessary to mention yet another pathology that extracorporeal artificial fertilization attempts to overcome: that of the uterus. In this case, a woman aspiring to motherhood may prove capable of fertilization but not gestation. What is proposed here is the use of a so-called *surrogate mother*. This introduces a further complication not only to the technical and medical problem but also to the ethical and legal problem, as will be seen in the following pages.

Another practice is that of transferring the embryos obtained from the gametes of another couple into the uterus. This is so-called *embryo donation*; the embryos can be obtained either from a previous IVF or by means of the *washing out* technique, and then used after having been frozen.[92] Washing out refers to the removal of an embryo from the uterus before implantation in order to transfer it to the uterus of the woman requesting it. Embryo donation is usually used in cases of anovularity or failure of preceding attempts at heterologous artificial insemination or IVF.

Complex legal cases have arisen alongside the complex medical cases, not the least of which is the problem of a multiple pregnancy. The practice of transferring several embryos in an attempt to increase the success rate of the technique involves— as already mentioned—the possibility that several embryos may implant and that a multiple pregnancy will begin. The selective reduction of embryos has been proposed in these cases, and it has also been legalized in the United Kingdom.[93]

The field is still wide open on the technological level, however, and it becomes difficult to follow the development of the individual techniques in all their particulars.

In intending to examine the set of ethical problems with heterologous IVF-ET and presuming the validity of the reservations expressed with regard to the homologous variety, the considerations can be summarized as follows:

---

[90] T. L. Toth et al., Fertilization and In Vitro Development of Cryopreserved Human Prophase I Oocytes," *Fertility and Sterility* 61, no. 5 (1994): 891–894; S. G. Baka et al., "Evaluation of the Spindle Apparatus of In Vitro Matured Human Oocyte following Cryopreservation," *Human Reproduction* 10, no. 7 (1995): 1816–1820.

[91] R. Fabbri et al., "Technical Aspects of Cryopreservation," *Molecular and Cellular Endocrinology* 169 (2000): 39–42; L. Kuleshova and A. Lopata, "Vitrification Can Be More Favorable than Slow Cooling," *Fertility and Sterility* 78 (2002): 449–454.

[92] P. Quinn, "Cryopreservation of Embryos and Oocytes," in *Infertility*, Keye et al., 821–840.

[93] Parliament of the United Kingdom, *Human Fertilisation and Embryology Act 1990* (London: HMSO, 1990), http://www.legislation.gov.uk/ukpga/1990/37/pdfs/ukpga _19900037_en.pdf.

- The repercussions on marital and parental unity
- The biological, psychological, and legal identity of the unborn child
- The eugenic incentive
- The consequences of involving a surrogate mother

REPERCUSSIONS ON MARITAL AND PARENTAL UNITY. It is easy to see that when there is a "donation" of the sperm or egg or both, it creates a divide between the figures of the "parents" and the figures of the spouses: being parents is not the same—or is the same only in part—as being spouses.

The unitary structure of marriage becomes compromised and fractured, as already noted in the case of heterologous artificial insemination, and furthermore fertilization usually takes place extracorporeally.

Thus there is a twofold breach of the unity of marriage: the unity between parenthood and conjugality is broken, and the union between the unitive and procreative dimensions of the conjugal act is broken.[94] In this context, "*family* represents a way of interpreting and ordering a set of relationships, regardless of any legal ties among adults or any consanguineal connections."[95]

I maintain that this new infraction against familial unity must be rejected on the ethical and rational level. Nor is the comparison with adoption valid: adoption does not entail any breach of the marital relationship. The adopted child has two more or less legitimate parents who together gave life to the child and between them had a relation of mutual self-giving. In the case of adoption, the task of bringing up the child is entrusted to another couple.

Nor can we invoke the concept of *donation* (in the sense of charity and solidarity) when speaking about gametes, as though it were a matter of donating an organ or giving blood. In the case of donating a kidney for a transplant operation or blood for a transfusion, no unity is violated, nor is life given to a new human being.

If this distinction were not presupposed, and if the concept of charity and altruism had to be extended to such procedures, then we would also have to justify every adulterous and extramarital relationship by the same principle. It really would not be worth the trouble of considering such shortcuts if they had not been brought up by the literature—not by medical or legal publications, but usually by journalism and opinion in the mass media.

The discrepancy between conjugality and parenthood becomes even more serious and complex when associated with a surrogate mother.

---

[94] "Heterologous artificial fertilization is contrary to the unity of marriage, to the dignity of the spouses, to the vocation proper to parents, and to the child's right to be conceived and brought into the world in marriage and from marriage. ... Recourse to the gametes of a third person, in order to have sperm or ovum available, constitutes a violation of the reciprocal commitment of the spouses and a grave lack in regard to that essential property of marriage which is its unity." CDF, *Donum vitae*, part II, no. 2.

[95] J. N. Edwards, "New Conceptions: Biosocial Innovations and the Family," *Journal of Marriage and the Family* 53, no. 2 (1991): 357; P. Donati, "Trasformazioni socio-culturali della famiglia e comportamenti relativi alla procreazione," *Medicina e Morale* 43, no. 1 (1993): 117–163; G. Rossi Sciumé, "Problemi sociologici emergenti nel merito del dibattito sulla procreazione assistita," *Medicina e Morale* 43, no. 1 (1993): 165–181.

THE IDENTITY OF THE UNBORN CHILD. The most grave and sometimes strange consequences are found at this level. First, the biological and social identities of the unborn child do not coincide. The fact is that every one of us has the right to know whose child he or she is. But once the origin of someone born through heterologous IVF-ET is explained to that person, it cannot help but cause difficulties in the relationship with his or her parents, one of whom is only a putative parent while the donor remains unknown.[96]

This requirement and this right are also recognized at the legal level, inasmuch as most countries that have legislated on the matter require the establishment of centralized registries (while respecting the anonymity of the donors) from which generic information can be obtained in case the child requests it. Some of those laws intend to prescribe secrecy concerning the names of donors, but this does not eliminate the difficulty; nor can the difficulties be eliminated by proposed norms concerning the choice of the donor so that, in addition to being healthy, the donor may resemble the putative parent physically and psychologically.[97] These realities are even more serious in the absence of legal guidelines.

Strange and legally intricate cases arise when clients turn not to a sperm bank but to an embryo bank, when embryos are transferred to a woman's uterus after the death of the husband donor (in a case of homologous fertilization), or after both commissioning individuals have died in a fatal accident (as has occurred), perhaps even leaving a considerable inheritance to the unborn child. This is also how the cases of children "from beyond the grave" come about: when the donor father dies and the embryos become orphans before being transferred to a woman's womb or when a widow wants a child by the semen of the deceased husband after the sperm was deliberately obtained during an illness that was diagnosed as terminal.

The possible public health consequences of heterologous IVF resulting from the freezing process and the successive phases are not yet foreseeable and will have to be ascertained over the course of time. Another possibility that has been examined in legal contexts and in discussions about embryo banks must also be mentioned. Since many ova can be fertilized with the ejaculate of just one man, and the embryos thus obtained come to be implanted in different women, and since the "paternity" of the donor is supposed to remain unknown and in any case would be difficult to trace, theoretically there could be a certain number of persons in a given population who are related by blood but unaware of their consanguinity.[98]

At this point, as already mentioned in the case of heterologous artificial insemination, marriages between blood relations may occur. This matter is important not

---

[96] "Heterologous artificial fertilization violates the rights of the child; it deprives him of his filial relationship with his parental origins and can hinder the maturing of his personal identity." CDF, *Donum vitae*, part II, no. 2. On this subject, see B. Z. Sokoloff, "Alternative Methods of Reproduction," *Clinical Pediatrics* 26, no. 1 (1987): 11–17; P. Vercellone, "Children's Rights and Artificial Procreation," *Medicine and Law* 14 (1995): 13–22; A. McWhinnie, "A Study of Parenting of IVF and DI Children," *Medicine and Law* 14 (1995): 501–508.

[97] Wattiaux, "Insémination artificielle."

[98] A. Jacquard and D. Schoevaert, "Artificial Insemination and Consanguinity," in *Human Artificial Insemination and Semen Preservation*, ed. G. David and W. S. Price (New York: Plenum Press, 1980).

only for legal reasons but also for public health, because marriage between blood relations is known to increase the risk of genetic disorders. The fact that the statistical probability of this eventuality would be low does not alter the ethical judgment.

The case of a child commissioned by a homosexual couple is aberrant with respect to the parental concept of procreation and the structure of marriage to such a degree that no further comment is necessary to demonstrate its unethical nature. For the sake of technical completeness, it must be added that for two homosexual men it would obviously be necessary to resort to a surrogate mother, whereas two homosexual women could, at least theoretically, have a child with the biological patrimony of that couple by means of a cloning procedure, as mentioned in the chapter about genetic manipulation, or by means of heterologous IVF-ET, which is a possibility currently being examined in for legislation.

THE EUGENIC INCENTIVE. It is not difficult to understand that some techniques of artificial fertilization, especially heterologous artificial insemination, involve additional aims other than the "treatment" of infertility. This refers in particular to the *eugenic* aim that has arisen with the possibility of *donor selection* through sperm banks. The goal of this selection was and is to give the couple making the request not only the desired child but also the *perfect child* or, alternatively, a child resembling the legal father as much as possible.

The problem is not new: Jacques Testart[99] recalls that this eugenic aim was proposed (though with other methods) as early as the turn of the twentieth century by a eugenics movement headed by Francis Galton, an Englishman.

In his book *La selection humaine* (1919), Charles Richet, a Nobel Prize winner in medicine, described this sort of selection by parents as *"tranquil racism"* when it was not yet being pursued through techniques of artificial reproduction: such selection, he said, "is a monument of tranquil racism, in which thinking ceaselessly drifts from incontestable biological findings to an ideology that can be described as totalitarian."[100]

Yet some geneticists, including Nobel Prize winner Hermann Müller, maintained even then that the road to *true and radical eugenics* had to pass by way of "new technical inventions consisting of artificial insemination and associated biological inventions."[101]

It is understandable that once the eugenic idea is affirmed, even if it is only pursued through donor selection (*first-level eugenics*), the procedure surpasses even the limits of "infertility treatment" and embarks on a sort of ideological aim, driven by technology, in which conjugal love loses its meaning and its interpersonal, self-giving character.

The eugenic idea was given new momentum by in vitro fertilization and the possibility of selecting egg cells before fertilization and selecting embryos using *preimplantation diagnosis* techniques later. Testart's *Le desir du gene* analyzes this

[99] J. Testart, *Le désir du gène* (Paris: Francois Bourin, 1992), 44 ff. Recall that there is also a sperm bank for Nobel Prize winners that was founded by Robert Graham.

[100] Ibid., 41.

[101] Ibid., 48–49.

problem with extraordinary clarity:[102] this is the *second level of eugenics*, which is pursued not through the selection of donors but through the selection of gametes and embryos and the subsequent elimination of embryos that do not possess the desired characteristics.[103]

Armed with techniques for selecting egg cells and embryos before implantations, negative eugenics could take aim not only at those subjects who are afflicted with hereditary illnesses but also at the carriers. It should be recalled that negative eugenics has been put into practice for more than twenty years now through the improper and selective use of postimplantation prenatal diagnosis techniques that often lead to abortion: this is another way of selectively shaping future generations. While in postimplantation prenatal diagnosis the selection constitutes an abuse and a corruption of a technique that is licit per se for other purposes, there is an intrinsic logic in extracorporeal artificial reproduction whereby, through observation and diagnosis before implantation, the desired child is selected through a technique that is already illicit per se.

In vitro fertilization ultimately offers the possibility of implementing yet a third level of eugenics, known as *positive eugenics*, through eventual recourse to genetic engineering techniques, which give man the utmost power over other human beings and against which the interested parties cannot defend themselves. Testart writes, "It is theoretically possible to obtain a genetically modified ovum by incorporating new genes (transgenesis) either into the ovum itself or right after fertilization."[104]

As a result of pharmacologically induced superovulation and multiple fertilization—in vivo and in vitro—another type of embryo selection takes place. This selection is guided by reasons that are not qualitative but quantitative: as already noted, the purpose of this *embryo reduction* or *fetal reduction* is supposedly to avoid maternal risk and the complications of a multiple pregnancy. Fetal reduction, which terminates the life of the tiny and defenseless embryos, is performed in utero in order to reduce technical errors.

SURROGATE MOTHERS. Surrogate mothers have appeared under many names and nicknames in magazines and on television. Receiving a paid commission and through the mediation of agencies, these women gestate and bring to term—on behalf of third parties—embryos that have been fertilized using the ovum and sperm of other consigning persons.[105] There have been various cases in which a mother,

---

[102] Ibid., 90 ff.

[103] See M. L. Di Pietro, A. Giuli, and A. Serra, "La diagnosi preimpianto," *Medicina e Morale* 54, no. 3 (2004): 1–33.

[104] Testart, *Le désir du gène*, 143.

[105] Some writers distinguish the child-bearing mother from the surrogate mother: in the first case the woman who is called on to offer her womb for the gestation receives the implantation of an embryo that is completely foreign to her; in the second case the pregnancy comes about with the participation of the surrogate mother through the donation of her own ovum as well, a donation made on account of a consigning mother—but these two terms are not clearly differentiated. For the purposes of ethical assessment, the instruction *Donum vitae* understands "surrogate mother" to mean both the woman who carries in pregnancy an embryo implanted in her womb and who is genetically a stranger to it and also the woman who contributes to the pregnancy by contributing her own ovum as well, which is fertilized through insemination with the sperm of a man other than her husband, whereby it is foreseen

despite having her own children, has performed this duty so as to give a child to her own infertile sister; there have also been reports of "rental mothers" who, after carrying a commissioned child, have refused to surrender the child based on a feeling that it is their own. In such a case involving a surrogate womb, it is evident that the commissioning couple is unconnected to the outside mother, who nevertheless manages to be intimately connected with the child through the close ties of biological communication during gestation. This constitutes a manipulation of the corporality of the child, who receives his genetic patrimony from two persons while receiving his blood, nourishment, and vital intrauterine communication (with psychological consequences as well) from another person: the surrogate mother. All this constitutes a series of abuses not only with respect to the marriage but also to the child, who thereby comes to be treated as an animal specimen and not as a person with the right to know his own parents and identify with them.

The instruction *Donum vitae* considers surrogate motherhood ethically unacceptable for the very same reason it cites in rejecting heterologous artificial fertilization: "It is contrary to the unity of marriage and to the dignity of the procreation of the human person."[106] Moreover it damages not only conjugal unity but also parental unity, the close relationship between parents and children: "It sets up, to the detriment of families, a division between the physical, psychological and moral elements which constitute those families."[107]

Some have tried to compare surrogate motherhood with wet-nursing and maintain that the practice is not only licit but also the expression of altruism. It seems that the difference of intensity in the relationship between surrogate mother and fetus in comparison to the relationship of wet nurse to baby needs to be underscored. Others emphasize the risk of exploiting the uterus of the woman and starting a new profession, namely, that of the "surrogate mother."

One aspect of the problem that is often downplayed is that the object of the contract and of the commercial bargain is not only the mother's womb but also and especially the baby. Indeed, if the object of the bargain in surrogacy contracts were solely gestational maternity, then the sum agreed upon ought to be handed over to the child-bearing mother from the start of her pregnancy as "security" for the baby's stay in her womb. Instead, part of the money is paid at the end of the pregnancy, after birth, and this is indicative of the fact that the contract is designed to ensure delivery of the product, and therefore is a deed of sale for a child; if the child is deformed, for example, the commissioning couple does not want it. Why pay the whole sum

---

that the newborn will be surrendered to the commissioning couple (CDF, *Donum vitae*, part II, no. 3). For a detailed analysis of the problem and of the types of surrogate motherhood, see M. L. Di Pietro, "Fecondazione artificiale e frammentazione della maternità: Considerazioni etiche e giuridiche." *La Famiglia* 154 (1992): 5–19; Di Pietro and Sgreccia, *Procreazione assistita e fecondazione*. For a survey of the debate on this topic, see G. J. Annas, "Baby M.: Babies and Justice for Sale," *Hastings Center Report* 17 (1987): 13–15; W. F. May, "Maternità surrogata e mercato: Un punto di vista," *KOS* 75 (1991): 34–38; E. Shuster, "When Genes Determine Motherhood: Problems in Gestational Surrogacy," *Human Reproduction* 7 (1992): 1029–1033; A. van Niekerk and L. van Zyl, "Commercial Surrogacy and the Commodification of Children: An Ethical Perspective," *Medicine and Law* 14 (1995): 163–170.

[106] CDF, *Donum vitae*, part II, no. 3.

[107] Ibid.

that was agreed upon for a "product" that does not satisfy the demand? Considering the ethical and social problems inherent in recourse to surrogate motherhood, many countries prohibit surrogacy contracts or consider them legally invalid.[108]

If the transaction were to take place in Italy under the current law, including the explicit prohibition in law no. 40/2004, then woman who should be registered as "mother" on the birth certificate is the one who gave birth to the child, not the woman who contributed the ovum for fertilization or who commissioned the gestation, unless the mother who gave birth explicitly requested that the child not be recognized as her own, in which case the child would be officially put up for adoption with precious little chance of being entrusted to the couple who commissioned the transaction.

In vitro fertilization and experimentation

This section will include all those experimental procedures which, through IVF, are performed with the intention of gaining knowledge about human DNA, immunological compatibility, the effectiveness of drugs, and so forth, or even for the purpose of concluding research into further combinations involving the cloning of embryonic cells or interspecies fertilization.[109]

Although these experiments have also been conducted on spare embryos and fetuses obtained from spontaneous or procured abortion (a topic that will have to be discussed separately), some regulations nevertheless provide for the possibility of "manufacturing" embryos for experimental purposes.[110] The specified limit is two weeks of development, which is the time when formation of the *embryonic primitive streak* is said to begin and the embryo is supposed to have completed the implantation phase. Other regulations, such as the German law and the law of the state of Victoria (Australia), allow experimentation until *syngamy*, i.e., until 21 to 22 hours after fertilization.[111]

The types of experiments that can be conducted on in vitro embryos during the first days of development include the study of differentiation and morphogenesis mechanisms in the human embryo, research into the practical possibility of preimplantation diagnosis of genetic disorders for the purpose of selecting only genetically healthy embryos for transfer to the uterus, the study of the "efficacy" of new abortion techniques, research into the properties of embryonic stem cells and the possibility of

---

[108] Di Pietro, "Analisi comparata."

[109] On this subject, see Dyson and Harris, *Experiments on Embryos*; Dunstan and Seller, *Status of the Human Embryo*; Spagnolo and Sgreccia, "Il feto umano"; Serra, "La sperimentazione sull'embrione"; Walters, "Fetal Research"; E. Sgreccia, "Interventi su embrioni e feti umani," in *Commento interdisciplinare*, Sgreccia and Lucas Lucas, 617–635.

[110] See the UK law on this topic, the Human Fertilization and Embryology Act 1990, which at article 3, clause 2, reads as follows: "A licence cannot authorize ... keeping or using an embryo after the appearance of the primitive streak."

[111] See Federal Republic of Germany, Embryonenschutzgesetz–ESchG, December 13,1990, http://www.gesetze-im-internet.de/bundesrecht/eschg/gesamt.pdf; State of Victoria (Australia), Infertility (Medical Procedures) Act 1984, law no. 10163, http://www.austlii.edu.au/au/legis/vic/hist_act/ipa1984311.pdf; idem, Infertility (Medical Procedures) (Amendment) Act 1987, law no. 86, http://www.austlii.edu.au/au/legis/vic/hist_act/ipa1987391.pdf. The issues related to experiments on aborted fetuses and on tissues obtained from fetuses in utero or after abortion will be also addressed in the chapter on experimentation.

manipulating them with a view to their use in transplants, and attempts at embryonic gene therapy by insertion into the genome of the embryo a gene that is supposed to prevent the manifestation of a pathological condition. Additionally, embryonic and fetal tissues are used to isolate viruses and prepare vaccines, extract hormones, prepare messenger RNA, isolate oncofetal antigens, and study the mechanisms of sexual differentiation.

What reasoning is proposed to justify these types of experimentation? Two arguments are usually provided: one is the denial of the embryo's humanity at that phase, which generally takes the form of a statement that the embryo should be considered a "potential" human person, at least until the formation of the nervous system; the other argument, which is of a scientific and "therapeutic" sort, is the claim that these experiments are necessary for the advancement of science and the treatment of illness, especially of genetic disorders that are otherwise incurable. It is said that medical science makes no progress without experiments, that animal experiments cannot always be validated, especially in matters pertaining to the genetic code, and that it is therefore necessary to resort to this type of experimentation.

In the interest of briefly explaining these two arguments while referencing ethical concepts presented in other chapters, it must first be said that the human embryo is not just potentially but actually an individual human being from the ontological and ethical viewpoint; only its development is potential, as it will continue to be even after birth. Even if there were doubts concerning the philosophical concept of personhood, there is still an ethical prohibition against terminating a life that, if allowed to continue, has within it the *real* capacity to mature as a human person, for it is on the basis of physical life that all other types of values can be realized. The obligation to defend human life becomes fundamental and takes priority over the other values, including the acquisition of new scientific knowledge. This obligation holds even for the "spare" embryos that have been "abandoned" or deemed unsuitable for transfer into a woman's genital tract.[112]

The Italian National Bioethics Committee has not expressed a unanimous opinion on this topic; one part of the committee, in the document *Identity and Status of the Human Embryo*, writes the following:

> 9.2.2 Other Committee members, having as a moral as well as legal criterion the values of the protection of health and scientific research, the meaning of which is individual as well as social, believe that we can *morally accept* experimentation for basic or applicative research on:
>
> 9.2.2.1 fresh embryos objectively judged to be unsuitable for transferral, applying to their use the conditions in force for the removal of organs from brain-dead individuals. In effect, it would be about considering these embryos as "cell donors," instead of organs;
>
> 9.2.2.2 cryopreserved embryos, as long as the term fixed for their preservation for reproductive purposes has passed, and as long as their further

---

[112] See Bioethics Center of the Catholic University of the Sacred Heart, "Contro la sperimentazione."

development is not protracted beyond the time in which, as in the case of normal development, such embryos could have been implanted.[113]

Furthermore, the therapeutic principle cannot be applied in vivo when the death of the subject on whom the experiment is performed is highly probable—or in these cases, certain.[114] The ethics of human experimentation is quite firm on this point, which cannot be replaced by reasons of convenience or utility. Only in the Nazi concentration camps were in vivo experiments accepted and undertaken in which the death of the subjects was foreseen, without their consent. Man is worth more than science and, above all, he is worth more than the aspirations of scientists: the Declaration of Helsinki is also explicit on this point regarding the regulation of human experimentation.

Aberrant forms of fertilization and gestation

Here we will mention certain aberrant forms of fertilization that have been proposed, and to some extent implemented, by unscrupulous researchers. One of these is *cloning*, which was already discussed at length in the chapter on genetics. Another possibility is interspecies fertilization and gestation. Interspecies fertilization, or *hybridization*, means the possibility of fertilization by combining human and animal gametes. Interspecies gestation, on the other hand, is the possibility of implanting embryos of one species into the uterus of animals belonging to another species.

These techniques have already implemented in animals (guinea pig–rabbit, goat–sheep, etc.) and there has been speculation about applying them to man as

---

[113] Italian National Bioethics Committee, *Identity and Status of the Human Embryo*, June 27, 1996 (Rome: Presidenza del Consiglio dei Ministri Dipartimento per l'Informazione e l'Editoria, 1996), 19, http://www.governo.it/bioetica/eng/pdf/human_embryo_19960622.pdf.

[114] Concerning experimentation on embryos, the instruction declares that "medical research must refrain from operations on live embryos, unless there is a moral certainty of not causing harm to the life or integrity of the unborn child and the mother, and on condition that the parents have given their free and informed consent to the procedure. It follows that all research, even when limited to the simple observation of the embryo, would become illicit were it to involve risk to the embryo's physical integrity or life by reason of the methods used or the effects induced. As regards experimentation, and presupposing the general distinction between experimentation for purposes which are not directly therapeutic and experimentation which is clearly therapeutic for the subject himself, in the case in point one must also distinguish between experimentation carried out on embryos which are still alive and experimentation carried out on embryos which are dead. *If the embryos are living, whether viable or not, they must be respected just like any other human person; experimentation on embryos which is not directly therapeutic is illicit.* No objective, even though noble in itself, such as a foreseeable advantage to science, to other human beings or to society, can in any way justify experimentation on living human embryos or foetuses, whether viable or not, either inside or outside the mother's womb. The informed consent ordinarily required for clinical experimentation on adults cannot be granted by the parents, who may not freely dispose of the physical integrity or life of the unborn child. Moreover, experimentation on embryos and foetuses always involves risk, and indeed in most cases it involves the certain expectation of harm to their physical integrity or even their death." CDF, *Donum vitae*, part I, no. 4. See also *Catechism of the Catholic Church*, no. 2275; Bioethics Center of the Catholic University of the Sacred Heart, "Contro la sperimentazione."

well (e.g., inseminating a female chimpanzee with human sperm) for the purpose of "producing" subhuman beings designed to perform repetitive or unpleasant tasks or for use as "reservoirs" of organs for transplantation.

The ethical judgment is clearly negative, not only with regard to the "viability" of these embryos but also in relation to the experimentation itself: human identity and the dignity of the individual and of the family would be violated in the most blatant and monstrous way. This is truly an opportunity to repeat the quotation from Chesterton: "The madman is not the man who has lost his reason. The madman is the man who has lost everything except his reason."[115]

The instruction *Donum vitae* states that "attempts or plans for fertilization between human and animal gametes and the gestation of human embryos in the uterus of animals ... are contrary to the human dignity proper to the embryo, and at the same time they are contrary to the right of every person to be conceived and to be born within marriage and from marriage."[116]

Artificial fertilization changes the meaning of human relations

Biomedicine makes possible the nonfertile exercise of sexuality and a fertility that is not mediated by sexual relations; it seems that we are headed for the reign of *artificial breeding* depicted by Aldous Huxley in his famous novel *Brave New World*.[117]

For technological man, who decides to reproduce himself by means of assisted procreation, the presence of the sexual urge is irrelevant, because the *plan* has replaced the *desire*, and this is fraught with consequences: "Indeed, desire expresses a dimension of *humanity*, which can extend it as an indefinite longing to the very ends of the earth; yet desire is irremediably absent in the coldness of the plan, confined within the very concrete limits of real feasibility. Goethe knew this well ... ; he places the following prophetic words, at once trivial and tragic, in the mouth of Homunculus, who was artificially procreated in a vial: 'Things natural find all the world scant space, / While things synthetic want a sheltered space.'"[118]

Scientific technologies place the cognitive and operational instruments at our disposal that transform generation into a planned, controlled process, but this affects our anthropological identity and our capacity for interpersonal relations. These techniques cause an involution of human generation itself, which is analogous to that of all mammals from the biological and functional viewpoint yet requires emotional and moral integration in the case of man.

It is important to clarify the ambiguous uses of the term *nature*:

If by nature is meant the biological element, then any type of fertilization is natural inasmuch as it uses biological material. But if by nature is meant that which is specific and proper to a certain act or behavior, then it is not enough to consider the biological element; a rape, for example, is not an

---

[115] G. K. Chesterton, *Orthodoxy*, 6th ed. (London: John Lane, 1919), 30.

[116] CDF, *Donum vitae*, part I, no. 6, emphasis removed.

[117] See A. Huxley, *Brave New World* (New York: Chelsea House, 2004).

[118] F. D'Agostino, "Fertilità," in *Parole di bioetica* (Turin: Giappichelli, 2004), 94, citing J. W. von Goethe, *Faust: A Tragedy*, ed. C. Hamlin, trans. W. Arndt (New York: Norton, 1976), p. 173.

artificial act, yet it is still, for someone who uses the term "nature" in the second sense, an *unnatural* act (i.e., not in conformity with the specificity of human, interpersonal sexual relations). Therefore "artificial" cannot be adopted as a synonym for "unnatural" (in the normative sense of the word). The use of eyeglasses does not make vision unnatural.[119]

Techniques of artificial fertilization are not only "artificial" but also "unnatural."

Artificial fertilization is *qualitatively different* from natural procreation, not-withstanding the fact that the two processes both lead to the birth of a baby. Some attempt to conceal the difference by verbal engineering and therefore talk about "assisted" procreation, disguising the fact that the subject who is being assisted acts as the primary subject when it is true assistance, while the assistance is limited to cooperation with that primary act.

The linguistic ambiguity is also evident when those who make their semen available for the purposes of heterologous fertilization, without any interest whatsoever in their children, are described as "donors." A "donation" or a "gift" is a personal, free, intentional act directed toward someone who stands before us, whose face we see; yet the term is used to describe making one's sperm available in a way that separates the animal-biological dimension of fertilization from the fundamental dimension of the encounter between a man and a woman in an act that not only is biological but also involves the complete personal and interpersonal dynamics of human procreation.

The reductive, mechanistic mentality conceives of the human being as a machine, as though there were neither an "intrinsic nature" to respect nor "natural ends" to consider, as if his whole being and all his actions depended on the will of the planner and on the technical skills of the craftsman, and thus anything *can* be done with these machines so long as one has the technical skill to do so.

The technological culture tends to try to dominate natural dynamics in the hope of successfully subordinating them to man's desires. According to this mentality, children are planned on the drawing board; they become less of a "gift" and more of a meticulously planned "product" that is sometimes "manufactured" quite literally through biotechnological methods. This cultural phenomenon is certainly not without consequences.

It shows no respect for the dignity of human persons to cause or allow them to enter life by being "manufactured" in a laboratory (and not in their mother's womb) in order to perhaps be transferred to a uterus. With this are significant risks that the transfer will result in failed implantation, not to mention that the less fortunate "spare embryos" can be frozen in an "ice prison" before being subjected to experimentation.

Those embryos who manage to see the light of day may not be aware of what has happened, but that does not change the reality: offending someone who cannot perceive or react to the offense is always deplorable.

The child is a *subject* and not an object, and must be immediately considered as such, even with reference to way he or she is called into existence. A context worthy of a child is one in which he or she is loved: it is the context of mutual conjugal love, which is also the impetus leading to sexual union, which in turn is the physiological prerequisite for the conception of a new human being. Only a context of love can respect the natural modalities by which this is expressed, recognizing the truth of

---

[119] A. Pessina, *Bioetica: L'uomo sperimentale* (Milan: Bruno Mondadori, 1999), 118.

conjugal love (inseparability of the unitive dimension from the procreative dimension), in which the spouses are disposed to love anyone who might be called into existence by virtue of the sexual expression of that love.[120]

Artificial fertilization clearly detaches the procreative outcome (conception) from the conjugal act (sexual relations) and inaugurates a sort of despotic rule over nature.

It is evident that in certain cases exercising dominion over the nature of one's acts poses no problems from the ethical standpoint based on the mere fact that there has been an integration or replacement of some aspects of biological nature. It is a different matter when the *natural moral law* is violated, that is, when one offends against the dignity of the person by denying his or her right to life, health, psychophysical integrity, knowledge of the truth, and free and peaceful life in society.

In human sexuality, first and foremost the dignity of the person is called into play. As the corporeal expression of the total self-gift of two human beings united by a spiritual bond of love, it does not tolerate any form of manipulation and falsification such as the dissociation of its unitive and procreative elements.

Artificial fertilization attacks the dignity of the child to be conceived or already conceived by reducing the child to an object. The event of conception (the true origin of the human adventure of the child, who is a person) takes place not in the womb of a woman who is in love but in a test tube, in the presence of persons who treat the child as an object in a relationship of mastery and dominion (as in the relationship between producer and product), to the point that they are willing to destroy it if it exhibits some "production defect."

The cold and inhuman "relational context" in which conception itself often occurs is an offense against the person of the unborn child. It is an offense inherent in the very use of procreative technologies and cannot be eliminated even if the quality of the techniques and their success rates improve.

If the conceptus is a person, he has the right to have his dignity as a person respected from the outset, beginning with the right to be generated as a person and not manufactured as an item or bred as an animal.

Adriano Pessina[121] underscores how the expression of a negative judgment on artificial fertilization does not mean underestimating the suffering that any type of renunciation involves; rather, it means appreciating what is renounced and why it must be renounced. What is being given up is not a child (indeed, the couple could adopt) but rather the plan of delegating the generation of a child to health care institutions. Some object that generation cannot be reduced to its biological component and that the parents' love is precisely what redeems this expropriation, which is remedied by their ability to fully love the child who was so desired and therefore so planned. Yet if it were not essential for these parents to generate a child physically, if the moral and affective component were enough to transform a couple into parents, then there would be no reason to resort to artificial fertilization and they would pursue the course of adoption.

---

[120] See M. Rhonheimer, *The Ethics of Procreation and the Defense of Human Life: Contraception, Artificial Fertilization, and Abortion* (Washington, DC: Catholic University of America Press, 2010).

[121] See Pessina, *Bioetica*, 124–129.

Regarding a child as a need means returning to anthropological conceptions that the West has tried to overcome through the categories of *gift*, *guest*, and *other*. These are the categories that have prevented people from seeing adult others in terms of their usefulness alone, and they have lent consistency to the thesis of equal dignity and equality among all human beings. Changing the meaning of generation therefore means altering the meaning of human relations.

What should be the fate of frozen embryos: adoption for birth?

The question about the fate of frozen embryos arises from the practice of artificial fertilization, which is intrinsically immoral, as already explained.

Even when legislation tends to limit the number of embryos and to discourage cryopreservation, the freezing of embryos can never be completely ruled out: the woman can refuse to allow implantation. Some think that these embryos should be used for research; according to others they must be thawed and allowed to die or else preserved for an indefinite period of time; finally, as the Italian National Bioethics Committee (NBC) has proposed,[122] they can be entrusted to someone who wishes to adopt them.

Adoption is favored both by advocates of artificial fertilization, who recommend the *donation* of the embryos to adopting couples simply as a means of moral validation for artificial fertilization, as well as by sincere defenders of innocent human life who see adoption as the embryos' only chance at life, however remote. Adoption for birth would give witness to the humanity of the embryo and hasten the day when that humanity is recognized legally.

In the foreword to the NBC document, Professor Francesco D'Agostino writes that "the right to be born must prevail on any ethical and legal consideration against it, although they highlight the not small problems arising from this solution."[123]

The question is being debated among Catholic moral theologians, and the Magisterium has not yet made a clear pronouncement on the matter.

Adoption for birth could justify the procedures that produce spare embryos. For these reasons, together with Professors Adriano Bompiani and Maria Luisa Di Pietro, I signed the following remark at the end of the NBC document: "With this personal remark we express our *abstention from vote* because we believe that the debate on the issue of the [adoption for birth] has not yet given sufficient elements for an adequate ethical evaluation. In addition, although it has the best intentions, which coincides with our vision and our commitment, the solution presented by the document approved in the plenary meeting appears theoretical and imperfect, and it does not fit in a context of real protection of life of all conceived embryos."[124]

Technology, which has fragmented the reproductive process and parental figures, forces us to go back to reflect on the meaning of our coming into and being in the world. The right to life must undoubtedly be emphasized, but so should the

---

[122] Italian National Bioethics Committee, *Adoption for the Birth of Cryopreserved and Residual Embryos Obtained by Medically Assisted Procreation (MAP)*, November 18, 2005 (Rome: Presidenza del Consiglio dei Ministri Dipartimento per l'Informazione e l'Editoria, 2005), http://www.governo.it/bioetica/eng/pdf/ADOPTION_18112005.pdf.

[123] Ibid., 2.

[124] Ibid., 12–13.

fact that human procreation is an integral process that binds a father, a mother, and a child into an *existential, moral, and corporeal relationship*.

This form of *gestation and birth* is proposed because artificial fertilization produces spare embryos that, as the NBC document states, would have to "die without ever being born";[125] however, it is a stretch to describe this as "adoption." Every human being has the right to be born, which involves a corresponding duty of gestation and delivery, but only on the part of the natural mother. There is no duty for third parties to respond to this right because maternal gestation cannot be carried out by any woman other than the mother. Certainly it is a noble intention to permit an abandoned embryo to be born, but one cannot help but reflect on the manner of doing so. Can there be a right for a woman to welcome the child of another mother into her own womb? These considerations aside, it seems to us that this type of adoption could not be proposed unless the further freezing of embryos had already been forbidden, so as to avoid justifying cryopreservation itself and its continuation. Moreover, this proposal could never be presented as a moral obligation given the health risks and the "unnaturalness" of such a gestation.

Artificial fertilization in ethics and law

The anthropological vision and ethical indications provided thus far involve a negative judgment on artificial fertilization techniques that do not assist the conjugal act, with greater gravity in the case of techniques involving the loss of human embryos and the presence of a third party outside the unity of the couple. Conversely, all treatments aimed at truly healing or preventing a couple's infertility are approved and recommended.

In proceeding to the judicial and legislative level while taking into account the plurality of ethical convictions, it is difficult for the personalist ethical position on artificial fertilization to be accepted in its entirety. I maintain, however, that while the rule of law cannot adopt a single school of thought as its own, neither can it renounce certain values that are foundational for the state itself. These values are identified first in those principles upon which the citizens have agreed to base their social life, that is, in the principles contained in the constitution.

In the Constitution of the Republic of Italy, for example, the principles in question can be summarized as follows: (1) *the personalist principle*, which entails the affirmation of man as an ethical value in himself and the inviolability of the right to life along with its legal protection from the beginning and respect for the dignity of the human person (arts. 2, 13, and 24; see also the incisive language of law no. 194/78 itself in art. 1); (2) *the principle of equality*, which entails the prohibition against discrimination for any motive whatsoever, including personal and social conditions (art. 3); (3) *the familial principle*, whereby the family, being "a natural society founded on marriage," is the fundamental human relational structure through which every subject acquires his or her own personal and subjective identity (arts. 29 and 31); and (4) *the principle of freedom of conscience*, which means acknowledging the

---

[125] Ibid., *Adoption for the Birth*, 4.

individual's freedom to refuse to act according to the provisions of the law when it conflicts with his conscience (art. 19).[126]

Several conclusions can be drawn from the perspective of civil law in light of these principles: heterologous artificial fertilization should be absolutely forbidden, whereas one might allow, among the techniques of artificial fertilization properly speaking, artificial insemination and homologous GIFT and IVF-ET only in cases of necessity, provided that the techniques are applied gradually (i.e., less invasive first) and that no more embryos will be fertilized than will be transferred to the uterus (no more than three, according to the Italian and German laws). Artificial fertilization should likewise be forbidden for a single woman, a homosexual couple, a woman in menopause, and a surrogate mother; for any form of experimentation on the embryo unless it is intended for the well-being and treatment of the embryo on which it is performed; and for ectogenesis, cloning, twin fission, and interspecies twin fertilization.

These principles—the defense of the embryo's life, the defense of family unity (in both the biological and emotional sense), and the defense of the identity of the conceptus—are therefore indispensable, even from a legal perspective, as also noted in the documents that the Catholic Magisterium has directed to scientific societies and governing authorities:[127]

> The intervention of the public authority must be inspired by the rational principles which regulate the relationships between civil law and moral law. The task of the civil law is to ensure the common good of people through the recognition of and the defense of fundamental rights and through the promotion of peace and of public morality. In no sphere of life can the civil law take the place of conscience or dictate norms concerning things which are outside its competence. It must sometimes tolerate, for the sake of public order, things which it cannot forbid without a greater evil resulting. However, the inalienable rights of the person must be recognized and respected by civil society and the political authority. These human rights depend neither on single individuals nor on parents; nor do they represent a concession made by society and the state: they pertain to human nature and are inherent in the person by virtue of the creative act from which the person took his or her origin. Among such fundamental rights one should mention in this regard: a) every human being's right to life and physical integrity from the moment of conception until death; b) the rights of the family and of marriage as an institution and, in this area, the child's right to be conceived, brought into the world and brought up by his parents. ... The law cannot tolerate—indeed it must expressly forbid—that human beings, even at the embryonic stage, should be treated as objects of experimentation, be mutilated or destroyed with the excuse that they are superfluous or incapable of developing normally. ... Civil law cannot legalize the donation of gametes between persons who are not legitimately united in marriage. ... [It] must also prohibit, by

---

[126] This subject was already addressed in Italian National Bioethics Committee, *La fecondazione assistita*, 171 ff

[127] John Paul II, Address to Members of the World Medical Association, October 29, 1983, in *Insegnamenti di Giovanni Paolo II* (Vatican City: Libreria Editrice Vaticana, 1983), 6.2:917–923 [Italian]; Catholic Bishops' Joint Committee on Bioethical Issues, *In Vitro Fertilisation: Morality and Public Policy* (London: Catholic Media Office, 1983).

virtue of the support which is due to the family, embryo banks, *post mortem* insemination and "surrogate motherhood."[128]

Italian guidelines on assisted fertilization, despite several lacunae, take these points into account. Law no. 40/2004[129] deals with medically assisted procreation techniques and the limits on experimentation with human embryos. The fundamental hermeneutic criterion adopted and formally enunciated by the lawmakers is the defense of "the rights of all subjects involved, including the conceptus" (art. 1, § 1).

Medically assisted procreation is understood by lawmakers solely as an instrument for solving reproductive problems due to human sterility or infertility when there are no other effective therapeutic methods to remove the causes thereof (art. 1, §§ 1–2). One may not resort to these techniques as a means of allowing subjects who carry genetically transmissible diseases to procreate healthy children through an intervention upon their genetic inheritance that is transmitted, which would essentially clear the way to eugenic practices that are explicitly forbidden by the same law (art. 13, § 3.b).

From another standpoint, through the application of the techniques provided for by the law, separation between the sexual act and procreation is allowed that results in a conflict with the natural moral law and the Magisterium of the Catholic Church. Law no. 40/2004 cannot be described as Catholic for this reason, even though many Catholic members of parliament supported it so as to "limit the harm" of a worse law, in keeping with paragraph 73 of the encyclical *Evangelium vitae*.

One mitigating aspect of the law is its defense of the conceptus, which the law expressly recognizes as a subject and not a mere object. From the moment the ovum is fertilized, a new subject is formed—the embryo—and the law protects its anticipation of life and its original genetic integrity against the will of the couple and the physician. After this point, in fact, the couple can no longer formally revoke, either jointly or singly, the consent that they have given (art. 6, § 3), thus signaling that this new biological reality is outside the couple's control and under the exclusive protection of the law. However, it is up to the woman to consent to the implantation procedure (i.e., it cannot be carried out coercively).

---

[128] CDF, *Donum vitae*, part III.

[129] See C. Casini, M. Casini, and M. L. Di Pietro, *La legge 19 febbraio 2004, n. 40 "Norme in materia di procreazione medicalmente assistita": Commentario* (Turin: Giappichelli, 2004); C. Casonato and T. E. Frosini, eds., *La fecondazione assistita nel diritto comparato* (Turin: Giappichelli, 2006); A. Bucelli, ed., *Produrre uomini: Procreazione assistita; Un'indagine multidisciplinare* (Firenze: Firenze University Press, 2006); A. Martini, *Profili giuridici della procreazione medicalmente assistita* (Naples: Editoriale Scientifica, 2006); F. A. Nastasi, *La fecondazione artificiale nella prospettiva antropologica del diritto canonico, del matrimonio e della famiglia* (Rome: Università della Santa Croce, 2005); A. Socci and C. Casini, *In difesa della vita: Legge 40, fecondazione assistita e mass media* (Casale Monferrato, AL: Piemme, 2005); R. Balduzzi, C. Cirotto, and I. Sanna, *Le mani sull'uomo: Quali frontiere per la biotecnologia?* (Fondazione Apost. Actuositatem, 2005); A. Gentilomo, A. Piga, and S. Nigrotti, *La procreazione medicalmente assistita nell'Europa dei quindici: Uno studio comparatistico* (Milan: Giuffrè, 2005); M. Fortino, ed., *La procreazione medicalmente assistita: Atti del convegno internazionale; Messina, 13–14 dicembre 2002* (Turin: Giappichelli, 2005); R. Rossano and S. Sibilla, eds., *La tutela giuridica della vita prenatale* (Turin: Giappichelli, 2005).

Only "couples who have reached the age of majority and are of different sexes, married or cohabiting, at a potentially fertile age, and both living" (art. 5) can resort to these techniques; the intervention of a gamete donor outside of the couple is expressly forbidden (prohibition of heterologous fertilization, art. 4, § 3).

Those born as a result of such techniques acquire the legal status of legitimate children or acknowledged children of the couple depending on whether the couple is married or cohabiting (art. 8). The couple is expressly forbidden to disacknowledge paternity or to maintain the anonymity of the mother in the event of heterologous fertilization in violation of the law (art. 9, § 1–2).

With respect to the applications of these techniques and in keeping with the principle of protecting the conceptus, the law first formulates a general prohibition against the cryopreservation and destruction of embryos excepting what the voluntary termination of pregnancy law provides for once the embryo has implanted in the woman's uterus (art. 14, § 1). A central provision in the logic of the law is that "they must not create a number of embryos greater than what is strictly necessary for a single and simultaneous implantation, in any case not more than three" (art. 14, § 2). The intention of the lawmakers is to avoid as much as possible the production of so-called spare embryos in excess of those destined for implantation and hence deprived of all prospects of development, also avoiding in this way their possible utilization for research purposes. The following are therefore expressly forbidden by article 13, § 3:

- The production of human embryos for purposes other than those permitted by the law

- Any form of eugenically motivated selection of embryos and gametes and any type of direct intervention to "alter the genetic inheritance of the embryo or of the gamete or to predetermine their genetic characteristics," except for interventions having diagnostic and therapeutic purposes

- Any form of human cloning

- The fertilization of a human gamete with the gametes of a different species and the production of hybrids and chimeras

The violation of each of these prohibitions is punishable by imprisonment and heavy fines and, in the case of a health care worker, by suspension from professional practice (art. 13, § 4).

Some legal inconsistencies also emerge in the text under consideration. One specifically results from the provision that the couple may be informed about the state of health of the embryo that is produced in vitro and, on the basis thereof, may subsequently refuse to implant it, which partly contradicts the declared aims of the law.

It has happened that members of parliament have found themselves discussing and voting on a proposed law on artificial fertilization that does not correspond to their personal ethical convictions.[130] We can draw some guidance on this matter from what the encyclical *Evangelium vitae* says in no. 73 about abortion and laws that put human

---

[130] See A. Rodríguez Luño, "Il parlamentare cattolico di fronte ad una legge gravemente ingiusta: Una riflessione sul n. 73 di *Evangelium vitae*," *L'Osservatore Romano*, September 6, 2002, 8; idem, "Leggi imperfette e inique," in *Lexicon: Termini ambigui e discussi su famiglia, vita e questioni etiche* , ed. Pontifical Council for the Family (Bologna: EDB, 2003), 523–527.

life at risk in response to the question as to whether one may vote for a law that, while more restrictive than a law already passed, is still permissive. The answer is in the affirmative, but under precise conditions: (1) the alternative of abrogating the existing law or preventing the introduction of a law promoting abortion is truly impossible, (2) the vote in question would have decisive weight for passing a less permissive law or modifying the current law so as to make it more restrictive, and (3) the lawmaker has clearly and publicly affirmed his opposition to any law allowing abortion:

> In a case like the one just mentioned, when it is not possible to overturn or completely abrogate a pro-abortion law, an elected official, whose absolute personal opposition to procured abortion was well known, could licitly support proposals aimed at *limiting the harm* done by such a law and at lessening its negative consequences at the level of general opinion and public morality. This does not in fact represent an illicit cooperation with an unjust law, but rather a legitimate and proper attempt to limit its evil aspects.[131]

In reality, already existing laws overlook or disregard these key points. The Norwegian and the German laws on artificial fertilization can be counted among the laws that take these principles into account at least in part.

It is not easy, however, for political authorities to resist pressure from couples who want a child "at all costs," pressure from researchers, and ideological claims. But it ought to be the state's responsibility to nudge scientific research toward the goals of true and proper treatments for infertility, the defense of human life both born and unborn, and the defense of family unity: "Scientists therefore are to be encouraged to continue their research with the aim of preventing the causes of sterility and of being able to remedy them so that sterile couples will be able to procreate in full respect for their own personal dignity and that of the child to be born."[132]

Concluding considerations

In concluding this section, we return to the remark made near the beginning about the *de-escalation* of human value as these sorts of experiments in the biomedical field continue to advance.

But another observation also needs to be added, namely, that it is necessary to connect this degradation with a model of medicine that is gaining ground in this way and has even been defended theoretically and epistemologically: I am talking about a reductionist understanding of medicine, not only as a method but also as a purpose.[133] This sort of medicine becomes reductive and ideological in the sense that it prescinds from the human person and from his ontological value, setting human biology on the same level as animal biology.

---

[131] John Paul II, *Evangelium vitae*, no. 73, emphasis added. On this subject, see E. Sgreccia, "L'Enciclica *Evangelium vitae*: Quali novità?" in *"Evangelium vitae" e bioetica*, 13–33; L Eusebi, "Corresponsabilità verso le scelte giuridiche della società pluralista e criteri di intervento sulle c.d. norme imperfette," in *"Evangelium vitae" e diritto*, ed. A. López Trujllo, J. Herranz, and E. Sgreccia (Vatican City: Libreria Editrice Vaticana, 1997), 389–406.

[132] CDF, *Donum vitae*, part II, no. 8.

[133] E. Ciaranfi, "L'evoluzione della medicina ed i problemi che ne derivano," *Federazione Medica* 35 (1982): 292–295; L. Lombardi Vallauri, "Bioetica, potere, diritto," *Jus* 31 (1984): 42.

Sex Selection or Determination

The terms *sex predetermination* and *sex selection* refer to procedures that allow parents to predetermine and choose the sex of their unborn children.[134] The issue is discussed in conjunction with IVF-ET and the manipulation of embryos because, while there is not yet a precise technique or practical experience in the matter, research is carried out primarily with these methods (there is also a "natural" method that will be seen), and this aim is pursued as well through a form of technological manipulation of genes and embryos.

Given the poorly defined situation, the brief reflections in this section are limited to the following: (a) the reasons that motivate this type of request, (b) the methods considered up to this point, and (c) the admissibility of the sex predetermination by the parents.

Motives for sex selection or sex predetermination

The motives can be summed up as follows:

• *Prevention of sex-linked disorders* (hemophilia, Parkinson's disease, etc.). This can be described as a therapeutic and preventive reason.

• *Birth control.* It is common knowledge that some couples who want to have at least one child of a different sex from that of their already-born children seek to conceive again and again until they end up with a large family, which could prove burdensome; the ability to control select the sex would help limit the number of births.

• *Experimentation.* To find out how sex is determined by the meeting of female with male gametes carrying the X or the Y chromosome, respectively, experiments must be conducted. These experiments are also necessary to determine whether concomitant factors other than the X and Y chromosomes should be considered, such as the H-Y antigen.

• Plain and simple parental *preference.*

• *Demographic correction* of a disproportion in the male-to-female ratio of a particular population.

• *Ideological bias* toward the sex that is thought to be superior.

---

[134] G. Largey, "Reproductive Technologies: Sex Selection," in *Encyclopedia of Bioethics*, Reich, 1439–1443; L. De Marinis, A. Barbarino, and A. Serra, "Biologia della differenziazione sessuale," *Medicina e Morale* 34, no. 2 (1984): 155–165; L. McSweeney, "Preselection of the Sex of Baby in Nigeria Using Billings Method," in *Report to the Sixth International Institute of the Ovulation Method* (Los Angeles: 1980); J.C. Fletcher, "Is Sex Selection Ethical?" *Progress in Clinical and Biological Research* 28 (1983): 333–334; R.B. Diasco and R.H. Glass, "Effects of pH on the Migration of X and Y Sperm," *Fertility and Sterility* 22 (1971): 303 ff; J.T. France et al., "A Prospective of the Preselection of the Sex of Offspring by Timing Intercourse Relative to Ovulation," *Fertility and Sterility* 42 (1984): 894–900; S. Harlap, "Gender of Infants Conceived on Different Days of the Menstrual Cycle," *New England Journal of Medicine* 300 (1979): 1145 ff; P. Zarutskie et al., "The Clinical Relevance of Sex Selection Techniques," *Fertility and Sterility* 6 (1989): 891–904.

Since the morality of an action is also judged by the end pursued by the acting person, these various aims must also be kept in mind in order to form an ethical judgment.

Hypothetical and experimental methods before conception

The morality of an act also depends on the method and the means employed. Methods for controlling or establishing the sex of an unborn child can be used at various times with respect to human conception.

First, control of the determining factors may be sought by intervening in a phase prior to conception.

The fact that the sex of the unborn child is due to the type of spermatozoon that fertilizes the ovum—whether it contains an X chromosome (female sperm) or a Y chromosome (male sperm)—has prompted researchers to deepen their understanding of what determines the presence of the desired chromosome and, furthermore, to control it so that the spermatozoa desired to fertilize the ovum are among the many present in a given ejaculate.

However, it is reasonably certain now that sex determination depends not only on the presence of the X or Y chromosome but also on the presence of one or more specific genes that encode the expression of an antigen known as the H-Y antigen; this, in turn, is said to interact with specific and nonspecific receptors of the cells. It will nonetheless be necessary to localize the gene that codes for this antigen along with its repressor genes and to determine the mechanism by which the antigen acts during the development of the undifferentiated gonads. It seems premature, however, to propose a genetic model for controlling the expression of the H-Y antigen.

Another research method focuses on spermatogenesis in order to ascertain which factors (biochemical, temperature, stress) could favor the maturation of male sperm and female sperm: determining those conditions could make it possible to produce the desired type of sperm in greater abundance, but it would not be easy, because there are differences among male subjects and in the same subject at different times.

Other methods that have been examined set out to control the biochemical conditions of the female genital tract which, in relation to the presence of an acidic or alkaline environment, can favor greater mobility in either the male sperm or the female sperm. It appears that several studies have yielded some results, but they are not yet conclusive.

Another hypothesis supposes that the male sperm are smaller than the female sperm, and the idea is to equip the woman with a diaphragm to filter them. This hypothesis has not been confirmed. Another hypothetical method is to use spermatozoa that have been selected by means of centrifugation, electrophoresis, and so on for IVF or artificial insemination.

More advanced studies have shown positive results with the use of the Lizuka method: the seminal fluid is centrifuged and the female sperm, unlike the male sperm, still retain the ability to fertilize the egg cell; in this way it is possible to conceive subjects of female sex alone.

Electrophoresis, on the other hand, allows separation of the male sperm from the female sperm without altering their ability to fertilize, making it possible to utilize either for the desired fertilization.

Methods proposed for after conception

These methods are more serious from an ethical standpoint and also more costly from a human perspective because they are more invasive.

One method, devised by Gardner and Edwards, is microsurgery: it consists of removing the blastocyst from the mother during the developmental period suited to sex determination, observing it to ascertain whether it is of the desired sex, then reimplanting or eliminating it accordingly. Only 20 percent of reimplantations succeed, however, and it must also be considered that the blastocysts of the opposite and undesired sex are destroyed; the trial was conducted on rabbits.

Another sex-selection technique is preimplantation diagnosis of embryos that have been fertilized in vitro or in vivo. In the latter case the embryo is removed by means of washing out, and is reintroduced into the maternal uterus only if it is of the desired sex.[135]

Another technique uses amniocentesis to determine the sex from the twelfth week on. Once the sex has been determined, the problem of selecting the desired sex remains: it is invariably done by means of abortion.

Natural methods and sex selection

It is a different matter to *predict*, and not select, the sex of offspring by means of natural conditions such as the production of cervical mucus with specific characteristics in relation to the various phases of the menstrual cycle and the time of ovulation.

According to several studies of the Billings method, which is based precisely on observation of the cervical mucus and the sensations it produces at the level of the external genitals during the menstrual cycle, it appears possible to favor the birth of a child of the desired sex. Indeed, when intercourse takes place on the days when the cervical mucus is thicker and less fluid, and therefore on the days further from the "peak" and from the time of ovulation, it is said to favor the birth of girls because the male sperm are thought to be weaker and less mobile and not as well-suited to reaching the fallopian tubes. The probability of the birth of a boy, on the other hand, is said to increase during the period in which cervical mucus is most fluid. The experiments that have been reported offer some initial statistical confirmation.[136]

---

[135] For a discussion of preimplantation diagnosis, refer to the chapter on bioethics and genetics. See also Italian National Bioethics Committee, *Diagnosi prenatali* [*Prenatal diagnoses*], July 18, 1992 (Rome: Presidenza del Consiglio dei Ministri Dipartimento per l'Informazione e l'Editoria, 1992), http://www.governo.it/bioetica/pdf/10.pdf, English abstract at http://www.governo.it/bioetica/eng/pdf/abstract_prenatal_diagnoses.pdf; I. Soussis et al., "Obstetric Outcome of Pregnancies Resulting from Embryos Biopsied for Pre-Implantation Diagnosis of Inherited Disease," *British Journal of Obstetrics and Gynaecology* 103 (1996): 784–788; A.H. Handyside, "In Vitro Fertilization and Preimplantation Genetic Diagnosis for Prevention of Inherited Disease," in *Infertility*, Keye et al., 859–867. For a review, see Di Pietro, Giuli, and Serra, "La diagnosi preimpianto."

[136] See the Nigerian study coordinated by L. McSweeney, which reports a success rate of 94 percent. The same results were obtained by Dedè with a success rate of 83.5 percent: A. Dedè, "Il metodo Billings: Marcatore della fertilità per la predeterminazione del sesso del concepito," in *La regolazione naturale della fertilità oggi: Certezze e dubbi; Congresso internazionale; Milano 9–11 dicembre 1988* (Rome: CIC Edizioni Internazionali, 1989),

Ethical assessment

There are many ethical problems involved, above all with reference to the legitimacy of predetermining the sex of offspring and therefore to the parents' motives. Furthermore, one has to wonder whether this might inaugurate a sexist ideology that could be cultivated in the social mentality or even exploited by the political authorities. Finally, one must ask about the possible social consequences of a disproportionate male-to-female ratio and the health care consequences resulting from the employment of certain selective techniques (e.g., microsurgery).

Finally, the morality of the methods varies according to their power of determination, the risks they pose for the embryo, and the extent to which they manipulate the genetic patrimony.

Analytical ethical judgments cannot be given in without definitive confirmation of the individual methods. Nevertheless, it seems that some informal ethical guidelines can be provided regarding the ends and means employed.

THE ENDS OR MOTIVES. Assuming the ethical validity of the method employed, treatment with a view to preventing sex-linked genetic disorders appears to be the most legitimate and ethically acceptable end by virtue of the therapeutic principle understood in preventive terms. At any rate it does not seem to me, even in this case, that society could ever force the parents to choose the sex that is thought to not carry the genetic anomaly. The parents are responsible for the procreative decision and for the number of children that they have; state interference would therefore be inappropriate if it ordered a limit to the number of children or abstention from conception. I maintain that the state would also be overstepping its bounds with regard to determining the sex of offspring, even for eugenic reasons. The state may propose health care education but cannot impose the selection of offspring.

As for the ideological aim, it can be affirmed today that, after the campaigns for equal dignity and equality between men and women, the sense of superiority of one sex over the other is found less often; yet in certain social contexts or in times of war, ideological pressures or justifications for the preferred sex may arise. In such a case, this ideology undermining the respect due to the human person, who has equal dignity as an individual regardless of sex, would invalidate the choice and make it detrimental to moral values. It would be even more serious if the ideology were driven by political authorities and aimed at racist domination.

Concerning the *motive of education and balanced families*, it seems reasonable to state that the simultaneous presence of children of both sexes in a family is of greater educational value and fosters better psychological equilibrium. This would clearly not constitute an illicit motive. To want a boy after the birth of one or more girls, or vice versa, seems blameless so long as the good is being measured in terms of the children, rather than in terms of the parents' tastes and desires, and provided that the couple nevertheless remains open to accepting the child that arrives, whatever the sex, without psychological rejection or rejection by abortion.

As for the *motive of regulating births*, it seems that a fair regulation of the number of children is admissible in accordance with the principle of responsible procreation, and therefore so is seeking to have offspring of both sexes. What becomes

---

485–487. See E. L. Billings and A. Westmore, *The Billings Method* (New York: Random House, 1980), 70–71.

the decisive point in this matter, however, is the absence of egotism and therefore a proportionately grave reason for limiting the number; what becomes most important of all is the choice of what means to use. Purely experimental motives are obviously ruled out when dealing with human beings, especially if risks are involved.

THE ADMISSIBILITY OF METHODS AND TECHNIQUES. This aspect of the problem seems more difficult and complex. It goes without saying that techniques presupposing the destruction of embryos or blastocysts of the "wrong" sex must be ruled out (e.g., the proposed case of employing microsurgery and selective preimplantation diagnosis). Techniques for embryonic or genetic manipulation are also to be ruled out in the absence of a strictly therapeutic purpose (as in this case), especially if there are risks to the life of the embryo.

Techniques of absolute technological determinism must be excluded. The European Council has expressly forbidden sex predetermination through biological alterations, and this prohibition is found also in the instruction *Donum vitae*, which considers such techniques "contrary to the personal dignity of the human being and his or her integrity and identity." [137] What is licit is "establishing the conditions" for a subject of the desired sex to be born when there are valid reasons, assuming that there is an openness to welcoming that subject regardless of the actual sex. Techniques that are based on respect for the embryo, for the spouses as persons, and for their sexual intercourse in its dignity, which is not just physical but also emotional and spiritual, therefore seem acceptable. The only methods that do so are the ones that take into account the appropriate biochemical environment—the physiology of the procreative act—and the ones connected with the ovulation method.

In conclusion, since the ends and means employed must be licit if these procedures, like any others, are to be morally acceptable, it can be concluded that the pursuit of offspring of a desired sex is licit (1) when natural methods are used, that is, when establishing certain favorable conditions for the conjugal act, (2) when the choice is made for therapeutic reasons or for the sake of the psychological and pedagogical equilibrium of the family and, finally, (3) when this pursuit is not imposed on the couple from outside.

From Child as Therapy to Therapeutic Cloning

On February 27, 1997, the magazine *Nature* announced the successful cloning of a female sheep named Dolly. This fact reignited the discussion on human cloning, which could broaden the prospects of IVF and allow new reproductive methods. Moreover, it introduced the possibility of new prospective therapies as a source of cells and tissues for transplantation. "It was a quick step from the child as therapy to therapeutic cloning." [138] This gave support to the "slippery slope" theory on this issue, which predicts that anything becomes possible once the axiological barrier between man and animal has been broken down.

Human cloning for therapeutic purposes, which entails the death of the generated individual, is morally much worse than human cloning in order to have children (which, it should be emphasized, offends human dignity by violating each individual's right to have his own identity). D'Agostino aptly recalls that "human life has *dignity*

---

[137] CDF, *Donum vitae*, part I, no. 6, emphasis removed.

[138] A. Pessina, *Bioetica. L'uomo sperimentale* (Milan: Bruno Mondadori, 1999), 138.

because man is the only natural subject who possesses an identity that is not reducible to his strictly biological constitution; because, *as a natural subject*, he is the bearer of a *surpassingness* with respect to his natural endowment."[139] Cloning is performed with a view to an "ideal model," and the person who is brought to life in this way suffers the violence of *alienation* because he is deprived of the right to be valued and accepted as he is in himself, as an external standard for measuring his identity is instead imposed on him.[140]

With therapeutic cloning, the biotechnology industry instead intends to achieve the generation of human individuals in order to obtain the cells necessary for future transplants (e.g., stem cells). It is not a question using a somatic cell to reproduce tissues that are destined for transplantation (skin, bone, cartilage), as with the in vitro cellular proliferation technique, but rather of cloning human beings whose development will be terminated.

Hence we cannot speak in terms of bad cloning (reproductive) and good cloning (therapeutic) understood as a source of *clumps of cells* (stem cells) taken from embryos whose development is inhibited. This way of thinking assumes that human life is a *thing*, to be utilized as a *thing*.

Stem cells

Stem cells are cells that are found in every organism, but they are distinguishable from other types of cells in that they are undifferentiated or unspecialized cells. Stem cells can reproduce in an almost unlimited manner, simultaneously giving rise to other stem cells and to cells that are *precursors* to a cell line destined for differentiation and the production of tissues and organs, such as muscles, heart, liver, bones, and so forth.

Stem cells can be *totipotent*, coming from the embryo before the blastocyst stage and able to give rise to the whole organism; *pluripotent* (or multipotent), or able to generate only some tissues; or *unipotent*, able to give rise to only one type of cell.

- *Embryonic* stem cells are found in the *inner cell mass* of the embryo, and to extract them is to necessarily destroy the embryo.

- The stem cells present in *umbilical cord* blood are attracting much attention because of their potential use in creating suitable banks of (personal) autologous cells.

- Adult stem cells (or, more precisely, *somatic* stem cells) provide for the maintenance of tissues and their eventual repair. Recent studies seem to indicate that they have a unique plasticity, whereas it was once thought that adult stem cells were capable of differentiating only into the tissue that hosts them.

As the Pontifical Academy for Life testifies, the use of embryonic stem cells is not licit for the following reasons:

1. On the basis of a complete biological analysis, the living human embryo is—from the moment of the union of the gametes—a *human subject* with

---

[139] F. D'Agostino, *Parole di bioetica* (Turin: Giappichelli, 2004), 187.
[140] See ibid., 62.

a well-defined identity, which from that point begins its own *coordinated, continuous and gradual development*, such that at no later stage can it be considered as a simple mass of cells.

2. From this it follows that as a *"human individual"* it has the *right* to its own life; and therefore every intervention which is not in favour of the embryo is an act which violates that right. ...

3. Therefore, the ablation of the inner cell mass (ICM) of the blastocyst, which critically and irremediably damages the human embryo, curtailing its development, is a *gravely immoral* act and consequently is *gravely illicit*.

4. *No end believed to be good*, such as the use of stem cells for the preparation of other differentiated cells to be used in what look to be promising therapeutic procedures, *can justify an intervention of this kind.* A good end does not make right an action which in itself is wrong.

5. ... The Church has always taught and continues to teach that the result of human procreation, from the first moment of its existence, must be guaranteed that unconditional respect which is morally due to the human being in his or her totality and unity in body and spirit: "The human being is to be respected and treated as a person from the moment of conception; and therefore from that same moment his rights as a person must be recognized, among which in the first place is the inviolable right of every innocent human being to life" (*Donum vitae*, I, no. 1).[141]

Adult stem cells are the privileged field for research, and not just for ethical reasons. Experimentation on animals has shown that stem cells, precisely by reason of their higher replicative and epigenetic potential, give rise to uncontrolled neoplastic (tumor-forming) proliferation after transplantation.[142]

Therapeutic cloning

With "therapeutic" cloning,[143] it is technically possible to create an embryonic "clone" of the patient, from which embryonic stem cells that are perfectly and genetically compatible with those of the patient can be extracted.

The procedure is as follows: the nucleus of a cell from a given subject is transferred into an enucleated human oocyte; then the embryo develops until the blastocyst stage. The cells of its inner cell mass are then used to obtain stem cells and, from them, the desired differentiated cells. This kind of cloning is described as therapeutic because the stem cells thus produced are supposed to serve for the treatment of certain pathologies. If the cloning process were called "reproductive," the embryo would be transferred to a uterus.

---

[141] Pontifical Academy for Life, *Declaration on the Production and the Scientific and Therapeutic Use of Stem Cells*, August 25, 2000, http://www.vatican.va/roman_curia/pontifical_academies/acdlife/documents/rc_pa_acdlife_doc_20000824_cellule-staminali_en.html. See J. de D. Vial Correa and E. Sgreccia, "Cellule staminali autologhe e trasferimento di nucleo," *L'Osservatore Romano*, January 5, 2001.

[142] See S. Wakitani et al., "Embryonic Stem Cells Injected into the Mouse Knee Joint Form Teratomas and Subsequently Destroy the Joint," *Rheumatology* 42 (2003): 162–165.

[143] See "Documento della Santa Sede sulla clonazione umana," *L'Osservatore Romano*, October 17, 2004.

From an ethical standpoint,[144] any form of therapeutic cloning—which necessarily entails the production and subsequent destruction of human embryos—for the purpose of obtaining stem cells is illicit.

The ANT technique

The system of altered nuclear transfer [145] (ANT) has been proposed to overcome the ethical concerns involved in the production of human embryonic stem cells. This method supposedly makes possible a sort of cloning that gives rise not to a human embryo but to an "embryo-like" mass.

Indeed, the "clone" thus produced would develop in a way that lacks the coordination, continuity, and gradualness that, from a biological viewpoint, always accompany the life of an organism such as a human being. For lack of organized and purposeful growth, the resulting group of cells would be an entity that develops in a disordered and confused way, maintaining the undifferentiated character typical of embryonic stem cells but lacking true totipotency, i.e., the capacity to form an entire organism, owing to the absence of suitable genetic programming toward that end.

The question is currently being studied by Catholic moralists as well. One would have to be sure that the ANT procedure does not give rise to an embryo; only then could it be characterized as good. However, it seems that while intending to achieve the production of nonembryonic stem cells, the technique actually produces a minuscule sick person who is programmed to die, an embryo with no possibility of continuing to live (due to artificially induced abnormalities). For this reason, the judgment can only be negative, because the procedure would constitute the use of a seriously handicapped human being. The fact that the alteration of the process is intentional and provoked and has as its purpose the production of an embryo is a burden that necessarily weighs on the ethical responsibility of the experimenter.

---

[144] R. Colombo, "La clonazione umana: Aspetti di diritto internazionale," *L'Osservatore Romano*, August 27, 2003; L. Eusebi, "La clonazione come problema giuridico," *L'Osservatore Romano*, August 30, 2003.

[145] See U.S. President's Council on Bioethics, *Alternative Sources of Pluripotent Stem Cells* (Washington, DC: Council, 2005), http://bioethics.georgetown.edu/pcbe/reports/white _paper/; Ethics and Public Policy Center, "Production of Pluripotent Stem Cells by Oocyte Assisted Reprogramming," June 20, 2005, http://www.eppc.org/publications/pubID.2374/ pub_detail.asp; I. L. Weissman, "Medicine: Politic Stem Cells," *Nature* 439 (2006): 145–147; A. Meissner and R. Jaenisch, "Generation of Nuclear Transfer-Derived Pluripotent ES Cells from Cloned Cdx2-Deficient Blastocysts," *Nature* 439 (2006): 212–215.

Summary Outline for Chapter 11

Artificial fertilization

"Artificial fertilization" or "artificial reproduction" refers to techniques used to bring about conception outside of sexual relations. It can be

- *Homologous*, if it uses gametes from the couple, or
- *Heterologous*, if at least one gamete does not come from the couple; this leads to the expedience of using semen that is always available, and hence the necessity of preserving semen (semen banks).

A further distinction can be made between artificial fertilization that is

- *Intracorporeal*, occurring within the female genital tract, or
- *Extracorporeal*, occurring outside the body, in the laboratory, in vitro

Two main methods can be emphasized:

- IVF-ET (*in vitro fertilization and embryo transfer*): fertilization occurs in vitro, where an ovum has been placed with sperm that have been treated appropriately. Here the process of fertilization is quite similar to the one that takes place in the woman's body following an act of sexual union.
- ICSI (*intracytoplasmic sperm injection*): the introduction of one sperm cell directly into the ovum. In this case the process of fertilization takes place more rapidly than it does intracorporeally.

All these techniques involve *ovarian hyperstimulation* and can lead to the production of a surplus of embryos, which may be preserved or implanted (followed almost always by so-called *embryo reduction*), possibly after embryo selection.

Italian law no. 40/2004 concerning medically assisted procreation

Italian law no. 40 of 2004 deals with medically assisted techniques for procreation and limits to experimentation on human embryos. The fundamental hermeneutic criterion adopted and formally enunciated by the lawmakers is the defense of "the rights of all subjects involved, including the conceptus" (art. 1, § 1).

Medically assisted procreation is understood by lawmakers solely as an instrument for solving reproductive problems due to human sterility or infertility when there are no other effective therapeutic methods to remove the causes thereof (art. 1, §§ 1–2). One may not resort to these techniques as a means of allowing subjects who carry genetically transmissible diseases to procreate healthy children through an intervention upon their genetic inheritance that is transmitted, which would essentially clear the way to eugenic practices that are explicitly forbidden by the same law (art. 13, § 3.b).

From another standpoint, through the application of the techniques provided for by the law, separation between the sexual act and procreation is allowed that results in a conflict with the natural moral law and the Magisterium of the Catholic Church.

One mitigating aspect of the law is its defense of the conceptus, which the law expressly recognizes as a subject and not a mere object. From the moment the ovum is fertilized, a new subject is formed—the embryo—and the law protects its anticipation of life and its original genetic integrity against the will of the couple and the physician. After this point, in fact, the couple can no longer formally revoke, either jointly or singly, the consent that they have given (art. 6, § 3), thus signaling that this new biological reality is outside of the couple's control and under the exclusive protection of the law. However, it is up to the woman to consent to the implantation procedure (i.e., it cannot be carried out coercively). Only "couples who have reached the age of majority and are of different sexes, married or cohabiting, at a potentially fertile age, and both living" (art. 5) can resort to these techniques; the intervention of a gamete donor outside the couple is expressly forbidden (prohibition of heterologous fertilization, art. 4, § 3).

Those born as a result of such techniques acquire the legal status of legitimate children or acknowledged children of the couple depending on whether the couple is married or cohabiting (art. 8). The couple is expressly forbidden to disacknowledge paternity or to maintain the anonymity of the mother in the event of heterologous fertilization in violation of the law (art. 9, § 1–2).

With respect to the applications of these techniques and in keeping with the principle of protecting the conceptus, the law first formulates a general prohibition against the cryopreservation and destruction of embryos excepting what the voluntary termination of pregnancy law provides for once the embryo has implanted in the woman's uterus (art. 14, § 1). A central provision in the logic of the law is that "they must not create a number of embryos greater than what is strictly necessary for a single and simultaneous implantation, in any case not more than three" (art. 14, § 2). The intention of the lawmakers is to avoid as much as possible the production of so-called spare embryos in excess of those destined for implantation and hence deprived of all prospects of development, also avoiding in this way their possible utilization for research purposes.

The following are therefore expressly forbidden by article 13, § 3:

- The production of human embryos for purposes other than those permitted by the law

- Any form of eugenically motivated selection of embryos and gametes and any type of direct intervention to "alter the genetic inheritance of the embryo or of the gamete or to predetermine their genetic characteristics," except for interventions having diagnostic and therapeutic purposes

- Any form of human cloning

- The fertilization of a human gamete with the gametes of a different species and the production of hybrids and chimeras

The violation of each of these prohibitions is punishable by imprisonment and heavy fines and, in the case of a health care worker, by suspension from professional practice (art. 13, § 4).

Nevertheless, some legal inconsistencies also emerge in the text under consideration. One specifically results from the provision that the couple may be informed about the state of health of the embryo that is produced in vitro and, on the basis thereof, may subsequently refuse to implant it, which partly contradicts the declared aims of the law.

Artificial fertilization

The children of technological science

Artificial fertilization is *qualitatively different* from natural procreation, notwithstanding the fact that the two processes both lead to the birth of a baby. Some advocates attempt to conceal the difference by verbal engineering and therefore talk about "assisted" procreation, disguising the fact that the subject who is being assisted acts as the primary subject when it is true assistance, while the assistance is limited to cooperation with that primary act.

The linguistic ambiguity is also evident when those who make their semen available for the purposes of heterologous fertilization, without any interest whatsoever in their children, are described as "donors." A "donation" or a "gift" is a personal, free, intentional act directed toward someone who stands before us, whose face we see; yet the term is used to describe making one's sperm available in a way that separates the animal-biological dimension of fertilization from the fundamental dimension of the encounter between a man and a woman in an act that is not only biological but also involves the complete personal and interpersonal dynamics of human procreation.

The reductive, mechanistic mentality conceives of the human being as a machine, as though there were neither an "intrinsic nature" to respect nor "natural ends" to consider, as if his whole being and all his actions depended on the will of the planner and on the technical skills of the craftsman, and thus anything *can* be done with these machines so long as one has the technical skill to do so.

The technological culture tends to try to dominate natural dynamics in the hope of successfully subordinating them to man's desires. According to this mentality, children are planned on the drawing board; they become less of a "gift" and more of a meticulously planned "product" that is sometimes "manufactured" quite literally through biotechnological methods. This cultural phenomenon is certainly not without consequences.

It shows no respect for the dignity of human persons to cause or allow them to enter into life by being "manufactured" like items in a laboratory (and not in their mother's womb), in order to perhaps be transferred in utero with significant risks that the transfer will result in failed implantation. Not to mention that the less fortunate "spare embryos" can be frozen in an "ice prison" before being subjected to experimentation. Those who manage to see the light of day may not be aware of what has happened, but that does not change the reality: offending someone who cannot perceive or react is always deplorable.

The ethical perspective

The child is a *subject* and not an object and must be immediately considered as such, even with reference to way that child is called into existence. A context worthy of a child is one in which he or she is loved: it is the context of mutual conjugal love, which is also the impetus leading to sexual union, which in turn is the physiological prerequisite for the conception of a new human being. Only a context of love can respect the natural modalities by which this is expressed, recognizing the truth of conjugal love (inseparability of the unitive dimension from the procreative dimension), in which the spouses are disposed to love anyone who might be called into existence by virtue of the sexual expression of that love.

Artificial fertilization clearly detaches the procreative outcome (conception) from the conjugal act (sexual relations) and inaugurates a sort of despotic rule over nature.

It is evident that in certain cases exercising dominion over the nature of one's acts poses no problems from the ethical standpoint based on the mere fact that there has been an integration or replacement of some aspects of biological nature. It is a different matter when the *natural moral law* is violated, that is, when one offends against the dignity of the person by denying his or her right to life, health, psychophysical integrity, knowledge of the truth, and free and peaceful life in society.

In human sexuality, first and foremost the dignity of the person is called into play. As the corporeal expression of the total self-gift of two human beings united by a spiritual bond of love, it does not tolerate any form of manipulation and falsification such as the dissociation of its unitive and procreative elements.

Artificial fertilization attacks the dignity of the child to be conceived or already conceived by reducing the child to an object. The event of conception (the true origin of the human adventure of the child, who is a person) takes place not in the womb of a woman who is in love but in a test tube, in the presence of persons who treat the child as an object in a relationship of mastery and dominion (as in the relationship between producer and product), to the point that they are willing to destroy it if it exhibits some "production defect."

The cold and inhuman "relational context" in which conception itself often occurs is an offense against the person of the unborn child. It is an offense inherent in the very use of procreative technologies and cannot be eliminated even if the quality of the techniques and their success rates improve.

If the conceptus is a person, he has the right to have his dignity as a person respected from the outset, beginning with the right to be generated as a person and not manufactured as an item or bred as an animal.

Documents on biotechnology (stem cells, fertilized ova, cloning)

PDL no. 6106—Regulations concerning the donation of fetal stem cells, umbilical cord stem cells and adult stem cells and their use for therapeutic purposes and research (September 30, 2005).

NBC, *Bioethical Considerations on the So-called "Ootide"* (July 15, 2005).

NBC, *Cellular Therapy of Huntington's Disease through the Implantation of Fetal Neurons* (May 20, 2005).

Document on stem cells by the Accademia dei Lincei (April 22, 2005).

"Biotechnology and Public Opinion in Italy 2004," in *Observa* (March 14, 2005).

Italian National Bioethics Committee (NBC) opinion on the use of embryonic cell lines for research purposes (October 27, 2000).

"Document of the Holy See on Human Cloning," *L'Osservatore Romano* (October 17, 2004).

Institutional declaration in favor of research on embryonic stem cells, signed by seven Spanish autonomous regions (Granada, January 23, 2004).

Proposal for a referendum abrogating the Italian law on assisted fertilization (December 12, 2003).

Declaration on human cloning by the InterAcademy Panel on International Issues (IAP) (September 22, 2003).

NBC opinion on the use of spare embryos (April 11, 2003).

Report on spare embryos by the EU commission for research (April 7, 2003).

Report on the proposed directive of the European Parliament and of the Council on tissues and cells of human origin (April 9, 2003).

Documentation of the Italian Society for Medical and Reproductive Sciences concerning proposed law 1514-1 (October 2002).

Manifesto "The Embryo as Patient" (February 2, 2002).

World Medication Association, Declaration of Helsinki (June 1964; October 2008).

Intervention of Tony Blair at the Royal Academy of Sciences (May 28, 2002).

Guidelines for research on embryos issued by the Canadian Institute of Health Research (March 4, 2002).

"Embryo and Research" manifesto presented by sixteen Italian scientists denying the scientific basis for equating an embryo with a human individual (March 4, 2002).

Pontifical Academy for Life, "Declaration on the Production and the Scientific and Therapeutic Use of Human Embryonic Stem Cells" (August 25, 2000).

Pontifical Academy for Life, "Prospects for Xenotransplantation: Scientific Aspects and Ethical Considerations" (September 26, 2001).

Ordinance of the Italian National Commission on Stem Cells (November 27, 2001).

Council of Europe, Convention for the protection of human rights and the dignity of the human being in view of applications of biology and medicine: Convention on human rights and biomedicine (Oviedo, April 4, 1997).

Additional Protocol to the Convention for the Protection of Human Rights and Dignity of the Human Being with regard to the Application of Biology and Medicine, on the Prohibition of Cloning Human Beings (January 1, 1998).

Manifesto of 1,150 students led by Nobel laureate Renato Dulbecco against the ban on research intended by the Italian government (February 2001).

Report of the study commission on the use of stem cells for therapeutic purposes (part 1) (December 28, 2000).

Report of the study commission on the use of stem cells for therapeutic purposes (part 2) (December 28, 2000).

NBC, *Therapeutic Use of Stem Cells* (October 27, 2000).

Resolution of the European Parliament on human cloning (September 7, 2000).

Report by Liam Donaldson [UK chief medical officer], "Stem Cell Research: Medical Progress with Responsibility" (Great Britain, June 2000).

NBC declaration on the possibility of patenting cells of human embryonic origin (February 25, 2000).

Declaration of Professor Giovanni Berlinguer, President of the NBC (February 22, 2000).

Declaration of Professor Leonardo Santi, President of the Italian National Committee for Biosecurity and Biotechnology (February 22, 2000).

Document on human and animal cloning composed by the working group on cloning (June 16, 1999).

NBC, *Cloning* (October 17, 1997).

Stem cells

Stem cells are cells that are found in every organism, but they are distinguishable from other types of cells in that they are undifferentiated or unspecialized cells. Stem cells can reproduce in an almost unlimited manner, simultaneously giving rise to other stem cells and to cells that are *precursors* to a cell line destined for differentiation and the production of tissues and organs, such as muscles, heart, liver, bones, and so forth.

Stem cells can be *totipotent*, coming from the embryo before the blastocyst stage and able to give rise to the whole organism; *pluripotent* (or multipotent), or able to generate only some tissues; and *unipotent*, able to give rise to only one type of cell.

- *Embryonic* stem cells are found in the *inner cell mass* of the embryo and to extract them is to necessarily destroy the embryo.

- The stem cells present in *umbilical cord* blood are attracting much attention because of their potential use in creating suitable banks of (personal) autologous cells.

- *Adult* stem cells (or, more precisely, *somatic* stem cells) provide for the maintenance of tissues and their eventual repair. Recent studies seem to indicate that they have a unique plasticity, whereas it was once thought that adult stem cells were capable of differentiating only into the tissue that hosts them.

As the Pontifical Academy for Life's Declaration on the Production and the Scientific and Therapeutic Use of Stem Cells (August 25, 2000) testifies, the use of embryonic stem cells is not licit for the following reasons:

1. On the basis of a complete biological analysis, the living human embryo is—from the moment of the union of the gametes—a *human subject* with a well-defined identity, which from that point begins its own *coordinated, continuous and gradual development*, such that at no later stage can it be considered as a simple mass of cells.

2. From this it follows that as a *"human individual"* it has the *right* to its own life; and therefore every intervention which is not in favour of the embryo is an act which violates that right. ...

3. Therefore, the ablation of the inner cell mass (ICM) of the blastocyst, which critically and irremediably damages the human embryo, curtailing its development, is a *gravely immoral* act and consequently is *gravely illicit.*

4. *No end believed to be good,* such as the use of stem cells for the preparation of other differentiated cells to be used in what look to be promising therapeutic procedures, *can justify an intervention of this kind.* A good end does not make right an action which in itself is wrong.

5. ... The Church has always taught and continues to teach that the result of human procreation, from the first moment of its existence, must be guaranteed that unconditional respect which is morally due to the human being in his or her totality and unity in body and spirit: "The human being is to be respected and treated as a person from the moment of conception; and therefore from that same moment his rights as a person must be recognized, among which in the first place is the inviolable right of every innocent human being to life" (*Donum vitae*, part I, no. 1).

Adult stem cells are the privileged field for research, and not just for ethical reasons.

Therapeutic cloning

With "therapeutic" cloning it is technically possible to create an embryonic "clone" of the patient, from which embryonic stem cells that are perfectly and genetically compatible with those of the patient can be extracted.

The procedure is as follows: the nucleus of a cell from a given subject is transferred into an enucleated human oocyte, then the embryo develops until the blastocyst stage. The cells of its inner cell mass are used to obtain stem cells and, from them, the desired differentiated cells. This kind of cloning is described as therapeutic because the stem cells thus produced are supposed to serve for the treatment of certain pathologies. If the cloning process were "reproductive," the embryo would be transferred to a uterus.

From an ethical standpoint, any form of therapeutic cloning—which necessarily entails the production and subsequent destruction of human embryos—for the purpose of obtaining stem cells is illicit.

The ANT technique

The system of altered nuclear transfer (ANT) has been devised to overcome the ethical concerns involved in the production of human embryonic stem cells.[1] This

---

[1] See United States President's Council on Bioethics, *Alternative Sources of Pluripotent Stem Cells* (Washington, DC: US President's Council on Bioethics, 2005), http://bioethics .georgetown.edu/pcbe/reports/white_paper/; Ethics and Public Policy Center, "Production of Pluripotent Stem Cells by Oocyte Assisted Reprogramming," June 20, 2005, http://www. eppc.org/publications/pubID.2374/pub_detail.asp; I. L. Weissman, "Medicine: Politic Stem

method supposedly makes possible a sort of cloning that gives rise not to a human embryo but to an "embryo-like" mass.

Indeed, the "clone" thus produced would develop in a way that lacks the coordination, continuity, and gradualness that, from a biological viewpoint, always accompany the life of an organism such as a human being. For lack of organized and purposeful growth, the resulting group of cells would be an entity that develops in a disordered and confused way, maintaining the undifferentiated character typical of embryonic stem cells but lacking true totipotency, i.e., the capacity to form an entire organism, due to the absence of suitable genetic programming toward that end.

The question is currently being studied by Catholic moralists as well. One would have to be sure that the ANT procedure does not give rise to an embryo; only then could it be characterized as good. However, it seems that while intending to achieve the production of nonembryonic stem cells, the technique actually produces a minuscule sick person who is programmed to die, an embryo with no possibility of continuing to live (due to artificially induced abnormalities). For this reason the judgment can only be negative, because the practice would constitute the use of a seriously handicapped human being.

Cells," *Nature* 439 (2006): 145–147; A. Meissner and R. Jaenisch, "Generation of Nuclear Transfer-Derived Pluripotent ES Cells from Cloned Cdx2-Deficient Blastocysts," *Nature* 439 (2006): 212–215.

Chapter 12

BIOETHICS AND
STERILIZATION

Historical Notes
and Various Forms of Sterilization

There have been many different reasons and applications for the acts of castration and sterilization over the course of history, but only in recent decades have they assumed such an unprecedented vastness and voluntary character.[1]

Past forms of sterilization will not be examined at length but will only be briefly mentioned. Its more recent forms, particularly contraceptive sterilization, will be examined in terms of their quantitative dimensions, the motivations behind them, and the ethical and social assessments of them.[2]

History records a form of castration with a *pseudoreligious* motivation. Origen, a famous teacher of biblical exegesis in the early Church at the school of Alexandria, despite the fact that he followed an allegorical hermeneutic, took the following passage from the Gospel too literally: "There are eunuchs who have made themselves eunuchs for the sake of the kingdom of heaven" (Mt 19:12). The medieval philosopher

---

[1] I. Riquet, *La castration* (Paris: Lethielleux, 1948); A. Gennaro, "La sterilizzazione sessuale volontaria o coatta," in *Questioni matrimoniali: IIa "Tre giorni" di teologia morale* (Turin: LICE, 1950); J. J. Speidel and R. T. Ravenholt, eds., *Atti del Workshop sulla Sterilizzazione* (San Francisco: December 4–6, 1977); Y. Grenier, "La stérilization volontaire, involontaire ou obligatoire," *Éthique* 5 (1992): 41–56.

[2] For an ethical treatment of the subject, see D. Tettamanzi, *La sterilizzazione anticoncezionale* (Brezzo di Bedero: Salcom, 1981); L. Ciccone, "La sterilizzazione antiprocreativa," in *Non uccidere: Questioni di morale nella vita fisica* (Milan: Ares, 1984), 311–360; C. Caffarra, "Il problema morale della sterilizzazione," *Medicina e Morale* 29, no. 2 (1979): 201–206; F. Mantovani, "La sterilizzazione consensuale ed irreversibile alla luce del principio personalistico," *Medicina e Morale* 29, no. 2 (1979): 191–198; G. Perico, "La sterilizzazione umana," in *A difesa della vita* (Milan: Centro Studi Sociali, 1965), 7–42; idem, "La sterilizzazione a scopo contraccettivo," in *Problemi di etica sanitaria* (Milan: Ancora, 1992), 361–372; L. Rossi, "Sterilità e sterilizzazione," in *Dizionario enciclopedico di teologia morale* (Rome: Paoline, 1981), 1055–1062; E. Sgreccia, "Sterilizzazione volontaria e mentalità contraccettiva," *La Famiglia* 75 (1979): 227–240; S. Leone, "Sterilizzazione," in *Nuovo dizionario di bioetica*, ed. S. Leone and S. Privitera (Acireale–Bologna: EDB-ISB, 1994), 945–947; D. Tettamanzi, "La sterilizzazione antiprocreativa," in *Bioetica: Difendere le frontiere della vita* (Casale Monferrato: Piemme, 1996), 273–285. Each of these works includes an extensive bibliography.

Abelard was castrated to end an amorous relationship. For a long time during the Renaissance period, Roman choirs practiced the castration of preadolescent male singers in order to have male soprano voices, until Pope Sixtus V felled the ethical validity of the motivation "laudes Domini suavius canere" (so that they might sing the Lord's praises more pleasingly) in his constitution *Cum frequenter* (1587).

History has also witnessed *penal or punitive castration* in every era, used against prisoners who were responsible for particularly serious and repeated sexual crimes; moralists have not always been in agreement as to whether this procedure is licit when the penalty is imposed by state tribunals.[3]

In recent German history, racism has made notorious the *eugenic sterilization* practiced on individuals and at the organizational level for the extinction of hereditary diseases and in pursuit of racial purification.[4] The condemnation came from Pope Pius XI in the encyclical *Casti connubii* (December 31, 1930). It can be mentioned simply here that the eugenic abortion practiced today basically obeys the same logic but is even more serious: in order to prevent the birth of a person afflicted with a serious defect or deformity, instead of an attack on the physical integrity of the parents who might transmit it, the fetus (and thus the carrier of the defect) is eliminated. It must be stated that racist Germany was not the only one to practice eugenic sterilization: nineteen nations practiced it in 1920 and twenty-five in 1930.[5] And it seems that the clandestine practice of forced sterilization of the handicapped and of ethnic minorities has never ceased: complaints of sterilizations for eugenic purposes have been filed in France and Sweden.[6]

Therapeutic or curative sterilization is practiced to this day and, when properly understood, does not raise or has never raised moral difficulties. It occurs when action must be taken *directly* to remove a tumor, a diseased organ, or the direct cause of an illness and when the sterilization is caused *indirectly*. According to the therapeutic principle or principle of totality, which states that it is licit to sacrifice a diseased part of the body to heal the entire organism, the intervention remains justified just like any other surgical intervention. It is described as *indirect* sterilization because the intention of the surgical intervention is not sterilization but the removal of the diseased organ or tissue. The so-called *indirect voluntary* principle, or principle of double effect, justifies the therapeutic action for a proportionately serious reason. The correct application of the principle of the double effect is discussed in the chapter on bioethics and abortion.

The kind of sterilization that has gained the upper hand today is *contraceptive* or *antiprocreative* sterilization in both its compulsory or coerced and voluntary forms. It is on this frontier of human manipulation that we must focus our attention at greater length.

---

[3] A. Boschi, *Questioni morali sul matrimonio* (Turin: Marietti, 1963), 479–483; Tettamanzi, *La sterilizzazione anticoncezionale*, 12.

[4] See the facts reported in R. J. Lifton, *I medici nazisti* (Milan: Rizzoli, 1988).

[5] Perico, "La sterilizzazione umana," 4.

[6] See K. Ettershank, "Report Reveals Australia's Illegal Sterilisations," *Lancet* 351, no. 9095 (1998): 44; A. Dorozynski, "France to Investigate Illegal Sterilization of Mentally Ill Patients," *British Medical Journal* 315, no. 7110 (1997): 697; C. Armstrong, "Thousands of Women Sterilized in Sweden without Consent," *British Medical Journal* 315, no. 7108 (1997): 563.

Contraceptive Sterilization in the World

The contraceptive push was very powerful and exploded in the 1960s and 1970s, and it still continues today throughout the world.

The governments of developing nations got involved after the sixties, frightened by the so-called population explosion, and were supported in this policy by the interventions of developed nations, which went so far as to make economic aid conditional on a drastic reduction in the number of births. The case of India during the first presidency of Indira Gandhi is well known. Concerning Mrs. Gandhi's population policies in India, Emérentienne de Lagrange and colleagues state, "This strategy was followed effectively by the Indian government: it moved from methods of positive encouragement to repressive measures, and then, after proclaiming a so-called *state of emergency*, a law on compulsory sterilization was promulgated and applied on a large scale, with 7.5 million persons sterilized in less than ten months; this is one of the main reasons for the condemnation of Mrs. Gandhi's government."[7]

After the seventies, the strategy focused on persuasive power to promote voluntary sterilization for both men and women. The efficacy of this sort of contraceptive method, the harmful and less effective results of other chemical, mechanical, and hormonal contraceptives, the dangerousness of repeated abortions, and above all the propaganda of ideological and heavily financed political movements led to a significant expansion of voluntary sterilization. Associations (complete with emblems and badges) and movements were established to support the practice and legalization of sterilization, especially in America.

From the legal perspective, sterilization can be performed without legal impediments in Eastern Europe[8] and in developed nations in general. In some countries, like England, the procedure is even free of charge.

In Sweden between 1935 and 1996, around 230,000 handicapped and "asocial" individuals were sterilized, 90 percent of them women, as a result of a law proposed in 1922 by the Social Democrats for "humanitarian reasons"; it remained in force until 1976. Sterilization laws were also enforced in Norway (1934) and Denmark (1929) for economic reasons: there were 40,000 sterilizations in Norway and 6,000 in Denmark. In Switzerland compulsory sterilization was approved by law in 1928 and remained in force until 1970. In Finland, too, compulsory sterilization was performed on 11,000 persons.

In France, at least 15,000 women afflicted with mental illness were sterilized, while in Austria, in the absence of pertinent regulations, 70 percent of mentally handicapped women are currently sterilized; in Spain, furthermore, the Constitutional Court itself allowed for compulsory sterilization of the mentally ill in 1994. In Italy, from 1985 to the present, at least 6,000 disabled persons have been sterilized, given that between 400 and 600 sterilization operations have been performed annually on patients in psychiatric hospitals since the Basaglia reform (law no. 180/1978).

In Asian countries, sterilization is still prohibited by law, but the political mentality is changing so much that the practice is quite widespread in some nations, such

---

[7] E. de Lagrange, M. M. de Lagrange, and R. Bel, *Il complotto contro la vita* (Milan: Ares, 1979), 80–81.

[8] See A. Romano, *Sterilizzazione umana e legalità costituzionale* (Naples: Edizioni Scientifiche Italiane, 2000), 16–18.

as China, and is carried out coercively in some, such as Bangladesh. According to 1994 estimates, the number of women sterilized—out of the female population—is said to be 39 percent in the Dominican Republic, 35 percent in Korea, 34 percent in China, 31 percent in Canada, and 23 percent in the United States. As for male sterilization, the number of sterilized men is said to be 16 percent in the United Kingdom, 13 percent in the United States, 13 percent in Canada, 10 percent in Australia, 12 percent in Korea, and 10 percent in China.[9]

After the fall of the dictatorships in central and eastern European nations, the practice of forced sterilization in the female Roma population and also in the former Czechoslovakia was revealed to the international community.[10]

Female sterilization is predominant almost everywhere at present, but male vasectomy is becoming more widespread as well.

From the quantitative perspective, it is not easy to obtain up-to-date, definitive statistics; the World Health Organization declared, however, that sterilization has become one of the chief methods of fertility control in the world.[11]

Some data are significant and shocking because of the trends they illustrate. In the United States, 942,000 cases of voluntary sterilization were counted in 1970; by 1975, the Association for Voluntary Sterilization estimated the number at 8,026,000, of which 4,499,000 were on men and 3,527,000 on women.[12]

At the beginning of 1976, there were 65 million sterilized persons worldwide, distributed as follows: 30 million in China, 17 million in India (by procedures that were for the most past coercive), 2 million in other Asian countries, 8 million in the United States, 4.5 million in Europe, 2 million in Latin America, 1 million in Canada, and 500,000 in Africa. In 1980, according to various authors, there were 102.5 million sterilized persons in the world: 42 million in China, 27 million in India, 12.5 million in the United States, 7 million in Europe, 2 million in Latin America and 1 million in Africa. These statistics were confirmed by studies published in the mid-1980s, which claimed that sterilization was used by 30 percent of couples who resorted to "contraception" during this period, involving at least 99 million couples worldwide.[13]

The report of the United Nations shows, moreover, that the countries with the highest rates of sterilization are the Republic of Korea, Puerto Rico, and Canada.

---

[9] United Nations Department for Economic and Social Information and Policy Analysis, Population Division, *World Contraceptive Use* (New York: United Nations, 1994).

[10] K. Krosnar, "Report Says 100 Roma Women Have Been Forcibly Sterilised in Slovakia," *British Medical Journal* 326 (2003): 302; E. Holt, "Roma Women Reveal That Forced Sterilisation Remains," *Lancet* 365 (2005): 927–928.

[11] See World Health Organization Special Programme of Research Development and Research Training in Human Reproduction, 1977.

[12] "$100 Million over 5 Years Will Bring Sterilization to 80–100 Million Couples in Developing Lands," *International Family Planning Digest* 2, no. 1 (1976): 3–5; R. T. Ravenholt, "World Epidemiology and Potential Fertility Impact of Voluntary Sterilization," presented at the Third International Conference on Voluntary Sterilization, February 1, 1976. See Tettamanzi, *La sterilizzazione anticoncezionale*, 15.

[13] United Nations, *World Population Trends and Policies* (New York: United Nations, 1988), 80 ff.

The Situation in Italy

Over the course of the years and of legislative sessions, various bills have been presented to the Italian Parliament for the purpose of depenalizing or legalizing sterilization in Italy. We cite as an example the Del Pennino–Agnelli Bill from 1979 and the one introduced by Testa in 1986. The first, introduced several times as the assemblies reconvened, contained the following language in article 1: "Within the context of conscientious and responsible procreation it is acknowledged that everyone shall have the option to avoid procreation through apt sterilization procedures performed on one's own person." For the moment, let us forgo commentary on the words "conscientious and responsible procreation."

The second bill essentially comprised only two articles: the first established that one can legitimately regulate one's own procreative capacity; the second specified that acts performed on consenting persons and aimed at permanently impeding the ability to procreate are not punishable. Both bills took their cue from the well-known fact that article 22 of law no. 194/1978 on the Social Protection of Motherhood and the Voluntary Termination of Pregnancy entirely abrogated Title X of book 2 of the penal code, including article 552, which punished "procured impotence to procreate." On the other hand, according to the same proponents, one cannot invoke article 5 of the civil code, which declares one's own body to be nondisposable, or article 583 of the penal code, which punishes injuries and impairments of physical integrity (as the Hon. Orsini recalled in an interpellation on September 28, 1978); instead it references a fundamental article of the Italian constitution, article 32, which is said to leave any personal option completely open in matters of health.

The Del Pennino–Agnelli Bill did not conceal the desire to adapt to "changes that have occurred in society and in customs and to the need for increasingly widespread use to modern contraceptive techniques so as to guarantee deliberate and responsible procreation." The legislators were even glad that this practice was increasing in Italy; not only that, but they also declared that given the ideological conflict dividing the world, it was comforting to note that voluntary sterilization had become widespread in both communist and capitalist countries.[14]

In the same report asserted that at the moment male sterilization was predominant, with 540 cases of vasectomy publicly reported by ASSTER, AIECS, ANCED and AIED,[15] as opposed to seventy cases of female sterilization in two months (September–October 1978), which were added to the 5,000 men who had undergone vasectomy in Switzerland and England while the procedure was illegal

---

[14] United Nations, *World Population Trends and Policies.*

[15] The association that offered its services specifically to promote and support sterilization in Italy was the Association for Male and Female Sterilization (Associazione per la Sterilizzazione Volontaria Maschile e Femminile, ASSTER), headquartered in Milan, which proposed to increase the number of sterilized couples to 350,000 by 1983. The other associations mentioned—Associazione Italiane per l'Educazione Contraccettiva Sessuale (AIECS), Associazione Nazionale Consultori Educazione Demografica (ANCED), and Associazione Italiana per l'Educazione Demografica (AIED)—have the additional purpose of promoting contraception and abortion, as is well known.

in Italy. Female sterilization was said to be sure to increase gradually as concerns about hormonal contraception increased.[16]

Other political groups have prepared and continue to promote bills that would restrict the scope of more permissive proposals, and parliamentary debate is ongoing.

An important fact that makes it even more urgent to fill the legislative gap left open by Italian law no. 194/1978 was a ruling by the Supreme Court of Cassation on June 18, 1987. The Court of Appeals of Florence had condemned a surgeon for having performed vasectomies on around fifty men with their consent, causing the irreversible loss of their ability to procreate, but this ruling was overturned "because the deed did not constitute a crime."[17] The Supreme Court of Cassation considered the surgeon's reason for appealing well founded, since law no. 194/1978 effected a true *abolitio criminis* (decriminalization), such that the procedure had to be considered licit in the case of the sterilization of a consenting party. The question about the nondisposability of one's own body raised by article 5 of the civil code was irrelevant.

According to the Court of Cassation, the generic prohibition against disposing of one's own body contained in the abovementioned article—a prohibition that would pertain to commercial acts of disposal (e.g., selling body parts for transplantation), among others—does not rule out the possibility of lawmakers deviating from it, as they did in abrogating article 552 of the penal code, with regard to "certain acts" that order procreative function.

Again according to the Supreme Court of Cassation, such a repeal would not conflict with article 32 of the constitution, which safeguards physical and mental health as a fundamental right of the individual and a matter of public interest. In fact, that very norm would instead be the key for interpreting the same article 5 of the civil code.

The Supreme Court of Cassation therefore determined that within the sphere of safeguarding health—even mental health alone—sterilization would contribute "to a greater ease and serenity in relations with one's own spouse or partner" for socially relevant or even indifferent purposes. In conclusion, according to the Supreme Court of Cassation, the present legal system does not rule out purely hedonistic sterilization.[18]

The Italian National Bioethics Committee (NBC) addressed the issue of nonvoluntary sterilization only in its document titled *Il problema bioetica della sterilizzazione non volontaria* (*The Bioethical Issue of Non-Voluntary Sterilization*), published on November 20, 1998. The conclusion of this document asserts that "in the present state of things, a broad interpretation of the law in force cannot include coerced sterilization of mentally handicapped individuals by third parties."[19]

---

[16] United Nations, *World Population Trends and Policies*.

[17] Italian Court of Cassation—V Sezione Penale, *Conciani v. Corte d'Appello di Firenze*, decision no. 438 on March 18, 1987. On this subject see also C. Calcagni and E. Marziano, *Ostetricia e ginecologia forensi* (Rome: Società Editrice Universo, 1996), 375–380.

[18] For a full and detailed review of the legal doctrine and penal provisions concerning human sterilization, see V. Scordamaglia, "La rilevanza penale della sterilizzazione umana," in *La sterilizzazione come problema biogiuridico*, ed. F. D'Agostino (Turin: G. Giappichelli Editore, 2002), 25–94.

[19] Italian National Bioethics Committee, *Il problema bioetica della sterilizzazione non volontaria* [*The bioethical issue of nonvoluntary sterilization*] (Rome: Presidenza del Consiglio dei Ministri Dipartimento per l'Informazione e l'Editoria, 1998), 31, http://www

The fact remains that one thing making the spread of sterilization more attractive is the refinement of the techniques; people speak about "reversible" sterilization and "recanalization," whether to reassure spouses with regard to second thoughts they may later have or to prevent conflicts with articles 2 and 32 of the Italian constitution, which protect the integrity of the human person.

The Techniques

In this regard, it seems opportune to describe some technical concepts representing the concrete possibilities of sterilization procedures.[20]

There are various methods of female sterilization, but almost all of them act upon the fallopian tubes. The modalities of tubal sterilization have changed quickly over the years: indeed, they have gone from a transabdominal approach by laparotomy, minilaparotomy, or laparoscopy to a vaginal approach through the posterior vaginal fornix or a transcervical or hysteroscopic approach.

The methods utilized to bring about tubal occlusion include excision and suture of the tubes, electrocauterization, the use of clips or rings, the use of occlusive intratubal instruments (made of silicone, ceramics, nylon, or polyethylene), or chemical substances that produce scar tissue (based on phenol, atabrine, tetracycline, or quinacrine).[21] Some of these techniques are still at the experimental stage.

Sterilization through surgical intervention on the fallopian tubes is euphemistically called "uterine isolation" as well since, by manipulating the passage of the female gametes (oocytes) from the ovary to the uterus, it is said to isolate the latter from the gonads.

The more radical and irreversible techniques such as oophorectomy (the removal of the ovaries) and hysterectomy (removal of the uterus) are performed for therapeutic purposes (treating tumors or hemorrhages) and not for simple contraceptive purposes.

The methods for sterilizing a man can, for the most part, be reduced to surgical or mechanical vasectomy, that is, cutting and tying the vas deferens in the first case or inserting little polyurethane stoppers in said ducts in the second case.[22]

.governo.it/bioetica/pdf/35.pdf, English abstract at http://www.governo.it/bioetica/eng/pdf/abstract_the_bioethical_issue.pdf.

[20] R. M. Richart and D. J. Prager, eds. *Human Sterilization* (Springfield, IL: C.C. Thomas, 1972); M. Policar, "Fertility Control: Medical Aspects," in *Encyclopedia of Bioethics*, ed. S. G. Post, 3rd ed. (New York: Macmillan Reference USA, 2004), 891–901; G. B. Candiani, V. Danesino, and A. Gastaldi, eds., *La clinica ostetrica e ginecologica* (Milan: Masson, 1996), 2:1158–1159, 1162.

[21] E. W. Wilson, "The Evolution of Methods for Female Sterilization," supplement 1, *International Journal of Gynecology and Obstetrics* 51 (1995): S3–S13; D. C. Sokal, J. Zipper, and T. King, "Transcervical Quinacrine Sterilization: Clinical Experience," supplement 1, *International Journal of Gynecology and Obstetrics* 51 (1995): S57–S59; R. Kulier et al., "Minilaparatomy and Endoscopic Techniques for Tubal Sterilization," *Cochrane Database of Systematic Reviews* 3 (2004): CD001328.

[22] F. H. Comhaire, "Male Contraception: Hormonal, Mechanical and Other," *Human Reproduction* 9, no. 4 (1994): 586–590; P. Romero Perez et al., "Vasectomy: Study of 300 Interventions: Review of the National Literature and of Its Complications," *Actas Urológicas Españolas* 28, no. 3 (2004): 175–214.

As for reversibility, this varies according to the sterilization technique. In men, the success rate for recanalization of the vas deferens by microsurgery varies from 40 to 70 percent;[23] for women, since the application of microsurgical techniques, it seems that full-term pregnancies can be achieved in more than 60 percent of cases. These results depend more on the remaining length of the tube than on the site of anastomosis. It should always be considered that some of these techniques require special equipment and highly trained technicians that few hospitals may have at their disposal. Moreover, the reversal procedure is not without complications.[24]

Another point to consider, which is usually not mentioned in the favorable publicity for these procedures, just as in the early years of widespread contraception, is the possibility of complications.[25] Researchers have noted, for example, an increased risk of ectopic pregnancy after sterilization. Indeed, one-third of pregnancies occurring after tubal sterilization (0.1–0.8) are ectopic and hence make it impossible for the embryo to survive without exposing the mother to a lethal danger.[26]

The risk of complications also seems significant in male sterilization: granulomas, fistulas, hematomas, and infections can occur in as many as 5 to 6 percent of cases. The most serious complication, however, which is still being studied, is the onset of an immunological reaction leading to the production of antibodies capable of causing permanent infertility even after the recanalization of the vas deferens[27] (sperm-agglutinating antibodies in 50–60 percent of the cases, sperm-immobilizing antibodies in 20–40 percent of the cases). The more recent debate about the complications of vasectomy centers on the possible increased incidence of prostate tumors and the emergence of certain forms of testicular cancer from a latent, precancerous stage.[28]

---

[23] W. F. Hendry, "Vasectomy and Vasectomy Reversal," *British Journal of Urology* 73 (1994): 337–344; C. P. Pavlovich and P. N. Schlegel, "Fertility Options after Vasectomy: A Cost-Effectiveness Analysis," *Fertility and Sterility* 67, no. 1 (1997): 133–141.

[24] E. Hardy et al., "Risk Factors for Tubal Sterilization Regret, Detectable Before Surgery," *Contraception* 54, no. 3 (1996): 159–162.

[25] H. B. Peterson and L. S. Wilcox, "Female Sterilization," in *Reproductive Health Care for Women and Babies*, ed. B. P. Sachs, R. Beard, E. Papiernik, and C. Russel (New York: Oxford, 1995), 161–169; V. Phupong, S. Taneepanichskul, and T. Rungruxsirivorn, "Bilateral Tubal Pregnancies After Tubal Sterilization in a Human Immunodeficiency Virus Seropositive Woman," *Journal of the Medical Association of Thailand* 85, no. 11 (2002): 1236–1239.

[26] H. B. Peterson et al., "The Risk of Ectopic Pregnancy after Tubal Sterilization: U. S. Collaborative Review of Sterilization Working Group," *New England Journal of Medicine* 336, no. 11 (1997): 762–767; T. Tulandi, "Tubal Sterilization," *New England Journal of Medicine* 336, no. 11 (1997): 796–797.

[27] I. P. Sardon, "La stérilisation dans le monde: Aperçus médicaux et legislatifs," *Population* (March–April 1977): 415; Hendry, "Vasectomy"; C. C. Coddington et al., "Hemizona Assay: Evaluation of Fertility Potential in Patients with Vasectomy Reversal," *Archives of Andrology* 38 (1997): 143–150; N. S. Awsare et al., "Complications of Vasectomy," *Annals of The Royal College of Surgeons of England* 87, no. 6 (2005): 406–410.

[28] S. Rohrmann et al., "Association of Vasectomy and Prostate Cancer among Men in a Maryland Cohort," *Cancer Causes Control* 16, no. 10 (2005): 1189–1194; N. Jorgensen et al., "Testicular Cancer after Vasectomy: Origin from Carcinoma In Situ of the Testis," *European Journal of Cancer* 29A, no. 7 (1993): 1062–1064; E. Lynge, L. B. Knudsen, and H. Muller, "Vasectomy and Testicular Cancer: Epidemiological Evidence of Association," *European Journal of Cancer* 29A, no. 7 (1993): 1064–1066; E. Giovannucci et al., "A Retrospective

According to studies of various degrees of significance, as more extensive research is awaited, these complications in both women and men have psychological and psychosexual consequences.[29]

It appears it can be said, however, that the very choice of sterilization, induced to a greater or lesser extent by external pressures or increased anxieties, is already a symptom and contributing factor in a self-destructive dynamic. Psychological complications are inevitable to the extent that the impairment and the uncertainty about its temporary character are felt and experienced both by the subject personally and in the relationship of the couple. Such psychological repercussions will be all the more profound to the extent that functional repercussions or disturbances arise in connection with at least some techniques.

From this brief presentation on the state of affairs in Italy, it seems that several facts can be underscored:

a. The motivation prompting people today to practice voluntary sterilization and to propose laws concerning it is clearly a contraceptive one: society has moved from mechanical contraception to increasingly abortifacient forms of hormonal contraception, and then on to surgical contraception or sterilization, in its search for a more radical and more reliable technical means. One factor in the growing trend toward sterilization has certainly been the widespread awareness, by now well documented, of the harms and pathologies caused by hormonal contraceptives; the legal opportunity arose, however, with the repeal of article 552 of the Italian penal code.

    What is striking is the fact that this contraceptive mentality is called "the pursuit of conscientious and responsible procreation," when it is obvious that the sterilized person is deprived of the ability to procreate responsibly, i.e., to choose between procreating and not procreating at every moment of his or her fertile life. This is sophistry resembling the hypocrisy with which Law no. 194/1978 refers to the liberalization of abortion as "safeguarding maternity." The reasoning adopted by some that voluntary sterilization would limit the spread of voluntary abortion and thus constitute a lesser evil is ethically untenable, because there are other licit ways to avoid abortion and because the lesser evil cannot be chosen voluntarily, although it must sometimes be accepted when imposed by others.

b. In this determination to sterilize couples, which seems to reflect a coercive and propagandistic climate produced by pressure groups, there is a search for methods that might offer some chance of reversibility, either to make

---

Cohort Study of Vasectomy and Prostate Cancer in US Men," *Journal of the American Medical Association* 269 (1993): 878–882; E. Giovannucci et al., "A Prospective Cohort Study of Vasectomy and Prostate Cancer in US Men," *Journal of the American Medical Association* 269 (1993): 873–877; K. P. Dieckmann, "Vasectomy and Testicular Cancer," *European Journal of Cancer* 30A, no. 7 (1994): 1040–1041; L. Rosenberg et al., "The Relation of Vasectomy to the Risk of Cancer," *American Journal of Epidemiology* 140, no. 5 (1994): 431–438.

    [29] K. L. Kohll and A. J. Sobrero, "Vasectomy: A Study of Psychosexual and General Reaction," *Social Biology* 20 (1973): 298–302; W. Neuhaus and A. Boltea, "Prognostic Factors for Preoperative Consultation of Women Desiring Sterilization: Findings of a Retrospective Analysis," *Journal of Psychosomatic Obstetrics and Gynaecology* 16, no. 1 (1995): 45–50.

the procedure more acceptable psychologically or else to elude legal prescriptions against permanent impairment of physical integrity.

c. What is already known about later psychological, biological, or psychosexual complications is kept silent. There is a failure to recommend and take precautions, before making the practice more widespread, that would allow adequate research to ascertain what has already been done in other nations and what the general consequences have been in relation to the health and emotional stability of individuals and couples.

Moral Assessment

After what has just been explained, there is hardly any need to recall that the issue of voluntary sterilization, aimed at causing infertility simply and directly, is different from two other types of sterilization: *forced* and *therapeutic*.

The first has been condemned by popular opinion as well as by the Catholic Church for the twofold reason that it harms physical integrity and violates the person's freedom.[30] We should ask ourselves whether certain propagandistic campaigns, aided and abetted by the mass media and based on incomplete, distorted information with the support of political movements, might not end up exerting a similar coercive pressure. For example, everyone today recognizes the health risks involved in hormonal contraception, but the dissenting voices and those urging caution are stifled before they can influence the marketplace. Governments and legislatures hasten to liberalize sterilization procedures, while political movements with misleading libertarian slogans and financial organizations have put pressure on populations, especially those in underdeveloped countries with fewer health care resources to begin with.

All that remains of sterilizations performed on individuals as a penalty for particular sexual crimes is the historical memory of excessive, inhumane punishment.

Therapeutic or *curative* sterilization, which has been practiced for ages in hospitals, does not raise any particular moral problems, as already mentioned. This procedure, often motivated by the occurrence of tumors or else by pathological processes that cannot be cured effectively except by the removal of the reproductive organs, remains morally licit (just like any surgical intervention involving an impairment) on these conditions: it must be ordered to the good of the whole organism, it must be necessary for the soundness of the body, this soundness must not be attainable otherwise, and it must address an urgent necessity.[31]

---

[30] The condemnation by the Catholic Church was pronounced not only in *Casti connubii* by Pius XI but also in various interventions of the Holy Office in 1931 and 1940, in the address by Pius XII to midwives on October 29, 1951, and in other interventions on the occasions of medical conferences (October 8, 1953, and September 12, 1958). The policy in India that was imposed during the Gandhi administration for demographic reasons was included in the condemnation by the Congregation for the Doctrine of the Faith (CDF) on March 13, 1975 (*L'Osservatore Romano*, September 12, 1976), and was reiterated by the Indian Bishops Conference in a declaration on November 22, 1977. It is well known that this policy was a major factor in the fall of that government. See also A. Fonseca, "Sterilizzazione obbligatoria in India," *La Civiltà Cattolica* 3 (1976): 153–162.

[31] I. Paquin, *Morale e Medicina* (Rome: Orizzonte Medico, 1962), 239–241; D. Tettamanzi, "La sterilizzazione: Problemi morali oggi," *Rivista del Clero Italiano* 2 (1979): 120; idem, *Bioetica: Difendere le frontiere della vita* (Casale Monferrato: Piemme, 1996), 274.

It should be noted that the reasons supporting the moral judgment of the illiceity of coerced sterilization and the liceity of curative sterilization are evident even in the light of ethical reason: in the first case, the *physical integrity of the person* is harmed without any objective need of a therapeutic kind, and the person's freedom is violated in the fundamental and inalienable right of the *possibility of procreating in marriage*; in the second case, the soundness of the organism, according to the principle of the totality of personal good, is what requires and justifies the intervention. Obviously, the patient must be notified in order to grant his or her consent. The therapeutic principle or principle of totality is discussed in greater depth later on.

The ethical problem presents itself more acutely in cases of voluntary sterilization for contraceptive purposes, that is, when performed with the consent of the person it affects.[32] The position of the Magisterium of the Catholic Church on this precise point is quite clear. Suffice it to cite several of the most significant and explicit documents; the arguments advanced are worth attention even in the light of human reason alone.

In the encyclical *Humanae vitae*, Pope Paul VI declares, "Equally to be condemned, as the magisterium of the Church has affirmed on many occasions, is direct sterilization, whether of the man or of the woman, whether permanent or temporary."[33] More elaborate and interesting, also because of its references to the opinion of certain theologians, is the document on sterilization in Catholic hospitals by the Congregation for the Doctrine of the Faith on March 13, 1975, titled *Quaecumque sterilizatio*, cited here verbatim:

> 1. Any sterilization which of itself, that is, of its own nature and condition, has the sole immediate effect of rendering the generative faculty incapable of procreation, is to be considered direct sterilization, as the term is understood in the declarations of the pontifical Magisterium, especially of Pius XII. Therefore, notwithstanding any subjectively right intention of those whose actions are prompted by the care or prevention of physical or mental illness which is foreseen or feared as a result of pregnancy, such sterilization remains absolutely forbidden according to the doctrine of the Church. And indeed the sterilization of the faculty itself is forbidden for an even graver reason than the sterilization of individual acts, since it induces a state of sterility in the person which is almost always irreversible.

> Neither can any mandate of public authority, which would seek to impose direct sterilization as necessary for the common good, be invoked, for such sterilization damages the dignity and inviolability of the human person. Likewise, neither can one invoke the principle of totality in this case, in virtue of which principal interference with organs is justified for the greater good of the person; sterility intended in itself is not oriented to the integral good of the person as rightly pursued "the proper order of goods being preserved" inasmuch as it damages the ethical good of the person, which is the highest good, since it deliberately deprives foreseen and freely chosen sexual activity of an essential element. ...

---

[32] On this subject, see also Tettamanzi, *Bioetica*, 275–277.

[33] Paul VI, Encyclical Letter *Humanae vitae*, July 25, 1968, no. 14.

2. The Congregation, while it confirms this traditional doctrine of the Church, is not unaware of the dissent against this teaching from many theologians. The Congregation, however, denies that doctrinal significance can be attributed to this fact as such, so as to constitute a "theological source" which the faithful might invoke and thereby abandon the authentic Magisterium, and follow the opinions of private theologians which dissent from it.[34]

Contraceptive sterilization, which has become a method of birth control,[35] is the increasingly widespread "fruit" of an anti-life mentality that "presents recourse to contraception, sterilization, abortion, and even euthanasia as a mark of progress and a victory of freedom."[36]

I believe it unnecessary to cite other documents of the Magisterium in addition to these, although there are numerous examples; it instead seems to opportune to offer some explanatory notes on the document by the Congregation for the Doctrine of the Faith, which is already clear and thoroughly reasoned in itself.

For this purpose it is helpful to recall two principles that are underscored in the documents of the Magisterium concerning this topic. These principles are of great importance and have universal applications on the anthropological and ethical levels: (1) the nondisposability or inviolability of the human person as the supreme good in the world—for believers, the human person is protected by the Creator and redeemed by Christ; and (2) the principle of the unity and totality of the person (unitotality), which was already discussed in the chapter on the general principles of bioethics.

For these reasons a health care worker's right to conscientious objection must be defended. He or she must have the option not to tolerate cooperation with an impairment of the physical integrity of a subject for contraceptive purposes.[37]

The Inviolability of the Person and of His Physical Integrity

The principle of the inviolability or nondisposability of the human person, even with respect to the subject's own will, is a clear teaching of the Church but also of sound philosophy: it constitutes the ontological and ethical benchmark for every ethical and legal norm. This supreme principle is reiterated in *Humanae vitae*:

> Consequently, unless we are willing that the responsibility of procreating life should be left to the arbitrary decision of men, we must accept that there

---

[34] CDF, *Quaecumque sterilizatio*, March 13, 1975, nos. 1–2. Many papal documents deal with sterilization, including Pius XII's addresses on October 29, 1951, September 7, 1953, and September 12, 1958, available in *AAS* 43 (1951): 843–844; (1953): 606 and 675; and 56 (1958): 743–745, respectively. For further confirmation of these pronouncements, see also CDF, Responses to Questions Proposed concerning "Uterine Isolation" and Related Matters, July 31, 1993.

[35] "Contraception, sterilization and abortion are certainly part of the reason why in some cases there is a sharp decline in the birthrate. It is not difficult to be tempted to use the same methods and attacks against life also where there is a situation of 'demographic [population] explosion.'" John Paul II, *Evangelium vitae*, March 25, 1995, no. 16.

[36] Ibid., no. 17.

[37] M.L. Di Pietro et al., "Obiezione di coscienza e sterilizzazione contraccettiva," in *Obiezione di coscienza in sanità: Nuove problematiche per l'etica e per il diritto* (Florence: Cantagalli, 2005), 214–217.

are certain limits, beyond which it is wrong to go, to the power of man over his own body and its natural functions—limits, let it be said, which no one, whether as a private individual or as a public authority, can lawfully exceed. These limits are expressly imposed because of the reverence due to the whole human organism and its natural functions, in the light of the principles We stated earlier, and in accordance with a correct understanding of the 'principle of totality' enunciated by Our predecessor Pope Pius XII.[38]

The pastoral constitution *Gaudium et spes* of Vatican Council II says this: "Though made of body and soul, man is one. Through his bodily composition he gathers to himself the elements of the material world; thus they reach their crown through him, and through him raise their voice in free praise of the Creator. For this reason man is not allowed to despise his bodily life, rather he is obliged to regard his body as good and honorable since God has created it and will raise it up on the last day."[39]

The basis or ultimate reason for this nondisposability of the human person is the fact that man is a creature of God, and therefore his entire personal reality is a gift of God belonging to God. Man has a responsibility for himself, but not a dominion over or an arbitrary possession of himself. On the purely rational level, this foundation consists in the fact that the person is the primary and transcendent value; if this foundation fails, it leads to complete relativism, as already mentioned several times. This foundation is ontological and ethical; if it is taken away, all of ethics falls and the foundation of human civilization collapses. If man had arbitrary dominion over himself as his own despotic master, why should he not have the same dominion over others? Perhaps because another person's life should be worth more than one's own and have greater protection—but from whom?

As made apparent, there are no other alternatives: either one accepts a creational or at least personalistic vision of man, and man's value is therefore absolute and inviolable, or else one ends up with an immanentistic vision (man as the master of man / the state as the master of man) that opens the door not only to sterilization but also to a "right" to suicide and euthanasia and to the justification of voluntary homicide, abortion, and all forms of violence. This is the dilemma posed by J. Maritain:

> We are led to distinguish two types of humanism: a theocentric or truly Christian humanism; and an anthropocentric humanism ...
>
> The first kind of humanism recognizes that God is the center of man; it implies the Christian conception of man, sinner and redeemed, and the Christian conception of grace and freedom, of which we have noted the principles.
>
> The second kind of humanism believes that man himself is the center of man, and therefore of all things. It implies a naturalistic conception of man and of freedom.
>
> If this conception is false, one understands that anthropocentric humanism merits the name of inhuman humanism, and that its dialectic must be regarded as the *tragedy of humanism*.[40]

---

[38] Paul VI, *Humanae vitae*, no. 17.

[39] Vatican Council II, Pastoral Constitution *Gaudium et spes*, December 7, 1965, no. 14.

[40] J. Maritain, *Integral Humanism: Temporal and Spiritual Problems of a New Christendom*, trans. J. W. Evans (New York: C. Scribner's Sons, 1968), 27–28.

Acknowledging oneself as a responsible steward and not the arbitrary master of one's own person means respecting the needs, potential, and integrity of that person; in other words, it demands that the moral order be conformed to the ontological order. This is what is meant by talking about the natural law.

Natural law, as mentioned in the chapter on bioethics and artificial fertilization, should not be understood merely in the sense of a biological law but rather as an ontological law in which biology is involved in an essential way, since the body is not an accessory but an essential component of man—in fact, it is the manifestation and demonstrative instrument of the whole, entire person. Following the natural law does not mean conforming to a static demand but rather following the essential and structural requirements of the growth and development of the person, a process which is open to the expansive and realistic dynamic of growth in being and love. One cannot cite the dynamic of the natural law in order to legitimize the voluntary and arbitrary suppression of any essential dimension of one's own humanity and being. Morality is respect for the totality and unity of the human being.

Sexuality, as recalled earlier, is an essential physical, psychological, and spiritual dimension of the totality of the person; in the exercise of sexuality the unitive and procreative dimensions are distinct yet so united (unitotality) that, in terms of the structural order, one cannot suppress one without deadening the other. The two are ontologically united and must remain united, from the perspective of their existential openness, both in the performance of a single sexual act and in terms of the integrity of the reproductive faculties, regardless of the procreative outcome, which itself is regulated by laws that are intrinsic to natural dynamics. This topic was examined in the chapter on responsible procreation.

This is the problem addressed by *Humanae vitae*, and the recognition of an objective and ontological order at the foundation of the ethical order is valid in the light of sound reason. The biblical norm is also applicable here: "What God has joined together, let no one separate" (Mk 10:9).

Respect for the ontological or natural order—in other words, the observance of the moral law and the moral order—is man's supreme good: it is like the head of the human body. No opportunistic calculations or human problems, however distressing, are worth abandoning this rectitude.

Christians learn that they must lose even their physical life rather than fall short of the supreme good of moral rectitude. Any other morality is subjectivistic and arbitrary, a result of hedonistic stupefaction or of a materialistic concept of life.

It should be added that in spite of the aforementioned interpretation by the Supreme Court of Cassation, it seems that the Italian constitution has accepted this concept of the transcendence and inviolability of the human person, precisely in articles 2 and 32, and is entirely founded on this concept. A law such as law no. 194 on abortion, in order to offer a window for the legalization of voluntary sterilization, also has to ignore the fundamental meaning of the Constitution of the Italian Republic.[41]

It is the widespread opinion of many legal scholars, however, that the traditional legal rules, which are so reasonable that they once appeared nearly unquestionable, are too generic in the present context and especially with respect to advances in

---

[41] Mantovani, "La sterilizzazione consensuale."

biomedicine; they hence anticipate the need to develop a new set of fundamental laws regarding the human body.[42]

The Unitotality of the Human Being

The human person is a unifying existence: he is a plurality and diversity of faculties and vital expressions in which all unity refers to the plurality and all plurality and diversity calls upon unification as an essential reference, aim, and structural dynamism. Hence, this structure implies a hierarchy of personal goods that is ordered to the whole and to the supreme good.

This means that while it is permissible to surgically remove an organ or one of its parts *when there is no other remedy* for the purpose of safeguarding physical life as a whole (as a good that is understood and sought objectively)—as may happen in the case of curative or therapeutic sterilization—it is not permissible to destroy a physical good simply as an act of the individual will or as a psychological accommodation to the detriment of the moral good of the whole, when there are still other possibilities that could feasibly solve the problem. And this is the case with contraception in general and contraceptive sterilization in particular: the immorality of sterilization is considered more serious in comparison with contraception because in sterilization there is not only a deadening of the act, but also a mutilation or a lasting suspension of the procreative faculty. In this area, which clearly presupposes the option for the objective and ontological value of the person (already discussed at length in the chapter dedicated to the general principles of bioethics), the ideas of those theologians who are described as "dissenting" in the above-cited document by the Congregation for the Doctrine of the Faith appear unfounded, and the CDF has declared that the faithful may not take them as authoritative. Those theologians appeal in both cases (contraception and voluntary sterilization) to two so-called principles according to which the teaching of *Humanae vitae* is supposedly surpassed and essentially gutted: the principle of *optionality* among the ends to be pursued and an overextended and improperly understood principle of *totality*,[43] understood as obliging the moralist to consider the whole set of the subject's existential psychological conditions.

For the principle of totality to be legitimately and licitly applied, certain essential conditions are required. They are recalled here because the principle of totality forms the basis of a derivative principle, the so-called therapeutic principle. The intervention must fulfill these conditions:

1. It must be a necessary intervention for which there are no valid alternatives.

2. The direct intervention must be upon the diseased part so as to remove it; if sterilization results, it must be indirect. One can remove the healthy part only if it is the real cause of a pathology that cannot be otherwise eliminated.

---

[42] F. D'Agostino, "Verso uno statuto giuridico del corpo umano," in *La sterilizzazione come problema giuridico* (Turin: Giappichelli, 2002), 1–9.

[43] Rossi, "Sterilità e sterilizzazione," 1059–1060; B. Häring, *Medical Ethics* (Notre Dame, IN: Fides, 1973), 62–63; E. Chiavacci, *Morale della vita fisica* (Bologna: Dehoniane, 1976), 72–73. For a more exhaustive response to these ideas, see Tettamanzi, "La sterilizzazione: Problemi," 126–127.

3. The necessity must be present at the time of intervention.

4. It must be ordered to the good of the same organism on which the intervention is performed; at the very least, it must consider and include the good of the totality of the organism on which the intervention is performed with a good probability of success.[44]

5. It must have the patient's consent.

The effort made by some moralists to go beyond these "objective" conditions and shift the assessment to the "personalistic" level actually tends to eliminate the assessment of the biological component within the context of the whole person, and therefore ends up introducing a subjectivistic criterion.

It is needless to explain that, in talking about the whole or about totality, it must be understood to include the physical organism in its integrity *as well*. If one is speaking about the understanding of the person, then the person comprises not just total freedom and social needs but also a body in which the person is embodied and manifests himself and for which his freedom is responsible.

The requirement of considering the body in the totality of the person must be understood not in the exclusive sense (the body alone) but in the inclusive sense (the body as well). Thus, the total good of the person is not achieved when the person is satisfied in his social and utilitarian or psychological choices (however well motivated) yet at the expense of his corporeity and without any benefit to corporeity itself.

Furthermore, let us repeat that it is not permissible to appeal to the principle of totality when sterilization is solicited to prevent a pregnancy *that has free sexual activity as its only true cause*. If one wishes to remove the cause of pregnancy, and there may be valid reasons for not wishing to initiate one, then said cause is not to be identified in the functioning of a healthy organ but rather in the unregulated use of sexuality and procreative acts, which are the true "cause" of the pregnancy.

The Principles of Optionality and Totality

The principle of *optionality* is formulated by some theologians as the choice of the lesser evil or of the superior end out of two possibilities that cannot be put into practice simultaneously—for example, fertility and conjugal union; in order to save the second, it is said to be necessary and permissible to eliminate the first, either by way of contraception or even by sterilization. It is usually assumed that this dilemma does not admit of a third way, that of naturally regulating the procreative act.

Again it should be kept in mind that what may cause difficulty for the union of spouses is not the generative faculty but the exercise of it; therefore it is not reasonable to destroy the generative faculty because its exercise is difficult.

Concerning this argument, it should be added that the method of "natural" regulation, in keeping with the ethical and medical-psychological good of the person,

---

[44] This extension of the principle of totality is found in Häring, *Medical Ethics*, 62–63; idem, *Free and Faithful in Christ* (New York: Crossroad, 1981), 3:20–21; F. Böckle, "Ethische Aspekte der freiwilligen operativen Sterilisation," *Stimmen der Zeit* 99 (1974): 755; Chiavacci, *Morale*, 72–73. For a critical examination of these positions, see M. Zalba, "Totalità (principio di)," in *Dizionario enciclopedico di teologia morale* (Rome: Paoline, 1981), 1141–1150, with an exhaustive bibliography.

is increasingly feasible and reputable given the scientific progress in the field of "natural methods." In the second place, since the two ends of marriage—openness to fertility and affective union—are combined in unity, the *voluntary* suppression of one impoverishes and distorts the meaning of the other as well. Finally, this "optionality" does not consider the higher ethical good that consists in accepting and promoting the ontological order of the person and of his expressive totality, where no curative motivation involves the suppression of a faculty that is at once physical and spiritual. It is easy for subjectivism to hide under the guise of optionality and, even with the best intentions and the most attentive sympathy for conjugal difficulties, it leads to a kind of ethical and ontological decay of love by stripping the person of responsibility.

The alleged principle of the *totality* of personal good means the totality of conjugal, familial, or social wellbeing which, according to these authors, would require contraceptive intervention in certain cases or, if that method proved to be impractical, sterilization.

The extension of the principle of totality to the psychological and social dimension, upheld even for abortion and contraception, becomes an absurdity: there is no longer any totality when one begins to destroy the physical dimension. Moreover, ethics loses its objective foundation when it legitimizes subjectivism and political exploitation.

It has now been clarified that the term "totality" can conceal ambiguity or deception: the totality of the good of the person and consequently of the family and of society cannot prescind from the real, essential order of the person; therefore, in the absence of a physical ailment, one cannot proceed to a surgical intervention that is physically harmful. Rather, one is called to align the physical, biological, and affective goods in harmony with the total and supreme good of the person according to the moral order of goods. The body cannot be excluded; it must be included in the totality of the ordered good of the person.

And this is what the pertinent document by the Congregation for the Doctrine of the Faith declares: "Neither can one invoke the principle of totality in this case, the principle which would justify interference with organs for the greater good of the person. Sterility induced as such does not contribute to the person's integral good, properly understood, 'keeping things and values in proper perspective.' Rather it does damage a person's ethical good, since it deprives subsequent freely chosen sexual acts of an essential element."[45]

The Request for Voluntary Sterilization as a Symptom

In an analysis of the relationship between social transformations and the values system in Italy, it has been written that "beyond the various and insightful interpretations and diagnoses that have been offered, there is reason to think that our time is marked by a profound and widespread uncertainty about models of behavior, norms, and of course values; the last-mentioned, to cite a famous definition by Linton, constitute the heart of culture."[46]

The proposal to legalize voluntary sterilization and the desire to make the practice more widespread appear to be symptoms of uncertainty and of the col-

---

[45] CDF, *Quaecumque sterilizatio*, no. 1.

[46] V. Cesareo, "L'epoca dell'incertezza," *Vita e Pensiero* 2 (1979): 36.

lapse of ethical values—the higher values concerning human life and the person. Contraception, sterilization, and abortion, together with many other acts and crimes of violence and oppression, are signs of an anomy: a void left by the missing values of life and of the person. As far as seeking voluntary sterilization is concerned, it is probably a psychological symptom of a destructive sort that ought to cause alarm and spur appropriate treatment; it should not spur encouragement, which is what proponents of the law suggest. It is no surprise that this is happening in economically advanced societies in which the economy is considered more important than the person. This gives reason to think that *homo oeconomicus* (economic man) or *homo ludicus* (leisurely man) is not *homo humanus* (human man); he is not the whole man but a sick man, afflicted with a sort of destructive and aggressive disease due to a lack of personal values.

The author of the encyclical *Humanae vitae* is aware that contemporary man succumbs to alienation from his own humanity in a way that may be imperceptible to him; in the name of progress, he often becomes merely *homo oeconomicus* or *homo technicus* (technological man). Paul VI is obliged, with all humility and firmness, to recall the integral vision of man, a vision expressed so well by Pope John Paul II, in which man rediscovers and affirms himself as *homo humanus*:[47]

> The truth that we owe to man is, first and foremost, a truth about man. … Perhaps one of the most obvious weaknesses of present-day civilization lies in an inadequate view of man. Without doubt, our age is the one in which man has been most written and spoken of, the age of the forms of humanism and the age of anthropocentrism. Nevertheless it is paradoxically also the age of man's deepest anxiety about his identity and his destiny, the age of man's abasement to previously unsuspected levels, the age of human values trampled on as never before.[48]

A renewal of philosophical and theological meditation on man that is closely wedded to serious scientific data, without hidden agendas, has therefore become urgent for Catholic culture and for the promotion of human and social progress. Teilhard de Chardin has written, "Like the meridians as they approach the poles, science, philosophy and religion are bound to converge as they draw nearer to the whole."[49] By drawing near to the holistic vision of man as person, the image of God, the points where these great meridians intersect can be rediscovered.

As for the marital and family problems that are invoked to justify the destructive procedure of voluntary sterilization, it seems that the time has come to confidently and courageously show the way—the only way available to the upright conscience, namely, learning responsible procreation and the approved use of natural methods. Morality has provided valid guidance for many years, and science now offers increasing reliability.

---

[47] K. Wojtyla, "La visione antropologica della *Humanae Vitae*," *Lateranum* (1978): 129.

[48] John Paul II, Address to Members of the Third General Conference of the Latin American Episcopate, January 28, 1979.

[49] P. Teilhard de Chardin, *The Phenomenon of Man*, trans. B. Wall (New York: Harper and Brothers, 1959), 30.

Corollaries and Specific Cases of Particular Gravity

There are certain specific cases that are dramatic and pose ethical difficulties not only for those directly concerned but also for physicians: the case in which a married couple faces the alternative between sterilization and total abstinence (the so-called *extreme case*); the case of a mentally ill person (generally the woman) who may be the object of sexual abuse; and the case of sexual violence (whether intra- or extramarital).[50]

The extreme case

To explain the extreme case, it is necessary to refer to documents of the Catholic Magisterium, which are cited often and by many parties.

There may be medical situations in which a new pregnancy is absolutely inadvisable and risky (serious cardiac, circulatory, or renal diseases; multiple previous cesarean sections). The passage from *Gaudium et spes* is usually cited which underscores that "where the intimacy of married life is broken, it often happens that faithfulness [*bonum fidei*] is [*or:* can be] imperiled and the good of the children [*bonum prolis*] suffers: then the education of the children as well as the courage to accept more children are both endangered."[51] From this, some draw the conclusion that if conjugal intimacy should not be interrupted yet cannot be exercised without the risk of causing a pregnancy that would be lethal, then sterilization must be justified as a last resort.

Three observations can be made about this proposal. First of all, the passage from the Vatican Council II document does not have the meaning that is sought in it and does not allow us to declare that total abstinence for serious and proportionate reasons is illicit. (Much less does it permit recourse to sterilization.) Second, sterilization is not the only way to avoid pregnancy. Third, the application here of the principle of the lesser evil is arbitrary and strained.

Let us explain each of these three statements. The paragraph from the conciliar document (*Gaudium et spes*, no. 51) means to underscore the need to harmonize the demands of conjugal love with those of responsible fertility regulation. One can conclude from the passage that in the normal course of events it is not good to interrupt conjugal relations without just reasons, because in the long run it can be detrimental to the harmony between the spouses and to the education of their children. But the passage must not be pushed to the point of making it say what it does not say, namely, that there cannot be situations so grave that they require deliberate abstinence for good reasons. The language in which the passage is expressed—"*it often happens* that faithfulness is imperiled" (emphasis added)—indicates a typical situation and not an absolute prohibition on interrupting conjugal relations. Besides, the Magisterium itself does not conceal from spouses the fact that their vocation, like any other, may in some situations require exceptional virtue and even heroism.

The same council declares that "outstanding courage is required for the constant fulfillment of the duties of this Christian calling: spouses, therefore, will need grace

---

[50] The cases are clearly described and analyzed in the volume by Ciccone, "La sterilizzazione," 346–352.

[51] Vatican Council II, *Gaudium et spes*, no. 51.

for leading a holy life; they will eagerly practice a love that is firm, generous, and prompt to sacrifice and will ask for it in their prayers."[52]

Think of what can happen to the will of the couple because of events beyond their control, such as the husband having a spinal cord injury from accidental trauma that prevents him from physically achieving marital union. This is the case of paraplegics, who can only sometimes regain the ability to achieve sexual union through special therapy and rehabilitation.

The Magisterium, however, does not intend to indirectly advocate voluntary sterilization in these cases; this can be proved by the following passage from the above-cited document by the Congregation for the Doctrine of the Faith: "[Direct, voluntary sterilization] is absolutely forbidden, therefore, according to the teaching of the Church, even when it is motivated by a subjectively right intention of curing or preventing a physical or psychological ill effect which is foreseen or feared as a result of pregnancy."[53]

The second observation is that sterilization is not the only way to avoid pregnancy: aside from continence, there are possibilities connected with the use of natural fertility regulation methods, which are mentioned and discussed in another chapter.

Finally, the principle of the lesser evil is not applicable in this case, both because an ironclad dilemma does not exist and because the lesser evil, though it may be tolerated when it is inflicted, can never be chosen as a means to an end.

Having clarified these arguments, some practical conclusions can be drawn from them. Foreseeing that a subsequent pregnancy will be difficult or high-risk (because of kidney or heart disease, for example) is not sufficient reason for tying a woman's tubes during a cesarean section. The reasoning that more than one or two cesarean sections have already been performed should not be adopted either, because it is not the number of cesarean sections per se that causes disease. The generative faculty should therefore not be taken away in such cases; rather, the exercise of procreative activity should be regulated.

Only the presence of a lacerated or diseased uterus or one that has become the source of serious pain and harm can justify hysterectomy, with the resulting infertility, because the diseased part must be removed when there are no other treatments. In this case, the infertility is only an indirect consequence (principle of double effect). The damage to and alteration of the uterus can also be the result of multiple cesarean sections, but its state of pathological alteration is what can justify the removal of the uterus; this supposes that such alteration truly exists and that action is not taken based simply on the number of cesarean sections that the woman has undergone.

What has been said is also valid when tubal sterilization is requested to counteract the syndrome of recurrent spontaneous abortion.[54] There are actually some who have proposed sterilization as a preventive measure. The uterus and the fallopian tubes are certainly not the cause of the pathology in recurrent spontaneous abortion, which would actually benefit from quite different types of medical intervention.

---

[52] Vatican Council II, *Gaudium et spes*, no. 49.

[53] CDF, *Quaecumque sterilizatio*. See Ciccone, "La sterilizzazione," 346.

[54] B. Cinque et al., "Aborto ripetuto spontaneo: Aspetti scientifici e obbligazioni morali," *Medicina e Morale* 42, no. 5 (1992): 889–910

Likewise, it is not permissible to remove healthy ovaries to prevent pregnancies in women who may have renal or coronary complications. The reason, once again, is that the ovaries are not responsible for the woman's possible infirmity; rather, it is the exercise of procreative activity by the couple, combined with renal or cardiac difficulties, that poses a risk. It is the heart or the kidneys that must be treated, and free procreative activity should be appropriately regulated. In confirmation of this teaching, let us cite what Pope Pius XII had to say on this subject:

> In order to justify the removal of the ovaries the above-cited principle (the principle of totality) is alleged, and it is said to be morally permissible to intervene upon healthy organs when the good of the whole demands it. But here reference is made erroneously to this principle. Because in this case the risk that the mother runs does not proceed, directly or indirectly, from the presence or from the normal functioning of the ovaries nor from their influence upon diseased organs, kidneys, lungs or heart. The danger appears only if free sexual activity causes a pregnancy that could endanger the above-mentioned weakened or diseased organs. The conditions that allow one to dispose of a part for the sake of the whole in virtue of the principle of totality are lacking. Therefore it is not morally permitted to intervene on healthy ovaries.[55]

The mentally ill woman and the case of intra- or extramarital violence

All that has been said thus far refers to the context in which sexual activity is freely chosen and foreseen. What is to be said when procreation is imposed by force, or when sexual violence is committed, with the risk of potentially serious contraindications to maternity?

It is often noted that the magisterial documents themselves, in declaring sterilization to be illicit, constantly refer to the free and conscious exercise of sexuality.[56] It is also pointed out that even paragraph 13 of the encyclical *Humanae vitae* declares illicit "a conjugal act imposed upon one's partner without regard for his or her condition and lawful desires."[57] Besides, this is consistent with what the encyclical teaches about the need to keep united the two meanings of the conjugal act: the procreative and unitive meanings. In cases of violence, the unitive aspect is lacking.

---

[55] Pius XII, Address to Participants in the Twenty-sixth Congress of the Italian Urology Association, October 8, 1953, in *Discorsi e Radiomessaggi*, 15:371–379 [Italian]; see Tettamanzi, *La sterilizzazione anticoncezionale*, 50–52; G. B. Guzzetti, *Sterilizzazione a scopo contraccettivo* (Milan: Massimo, 1981), 17.

[56] The document on sterilization by the CDF dated March 13, 1975, declares, "[Sterilization] damages the ethical good of the person, which is the highest good, since it deliberately deprives foreseen and freely chosen sexual activity of an essential element" (no. 1). And further on, "Any cooperation which involves the approval or consent of the hospitals to actions which are in themselves, that is, by their nature and condition, directed to a contraceptive end, namely, in order that the natural effects of sexual actions deliberately performed by the sterilized subject be impeded, is absolutely forbidden" (no. 3). For a commentary on these passages, see Ciccone, "La sterilizzazione," 348–357.

[57] See M. Zalba, "Principia ethica in crisim vocata intra (propter)? Crisim morum," *Periodica de re morali, canonica et liturgica* 71 (1982): 25–63, 319–357.

Some arguments are also based on a passage from an address by Pius XII in which he said verbatim, "When the carrier of a hereditary defect is incapable of behaving humanly and consequently of contracting marriage, or when later on he has become unable to claim by a free act the right acquired by a valid marriage, he can be legitimately prevented from procreating a new life."[58]

In conclusion, when there is a simultaneous occurrence of the conditions of physical violence (or mental incapacity to give free consent) and, furthermore, a justified need not to procreate for proportionately serious reasons, and finally, if there is no other way to prevent conception, some moralists maintain that recourse to sterilization is licit *by virtue of the principle of legitimate defense against unjust aggression.*

The principle of legitimate defense can allow for a harm inflicted on the aggressor; therefore in these hypothetical cases, recourse to the law or to the use of force *against the aggressor* would be automatically legitimized. The question then arises: Can the same principle authorize a woman to have herself sterilized, and thereby to cause harm to herself, in order to prevent not the violence, but the result thereof—namely, conception? Can a mentally handicapped woman who is subjected to sexual abuse, and thereby forced to become pregnant, be sterilized? And who takes the initiative or responsibility?

One also has to wonder whether the sterilization of such persons is not perhaps a quick, safe, and inexpensive way to deal with conditions that would otherwise involve a greater investment of economic and human resources—implementing medical, psychological, educational, and family assistance—in order to bring about appropriate social integration.[59]

There is in fact a tendency on the part of courts to authorize sterilization procedures upon mentally handicapped subjects. It is well known that as early as March 1987 the Court of Appeals in Great Britain[60] authorized this procedure on a minor known as Janette who was afflicted with a serious mental handicap. The legal guardian contested the ruling and submitted the case to the House of Lords, which made the practice legal in April 1987. Consequently, the judiciary authority, in exercising legal guardianship, can authorize the sterilization of a minor with a serious mental handicap whenever this procedure proves necessary to protect the minor herself from serious and certain psychological and physical harm in the event of pregnancy.[61]

---

[58] Pius XII, Address to the Italian Medical-Biological Association of St. Luke, November 12, 1944, in *Discorsi e Radiomessaggi*, 6:192 [Italian].

[59] J.C. Truffino, "La esterilización en los enfermos mentales: Casos clínicos y consideraciones éticas," *Cuadernos de Bioética* 5, no. 22 (1995): 170–172.

[60] "In Re B (A Minor) (Sterilization): Law Report," *The Times*, March 17, 1987, p. 35 (col. 1); C. Dyer, "Sterilization of Mentally Handicapped Woman," *British Medical Journal* 294 (1987): 825; D. Chakraborti, "Sterilisation and the Mentally Handicapped," *British Medical Journal* 294 (1987): 794.

[61] K. Petersen, "Private Decision and Public Scrutiny: Sterilization and Minors in Australia and England," in *Contemporary Issues in Law, Medicine and Ethics*, ed. S.A.M. McLean (Brookfield, VT: Dartmouth Publishing, 1996), 57–77.

The sterilization of mentally handicapped women is currently possible in Spain as well: in 1994, the Constitutional Court rejected an appeal, filed at the lower court in Barcelona, challenging the constitutionality of article 428 of the 1989 penal code.[62]

This trend has been confirmed by resolution A3-0231/92, approved by the European Parliament on September 16, 1992, concerning the rights of the mentally handicapped.[63] With respect to the problem of sterilization and civil rights, the European Parliament

> Calls for sterilization to be considered as the last resort and for steriliza-tion to be carried out only if other methods or means of contraception are impracticable or unreliable (art. 6);

> Calls for the sterilization of persons incapable of giving consent to be allowed only after thorough examination by at least two doctors and on presentation of a favourable written opinion, with the consultation of the parents or the legal representatives and of the persons concerned; calls for a final decision on sterilization to be made only by the competent court in accordance with the procedures provided for in the Member State in question; a representa-tive of the state must be involved in this procedure if so provided for in the Member State (art. 7);

> Calls for sterilization to be carried out no sooner than two weeks after the court delivers its ruling and for it to be carried out in accordance with medical procedures which enable the operation to be reversed if possible (art. 8).[64]

Interpreting these three points proves quite problematic. Indeed, what does "calls for" mean when it says that sterilization should be considered the "last resort," to be practiced when other methods are "impracticable or unreliable"? Perhaps it means to say that, instead of being a "last resort," sterilization ought to be practiced very frequently. Indeed, aside from the immorality of practicing contraception without valid consent (among other considerations), if one considers the difficulties with ensuring the proper administration or application of most contraceptive methods, they could be classified as "impracticable or unreliable"; one could therefore conclude that sterilization should be practiced on a large scale instead of being the "last resort."

Moving forward, what is article 7 supposed to mean? That the physicians must certify that the individual is incapable of understanding and making decisions and should therefore be sterilized? And on the basis of what criteria should one decide who should be sterilized and who should not? Or does it perhaps only mean that certain mental illnesses must be identified that, unlike others, call for forced sterilization? The resolution also calls for consultation with parents and legal representatives,

---

[62] See F. Muñoz Conde, "Sterilization of the Mentally Handicapped: Comments on the Ruling of Spain's Constitutional Court (July 14, 1994)," *Law and Human Genome Review* 2 (1995): 175–196; F. C. Fernández Sánchez, "La esterilización de incapacitados mentales y su calificación moral objetiva," *Cuadernos de Bioética* 5, no. 20 (1994): 361–367.

[63] European Parliament, Resolution A3-0231/92 on the Rights of the Mentally Handicapped, nos. 6–8, *Official Journal of the European Communities* C 284 (1992): 18–22. See A. Romano, *Sterilizzazione umana*, 52–77.

[64] European Parliament, "Rights of the Mentally Handicapped," 20.

while leaving the final decision to the judicial authority and, where provided for by law, to a representative (attorney) for the public's interest. Everything seems fine, yet there is a major problem: everyone has a say except the interested party, who must meekly accept the fate decided by others.

Article 8 calls for forced sterilization to be performed with the utmost caution, which means waiting a good two weeks after the date of the decision (it is not clear what happens during these fourteen days of reflection or who is supposed to be reflecting) and using, if possible, techniques that best allow for the restoration of fertility. Then, if these techniques prove unsuccessful or unreliable and therefore require a second surgical intervention with local anesthesia, it matters little: the mentally handicapped person—according to the resolution—has less dignity and is less of a person than those who are healthy, and can therefore put up with that, too![65]

The French National Ethics Advisory Committee for Life and Health Sciences (Comité Consultatif National d'Éthique pour le Sciences de la Vie et de la Santé, CCNE) has intervened on this matter as well with two opinions, no. 49 and no. 50, both dated April 3, 1996.[66] The conclusions reached in the two documents can be summarized as follows:

1. The subject's informed consent is indispensable, even though the possibility of allowing sterilization of the mentally handicapped under certain circumstances (therefore without consent) is not ruled out. In the latter case certain conditions are required: a thorough assessment of the subject's state of incompetence, the individual's interest in having sexual activity and in preserving fertility or not, the proven impossibility of resorting to other contraceptive practices, and the possibility of performing a reversible procedure. The decision is made by a panel made up of legal scholars, physicians, and others.

   It must be recognized that very restrictive rules have been devised in this case, yet they remain questionable from an ethical standpoint since no effort is made to protect the mentally handicapped women from possible sexual aggression.

2. Sterilization for contraceptive purposes—and hence at the individual's request—cannot be classified as sterilization for therapeutic purposes, according to the provisions of article 16-3 of the French civil code.

3. Since the reversibility of the sterilization procedure cannot be guaranteed in any case, it is necessary that anyone requesting it be informed about the

---

[65] See also M. Scalabrino Spadea, "La tutela del malato di mente nel diritto internazionale dei diritti dell'uomo: Documenti vecchi e nuovi," *Medicina e Morale* 42, no. 6 (1992): 1105–1118. On the same topic, see the standards established by the courts in a growing number of states in the United States as presented in J. Areen, "Limiting Procreation," in *Medical Ethics*, ed. R. M. Veatch, 2nd ed. (Boston: Jones & Bartlett Publishers, 1997), 115–117.

[66] French National Ethics Advisory Committee for Life and Health Sciences, "La contraception chez les personnes handicapées mentales, " opinion no. 49 dated April 3, 1996, *Les Cahiers du CCNE* 8 (1996): 3–5; idem, "La stérilisation envisagée comme mode de contraception définitive," opinion no. 50 dated April 3, 1996, *Les Cahiers du CCNE* 9 (1996): 3–19.

consequences of the intervention (risk of failure, uncertain reversibility, surgical risks) and provide her own consent orally or in writing. Health care workers should likewise prudently evaluate the conditions under which contraceptive sterilization can be performed or simply proposed.

4. The possibility should be kept in mind that a health care worker may refuse to participate in sterilization procedures through conscientious objection.

Finally, the CCNE recommends particular care in taking suitable measures to protect the rights and interests of the persons whose request for sterilization is presented by others.

Now the fact that there are some—even moralists—who consider sterilization licit in these cases does not dispel the serious doubts about the efficacy of such measures against violence. Nor does it exempt from the duty to seek more legitimate and less harmful means of self-defense that are more respectful of human dignity and can be employed instead—in the event of necessity—against the aggressor.

The debate is ongoing not only in the context of civil law but also among ethicists. It should be noted that even among those who are outside the Catholic tradition and generally allow sterilization for contraceptive purposes, the prevalent opinion is that sterilization without explicit consent is unacceptable.[67]

Assuming therefore that the sterilization of a handicapped woman is a doubly illicit act (a human being is deprived of the faculty of procreation, and this is done without her consent), I maintain that a law against sexual violence can be the most instructive remedy at the social level. This is the present position held by Italian law, which does not provide, however, for any particular protection in the case of sexual violence against an incompetent woman.[68] Among other things, the condition likewise stipulated by moralists who favor sterilization in these cases should be emphasized: namely, there must be no other suitable method of preventing the violence. On the other hand, it seems absurd that, in order to prevent an abuse, physical harm should be done not to the person who commits the abuse but to the one who endures it and who, in the case of the mentally incompetent, is already burdened by a handicap. Furthermore, we do not know what the psychological repercussions of such a mutilation might be for a subject who is already mentally vulnerable.

The real solution that will protect the mentally handicapped person from sexual abuse and from forced pregnancy consists of assistance to the mentally infirm subject, including through state-subsidized caregivers, as already occurs in many countries in particularly serious cases.

Sterilization and the Law

There are various possible situations for the relationship between the law and sterilization: the depenalization of voluntary sterilization, the legalization of voluntary sterilization, and making sterilization mandatory by law in some situations.

---

[67] R. Gillon, "On Sterilising Severely Mentally Handicapped People," *Journal of Medical Ethics* 13 (1987): 59–61.

[68] See Italian Parliament, Norms against Sexual Violence, law no. 66, February 15, 1996, art. 609 *bis*, in *Medicina e Morale* 46, no. 6 (1996): 1190–1195 [Italian].

Let us begin with the last possibility, which has already been mentioned. It occurred in Germany under Hitler for eugenic purposes and was condemned in Catholic circles by Pope Pius XI in *Casti connubii* on December 31, 1930, and then repeatedly by decrees of the Holy Office in 1931 and 1940; the condemnation was reiterated by Pius XII in various addresses (October 29, 1951; September 7, 1953; September 12, 1958). Population control legislation was passed in India in 1973, resulting in the sterilization of thirteen million Indians. The reaction of the Indian bishops was expressed and explained in a document signed by Lawrence Cardinal Picachy (March 19, 1976) and again in a document of the Standing Committee of the Catholic Bishops Conference of India in 1976, and it was repeated on the tenth anniversary of *Humanae vitae*. Paul VI made a statement along the same lines on the occasion of the World Population Conference held in Bucharest. The condemnation was sanctioned in the concluding document of the Synod on the Family in 1980.[69]

The reasoning behind this condemnation is intuitive and based on the fact that sterilization violates two fundamental rights that are recognized not only by rational morality but also by international law: the right to physical integrity and the right to the fundamental freedom to have a family and to practice responsible procreation within it. Procreation remains a responsibility of spouses. This type of forced sterilization is instinctively rejected by populations themselves, which is the reason population controllers now resort to other sorts of procedures.

Although the idea of forced sterilization for the purpose of birth control has been rejected, no such unanimity exists with regard to compulsory sterilization of the mentally ill. A comparative study conducted on behalf of the European Parliament finds that in the individual European member states, there is a consensus that the practice of forcibly sterilizing the mentally ill should be continued with the prior consent of the legal guardian.[70]

Very clear guidelines on this matter were contained in the draft of the civil code of the Federal Republic of Germany, which listed several conditions for sterilizing the mentally handicapped in article 1095: the patient must be incapable of forming a lasting intention, there must be danger of a pregnancy unless the sterilization is performed, no other contraceptive is suitable, and some danger to the life of the pregnant woman or the danger of a serious impairment of her physical and psychological health must be a foreseeable consequence of pregnancy.

The depenalization of voluntary sterilization is already under way in many countries. In effect, this is the case in Italy as well since the repeal of article 552 of the penal code, notwithstanding the fact that serious legal scholars invoke articles 582 and 583 of the code (concerning fraudulent personal injuries), which are still in

---

[69] See "Contre un projet de loi sur la stérilisation obligatoire en Inde," *La Documentation Catholique* (1976): 420–421; *L'Osservatore Romano*, May 30, 1976, p. 2, and March 1, 1978; Paul VI, Address to Participants in the Twenty-fifth General Assembly of the International Pharmaceutical Federation and the Thirty-fourth International Conference on Pharmaceutical Sciences, September 7, 1974, in *Insegnamenti di Paolo VI* (Vatican City: Tipografia Poliglotta Vaticana, 1974), 12:800 [Italian]; Synod of Bishops, Message to Christian Families, October 24, 1980, in *Enchiridion Vaticanum* (Bologna: Dehoniane, 1982), 7:743–759 [Italian].

[70] B. Schmidbauer, *Relazione della Commissione per le petizioni sui diritti dei minorati mentali: Studio comparato sulla situazione giuridica dei minorati psichici e sull'attuazione pratica delle norme giuridiche degli Stati membri* (Strasbourg: 1992), in offset.

force, and article 50 of the same code, as well as articles 2 and 32 of the Italian constitution.[71] In practice, however, since the above-cited ruling of the Supreme Court of Cassation, the Italian courts are no longer in a position to intervene in matters of voluntary sterilization.

Depenalization nevertheless allows pro-sterilization groups and associations to carry on propaganda and prepares the way for actual legalization. The legalization of voluntary sterilization is sought as an effective form of birth control and is claimed as a free service. It is obvious that such a law, which purports to legitimize real harm to the integrity of the human person and to the procreative faculty of the family, is characterized as objectively and morally unjust and as a perversion of the moral law. It is therefore up to the citizens, whether believers or not, to oppose such a law by all legally permissible means before it is formulated and approved.[72]

In the event, however, that the law succeeds in gaining a majority consensus, many physicians, paramedics, and other collaborators will have to become *conscientious objectors*, as in the case of a request for abortion. Anyone who acts in accordance with the provisions of such a law shares in conscience in its intrinsic violence and injustice. Catholic hospitals, obviously, will be obliged by the Church's moral teaching to refrain from and oppose this practice.

It is not unlikely that a law legalizing voluntary sterilization will be passed in Italy; if so, this will be another collision of traditional ethics and contemporary ideology within hospitals and in the world of health care.

Voluntary Sterilization and Libertarian Neo–Natural Law Theory

Francesco D'Agostino correctly points out that we are faced with a renewed *appeal to the symbolic code of nature*, with an *unexpected variation on the natural law paradigm*: *nature* is identified with *subjective desire* or with *autonomy*. Even the request to recognize "nontherapeutic voluntary sterilization as a private and incontrovertible decision is ultimately rooted in a curious reasoning of a *naturalistic* sort: human sexuality can be fully expressed in its most authentic and hence its only truly 'natural' dimension only if it is (technologically!) freed from all procreative risk. 'Deceiving' nature, that is, bringing about an 'artificial' sterility, would finally allow nature to conquer itself."[73] This mentality is rooted in modern philosophy, with its emphasis on self-determination, as well as in sadist libertinism.

Theoretical reflection on *nature* is necessary for confronting this *libertarian neo–natural law theory*, thereby distinguishing *what can be discovered empirically* from *what can be reasoned metaphysically*. But to appeal to the contemporary mentality it is not enough to develop new and improved speculative paradigms;

---

[71] Mantovani, "La sterilizzazione consensuale," 191.

[72] See Caffarra, "Il problema della sterilizzazione," 201–206; Ciccone, "La sterilizzazione," 352–359; Tettamanzi, *La sterilizzazione anticoncezionale*, 106–110.

[73] F. D'Agostino, "Il diritto naturale, il diritto positivo e le nuove provocazioni della bioetica," in *Natura e dignità della persona umana a fondamento del diritto alla vita: Le sfide del contesto culturale contemporaneo; Atti dell'Ottava Assemblea Generale della Pontificia Accademia per la Vita; Città del Vaticano, 25–27 febbraio 2002*, ed. J. de D. Vial Correa and E. Sgreccia (Vatican City: Libreria Editrice Vaticana, 2003), 183.

symbolic codes must be developed that are capable of affecting personal and shared sensibilities. This effort, which is at once theoretical, cultural, and spiritual, allows the development of a code of *comfort*, which is "a synthesis of understanding, solidarity, closeness, solicitude, care, sharing, support, helping, and selfless friendship." [74]

Comfort can provide guidance in situations, such as those connected with mental illness, in which we cannot make *utility* the decisive criterion (since that is the code for things and not persons). Mental illness cannot justify sterilization, which is the imposition of a physical impairment that is supposed to free someone from the possible consequences of unconscious or violent sexual relations. All human frailty, even mental, is nothing but a cry for help that does not demand an anxious or objectivizing response attempting to eliminate the immediate problems, but rather a response of partaking and caring.

---

[74] D'Agostino, "Il diritto naturale," 192.

Summary Outline for Chapter 12

Sterilization

Definition

Sterilization is understood to be a medical act that will cause sterility in a subject who is capable of procreation.

A distinction can be made between reversible and irreversible sterilization:

- Sterilization can be described as *total* when the reproductive organs are altered in a generally *irreversible* way.

- On the other hand, it can be described as *functional* or *temporary* when it respects the integrity of the organs and is limited to preventing their normal functioning. Functional sterilization is generally reversible and corresponds to a suspension of functions.

- Sterilization is described as *therapeutic* or *indirect* when it is brought about as a consequence of safeguarding the patient's own health which is threatened by a pathology that affects the organs necessary for procreation.

- In contrast, *anti-procreative* or *direct* sterilization has the sole purpose of rendering the person objectively incapable of procreation.

Motivations

There are various motives for anti-procreative sterilization:

- Private (contraceptive purposes)
- Social (ritual and/or eugenic sterilizations)
- Public (demographic and/or penal sterilizations following a criminal sentence)

Sterilization is *voluntary*, whether therapeutic or anti-procreative, when the subject requests it and give his or her own adequately informed consent.

It is *nonvoluntary* when it dispenses with the consent of the subjects who are subjected to it:

- Minors

- Incapacitated persons who cannot adequately express their consent or lack understanding and decision-making capacity

(See NBC, *The Bioethical Issue of Non Voluntary Sterilization*, November 20, 1998 [Italian].)

Bioethical issues

1. With respect to temporary and permanent *voluntary sterilizations* performed on individuals who have reached majority and are adequately informed, *there are legitimate objections* of both a medical and ethical nature, because the integrity of the person is violated.

2. *Forced sterilizations* are *illicit*, both legally and ethically, regardless of the person who decides to have them performed (parents or guardians, physicians, judges, the State) or the reasons (particularly those of a psychological and social nature) adduced to justify them.

    In particular, the following are unacceptable:

    • Ritual sterilizations

    • Penal sterilizations performed on individuals who have committed sexual crimes

    • Demographic sterilizations

    • Eugenic sterilizations, particularly when performed on the disabled

3. The case of sterilizations performed on the mentally disabled has been examined very carefully, including by Catholic moralists. *Any sterilization of a disabled person motivated by the intention to make caring for the sick person easier or more economical* for the family or the institution in charge *should be condemned.*

The Magisterium of the Church and sterilization

The pontifical Magisterium, especially of Pius XII, has defined as "direct sterilization" any sterilization that "has the sole immediate effect of rendering the generative faculty incapable of procreation."

According to some Catholic moralists, sterilization could be licit in a few special cases (the mentally ill, women subject to the risk of violence), but the official documents always stress that "notwithstanding any subjectively right intention of those whose actions are prompted by the care or prevention of physical or mental illness which is foreseen or feared as a result of pregnancy, such *sterilization remains absolutely forbidden according to the doctrine of the Church.* And indeed the sterilization of the faculty itself is forbidden for an even graver reason than the sterilization of individual acts, since it induces a state of sterility in the person which is almost always irreversible. Neither can any mandate of public authority, which would seek to impose direct sterilization as necessary for the common good, be invoked, for such sterilization damages the dignity and inviolability of the human person. Likewise, neither can one invoke the principle of totality in this case, in virtue of which principal interference with organs is justified for the greater good of the person; sterility intended in itself is not oriented to the integral good of the person as rightly pursued" (Congregation for the Doctrine of the Faith, *Quaecumque sterilizatio*, March 13, 1975, no. 1).

Voluntary sterilization and libertarian neo–natural law theory

We are faced with a renewed *appeal to the symbolic code of nature*, with an *unexpected variation on the natural law paradigm: nature* is identified with *subjective desire* or with *autonomy.* Even the request to recognize "nontherapeutic voluntary sterilization as a private and incontrovertible decision is ultimately rooted in a curious reasoning of a *naturalistic* sort: human sexuality can be fully expressed in its most authentic and hence its only truly 'natural' dimension only if it is (technologically!) freed from all procreative risk" (F. D'Agostino).

Theoretical reflection on *nature* is necessary for confronting this *libertarian neo–natural law theory*, thereby distinguishing *what can be discovered empirically* from *what can be reasoned metaphysically*. But *symbolic codes* must be developed that are capable of affecting personal and shared sensibilities. The code of *comfort* should be highlighted.

All human frailty, even mental, is nothing but a cry for help, which does not demand an anxious or objectivizing response attempting to eliminate the immediate problems but rather a response of partaking and caring.

Italian law on matters of sterilization

Following the abrogation of article 552 by Italian law no. 194 on May 22, 1978, which legalized abortion, the problem of irreversible consensual sterilization has remained unresolved.

In Italian law, the chief general norms relating to how the human body may be treated are articles 2, 13, and 32 of the constitution and article 5 of the civil code. Article 2 of the Constitution recognizes and guarantees fundamental human rights, article 13 declares the inviolability of personal freedom, while article 32 safeguards health as a right of the individual and a collective interest; finally, article 5 of the civil code regulates acts of disposal over one's body, which are forbidden when they cause a permanent harm to physical integrity or when they are otherwise contrary to the law, public order, or good custom.

There is currently a trend of absolute liberalization after a 1987 decision by the Supreme Court of Cassation that dismissed charges of personal injury (art. 583, penal code) filed against a physician who had performed a consensual sterilization for reasons of "convenience."

Yet the legal situation relative to the problem of sterilization, which is managing to reenter the scene under the common discipline provided in articles 582 and 583 of the penal code (regarding personal injury), by article 50 of the penal code (consent by the bearer of a right) and by article 5 of the civil code (limits on acts of disposal over one's own body), is not clear and exhaustive and requires a more careful legislative definition.

Sterilization of particular categories of persons

Resolution A3-0231/92 adopted on September 16, 1992 by the European Parliament—following a trail already blazed by recommendation R (83) 2 adopted by the Committee of the Ministers of the Council of Europe on February 22, 1983—titled "Resolution on the Rights of the Mentally Handicapped," provides for the right of the mentally handicapped, like all human beings, to satisfy their sexual needs, calling for an "improved emotional and sex education" (article 5); whereas in article 6 it calls for "sterilization to be regarded as the last resort" and that it be performed "only if other methods or means of contraception are impracticable or unreliable."

Article 7 proposes "the sterilization of persons incapable of giving consent to be allowed only after thorough examination by at least two doctors and on presentation of a favourable written opinion," and it also declares that "a final decision on sterilization to be made only by the competent court in accordance with the procedures provided for in the Member State in question."

Some data on sterilization practices

Nazism was not alone in enacting sterilization policies. In Sweden between 1935 and 1996, around 230,000 handicapped and "asocial" individuals were sterilized, 90 percent of them women, as a result of a law proposed in 1922 by the Social Democrats for "humanitarian reasons"; it remained in force until 1976.

Sterilization laws were also enforced in Norway (1934) and Denmark (1929) for economic reasons: there were 40,000 sterilizations in Norway and 6,000 in Denmark.

In Switzerland, compulsory sterilization was approved by law in 1928 and remained in force until 1970. In Finland, too, compulsory sterilization was performed on 11,000 persons.

In France, at least 15,000 women afflicted with mental illness were sterilized, whereas in Austria, in the absence of pertinent regulations, 70 percent of mentally handicapped women are currently sterilized; in Spain, furthermore, the Constitutional Court itself allowed for compulsory sterilization of the mentally ill in 1994.

In Italy, from 1985 to the present, at least 6,000 disabled persons have been sterilized, given that between 400 and 600 sterilization operations have been performed annually on patients in psychiatric hospitals since the Basaglia reform (law no. 180/1978).

Since 1960, moreover, sterilization practices have been making a comeback as instruments of international *family planning* policies. Since the International Conference on Population and Development in Cairo (1994), it has been justified as part of "reproductive health" programs along with various forms of contraception and even some forms of abortion.

Chapter 13

BIOETHICS AND
HUMAN EXPERIMENTATION

Clinical Drug Trials

Experimentation is necessary

Because of an intrinsic epistemological requirement, experimental science cannot do without experimentation: empirical sciences are distinguished from nonempirical sciences precisely on this frontier of experimentation.[1] The history of experimental science since the time of Galileo has seen growth not only in refinements of the experimental method but also in its scope of application and its potential.[2]

Experimentation has expanded from the objective, external world into the subjective, interior, and social world of psychology and sociology. The potential for experimentation, furthermore, has increased enormously with the refinement and advancement of technology: in the fields of molecular biology, genetics, and embryology, as well as in the domain of chemistry, biochemistry, and physics, experimentation can now penetrate—by means of the most refined instruments—into the primordial fibers of the living being, of corporality, and of human life.

---

[1] C. Huber, "Limiti della validità del sapere scientifico," in *Lineamenti di etica della sperimentazione clinica*, ed. A.G. Spagnolo and E. Sgreccia (Milan: Vita e Pensiero, 1994), 29–38; B. Bleidt, *Clinical Research in Pharmaceutical Development* (New York: Dekker, 1996); P. Rossi, *La nascita della scienza moderna in Europa* (Rome: Laterza, 1997); C. Weijer, B. Dickens, and E.M. Meslin, "Bioethics for Clinicians: 10. Research Ethics," *Canadian Medical Association Journal* 156, no. 8 (1997): 1153–1157; S.F. Palter, "Ethics of Clinical Trials," *Seminars in Reproductive Endocrinology* 14, no. 2 (1996): 85–92; M.T. Claessens, J.L. Bernat, and J.A. Baron, "Ethical Issues in Clinical Trials," supplement 2, *British Journal of Urology* 76 (1995): 29–36; A. Bompiani, "La sperimentazione clinica dei farmaci: Stato attuale del problema e proposte di riforma," *Medicina e Morale* 32, no. 2 (1982): 95–135; C. Hempel, *Filosofia delle scienze naturali* (Bologna: Il Mulino, 1968); L. Villa, "Etica e deontologia della sperimentazione sull'uomo," *Minerva Medica* 57, no. 89 (1966): 3733–3739; E. Agazzi, *Il concetto di progresso della scienza* (Milan: Feltrinelli, 1976).

[2] On this subject, see E. Sgreccia, "Autonomia e responsabilità della scienza," in *Lineamenti di etica*, A.G. Spagnolo and E. Sgreccia, 39–50; idem, "Potenzialità e limiti del progresso scientifico e tecnologico," *Dolentium Hominum* 37, no. 1 (1998): 137–144; A.G. Spagnolo, "La sperimentazione e i soggetti di sperimentazione," in *Bioetica nella ricerca e nella prassi medica* (Turin: Camilliane, 1997), 453–482.

This expansion and intensification of the experimental method has established an enormous potential for domination and manipulation by man, especially by scientists, at the expense of human nature itself. This state of affairs, which has reached the frontiers of genetic engineering, has introduced the possibility of "changing" nature, thereby posing philosophic and ethical problems of great import.

Some speak about a "subjective motivation"[3] that has developed along with experimental science and technology and is aimed at understanding the world with the precise purpose of dominating it; it is knowledge that turns itself into power. This is said to be in contrast with the "objective motivation" proper to classical medieval philosophy, which is characterized by a theoretical purpose and knowledge for sake of the joy of contemplating reality. There are those who perceive the distinguishing characteristic of the modern era, as well as one of the reasons for its ethical decline, in this desire to manipulate and dominate nature.[4] One final observation should be made in introducing this topic: when we talk about the experimental manipulation of a living being and in particular of a human being, we tend to think solely of the manipulation of biological nature; yet today's society also manipulates human culture and, by that very fact, introduces many changes that are not always positive and are perhaps even more serious than the ones caused by biological manipulation.[5]

These observations simply intend to underscore that, as the urge to dominate and manipulate that is inherent in scientific knowledge per se has grown, the impulse to experiment has increased. Experimentation is indeed one phase in "experimental" knowledge and in the conquests of the technology based on that knowledge. All this should not make us forget the positive aspects of experimentation that are also present, which do not aim at negative and exploitative manipulation but rather at treatment to restore the health, functional integrity, and social interaction of a human being.

One important matter is defining the end, the subject, and the conditions of experimentation: for what purpose (therapeutic or not), on what subjects (the sick, fetuses, prisoners), and under what conditions (freedom, informed or presumed consent) will the experimentation be conducted?

Experiments, including drug trials, prove necessary for the progress of science in general and of medical science in particular. Within science and within scientific research the dimensions of knowledge and utility are present and interdependent; it is obvious, however, that the tendency for dominion is pervasive in both and can tend to pervert their ends, methods, and procedures. Yet it is true, as the first pages

---

[3] J. de D. Vial Correa and E. Sgreccia, eds., *Etica della ricerca biomedica; Per una visione cristiana; Atti della nona assemblea generale della Pontificia Accademia per la Vita; Città del Vaticano, 24–26 febbraio 2003* (Vatican City: Libreria Editrice Vaticana, 2004); M. Horkheimer, *Eclipse of Reason* (New York: Oxford University Press, 1947); see G. Reale and D. Antiseri, *Il pensiero occidentale dalle origini ad oggi* (Brescia: La Scuola, 1983), 3:629.

[4] R. Guardini, *La fine dell'epoca moderna* (Brescia: Morcelliana, 1979), 53–58; idem, *Etica* (Brescia: Morcelliana, 2001), 950–1019; A. Pessina, "La relazione fra la ricerca biomedica, l'antropologia e l'etica filosofica," in *Etica della ricerca*, Vial Correa and Sgreccia, 144–158.

[5] M. Horkheimer and T. W. Adorno, *Dialectic of Enlightenment*, trans. J. Cumming (New York: Herder and Herder, 1972).

of the Bible attest, that knowing and having dominion over the world can also correspond to the plan and will of the Creator.[6]

All knowledge will involve a path toward good or evil: every form of dominion over the world can either serve or enslave man depending on the ethics applied in the processes and goals of man's knowledge and power. This illustrates once again that ethics is essential as a dimension of equilibrium between nature and the person, between technology and human life.

Hippocratic medicine had already set an objective ethical limit on medical intervention with the well-known aphorism *primum non nocere* (first do no harm). Gradually, as experimentation has assumed its indispensable place in medical science and has performed the job of driving further research, the relationship between the experimenting subject and the subject on whom the experiment is performed has become increasingly well defined.

As pharmaceuticals in particular have gradually lost their magical, suggestive character and acquired their true value as biochemical agents aimed at influencing the mechanisms of the subject's molecular structure and the function of the body, they have been increasingly subjected to experimental verification both before their use and during administration. Thus, in pharmacological experimentation—although it starts with a laboratory procedure, which in turn is preceded by a scientific theory—and in experimentation on animals, the validation process is ultimately completed through experimentation on the recipient for whom it is ultimately intended: man himself.

The technical meaning of pharmacological experimentation

The word *experimentation* is not always understood in the same way: it is necessary to define it more precisely in relation to this topic.

To experiment can have a "subjective" meaning in the sense that the subject masters a skill he did not previously have by repeating an action whose ordinary modalities and outcomes are known; in this case, *experimentation* is equivalent to *learning*. Even this type of experimentation can have applications in medicine, such as when a young surgeon first performs an operation that he had simply studied theoretically or seen others perform until then. It will also be technically and ethically necessary to have an apprenticeship for this necessary form of "experimentation," reducing the risk to the patient's life through the gradual refinement of skill.

But the meaning given to the word in the medical field and in particular in the field of pharmacological research is the "objective" meaning. In this latter sense, "to experiment" means *to subject to testing, through direct use*, procedures or means (drugs) that are new or else approved but with unknown effects, whether direct or indirect, immediate or long-term.[7] This "newness" or "lack of knowledge" can be

---

[6] See Gen 1:26–3:23.

[7] The instruction *Donum vitae* by the Congregation for the Doctrine of the Faith (CDF) (February 22, 1987) rightly distinguishes between *experimentation* and *research*. Since the terms are often used synonymously and ambiguously, it made sure to define the meanings attributed in the document itself: "(1) By *research* is meant any inductive–deductive process which aims at promoting the systematic observation of a given phenomenon in the human field or at verifying a hypothesis arising from previous observations. (2) By *experimentation* is meant any research in which the human being (in the various stages of his existence: embryo, foetus, child or adult) represents the object through which or upon which one intends

absolute or partial, or the "means" may not be considered fully usable until more rigorous tests are conducted.[8]

Since the effectiveness of a drug can be confirmed only in relation to its impact on the preexisting, simultaneous, or subsequent causes (whether physiological, pathological, or pharmacological—by combining or interfering with other drugs or even counteracting them) in the organism for which it is intended, a drug designed for human beings will never have scientific and clinical confirmation and validity until it has been used experimentally on human being themselves. The first phases of study and experimentation—in the laboratory and on animals—therefore have a preparatory and necessary character but are neither definitive nor final.

It is appropriate in this regard to make a few additional clarifications concerning experimentation on animals. Among the many problems that concern our age, the issue of experimentation on animals is indeed of lively interest, and public opinion is largely divided between those who consider experimentation on animals, and vivisection in particular, absolutely cruel and inhumane and hence unacceptable, and those who hold that the use of animals for research is indispensable, provided that it is closely regulated and avoids any sort of cruelty.

Several associations have been founded within the first category, such as the Italian Antivivisectionist Union (Unione Antivivisezionista Italiana, UAI), which tend to confer certain "rights" upon animals. They apply to animals the maxim "do not do to others what we would not want done to us." In support of this position they give reasons that are either scientific (such as disputing the scientific validity of animal experimentation or pointing out the existence of alternative methods such as culturing cells or organs in vitro) or religious (appealing to respect for creation or referencing the figure of Saint Francis of Assisi).

Research centers favoring animal experimentation declare that all the fundamental scientific discoveries that have allowed mankind to cure and conquer many human diseases have been possible through preliminary experimentation on animals and that discoveries about evolution allow the extrapolation to human beings, with very little adjustment, of the data from these experiments.[9]

---

to verify the effect, at present unknown or not sufficiently known, of a given treatment (e.g., pharmacological, teratogenic, surgical, etc.)." *Donum vitae*, part I, no. 4.

[8] G. De Vincentiis and P. Zangani, "Sulla liceità e sui limiti della sperimentazione sull'uomo," *Giustizia Penale* 1 (1968): 332–334; Bompiani, "La sperimentazione clinica," 101–102; Italian National Bioethics Committee, *La sperimentazione dei farmaci* [*Drug experimentation*], November 17, 1992 (Rome: Presidenza del Consiglio dei Ministri Dipartimento per l'Informazione e l'Editoria, 1993), http://www.governo.it/bioetica/pdf/11 .pdf, English abstract at http://www.governo.it/bioetica/eng/pdf/abstract_drug_trials.pdf.

[9] For a more in-depth treatment of the subject, see Italian National Bioethics Committee, *Sperimentazione sugli animali e salute dei viventi* [*Animal testing and health of living beings*], April 17, 1997 (Rome: Presidenza del Consiglio dei Ministri Dipartimento per l'Informazione e l'Editoria, 1997), http://www.governo.it/bioetica/pdf/27.pdf, English abstract at http://www .governo.it/bioetica/eng/opinions/abstracts/abstract_ANIMAL_TESTING.pdf; P. Preziosi, "Valore predittivo della sperimentazione pre-clinica sull'animale," in *Lineamenti di etica*, Spagnolo and Sgreccia, 71–84; S. Castignone, ed., *I diritti degli animali* (Bologna: Il Mulino, 1985); R. G. Frey, "Vivisection Morals and Medicine," *Journal of Medical Ethics* 9 (1983): 94–97; M. Midgley, *Animals and Why They Matter* (Harmondsworth, UK: Penguin Books, 1983); W. Paton, "Vivisection, Morals, Medicine: Commentary from a Vivisecting Professor

I maintain that there are certain boundaries separating man from the rest of living beings: they are called spirit, person, and transcendence. Therefore, man cannot be reduced to the status of a "higher animal," nor can it be claimed that an animal has a dignity equal to man's. The personalist ethic does not place every living being on the same level; rather, in the truth of each creature, it respects the truth of the hierarchical ordering of the subhuman world toward man and his true good.

Hence it can be acceptable to utilize an inferior organism for the true good of man; no one objects, for example, to the use of sheep's wool for protection from the cold or the use of animal meat for nourishment. Obviously, the animal world is not to be treated frivolously or as a way of exercising power without a real motive for the benefit of man and without respect for the equilibrium among the species. Therefore, if there are no real alternatives to the use of animals in experimentation, if the end is man's true well-being, and if all possible analgesic and anesthetic measures are taken so as to prevent unnecessary cruelty to the animals, then experimentation can and must be ethically accepted.[10]

The phases of pharmacological experimentation on human beings have been described in various ways. The following paragraphs cite the information contained in the Italian ministerial decree dated April 27, 1992, which accepted the classification reported in the 1991 edition of the norms of good clinical practice (GCP) issued by the European Economic Community (EEC) in directive 91/507/EEC:

> Clinical trials are generally classified into phases: I through IV. It is not possible to draw precise boundaries between the individual phases, and there are differing opinions in this regard about the details and the methodology. The individual stages are briefly described in the following lines, based on aims pertaining to the clinical development of medicinal products.
>
> *Phase I:* Initial studies of a new active pharmaceutical ingredient conducted on human subjects, often on healthy volunteers. The purpose is to provide a preliminary assessment of its safety and an initial profile of the pharmacokinetics and the pharmacodynamics of the active ingredient in man.
>
> *Phase II:* Pilot therapeutic studies. The purpose is to prove the activity and to evaluate the short-term safety of an active ingredient in patients afflicted

---

of Pharmacology," *Journal of Medical Ethics* 9 (1983): 102–104; idem, *Man and Mouse: Animals in Medical Research* (New York: Oxford University Press, 1984); G. Perico, "La sperimentazione scientifica sugli animali: La vivisezione," in *Problemi di etica sanitaria* (Milan: Ancora, 1992), 123–135; L. S. Sprigge, "Vivisection, Morals, Medicine: Commentary from Anti-vivisectionist Philosopher," *Journal of Medical Ethics* 9 (1983): 98–101. Volume 39, no. 6 (1989) of *Medicina e Morale* was also dedicated to this topic.

[10] "If it be useful and necessary, the *testing* of new pharmaceutical products or of new techniques should first be done *on animals* before they are tried on humans. 'It is certain that the animal is for the service of man and can therefore be the object of experimentation. However, it should be treated as one of God's creatures, meant to cooperate in man's good but not to be abused.' It follows that all experimentation 'should be carried out with consideration for the animal, without causing it useless suffering.'" Pontifical Council for Pastoral Assistance to Health Care Workers, *Charter for Health Care Workers* (Boston: Pauline Books, 1995), no. 79; see J. de D. Vial Correa, "L'etica della sperimentazione sugli animali," in *Etica della ricerca*, Vial Correa and Sgreccia, 270–277.

with a disease or a clinical condition for which the active ingredient is proposed. The studies are conducted on a limited number of subjects and often, at a more advanced stage, according to a comparative scheme (e.g., control experiment with a placebo). This phase also aims to determine an appropriate interval for doses and/or therapeutic programs and (if possible) to identify the relation between dosage and response, for the purpose of providing the best preliminary data in order to plan more extensive therapeutic studies.

Phase III: Studies on larger (and possibly diversified) groups of patients for the purpose of determining the short- and long-term relation between the safety and efficacy of the formulas of the active principle, and also of assessing its absolute and relative therapeutic value. The course and characteristics of the most frequent adverse reactions must be investigated and the specific characteristics of the product must be examined (e.g., clinically important interactions between drugs, factors that induce different responses, such as age, etc.). The experimental program should preferably be a randomized double-blind procedure, but other designs may be acceptable, for example, in the case of long-term studies on safety. Generally the conditions of the studies should be as near as possible to the normal conditions of use.

Phase IV: Studies conducted after the marketing of the medicinal product(s), although there is not complete agreement as to the definition of this phase. Phase IV studies are conducted on the basis of the information contained in the summary of the characteristics of the product relative to the authorization to distribute it commercially (e.g., drug reviews or assessment of the therapeutic value). Depending on the case, phase IV studies require experimental conditions (including at least a protocol) similar to those described above for the pre-marketing studies. Once a product has been placed on the market, clinical studies aimed at examining, for example, new guidelines, new methods of administration, or new interactions are to be considered studies on new medicinal products.[11]

The experiment in fact concludes when knowledge about all the above-cited aspects has reached the point where the discrepancies between short- and long-term results, if present at all, are statistically insignificant.

Once this phase of the "objective" experimentation with the drug has been concluded, the "subjective" experimentation begins—or better, the learning process whereby the researchers discover the properties, doses, and modalities of application. Given the relation of trust and competence among professionals and the documen-

---

[11] Italian Ministry of Health, ministerial decree of April 27, 1992, "Disposizioni sulle documentazioni tecniche da prestare a corredo delle domande di autorizzazione all'immissione in commercio di specialità medicinali per uso umano, in attuazione della direttiva n. 91/507/CEE" [Regulations on the technical documentation to be submitted with requests for permission to market medicinal products for human use, pursuant to directive no. 91/507/EEC], published in *Gazzetta Ufficiale della Repubblica Italiana* 139 (June 15, 1992). As will be explained later, the European guidelines for good clinical practice were updated in 1996 and harmonized with those of the United States and Japan, which have varying regulations, although from the conceptual point of view the classification of the experimental phases is still valid. For further discussion of the methodology for developing pharmaceuticals and clinical trials, see C. Bascarin et al., *Manuale pratico di sperimentazione clinica* (Fidenza: Mattioli 1885, 2002).

tation regarding the drug itself, this phase does not present difficulties nor does it require every physician to repeat the sequence of experiments carried out by those who conducted the drug trial leading to approval.

It is clear, however, that the position and the degree of ethical and legal responsibility of the researcher conducting the experiment for initial approval of the drug are quite different from the position and degree of responsibility of someone prescribing a drug that is already in general use. In the second case the responsibility is equal to that of all other professional and therapeutic activities.

The level of ethical and legal responsibility that must now be examined is that of the researcher conducting experimentation of an innovative sort. But before addressing the legal and ethical issues, several more fundamental distinctions must be made.

The first distinction that must be made is the one already mentioned between experiments in the *laboratory* and/or on *animals* and experiments on *human beings*. The transition to experimentation on human subjects must be preceded and justified by the attainment of all possible elements of technical certainty in such a way that the transition to human trials is free of risks insofar as the risks are foreseeable in the laboratory.

On the other hand, the transition to human trials will always introduce new factors and will inevitably involve risk. The ethical problem arises at this precise moment; it consists of establishing what values are at stake and the conditions for initiating the various phases of experimentation with minimal risk.[12]

Within the scope of experimentation on humans, since the risk factor is the decisive element, a distinction must be made between experimentation on sick persons, on healthy volunteers, on the researchers themselves, on detainees or prisoners, and on those who have been sentenced to death. In the case of administering a drug to a sick person, a further distinction is made between *experimentation for therapeutic purposes*, that is, done with the intention of curing the patient himself, and *purely clinical experimentation*, which is conducted simply for the purpose of ascertaining the functionality of the drug (or of a surgical technique).

These various possibilities have had a history of their own and have also prompted laws and pronouncements that warrant brief examination.

---

[12] A. G. Spagnolo, "Principi etici e metodologie di sperimentazione clinica," in *Lineamenti di etica*, Spagnolo and Sgreccia, 51–70; Italian National Bioethics Committee, *La sperimentazione dei farmaci*; G. Perico, "Sperimentazione clinica," in *Dizionario enciclopedico di teologia morale* (Rome: Paoline, 1981), 1025–1034; idem, "La sperimentazione clinica," in *Problemi di etica sanitaria* (Milan: Ancora, 1992), 107–121; idem, *A difesa della vita* (Milan: Centro Studi Sociali, 1975), 203 ff.; A. Fiori, "La sperimentazione sull'uomo pone gravi interrogativi," *Orizzonte Medico* (July–August 1971): 2; A. Valsecchi, "Principi etici generali sulla sperimentazione clinica," in *Medicina e Morale* (Rome: Orizzonte Medico, 1970), 27 ff.; A. Bompiani, "La ricerca sperimentale in ambito biomedico: Ambiti, metodologie, criteri di validità dei progetti di ricerca," in *Etica della ricerca biomedica*, Vial Correa and Sgreccia, 95–127.

History, practice, and legislation [13]

EXPERIMENTS ON ONESELF. Moses Maimonides, a Jewish physician and philosopher of the twelfth century, instructed his students to test the medications then available on themselves to see whether they were effective. Let us recall, more recently, Eusebio Valli (1755–1816), who injected himself with a mixture of pus from patients with smallpox; Lazzaro Spallanzani (1729–1799), who repeatedly ingested small tubes containing foods prepared in different ways to study the mechanism of digestion; René-Nicolas Desgenettes (1762–1837), who inoculated himself with the contents of a bubo caused by the plague; Max Joseph Pettenkofer (1818–1901), who ingested cholera bacilli; and a German physician, Dr. Lindeman, who in 1850 and 1851 injected himself with the spirochete that causes syphilis. It is also recorded that the young physician Werner Forssmann (1904–1979) performed the first cardiac catheterization on himself.

EXPERIMENTS ON VOLUNTEER SUBJECTS. Among early clinical experiments on consenting persons in the West was the famously successful attempt at a smallpox vaccination by Edward Jenner at the end of the 1700s. It was often professors of medicine, researchers in the field of pharmacology, and their associates who volunteered for pharmacological experimentation. Recall, for example, the discovery of adrenaline following experiments done in 1895 by English physician George Oliver on his consenting son. Much earlier, in 1667, a young patient of Jean-Baptiste Denys agreed to submit to a transfusion of blood from a lamb. Many experiments have been conducted for research on syphilis, gonorrhea, and others on volunteers who remained anonymous.

EXPERIMENTS ON THOSE SENTENCED TO DEATH. Herophilos, a Greek physician born in 335 BC, practiced vivisection on human beings who had been sentenced to death, so as to study the anatomy and physiology of the body. Every now and then during the Renaissance there is mention of the use—authorized by public authorities—of convicts who had been sentenced to death in the study of certain diseases. (In the 1400s, King Louis XI of France allowed the use of condemned prisoners for the study of kidney stones; in the following century, Leonardo Fioravanti studied the contagiousness and symptoms of the plague on condemned convicts whom he sought to anesthetize first.) Yet it is common knowledge that high-risk experiments and treatments have been conducted in certain countries even in recent times on prisoners condemned to death, who perhaps consent in the hope of having their sentence commuted.

EXPERIMENTS ON PRISONERS IN THE NAZI CONCENTRATION CAMPS. [14] According to Nuremberg trial documents, Jewish, Polish, Russian, and Italian prisoners were subjected to cruel experiments with drugs, gas, and poisons. Many experiments involved atrocious sufferings and death. Lethal experiments were carried out with decompression chambers to study the effects of flight at high altitudes, and studies at artificially produced polar temperatures were performed on both naked and clothed prisoners to

---

[13] Perico, "Sperimentazione clinica"; D. J. Rothmas, "Research, Human: Historical Aspects," in *Encyclopedia of Bioethics*, ed. S. G. Post, 3rd ed. (New York: Macmillan Reference USA, 2004), 2316–2326. For a historical review of unethical experimentation, see P. McNeill, *The Ethics and Politics of Human Experimentation* (Cambridge: Cambridge University Press, 1993).

[14] A detailed description of these experiments is provided in R. J. Lifton, *I medici nazisti* (Milan: Rizzoli, 1988). [In English: *The Nazi Doctors: Medical Killing and the Psychology of Genocide* (New York: Basic Books, 1986).]

analyze the effects of freezing. Other experiments in German prison camps studied burns inflicted with mustard gas; resected bones, muscles, and nerves; and injections with vaccines, supposed anti-cancer serums, hormones, and other chemicals. The whole business was justified by an alleged supremacy of science, behind which was hidden only the will of the state.

Even after World War II, abominable experiments were conducted on prisoners and populations in countries experiencing conflict—experiments with vaccines, drugs, bacteria and viruses (which were then used as biological weapons), and forms of radiation.[15]

It is understandable from all of this how the historical circumstances of recent decades, and especially the Nazi experiments, were able to bring about the rise of both international and national norms. At the international level, the most significant documents will be listed along with the normative force proper to each of these acts.

THE NUREMBERG CODE. Precisely in response to the crimes perpetrated by Nazi physicians, including experimentation on human subjects conducted with contempt for the most elementary norms of respect for individual dignity, a document was issued in 1947 that was intended to place explicit limitations on any experimental intervention on human beings. This document served as the point of departure for the subsequent elaboration of specific guidelines.

THE DECLARATION OF HELSINKI. The World Medical Association (WMA), on the occasion of its Fifty-ninth General Assembly held in Seoul, South Korea, in October 2008, approved a new version of the Declaration of Helsinki,[16] the most important and widely accepted set of guidelines in the field of medical research on human subjects. The Declaration of Helsinki, first adopted in June 1964 by the Eighteenth General Assembly of the WMA, had been amended by the Twenty-ninth General Assembly in Tokyo in October 1975; the Thirty-fifth General Assembly in Venice in October 1983; the Forty-first General Assembly in Hong Kong in September 1989; the Forty-eighth General Assembly in Somerset West (South Africa) in October 1996; the Fifty-second General Assembly in Edinburgh in October 2000; the Fifty-third General Assembly in Washington, DC, in October 2002; and the Fifty-fifth General Assembly in Tokyo in 2004.

The recommendations added in the year 2000[17] pertained to some crucial aspects of clinical research that were already present in part in the preceding versions; they were examined here in greater depth and also clarified in light of the new problems posed by research. Four topics were addressed:

1. Once again it was forcefully and unequivocally emphasized that it is necessary to obtain informed consent, preferably in writing, from all subjects participating in the research, and that participation in research by individuals incapable of giving consent must be a rare exception to the rule.

---

[15] See A. Goliszek, *In the Name of Science: A History of Secret Programs, Medical Research, and Human Experimentation* (New York: St. Martin's Press, 2003).

[16] World Medical Association, Declaration of Helsinki: Ethical Principles for Medical Research Involving Human Subjects, June 1964, last amended 2008, http://www.wma.net/en/30publications/10policies/b3/.

[17] The entire English version was published in *Medicina e Morale* 50, no. 5 (2000), the journal of the Bioethics Center of the Catholic University of the Sacred Heart in Rome.

2. It was established that all research must guarantee the possibility of a real benefit to those who participate. It was stated explicitly that a research project is justified only if the population that is the object of study will be able to enjoy eventual benefits resulting from the research itself. This is a clear signal, aimed at all research institutions, of the intention to avoid what has happened in the past, namely, that wealthier populations exploit poorer ones by testing new drugs on them while the pharmaceutical products will then be available only to those who have the money with which to buy them. The population involved in the study of new drugs must receive a guarantee of its right to benefit from the results of the research regardless of economic availability or obstacles of any other kind.

3. It was stated forcefully that any study aimed at documenting the efficacy of a treatment must guarantee that the participants receive the best available therapy. Only in the case of a lack of an effective treatment can there be recourse to the use of a placebo. This amounts to a big step forward in protecting individuals who participate in research, hopefully putting an end to the clearly unethical requests by some regulatory institutions and pharmaceutical companies, often without justification, to conduct controlled studies with a placebo even in instances where effective treatments are well established.

4. Finally the recommendations required that all experimenters participating in the study declare any economic interests or other potential conflicts of interest in order to allow objective assessment of the potential risk of interpreting research data in light of other interests that are not just scientific.

The Declaration of Helsinki (2008) is divided into three parts: (A) introduction (points 1–10); (B) principles for all medical research (points 11–30); and (C) additional principles for medical research combined with medical care (points 30–35).

INTERNATIONAL ETHICAL GUIDELINES FOR BIOMEDICAL RESEARCH INVOLVING HUMAN SUBJECTS. This document was first promulgated by the Council for International Organizations of Medical Sciences in collaboration with the World Health Organization in 1982; it was revised in 1993 and again in 2002. It contains twenty-one guidelines and provides extensive commentary on them. They address the fundamental issues in human experimentation: individual informed consent, risk-benefit assessment, research subject selection, research in populations and communities with limited resources, experimentation involving minors and incompetent subjects, involvement of women of fertile age and pregnant women, protection of confidentiality, compensation for subjects involved in research in the event of accidental injury, review procedures, and externally sponsored research.[18]

EUROPEAN DIRECTIVES. Since 1965 several directives have been issued for the purpose of overcoming the procedural differences among the various states in the European Union. The standardization of laws, however, has proceeded very slowly.

---

[18] Council for International Organizations of Medical Sciences and World Health Organization, *International Ethical Guidelines for Biomedical Research Involving Human Subjects* (Geneva: CIOMS, 2002), 3–4, http://www.cioms.ch/publications/layout_guide2002.pdf.

The the need to develop guidelines for experimentation with new pharmaceutical products as quickly as possible in Europe, by analogy with those that existed in the United States, prompted the pharmaceutical industries themselves to become promoters of a working group that would develop some guidelines that would subsequently involve the European Economic Community itself.

The objective for these guidelines was to harmonize the requirements for conducting trials in various European research centers, making them uniform with the international requirements; to make the trials themselves credible so as not to have to repeat them in the various countries where registration of the drug would be requested; to ensure the highest ethical standard to protect the subjects of the experimentation; and to guarantee a high scientific standard in general.

Every member state appointed a delegate to participate in the work of the Working Party on the Efficacy of Drugs, one of the study groups formed by the Committee for Proprietary Medicinal Products (CPMP), which was responsible for drawing up the document. Thus, starting in 1985, they prepared a first draft of *Good Clinical Practice for Trials on Medical Products in the European Community* (GCP), i.e., the guidelines for good clinical practice mentioned earlier. Aimed at establishing standard principles for conducting trials for new pharmaceutical products through experimentation on human subjects, the text was completed by 1988. After various revisions of the text, the document became directive 91/507/EEC, which was ratified in Italy by the ministerial decree on April 27, 1992.

Taking into account the preceding EEC directives on the subject, the GCP guidelines were devised in such a way as to primarily address the pharmaceutical industries' design of experiments for new products; in practice, they also involved all those taking part in the production of the clinical data used in the registration and sale of a drug. Hence, they immediately affect the physicians and the researchers administering the drug as well as all personnel directly or indirectly collaborating in every single phase of experimentation.[19]

In July 1996, the Committee for Proprietary Medicinal Products (CPMP, replaced in 2004 by the Committee for Medicinal Products for Human Use, CHMP) of the European Agency for the Evaluation of Medicines (EMEA) updated the GCP—to which every country belonging to the European Union is expected to conform its legislation—with the document *Good Clinical Practice: Consolidated Guidance* (CPMP/ICH/135/1995).[20] The directive was adopted in Italy by decree of the Ministry

---

[19] For general information about the 1991 edition of the GCP, see A.G. Spagnolo, "'Norme di buona pratica clinica': Il documento della Comunità Europea sulla sperimentazione di nuovi prodotti farmaceutici," *Medicina e Morale* 41, no. 2 (1991): 201–227; L.M. Fuccella, "Come sorgono e cosa sono le GCP della Comunità Europea," in *Lineamenti di etica*, Spagnolo and Sgreccia, 95–104; D. Criscuolo, "Definizione dei termini utilizzati nella GCP," in *Lineamenti di etica*, Spagnolo and Sgreccia, 105–112; A.A. Bignamini and D.R. Hutchinson, "Guida pratica alle GCP per i ricercatori," in *Lineamenti di etica*, Spagnolo and Sgreccia, 181–194. Concerning clinical trials in light of the 1991 GCP, see also L.M. Fuccella and M. Frascio, *Guida pratica alla sperimentazione clinica dei farmaci* (Milan: OEMF, 1995).

[20] The English text of the 1996 International Conference on Harmonisation (ICH) GCP was published, among other places, in the supplement to *Good Clinical Practice Journal* 3, no. 4 (1996).

of Health on July 15, 1997.[21] It should also be noted that this second edition of the GCP serves as a reference not only for the European Union (EU) but also for the United States and Japan. A pertinent committee called the International Conference on Harmonization (ICH), made up of representatives from the three geographical areas (EU, USA, and Japan), worked jointly to issue a single standard text.

Various modifications have been made to the 1991 edition of the GCP. Since it is not possible in this chapter to examine in detail the features differentiating the two documents, suffice it to say in general that the differences are in both the vocabulary and the instructions for protocol design.

Regarding the first aspect, the following concepts are defined more precisely: adverse drug reaction, adverse event, audit, ethics committee, types of controlled clinical trials, investigators, data monitoring, trial subjects, and many more. The definitions of confidentiality and quality control are narrowed down, and the term "good manufacturing practice" (GMP) is abolished.

As for the second aspect, concerning protocol design, the general information necessary for designing a clinical experimentation protocol is illustrated with a greater wealth of particulars, while the part concerning experimental design is presented with greater statistical elaboration. Clear reference to the concept of "ethics" is instead removed.

The text contains other elements that could be criticized.[22] By way of example, consider that in the case of multicenter experimentation the opinion of the ethics committee of the health care institution coordinating the study would be binding for the committees of the other institutions participating in the trial, which would no longer be allowed to add "local" amendments to the protocol in question but could only reject or accept it in full. This entails foreseeable disadvantages in terms of the practicability of pharmacological research in a given country or with regard to non-negligible ethical problems that multicenter studies sometimes reveal when examined by local ethics committees, especially in relation to informed consent documents for the subjects of the experimentation.

In the context of the member states, additional details and essential clarifications regarding the duties and procedures to be adopted by ethics committees, research sponsors, and investigators were provided by directive 2001/20/EC,[23] concerning the

---

[21] The decree was titled "Recepimento delle linee guida dell'Unione Europea di buona pratica clinica per la esecuzione delle sperimentazioni cliniche dei medicinali" [Ratification of the European Union guidelines for good clinical practice for conducting clinical trials of medications] and was published in supplement 162 to *Gazzetta Ufficiale della Repubblica Italiana* 191 (August 18, 1997).

[22] For a critical analysis of the 1996 GCP, see A. G. Spagnolo, A. A. Bignamini, and A. de Franciscis, "I Comitati di Etica fra linee-guida dell'Unione Europea e decreti ministeriali," *Medicina e Morale* 47, no. 6 (1997): 1059–1088.

[23] European Parliament and Council of Europe, directive 2001/20/EC, "On the Approximation of the Laws, Regulations and Administrative Provisions of the Member States Relating to the Implementation of Good Clinical Practice in the Conduct of Clinical Trials on Medicinal Products for Human Use," *Official Journal of the European Communities* L 121, no. 34 (2001), http://www.eortc.be/services/doc/clinical-eu-directive-04-april-01.pdf, ratified in Italy by legislative decree no. 211 of June 24, 2003, "Attuazione della direttiva 2001/20/ CE relativa all'applicazione della buona pratica clinica nell'esecuzione delle sperimentazioni

application of good clinical practice in the performance of clinical pharmacological experiments, including multicenter studies, and by directive 2005/28/EC of the Commission of the European Community on April 8, 2005, which establishes principles and detailed guidelines for medicines used in experiments on human beings and the requirements for authorizing the manufacture and importation of medicines.[24]

The Convention for the Protection of Human Rights and the Dignity of the Human Being with regard to the Application of Biology and Medicine: Convention on Human Rights and Biomedicine, by the Council of Europe,[25] addresses the issues relating to scientific research on human beings. It lists the conditions under which a person can legitimately undergo scientific research (article 16) and the criteria for safeguarding persons who are unable to express valid, informed consent (article 17). It also takes a position on research involving in vitro embryos (article 18), which we will return to later on.

Let us now take a look at examples of Italian legislation.[26] The *unified text of health laws*,[27] in article 162, granted the Minister of Health authority to examine the qualitative composition of medicines, which were all required to be registered. The head provincial physician was responsible for oversight in matters pertaining to

---

cliniche di medicinali per uso clinico," ordinary supplement to *Gazzetta Ufficiale della Repubblica Italiana* 184, no. 9 (August 2003).

[24] Published in *Official Journal of the European Union* L 91, no. 3 (2005), http://eur-lex .europa.eu/LexUriServ/LexUriServ.do?uri=OJ:L:2005:091:0013:0019:en:PDF.

[25] The document, also known as the European Bioethics Convention or the Oviedo Convention, can be found in English at http://conventions.coe.int/Treaty/en/Treaties/html/164 .htm. In summary, the conditions under which a human being can legitimately be subjected to experimentation are the following: (i) no alternative research method of comparable efficacy; (ii) proportionality of the associated risks; (iii) approval of the document by a "competent body"; (iv) information about the subject's rights and the safeguards for his protection; (v) voluntary and written informed consent. With regard to the protection of persons who are not able to consent, scientific research cannot normally be performed on them, given that valid informed consent cannot be obtained, unless (i) the first four conditions listed above are satisfied; (ii) the results expected from the research involve a real and direct benefit to the health of the person; (iii) the research cannot be conducted with comparable efficacy on competent subjects; (iv) authorization is granted by the patient's legal representative; (v) the person directly concerned does not refuse. For a commentary on the document, see A. Bompiani, "Una valutazione della 'Convenzione sui diritti dell'uomo e la biomedicina' del Consiglio d'Europa," *Medicina e Morale* 47, no. 1 (1997): 37–56.

[26] See "Testo aggiornato del decreto legislativo 30 dicembre 1992, n. 502, recante 'Riordino della disciplina in materia sanitaria, a norma dell'art. 1 della legge 23 ottobre 1992, n. 421'" [updated text of legislative decree no. 502 of December 30, 1992, pertaining to "restructuring health care regulations in accordance with article 1 of law no. 421 of October 23, 1992], ordinary supplement to *Gazzetta Ufficiale della Repubblica Italiana* 4 (January 7, 1994). This updated text also provides the changes made to law no. 502/1992 by legislative decree no. 517 of December 7, 1993, published in *Gazzetta Ufficiale della Repubblica Italiana* 293 (December 15, 1993). It is well known that those legislative measures in Italy introduced new aspects to the public management of health care such as the shift to a business model of health care, the pay-for-service system, and mechanisms allowing economic competition among service providers offering health care on behalf of the SSN.

[27] Royal decree no. 1265 of July 27, 1934, published in *Gazzetta Ufficiale* 186 (August 9, 1934).

vivisection. Royal decree no. 27 of June 11, 1934, established the Greater Institute of Health (Istituto Superiore di Sanità, ISS). Law no. 519 of August 7, 1973, defined and modified the characteristics of the ISS: among other tasks, it is charged with verifying the composition and safety of pharmaceutical products prior to clinical experimentation on human beings.

DECREES AND CIRCULARS OF THE MINISTRY OF HEALTH. The circulars of the Ministry of Health on April 9, 1975, and on July 28, 1977, already prescribed a sort of experimentation program that had to be presented in connection with requests for authorization to proceed to human applications.

Law no. 877/1978, which established the Italian National Health Care System (Sistema Sanitaria Nazionale, SSN), outlined the objectives of its services. This included the objective of "establishing regulations for experimentation, production, commercialization, and distribution of drugs and scientific information pertaining to those drugs, with the aim of ensuring a therapeutic approach and the familiarity and economic availability of the products." The same law entrusted and reserved to the state the administrative regulations concerning "production, registration, research, experimentation, commerce, and information pertaining to the chemical products used in medicine, pharmaceutical compounds, galenic compounds, and medicinal specialities." Regarding methods of experimentation, article 29 established that the authority to establish regulations is held by the state.

In addition to the previously cited ministerial decree on July 15, 1997, which implemented the European Union's guidelines for GCP in the conduct of clinical pharmacology trials and affirmed that "no clinical drug trials may be conducted without the approval of the competent ethics committee" (article 3, §4), the following warrant mention: the ministerial decrees on March 18, 1998, and March 19, 1998, regarding the institution and function of ethics committees and the certification of the suitability of centers for clinical drug trials, respectively. A significant change in the evaluation procedure for authorization to conduct multicenter experiments was made by legislative decree no. 211 of 2003, which implemented European directive 2001/20/EC: with particular regard to multicenter experiments, the ethics committee of the coordinating center was given a key role in having to issue a "unified opinion" regarding the entire experimentation protocol, including even the pharmacological assessment of the safety of the newly instituted active ingredient (except for certain classes of drugs that must be screened by the ISS), that—once issued—could only be either accepted or rejected by the ethics committees of the other centers involved in the experimentation.

Additional fundamental regulations on this topic were approved by two other decrees. The ministerial decree on May 8, 2003, regarding the therapeutic use of medicines subjected to clinical trials allows physicians to use experimental drugs— which must be requested directly from the manufacturer outside the context of clinical trials—in a limited capacity when the proper conditions have been met, such as the lack of a viable therapeutic alternative for the treatment of serious pathologies or rare diseases or conditions of illness endangering the life of the patient. Furthermore, the ministerial decree on December 17, 2004, regarding general prescriptions and conditions affecting the conduct of clinical drug trials, especially those whose purpose is the advancement of clinical practice as an integral part of health care, recognized a request from a large number of clinical researchers by regulating unsponsored clinical research not directed toward industrial development, recognizing it as an integral part of health care and, as such, allowing it to be covered by the SSN in terms of the cost of drugs already on the market to be administered to patients in the context of clinical trials.[28] This was an innovative law in the European context that could give a strong impetus to unsponsored biomedical research, especially in clinical areas that are not of interest to the pharmaceutical industry—and therefore scarcely investigated—yet involve groups of patients who are rightfully awaiting the attention and concrete commitment from public institutions.

The Italian *Codice di deontologia medica* (Code of medical ethics), approved by the national council of the Italian National Federation of Medical and Dental Associations (Federazione Nazionale degli Ordini dei Medici Chirurghi e degli Odontoiatri, FNOMCeO) in 2006,[29] expresses its position on the topic under chapter 8 ("Experimentation") of title 3 ("Relations with Citizens").[30]

---

[28] The ministerial circulars and decrees that are currently in force regarding clinical experimentation in Italy can be readily obtained in Italian at www.agenziafarmaco.gov.it under the heading "Normativa."

[29] Italian National Federation of Medical and Dental Associations (FNOMCeO), *Codice di deontologia medica*, rev. ed., December 16, 2006, http://portale.fnomceo.it/PortaleFnomceo/showVoceMenu.2puntOT?id=5.

[30] For the sake of completeness, here is the full text of each article referenced above. "*Article 45: Interventions on the genome*—Any intervention on the human genome must be intended to prevent and/or correct pathological conditions. *Article 47: Scientific experimentation*—The advancement of medicine is based on scientific research, which also avails itself of experimentation on animals and on human beings. *Article 48: Biomedical research and experimentation on human beings*—Biomedical research and experimentation on human beings must be informed by the inviolable principle of the protection of the psychophysical integrity and of the life and dignity of the person. They are subordinated to the consent of the subject undergoing experimentation, which must be expressed freely and consciously in writing after having been provided with specific information about the objectives, methods, foreseen benefits, and potential risks to the subject as well as about the subject's right to withdrawal from experimentation at any time. In the case of subjects who are minors or disabled or otherwise incompetent or under the legal responsibility of others, only experimentation for the preventive and therapeutic benefit of the subject is allowed. Consent must be expressed by the subject's legal representative(s), but the medical researcher must inform the person and record his or her wishes and is nonetheless required to take those wishes into consideration. All type of experimentation, including clinical experimentation, must be planned and conducted in accordance with appropriate protocols in the context of the laws in force and after having received prior approval from an independent ethics committee. *Article 49: Clinical experimentation*—Experimentation can be conducted in the context of diagnostic and/or therapeutic treatments only insofar as it is open to diagnostic or therapeutic utility for the citizens concerned. In all cases of clinical study, the patient cannot be deliberately deprived of confirmed diagnostic and therapeutic means that are indispensable to the maintenance and/or restoration of health. The aforementioned principles adopted on the topic of experimentation are also applicable to healthy volunteers. *Article 50: Experimentation on animals*—Experimentation on animals must be configured by the needs and purposes of garnering knowledge that is not otherwise obtainable and without the aim of profit, and it must be conducted with appropriate methods and means for avoiding useless suffering, and the protocols must have received the prior approval of an independent ethics committee."

Finally, the World Health Organization issued its *Operational Guidelines for Ethics Committees That Review Biomedical Research* in 2000 to call attention to the need to establish adequate procedures for biomedical research committees (Table 1).[31]

Ethics of human experimentation[32]

THE FUNDAMENTAL ETHICAL VALUES. Before going on to the specific assessment of individual cases it is helpful to recall some of the fundamental values and principles that are relevant here, as in other fields of applied medicine.

---

[31] World Health Organization, *Operational Guidelines for Ethics Committees That Review Biomedical Research* (Geneva: WHO, 2000), http://www.searo.who.int/LinkFiles/ RPC_Operational_Guidlines_Ethics.pdf. See also A.G. Spagnolo, "Comitati di etica per la ricerca: Procedure e qualità della revisione etica," in *Etica della ricerca*, Vial Correa and Sgreccia, 245–269.

[32] It must be noted that in recent years the quantity of literature on the ethical aspects of human experimentation has increased significantly. Here are several works that take different approaches with regard to their ethical standards: Spagnolo and Sgreccia, eds., *Lineamenti di etica*; McNeill, *Ethics and Politics*; R.J. Levine, *Ethics and Regulation of Clinical Research*, 2nd ed. (New Haven: Yale University Press, 1988); B.A. Brody, *Ethical Issues in Drug Testing, Approval, and Pricing* (New York: Oxford University Press, 1995); H.Y. Vanderpool, *The Ethics of Research Involving Human Subjects. Facing the 21st Century* (Frederick, MD: Univ. Publishing Group, 1996); E.J. Emanuel et al., *Ethical and Regulatory Aspects of Clinical Research* (Baltimore–London: Johns Hopkins University Press, 2003); on more specific topics, see E. Kodish, ed., *Ethics and Research with Children: A Case-Based Approach* (New York: Oxford University Press, 2005); M.J. Field and R.E. Behrman, eds., *Ethical Conduct of Clinical Research Involving Children* (Washington, DC: National Academies Press, 2004); M.A. Grodin and L.H. Glantz, *Children as Research Subjects* (Oxford: Oxford University Press, 1993); A.C. Mastroianni, R. Faden, and D. Federman, eds., *Women and Health Research: Ethical and Legal Issues of Including Women in Clinical Studies*, 2 vols. (Washington, DC: National Academy Press, 1994); B. Minogue, "Medical Research Involving Persons," in *Bioethics: A Committee Approach* (Boston: Jones and Bartlett, 1996); D. Evans and M. Evans, *A Decent Proposal: Ethical Review of Clinical Research* (Chichester: John Wiley, 1996); P. Cattorini, "Sperimentazione clinico-farmacologica: Problemi bioetici di attualità," in *A sua immagine e somiglianza?* ed. A. Mazzoni (Rome: Citta Nuova, 1997), 236–251; A. Fagot-Largeault, "Experimentation humaine," in *Les mots de la bioéthique*, ed. G. Hottois and M.H. Parizeau (Brussels: De Boeck Université, 1993), 219–228; F. Enia and E. Geraci, "Protocolli terapeutici," in *Nuovo dizionario di bioetica*, ed. S. Leone and S. Privitera (Palermo: EDB-ISB, 2004), 941–945; L. Pagliaro, "Trials," in *Nuovo dizionario di bioetica*, Leone and Privitera, 1211–1214; C. Foster, *The Ethics of Medical Research on Humans* (Cambridge University Press, 2001); E. Sgreccia, ed., *Storia della medicina e storia dell'etica medica verso il terzo millennio* (Soveria Mannelli: Rubbettino, 2000); B. Pitt et al., *La sperimentazione clinica* (Rome: Il Pensiero Scientifico Ed., 2000); S. Galbraith, M. Stat, and I. Marschner, "Guidelines for the Design of Clinical Trials with Longitudinal Outcomes," *Controlled Clinical Trials* 23 (2002): 257–273; L. Curtis and T. Meiner, "Clinical Trials and Treatment Effects Monitoring," *Controlled Clinical Trials* 19 (1998): 515–522; R. Horton, "The Clinical Trials: Deceitful, Disputable, Unbelievable, Unhelpful, Shameful: What Next?" *Controlled Clinical Trials* 22 (2001): 593–604; McKibbon A. et al., *Guida alla evidence based medicine* (Rome: Il Pensiero Scientifico, 2000); D. Pullman and X. Wang, "Adaptive Designs, Informed Consent and the Ethics of Research," *Controlled Clinical Trials* 22 (2001): 203–210.

Table 1. *Elements of the ethical review of a study*

Scientific design and conduct of the study—the appropriateness of the study design in relation to the objectives of the study, the statistical methodology (including sample size calculation), and the potential for reaching sound conclusions with the smallest number of research participants; the justification of predictable risks and inconveniences weighed against the anticipated benefits for the research participants and the concerned communities; the justification for the use of control arms; criteria for prematurely withdrawing research participants; criteria for suspending or terminating the research as a whole; the adequacy of provisions made for monitoring and auditing the conduct of the research, including the constitution of a data safety monitoring board (DSMB); the adequacy of the site, including the supporting staff, available facilities, and emergency procedures; the manner in which the results of the research will be reported and published;

Recruitment of research participants—the characteristics of the population from which the research participants will be drawn (including gender, age, literacy, culture, economic status, and ethnicity); the means by which initial contact and recruitment is to be conducted; the means by which full information is to be conveyed to potential research participants or their representatives; inclusion criteria for research participants; exclusion criteria for research participants;

Care and protection of research participants—the suitability of the investigator(s)'s qualifications and experience for the proposed study; any plans to withdraw or withhold standard therapies for the purpose of the research, and the justification for such action; the medical care to be provided to research participants during and after the course of the research; the adequacy of medical supervision and psychosocial support for the research participants; steps to be taken if research participants voluntarily withdraw during the course of the research; the criteria for extended access to, the emergency use of, and/or the compassionate use of study products; the arrangements, if appropriate, for informing the research participant's general practitioner (family doctor), including procedures for seeking the participant's consent to do so; a description of any plans to make the study product available to the research participants following the research; a description of any financial costs to research participants; the rewards and compensations for research participants (including money, services, and/or gifts); the provisions for compensation/treatment in the case of the injury/disability/death of a research participant attributable to participation in the research; the insurance and indemnity arrangements;

Protection of research participant confidentiality—a description of the persons who will have access to personal data of the research participants, including medical records and biological samples; the measures taken to ensure the confidentiality and security of personal information concerning research participants;

Informed consent process—a full description of the process for obtaining informed consent, including the identification of those responsible for obtaining consent; the adequacy, completeness, and understandability of written and oral information to be given to the research participants, and, when appropriate, their legally acceptable representative(s); clear justification for the intention to include in the research individuals who cannot consent, and a full account of the arrangements for obtaining consent or authorization for the participation of such individuals; assurances that research participants will receive information that becomes available during the course of the research relevant to their participation (including their rights, safety, and well-being); the provisions made for receiving and responding to queries and complaints from research participants or their representatives during the course of a research project;

Community considerations—the impact and relevance of the research on the local community and on the concerned communities from which the research participants are drawn; the steps taken to consult with the concerned communities during the course of designing the research; the influence of the community on the consent of individuals; proposed community consultation during the course of the research; the extent to which the research contributes to capacity building, such as the enhancement of local health care, research, and the ability to respond to public health needs; a description of the availability and affordability of any successful study product to the concerned communities following the research; the manner in which the results of the research will be made available to the research participants and the concerned communities.

SOURCE: World Health Organization, *Operational Guidelines for Ethics Committees That Review Biomedical Research* (Geneva: WHO, 2000).

The values that come into play in this field are the following, in hierarchical order: the defense of life and of the human person, the legitimacy of the therapeutic principle, and the social issue connected with scientific progress.

The inviolability of the human person—the primacy of the person—is the most pivotal principle of ethics as a whole. It has been recalled several times that the person is a "transcendent" value in comparison with all the rest of created reality. Even for someone who does not share the Christian faith, the value of the human person is not instrumental and cannot be turned instrumental.[33] Those who then discover the image of the Creator therein, the subject whom Christ has made a participant in divine life itself, understand even more clearly this primacy of the value that the person is.

Yet it is not enough to declare this, because a distinction could be made between physical life and personal life. This alleged distinction was already encountered in the discussion on abortion and will arise again with the topic of euthanasia. Some would say, "Yes, defend human life, but only when it is personal life, when it is worthy of the person." Clarity is necessary here: the human person does not exist in his earthly concreteness without physical life; physical life, while it does not exhaust the totality of personal life, is the foundation of personal life. Hence every offense against physical life (or the destruction of it) is an offense against the person. The body is co-essential to the person, and the life of the body must be defended as the first duty of every human being; the inviolability of human life and the defense of its integrity constitute the first duty and the first right. This is not to say that the physical life of a human being is the all-encompassing value of the person as a whole; higher moral demands and religious values can even require putting life itself at risk (exposure to risk, not the direct taking of life). Yet physical life is the *fundamental value*, because all other successive developments of the person and all other potential for growth and freedom are "founded" upon it.

The other value that provides orientation for the ethical evaluation under discussion is the value related to the enunciation of the *principle of totality,* or *therapeutic principle*. Precisely in order to defend physical life as the fundamental value of the person, it is permissible to sacrifice one part of the organism in order to save that organism. The limits and applications of this principle were addressed in the preceding chapters; here it must be said that in administering a drug there is an inevitable and justifiable minimum of risk weighed against other risks and advantages for the physical integrity of the subject upon whom the experiment is performed. Within the scope of this principle, "a prudent administration, however, requires that the damage done to the body ... be a lesser evil than what is sought to be avoided by means of this sacrifice. Otherwise it would amount to the absurd act by which a greater evil is committed in order to avoid a lesser one."[34] Clearly the limits already mentioned may slow scientific research, but "it goes against all common sense to violate fundamen-

---

[33] No. 75 of the *Charter for Health Care Workers* by the Pontifical Council for Pastoral Assistance to Health Care Workers reads, "[The human person], because of his unique dignity, can be the subject of research and clinical experimentation with the safeguards due to a being with the value of a subject and not an object. For this reason, biomedical sciences do not have the same freedom of investigation as those sciences which deal with things. 'The ethical norm, founded on respect for the dignity of the person, should illuminate and discipline both the research stage and the application of the results obtained from it.'"

[34] Perico, "Sperimentazione clinica."

tal human values in order to heal human beings,"[35] and it is contrary to reasonable morality that one should do evil to attain good. What has just been delineated is the principle of proportionate risk.[36]

The principle or value of *social solidarity* is within the scope of this principle of making a particular sacrifice for the sake of a higher good. By virtue of this principle it is permissible to ask the individual (except for what was said about the two preceding values that must be safeguarded) to accept an amount of sacrifice or risk for the good of the whole society of which everyone is a member. This principle must be understood correctly so as not to fall into a collectivist understanding of health. The notions of "the part" and "the whole" cannot be applied to the individual person and society, respectively: the value of a whole and the fundamental reason for the very existence of society are found in the individual person.[37] Hence, except in the case of legitimate defense, one may never expose an individual to a fatal risk for the advantage of society. Very clear limits must therefore be set to this social principle; there are two such limits of particular importance:

1. Society (or the public authorities or the researcher) cannot demand a greater risk of a subject than what the individual subject can demand of himself. Thus, neither medical science nor the physician has an amount of control over the patient's life that is greater than what the patient himself possesses. Just as the subject must respect the intrinsic *order of values* in his moral behavior, so also should the physician with regard to the patient: survival, substantial (mental and physical) integrity, and personal and family morality are factors and values that surpass the value of health and healing. Medical science exists and functions to safeguard these values, not to damage them.

2. Since the physician researcher is acting upon another person, while always observing the objective limits described previously, he must furthermore obtain prior *informed consent*.[38]

---

[35] Perico, "Sperimentazione clinica."

[36] "Of its nature, every experimentation has risks. Hence, 'it cannot be demanded that all danger and all risk be excluded. This is beyond human possibility; it would paralyze all serious scientific research and would quite often be detrimental to the patient. ... But there is a level of danger that the moral law cannot allow.' A human subject cannot be exposed to the same risk as beings which are not human. There is a threshold beyond which the risk becomes humanly unacceptable. This threshold is indicated by the inviolable good of the person, which forbids him 'to endanger his life, his equilibrium, his health, or to aggravate his illness.' ... Experimentation on the human person must be in accord with the principle of *proportionate risk*, that is, of due proportion between the advantages and foreseeable risks." Pontifical Council for Pastoral Assistance to Health Care Workers, *Charter*, nos. 78–79.

[37] This concept was explained very clearly in J. Maritain, *The Rights of Man and Natural Law*, trans. D.C. Anson (New York: C. Scribner's Sons, 1943), 11–17.

[38] The patient "should be informed about the experimentation, its purpose and possible risks, so that he can give or refuse his consent with full knowledge and freedom. In fact, the doctor has only that power and those rights which the patient himself gives him." Pontifical Council for Pastoral Assistance to Health Care Workers, *Charter*, no. 77.

Invoking, as some do, the so-called *teleological principle* to justify harmful or immoral experiments on human subjects with a view to scientific progress or the good of society is tantamount to reintroducing, under another guise, the false justifications that led to the crimes in the concentration camps during World War II and to inverting the relation between person and society.

Finally, the values just cited prompt measures for *structural, planning, and procedural correctness* in clinical experimentation,[39] which involve the necessary safety precautions that are required in the research phase; this means condemning all abuses and crimes, such as the Nazi atrocities.[40] More recently the problem of ethical protocols has expanded to include clinical experimentation conducted in so-called developing nations.[41]

In light of all these principles it is understandable that Pius XII declared, "Medical research will seek to practice upon its immediate object, a living human being, in the interests of science, in the interests of the sick person, and in the interests of the community. This should certainly not be condemned, so long as it stops at the limits defined by moral principles."[42]

---

[39] See, for example, K. F. Schaffner, "Research Methodology: I. Conceptual Issues," in *Encyclopedia of Bioethics*, ed. Post, 2326–2334; L. M. Kopelman, "Research Methodology: II. Clinical trials," in *Encyclopedia of Bioethics*, ed. Post, 2334–2343; C. Levine and T. W. Ogletree, "Research Methodology: III. Subjects," *Encyclopedia of Bioethics*, ed. Post, 2343–2347; S. Lock and F. Wells, *Fraud and Misconduct in Medical Research* (London: British Medical Journal Publishing Group, 1996); Spagnolo, "Principi etici."

[40] See B. Freedman, "Research, Unethical," in *Encyclopedia of Bioethics*, ed. Reich, rev. ed., 2258–2261.

[41] See, for example, C. A. Pace and E. J. Emanuel, "The Ethics of Research in Developing Countries: Assessing Voluntariness," *Lancet* 365, no. 9453 (2005): 11–12; A. A. Hyder et al., "Ethical Review of Health Research: A Perspective from Developing Country Researchers," *Journal of Medical Ethics* 30, no. 1 (2004): 68–72; N. E. Kass, L. Dawson, and N. I. Loyo-Berrios, "Ethical Oversight of Research in Developing Countries," *IRB* 25, no. 2 (2003): 1–10; D. Resnik, "Research Subjects in Developing Nations and Vulnerability," *American Journal of Bioethics* 4, no. 3 (2004): 63–4; M. C. Ruof, "Vulnerability, Vulnerable Populations, and Policy," *Kennedy Institute of Ethics Journal* 14, no. 4 (2004): 411–25; P. Lurie and S. M. Wolfe, "Unethical Trials of Intervention to Reduce Perinatal Transmission of the Human Immunodeficiency Virus in Developing Countries," *New England Journal of Medicine* 337, no. 12 (1997): 853–856; M. Angell, "The Ethics of Clinical Research in the Third World," *New England Journal of Medicine* 337, no. 12 (1997): 847–849.

[42] Pius XII, An Address to the First International Congress on the Histopathology of the Nervous System, September 14, 1952, http://www.ncbcenter.org/Page.aspx?pid=388. Papal teaching, especially that of Pius XII and John Paul II, serves as an important reference point on the topic of human experimentation. See in particular Pius XII, Address at the Sixteenth Session of the International Office of Documentation on Military Medicine, October 19, 1953, in *Discorsi e Radiomessaggi* (Vatican City: Tipografia Poliglotta Vaticana, 1969), 11:415–428 [Italian]; idem, Address to Participants in the Eighth Assembly of the World Medical Association, September 30, 1954, in *Discorsi e Radiomessaggi*, 17:167–179 [Italian]; Paul VI, Address at the Eleventh National Congress of the Italian Society of Pathology, October 31, 1968, in *Insegnamenti di Paolo VI* (Vatican City: Tipografia Poliglotta Vaticana, 1969), 7:717–720 [Italian]; John Paul II, Address to Participants in Two Congresses on Medicine and Surgery, October 27, 1980, in *Insegnamenti di Giovanni Paolo II* (Vatican City: Libreria Editrice Vaticana, 1980–1987), 3.2:1005–1010 [Italian]; idem, Address to the

It is necessary to add, however, that even the good of science must be understood not in an absolute but rather in a relative sense: science is of man and for man and cannot require sacrifices or impose violations on a human being beyond what the good of the human person, correctly understood, can require of individuals.

In the same way, the authority of the state cannot go beyond the limits established by the preceding principles, not even in the name of an alleged good for individual persons. The state should issue clear norms regarding pharmacological research so as to safeguard the human person. We know how easily the eagerness of researchers on the one hand and economic considerations by the pharmaceutical industries on the other can cross the limits of personalist morality.[43]

PRACTICAL ETHICAL GUIDELINES REGARDING HUMAN EXPERIMENTATION. The guidance arising from the ethical values described above can be outlined as follows:

a. As a necessary means in the fight against disease, clinical drug trials are not only permissible but actually constitute a service to mankind and share in the goodness and *value of science*, provided they are conducted properly and under morally acceptable conditions.

b. It is in fact the duty of the state and the community to encourage, finance, and regulate scientific research for the relief of human suffering and to prevent abuses and profiteering at the expense of individuals and the community. Cutting off funds for scientific research is as serious as depriving children of food and sick persons of medicine; to neglect regulation in this sector would mean promoting the subordination of science to profit and betraying the expectations and confidence of the sick.

c. The preclinical phase of experimentation should be conducted with all possible diligence so as to acquire a maximum of valid findings and minimize risks in the trials on human subjects. In light of this, animal experimentation—conducted with due respect for protective legislation—should not be so narrowly restricted by law as to cause increased risk during the phase involving human subjects.

d. Experimentation must be conducted by a competent person and supervised by a clinic that is likewise competent and conscientious.

e. The *risk* posed during the clinical phase of experimentation, whether therapeutic or nontherapeutic, must be *proportionate* to the purpose and guarantee the

---

Fifteenth Congress of Catholic Physicians, October 4, 1982, in *Insegnamenti di Giovanni Paolo II*, 5.3:669–677 [Italian]; idem, Address to Members of the Pontifical Academy of Sciences, October 23, 1982; idem, Address to Participants in the Conference of the Pro-Life Movement, December 3, 1982, in *Insegnamenti di Giovanni Paolo II*, 5.3:1509–1513 [Italian]; idem, Address to Participants in a Study Course on Human Preleukemias, November 15, 1985, in *Insegnamenti di Giovanni Paolo II*, 8.2:1265 [Italian]; idem, Address to Participants in the International Congress on Cancer and Hormones, April 26, 1986; idem, Address to a Conference on Pharmaceuticals in the Hall of the Synod, October 24, 1986, in *Insegnamenti di Giovanni Paolo II*, 9.2:1183 [Italian]; idem, Address to Participants in a Congress on Surgery, February 19, 1987, in *Insegnamenti di Giovanni Paolo II*, 10.1:376 [Italian]; idem Address to Scientists and Health Care Workers, November 12, 1987, in *Insegnamenti di Giovanni Paolo II*, 10.3:1086–1087 [Italian]; *Catechism of the Catholic Church*, nos. 2292–2295; Pontifical Council for Pastoral Assistance to Health Care Workers, *Charter*, nos. 75–82.

[43] G. Gismondi, *Etica fondamentale della scienza* (Rome: Citta Nuova, 1997); idem, *Critica ed etica nella ricerca scientifica* (Turin: Marietti, 1978); various authors, *Manipolazione e futuro dell'uomo* (Bologna: Dehoniane, 1972).

integrity of the subject upon whom the experiment is being conducted. The trials must be suspended as soon as it becomes evident that this proportion and guarantee no longer exist. The risk factor must also be taken into account in the case of volunteers and of experimentation conducted by the physician upon himself. In particular, extreme care should be taken before including women of childbearing age in clinical trials, even when there is only a remote possibility of a harmful effect on a fetus. Indeed, in the case of a pregnant or nursing mother, the prospective subject must be excluded from experimentation as a necessary safeguard for the health of both the woman and the child.

Within the category of women of childbearing age, who are therefore potentially open to becoming pregnant before or during the experimentation, it is necessary to make distinctions: If a drug has the potential to cause deformities, in a case where there are standard treatments other than the experimental one or in a case of experimentation on healthy volunteers, women of childbearing age must be excluded. If on the other hand there are no standard therapies available and the seriousness of the pathology makes treatment necessary, then the woman who subjects herself to the experimental study must be advised not to become pregnant during the trial. Finally, similar precautions must be taken with an experimental drug that may cause harmful side effects for a fetus through its action on the genetic inheritance of the parents through their germ cells, meaning that careful attention must also be given to the role of potential fathers.[44]

f. *Informed consent* is absolutely necessary in the case of a nontherapeutic experimentation, the advantage and purpose of which have no immediate connection with the subject on whom the experiments are conducted; all subjects incapable of giving completely free and fully conscious consent must therefore be excluded. With regard to informed consent in medicine[45] and in particular with a view to experimentation,[46] it is necessary to insist that this consent, when requested, be truly

---

[44] For further details, consult J. C. Bennet et al., "Inclusion of Women in Clinical Trials: Policies for Population Subgroups," *New England Journal of Medicine* 329, no. 4 (1993): 288–292; R. B. Merkatz, "Women in Clinical Trials of New Drugs: A Change in Food and Drug Administration Policy," *New England Journal of Medicine* 329, no. 4 (1993): 292–296; A. G. Spagnolo, "L'inclusione delle donne nelle sperimentazioni farmacologiche," in *Bioetica nella ricerca*, 470–472; and the "Raccomandazioni riguardo alla inclusione delle donne in età fertile nei protocolli di sperimentazione clinica" (Recommendations regarding the inclusion of women of childbearing age in clinical trial protocols) drawn up and adopted by the ethics committee of the A. Gemelli Polyclinic of the Catholic University of the Sacred Heart, published in *Medicina e Morale* 53, no. 6 (2003): 1271–75.

[45] For a general overview of the ethical and legal aspects, see Italian National Bioethics Committee, *Information and Consent to Medical Treatment*, June 29, 1992 (Rome: Presidenza del Consiglio Dipartimento per l'Informazione e l'Editoria, 1992), http://www.governo.it/bioetica/eng/pdf/Informazione_e_consenso_all%27atto_medico_2.pdf; A. Santosuosso, ed. *Il consenso informato: Tra giustificazione per il medico e diritto del paziente* (Milan: Cortina, 1996); A. Fiori, "Problemi attuali del consenso informato," *Medicina e Morale* 43, no. 6 (1993): 1123–1138; L. Eusebi, "Sul mancato consenso al trattamento terapeutico: profili giuridico-penali," *Rivista Italiana di Medicina Legale* 17 (1995): 727–740.

[46] On this topic, see P. Casali and A. Santosuosso, "Il consenso informato nella sperimentazione clinica," in *Il consenso informato*, Santosuosso, 169–184; R. J. Levine, "Informed Consent: III. Consent Issues in Human Research," in *Encyclopedia of Bioethics*,

informed; that is, it must be based on complete, understandable, and duly received information. It must also be truly free; that is, it must be requested of persons who have no form of moral dependence or physical constraint (children, detainees, prisoners, employees, or individuals who have been sentenced to death). This informed consent must be obtained for therapeutic experimentation as well, although it can be presumed or omitted for very serious reasons in order to save the patient's life. This may occur in an emergency when an experimental drug is used in a final effort to save the life of a patient for whom all other therapeutic measures have failed (the patient does not have the ability to understand and decide and it impossible to wait any longer). Moreover, the request for informed consent should not be the time or occasion to provide the patient with information about the state of his illness that might make him even more dejected, aggravate his depression, or risk inciting him to perform acts that might endanger his life, as in the case of patients in the final stages of a terminal illness. The case of consent for experimentation on minors will be considered separately.

g. *The protection of personal information*, in other words, information about subjects who participate in clinical experimentation must be treated with the appropriate confidentiality. The 1996 revision of the GCP states in this regard, "The confidentiality of records that could identify subjects should be protected, respecting the privacy and confidentiality rules in accordance with the applicable regulatory requirement(s)."[47]

h. Attention also needs to be paid to so-called *orphan drugs*. Briefly, these are pharmacological substances that are useful for the treatment or cure of diseases with extremely low rates of epidemiological incidence or of endemics afflicting popula-

---

Post, 1280–1290; S. Hewlett, "Consent to Clinical Research: Adequately Voluntary or Substantially Influenced?" *Journal of Medical Ethics* 2 (1996): 232–237; N.E. Kass et al., "Trust: The Fragile Foundation of Contemporary Biomedical Research," *Hastings Center Report* 26, no. 5 (1996): 25–29; B. MacKinnon, "How Important Is Consent for Controlled Clinical Trials?" *Cambridge Quarterly of Healthcare Ethics* 5, no. 2 (1996): 221–227; J.P. Porter, "Informed Consent Issues in International Research Concerns," *Cambridge Quarterly of Healthcare Ethics* 5, no. 2 (1996): 237–243. There have been attempts to improve public awareness of informed consent and to make the documents more "comprehensible" for patients who sign them; see in particular D.J. Willison et al., "Patients' Consent Preferences for Research Uses of Information in Electronic Medical Records: Interview and Survey Data," *British Medical Journal* 326, no. 7385 (2003): 373–77; European Organization for Research and Treatment of Cancer (EORTC), *What Are Cancer Clinical Trials All About?* (Brussels: 1997); J.J. Giffels, *Clinical Trials: What You Should Know before Volunteering to Be a Research Subject* (New York: Damos Vernande, 1996).

[47] European Medicines Agency, *Note for Guidance on Good Clinical Practice* (London: EMEA, 2006), §2.11, p. 12, http://www.ich.org/fileadmin/Public_Web_Site/ICH_Products/Guidelines/Efficacy/E6_R1/Step4/E6_R1__Guideline.pdf. Currently in force in Italy is the related legislative decree no. 196 of June 30, 2003, on the protection of personal data (published in *Gazzetta Ufficiale della Repubblica Italiana* 174 [July 29, 2003]); this document deals extensively with protection of sensitive data in the field of scientific and health care research. On this topic, the reader may consult D. Kalra et al., "Confidentiality of Personal Health Information Used for Research," *British Medical Journal* 333, no. 7560 (2006): 196–8; R.M. Veatch, "Consent, Confidentiality, and Research," *New England Journal of Medicine* 336, no. 12 (1997): 869–870; R. Minacori and A.G. Spagnolo, "Consenso, 'privacy' e ricerca medica," *Medicina e Morale* 47, no. 4 (1997): 811–814.

tions that are economically indigent and hence incapable of paying for appropriate treatment.[48] In either case it is difficult for the pharmaceutical industries to expend resources and time on research that, for various reasons, would not bring them a return in the form of profits. Obligating pharmaceutical firms to produce drugs at a net loss is not feasible; to get around this situation, plans have been undertaken in some countries, such as the United States, to provide financial incentives for industries that pledge to develop orphan drugs to treat rare diseases. For diseases in developing countries, on the other hand, an international collaborative initiative coordinated by the United Nations has been started.

i. *Drug monitoring* can be defined as the process of monitoring possible adverse reactions to drugs[49] and notifying the public health authorities. The topic has no short supply of specific ethical implications;[50] some examples include (1) the recognition and communication of adverse events and (2) the types of clinical drug monitoring studies. In the first case, it is ethically obligatory for the physician to report to the public health authority in a timely manner any adverse events presumably connected with the taking of a certain drug and for the authorities to draw from them the appropriate conclusion in practical terms. At a more general level, international data banks on drug reviews prove to be extremely useful, since they allow rapid and efficient exchange of information in real time. In the second case, it must be said that even the 1996 GCP guidelines make no reference to phase IV studies, which concern both comparisons with other drugs already on the market and studies of drug reviews. Ethical standards must be upheld in the latter type of studies by maintaining a satisfactory physician-patient relationship and by avoiding the introduction of conditions that could alter that relationship. On the other hand, again for ethical reasons, it is desirable to have the collaboration of other health care professionals (pharmacists, medical specialists) and adequate education and training for the primary physician in order to achieve efficient collaboration in phase IV studies.

j. The correct use of the *placebo* must be considered. A placebo is an inert substance devoid of pharmacological effects that is called for in various controlled clinical experiments. It is administered to so-called experimental control groups and compared with the study "arm" to which the drug being tested is given. The general use of a placebo in therapeutic clinical experimentation on humans is debatable; three reasons are given in support of it: (1) to eliminate all possible interference in interpreting the effects attributed to the new drug, (2) to avoid the difficulty of choosing a treatment with which to compare the experimental therapy, and (3) to take advantage of the presumed greater ease of demonstrating the statistical significance of an innovative treatment in comparison with a placebo, as opposed to a comparison between two therapies (the experimental one and a standard treatment).

---

[48] Italian National Bioethics Committee, *La sperimentazione dei farmaci,* chap. 5, "I farmaci orfani" (orphan drugs), 43–44.

[49] In the case of medicinal products already on the market, the above-cited 1996 revision of the GCP defines an adverse drug reaction as "a response to a drug which is noxious and unintended and which occurs at doses normally used in man for prophylaxis, diagnosis, or therapy of diseases or for modification of physiological function" (§1.1, p. 5).

[50] Italian National Bioethics Committee, *La sperimentazione dei farmaci*, chap. 7: "Farmacovigilanza" (drug monitoring), 47–55.

These reasons, however, do not hold up to the following considerations of methodology and professional ethics:[51] (1) The purpose of medicine is to take into account what already exists and to find something better. (2) The physician's primary obligation with respect to the sick person is therapeutic, even before his obligations to science or to future patients. (3) It is illusory to think that a study which is still small-scale (in comparison to a generalized use of the substance once it is marketed) could have absolute scientific validity, given that measurements are inherently and demonstrably susceptible to appreciable statistical error, meaning that the real value of the new product, with regard to its efficacy in comparison to proven treatments, is still not at all certain. Moreover, again from the perspective of professional ethics and law, other problems arise by virtue of the following additional factors: (4) Obtaining valid informed consent for experimentation with a placebo does not relieve the physician of the duty to administer the best treatment available to the patient. And (5) if, on the other hand, it happens that the patient does not receive adequate information, then it constitutes an actual act of "deception" and the consent obtained is not valid.

Ethical assessment and specific cases

Let us first take the more common case of *experimentation on a sick person.* This experimentation can be conducted for therapeutic purposes, as pure experimentation for the sake of knowledge, or as a combination of the two.

We will examine the case in which the drug is used because it is therapeutically justified, has already proved innocuous in the context of basic and laboratory experimentation, has been authorized for use on human subjects, but still needs clinical confirmation and further certification. In this case the physician can use the drug on the patient with the latter's specific consent or, in the event that he is not in a position to give it, with the consent of family members or of a legal representative.

If the drug is being used not primarily for the good of that patient but rather to test the drug itself or for the benefit of the researcher (nontherapeutic purpose), then the patient's personal and explicit consent must absolutely be given if it is to be used.

There could also be a case in which the use of a method or drug that has not yet been proven effective is the *last chance* for an extreme attempt to save a life that will otherwise be lost. In this case the experimental character happens to coincide with the therapeutic character, because the drug is the only treatment possible; consent itself, which sometimes cannot be obtained directly from the patient or from the

---

[51] For an in-depth treatment of the topic, see G. Torlone, "Le sperimentazioni internazionali: Il dibattito sugli studi placebo-controllati," *Medicina e Morale* 54, no. 3 (2004): 555–588; S.J. Ellis and R.F. Adams, "The Cult of the Double-Blind Placebo-Controlled Trial," *British Journal of Clinical Practice* 51 (1997): 36–39; P.P. De Deyn and R. D'Hooge, "Placebos in Clinical Practice and Research," *Journal of Medical Ethics* 22 (1996): 140–146; G. Folli, "L'uso del placebo nei trials clinici: Significato scientifico e valore sperimentale," in *Lineamenti di etica*, Spagnolo and Sgreccia, 85–90; L. Candia, "L'uso del placebo nei trials clinici: Considerazioni etico-deontologiche," in *Lineamenti di etica*, Spagnolo and Sgreccia, 91–94; K.J. Rothman and K.B. Michels, "The Continuing Unethical Use of Placebo Controls," *New England Journal of Medicine* 331, no. 6 (1994): 394–398; A.G. Spagnolo, "L'uso del placebo nelle sperimentazioni farmacologiche," in *Bioetica nella ricerca*, 464–468; R. Minacori and A.G. Spagnolo, "Il placebo nella pratica clinica e nella ricerca," *Medicina e Morale* 47, no. 2 (1997): 444–447; S. Bok, "The Ethics of Giving Placebos," *Scientific American* 15 (1974): 17–23.

parents (in the case of minors) or from the legal representative, can be considered implicitly and tacitly given in the general intention of the patient and parents that everything possible be done to save the life of the patient. In this way one could justify, for example, Pasteur's use of an experimental rabies vaccine on a boy who had been bitten by a rabid dog and could not otherwise be restored to health.

In this regard Pius XII declares, "In desperate cases when the sick person is lost unless an intervention is made, if there is a medication, a measure, an operation which, although not entirely without danger, still has some possibility of success, an upright and thoughtful mind at once admits that the physician may, with the explicit or tacit consent of the patient, proceed to the application of such a treatment." [52]

Laws protecting the inviolability of the person and forbidding exploitation do not change when the person in question is handicapped or mentally ill or dying (the case of fetuses will also be examined); since they are *incapable of consent, they cannot be made subjects of experiments* that require informed consent. Therefore, *only therapeutic experimentation* can be conducted on those who are incapable of consent. [53]

---

[52] Pius XII, Address to the World Medical Association, September 30, 1954.

[53] Ibid. As for the participation of these subjects in medical research, including research for nontherapeutic purposes, notwithstanding suitable and rigorous precautions concerning the definition and control of risk, it is necessary to emphasize a clear-cut change from the past in the international debate, in which there is a much greater orientation toward the possibility. Proof of this is the appearance, in recent revisions of both the Declaration of Helsinki (2008) and the *International Ethical Guidelines for Biomedical Research Involving Human Subjects* (2002), of the distinction between therapeutic and nontherapeutic purposes of research. The 2008 revision of the Declaration of Helsinki, at no. 27, which deals with subjects incapable of giving consent, whether adults or minors, declares that "for a potential research subject who is incompetent, the physician must seek informed consent from the legally authorized representative. These individuals must not be included in a research study that has no likelihood of benefit for them unless it is intended to promote the health of the population represented by the potential subject, the research cannot instead be performed with competent persons, and the research entails only minimal risk and minimal burden." Guideline 15 of the CIOMS *International Ethical Guidelines* has similar wording: "Before undertaking research involving individuals who by reason of mental or behavioural disorders are not capable of giving adequately informed consent, the investigator must ensure that: such persons will not be subjects of research that might equally well be carried out on persons whose capacity to give adequately informed consent is not impaired; the purpose of the research is to obtain knowledge relevant to the particular health needs of persons with mental or behavioural disorders; the consent of each subject has been obtained to the extent of that person's capabilities, and a prospective subject's refusal to participate in research is always respected, unless, in exceptional circumstances, there is no reasonable medical alternative and local law permits overriding the objection; and, in cases where prospective subjects lack capacity to consent, permission is obtained from a responsible family member or a legally authorized representative in accordance with applicable law." On the other hand, in guideline 9 ("Special Limitations on Risks When Research Involves Individuals Who Are Not Capable of Giving Consent") the CIOMS defines additional elements to protect these subjects in a particular way given the possibility that there may be no direct benefit from experimentation: "When there is ethical and scientific justification to conduct research with individuals incapable of giving informed consent, the risk from research interventions that do not hold out the prospect of direct benefit for the individual subject should be no more likely and not greater than the risk attached to routine medical or psychological examination of such persons. Slight or minor increases above such risk may be permitted when there is

EXPERIMENTATION ON HEALTHY VOLUNTEERS. The principle of social solidarity, which I attempted to explain earlier, applies to this matter and makes it permissible to experiment on a healthy volunteer under certain conditions.[54]

The first condition is that the consent is free, informed, explicit, and personal: since the person in question is healthy, there is no room for presumed consent. Experiments performed on persons who are dependent on the experimenter for employment or who act as his subordinates remain of dubious morality, because of the presumption of coercion.

The risk to which the volunteer is subjected cannot and must not in any way encroach upon the defense of life and of substantial integrity. This is the limit on the control that the subject has even over himself.

Even on this side of the limit, the *law of proportionality* between risks and benefits still holds; experiments lacking scientific value and validity or performed out of sheer curiosity are therefore to be ruled out. The experimenter must be ready to interrupt the experimentation at any moment: either when consent is withdrawn or when unforeseen risks arise. The organization of the research must therefore provide procedures for stopping the experiment at any stage.

Anything degrading or immoral and any dishonest intention, whether with or without the concurrence of the person undergoing the experiment, must be excluded from the experimentation.[55]

EXPERIMENTATION ON ONESELF. The intention of performing a heroic act in good faith for the benefit of science or the desire for the admiration of society does not change the objective moral assessment of experimentation on oneself; therefore the limits and conditions explained for healthy volunteer subjects still apply to this case.

It must be added that when experimentation is performed on oneself, the law should provide for the presence of another person capable of stopping the experiment in the event of unforeseen or serious risks.

EXPERIMENTATION ON DETAINEES. The objective limits of life and substantial integrity are valid in every case, even for prisoners. It is permissible to offer the opportunity to volunteer for experimentation, within the aforesaid limits, even when there may be rewards (for instance a commuted sentence), provided that the consent is well informed and free of any sort of deception. Certain experiments with carcinogenic substances— which have been performed even in the United States—or with other dangerous or harmful substances are absolutely inadmissible, even on prisoners of war or convicts. In the case of a prisoner who has been condemned to death (in those countries where the death penalty still exists), some would justify unsafe experimentation with lethal risks if the condemned person's consent is obtained; it is said that it would be a sort of substitute punishment (the possibility of dying from the drug rather than by the

---

an overriding scientific or medical rationale for such increases and when an ethical review committee has approved them."

[54] On this subject, see also Pontifical Council for Pastoral Assistance to Health Care Workers, *Charter*, no. 81.

[55] For an in-depth treatment of the ethical aspects of experimentation on volunteers, see R. Mordacci, "Disponibilità e disposizione: Riflessioni etiche sulla partecipazione di volontari sani alla ricerca biomedica," *Medicina e Morale* 41, no. 4 (1991): 585–611 I; P. Tiraboschi and A. Spagnoli, "Le indagini sull'uomo sano," *Federazione Medica* 44 (1991): 27–30.

electric chair), and it is also said that such experiments must not involve additional pain or degrading treatments. But in my opinion this possibility of experimentation on prisoners condemned to death does not take into account the impossibility of free, informed consent under those conditions. In any case this possibility was not taken into consideration in the Declaration of Helsinki precisely because of the difficulty in reaching an agreement on the subject.[56] Such experimentation with that degree of risk would appear to be unacceptable from an ethical perspective.

EXPERIMENTATION ON CHILDREN AND MINORS. In articulating general principles it was already explained that these subjects are included among those who are unqualified or unable to give valid consent; experimentation on these subjects is therefore described as licit only if it is conducted for therapeutic purposes, for the benefit of the subjects themselves taken individually, and with the consent of parents or of legal representatives.

In recent years, however, the ethical literature has proposed more nuanced responses. First of all, a distinction is introduced between real mental capacity and the legal capacity to give informed consent: with regard to consent, a five-year-old child is quite different from a fifteen-year-old youth, who can even contract a valid marriage in some countries.

Recall, moreover, that there are experiments involving almost no risk at all: for example, one such experiment might involve only obtaining a small quantity of blood. Therefore, according to some authors, not only age but also risk should be considered: a dermatologist's experiment with a compound that may cause irreversible hair loss is one thing, while the blood sample mentioned above, when taken in the interests of research, is another.

Others introduce a consideration with utilitarian or "teleological" overtones based on the great progress that certain experiments, which were illicit in themselves, have brought about in curing diseases that were untreatable with previously known drugs. An example is the research on the Salk vaccine that led to the eradication of poliomyelitis after experiments on children.

Finally, some authors emphasize the possibility of the combination ("identification") of the consent given by the parents or legal representatives with the explicit consent (assent) of a minor who has the use of reason. Since the latter is capable of acts of generosity he can also accept, if properly instructed, an altruistic agreement that involves some risk.[57]

The Declaration of Helsinki, updated by the World Medical Association in 2008, effectively combines the case of research involving a minor with that of the incompetent adult subject, making no explicit mention of minors. Concerning consent, the document reads as follows in paragraphs 27 and 28: "For a potential research subject who is incompetent, the physician must seek informed consent from the legally authorized representative. . . . When a potential research subject who is deemed incompetent

---

[56] Guideline 13 of the above-cited CIOMS *International Ethical Guidelines* (2002), which refers to research involving vulnerable persons, recommends requiring a special justification for inviting vulnerable individuals to participate as subjects of research, and if they are included, rigorously applying measures for the protection of their rights and well-being.

[57] R. H. Redmon, "How Children Can Be Respected as 'Ends' Yet Still Be Used as Subjects in Non-therapeutic Research," *Journal of Medical Ethics* 12 (1986): 77–82.

is able to give assent to decisions about participation in research, the physician must seek that assent in addition to the consent of the legally authorized representative."[58]

I maintain that nontherapeutic experimentation is morally inadmissible in subjects who are really incapable of consent, inasmuch as they do not yet have or will no longer be able to have the use of reason.[59] In this category one may only consider the use of an experimental drug as the last chance to save the life of a minor who could not be saved with known drugs. Strictly speaking, this would be a case of therapeutic experimentation, even though one might be able to derive from it, fortuitously, valuable findings for science and society in the event of a positive outcome.[60]

On the other hand, if it were a question of legal incompetence, but with the real ability to give free and informed consent, that consent, reinforced by that of the legal representative, could be considered valid for nontherapeutic experimentation, provided that there were no serious risks either to the life or the physical integrity of the subject.[61] The fulfillment of these conditions, however, would have to be rigorously assessed by an ethics committee.

The fact that the minor subject, who is really or legally incapable of consent, is destined for certain death because of a disease does not change either the nature or the extent of the moral protection due to him or her.

EXPERIMENTATION BY MEANS OF RANDOMIZED CLINICAL TRIALS. It is well known that the experimental path, in order to verify the effects of a particular drug, usually requires the control method, which involves two groups of patients receiving two types of drugs (the experimental drug and a standard drug) or else one group that receives the experimental drug and another to which that drug is not administered. Since the selection of subjects for both groups is often made at random, these are called randomized clinical trials.

In certain protocols, a *placebo* is administered to the control group to ensure that the research subjects do not know which group they belong to; this is an effort to avoid the psychological factor (the expectation of an effect). This theme was discussed earlier.

Sometimes, besides the subjects on whom the experiments are performed, some of those who are conducting the research (those administering the drugs and gathering the data concerning the effects) do not know which subjects belong to which group, again to avoid any possibility of subjective influence. Only the person overseeing the

---

[58] World Medical Association, Declaration of Helsinki, nos. 27–28.

[59] As in the previously discussed case of the incompetent adult subject, so too with the category of the minor who is involved in research: guideline 14 of the *International Ethical Guidelines for Biomedical Research* (2002) of the CIOMS explicitly talks about the involvement of pediatric subjects in research in general without distinguishing the case of research for a nontherapeutic purpose: "Before undertaking research involving children, the investigator must ensure that: the research might not equally well be carried out with adults; the purpose of the research is to obtain knowledge relevant to the health needs of children; a parent or legal representative of each child has given permission; the agreement (assent) of each child has been obtained to the extent of the child's capabilities; and a child's refusal to participate or continue in the research will be respected."

[60] Pius XII, Address to the World Medical Association, September 30, 1954.

[61] Redmon, "How Children," 80.

execution of the experiment knows who in fact receives the drug and who receives the placebo. Under these conditions it is called a *double-blind* experiment.

In using this method (randomization, placebo, double-blind), ethical problems can arise within the technical aspects. They can be summarized with the following observations:[62]

a. When proceeding to a drug trial for nontherapeutic research on patients who are already being treated, there is an ethical obligation to ensure that suspension of the ordinary therapy does not expose the patients' health or life to harm or risk. Moreover, it is necessary to obtain the consent of the subjects even when applying the double-blind method, although it will be limited to the fact that experimentation will be conducted and to the method that will be used.[63] Whenever this would prove impossible to propose (because of a refusal to subject oneself to experimentation under those conditions, or a strong emotional response from the subjects), all of these methodologies become ethically inadmissible. In such cases there is no ethical justification for deception, which cannot be legitimized even for reasons of the patient's health and would end up damaging the credibility of the medical profession and of hospitals.

b. When the experimentation is conducted for therapeutic purposes, it is necessary—in addition to obtaining informed consent—to ensure that the suspension of ordinary treatment does not involve risk or harm to the patients and that the experimentally administered drug is at least as effective as the already known treatment.

Experimentation on Human Fetuses and Embryos

The novelty of the issue

The matter of experimentation on human embryos was already assessed during our discussion of in vitro fertilization. It was declared ethically inadmissible, regardless of the type and purposes of the experimentation. Whether it is genetic experimentation (recombinant DNA) or chromosome substitutions, whether pharmacological research or an experimental study in embryonic biology, the judgment from the ethical viewpoint is still negative, because the experimentation often involves bringing a human being to life with the sole purpose of making him or her an "object" of the experiments. The depravity of the end infects the morality of the entire act. Furthermore, such experimentation normally requires the loss and destruction of

---

[62] P. Arpaillange and S. Dion, "Considérations sur l'éthique de la randomisation," *Biomedicine and Pharmacotherapy* 38 (1984): 426–429.

[63] The fact that the subjects invited to participate in an experiment understand the meaning of randomization may be insufficient from their perspective; it might also be necessary to provide them with a justification for such a choice, even if it is scientifically correct. On this matter, see C. Kerr et al., "Randomisation in Trials: Do Potential Trial Participants Understand It and Find It Acceptable?" *Journal of Medical Ethics* 30, no. 1 (2004): 80–84; K. Featherstone and J. L. Donovan, "Random Allocation or Allocation at Random? Patients' Perspectives of Participation in a Randomised Controlled Trial," *British Medical Journal* 317, no. 7167 (1998): 1177–1180.

the embryo, which is an aggravating result. I therefore maintain that the conditions required for human experimentation in general must be applied here as well.[64]

The problem of experimentation has been posed more acutely in recent times by public condemnations of trafficking in aborted fetuses, whether for the purposes of industrial use in the production of cosmetics or as a way of providing experimental material.[65]

During the preparation of the European Convention on Human Rights and Biomedicine, there was a lengthy discussion of this issue, and a distinction was made between the production of embryos for the sole purpose of experimenting on them and the use of embryos—again for experimental purposes—that are "left over" following the applications of reproductive technologies. The conclusion was a compromise, formulated in article 18,[66] which does not explicitly forbid research on embryos produced in vitro if it is permitted by national law, while recalling the obligation to provide adequate protection for the embryo.

Subsequently, there was an announcement by researchers that they had obtained fetal brain tissue or tissues from fetal suprarenal glands for the purposes of preparing an injection for individuals afflicted with Parkinson's disease.[67]

---

[64] Pontifical Council for Pastoral Assistance to Health Care Workers, *Charter*, no. 82: "Since the human individual, in the prenatal stage, must be given the dignity of a human person, *research and experimentation on human embryos and fetuses* is subject to the ethical norms valid for the child already born and for every human subject" (original emphasis); Perico, "Sperimentazione clinica"; Villa, "Etica e deontologia"; Häring, *Medical Ethics*; *L'expérimentation humaine*.

[65] D. Tettamanzi, "Feti umani e sperimentazione biomedica: Problemi etici," *Anime e Corpi* 111 (1984): 37–50; idem, "Problemi morali circa alcuni interventi sui feti ed embrioni umani," *Medicina e Morale* 35, no. 1 (1985): 23–43. For experimentation on embryos, see the report by the Warnock Committee presented to the Parliament of the United Kingdom: *Report of Inquiry into Human Fertilisation and Embriology* (London: Her Majesty's Stationery Office, 1984), http://www.hfea.gov.uk/docs/Warnock_Report_of_the_Committee_of_Inquiry_ into_Human_Fertilisation_and_Embryology_1984.pdf. In two of its opinions (May 22, 1984, and December 15, 1986), the French National Ethics Advisory Committee expressed its own position on the topic of experimentation on embryos. With regard to experimentation on and trafficking in fetuses, see also M. Lichfield and S. Kentish, *Bambini da bruciare* (Catania: Paoline, 1976), 175–181; C. Jacquinot and J. Delaye, *Le trafiquants des bébés à naître* (Paris: P. M. Faure, 1984), 158; S. Udo, "Protecting the Vulnerable: Testing Times for Clinical Research Ethics," *Social Science and Medicine* 51 (2000): 969–977; R. Nicholson, "Who Is Vulnerable in Clinical Research?" *Bulletin of Medical Ethics* 181 (2002): 19–24; B. Steinbock, J. D. Arras, and A. J. London, *Ethical Issues in Modern Medicine* (Boston: McGraw-Hill, 2003); S. Eckstein, ed., *Manual for Research Ethics Committees* (New York, NY: Cambridge University Press, 2003); R. Colombo, "La vulnerabilità nella ricerca biomedica: Il caso dell'embrione umano," in *Etica della ricerca*, Vial Correa and Sgreccia, 217–244.

[66] Indeed, the article reads, "1. Where the law allows research on embryos in vitro, it shall ensure adequate protection of the embryo. 2. The creation of human embryos for research purposes is prohibited." The assembly of the European Council, nevertheless, had proposed replacing the first part of the article with the following language: "Research on embryos in vitro shall not be permitted except in the interest of their development. It can however be conducted for the diagnosis of more serious illnesses."

[67] M. J. Perlow, "Brain Grafting as a Treatment for Parkinson's Disease," *Neurosurgery* 20 (1987): 335–342; M. B. Mahowald et al., "Trasplantation of Neural Tissue from Fetuses,"

There is no distinction between embryos and fetuses as far as the ethical aspect is concerned, given that it is always a matter of one and the same life of a human being, whose dignity as a human person must be recognized.[68] This has been recalled and demonstrated several times.

The instruction *Donum vitae* by the Congregation for the Doctrine of the Faith also stressea the identical ethical weight of the terms "zygote," "preembryo," "embryo," and "fetus," which only indicate successive stages in the development of a human being, since they all designate the fruit (whether visible or not) of human generation from the first moment of its existence until birth.[69]

The more relevant distinctions from an ethical viewpoint are those concerning the *end* of the interventions on the fetus. The end could be pure biomedical research or experimentation that is therapeutic or has therapeutic aims (e.g., obtaining tissues in the hope of curing other sick persons). Moreover, and even more importantly, there are distinctions regarding the condition of the fetus: if alive, viable or nonviable; if dead, dead as a result of spontaneous abortion or voluntary termination of pregnancy.

Finally, it is necessary to keep in mind the *risk factor* for live embryos, that is, whether the intervention or experimentation can cause death or compromise the substantial integrity of the embryo's life in the future.

Let us consider the cases individually.

Therapeutic intrauterine interventions

In addition to blood transfusions to save the life of the fetus, which have helped save many unborn children, the range of fetal surgical interventions continues to expand with the progress of science and of prenatal care; for example, there are surgeries for malformations or obstructions of the urinary tracts and for cases of hydrocephaly. Nevertheless, some maintain that some of these surgeries can be performed with better results in the immediate postnatal phase.[70] *Diagnostic intervention*, also termed "prenatal diagnosis," is also a topic included in this field; it must be considered not only in terms of the risk to the fetus at the moment when the diagnostic technique is employed but also in terms of the link to *selective abortion*, which was discussed in a separate section.[71]

Science 25 (1987): 1307–1308; G. Pezzoli et al., "Human Fetal Andrenal Medulla for Transplantation in Parkinsonian Patients," *Annals of the New York Academy of Sciences* 495 (1987): 771–773.

[68] A. Serra, "Pari dignità all'embrione umano nell'enciclica *Evangelium vitae*," in *"Evangelium vitae" e bioetica: Un approccio interdisciplinare*, ed. E. Sgreccia and D. Sacchini (Milan: Vita e Pensiero, 1996), 147–173.

[69] CDF, *Donum vitae*, foreword.

[70] S. Calisti, "Il feto, paziente chirurgico," *Medicina e Morale* 33, no. 1 (1983): 49–58.

[71] See the chapter on genetic diagnosis. The reader may also consult A. G. Spagnolo, "Diagnosi e terapie fetali," in *Bioetica nella ricerca*, 159–192; the articles on prenatal diagnosis by A. Serra, C. Caffarra, and L. Leuzzi in *Medicina e Morale* 34, no. 4 (1984): 433 ff., 449 ff., and 458 ff.; A. Serra, "Problemi etici della diagnosi prenatale," *Medicina e Morale* 32, no. 1 (1982): 52–61; D. Tettamanzi, "Diagnosi prenatale e aborto selettivo: Problemi etici," *Anime e Corpi* (1983): 339–360.

Therapeutic intervention on a fetus, as on any individual human being, is licit where there is a good possibility of success; it is a special case, however, since the subject is one who cannot provide consent.

Any therapeutic intervention (diagnostic or curative, medical or surgical) must therefore be guaranteed by three essential conditions: it must be required by a serious and proportionate reason and therefore cannot be postponed until after birth with the prospects of better conditions for executing the procedure; there must be no serious risk either for the life or for the physical integrity of the fetus or the mother (i.e., serious impairment to fetus or mother must be avoided); and the parents must provide informed, truly free consent that substitutes and represents—from the legal perspective as well—the consent of the concerned party.

The instruction *Donum vitae* offers clarification in this regard:

> As with all medical interventions on patients, one must uphold as licit procedures carried out on the human embryo which respect the life and integrity of the embryo and do not involve disproportionate risks for it but are directed towards its healing, the improvement of its condition of health, or its individual survival. Whatever the type of medical, surgical or other therapy, the free and informed consent of the parents is required, according to the deontological rules followed in the case of children. The application of this moral principle may call for delicate and particular precautions in the case of embryonic or foetal life. [72]

It continues further on, "In the case of experimentation that is clearly therapeutic, namely, when it is a matter of experimental forms of therapy used for the benefit of the embryo itself in a final attempt to save its life, and in the absence of other reliable forms of therapy, recourse to drugs or procedures not yet fully tested can be licit." [73]

In its document on the identity and status of the embryo, the Italian National Bioethics Committee (NBC) lists the following in the category of *morally acceptable*: "Eventual therapeutic interventions, still at the experimental phase, carried out on the embryo, when they are finalised to the protection of the embryo's life or health. In this case, the deontological [professional ethics] rules for children apply, in particular with regards to the [sic] free and informed consent." [74]

Interventions for purely scientific research

Recalling that the fetus, like the embryo, has the dignity of a human person and that, on the other hand, he or she is a subject incapable of consent, it is necessary to mention again that only therapeutic interventions can be made on fetuses. Therefore, experiments performed solely for the purpose of scientific research should be considered immoral, even when no lethal outcome is foreseen.

---

[72] CDF, *Donum vitae*, part I, no. 3, emphasis removed.

[73] Ibid., no. 4. Even the Italian *Codice di Deontologia Medica* in article 43 forbids "genetic modification of the embryo that is not for the purpose of preventing and correcting pathological conditions."

[74] Italian National Bioethics Committee, *Identity and Status of the Human Embryo*, June 22, 1996 (Rome: Presidenza del Consiglio dei Ministri Dipartimento per l'Informazione e l'Editoria, 1996), point 8.2.2, p. 17, http://www.governo.it/bioetica/eng/pdf/human_embryo _19960622.pdf.

The Declaration of Helsinki itself prohibits nontherapeutic experimentation on subjects who are unable to consent, given that participation in such scientific undertakings must be voluntary. The fetus must not be treated like a guinea pig, and for this reason purely investigative experimentation on living human fetuses must be declared immoral. Nor can we accept the justification that this might bring about advances in research that would help cure other human subjects, because a good purpose cannot be attained by an evil means (in this case by the destruction of a human individual). Another invalid argument is that the part must be sacrificed for the whole (whereby the individual fetus is understood as the "part" and the "whole" is the good of humanity), because the individual subject is not a part of the social organism; rather, he is the aim of society. The common good must be understood as the good of individuals.

In this regard, the encyclical *Evangelium vitae* declares that "the use of human embryos or fetuses as an object of experimentation constitutes a crime against their dignity as human beings who have a right to the same respect owed to a child once born, just as to every person." [75]

Even when it has already been decided that the fetus shall die by abortion, regardless of the input of the researcher, it is not permissible to perform an experiment on the still-living fetus. The reason is that "if the embryos are living, whether viable or not, they must be respected just like any other human person." [76]

In this regard the above-mentioned European Convention on Human Rights and Biomedicine displays some omissions and inadequacies. As for the former, nothing is said about the use of fetuses after procured or spontaneous abortion, nor about the right to conscientious objection (on the part of the researcher) to the practice of in vivo and in vitro research in the area of procreation, nor, finally, about the right to life that ought to constitute the source of all rights for every individual human being.

On the level of textual inadequacies, the convention forbids discrimination only against "persons" and not against "human beings," terms that are distinguished speciously in the document. [77]

The above-cited document by the NBC on the human embryo reaches several unanimous conclusions, one of which is the judgment that many forms of purely scientific research are morally illicit, including "the in vitro creation of human embryos

---

[75] John Paul II, Encyclical Letter *Evangelium vitae*, March 25, 1995, no. 63.

[76] CDF, *Donum vitae*, part I, no. 4.

[77] This article manifests primarily anthropological shortcomings in that it neither gives a definition of the embryo nor specifies when its formation and identity begin. Consequently, the door is left open to conflicting ideas (that of the "preembryo" being one possibility). Relegating a topic of this sort to an additional paper seems to indicate the weakness of the entire document. In my opinion, a third clause should have been added to the article to forbid the production of so-called spare embryos with reproductive technologies, which are produced with a view to possible authorization to experiment on them. And again, what would happen to the solemn affirmation of the principle of assuring their individual dignity, identity, and integrity (article 1)? Moreover, without this third clause, not even the first clause logically governs if it declares an assurance of adequate protection for the embryo. Not to mention that the same first clause in itself proves vacuous in the absence of a declared right to life, given that the exact parameters of the stated "adequate protection" of the embryo are lost. If a third clause were included, it would assure adequate protection for the life and integrity of the embryo.

for the only purpose of using them for experimental research, or for commercial or industrial uses."[78] On the other hand, the committee is internally divided in its ethical judgment about the illicitness of "the suppression and any form of harmful manipulation of embryos, even at the pre-implantation stage of development" (8.1.6), about "experimentation on embryos, as in effect it implies their suppression" (8.1.18), and about "the in vitro creation of embryos for whom we have no intention of proceeding to the implantation into the maternal uterus" (8.1.9).

The perspective of personalist ethics affirms the speciousness of the distinction between human individuality and personhood; the fundamental value of individual physical human life, which is to be protected at every stage of its development; and the dignity of the human being, which confers upon him a higher value and makes him deserving of respect. The following are therefore considered illicit: (1) "the suppression, including embryonic reduction and any form of embryonic manipulation even before the 14th day of development"; (2) "any nontherapeutic trial on embryos from the first moment of their formation, namely, also on the so-called 'early embryos' "; (3) "the suppression and any form of manipulation of frozen and/or residual embryos, who must be considered alive although not suitable for implantation"; and (4) "the in vitro creation of embryos, especially when there [is] no certainty that they will be transferred in a woman's uterus."[79]

Therefore the warning to the members of the Pontifical Academy of Sciences by Pope John Paul II is fully justified: "Consequently, I have no reason to be apprehensive for those *experiments in biology* that are performed by scientists who, like you, have a profound respect for the human person, since I am sure that they will contribute to the *integral well-being of man*. On the other hand, I condemn, in the most explicit and formal way, experimental manipulations of the human embryo, since the human being, from conception to death, cannot be exploited for any purpose whatsoever. Indeed, as the Second Vatican Council teaches, man is 'the only creature on earth which God willed for itself.'"[80] To the participants in the conference in Rome organized by the Italian Pro-Life Movement on the subject of prenatal diagnosis and surgical treatment of congenital deformities, held on December 3, 1982, the John Paul II had already declared unacceptable "any form of experimentation on fetus that could harm its integrity or worsen its conditions, unless it is an attempt to save the fetus from certain death."[81]

Experiments on aborted human fetuses

The discussion about aborted human fetuses refers to the fetus outside the womb, regardless of whether the fetus is alive (viable or not) or dead—a condition that must always be ascertained. For example, if an abortion is performed by expulsion or by cesarian section, some fetuses survive the procedure and are sometimes still alive. If the aborted fetuses are alive, whether viable or not, then the above-mentioned rules for intrauterine interventions and for minors therefore apply. The practice of keeping

---

[78] Italian National Bioethics Committee, *Identity and Status*, point 8.1.1, p. 17.

[79] The passage is taken from the "Additional Declaration by Some NBC Members" in the Italian National Bioethics Committee's *Identity and Status*, 27.

[80] John Paul II, Address to the Pontifical Academy of Sciences, October 23, 1982.

[81] John Paul II, Address to the Pro-Life Movement, December 3, 1982.

human fetuses or embryos alive, in vivo or in vitro, for experimental or commercial purposes is utterly contrary to human dignity.[82]

As for fetuses that are already dead—which, I repeat, must be ascertained in a timely fashion—it is necessary to make a distinction between a spontaneously aborted fetus and a voluntarily aborted fetus.

When fetuses come from elective abortions, and this fact is known to the researchers, it could be licit in itself to use them as long as there is no connection by way of previous agreements, participation in the abortion, or sharing of the intention of abortion; in practice, this connection is always established directly or indirectly—even through ethics committees, as some would have it—insofar as the abortion is often performed ad hoc for this use and hence the fetus may still be alive at the end of the procedure, albeit not viable. Furthermore, there is a danger of encouraging the procedure in this way (*ratio scandali*), given the need to request the mother's consent.[83] Regardless of this specific circumstance, the eventual connection nonetheless implies collaboration.

Even the French regulations concerning the extraction of fetal tissues following an elective abortion mention this ethical problem and declare it important enough to admit the possibility of conscientious objection for those who might be called upon to collaborate in this field and might not be able in conscience to conduct experiments on fetuses derived from voluntary terminations of pregnancy.[84]

In the case of fetuses from spontaneous abortions, the consent of the parents or mother is required in addition to certification of death, and any form of speculation or recompense of a commercial sort must be ruled out. There must be a reasonable and foreseeable advantage from such research at the practical human level, with a view to studying a disease.

The NBC, in its document on the human embryo, considers morally permissible "experimentation for nontherapeutic purposes on dead embryos, obtained from spontaneous or induced abortions, as long as the parents give their free and informed consent and the independence between the medical personnel and/or the institutions that practice voluntary abortion and those that carry out the experimentation, is ascertained."[85]

Finally, the instruction *Donum vitae* recalls that "the corpses of human embryos and foetuses, whether they have been deliberately aborted or not, must be respected just as the remains of other human beings."[86]

---

[82] John Paul II, Address to the Pro-Life Movement, December 3, 1982.

[83] On this question, see A.G. Spagnolo and E. Sgreccia, "Prelievi di organi e tessuti fetali a scopo di trapianto: Aspetti conoscitivi e istanze etiche," in *Trapianti d'organo*, ed. A. Bompiani and E. Sgreccia (Milan: Vita e Pensiero, 1989), 47–84; A.G. Spagnolo, "L'inevitabile complicità nel trapianto di tessuti fetali da aborti volontari," *L'Osservatore Romano*, January 28, 1995.

[84] Tettamanzi, "Feti umani e sperimentazione," 46–47; idem, "Interventi su embrioni/feti umani," *La Famiglia* 120 (1986): 31–53. The regulation issued by the French National Ethics Advisory Committee is titled *Avis sur les prélèvements de tissus d'embryons ou de foetus humains morts, à des fins thérapeutiques, diagnostiques et scientifiques* (May 22, 1984).

[85] Italian National Bioethics Committee, *Identity and Status*, point 8.2.2, p. 17.

[86] CDF, *Donum vitae*, part I, no. 4.

The extraction of fetal tissues for the purpose of transplantation

Tissues may be obtained by fetal extraction—and the procedure has in fact been carried out[87]—for the purpose of transplanting them into persons suffering from diseases that are otherwise untreatable: it might be a transplantation of bone marrow (to cure leukemia), nerve cells, or liver or pancreatic tissues. The advantage of fetal tissues over those from adults seems to be that, since fetal tissues are in a stage of rapid cellular development, they contribute to a better therapeutic outcome and are less likely to cause reactions from the recipient's immune system (rejection or incompatibility).

Here too it is necessary to distinguish among various cases: if it is a question of obtaining tissues from live fetuses, with the risk of causing serious harm or the death of the fetus, the procedure is plainly illicit. If the fetuses in question are dead (as a result of spontaneous abortion or voluntary termination of pregnancy), the tissues are normally not suitable for transplantation unless they were obtained in the very first moments afterward. In any case, the observations and stipulations made in the preceding section are valid. But the danger is that researchers will seek to obtain tissues that are as intact as possible and capable of regenerating themselves and hence will proceed to conduct research on living fetus through contacts with women who are predisposed to abortion or with clinics where elective abortions are performed. An intention of this sort has been denounced as "trafficking in human fetuses."[88]

Whenever practices of this kind have been implemented in laboratories or hospitals under ethically unacceptable conditions, conscientious objection is obligatory for those who perceive the illicit character thereof.

The utilization of live but nonviable fetuses to obtain tissues in the period preceding or immediately following abortion must be ruled out morally, because foreseeing an intended death does not change the illicit character of the harmful act that is performed on the fetuses while they are still alive, which thereby becomes the direct cause of death.

The protection of embryos must also be promoted by regulations and laws. It is strange that there is no legislation in this area, whereas there are laws defending experimental animals and regulating the performance of autopsies on human cadavers. It is encouraging that at the international level a charter of embryonic rights is being formulated, similar to the one drawn up for children in keeping with the rights of the family concerning the protection and defense of newborn and unborn human life, as also declared in the *Charter for Family Rights* published by the Catholic Church.[89]

The Parliamentary Assembly of the European Council also issued recommendation 1046 (1986) on the use of human embryos and fetuses for diagnostic, thera-

---

[87] M. J. Perlow, "Brain Grafting as a Treatment for Parkinson's Disease," *Neurosurgery* 20 (1987): 335–342; M. B. Mahowald et al. "Transplantation of Neural Tissue from Fetuses," *Science* 25 (1987): 1307–1308; G. Pezzoli et al., "Human Fetal Andrenal Medulla for Transplantation in Parkinsonian Patients," *Annals of the New York Academy of Sciences* 495 (1987): 771–773.

[88] Jacquinot and Delaye, *Les trafiquants*.

[89] Pontifical Council for the Family, *Charter of the Rights of the Family*, October 22, 1983.

peutic, scientific, and industrial purposes. It is significant that even this document, issued by a secular organization, acknowledges the existence of a human life from the moment of fertilization; human embryos and fetuses must therefore be treated in all circumstances with the respect due to human dignity (no. 10). Consequently, any intervention on live fetuses, even if they are not viable, was prohibited by the recommendation in the hope that national legislatures would accept those guidelines.

Subsequently, the European Council itself confirmed this line of thought concerning the use of human embryos for research purposes, in recommendation 1100 (1989).[90] On the other hand, the European Convention on Human Rights and Biomedicine, which has greater weight from the legal perspective, weakened this defense with its article 18, which was cited above.

"Proposal of an Ethical Commitment for Researchers in the Biomedical Field"

Among the contributions by the Catholic Church, the "Proposal of an Ethical Committment for Researchers in the Biomedical Field" was published at the conclusion of the Ninth General Assembly of the Pontifical Academy for Life in 2003, which was dedicated to the theme "Biomedical Research Ethics: For a Christian Vision." The "manifesto" was published in the proceedings of the conference:

Premise

The scientific developments of recent decades have brought about important cultural and social transformations, modifying in a qualitative way many aspects of human life. Indeed, the advance of scientific progress in many sectors has given rise to great hopes of concrete improvements for the life and future of the human person. However, in certain sectors of scientific research problems and/or doubts of an ethical and religious nature have arisen; they have demonstrated unequivocally the real need for constant dialogue/ integration between the experimental sciences and the broader human sciences and philosophy in terms of operating in a more ample perspective so that the acquisition of greater knowledge may effectively serve the true good of the human person.

Human life and human nature appear to be realities too complex to be exhaustively evaluated from a single perspective; a multidisciplinary approach therefore appears indispensable for a better understanding of the human being in his integrity and contribute to a meaningful growth of a science that would truly be *for* the human being.

Moreover, such an interdisciplinary dialogue, by re-focusing attention on the centrality of the human person, would make the scientists more aware of the ethical implications of their work, and, conversely, would incite those involved in philosophical and theological anthropology to assume toward the scientists a mission of dialogue, collaboration and practical support, with

---

[90] Parliamentary Assembly of the Council of Europe, recommendation 1100, "On the Use of Human Embryos and Foetuses in Scientific Research," February 2, 1989, http:// assembly.coe.int/Main.asp?link=/Documents/AdoptedText/ta89/EREC1100.htm.

the mutual intention of developing cognitive and applied tools for the service of the human community.

In this perspective, the reference to human values, and finally, to an anthropological and ethical vision, is an indispensable premise for a correct scientific research, that recognises the person's responsibility to himself and to others.

In fact, without reference to ethics, science and technology can be used either to kill or to save human lives, to manipulate or to promote, to destroy or to build. It is therefore necessary that, through responsible management, research be addressed toward the true common good, a good that transcends any merely private interest, going beyond the geographical and cultural boundaries of nations and keeping one's vision directed toward the good of future generations.

For science to be really placed at the service of the human being, it is necessary that it goes "beyond matter," intuiting in the corporeal dimension of the individual the expression of a greater spiritual good. Scientists should understand the human body as the tangible dimension of a unitary personal reality, which is at the same time corporeal and spiritual. The spiritual soul of the human being, although not in itself tangible, it is always the root of his existential and tangible reality, of his relationship with the rest of the world, and consequently, of his specific and inalienable value.

Only such a vision can make scientific research effectively respectful of the human person, considered in his complex corporeal–spiritual unity, every time he/she becomes the object of investigation, with particular reference to those events that constitute the beginning and the end of the individual human life.

For this reason, emerges a strong need to offer to young researchers formative programmes that put the accent not only on the scientific preparation, but also on the acquisition of the fundamental notions of anthropology and ethics. The expression of such programmes could, then, crystallize in the elaboration of a true and proper *Deontological Code* for researchers, to which each researcher could safely refer in his work, and which, at the same time, would represent a sign of hope and commitment for a truly "humanized" medicine in the new millennium.

A first indication of the way to take, might concern the manner in which the researchers should behave and the norms they should observe in order to direct their research towards the objective just recalled above. It is our desire to propose such ethical indications, to which we firmly adhere, to all others who are involved in the world of biomedical research; somehow, they delineate the principal features of the researcher's "moral personality."

Commitment

I commit myself to adhere to a methodology of research characterized by scientific rigour and a high quality of the information that is furnished.

I will not take part in research projects in which I could be subject of a conflict of interests, from the personal, professional or economic point of view.

I recognise that science and technology must be at the service of the human person, fully respecting his dignity and rights.

I recognise and respect all researches and their applications which are based on the principle of "moral goodness" and referring to the correct vision of the corporeal and spiritual dimensions of the human being.

I recognise that every human being, from the first moment of his existence (process of fertilization) up to the moment of his natural death, is to be guaranteed the full and unconditional respect due to every human person by virtue of his peculiar dignity.

I recognise, because of my duty to safeguard human life and health, the usefulness and the obligation of a serious and responsible experimentation on animals, carried out according to determined ethical guidelines, before applying new diagnostic and therapeutic methodologies to human beings. I also recognise that the passage from the experiments with animals to the clinical experimental stage (on man) should take place only when the evidences resulting from the experiments with animals sufficiently demonstrate the harmlessness or the acceptability of the possible harms and risks that such experiments might involve.

I recognise the legitimacy of clinical experiments on the human being, but only under precise conditions, including, in the first place, the safeguarding of the life and physical integrity of human beings who are involved. Then, there is the need that the experiments be always preceded by proper, correct and complete information regarding the significance and developments of the same experiments. I will treat each person who submits to an experiment as a free and responsible subject and never as a mere means to achieve other ends. I will never let a person be involved in an experiment unless he/she has given his/her free and informed consent.[91]

---

[91] Pontifical Academy for Life, "Proposal of an Ethical Committment for Researchers in the Biomedical Field," in *Ethics of Biomedical Research in a Christian Vision: Proceedings of the Ninth Assembly of the Pontifical Academy for Life, Vatican City, 24–26 February 2003*, ed. Juan de Dios Vial Correa and Elio Sgreccia (Vatican City: Libreria Editrice Vaticana, 2004), http://www.academiavita.org/index.php?view=article&catid=54%3Aatti-della-ix-assemblea-della-pav-2003&id=242%3Aproposta-di-impegno-etico-per-i-ricercatori-in-ambito-biomedico&format=pdf&option=com_content&Itemid=66&lang=en.

Summary Outline for Chapter 13

Norms concerning experimentation

The primary reason for the passage of laws regulating research in various parts of the world was the news about the research conducted during World War II by the Nazis: sterilization techniques, methods of mass extermination, experiments in hypothermia, etc. But Nazi experimental practices were not the only evidence of the need for moral norms. The most infamous illicit experiment conducted in the United States was the Tuskegee syphilis experiment, part of a project prepared by the Public Health Service: U.S. citizens of color were subjected to experimentation, and of these 20 percent died following invasive treatments. Even when penicillin was available, the experimentation continued.

In 1947, the Nuremberg Code was published; among its ten principles was a prohibition on conducting experiments without the subject's consent. The same principle of consent for physical interventions on and treatments of an individual is also included in the Universal Declaration of Human Rights (1948).

The most famous and still current code of professional ethics is the Declaration of Helsinki of 1964, which has been revised several times (the last revision was in 2008) to address new questions that arose. In this code, besides the requirement of consent, we find the principle of avoiding harm to the patient. The document was issued by the World Medical Association and is still taken as the basis for clinical trials.

Another important international document concerning experimentation is the International Ethical Guidelines for Biomedical Research involving Human Subjects, published by the Council for International Organizations of Medical Sciences and revised most recently in 2002. The same organization published a document on experimentation in the field of epidemiology titled *International Guidelines for the Review of Epidemiological Studies* (1999).

In the European context, we should recall the guidelines *Good Clinical Practice for Trials on Medical Products in the European Community* (1991 and 1996; the final version dates from 2006), which were accepted by the member states of the European Union. Norms regarding experimentation are contained in the European Convention on Human Rights and Biomedicine (1997), and as far as genetic experimentation is concerned, we should keep in mind the Universal Declaration on the Human Genome and Human Rights by UNESCO (1997).

On the national level, it can be said that many government ministries, medical associations, and organizations of specialists have produced their own normative documents (national codes of professional ethics). Still historically important is the Belmont Report in the United States, issued in 1978 by the national commission formed specifically in response to revelations about savage experiments: this document underscores the principles of respect for persons, beneficence ("Do no harm"), and justice.

In the United States in 1980, the President's Commission for the Study of Ethical Problems in Medicine and Biomedical and Behavioral Research was established, which published a set of documents providing guidance in the most advanced fields of biomedical research: genetic engineering, artificial procreation, and organ transplantation. This commission can be considered the first national ethics committee; later, in 1984, came the initiative by President Mitterand in France with the formation of the National Ethics Advisory Committee (CCNE). (See J. Vial Correa and E. Sgreccia, *Etica della ricerca biomedica: Per una visione cristiana; Atti della nona assemblea generale della Pontificia Accademia per la Vita; Città del Vaticano, 24–26 febbraio 2003* [*Ethics of biomedical research: for a Christian vision; acts of the ninth general assembly of the Pontifical Academy for Life; February 24–26, 2003*].)

Patient health is the highest law (Salus aegroti suprema lex)

Kant formulated the categorical imperative, declaring that human beings must never be used as mere means. And every person on whom experiments are conducted is instrumentalized, in other words, used as a means to an end. But the limits of this use are what is decisive. They require first that no one be "used" without his consent. This implies, for example, that generally children and mentally handicapped persons cannot be used as guinea pigs in an experiment if it involves any harm to them. This naturally implies that the life and health of any person whatsoever cannot be sacrificed for the benefit of the life and health of others. The limits on the use of persons as means also forbid any acquisition of knowledge derived from experiments that involve the destruction of human embryos. But even in everyday experimental practice this problem comes up with regard to experimentation with new drugs. Sometimes it happens that before the planned series of experiments is completed the physician becomes convinced that the medicine in question is indeed very effective in curing an illness. The moment he is convinced of this, he must interrupt the experiment and administer that drug to all the patients, even to the control group that until then had received a placebo. The health of a specific medical patient cannot be sacrificed to the *salus* of an indistinct mass of future patients. (See Spaemann, in *Etica della ricerca biomedica: Per una visione cristiana*.)

The four phases of drug trials

- *Phase I trials*. Trials in phase I are the first time that human subjects are exposed to potential new medicines. The purpose of these trials is to study pharmacodynamics and response to dosage—and the immune response in vaccine tests—and to determine the maximum dose that can be tolerated by the participants.

  With most new medicines, these trials are conducted on a small number of healthy volunteers. It is not expected that the effectiveness of the medicine will be made evident in phase I studies.

- *Phase II trials*. Using the information regarding safe doses provided by phase I trials, researchers administer the drug to patients who suffer from the targeted illness, and now a significant number of individuals are involved in the study. These studies are almost always conducted in various clinical centers. The purpose of the phase II study is to prove the efficacy of the medicine against a specific disease. Further information about the safety of the cure will be made evident by these studies, because greater numbers of individuals are exposed to the cure. In the phase II trial, the patients are often randomly assigned to the group that will receive the new cure or to the group that will receive either a placebo (a substance that has no therapeutic effect) or, more often, the conventional treatment that has already been approved.

- *Phase III trials.* If a drug has proved to be effective and without significant side effects, it will enter into phase III, in which many hundreds—sometimes several thousands—of patients will be enrolled. This phase is generally conducted not only to confirm the clinical effect of the drug, but also to establish its efficacy in comparison with traditional treatments. These studies often take place in multiple centers and are often conducted on an international basis. Once again great attention is given to the possible side effects inasmuch as large numbers of patients are exposed to the cure. The endpoints of phase III include the demonstration of a statistically significant improvement in the efficacy of the new medicine in comparison with already proven therapies, if there are any.

- *Phase IV trials.* Once the new medicine hits the market, it will be subject to post-marketing monitoring to identify the side effects and other adverse effects made evident as many more individuals are treated with it. Moreover, formal clinical studies will continue to be performed to give researchers a better understanding of the drug and its effects in a wider clinical context, and also to extend its use to other clinical indications or to different groups of patients, such as children or older persons. Specially designed studies can be used, depending on the objectives of the study for the evaluation of efficacy and safety.

Placebo (substance without therapeutic effect)

According to the Declaration of Helsinki (2008), "The benefits, risks, burdens and effectiveness of a new intervention must be tested against those of the best current proven intervention. . . . The use of placebo, or no treatment, is acceptable in studies where no current proven intervention exists." The use of a placebo can open the door to abuses. (See some of the therapeutic research trials on AIDS conducted in some African or Asian nations.)

Double blind

Patients are randomly assigned to the group receiving a placebo or the standard drug and to the group receiving the drug being studied. This is necessary to prevent the physician, even unconsciously, from choosing patients to assign to one group or the other on the basis of the seriousness of their disease or other criteria, such that the groups being studied are not comparable. This technique ensures a distributive "justice," so that all the patients have the same advantages and risks a priori with respect to the treatments in question. (See Serrão, in *Etica della ricerca biomedica: Per una visione cristiana.*)

Principles of the Declaration of Helsinki for medical research

- The physician has the duty to protect the life, health, privacy, and dignity of the subjects.

- The research must be in keeping with universally accepted scientific principles and based on the literature and preclinical studies.

- A design for the research project is necessary, described in a protocol to be submitted to an ethics committee.

- The responsibility always falls on the researcher (physician).

- The importance of the objective must outweigh the risks for the subject, whose integrity must be safeguarded.

Declaration of Helsinki: Points from the 2008 revision

- "Research involving human subjects [includes] research on identifiable human material and data" (no. 1).

- "Even the best current interventions must be evaluated continually through research for their safety, effectiveness, efficiency, accessibility and quality" (no. 7).

- "Medical research involving a disadvantaged or vulnerable population or community is only justified if the research is responsive to the health needs and priorities of this population or community and if there is a reasonable likelihood that this population or community stands to benefit from the results of the research" (no. 17).

- "Research involving subjects who are physically or mentally incapable of giving consent, for example, unconscious patients, may be done only if the physical or mental condition that prevents giving informed consent is a necessary characteristic of the research population. In such circumstances the physician should seek informed consent from the legally authorized representative" (no. 29).

Ethics of biomedical research

From the Pontifical Academy for Life, "Final Communiqué of the Ninth General Assembly of the Pontifical Academy for Life on the Ethics of Biomedical Research: For a Christian Vision," February 26, 2003:

> Biomedical research has been instrumental in the progress of knowledge in this sector of medicine. ... (no. 1).

> In principle, therefore, there are no ethical limits to the knowledge of the truth, that is, there are no "barriers" beyond which the human person is forbidden to apply his cognitive energy. ... On the other hand, precise ethical limits are set out for the manner the human being in search of the truth should act, since *"what is technically possible is not for that very reason morally admissible."* ... (no. 3).

> The ultimate aim of every research activity in this field must be the integral good of man. The means it uses, must fully respect every person's inalienable dignity as a person, his right to life and and his substantial physical integrity. ... (no. 4)

Another theme of great importance in the context of biomedical research is certainly that of therapeutic and non-therapeutic experimentation, considered from the perspective of its application to the human being. It involves many problematic aspects, both of a scientific and ethical nature. It is indispensable, for example, to demand a high professional standard from the researchers involved in the experimental project, and to adopt a methodology that is rigorous in determining and applying procedural criteria. Moreover, It is also ethically necessary that the person conducting the experiment, with his collaborators, maintain total personal and professional independence with regard to possible interests (financial, ideological, political, etc.) unrelated to the goal of the research, for the good of the subjects involved and the genuine progress of humanity (no. 7).

Besides, we want to reaffirm the need to do sufficient experimentation on animals prior to the clinical experimental phase (the application on human beings) that will enable researchers to acquire advance knowledge of the possible harm and risks that this experimentation could have in order to guarantee the safety of the human subjects involved. Naturally, experimentation on animals also has to be carried out with the observance of precise ethical norms to safeguard, as far as possible, the well-being of the specimens used (no. 8).

Special attention must also be paid to the treatment of human subjects who undergo research who are especially "vulnerable" because of their state of life, as the example of human embryos clearly illustrates. Because of the delicate stage of their development, possible experimentation on them in the light of current technological advances would involve a very high—and therefore ethically unacceptable—risk of causing them irreversible damage and even death. … (no. 9).

Given the growing limitation of the resources that are available for the development of biomedical research, it is in fact necessary to pay great attention to achieving a just distribution between the different countries, taking into account the living conditions in the various parts of the world and the emergence of the primary needs of the poorest and most harshly tried peoples. That means that all should be guaranteed the conditions and minimal means so that they can enjoy the benefits deriving from research, and develop and support an endogenous capacity for research (no. 10).

Chapter 14

BIOETHICS AND
ORGAN TRANSPLANTS

Scientific and Technological Advances

The field of transplantation is one area in which medical science and surgical techniques are making continuous and rapid progress. In just a few decades, increasingly important organs have been transplanted successfully: kidney, heart, liver, pancreas, and so on. Even though surgeons have not yet gone so far as to perform torso or head transplants in human beings, that technical possibility nevertheless is foreseen. In recent years there have also been attempts at heterologous transplantations of a baboon heart and a chimpanzee liver.[1]

---

[1] The scientific as well as legal and ethical bibliography on this topic is enormous, and I do not intend to provide a complete review of it. I limit the following list to works of an ethical or legal character and, among those, I cite the works that are most in line with the personalist approach: G. Perico, "Trapianti umani," in *Nuovo dizionario di teologia morale*, ed. F. Compagnoni, G. Piana, and S. Privitera (Cinisello Balsamo: Paoline, 1990), 1383–1391; B. Häring, *Free and Faithful in Christ* (New York: Crossroad, 1981), 3:91–94; A. Fiori, "I trapianti d'organo ed i costi-benefici del progresso medico," *Medicina e Morale* 34, no. 1 (1984): 16–26; A. Bompiani and E. Sgreccia, eds., *Trapianti d'organo* (Milan: Vita e Pensiero, 1989); and *Trapianto di cuore e trapianto di cervello* (Rome: Orizzonte Medico, 1983), especially the following articles: C. Caffarra, "Scienza e tecnologia a servizio della vita con il trapianto di cuore artificiale," 97–101; P. K. Demmer, "Liceità dell'ardita sperimentazione del trapianto cerebrale," 150–169; P. A. Marino, "Sviluppi della medicina e progresso umano," 236–246; and R. Cortesini, "Il cuore artificiale: Esperienze scientifiche e riflessioni etico-morali," 52–86. For magisterial teachings, see Pius XII, Address to Managers and Associates of the Italian Society of Cornea Donors, May 14, 1956, in *Discorsi e Radiomessaggi* (Vatican City: Tipografia Poliglotta Vaticana, 1969), 18:192–201 [Italian]; *Catechism of the Catholic Church*, no. 2296; John Paul II, Encyclical Letter *Evangelium vitae*, March 25, 1995; Pontifical Council for Pastoral Assistance to Health Care Workers, *Charter for Health Care Workers* (Boston: Pauline, 1995), nos. 83–91; I. Carrasco De Paula, "Il problema filosofico ed epistemologico della morte cerebrale," *Medicina e Morale* 43, no. 5 (1993): 889–902; C. Manni, R. Proietti, and F. Della Corte, "La morte cerebrale aspetti diagnostici," *Medicina e Morale* 43, no. 5 (1993): 903–918; S. Mazza, "Aspetti neurologici della morte cerebrale," *Medicina e Morale* 43, no. 5 (1993): 919–932; A. Fiori and M. L. Di Pietro, "Accertamento della morte: Normativa vigente e prospettive future," *Medicina e Morale* 43, no. 5 (1993): 945–965; A. Puca, *Trapianto di cuore e morte cerebrale: Aspetti etici* (Turin: Camilliane, 1993); idem, "I trapianti d'organo," in C. Romano and G. Grassani, eds., *Bioetica* (Turin: UTET, 1995), 487–504; A. Bondolfi,

Aside from factors connected with the development of surgery, what has contributed most to the progress in this field has been the possibility of experimenting on animals and thereby deducing effective techniques for human applications. Another decisive factor has been the progress made in our knowledge of immunology. Another contribution was made by studies on blood transfusion and blood types after the discoveries by Landsteiner in 1900: even if a blood transfusion cannot be described as a transplant, strictly speaking, inasmuch as the tissue introduced into the host organism survives only for a short time, such experiments provided a basis and an incentive for the study of histocompatibility, and which in turn has favored the development of transplantation techniques. The confidence of success in the field of transplantation has fluctuated: there have been moments of euphoria followed by moments of greater caution; the years 1968 to 1970 were a time of enthusiasm about heart transplants, for example, which was followed by a period of greater prudence.

We have more recently witnessed renewed optimism pushing the limits of what is human in, for instance, the widely discussed idea of head/torso transplantation.[2] Technical potential is also expanding toward new objectives: the transplantation of portions of the brain (e.g., pituitary gland, hypothalamus), gonads, and—in the area of genetic engineering—even the transplantation of genes and fetal tissues (which was addressed in a separate section). The increased confidence is due to the availability of more effective drugs to combat the phenomenon of rejection.

Generally, and by way of introduction, it must also be stated at the outset that while these trends seemed to be simplifying the ethical problem, it is now becoming ever more complex. At first glance it would seem that, since the purpose of transplantation is clearly therapeutic, the ethical question becomes simple and can be answered in the affirmative, provided that death is certified before the tissue is obtained from a cadaver or that the survival and good health of a living donor has been assured.

In reality, given the progress of the techniques, the greater number of requests, the scarcity of donors, and the quality of organs needed for transplantation suitability, the ethical problems have grown more complex. The informed consent of the

---

"Per una considerazione etica globale della medicina dei trapianti," *Bioetica e Cultura* 3, no. 5 (1994): 77–92; L. Mingone, "I trapianti di organi nei dibattiti dell'etica contemporanea," *Medicina e Morale* 44, no. 1 (1994): 11–37; "Osservatorio sulla donazione di organi," *Bioetica e Cultura* 9 (1996): 27–61; M. Cozzoli, "Il trapianto di organi nella prospettiva valoriale del dono," *Medicina e Morale* 47, no. 3 (1997): 461–473; A. Anzani, *Trapianti d'organo: Problemi etici ed aspetti sociali* (Milan: Lauri, 1996), 154. As for the legal aspects, see G. Biscontini, *La morte e il diritto: Il problema dei trapianti d'organo* (Naples: ESI, 1994); M.B. Fisso, "Trapianti (diritto)," *Enciclopedia italiana*, appendix V, updated 1978–1992, 551–553. For the social and psychological aspects, see E. Soricelli, *Il trapianto d'organi* (Genoa: Pantograph, 1994). With regard in particular to the problem of commercialization, see G. Berlinguer and V. Garrafa, *La merce finale* (Milan: Baldini and Castoldi, 1996); R.P. Baker and V. Hargreaves, "Organ Donation and Transplantation: A Brief History of Technological and Ethical Developments," in *The Ethics of Organ Transplantation*, ed. W. Shelton and J. Balint (Oxford: Oxford University Press, 2001), 1–42.

[2] Fiori, "I trapianti d'organo"; R.J. White et al., "The Isolation and Transplantation of the Brain: An Historical Perspective Emphasizing the Surgical Solutions to the Design of These Classical Models," *Neurological Research* 18 (1996): 194–203. Pontifical Academy for Life, "Prospects for Xenotransplantation: Scientific Aspects and Ethical Considerations," September 26, 2001. This issue will be examined further on.

recipient, the freedom of the donor and of the relatives, the right of society to obtain organs from cadavers without explicit consent, the permissibility of recompense, the permissibility of certain transplants that can affect the identity of the recipient person, the legitimacy of experimental transplantation, determining death in the case of transplants from "beating-heart" cadavers and, finally, determining the criteria by which to assign transplantation organs among the various persons awaiting them[3]—all these are complex and multifaceted problems.[4]

Aside from the individual cases and particular problems, there are always the fundamental problems of all bioethics: the person's relationship of respectful dominion over his corporeal nature and the relation between technology and ethics.[5]

The implantation of artificial organs, such as an artificial heart, should be mentioned as a corollary to this transplantation technology; it does not pertain to our discussion of the transplantation of human organs per se but often constitutes one of its phases or an alternative to it.

Terminology and Historical-Descriptive Notes

Transplant and *graft* are terms for a surgical operation whereby an organ or tissue obtained from a donor is inserted into the host organism. When the tissues are dead or preserved, it is more properly called an *implant* or an *inclusion*. As a result of the transplant, the vital phenomena known as survival, adaptation, and thriving take place between the grafted tissues and the host organism. In the case of the cornea, even though it does not have a vascular system, it is nevertheless a living tissue and is therefore still described as a transplant.

When the tissue is transferred from one site to another in the same organism, it is called an *autologous transplant* or *autograft*. The term *homograft* or *homologous transplant* denotes the transfer of tissues from an individual donor to a host belonging to the same species; otherwise, when the procedure involves individuals of different species, it is called a *heterograft* or *heterologous transplant*. Depending on the technique used, it is called a *pedunculated* transplant when it involves the transfer of pieces of tissue not entirely detached from the donor organism; it is described as *free* when the transplanted tissue is detached from its natural supports. Finally, when the transplant requires the reconstruction of one or more ducts (vessels, bronchia, or sections of intestine), it is referred to as a transplant *with anastomosis*.

Obviously, in addition to the transplantation of tissues, one can speak about the transplantation of entire organs (e.g., kidney, heart). The organs involved in transplantation can be of a functionally directed type (e.g., joint, kidney, heart) or organs with functions of central coordination or individual differentation (e.g., brain, gonads).

Obviously, one must also consider whether the organ or tissues are obtained from a dead donor (*ex mortuo*) or from a live donor (*ex vivo*). This is not only because of the relevant ethical assessments but also to determine the feasibility of surgical

---

[3] G. Nanni, "Microallocazione delle risorse: Il punto di vista del chirurgo," in *Etica e allocazione delle risorse in sanità*, ed. E. Sgreccia and A.G. Spagnolo (Milan: Vita e Pensiero, 1996), 135–136.

[4] J. Dausset, "I successi ed i limiti dei trapianti effettuati sull'uomo," in *Trapianto di cuore*, 12–42.

[5] Caffarra, "Scienza e tecnologia."

extraction without causing harm to the donor.[6] We also need to remember the facts that follow or accompany the transplant. *Thriving* is the vital phenomenon whereby most of the grafted tissue survives and adapts to the new living conditions; tissue, blood vessel, and nerve connections are reestablished in such a way that the grafted tissue participates permanently in the life of the host organism.

When the transplant is not successful, the grafted tissue gradually dies and disappears by inclusion and reabsorption. It is essential to reestablish the circulation of blood in order for a transplant to succeed—except in the case of a cornea, given its particular histological structure without blood vessels.

The phenomenon of *incompatibility* occurs through a reaction of the host organism itself to the grafted tissue due to a complex set of factors. There are identifying structures in the cells of every organism called tissular *antigens*, which serve to detect tissues that are foreign to the organism in which they are transplanted and with which they are incompatible. A complex sequence of biochemical and anatomical-histological phenomena result from the immunological reaction of antibodies that are drawn to the grafted tissue, even after initial acceptance and reconstruction of blood vessels (so-called rejection). The study of various groups of related *tissular antigens* is a necessary requirement before transplantation if the procedure is to be successful.

Another fundamental technical problem is the preservation of the organs destined for transplantation. While blood for transfusions can be preserved for a long time under certain conditions without undergoing substantial changes, the tissues of an organ can be preserved unchanged, in a manner suitable for transplantation, for only a short time. Planning to obtain an organ surgically therefore requires a set of considerations and complex preparations, whether it is done *ex vivo* (e.g., a kidney donation) or *ex mortuo*; in the latter case, the clinical death of the donor subject must be confirmed and the transplant must be organized within a very short interval of time. "Centers" for rapid communication and transportation are therefore organized both at the national and international levels. It should be kept in mind that, prior to transport and transplantation, histocompatibility tests must be conducted to best assign the organ along with all other tests to rule out the presence of infectious diseases (e.g., viral hepatitis and AIDS).

Autologous transplants, for which histological incompatibility is a nonissue, include transplants that are reparative (for burns) or aesthetic in nature; in this case, the operation takes place in three phases: tissue preparation, transplantation, and modeling/shaping.

Tissues that can theoretically be considered for *homologous transplantation* include endocrine glands such as the ovaries, pituitary, thyroid, thymus, pancreas, and testicles; other possible tissues are blood vessels, bone tissues, bone marrow (for the cure of various forms of leukemia), nerves, tendons, patches of skin, cornea, and liver.

The transplantation of entire organs has taken on particular importance; examples include the kidney, which can be obtained from a live donor, and the heart, which obviously can be taken only *ex cadavere* (from a cadaver). Obtaining an organ from a cadaver involves the complex question of determining clinical death, which is both an ethical and a legal problem; this matter will therefore have to be treated separately.

---

[6] Perico, "Trapianti umani," 1162–1163.

From a historical perspective, suffice it to recall that the first *experimental kidney transplants* were performed by Emerich Ullmann in 1902. In the same year in France, Alexis Carrel perfected the technique of anastomosis for kidney transplants; however, the first scientific study on kidney transplants in humans was published in 1955 by David Hume in the United States: he described transplants performed without immunosuppressive therapy and based only on the prior study of ABO blood type compatibility, with survival ranging from 37 to 180 days.[7] The first successful kidney transplant in Italy was performed by a group headed by Carlo Casciani in Rome on May 20, 1966, and the patient enjoyed good health for a long time.

Significant progress has been made since then: by 1990 more than 200,000 kidney transplants had been performed worldwide, and kidney transplants have now become routine in large hospitals. In Italy, around 800 kidney transplants were performed in 1990; in 1995 there were 541 kidney transplants, some of them involving multiple organs, and 236 liver transplants.[8] A campaign was started for voluntary postmortem organ donation through associations such as the Italian Organ Donors Association (Associazione Italiana Donatori d'Organo, AIDO). Another voluntary association is the National Hemodialyzed, Dialysis, and Transplantation Association (Associazione Nazionale Emodializzati Dialisi e Trapianto, ANED).

Heart transplants began in Cape Town, South Africa, in 1967, with the work of Christiaan Barnard. For a long time his attempts were followed by unsatisfactory results and involved large areas of uncertainty. Beginning in 1985, the Italian Ministry of Health authorized several centers to perform heart transplants; since then various Italian hospitals have quickly begun to perform heart transplants on adults and on children, for the most part successfully. The data indicate that there have been several hundred heart transplants, some of which were heart-lung transplants.[9] The difficulty of the surgical procedure and the scarcity of donors are reasons why research is being done into various forms of an "artificial heart," which is a machine that is supposed to replace the heart entirely. Models of artificial hearts are designed to be increasingly manageable and portable; the purpose of this technology is also to provide an interim solution for patients awaiting the transplantation of a compatible human heart. But research in this area is still in the experimental phase.[10]

Liver Transplants from Living Donors

A liver transplant from a living donor is a surgical procedure designed to remove part of the liver of a volunteer donor and then transplant it into a patient suffering

---

[7] David M. Hume et al., "Experiences with Renal Homotransplantation in the Human: Report of Nine Cases," *Journal of Clinical Investigation* 34 (1955): 327–382.

[8] Nord Italia Transplant, "Dati sull'attività di prelievo e trapianto in Italia e all'estero," in *Servizio Studi Camera dei Deputati: XII Commissione - Indagine Conoscitiva sui Trapiant, n. 16*, XII Legislatura, ed. D. Chiassi (February 1996), 1–10.

[9] Nord Italia Transplant, "Dati sull'attività"; R. Bernardini, "In lista d'attesa con poche speranze," *Missione Salute* 4 (1991); R. W. Evans, "Organ Transplantation and Inevitable Debate as to What Constitutes a Basic Health Care Benefit," *Clinical Transplantation* (1993): 359–391; Task Force on Heart Transplantation, *Journal of American College of Cardiology* 22 (1993): 1–64.

[10] Cortesini, "Il cuore artificiale."

from a chronic liver disease. The right part of the liver is the one more commonly removed from the donor to perform the transplant; this is possible thanks to the liver's capacity to regenerate after a portion has been removed.

The technique, carried out the first time in the late 1980s, met with great success in Japan, where organ donation is extremely rare.

Liver transplants from living donors are regulated in Italy by law no. 483 of December 16, 1999, "Norms for allowing partial liver transplants," published in the *Gazzetta Ufficiale della Repubblica Italiana*, issue 297, on December 20, 1999.[11]

"Non-life-saving" Transplants of Visible Parts [12]

It is necessary to distinguish so-called *life-saving transplants* (e.g. of liver, heart and lungs), in which the patient is doomed to die if he does not receive a transplant, from *non-life-saving* transplants (e.g., of the hand or face), which are supposed to improve the patient's quality of life.

There are many reservations from the ethical perspective because the face and hands also represent a person's identity, and these techniques are therefore an assault on its uniqueness. The first hand transplant was completely successful, but after a while the patient had the new "piece" removed, because he never managed to adapt to the new hand.

Furthermore, it is debatable whether it is permissible to transform a mutilated but "healthy" person into a "reconstructed" but sick person. A French patient from Amiens who underwent a face transplant in December 2005 is now severely immunodeficient and in practice can hardly leave the house because it would expose her to infections (the skin is the most dangerous tissue from this perspective). It might well be asked whether she is able to have a better life than she had before.

"Crossover" Transplants

The "crossover" transplant is a special case of transplant from a living donor. The procedure is performed in the case of a kidney donation, although it could potentially be applied in a partial liver transplant as well. It involves two or more couples of subjects; each couple consists of blood relatives, spouses, or friends, one of whom needs a transplant while the other is willing to make an organ donation, although they are immunologically incompatible. If another couple in a similar situation is found and it is ascertained that there is a crossover biological compatibility (i.e., between donor A and recipient B and between donor B and recipient A), a crossover transplant can be performed, if the two couples give consent. The technique thus makes it possible to overcome the barrier of biological incompatibility

---

[11] "Art. 1. *(Partial liver transplant)*: 1. Notwithstanding the prohibition in article 5 of the civil code, it is permissible to give away part of one's liver without remuneration for the exclusive purpose of transplantation among living persons. 2. To the extent that they are compatible, the provisions of law no. 458 of June 26, 1967, apply to the purposes described in section 1."

[12] U. Genovese, R. Stucchi, and A. Farneti, "I trapianti 'non salvavita': aspetti medico-legali," *Rivista Italiana di Medicina Legale* 22, no. 2 (2000): 417–430.

that prevented a direct transplant within each couple. The procedure was proposed for the first time in 1986.[13]

There have also been proposals for so-called indirect modes of crossover transplantation. These are carried out when a living donor is available for a recipient but the donor and recipient are biologically incompatible. The donor can then offer to donate his organ to a compatible patient who is near the head of the waiting list for cadaver transplants. In this way the recipient originally associated with the donor takes a position near the head of the waiting list for cadaver transplants.[14]

Crossover transplantation is practiced in several countries, including South Korea,[15] Israel,[16] Mexico,[17] and the United States.[18] In Europe, crossover transplants are performed in Germany,[19] Great Britain,[20] the Netherlands (where there is a pertinent national program with a centralized system that identifies matches with the best compatibility and guarantees anonymity),[21] Romania,[22] and Switzerland.[23]

In 2006, a new, so-called domino plan for indirect crossover transplants was proposed, in which a single donor sets off a "cascade" of transplants. The plan entails a living donor (donor 1) who donates an organ to a biologically compatible recipient (recipient 1), who has a blood relative, spouse, or friend (donor 2) who is biologically incompatible but willing to donate an organ. Donor 2 (who is associated

---

[13] F. T. Rapaport, "The Case for a Living Emotionally Related International Kidney Donor Exchange Registry," supplement 2, *Transplantation Proceedings* 18 (1986): 5–9; idem, "Exchange Donor Program in Kidney Transplantation," *Transplantation Proceedings* 19, no. 1, part 1 (1987): 169–173.

[14] F. L. Delmonico et al., "Donor Kidney Exchanges," *American Journal of Transplantation* 4, no. 10 (2004): 1628–1634.

[15] K. Park et al., "Exchange Donor Programme in Kidney Transplantation," *Transplantation* 67, no. 2 (1999): 336–338.

[16] T. Klein et al., "Successful Interspousal Swap of ABO Incompatible Living Donor Kidneys," *Transplantation Proceedings* 32, no. 4 (2000): 688–689.

[17] F. Juarez et al., "Domino (Crossover) Kidney Transplantation Using Low Doses of Neoral," *Transplantation Proceedings* 30, no. 5 (1998): 2289–2290.

[18] F. McLellan, "US Surgeons Do First 'Triple-Swap' Kidney Transplantation," *Lancet* 362, no. 9382 (2003): 456; A. Spital, "Should People Who Donate a Kidney to a Stranger Be Permitted to Choose Their Recipients? Views of the United States," *Transplantation* 76, no. 8 (2003): 1252.

[19] A. Tuffs, "Surgeons Perform Germany's First Crossover Kidney Transplantation," *British Medical Journal* 331, no. 7520 (2005): 798–799.

[20] C. Dyer, "Paired Kidney Transplants to Start in the United Kingdom," *British Medical Journal* 332, no. 7548 (2006): 989.

[21] L. Kraneburg et al., "Starting a Crossover Kidney Transplantation Program in the Netherlands: Ethical and Psychological Considerations," Transplantation 78, no. 2 (2004): 194–197; Academisch Medisch Centrum Amsterdam et al., *National Protocol Crossover Kidney Transplantation*, 2004; M. De Klerk et al., "The Dutch National Living Donor Kidney Exchange Program," *American Journal of Transplantation* 5, no. 10 (2005): 2302–2305.

[22] M. Lucan et al., "Kidney Exchange Program: A Viable Alternative in Countries with Low Rate of Cadaver Harvesting," *Transplantation Proceedings* 35, no. 3 (2003): 933–934.

[23] G. Thiel et al., "Crossover Renal Transplantation: Hurdles To Be Cleared!" *Transplantation Proceedings* 33, no. 1–2 (2001): 811–816.

with recipient 1 but incompatible) can, in turn, agree to donate the organ to the next compatible patient on the waiting list (recipient 2), producing a "domino effect." The proponents of this plan, employing a mathematical simulation, have calculated that if the model had been applied in the United States from 1998 (when the first indirect transplant from a living donor was performed) until 2006, then 583 kidney transplants could have been performed instead of the 302 that were effectively performed (according to data from the United Network of Organ Sharing, UNOS). According to the same authors, the "domino" model satisfies the ethical requirements of the three principal ethical models that are used to evaluate organ donation by a living donor. The donor-centered model, which emphasizes the likelihood of a successful intervention, is satisfied inasmuch as the probability of a favorable outcome is increased. The recipient-centered model, which states that priority should be given to the patient in greatest need, is satisfied because the waiting lists are shortened. Finally, the society-centered model, which emphasizes the criterion of equal weight for the criteria of success and need, is satisfied because the organs are allocated to those at or near the head of the waiting list.[24]

Of all national ethics committees, only the Swiss National Advisory Commission on Biomedical Ethics (Nationale Ethikkommission-Commission Nationale d'Éthique, NEK-CNE) has dealt with crossover transplants; it did so in an opinion on living donor transplants.[25] It expressed a favorable opinion and specifically stated that it "sees no reason to introduce special restrictions on crossover transplantation in the Law."[26]

From the ethical perspective, crossover transplantation poses all the ethical problems typical of transplants from living donors (see the discussion above), with some peculiarities that are especially pronounced and deserve particular attention. These include informed consent (which involves a condition concerning not so much the "donation" typical in living donor transplantation as a consignment for the purpose of an exchange), an evaluation of the psychological characteristics and motives of the donor, and the risk of commercialization or exploitation.[27] The importance of these problems suggests that the practice of crossover transplantation should linked with anonymity between the two couples. For example, according to the NEK-CNE, this aspect "should be disclosed to the donors and recipients before the final decision is taken."[28]

Crossover transplantation requires precise regulations aimed at guaranteeing respect for ethical requirements, as well as the correct execution of the procedures, which in some practical aspects are more complex than in typical live-donor trans-

---

[24] N. R. Brook and M. L. Nicholson, "Non-Directed Live Kidney Donation," *Lancet* 368, no. 9533 (2006): 346–347; R. A. Montgomery et al., "Domino Paired Kidney Donation: A Strategy to Make Best Use of Live Non-Directed Donation," *Lancet* 368, no. 9533 (2006): 419–421.

[25] Swiss National Advisory Commission on Biomedical Ethics (NEK-CNE), "On the Regulation of Living Donation in the Transplantation Law," opinion no. 6/2003, November 17, 2003, http://www.bag.admin.ch/nek-cne/04229/04232/index.html?lang=de.

[26] Swiss Advisory Commission, "Living Donation," 33.

[27] P. E. Morrissey et al., "Good Samaritan Kidney Donation," *Transplantation* 80, no. 10 (2005): 1369–1373.

[28] Swiss Advisory Commission, "Living Donation," 33.

plants (for example, it is necessary to coordinate the two interventions efficiently and to perform them simultaneously, while realizing that the two interventions could have different outcomes).[29]

Legislative Prospects and International Cooperation

The Council of Europe, according to the inspiring principles of the Convention on Human Rights and Biomedicine, which is its fundamental and "constitutional" charter, has fostered intensive cooperation among the member states in the area of health care legislation.

In the European sphere, again on the level of enunciating general principles, the European Social Charter should first be recalled. In article 11, it declares that nations are obliged to safeguard and protect the health of their citizens. Also to be recalled is the European Code of Social Security, which guarantees preventive, curative, and rehabilitative measures.

The guiding principles that inspired the legislation of the Council of Europe pertaining to blood transfusions (Reykjavik Code of 1975), the exchange of tissue compatibility tests (agreement no. 84 of 1974), and problems concerning transplants, grafts, and obtaining tissues and organs of human origin (resolution [78] 29 and recommendation R [79] 5), are designed to promote health care collaboration and security and to prevent commercialization.[30]

To guarantee greater legislative uniformity among the various European countries, resolution (78) 29, approved by the Council of Europe's Committee of Ministers on May 11, 1978, contains a series of rules, definitions, and invitations to the member states for their respective national legislations.

Particularly noteworthy are the guidelines about consent for obtaining organs from living donors and about precise sanctions for false declarations of death in order to obtain an awaited transplant from a cadaver. In the same resolution, governments are invited to provide the opportunity for citizens to declare in advance their willingness to make a postmortem organ donation on state-issued identification documents or on driver's licenses. The general slant and drift of the document favors the possibility of making use in the member states of "presumed consent" to obtain organs post mortem, unless the subject has explicitly "opted out" during his lifetime. This position was also reiterated by the Conference of European Health Ministers held in Paris on November 16–17, 1987.[31] The tendency is therefore to consider the cadaver *res communitatis* (community property); nonetheless, the document also encourages

---

[29] F. L. Delmonico, "Exchanging Kidneys: Advances in Living-Donor Transplantation," *New England Journal of Medicine* 350, no. 18 (2004): 1812–1814; R. Veatch, *Transplantation Ethics* (Washington, DC: Georgetown University Press, 2000), 186–188.

[30] F. Marziale, "Cooperazione europea nell'ambito dei trapianti e problemi etici," in *Trapianto di cuore*, 184–207; M. Torrelli, *Le médecin et les droits de l'homme* (Paris: Berger-Lévrault, 1983). With regard to the complex problems associated with buying and selling human organs, see also Anzani, *Trapianti d'organo*, 49–74; Berlinguer and Garrafa, *La merce finale*.

[31] Conference of European Health Ministers, "Organ Transplantation," in the appendix to Bompiani and Sgreccia, *Trapianti d'organo*, 295–305.

respect for the religious convictions of the living person whenever such convictions would be opposed to postmortem organ donation.

As for ascertaining clinical death, it is forbidden to remove organs or tissues unless the donor has lost brain function totally and irreversibly: the document explains, however, that this does not rule out the possibility that certain vital functions of other organs may be maintained artificially.[32]

Articles 9 and 14 of the same resolution prescribe that the donation of organs and tissues should be done free of remuneration, except for the rights to reimbursement for expenses and the guarantee of social security for injuries resulting from a donation by a living donor. The other points deal with preliminary examinations and the location and conditions for surgically obtaining and transplanting the tissue or organ.

Recommendation R (79) 5 concerns safety measures and customs regulations for the rapid and expedited international transport of substances destined for transplantation. The goods travel with a European "label."

Thanks to these provisions, which are incorporated into the rules established by resolution no. 24/79 of the European Parliament on organ banks,[33] "data banks" are organized along with programs for the exchange of related information at the international level to make it possible to locate an organ for transplant, determine its type, and ship organs by air freight, along with blood for transfusions and other substances destined for transplantation in the event of a catastrophe.

Lastly, we also cite the provisions of the so-called Convention on Bioethics by the Council of Europe.[34] Although this document does not address at length the topic of postmortem organ donation—the topic is reserved for a later, more specific protocol—it reiterates the prohibition against commercializing human body parts.

The Italian Legal Situation

Law no. 91 of April 1, 1999, which deals with regulations on organ and tissue procurement and transplantation, was approved after eleven years of debate.[35] Accord-

---

[32] World Medical Association, Declaration of Sydney on the Determination of Death and the Recovery of Organs, August 1968, updated October 2006, http://www.wma.net/en/30publications/10policies/d2/.

[33] European Parliament, "Resolution on Organ Banks," May 21, 1979, *Official Journal of the European Communities* C 127 (1979): 71.

[34] Council of Europe, Convention for the Protection of Human Rights and the Dignity of the Human Being with Regard to the Application of Biology and Medicine: Convention on Human Rights and Biomedicine, April 4, 1997, http://conventions.coe.int/Treaty/en/Treaties/html/164.htm.

[35] F. Panarese et al., "Trapianti, donazione e normativa: Risultati di una legge controversa; Quattro anni dopo," *Difesa sociale* 82, no. 4–5 (2003): 11–36; P. Stanzione, ed., *La disciplina giuridica dei trapianti: Legge 1 aprile 1999 n. 91* (Milan: Giuffré, 2000); V. Palermo and E. Ravera, "Note sulla legge 1 aprile 1999, n. 91 'Disposizioni in materia di prelievi e di trapianti di organi e tessuti'," *Rivista di Diritto delle Professioni Sanitarie* 2, no. 2 (1999): 104–117; E. Palmerini, "La nuova legge sui trapianti d'organo: Prime notazioni," *Studium Iuris* 12 (1999): 1311–1321; P. Becchi and P. Donadoni, "Informazione e consenso all'espianto di organi da cadavere: Riflessioni di politica del diritto sulla nuova legislazione," *Politica del Diritto* 32, no. 2 (2001): 257–287; Camera dei Deputati Servizio Studi, ed., *Legge 1 aprile 1999, n. 91—Disposizioni in materia di prelievi e di trapianti di organi e di*

ing to the law, brain death—or the death of the encephalon—must have occurred in order to proceed with organ removal; the body must also still be "viable" in order to allow heart-beating organ removal while the organ perfusion devices are operating.

The law accepts what the Ad Hoc Commission of the Harvard Medical School established in the United States in 1968,[36] according to which an individual is brain dead when the following conditions are met: lack of brain activity, lack of spontaneous or induced movement, lack of spontaneous breathing, and lack of central brainstem reflexes. This definition of brain death[37] is adopted as a justification for the official declaration of death, the suspension of resuscitation efforts and, lastly, the removal of organs.

According to article 4 of law no. 91/99, every citizen shall be called to express his or her willingness to donate or not in written form; if the citizen's will is not so manifested, it shall be considered assent to organ removal.

This effort to increase the number of donations, which is good in itself, could deprive individuals of the protection to make decisions regarding their own bodies. To avoid this danger, lawmakers introduced guarantees against the organ "predation" that might ensue, according to some, from this form of "tacit assent,"[38] which is nevertheless different from presumed consent.

Article 4, section 5, in fact prohibits proceeding with removal if there is "a written declaration of unwillingness to undergo removal, authored by the subject whose death has been confirmed." The relatives or legal representative of the potential donor can therefore prevent organ removal by presenting a document written by the potential donor that manifests his or her refusal of organ donation.

Taking away from the family the possibility of expressing consent for organ removal from a brain-dead relative, which was the case under law no. 644 of 1975, lifted the burden of a difficult decision from the family but created a problem that could be considered just as challenging as the one it resolved. In cases involving the valid application of tacit assent, the family is charged with an unpleasant task: searching for and finding—in the limited time required for confirming brain death—a written declaration *authored by the subject* expressing his or her refusal to donate.

The time needed to fully implement law no. 91/99 will necessarily be lengthy in virtue of the creation of the information system required by the law (article 7).

---

tessuti: Iter parlamentare; Dossier provvedimento n. 370/3: parti I, II e III (Rome: Camera dei Deputati, 1999); M. Pappalardo and F. Martinelli, "Qualche osservazione sull'assenso alla donazione post mortem di organi e tessuti e sull'art. 14, punto 2 della legge 1 aprile 1999, n. 91," *Sanità Pubblica* 10 (2002): 1215–1224.

[36] See "A Definition of Irreversible Coma: Report of the Ad Hoc Committee of the Harvard Medical School to Examine the Definition of Brain Death," *Journal of the American Medical Association* 205, no. 6 (1968): 337–340; J. L. Bernat, C. M. Culver, and B. Gert, "On the Definition and Criterion of Death," *Annals of Internal Medicine* 94 (1981): 389–394; J. Korein, "Terminology, Definitions, and Usage," *Annals of the New York Academy of Sciences* 315 (1978): 6–10.

[37] R. Lucas Lucas, "Morte encefalica e morte umana," in *Antropologia e problemi bioetici* (Cinisello Balsamo: Edizioni San Paolo, 2001), 119–158.

[38] I. Nicotra Guerra, "Il silenzio legale informato nella recente legge sui trapianti dalla regola coercitiva alla norma pedagogica," *Rivista Trimestrale di Diritto Pubblico* 3 (1999): 829–848.

Provisional regulations were put in place that would remain in force until the completion of the information system. The precedence of the family's wishes over the wishes of the deceased, which was sanctioned by law no. 644 of 1975, is abolished if the subject expressed a favorable opinion about donation while alive. Organ and tissue removal is permitted unless the subject explicitly refused to give his or her assent, or unless the non-separated spouse or cohabitating partner *more uxorio* (as though husband and wife)—or the adult children (in the absence of the preceding possibility) or the parents or legal representative (in the absence of both preceding possibilities)—presents written opposition within the observation period required for the confirmation of brain death.

In general, the legal recognition afforded to the explicit manifestation of wishes during life with respect to organ donation after death takes up what was already anticipated in the document by the Italian National Bioethics Center (NBC), *Donazione d'organo a fini di trapianto* (Organ donation for transplants),[39] published on October 7, 1991, which hoped for "the transition toward a legal framework of consent in which the personal choice to donate one's organs and the qualified presumption of consent from those who remain silent can be more reasonably recognized, though with a norm requiring the expression of one's wishes with regard to the acceptance of organ removal."[40] Once the situation is established that would allow all citizens to express their wishes, even the lack of explicit wishes to the contrary could be considered sufficient for eventual organ removal.

According the NBC, it is essential that the subject be solicited to express his opinion in an official manner and that the lack of an explicit pronouncement by the subject should be only considered equivalent to the manifestation of assent after a certain amount time has elapsed. In this manner the limits dictated by the religious denomination to which the subject belongs or the subject's personal convictions can be respected: through the explicit request of assent, the subject is provided with all of the necessary information that would allow the subject, if he considered it appropriate, to refuse his consent to organ removal. This would ensure his right to be informed of the legal significance of his gesture (silence), after which the lack of an explicit declaration could be attributed with the positive significance of consent to donation.

On the other hand, in discussing an appropriate legal framework for tacit assent in its document, the NBC gave the sense that the introduction of this practice should be surrounded by safeguards. A law that would require the subject to take the initiative in making known his wishes against postmortem organ donation would be unacceptable from both the ethical and the legal perspectives. It is unacceptable because it would be embodied in the imposition of an obligation on the subject to concern himself with declaring his refusal of organ donation in the forms and manners established by the law. The law should provide for ways in which the subject is repeatedly invited to make his wishes on the subject known; it is only in this way

---

[39] See M. Baraldo et al., "Riflessioni sulla disponibilità del cadavere da parte dei familiari," *Rivista Italiana di Medicina Legale* 22, no. 4–5 (2000): 947–953.

[40] Italian National Bioethics Committee, *Donazione d'organi ai fini di trapianto*, October 7, 1991 (Rome: Presidenza del Consiglio dei Ministri Dipartimento per l'Informazione e l'Editoria, 1991), 5, http://www.governo.it/bioetica/pdf/donazione_d_organo_a_fini_di_trapianto_ok.pdf, English abstract at http://www.governo.it/bioetica/eng/opinions/abstracts/abstract_organ_donation.pdf.

that a positive interpretation of his silence, equated with consent, can be admissible. Vice versa, the lack of an explicit legal provision for inviting the subject to declare his wishes on the question of donation would give silence, even if it is due simply to the ignorance of the subject, a positive significance that is does not have, illegitimately overriding what could be the actual wishes of the individual.

The real solution to the problem should be sought in a major awareness effort directed toward public opinion to help spread a new culture, helping organ donation become an increasingly spontaneous act. Our hope must therefore be to overcome the tacit assent regime through the dissemination of a *gift culture*, such that postmortem organ donation may become a habitual form of solidarity—without losing its high moral value—practiced by all, regardless of the personal convictions of individuals.

Before considering the ethical aspects of the issue in greater detail, I would like to mention several documents of a deontological character regarding this topic. The *Principles of European Medical Ethics*, adopted by the International Conference of Medical Professional Associations of the European Economic Community (EEC) on January 6, 1987, and amended on February 6, 1995, makes reference to transplants in articles 13, 14, and 15. The Thirty-ninth Assembly of the World Medical Association (WMA) also approved its Declaration on Organ Transplants in October 1987. The *Codice italiano di deontologia medica* of 2006, in articles 40 and 41, considers the promotion of a culture of organ donation, which is an inescapable reference to the regulations in force on transplantation, determination of death, and organ and tissue procurement from living persons.[41]

The Ethical Aspect of the Problem:
General Principles

There are three general principles that define the ethical problem of transplants: defending the lives of the donor and recipient, protecting personal identity, and obtaining informed consent. Let us now briefly reflect on each of these.

Defending the lives of the donor and the recipient

It may seem that the question about the permissibility of a transplant performed for the purpose of prolonging the life of a seriously ill and otherwise incurable patient has already been settled; in reality, more in-depth bioethical reflection is required.

It should be considered that even when there is an effective benefit for the patient receiving the organ, some harm to a living donor is at times required; moreover, the prolonged life of the patient receiving the transplant organ is not always of satisfactory quality. Even in the case of an *ex cadavere* organ, offenses against life can be committed when death has not actually occurred or has not been properly certified. Finally, the surgical practice can be infused by a desire for success at all costs and by experimental aims at the expense of the patient, who may function as a guinea pig for the progress of surgical techniques. Ethical reflection must therefore be attentive and cautious while remaining open to genuinely serving the lives of patients.

---

[41] Italian National Federation of Medical and Dental Associations (FNOMeO), *Codice di deontologia medica*, rev. ed., December 16, 2006, http://portale.fnomceo.it/PortaleFnomceo/showVoceMenu.2puntOT?id=5.

The principle of respect for a person's physical life logically implies and includes the consequent obligation of the "non-disposability" of one's own body except for some greater good of the body itself (principle of totality) or for a greater moral good relative to the same person. In the case under consideration, the principle of totality or the therapeutic principle alone justifies *autoplastic* transplants (even those of an aesthetic or corrective sort); however, the principle of totality must join forces with the *principle of solidarity and sociality* to justify homoplastic transplants (kidney, heart, etc.).

In other words, a *homoplastic transplant* is permissible under the following two conditions: (a) The donor (if living) must not suffer substantial and irreparable harm to his own life and functionality. This is the case in the donation of a paired organ (kidney) or part of an organ (liver, lung) that will normally allow the donor, especially as the techniques are perfected, to continue living and working. It is understandable that Catholic moral teaching, which at the beginning of transplant experiences was rather reluctant to admit their permissibility, has gradually broadened its outlook to the point of recommending postmortem organ donation: the attitude toward permissibility depends on the success of the technique, most importantly in preserving the life of the donor. (b) There must be a high and confirmed probability of success for the transplant in the patient recipient: there must be a certain proportionality between the donor's sacrifice and the probability of a real improvement in the life of the beneficiary. Furthermore, the recipient's life is also sacred and may only be subjected to a risky, invasive treatment if there are well-founded hopes of a real prolongation of life. The *Catechism of the Catholic Church* also takes this position, considering organ transplantation a meritorious act in conformity with the moral law, provided that "the physical and psychological dangers and risks to the donor are proportionate to the good sought for the recipient. ... It is not morally admissible to bring about the disabling mutilation or death of a human being, even in order to delay the death of other persons" (no. 2296).

Hence, transplants that are risky enough to be considered primarily experimental must be condemned morally. Consider, for example, the adventurous risks run by those who have performed heart transplants at certain times in the past: the transplant must prove the only valid remedy that can prolong the patient's life. In the case of a kidney transplant, for example, the transplant procedure is appropriate only if dialysis can no longer be performed successfully in the long term; moreover, there must be clear proof of the donor's death when the organ is obtained from a cadaver. We will return to this point later.

To conclude, we will pause briefly to analyze the criteria that can be used in selecting the patients who will be granted access to the surgical procedure. It seems to me that there are at least three different criteria: a utilitarian criterion, a random criterion, and a therapeutic criterion.

According to the *utilitarian* criterion, organs should be allocated based on the criterion of *social productivity*, whereby it is preferable to choose as a recipient a patient who, after being healed, will return to work. The *random* criterion is based on the principle of impartiality and leaves the allocation of organs for transplantation to chance, on a first-come, first-served basis. Lastly, the *therapeutic* criterion, which I tend to favor, takes into account several clinical factors such as urgency, the probability of successful transplant surgery given the patient's condition, the prospect that the organ will thrive, and, as a final consideration, the order priority of the request.

The choice of the patient who will undergo the procedure must be informed, moreover, by the principle of *nondiscrimination*: there must be no social or racial reason for denying access to anyone for whom the procedure is truly necessary and medically indicated.[42]

Protecting the personal identity of the recipient and the recipient's descendents

The moral problem arises with the transplantation of organs that are not just functionally directed but rather structurally connected with the patient's thinking and biological-procreative identity. This is the case in the daring hypothesis of a head/torso transplant, or of the technically less complex transplantation of genital organs (ovaries, testicles) or even of glands that are very important for the subject's biopsychological equilibrium (e.g., pituitary).

When quality of life and personal identity are severely threatened and compromised, when the result of the transplantation threatens and upsets the subject so profoundly, this raises the question of the permissibility of the transplant and of whether the sole purpose is to achieve a kind of survival that would be merely biological.

The case of a head/torso transplant, which has been carried out on dogs and apes[43] and is being considered for human beings, would involve a personal identity linked to the brain. The new individual would be the one identified with the head, because "personal memory" is stored in the brain; this is why it is preferable to speak more about a torso transplant rather than a head transplant. This does not seem to be enough, however, to protect an overall identity of the person: indeed, it would be very difficult to connect the torso at the level of sensation, innervation, and motor function, which means it could not be "sensed" by the subject either as a sense organ or as a means of expression and action. But even more problematic is the fact that the brain's memory would continue to store the sensations and experiences from the previous body. The mechanistic assumptions of those proposing such transplants are an inadequate perspective for describing corporality, which is an integral part of the person and the embodiment of the self, of which it is the identity, manifestation, and language. From the psychological and sensorial perspective, the self is constructed through bodily experience: memory stores up the self-image and identity is reflected in corporality: How could a mind that belongs to a body which no longer exists recognize itself in another body?

For this reason I venture to declare that such a "surgical construction" would not have a limited human character and would end up disturbing the subject's personal identity.[44]

---

[42] P. J. Heid et al., "Access to Kidney Transplantation: Has the United States Eliminated Income and Racial Differences?" *Archives of Internal Medicine* 148 (1988): 2594–2600; Nanni, "Microallocazione delle risorse."

[43] R. J. White, "Individualità e trapianto cerebrale," in *Trapianto di cuore*, 102–131; idem, "Isolation and Transplantation." The author says he is in favor of this transplant procedure.

[44] Demmer, "Liceità." The author does not rule out a moral justification for a head/torso transplant.

As for the gonads and the organs connected with reproduction and the secretion of particular hormones, they are not connected with vital functions; hence, a therapeutic transplant (i.e., one that aims to save the life of the recipient) is not required due to a need that cannot otherwise be met. In general, it seems that such procedures should be considered as threat to the biological and psychological identity of the recipient and the recipient's descendents. In particular, the prohibition is even clearer when the transplant is performed merely to cure the organ and not the individual, that is, simply in order to have normal ovaries or testicles instead of diseased ones and thus to assure fertility. In this case the point of the transplant would not be to save the subject's life—as in the case with a kidney, a heart, etc.—but rather simply to treat infertility. This reason does not justify a transplant that would then entail disrupting the biological identity of the descendants—in the same manner that the moral judgment on heterologous in vitro fertilization is negative. The presence of a threat to the patient's life through a disease process (e.g., a tumor) in the genital organs can justify and involve the removal of the diseased organ, but in my opinion it cannot require replacement by means of transplant with a procreative aim.

This principle has been reiterated clearly in the *Charter for Health Care Workers*, which declares that "ethically, not all organs can be donated. The brain and the gonads may not be transplanted because they ensure the personal and procreative identity respectively. These are organs which embody the characteristic uniqueness of the person which medicine is bound to protect."[45]

One thing that remains to be discussed is the idea that the transplantation of gonads (e.g., ovaries) might be more ethically acceptable—and more therapeutically appropriate—than the administration of exogenous hormones in pathologies where the removal of diseased ovaries could result in complications because of a subsequent lack of the hormones they produce (estrogens and progestogens), causing premature iatrogenic menopause and a host of associated bone and cardiovascular disorders.[46] In such a case, however, it would be necessary to take preventive measures to ensure that gonad transplantation does not later lead to procreation.

Informed consent

With regard to the problem of consent, the following two scenarios must be considered: when the organ or tissue is obtained *ex vivo* and when it is obtained *ex cadavere*. But in any case, and most importantly, the obligation to provide precise and complete information about the risks, consequences, and difficulties must also be fulfilled *with respect to the recipient* of the organ or tissue.

Since this type of surgery sometimes involves very high risks (e.g., heart transplant) and other times requires the patient to face uncertainties and negative

---

[45] Pontifical Council for Pastoral Assistance to Health Care Workers, *Charter*, no. 88.

[46] The state of the art of these procedures has recently been reviewed in the scientific literature. See Y. Aubard et al., "Greffes et transplantations ovariennes chez la femme: Le point," *Revue Française de Gynécologie et d'Obstétrique* 88 (1993): 583–590; E. J. Barten and D. W. Newling, "Transplantation of the Testis: From the Past to the Present," *International Journal of Andrology* 19 (1996): 205–211. A legal perspective is given in M. Gennari and C. Moreschi, "Un divieto irrazionale il trapianto delle ghiandole della sfera genitale e della procreazione," *Rivista Italiana di Medicina Legale* 14, no. 4 (1992): 805–813; M. L. Faggioni, "Il trapianto di gonadi: Storia e attualità," *Medicina e Morale* 48, no. 1 (1998): 15–46.

consequences such as possible rejection or special precautions and care for the rest of his life (e.g., kidney, liver, and pancreas transplants), the information must be accurate and consent must be explicit and formal before implantation of a new organ.

When the extraction is performed on a living donor, as in the case of tissues, parts of organs, or paired organs (e.g., kidney), the obligation of informed consent also applies to the donor and concerns all the consequences to the donor's health and future ability to work. There could be no act of donation, no expression of solidarity, if there were no well-founded awareness of all the consequences of the gesture.

When the extraction is performed on a cadaver, the legal trend is to consider the cadaver *res communitatis*, as mentioned earlier, and to favor its use for the common good whenever a social need arises and the donor subject did not "opt out" by voicing a contrary intention while living.

Not all authors agree with this criterion from an ethical perspective: they recall that the cadaver, while still a *substance* and *no longer a person*, retains a kind of *sacred* character through the phenomenological and psychological reference it becomes for the survivors. For this reason, while it is still true that usefulness for the common good can justify certain hygienic or sanitary measures, emotional ties on the part of the survivors should not be completely excluded from consideration. While it remains true that the common good can justify certain measures for health and hygiene, the connection of psychological and emotional belonging of those still living should not be disregarded. Therefore, respect for the subject's own intentions and, whenever possible, information about and respect for the preferences of surviving relatives still maintain some ethical weight. Public utility, which can require sacrifices even from the living, can require some surgical procedures and the extraction of tissues or organs from cadavers without causing any harm to human life—yet respect is still due to this *substance* that has a psychological association with the person. In the above-cited document *Donazione d'organo a fini di trapianto*, the Italian National Bioethics Committee (NBC) also mentions this "respect for the values expressed by persons connected with the deceased person by a bond of family or companionship."[47]

The spread of an "authentic culture of donation" is therefore particularly important from the ethical perspective: in such a context everyone, from childhood on, would be aware of the need to *explicitly* give one's personal consent to this act of profound human solidarity and great social value.[48]

The particular connotations assumed by the problem when dealing with organ procurement from a minor must not be forgotten, either. The NBC thought it appropriate to return to the subject again, dedicating an entire document to pediatric transplants,[49] in which all the special problems of transplantation in that age group are dealt with comprehensively. In this case, indeed, health care personnel deal with quite a range of different situations, depending on whether the minor recipient is

---

[47] Italian National Bioethics Committee, *Donazione d'organo*.

[48] The *Catechism of the Catholic Church* (1992) has this to say on the subject: "Donation of organs … is not morally acceptable if the donor or those who legitimately speak for him have not given their explicit consent" (no. 2296).

[49] Italian National Bioethics Committee, *Organ Transplants in Childhood*, January 21, 1994 (Rome: Presidenza del Consiglio dei Ministri Dipartimento per l'Informazione e l'Editoria, 1994), http://www.governo.it/bioetica/eng/pdf/PCM_trapianti.pdf.

a little child or an adolescent capable of understanding the significance of what is happening. In the latter instance a special effort to communicate is necessary, partly because of the difficulty of the message to be conveyed, which essentially involves information about chances of survival.

The problem of determination of death in transplants from a cadaver

This is the most delicate problem of all in the matter of human transplants. We can proceed by reaffirming several principles that are based on respect for life.

Above all, death must not be hastened by performing any intervention whatsoever on the patient for the purpose of obtaining an organ. The reason is that one may not do evil for the sake of a good result; in this case the evil would be the killing or the culpable omission of medical assistance.

It is therefore understandable why the law prescribes, in the articles mentioned above, a set of "objective" conditions that provide certainty as to the parameters of clinical brain death.

The second point to be clarified is that the determination of death should be left to the scientific knowledge and experience of competent individuals and is not per se or primarily a philosophical or theological problem. This means that signs of the cessation of human life are to be deduced through the means and methods that science progressively develops and perfects. At one time, a person whose heartbeat and breathing had stopped was considered to be dead on the basis of empirical methods alone; now, with the development of technology and resuscitation methods, science uses more refined means and methods to avoid errors in certain situations. This has led to the definition of "clinical death," which consists of the determination that circulatory, respiratory, and nervous activity has ceased in a manner that is irreversible and not temporary.

The introduction of resuscitation techniques allows the continuation of respiratory and cardiocirculatory function for some time; however, if in the meantime there has been irreversible degeneration of the higher nervous centers, which is evidenced definitively by the cessation of all brain activity, then that human life cannot be reactivated despite a modicum of biological activity that may continue with the help of machines.

Once the state of "clinical death" has been defined as "brain death," it is considered permissible to obtain organs that continue to be viable only because they are maintained by means of instruments.

In order to guarantee that this cessation of life is irreversible and not momentary, a certain waiting period must also be required; it can be specified by science and law. I maintain that the provisions of the above-cited Italian law are sufficient guarantees, from the ethical perspective as well, for the determination of death.

In the earlier Italian law, particular difficulties could be caused by the fact that different conditions were prescribed for declaring a person's death based on whether that person was a candidate for supplying organs for transplantation ("brain death") or not ("cardiac death"). The NBC,[50] claiming that this lack of uniformity and clarity

---

[50] Italian National Bioethics Committee, *Definizione e accertamento della morte nell'uomo* [*Definition and ascertainment of death in man*], February 17, 1991 (Rome: Presidenza del Consiglio dei Ministri, Dipartimento per l'Informazione e l'Editoria, 1991),

tended to instigate debates over the reliability of diagnostic systems and about the very idea of identifying brain death with the death of the entire organism, hoped for and encouraged a single definition of death in the legislative arena, regardless of the destination of the cadaver. As noted, this is the position taken by law no. 578/1993, which declares the following in article 1: "Death is identified with the irreversible cessation of the functions of the entire brain."

Some authors maintain that the six-hour period of constant electroencephalo-gram monitoring is too short an interval to be the basis for a declaration of death. The criteria for assessment actually provide that a specialist verify not only the cessation of cerebral cortex activity, but also that the deeper nuclei of the brain (which coordinate vital functions) have been irreversibly compromised.[51] Therefore, it is not enough, as some maintain,[52] for there to be merely a loss of relational functions because the cerebral cortex has been compromised, even irreversibly; the entire brain must be dead. One cannot introduce a distinction between "biological life" (organic functions) and "personal life" (life of consciousness and relating): there is one life force in man, and as long as there is life, it is to be considered the life of a person.[53]

This is why specialists must verify that those vital functions dependent on the inner centers of the brain have ceased, in keeping with the precepts of the law, before proceeding with organ removal, continuing artificial respiration to maintain heartbeat, and infusing the organ to be transplanted. This artificial respiration is

---

http://www.governo.it/bioetica/pdf/2.pdf, English abstract at http://www.governo.it/bioetica/eng/opinions/abstracts/abstract_definition_and_ascertainment.pdf.

[51] Notwithstanding these criteria, the concept of brain death has been strongly criticized by some authors: see R. D. Truog, "Is It Time to Abandon Brain Death?" *Hastings Center Report* 27, no. 1 (1997): 29–37. Truog maintains that the use of such criteria is inappropriate and a source of confusion, even though in practice it makes it easier to obtain organs for transplantation. For this reason he looks forward to the abandonment of those criteria, maintaining that the need to procure organs for transplantation may be a sufficient reason to hasten a person's death. Recourse to the principles of consent and nonmaleficence could avoid abuses. Regarding the main points of the debate about brain death and its ambiguities, see R. J. White, H. Angstwurm, and I. Carrasco de Paula, eds., *Working Group on the Determination of Brain Death and Its Relationship to Human Death* (Vatican City: Pontifical Academy of Sciences, 1992). In particular, see the essays that are critical of the definition of human death in terms of brain death: D. A. Shewmon, "'Brain Death': A Valid Theme with Invalid Variations, Blurred by Semantic Ambiguity," 23–51; J. Seifert, "Is 'Brain Death' Actually Death? A Critique of Redefining Man's Death in Terms of 'Brain Death,'" 95–143. See also D. A. Shewmon, "Recovery from 'Brain Death': A Neurologist's Apologia," *Linacre Quarterly* 64, no. 1 (1997): 30–96. For an essay on the philosophical aspects of clinical death, see A. Rodríguez Luño, "Rapporti tra il concetto filosofico e il concetto clinico di morte," *Acta Philosophica* 1, no. 1 (1992): 54–68.

[52] See, for example, C. A. Defanti, "I concetti di morte dell'organismo, morte cerebrale, morte corticale," in *Atti del II incontro di aggiornamento in neurologia: Attualità in tema di morte cerebrale; Perugia 20 febbraio 1993*, ed. A. Ferroni, *Annali di Neurologia e Pischiatria* 87 (1993): 21–29.

[53] E. Sgreccia, "La persona e la vita," *Dolentium Hominum* 2 (1986): 38–41.

initiated insofar as spontaneous breathing has ceased because of irreversible damage to the inner nerve centers of the brain, on which it depends.[54]

Particular problems can arise in diagnosing brain death in the case of a pediatric patient. Various laws, including the Italian one, provide different criteria from those applied to an adult donor or require additional evaluations. Indeed, in the document mentioned previously, the NBC expressed the hope that the criteria developed by an ad hoc task force for the determination of death in children[55] would be introduced in Italy as well. The NBC had emphasized in particular the need to use different criteria and specific examinations, since the generally prescribed criteria are valid only for children from the age of 5 years and up. Indeed, other parameters may be lacking or modified owing to various causes, ranging from the incomplete maturation of organs to the particular conditions of children who are kept in intensive care, especially newborns.[56] This information was in fact incorporated, albeit partially, into the determination-of-death law.

In conclusion, however, it is appropriate to emphasize that to obtain organs for transplantation, it is *always* necessary to be certain, duly taking into account the latest medical knowledge, that one is truly in the presence of a cadaver. This is because it is certainly not permissible to directly cause the death of a human being, even to delay the death of another.

The encyclical *Evangelium vitae* notes that in contemporary culture it often happens that "organs are removed without respecting objective and adequate criteria which verify the death of the donor," and describes this as one of the "more furtive, but no less serious and real, forms of euthanasia."[57] This principle has also been reiterated in the *Charter for Health Care Workers*, which declares that "the removal of organs from a corpse is legitimate when the certain death of the donor has been ascertained. Hence the duty of 'taking steps to ensure that a corpse is not considered and treated as such before death has been duly verified.'"[58]

The Current Debate on Brain Death [59]

Within a month of the publication of the Harvard report, Hans Jonas declared his determined opposition to the notion of "brain death" in a talk that he gave at a convention, citing the fact that we do not know with certainty the boundary line

---

[54] Recall that this procedure is not followed in obtaining corneas, since that organ scarcely has any vasculature and is not subject to rapid necrosis; it is therefore possible to obtain and use it even after ascertaining the death of an individual by cardiac arrest. See law no. 301 of August 18, 1993, on regulations regarding cornea procurement and grafting.

[55] American Academy of Pediatrics Special Task Force on Brain Death in Children, "Guidelines for the Determination of Brain Death in Children," *Pediatrics* 80, no. 2 (1987): 298.

[56] Italian National Bioethics Committee, *Transplants in Childhood*, 11–17.

[57] John Paul II, *Evangelium vitae*, no. 15.

[58] Pontifical Council for Pastoral Assistance to Health Care Workers, *Charter*, no. 87.

[59] See P. Becchi, *La morte nell'età della tecnica: Lineamenti di tanatologia etica e giuridica* (Genoa: Compagnia dei librai, 2002); see also R. Barcaro and P. Becchi, eds., *Questioni mortali: L'attuale dibattito sulla morte cerebrale e il problema dei trapianti* (Naples: Edizioni Scientifiche Italiane, 2004); Truog, "Is It Time," 29–37.

between life and death. Subsequently he has maintained[60] that, when the brain has irreversibly ceased to function, we can suspend treatments that sustain life artificially, not because the patient is dead, but because there is no sense in prolonging his life, given the patient's condition. Albert Jonsen[61] also wonders whether we should discontinue life support to allow a patient to die or whether we should turn off the ventilator when the body is already dead.

In Germany, the debate was reopened in October 1992 when, as a result of an automobile accident, a young woman went into a coma and, after the necessary verifications, was declared brain dead. With her parents' consent, the doctors intended to proceed with organ extraction, but they discovered the woman was pregnant. This fact influenced the doctors' decision not to remove the organs but instead to do everything possible to allow the pregnancy to continue. This sparked discussion on brain death in Germany, and many people then asked themselves how it was possible for a "cadaver" to continue a pregnancy.

Jonas, a philosopher of Jewish extraction, is not the only one to dispute the definition of brain death. Others include the Catholic philosophers Josef Seifert[62] and Robert Spaemann.[63] More surprisingly, the utilitarian philosopher Peter Singer[64] is also among the ranks of those who contest the definition of the Harvard committee. In *Rethinking Life and Death*,[65] published in 1994, he maintains that the very notion of brain death is in crisis. But while Jonas, Seifert, and Spaemann hold that one cannot proceed with organ removal if the brain-dead patients are still alive at the time of surgery, the intervention is licit for Singer because he does not regard life as a sacred or inviolable value.

I am certainly far from the positions of Singer, but I agree with the U.S. President's Commission that death is "that moment in which the physiological system

---

[60] See H. Jonas, *Tecnica, medicina ed etica: Prassi del principio responsabilità*, 2nd ed. (Turin: Einaudi, 1999), 202 [Italian]; originally published as *Technik, Medizin und Ethik: Zur Praxis des Prinzips Verantwortung* (Frankfurt am Main: Suhrkamp, 1985).

[61] See A. R. Jonsen, *The Birth of Bioethics* (New York: Oxford University Press, 1998).

[62] See J. Seifert, *Leib und Seele: Ein Beitrag zur philosophischen Anthropologie* (Salzburg: Pustet, 1973); idem, *Das Leib–Seele Problem und die gegenwärtige philosophische Diskussion: Eine kritisch-systematische Analyse* (Darmstadt: Wissenschaftliche Buchgesellschaft, 1989); idem, *What is Life? On the Originality, Irreducibility and Value of Life*, ed. H. G. Callaway (Amsterdam: Rodopi, 1997); idem, "Is 'Brain Death' Actually Death?" 95–143; idem, "Is 'Brain Death' Actually Death?" *Monist* 76 (1993): 175–202.

[63] R. Spaemann, "La morte della persona e la morte dell'essere umano," in "Ai confini della vita," special issue, *Lepanto* 21, no. 162 (2002); idem, "Ars longa vita brevis," in *Etica della ricerca biomedica: Per una visione cristiana; Atti della nona assemblea generale della Pontificia Accademia per la Vita; Città del Vaticano, 24–26 febbraio 2003*, ed. J. Vial Correa and E. Sgreccia (Vatican City: Libreria Editrice Vaticana, 2004), 159–174.

[64] See P. Becchi, "Un passo indietro e due avanti: Peter Singer e i trapianti," *Bioetica* 10, no. 2 (2002): 226–247.

[65] See P. Singer, *Rethinking Life and Death: The Collapse of Our Traditional Ethics* (New York: St. Martin's Press, 1994); see Singer, "Morte cerebrale ed etica della sacralità della vita," trans. Stephano Rini, *Bioetica* 8, no. 1 (2000): 31–49.

of the organism ceases to constitute an integral whole"[66] and that the brain is the critical organ in bodily integration. The error is in combining two things that should be kept distinct: the problem of ascertaining death and the problem of what to do about those who can be considered brain dead.

On the occasion of the Eighteenth International Congress of the Transplantation Society (August 29, 2000), Pope John Paul II posed the question

> When can a person be considered dead with complete certainty?

> In this regard, it is helpful to recall that *the death of the person* is a single event, consisting in the total disintegration of that unitary and integrated whole that is the personal self. It results from the separation of the life-principle (or soul) from the corporal reality of the person. The death of the person, understood in this primary sense, is an event which *no scientific technique or empirical method can identify directly*.

> Yet human experience shows that once death occurs, *certain biological signs inevitably follow*, which medicine has learnt to recognize with increasing precision. In this sense, the "criteria" for ascertaining death used by medicine today should not be understood as the technical-scientific determination of the *exact moment* of a person's death, but as a scientifically secure means of identifying *the biological signs that a person has indeed died*.

> It is a well-known fact that for some time certain scientific approaches to ascertaining death have shifted the emphasis from the traditional cardio-respiratory signs to the so-called *"neurological" criterion*. Specifically, this consists in establishing, according to clearly determined parameters commonly held by the international scientific community, the complete and irreversible cessation of all brain activity (in the cerebrum, cerebellum and brain stem). This is then considered the sign that the individual organism has lost its integrative capacity.

> With regard to the parameters used today for ascertaining death—whether the "encephalic" signs or the more traditional cardio-respiratory signs—the Church does not make technical decisions. She limits herself to the Gospel duty of comparing the data offered by medical science with the Christian understanding of the unity of the person, bringing out the similarities and the possible conflicts capable of endangering respect for human dignity.

> Here it can be said that the criterion adopted in more recent times for ascertaining the fact of death, namely the *complete* and *irreversible* cessation of all brain activity, if rigorously applied, does not seem to conflict with the essential elements of a sound anthropology.[67]

---

[66] United States President's Commission for the Study of Ethical Problems in Medicine and Biomedical and Behavioral Research, *Defining Death: A Report on the Medical, Legal and Ethical Issues in the Determination of Death* (Washington, DC: 1981), 33, http://bioethics .georgetown.edu/pcbe/reports/past_commissions/defining_death.pdf.

[67] John Paul II, "Address to the Eighteenth International Congress of the Transplantation Society," August 29, 2000, nos. 4–5.

It is necessary to recall that a precise definition of brain death[68] must apply to *the entire brain* and must be ascertained through various procedures over a period of time that allows for comparison of the data accumulated. Yet even though the brain plays a central role in the organism, this does not mean that a human being can be identified with it; indeed, when death used to be verified by the cessation of cardiac and pulmonary activity, the person was not identified with his heart or lungs.

The Case of Heterologous Transplant[69]

The case of a heart transplant from an ape to a baby (Baby Fae) in 1984 and cases of kidney and liver transplants from animals (pig, chimpanzee) to adult human subjects have raised questions in the press as to the permissibility and practicability of similar interventions.

There are two aspects to the ethical problem: whether the uncertainty of success and the extremely high risk of rejection present obstacles serious enough to make such attempts unreasonable, and whether the introduction of an animal organ—or of an artificial organ, for that matter—might create a change of personality.

The first objection must still be considered well founded, inasmuch as there seems to be no way of ascertaining a well-founded hope of successful transplantation beforehand. Proceeding simply on the basis of the impulse to experiment is not allowed.

Some authors say that there are already necessary and sufficient conditions to justify the use of animals in medical treatments and in research on human beings.[70] In reality, the problem is not so simple and does not concern the mere use or non-use of animal organs for the benefit of a human being—which in itself could be ethically acceptable. The unacceptability of the experiment on Baby Fae, for example, was not due to the fact that a baboon heart was used but to the fact that the theoretical premises and the state of the art and of research up to that point offered no prospect of success. (There was absolutely an issue of blood type incompatibility.) Hence, it was not a therapeutic experiment but only an experiment as an end in itself. In fact, the first kidney transplants from chimpanzees and baboons to humans, performed in the 1960s by Keith Reemtsma and Thomas Starzl, failed: rejection occurred quickly.[71]

Attempts have been made to use pig organs and to treat recipients with immuno-suppressive drugs. In the former case, however, when the organ from the animal

---

[68] A. Pessina, "Filosofia e scienza al capezzale dell'uomo: La morte cerebrale," in *Bioetica: L'uomo sperimentale* (Milan: Bruno Mondadori, 1999), 159–171.

[69] M. Pennacchini, *Il trapianto eterologo: Storia, problemi etici e impatto psicologico nei pazienti in lista d'attesa* (Rome: Aracne, 2004); G. Cordini, "Xenotrapianti: I profili costituzionali," *Rassegna Ammnistrativa della Sanità* 40, no. 1 (2001): 5–13; L. Ravarotto and R. Pegoraro, *Transgenesi, clonazione, xenotrapianto: Analisi scientifica, giuridica ed etica sull'impiego degli animali* (Padua: Piccin-Nuova Libraria, 2003); D. Palombo, A. Ramello, and P. Tappero, eds., *Trapianti e xenotrapianti: Aspetti etici e giuridici; Atti del Convegno; Agliè, 12 ottobre 2002* (Turin: Selcom, 2003); Pontifical Academy for Life, "Prospects for Xenotransplantation."

[70] T. Kushner and R. Belliotti, "Baby Fae: A Beastly Business," *Journal of Medical Ethics* 11 (1985): 178–183.

[71] R. Calner, "Organs from Animals: Unlikely for a Decade," *British Medical Journal* 307 (1993): 306–307.

comes into contact with human blood, it immediately sets off an antigen-antibody reaction with complement activation resulting in the destruction of the vascular endothelium. Experiments aimed at blocking this destructive reaction have not yet been fully corroborated, which is why many authors describe those predicting the forthcoming success of this type of transplantation as optimists and instead agree with others who expect concrete and lasting results to require much more time.

Ethical reflection is therefore necessary and is connected precisely with the intervention's uncertainty of success (as noted in the literature) and the extremely serious risk of rejection. These elements could indeed constitute an ethical obstacle against continuing along this path unless a well-founded hope of success for the transplant itself is first acquired.

Among the questions concerning the permissibility of heterologous transplants, there is also the one posed by those who conjecture that a human subject might undergo a personality change owing to the introduction of an animal organ. The answer here is actually simple, given that organs such as the liver or heart are just functionally directed and are not the seat of functions that characterize the person in an essential way.

Obviously, the recipient must also be psychologically prepared for the prospect of living with an organ derived from an animal and must be supported in that decision, because he might feel a sense of rejection and conflict toward the transplant, perhaps even for cultural reasons, as sometimes happens with other handicaps that are remedied by prostheses. Indeed, the question of the social acceptability of these transplants should also be considered: Paula Mohacsi and colleagues,[72] on the basis of a survey conducted among 1,728 nurses at fifty-nine Australian public hospitals, reported that the majority of those surveyed were strongly opposed to accepting a xenotransplant; moreover, many cited the risk of viral agent transmission from donors.

In recent years, new hopes have been nurtured thanks to biotechnological advances that have made it possible to transplant organs from transgenic animals, theoretically with less likelihood of rejection.[73] Quite a few doubts still remain, however, and the opinions of researchers are still divided, especially regarding the possibility that other factors besides immunological compatibility are involved in the mechanism that "rejects" an organ. It would therefore be premature to initiate experimentation on human beings, and at any rate it is important to verify first the real scientific validity of such results.[74]

The document by the Pontifical Academy for Life cited earlier, which was authored by a group of international specialists, emphasizes the risk of transmitting pathogenic agents from animals to humans in the scenario of xenotransplantation. Here are some excerpts from the document regarding the transplantation of pig livers into humans:

> Over sixty porcine infectious agents with a potential to cause disease in humans have been identified. Development of "clean" lines of source animals,

---

[72] See P. J. Mohacsi et al., "Aversion to Xenotransplantation," *Nature* 378 (1995): 434.

[73] D. Dickson, "Pig Heart Transplant 'Breakthrough' Stirs Debate over Timing of Trials," *Nature* 377 (1995): 185–186.

[74] On the ethical aspects, see also M. J. Hanson, "The Seductive Sirens of Medical Progress: The Case of Xeno-Transplantation," *Hastings Center Report* 25, no. 5 (1995): 5–6.

with a certified health status, is under way. Control measures include the birth of pigs by hysterotomy (caesarean derived), carefully controlled environments and *routine* monitoring of pigs and their handlers. These steps appear to have excluded almost all known infectious agents of concern. However, it cannot be ruled out that an unknown porcine virus might exist which causes no pathology in pigs but which may cause disease in humans.

As is true for all other mammalian species, pigs have sequences in their DNA that encode retroviruses (PERV, *porcine endogenous retroviruses)*. Weiss and colleagues showed that pig retroviruses could infect human cells in vitro. There are no satisfactory animal models to test the pathogenicity of these agents.[75]

On the ethical level, the document concludes with an appeal to consider

the possible transmission to the recipient of infections arising from the xenotransplant (*zoonoses)* by known or unknown pathogenic agents which are not harmful to the animal but which are possibly dangerous for man. Such infections could escape detection, with the consequent possibility of the spread of the infection to those having close contacts with the patient, leading eventually to its being spread to the entire population.

Since clinical experience of xenotransplantation is quite limited and certainly insufficient to provide reliable statistics on the real probability of occurrences and spread of infections, any decision concerning clinical development of the new therapy can only be based on hypothesis. There is, therefore, an ethical requirement to proceed with the greatest caution.[76]

There is one last ethical problem connected with the use of animals that I would like to mention: the objection raised by groups of animal rights activists who find this solution to the shortage of organs unacceptable. It seems to me that, in view of the truth of the ontological subordination of the animal world to the human person, there are no difficulties in accepting the procedure of transplanting animal organs into a human being. Refraining, as much as possible, from all senseless cruelty to the animal, it is therefore ethically acceptable and justifiable to use animals experimentally if necessary to save and heal the lives of human persons.

Tissue Grafts

The practice of obtaining and grafting tissues from human fetuses for the purposes of therapeutic transplantation was discussed in the preceding chapter on experimentation on fetuses and embryos.

This section will deal exclusively with the ethical problems relating to *grafting or transplanting bone marrow*, which is the tissue found inside bones that is responsible for hematopoiesis, or the production of the cellular components present in the blood as it circulates (e.g., red cells, white cells, platelets).

This sort of transplant is now performed with greater success as a result of research developments, such as a better definition of the major histocompatibility

---

[75] Pontifical Academy for Life, "Prospects for Xenotransplantation," no. 4.
[76] Ibid., no. 14.

complex (MHC), recent progress in chemotherapy and radiation therapy, the possibility of preserving hematopoietic precursors (the fundamental stage for autologous transplantation) for years in liquid nitrogen, as well as the possibility of identifying and eradicating residual neoplastic cells in the marrow. Bone marrow transplants on neonates, and the centers that perform them, are now countless.

At one time this procedure was only indicated in cases of acute leukemia and aplastic anemia, but indications now include various other hematological diseases—even of a noncarcinogenic nature—as well as non-hematological diseases (neuroblastoma, melanomas, and other solid tumors).

The marrow from a healthy donor can be used to replace marrow that is damaged and incapable of normal hematopoeisis or of producing immunocompetent cells, as occurs in aplastic anemia, various forms of leukemia, immunodeficiency syndromes, certain congenital defects affecting metabolism, and thalassemia major. In neoplasms of the hematopoietic system and other solid tumors, the marrow of the healthy donor (allogeneic or syngeneic transplant) or the cryopreserved marrow of the patient himself (autologous transplant or autograft) offers the possibility of hematopoietic "rescue," and specifically of increasing the doses of cytostatic drugs or radiation treatment beyond the commonly accepted limits, in the hope of definitively eradicating the neoplastic disease.

Acute, non-lymphatic types of leukemia have a better prognosis with a marrow transplant than do acute lymphoid forms of leukemia, which present more difficult problems. For acute lymphatic leukemia in infancy, except in the rarest cases, a marrow transplant is indicated in the second remission, i.e., after an initial attempt to treat it with chemotherapy alone; indeed, chemotherapy currently seems to be the best path for successful treatment of this disease. In adults, on the other hand, generally until the age of 45 to 50 years, a bone marrow transplant is indicated in cases of acute lymphatic leukemia and acute myeloid leukemia in the first remission, i.e., in connection with the first chemotherapy treatment.[77]

In recent years, the number of marrow transplants for chronic myeloid leukemia has increased worldwide. The idea is being confirmed that the sooner the transplant is performed, the better the prognosis.

Attempts by Italian researchers[78] to cure thalassemia major by bone marrow transplant have met with great scientific interest. In late 1981, these researchers started a project, in collaboration with the Freud Hutchinson Cancer Research Center in Seattle, to evaluate bone marrow transplant as a possible treatment for thalassemia major. The first transplant was performed in Seattle, on a thalassemic child who had never had a blood transfusion because he was the son of Jehovah's

---

[77] A. Marmont, "Stato attuale del trapianto del midollo osseo allogenico nelle leucemie," in *Atti del convegno internazionale: Trapianto di midollo osseo in pediatria; Problematiche etico-giuridiche e scientifiche; Pavia, 15–16 novembre 1986,* ed. Direzione Medico-Scientifica della Immuno (Pisa: Pacini, 1987), 17–22.

[78] G. Lucarelli et al., "Marrow Transplantation for Thalassaemia following Busulphan and Cyclophosphamide," *The Lancet* 1 (1985): 1335–1357; idem, "Marrow Transplantation in Patients with Advanced Thalassaemia," *New England Journal of Medicine* 316 (1987):1050–1055; idem, "Allogeneic Marrow Transplantation for Thalassaemia," *Experimental Haematology* 12 (1984): 676–681.

Witnesses. One week later the first Italian transplant for thalassemia was carried out on a 14-year-old boy who had been transfused repeatedly. While the project was suspended in Seattle after several transplants with favorable outcomes and did not recommence until several years later, the Lucarelli group in Italy moved forward, gradually accumulating many case studies. The practice has since become firmly established with satisfactory results.

The problems of a scientific and ethical sort are in connection with disagreements among researchers regarding their assessments of the efficacy of this type of transplantation as compared to traditional treatments, especially in the case of thalassemia.[79] Indeed, some maintain that the survival rate decreases among subjects who undergo a marrow transplant at an advanced stage of the disease. Lucarelli's group, in contrast, asserts that transplantation still remains the sole treatment possibility, even in advanced thalassemia.

This second hypothesis has been roundly criticized by some pediatricians and hematologists. The traditional therapy involving periodic transfusions, which currently contain iron chelates and young red blood cells, would actually reduce the side effects of such a treatment and would allow for longer survival than in the past. In practice, if transfusion therapy is conducted successfully, a patient who was thalassemic at birth would have a chance of almost normal development until puberty and an average life expectancy of around thirty years. Moreover, it is reasonable to predict that with subsequent innovations in the biomedical field there will be further improvement in this type of therapeutic approach.

Again, if we compare the mortality rates of the two types of therapy, transfusion and transplant, we see that they are higher for transplantation. This treatment certainly offers the hope of a lasting cure, then, but for a limited number of persons.

Hence, while anticipating the possibility of a definitive cure by transplantation, which again requires confirmation from a longer follow-up, the traditional therapy still remains the more effective approach to treating thalassemia, inasmuch as it makes possible a longer period of survival.[80]

It is a different matter with some types of acute leukemia or chronic myeloid leukemia and aplastic anemia; in such cases, if conditions permit the procedure, a bone marrow transplant seems to be the most effective solution.[81]

---

[79] W. Krivit and C. B. Whitley, "Bone Marrow Transplantation for Genetic Diseases," *New England Journal of Medicine* 316 (1987): 1085–1087; S. Piomelli et al., "Bone Marrow Transplantation for Thalassaemia," *New England Journal of Medicine* 317 (1987): 964.

[80] Several authors have insisted on the need for careful, long-term supervision of subjects who have received bone marrow transplants, having discovered in a study of around 20,000 transplant patients an increased risk of developing a solid tumor long after the transplant, with a greater risk among young subjects. See R. E. Curtis et al., "Solid Cancer after Bone Marrow Transplantation," *New England Journal of Medicine* 336 (1997): 897–904.

[81] See G. Fasanella and E. Sgreccia, "Il trapianto di midollo osseo: Aspetti etici," *Medicina e Morale* 38, no. 3/4 (1988): 397–409; E. Sgreccia, M. L. Di Pietro, and G. Fasanella, "I trapianti d'organo e di tessuti nell'uomo: Aspetti etici," in *Trapianti d'organo*, Bompiani and Sgreccia, 150–154.

There is no doubt that the decision to undertake a bone marrow transplant is always a grueling decision, which opens up new horizons for medicine but at the same time reveals its inconsistencies and failures.

In the contemplation of a bone marrow transplant, it is also necessary to reference the ethical values already mentioned in relation to transplants in general. In particular, one must respect the principle of defending life as well as the principle of informed consent, which is closely connected with the first and is a necessary part of respect for the dignity and freedom of the human person. Both principles must be applied to the donor and the recipient. The donor must not suffer irreparable damage to his own life or functionality, and there must be some proportionality between his sacrifice and the possibility of real benefit to the life of the recipient. Both the donor and the recipient must give their own informed consent.

As for the donor's physical integrity, the removal of bone marrow in fact only causes it temporary harm, since the bone marrow reconstitutes itself in a short time. The only real danger to his life is associated with the general anesthesia administered, since multiple needle punctures are necessary to aspirate a sufficient quantity of cells without too much admixture of blood. It is nonetheless a very low and therefore acceptable risk. The danger of infection connected with the punctures also seems more theoretical than practical. Because of this lack of significant physical harm, obtaining marrow from minor donors who are incapable of consent is also justified, with qualifications that will be spelled out a little further on.

From the legal perspective, removing bone marrow is not contrary to article 5 of the Italian civil code inasmuch as it does not cause permanent damage to physical integrity. (The sole exception allowed for by the Italian ordinance involves the removal of a kidney from a living donor, pursuant to law no. 458 of June 26, 1967.)

With regard to consent, on the other hand, there are some problems because the suitable marrow donor is often a minor. With law no. 107 of December 4, 1990 regulating transfusions, the Italian legislation filled the lacuna that existed in the matter of obtaining marrow from a minor. Indeed, while providing in article 3 that whole blood or plasma may be taken only from consenting persons no younger than 18 years of age, it admits that one may, on the other hand, obtain platelets and leukocytes by means of hemapheresis and medullar stem cells and peripheral cells even from subjects younger than 18 years of age with the prior consent of those exercising legal authority in the capacity of parent, guardian, or supervising judge.

There is already legislation on this matter in other countries as well. For example in Switzerland, Germany, and England, the written informed consent of the parents or whoever exercises the authority of legal guardianship is sufficient. In France, in addition to the parents' consent, it is necessary to obtain the approval of three experts, including two physicians who are not members of the team that will carry out the removal or the transplant.

Regardless of the legislation currently in force, I maintain that the principle of informed consent must be preserved and respected, to the extent possible, by simultaneously introducing a criterion for making age distinctions, just as already noted for minor consent to nontherapeutic experimentation.

Indeed, the case of a 2-year-old baby who is just beginning to put together his first words is quite different from the case of a 14-year-old youth who, although he has not yet acquired "the capacity to perform all acts for which a different age has not been established" (article 2 of the civil code), can nevertheless sense and comprehend, at least partially, the significance and the value of a donation. The parents can guide their child, without coercion or intimidation, to express valid consent.[82]

The parents and child in this case would constitute "the unit that expresses consent." Incidentally the revision of the Declaration of Helsinki in 2000 provided that whenever a minor is capable of expressing consent, it must be obtained together with that of the parents. In the case of obtaining tissue from a minor who is incapable of any consent whatsoever, the European Convention on Human Rights and Biomedicine decided that, in view of a general prohibition against obtaining organs or tissues from a subject who is incapable of expressing consent (article 20, section 1), the removal of replaceable tissues (such as bone marrow) from a subject incapable of consent can be permitted under certain specific conditions (article 20, section 2).[83]

As for the recipient, the parents' consent can be sufficient for minors provided the therapy in question benefits the subject who is directly involved.

Finally, with regard to transplants of fetal livers, which are also used for hematopoietic tissue transplants, the guidelines already stated concerning the extraction of embryonic and fetal tissues in the chapter on experimentation are valid in this case as well.

The Anencephalic Newborn as Organ Donor

The difficulty of finding small organs to use for transplantations in children has suggested recourse to removing organs, especially kidneys, from anencephalic fetuses or newborns.[84] Before examining the ethical problems related to the removal of organs from an anencephalic fetus or newborn, it should be noted that they are generally not "good donors," whether because of the frequent occurrence of premature birth (53 to 58 percent) with resulting organ immaturity or because of the higher incidence of malformations associated with the underlying condition (around 30 percent have urinary tract malformations).[85]

From an ethical perspective there are several practical questions that arise when considering the possibility of using anencephalic fetuses or newborns as organ donors: (1) Is it permissible to use these subjects as "reservoirs" of tissues and organs? (2) Is it acceptable to plan and carry out the "resuscitation" of an anencephalic newborn with the sole purpose of continuing to perfuse the organs that are to be extracted later for transplantation? (3) What are the criteria for determining brain death in an

---

[82] This problem was mentioned earlier and is highlighted in the NBC document. See Italian National Bioethics Committee, *Organ Transplants in Childhood*, 18–20.

[83] Specifically, "(i) there is no compatible donor available who has the capacity to consent; (ii) the recipient is a brother or a sister of the donor; (iii) the donation must have the potential to be life-saving for the recipient; (iv) the authorisation provided for under paragraphs 2 and 3 of Article 6 has been given specifically and in writing, in accordance with the law and with the approval of the competent body; (v) the potential donor concerned does not object. "

[84] K. Iitaka et al., "Transplantation of Cadaver Kidneys from Anencephalic Donors," *Journal of Pediatrics* 93, no. 2 (1978): 216–220.

[85] P. A. Baird and A. D. Sadovnick, "Survival in Infants with Anencephaly," *Clinical Pediatrics* 23, no. 5 (1984): 268–271; D. A. Shewmon, "Anencephaly: Selected Medical Aspects," *Hastings Center Report* 18, no. 5 (1988): 11–18.

anencephalic patient? Underlying these questions, however, is another and perhaps the most important one: What is the *identity of the anencephalic subject?*

It seems to me an indisputable fact that the anencephalic patient is the product of a human conception, with a human form, who from the moment of conception has been teleologically directed by his or her own vital principle.[86] There should consequently be no doubt that this is an individual of the human species, to be respected as a person just like any other embryo. The reflections in response to the questions will be divided into three points: (1) removing organs from an anencephalic subject who is alive, (2) keeping anencephalic newborns alive, and (3) removing organs after death.

The procedure of removing organs from an anencephalic subject who is alive is generally accepted by those who assert that the anencephalic subject is not a human individual or, more often, that "it" is in a situation resembling brain death given the absence of a large portion of cerebral mass. This hypothesis is neither scientifically correct nor ethically acceptable: indeed, the term *brain death* designates[87] damage to the entire brain, both cortex and brain stem, whereas in the anencephalic fetus or newborn the brain damage is partial, i.e., it does not involve the structures of the brain stem, which are therefore capable of autonomously maintaining vital functions for a certain period of time.

As for the resuscitation of an anencephalic newborn, there are four distinguishable ways of approaching the matter:[88] (a) As soon as he is born, the anencephalic child is intubated and put on a ventilator; all vital signs are maintained for the purpose of proceeding to remove organs as the need may arise, regardless of whether or not brain stem activity is present. This approach, as already explained in point 1, *is ethically unacceptable.* (b) As soon as he is born, the anencephalic child is resuscitated and monitored until brain stem activity disappears; this approach seems *disproportionate* with respect to the prognosis and nature of the pathology, thus constituting therapeutic obstinacy in the strict sense. (c) The newborn is monitored and ordinary care alone is provided until either hypertension or bradycardia occurs, at which time the subject is resuscitated and the transplantation team waits for the death of the brain stem: this approach seems to be aimed exclusively at obtaining organs and is therefore a form of *exploitation* of a human being. (d) The anencephalic newborn is monitored and ordinary care alone is provided until cardiorespiratory arrest, after which the team proceeds with organ removal. This approach is the one that most *respects the anencephalic subject's value as a person*, avoiding the unacceptable therapeutic obstinacy, exploitation of the person, and actual vivisection—a strongly disparaged practice when performed on animals—involved in the other three approaches.

Ultimately, there are no ethical impediments to removing organs from anencephalic fetuses or newborns after death—once the problem of determining death has been solved, given the objective difficulty of using the criteria for adults or children.

---

[86] A.G. Spagnolo and E. Sgreccia. "Prelievi di organi e tessuti fetali a scopo di trapianto," in *Trapianti d'organo*, Bompiani and Sgreccia, 49 ff.

[87] Task Force for the Determination of Brain Death in Children, "Guidelines for the Determination of Brain Death in Children," *Neurology* 37 (1987): 1077–1078.

[88] Medical Task Force on Anencephaly (MTFA), "The Infant with Anencephaly," *New England Journal of Medicine* 322 (1990): 669–673.

The report by the Medical Task Force on Anencephaly[89] considers it superfluous to order an electroencephalogram (EEG) or to monitor brain waves given the intrinsic anatomical condition of the anencephalic patient. The report also underscores the difficulty of assessing brain stem function by checking reflexes, because the condition is frequently associated with abnormalities of the cranial nerves. Consequently, the death of the brain stem can only be determined by observing the disappearance of reflexes that were previously exhibited; if there was never any evidence of reflexes, their absence from the start does not necessarily indicate the death of the brain stem.

The NBC examined the problem of anencephalic newborns as potential organ donors in one of its documents, bringing to light the fact that even if it is an extreme case, "we are not thereby authorized to devise for these subjects a special biological or legal category, as the case may be," in order to use the organs of these newborns.[90]

---

[89] Ibid.

[90] Italian National Bioethics Committee, *The Anencephalic Newborn and Organ Donation*, June 21, 1996 (Rome: Presidenza del Consiglio dei Ministri Dipartimento per l'Informazione e l'Editoria, 1996), available at http://www.governo.it/bioetica/eng/opinions/abstracts/newborn_21.pdf.

Summary Outline for Chapter14

Organ transplants

Definition

A "transplant" is a surgical operation whereby an organ or a tissue obtained from a donor is inserted into the host organism. [The donor may be]

- *A cadaver*
- *A living person*

> From cadavers: skin, corneas, bone marrow, kidneys, heart, lungs, liver; pancreas, intestines (but also hands and face).

> From living persons: kidney and liver (partial transplant).

The transplant can be

- *Autologous*, from one site to another in the same individual (skin)
- *Homologous*, between subjects *of the same species* (there is a problem of *histocompatibility* and rejection)
- *Heterologous*, between subjects *of different species* (xenotransplant; experimental only)

Factors that have encouraged transplants

Various factors have encouraged transplants:

- *The discovery of cyclosporin*, a well-tolerated drug that depresses the immune system, which is capable of quickly detecting protein structures within the body "different from itself" and of destroying them.

- *The multiplication of "intensive care units,"* which, in the case of death, make possible a prompt diagnosis of death, allowing the removal of organs *before* the beginning of regressive phenomena that are incompatible with successful transplantation.

- *Advances in surgical techniques* entailing "extracorporeal circulation."

Procurement from a living donor

The donation of one's own organ (kidney, part of the liver) certainly causes harm to the organism but enriches the person on the level of moral values: solidarity, altruism, charity.

The buying and selling of human organs is universally condemned.

Procurement from a cadaver

The removal of organs not only shows respect for the cadaver but also increases its value, because it makes possible a new form of solidarity: preventing the premature death of other persons.

Determination of death

The determination of the *death* of a person can be made by noting the complete and irreversible cessation of all brain activity.

The situation of *brain death* is a state of complete cerebral infarct, with the subsequent impossibility of any flow of blood to the brain and, therefore, *the decay of the brain cells*. The mechanical tests detecting such a situation are, by Italian law, protracted over a period of hours. The law establishes an obligation that these tests be conducted by a committee of specialists appointed by the hospital management and made up of persons who are altogether separate from the transplant team. *The practice is the same whether or not the removal of an organ is planned.*

"Non-life-saving" transplants

It is necessary to distinguish so-called *life-saving transplants* (e.g., of liver, heart, and lungs), in which the patient is doomed to die if he does not receive a transplant, from *non-life-saving* transplants (e.g., of the hand or face), which are supposed to improve the patient's quality of life.

There are many reservations from the ethical perspective because the face and hands also represent a person's identity, and these techniques are therefore an assault on its uniqueness. The first hand transplant was completely successful, but after a while the patient had the new "piece" removed, because he never managed to adapt to this new hand.

Furthermore, it is debatable whether it is permissible to transform a mutilated but "healthy" person into a "reconstructed" but sick person, owing to a suppressed immune system.

"Crossover" transplant

The "crossover" transplant is a special case of transplant from a living donor. The procedure is performed in the case of a kidney donation, although it could potentially be applied in a partial liver transplant as well. It involves two or more couples of subjects; each couple consists of blood relatives, spouses, or friends, one of whom needs a transplant while the other is willing to make an organ donation, although they are immunologically incompatible. If another couple in a similar situation is found and it is ascertained that there is a crossover biological compatibility (i.e., between donor A and recipient B and between donor B and recipient A), a crossover transplant can be performed if the two couples give consent. The technique thus makes it possible to overcome the barrier of biological incompatibility that prevents a direct transplant within each couple.

Transplants of gonads or brain

Transplants of the gonads or of the brain are ethically unacceptable.

With regard to *the gonads*, their decisive role in constructing the biological identity of every child that the person might generate must be kept in mind. The haploid genome of every ovum and every sperm cell is completely derived from the genome of the organism to which the ovary or the testicle in question belongs from the beginning, and this continues to be true even after the transplant to another organism. This causes a profound change in the biological identity of any child generated by a person to whom the gonad of another person has been transplanted, for he is genetically the child of the latter.

A *brain* transplant would be an attempt to give a healthy body to a person with a perfectly functioning brain but a body that has sustained massive trauma. Therefore it is a transplant of a torso rather than of a brain.

Even if such a thing were technically possible, it would be ethically unacceptable because it would constitute an attack on individual identity.

Xenotransplants

"Xenotransplantation, the transplantation of organs, tissues or cells from one species to another, if applied to man, would offer the possibility of a huge supply of organs, tissues and cells for transplantation thereby relieving the 'chronic' shortage of human donors. However, before xenotransplantation becomes a clinical reality, there are practical challenges that must be overcome. One is rejection, the process by which the body of the transplant recipient attempts to rid itself of the transplant. Another is to ensure the correct functioning, across species barriers, of the transplant in its new host. Also, there is the need to minimize the likelihood of the introduction of new infectious agents into the human population via the transplant" (Pontifical Academy for Life, "Prospects for Xenotransplantation," 2001).

The first attempts were made in the 1960s and the early 1970s. One man managed to survive for nine months with a chimpanzee kidney.

In the 1980s, a chimpanzee heart was transplanted into an infant girl (Baby Fae), who survived for a short time. It was a purely experimental attempt.

The Italian law concerning transplants

According to Italian law no. 91/99 ("Regulations pertaining to the procurement and transplantation of organs and tissues"), brain death—or the death of the encephalon—must have occurred in order to proceed with organ removal; the body must also still be "viable" in order to allow heart-beating organ removal while the organ perfusion devices are operating.

The law accepts what the Ad Hoc Commission of the Harvard Medical School established in the United States in 1968, according to which an individual is brain dead when the following conditions are met: lack of brain activity, lack of spontaneous or induced movement, lack of spontaneous breathing, and lack of central brainstem reflexes.

According to article 4 of law no. 91/99, every citizen shall be called to express his or her willingness to donate or not in written form; if the citizen's will is not so manifested, it shall be considered assent to organ removal.

The time needed to fully implement law no. 91/99 will necessarily be lengthy in virtue of the creation of the information system required by the law (article 7).

Provisional regulations were put in place that would remain in force until the completion of the information system. The precedence of the family's wishes over the wishes of the deceased, which was sanctioned by law no. 644 of 1975, is abolished if the subject expressed a favorable opinion about donation while alive.

Organ and tissue removal is permitted unless the subject explicitly refused to give his or her assent, or unless the non-separated spouse or cohabitating partner *more uxorio* (as though husband and wife)—or the adult children (in absence of the preceding possibility) or the parents or legal representative (in absence of both preceding possibilities)—presents written opposition within the observation period required for the confirmation of brain death.

In general, the legal recognition afforded to the explicit manifestation of wishes during life with respect to organ donation after death takes up what was already anticipated in the document by the Italian National Bioethics Center (NBC), *Donazione d'organo a fini di trapianto* (Organ donation for transplants), published on October 7, 1991.

The current debate about brain death

Hans Jonas, a philosopher of Jewish extraction, is not the only one to dispute the definition of brain death. Others include the Catholic philosophers Josef Seifert and Robert Spaemann. More surprisingly, the utilitarian philosopher Peter Singer is also among the ranks of those who contest the definition of the Harvard committee. In *Rethinking Life and Death*, published in 1994, Singer maintains that the very notion of brain death is in crisis. But while Jonas, Seifert, and Spaemann hold that one cannot proceed with organ removal if the brain dead patients are still alive at the time of surgery, the intervention is licit for Singer because he does not regard life as a sacred or inviolable value.

I agree with the U.S. President's Commission that *death is "that moment in which the physiological system of the organism ceases to constitute an integral whole"* and that the brain is the critical organ in bodily integration. The error is in combining two things that should be kept distinct: the problem of ascertaining death and the problem of what to do about those who can be considered brain dead.

It is necessary to recall that a precise definition of brain death must apply to *the entire brain* and must be ascertained through various procedures over a period of time that allows for comparison of the data accumulated.

Transplantation of organs from anencephalic newborns

It is difficult to determine brain death in an anencephalic neonate, and therefore the absence of spontaneous respiration and cardiorespiratory activity (regulated by the brain stem) is the primary element in identifying the cessation of function of the brain stem.

There are ethical questions about the resuscitation of an anencephalic neonate. It is not permissible to consider the newborn instrumentally as an organ deposit and to subject him to resuscitation solely to ensure good perfusion.

It is necessary to tend the neonate, resorting only to ordinary and proportionate treatments (aspiration of bronchial secretions, nutrition and hydration) until cardiorespiratory arrest, barring euthanizing interventions.

After cardiorespiratory arrest it is possible to proceed with organ procurement.

Chapter 15

Bioethics, Euthanasia, and Death with Dignity

Definition of Terms
and History of the Problem

The scope of this chapter follows a twofold purpose: to recall the essential reference points of ethics and in particular of Catholic moral teaching, which has examined this topic extensively, and at the same time to clarify the implications of the related theme that goes by the name of "death with dignity" or "humanization of death." These are two aspects connected with assistance for the dying and connected with one another, but they are not identical. As the discussion will demonstrate, euthanasia must be condemned because it involves the premature—even if "compassionate"—killing of a dying person, whereas the humanization of death must be promoted with a whole array of means and provisions. I once again mention at the outset that reference will be made in this chapter to the documents of the Magisterium of the Catholic Church, both because its ethical doctrine in this area has in fact been extensively developed and also because the reasons provided are often valid for nonbelievers as well.[1]

It would be interesting to preface this chapter with a historical survey of its specific topic, identifying behaviors that can be traced among primitive peoples, in classical antiquity, in the Middle Ages, in the Renaissance period, and in the modern era. Historical research on euthanasia has ethical relevance especially if it aims to bring to light the reasons and concepts about life that undergird the practice. Such a treatment has been outlined, and there is still an active interest in historical research, especially regarding the concept of death among various peoples and in various civilizations, thanks to the efforts of specialists in ethnology, cultural anthropology, and the history of customs;[2] we will therefore be limited to a rapid review of them here. After a comparative historical study, the anthropologist Louis-Vincent Thomas drew this rather paradoxical conclusion: "There is one society that respects man and accepts death: African society; there is another that is mortiferous, thanatotic,

---

[1] See J. de D. Vial Correa and E. Sgreccia, eds., *The Dignity of the Dying Person* (Vatican City: Libreria Editrice Vaticana, 2000); Pontifical Academy for Life, "Respect for the Dignity of the Dying," December 9, 2000; C. Zuccaro, *Il morire umano: Un invito alla teologia morale* (Brescia: Queriniana, 2002).

[2] G. Pelliccia, "L'eutanasia ha una storia?" in *Morire sì, ma quando?* ed. P. Beretta (Rome: Paoline, 1977), 68–96.

obsessed with and terrified by death: Western society."[3] Obviously it is in this latter society that the demand for legalized euthanasia is promoted.[4]

Yet among primitive peoples one encounters practices analogous to euthanasia and even the performance of human sacrifice against a religious background. Among the Battaki of Sumatra an aged father, upon inviting his children to eat his flesh, allows himself to fall from a tree like a ripe fruit, whereupon his relatives kill him and eat his flesh. Practices of killing the aged have been discovered in some tribes of Arakan (India) and of lower Siam (Thailand), among the Cachibas and the Tupi of Brazil, in Europe among the ancient Wends (a Slavic people), and finally in twentieth-century Russia in the pseudoreligious sect of "stranglers." Human sacrifices of young persons or firstborn children are found among the peoples of ancient civilizations on all the continents.[5]

It would be more interesting, with respect to intellectual history as well, to examine this topic in terms of the Western world. Everyone knows the fate in store for deformed newborns in Sparta, and we know that Aristotle (*Politics*, VII, 1335b) approved that practice as a matter of policy for utilitarian reasons. Plato, however, in a controversial passage of the *Republic*, declares that human beings with incurable physical illnesses should be left to die and that those who are mentally depraved (without any possibility of correction) should be put to death (409e–410a).

In Rome, in addition to the custom of exposing deformed newborns, which persisted until the days of Emperor Valens, we find the practice of suicide, which is described sympathetically by many writers especially during the period of the Empire: Tacitus eulogizes the suicide of Petronius (*Annales*, XVI, 18–19); Valerius Maximus delights in mentioning that the Senate of Marseilles has custody of the "state poison"; and Silius Italicus, who performed euthanasia on himself, praises the customs of the Celts who are "quite ready to hasten the death" of their aged, sick, and war-wounded. In Rome the exaltation of power, youth, and physical strength—which inculcated a genuine repugnance for old age and sickness—was also combined with Stoic teaching, which praised and commemorated many suicides by well-known personas of the culture: Seneca, Epictetus, and Pliny the Younger.

Yet even in the Greco-Roman world there were some who opposed such practices and theories: among the Greeks, Pythagoras and especially Hippocrates and Galen. On this subject, the famous Oath of Hippocrates reads as follows: "I will not give a poison to anyone though asked to do so, nor will I suggest such a plan."[6] Among the Romans, we should recall what Cicero writes in *Somnium Scipionis*: "Therefore, Publius, both you and all righteous men must let the soul be kept in the custody of the body and not allow it to emigrate from human life unbidden by him

---

[3] L.-V. Thomas, *Anthropologie de la mort* (Paris: Payot, 1975).

[4] For the history of law regarding euthanasia, see F. D'Agostino, "Eutanasia e diritto," in *Morire sì*, Beretta, 164–178.

[5] Thomas, *Anthropologie de la mort*.

[6] W. H. S. Jones, *The Doctor's Oath: An Essay in the History of Medicine* (New York: Macmillan, 1924), 11. A more recent translation is provided in S. H. Miles, *The Hippocratic Oath and the Ethics of Medicine* (New York: Oxford, 2004), p. xiv: "I will never give a drug that is deadly to anyone if asked [for it], nor will I suggest the way to such a counsel."

who has endowed you with this soul, or you will seem to have shunned the gift of humanity allotted you by god."[7]

Legal historians agree that the advent of Christianity in the Western world was a turning point in customs and thinking about euthanasia as well. The custom was revived somewhat in the modern era due to the influence of Stoic and utilitarian thought, as might be deduced from some of the statements by Saint Thomas More, Francis Bacon, and John Locke (statements that are variously interpreted[8]), but apart from that it was not until Nazism that the practice burst onto the scene again in organized form. As Francesco D'Agostino notes, "Since the advent of Christianity, euthanasia never experienced a true moment of reemergence—until our age."

I believe it is necessary, however, to stop dwelling on the history of this subject, because it would require too much space. While referring the reader to studies and summaries that are available to all,[9] I must concentrate on a discussion of the topic in relation to current cultural trends from an ethical and theological perspective.

Furthermore the pro-euthanasia opinion[10] that is currently active has its own characteristic features and reasons that are not identical to those that supported mercy killing in other historical periods. The present movement is not limited to the attitude of humanitarian understanding of the reality when so-called mercy killing occurs; it aims to legalize it. This is therefore the movement that must be discussed in order to grasp the underlying ideology and also to examine the ethical and cultural context from which it springs and on which it feeds.

There is frequently a "natural" alliance with the intellectual movement that has led to the legalization of voluntary abortion in many countries; indeed, it is not difficult to understand the cultural background common to both instances of legalizing

---

[7] Cicero, "The Dream of Scipio," in *Nine Orations and The Dream of Scipio*, trans. P. Bovie (New York: New American Library, 1967), 299.

[8] For an in-depth analysis of the true meaning of the expressions used by Bacon, see M. Cuyás, *Eutanasia: L'etica, la libertà e la vita* (Casale Monferrato: Piemme, 1989). The author shows that in reality Bacon used the term "euthanasia" to designate an easy death supported actively by the physicians, who do not abandon the sick patient when they can no longer help cure him.

[9] For the history of euthanasia, see, in addition to the work by G. Pelliccia cited above, F. D'Agostino, "Eutanasia, diritto e ideologia," *Iustitia* 30, no. 3 (1977): 285–306; E. Volterra, "Esposizione dei nati: diritto greco e romano," *Novissimo Digesto Italiano*, ed. A. Azara and E. Eula, 3rd ed. (Turin: UTET, 1960), 6:878–879; A. Oddone, "L'uccisione pietosa," *La Civiltà Cattolica* 1 (1950): 248 ff; H. J. Rose, "Euthanasia," *Encyclopedia of Religion and Ethics*, ed. J. Hastings, J. A. Selbie, and L. H. Gray, vol. 5 (Edinburgh: T. & T. Clark, 1969–71); see A. W. Maier, ed., "Suicide," in *Encyclopedia of Religion*, Hastings, Selbie, and Gray, 12:32. A useful reference work for the Greco-Roman era is M. Carpitella, ed., *Dizionario delle antichità classiche*, 3 vols. (Rome-Alba: Paoline, 1963); G. J. Gruman, "Death and Dying: Euthanasia and Sustaining Life; Historical Perspectives," in *Encyclopedia of Bioethics*, ed. W. T. Reich (New York: Free Press, 1978), 261–268; E. Sgreccia and M. L. Di Pietro, "Storia del fenomeno dell'eutanasia dall'antichità ai nostri giorni," in *Eutanasia, diritto alla vita*, ed. A. Tarantino and M. L. Tarantino (Lecce: Edizioni del Grifo, 1994), 13–46; D. Gracia Guillén, "Historia de la eutanasia," in S. Urraca Martinez, *Eutanasia hoy: Un debate abierto* (Madrid: Editorial Noesis, 1996), 67–91.

[10] S. Agostini and F. Perazza. *Eutanasia: Problematiche etiche, medico-legali, giuridiche* (Turin: Minerva Medica, 2004).

"inflicted death," which entails disregarding the value of the human person. Another similarity is the strategy adopted by the supporters of both instances of death: it begins with sensitizing public opinion concerning the "compassionate cases," then praises the mitigated sentences handed down by the courts that have held criminal proceedings in such cases, until finally there are requests for legalization, once public opinion has been suitably sensitized by the mass media and public debates.[11]

But there is a new, peculiar, and more frightening aspect, if you will, to the campaign in support of legalizing euthanasia—namely, the much greater potential for social and personal involvement than with the legalization of abortion, which may have appeared to be of immediate concern to only a few; abortion can happen to some, but everyone is destined to die.

I think it is appropriate, in this introductory section, to present first of all a precise definition of euthanasia and to distinguish this type of procedure from every other medical practice aimed at alleviating pain or avoiding therapeutic treatments that are unnecessary or disproportionate to the desired effect of prolonging life.

Let us take V. Marcozzi's definition of euthanasia, which is cited by other well-qualified legal scholars and moralists: euthanasia is "the painless or compassionate killing of someone who is suffering or who claims that he is suffering and that he might suffer unbearably in the future."[12]

This definition is substantially identical to the one provided by the Congregation for the Doctrine of the Faith in its declaration on euthanasia, *Iura et bona* (May 5, 1980), which is formulated in even more analytical terms: "By euthanasia is understood an action or an omission which of itself or by intention causes death, in order

---

[11] R. F. Esposito, "L'eutanasia nella stampa di massa italiana," in *Morire sì*, Beretta, 17–35.

[12] V. Marcozzi, "Il cristiano di fronte all'eutanasia," *La Civiltà Cattolica* 4 (1975): 322. This definition is also provided by S. Lener in an essay in which he examines the recommendation of the European Council on the rights of the sick and dying (January 29, 1976), titled "Sul diritto dei malati e dei moribondi: È lecita l'eutanasia?," *La Civiltà Cattolica* 2 (1976): 217–232. See also B. Häring, "Eutanasia e teologia morale," in *Morire sì*, Beretta, 221–232; idem, *Free and Faithful in Christ* (New York: Crossroad, 1978–81), 3:84–89; idem, *The Law of Christ: Moral Theology for Priests and Laity* (Westminster, MD: Newman Press, 1966), 3:213; idem, *Medical Ethics* (Notre Dame, IN: Fides, 1973), 144–148; G. Perico, *Difendiamo la vita*, 2nd ed. (Milan: Centro studi sociali, 1962), 465 ff; A. Boschi, *L'eutanasia* (Turin: 1950); P. Palazzini, *Dictionarium Canonicum et Morale* (Rome: Officium Libri Catholici, 1965); S. Bok, "Death and Dying: Euthanasia and Sustaining Life; Ethical Views," in *Encyclopedia of Bioethics*, Reich, 268–277. In the following work, no precise definition is found, but an attempt is made to draw a line of demarcation between what is responsible deliberate killing and what is not: E. Chiavacci, "Promozione dei diritti del malato posto di fronte alla prospettiva della morte," in *Morire sì*, Beretta, 253–266; L. Rossi, "Eutanasia," in *Dizionario enciclopedico di teologia morale* (Rome: Paoline, 1981), 380–386; G. Spagnolo, "L'eutanasia: Aspetto etico del problema," *Scienza e fede* 8 (1983): 1–38; G. Perico, *Problemi che scottano* (Milan: Ancora, 1976), 229 ff; Oddone, "L'uccisione pietosa," 248 ff; A. Günthor, *Chiamata e risposta* (Alba: Paoline, 1977), 3:602; Chiavacci, "Morale della vita fisica: Eutanasia e diritto di morire con dignità," *La Civiltà Cattolica* 4 (1983): 313–329.

that all suffering may in this way be eliminated." [13] The encyclical *Evangelium vitae* by Pope John Paul II repeats this definition (no. 65).[14] *Iura et bona* distinguishes this definition from other meanings that are often attributed to the word, such as the generic etymological sense of "painless death," which can also refer to natural death, or else "some intervention of medicine whereby the suffering of sickness or of the final agony are reduced, sometimes also with the danger of suppressing life prematurely." [15]

To avoid potential confusion I will use the term *euthanasia* only in the true and proper sense defined by the document and by moral theologians, whereas the term "treatment of pain" or more technical medical terminology will be used in other cases.

To complete our survey of definitions it is necessary to add that today people talk about euthanasia not only in reference to patients who are seriously and terminally ill, but also in other situations. Take the case of a newborn child with serious defects ("wrongful life"), who some suggest should be abandoned by the withholding of food so as to prevent his suffering—so they say—and avoid the burden to society; this situation is referred to as *neonatal euthanasia*. This was mentioned in the chapter on prenatal diagnosis. Yet another meaning of euthanasia is now gaining popularity: so-called *social euthanasia*, which takes the form of a choice not by a single individual but by society and results from the idea that health care budgets can no longer sustain the financial burden required to care for patients with a prognosis of long-term illness that is very costly to treat. In this way, economic resources are reserved for patients who are able to be cured and return to a productive life of work; this is one of the dangers of an economy that takes the cost-benefit criterion as its sole guide.

The Present Cultural Context

The Nazi practice of programmed euthanasia has already been mentioned; it was the first political euthanasia program to be devised and put into action. According to studies that drew on the transcripts of the Nuremberg trials, from 1939 to 1941, more than seventy thousand lives defined as "not worth living" were eliminated.[16]

The reason behind this program—as well as the one for eliminating Jews and prisoners in concentration camps—was connected with racism and absolute statism, which was made to coincide with the utterly cynical calculus of reducing state expenses for the purpose of directing economic resources to the costs of war. It has been rightly noted that the ideology currently pushing societies toward the

---

[13] Congregation for the Doctrine of the Faith (CDF), Declaration *Iura et bona*, May 5, 1980, part II.

[14] To be precise, *Evangelium vitae* says "of itself *and* by intention" rather than "of itself *or* by intention" [emphasis added]. Since the declaration *Iura et bona* uses the conjunction "vel ... vel," the meaning is the same as "et ... et"; hence, there is no contradiction between the two pontifical documents.

[15] CDF, *Iura et bona*, II. For commentary on this declaration, see J. V. Visser, "Pronunziamento ufficiale della S. Sede sull'eutanasia," *Medicina e Morale* 31, no. 3 (1981): 358–372.

[16] D'Agostino, "Eutanasia, diritto e ideologia," 298. The data on this subject are in A. Mitscherlich and F. Mielke, *Medizine und Menschheit: Dokumenten des Nürnberger Ärzteprozesses* (Frankfurt am Main: 1960); S. Cotta, *Vita fisica e legislazione* (Rome: 1985).

legalization of euthanasia is not the same, and it would be a sociological and histori-cal error to liken it with Nazism in order to combat it.[17]

The reasons adduced by supporters of euthanasia today are certainly not the same ones, and the analysis should be conducted objectively and dispassionately. Nazi theories and today's pro-euthanasia ideology do have one point in common, however, which is their lack of a concept of the transcendence of the human person. Once society betrays this value, which is closely connected with the affirmation that there is a personal God, arbitrary human control over human beings is eventu-ally claimed by the political leader of an absolutist regime or else by the demands of individualism. If human life is not valuable for its own sake, then someone can always exploit it,[18] using it as a means toward some contingent end. Even though there is no systematized sociology of the phenomenon under analysis here, the con-clusions of scholars, jurists, and sociologists can be summarized in the following three components, or "matrices," of the pro-euthanasia movement.

The secularization of thought and life

This prevents people from understanding the significance of death and the value of suffering. Everyone knows that there are various gradations of the secu-larized mindset: it can express itself as a fair assessment of relative autonomy and the value of temporal realities; it can also express itself as an exclusive interest in worldly realities and, furthermore, as a rejection of any dependence of man upon God or the moral law. In these last two attitudes, secularization reveals its inability to give meaning to suffering and death. The inability to give meaning to death leads to two interconnected attitudes: on the one hand death is ignored and banished from consciousness, culture, and life, and it is excluded above all as a criterion for verify-ing and assessing everyday life; on the other hand, death is hastened so as to avoid a head-on collision between it and the awareness of it.[19] G. Campanini observes that

---

[17] It should be noted, however, that according to Brian Pollard, the modern history of euthanasia can be said to have started in 1895 in Germany with the publication of a book written by A. Jost, *Das Recht auf den Tod* (The right to death). In 1920, another German volume appeared with the title *Die Freigabe der Vernichtung Lebensunwerten Lebens* (Allowing the destruction of life unworthy of life), written by the lawyer K. Binding and the psychiatrist A. Hoche. This book had remarkable influence and can be considered one of the key elements of the euthanasia phenomenon in Germany in the 1920s and 1930s. It is interesting to note that the original idea proposed in the latter book was not based on racist reasons (in fact it was proposed that euthanasia be applied to individuals from the German people, too). Rather, the reasons given for euthanasia were compassion, the meager quality of life in certain cases, and the need to contain social costs. See B. Pollard, *The Challenge of Euthanasia* (Crows Nest, NSW: Little Hills Press, 1994).

[18] On this point, I would certainly agree with the thesis of E. Levinas found in *Etica ed infinito* (Rome: Città Nuova, 1984).

[19] P. Ariès, *Essais sur l'histoire de la mort en Occident da Moyen age à nos jours* (Paris: Editions du Seuil, 1975); idem, "La mort inversée," *La Maison Dieu* 101 (1970): 57–88; E. Morin, *L'homme et la mort devant l'histoire* (Paris: Éditions Corrêa, 1951); S. Spinsanti, "Psicologi incontro ai morenti," *Medicina e Morale* 26, no. 1/2 (1976): 79–96; idem, ed., *Umanizzare la malattia e la morte: Documenti pastorali dei vescovi francesi e tedeschi* (Rome: Paoline, 1980); idem, "Salute, Malattia, Morte," in *Nuovo dizionario di teologia morale*, ed. F. Compagnoni, G. Piana, and S. Privitera (Cinisello Balsamo: Paoline, 1990),

"euthanasia is part of the process of secularization that pervades our society and finds expression above all as the supreme form of man's claim to independence, even—and especially—from God; consequently it views suffering as futile and rejects the religious symbolism of death."[20]

For a believer, death shows one's own contingency and dependence on God in the first place; it places one's life in God's hands in an act of total obedience. Euthanasia and—similarly—suicide are signs of man's claim to have the full right of control over himself, his own life, and his own death. The trend of secularization was also reinforced in the Industrial Age by a kind of utilitarianism based on economic productivism and consequently by the hedonistic ethic, which is immensely unsettled by death and pain. For a culture of this type, pain and suffering take on a negative value more than anything else and are rejected.

This is the origin of the "taboo" about death and everything associated with it; this is the origin of the social demand for a medical profession that would assure "complete physical, mental, and social well-being"[21] and even a painless death. Death has become a "taboo," an unmentionable thing, and, as once was the case with sex, it must not be mentioned in public. Philippe Ariès notes,

> In the twentieth century death has replaced sex as the principal forbidden subject. Once people used to tell children that they were born under a cabbage, but they would be present at the great scene of bidding final farewells to a dying person in his room and at his bedside. Today children from the most tender age are initiated into the physiology of love and birth, but when they no longer see their grandpa and ask why, they are told in France that he has gone on a journey and is far, far away; in England they are told he is resting in a beautiful garden where honeysuckles are blooming. It is no longer the children who are born beneath cabbages; instead it is the dead who disappear amid the flowers.[22]

Euthanasia, as a flight from suffering and agony, occurs first in the spirit and then in society and in law.

To verify all of this, suffice it to observe the countries and cultural contexts from which the request for euthanasia comes: it comes from the countries with industrialized and secularized societies. It started with the state of California, which passed the Natural Death Act in 1976 and effectively paved the way for the decriminalization of euthanasia when performed at the request of the patient, as expressed in the form of a living will. Six other states passed similar bills the following year. These laws recognize the right of any adult to sign a directive, valid for a five-year period, that instructs the personal physician to not apply or to interrupt "life-sustaining treatments" if the patient is drawing near the end of life.

1134–1144: M. Petrini, "L'assistenza al morente: Orientamenti e prospettive," *Medicina e Morale* 35, no. 2 (1985): 365–398.

[20] G. Campanini, "Eutanasia e società," in *Morire sì*, Beretta, 62.

[21] World Health Organization, Constitution of the World Health Organization, in *Basic Documents*, 45th ed., supplement (October 2006), 1, http://www.who.int/governance/eb/who_constitution_en.pdf.

[22] Ariès, *Essais sur l'histoire*, 186.

On September 27, 1977, the canton of Zurich (Switzerland) approved a euthanasia law by referendum. The debate flared up again in London even after proposed bills in favor of euthanasia had been rejected by the House of Lords in previous years (1969 Voluntary Euthanasia Bill). The pressure of public opinion polls has been strong in Germany and Belgium and intensifies every time hard cases are reported by the media.[23] In Italy there was a sense that the clandestine practice preceded the public debate.

Contributing to the pressure for legalization of euthanasia are associations such as the Euthanasia Society of America, which submitted to the United Nations a petition to include the "right" to euthanasia in the Universal Declaration of Human Rights. This cultural pressure is reinforced by the influence of propaganda groups and movements in favor of suicide, understood as "self-deliverance." Throughout the world, associations and societies dedicated to the spread of euthanasia have multiplied and tried to bring about legislative change in favor of that practice. The World Federation of Right to Die Societies comprises thirty-two societies in twenty-one countries. Among their other activities, these groups conduct propaganda campaigns and provide practical "services" through brochures or books dedicated to teaching the various techniques for committing a "dignified suicide." Arthur Koestler, who suffered from Parkinson's disease and leukemia and killed himself together with his wife Cynthia in 1983, had previously written the preface to a manual titled *A Guide to Self Deliverance*; it was then distributed to thousands of members of Exit, the British society for voluntary euthanasia, of which Koestler was the vice-president. It is reported that numerous groups similar to this society exist in various European countries. Thus copies of a similar how-to book in France, *Suicide: Mode d'emploi*, have circulated in the hundreds of thousands. The suicide manual *Final Exit*, by Derek Humphry, founder of the Hemlock Society in the United States, is now known throughout the world.[24]

Major periodicals such as *Le Monde, Time,* and *Lancet* often report "mercy killings" of children who were born with deformities. The practice has been proposed in England by some physicians for the purpose of avoiding the expense and risk of prenatal genetic diagnosis for defective neonates.[25]

The Royal College of Pediatrics and Child Health in England published a guidebook for physicians to instruct them on dealing with children born with serious conditions; some concerned voices were raised in protest, such as the voice of

---

[23] See Spagnolo, "L'eutanasia." The article recalls the so-called compassionate cases that fuel pro-euthanasia propaganda: the 1950 case of Dr. Herman Sanders, a physician who killed a 59-year-old woman suffering from cancer; the Belgian case of Corinne Vandelput, a little bow-legged girl who was killed by her parents; the French case of Luigi Faita, who killed his seriously ill brother and was acquitted by the Court of Assizes in Reno; the 1970 Italian case of Livio Davani, who threw his little deformed son into the Tiber River and escaped all punishment; and the widely reported case of Baby Jane Doe, who in 1983 was starved to death in the United States at her parents' request because she was born with hydrocephaly and spina bifida. In *The Challenge of Euthanasia*, Pollard presents more than forty cases since 1989; the judicial sentence was lenient in many of them.

[24] D. Humphry, *Final Exit: The Practicalities of Self-Deliverance and Assisted Suicide for the Dying* (Eugene, OR: Hemlock Society, 1991).

[25] See Chiavacci, "Morale della vita fisica."

the anti-euthanasia group Alert.[26] In France, the Association for the Prevention of Handicapped Childhood (Association pour la prevention de l'enfance handicapée, APEH) proposed a bill whose first article states, "A physician will commit neither a crime nor a misdemeanor if he refrains from administering to a three-day-old newborn the treatments necessary for its survival when the infant has an incurable infirmity making it foreseeable that it will never be able to have a life worth living."[27]

The secularized view of life and of the human being has been the soil in which this mentality has flourished—a cultural vision that John Paul II described as a "culture of death."[28] Having lost sight of the transcendent meaning of the human person, this vision is no longer capable of recognizing the inviolable value of a person's life and therefore ends up proposing as a good the elimination of human life in certain circumstances.[29]

The well-known kick-off for this type of pro-euthanasia ideology, however, was "A Plea for Beneficent Euthanasia," which was published in *The Humanist* in July 1974 and signed by around forty notable figures, including Nobel Prize winners Jacques Monod, Linus Pauling, and George Thomson. This manifesto deserves some comment, because it brings to light an analogous component of the pro-euthanasia mentality.

Rationalistic and humanitarian scientism

Scientism, of which Monod is one of the chief proponents, proceeds from the assumption that objective knowledge is only possible in the field of experimental science; the latter is said to be incompatible with every type of knowledge that it calls "subjective," and would therefore exclude ethical values, which Monod relegates to the realm of myth and imagination. Man, who emerged by chance in a universe that emerged from "chance" and "necessity," is his own judge and has no other point of reference outside of his own being: "scientific" reason is his sole guide, and he does not have to answer to anyone else about his own destiny. "Man knows at last that he is alone in the universe's unfeeling immensity, out of which he emerged only by chance."[30]

From these premises, the plea or "manifesto" for euthanasia more or less logically deduces the following:

We deplore moral insensitivity and legal restrictions that impede and oppose consideration of the ethical case for euthanasia. We appeal to an enlightened

---

[26] See *La Repubblica*, September 22, 1997.

[27] *Corriere della Sera*, November 7, 1987.

[28] John Paul II, Encyclical Letter *Evangelium vitae*, March 25, 1995, no. 12.

[29] For an analysis of the culture of death concept, see G. Miranda, "'Cultura della morte': Analisi di un concetto e di un dramma," in *Commento interdisciplinare alla "Evangelium vitae,"* ed. E. Sgreccia and R. Lucas Lucas (Vatican City: Libreria Editrice Vaticana, 1997); see also G. F. Morra, "¿Por qué la cultura contemporánea no respeta la vida?" *Ecclesia* 1, no. 1 (1987): 53–67; P. Donati, *La cultura della vita: Dalla società tradizionale a quella postmoderna* (Milan: F. Angeli, 1989).

[30] J. Monod, *Chance and Necessity: An Essay on the Natural Philosophy of Modern Biology*, trans. A. Wainhouse (New York: Knopf, 1971), 180. See commentary in G. Giusti, *L'eutanasia, diritto di vivere, diritto di morire* (Padua: CEDAM, 1982).

public opinion to transcend traditional taboos and to move in the direction of a compassionate view toward needless suffering in dying. . . .

We hold that the tolerance, acceptance, or enforcement of the unnecessary suffering of others is immoral.

We believe in the value and dignity of the individual person. This requires respectful treatment, which entails the right to reasonable self-determination.[31]

In other words, it is necessary to provide to those who are afflicted with incurable diseases and have reached the terminal stage the means by which to die gently and easily. Furthermore, "every individual has the right to live with dignity . . . [and] every individual has the right to die with dignity" and

it is beneficent euthanasia if, and only if, it results in a painless and quick death, and if the act as a whole is beneficial to the recipient. . . . To require that a person be kept alive against his will and to deny his pleas for merciful release after the dignity, beauty, promise, and meaning of life have vanished, when he can only linger on in stages of agony or decay, is cruel and barbarous. The imposition of unnecessary suffering is an evil that should be avoided by civilized society. . . .

We recommend that those individuals who believe as we do sign a "living will," preferably when they are in good health, stating unequivocally the expectation that the right to die with dignity will be respected.[32]

The inherent contradiction in the document can be noted: it starts by condemning morality and the law for demanding that society support suffering, describing it as cruel, and ends by invoking the "ethical" demand for a euthanasia law that involves the hastened taking of another's life.

The cultural perspective of the document is clear, however: against the background of materialistic atheism, science claims to transform death from an "occurrence" into a calculated and planned "outcome."[33] It has been rightly noted that behind these notions there is not only a lack of faith in God and in the eternal life that awaits man, but also—perhaps even prior to that and more radically—the death of metaphysics and of the ontology of the person. Since the "objective" value of the person has disappeared from Western thought through the triumph of philosophies of immanence and subjectivism, the death of the transcendent value of man has already been impressed upon the modern consciousness; the rest—euthanasia, suicide, and violence—has logically followed. "Humanitarianism is the metaphysics of subjectivity"; in other words, it is "the unchallenged primacy that the 'self' has commandeered in philosophy, ethics, art, and politics."[34] Theodor Adorno observes that "in socialized societies, in the thick and inescapable fabric of immanence, human beings now perceive death only as something strange and external. They are unable

---

[31] "A Plea for Beneficent Euthanasia," *The Humanist* 34, no. 4 (1974): 4, available at http://users.rcn.com/pknyc/articles/doc68.pdf.

[32] "Plea for Euthanasia," 4; see the first commentary in Marcozzi, "Il cristiano."

[33] Campanini, "Eutanasia e società," 65.

[34] D'Agostino, "Eutanasia, diritto e ideologia," 301.

to realize that they must die. … Due to the fact that it [death] literally transforms them into things, they reap their permanent death: reification." [35] This is as if to say that when man no longer heeds the transcendent value of the person, there is nothing left for him to do but to feel that he is a thing.

The personalist understanding of man, while accepting the fact that every person is limited by time and mortality, nevertheless surpasses the earthly perspective of individualism by recognizing the objective, transcendent value of the person and his otherworldly destination. We should take in the lesson from Heidegger, who sees death inscribed in life as a whole as the light that unveils the limit, and combine it with the metaphysics of Saint Thomas Aquinas, which opens man's personal being to life beyond this world.

The imbalance in medicine between technology and humanization

Developments in medicine have made the problem of euthanasia more acute or, at the very least, have underscored the problem of "death with dignity." This has come about in two ways: through technological progress in care for the dying and through the so-called socialization of medicine.

The Pontifical Council Cor Unum observes that "recent advances in science increasingly have repercussions on medical practice, in particular with regard to care for the seriously ill and the dying." [36] The debates in 1975 about the Karen Ann Quinlan case made it clear that medical advances render it increasingly difficult to define the boundaries between life and death, between irreversible and reversible comas. Resuscitation techniques make possible prodigious and complete recovery for many, but often condemn some individuals to treatments that prolong their agony rather than their lives. [37] Technological efforts in intensive care units are often accompanied by isolation and solitude for the patient: isolation from relatives even at the point of death, and solitude even with respect to the medical personnel, who are busy with the machines.

These extreme situations pose ethical questions: Beyond a certain point, are some resuscitation procedures permissible? If so, are they obligatory? The situations also pose the ethical problem of the obligation to provide humane psychological care to dying patients of this type. [38]

Other ethical problems for contemporary medicine result from the so-called socialization of medicine. The demand for health, driven by the quest for individual and social well-being, leads to overcrowding in hospitals and therefore to the depersonalization of health care and the isolation of dying patients in the ward; all this

---

[35] T. W. Adorno, *Dialettica negativa* (Turin: Einaudi, 1970), 334.

[36] Pontifical Council Cor Unum, "Ethical Questions Regarding the Seriously Ill and Dying," June 27, 1981, in *Enchiridion Vaticanum* (Bologna: Dehoniane, 1982), 7:1133–1173 [Italian].

[37] Secretariat of the French Bishops Conference, "Problème étiques posés aujourd'hui par la mort et le mourir," March 6, 1976, *Bulletin du Secrétariat de la Conférence Episcopale Française* (1976): 6, published in Italian as "Problemi etici posti oggi dalla morte e dal morire," in *Umanizzare la malattia*, Spinsanti, 43–44.

[38] A. Pessina, ed., *Scelte di confine in medicina: Sugli orientamenti dei medici rianimatori* (Milan: Vita e Pensiero, 2004).

makes it very difficult for the health care personnel to shift from simple technical treatment to human care.[39]

The Teaching of the Church's Magisterium

Going now to the heart of the ethical discussion, let us begin with a summary review of the documents of the Church's Magisterium on this topic, illustrating the progressive stages of its enrichment and clarification. If examined chronologically, the lines of development in the Church's teaching are ordered in such a way that they reveal (a) the progressive clarification and distinction of concepts, which has led to the true and proper definition of euthanasia, the notion of palliative care, which can involve the indirect shortening of life, the concept of "extraordinary" or "disproportionate" therapeutic means, and the rejection of therapeutic obstinacy or artificially prolonged death; as well as (b) an ever more widely ranging discussion about euthanasia, the identification of a connection with other anti-life cultural forms and attitudes, and an increasingly well-defined duty of the Christian community to prevent euthanasia by offering appropriate care to the dying.

In this gradual development of guidelines, which start as doctrinal statements and become increasingly pastoral and cultural, the Church has clearly followed both the progress of medical science and the development of social customs, which are not always positive.

In examining the magisterial teaching of the Church, I will take into consideration not only pontifical and conciliar documents but also documents issued by national bishops conferences.

The doctrine of the Church starts from several fixed points of reference: the acknowledgment of the sacred character of human life as created by God, the primacy of the human person over society, the consequent duty of those in authority to respect innocent life (which does not prejudice the question of the death penalty): these are the key points on which the Church's thinking will never be revised.[40]

Pius XII was called to intervene on the matter by two simultaneous and interconnected circumstances: Nazi doctrine and practice, which had been asserted as of 1939, and the request for clarification posed to the Pontiff by the medical corps.[41]

In citing the principal addresses and the most important passages from the teaching of Pius XII, I will recall his statements in the encyclical *Mystici Corporis* (June 29, 1943), which repeated an official response given in a decree by the Holy Office (December 2, 1940) in direct reference to the Nazi practices of "compulsory killing" due to "psychological and physical defects":

---

[39] L. Villa, *Medicina oggi: Aspetti di ordine scientifico, filosofico, etico-sociale* (Padua: Piccin, 1980).

[40] T. Jorio, *Theologia moralis*, II (Naples: M. D'Auria, 1939), 143; P. Palazzini, *Dictionarium canonicum et morale* (Rome: Officium Libri Catholici, 1965), 2:304–305.

[41] For a collection of the documents of the papal and episcopal Magisterium (excluding several more recent documents), see G. Caprile, "Il Magistero della Chiesa sull'eutanasia," in *Morire sì*, Beretta, 192–220, where the reader can find the citations from the official collections; Pius XII, *Discorsi e Radiomessaggi* [Addresses and radio messages], collected in numerous volumes by Tipografia Poliglotta Vaticana from *L'Osservatore Romano* and *Acta Apostolicae Sedis*.

674

To Our profound grief We see at times the deformed, the insane, and those suffering from hereditary disease deprived of their lives, as though they were a useless burden to Society; and this procedure is hailed by some as a manifestation of human progress, and as something that is entirely in accordance with the common good. Yet who that is possessed of sound judgment does not recognize that this not only violates the natural and the divine law written in the heart of every man, but that it outrages the noblest instincts of humanity? The blood of these unfortunate victims who are all the dearer to our Redeemer because they are deserving of greater pity, "cries to God from the earth."[42]

There are many interventions by Pius XII concerning mercy killing on the physician's initiative or at the patient's request, starting with the address given on November 12, 1944, to Italian Medical-Biological Association of St. Luke, in which he summarizes the traditional teaching of the Church: "God is the only master of the life of a human being who is not guilty of a crime punishable by death. ... No one in the world, no private person, no human power can authorize [the physician] to destroy it directly. His duty is not to destroy life but to save it."[43]

A remarkable development of medical ethics, which certainly ought to be kept in mind today, is offered by the same Pontiff in his address to the participants in the First International Congress on the Histopathology of the Nervous System (September 14, 1952): "In the first place it must be assumed that, as a private person, the doctor can take no measure or try no course of action without the consent of the patient. The doctor has no other rights or power over the patient than those which the latter gives him, explicitly or implicitly and tacitly. On his side, the patient cannot confer rights he does not possess. ... As for the patient, he is not absolute master of himself, of his body or of his soul. He cannot, therefore, freely dispose of himself as he pleases."[44]

The same idea is repeated with reference to legal representatives, who "have no other rights over the body and life of those they represent than those people would have themselves if they were capable. And they have those rights to the same extent."[45] This position regarding the physician's duty was repeated to participants in the Seventh Congress of Catholic Physicians on September 11, 1956.

With regard to the duties of public authorities, the same Pontiff repeated the same directive in the above-cited address to participants in the First International Congress on the Histopathology of the Nervous System, in which he spoke about "the personal right of the patient to the life of his body and soul in its psychic and moral integrity"[46] and declared, "It must be noted that, in his personal being, man is

---

[42] Pius XII, Encyclical Letter *Mystici Corporis*, June 29, 1943, no. 94.

[43] Pius XII, Address to the Italian Medical-Biological Association of St. Luke, November 12, 1944, in *Discorsi e Radiomessaggi* (Vatican City: Tipografia Poliglotta Vaticana, 1960), 6:181–196 [Italian].

[44] Pius XII, Address to the First International Congress on the Histopathology of the Nervous System, September 14, 1952, nos. 12–13, http://www.ncbcenter.org/Page.aspx?pid=388.

[45] Pius XII, Address on Histopathology, September 14, 1952, no. 19.

[46] Ibid., no. 8.

not finally ordered to usefulness to society. On the contrary, the community exists for man."[47]

An explicit condemnation is repeated by the same Pontiff with regard to euthanasia in his aforementioned address to the International Congress of Catholic Physicians and in his address to participants in the congress of the Italian Catholic Union of Midwives (October 29, 1951).[48] An important clarification is offered as well by the same Pontiff on the subject of what some inaccurately call "indirect euthanasia": in reality, it should be described as "pain management" or "palliative care." This applies to cases where the administration of analgesics may have two indirect consequences: the loss of consciousness or the shortening of life. The Pontiff dealt with the problem of anesthesia in terminal patients afflicted with an otherwise untreatable disease in his address to the Ninth Congress of the Italian Society of Anesthesiology (February 24, 1957), declaring, among other things, "You ask Us: Is the suppression of pain and consciousness by means of narcotics, when medically indicated, permitted by religion and morality to the doctor and the patient, even when death draws near and it is foreseen that the use of narcotics will shorten life? It is necessary to answer yes, if no other means exist and if, in the given circumstances, this does not prevent the carrying out of other religious and moral duties."[49] In a subsequent address to the Collegium Internationale Neuro-Psychopharmacologicum (September 9, 1958), he further defines the instruction by requiring the patient's consent.[50]

Several times during his pontificate, Pope Paul VI had occasion to repeat the condemnation of euthanasia in the strict sense, and he constantly connected these teachings with the theme of respect for human life in its totality and, in particular, linked the condemnation of euthanasia to the condemnation of abortion. Note, moreover, that Paul VI replaced natural law language with the expression "human rights" or "rights of the human person."

In his address to a committee of the United Nations on the subject of racial discrimination, he declared, "Above all the precious right to life—that most fundamental of all human rights—must be affirmed anew, together with the condemnation of that massive aberration which is the destruction of innocent human life, at whatever stage it may be, through the heinous crimes of abortion or euthanasia."[51] The declaration by the Congregation for the Doctrine of the Faith on the topic of procured abortion (November 18, 1974) repeats, "The right to life remains complete in an old person, even one greatly weakened; it is not lost by one who is incurably sick."[52]

---

[47] Ibid., no. 28.

[48] Pius XII, Address to Conference Participants of the Italian Catholic Union of Midwives, October 29, 1951, in *Discorsi e Radiomessaggi*, 13:336 [Italian]; idem, Radio Message to the Seventh International Congress of Catholic Physicians, September 11, 1956, in *Discorsi e Radiomessaggi*, 18:425 [Italian].

[49] Pius XII, Response to Questions of the Ninth Congress of the Italian Society of Anesthesiology, February 24, 1957, in *Discorsi e Radiomessaggi*, 18:794, 797–798 [Italian].

[50] Pius XII, Address to Conference Participants of the Collegium Internationale Neuro-Psychopharmacologicum, September 9, 1958, in *Discorsi e Radiomessaggi*, 20:331 [Italian].

[51] Paul VI, Address to the Committee of the United Nations on Apartheid, May 22, 1974.

[52] CDF, Declaration *Quaestio de abortu*, November 18, 1974, no. 12.

The topic is also mentioned in other papal speeches. We should note the address to the International College of Psychosomatic Medicine (September 18, 1975), in which Paul VI introduces the concept of the "dignity" of death in these words: "Keeping in mind the value of every human person, we wish to recall that it is the physician's duty always to be at the service of life and to assist it until the end, without ever accepting euthanasia or renouncing the exquisitely human duty to help it complete its earthly course with dignity." [53] We should recall, finally, a further clarification introduced by Paul VI through a letter by Jean-Marie Cardinal Villot, dated October 3, 1970, to the secretary general of the International Federation of Catholic Medical Associations concerning the rejection of what would be described as *therapeutic obstinacy*: "In many cases would it not be futile torment to order vegetative resuscitation in the final stage of an incurable illness? The physician's duty consists rather of endeavoring to soothe the suffering, instead of prolonging as long as possible by any means and under any conditions a life that is naturally approaching its conclusion." [54]

During the pontificate of Paul VI the condemnation of euthanasia was solemnly reaffirmed by Vatican Council II in the pastoral constitution *Gaudium et spes*, again in connection with other crimes against human life: "Whatever is opposed to life itself, such as any type of murder, genocide, abortion, euthanasia or willful self-destruction ... all these things and others of their like are infamies indeed. They poison human society, but they do more harm to those who practice them than those who suffer from the injury. Moreover, they are supreme dishonor to the Creator." [55] Similarly, the statement by the 1974 Synod of Bishops on the subject of human rights and reconciliation reaffirmed that "the right to life is a fundamental, inalienable right; today it is subject to serious violations: contraception, abortion, euthanasia." [56] Many documents have been issued by conferences of bishops, during and after the pontificate of Paul VI, that have broadened the doctrinal perspective on the ethical level and especially at the pastoral level.

Various documents by the bishops' conferences deal fully with this theme, while some of them treat it at greater length (the letter on public morality by the bishops of England and Wales on December 31, 1970; the communiqué of the bishops of Panama on November 23, 1974; the declaration on respect for human life by the bishops of Mexico on September 8, 1975; the letter by the bishops of Rwanda on May 31, 1975; the declaration by the bishops of West Germany on June 1, 1975; the pastoral letter by the Irish bishops on May 1, 1975).[57] Other episcopal documents are dedicated entirely to the problem of care for the dying, such as the document by the secretariat of the French Bishops Conference in March 1976 and the declaration of the West German Bishops Conference on November 20, 1978, on the theme of death

---

[53] Paul VI, Address to Participants in the Third World Congress of the International College of Psychosomatic Medicine, September 18, 1975, in *Insegnamenti di Paolo VI* (Vatican City: Tipografia Poliglotta Vaticana, 1975), 13:953–956 [Italian].

[54] Printed in *La Civiltà Cattolica* 4 (1970): 275–277 [Italian].

[55] Vatican Council II, Pastoral Constitution *Gaudium et spes*, December 7, 1965, no. 27.

[56] G. Caprile, *Il Sinodo dei Vescovi: Terza Assemblea Generale; 27 settembre–26 ottobre 1974* (Rome: La Civiltà Cattolica, 1975), 708 [Italian].

[57] G. Caprile, "Il Magistero della Chiesa," 202–217.

worthy of man and Christian death.[58] The Bishops Committee for the Defense of Life of the Spanish Bishops Conference published an interesting document in 1993 on the topic—it took the format of a catechism made up of questions and answers.[59]

Because of space limitations I will not cite letters and documents by individual bishops, although Msgr. Léon-Arthur Elchinger, Bishop of Strasbourg, wrote a very important one during the studies on the rights of the sick and dying conducted by the Committee on Social and Health Questions of the Council of Europe.[60] Finally, it is necessary to add that the position of other Christian denominations is very close to that of the Catholic Church with regard to euthanasia, properly speaking. The bishops' documents develop papal teaching and also incorporate the contributions of moral theologians. It is frequently noted that "care for the dying does not mean keeping a person alive by extraordinary means when there is no hope of recovery" (bishops of England and Wales, 1970 letter).[61] The criterion for "ordinary and extraordinary means," already enunciated by Pius XII, is common in those documents that also develop the concept of "death with dignity" (bishops of West Germany, 1975 declaration).

The 1976 document by the French bishops presents a unique development in the topic of extraordinary, intensive treatments aimed at prolonging life at all costs. This document addresses the problem of the technical difficulty of determining death and expresses doubts about the criteria used by medical personnel, which it considers too "biological." But the documents of the bishops conference devote space above all to the theme of *humanizing death*, understood as the need and duty on the part of community and of medical personnel to make the patient feel that someone is near, to keep communication lines open, and to spare the dying or sick person loneliness

---

[58] Spinsanti, *Umanizzare la malattia*. Other documents of the Catholic Magisterium have addressed the topic of death and dying: see Florida Catholic Conference, "Statement on the Life, Death and Treatment of Dying Patients," April 27, 1989, http://www.flacathconf .org/statements/1989/dyingtreatment.pdf; Standing Council of the French Bishops Conference, "Respecter l'homme proche de sa mort," *Medicina e Morale* 42, no. 1 (1992): 124–133; John Paul II, Address to the International Congress on Assistance to the Dying, March 17, 1992, in *Medicina e Morale* 42, no. 3 (1992): 419–422 [Italian]; Pennsylvania Catholic Conference, "Nutrition and Hydration: Moral Considerations; A Statement of the Catholic Bishops of Pennsylvania," rev. ed., 1999, http://www.pacatholic.org/bishops -statements/nutrition-and-hydration-moral-considerations; U.S. Conference of Catholic Bishops Committee for Pro-Life Activities, "Nutrition and Hydration: Moral and Pastoral Considerations," 1992, http://old.usccb.org/prolife/issues/euthanas/nutindex.shtml.

[59] Spanish Bishops Conference Committee for the Family and the Defense of Life, *La Eutanasia: 100 cuestiones y respuestas sobre la defensa de la vida y la actitud de los católicos* (Madrid: Paulinas, 1993), available online at http://www.conferenciaepiscopal. nom.es/archivodoc/jsp/system/win_main.jsp.

[60] *L'Osservatore Romano*, January 1, 1976. The aforementioned committee of the Council of Europe carried out the preparatory work for Recommendation 779/1976 of the Parliamentary Assembly of the Council of Europe on the rights of the sick and dying (http:// assembly.coe.int/main.asp?Link=/documents/adoptedtext/ta76/erec779.htm), which was discussed and approved by the same assembly January 26–29, 1976.

[61] Bishops Conference of England and Wales, "Moral Questions of Today," December 31, 1970, *L'Osservatore Romano*, January 28, 1971, excerpt here taken from http://catholicinsight. com/online/church/vatican/article_468.shtml.

and a sense of abandonment. The 1978 document by the German bishops gives some pastoral guidelines and even draws on the findings of several psychological studies on death and care for the dying.[62]

The teaching recalled thus far was repeated in *Iura et bona*.[63] Then there were further doctrinal refinements in the 1995 *Charter for Health Care Workers* by the Pontifical Council for Pastoral Assistance to Health Care Workers.[64] That same year John Paul II, in his encyclical *Evangelium vitae*, treated the topic in even greater depth and forcefully and solemnly pronounced that euthanasia is morally unacceptable. These documents will guide us in the attempt to summarize the Church's moral teaching.

Summary of Moral Teaching on Euthanasia

Having examined this series of statements by the Magisterium,[65] we can now offer a doctrinal summary.

The rejection of euthanasia properly speaking

The definition of euthanasia provided by the Congregation for the Doctrine of the Faith[66] has already been recalled, as well as the fact that *Evangelium vitae* adopts this definition.[67]

Several comments must immediately be made on this definition to point out the refinement of terminology common to theologians and medical practitioners. The distinction between *direct* and *indirect* euthanasia is omitted; in the previous language, which was still used by Pius XII, indirect euthanasia was understood to mean "pain management" or "palliative care," which was considered licit under certain conditions, even when as a consequence it could shorten life. In reality, neither the action per se nor the intention is aimed at ending life or anticipating death in this case, and therefore, to avoid confusion, it should not be referred to as "euthanasia" at all. Further on, the same document more appropriately uses the expression "use of analgesics."

[62] E. Kübler-Ross, *On Death and Dying* (New York: Macmillan, 1969). See also Spinsanti, "Psicologi incontro ai morenti."

[63] CDF, *Iura et bona*. See also the commentary by Visser, "Pronunziamento ufficiale."

[64] Pontifical Council for Pastoral Assistance to Health Care Workers, *Charter for Health Care Workers* (Boston: Pauline, 1995).

[65] See also John Paul II, Apostolic Letter *Salvifici doloris*, February 11, 1984; *Catechism of the Catholic Church*, nos. 2276–2283.

[66] CDF, *Iura et bona*, II. See for example an analysis of the definition in G. Miranda, "Riflessioni etiche intorno alla fine della vita," in *A sua immagine e somiglianza?*, ed. Mazzoni (Rome: Città Nuova, 1997), 180–202.

[67] On this point see G. Miranda, "I problemi etici dell'eutanasia nell'enciclica *Evangelium vitae*," *Medicina e Morale* 45, no. 4 (1995): 719–738; L. Ciccone, "L'eutanasia ed il principio dell'inviolabilità assoluta di ogni vita umana innocente," in *Commento interdisciplinare*, Sgreccia and Lucas Lucas, 453–465; M. Calipari, *Curarsi e farsi curare* (Turin: Edizioni Paoline / San Paolo, 2006).

Also avoided is a distinction frequently used in medical language between *active* euthanasia and *passive* euthanasia, where the adjective "passive" is supposed to indicate the omission of treatments and medical procedures. However, the word "passive" has a much broader meaning and could therefore give rise to ambiguities; euthanasia is always passive in a certain sense, from the perspective of the sick person, and always active on the part of those who instigate it, whether by action or by omission.

The document of the Holy See expresses its moral judgment on euthanasia, with the understanding and clarifications just described, as follows:

> It is necessary to state firmly once more that nothing and no one can in any way permit the killing of an innocent human being, whether a fetus or an embryo, an infant or an adult, an old person, or one suffering from an incurable disease, or a person who is dying. Furthermore, no one is permitted to ask for this act of killing, either for himself or herself or for another person entrusted to his or her care, nor can he or she consent to it, either explicitly or implicitly, nor can any authority legitimately recommend or permit such an action. For it is a question of the violation of the divine law, an offense against the dignity of the human person, a crime against life, and an attack on humanity.[68]

Evangelium vitae expresses this in language that is even more peremptory and formal, approaching the solemnity of dogmatic formulas: "In harmony with the Magisterium of my Predecessors and in communion with the Bishops of the Catholic Church, I confirm that euthanasia is a grave violation of the law of God, since it is the deliberate and morally unacceptable killing of a human person. This doctrine is based upon the natural law and upon the written word of God, is transmitted by the Church's Tradition and taught by the ordinary and universal Magisterium."[69]

These condemnations reaffirm, as Pope John Paul II explicitly notes, all the preceding pronouncements of the Magisterium and the constant teaching of moral theology. The condemnation is also extended to suicide by those same documents. The declaration *Iura et bona* states:

> Intentionally causing one's own death, or suicide, is therefore equally as wrong as murder; such an action on the part of a person is to be considered as a rejection of God's sovereignty and loving plan. Furthermore, suicide is also often a refusal of love for self, the denial of a natural instinct to live, a flight from the duties of justice and charity owed to one's neighbor, to various communities or to the whole of society—although, as is generally recognized, at times there are psychological factors present that can diminish responsibility or even completely remove it. However, one must clearly distinguish suicide from that sacrifice of one's life whereby for a higher cause, such as God's glory, the salvation of souls or the service of one's brethren, a person offers his or her own life or puts it in danger.[70]

Evangelium vitae notes that euthanasia involves the malice proper to suicide or homicide and declares that "suicide is always as morally objectionable as murder,"

---

[68] CDF, *Iura et bona*, II.

[69] John Paul II, *Evangelium vitae*, no. 65.

[70] CDF, *Iura et bona*, I.

adding shortly afterward that "to concur with the intention of another person to commit suicide and to help in carrying it out through so-called 'assisted suicide' means to cooperate in, and at time to be the actual perpetrator of, an injustice which can never be excused, even if it is requested."[71]

The encyclical analyzes euthanasia in terms of three different "steps" of increasing gravity. It first refers to euthanasia committed with the motivation of compassion: "Even when not motivated by a selfish refusal to be burdened with the life of someone who is suffering, euthanasia must be called a false mercy, and indeed a disturbing 'perversion' of mercy. True 'compassion' leads to sharing another's pain; it does not kill the person whose suffering we cannot bear." Then it speaks about the act committed without the request or consent of the sick person (so-called involuntary or nonvoluntary euthanasia[72]): "The choice of euthanasia becomes more serious when it takes the form of a murder committed by others on a person who has in no way requested it and who has never consented to it." Finally, it adds that "the height of arbitrariness and injustice is reached when certain people, such as physicians or legislators, arrogate to themselves the power to decide who ought to live and who ought to die."[73]

We certainly know that the laws regulating these practices today usually require the explicit expression of consent by the interested party; yet we also know that various court rulings, even in places where such laws exist (e.g., Holland[74]), have declared such acts unpunishable when committed against the life of a newborn child or at the request of persons in a state of serious psychological depression. Holland quickly passed from a law that exempted physicians from punishment for performing euthanasia on infirm adults who made explicit, reasoned, and repeated requests for it—including adolescents between the ages of 16 and 18 and even, if accompanied by an additional request from the parents, children between the ages of 12 and 16 (April 1, 2002)—to allowing euthanasia for children younger than 12, including newborns, through an agreement between the pediatric department and a local court (September 2004).[75] We know that there have been laws imposing involuntary eutha-

---

[71] John Paul II, *Evangelium vitae*, no. 66.

[72] A distinction between voluntary and nonvoluntary euthanasia can be found, for example, in J. Harris, who writes that "If that decision coincides with the individual's own wishes and he or she has consciously and expressly approved of the decision, I will call this *voluntary euthanasia*. Where the individual concerned does not know about the decision and has not consciously and expressly approved it in advance, I will call this *non-voluntary euthanasia*, even where he or she is believed or presumed to be in accord." J. Harris, "Euthanasia and the Value of Life," in *Euthanasia Examined: Ethical, Clinical and Legal Perspectives*, ed. J. Keown (New York: Cambridge University Press, 1995), 6–7.

[73] All of the passages cited in this paragraph are found in John Paul II, *Evangelium vitae*, no. 66.

[74] A regulation went into effect on November 28, 2000 that not only declares physicians who perform euthanasia unpunishable, but also affirms the licitness of acts of euthanasia. See G. Bognetti, "La legge olandese su eutanasia e suicidio assistito," *Corriere giuridico* 1 (2001): 705; P. Ricca, ed., *Eutanasia: La legge olandese e commenti* (Turin: Claudiana, 2002).

[75] See W. Eijk, "Is the Dutch Euthanasia Regulation Compatible with *Evangelium Vitae*?" *Medicina e Morale* 46, no. 3 (1996): 469–481; E. Sgreccia, "L'eutanasia in Olanda: Anche per i bambini!," *L'Osservatore Romano*, September 3, 2004, p. 8.

nasia (see what was said above on Nazism), and that there is always the possibility of a return to similar situations, which must be condemned forcefully to prevent such a recurrence. On the other hand, when legislators or medical associations decide that euthanasia or assisted suicide is permissible only in the case of persons in certain circumstances, is this not perhaps an instance of how they "arrogate to themselves the power to decide who ought to live and who ought to die"?

For the purpose of dialogue with the secular world, I think it is important to note what *Iura et bona* states about the reasons for this rejection. The document teaches that "the considerations set forth in the present document concern in the first place all those who place their faith and hope in Christ, who, through His life, death and resurrection, has given a new meaning to existence and especially to the death of the Christian, as St. Paul says: 'If we live, we live to the Lord, and if we die, we die to the Lord.' As for those who profess other religions, many will agree with us that faith in God the Creator, Provider and Lord of life—if they share this belief—confers a lofty dignity upon every human person and guarantees respect for him or her." The passage continues, making an appeal to "many people of good will, who, philosophical or ideological differences notwithstanding, have nevertheless a lively awareness of the rights of the human person. ... And since it is a question here of fundamental rights inherent in every human person, it is obviously wrong to have recourse to arguments from political pluralism or religious freedom in order to deny the universal value of those rights."[76]

This passage makes an important reference to the rational, secular, and universal basis for defending human life and rejecting euthanasia: out of respect for the truth, even more than for strategic reasons, one should avoid basing arguments against euthanasia on reasons of faith alone, as though defending the life of the sick and dying were a duty of believers alone. Human life is a secular good and a secular value, recognizable by all who intend to follow right reason and objective truth.

What Pius XII called "natural law" is described in the document here examined as a "fundamental right" of man, the first of human rights: it is defined as fundamental because all other human rights are based on it. "Human life," the declaration affirms, "is the basis of all goods, and is the necessary source and condition of every human activity and of all society."[77]

The foundation of ethics is respect for the truth of man, respect for the human person as he is: there can be no other real foundation for ethics. Ethics guides man from *is* to *ought*; other criteria would be the benefit of one to the detriment of another, or the power of some over others, or the effectiveness of that power, which becomes increasingly vast for some and more and more oppressive for others.

Respecting the truth of the person from the moment of conception means respecting God who creates the human person just as He creates him or her; respecting man in his final phase of life means respecting the encounter of man with God, his return to the Creator, to the exclusion of any other human power, both the power to anticipate that death (euthanasia) and the power to prevent that encounter by a form of biological tyranny (therapeutic obstinacy). The boundary between 'euthanasia' and "death with dignity" is established in this perspective.

---

[76] CDF, *Iura et bona*, intro.

[77] Ibid., I.

682

Doing away with this boundary means doing away with any objective foundation for law, for ethics itself, and, at the same time, for the identity of the medical profession.[78] The following principles elucidate the criterion of death with dignity.

The proportionate use of therapeutic means

Morality cannot ignore the problem and the obligation of making death worthy of man and worthy of believers: the expression "death with dignity," when it is not meant to suggest disguised forms of euthanasia, expresses a criterion that is ethically acceptable and proper. It is true that many people die peacefully and, as the CDF document warns, we should not "think only of extreme cases":

> Nevertheless the fact remains that death, often preceded or accompanied by severe and prolonged suffering, is something which naturally causes people anguish. ... Today it is very important to protect, at the moment of death, both the dignity of the human person and the Christian concept of life, against a technological attitude that threatens to become an abuse. Thus some people speak of a "right to die," which is an expression that does not mean the *right to procure death* either by one's own hand or by means of someone else, as one pleases, but rather *the right to die peacefully with human and Christian dignity*.[79]

In light of this consideration, the declaration introduces new technical language as required by medical advances and as previously noted by some theologians. Since the time of Pius XII moralists have spoken about "ordinary" and "extraordinary" therapeutic means: the use of ordinary means to sustain the dying person is obligatory, but one may licitly refuse extraordinary means, with the patient's consent or pursuant to his request, even when that refusal causes death to occur sooner. The "extraordinary" character was defined in relation to the increase of suffering that could result from such means, or else in terms of the expense or even of the difficulty of providing access to all who might request it. Advances in medicine have made it difficult to apply this distinction, inasmuch as many means that were considered extraordinary in the past have now become ordinary and, furthermore, as shown by renowned clinics,[80] new techniques of resuscitation and intensive care have saved many lives. Hence the need to find another standard criterion, no longer based on the "therapeutic means" but rather on the "therapeutic result" expected from it. As *Iura et bona* puts it,

> In the past, moralists replied that one is never obliged to use "extraordinary" means. This reply, which as a principle still holds good, is perhaps less clear today, by reason of the imprecision of the term and the rapid progress made in the treatment of sickness. Thus some people prefer to speak of "proportionate" and "disproportionate" means. In any case, it will be possible to

---

[78] D'Agostino, "Eutanasia e diritto."

[79] CDF, *Iura et bona*, III–IV (emphasis added). On this point of the proportionate use of therapeutic means, see Visser, "Pronunziamento ufficiale," 369–370; Häring, "Eutanasia e teologia morale," 164–178. Concerning the so-called right to die, see the fine study by L. R. Kass, "Is there a right to die?" *Hastings Center Report* 23, no. 1 (1993): 34–43.

[80] C. Manni, "Considerazioni mediche sull'eutanasia: Il pensiero di un medico anestesista-rianimatore," in *Morire sì*, Beretta, 103–120.

make a correct judgment as to the means by studying the type of treatment to be used, its degree of complexity or risk, its cost and the possibilities of using it, and comparing these elements with the result that can be expected, taking into account the state of the sick person and his or her physical and moral resources.[81]

The declaration deduces four very useful guidelines from this distinction:

> If there are no other sufficient remedies, it is permitted, with the patient's consent, to have recourse to the means provided by the most advanced medical techniques, even if these means are still at the experimental stage and are not without a certain risk ... .

> It is also permitted ... to interrupt these means, where the results fall short of expectations. But for such a decision to be made, account will have to be taken of the reasonable wishes of the patient and the patient's family, as also of the advice of the doctors who are specially competent in the matter ... .

> It is also permissible to make do with the normal means that medicine can offer. Therefore one cannot impose on anyone the obligation to have recourse to a technique which is already in use but which carries a risk or is burdensome. ...

> When inevitable death is imminent in spite of the means used, it is permitted in conscience to take the decision to refuse forms of treatment that would only secure a precarious and burdensome prolongation of life, so long as the normal care due to the sick person in similar cases is not interrupted.[82]

The most recent literature makes a more precise distinction between the concepts *ordinary/extraordinary*, on the one hand, and *proportionate/disproportionate* on the other. Maurizio Calipari proposes a new synthesis of moral doctrine concerning the use of life-sustaining means in continuity with the teachings of the moral tradition, but taking into account the new ethical demands made by the constant development of the medical sciences and of the technologies applied to them. He outlines a new systematic framework in which he dynamically combines the concept pairs "proportionate/disproportionate" and "ordinary/extraordinary," proposing *the principle of ethical appropriateness regarding the use of life-sustaining means.*[83]

Calipari maintains that the process of evaluating is divided into three phases according to a logic that examines the medical and technical elements; that is, first one examines all the aspects lending themselves to an objective evaluation (phase 1); next one takes into consideration the more subjective aspects of the patient (phase 2); finally one arrives at an ethical judgment that takes those two phases of evaluation into account and is translated into a morally appropriate, practical decision (phase 3).

The *proportionality* or *disproportionality* of a means of preserving life designates the *medical and technical* appropriateness or inappropriateness of its use, in relation to attaining a definite objective (restoring the patient to health or keeping

---

[81] CDF, *Iura et bona*, II.

[82] Ibid., IV.

[83] See Calipari, *Curarsi e farsi curare*, esp. 151–170.

him alive). *Medical effectiveness* designates the objective salutary effects produced by a means of preserving life in relation to a precise medical objective, whereas *global effectiveness* refers to "the attainment of salutary effects that prove to be really significant for the life of the patient, according to his personal assessment, in the complex context of his existence and based on the axiological scale he has adopted." [84] Obviously medical effectiveness is the minimum prerequisite for global effectiveness, i.e., it is necessary but not sufficient component of global effectiveness.

A medical intervention is proportionate if it proves to be appropriate in attaining a precise medical objective. This judgment evaluates "(a) the concrete availability or likelihood of being able to find the means; (b) the actual technical possibility of using the means appropriately; (c) the reasonable expectation of real medical efficacy; (d) the possible harmful side effects; (e) the foreseeable risks for the patient's health or life that may be involved in using that means; (f) the actual possibility of recourse to alternative therapies that are equally or more effective; (g) the allocation of health care resources, both technical and financial, that are necessary in order to employ the means." [85]

With regard to the final criterion, which is economic, Calipari makes it clear that even though it is necessary to ration economic resources as well as possible, ambiguous interpretations must be avoided, keeping in mind that the fundamental good of human life is not quantifiable in merely economic terms and cannot be measured against dissimilar and inferior goods.

Phase 2 takes into consideration those aspects which are strictly dependent on the subjectivity of the patient: what proves to be ordinary (or extraordinary) for a given person in a given clinical situation might not be so for another, or for the same person in a different situation.

Given the importance of the value at stake, physical human life, any means aimed at preserving it must always be considered *ordinary* "unless its use *in that situation* involves for the patient, according to his own prudent judgment, at least one significant element that characterizes it as extraordinary." [86]

Criteria for classifying a means as extraordinary may include (a) excessive effort required to find and/or use the means; (b) the experience of very great or unbearable pain that cannot be alleviated adequately; (c) economic costs connected with the means that are very burdensome for the patient or for his relatives; (d) the experience of tremendous fear or strong repugnance in relation to the application of the means; (e) a reasonably high probability of the occurrence of serious risks to the life or health of the patient associated with the use of the means, as assessed by the patient himself in relation to the seriousness of his present clinical condition; (f) a low rate of "global effectiveness" of the means in relation to the benefits reasonably expected by the patient, evaluated according to the axiological scale he has adopted; and (g) the permanence of clinical conditions, following use of the means, that would prevent the patient from fulfilling moral obligations that cannot be deferred.

I maintain that it is necessary for the patient to be able to make these evaluations personally, with the awareness, however, of the support of the community,

---

[84] Ibid., 153.

[85] Ibid., 154.

[86] Ibid., 159.

including financial support. Respecting another's will does not mean disregarding the duty to practice solidarity.

By combining the objective and subjective variables, it is possible to deduce the following theoretical classification of means for preserving human life: (a) proportionate and ordinary means, (b) proportionate and extraordinary means, (c) disproportionate and ordinary means, and (d) disproportionate and extraordinary means.

As far as moral duty is concerned, the application of any means for preserving life can be *obligatory*, *optional*, or *illicit*. "Whenever the use of a certain means for preserving life, which is deemed proportionate, proves to be 'ordinary' as well for the patient, recourse to that means should be considered *obligatory* for him; whenever the same means proves, on the other hand, to be 'extraordinary' for the patient, recourse to it would be *optional* for him, in principle. Indeed, there can be particular circumstances in which even the use of a proportionate and extraordinary means could prove to be *obligatory* for the patient so that he can carry out more serious duties (of charity or justice, toward God or toward neighbor)."[87]

In contrast, a disproportionate means for preserving life is *illicit* even if it should prove to be ordinary for the patient (unless it is the only way for the patient to fulfill very serious moral obligations). In this case the disproportionate procedure would be an intervention that procures a certain benefit, but to a degree insufficient to outweigh the possible harmful side effects associated with it. A disproportionate means that is incapable of procuring any benefit or is simply harmful would always be illicit.

Calipari appropriately makes clear that all moral reasoning about preserving life and caring for health must preserve the "centrality" of the human person in his integral truth.

The administration of normal care and palliative care

The moment eventually comes when no therapeutic intervention, properly speaking, can be undertaken for the purpose of arresting or reversing the course of an illness. In this situation, any true and proper therapeutic intervention runs the risk of being disproportionate.

Medicine still has resources to employ at this point, and therefore it has the obligation to use them within the limits of what is possible. Such an act per se is no longer aimed at healing or prolonging life, but shows the respect due to the patient and seeks to maintain a certain quality of life. These resources are *normal care* and *palliative care*. The two concepts are not automatically equivalent.

Normal care is understood to include nutrition and hydration (whether artificial or not), the aspiration of bronchial secretions, and the cleansing of bedsores. Along these lines, for example, the *Charter for Health Care Workers* declares, "The administration of food and liquids, even artificially, is part of the normal treatment always due to the patient when this is not burdensome for him: their undue suspension could be real and properly so-called euthanasia" (no. 120).

I have deliberately expressed myself using the very words of the magisterial documents because of the precision that this delicate topic demands. It would be negligent not to emphasize the humane feeling contained in the guidelines and the profound respect due to the dying person.

---

[87] Ibid., 165.

A controversy over normal care has been ongoing, especially in the United States, because some medical centers tend to consider artificial nutrition and hydration therapeutic interventions that are, furthermore, extraordinary in character and therefore not due to the patient. Several local bishops' conferences in the United States[88] have intervened in the debate to reaffirm that such support certainly does not amount to torture and can contribute in many cases not so much to prolonging life but to making death less distressing. They also recall that food and hydration cannot constitute a medical act but rather are normal care, even when the means of administration is artificial. When the organism is no longer capable of receiving that support or of benefiting from it, then it obviously ceases to be care and the obligation to administer it ceases.

Palliative care is a broader category than normal care because it is aimed first of all at reducing the symptoms of the illness and relieving pain (but not only pain); this will be discussed in a moment. Methods of palliative care are commonly understood to mean treatments for the benefit of patients suffering from illnesses that are no longer curable, aimed at controlling the symptoms more than the underlying pathology, through the application of procedures that allow the patient to enjoy a better quality of life.[89]

Palliative care today in fact comprises (a) palliative oncological therapy, i.e., the whole set of "classic oncological therapies (surgery, radiation therapy, chemotherapy) applied to patients for the purpose of treating their symptoms,"[90] and (b) so-called support care, which includes "analgesic, non-causal therapies aimed at reducing or eliminating pain; nutritional evaluation and electrolytic regulation; the treatment of opportunistic infections; physical therapy and rehabilitation procedures; psychological support, which is a particularly important part of the support offered to patient and family; and psychological supervision of the team of caregivers, whose good emotional performance is fundamental to optimizing the therapeutic outcome in this delicate phase of chronic illness."[91]

The encyclical *Evangelium vitae* says the following about palliative care: "In modern medicine, increased attention is being given to what are called 'methods of

---

[88] See the essays by the Pennsylvania Catholic Conference and the Committee for Pro-Life Activities of the United States Conference of Catholic Bishops (see note 58 above) and the commentary by D. Tettamanzi, "Nutrizione e idratazione medicalmente assistite nel paziente in stato di incoscienza: Problemi morali," *L'Osservatore Romano*, December 11, 1992. For a summary analysis of the problems surrounding this point, see also E. P. Flynn, *Hard Decisions, Forgoing and Withdrawing Artificial Nutrition and Hydration* (Kansas City, MO: Sheed and Ward, 1990); Miranda, "Riflessioni etiche."

[89] For an in-depth study, see V. Ventafrida, "Cure palliative," in *Dizionario di teologia pastorale sanitaria* (Turin: Camilliane, 1997), 325–326; G. Di Mola, ed., *Cure palliative: Approccio multidisciplinare alle malattie inguaribili* (Milan: Masson, 1994); N. Cellini et al., "Unità di Cura Continuativa: esperienza in tema di assistenza domiciliare a pazienti oncologici in fase avanzata," *Anziani Oggi* 1 (1996): 59–65; A. G. Spagnolo and D. Sacchini, "Etica della gestione delle risorse per malattie terminali," *Quaderni di cure palliative* 5, no. 3 (1997): 231–236.

[90] N. Cellini, presentation to *Terapie palliative e cure di supporto in oncologia*, by A. Ciabattoni and M. Pittiruti (Rome: SEU, 1996), xi.

[91] Ibid., ix.

palliative care,' which seek to make suffering more bearable in the final stages of illness and to ensure that the patient is supported and accompanied in his or her ordeal." [92]

The implementation of this strategy has given rise to the hospice movement, especially in the English-speaking world, and in Italy to various initiatives for home care that make it possible to provide palliative care even at home, with the collaboration of psychologists, pastors, and volunteers who support the family. [93]

The refusal of therapeutic obstinacy and dysthanasia

The last of the four guidelines from the declaration *Iura et bona* mentions the refusal to cease treatment, or so-called *therapeutic obstinacy*, which goes to the opposite extreme of causing a painful death by trying to prolong life at all costs. In order to define this concept is it necessary to recall first of all the criteria for determining death.

We know that the problem of determination of death is the topic of various international charters that set the parameters within which the physician can sign a death certificate. The 1968 Geneva Charter defines the "state of death" as when the following indications are all observed (present together): the cessation of any sign of interaction, the absence of spontaneous breathing, muscular atony and absence of reflexes, drop in blood pressure as soon as it is no longer pharmacologically sustained, and absence of brain waves on an electroencephalogram (EEG). As explained in the chapter on transplantation, thanks to advances in both the neurological sciences and diagnostic techniques, it is now recognized almost universally that the clinical death of an individual can be licitly determined upon certifying the state of *complete brain death*, i.e., encephalic death. In this area of medical morality, nevertheless, it will be helpful to go a step further and consider several delicate cases of patients in a coma while citing documents by several bishops' conferences, especially with regard to what was declared in the document by the Secretariat of the French Bishops Conference:

a. In the case of a coma that is considered "reversible," there is an obligation to use all available means, because the possible or probable recovery of a human life is worth any sort of economic sacrifice. This seems all the more necessary inasmuch as the comatose patient cannot express himself and give consent; therefore his relatives and the medical team are obliged to do everything possible to resuscitate the patient, even using extraordinary means if they are available.

b. When the coma appears in the opinion of experts to be "irreversible," there is still an obligation to provide ordinary care (including, as noted, hydration and parenteral nutrition), but one is not obliged to employ means that are particularly debilitating or burdensome for the patient, thus condemning him to prolongation of an agony experienced under conditions that rule out any possibility of regaining consciousness or the capacity for interaction. In that case it would be a case of undue *therapeutic obstinacy*. It is not easy to determine the irreversibility of a coma or assess a patient's chances of ever

---

[92] John Paul II, *Evangelium vitae*, no. 65.

[93] N. Cellini, "Problemi etici dell'assistenza al morente," in *Corso di Bioetica*, ed. E. Sgreccia (Milan: F. Angeli, 1986), 153–163; M. Petrini, *Accanto al morente* (Milan: Vita e Pensiero, 1990); Cellini et al., "Unità di Cura Continuativa."

regaining consciousness; these judgments must be left to the evaluation of competent and conscientious health care professionals.

c. To prolong life that is merely apparent and totally artificial, after brain function has completely and irreversibly ceased (as indicated by a flat electroencephalogram [EEG] and signs of death in all sectors of the brain), would be an offense to the dying person and to his death, as well as a deception at the expense of his relatives.[94]

A flat EEG in repeated tests over a certain time is a sign of the irreversible failure of cortical functions. As already mentioned in the chapter on transplants, in order to obtain organs for transplantation, Italian law[95] requires that at least six hours elapse with a flat EEG. The other above-mentioned parameters relative to whole brain death must be considered together with this sign.

But the term "therapeutic obstinacy," which sometimes seems deliberately overdramatized, applies not only when technical measures are employed on someone who is practically dead, in other words after brain death, but also when intervention with medical or surgical treatments (except for ordinary treatments) is done in a way that is "disproportionate" to the foreseeable effects.

It must be recognized that, despite these detailed guidelines, there are some cases not only of profound and irreversible coma but also of *prolonged* coma: the patient continues to remain in a coma *even with ordinary care* alone. There have been cases in which this irreversible comatose state, of apparently biological life alone, has lasted for several months or years (persistent vegetative state). Such perhaps was the situation of Karen Ann Quinlan, the young American woman who was in the news for about ten years. A similar and famous case involved the young Nancy Cruzan, who received nutrition for about eight years by means of a feeding tube while in a persistent vegetative state; after various court rulings it was decided to discontinue feeding—causing her death around ten days later—on the assumption, supported by testimonies, that this would have been her decision.[96] Similarly, the death of Terri Schiavo occurred in 1995 at the behest of her husband and through a decision of the Supreme Court of Florida; she had been in a vegetative state since 1990.

The question arises, Are we dealing with real life when there is almost complete certainty as to the irreversibility of the coma, the state of unconsciousness, and the absence of interaction, and when the EEG has been flat for the appropriate number of hours despite the persistence of certain vital functions? The document by the French bishops affirms that it should not be considered euthanasia if all treatments were stopped in certain cases of extended coma, because it would not be hastening but rather verifying death.[97] The same document, however, states that this is a theoretical problem, because even after weeks of observation it is still difficult for physicians to define a coma as irreversible and hence to declare the impossibility of recovering consciousness and interacting. Therefore, in practice, it seems difficult to

---

[94] Secretariat of the French Bishops Conference, "Problemi etici," 42–44; Manni, "Considerazioni mediche," 112–114.

[95] Law no. 578 of December 29, 1993, and ministerial decree no. 582 of August 22, 1994.

[96] See A. Puca, "Il caso di Nancy Beth Cruzan," *Medicina e Morale* 42, no. 5 (1992): 911–932.

[97] Secretariat of the French Bishops Conference, "Problemi etici," 43–44.

justify the withdrawal of ordinary care, even in similar cases that are truly saddening. It is necessary to keep in mind that there is one single existential and personal act in a human being that sustains vegetative, sensitive, and intellectual-relational life. It is therefore my opinion that it is not possible to introduce a distinction between "human life," understood as biological life, and "personal life," understood as a life of interaction with one's surroundings.

Some indications regarding the vegetative state (VS) are relevant here. A congress organized by the Pontifical Academy for Life and the World Federation of Catholic Medical Associations in 2004 took a position on this topic, which will be summarized here by citing some of the more important passages from their final document:

> Vegetative state (VS) is a state of unresponsiveness, currently defined as a condition marked by: a state of vigilance, some alternation of sleep/wake cycles, absence of signs of awareness of self and surroundings, lack of behavioural responses to stimuli from the environment, maintenance of autonomic and other brain functions.

> VS must be clearly distinguished from: encephalic death, coma, "locked-in" syndrome, minimally conscious state. VS cannot be simply equated with cortical death either, considering that in VS patients islands of cortical tissue which may even be quite large can continue functioning.

> In general, VS patients do not require any technological support in order to maintain their vital functions.

> VS patients cannot in any way be considered terminal patients, since their condition can be stable and enduring. ...

> No single investigation method available today allows us to predict, in individual cases, who will recover and who will not among VS patients.[98]

In ethical and deontological terms, the document makes the following affirmations:

> We acknowledge that every human being has the dignity of a human person, without any discrimination based on race, culture, religion, health conditions or socio-economic conditions. Such a dignity, based on human nature itself, is a permanent and intangible value that cannot depend on specific circumstances of life and cannot be subordinated to anyone's judgment.

> We recognize the search for the best possible quality of life for every human being as an intrinsic duty of medicine and society, but we believe that it cannot and must not be the ultimate criterion used to judge the value of a human being's life.

---

[98] Pontifical Academy for Life (PAV) and World Federation of Catholic Medical Associations (WFCMA), "Joint Statement on the Vegetative State," nos. 1–4 and 7, International Congress on Life-Sustaining Treatments and Vegetative State: Scientific Advances and Ethical Dilemmas, March 17–20, 2004, http://www.vatican.va/roman_curia/pontifical_academies/acdlife/documents/rc_pont-acd_life_doc_20040320_joint-statement-veget-state_en.html.

We acknowledge that the dignity of every person can also be expressed in the practice of autonomous choices; however, personal autonomy can never justify decisions or actions against one's own life or that of others: in fact, the exercise of freedom is impossible outside of life.

Based on these premises, we feel the duty to state that VS patients are human persons, and as such, they need to be fully respected in their fundamental rights. The first of these rights is the right to life and to the safeguard of health. In particular, VS patients have the right to

- Correct and thorough diagnostic evaluation, in order to avoid possible mistakes and to orient rehabilitation in the best way;
- Basic care, including hydration, nutrition, warming and personal hygiene;
- Prevention of possible complications and monitoring for any possible signs of recovery;
- Adequate rehabilitative processes, prolonged in time, favouring the recovery and maintenance of all progress achieved;
- Be treated as any other patients with reference to general assistance and affective relationships.

This requires that any decision of abandonment based on a probability judgment be discouraged, considering the insufficiency and unreliability of prognostic criteria available to date. The possible decision of withdrawing nutrition and hydration, necessarily administered to VS patients in an assisted way, is followed inevitably by the patients' death as a direct consequence. Therefore, it has to be considered a genuine act of euthanasia by omission, which is morally unacceptable.

At the same time, we refuse any form of therapeutic obstinacy in the context of resuscitation, which can be a substantial cause of post-anoxic VS.

To the rights of VS patients corresponds the duty of health workers, institutions and societies in general to guarantee what is needed for their safeguard, and the allocation of sufficient financial resources and the promotion of scientific research aimed to the understanding of cerebral physiopathology and of the mechanisms on which the plasticity of the central nervous system is based.

Particular attention has to be paid to families having one of their members affected by VS. We are sincerely close to their daily suffering, and we reaffirm their right to obtain help from all health workers and full human, psychological and financial support, which enables them to overcome isolation and feel part of a network of human solidarity.[99]

---

[99] Ibid., nos. 9–12.

The substance of these ethical positions was taken up in the address by John Paul II to congress participants,[100] and the acts of the congress were published in the journal *NeuroRehabilitation*.[101]

The use of analgesics

This topic was partly covered in the discussion of palliative care, but there are some specific ethical aspects of it to consider. One problem in particular found an ethical solution in the teaching of Pope Pius XII, which was essentially confirmed by the CDF declaration *Iura et bona*: the use of analgesics or painkillers is licit even when it could involve the risk of shortening life, provided there is no other means of alleviating the pain; the use of painkillers that render the patient unconscious is also licit, provided he has had the time to fulfill his religious and moral duties toward himself, his family, and society: "It is not right to deprive the dying person of consciousness without a serious reason." [102]

The CDF document offers further clarification by insisting on the patient's consent: he could legitimately refuse the use of analgesics, as a whole or in part, in order to be able to lend his own pain the fuller meaning of "sharing in Christ's Passion" and of "union with the redeeming sacrifice which He offered in obedience to the Father's will." [103]

It should be added that this problem, concerning the consequences attendant upon the use of painkillers, is perceived as less dramatic today because medical science, especially in the field of cancer care, has made remarkable progress and now offers regimens of pain management that limit or even eliminate such adverse consequences, especially the loss of consciousness. On the other hand one must avoid the deliberate commission of euthanasia, properly speaking, through massive doses of painkillers (opiates). The dose of analgesics must therefore be proportionate to the pain.[104]

Telling the truth to a terminally ill patient

This problem is not treated explicitly in *Iura et bona*; however, it is implicitly presupposed in this and other documents of the Magisterium whenever the patient's consent is referenced: such consent requires that the patient in question be informed about his real situation. The problem is considered more explicitly in the document by the Secretariat of the French Bishops Conference and in the declaration by the

---

[100] John Paul II, Address to Participants in the International Congress on Life-Sustaining Treatment and Vegetative State: Scientific Advances and Ethical Dilemmas, March 20, 1994.

[101] Pontifical Academy for Life and World Federation of Catholic Medical Associations, "Life-Sustaining Treatments in Vegetative State: Scientific Advances and Ethical Dilemmas," ed. G. L. Gigli and N. D. Zasler, *NeuroRehabilitation* 19, no. 4 (2004).

[102] CDF, *Iura et bona*, III, citing Pius XII, Address to the Collegium Internationale, September 9, 1958.

[103] CDF, *Iura et bona*, III; Pius XII, Response to the Society of Anethesiology, February 24, 1957; idem, Address to the Collegium Internationale, September 9, 1958.

[104] See A. G. Spagnolo, "Ai confini tra atteggiamento eutanasico e terapia palliativa," *Quaderni di cure palliative* 1 (1994): 49–51.

German Bishops Conference on November 20, 1978;[105] moreover, this problem is widely debated in the field of medical ethics.

The following is a summary of the ethical criteria that may be applicable:

1. The truth remains a fundamental requirement in order for a moral act to be objectively positive. It is therefore necessary to avoid deliberately misleading behavior on the part of relatives and caregivers. Systematic falsehood does not benefit the patient, who has the right to information as well as the right to prepare for a dignified death, and it may become useless and counterproductive when—as readily and frequently happens—the patient comes to discover it anyway. Published studies confirm that, when the truth has been opportunely offered and accepted, it leads to a psychologically and spiritually positive response on the part of both patients and family members.[106] The right to information is therefore included in all proposed lists of patients' rights and should be considered, albeit with certain qualifications, in dealing with terminally ill patients as well.[107]

2. In the case of seriously or terminally ill patients, this information must be offered within the context of broader and interpersonal "human communication," which is not limited to providing diagnostic data and prognoses for the condition. There is an obligation, above all, to *listen to the patient*; after this, it will be possible to talk to the sick person about the seriousness of his condition. What the sick—and especially the dying—person is looking for from the caregiver is solidarity, not being left alone, being able to communicate, and feeling a sense of sharing;

3. While lying should not be taken as an approach and one's goal should always be to communicate the truth, it is still necessary to recall that the truth to be conveyed must be *commensurate with the subject's capacity* to receive it in a healthy manner. It is therefore necessary to prepare a suitable state of mind: one should know the various psychological phases of a dying person (especially of a cancer patient) so as not to aggravate the depressive phases; one should also present the facts gradually and, if necessary, know how to stop at the right moment. It is never necessary to rule out all hope, given the reality that there are no absolutely certain predictions in medicine. In this regard, the 2006 revision of the Italian *Codice di deontologia medica* states the following in article 33:

> The doctor must provide the patient with the most appropriate information about the diagnosis, the prognosis, the prospects and possible diagnostic/therapeutic alternatives, and the foreseeable consequences of the choices made.
>
> The doctor must take into account the subject's ability to comprehend when communicating with him, in order to promote the highest possible

---

[105] See Spinsanti, *Umanizzare la malattia*, 59–60 and 94–96; Pontifical Council Cor Unum, "Ethical Questions."

[106] R. Zorza and V. Zorza, *Un modo di morire*, 2nd ed. (Rome: Paoline, 1983).

[107] R. Ziglioli, "I diritti del malato," *Anime e Corpi* 109 (1983): 481–498.

participation in decision making and compliance with the diagnosis and treatment proposals. ...

Information regarding prognoses that are serious, bleak, or otherwise sufficient to cause the person concern and suffering, should be provided prudently, with terminology that is not traumatizing and without ruling out hope.[108]

4. While keeping in mind the preceding methodological norm, I think we should emphasize the obligation not to conceal the essential gravity of the situation, especially when the patient has the duty to make important decisions before his death and has, like any one of us, the duty and right to prepare for a good death.

Yet these directives, which have an ethical character, are also part of the broader topic of care for the sick and dying. Even from a purely deontological viewpoint, in addition to the ethical perspective, physicians are appointed not only to administer treatments to patients but also to care for the dying.

Review of Some Deontological and Legal Documents

Since a physician has both his own conscience, which is presumably well formed, and deontological codes and guidelines (whether international or national) as standards for his behavior, it seems necessary to examine some of these documents from the ethical perspective. This will help also provide further clarifications with regard to the subject of euthanasia itself.

It is obvious that these codes and guidelines show signs of contemporary cultural debates and by nature suggest general courses of action into which it is not always possible to fit the individual case; therefore the decision ultimately remains up to the ethically formed and informed conscience.[109]

Recommendation of the Council of Europe

We will begin by examining Recommendation 779/1976 of the Parliamentary Assembly of the Council of Europe on the rights of the sick and dying.[110]

The first six articles, after recalling the reasons that prompted the assembly the draw up the recommendation (continual advances in medicine can create problems for and conceal threats to fundamental human rights and the dignity and integrity of the patient), state the rights of the sick: the right to respect for the patient's will

---

[108] Italian National Federation of Medical and Dental Associations, *Codice di deontologia medica*, rev. ed., December 16, 2006, art. 33, p. 9, http://portale.fnomceo.it/PortaleFnomceo/showVoceMenu.2puntOT?id=5.

[109] For an overview of these texts, see Lener, "Sul diritto dei malati"; Giusti, *L'eutanasia*; L. Ciccone, *Non uccidere: Questioni di morale della vita fisica* (Milan: Ares, 1984), 1:257–294; A. Bompiani, "Eutanasia e diritti del malato 'in fase terminale': Considerazioni giuridiche," *Presenza Pastorale* 5, no. 6 (1985): 76–119.

[110] Council of Europe, Recommendation 779, "On the Rights of the Sick and Dying," January 21, 1976, http://assembly.coe.int/main.asp?Link=/documents/adoptedtext/ta76/erec779.htm.

concerning what treatment to apply, the right to dignity and integrity, the right to information, the right to adequate care, and the right to avoid undue suffering.

Article 7 rules out euthanasia with these words: "Considering that the doctor must make every effort to alleviate suffering, and that he has no right, even in cases which appear to him to be desperate, intentionally to hasten the natural course of death." Nevertheless, the document means to underscore that "the prolongation of life by artificial means depends to a large extent on factors such as the availability of efficient equipment, and that doctors working in hospitals where the technical equipment permits a particularly long prolongation of life are often in a delicate position as far as the continuation of the treatment is concerned, especially in cases where all cerebral functions of a person have irreversibly ceased" (article 8). It also highlights the fact that "doctors shall act in accordance with science and approved medical experience, and that no doctor or other member of the medical profession may be compelled to act contrary to the dictates of his own conscience in relation to the right of the sick not to suffer unduly" (article 9).

In these last articles it is quite plain that the recommendation is positing the premises of facts (when the cessation of a person's brain functions is irreversible: this is the case of an irreversible coma), of rights, and of competences (the physician cannot be forced to act against his conscience when he means to safeguard the sick person's right to avoid undue suffering). This is done to accommodate the instances of suspending treatment that are spelled out in section 1 of the following article, which states that the assembly

> recommends that the Committee of Ministers invite the governments of the member states:
>
> I. *a.* to take all necessary action, particularly with respect to the training of medical personnel and the organisation of medical services, to ensure that all sick persons, whether in hospital or in their own homes, receive relief of their suffering as effective as the current state of medical knowledge permits;
>
> *b.* to impress upon doctors that the sick have a right to full information, if they request it, on their illness and the proposed treatment, and to take action to see that special information is given when entering hospital as regards the routine, procedures and medical equipment of the institution;
>
> *c.* to ensure that all persons have the opportunity to prepare themselves psychologically to face the fact of death, and to provide the necessary assistance to this end.[111]

And here is the text on which the debate and the problem hinge: in section II of the same article the invitation to governments is further specified to mean that each individual state should decide to

> establish national commissions of enquiry, composed of representatives of all levels of the medical profession, lawyers, moral theologians, psychologists and sociologists, to establish ethical rules for the treatment of persons approaching the end of life, and to determine the medical guiding principles for the application of extraordinary measures to prolong life, thereby considering inter alia

---

[111] Ibid., art. 10.

the situation which may confront members of the medical profession, such as legal sanctions, whether civil or penal, when they have refrained from effecting artificial measures to prolong the death process in the case of terminal patients whose lives cannot be saved by present-day medicine, or have taken positive measures whose primary intention was to relieve suffering in such patients and which could have a subsidiary effect on the process of dying, and to examine the question of written declarations made by legally competent persons, authorising doctors to abstain from life-prolonging measures, in particular in the case of irreversible cessation of brain function.

The rest of the article (sections 2 and 3) concerns the establishment of national commissions responsible for examining complaints made against physicians and an invitation to communicate analysis results and conclusions drawn by said commissions to the Council of Europe.

Practically speaking, the recommendation frames the problem of refusing treatments that would prolong life to the bitter end by dividing it into three distinct cases: (a) the case of a patient not yet in the throes of death; (b) the case determined by the use of so-called *invasive analgesia*; (c) the case of a written declaration by a patient along the lines of a living will, which originated in California.

Since the recommendation identifies points that need clarification in relation to charges that could be brought against physicians but does not define behavior in concrete cases, it cannot be said that it expresses a favorable opinion toward interrupting the means of prolonging life in these cases. It should be observed, however, that while case (a) talks about refusing the use of "artificial means" for postponing death in patients in irreversible conditions where death has begun or is imminent, point (c) simply talks about refusing means of prolonging life "in particular [*author's note*: therefore not only?] in the case of irreversible cessation of brain function"[112]—a refusal motivated by the patient's written will.

Even the use of analgesics mentioned in point (b), which could contribute to the patient's death as a concomitant and indirect effect, can prove to be ambiguous and lend itself to abuses. The analgesic must be "proportionate" to the mitigation and tolerability of the pain and is licit only in that case; moreover, one must choose from among the suitable analgesics the one with the fewest risks of shortening life. The point concerning the patient's written declaration or *living will* requires examining the term in the law of the state of California, which was later adopted by other states.

In conclusion, the recommendation notes the onset of problems caused by the principle of the living will, which ends up limiting medical action to sustain the dying person.

Advance directives

On December 18, 2003, the Italian National Bioethics Committee (NBC) issued a document on advance directives titled *Advanced Treatment Statements*. The introduction states the following:

> This document concerns advance treatment directives, the importance of which has grown constantly in recent years and which are frequently referred to in the national and international literature on bioethics with the

---

[112] Ibid., 10(II).

expression *living will*. Other expressions used include biological will, life will, prior treatment wishes, etc. These expressions all refer to a document through which individuals, in full possession of their mental faculties, express their wishes regarding the treatments that they would or would not want to undergo to [*sic*] in the event that, in the course of an illness or as a result of sudden trauma, they were no longer able to express their informed consent or dissent. The different forms that an advance treatment directive can take (some of which have been legally recognised in a number of countries) have been discussed in the literature.

To give these documents public (although not necessarily legal) status, they are required to be drawn up in writing in such a way that there can be no doubt as to the identity and capacity of the person signing them, their authenticity and the date of signing. They should if possible be countersigned by a physician, who should guarantee that the signer has been fully informed of the possible consequences of the decisions he is taking in the document. It is desirable for the signer to indicate a date for the confirmation and/or renewal of the directive, without prejudice to their right at any time to withdraw or amend the instructions it contains.[113]

The ethical and legal implications of the topic raise many questions.[114] First of all, how can one avoid the inevitably abstract character of these living wills in comparison with the real situation of sickness in which they would have to be applied? What legal status is to be attributed to a potential "proxy" who is expected to act according to the instructions contained in the "will" and in the exclusive interest of the incapacitated person? How can one resolve possible contradictions between living wills, positive law, and the norms of good clinical practice and medical ethics? Can the withdrawal of artificial nutrition and hydration be considered a simple suspension of treatment? Do such directives have a binding character, or are they simply guides for health care personnel? Can advance directives promote a bureaucratic acceleration of death?

The NBC expressed itself on the subject in this manner: "The NBC deems it essential to eliminate any ambiguity and emphasises that the right being proposed—for patients to influence the treatment to which they might be subjected in the event of their being considered incompetent—is not a right to euthanasia ... It is, rather, a right solely to ask physicians to interrupt or not to undertake therapeutic actions."[115]

As Adriano Pessina rightly observes, if the purpose of the advance directive or living will is not to introduce euthanasia but to prevent therapeutic obstinacy, then we must ask why that purpose should have to be enacted by choice—by an explicit request—and why it could not be established as a norm of medical practice.[116]

---

[113] Italian National Bioethics Committee, *Advanced Treatment Statements*, December 18, 2003 (Rome: Presidenza del Consiglio Dipartimento per l'Informazione e l'Editoria, 2003), 2, available at http://www.governo.it/bioetica/eng/opinions/advance_treatment_directives.pdf.

[114] See F. Turoldo, ed., *Le dichiarazioni anticipate di trattamento: Un testamento per la vita* (Padua: Fondazione Lanza / Gregoriana Libreria Editrice, 2006).

[115] Italian National Bioethics Committee, *Advanced Treatment Directives*, 12.

[116] See A. Pessina, *Bioetica: L'uomo sperimentale* (Milan: Bruno Mondadori, 1999), 153.

Principles of European Medical Ethics

The document titled *Principles of European Medical Ethics* was approved in Paris by the International Conference of Medical Professional Associations on January 6, 1987,[117] modifying the preceding text that had been issued in Brussels with the title *European Guide for Medical Ethics*, and amended on February 6, 1995. The original *Principles* comprise thirty-seven articles that are meant to offer physicians in the European Community deontological guidelines for solving problems such as independence from political authorities, professional confidentiality, organ transplantation, human reproduction, abortion and conscientious objection, human experimentation, assistance to the dying, and so on.

The article arousing the greatest interest and controversy is the twelfth, which to some seemed open to a form of euthanasia improperly described as "passive." The article is worded as follows: "In all circumstances, medicine implies constant respect for life, moral autonomy and the patient's free choice. However, in the case of incurable and terminal conditions, the doctor may limit himself or herself to relieving the physical and moral suffering of the patient by giving appropriate treatment and by maintaining, as far as is possible, the quality of a life nearing its end. It is essential to assist a dying person until the end and to act in such a way as to maintain that person's dignity."

The expression "passive euthanasia" does not appear in the text, and the principle of respect for life "in all circumstances"—therefore in the terminal phase as well—does not allow room for an interpretation that would authorize terminating a life.

The controversial point, however, is the second part of the first sentence and the explanation provided in the following statement; it concerns the physician's obligation to respect the patient's "moral autonomy" and "free choice" and the resulting authorization for the physician to limit his or her intervention, in the terminal phase of an incurable illness, to alleviating the physical and moral sufferings of the patient with treatments proportionate to that purpose. The question is whether these guidelines create the expectation that the physician will dutifully accompany the patient to a dignified death or whether they pave the way for the concept of passive euthanasia with the patient's agreement.

Examining the text and keeping to the literal sense, the following conclusions can be drawn:

- The first sentence of article 12 states a principle that has the weight of a premise: "In all circumstances, medicine implies constant respect for life, moral autonomy and the patient's free choice." One can only agree with and celebrate the first part of the sentence, which refers to "constant respect for life" in all circumstances.

- The second part of the phrase is incomplete in my opinion; it is true but insufficient and requires further clarification. It is clear that the patient's

---

[117] International Conference of Medical Professional Associations, *Principles of European Medical Ethics*, January 6, 1987, amended February 6, 1995, http://www.ceom -ecmo.eu/sites/default/files/documents/european_medical_ethics_principles-1987–1995 _ceom_cio.pdf.

moral autonomy and free choice must be respected, but there is also an obligation for the patient to respect his own life and to respect the moral autonomy of the physician as well. In other words, this part of the sentence must be subordinated to the first part and not simply juxtaposed with it as though equivalent.

It is clear that the patient cannot impose on the physician a decision to refuse legitimate treatments and thereby hasten death.

The relation of freedom and responsibility between patient and physician certainly must not be understood to mean that the physician substitutes for the patient's responsibility (medical paternalism), nor that the physician must simply execute the patient's will. The relationship must therefore be triangular: at the vertex is constant respect for life, and this respect must be valid both for the patient and the physician. This constant respect for life must not be understood to mean simply that one must not perform "active" euthanasia, whereas all other matters can be regulated by the patient's free choice, because that attitude would be tantamount to admitting the possibility of passive abstention even from efficacious and proportionate treatments whenever the sick person refused them, and this would in effect suggest passive euthanasia.

It must be added that the following sentence seems to rule out this pessimistic interpretation favorable to euthanasia: "However, in the case of incurable and terminal conditions, the doctor may limit himself or herself to relieving the physical and moral suffering of the patient by giving appropriate treatment and by maintaining, as far as is possible, the quality of a life nearing its end." The article then concludes, "It is essential to assist a dying person until the end and to act in such a way as to maintain that person's dignity."

We are therefore dealing with the terminal stage and irreversible course of a pathology, and the physician is authorized to limit his intervention to the commitment of alleviating physical and moral suffering with suitable means. These "proportionate treatments" ought to include appropriate pain management and normal care as well as human assistance. On these conditions, the commitment could be considered fulfilled.

Yet, in this second sentence of article 12, it seems there is not much emphasis on the active use of proportionate therapeutic means to safeguard respect for life, leaving room for the patient's moral autonomy and free choice as enunciated in the first part.

In conclusion, it can be said that this text, which is neither very complete nor perfectly consistent in terms of its internal logic, could lead the reader—regardless of its true intentions—to interpret "in all circumstances ... respect for life" as a rejection of active euthanasia alone, meaning that the physician can and must be guided by the free choice of the patient with regard to the remaining treatments and, obviously, must limit himself to relieving physical and moral suffering in the terminal phase. This interpretation would truly diminish or eliminate the duty to administer and receive effective and proportionate treatments and would constitute a subtle opening toward euthanasia by omission.

Another reason for the interpretive difficulties presented by the text is the fact that the Plenary Assembly of the Standing Committee of Doctors of the European Economic Community, held in Berlin on November 20–21, 1987, approved the amendment proposed by the Italian delegation, which reads, "Every act aimed at deliberately causing the death of a patient is contrary to medical ethics,"[118] thereby clarifying the meaning of article 12 of the *Principles of European Medical Ethics*.

A proposed resolution in the European Parliament

On April 30, 1991, a resolution was proposed to the European Parliament by the Committee on the Environment, Public Health and Consumer Protection concerning care for terminal patients (presented by the Honorable Léon Schwartzenberg). This proposal explicitly requested that "whenever a fully conscious sick person insistently and continuously asks to end a life that for him is now deprived of all dignity and a panel of physicians established for this purpose ascertains that it not possible to administer specific new treatments, this request must be fulfilled" (article 8).

It is clear that on the deontological level this item in the proposal represents truly active euthanasia, properly speaking, and not just the rejection of therapeutic obstinacy. This conflicts with the previously mentioned *Principles of European Medical Ethics*, with many codes of medical ethics, and also with human rights and the constitutions of many nations.

But even on the anthropological level there are many problems with affirming that the foundation of human life is its spirituality (preamble A): in effect, bodily life is not considered a fundamental value of human life, which is to say that the status of corporality is ignored and denied. Moreover, the concept of "natural functions" cannot be identified solely with the concept of "functions of vegetative life." Furthermore, by affirming in preamble C that "the cerebral functions determine the level of consciousness and that the level of consciousness defines a human being," the proposed resolution confuses the distinctive characteristics of the human being—rationality and consciousness—with their exercise, i.e., the present and acting capacity of reasoning and consciousness. The logical consequence of this would be that where the exercise of rationality and consciousness does not exist or no longer exists, there would be no individual human being to protect and no human life to defend. Both preambles disagree with other previous recommendations issued by the Council of Europe concerning the terminally ill (for example, point 8 of the Recommendation 779/1976).

Finally, on the medical and neurological level as well, the combination of preamble B ("considering that the death of an individual is defined in terms of the cessation of cerebral functions, even when biological functions continue") and preamble C intentionally confuses *cerebral* functions and *cortical* functions. It must be reiterated clearly that total brain death corresponds to the death of the individual, a criterion which has been accepted by all the reports of the international ad hoc committees since the Harvard Committee.

Based on these objections, therefore, the proposal cannot be accepted, which is in fact what the European Parliament subsequently declared. In a document of its own, the Italian National Bioethics Committee had also expressed its negative assessment of the proposed recommendation.[119]

---

[118] "Il Medico d'Italia," December 8, 1987.

[119] Italian National Bioethics Committee, *Parere sulla proposta di risoluzione sull'assistenza ai pazienti terminali* [*Opinion on the resolution proposal concerning assistance to terminally ill patients*], September 6, 1991 (Rome: Presidenza del Consiglio

Initiatives for the legalization of euthanasia

We will now take into consideration several initiatives, in various countries, that align themselves with the push for legalized euthanasia.

In the United States, the Natural Death Act, approved by the state of California and later extended in virtually identical terms to other states, goes back to 1976. The specific provisions of the law recognize the right of every adult to arrange for the non-application and interruption of "life-sustaining procedures," should he find himself in an "end-stage condition."

The living will must be signed by the author in the presence of two witnesses, who must not be related to him by ties of blood or affinity, nor be heirs to his estate, nor the treating physician or his subordinate, nor an employee of the health care facility.[120] The living will, put into writing in a very precise format, also provides for its own invalidation in the event that the patient is pregnant. The document is considered valid for five years.

An "end-stage condition" means the final phase of an incurable condition, in which the application of treatments might postpone death but could not restore life. "Life-sustaining procedures" include any medical means or intervention that uses mechanical or artificial devices in order to sustain, reactivate, or replace a natural vital function and that, if applied, would serve only to postpone the time of death. The patient must already have received an unfavorable prognosis signed by two physicians.

It should be noted that, at first glance, this protocol might correspond to the one described in the declaration on euthanasia by the Congregation for the Doctrine of the Faith, where it states, "It is also permissible to make do with the normal means that medicine can offer. Therefore one cannot impose on anyone the obligation to have recourse to a technique which is already in use but which carries a risk or is burdensome."[121] It might also correspond to the preceding passage, which states, "It is permitted, with the patient's consent, to have recourse to the means provided by the most advanced medical techniques, even if these means are still at the experimental stage and are not without a certain risk. ... But for such a decision to be made, account will have to be taken of the reasonable wishes of the patient and the patient's family, as also of the advice of the doctors who are specially competent in the matter."[122]

The U.S. Catholic Health Association, as early as 1974, had also issued a document ("Christian Affirmation of Life") that included this statement: "I request that, if possible, I be consulted concerning the medical procedures which might be used to prolong my life as death approaches. If I can no longer take part in decisions concerning my own future and if there is no reasonable expectation of my recovery from physical and mental disability, I request that no extraordinary means be used to

---

dei Ministri Dipartimento per l'Informazione e l'Editoria, 1991), http://www.governo.it/bioetica/pdf/5.pdf; English abstract, http://www.governo.it/bioetica/eng/opinions/abstracts/abstract_opinion_on_the_draft_resolution.pdf.

[120] Bompiani, "Eutanasia e diritti."

[121] CDF, *Iura et bona*, IV.

[122] Ibid., IV.

prolong my life."[123] Other similar initiatives were issued later, such as the "living will" recommended by the Respect Life Committee of the Spanish Bishops Conference.[124]

Nevertheless, significant procedural problems remain with the living will,[125] especially with regard to the legal and moral validity of a will and testament expressed in advance, rather than in the specific conditions of an illness, concerning a good that is not a thing but rather life itself. There is also the fundamental problem of interpreting what procedures are to be considered life-sustaining in a specific case and determining when a condition is irreversible: Do we understand "life-sustaining procedures" also to include assisted respiration, nutrition, personal hygiene, or hydration? Are these really the means that the CDF declaration is talking about, or are they the extraordinary means described in "Christian Affirmation of Life"? Furthermore is it permissible to "exempt" the physician from making his own assessment, even against the will of his patient? Under those conditions, how could the physician remain autonomous in his own conscience and in his capacity as an "intellectual service provider," whereby he himself must intelligently evaluate the means suited to assisting a terminally ill patient? One author has observed that the living will starts from the assumption that the physician wants to employ heroic measures always and at all costs.

In 1990, after the famous case of Nancy Cruzan was decided, the Patient Self-Determination Act was passed in the United States; it is meant to promote the patient's role in making decisions concerning his own life, especially in the final stage of illness.[126] A referendum in the state of Washington in 1991, however, which passed by a very small margin, gave additional cause for concern to those who oppose the legalization of euthanasia.

In Canada, the so-called "do-not-resuscitate policy" was declared ethical by the General Council of the Canadian Medical Association in 1974 and goes beyond even the limits foreseen by living wills. Practically speaking, it consists in the refusal or non-use of resuscitation techniques on patients for whom they are considered useless or burdensome, even when this involves the hastening of death.

The application of this policy varies greatly in practice: there are hospitals where the opinion of the ethics committee is required; there are physicians who agree in advance with the patient on the course of action to be taken, after explaining to him honestly his condition and prognosis; there are physicians who reserve for themselves the judgment as to the whether or not it is appropriate to use resuscitation techniques.

Clearly, in the absence of objective criteria that can be physically measured by instrumentation, and given the variety of cases, which prohibits generalizations, this

---

[123] K. O'Rourke, "The Christian Affirmation of Life," *Hospital Progress* 55 (1974): 67.

[124] On this subject see, for example, G. Perico, "Testamento biologico e malati terminali," *Aggiornamenti Sociali* 43, no. 11 (1992): 677–692; Pennsylvania Catholic Conference, *Living Will and Proxy for Health Care Decisions* (Harrisburg, PA: Pennsylvania Catholic Conference, 2004), http://www.pacatholic.org/wp-content/uploads/livingwill.pdf.

[125] See J. R. Wernow, "The Living Will," *Ethics & Medicine* 10 (1994): 27–35.

[126] For an in-depth treatment, see F. Rouse et al., "Practicing the PSDA," supplement, *Hastings Center Report* 21, no. 5 (1991): 15–165. On the topic of living wills and advance patient directives, see A. G. Spagnolo, "Il bene del paziente e i limiti dei testamenti di vita," *L'Osservatore Romano* 138 (June 17–18, 1996); idem, "Testamenti di vita," in *Bioetica in medicina*, ed. A. Bompiani (Rome: CIC Edizioni Internazionali, 1996), 340–355.

procedure presents more serious ethical difficulties. Moreover there is a broad and almost unlimited scope for subjectivity on the part of patients or physicians, both in determining the unfavorable prognosis and in applying the norm.

The Bioethics Council (Consulta di Bioetica), an Italian association with a secular orientation, also proposed a "self-determination charter" that adopts the logic of the living will.[127] Before proceeding to examine the bill proposed in the Italian Parliament, the case of Holland and the law that passed by the Dutch Parliament should be mentioned. Notoriously, separate surveys indicated that the medical practice of euthanasia was quite widespread in that country,[128] which is why the law (November 28, 2000) intended—controversially—to welcome that practice, as if it were the job of the law to legalize what already happens in practice rather than to establish dispositions that practice should follow. In effect that law does not legalize but rather depenalizes euthanasia to prevent lawsuits against physicians who, while complying with certain well-codified requirements, end the life of a patient. Nevertheless, this does not change the substance of an action that is gravely illicit on the ethical level, and it paves the way for other forms of euthanasia, including involuntary euthanasia.

The bill proposed by the Health and Hygiene Committee of the Italian Senate[129]

On July 13, 2005, the Twelfth Committee of the Senate of the Italian Republic (Health and Hygiene) approved a bill, presented by Senator Antonio Tomassini, comprising sixteen articles on regulations for informed consent and advance directives. Several newspapers described it as "the first step toward euthanasia" (*Libero*, July 15, 2005).

Nonetheless, talking about "the first step toward euthanasia" in reference to advance directives is not automatic: there can be instructions that are compatible with respect for life, even aiming to provide adequate care to the dying person in accepting death with the peacefulness hoped for. Yet it must be recalled that there is another type of "advance directive," historically and ethically speaking, inspired by the claim to a *right to end life* based on the so-called principle of autonomy, which is understood as the patient's right to autonomously decide to end a life he considers meaningless.

---

[127] For the Italian text of the "charter" and commentary on it, see Consulta di Bioetica, "Carta dell'autodeterminazione," *Notizie di Politeia* 24 (1991): 4–5. See also A. G. Spagnolo, "Il dibattito sull'eutanasia in Italia," *Le Scienze* 297 (1993): 9–10.

[128] See Dutch Committee to Study the Medical Practice concerning Euthanasia (Remmelink Committee), *Medische Beslissingen rond het Levenseinde: Rapport van de Commissie Onderzoek Medische Praktijk inzake Euthanasie* [Medical decisions about the end of life: Report of the committee to study the medical practice concerning euthanasia], 2 vols. (The Hague: SDU Publishing, 1991); a summary in English can be found in P. J. Van Der Maas et al., "Euthanasia and Other Medical Decisions concerning the End of Life," *Lancet* 338 (1991): 669–674. See also Holland, "Introduzione di una disciplina giuridica per la procedura di notifica degli interventi di eutanasia," *Medicina e Morale* 43, no. 2 (1993): 446–448.

[129] E. Sgreccia, "Le 'Disposizioni anticipate di trattamento,'" *L'Osservatore Romano*, July 25–26, 2005, p. 8.

The first observation that arises from the analysis of the document is the nearly total lack of indications regarding what the content of these declarations should be. In fact, it limits itself to describing the formal, procedural, and legal aspects (especially regarding direct and proxy consent), offering many indications regarding the guarantees that should be provided—through legal representatives or proxies—to minors or those incapable of consent, but greater specificity was hoped for in terms of content. As it was presented in this draft form, the bill showed two major shortcomings: no explicit prohibition against euthanasia and minimal consideration of the viewpoint of the physician, who is called to apply the advance directives.

The absence of such an explicit rejection of the practice of euthanasia in the text of the bill on advance directives, precisely because of the affinity and implications of its topic, could induce a surreptitious or tacit acceptance of directives aimed at ending human lives, whether actively or passively. In fact, while it explicitly elaborates an absolute condemnation of therapeutic obstinacy, which consists in using therapeutic means that are futile and devoid of benefits—both at present and later on—or gravely disproportionate, no similarly clear statements are made to rule out euthanasia.

The other shortcoming is the lack of mention of the physician's autonomy of conscience. Article 13 of the bill establishes that the physician may disregard the instructions or directives of the patient only if he finds them "outdated or inappropriate from a scientific and therapeutic perspective"; there is no mention of the physician's conscience and professional ethics, yet the 2006 *Codice di deontologia medica* (Code of medical ethics) explicitly discusses these matters in articles 37–39. Finally, it did not state whether citizens would be obliged to provide these declarations or simply urged or authorized to present them upon entering a hospital.

The Italian Code of Medical Ethics of 2006

Article 17 of the code clearly prohibits any act of euthanasia: "The physician, even upon the request of the patient, must not carry out or enable treatments whose purpose is to cause the patient's death." Article 39 provides guidelines for the care of patients with incurable conditions: "In the case of illnesses with a definitively unfavorable prognosis or having reached the terminal stage, the physician must direct his work toward actions and conduct that will spare the patient unnecessary psychophysical suffering and provide the person with treatments suitable to safeguarding, to the extent possible, the quality of life and dignity of the person. If the patient's state of consciousness is compromised, the physician must continue life support therapy as long as it is considered reasonably useful, avoiding any form of therapeutic obstinacy."

Article 16 of the code instructs the physician, "taking into account the wishes of the patient if expressed," to "abstain from obstinacy in diagnostic and therapeutic treatments offering no founded expectation of benefit to the health of the patient and/or improvement in quality of life."

At the end of this presentation of various legal or deontological documents that focus on the crucial point of whether the use of means for resuscitation or artificial life support are always necessary, I believe I can conclude by saying this: It is possible that a physician, in conscience and out of respect for life, may in certain cases judge the application of these technical means to be useless or a serious disturbance to the dying process for a patient who, let us suppose, would have to be transported from his room to the resuscitation area when at this point there is no longer hope of a recovery and he would not be helped by the treatment. On the contrary, it would actually render the process of dying more painful. Imagine a sudden heart attack in a cancer patient who has reached the limit of his remaining strength and life. But judgment not to begin resuscitation with technological means in these cases will have to be made by the physician in accordance with the criterion of proportionate therapeutic means.

I believe the case of someone who is already under resuscitation treatment is different, and the means for resuscitation can be considered life sustaining as well as pain reducing. Disengaging them could render the dying process and death itself more painful as well as hastening both. In these cases, it seems to me that the duty to use these means up to the time of clinical death should be reaffirmed.

The Present Legislative Situation Worldwide [130]

The problems surrounding euthanasia and assisted suicide have been complicated by initiatives and pronouncements based on various standards. In addition to the Dutch law briefly noted earlier, a euthanasia law was also approved in Belgium on April 10, 2001.

In the Northern Territory in Australia, the Rights of the Terminally Ill Act was approved in May 1995 and went into effect on July 1, 1996. It was the first law in the world to approve euthanasia in our time, considering it the right of a citizen in certain conditions. The tremendous debate over this law that ensued in Australia led to the passage of a federal law revoking the law of the state of the Northern Territory (by vote of the federal senate on March 24, 1997).

On June 26, 1997, the Supreme Court of the United States ruled on two decisions on this matter handed down by the Court of Appeals of the Ninth and Second Circuits. The two lower courts had annulled laws prohibiting assisted suicide in the states of Washington and New York. Both had declared that prohibition to be unconstitutional and claimed that citizens had a constitutional right to choose the manner and timing of their own death. The Supreme Court reversed those decisions in a rare unanimous ruling (9–0), declaring that the alleged right cannot be included among the rights recognized by the constitution and that the separate states therefore have the power to regulate the matter by law. Indeed, in his reasoning for the decision, Chief Justice William Rehnquist declared that state laws prohibiting assisted suicide are "expressions of the States' commitment to the protection and preservation of all human life," [131] adding that "the State's assisted suicide ban reflects and reinforces its policy that the lives of terminally ill, disabled, and elderly people must be no less valued than the lives of the young and healthy, and that a seriously disabled person's suicidal impulses should be interpreted and treated the same way as anyone else's." [132]

---

[130] P. Lettellier and G. Gambino, eds., *L'eutanasia*, vol. 1, *Aspetti etici e umani* (Rome: Sapere 2000, 2004); G. Gambino, ed., *L'eutanasia*, vol. 2, *Diritto e prassi in Italia, Europa e Stati Uniti* (Rome: Sapere 2000, 2005).

[131] U.S. Supreme Court, *Washington v. Glucksberg*, 117 S. Ct. 2258 (1997), §I, available at http://www.law.cornell.edu/supct/search/display.html?terms=assisted%20suicide&url=/supct/html/96–110.ZO.html.

[132] Ibid., §II; see International Anti-Euthanasia Task Force (IAETF), "The Courts Have Spoken: No Constitutional Right to Assisted Suicide," *IAETF Update* 11, no. 3: 2 ff.

As of 1995, more than thirty states in the United States considered assisting a suicide a criminal offense.[133] There have been different attempts to change this situation on the part of various pro-euthanasia groups and associations. After the failure of some attempts in various states, they succeeded in November of 1994 in having a law passed in the state of Oregon that allowed assisted suicide for terminally ill patients under very restricted conditions (e.g., it is forbidden to help someone to die with an injection, even though it is well known that the approved method of administering a lethal dose of pills is ineffective in many cases). Following the referendum that approved the law, a judge of the District Court of Oregon blocked the law, considering it contrary to the United States constitution, which obliges the states to protect all their citizens, even those who are terminally ill. In October 1997, the Supreme Court, in keeping with its decision in June, decided not to hear the appeal against the law approved in 1994, thereby allowing such laws to be discussed and passed in the various states. On October 27, 1997, a second referendum on assisted suicide was put to a vote. This time it had been proposed by the opponents of euthanasia, asking the citizens to vote to outlaw it. Sixty percent of those who voted declared themselves against outlawing assisted suicide, and thus the law approved by the 1994 referendum returned to the spotlight.

Finally, in a very surprising development, a ruling by the Constitutional Court of Columbia on May 20, 1997, approved euthanasia for the terminally ill, so long as they provide personal consent.[134] The implications and consequences of that ruling were not fully appreciated at the time, since it was handed down before there was any real debate on the subject in that Latin American country.

In China in 1998, the government authorized hospitals to perform euthanasia on the incurable and terminally ill.

Pediatric Euthanasia: The Gröningen Protocol[135]

Based on an agreement reached between the clinic at the University of Gröningen in Holland and the Dutch judicial authorities, the possibility of euthanasia was extended to children under the age of twelve, including newborns. The protocol established detailed procedures that physicians must follow in order to deal with the problem of seriously ill children by "freeing them from pain" and subjecting them to euthanasia.[136]

Requirements that must be met are as follows:

- The diagnosis and the prognosis must be certain.

- The patient's suffering must be unbearable and hopeless.

---

[133] R. L. Worsnop, "Assisted Suicide Controversy," *CQ Researcher* 5, no. 17 (1995): 393–416.

[134] IAETF, "Colombian Constitutional Court OK's Euthanasia," *IAETF Update* 11, no. 2 (March–May 1997): 12–13.

[135] E. Sgreccia, "L'eutanasia in Olanda."

[136] This protocol was formulated in 2002. See E. Verhagen and P.J.J. Sauer, "The Gröningen Protocol: Euthanasia in Severely Ill Newborns," *New England Journal of Medicine* 352 (2005): 959–962; S.A. Bondi, D. Gries, and K. Faucette, "Neonatal Euthanasia?" *Pediatrics* 117, no. 3 (2006): 983; G. Dworkin, R.G. Frey, and S. Bok, *Euthanasia and Physician-Assisted Suicide: For and Against* (Cambridge: Cambridge University Press, 1998).

- The diagnosis, prognosis, and unbearable suffering must be confirmed by at least one independent physician.

- Both parents must provide informed consent.

- The procedure must be performed according to the standards accepted by the medical profession.

Information necessary to support and explain the decision for euthanasia is as follows:

Diagnosis and prognosis

- Describe all the relevant medical data and the results of the diagnostic tests on which the diagnosis is based.

- List all participants in the decision making process, all opinions expressed, and the agreement reached.

- Describe how the long-term prognosis concerning the patient's health was reached.

- Describe how the degree of suffering and hope for survival were assessed.

- Describe the availability of alternative treatments, alternative methods of pain relief, or both.

- Describe the treatments and results of treatments that preceded the decision for euthanasia.

The decision for euthanasia

- Describe who initiated the discussion on the possibility of euthanasia and the circumstances of the discussion.

- List the considerations that led to the decision.

- List all participants in the decision making process, all opinions expressed, and the agreement reached.

- Describe the way in which the parents were informed and their opinions.

Consultation

- Describe the physician(s) who provided a second opinion (name, specialization).

- List the results of the visits and recommendations made by physicians or specialists.

Implementation

- Describe the actual euthanasia procedure (time, place, participants, and drugs administered).

- Describe the reasons why that method of euthanasia was selected.

After death

- Describe the assessment determined by the coroner or medical examiner.

- Describe how the case of euthanasia was reported to the public prosecutor or district attorney.

- Describe how the parents are being attended and assisted.

707

- Describe the review plan, such as the reexamination of the case, the post-mortem examination, and genetic evaluation.

What is at stake in this protocol is no longer even the principle of autonomy, but rather an "external" decision, an imposition of a conscious adult upon someone who has no autonomy. The adults are the ones who deem the children's sufferings "unbearable." This step in the Dutch euthanasia law moves decisively in the direction of "disguised eugenics."

Euthanasia as a Legal Problem [137]

Reflecting on euthanasia as a legal problem entails reflecting on the legitimacy of euthanasia legislation in principle, that is, making evaluations in terms of *justice*. According to D'Agostino,

> any case in which the possibility of recourse to euthanasia is discussed necessarily has a tragic, heart-rending, and distressing character, which deserves careful attention and deep respect; all that notwithstanding, I am convinced that dealing with similar cases by means of a generalized legal instrument, that of the *law*, and in particular of a *permissive* law, is tantamount to giving a *wrong* answer to *real* problems. The reason for this statement can be summarized quickly: problems involving euthanasia are always *extreme* problems, but the law is not suited to resolving such problems. The law exists to govern and regulate ordinary situations, not exceptional situations; much less is it designed to socially regulate a value like *compassion*, which in itself is highly noble yet cannot be reduced to any sort of legal formula. And situations involving euthanasia are typically exceptional situations calling for compassion, because everyone is endowed with an individual, unrepeatable emotional profile that cannot be compared to any other.[138]

It is first of all necessary to make a linguistic and conceptual clarification. *Euthanasia, from the legal perspective, is not the refusal of treatments made by a competent patient*, even though such a refusal could be the equivalent of suicide in terms of a moral analysis. In no modern legal system can a competent patient be coercively subjected to a treatment (except for reasons of public health).

- *Euthanasia is not the abandonment of therapeutic obstinacy*, that is, of all medical practices that are not proportionate to the patient's real health care situation.

- *Euthanasia is not equivalent to palliative care*, which is capable of controlling pain in terminally ill patients and assuring them a more than acceptable quality of life, though in some cases the use of certain substances in palliative care produces the double effect of sedating the patient's pain yet hastening his death.

- *The legal discussion of euthanasia should not include eugenic euthanasia*, that is, the killing of deformed newborns shortly after birth or infanticide

---

[137] See F. D'Agostino, "Eutanasia," in *Parole di bioetica* (Turin: Giappichelli, 2004), 77–88.

[138] Ibid., 77–78.

committed upon handicapped newborn children, which is connected with the problems of how to cope with the handicap.

Having eliminated these forms of pseudo-euthanasia, how do we define euthanasia in the true and proper sense? "It is euthanasia *if and only if* a physician puts an end to the life of a terminally ill patient at his request."[139]

If we analyze the Dutch experience, we find that the law defines *euthanasia* as the killing of a patient *at his explicit request*, but the numbers show that in approximately 40 percent of cases the patient's life is ended *without his explicit request*. Those cases are justified hypocritically by the fact that there are situations in which all vital functions begin to be impaired irreversibly, and hence there can be no explicit request by the patient.

Euthanasia is always proposed as a humanitarian solution in terrible cases where the patients are the victims of agonizing pain. Palliative care is systematically ignored. The argument inevitably drifts from *physical pain* to *existential suffering*, and consequently "aid in dying" is given even to a woman who is lonely and depressed but not physically sick.

In conclusion, it is necessary to recall that "anyone who asks to die is really asking for help to live *differently*: sometimes it is enough to learn that one's pain is shared by others in order to cope with it positively (and therefore rigorous empirical studies recommend that terminal patients in hospitals *not be isolated*, because just the experience of a ward in which all the patients are suffering in a similar way is capable of arousing in each of them an incredible ability to resist depression and pain). Selling lethal pills in a pharmacy is an insidious way of inviting the elderly not to burden their families or public assistance programs with all their needs."[140]

Euthanasia and Suicide

It was not our intention to deal at length with the topic of suicide because it is not part of medical practice, except for the obligation to care for those who have attempted suicide and the activities of forensic medicine related to determining the cause of death. The topic belongs more specifically to the ethical aspects of social life and psychiatry, to which ample and appropriate attention should be given.[141]

Here I simply wish to explain that from an objective ethical perspective, suicide must be condemned because it is contrary to the inviolability of life, to the dignity of the person, and also to respect for the goodness of the Creator. There is, however, the complex and delicate problem of assessing both the actual degree of the subject's responsibility and, on the other hand, how much responsibility must be borne by society and by the ideological or cultural or familial environment in which the suicidal tendency developed. Defining suicide as "the ultimate freedom of life" means situating it within an ideology of freedom without responsibility in

---

[139] Ibid., 81.

[140] Ibid., 85.

[141] See G. Masi, "Suicidio," in *Enciclopedia Filosofica*, 2nd ed. (Rome, 1979), vol. 8, cols. 26–28; E. Durkheim, *Il suicidio: Studio di sociologia* (Milan: Rizzoli, 1989); G. Girgenti, ed., *Il suicidio: Follia o delirio di libertà* (Cinisello Balsamo: Paoline, 1989); M. Bertolino, "Suicidio (istigazione o aiuto al)," in *Digesto, Sezione Penale* (Turin: UTET, 1998): 14:113.

which the values of freedom and life are reversed, which fuels both the campaign to legalize euthanasia and the one that encourages suicide.

A complete study of the phenomenon from a historical, sociological, and cultural perspective cannot be undertaken in this work, which intends to focus on the principal problems involving the direct responsibility of physicians for medical actions. As far as direct participation is concerned, the ethical-deontological problem for the physician would be simply in the possibility of prescribing or knowingly administering lethal drugs to a consenting suicide who requests them. And this aspect concerns the pharmacist as well whenever the composition and quality of the drug make it suitable solely for the purpose of committing suicide.

But this type of responsibility is very easy to understand, from the legal perspective as well, for those who maintain that suicide is per se illicit and inhumane. As for the refusal of the truth and the temptation of suicide, there is extensive literature from a strictly anthropological viewpoint that accompanies and includes a scientific approach.[142]

At the conclusion of this chapter it is hard to miss a paradox: in an age when medical advances have made pain control easier and more readily available and, generally speaking, have made life more comfortable, especially in the Western world, we hear arguments from more quarters in favor of euthanasia and suicide. Perhaps we are to understand that it is not the pain that has become unbearable; in reality, our reasons for living have been lost along with much of the meaning afforded to suffering and death.

---

[142] Some aspects can be found in E. Sgreccia, "Il rifiuto della verità e la tentazione del suicidio," *Prospettive nel mondo* 75, no. 76 (1982): 35–41; M. L. Di Pietro and A. Lucattini, "Condotte suicidiarie e adolescenza nel dibattito attuale," *Medicina e Morale* 44, no. 4 (1994): 667–690. See also E. Fizzotti, "L'onda lunga del suicidio tra vuoto esistenziale e ricerca di senso," *Anime e Corpi* 161 (1992): 273–294; L. Cantoni, "Su alcune dimensioni del suicidio: Il caso dell'Emilia Romagna," *Medicina e Morale* 44, no. 6 (1994): 1143–1160.

Summary Outline for Chapter 15

Respect for the dignity of the dying:
Ethical considerations on euthanasia

From the 1970s on, starting in the most developed countries of the world, an insistent campaign has been spreading in favor of euthanasia, understood as a deliberate action or omission that of itself and by intention cuts short the life of a seriously sick patient or of a deformed newborn. The motive that is usually cited is the desire thus to spare the patient himself useless suffering.

The case of Holland, where for some years there had already been a sort of regulation considering unpunishable a physician who practices euthanasia at the patient's request, is presently an instance of the true and actual legalization of *euthanasia on demand*, although limited to cases of serious, irreversible illness accompanied by suffering, and on the condition that this situation be verified by a medical process that claims to be rigorous.

The justification it attempts to put forward and have accepted by public opinion essentially hinges on two fundamental ideas:

- the *principle of personal autonomy*: the individual is said to have the absolute right to dispose of his own life

- the more or less explicit conviction that the pain sometimes accompanying death is *unbearable* and *useless*

The Church in her reflections has constantly kept in contact with medical workers and specialists, striving for fidelity to the principles and values of humanity shared by the majority of mankind, in the light of reason enlightened by faith:

- Declaration *Iura et bona* (1980) on euthanasia, published by the Congregation for the Doctrine of the Faith

- The document by the Pontifical Council Cor Unum, "Ethical Questions Regarding the Seriously Ill and Dying" (1981)

- The encyclical letter *Evangelium vitae* (1995) by John Paul II, in particular nos. 64–67

- The *Charter for Health Care Workers* drawn up by the Pontifical Council for Pastoral Assistance to Health Care Workers (1995)

These documents of the Magisterium do not limit themselves to describing euthanasia as morally unacceptable "since it is the deliberate ... killing of a human person" (*Evangelium vitae*, no. 65), but also offer a way to assist the seriously ill or dying patient which is, both in terms of medical ethics and in spiritual and pastoral terms, inspired by the dignity of the person, respect for life, and the values of fraternity and solidarity, calling on persons and institutions to respond with concrete witness to the current challenges of a spreading culture of death.

In reference to the documents just cited, it is worth recalling that the patients' pain, which some would argue makes euthanasia and/or assisted suicide somehow justifiable or even obligatory, is today more than ever a pain that is "treatable" with appropriate analgesic methods and palliative treatments proportioned to the pain itself; this pain, if accompanied by suitable human and spiritual assistance, can be mitigated and comforted in an atmosphere of psychological and emotional support.

Consequently, the consideration proposed by the *Charter for Health Care Workers* is all the more true: "The sick person who feels surrounded by a loving human and Christian presence does not give way to depression and anguish as would be the case if one were left to suffer and die alone and wanting to be done with life. This is why *euthanasia is a defeat* for the one who proposes it, decides it and carries it out" (no. 149).

The *principle of autonomy*, which is sometimes used to exaggerate the concept of individual freedom, pushing it beyond its rational limits, certainly cannot justify putting an end to one's own life or the life of another. Indeed, personal autonomy presupposes first of all *being alive* and requires the responsibility of the individual, who is *free to* do the good according to the truth; he will succeed in affirming himself, without contradictions, only by recognizing (again from a purely rational perspective) that he has received his life *as a gift* and therefore cannot be the "absolute master" of it; putting an end to life, after all, means destroying the very roots of personal freedom and autonomy.

Justifying euthanasia involves a perverse complicity on the part of the physician, who, by his professional identity and by virtue of the strict ethical demands connected with it, is always called to sustain life and to alleviate pain, but never to cause death "though asked to do so" (Oath of Hippocrates).

The treatment of a seriously ill, dying patient should be inspired by respect for the life and dignity of the person; it should aim to make appropriate therapies available, without indulging in any sort of "therapeutic obstinacy"; it should consult the will of the patient when dealing with extraordinary or risky treatments (to which one is not morally obliged to consent); it should always provide ordinary care (including nutrition and hydration, even if by artificial means) and be committed to palliative care, above all in treating pain adequately, always encouraging dialogue and informing the patient himself.

When death appears to be inevitable and imminent, "it is permitted in conscience to take the decision to refuse forms of treatment that would only secure a precarious and burdensome prolongation of life" (CDF, *Iura et bona*, part 4), since there is a big ethical difference between "causing death" and "allowing to die": the first attitude rejects and denies life, while the second accepts the natural conclusion of it. (See Pontifical Academy for Life, "Respect for the Dignity of the Dying," December 9, 2000.)

Euthanasia as a legal problem

It is necessary to make a linguistic and conceptual clarification:

- *Euthanasia, from the legal perspective, is not the refusal of treatments made by a competent patient*, even though such a refusal could be the equivalent of suicide in terms of a moral analysis.

- *Euthanasia is not the abandonment of therapeutic obstinacy*, that is, of all medical practices that are not proportionate to the patient's real health care situation.

- *Euthanasia is not equivalent to palliative care*, which is capable of controlling pain in terminally ill patients and assuring them a more than acceptable quality of life, though in some cases the use of certain substances in palliative care produces the double effect of sedating the patient's pain yet hastening his death.

- *The legal discussion of euthanasia should not include eugenic euthanasia*, that is, the killing of deformed newborns shortly after birth or infanticide committed upon handicapped newborn children, which is connected with the problems of how to cope with the handicap.

Having eliminated these forms of pseudo-euthanasia, how do we define euthanasia in the true and proper sense? "It is euthanasia *if and only if* a physician puts an end to the life of a terminally ill patient at his request."

If we analyze the Dutch experience, we find that the law defines *euthanasia* as the killing of a patient *at his explicit request*, but the numbers show that in approximately 40 percent of cases the patient's life is ended *without his explicit* request.

Those cases are justified hypocritically by the fact that there are situations in which all vital functions begin to be impaired irreversibly, and hence there can be no explicit request by the patient.

Euthanasia is always proposed as a humanitarian solution in terrible cases where the patients are the victims of agonizing pain. Palliative care is systematically ignored.

The argument inevitably drifts from *physical pain* to *existential suffering*, and consequently "aid in dying" is given even to a woman who is lonely and depressed but not physically sick.

In conclusion, it is necessary to recall that "anyone who asks to die is really asking for help to live *differently*: sometimes it is enough to learn that one's pain is shared by others in order to cope with it positively (and therefore rigorous empirical studies recommend that terminal patients in hospitals *not be isolated*, because just the experience of a ward in which all the patients are suffering in a similar way is capable of arousing in each of them an incredible ability to resist depression and pain). Selling lethal pills in a pharmacy is an insidious way of inviting the elderly not to burden their families or public assistance programs with all their needs." (D'Agostino, "Eutanasia," in *Parole di bioetica*.)

The legalization of euthanasia: International overview

In *Holland*, euthanasia was legalized in late 2000. The process leading up to that step was gradual. In the first phase there were guarantees that a physician committing an act of euthanasia would not be punishable. This mechanism went into effect on June 1, 1994; it was based on the premise that the physician involved in euthanasia would issue a certificate of nonnatural death, together with a report on the incident (a sort of self-accusation), drafted in accordance with guidelines that had already been developed in 1981 by the Court of Rotterdam and then revised by the Royal Dutch Medical Association in 1985. The guidelines specified the following criteria for a legitimate euthanasia request:

- The patient must be suffering from intolerable pain.
- The patient must be conscious.
- The request for death must be voluntary.
- Other alternatives must be offered to the patient along with time to consider them.

- There must be no other reasonable solutions to the problem.
- The death of the patient cannot cause useless suffering to others.
- There must be more than one person involved in the decision.
- Only a physician can perform euthanasia on the patient.
- It is necessary to take great care in making the decision.

It was later determined that not only physical suffering justified the request for euthanasia, but also psychological suffering and possible personality disorders. This meant that even a subject who might not be described as clinically ill or suffering would have access to euthanasia.

In *Italy*, euthanasia is not addressed by the law. If someone kills a person who made an explicit request for death, he faces the consequences put forth in article 579 of the penal code (homicide of a consenting person), provided that the victim was at least 18 years old, was not mentally ill, was not in a mentally deficient state, and was not influenced or persuaded to ask for death by threats or deception. In the absence of these circumstances, or in the absence of any request whatsoever—that is, in the case of euthanasia for pious motives—the deed must be treated as an instance of voluntary homicide (article 575 of the penal code).

In *Belgium*, the law permitting euthanasia was approved on April 10, 2001.

In the *United States*, federal law forbids euthanasia. Oregon was the first state to legalize physician-assisted suicide (in 1994) for terminally ill patients who have made a formal request. A state legislative challenge failed in 1997, and the Supreme Court upheld the law in *Gonzales v. Oregon* in 2006. Physician-assisted suicide is also legal in the states of Washington and Montana.

In *Australia*, a pro-euthanasia law was passed in 1996 by the parliament of the Northern Territory, but nine months later, in March 1997, it was repealed by the federal Parliament.

In *China* in 1998, the government authorized hospitals to practice euthanasia on patients in the terminal phase of an incurable illness.

The Gröningen Protocol (Holland) for neonatal euthanasia

The following requirements must be satisfied:

- The diagnosis and the prognosis must be certain.
- The patient's suffering must be unbearable and hopeless.
- The diagnosis, prognosis, and unbearable suffering must be confirmed by at least one independent physician.
- Both parents must provide informed consent.
- The procedure must be performed according to the standards accepted by the medical profession.

Information necessary to support and explain the decision for euthanasia:

Diagnosis and prognosis

- Describe all the relevant medical data and the results of the diagnostic tests on which the diagnosis is based.

- List all participants in the decision making process, all opinions expressed, and the agreement reached.
- Describe how the long-term prognosis concerning the patient's health was reached.
- Describe how the degree of suffering and hope for survival were assessed.
- Describe the availability of alternative treatments, alternative methods of pain relief, or both.
- Describe the treatments and results of treatments that preceded the decision for euthanasia.

The decision for euthanasia

- Describe who initiated the discussion on the possibility of euthanasia and the circumstances of the discussion.
- List the considerations that led to the decision.
- List all participants in the decision making process, all opinions expressed, and the agreement reached.
- Describe the way in which the parents were informed and their opinions.

Consultation

- Describe the physician(s) who provided a second opinion (name, specialization).
- List the results of the visits and recommendations made by physicians or specialists.

Implementation

- Describe the actual euthanasia procedure (time, place, participants, and drugs administered).
- Describe the reasons why that method of euthanasia was selected.

After death

- Describe the assessment determined by the coroner or medical examiner.
- Describe how the case of euthanasia was reported to the public prosecutor or district attorney.
- Describe how the parents are being attended and assisted.
- Describe the review plan, such as the reexamination of the case, the postmortem examination, and genetic evaluation.

Advance directives

On December 18, 2003, the Italian National Bioethics Committee (NBC) issued a document on advance directives titled *Advanced Treatment Statements*. The introduction states the following:

> This document concerns advance treatment directives, the importance of which has grown constantly in recent years and which are frequently referred to in the national and international literature on bioethics with the expression *living will*. Other expressions used include biological will, life will, prior treatment wishes, etc. These expressions all refer to a document through which individuals, in full possession of their mental faculties, express their wishes

regarding the treatments that they would or would not want to undergo to [*sic*] in the event that, in the course of an illness or as a result of sudden trauma, they were no longer able to express their informed consent or dissent. The different forms that an advance treatment directive can take (some of which have been legally recognised in a number of countries) have been discussed in the literature.

To give these documents public (although not necessarily legal) status, they are required to be drawn up in writing in such a way that there can be no doubt as to the identity and capacity of the person signing them, their authenticity and the date of signing. They should if possible be countersigned by a physician, who should guarantee that the signer has been fully informed of the possible consequences of the decisions he is taking in the document. It is desirable for the signer to indicate a date for the confirmation and/or renewal of the directive, without prejudice to their right at any time to withdraw or amend the instructions it contains.

The ethical and legal implications of the topic raise many questions. First of all, how can one avoid the inevitably abstract character of these living wills in comparison with the real situation of sickness in which they would have to be applied? What legal status is to be attributed to a potential "proxy" who is expected to act according to the instructions contained in the "will" and in the exclusive interest of the incapacitated person? How can one resolve possible contradictions between living wills, positive law, and the norms of good clinical practice and medical ethics? Can the withdrawal of artificial nutrition and hydration be considered a simple suspension of treatment? Do such directives have a binding character or are they simply guides for health care personnel? Can advance directives promote a bureaucratic acceleration of death?

The NBC expressed itself on the subject in this manner:

The NBC deems it essential to eliminate any ambiguity and emphasises that the right being proposed—for patients to influence the treatment to which they might be subjected in the event of their being considered incompetent—is not a right to euthanasia. ... It is, rather, a right solely to ask physicians to interrupt or not to undertake therapeutic actions.

The more debatable aspects are as follows:

- There is a danger of implicitly recognizing the principle that human life is disposable.

- The physician's position of indemnity is weakened.

- Importance is given to the patient's explicitly expressed wishes (written), but not necessarily to the patient's present wishes.

Chapter 16

Bioethics
and Technology

The Insatiability of History

The expression "the insatiability of history" was used by a philosopher of law to describe the perpetual expanse existing in man between an insuppressible yearning of the will for a complete yet unfulfilled happiness and his concrete actions circumscribed by limits and negativity; between a yearning to conquer that "something more," that fullness, and a falling back to the negative and the "no" that accompanies every life experience.[1] A tension arises from this expanse because the human will, since it is called by the eternal and the infinite, never ceases to struggle against limitation and to push itself beyond in an experience of conquering the cosmos that surrounds life and the mystery that envelops the mind. Science and technology have always been interlaced in learning more, having dominion over more, and being more.

There is an incoercible will beneath this; hence there is an ethical dimension in the thirst for knowledge and for action. Whether this ethicality is properly ordered and positive depends not only on the ends sought by the conquest but also on the modalities and means enacted along the journey and in the conquest itself; it depends above all on the project for man and for humanity that is gradually projected on the horizon: teleology and deontology must join together along the journey. But in the meantime it is certain that the journey has been and will continue to be real, because man is made for journeying: he is a pilgrim in search of greatness beyond himself and unexplored horizons. No one and nothing other than the final eschaton will halt the first command of Genesis to "fill the earth and subdue it" (Gen. 1:28), even at the cost of an occasionally foolish hurdling toward an unknown that evades exploration and understanding.

But the question is why the problem has become even more acute today and why it is the subject of ethical investigation. No ethical problem was posed when humans domesticated horses, although they were later used for war, among other things. No ethical problem was posed when humans invented the wheel or the plough, although these conquests served—in addition to working the fields and transporting goods—to reinforce armies and whet human appetites for military conquest. The ethical problem is now posed in more acute terms for various reasons: the increased power of explosives in human hands, capable of destroying man himself and all

---

[1] G. Capograssi, *Introduzione alla vita etica* (Rome: Studium, 1976).

717

humanity; the need for a new project for humanity that integrates these conquests with perennial and profound human values; and the conviction that the will of a few is insufficient to achieve this, that a few laws alone are insufficient, and that worldwide human growth is needed.[2]

Let us first attend to defining the role of technology in relation to the person. Technology, as practice, shapes the instrumental relationship of the person with the world. It is *an extension and enhancement of the body*. Man has sought instruments with which to enhance his muscular power for millennia; technology later amplified the senses: sight and hearing. Just think of what has become possible to see and hear through ultrasounds, x-rays, radio, and television. Finally, in own age and therefore marking the transition into a new era, man has invented means of enhancing the central nervous system—no longer just muscular power, but now mental power. As will be seen, not all thought is a part of this instrumental plan—only thoughts that are quantifiable and calculable, reducible to formulas. Nonetheless, a new era has begun: the computer science and robotics era, in which the power of dominion over the world has grown and the human brain has been enhanced by machines.

But let us rapidly review this development of technology. In reconstructing the various stages of technological progress and the relationship between man and nature, scholars of the history of civilization affirm that there has been a "mutation" of the society and culture at every phase of technological advancement; in other words, there has been a change in the human project. This fact is important for pointing out that technology is not just an instrumental factor but also a cultural factor, and therefore it has an ethical dimension.

Under the category of the history of technological development, humanity has experienced four eras. They are sequential stages, but the previous stage does not immediately vanish with the appearance of the next one. The first is the *primitive* age, also known as the "hunter-gatherer" stage: man gathers what the earth spontaneously produces. His instruments are those of immediate contact: stones, clubs, arrows, and hatchets involve direct contact with the object. This phase entails a very simple form of social organization: the family and the tribe. Habitations are temporary, in the form of huts or natural places of refuge like caves.

The *agricultural and shepherding* age is next, characterized by working the land to reap its fruits. It includes the invention of the wheel, house building, metal working, product exchange, currency, writing, and defense building; in sum, this is the great agricultural civilization known in the East (Egypt, Mesopotamia, and central Asia), pre-Columbian America, Greece, and Rome. The agricultural phase continued up to the invention of machines and has continued alongside industrial civilization. Traces and areas of primitive culture have also continued to coexist with the later stages. In the agricultural society, civilization attained a level of development in the arts, literature, and philosophical thought to which we are still indebted. Social organization achieved the stage of law and experienced what our history knows up until the eighteenth century.

---

[2] *Proceedings of the 11th International Congress of Philosophy; Brussels, August 20–26, 1953*, vol. 8, *Philosophy of Culture* (Amsterdam: North-Holland, 1953); B. Chiarelli, "Storia naturale del concetto di etica e le sue implicazioni per gli equilibri naturali attuali," *Federazione Medica* 37, no. 6 (1984): 542–547.

The *industrial age* is characterized by machines, which magnify the power of man and profoundly transform nature—the raw materials man finds in his environment. These machines are increasingly powerful, capable of using ever newer forms of energy (e.g., vapor, electricity, petroleum products). The unique attributes of industrialized civilization arise here: the social question, urbanization, fragmentation of families, productivist consumerism, hedonism, demographic problems, and political and social revolutions intersecting one another. Machines have transformed not only raw materials but also society and values.

Cultural anthropologists perceive that we have reached the fourth age of the world,[3] which the majority defines as the *technological age*. Technological progress is what characterizes this era, in which machines guide other machines, and this is made possible primarily through computer science. Computer science has allowed the imitation and enhancement not only of human muscle power but also of human mental power. This is not to say that man does not need to use mental resources to build machines of the first type: science and technology have always been intertwined throughout human history; but the purpose of information machines (robots and computers) is to use calculations, information, and mental data supplied by humans and proceed with autonomous combination, simulation, and calculation. This is how cybernetics[4] was possible or, rather, how it became possible to transfer cybernetics from within man into cybernetics external to man.

The creator of the field of cybernetics was Norbert Wiener,[5] an esteemed mathematician who lived during the first half of the twentieth century and whose scientific contributions ranged from theoretical reflection to technological applications such as the first calculators, to cite just one example. Cybernetics is based on three elements: the use of probabilistic calculations and functional analysis through information and control technologies, the similarities between biological systems and structural and functional aspects of electromechanical systems, and a cybernetic image of the world.[6]

It follows that this scientific theory, like many others throughout the history of science in fact, quickly passed beyond its highly specialized confines and began

---

[3] See S. Cotta, *L'era tecnologica* (Bologna: Il Mulino, 1981); J. Ladrière, *I rischi della razionalità* (Turin: SEI, 1987); C. Angelini, "Informatica e morale," *Vita e Pensiero* 9 (1983): 18–25, and *Vita e Pensiero* 10 (1983): 23–33; M. Vidal, "Ética de la actividad cientifico-técnica," *Moralia* 5 (1983): 419–443; E. Sgreccia, "Il progresso scientifico-tecnologico di fronte all'etica," *Medicina e Morale* 33, no. 4 (1983): 335–341; various authors, "L'appello del futuro e l'intelligenza dell'uomo: Lavoro e cultura nella nuova età tecnologica," in "Atti del II Congresso del Movimento Ecclesiale di Impegno Culturale (MEIC)," *Coscienza* 5–6 (1985).

[4] *Cybernetics* is defined as a "theory of control systems that uses especially analogies between machines and the central nervous system of animals and human beings." N. Zingarelli, "Cibernetica," in *Il nuovo Zingarelli: Vocabolario della lingua italiana*, ed. M. Dogliotti and L. Rosiello (Bologna: Zanichelli, 1991). The etymological root of the word comes from the Greek verb *kybernao* ("to govern a ship") or the noun *kybernetike* ("the art of navigating").

[5] N. Wiener, *Cybernetics*, 2nd ed. (New York: MIT Press, 1961).

[6] For an overview of the origins of cybernetics and the debate surrounding it, see P. A. Rossi, "La cibernetica e la teoria dell'informazione nel XX secolo," in *Storia delle scienze*, ed. E. Agazzi (Rome: Città Nuova, 1984), 2:423–435.

to acquire nonscientific connotations on various levels. In fact, in light of the use of automatons—complex machines able to execute sophisticated tasks—cybernetics maintains that neuropsychological disturbances on the biophysical level can be considered analogous to deficits in data processing and control in machines. It affirms the reproducibility of the physical foundations of life and intelligence on the biological level and—through robots—the possibility of reproducing any human behavior on the psychological level. In the anthropological realm, it affirms the capacity in principle of automatons to carry out any human activity. Finally, cybernetic theory attempts a metaphysical analysis of reality based on the dialectic between *positive entropy* (disorder) and *negative entropy* (order).

Given all this, it appears that the implications of this new age are immense and not yet entirely predictable and balanced. First and foremost atomic fission has been achieved in this manner, which is why this is also called the atomic age; space travel has been made possible, which is why it has also been called the space age. Simulated design has been made possible, which is why it has become the age of planning and of the future. The medical field has witnessed the creation not only of machines that make use of light rays, atomic radiation, and nuclear energy detection in the body but also of telemedicine and computerization in patient management and in the storage and study of nosological, operational, and predictive data. Other consequences of the technological age are predicted in the relationship between man and nature due to the growing consequences of exploitation, pollution (at times radioactive), and the potential biological and genetic modifications achieved through genetic engineering technologies.[7] This insatiability of history and of human progress therefore relates to anthropology and ethics (and law), and in many different ways.

Technological and Scientific Progress, Anthropology, and Ethics

This is precisely the reality: there is an interdependence between technological progress and human sociocultural changes on the planet. Yet two additional questions then arise: (1) Does "progress," which is linear in terms of technology, automatically entail a linear process of anthropological perfection? (2) Is the change affecting man's lifestyle as a result of scientific progress the sort of change that man himself can control? The philosophy of science,[8] with some of its widely shared conclusions, can be of assistance in responding.

The first observation in this philosophical reflection is that scientific and technological progress is linear, and therefore progressive, yet reductive at the same time. These two characteristics depend on the fact that experimental science considers the quantitative aspect to be measurable and reducible to mathematical formulas

---

[7] A.K. Tanswell and H.M. O'Brodovich, "The Present and Future Role of Gene Therapy in the Newborn," *Current Opinion in Pediatrics* 9, no. 2 (1997): 141–145.

[8] A specific sector within the philosophy of science deals with the theoretical prospects of technology itself. Different ideas have been formulated on the matter: *technology as object* (tool), *technology as knowledge* (epistemology), *technology as activity*, and *technology as individual volition*. For a systematic discussion of the topic, see C. Mitcham, "Philosophy of Technology," in *Encyclopedia of Bioethics*, ed. W. T. Reich, rev. ed. (New York: Macmillan, 1995), 5:2477–2484.

describing reality, both in research and in applied studies. Any conclusion rendered through a descriptive formula therefore becomes the end point for a study and the starting point for the next phase. There is progress in this fact, whereby subsequent theories preserve the previous ones at least in part: in the verified, "not falsifiable" experimental part. The subsequent theory can modify it in terms of completion, correction, and reinterpretation, but cannot help but be based upon the knowledge acquired previously, even if only to correct it. Newton presupposes Galileo, Einstein cannot ignore Newton, and those who build nuclear power plants cannot ignore the laws of thermodynamics and atomic studies.

To be clear, this does not mean that all scientific research is devoid of ethicality; indeed, it arises precisely from an ethical impulse: the insatiability of the human spirit, which is open to knowledge and also to the will. The insatiability of history is the insatiability of man, of science, and of progress.[9]

But this ethicality at the source requires a mouth and an outlet that are equally ethical: Where should we go and how should we get there? The ethics of the impulse is present in science and technology, but it does not know the ethics of means and ends: deontology and teleology. This is why the various philosophical currents—existentialism, phenomenology, personalism, and the schools of epistemology themselves—have demanded and demand a new synthesis between technological culture and humanism, between the experimental sciences and the human sciences, invoking ethics in particular to take hold of civilization's sense of direction and safeguard its journey.

This ethical demand becomes more acute—and here the second point mentioned will be examined—because scientific and technological "power" has reached a breaking point. The break is at the level of the "technical" possibility of destroying all of humanity, on the one hand, through atomic weapons or environmental pollution; and at the level of potentially inducing a genetic "mutation" in man, on the other hand. The ethical question today can be formulated as follows: What must be done or not done so that man survives and remains man? This question presupposes and recalls two others: What is it that characterizes man as such? Is it necessary for man to exist in the world? The end of the world was proclaimed by structuralists like Michel Foucault[10] but may be brought about by modern technology.

The ethics of technology should not be considered only in terms of the applied phase, however, but also in its radical insufficiency, its teleological ambivalence, and its ever increasing dynamic of knowledge and power, and therefore in its explanatory phase as well.[11] In other words, technology demands to be completed and to have a

---

[9] Capograssi, *Introduzione.*

[10] H. Jonas, "Technique, morale et génie génétique," *Communio* 9, no. 6 (1984): 46–65; J.L. Ruiz de la Peña, "Anthropologie et tentation biologiste," *Communio* 9, no. 6 (1984): 66–79. For Foucault's thinking, see G. Reale and D. Antiseri, *Il pensiero filosofico occidentale dalle origini ad oggi* (Brescia: La Scuola, 1983), 3:697–699. On this topic, see also A.P. Iannone, *Contemporary Moral Controversies in Technology* (New York: Oxford University Press, 1987); A. Bausola, ed. *Etica e trasformazioni tecnologiche: Atti del 57° corso di aggiornamento culturale dell'Università Cattolica; Arezzo, 20–25 settembre 1987* (Milan: Vita e Pensiero, 1987).

[11] A contribution to an overall improvement in assessing technology could come through technology assessment (TA), which is the multidisciplinary assessment of the impact

reference within an integral anthropology, in which it can find its place alongside man's other dimensions.[12] This presumes a project for man that integrates technological development without dehumanizing or absolutizing it.

Ultimately, then, to say it with Hans Jonas,[13] technology harbors two basic elements within itself: a *formal dynamic*, understood as an "uninterrupted collective enterprise that advances according to its own 'laws of movement,'" and a *substantial content* "made up of the things it puts at the service of man, the possibilities and powers it confers, the new objectives it presents or imposes, and the changed modes of action and human behaviors."[14] Both factors inhere in the ethical dimension, actually constituting a new and unique case to be morally analyzed in light of a "robust" reference anthropology—an anthropology centered on the ontological dimension of man.

Further observations and analyses of an anthropological nature can be pursued with regard to computer science, this unique and recent element in the age of technology.

Computer Science and Anthropology

Faced with the active process caused by the perfusion of computer science[15] into all aspects of life (from family to elementary school, the workplace, and the economy), it is necessary clarify what the presently active or foreseeable consequences are at the human level, what positive and negative factors there are for human development and autonomy, and on what conditions this technological progress can become human progress.

Reassuring consequences of computer science at the anthropological and cultural level[16]

Many scholars are critically and guardedly optimistic about man's potential to absorb this new impact based on his capacity for adaptation and synthesis as revealed by the history of human culture in the previous phases of his millenarian development. Virtually all of them underscore first of all an anthropological and logical limit to computer science: it works at the level of formal logic and can be considered artificial intelligence insofar as that intelligence works with elements that are reducible to combinations, identity, and unity. In other words, computers and robots can never be attributed with the most typical function of the human intellect: intuition or creativity or symbolic language. Man—individual men and women—can therefore never be replaced. This is not only because man will always be the one to provide

---

of technological processes, substance, programs, and methods. See K. Shrader-Frechette, "Technology Assessment," in *Encyclopedia of Bioethics*, ed. Reich, 5:2484–2489.

[12] H. Jonas, *The Imperative of Responsibility: In Search of an Ethics for the Technological Age* (Chicago: University of Chicago Press, 1984).

[13] H. Jonas, *Tecnica, medicina ed etica* (Turin: Einaudi, 1997), 7–36, originally published as *Technik, Medizin und Ethik: Zur Praxis des Prinzips Verantwortung* (Frankfurt am Main: Suhrkamp, 1985).

[14] Ibid.

[15] A. Ruberti, ed., *Tecnologia domani: Utopie differite e transizione in atto* (Rome: Laterza, 1987).

[16] "L'appello del futuro."

the initial program and laws for the computer, but also because the computer or robot will never be capable of receiving the profoundness of the creative intellect. Poetry, art, moral values, and religion are fruits that can be memorized after they have been produced, but can never be suggested and developed by a machine.

Another preliminary observation delimiting artificial thought is the fact that beneath a program there are not only logical thoughts and data provided by man, but also decisions made in light of a design. Hence there is an ethical aspect not only at the conclusion of the process—during the applied phase of what has been developed—but also at the beginning of a designed computer operation. This is especially evident in the processes of "simulation" and predictive programs: everything has already been "directed" in accordance with an initial option.

The link between computer science and ethics therefore cannot be lost; it is present at the beginning and the end of every form of design. Computer science cannot therefore escape the influence of ethics. Other scholars who aim to shed light on the reassuring aspects of the emergence of computer science note another fact or, rather, another possibility: the delegation of operational aspects of thought to computerization will allow man to become freer, creating more room for creative and meditative moments.

By delegating to computers his attention to multiplicities, man will be able to devote himself to that which is unifying in life: the questions that restore *meaning* to life. Since moral formation must tend not toward the *multa* (many) but rather toward the *multum* (much), he will be at an advantage. The foreseeable increase in free and "unoccupied" time will obviously entail the need to fill those human spaces with free and liberating activities. A further reassuring prediction has also been offered: the possibility of overcoming the phenomenon of urban migration and urbanization as more distant locations are sought for many working sectors currently concentrated in cities, where living space and environmental resources no longer seem sufficient.

Problematic aspects of the emergence of computer science

Of all the risks highlighted—and it is indeed a matter of risk—the one most discussed is the potential impoverishment of the perception of reality. It is said that reality will be reduced to information, formal data, and consequently the intellect, accustomed to formal exercises, will experience an attenuation of its creative and imaginative faculties. Even language will risk becoming impoverished to a merely descriptive structure to the detriment of quality and value.

A more specific form of impoverishment is feared and signaled in the realm of values, since these are not quantifiable and descriptive aspects. Human and social relations would be impoverished because computers are not suited to providing judgments of meaning and metaphysical significance. Some also predict an identity crisis and a loss of the meaning of corporality and of natural cosmic reality as a result of formal reduction.

But the greatest fear is the change in society due to the necessary concentration of the power of knowledge in the hands of a few: programs, which require specialized knowledge and rapid decisions, can only be created and studied by a few. The specter of a new technocratic oligarchy programming and controlling development looms on the horizon; this is coupled with the risk that society will let itself be trained to assess realities in terms of their quantitative and functional aspects, thereby becoming further disposed to the loss of freedom. There would be those who do the programming for everyone by institutional right, and those who let themselves be programmed and who could not, in any event, prevent the execution of programs and of control over the masses.

Ethical Requirements vis-à-vis the Computer Science Culture

The position proposed first and foremost is neither a plain and simple rejection nor an enthusiastic acceptance. Faced with the facts of scientific progress, positions of exorcism or fideism are inadequate and ineffective in terms of a humanization process for what is a part of the realm of means and culture. But the criterion of plain and simple complementarity between technological civilization and humanistic civilization is also inadequate. Technology is also a product of man.

What is necessary is a new project for man and culture that achieves a new synthesis and not a mere addition of a new patch on a piece of old clothing.

It is not easy to delineate the new man, the human man who is not merely economic man or technological man, who is not resigned to the disembodiment of what is real and the preservation of its functional, informational, or cybernetic aspects alone. This new humanism will surely emerge through a painful and difficult transition. The thoughts of specialists and philosophers that appear relatively well defined at present can be summarized by the following requirements.

The first requirement is the recovery of what is irreducible to informational, formal, and quantitative rationality in the human soul. This means intensifying creative activity, artistic and symbolic language, and the experience or understanding of mystery and being in the educational process. The "civilization of being" is beyond any computerization and is manifested in the contemplation, the search for what lies beyond the "sign" and is found within what is "signified." The ethical requirement emerges in the same way as not only an ethics of obligation but also an ethics of design, a teleological ethics grafted together with deontological ethics and both rooted in ontology. Any design or project requires meaning, any "meaning" requires a goal, or *télos*, and the *télos* elicits and demands a fullness of being and a structure of essence.

Computer programmers make decisions and work in functions of projects: the meaning and the project require an end and a path and the discovery and clarification of values. These values constitute the goal providing direction and meaning to reality; their foundation must be objective and rational and not merely emotional to avoid instilling a dualism in man between the knowledge process and the decisional process. A value must have its own ontology and rational justification.

Another dimension this new culture should cultivate for anthropological equilibrium is sociality, primarily understood as the abundance of close social relations. The integration of the civilization of being with the civilization of numbers and data also entails the civilization of love and interpersonal relations.[17] Yet the requirement of sociality also implicates the need for participation: computer science will either become social or it will create privileged classes and a form of technocracy with some who design and others who submit to the design.

---

[17] On this question, see H. Marcuse, *One-Dimensional Man: Studies in the Ideology of Advanced Industrial Society* (Boston: Beacon Press, 1964); E. Fromm, *To Have or To Be?* (New York: Harper and Row, 1976).

Culture creates humanism, in fact, only when it becomes social. This is what Pope Paul VI said in *Populorum progressio* in coining the expression "man's complete development and the development of all mankind,"[18] a concept that he also defined as "full-bodied humanism."[19] When Jacques Maritain spoke of "integral humanism," he envisioned harmony between the natural and supernatural values, in contrast with both secularistic naturalism and fideistic supernaturalism. With the expression "full-bodied humanism," Paul VI sought to underscore both the attainment of the inner fullness of the person through the educational development (such development also entails harmony) of all man's capacities and dimensions, and the extension of human goods to all people throughout the earth. In this perspective, it is necessary to pursue a design and a new humanism in which the values of technology and computer science are harmonized with those of creativity and freedom at both the microsocial level as well as the macrosocial and global level.

Another ethical requirement takes shape for this type of civilization by way of contrast with technicism and formal rationality: the need for symbolic and prophetic witness. In order to overcome the risk of conformist uniformity and to enliven the mass-mentality society which is its most natural fruit, moral values must not be expressed through verbal and rational language alone; they must also be expressed through symbolic and emblematic signs and embodied in facts. This is what theological language defines as "witness."

Some Specific Ethical Frontiers of Applied Technology

Returning to technological development considered in its entirety, several unique problems are recognized as "frontiers" beyond which the use of technology without guidelines and moral finality could lead to a rupture in human equilibrium in a cosmic and social sense. There are primarily three such frontiers affected by technological development: the accumulation of atomic power, environmental pollution, and technology applied to genetics. Each one deserves its own separate treatment.

All three have implications for the medical and biological field and closely touch the aspects of medicine and medical professionalism. Moreover, they also have potential relations of interdependence: atomic energy can cause pollution—just as it can be used for medical applications—and can influence the genetic inheritance of future generations. These arguments are not the exclusive prerogative of politics; they also involve medical morality and ethics.

Genetic modification

Suffice it to briefly mention here the problem of applying scientific and technological progress to the field of genetics; the reader is invited to consult the chapter on this subject for a more detailed treatment. It is nonetheless important to recall that the moral assessment changes depending on whether it is a question of gene therapy—which is licit in itself with certain conditions (e.g., no embryos are destroyed, there are no serious risks of causing errors or aberrations that are worse

---

[18] Paul VI, Encyclical Letter *Populorum progressio*, March 26, 1967, no. 5.
[19] Ibid., no. 42.

than the ones to be corrected, and there is a certain probability of success)—or a question of alteration or eugenics.[20]

The use of atomic energy

Atomic energy can be used for the purposes of war, and we know from studies conducted by the Pontifical Academy of Sciences that the potential of atomic warheads in 1982 was equivalent to about three tons of trinitrotoluene (TNT) for every individual inhabitant of the earth.[21] The biological consequences of sporadic explosions were also studied, indicating potential harm to the health of survivors.

The Declaration on Prevention of Nuclear War formulated by the Pontifical Academy of Sciences in 1982 states the following:

> For the first time, it is possible to cause damage on such a catastrophic scale as to wipe out a large part of civilisation and to endanger its very survival. The large-scale use of such weapons could trigger major and irreversible ecological and genetic changes, whose limits cannot be predicted.
>
> Science can offer the world no real defense against the consequences of nuclear war. ...
>
> There are now some 50,000 nuclear weapons, some of which have yields a thousand times greater than the bomb that destroyed Hiroshima. The total explosive content of these weapons is equivalent to a million Hiroshima bombs, which corresponds to a yield of some three tons of TNT for every person on earth. Yet these stockpiles continue to grow. ...
>
> Not only the potentialities of nuclear weapons, but also those of chemical, biological and even conventional weapons are increasing by the steady accumulation of new knowledge. ... It is the duty of scientists to help prevent the perversion of their achievements and to stress that the future of mankind depends upon the acceptance by all nations of moral principles transcending all other considerations. ...
>
> IV. Finally, we appeal:
>
> 1. To national leaders, to take the initiative in seeking steps to reduce the risk of nuclear war. ...
>
> 2. To scientists, to use their creativity for the betterment of human life, and to apply their ingenuity in exploring means of avoiding nuclear war and developing practical methods of arms control.

---

[20] On this subject, see Italian National Bioethics Committee, *Terapia genica* [*Gene therapy*], February 15, 1991 (Rome: Presidenza del Consiglio dei Ministri Dipartimento per l'Informazione e l'Editoria, 1991), http://www.governo.it/bioetica/pdf/1.pdf, English abstract at http://www.governo.it/bioetica/eng/opinions/abstracts/abstract_gene_therapy.pdf; idem, *Documento sulla sicurezza delle biotecnologie* [*Document on biotechnology safety*], May 28, 1991 (Rome: Presidenza del Consiglio dei Ministri Dipartimento per l'Informazione e l'Editoria, 1991), http://www.governo.it/bioetica/pdf/4.pdf, English abstract at http://www.governo.it/bioetica/eng/opinions/abstracts/abstract_document_on_biotechnlogy.pdf.

[21] Pontifical Academy of Sciences, "Declaration on Prevention of Nuclear War," September 24, 1982, 14, http://www.vatican.va/roman_curia/pontifical_academies/acdscien/2010/documenta_3_4_7_11_pas.pdf.

3. To religious leaders and other custodians of moral principles, to proclaim forcefully and persistently the grave human issues at stake so that these are fully understood and appreciated by society.

4. To people everywhere, to reaffirm their faith in the destiny of humankind, to insist that the avoidance of war is a common responsibility, to combat the belief that nuclear conflict is unavoidable, and to labor unceasingly towards ensuring the future of generations to come.[22]

Faced with this tragic possibility that envisions the end of humanity, there is also a possibility of peaceful uses for nuclear energy and even medical applications, especially in diagnostic radiology. In the latter case, however, there is a "risk" of bearableness; safety regulations and dosage prescriptions are set in place in light of this. These regulations are also ethically binding in the name of defending the physical lives of individuals. Protection should be considered in relation to patients, health care workers, and the social and ecological environment. The radiological risk should also be considered in relation to other types of radiation, which is why there are legal regulations in this area as well that are morally binding with regard to effective dangerousness and harmfulness.

The harmfulness and dangerousness of radiation should also be considered in relation to the fetus if a woman is pregnant, meaning that potentially harmful radiation should be avoided if pregnancy is certain or even just suspected. If a pregnancy is discovered after the pregnant woman and her child were already exposed to radiation, voluntary termination of pregnancy is unacceptable from the perspective of personalist ethics, regardless of what the present law allows. Prenatal diagnosis can contribute to dispelling doubt in certain cases, but termination of pregnancy does not become licit even if a diagnosis of malformation is confirmed, as has been explained. But this precise aspect of the problem—the peaceful use of radioactivity, which has primarily medical applications—requires a more extensive discussion that cannot be addressed here in its entirety and must be saved for a treatment of more particular nature.

This chapter, which is general in nature, will clarify certain concepts or at least highlight some ethical questions. Remaining on the subject of the peaceful use of nuclear energy, it must first be underscored that there is no "threshold" of safety below which there is no risk: there is always a risk, and this is why the concept of "acceptable risk" has been proposed and the principle of conduct in terms of a "risk-benefit assessment" has been advanced. These concepts have been developed and quantified even in precise reports by the International Commission on Radiological Protection (ICRP)—established by the treaty that instituted the European Atomic Energy Community (Euratom)[23]—and by the United States Advisory Committee on Human Radiation Experiments (ACHRE).[24]

---

[22] Pontifical Academy of Sciences, "Prevention of Nuclear War," 14–15.

[23] European Council directives 80/836/Euratom, 84/466/Euratom, and 84/467/Euratom were accepted into Italian law by legislative decree no. 230/95.

[24] The committee published its results in U.S. Advisory Committee on Human Radiation Experiments, *Final Report* (Washington, DC: US Government Printing Office, 1995), http://hss.energy.gov/healthsafety/ohre/roadmap/achre/report.html.

But what is meant by this problematic "risk-benefit" principle? The tendency, or at least the danger, is that these two terms are given either a purely economic meaning or noncomparable meanings.[25]

The purely economic meaning entails a calculation in monetary terms of the human lives that might be lost and compensated or the economic cost of diseases that could be induced, and this factor is placed on one side of the scale. The other side of the scale might include the economic benefits (jobs, machine production, etc.) that could derive from these applications of nuclear energy. The calculation is done in statistical terms just like an insurance contract; the problem is different, however, because with insurance contracts the cause of injury is external, accidental, and surely not internal and intrinsic to the use of the means or method. This sort of understanding of "risk-benefit" must therefore be rejected.

Another equivocation occurs when the values calculated are not comparable, as when the risk weighed is a truly inflicted injury to health and therefore an anthropological good but the other side of the scale weighs socioeconomic advantages, that is, economically necessary and socially useful activities. The risk-benefit concept, in summary, is properly applied in a personalistic sense when it is applied in reference to the same subject and the same order of goods.

In each of the two scenarios the calculation is always unfavorable toward human value. A terminological and ethical clarification therefore needs to be introduced within this well-known principle, which is taking a role of excessive importance in the context of industrial productivity and in a society based on output and economic profit. It maintains its validity only when calculating and comparing economic goods with one another; yet even in this case, it must be subordinated to higher ethical principles and defined as a cost-benefit relationship.

In the specific case of using nuclear energy for peaceful purposes,[26] I will limit myself to mentioning the following criteria for guidance, which are being proposed with great difficulty in international arenas as well.

First of all, the principle of *therapeutic proportionality* should apply in terms of *patient application*: the harm inflicted on the patient's health and the therapeutic result for the same person must be calculated, seeking a positive proportionality. It is pointless and illicit to use a therapy if it does not have a proportionately equal or greater result with respect to the harm caused. If the result is positive and recovery is expected, the economic cost should not interfere because economics serve the person. The fact that the patient is elderly or handicapped should not have any selective and exclusionary bearing either.

With regard to the harms to workers (doctors, technicians, etc.) and the population or environment, the concept of "acceptable" risk should not be employed: it

---

[25] L. Failla, "Radiazioni ionizzanti: Rischi/benefici o cost/benefici?" *Medicina e Morale* 39, no. 3 (1989): 505–513; Pontifical Academy of Sciences, *Biological Implications of Optimization in Radiation Procedures: Conclusions; May 2–5, 1983* (Vatican City: Pontifical Academy of Sciences, 1985); E. Sgreccia, "Bioeticità delle metodiche strumentali in geriatria," *Giornale di Gerontologia* 36 (1988): 545–554.

[26] On the topic of using ionized radiation in medicine, see the article R. Nicholson, "Ethics and the Use of Ionising Radiation in Research on Humans," *Bulletin of Medical Ethics* 132 (1997): 13–24.

is a relativistic criterion (What could make death due to a radiation-caused tumor acceptable?), and the concept of "optimization" should be employed instead. In other words, the harm inflicted on workers and on the population should tend toward the minimal, since risk is intrinsic in this area and is not caused by accident. The criterion of sociality could make acceptable a risk that is deliberately foreseen and not otherwise reducible in order to help one's neighbor, so long as the axiom *non sunt facienda mala ut veniant bona* is not violated.

These brief references help us understand the complexity of the relationship between technology and morality, and how utilitarian criteria can be introduced too empirically in the management of a technological means when it is used on human health.

Environmental pollution

It has already been explained that the use of nuclear power can ultimately produce—if the doses exceed a certain risk level—a pollution level dangerous to life. When discussing pollution, in fact, a distinction must be drawn between forms of environmental pollution and forms of biological pollution, though they are often tied to one another.

The applications of technologies for harvesting natural resources and manufacturing products, the use of agricultural products, and the dumping of factory waste can lead to environmental degradation and harm to life. Human life can be indirectly compromised and harmed by pollutants in the atmosphere, local environment, and food. But there are forms of compromise of the biological equilibrium that more deliberately and severely compromise the relation between person and nature. Anything resulting from genetic and embryonic manipulation or the separation of human love from physical life (e.g., contraception, sterilization, abortion) is part of this wave of biological pollution in which man's physical life is comprised by technology.

On this topic, the declaration adopted in São Paulo (Brazil) in October 1976 by the Thirtieth Assembly of the World Medical Association should be kept in mind. This document underscores the fundamental importance of the relationship between man and the environment and the need to rethink socioeconomic development in qualitative terms. The text affirms that this danger of pollution threatens the future of the human race and has become the primary factor threatening life.[27] The role of physicians is considered essential since many physical and chemical pollutants can have consequences at the individual biological level and genetic level. It makes specific reference to foods, the use of pesticides and additives in agriculture, the factory setting, and the workplace as sources of disease. The task of the physician and of medical associations is to point out, at the preventive and educational level, the connection between quality of environment and quality of life.[28]

---

[27] World Medical Association, Declaration of São Paulo on Pollution, October 1976, revised in 1984 and rescinded in 2005, http://www.wma.net/en/30publications/10policies/20archives/p22. The threat is described in less extreme terms in the more recent version: "The problem of pollution affects not only the viability and beauty of the environment, but constitutes a growing threat to the very health of the humans who occupy it."

[28] M. Torrelli, *Le médecin et les droits de l'homme* (Paris: Berger-Lévrault, 1983), 410–412; see also E. Sgreccia, "Cumbre de la tierra + 5," *Medicina e Morale* 47, no. 4 (1997): 659–663.

The United Nations Conference on Environment and Development in Rio de Janeiro in 1992 (the Earth Summit) and the more recent conference in Kyoto in 1997 also brought attention to the same topic, especially atmospheric pollution.

What seems obvious from what has been presented here with regard to the use of technology in the biological and genetic sector, the nuclear sector, and the transformation of the environment is that the ethical responsibility[29] of individual persons—citizens and physicians—should be linked to legislative, political, and planning decisions at the local, national, and international levels, in order to achieve a positive outcome.[30] The technological revolution is centralized by its own nature and tends to become massive and above the will and choices of individuals. On the other hand, the formation of individual consciences is indispensable for creating a thought movement and for everyone's collaboration in defending the environment.

Incidents in Seveso, Bhopal, and Chernobyl show how environmental disaster can occur because of a lack of sufficient safety regulations in industries and factories. In other areas, it has been observed how entire coastlines and bodies of water themselves have been harmed by environmental degradation resulting from industries and the use of chemicals in agriculture. The equilibrium of the world and of the biosphere must be considered two fundamental tasks of society from now on in terms of education and legislation, with a strong emphasis on environmental ethics.

We have merely scratched the surface of just a few topics in order to underscore the significant impact of technological advancements in medicine and health care. This impact is no less relevant in the context of culture and society. Engaging ethical values is becoming increasingly urgent and salutary with respect to the human meaning of these same technological advancements.

Technology as the Sole Horizon
Transforms Man into an Automaton

As Martin Heidegger affirmed,[31] we need a form of meditative thought capable of engaging with the complete domination of technology, the characteristic of our age. This is the horizon, the origin, and the meaning of bioethics, which Adriano Pessina calls the critical conscience of technological man: "Bioethics is the expression of a 'critical' moment, in fact: the collapse of trust in the self-regulating capacity of technological processes and the dissatisfaction with certain moral criteria that have become the background for scientific research and practice."[32]

---

[29] For a review of the literature on the topic, see Italian National Bioethics Committee, *Bioetica e ambiente* [*Bioethics and the environment*], September 21, 1995 (Rome: Presidenza del Consiglio dei Ministri Dipartimento per l'Informazione e l'Editoria, 1995), http://www.governo.it/bioetica/pdf/19.pdf [Italian], especially chapter 3, "Interventi sull'uomo e stabilità degli ecosistemi" (Interventions on man and the stability of ecosystems); E. Sgreccia and M. B. Fisso, "Etica dell'ambiente," supplement, *Medicina e Morale* 1947, no. 3 (1997); L. A. Schokel, "Tecnologia, ecologia e contemplazione," *Civiltà Cattolica* 3 (1987): 105–114.

[30] G. B. Marini Bettolo, ed., *A Modern Approach to the Protection of the Environment* (Vatican City: Pontifical Academy of Sciences, 1989); G. Sciaudone, ed., *L'umana dimora già e non ancora* (Rome: Istituto Italiano Medicina Sociale, 1989).

[31] See M. Heidegger, *The Question concerning Technology, and Other Essays* (New York: Harper and Row, 1977).

[32] A. Pessina, *Bioetica: L'uomo sperimentale* (Milan: Mondadori, 1999), 3.

According to Arnold Gehlen [33] and Helmuth Plessner,[34] technical operation is man's response to the solicitations of an environment that is not familiar to him.[35] Jacques Ellul [36] specifies that Christianity, with its reference to otherworldly reality, introduces a criterion that is extraneous to the selection of the technical means; he maintains that technique [37] develops if it proceeds according to this logic. It requires the unification and rationalization of knowledge, which historically occurred in the seventeenth century but has its apex in the twenty-first century. The development of technique is characterized by the quest for the greatest possible efficiency ("the one best way in the world").

Mariella Lombardi Ricci [38] appropriately examines the reasons for the uneasiness technology elicits—the same reasons for the emergence of bioethics—and maintains the importance of comparing traditional technology with present-day technology. Past technologies did not take up much space and the means available were limited; hence greater efficiency was sought by modifying not so much the instrument as the manual ability of man. Technology was always local and did not have issues with adaptation: since there was a variety of models and no tool had an absolute value, man could break the connection with a technology, using it for the aim he sought.

---

[33] See A. Gehlen, *Man in the Age of Technology* (New York: Columbia University, 1980).

[34] See H. Plessner, *Die Stufen des Organischen und der Mensch: Einleitung in die philosophische Anthropologie* (Berlin: Walter de Gruyter, 1965).

[35] Technique has always aided man, who, as Plato noted (*Protagorus*, 321 c), is the animal that nature left most unequipped.

[36] See J. Ellul, *The Technological Society*, trans. J. Wilkinson (New York: Knopf, 1964), originally published as *La Technique ou l'enjeu du siècle* (Paris: A. Colin, 1954).

[37] TRANSLATOR'S NOTE: The author generally uses the Italian terms *tecnica* and *tecnologia* interchangeably throughout most of this chapter. "Technology" is the English term typically most fitting for both concepts. At this point in the chapter, however, he begins to cite J. Ellul (and others who in turn cite Ellul's work), who applies a unique concept to each of the two terms. Ellul uses the term "technique" to discuss a concept including but not limited to technology, of which technology is an example: "Technique refers to any complex of standardized means for attaining a predetermined result." R. K. Merton, foreword to *The Technological Society*, by J. Ellul (New York: Knopf, 1964). In Ellul's words, "The term *technique*, as I use it, does not mean machines, technology, or this or that procedure for attaining an end. In our technological society, *technique* is the *totality of methods rationally arrived at and having absolute efficiency* (for a given stage of development) in *every* field of human activity" (p. xxv). Furthermore, "technique is not an isolated fact in society (as the term *technology* would lead us to believe) but is related to every factor in the life of modern man; it affects social facts as well as all others. Thus technique is a sociological phenomenon, and it is in this light that we shall study it" (p. xxvi). Other authors whose works are published in English do not necessarily use the term "technique" but still discuss "technology" in its sociological or more abstract dimensions; Italian authors (including the author of this work and the Italian authors he cites) also use *tecnica* and *tecnologia* in interchangeable ways, decipherable primarily by context. Hence, as in the original Italian, the terms still maintain a certain degree of interchangeability in the discussion presented here.

[38] See M. Lombardi Ricci, *Il cantiere della vita: Risvolti culturali delle biotecnologie* (Bologna: Pardes, 2006).

Then the need arose to organize all of the individual techniques in order to improve and maximize their output, leading to the rise of what Ellul calls the "technical phenomenon," with two impacts. The negative impact was losing the relationship between technique, society, and individual; the positive impact was gaining worldwide geographical distribution and rapid transmission (though the differences between individual civilizations were eliminated).[39]

In this context, *technical automatism* reigns: technical direction and decisions happen on their own.[40] "Man is there like a device registering effects, in the sense that he decides according to the criterion of maximum effectiveness and efficiency. Technical automatism does not tolerate being questioned; in fact, the harshest reprobation in our world is the accusation that someone or something is impeding it—an act that is interpreted as putting the brakes on progress."[41] Anything that is possible becomes obligatory: the message is "we have to try," "science should not be impeded," "all obstacles must be overcome."

It is important to point out that *technical self-augmentation* happens anonymously and collectively; man's inventiveness is no longer important as connectivity counts even more. Technique repairs its own negative effects, even when they are not of a technical sort, through technical instruments: the answer to all problems is sought in technique.[42] Technical progress is difficult to reverse, and we do not reject what has become a part of our daily lives. In this manner, technological growth takes on the contours of a geometric progression.

Uniqueness is a fundamental characteristic of the technical phenomenon: individual technologies share only essential traits. According to Ellul, it is impossible to separate the technology from its use, what is good from what is bad and therefore in need of improvement, because modern technique cannot be reconnected to the technique-instrument of centuries past: its use is now ontologically linked to its being.

Based on this, it is impossible to give it a direction: that would be subjecting it to a nontechnical judgment. *Modern-day technique*, according to Ellul, *is rigorously independent of morality*: "It does not act in sight of a recognized and critically planned aim; rather it is the combination of prior elements that creates new technical elements. Proposing a direction amounts to denying the nature of technique and limiting its range."[43]

In this framework, it is impossible to distinguish the technical means from its use: there is only one effective and efficient use. In this manner, man closes himself

[39] Ellul, *Technological Society*, 77–78.

[40] Ibid., 79–80.

[41] Lombardi Ricci, *Il cantiere della vita*, 29, referencing Ellul, *Technological Society*, 80: "The worst reproach modern society can level is the charge that some person or system is impeding this technical automatism."

[42] According to P. Sloterdijk only technique can "save us," thanks to the intelligent technologies constituting the new artificial humanism. He is trusting toward "anthropotechnology," which will ultimately achieve an explicit programming of genetic characteristics so that humanity can move beyond "the fatalism of birth" through optional birth with prenatal selection. See P. Sloterdijk, *Nicht gerettet: Versuche nach Heidegger* (Frankfurt am Main: Suhrkamp, 2001).

[43] Lombardi Ricci, *Il cantiere della vita*, 31.

into a closed universal language and proper technical operation, which is objectively measurable and verifiable.

Lombardi Ricci rightly asks whether a free man can be a good technician: "Neither right nor good can be a criterion for evaluating technical operation. Independence from morality and spiritual values is therefore an indispensible requirement; good and right are actually kept but eviscerated of their original meaning and set in relation to the principles of technical operation. In this manner, technique creates an internal ethics which is the indispensible condition for its development; the consequence is man's adherence to machine."[44]

If technique proceeds only by calculation, if it transforms means into ends and only accepts assessments of a technical nature, then it is clear that human spirituality and the yearning for transcendence are stifled. A world in which machines dominate is standardizing and stifling: it is a world in which quantity takes the place of quality and spiritual values are replaced with instrumental and utilitarian values.

Thinkers like Ellul, Emanuele Severino, and Jonas all underscore the self-referential structure of technique. According to Severino,[45] technique is not aimed at a specific goal but at an increase in the power itself to produce objectives and technicality is the very essence of man. This is contested by Jonas,[46] though he also points out the self-referential aim of technique at present.

In the current technical context, truth is no longer the profound essence of reality and the aim of knowledge; rather, it is the effective procedure for obtaining a result—it is something merely operative. Reason clearly reaches no longer toward the end and toward the good; it is procedural and instrumental.

In the classical understanding, *episteme* (knowledge) can be considered "pure" to the extent that it indicates the essence of things, the truth inscribed in the natural order of things. Theory is separate from practice and the disposition of assessment can be assumed at the moment of application.

The experimental method changed the way that knowledge is understood: experimentation requires isolating phenomena from the whole context and controlling, measuring, and quantifying them with the appropriate instruments. With the introduction of mathematics and experimentation, reality ceased to be a source of meaning: "The world loses its meaning, vanishing as a language that speaks to us and reveals the essence of things. Nature is no longer seen in its totality as a manifestation, but as a mosaic of elements to be used for the purpose of improving the material conditions of humanity."[47] Contemplation is replaced by manipulation of reality, and so "truth as the revelation of the essence of things is replaced by the need to verify the effectiveness of provisional procedures in view of an outcome to be obtained. The world, nature itself, is changed through the hypotheses advanced by

---

[44] Ibid., 32.

[45] See E. Severino, *Il destino della tecnica* (Milan: Rizzoli, 1988).

[46] See H. Jonas, *Imperative of Responsibility*; idem, *Tecnica*; idem, *Sull'orlo dell'abisso: Conversazioni sul rapporto uomo e natura* [At the edge of the abyss: Conversations on the relationship between man and nature] (Turin: Einaudi, 2000); E. Colombetti, "La tecnologia tra strumento e paradigma," in *Vita e Pensiero* 2 (2002): 135–151.

[47] Lombardi Ricci, *Il cantiere della vita*, 40.

man, insofar as man cuts from the whole a fragment on which he intends to work, and configures it—by the instruments he uses—in the ways he finds most suitable."[48]

The old science investigated reality; technology would construct it to our liking. Yet we no longer know how to use metaphysical reason for grasping ends and goods, so we become automatons at the mercy of mechanical causes, incapable of grasping life, which is teleology. The contemporary technological mentality trumpets the insignificance of limits,[49] not admitting the reality of constitutive limits that can be seen as good insofar as they correspond to the being that is proper what is human. The ontological and axiological evisceration of reality, which is transformed into a collection of elements open to manipulation at man's pleasure, brings to light a new existential angst: the lack of any meaning to be recognized and developed.

The attempt to "construct" human reality cannot help but fail. It is difficult for us to talk about *human nature*, which is to say our essence, which is distinct from the *human condition*, which is the *setting* in which we actually live. As Elena Colombetti observes, "The two are not unrelated, of course: the human *condition* must answer to what human *nature* requires for survival and development in the totality of somatic unity."[50] Some changes are peripheral, but others affect our identity. This is precisely why technological practice raises the question about man, the meaning of his existence, and the meaning of his actions, and has led to the emergence of bioethics, which reflects primarily on the intrusion of technology into organic life.

---

[48] Ibid., 40–41.

[49] A. Pessina, "Il 'senso' del possibile e l'orizzonte del limite nella civiltà tecnologica," in "Domande di etica," *Hermeneutica* (2001): 41–64.

[50] E. Colombetti, *Incognita uomo: Corpo, tecnica, identità* (Vita e Pensiero, Milano 2006), 139.

Summary Outline for Chapter 16

Bioethics and technology

What is possible is obligatory

Technology transforms spatiotemporal perceptions and allows the human subject to broaden the scope of his planning, his human relations, and his very rationality.

Technological science is not only a sensorial *enhancement* (the microscope or the webcam), but also the grafting of mechanical parts into our body (artificial heart valves) or the *replacement*, as in IVF–embryo transfer, of human acts of procreation through the mechanical operations of scientific technology.

Every development is presented as *progress*, and the limits of human existence are always viewed as plain and simple "obstacles" or something else that should inherently be overcome.

Technology is proposed as *good* precisely because *it frees us from many physical limitations*. The advent of biotechnology, from this perspective, is the sign of a momentous change: for the first time in history man is capable of reproducing outside of a mother's womb. What emerges first of all from the experience of technology is the notion that *what is possible is obligatory*. For experimental science, truth is the effectiveness of the procedure, in other words, it is found at the end of the experimental process and not at the beginning. Hence, for technology, a limit is always thought of as an obstacle, that is, as something that must be overcome. "In ethics, by contrast, there are obstacles (which can be surmounted) that must be taken as limits, because they call on us to respect the good." (A. Pessina)

Technological man has lost the capacity to recognize in nature an immanent dynamic that can nonetheless be guided by the human will, first and foremost through imitation of it.

The advent of genetic engineering involves the conviction that human planning is the only horizon of meaning. The fear of the collapse of the ecosystem is not enough to put an end to the illusion of those who think that every problem raised by the expansion of biotechnology can in turn be corrected by new technological advances. Biological nature has been deprived of all intrinsic finality, becoming mere raw material.

Three epistemological models, three approaches to reality

Pessina proposes three epistemological models:

- The first belongs to the Western *metaphysical tradition* and is typical of the Judeo-Christian culture, which *identifies the source of the intelligibility of reality in the doctrine of creation*. It affirms *the equivalence between the truth and being of things*.

- The second belongs to the *modern scientific revolution* (first stage of scientific development). In contrast, it identifies *the locus of truth in what is made*.

- The third characterizes the present scientific situation, in which *a clear-cut separation between technology and science occurs*, at least *at the level of the experimental sciences.* This technological culture attributes the connotation of *truth* to *what is feasible or makeable.*

It can also be said that

- A *speculative model* that acknowledges the truth in reality (intelligibility belongs to what is real and is not the product of human cognitive activity) is gradually being replaced by

- An *operative model*, which claims to be able to guarantee the value of knowledge wherever man deals with something that has been constructed or that can be constructed (intelligibility is the operation by which man attributes meaning to the reality that surrounds him because he is capable of constructing theoretical and operative models that explain reality).

- Finally, *intelligibility is identified with planning*, and this has no predetermined models because it is enmeshed in the *possibilities that continually open up* through increased experimentation.

Is technological man still free?

The uneasiness generated by technology is at the origin of bioethics, and it is important to contrast traditional technology with modern technology.

Past technologies did not take up much space and the means available were limited; hence greater efficiency was sought by modifying not so much the instrument as the manual ability of man. Technology was always local and did not have issues with adaptation: since there was a variety of models and no tool had an absolute value, man could break the connection with a technology, using it for the aim he sought.

Then the need arose to organize all of the individual techniques to improve and maximize their output, leading to the rise of what Ellul calls the "technical phenomenon," with two impacts. The negative impact was losing the relationship between technique, society, and individual; the positive impact was gaining worldwide geographical distribution and rapid transmission (though the differences between individual civilizations were eliminated).

- In this context, *technical automatism* reigns: technical direction and decisions happen on their own.

- It is important to point out that *technical self-augmentation* happens anonymously and collectively; man's inventiveness is no longer important, since connectivity counts even more.

- *Technique repairs is own negative effects*, even when they are not of a technical sort, *through technical instruments*: the answer to all problems is sought in technique.

- *Technical progress is difficult to reverse*, and we do not reject what has become a part of our daily lives. In this manner, technological growth takes on the contours of a geometric progression.

- *Uniqueness is a fundamental characteristic of the technical phenomenon*: individual technologies share only essential traits.

- Based on this, *it is impossible to give it a direction*: that would be subjecting it to a nontechnical judgment. *Modern-day technique*, according to Ellul, *is rigorously independent of morality.*

In this framework, it is impossible to distinguish the technical means from its use: there is only one effective and efficient use. In this manner, man closes himself into a closed universal language and proper technical operation, which is objectively measurable and verifiable. But is he still free?

Return to meditative thought

Reality, transformed into a set of elements that can be manipulated at man's whim, causes the emergence of an existential unease connected with the lack of a meaning that can be recognized and developed.

It must be stressed that technology is not self-referential and that one must return to thinking and operating within the truth of being by once again appropriating ontology and metaphysics.

As Heidegger affirmed, we need a form of meditative thought capable of engaging with the complete domination of technology, the characteristic of our age.

Christian thought recognizes that technology opens up new paths, contributes to improved living conditions, and spreads culture; it emphasizes, however, that it does not always pursue true human values. The balance between technological-scientific development and these values, therefore, is one of the most urgent tasks in contemporary technological cultures. Science and technology can improve societies and cultures, make nature better known, and transform it.

Scientific research and technological transformations can make possible better social coexistence and greater responsibility. However, they also lead to agnosticism, the loss of transcendence, and illusions of self-sufficiency. (See *Gaudium et spes*, nos. 54, 56–57; see John Paul II, "Address to the Pontifical Academy of Sciences," October 29, 1990, no. 6.)

With regard to "technological cultures," the Christian faith will have to

- Elucidate and orient their planning according to the purposes, meanings, and values of the Gospel

- Acknowledge the value and originality of technology as a glorification of God in creation, the transformation of reality for service of the common good and our neighbor's good, and the actualization and growth of human persons

- Guide those who are responsible for overseeing technologies and innovations to respect, conserve, and develop creation for the service of mankind

- Address with suitable measures the constant decrease in jobs resulting from technological innovations and developments

- Oppose the loss of human abilities resulting from technological developments

With regard to the "technologies," the faith should lead to

- Aiding labor, freeing it from the most harmful, dangerous, tiring, and frustrating aspects of technology

- Planning beneficial and reasonable transformations of creation
- Monitoring the negative consequences
- Keeping in mind the needs of future generations
- Monitoring and opposing the negative consequences of technological developments.

(See G. Gismondi, "Tecnologia," in *Dizionario interdisciplinare di scienza e fede.*)

Bibliography
and Indexes

Bibliography

Aancsen, A., and F. Flam. "Bilateral Tubal Pregnancy following In Vitro Fertilization and Transfer of Two Embryos." *European Journal of Obstetrics and Gynecology and Reproductive Biology* 64 (1996): 235–236.

Abbagnano, N. "Addio cicogna, ora avremo il neonato ordinato per telefono." *Corriere della Sera*, October 15, 1984.

Abel, F. "Bioethics in Spain: 1991–1993." In *Bioethics Yearbook: Regional Developments in Bioethics, 1991–1993*, edited by B.A. Lusting, 4:269–283. Dordrecht: Kluwer, 1995.

Abraham, B., and R. Perto. *Psicanalisi e terapie sessuologiche*. Milan: Feltrinelli, 1979.

Academisch Medisch Centrum Amsterdam, Academisch Ziekenhuis Groningen, Leids Universtair Medisch Centrum, et al. *National Protocol Crossover Kidney Transplantation*. 2004.

Adinolfi, M., C. Camporese, and T. Carr. "Gene Amplification to Detect Fetal Nucleated Cells in Pregnant Women." *Lancet* 2 (1989): 328–329.

Adinolfi, M., A. Davies, S. Sharif, P. Soothill, and C. Rodeck. "Detection of Trisomy 18 and Y-Derived Sequences in Fetal Nucleated Cells Obtained by Transcervical Flushing." *Lancet* 342 (1993): 403–404.

Adinolfi, M., J. Sherlock, P. Soothill, and C. Rodeck. "Molecular Evidence of Fetal Derived Chromosome 21 Markers (STRs) in Transcervical Samples." *Prenatal Diagnosis* 15 (1995): 35–39.

Adinolfi, M., P. Soothill, and C. Rodeck. "A Simple Alternative to Amniocentesis?" *Prenatal Diagnosis* 14 (1994): 231–233.

Adorno, T.W. *Dialettica negativa*. Turin: Einaudi, 1970.

Agazzi, E. *Il bene, il male e la scienza: Le dimensioni etiche dell'impresa scientifico-tecnologica*. Milan: Rusconi, 1992.

———. *Il concetto di progresso della scienza*. Milan: Feltrinelli, 1976.

———, ed. *Manipolazioni genetiche e diritto: Atti del XXXV Convegno di studio; Roma, 7–9 dicembre 1984*. Quaderni di Iustitia 34. Rome: Giuffrè, 1986.

———, ed. *Storia delle scienze*. Rome: Città Nuova, 1984.

Agostini, S., and F. Perazza. *Eutanasia: Problematiche etiche, medico-legali, giuridiche*. Turin: Minerva Medica, 2004.

Aiken, H.D. "Life and the Right to Life." In *Ethical Issues in Human Genetics*, edited by B. Hilton, 173–183. New York: Plenum Press, 1973.

Ales Bello, A. "L'analisi della corporeità nella fenomenologia." In *Corporeità e pensiero: Atti dell'VIII Convegno Studium; Roma, 21–23 ottobre 1999*. Rome: Studium, 2000.

Allen, P., and W.E. Waters. "Attitudes to Research Ethics Committees." *Journal of Medical Ethics* 9 (1983): 61–65.

Allport, G.W. *Personality: A Psychological Interpretation*. New York: Holt, 1937.

American Academy of Pediatrics Special Task Force on Brain Death in Children. "Guidelines for the Determination of Brain Death in Children." *Pediatrics* 80, no. 2 (1987): 298–300.

American College of Obstetricians and Gynecologists. *Ultrasound in Pregnancy: Technical Bulletin 116.* Washington, DC: ACOG, 1988.

American College of Occupational Medicine. "Code of Ethical Conduct for Physicians Providing Occupational Medical Services." *Journal of Occupational Medicine* 18, no. 8 (1976): 1.

Ancona, L. "Prospettive psicologiche in tema d'aborto." *Medicina e Morale* 24, no. 3 (1974): 377–397.

Ancora, G., ed. *Biotecnologie animali e vegetali: Nuove frontiere e nuove responsabilità.* Vatican City: Libreria Editrice Vaticana, 1999.

Andersen, A.N., L. Gianaroli, R. Felberbaum, J. de Mouzon, K.G. Nygren, European IVF-Monitoring Programme (EIM), and European Society of Human Reproduction and Embryology (ESHRE). "Assisted Reproductive Technology in Europe, 2001: Results Generated from European Registers by ESHRE." *Human Reproduction* 20, no. 5 (2005): 1158–1176.

Anderson, C., and P. Aldhous. "Human Genome Project: Still Room for HUGO?" *Nature* 355 (1992): 4–5.

———. "Secrecy and the Bottom Line." *Nature* 354 (1991): 96.

Anderson, W.F. "Prospects of Human Gene Therapy." *Science* 226 (1984): 401–409.

Andorno, R. "The Oviedo Convention: A European Legal Framework at the Intersection of Human Rights and Health Law." *Journal of International Biotechnology Law* 2 (2005): 133–143.

Andrews, L.B., J.E. Fullarton, N.A. Holtzman, and A.G. Motulsky, eds. *Assessing Genetic Risks: Implications for Health and Social Policy.* Washington, DC: National Academy Press, 1994.

Angelini, F., ed. *Pio XII: Discorsi ai medici.* Rome: Orizzonte Medico, 1959.

Angell, M. "The Ethics of Clinical Research in the Third World." *New England Journal of Medicine* 337, no. 12 (1997): 847–849.

Annas, G.J. "Baby M. Babies and Justice for Sale." *Hastings Center Report* 17 (1987): 13–15.

———. "Reforming Informed Consent to Genetic Research." *Journal of the American Medical Association* 286, no. 18 (2001): 2326–2328.

Annuaire européenne de bioéthique. Paris: Lavoiser, 1996.

Antico, L., and E. Sgreccia, eds. *Anzianità creativa.* Milan: Vita e Pensiero, 1989.

Antiseri, D. *Cristiano perché relativista, relativista perché cristiano: Per un razionalismo della contingenza.* Soveria Mannelli: Rubbettino, 2003.

———. *Trattato di metodologia delle scienze sociali.* Turin: UTET, 1996.

Anzani, A. *Trapianti d'organo: Problemi etici, aspetti sociali.* Milan: Lauri, 1996.

Apel, K.O. *Comunità e comunicazione.* Turin: Rosenberg and Sellier, 1977.

Aquinas, T. *De Anima.*

———. *Quaestiones disputatae: De potentia.* Vol 3. Casale Monferrato: Marietti, 195.

———. *Quaestiones disputatae: De spiritualibus creaturis*, edited by L. Keeler. Rome: Gregorianum, 1946.

———. *Sententia libri ethicorum Aristotelis*. Casale Monferrato: Marietti, 1949.

———. *Summa contra gentiles*. Notre Dame: University of Notre Dame Press, 1975.

———. *Summa theologiae*. Bologna: ESD, 1984.

Areen, J. "Limiting Procreation." In *Medical Ethics*, edited by R. M. Veatch. London: Jones and Bartlett, 1997.

Arendt, H. *The Human Condition*. Garden City, NY: Doubleday, 1959.

Ariès, P. *Essais sur l'histoire de la mort en Occident du Moyen Age à nos jours*. Paris: Éditions du Seuil, 1975.

Aristotle. *The Complete Works: The Revised Oxford Translation*, edited by J. Barnes. Princeton: Princeton University Press, 1984.

Armstrong, C. "Thousands of Women Sterilized in Sweden without Consent." *British Medical Journal* 315, no. 7108 (1997): 563.

Arpaillange, P., and S. Dion. "Considérations sur l'éthique de la randomisation." *Biomedicine and Pharmacotherapy* 38 (1984): 426–429.

Artigas, M. *Le frontiere dell'evoluzionismo*. Milan: Ares, 1993.

Asch, R. H., J. P. Balmaceda, L. R. Ellsworth, and P. C. Wong. "Gamete Intra-Fallopian Transfer (GIFT): A New Treatment for Infertility." *International Journal of Fertility* 30 (1985): 41–45.

Asch, R. H., J. P. Balmaceda, L. R. Ellsworth, and P. C. Wong. "Preliminary Experiences with Gamete Intra-Fallopian Transfer (GIFT)." *Fertility and Sterility* 45 (1986): 366–371.

Ascone, G. B., and L. Rossi Carleo. *La procreazione artificiale: Prospettive di una regolamentazione legislativa nel nostro Paese*. Naples: Edizioni Scientifiche Italiane, 1986.

Ashley, B. M. *Theologies of the Body: Humanist and Christian*. Braintree, MA: Pope John XXIII Medical-Moral Research Center, 1985.

"Assisted Reproductive Technology in the United States and Canada: 1994 Results Generated from the American Society for Reproductive Medicine/Society for Assisted Reproductive Technology Registry." *Fertility and Sterility* 66, no. 5 (1996): 697–705.

"Assisted Reproductive Technology in the United States: 2000 Results Generated from the American Society for Reproductive Medicine/Society for Assisted Reproductive Medicine/Society for Assisted Reproductive Technology Registry." *Fertility and Sterility* 81 (2004): 1207–1220.

Aubard, Y., M. P. Teissier, and J. H. Baudet. "Greffes et transplantations ovariennes chez la femme: Le point." *Revue Française de Gynécologie et d'Obstétrique* 88 (1993): 583–590.

Aubert, J. M. *Sexualité, amour et mariage*. Paris: Beauchesne, 1970.

Augustine. *Confessions*. Translated by F. J. Sheed. New York: Sheed and Ward, 1943.

Australian In Vitro Fertilization Collaborative Group. "High Incidence of Preterm Birth and Early Losses in Pregnancy after In Vitro Fertilization." *British Medical Journal* 291 (1985): 1160.

Awsare, N. S., J. Krishnan, G. B. Boustead, D. C. Hanbury, and T. A. McNicholas. "Complications of Vasectomy."*Annals of the Royal College of Surgeons of England* 87, no. 6 (2005): 406–410.

Aznar, J. "Multifetal Pregnancy Reduction." *Acta Obstetricia et Gynecologica Scandinavica* 75, no. 1 (1996): 90–91.

Azzone, F. G. *Il senso della vita*. Bari: Laterza, 1994.

Babolin, A. *Essere e alterità in M. Buber*. Padua: Gregoriana, 1965.

Baccarini, E. *La persona e i suoi volti: Etica e antropologia*. 2nd ed. Rome: Anicia, 2003.

Bacchus, C., and W. Buselmaier. "Blastomere Karyotyping and Transfer of Chromosomally Selected Embryos: Implications for the Production of Specific Animal Models and Human Prenatal Diagnosis." *Human Genetics* 80 (1988): 333–336.

Baier, K. *The Rational and the Moral Order*. Chicago: Open Court, 1995.

Baird, D. T., and A. F. Glasier. "Hormonal Contraception." *New England Journal of Medicine* 328, no. 21 (1993): 1543–1549.

Baird, P. A., and A. D. Sadovnick. "Survival in Infants with Anencephaly." *Clinical Pediatrics* 23, no. 5 (1984): 268–271.

Baka, S. G., T. L. Toth, L. L. Veeck, H. W. Jones Jr., S. J. Muasher, S. E. Lanzendorf. "Evaluation of the Spindle Apparatus of In Vitro Matured Human Oocytes following Cryopreservation." *Human Reproduction* 10, no. 7 (1995): 1816–1820.

Baker, R. P., and V. Hargreaves. "Organ Donation and Transplantation: A Brief History of Technological and Ethical Developments." In *The Ethics of Organ Transplantation*, edited by W. Shelton and J. Balint, 1–42. Oxford: Oxford University Press, 2001.

Balduzzi, R., C. Cirotto, and I. Sanna. *Le mani sull'uomo: Quali frontiere per la biotecnologia?* Rome: AVE, 2005.

Ballesteros, J. "La costruzione dell'immagine attuale dell'uomo." In *Immagini dell'uomo: Percorsi antropologici nella filosofia moderna*, edited by I. Yarza. Rome: Armando, 1997.

Balmaceda, J. P., A. Manzur, and R. H. Asch, "Gamete Intrafallopian Transfer." In *Infertility: Evaluation and Treatment*, edited by W. R. Keye, R. J. Chang, R. W. Rebar, and M. R. Soules, 772–779. Philadelphia: W. B. Saunders, 1995.

Bandeira, R. "Hospital Ethics Committees in Portugal." *HEC Forum* 3, no. 6 (1991): 347–348.

Baraldo, M., G. Fassina, A. Osculati, A. Sali, and G. Vandoni. "Riflessioni sulla disponibilità del cadavere da parte dei familiari."*Rivista Italiana di Medicina Legale* 22, no. 4–5 (2000): 947–953.

Barberi, P., and D. Tettamanzi, eds. *Matrimonio e famiglia nel magistero della Chiesa*. Milan: Massimo, 1986.

Barcaro, R., and P. Becchi, eds. *Questioni mortali: L'attuale dibattito sulla morte cerebrale e il problema dei trapianti*. Naples: Edizioni Scientifiche Italiane, 2004.

Barclay, W. *Ethics in a Permissive Society*. Glasgow: Collins, 1971.

Barten, E. J., and D. W. Newling. "Transplantation of the Testis: From the Past to the Present." *International Journal of Andrology* 19 (1996): 205–211.

Bartolomei, S. *Etica e ambiente*. Milan: Guerini, 1989.

———. *Etica e natura*. Bari: Laterza, 1995.

Bascarin, C., W. Bianchi, M. Costanza, et al. *Manuale pratico di sperimentazione clinica.* Fidenza: Mattioli, 1885, 2002.

Baschat, A.A., E. Schwinger, and K. Diedrich. "Assisted Reproductive Techniques: Are We Avoiding the Genetic Issue?" *Human Reproduction* 11 (1996): 926–928.

Bateson, G. *Steps to an Ecology of Mind.* San Francisco: Chandler Publishing, 1972.

Batzer, F.R., B. Gocial, S.L. Corson, S. Weiner, R.J. Wapner. "Multiple Pregnancies with Gamete Intrafallopian Transfer (GIFT): Complications of a New Technique." *Journal of In Vitro Fertilization and Embryo Transfer* 5, no. 1 (1988): 35–37.

Baulieu, E.E. "Contragestation by Antiprogestin: A New Approach to Human Fertility Control." In *Abortion: Medical Progress and Social Implications*, Ciba Foundation Symposium 115, edited by R. Porter and M. O'Connor, 192–210. London: Pitman, 1985.

Baumgärtel, F., R. Meyer, and E. Schweiser. "Sarx." In *Grande lessico del nuovo testamento*, edited by G. Kittel and G. Friedrich, translated by G. Torti and U. Argenti, 11:1265–1398. Brescia: Paideia, 1976.

Bausola, A. *Libertà e responsabilità.* Milan: Vita e Pensiero, 1980.

———. *Natura e progetto dell'uomo.* Milan: Vita e Pensiero, 1977.

Bayle, B. "L'activité antinidatoire des contraceptifs oraux." *Contraception Fertilité Sterilité* 22, no. 6 (1994): 391–395.

Beauchamp, T.L. "The Promise of the Beneficence Model for Medical Ethics." *Journal of Contemporary Health Law and Policy* 6 (1990): 145–155.

Beauchamp, T.L., and J.F. Childress. *Principles of Biomedical Ethics.* 4th ed. New York: Oxford University Press, 1994.

———. "Un passo indietro e due avanti: Peter Singer e i trapianti." *Bioetica* 10, no. 2 (2002): 226–247.

Beauvoir, S. de. *The Second Sex.* New York: Knopf, 1952. Originally published as *Le deuxième sexe*, 2 vols. (Paris: Gallimard, 1949).Becchi, P. *La morte nell'età della tecnica: Lineamenti di tanatologia etica e giuridica.* Genoa: Compagnia dei librai, 2002.

Becchi, P., and P. Donadoni. "Informazione e consenso all'espianto di organi da cadavere: Riflessioni di politica del diritto sulla nuova legislazione." *Politica del Diritto* 32, no. 2 (2001): 257–287.

Becker, E. *The Denial of Death.* New York: Free Press, 1973.

Beirnaert, L. "L'avortement est-il un infanticide?" *Études* (November 1970): 520.

Bell, J. "The New Genetics in Clinical Practice." *British Medical Journal* 316 (1998): 618–620.

Bellavite, P., P. Musso, and R. Ortolani, eds. *Il dolore e la medicina: Alla ricerca di senso e di cure.* Firenze: Società Editrice Fiorentina, 2005.

Benciolini, P. "Deontologia e obiezione di coscienza." In *Bioetica in medicina*, edited by A. Bompiani, 366–382. Rome: CIC Edizioni Internazionali, 1996.

———. "Obiezione di coscienza alla pillola?" *Bollettino dell'Ordine dei Medici di Padova* 1 (1994): 8–11.

Benciolini, P., and A. Aprile. *L'interruzione volontaria della gravidanza: Compiti, problemi, responsabilità.* Padua: CEDAM, 1990.

Bendall, C. *Standard Operating Procedures for Local Research Ethics Committees: Comments and Examples.* London: McKenna, 1994.

745

Bennet, J.C., for the Board on Health Sciences Policy of the Institute of Medicine. "Inclusion of Women in Clinical Trials: Policies for Population Subgroups." *New England Journal of Medicine* 329, no. 4 (1993): 288–292.

Bentham, J. *An Introduction to the Principles of Morals and Legislation.* 1779. Reprint, London: Athon Press, 1970.

Berkowitz, R.L. "From Twin to Singleton." *British Medical Journal* 313 (1996): 373–374.

Berlinguer, G. *Il corpo come merce o come valore*. Bari: Laterza, 1993.

———, ed. *La procreazione assistita: Aspetti scientifici, etici e giuridici; Atti della terza Giornata Europea di Bioetica; Milano, sabato 26 novembre 1988.* Turin: LDC, 1989.

———. *Questioni di vita: Etica scienza e salute*. Turin: Einaudi, 1991.

Berlinguer, G., and V. Garrafa. *La merce finale: Saggio sulla compravendita di parti del corpo umano.* Milan: Baldini e Castoldi, 1996.

Bernard, J. *De la biologiè à l'éthique: Nouveaux pouvoirs de la science, nouveaux pouvoirs de l'homme.* Paris: Buchet Chastel, 1990.

Bernardini, R. "In lista d'attesa con poche speranze." *Missione Salute* 4 (1991).

Bernat, J.L., C.M. Culver, and B. Gert. "On Definition and Criterion of Death." *Annals of Internal Medicine* 94 (1981): 389–394.

Berté, V., and A.A. Bignamini, eds. *Comitati di etica e farmacovigilanza.* Bologna: Health, 1994.

Bertolino, M. "Suicidio (istigazione o aiuto al)." In *Digesto, Sez. pen.*, 14:113. Turin: UTET, 1998.

Bianco, I., B. Graziani, M. Lerone, D. Ponzini, M.C. Aliquo, and E. Foglietta. "Updated Results of the Thalassemia Prevention Programme Carried Out in Latium." *Journal of Medicine and Genetics* 26 (1989): 667.

Biggers, J.D. "Arbitrary Partitions of Prenatal Life." *Human Reproduction* 5, no. 1 (1990): 1–6.

Bignamini, A.A., and D.R. Hutchinson. "Guida pratica alle GCP per i ricercatori." In *Lineamenti di etica della sperimentazione clinica*, edited by A.G. Spagnolo and E. Sgreccia, 181–194. Milan: Vita e Pensiero, 1994.

Billings, E., and A. Westmore. *The Billings Method: Controlling Fertility without Drugs or Devices.* New York: Random House, 1980.

Billings, P.R. "In Utero Gene Therapy: The Case Against." *Nature Medicine* 5 (1999): 255.

Binding, K., and A. Hoche. *Die Freigabe der Vernichtung Lebensunwerten Lebens.* Leipzig: Meiner, 1920.

Bioethics Center of the Catholic University of the Sacred Heart. "Contro la sperimentazione sugli embrioni umani." *Medicina e Morale* 46, no. 4 (1996): 802–809.

———. "Identità e Statuto dell'embrione umano." *Medicina e Morale* 39, no. 4, suppl. (1989): 665–666.

———. "Sulla cosiddetta 'contraccezione d'emergenza.'" *Medicina e Morale* 47, no. 3 (1997): 582–589.

———. "Sulla cosiddetta 'riduzione' embrionale." *Medicina e Morale* 47, no. 2 (1997): 374–381.

Birch, C., and P. Abrecht. *Genetics and the Quality of Life.* Oxford: Pergamon Press, 1975.

746

Biscaia, J., and W. Osswald. "Bioethics in Portugal: 1991–1993." In *Bioethics Yearbook: Regional Developments in Bioethics, 1991–1993*, edited by B.A. Lusting, 4:285–289. Dordrecht: Kluwer, 1995.

Biscontini, G. *La morte e il diritto: Il problema dei trapianti d'organo*. Naples: ESI, 1994.

Bishops Conference of England and Wales. "Moral Questions of Today: Statement." December 31, 1970. In *L'Osservatore Romano*. January 28, 1971.

Blandino, G. "L'argomentazione casualistica di Jacques Monod." *La Civiltà Cattolica* 2 (1978): 557–565.

Blandino, G. "Caso e finalità." *La Civiltà Cattolica* 2 (1977): 366–368.

Blazquez, N. *Bioética fundamental*. Madrid: Biblioteca Autores Cristianos, 1996.

Bleidt, B. *Clinical Research in Pharmaceutical Development*. New York: Dekker, 1996.

Blondel, M. "Le problème de l'immortalité de l'âme." *Supplément de la vie spirituelle* 61 (1939): 1–15.

Board of Directors of the American Society of Human Genetics. "ASHG Statement on Gene Therapy: April 2000." *American Journal of Human Genetics* 67 (2000): 272–273.

Böckle, F. "Ethische Aspekte der freiwilligen operativen Sterilisation." *Stimmen der Zeit* 99 (1974): 755–760.

———. *Morale fondamentale*. Brescia: Queriniana, 1979.

———. "Le pouvoir de l'homme sur l'homme." In *L'homme manipulé*, edited by C. Robert. Strasbourg: Cerdic, 1974.

Bognetti, G. "La legge olandese su eutanasia e suicidio assistito." *Corriere giuridico* 1 (2001): 705.

Boiardi, G. "Sessualità maschile e femminile tra natura e cultura." *Medicina e Morale* 33, no. 1 (1983): 12–24.

Bok, S. "Death and Dying: Euthanasia and Sustaining Life; Ethical Views." In *Encyclopedia of Bioethics*, edited by W.T. Reich, 268–277. New York: Free Press, 1978.

———. "The Ethics of Giving Placebos." *Scientific American* 15 (1974):17–23.

Bollen, N., M. Camus, C. Staessen, H. Tournaye, P. Devroey, and A.C. Van Steirteghem. "The Incidence of Multiple Pregnancy after In Vitro Fertilization and Embryo Transfer, Gamete, or Zygote Intrafallopian Transfer." *Fertility and Sterility* 55 (1991): 314–318.

Bompiani, A. *Bioetica in Italia: Lineamenti e tendenze*. Bologna: Dehoniane, 1992.

———. "Eutanasia e diritti del malato 'in fase terminale': Considerazioni giuridiche." *Presenza Pastorale* 5, no. 6 (1985): 76–119.

———. "Genomica funzionale e proteomica: Recenti sviluppi della ricerca nelle malattie poligeniche e considerazioni etiche." *Medicina e Morale* 53, no. 5 (2003): 797–840.

———. "Indicazioni all'aborto 'terapeutico': Stato attuale del problema." In *L'aborto: Riflessioni di studiosi cattolici*, edited by A. Fiori and E. Sgreccia, 191–215. Milan: Vita e Pensiero, 1975.

———. "Medico, servizio sanitario, economia." *Medicina e Morale* 35, no. 4 (1985): 691–716.

———. "Problemi biologici e clinici dell'ingegneria genetica." In *Bambini in provetta*, edited by G. Concetti, 43–96. Rome: Logos, 1986.

747

————. "La ricerca sperimentale in ambito biomedico: Ambiti, metodologie, criteri di validità dei progetti di ricerca." In *Etica della ricerca biomedica: Per una visione cristiana; Atti della nona assemblea generale della Pontificia Accademia per la Vita; Città del Vaticano, 24–26 febbraio 2003*, edited by J. de D. Vial Correa and E. Sgreccia, 95–127. Vatican City: Libreria Editrice Vaticana, 2004.

————. "La sperimentazione clinica dei farmaci: Stato attuale del problema normativo e proposte di riforma." *Medicina e Morale* 32, no. 2 (1982): 95–134.

————. *Le tecniche di fecondazione assistita: Una rassegna critica*. Milan: Vita e Pensiero, 2006.

————. "Una valutazione della 'Convenzione sui diritti dell'uomo e la biomedicina' del Consiglio d'Europa." *Medicina e Morale* 47, no. 1 (1997): 37–55.

Bompiani, A., and N. Garcea. "La fecondazione in vitro: Passato, presente, futuro." *Medicina e Morale* 36, no. 1 (1986): 79–102.

Bompiani, A., and E. Sgreccia, eds. *Trapianti d'organo*. Milan: Vita e Pensiero, 1989.

Bondi, S. A., D. Gries, and K. Faucette. "Neonatal Euthanasia?" *Pediatrics* 117, no. 3 (2006): 983.

Bondolfi, A. "Per una considerazione etica globale della medicina dei trapianti." *Bioetica e Cultura* 3, no. 5 (1994): 77–92.

Bonduelle, M., J. Legein, M. P. Derde, et al. "Comparative Follow-up Study of 130 Children Born after Intracytoplasmic Sperm Injection and 130 Children Born after In-Vitro Fertilization." *Human Reproduction* 10 (1995): 3327–3331.

Bongso, A., N. S. Chye, S. Ratnam, H. Sathananthan, and P. C. Wong. "Chromosome Anomalies in Human Oocytes Failing to Fertilize after Insemination In Vitro." *Human Reproduction* 3, no. 5 (1988): 645–649.

Bonjean, G., J. Bouchard, P. Forestier, N. Lery, N. Perrin, and G. Piot. "Le refus de soins: La dimension éthique du problème." *Médecine et Hygiène* 42 (1984): 1184–1189.

Bonnicksen, A. L. "In Vitro Fertilization and Embryo Transfer." In *Encyclopedia of Bioethics*, edited by W. T. Reich, 2221–2225. Rev. ed. New York: Macmillan, 1995.

Bontadini, G. *Saggio di una metafisica dell'esperienza*. Milan: Vita e Pensiero, 1995.

Bordignon, C., L. D. Notarangelo, N. Nobili, et al. "Gene Therapy in Peripheral Blood Lymphocytes and Bone Marrow for ADA-Immunodeficient Patients." *Science* 270, no. 5235 (1995): 470–475.

Borgonovo, G. "Essere e amore: Per un approccio personalista alla bioetica." *Annales Teologici* 17 (2003): 43–76.

Boschi, A. *Canonicum et Morale*. Rome: Officium Libri Catholici, 1965.

————. *L'eutanasia*. Turin: 1950.

————. *Questioni morali sul matrimonio*. Turin: Marietti, 1963.

Botturi, F. "Libertà e formazione morali." In *Alla ricerca delle parole perdute: La famiglia e il problema educativo*, edited by G. Borgonovo, 36–53. Casale Monferrato: Piemme, 2000.

————, ed. *Le ragioni dell'etica: Natura del bene e problema fondativo*. Milan: Vita e Pensiero, 2005.

Botturi, F., and C. Vigna, eds. "Affetti e legami." *Annuario di etica* 1 (2004).

Brandiff, B., L. Gordon, L. K. Ashworth, G. Watchmaker, and A. V. Carrano. "Detection of Chromosome Abnormalities in Human Sperm." *Progress in Clinical and Biological Research* 209B (1986): 469–476.

Brandt, R. *Ethical Theory.* Englewood Cliffs: Prentice Hall, 1959.

Brena, G. L., ed. *La libertà in questione.* Padua: Ed. Messaggero, 2002.

Brink, D. O. *Il realismo morale ed i fondamenti dell'etica,* edited by F. Castellani and A. Corradini. Milan: Vita e Pensiero, 2005.

Brody, B. A. *Ethical Issues in Drug Testing, Approval, and Pricing.* New York: Oxford University Press, 1995.

———. "The President's Commission: The Need to be More Philosophical." *Journal of Medicine and Philosophy* 14, no. 4 (1989): 369–383.

Brook, N. R., and M. L. Nicholson. "Non-Directed Live Kidney Donation." *Lancet* 368, no. 9533 (2006): 346–347.

Brouillard, H., and K. Barth. *Genèse et évolution de la théologie dialectique.* Paris: Montagnier, 1957.

Brovedani, E. "Il brevetto di organismi viventi ottenuti con l'ingegneria genetica: Aspetti scientifici, giuridici ed etici." *Aggiornamenti Sociali* 39, no. 4 (1988): 245–259.

———. "L'ingegneria genetica: Aspetti scientifico-tecnici." *Aggiornamenti Sociali* 7, no. 8 (1986): 517–534.

———. "Verso la terapia genica umana: Prospettive e implicazioni etiche." *Aggiornamenti Sociali* 9, no. 10 (1988): 591–610.

Brown, J. B., P. Harisson, M. A. Smith, et al. "Correlation between the Mucus Symptoms and the Hormonal Markers of Fertility throughout Reproductive Life." In *The Ovulation Method: The Achievement or Avoidance of Pregnancy by a Technique Which Is Reliable and Universally Acceptable,* J. J. Billings, 104 ff. Melbourne: Advocate Press, 1983.

Bruaire, C. *Filosofia del corpo.* Cinisello Balsamo: Paoline, 1975.

Buber, M. *I and Thou.* New York: Scribner, 1970.

———. *Werke.* Munich: Kösel-Verlag, 1962.

Bucelli, A., ed. *Produrre uomini: Procreazione assistita; Un'indagine multidisciplinare.* Firenze: Firenze University Press, 2006.

Bui, T. H., and H. Wramsby. "Micromanipulative Assisted Fertilization: Still Clinical Research." *Human Reproduction* 11 (1996): 925–926.

Burani, G. *Il passaggio dalla assistenza sanitaria alla tutela della salute.* Brezzo di Bedero: Salcom, 1985.

Buster, J. E., M. Bustillo, I. A. Rodi, et al. "Biologic and Morphologic Development of Donated Human Ova Recovered by Non-Surgical Uterine Lavage." *American Journal of Obstetrics and Gynecology* 135 (1985): 211–217

Butler, J. *Bodies That Matter: On the Discursive Limits of "Sex."* New York: Routledge, 1993.

Bynum, W. F., and R. Porter, eds. *Companion Encyclopedia of the History of Medicine.* 2 vols. London: Routledge, 1994.

Caferra, V. M. *Diritti della persona e stato sociale.* Bologna: Zanichelli, 1987.

Caffarra, C. "Aborto e obiezione di coscienza." *Medicina e Morale* 27, nos. 1–2 (1977): 101–109.

———. "Aspetti etici della diagnostica prenatale." *Medicina e Morale* 34, no. 4 (1984): 449–457.

———. *Non è bene che l'uomo sia solo.* Vol. 1, *Creati per amare.* Siena: Cantagalli, 2006.

———. "Il problema morale dell'aborto." In *L'aborto: Riflessioni di studiosi cattolici*, edited by A. Fiori and E. Sgreccia, 313–320. Milan: Vita e Pensiero, 1975.

———. "Il problema morale della sterilizzazione." *Medicina e Morale* 29, no. 2 (1979): 199–208.

———. "Scienza e tecnologia a servizio della vita con il trapianto di cuore artificiale." In *Trapianto di cuore e trapianto di cervello*, 97–101. Rome: Orizzonte Medico, 1983.

———. "La trasmissione della vita nella *Familiaris consortio.*" *Medicina e Morale* 33, no. 4 (1983): 391–400.

Calcagni, C., and E. Marziano. *Ostetricia e ginecologia forensi.* Rome: Società Editrice Universo, 1996.

Calguilhem, G. *Il normale and il patologico.* Turin: Einaudi, 1998, 65.

Calipari, M. *Curarsi e farsi curare.* Turin: Edizioni Paoline San Paolo, 2006.

Calisti, A. "Diagnosi prenatale e possibilità terapeutiche chirurgiche." *Medicina e Morale* 34, no. 4 (1984): 493–497.

———. "Il feto, paziente chirurgico." *Medicina e Morale* 33, no. 1 (1983): 49–58.

Callahan, D. *False Hopes: Why America's Quest for Perfect Health is a Recipe for Failure.* New York: Simon and Schuster, 1998.

———, ed. *The Hastings Center: A Short and Long 15 Years.* Hastings-on-Hudson, NY: Hastings Center, 1984.

———. "Religion and the Secularization of Bioethics." *Hastings Center Report* 6–7, suppl. (1990): 2–4.

———. *What Kind of Life: The Limits of Medical Progress*. New York: Simon and Schuster, 1990.

Calner, R. "Organs from Animals: Unlikely for a Decade." *British Medical Journal* 307 (1993): 306–307.

Camera dei Deputati Servizio Studi, ed. *Legge 1 aprile 1999, n. 91—Disposizioni in materia di prelievi e di trapianti di organi e di tessuti. Iter parlamentare; Dossier provvedimento n. 370/3: parti I, II e III.* Rome: Camera dei Deputati, 1999.

Campanini, G. "Eutanasia e società." In *Morire sì, ma quando?*, edited by P. Beretti, 58–67. Rome: Paoline, 1977.

Campbell, A.V. "Committees and Commissions in the United Kingdom. " *Journal of Medicine and Philosophy* 14, no. 4 (1989): 385–401.

Campbell, S., and J.M. Pearce. "Ultrasound Visualization of Congenital Malformations." *British Medical Bulletin* 39 (1983): 322–333.

Camus, A. "The Myth of Sisyphus." In *The Myth of Sisyphus and Other Essays*, translated by J. O'Brien. New York: Vintage Books, 1991. Originally published as "Le mythe de Sisyphe," in *Essais* (Paris: Bibliothèque de la Pléiade, 1965).

Candia, L. "L'uso del placebo nei trials clinici: Considerazioni etico-deontologiche." In *Lineamenti di etica della sperimentazione clinica*, edited by A.G. Spagnolo and E. Sgreccia, 91–94. Milan: Vita e Pensiero, 1994.

Candiani, G. B., V. Danesino, and A. Gastaldi, eds. *La clinica ostetrica e ginecologica*. 2 Vols. Milan: Masson, 1996.

Canepa, G. "Bioetica e deontologia medica: Aspetti problematici e conflittuali." *Rivista Italiana di Medicina Legale* 1 (1990): 3–6.

Canguilhem, G. *The Normal and the Pathological*. New York: Zone Books, 2007.

———. *Il normale e il patologico*. Turin: Einaudi, 1998.

Cantoni, L. *Il problema della popolazione mondiale e le politiche demografiche: Aspetti etici*. Piacenza: Cristianità, 1994.

———. "Su alcune dimensioni del suicidio: Il caso dell'Emilia Romagna." *Medicina e Morale* 44, no. 6 (1994): 1143–1160.

Caplan, A. L., and J. M. Wilson. "The Ethical Challenges of In Utero Gene Therapy." *Nature Genetics* 24 (2000): 107.

Capograssi, G. *Introduzione alla vita etica*. Rome: Studium, 1976.

Cappella, A., ed. *Liberi e responsabili collaboratori del Dio Creatore*. Rome: Centro Studi e Ricerche Regolazione Naturale della Fertilità, Facoltà di Medicina e Chirurgia, Università Cattolica del Sacro Cuore, 1980.

Cappella, A., E. Giacchi, G. Pompa, and C. Castagna. "Metodi naturali e cultura della vita: Valutazione di un'esperienza di insegnamento." *Medicina e Morale* 46, no. 4 (1996): 669–682.

Cappella, A., V. Navarretta, and E. Giacchi. "Il metodo dell'ovulazione Billings: Dati e valutazione dello studio multicentrico dell'OMS." *Medicina e Morale* 32, no. 4 (1982): 371–387.

Cappella, A. W., and A. Saporosi. "Impiego di moderne tecnologie nella regolazione naturale della fertilità: Il monitor ovarico di Brown." *Medicina e Morale* 47, no. 6 (1997): 1119–1128.

Caprile, G. "Il Magistero della Chiesa sull'eutanasia." In *Morire sì, ma quando?*, edited by P. Beretti, 192–220. Rome: Paoline, 1977.

———. *Non uccidere: Il Magistero della Chiesa sull'aborto*. Rome: La Civiltà Cattolica, 1973.

———. *Il Sinodo dei Vescovi: Terza Assemblea Generale; 27 settembre–26 ottobre 1974*. Rome: La Civiltà Cattolica, 1975).

Caprioli, A., and L. Vaccaro, eds. *Diritto morale e consenso sociale*. Brescia: Morcelliana, 1989.

Capuano, V. *Rilascio ambientale di organismi geneticamente modificati: La valutazione del rischio*. Rome: ENEA, 1991.

Carcaterra, G. *Il problema della fallacia naturalistica: La derivazione del dover essere dall'essere*. Milan: Giuffré, 1969.

Cardente, R. "La fecondazione artificiale." *Iustitia* 3 (1950): 21.

Carlson, J. W. "*Donum Vitae* on Homologous Interventions: Is IVF-ET a Less Acceptable Gift than GIFT?" *Journal of Medicine and Philosophy* 5 (1989): 523–540.

Carpitella, M., ed. *Dizionario delle antichità classiche di Oxford*. 3 vols. Rome: Paoline, 1963.

Carrasco de Paula, I. "Dal dono al Vangelo della vita: Per una lettura teologica dell'*Evangelium vitae*." In *"Evangelium vitae" e bioetica: Un approccio interdisciplinare*, edited by E. Sgreccia and D. Sacchini, 113–124. Milan: Vita e Pensiero, 1996.

———. "L'etica dell'intervento medico: Il primato dell'interesse del paziente." In *L'assistenza al morente*: *Atti del congresso internazionale "Care for Dying Persons"; 15–18 marzo 1992*, edited by E. Sgreccia, A.G. Spagnolo, and M.L. Di Pietro. Vatican City: Libreria Editrice Vaticana, 2000.

———. "Il problema filosofico ed epistemologico della morte cerebrale." *Medicina e Morale* 43, no. 5 (1993): 889–902.

Carse, A. "The Voice of Care: Implications for Bioethical Education." *Journal of Medicine and Philosophy* 16 (1991): 5–28

Carta dello Sviluppo sanitario della regione africana. Maputo: September 24, 1979.

Carter, B. "Large Number Coincidences and the Anthropic Principle in Cosmology." In *Confrontation of Cosmological Theories with Observational Data*, edited by M.S. Longair. Boston: Reidel, 1974.

Casali, P., and A. Santosuosso. "Il consenso informato nella sperimentazione clinica." In *Il consenso informato*, edited by A. Santosuosso, 169–184. Milan: Cortina, 1996.

Cascioli, R. *Il complotto demografico: Il nuovo colonialismo delle grandi potenze economiche e delle organizzazioni umanitarie per sottomettere i poveri del mondo*. Casale Monferrato: Piemme, 1996.

Casini, C. "Prospettive di riforma dell'attuale legislazione sull'aborto: Il dibattito italiano ed europeo." *La Speranza* suppl. (January 1995).

Casini, C., M. Casini, and M.L. Di Pietro. *La legge 19 febbraio 2004, n. 40 "Norme in materia di procreazione medicalmente assistita": Commentario*. Turin: Giappichelli, 2004.

Casonato, C., and T.E. Frosini, eds. *La fecondazione assistita nel diritto comparato*. Turin: Giappichelli, 2006.

Caspar, P. "Les fondements de l'individualité biologique." *Communio* 9, no. 6 (1984): 89–90.

———. "Individuazione genetica e gemellarità: L'obiezione dei gemelli omozigoti." *Medicina e Morale* 43, no. 3 (1994): 453–467.

Casson, F.F. "Dignità della professione medica." *Federazione Medica* 37 (1984): 936–941.

Castagnet, G. *Sexe de l'âme et sexe du corps*. Paris: Le Centurion, 1981.

Castignone, S., ed. *I diritti degli animali*. Bologna: Il Mulino, 1985.

———, ed. *Etica ambientale: Atti della giornata di etica ambientalista*. Naples: Guida, 1992.

Catholic Bishops' Joint Committee on Bioethical Issues. *Antenatal Tests: What You Should Know*. London: Incorporated Catholic Truth Society, 1989.

———. *In Vitro Fertilisation: Morality and Public Policy*. London: Catholic Media Office, 1983.

Catholic Bishops' Joint Committee on Bioethical Issues, Great Britain. *Genetic Intervention on Human Subjects*. London: Linacre Centre, 1996.

Cattorini, P. "Profilo della scuola di medicina e scienze umane: Educare ad un'intenzione antropologica." *Sanare Infirmos* 3 (1988): 19–23.

———. "Sperimentazione clinico-farmacologica: Problemi bioetici di attualità." In *A sua immagine e somiglianza?*, edited by A. Mazzoni, 236–251. Rome: Citta Nuova, 1997.

———. "Terapia e parola: Il rapporto medico-paziente come nucleo essenziale nella prassi medica." *Medicina e Morale* 35, no. 4 (1985): 781–799.

Cattorini, P., and V. Ghetti, eds. *La bioetica nelle facoltà di medicina.* Milan: F. Angeli, 1996.

Cavadi, A., N. Galantino, and E. Guarnieri. *Alla ricerca dell'uomo.* Palermo: Augustinus, 1988.

Cavicchi, I. *Ripensare la medicina: Restauri, reinterpretazioni, aggiornamenti.* Turin: Bollati Boringhieri, 2005.

———. *L'uomo inguaribile: Il significato della medicina.* Rome: Editori Riuniti, 1998.

Cellini, N. Presentation to *Terapie palliative e cure di supporto in oncologia*, by A. Ciabattoni and M. Pittiruti. Rome: SEU, 1996.

———. "Problemi etici dell'assistenza al morente." In *Corso di bioetica*, edited by E. Sgreccia, 153–163. Milan: F. Angeli, 1986.

Cellini, N., A. Ciabattoni, A. G. Morganti, et al. "Unità di Cura Continuativa: Esperienza in tema di assistenza domiciliare a pazienti oncologici in fase avanzata." *Anziani Oggi* 1 (1996): 59–65.

CENSIS Forum per la Ricerca Biomedica. *La ricerca biomedica in Italia.* Vol. 2, *Industria farmaceutica ed università verso l'integrazione europea.* Milan: F. Angeli, 1993.

Center for Bioethics and Human Rights of the University of Lecce. *I diritti del nascituro e la procreazione artificiale.* Vatican City: Libreria Editrice Vaticana, 1995.

Centro Italiano Femminile. *Uomo-Donna: Progetto di vita.* Rome: CIF/UECI, 1985.

Centro Studi e Ricerche sulla Regolazione Naturale della Fertilità, ed. *Procreazione responsabile: Quale realtà per la famiglia oggi? Atti del II Congresso Internazionale per la Famiglia d'Africa e d'Europa, Roma 12–15 marzo 1988.* Rome: Facoltà di Medicina e Chirurgia dell'Università Cattolica del S. Cuore, 1989.

Ceri, L., and S. F. Magni, eds. *Le Ragioni dell'Etica.* Pisa: Ets, 2004.

Cesareo, V. "L'epoca dell'incertezza." *Vita e Pensiero* 2 (1979).

Cesari, G. "Natura e interpretazione dei disorientamenti sessuali: L'omosessualità." In *Interrogativi per la bioetica*, edited by M. L. Di Pietro and E. Sgreccia, 78–96. Brescia: La Scuola, 1998.

Cesari, G., and M. L. Di Pietro. *L'educazione della sessualità.* Brescia: La Scuola, 1996.

Chadwich, R., M. Levitt, and D. Shickle. *The Right to Know and the Right Not to Know.* Chippenham: Antony Rowe, 1997.

Chakraborti, D. "Sterilization and the Mentally Handicapped." *British Medical Journal* 294 (1987): 794.

Chan, Y. F., and P. C. Ho. "Bilateral Tubal Pregnancies after Pronuclear Stage Embryo Tubal Transfer to One Tube." *Australian and New Zealand Journal of Obstetrics and Gynaecology* 33 (1993): 315–316.

Chandley, A. C., and T. B. Hargreave. "Genetic Anomaly and ICSI." *Human Reproduction* 11 (1996): 930–932.

Chang, J. C., T. T. Sun, and Y. C. Lin. "Simultaneous Ectopic Pregnancy with Intra-Uterine Gestation after In Vitro Fertilization and Embryo Transfer." *European Journal of Obstetrics and Gynecology and Reproductive Biology* 44, no. 2 (1992): 157–160.

Chang, M. C. "Fertilization of Rabbit Ova In Vitro." *Nature* 184 (1959): 466–467.

Changeux, J. P. *Neuronal Man: The Biology of Mind.* Translated by L. Garey. Princeton: Princeton University Press, 1997. Originally published as *L'Homme neuronal* (Paris: Fayard, 1983).

Charo, R.A. "Reproductive Technologies: Legal and Regulatory Issues." In *Encyclopedia of Bioethics*, edited by W.T. Reich, 2241–2248. New York: Macmillan, 1995.

Chen, C. "Pregnancy after Human Oocyte Cryopreservation." *Lancet* 1 (1986): 884–886.

Chesterton, G.K. *Orthodoxy.* 6th ed. London: John Lane, 1919.

Chevallier, M.D. *Rapport sur les applications des biotecnologies à l'agriculture et à l'industrie agroalimentaire.* Vol. 1. Paris: Office Parlamentaire des choix scientifiques et technologiques, Assemblée Nationale Francaise, Sénet, 1990–1991.

Chiarelli, B. *Problemi di bioetica nella transizione fra il II e il III millennio.* Florence: Il Sedicesimo, 1990.

———. "Storia naturale del concetto di etica e le sue implicazioni per gli equilibri naturali attuali." *Federazione Medica* 37, no. 6 (1984): 542–547.

Chiavacci, E. *Morale della vita fisica.* Bologna: Dehoniane, 1976.

———. "Morale della vita fisica: Eutanasia e diritto di morire con dignità." *La Civiltà Cattolica* 4 (1983): 313–329.

———. "Promozione dei diritti del malato posto di fronte alla prospettiva della morte." In *Morire sì, ma quando?* edited by P. Beretti, 253–266. Rome: Paoline, 1977.

Childress, J.F. "The Place of Autonomy in Bioethics." *Hastings Center Report* 20, no. 1 (1990): 12–17.

Choo, V. "Early Amniocentesis." *Lancet* 338 (1991): 750–751.

Ciaranfi, E. "L'evoluzione della medicina ed i problemi che ne derivano." *Federazione Medica* 35 (1982): 292–295.

Ciccone, L. "Etica della sessualità." In *Corso di bioetica*, edited by E. Sgreccia, 69–94. Milan: Angeli, 1986.

———. "L'eutanasia ed il principio dell'inviolabilità assoluta di ogni vita umana innocente." In *Commento interdisciplinare alla "Evangelium vitae,"* edited by E. Sgreccia and R. Lucas Lucas, 453–465. Vatican City: Libreria Editrice Vaticana, 1997.

———. *"Humanae vitae": Analisi e orientamenti pastorali.* Rome: Ed. Pastorali, 1970.

———. *Non uccidere: Questioni di morale della vita fisica.* Vol. 1. Milan: Ares, 1984.

———. "I problemi etici dell'aborto nell'enciclica *Evangelium vitae*." In *"Evangelium vitae" e bioetica: Un approccio interdisciplinare*, edited by E. Sgreccia and D. Sacchini, 59–76. Milan: Vita e Pensiero, 1996.

———. "La sterilizzazione antiprocreativa." In *Non uccidere: Questioni di morale della vita fisica*, 311–360. Milan: Ares, 1984.

Cicero. "The Dream of Scipio." In *Nine Orations and the Dream of Scipio*, 297–304. Translated by P. Bovie. New York: New American Library, 1967.

Cinque, B., M. Pelagalli, S. Daini, S. Dell'Acqua, and A.G. Spagnolo. "Aborto ripetuto spontaneo: Aspetti scientifici e obbligazioni morali." *Medicina e Morale* 42, no. 5 (1992): 889–910.

Cirotto, C., and S. Privitera. *La sfida dell'ingegneria genetica: Tra scienza e morale.* Assisi: Cittadella, 1985.

Cittadini, E. "Aspetti biomedici ed etici della fertilizzazione in vitro." In *La fertilizzazione in vitro: Atti del convegno della Societas Ethica; Palermo, 2–6 settembre 1985*, edited by S. Privitera, 9–34. Palermo: EDI OFTES, 1986.

Claessens, M. T., J.L. Bernat, and J.A. Baron. "Ethical Issues in Clinical Trials." *British Journal of Urology* 76, no. Suppl. 2 (1995): 29–36.

Clauser, F., P. Bailo, and P. Alfieri. *Problemi medico-morali.* Bergamo: Edizioni della Rotonda, 1958.

Clouser, K. D., and B. Gert. "A Critique of Principlism." *Journal of Medicine and Philosophy* 15, no. 2 (1990): 219–236.

Coccio, A. "Il problema dell'immortalità dell'anima nella Summa Theologica di S. Tommaso d'Aquino." *Rivista di Filosofia Neoscolastica* 38 (1946): 298–306.

Coddington, C.C., R. Demochowski, S. Oehninger, J. R. Auman, and G. D. Hodgen. "Hemizona Assay: Evaluation of Fertility Potential in Patients with Vasectomy Reversal." *Archives of Andrology* 38 (1997): 143–150.

Colombo, B. "Sulla diffusione degli aborti illegali in Italia." *Medicina e Morale* 26, nos. 1–2 (1976): 17–78.

Colombo, R. "La clonazione umana: Aspetti di diritto internazionale." *L'Osservatore Romano*, August 27, 2003.

———. "La vulnerabilità nella ricerca biomedica: Il caso dell'embrione umano." In *Etica della ricerca biomedica: Per una visione cristiana; Atti della nona assemblea generale della Pontificia Accademia per la Vita; Città del Vaticano, 24–26 febbraio 2003*, edited by J. de D. Vial Correa and E. Sgreccia, 217–244. Vatican City: Libreria Editrice Vaticana, 2004.

Comhaire, F. H. "Male Contraception: Hormonal, Mechanical and Other." *Human Reproduction* 9, no. 4 (1994): 586–590.

Commission of the European Communities. Directive 2005/28/EC. "Laying Down Principles and Detailed Guidelines for Good Clinical Practice as Regards Investigational Medicinal Products for Human Use, as well as the Requirements for Authorisation of the Manufacturing or Importation of Such Products." *Official Journal of the European Union* L 91, no. 3 (2005), http://eur-lex.europa.eu/LexUriServ/LexUri Serv.do?uri=OJ:L:2005:091:0013:0019:en:PDF

Committee to Consider the Social, Ethical and Legal Issues Arising from In Vitro Fertilization. *Report on the Disposition of Embryos Produced by In Vitro Fertilization.* Melbourne: Government Printer, 1984.

"Conception in a Waterglass." *New England Journal of Medicine* 217 (1937): 678.

Congregation for Catholic Education. *Educational Guidance in Human Love: Outlines for Sex Education.* November 1, 1983.

Congregation for the Doctrine of the Faith. Declaration *Iura et bona.* May 5, 1980.

———. Declaration *Persona humana.* December 29, 1975.

———. Declaration *Quaestio de abortu.* November 18, 1974.

———. Instruction. *Donum vitae.* February 22, 1987.

———. Letter *Homosexualitatis problema.* October 1, 1986.

———. Letter to Father Charles Curran. July 25, 1986.

———. *Quaecumque sterilizatio* [Sterilization in Catholic hospitals]. March 13, 1975. In *Vatican Council II: More Postconciliar Documents*, edited by A. P. Flannery, 454–455. Northport, NY: Costello Publishing Company, 1975.

———. *Quaestio de abortu* [Declaration on procured abortion]. November 18, 1974. In *Vatican Council II: More Postconciliar Documents*, edited by A. P. Flannery, 441–453. Northport, NY: Costello Publishing Company, 1975.

Congresso dei Biblisti e Moralisti dell'Italia Meridionale, ed. *Antropologia biblica e morale: Atti del 1° congresso dei biblisti e moralisti dell'Italia meridionale; Castellammare, 1–2 giugno 1971*. Naples: Dehoniane, 1972.

Consulta di Bioetica. "Carta dell'autodeterminazione." *Notizie di Politeia* 24 (1991): 4–5.

"Contre un project de loi sur la stérilisation obligatoire in Inde." *La Documentation Catholique* (1976): 420–421.

Cordini, G. "Xenotrapianti: I profili costituzionali." *Rassegna Ammnistrativa della Sanità* 40 (January–March 2001): 5–13.

Cortesini, R. "Il cuore artificiale: Esperienze scientifiche e riflessioni etico-morali." In *Trapianto di cuore e trapianto di cervello*, 52–86. Rome: Orizzonte Medico, 1983.

Cosmacini, G., and C. Crisciani. *Medicina e filosofia: Nella tradizione dell'Occidente*. Milan: Episteme, 1998.

Cosmacini, G., and R. Satolli. *Lettera a un medico sulla cura degli uomini*. Rome: Laterza, 2003.

Cotta, S. *Il diritto nell'esistenza: Linee di ontofenomenologia giuridica*. Milan: Giuffré, 1991.

———. *Giustificazione e obbligatorietà delle norme*. Milan: Giuffré, 1981.

Cotton, R. G., N. Rodrigues, and R. D. Campbell, "Reactivity of Cytosine and Thymine in Single-Base-Pair Mismatches with Hydroxylamine and Osmium Tetroxide and Its Application to the Study of Mutations." *Proceedings of the Natural Academy of Sciences of the USA* 85 (1988): 4397–4401.

Council for International Organizations of Medical Sciences (CIOMS) and World Health Organization (WHO). *International Ethical Guidelines for Biomedical Research Involving Human Subjects*. Geneva: CIOMS, 2002. http://www.cioms.ch/publications/layout_guide2002.pdf.

Council of Europe. Additional Protocol to the Convention on Human Rights and Biomedicine on the Prohibition of Cloning Human Beings. January 12, 1998. http://conventions.coe.int/treaty/en/treaties/html/168.htm.

———. Convention for the Protection of Human Rights and Dignity of the Human Being with regard to the Application of Biology and Medicine: Convention on Human Rights and Biomedicine. European Treaty Series No. 164. April 4, 1997. http://conventions.coe.int/Treaty/en/Treaties/html/164.htm.

———. "On Genetic Engineering." Recommendation 934. January 26, 1982. http://assembly.coe.int/Main.asp?link=/Documents/AdoptedText/ta82/EREC934.htm#1.

———. "On Genetic Testing and Screening for Health Care Purposes." Recommendation R (92) 3. *International Journal of Bioethics* 3, no. 4 (1992): 255–257.

———. "On the Rights of the Sick and Dying." Recommendation 779. January 21, 1976. http://assembly.coe.int/main.asp?Link=/documents/adoptedtext/ta76/erec779.htm.

———. "Preparation of a Convention on Bioethics. " Recommendation 1160. June 28, 1991. http://assembly.coe.int/Main.asp?link=/Documents/AdoptedText/ta91/EREC1160.htm.

———. "Protection of the Human Genome by the Council of Europe." Doc. 9002. March 19, 2001. http://assembly.coe.int/main.asp?Link=/documents/workingdocs/doc01/edoc9002.htm.

Council of Europe. Committee of Ministers. Convention for the Protection of Human Rights and the Dignity of the Human Being with Regard to the Application of Biology and Medicine: Convention on Human Rights and Biomedicine. CETS No. 164. April 4, 1997. http://conventions.coe.int/Treaty/en/Treaties/html/164.htm.

Coutelle, C., M. Themis, S. N. Waddington, et al. "Gene Therapy Progress and Prospects: Fetal Gene Therapy—First Proofs of Concept—Some Adverse Effects." *Gene Therapy* 12, no. 22 (2005): 1601–1607.

Cozzoli, M. "Il trapianto di organi nella prospettiva valoriale del dono." *Medicina e Morale* 47, no. 3 (1997): 461–473.

Cremaschi, S. "Il concetto di eros in *Le deuxième sexe* di Simone de Beauvoir." In *Amore e matrimonio nel pensiero filosofico e teologico moderno*, edited by V. Melchiorre, 296–316. Milan: Vita e Pensiero, 1976.

Cremer, T., P. Lichter, J. Borden, D. C. Ward, and L. Manuelidis. "Detection of Chromosome Aberrations in Metaphase and Interphase Tumor Cells by In Situ Hybridization Using Chromosome-Specific Library Probes." *Human Genetics* 80 (1988): 235–246.

Criscuolo, D. "Definizione dei termini utilizzati nella GCP." In *Lineamenti di etica della sperimentazione clinica*, edited by A. G. Spagnolo and E. Sgreccia, 105–112. Milan: Vita e Pensiero, 1994.

Cullen, M. T., E. A. Reece, J. Whethaam, and J. C. Hobbins. "Embryoscopy: Description and Utility of a New Technique." *American Journal of Obstetrics and Gynecology* 162 (1990): 82–86.

Curran, C. E. *Critical Concerns in Moral Theology*. Notre Dame: University Notre Dame Press, 1984.

Curson, R., and J. Parson. "Disappointing Results with Direct Intraperitoneal Insemination." *Lancet* 1 (1987): 112.

Curtis, L., and T. Meiner. *"Clinical Trials and Treatment Effects Monitoring." Controlled Clinical Trials* 19 (1998): 515–522.

Curtis, R. E., P. A. Rowlings, H. J. Deeg, et al. "Solid Cancers after Bone Marrow Transplantation." *New England Journal of Medicine* 336 (1997): 897–904.

Cuyás, M. *Eutanasia: L'etica, la libertà, la vita*. Casale Monferrato: Piemme, 1989.

———. "Problematica etica della manipolazione genetica." *Rassegna di Teologia* 5 (1985): 471–497.

Daffos F., M. Capella-Pavlowsky, and F. Forestier. "Fetal Blood Sampling During Pregnancy with Use of a Needle Guided by Ultrasound: A Study of 606 Consecutive Cases." *American Journal of Obstetrics and Gynecology* 153 (1985): 655–660.

———. "A New Procedure for Fetal Blood Sampling In Utero: Preliminary Results of Fifty-three Cases." *American Journal of Obstetrics and Gynecology* 146 (1983): 985–987.

D'Agostino, F. "L'approccio morale al diritto." In *Scritti in onore di Angelo Falzea*, 1:230. Milan: Giuffrè, 1991.

———. "Bioetica e diritto." *Medicina e Morale* 43, no. 4 (1993): 675–691.

———. *Bioetica nella prospettiva della filosofia del diritto*. Turin: Giappichelli, 1998.

———, ed. *Il corpo de-formato: Nuovi percorsi dell'identità personale*. Milan: Giuffré, 2002.

————. "Il diritto naturale e la fallacia naturalistica." In *Filosofia del diritto*. Turin: Giappichelli, 1996.

————. "Il diritto naturale, il diritto positivo e le nuove provocazioni della bioetica." In *Natura e dignità della persona umana a fondamento del diritto alla vita: Le sfide del contesto culturale contemporaneo; Atti dell'Ottava Assemblea Generale della Pontificia Accademia per la Vita; Città del Vaticano, 25–27 febbraio 2002*, edited by J. de D. Vial Correa and E. Sgreccia. Vatican City: Libreria Editrice Vaticana, 2003.

————. "Eutanasia, diritto e ideologia." *Iustitia* 30, no. 3 (1977): 285–306.

————. "Eutanasia e diritto." In *Morire sì, ma quando?* Beretta: Edizioni Paoline, 164–178.

————. *Una filosofia della famiglia*. Milan: Giuffrè, 1999.

————. *Parole di bioetica*. Turin: Giappichelli, 2004.

————. "Per una ermeneutica dell'*Evangelium vitae*: Legge morale e legge civile." *Bioetica* 3 (1995).

————. "La teologia del diritto positivo: Annuncio cristiano e verità del diritto." In *Evangelium vitae e diritto*, 121–131. Vatican City: Libreria Editrice Vaticana, 1997.

————. "Verso uno statuto giuridico del corpo umano." In *La sterilizzazione come problema giuridico*, 1–9. Turin: Giappichelli, 2002.

Dalla Serra, A. "Il congelamento di gameti ed embrioni." In *Attualità sulla procreazione medico assistita*, edited by C. Ragni and W. Vegetti, 99–106. Rome: CIC Edizioni Internazionali, 1997.

Dalton, M. E., and A. H. DeCherney. "Prenatal Diagnosis." *New England Journal of Medicine* 328, no. 2 (1993): 114–120.

Danish Folketing. Law no. 353. "Establishment of an Ethical Council and the Regulation of Certain Forms of Biomedical Experiments." June 3, 1987. http://www1.etiskraad.dk/sw297.asp.

Da Re, A. *L'etica tra felicità e dovere: L'attuale dibattito sulla filosofia pratica*. Bologna: Dehoniane, 1986.

Dausset, J. "I successi e i limiti dei trapianti effettuati sull'uomo." In *Trapianto di cuore e trapianto di cervello*, 12–42. Rome: Orizzonte Medico, 1983.

Davanzo, G. "L'aborto nella problematica etico-cristiana." *Anime e Corpi* 38 (1971): 550–551.

Davis, O. K., and Z. Rosenwaks. "In Vitro Fertilization." In *Infertility: Evaluation and Treatment*, edited by W. R. Keye, R. J. Chang, R. W. Rebar, and M. R. Soules, 759–771. Philadelphia: W. B. Saunders, 1995.

Dawson, K. "Fertilization and Moral Status: A Scientific Perspective." In *Embryo Experimentation*, edited by P. Singer, H. Kuhse, S. Buckle, K. Dawson, and P. Kasimba, 43–52. Cambridge: Cambridge University Press, 1990.

De Anna, G. *Realismo metafisico e rappresentazione mentale: Un'indagine tra Tommaso d'Aquino e Hilary Putnam*. Padua: Il Poligrafo, 2001.

Declaration of Alma-Ata. September 12, 1978.

Dedé, A. "Il metodo Billings: Marcatore della fertilità per la predeterminazione del sesso del concepito." In *La regolazione naturale della fertilità oggi: Certezze e dubbi; Congresso internazionale; Milano 9–11 dicembre 1988*, 485–487. Rome: CIC Edizioni Internazionali, 1989.

De Deyn, P. P., and R. D'Hooge. "Placebos in Clinical Practice and Research." *Journal of Medical Ethics* 22 (1996): 140–146.

Defanti, C.A. "I concetti di morte dell'organismo, morte cerebrale, morte corticale." In "Atti del II incontro di aggiornamento in neurologia: Attualità in tema di morte cerebrale; Perugia 20 febbraio 1993," edited by A. Ferroni, *Annali di Neurologia e Pischiatria* 87 (1993): 21–29.

Defaye, J., H. de Roissart, and P.M. Vignais, eds. *Risk Management in Biotechnology*. Paris: Technique et Documentation, 1990.

De Finance, J. *Etica generale*. Cassano Murge: Tipogr. Meridionale, 1982.

———. "L'ontologia della persona e della libertà in Maritain." In *Jacques Maritain oggi*, edited by V. Possenti, 156–173. Milan: Vita e Pensiero, 1983.

De Flora, A. "Strategie per la correzione delle malattie genetiche." *Medicina e Morale* 32, no. 1 (1982): 25–35.

de Klerk, M., K.M. Keizer, F.H. Claas, M. Witvliet, B.J. Haase-Kromwijk, and W. Weimar. "The Dutch National Living Donor Kidney Exchange Program." *American Journal of Transplantation* 5, no. 10 (2005): 2302–2305.

De Lagrange, E., M.M. De Lagrange, and R. Bel. *Un complot contre la vie: L'avortement*. Paris: Société de Production Littéraire, 1979.

———. *Il complotto contro la vita*. Milan: Ares, 1987. Originally published as *Un complot contre la vie: L'avortement* (Paris: Société de Production Littéraire, 1979).

Delfosse, M.-L. "I Comitati di Etica in Belgio." In *I Comitati di etica in ospedale*, edited by S. Spinsanti, 101–110. Cinisello Balsamo: Paoline, 1988.

Delgado, L. *Antropologia medica*. Milan: Paoline, 1991.

Delmonico, F.L. "Exchanging Kidneys: Advances in Living-Donor Transplantation." *New England Journal of Medicine* 350, no. 18 (2004): 1812–1814.

Delmonico, F.L., P.E. Morrissey, G.S. Lipkowitz, et al. "Donor Kidney Exchanges." *American Journal of Transplantation* 4, no. 10 (2004): 1628–1634.

Del Re, G. "Complessità." In *Dizionario interdisciplinare di scienza e fede*, edited by G. Tanzella-Nitti and A. Strumia, 259–265. Rome: Urbaniana University Press, 2002.

———. "Le tesi del filosofo Sloterdijk fanno rabbrividire i tedeschi." *Avvenire*. September 30, 1999. http://www.swif.uniba.it/lei/rassegna/990930b.htm.

De Marinis, L., A. Barbarino, and A. Serra, "Biologia della differenziazione sessuale." *Medicina e Morale* 34, no. 2 (1984): 155–165.

Demey, H.I., R. Daelemans, D. Galdermans, G.A. Verpooten, M.E. De Broe, and L.L. Bossaert. "Acute Oligo-Anuria during Ovarian Hyperstimulation Syndrome." *Acta Obstetricia et Gynecologica Scandinavica* 66 (1987): 741–743.

Demmer, K. "Liceità dell'ardita sperimentazione del trapianto cerebrale." In *Trapianto di cuore e trapianto di cervello*, 150–169. Rome: Orizzonte Medico, 1983.

———. "Opzione fondamentale." In *Nuovo dizionario di teologia morale*, edited by F. Compagnoni, G. Piana, and S. Privitera, 854–861. Cinisello Balsamo: Paoline, 1990.

De Paolis, V. "La protezione penale del diritto alla vita." In *Commento interdisciplinare alla "Evangelium vitae,"* edited by E. Sgreccia and R. Lucas Lucas, 501–520. Vatican City: Libreria Editrice Vaticana, 1997.

Department of Health and Human Services [U.S.], National Institutes of Health. *Guidelines for Research Involving Recombinant DNA Molecules*. May 2011. http://oba.od.nih .gov/oba/rac/Guidelines/NIH_Guidelines.Pdf.

―――. "Points to Consider in the Design and Submission of Human Somatic Cell Gene Therapy Protocols." *Recombinant DNA Technical Bulletin* 8, no. 4 (1985): 181–186.

Department of Health and Social Security Committee [U.K.]. *Report of Inquiry into Human Fertilisation and Embriology*. London: Her Majesty's Stationery Office, 1984. http:// www.hfea.gov.uk/docs/Warnock_Report_of_the_Committee_of_Inquiry_into _Human_Fertilisation_and_Embryology_1984.pdf.

Department of Health, Education, and Welfare Ethics Advisory Board [U.S.]. *HEW Support of Research Involving Human In Vitro Fertilization and Embryo Transfer*. Washington, DC: DHEW, 1979.

De Rosa, G. "Caso o finalismo nella evoluzione dei viventi?" *La Civiltà Cattolica* 2 (2006): 483–492.

―――. "L'evoluzione dei viventi: Il fatto e i meccanismi." *La Civiltà Cattolica* 3 (2006): 232–241.

Descartes, R. *Discourse on Method and Meditations on First Philosophy*. Translated by D.A. Cress. Indianapolis: Hackett Publishing Company, 1981.

De Vincentiis, G., and P. Zangani. "Sulla liceità e sui limiti della sperimentazione sull'uomo." *Giustizia Penale* 1 (1968): 332–334.

Diamond, M., C. Christianson, and J.F. Daniell. "Pregnancy following Use of the Cervical Cup for Home Artificial Insemination Utilizing Homologous Semen." *Fertility and Sterility* 4 (1983): 480 ff.

Diasco, R.B., and R.H. Glass. "Effects of pH on the Migration of X and Y Sperm." *Fertility and Sterility* 22 (1971): 303 ff.

Dickson, D. "Pig Heart Transplant 'Breakthrough' Stirs Debate over Timing of Trials." *Nature* 377 (1995): 185–186.

Diczfalusy, E., and M. Bygdeman, eds. *Fertility Regulation Today and Tomorrow*. New York: Serono Symposia, 1987.

Dieckmann, K.P. "Vasectomy and Testicular Cancer." *European Journal of Cancer* 30A, no. 7 (1994): 1040–1041.

Di Ianni, M. "Fecondazione artificiale." In *Dizionario enciclopedico di teologia morale*, 401–412. Rome: Paoline, 1981.

Dilella, A., W.M. Huang, and S.L.C. Woo. "Screening for Phenylketonuria Mutations by DNA Amplification with the Polymerase Chain Reaction." *Lancet* 2 (1988): 479–499.

Di Menna, R. "Umanizzazione e animazione del concepito umano." In *Scienza e origine della vita*, 36–72. Rome: Orizzonte Medico, 1980.

Di Pietro, M.L. "Analisi comparata delle leggi e degli orientamenti normativi in materia di fecondazione artificiale." *Medicina e Morale* 43, no. 1 (1993): 231–282.

―――. "Bioetica e demografia." In *A sua immagine e somiglianza?*, edited by A. Mazzoni, 323–344. Rome: Città Nuova, 1997.

―――. "Fecondazione artificiale: Sul parere del Comitato Nazionale per la Bioetica." *La rivista del clero italiano* 12 (1994): 837–853.

―――. "Fecondazione artificiale e frammentazione della maternità: Considerazioni etiche e giuridiche." *La Famiglia* 154 (1992): 5–19.

———. "La fecondazione extracorporea." *Medicina e Morale* 34, no. 4 (1984): 570–571.

———. "Identità sessuale e genere." *Anthopotes* 20, no. 1 (2004): 83–105.

———. "La lettera enciclica *Evangelium vitae* e l'aborto procurato: Nuovi elementi di riflessione nella continuità di un insegnamento." *Vita e Pensiero* 10 (1995): 653–676.

———. *Sexualidad y procreación humana*. Buenos Aires: Educa, 2005.

———. "Temi bioetici nelle Conferenze Internazionali delle Nazioni Unite." *Vita e Pensiero* 3 (1996): 175–188.

Di Pietro, M.L., M. Casini, R. Minacori, L. Romano, A. Fiori, and A. Bompiani. "Norlevo e obiezione di coscienza." *Medicina e Morale* 53, no. 3 (2003): 411–455.

Di Pietro, M.L., and M. Casini. "Il mifepristone." *Medicina e Morale* 52, no. 6 (2002): 1047–1079.

Di Pietro, M.L., C. Casini, M. Casini, and A.G. Spagnolo. *Obiezione di coscienza in sanità: Nuove problematiche per l'etica e il diritto*. Siena: Cantagalli, 2005.

Di Pietro, M.L., and S.M. Correale. "Valutazione delle terapie medico-chirurgiche e protesiche dell'impotentia coeundi nell'uomo ai fini della validità del matrimonio canonico." *Apollinaris* 66 (1993): 273–314.

Di Pietro, M.L., and M.B. Fisso. "La tutela dell'embrione umano in Germania: Dalla legge del 1990 alla Sentenza della Corte Costituzionale del 28 maggio 1993." *Vita e Pensiero* 4 (1994): 269–283.

Di Pietro, M.L., A. Giuli, and A. Serra. "La diagnosi preimpianto." *Medicina e Morale* 54, no. 3 (2004): 1–33.

Di Pietro, M.L., and A. Lucattini. "Condotte suicidiarie e adolescenza nel dibattito attuale." *Medicina e Morale* 44, no. 4 (1994): 667–690.

Di Pietro, M.L., and R. Minacori. "'Contraccezione d'emergenza': Problema medico, etico e giuridico." *Vita e Pensiero* 5 (1997): 353–361.

———. "Qual è il rischio delle tecniche di fecondazione artificiale?" *Medicina e Morale* 48, no. 3 (1998): 465–497.

———. "Sull'abortività della pillola estroprogestinica e di altri 'contraccettivi.'" *Medicina e Morale* 46, no. 5 (1996): 863–900.

Di Pietro M.L., and E. Sgreccia. *Procreazione assistita e fecondazione artificiale tra scienza, bioetica e diritto*. Brescia: La Scuola, 1998.

———. *La trasmissione della vita nell'insegnamento di Giovanni Paolo II*. Milan: Vita e Pensiero, 1989.

Di Pietro, M.L., and A.G. Spagnolo. "Sovrappopolazione come sottosviluppo? Alcune riflessioni per capire la posizione cattolica alla Conferenza de Il Cairo." *Bioetica* 2 (1995): 216–228.

Di Pietro, M.L., A.G. Spagnolo, and E. Sgreccia. "Meta-analisi dei dati scientifici sulla GIFT: Un contributo alla riflessione etica." *Medicina e Morale* 50, no. 1 (1990): 13–40.

Di Rocco, C. "Problemi decisionali relativi al trattamento chirurgico delle malformazioni congenite del Sistema Nervoso Centrale." *Medicina e Morale* 34, no. 4 (1984): 488–492.

"Documento della Santa Sede sulla clonazione umana." *L'Osservatore Romano*, October 17, 2004.

Doerfler, J.F. "Is GIFT Compatible with the Teaching of *Donum Vitae*?" *Linacre Quarterly* 64, no. 1 (1997): 16–29.

Dommergues, M., M. Lemerrer, G. Couly, A.L. Delezoide, and Y. Dumez."Prenatal Diagnosis of Cleft Lip at 11 Menstrual Weeks Using Embryoscopy in the Van der Woude Syndrome." *Prenatal Diagnosis* 15 (1995): 378–38.

Donati, P. "Il contesto sociale della bioetica: Il rapporto tra norme morali e norme di diritto positivo." In *Bioetica: Un'opzione per l'uomo; Atti del primo corso internazionale di bioetica; 15–16/29–30 aprile 1988*, 135–181. Milan: Jaca Book, 1988.

———. *La cultura della vita: Dalla società tradizionale a quella postmoderna*. Milan: F. Angeli, 1989.

———. "Trasformazioni socio-culturali della famiglia e comportamenti relativi alla procreazione." *Medicina e Morale* 43, no. 1 (1993): 117–163.

Donceel, J.F. "Immediate Animation and Delayed Hominization." *Theological Studies* 31 (1970): 76–106.

Dorozynski, A. "France to Investigate Illegal Sterilization of Mentally Ill Patients." *British Medical Journal* 315, no. 7110 (1997): 697.

Doyle, P. "The Outcome of Multiple Pregnancy." *Human Reproduction* 11, no. 4, suppl. (1996): 110–120.

Dulbecco, R. *La mappa della vita*. Milan: Sperling and Kupfer, 2001.

Dulbecco, R., and R. Chiaberge. *Ingegneri della vita: Medicina e morale nell'era del DNA*. Milan: Sperling and Kupfer, 1988.

Dumas, A. "Fondements bibliques d'une bioétique." *Le Supplément* 142 (1982): 153.

Dumez, Y., M. Dommergues, M.C. Gubler, et al. "Meckel-Gruber Syndrome: Prenatal Diagnosis at 10 Menstrual Weeks Using Embryoscopy." *Prenatal Diagnosis* 14, no. 2 (1994): 141–144.

Dunstan, G.R., and M.J. Seller, eds. *The Status of the Human Embryo*. London: King's Fund, 1988.

Durkheim, E. *Il suicidio: Studio di sociologia*. Milan: Rizzoli, 1989.

Dutch Committee to Study the Medical Practice Concerning Euthanasia. *Medische Beslissingen Rond Het Levenseinde: Rapport Van De Commissie Onderzoek Medische Praktijk Inzake Euthanasie* [Medical decisions about the end of life: Report of the committee to study the medical practice concerning euthanasia]. 2 vols. The Hague: SDU Publishing House, 1991 [Dutch].

Dworkin, G., R.G. Frey, and S. Bok. *Euthanasia and Physician-Assisted Suicide: For and Against*. Cambridge: Cambridge University Press, 1998.

Dyer, C. "Paired Kidney Transplants to Start in the United Kingdom." *British Medical Journal* 332, no. 7548 (2006): 989.

———. "Sterilization of Mentally Handicapped Woman." *British Medical Journal* 294 (1987): 825.

Dyson, A., and J. Harris. *Experiments on Embryos*. London: Routledge, 1990.

Eckstein, S., ed. *Manual for Research Ethics Committees*. 6th ed. New York: Cambridge University Press, 2003.

Edmonds, D.K., and K.D. Lindsay. "Early Embryonic Mortality in Women." *Fertility and Sterility* 38 (1982): 447.

Edwards, B.J., and A.M. Haddad. "Establishing a Nursing Bioethics Committee." *Journal of Nursing Administration* 18, no. 3 (1988): 30–33.

Edwards, J.N. "New Conceptions: Biosocial Innovations and the Family." *Journal of Marriage and the Family* 53, no. 2 (1991): 346–360.

Edwards, R.G. *Life Before Birth*. New York: Basic Books, 1989.

———. *A Matter of Life: The Story of the Medical Breakthrough*. London: Hutchinson, 1980.

———. "Maturation In Vitro of Human Ovarian Oocytes." *Lancet* 2 (1965): 926–929.

Edwards, R.G., D. Bavister, and P.C. Steptoe. "Early Stages of Fertilization In Vitro of Human Oocytes Matured In Vitro." *Nature* 221 (1969): 632–635.

Egidi, R., M. Dell'Utri, and M. De Caro, eds. *Normatività Fatti Valori*. Macerata: Quodlibet, 2003.

Egozcue, J. "The Ethical Committees in Spain." In *Funzione e funzionamento dei comitati etici*, edited by G. Gerin, 145–151. Padua: CEDAM, 1991.

Eijk, W. "Is the Dutch Euthanasia Regulation Compatible with *Evangelium Vitae*?" *Medicina e Morale* 46, no. 3 (1996): 469–481.

Elementi di medicina e psicologia pastorale. Brezzo di Bedero: Salcom, 1969.

Elliott, P.J. "La prospettiva etica nella Conferenza ONU di Pechino sulla donna." *Medicina e Morale* 45, no. 6 (1995): 1175–1182.

Ellis, S.J., and R.F. Adams. "The Cult of the Double-Blind Placebo-Controlled Trial." *British Journal of Clinical Practice* 51 (1997): 36–39.

Emanuel, E.J. "The Beginning of the End of Principlism." *Hastings Center Report* 25, no. 4 (1995): 37–38.

Emanuel, E.J., R.A. Crouch, J.D. Arras, J.D. Moreno, and C. Grady, eds. *Ethical and Regulatory Aspects of Clinical Research*. Baltimore: Johns Hopkins University Press, 2003.

Emanuel, E.J., and L.L. Emanuel. "Four Models of the Physician-Patient Relationship." *Journal of American Medical Association* 267, no. 16 (1992): 2221–2226.

Embury, S.H., S.J. Scharf, and R.K. Saiki, "Rapid Prenatal Diagnosis of Sickle Cell Anaemia by a New Method of DNA Analysis." *New England Journal of Medicine* 316 (1987): 646–651.

Engelhardt, D. von. "Storia dell'etica medica." In *Dizionario di bioetica*, edited by S. Leone and S. Privitera, 954–958. Bologna: Dehoniane / Istituto Siciliano di Bioetica, 1994.

Engelhardt, H.T. Jr. "La bioetica nell'era post-moderna." Interview by C. Botti. *Notizie di Politeia* 7, no. 24 (1991): 9–16.

———. *The Foundations of Bioethics*. 2nd ed. New York: Oxford University Press, 1996.

Engels, F. "Engels to F. Mehring: London, July 14, 1893." In *Marx Engels: Selected Correspondence*, 541. Moscow: Foreign Languages, 1953.

Enia, F., and E. Geraci. "Protocolli teraputici." In *Nuovo dizionario di bioetica*, edited by S. Leone and S. Privitera, 941–945. Palermo: EDB-ISB, 2004.

Erlich, H.A., and N. Arnheim. "Genetic Analysis Using the Polymerase Chain Reaction." *Annual Review of Genetics* 26 (1992): 479–506.

Esposito, R.F. "L'eutanasia nella stampa di massa italiana." In *Morire sì, ma quando?*, edited by P. Beretta, 17–35. Rome: Paoline, 1977.

Ethics and Public Policy Center. "Production of Pluripotent Stem Cells by Oocyte Assisted Reprogramming." June 20, 2005. http://www.eppc.org/publications/pubID.2374/pub_detail.asp.

Ethics Committee of the A. Gemelli Polyclinic of the Catholic University of the Sacred Heart. "Raccomandazioni riguardo alla inclusione delle donne in età fertile nei protocolli di sperimentazione clinica." *Medicina e Morale* 53, no. 6 (2003): 1271–1275.

Ethics Committee of the American Fertility Society. "Ethical Considerations of the New Reproductive Technologies." *Fertility and Sterility* 46, no. 3, suppl. 1 (1986): 345–385.

Ettershank, K. "Report Reveals Australia's Illegal Sterilisations." *Lancet* 351, no. 9095 (1998): 44.

European Economic Community. "Contained Use of Genetically Modified Micro-Organisms." Council Directive 90/219. April 23, 1990. http://eur-lex.europa.eu/LexUriServ/LexUriServ.do?uri=CELEX:31990L0219:EN:HTML.

———. "Deliberate Release into the Environment of Genetically Modified Organisms." Council Directive 90/220. April 23, 1990. http://eur-lex.europa.eu/LexUriServ/LexUriServ.do?uri=CELEX:31990L0220:EN:HTML.

———. "The Registration of Work Involving Recombinant Deoxyribonucleic Acid (DNA)." Recommendation 82/472. June 30, 1982. http://eur-law.eu/EN/82-472-EEC-Council-Recommendation-30-June-1982,96347,d.

European Forum for Good Clinical Practice. *European Guidelines for Auditing Independent Ethics Committees.* EFGCP, 2002. http://www.appliedclinicaltrialsonline.com/applied clinicaltrials/data/articlestandard//appliedclinicaltrials/232002/21190/article.pdf.

European Medicines Agency. *Note for Guidance on Good Clinical Practice (CPMP/ICH/135/95).* London: EMEA, 2006. http://www.emea.europa.eu/docs/en_GB/document _library/Scientific_guideline/2009/09/WC500002874.pdf.

European Organization for Research and Treatment of Cancer (EORTC). *What Are Cancer Clinical Trials All About?* Bruxelles: 1997.

European Parliament. "Resolution A3-0231/92 on the Rights of the Mentally Handicapped." *Official Journal of the European Communities* C 284 (1992): 18–22.

———. "Resolution on Cloning." Doc. B4-209/97. March 12, 1997. http://wfrt.org/humanrts/instree/cloning1.html.

———. "Resolution on Human Cloning." January 15, 1998. http://wfrt.org/humanrts/instree/cloning2.html.

———. "Resolution on Organ Banks." May 21, 1979. *Official Journal of the European Communities* C 127 (1979): 71.

———. "Resolution on the Ethical and Legal Problems of Genetic Engineering." Doc. A0327-88. March 16, 1990. In *Official Journal of the European Communities* C 96 (1989): 165–171. http://codex.vr.se/texts/EP-genetic.html.

European Parliament and Council. "Legal Protection of Biotechnological Inventions." Directive 98/44/EC. July 6, 1998. http://eur-lex.europa.eu/LexUriServ/LexUriServ.do?uri=CELEX:31998L0044:EN:HTML.

European Parliament and Council of Europe. "On the Approximation of the Laws, Regulations and Administrative Provisions of the Member States Relating to the Implementation of Good Clinical Practice in the Conduct of Clinical Trials on

Medicinal Products for Human Use." Directive 2001/20/EC. *Official Journal of the European Communities* L 121, no. 34 (2001). http://www.eortc.be/services/doc/clinical-eu-directive-04-april-01.pdf.

Eusebi, L. "La clonazione come problema giuridico." *L'Osservatore Romano*, August 30, 2003.

———. "Corresponsabilità verso le scelte giuridiche della società pluralista e criteri di intervento sulla c.d. norma imperfetta." In *"Evangelium vitae" e diritto*, edited by A. Lopéz Trujllo, J. Herranz, and E. Sgreccia, 389–406. Vatican City: Libreria Editrice Vaticana, 1997.

———. "Il dibattito nella bioetica: Obiezione di coscienza all'interruzione della gravidanza." *Medicina e Morale* 41, no. 6 (1991): 1068–1070.

———. "Sul mancato consenso al trattamento terapeutico: Profili giuridico-penali." In *Bioetica in medicina*, edited by A. Bompiani, 221–228. Rome: CIC Edizioni Internazionali, 1996.

———. "Sul mancato consenso al trattamento terapeutico: Profili giuridico-penali." *Rivista Italiana di Medicina Legale* 17 (1995): 727–740.

———. "Tutela giuridica dell'embrione ed esigenze irrisolte di prevenzione dell'aborto." In *Ingegneria genetica e biotecnologie nel futuro dell'uomo*, edited by E. Sgreccia and V. Mele, 329–358. Milan: Vita e Pensiero, 1992.

Evans, D., and M. Evans. *A Decent Proposal: Ethical Review of Clinical Research*. Chichester: John Wiley and Sons, 1996.

Evans, R.W. "Organ Transplantation and the Inevitable Debate as to What Constitutes a Basic Health Care Benefit." *Clinical Transplantation* (1993): 359–391.

Evans, W.E., and M.V. Relling. "Pharmacogenomics: Translating Functional Genomics into Rational Therapeutics." *Science* 286 (1999): 487–491.

Ewigman, B.G., J.P. Crane, F.D. Frigoletto, et al. "Effect of Prenatal Ultrasound Screening on Perinatal Outcome." *New England Journal of Medicine* 329 (1993): 821–827.

Fabbri, R., E. Porcu, T. Marsella, et al. "Technical Aspects of Oocyte Cryopreservation." *Molecular and Cellular Endocrinology* 169 (2000): 39–42.

Facchini, F. *Il cammino dell'evoluzione umana*. Milan: Jaca Book, 1985.

———. "Evoluzione e Creazione." *L'Osservatore Romano* (January 15–16, 2006), 4.

Faccini, J.M., P.N. Bennett, and J.L. Reid. "European Ethical Review Committee: The Experience of an International Ethics Committee Reviewing Protocols for Drug Trials." *British Medical Journal* 289 (1984): 1052–1054.

Faggioni, M.P. "Il trapianto di gonadi: Storia e attualità." *Medicina e Morale* 48, no. 1 (1998): 15–46.

Fagone, V. "Il problema dell'inizio della vita del soggetto umano." In *Aborto: Riflessioni di studiosi cattolici*, edited by A. Fiori and E. Sgreccia, 149–179. Milan: Vita e Pensiero, 1975.

———. "Il problema dell'inizio della vita del soggetto umano." *Medicina e Morale* 24, no. 2 (1974): 212–242.

Fagot-Largeault, A. "Experimentation humaine." In *Les mots de la bioéthique*, edited by G. Hottois and M.H. Parizeau, 219–228. Brussels: De Boeck Université, 1993.

Fasanella, G., and E. Sgreccia. "Il trapianto di midollo osseo: Aspetti etici." *Medicina e Morale* 38, nos. 3–4 (1988): 397–409.

Featherstone, K., and J. L. Donovan. "Random Allocation or Allocation at Random? Patients' Perspectives of Participation in a Randomised Controlled Trial." *British Medical Journal* 317, no. 7167 (1998): 1177–80.

Federal Republic of Germany. Embryonenschutzgesetz—EschG. December 13,1990. http://www.gesetze-im-internet.de/bundesrecht/eschg/gesamt.pdf.

Federazione Nazionale degli Ordini dei Medici Chirurghi e degli Odontoiatri (FNOMCeO), ed. *Guida all'esercizio professionale per i medici chirurghi e gli odontoiatri.* Turin: Edizioni Medico Scientifiche, 1994.

Feinberg, J. *The Problem of Abortion.* Belmont, CA: Wadsworth, 1984.

Fernández Sánchez, F. C. "La esterilización de incapacitados mentales y su calificación moral objetiva." *Cuadernos de Bioética* 5, no. 20 (1994): 361–367.

Ferraretti, P., L. Gianaroli, and C. Magli. "Sindrome da iperstimolazione: Prevenzione e trattamento." In *Attualità sulla procreazione medico assistita*, edited by C. Ragni and W. Vegetti, 47–61. Rome: CIC Edizioni Internazionali, 1997.

Field, M. J., and R. E. Behrman, eds. *Ethical Conduct of Clinical Research Involving Children.* Washington, DC: National Academies Press, 2004.

Fiksel, J., and V. T. Covello, eds. *Biotechnology Risk Assessment.* Oxford: Pergamon Press, 1986.

La filosofia dell'uomo: *Atti del congresso della Federazione Universitaria Cattolica Italiana.* Rome: FUCI, 1961.

"Final Regulations Amending Basic HHS Policy for the Protection of Human Research Subjects." *Federal Register* 46, no. 16 (January 26, 1981): 8366–8391.

Finfer, S. "Managing Patients Who Refuse Blood Transfusions: An Ethical Dilemma," *British Medical Journal* 308 (1994): 1423–1426.

Finnis, J. "Natural Law—Positive Law." In *"Evangelium vitae" e diritto*, edited by A. López Trujillo, G. Herranz, and E. Sgreccia, 199–212. Vatican City: Libreria Editrice Vaticana, 1997.

———. "Nature and Natural Law in Contemporary Philosophical and Theological Debates: Some Observations." In *The Nature and Dignity of the Human Person as the Foundation of the Right to Life: The Challenges of the Contemporary Cultural Context; Proceedings of the Eighth Assembly of the Pontifical Academy for Life: Vatican City, 25–27 February 2002*, edited by J. de D. Vial Correa and E. Sgreccia, 81–109. Vatican City: Libreria Editrice Vaticana, 2003.

Finnis, J. M. "Natural Law in *Humanae Vitae.*" *Law Quarterly Review* 84 (1968): 467–471.

———. "Personal Integrity, Sexual Morality and Responsible Parenthood." *Anthropos* 1 (1985): 43–55.

Fiori, A. "Bioetica laica e bioetica cattolica." *Medicina e Morale* 46, no. 2 (1996): 203–207.

———. "Contro la fecondazione eterologa." *Medicina e Morale* 47, no. 2 (1997): 241–266.

———. "Genetic Tests in Legal Medicine and Criminology." In *Human Genome: Human Person and the Society of the Future*, edited by J. de D. Vial Correa and E. Sgreccia. Vatican City: Libreria Editrice Vaticana, 1999.

———. "Medicina ippocratica, medicina ideologica, obiezione di coscienza." *Medicina e Morale* 47, nos. 1–2 (1977): 166–184.

———. "Problemi attuali del consenso informato." *Medicina e Morale* 43, no. 6 (1993): 1123–1138.

———. "La sperimentazione sull'uomo pone gravi interrogativi." *Orizzonte Medico* (July–August 1971): 2.

———. "I trapianti d'organo ed i costi-benefici del progresso medico." *Medicina e Morale* 34, no. 1 (1984): 16–26.

———. "Problemi attuali del consenso informato." *Medicina e Morale* 43, no. 6 (1993): 1123–1138.

Fiori, A., and L. C. Caprioli. "La fecondazione artificiale eterologa: Un altro ritorno al matrilineare?" *Medicina e Morale* 44, no. 2 (1994): 213–230.

Fiori, A., and M. L. Di Pietro. "Accertamento della morte: Normativa vigente e prospettive future." *Medicina e Morale* 43, no. 5 (1993): 945–965.

Fiori, A., and E. Sgreccia, eds. *Obiezione di coscienza e aborto*. Milan: Vita e Pensiero, 1978.

Firth, H. V., P. A. Boyd, P. F. Chamberlain, I. Z. MacKenzie, G. M. Morriss-Kay, and S. M. Huson. "Analysis of Limb Reduction Defects in Babies Exposed to Chorionic Villus Sampling." *Lancet* 343 (1994): 1069–1071.

Firth, H. V., P. A. Boyd, P. F. Chamberlain, I. Z. MacKenzie, R. H. Lindenbaum, and S. M. Huson. "Severe Limb Abnormalities after Chorion Villus Sampling at 56–66 Days' Gestation." *Lancet* 337 (1991): 762–763.

Fisichella, R. "Da credente in difesa della ragione." In *Cristiano perché relativista, relativista perché cristiano: Per un razionalismo della contingenza*. D. Antiseri. Soveria Mannelli: Rubbettino, 2003.

Fisso, M. B. "Trapianti (diritto)." *Enciclopedia italiana*, 551–553. Appendix V. Updated 1978–1992.

Fisso, M. B., and E. Sgreccia. "Etica dell'ambiente (I–II)." *Medicina e Morale* 46, no. 6 (1996): 1057–1082; 47, no. 1 (1997): 57–74.

Fizzotti, E. "L'onda lunga del suicidio tra vuoto esistenziale e ricerca di senso." *Anime e Corpi* 161 (1992): 273–294.

Flamigni, C., A. Masserenti, M. Mori, and A. Petroni. "Manifesto di bioetica laica." *Notizie di Politeia* 41/42 (1996).

Fletcher, J. C. "Is Sex Selection Ethical?" *Progress in Clinical and Biological Research* 28 (1983): 333–334.

Flick, M., and Z. Alszeghy. *Fondamenti di un'antropologia teologica*. Florence: Libreria Editrice Fiorentina, 1970.

———. *Il peccato originale*. Brescia: Queriniana, 1972.

Florida Catholic Conference. "Statement on the Life, Death and Treatment of Dying Patients." April 27, 1989. http://www.flacathconf.org/statements/1989/dyingtreatment.pdf.

Flynn, E. P. *Hard Decisions: Forgoing and Withdrawing Artificial Nutrition and Hydration*. Kansas City: Sheed and Ward, 1990.

———. *Human Fertilization In Vitro: A Catholic Moral Perspective*. Lanham, MD: University Press America, 1984.

Folli, G. "L'uso del placebo nei trials clinici: Significato scientifico e valore sperimentale." In *Lineamenti di etica della sperimentazione clinica*, edited by A. G. Spagnolo and E. Sgreccia, 85–90. Milan: Vita e Pensiero, 1994.

Fonseca, A. "Politiche demografiche e contraccezione oggi nel mondo." *La Civiltà Cattolica* 4 (1983): 134–148.

————. "Sterilizzazione obbligatoria in India." *La Civiltà Cattolica* 3 (1976): 153–162.

Food and Drug Administration, *Information Sheets for Institutional Review Boards and Clinical Investigators*. Rockville: FDA Office of the Associate Commissioner for Health Affairs, 1995.

Ford, N. M. *When Did I Begin? Conception of the Human Individual in History, Philosophy and Science*. Cambridge: Cambridge University Press, 1988.

Fortino, M., ed. *La procreazione medicalmente assistita: Atti del convegno internazionale; Messina, 13–14 dicembre 2002*. Turin: Giappichelli, 2005.

Fortuny, A., A. Borrell, A. Soler, et al. "Chorionic Villus Sampling by Biopsy Forceps: Results of 1580 Procedures from a Single Centre." *Prenatal Diagnosis* 15 (1995): 541–550

Fossum, G. T., A. Davidson, and R. J. Paulson. "Ovarian Hyperstimulation Inhibits Embryo Implantation in the Mouse." *Journal of In Vitro Fertilization and Embryo Transfer* 6 (1989): 7–10.

Fost, N., and R. E. Cranford. "Hospital Ethics Committees: Administrative Aspects." *Journal of the American Medical Association* 253 (1985): 2687–2692.

Foster, C. *The Ethics of Medical Research on Humans*. Cambridge: Cambridge University Press, 2001.

Foucault, M. *The Order of Things: An Archaeology of the Human Sciences*. New York: Vintage, 1973. Originally published as *Les Mots et les choses: Une archéologie des sciences humaines* (Paris: Gallimard, 1966).

Foulon, W., A. Naessens, L. de Catte, and J. J. Amy. "Detection of Congenital Toxoplasmosis by Chorionic Villus Sampling and Early Amniocentesis." *American Journal of Obstetrics and Gynecology* 103 (1990): 1511–1513.

France, J. T., F. M. Graham, L. Gosling, and P. I. Hair. "A Prospective Study of the Preselection of the Sex of Offspring by Timing Intercourse Relative to Ovulation." *Fertility and Sterility* 42 (1984): 894–900.

Franchini, A. "Le grandi scoperte della medicina nel XX secolo." In *Storia delle scienze*, edited by E. Agazzi, 2:387–400. Rome: Città Nuova, 1984.

Frankel, M. S. "Artificial Insemination." In *Encyclopedia of Bioethics*, edited by W. T. Reich, 1444–1448. New York: Free Press, 1978.

Frankena, W. *Ethics*. Englewood Cliffs: Prentice Hall, 1973.

————. "The Naturalistic Fallacy." *Mind* 48 (1939): 464–467.

Frankl, V. E. *The Doctor and the Soul: From Psychotherapy to Logotherapy*, 2nd ed. (New York, Knopf, 1965).

Frattallone, R. "Persona e atto umano." In *Nuovo dizionario di teologia morale*, edited by F. Compagnoni, G. Piana, and S. Privitera, 932–952. Cinisello Balsamo: Paoline, 1990.

Freedman, B. "Research, Unethical." In *Encyclopedia of Bioethics*, edited by W. T. Reich, 2258–2261. Rev. ed. New York: Macmillan, 1995.

French, C. V. "Against Biospherical Egalitarianism." *Environmental Ethics* 12 (1995): 41–57.

"French In Vitro Fertilization Registry: 1994 Report." *Contraception, Fertilité, Sexualité* 23, nos. 7–8 (1995): 490–493.

768

French National Ethics Advisory Committee for Life and Health Sciences. *Avis relatif aux recherches et utilisation des embryons humains in vitro à des fins médicales et scientifiques*. December 15, 1986. Paris: Centre de documentation et d'information d'éthique, 1991.

————. *Avis sur la non-commercialisation du corps humain*. December 13, 1990.

————. "La contraception chez les personnes handicapées mentales." Opinion no. 49. April 3, 1996. *Les Cahiers du CCNE* 8 (1996): 3–5.

————. *Rapport et Recommandations sur les Comités d'Éthiques Locaux*. November 7, 1988.

————. "La stérilisation envisagée comme mode de contraception définitive." Opinion no. 50. April 3, 1996. *Les Cahiers du CCNE* 9 (1996): 3–19

Freud, S. *The Standard Edition of the Complete Works of Sigmund Freud*, edited by J. Strachey. 24 vols. London: Hogarth, 1953–1974.

Frey, R.G. "Vivisection Morals and Medicine." *Journal of Medical Ethics* 9 (1983): 94–97.

Fuccella, L.M. "Come sorgono e cosa sono le GCP della Comunità Europea." In *Lineamenti di etica della sperimentazione clinica*, edited by A.G. Spagnolo and E. Sgreccia, 95–104. Milan: Vita e Pensiero, 1994.

Fuccella, L. M., and M. Frascio. *Guida pratica alla sperimentazione clinica dei farmaci*. Milan: OEMF, 1995.

Fujita, Y., K. Tasaka, A. Miyake, et al. "Methods for Monitoring Follicle Maturation Decrease: The Ovarian Hyperstimulation Syndrome during Gonadotropin Treatment." *European Journal of Obstetrics and Gynecology and Reproductive Biology* 32 (1989): 223.

"Future Offers Hope for Infertile Women." *Journal of the American Medical Association* 245, no. 6 (1981): 565.

Gadamer, H.-G. *The Enigma of Health: The Art of Healing in a Scientific Age*. Translated by J. Gaiger and R. Walker. Stanford: Stanford University Press, 1996. Originally published as *Über die Verborgenheit der Gesundheit* (Frankfurt: Suhrkamp, 1993).

Gainotti, S., and A.G. Spagnolo. "Test genetici: A che punto siamo in Europa? A margine del Rapporto e delle Raccomandazioni della Commissione Europea sugli aspetti etici, giuridici e sociali dei test genetici." *Medicina e Morale* 54, no. 4 (2004): 737–766.

Galantino, N. *Dire "uomo" oggi: Nuove vie dell'antropologia filosofica*. Milan: Paoline, 1993.

Galbraith, S., M. Stat, and I. Marschner. *"Guidelines for the Design of Clinical Trials with Longitudinal Outcomes." Controlled Clinical Trials* 23 (2002): 257–273.

Galli, A. "Nietzsche: Profeta dell'eugenetica." *Avvenire*. September 21, 2005.

Galvagni, L. *Bioetica e comitati etici*. Bologna: EDB, 2005.

Gambino, G., ed. *L'eutanasia*. Vol. 2, *Diritto e prassi in Italia, Europa e Stati Uniti*. Sapere 2000 Edizioni Multimediali, 2005.

Garcea, N. "La procreazione assistita." In *Trattamento della sterilità coniugale*, edited by S. Mancuso and E. Sgreccia, 145–159. Milan: Vita e Pensiero, 1988.

————. "Tecniche di procreazione assistita." *Medicina e Morale* 43, no. 1 (1993): 59–66.

Garcea, N., S. Campo, and P.L. Cannella. "Current Therapeutic Possibilities of GIFT." *Acta Europea Fertilitatis* 1 (1988): 315 ff.

Garcea, N., S. Campo, and R. D'Argenio. "Is GIFT a Possibility for Catholics in the Assisted Reproduction Field?" In *From Basics to Clinics*, edited by G.L. Capitanio. New York: Raven Press, 1989.

Garcia De Haro, R. "La legge morale e le norme civili." In *Persona, verità e morale: Atti del congresso internazionale di teologia morale; Roma, 7–12 aprile 1986*, edited by A. Ansaldo. Rome: Città Nuova, 1986.

Geddes, D.P. "New Light on Sexual Knowledge." In *About the Kinsey Report: Observations by 11 experts on "Sexual Behavior in the Human Male,"* edited by D.P. Geddes and E. Curie, 5–25. New York: New American Library, 1948.

Gemelli, A. *La fecondazione artificiale.* 2nd ed. Milan: Vita e Pensiero, 1949.

Gennari, M., and C. Moreschi. "Un divieto irrazionale: Il trapianto delle ghiandole della sfera genitale e della procreazione." *Rivista Italiana di Medicina Legale* 14, no. 4 (1992): 805–813.

Gennaro, A. "La sterilizzazione sessuale volontaria o coatta." In *Questioni matrimoniali: IIa "Tre giorni" di teologia morale.* Turin: LICE, 1950.

Genovese, U., R. Stucchi, and A. Farneti. "I trapianti 'non salvavita': Aspetti medico-legali." *Rivista Italiana di Medicina Legale* 22, no. 2 (2000): 417–430.

Gensabella Furnari, M. *Tra autonomia e responsabilità: Percorsi di bioetica.* Soveria Mannelli, CZ: Rubbettino, 2000.

Gentilomo, A., A. Piga, and S. Nigrotti. *La procreazione medicalmente assistita nell'Europa dei quindici: Uno studio comparatistico.* Milan: Giuffrè, 2005.

German Bishops' Conference. "The Future of Creation and the Future of Humanity." In *Medicina e Morale* 40, no. 2 (1990): 366–380 [Italian].

German Bundestag. "Report of the Commission of Enquiry on Prospects and Risks of Genetic Engineering." January 1987.

Germond, M., M.C. Gaillard, and A. Senn. "Syndrome d'hyperstimulation ovarienne." *Archives of Gynecology and Obstetrics* 246 (1989): S53–S64.

Gert, B., C.M. Culver, and K.D. Clouser. *Bioethics: A Return to Fundamentals.* New York: Oxford University Press, 1997.

Gevaert, J. *Il problema dell'uomo.* Turin: LDC, 1978.

Ghisalberti, A. "Anima e corpo in Tommaso d'Aquino." *Rivista di Filosofia Neoscolastica* 97, no. 2 (2005): 282–296.

Giannini, G. *Il problema antropologico.* Rome: Pontificia Università Lateranense, 1965.

Gidley-Baird, A.A., C. O'Neill, M.J. Sinosich, R.N. Porter, I.L. Pike, and D.M. Saunders. "Failure of Implantation in Human In Vitro Fertilization and Embryo-Transfer Patients: The Effects of Altered Progesterone-Estrogen Ratios in Humans and Mice." *Fertility and Sterility* 45 (1986): 69–74.

Giffels, J.J. *Clinical Trials: What You Should Know Before Volunteering to Be a Research Subject.* New York: Damos Vernande, 1996.

Gill, P., and D.J. Warret. "Exclusion of a Man Charged with Murder by DNA Fingerprinting." *Forensic Science International* 35 (1987): 145–148.

Gillon, R. "On Sterilising Severely Mentally Handicapped People." *Journal of Medical Ethics* 13 (1987): 59–61.

———. *Philosophical Medical Ethics*. Chichester: John Wiley and Sons, 1986.

Gindro, S., S. Mancuso, G. Astrei, R. Bracalenti, and E. Mordini, eds. *Aborto volontario: Le conseguenze psichiche*. Rome: CIC Edizioni Internazionali, 1996.

Giorlandino, C. *Trattato di diagnosi prenatale e terapia fetale*. Rome: CIC Edizioni Internazionali, 1997.

Giorlandino, C., P.M. D'Alessio, L. Mobili, O. Carcioppolo, L. Capolino, and A. Vizzone. "L'embrioscopia: Nuova tecnica di diagnosi prenatale invasiva." *Ultrasonica* 4 (1994): 103–106.

Giovannucci E, T.D. Tosteson, F.E. Speizer, A. Ascherio, M.P. Vessey, G.A. Colditz. "A Prospective Cohort Study of Vasectomy and Prostate Cancer in US Men." *Journal of the American Medical Association* 269, no. 17 (February 17, 1993): 873–877.

Girgenti, G., ed. *Il suicidio: Follia o delirio di libertà?* Cinisello Balsamo: Paoline, 1989.

Gismondi, G. *Critica ed etica nella ricerca scientifica*. Turin: Marietti, 1978.

———. *Etica fondamentale della scienza*. Rome: Citta Nuova, 1997.

Giunchedi, F. "Considerazioni morali sulla fecondazione artificiale." *La Civiltà Cattolica* 1 (1984): 223–241.

Giusti, G. *L'eutanasia, diritto di vivere, diritto di morire*. Padua: CEDAM, 1982.

Gnoth, G., D. Godehardt, P. Frank-Herrmann, and G. Freundl. "Time to Pregnancy: Results of the German Prospective Study and Impact on the Management of Infertility," *Human Reproduction* 18, no. 9 (2003): 1959–1966.

Gobel, P. "FDA Audit for Institutional Review Boards." *Applied Clinical Trials* (Oct. 1995): 54–59.

Goethe, J.W. von. *Faust: A Tragedy*, edited by C. Hamlin. Translated by W. Arndt. New York: W.W. Norton, 1976.

Goffi, T., and G. Piana, eds. *Corso di Morale*. Vol. 3, *Koinonia: Etica della vita sociale*. Brescia: Queriniana, 1984.

Goglia, G. "Jacques Monod." *L'Osservatore Romano*. August 28, 1976.

Goldenring, J.M. "The Brain-Life Theory: Towards a Consistent Biological Definition of Humanness." *Journal of Medical Ethics* 11 (1985): 198–204.

Goliszek, A. *In the Name of Science: A History of Secret Programs, Medical Research, and Human Experimentation*. New York: St. Martin's Press, 2003.

Goodman, M.F. *What Is a Person?* Clifton, N.J.: Humana Press, 1988.

Gracia, D. *Fundamentos de bioética*. Madrid: Eudema Universidad, 1989.

Gracia Guillén, D. "Historia de la eutanasia." In *Eutanasia hoy: Un debate abierto*, edited by S. Urraca Martinez, 67–91. Madrid: Editorial Noesis, 1996.

Grasse, P.P. *L'evoluzione del vivente*. Milan: Adelphi, 1979.

Grasso, P.G. "Personalità." In *Dizionario enciclopedico di pedagogia*, 3:680–682. Turin: SAIE, 1959.

Grenier, Y. "La stérilisation volontaire, involontaire ou obligatoire." *Éthique* 5 (1992): 41–56.

Griener, G.G., and J.L. Storch. "Hospital Ethics Committees: Problems in Evaluation." *HEC Forum* 4, no. 1 (1992): 5–18.

Griffith-Jones, M. D., D. Miller, R. J. Lilford, J. Scott, and J. Bulmer. "Detection of Fetal DNA in Transcervical Swabs from First Trimester Pregnancies by Gene Amplification: A New Route to Prenatal Diagnosis." *British Journal of Obstetrics and Gynaecology* 99 (1992): 508–511.

Grisez, G. *Contraception and Natural Law*. Milwaukee: Bruce, 1964.

Grmek, M. D. *Le malattie all'alba della civiltà occidentale*. Bologna: Il Mulino, 1985.

Grobstein, C. "Biological Characteristics of the Preembryo." *Annals of the New York Academy of Sciences* 541 (1988): 346–348.

Grodin, M. A., ed. *Meta-Medical Ethics: The Philosophical Foundations of Bioethics*. Dordrecht: Kluwer, 1995.

Grodin, M. A., and L. H. Glantz. *Children as Research Subjects*. Oxford: Oxford University Press, 1993.

Grose, C., O. Itani, and C. P. Weiner. "Prenatal Diagnosis of Fetal Infection: Advance from Amniocentesis to Cordocentesis; Congenital Toxoplasmosis, Rubella, Cytomegalovirus, Varicella Virus, Parvovirus and HIV." *Pediatric Infectious Disease Journal* 8 (1989): 459–468.

Gruman, G. J. "Death and Dying: Euthanasia and Sustaining Life; Historical Perspectives." In *Encyclopedia of Bioethics*, edited by W. T. Reich, 261–286. New York: Free Press, 1978.

Guardini, R. *Etica*. Brescia: Morcelliana, 2001.

———. *Fede, religione, esperienza*. Brescia: Morcelliana, 1984.

———. *La fine dell'epoca moderna*. Brescia: Morcelliana, 1979.

Guizick, D. S., C. Wilkes, and N. W. Jones. "Cumulative Pregnancy Rates for In Vitro Fertilization." *Fertility and Sterility* 46 (1986): 63 ff.

Gumby, J., S. Daya, and IVF Directors Group of the Canadian Fertility and Andrology Society. "Assisted Reproductive Technology (ART) in Canada: 2001 Results from the Canadian ART Register." *Fertility and Sterility* 84, no. 5 (2005): 590–599.

Günthor, A. *Chiamata e risposta*, 3 vols. Rome: Paoline.

Gura, T. "After a Setback, Gene Therapy Progresses...Gingerly." *Science* 291 (2001): 1692–1697.

Gutierrez, L. V., J. V. Gutierrez, and P. M. Baza. *Reprodución asistida en la comunidad europea: Legislación y aspectos bioéticos*. Valladolid: Universidad de Valladolid, 1993.

Guttmacher, A. E., and F. S. Collins. "Realizing the Promise of Genomics in Biomedical Research." *Journal of the American Medical Association* 294, no. 11 (2005): 1399–1402.

Guzmán, J. L. *Objeción de conçiencia farmacéutica*. Barcelona: Ediciones Internacionales Universitarias, 1997.

Guzzetti, G. B. *Sterilizzazione a scopo contraccettivo*. Milan: Massimo, 1981.

Haan, G., and R. Van Steen. "Costs in Relation to Effects of In Vitro Fertilization." *Human Reproduction* 7 (1992): 982–986.

Habermas, J. *Between Facts and Norms: Contributions to a Discourse Theory of Law and Democracy*. Cambridge, MA: MIT Press, 1996.

———. *The Future of Human Nature.* Malden: Polity, 2003. Originally published as *Die Zukunft der menschlichen Natur: Auf dem Weg zu einer liberalen Eugenik?* (Frankfurt: Suhrkamp, 2001).

———. *The Inclusion of the Other: Studies in Political Theory*, edited by C. Cronin and P. De Greiff. Cambridge, MA: MIT Press, 1998.

———. *Moral Consciousness and Communicative Action.* Translated by C. Lenhardt and S. Weber Nicholsen. Cambridge, MA: MIT Press, 1990. Originally published as *Moralbewusstsein und kommunikatives Handeln* (Frankfurt: Suhrkamp, 1983).

———. *Theory and Practice.* Boston: Beacon Press, 1973.

Habermas, J., and J. Ratzinger. *Ragione e fede in dialogo.* Venice: Marsilio, 2005.

Haddow, J.E., G.E. Palomaki, G.J. Knight, et al. "Prenatal Screening for Down's Syndrome with Use of Maternal Serum Markers." *New England Journal of Medicine* 327 (1992): 588–593.

Haddow, J.E., G.E. Palomaki, G.J. Knight, G.C. Cunningham, L.S. Lustig, and P.A. Boyd. "Reducing the Need for Amniocentesis in Women 35 Years of Age or Older with Serum Markers for Screening." *New England Journal of Medicine* 330 (1994): 1114–1118.

Haeckel, E. "Zellseelen und Seelenzellen." *Deutsche Rundschau* 16 (1978): 40–59.

Halvorson, H., ed. *Engineered Organism in the Environment: Scientific Issues.* Washington, DC: American Society for Microbiology, 1985.

Hampshire, S. "Fallacies in Moral Philosophy." In *Freedom of Mind.* Oxford: Clarendon Press, 1972.

Handyside, A.H. "In Vitro Fertilization and Preimplantation Genetic Diagnosis for Prevention of Inherited Disease." In *Infertility: Evaluation and Treatment*, edited by W.R. Keye, R.J. Chang, R.W. Rebar, and M.R. Soules, 859–867. Philadelphia: W.B. Saunders, 1995).

Handyside, A.H., J.K. Pattinson, R.J.A. Penketh, J.D.A. Delhanty, R.M.L. Winston, and E.G.D. Tuddenham. "Biopsy of Human Pre-Implantation Embryos and Sexing by DNA Amplification." *Lancet* 1 (1989): 347–349.

Haney, A.F. "Controlled Ovarian Hypersimulation and Intrauterine Insemination." In *Infertility: Evaluation and Treatment*, edited by W.R. Keye, R.J. Chang, R.W. Rebar, and M.R. Soules, 745–758. Philadelphia: W.B. Saunders, 1995.

Hanson, M.J. "The Seductive Sirens of Medical Progress: The Case of Xeno-Transplantation." *Hastings Center Report* 25, no. 5 (1995): 5–6.

Haraway, D. *Simians, Cyborgs, and Women: The Reinvention of Nature.* New York: Routledge, 1991.

Hardy, E., L. Bahamondes, J.J. Osis, R.G. Costa, and A. Faúndes. "Risk Factors for Tubal Sterilization Regret, Detectable Before Surgery." *Contraception* 54, no. 3 (1996): 159–162.

Hare, R.M. *Freedom and Reason.* Oxford: Oxford University Press, 1963.

———. *The Language of Morals.* London: Clarendon Press, 1952.

Häring, B. "Eutanasia e teologia morale." In *Morire sì, ma quando?*, edited by P. Beretta, 221–232. Rome: Paoline, 1977.

———. *Free and Faithful in Christ: Moral Theology for Clergy and Laity.* 3 vols. New York: Crossroad, 1978–1981.

———. *Healing and Revealing: Wounded Healers Sharing Christ's Mission*. Slough: St. Paul, 1984. Originally published as *Vom Glauben, der gesund macht: Ermutigung der heilenden Berufe* (Freiburg: Herder, 1984).

———. *The Law of Christ: Moral Theology for Priests and Laity*. Vol. 3, *Special Moral Theology*. Westminster, MD: Newman Press, 1966.

———. *Medical Ethics*, edited by Gabrielle L. Jean. Notre Dame, ID: Fides, 1973.

———. *Paternità responsabile*. Rome: Libreria Editrice Salesiana, 1970.

———. *Rapporti sessuali prematrimoniali e morale*. Rome: Paoline, 1973.

———. "Sessualità." In *Dizionario enciclopedico di teologia morale*, edited by L. Rossi and A. Valsecchi, 993–1006. Rome: Paoline, 1981.

Harlap, S. "Gender of Infants Conceived on Different Days of the Menstrual Cycle." *New England Journal of Medicine* 300 (1979): 1145 f.

Harris, J. "Euthanasia and the Value of Life." In *Euthanasia Examined: Ethical, Clinical and Legal Perspectives*, edited by J. Keown, 6–22. New York: Cambridge University Press, 1995.

———. *Wonderwoman and Superman: The Ethics of Human Biotechnology*. New York: Oxford University Press, 1992.

Harrison, M. R., M. S. Golbus, and R. A. Filly, *The Unborn Patient*: *Prenatal Diagnosis and Treatment*, 2nd ed. Philadelphia: W. B. Saunders, 1991.

Harsanji, G. C. "Rule Utilitarianism and Decision Theory." *Erkenntnis* 11 (1977): 25–53.

Harvard Medical School, "A Definition of Irreversible Coma: Report of the Ad Hoc Committee of the Harvard Medical School to Examine the Definition of Brain Death." *Journal of the American Medical Association* 205, no. 6 (1968): 337–340.

Hastings Center. "Gli scopi della medicina: Nuove priorità." *Notizie di Politeia* 45 (1997).

Hayes, G. J., S. C. Hayes, and T. Dykstra. "A Survey of University Institutional Review Boards: Characteristics, Policies, and Procedures." *IRB* 17, no. 3 (1995): 1–6.

Heaney, S. J. "Aquinas and the Presence of the Human Soul in the Early Embryo." *The Thomist* 56, no. 1 (1992): 19–48.

Held, P. J., M. V. Pauly, R. R. Bovbjerg, J. Newmann, and O. Salvatierra Jr. "Access to Kidney Transplantation: Has the United States Eliminated Income and Racial Differences?" *Archives of Internal Medicine* 148 (1988): 2594–2600.

Hellegers, A. E. "Fetal Research." In *Encyclopedia of Bioethics*, edited by W. T. Reich, 489–493. New York: The Free Press, 1978.

Hempel, C. *Filosofia delle scienze naturali*. Bologna: Il Mulino, 1968.

Hendry, W. F. "Vasectomy and Vasectomy Reversal." *British Journal of Urology* 73 (1994): 337–344.

Hens, L., M. Bonduelle, I. Liebaers, P. Devroey, and A. C. Van Steirteghem. "Chromosome Aberration in 500 Couples Referred for In Vitro Fertilization or Related Fertility Treatment." *Human Reproduction* 3, no. 4 (1988): 451–457.

Herranz, G. "Il Comitato Centrale di deontologia spagnolo." In *I Comitati di etica in ospedale*, edited by S. Spinsanti, 141–148. Cinisello Balsamo: Paoline, 1988.

———. "Scienze biomediche e qualità della vita." *Vita e Pensiero* 6 (1986): 415–424.

Hersch, J. "Nuovi poteri dell'uomo, senso della vita e salute." *Aut aut* 318 (2003): 180–187.

Hervada, J. *Introduzione critica al diritto naturale*. Milan: Giuffré, 1990.

Hewlett, S. "Consent to Clinical Research: Adequately Voluntary or Substantially Influenced?" *Journal of Medical Ethics* 2 (1996): 232–237.

Hilgers, T. W., K. D. Dailey, A. M. Prebil, and S. K. Hilgers. "Cumulate Pregnancy Rates in Patients with Apparently Normal Fertility: Focused Intercourse." *Journal of Reproductive Medicine* 37 (1992): 864–866.

Hodgkinson, K. A., L. Kerzin-Storrar, E. A. Watters, and R. Harris, "Adult Polycystic Kidney Disease: Knowledge, Experience, and Attitudes to Prenatal Diagnosis." *Journal of Medical Genetics* 27 (1990): 552–558.

Hoffman, E. P., R. H. Brown, and L. M. Kunkel. "Dystrophin: The Protein Product of the Duchenne Muscular Dystrophy Locus." *Cell* 51 (1987): 919–928.

Höffner, J. *La dottrina sociale cristiana*. Rome: Paoline, 1979.

Hogge, W. A., S. A. Schonberg, and M. S. Golbus. "Prenatal Diagnosis by Chorionic Villus Sampling: Lessons of the First 600 Cases." *Prenatal Diagnosis* 5 (1985): 393–400.

Holt, E. "Roma Women Reveal That Forced Sterilisation Remains." *Lancet* 365 (2005): 927–928.

Honings, B. "L'aborto e il momento della ominizzazione." In *Il diritto alla vita*, edited by G. Concetti, 23–41. Rome: Logos, 1981.

Horkheimer, M. *Eclipse of Reason*. New York: Oxford University Press, 1947.

Horton, R. *"The Clinical Trial: Deceitful, Disputable, Unbelievable, Unhelpful, and Shameful: What Next?" Controlled Clinical Trials* 22 (2001): 593–604.

Hosford, B. *Bioethics Committees*. Rockville: Aspen, 1986.

Housman, D., and F. Ledley. "Why Pharmacogenomics? Why Now?" *Nature Biotechnology* 16 (1998): 492–493.

Huber, C. "Limiti della validità del sapere scientifico." In *Lineamenti di etica della sperimentazione clinica*, edited by A. G. Spagnolo and E. Sgreccia, 29–38. Milan: Vita e Pensiero, 1994.

HUGO Ethics Committee. "Statement on DNA Sampling: Control and Access." February 1998. http://www.hugo-international.org/img/dna_1998.pdf.

Hull, M. G. R. "Efficacia dei trattamenti per l'infertilità: Scelta e analisi comparativa." *Giornale Italiano di Ostetricia e Ginecologia* (1995): 567–576.

———. "Infertility Treatment: Relative Effectiveness of Conventional and Assisted Conception Methods." *Human Reproduction* 7 (1992): 785–796.

Hume D. *A Treatise of Human Nature*. 2nd ed. Edited with analytical index by L. A. Selby-Bigge. Oxford: Clarendon Press, 1978.

Humphry, D. *Final Exit*: *The Practicalities of Self-Deliverance and Assisted Suicide for the Dying*. Eugene, OR: Hemlock Society, 1991.

Huxley, A., *Brave New World*. New York: Chelsea House Publications, 2004.

Iadecola, G. *Consenso del paziente e trattamento medico-chirurgico*. Padua: Liviana, 1989.

———. "I trattamenti sanitari obbligatori. " In *Potestà di curare e consenso del paziente*, 71–75. Padua: CEDAM, 1998.

Iandolo, C. *L'approccio umano al malato: Aspetti psicologici dell'assistenza*. Rome: Armando, 1979.

———. "Etica clinica e bioetica." *Giornale Italiano per la Formazione Permanente del Medico* 15 (1987): 88–103.

Iitaka, K., L.W. Martin, J.A. Cox, P.T. McEnery, and C.D. West. "Transplantation of Cadaver Kidneys from Anencephalic Donors." *Journal of Pediatrics* 93, no. 2 (1978): 216–220.

Immacolato, M., Maurizio Mori, and Søren Holm. "Carta di San Macuto." March 21, 2003. http://www.politeia-centrostudi.org/doc/CartaSMacuto.pdf.

Informal Working Party of Bioethics of the Section for Relations with States of the Vatican Secretariat of State. "Observations on the Universal Declaration on the Human Genome and Human Rights." November 11, 1997. http://www.vatican.va/roman_curia/pontifical_academies/acdlife/documents/rc_pa_acdlife_doc_08111998_genoma_en.html.

Ingelman-Sundberg, M. "Pharmacogenetics: An Opportunity for a Safer and More Efficient Pharmacotherapy." *Journal of Internal Medicine* 250 (2001): 186–200.

"In Re B (A Minor) (Sterilization): Law Report." *Times*. March 17, 1987. Col. 1, p. 35.

Institut International d'études des droits de l'homme. *Le médecin face aux droits de l'homme*. Padua: CEDAM, 1989.

International Anti-Euthanasia Task Force. "Colombian Constitutional Court OK's Euthanasia." *IAETF Update* 11, no. 2: 12–13.

———. "The Courts Have Spoken: No Constitutional Right to Assisted Suicide." *IAETF Update* 11, no. 3: 2 ff.

International Commission on Occupational Health. "International Code of Ethics for Occupational Health Professionals." *Bulletin of Medical Ethics* 82 (1992): 9.

International Conference on Bioethics. *International Conference on Bioethics: The Human Genome Sequencing; Ethical Issues; Roma, 10–15 April 1988*. Brescia: Clas International, 1989.

International Conference on Harmonization Steering Committee. "ICH Harmonised Tripartite Guideline for Good Clinical Practice." *Good Clinical Practice Journal* 3, no. 4, suppl. (1996): 2–27.

International Conference of Medical Professional Associations. *Principles of European Medical Ethics*. January 6, 1987. Amended February 6, 1995. http://www.ceom-ecmo.eu/sites/default/files/documents/european_medical_ethics_principles-1987-1995_ceom_cio.pdf.

International Human Genome Sequencing Corsortium. "Initial Sequencing and Analysis of the Human Genome." *Nature* 409 (February 15, 2001): 860–941.

International Theological Commission. "Communione e Servizio." *La Civiltà Cattolica* 4 (2004): 254–288.

Introna, F. "Legalizzazione dell'aborto: Riflessioni medico-sociali." *Medicina e Morale* 24, no. 3 (1974): 418–449.

Introna, F., M. Tantalo, and A. Coiafigli. *Il codice di deontologia medica correlato a leggi ed a documenti*. Naples: Liviana Medicina, 1992.

Isambert, F.A. "De la bioéthique aux comités d'éthique." *Études* 358 (1983): 671–683.

Isidori, A. "L'inseminazione artificiale omologa e eterologa nella sterilità maschile: Aspetti medici e psicologici." *Medicina e Morale* 43, no. 1 (1993): 75–96.

776

BIBLIOGRAPHY

Italian Court of Cassation—V Sezione Penale. Conciani v. Corte d'Appello di Firenze. Decision no. 438 of March 18, 1987.

Italian Ministry of Health. Ministerial Decree of April 27, 1992. "Disposizioni sulle documentazioni tecniche da prestare a corredo delle domande di autorizzazione all'immissione in commercio di specialità medicinali per uso umano, in attuazione della direttiva n. 91/507/CEE" [Regulations on the technical documentation to be submitted with requests for permission to market medicinal products for human use, pursuant to directive no. 91/507/EEC]. *Gazzetta Ufficiale della Repubblica Italiana* 139 (June 15, 1992).

———. "Relazione della Commissione di Studio per l'Ingegneria genetica." In *Documenti conoscitivi sulla riproduzione umana assistita, embriologia ed ingegneria genetica*, edited by F. Luzi. Rome: Senato della Repubblica Italiana, Servizio Studi, Ufficio Ricerche nel Settore Sociale, 1990.

Italian National Bioethics Committee. *Adoption for the Birth of Cryopreserved and Residual Embryos Obtained by Medically Assisted Procreation (MAP)*. November 18, 2005. Rome: Presidenza del Consiglio dei Ministri Dipartimento per l'Informazione e l'Editoria, 2005. http://www.governo.it/bioetica/eng/pdf/ADOPTION_18112005.pdf.

———. *Advanced Treatment Statements*. December 18, 2003. Rome: Presidenza del Consiglio dei Ministri Dipartimento per l'Informazione e l'Editoria, 2003. http://www.governo.it/bioetica/eng/opinions/advance_treatment_directives.pdf.

———. *The Anencephalic Newborn and Organ Donation*. June 21, 1996. Rome: Presidenza del Consiglio dei Ministri Dipartimento per l'Informazione e l'Editoria, 1996. http://www.governo.it/bioetica/eng/opinions/abstracts/newborn_21.pdf.

———. *Bioetica e formazione nel sistema sanitario* [*Bioethics and Professional Training in the Healthcare System*]. September 7, 1991. Rome: Presidenza del Consiglio dei Ministri Dipartimento per l'Informazione e l'Editoria: 1991. http://www.governo.it/bioetica/testi/070991.html. English abstract, http://www.governo.it/bioetica/eng/opinions/abstracts/abstract_bioethics_and_professional.pdf.

———. *La clonazione come problema bioetico*. March 21, 1997. In *Medicina e Morale* 47, no. 2 (1997): 360–362.

———. *Comitati Etici*. February 27, 1992. Rome: Presidenza del Consiglio dei Ministri Dipartimento per l'Informazione e l'Editoria: 1992. http://www.governo.it/bioetica/pdf/8.pdf. English abstract, http://www.governo.it/bioetica/eng/opinions/abstracts/abstract_the_ethics_committees.pdf.

———. *I Comitati Etici in Italia: Problematiche recenti* [*Ethics Committees in Italy: Recent Issues*]. April 18, 1997. Rome: Presidenza del Consiglio dei Ministri Dipartimento per l'Informazione e l'Editoria: 1997. http://www.governo.it/bioetica/testi/180497.html. English abstract, http://www.governo.it/bioetica/eng/opinions/abstracts/abstract_ETHICS_COMMITTEES.pdf.

———. *Definizione e accertamento della morte nell'uomo* [*Definition and Ascertainment of Death in Man*]. February 17, 1991. Rome: Presidenza del Consiglio dei Ministri Dipartimento per l'Informazione e l'Editoria, 1991. http://www.governo.it/bioetica/pdf/2.pdf. English abstract, http://www.governo.it/bioetica/eng/opinions/abstracts/abstract_definition_and_ascertainment.pdf.

————. *Diagnosi prenatali* [*Prenatal Diagnoses*]. July 18,1992. Rome: Presidenza del Consiglio dei Ministri Dipartimento per l'Informazione e l'Editoria, 1992. http://www.governo.it/bioetica/pdf/10.pdf.Englishabstract,http://www.governo.it/bioetica/eng/pdf/abstract_prenatal_diagnoses.pdf.

————. *Documento sulla sicurezza delle biotecnologie* [*Document on Biotechnology Safety*]. May 28, 1991. Rome: Presidenza del Consiglio dei Ministri Dipartimento per l'Informazione e l'Editoria, 1991. http://www.governo.it/bioetica/pdf/4.pdf. English abstract, http://www.governo.it/bioetica/eng/opinions/abstracts/abstract_document_on_biotechnlogy.pdf.

————. *Donazione d'organo a fini di trapianto* [*Organ Donation for Transplants*]. October 7, 1991. Rome: Presidenza del Consiglio dei Ministri Dipartimento per l'Informazione e l'Editoria, 1991. http://www.governo.it/bioetica/pdf/donazione_d_organo_a_fini_di_trapianto_ok.pdf. English abstract, http://www.governo.it/bioetica/eng/opinions/abstracts/abstract_organ_donation.pdf.

————. *La fecondazione assistita* [*Assisted Fertilisation*]. February 17, 1995. Rome: Presidenza del Consiglio dei Ministri Dipartimento per l'Informazione e l'Editoria, 1995. http://www.governo.it/bioetica/testi/feconfazione170295.pdf. English abstract, http://www.governo.it/bioetica/eng/opinions/abstract_assisted%2fertilisation.pdf.

————. *From Pharmacogenetics to Pharmacogenomics*. April 21, 2006. Rome: Presidenza del Consiglio dei Ministri Dipartimento per l'Informazione e l'Editoria, 2006. http://www.governo.it/bioetica/eng/opinions/Pharmacogenetics.pdf.

————. *Identity and Status of the Human Embryo*. June 22, 1996. Rome: Presidenza del Consiglio dei Ministri, Dipartimento per l'Informazione e l'Editoria, 1996. http://www.governo.it/bioetica/eng/pdf/human_embryo_19960622.pdf.

————. *Identity and Status of the Human Embryo*. June 27, 1996. Rome: Presidenza del Consiglio dei Ministri Dipartimento per l'Informazione e l'Editoria, 1996. http://www.governo.it/bioetica/eng/pdf/human_embryo_19960622.pdf.

————. *Information and Consent to Medical Treatment*. June 20, 1992. Rome: Presidenza del Consiglio dei Ministri Dipartimento per l'Informazione e l'Editoria, 1992. http://www.governo.it/bioetica/eng/pdf/Informazione_e_consenso_all%27atto_medico_2.pdf.

————. *Organ Transplants in Childhood*. January 21, 1994. Rome: Presidenza del Consiglio dei Ministri Dipartimento per l'Informazione e l'Editoria, 1994. http://www.governo.it/bioetica/eng/pdf/PCM_trapianti.pdf.

————. *Parere sulla proposta di risoluzione sull'assistenza ai pazienti terminali* [*Opinion on the Resolution Proposal Concerning Assistance to Terminally Ill Patients*]. September 6, 1991. Rome: Presidenza del Consiglio dei Ministri Dipartimento per l'Informazione e l'Editoria, 1991. http://www.governo.it/bioetica/pdf/5.pdf. English abstract, http://www.governo.it/bioetica/eng/opinions/abstracts/abstract_opinion_on_the_draft_resolution.pdf.

————. *Parere sulle tecniche di procreazione assistita: Sintesi e conclusioni* [*Opinion on Medically Assisted Procreation Techniques: Synthesis and Conclusions*]. June 17, 1994. Rome: Presidenza del Consiglio dei Ministri Dipartimento per l'Informazione e l'Editoria, 1994. http://www.governo.it/bioetica/pdf/16.pdf. English abstract, http://www.governo.it/bioetica/eng/opinions/NBC_opinion_on_medically_assisted_procreation_techniques_55.pdf.

778

———. *Il problema bioetico della sterilizzazione non volontaria* [*The Bioethical Issue of Non Voluntary Sterilization*]. Rome: Presidenza del Consiglio dei Ministri Dipartimento per l'Informazione e l'Editoria, 1998. http://www.governo.it/bioetica/pdf/35 .pdf. English abstract, http://www.governo.it/bioetica/eng/pdf/abstract_the_bioethical_issue.pdf.

———. *Problemi della raccolta e trattamento del liquido seminale umano per finalità diagnostiche* [*Problems of the Collection and Handling of Human Seminal Liquid for Diagnostic Purposes*]. May 5, 1991. Rome: Presidenza del Consiglio dei Ministri Dipartimento per l'Informazione e l'Editoria, 1991. http://www.governo.it/bioetica/ pdf/3.pdf. English abstract, http://www.governo.it/bioetica/eng/opinions/abstracts/ abstract_problems_of_the_collection.pdf.

———. *Progetto Genoma Umano* [*Human Genome Project*]. March 18, 1994. Rome: Presidenza del Consiglio dei Ministri Dipartimento per l'Informazione e l'Editoria, 1994. http://www.governo.it/bioetica/pdf/15.pdf. English abstract, http://www.governo .it/bioetica/eng/pdf/abstract_Human_genome_2.pdf.

———. *Rapporto sulla brevettabilità degli organismi viventi* [*Report on the Patentability of Living Organisms*]. November 11, 1993. Rome: Presidenza del Consiglio dei Ministri Dipartimento per l'Informazione e l'Editoria, 1993. http://www.governo .it/bioetica/pdf/12.pdf. English abstract, http://www.governo.it/bioetica/eng/pdf/ Report_patentability_2.pdf.

———. *Scopi, limiti e rischi della medicina* [*Aims, Limits and Risks of Medicine*]. December 14, 2001. Rome: Presidenza del Consiglio dei Ministri Dipartimento per l'Informazione e l'Editoria, 2001. http://www.governo.it/bioetica/pdf/51.pdf. English abstract, http://www.governo.it/bioetica/eng/opinions/abstracts/AIMS_RISKS_AND _LIMITS.pdf.

———. *Sperimentazione sugli animali e salute dei viventi* [*Animal Testing and Health of Living Beings*]. April 17, 1997. Rome: Presidenza del Consiglio dei Ministri Dipartimento per l'Informazione e l'Editoria, 1997. http://www.governo.it/bioetica/ pdf/27.pdf. English abstract, http://www.governo.it/bioetica/eng/opinions/abstracts/ abstract_ANIMAL_TESTING.pdf.

Italian National Federation of Medical and Dental Associations. *Codice di deontologia medica.* Rev. ed. December 16, 2006. http://portale.fnomceo.it/PortaleFnomceo/showVoce Menu.2puntOT?id=5.

Italian Parliament. Norms against Sexual Violence. Law no. 66. Feburary 15, 1996. Reprinted in *Medicina e Morale* 46, no. 6 (1996): 1190–1195 [Italian].

Italian Republic. Law No. 194. On the Social Protection of Motherhood and the Voluntary Termination of Pregnancy. May 22, 1978. Published in *Gazzetta Ufficiale della Repubblica Italiana* 140, May 2, 1978, Part 1, 3642–3646 [Italian].

Jackson, L. G., R. A. Warner, and M. A. Barr. "Safety of Chorionic Villus Biopsy." *Lancet* 1 (1986): 674.

Jacob, F. *The Logic of Living Systems: A History of Heredity.* London: Lane, 1974. Originally published as *La logique du vivant: Une histoire de l'hérédité.* Paris: Gallimard, 1970.

Jacquard, A., and D. Schoevaert. "Artificial Insemination and Consanguinity." In *Human Artificial Insemination and Semen Preservation*, edited by G. David and W. S. Price. New York: Plenum Press, 1980.

Jacquinot, C., and J. Delaye. *Les trafiquants des bébés à naitre*. Lausanne: P. M. Faure, 1984.

Jaspers, K. *Der Arzt im technischen Zeitalter*. Munich: R. Piper, 1986.

Jaspers, K., and R. Bultmann. *Myth and Christianity: An Inquiry into the Possibility of Religion without Myth*. New York: Noonday Press, 1958. Originally published as *Die Frage der Entmythologisierung*. Munich: Piper, 1981.

John Paul II. "Address to a Conference on Pharmaceuticals in the Hall of the Synod." October 24, 1986. In *Insegnamenti di Giovanni Paolo II*, 9.2:1183 [Italian]. Vatican City: Libreria Editrice Vaticana, 1986.

———. "Address to the Eighteenth International Congress of the Transplantation Society." August 29, 2000.

———. "Address to the Fifteenth Congress of Catholic Physicians." October 3, 1982. In *Insegnamenti di Giovanni Paolo II*, 5.3:669–677 [Italian]. Vatican City: Libreria Editrice Vaticana, 1982.

———. "Address to the International Congress on Assistance to the Dying." March 17, 1992. In *Medicina e Morale* 42, no. 3 (1992): 419–422 [Italian].

———. "Address to Members of the Third General Conference of the Latin American Episcopate—Puebla, Republic of Mexico." January 28, 1979.

———. "Address to Members of the World Medical Association." October 29, 1983. In *Insegnamenti di Giovanni Paolo II*, 6.2:917–923 [Italian]. Vatican City: Libreria Editrice Vaticana, 1983.

———. "Address to Midwives Participating in the Conference on Defense of Life and Family." January 26, 1980. In *Insegnamenti di Giovanni Paolo II*, 3.1:191–195 [Italian]. Vatican City: Libreria Editrice Vaticana, 1980.

———. "Address to Participants in a Congress on Surgery." February 19, 1987. In *Insegnamenti di Giovanni Paolo II*, 10.1:376 [Italian]. Vatican City: Libreria Editrice Vaticana, 1985.

———. "Address to Participants in the Congresses of the Italian Society of Surgery." October 27, 1980. In *Insegnamenti di Giovanni Paolo II*, 3.2:1005–1010 [Italian]. Vatican City: Libreria Editrice Vaticana, 1980.

———. "Address to Participants in the First International Medical Conference of the Pro-Life Movement on Prenatal Diagnosis and the Surgical Treatment of Congenital Malformations." December 3, 1982. In *Insegnamenti di Giovanni Paolo II*, 5.3:889–898 and 1509–1513 [Italian]. Vatican City: Libreria Editrice Vaticana, 1982.

———. "Address to Participants in the International Congress on Cancer and Hormones." April 26, 1986.

———. "Address to Participants in the International Congress on Life-Sustaining Treatment and Vegetative State: Scientific Advances and Ethical Dilemmas." March 20, 1994.

———. "Address to Participants in a Study Course on Human Preleukemias." November 15, 1985. In *Insegnamenti di Giovanni Paolo II*, 8.2:1265 ff [Italian]. Vatican City: Libreria Editrice Vaticana, 1985.

———. "Address to the Pontifical Academy for Sciences." October 23, 1982. In *Insegnamenti di Giovanni Paolo II*, 5.3:889–898 [Italian]. Vatican City: Libreria Editrice Vaticana, 1982.

———. "Address to Priests Participating in the Study Seminar on Responsible Procreation." September 17, 1983. In *Insegnamenti di Giovanni Paolo II*, 6.2:561–564 [Italian]. Vatican City: Libreria Editrice Vaticana, 1983.

———. "Address to Scientists and Health Care Workers." November 12, 1987. In *Insegnamenti di Giovanni Paolo II*, 10.2:1086–1087 [Italian]. Vatican City: Libreria Editrice Vaticana, 1985.

———. "All Demographic Policies Must Respect Man." In *Insegnamenti di Giovanni Paolo II*, 7.1:1625–1636 [Italian]. Vatican City: Libreria Editrice Vaticana, 1984.

———. Apostolic Exhortation *Familiaris consortio*. November 22, 1981.

———. Apostolic Letter *Mulieris dignitatem*. August 15, 1988.

———. Apostolic Letter *Salvifici doloris*. February 11, 1984.

———. Encyclical Letter *Evangelium vitae*. March 25, 1995.

———. Encyclical Letter *Fides et ratio*. September 14, 1998.

———. Encyclical Letter *Sollicitudo rei socialis*. December 30, 1987.

———. Encyclical Letter *Veritatis splendor*. August 6, 1993.

———. *Insegnamenti di Giovanni Paolo II*. 20 vols. Vatican City: Libreria Editrice Vaticana, 1979–2005.

———. *Letter to Families*. February 2, 1994.

———. *Man and Woman He Created Them: A Theology of the Body*. Translated and edited by M. Waldstein. Boston: Pauline Books and Media, 2006.

———. "Message for World Peace Day: Peace with God the Creator; Peace with all of Creation." January 1, 1990.

———. "Message to Participants in the Plenary Assemby of the Pontifical Academy of Sciences." October 22, 1996. *Orizzonte Medico* 5 (1996): 4–5 [Italian].

———. "Responsible Parenthood in the Light of *Humanae Vitae*." August 1, 1984. In *Insegnamenti di Giovanni Paolo II*, 7.2:144–151 [Italian]. Vatican City: Libreria Editrice Vaticana, 1984.

———. *The Theology of the Body: Human Love in the Divine Plan*. Boston: Pauline Books and Media, 1997.

John XXIII. *Discorsi, messaggi, colloqui del S. Padre Giovanni XXIII* [Speeches, messages, and talks by the Holy Father John XXIII]. 5 vols. Vatican City: Tipografia Poliglotta Vaticana, 1960–1964.

———. Encyclical Letter *Mater et magistra*. May 15, 1961.

———. Encyclical Letter *Pacem in terris*. April 11, 1963.

Jolif, J.Y. *Comprendre l'homme: Introduction à une anthropologie philosophique*. Paris: Cerf, 1967.

Jonas, H. *The Imperative of Responsibility: In Search of an Ethics for the Technological Age*. Chicago: University of Chicago Press, 1984. Originally published as *Das Prinzip Verantwortung* (Frankfurt: Insel Verlag, 1979).

———. *Organismo e libertà*, edited by P. Becchi. Turin: Einaudi, 1999.

———. *Philosophical Essays: From Ancient Creed to Technological Man*. Englewood Cliffs: Prentice Hall, 1974.

———. *Technik, Medizin und Ethik: Zur Praxis des Prinzips Verantwortung*. Frankfurt: Suhrkamp, 1985.

———. "Technique, morale et génie génétique." *Communio* 9, no. 6 (1984): 45–65.

———. *Tecnica, medicina ed etica*. Turin: Einaudi, 1997. Originally published as *Technik, Medizin und Ethik: Zur Praxis des Prinzips Verantwortung* (Frankfurt: Suhrkamp, 1985).

Jones, H. W. "History of In Vitro Fertilization." In *Infertility: Evaluation and Treatment*, edited by W. R. Keye, R. J. Chang, R. W. Rebar, and M. R. Soules, 736–744. Philadelphia: W. B. Saunders, 1995.

Jones, W. H. S. *The Doctor's Oath: An Essay in the History of Medicine*. New York: Macmillan, 1924.

Jonsen, A. R. *The Birth of Bioethics*. New York: Oxford University Press, 1998.

———. "Casuistry as Methodology and Clinical Ethics." *Theoretical Medicine* 12 (1991): 295–307.

Jonsen, A. R., A. L. Jameson, and A. Lynch. "Medical Ethics: History of North America in the Twentieth Century." In *Encyclopedia of Bioethics*, edited by W. T. Reich, 1:992–1001. New York: Free Press, 1978.

Jorgensen, N., A. Giwercman, S. W. Hansen, and N. E. Skakkebaek. "Testicular Cancer after Vasectomy: Origin from Carcinoma In Situ of the Testis." *European Journal of Cancer* 29A, no. 7 (1993): 1062–1064.

Jorio, T. *Theologia moralis*. Vol. 2. Naples: M. D'Auria, 1939.

Jost, A. "Das Recht auf den Tod." 1895.

Juarez, F., Y. Barrios, L. Cano, et al. "Domino (Crossover) Kidney Transplantation Using Low Doses of Neoral." *Transplantation Proceedings* 30, no. 5 (1998): 2289–2290.

Judicial Council of the Amerjcan Medical Association Chicago. "Guidelines for Ethics Committees in Health Care Institutions." *Journal of the American Medical Association* 253 (1985): 2698–2699.

Julien, C., A. Bazin, B. Guyot, F. Forester, and F. Dattus. "Rapid Prenatal Diagnosis of Down's Syndrome with In Situ Hybridization of Fluorescent DNA Probes." *Lancet* 2 (1986): 863–864.

Kahn, A. "Essais sur l'homme et essais d'hommes." *Le Cahiers du CCNNE* 3 (1995): 25–26.

Kalra, D., R. Gertz, P. Singleton, and H. M. Inskip. "Confidentiality of Personal Health Information Used for Research." *British Medical Journal* 333, no. 7560 (2006): 196–8.

Kaplan, H. S. *The New Sex Therapy: Active Treatment of Sexual Dysfunctions*. New York: Brunner Mazel, 1974.

Kardiner, A. *Sex and Morality*. Indianapolis, NY: Bobbs-Merill, 1954.

Kasimba, P., and P. Singer. "Australian Commission and Committees on Issues in Bioethics." *Journal of Medicine and Philosophy* 14, no. 4 (1989): 403–424.

Kass, L. R. "Is There a Right to Die?" *Hastings Center Report* 23, no. 1 (1993): 34–43.

———. "Organs for Sale? Propriety, Property and the Price of Progress." *The Public Interest* 107 (1992): 65–86.

Kass, N. E., L. Dawson, and N. I. Loyo-Berrios. "Ethical Oversight of Research in Developing Countries." *IRB* 25, no. 2 (2003): 1–10.

Kass, N. E., J. Sugarman, R. Faden, and M. Schoch-Spana. "Trust: The Fragile Foundation of Contemporary Biomedical Research." *Hastings Center Report* 26, no. 5 (1996): 25–29.

Kelly, M. J., and D. G. McCarthy, eds. *Ethics Committees: A Challenge for Catholic Health Care*. St. Louis, MO: The Pope John XXIII Medical-Moral Research Center, 1984.

Kelsen, H. *La dottrina pura del diritto.* Turin: Einaudi, 1966.

Kenny, D. T. "In Vitro Fertilization and Gamete Intrafallopian Transfer: An Integrative Analysis of Research 1987–1992." *British Journal of Obstetrics and Gynecology* 102 (1995): 317–325.

Kerr, C., E. Robinson, A. Stevens, et al. "Randomisation in Trials: Do Potential Trial Participants Understand It and Find It Acceptable?" *Journal of Medical Ethics* 30, no. 1 (2004): 80–84.

Kimura, R. "Ethics Committees for 'High Tech' Innovation in Japan." *Journal of Medicine and Philosophy* 14, no. 4 (1989): 457–464.

Kinsey, A. C., W. B. Pomeroy, and C. E. Martin. *Sexual Behavior in the Human Female*. Philadelphia: W. B. Saunders, 1953.

Klein, T., H. Or, C. Zakai, Z. Shapira, and A. Yussim. "Succesful Interspousal Swap of ABO Incompatible Living Donor Kidneys." *Transplantation Proceedings* 32, no. 4 (2000): 688–689.

Kleinknecht, H. "Pneuma." In *Grande lessico del nuovo testamento*, edited by G. Kittel, G. Friedrich, 10:767–849. Brescia: Paideia, 1975.

Kodish, E., ed. *Ethics and Research with Children: A Case-Based Approach*. New York: Oxford University Press, 2005.

Kohll, K. L., and A. J. Sobrero. "Vasectomy: A Study of Psychosexual and General Reaction." *Social Biology* 20 (1973): 298–302.

Komersaroff, P. A. *Troubled Body: Critical Perspectives on Postmodernism, Medical Ethics and the Body*. Durham: Duke University Press, 1995.

Kopelman, L. M. "Research Methodology: II. Clinical Trials." In *Encyclopedia of Bioethics*, edited by S. G. Post, 2334–2343. New York: Macmillan Reference USA, 2004.

Korein, J. "Terminology, Definitions, and Usage." *Annals of the New York Academy of Sciences* 315 (1978): 6–10.

Kovacs, G. T., P. Rogers, J. F. Leeton, A. O. Trounson, C. Wood, and H. W. Baker. "In Vitro Fertilization and Embryo-Transfer." *Medical Journal of Australia* 144 (1986): 682–683.

Kranenburg, L., T. Visak, W. Weimar, et al. "Starting a Crossover Kidney Transplantation Program in the Netherlands: Ethical and Psychological Considerations." *Transplantation* 78, no. 2 (2004): 194–197.

Krimsky, S. *Genetic Alchemy: A Social History of the DNA Controversy*. Cambridge, MA: MIT Press, 1982.

Krivit, W., and C. B. Whitley. "Bone Marrow Transplantation for Genetic Diseases." *New England Journal of Medicine* 316 (1987): 1085–1087.

Krosnar, K. "Report Says 100 Roma Women Have Been Forcibly Sterilised in Slovakia." *British Medical Journal* 326 (2003): 302.

Kübler-Ross, E. *On Death and Dying*. New York: Macmillan, 1969.

Kuleshova, L., and A. Lopata. "Vitrification Can Be More Favorable Than Slow Cooling." *Fertility and Sterility* 78 (2002): 449–454.

Kulier, R., M. Boulvain, D. Walker, et al. "Minilaparotomy and Endoscopic Techniques for Tubal Sterilization." *Cochrane Database of Systematic Reviews* 3 (2004): CD001328.

Kushner, T., and R. Belliotti. "Baby Fae: A Beastly Business." *Journal of Medical Ethics* 11 (1985): 178–183.

Kvistgaard, M., and A.M. Olsen. *Biotechnology and Environment: A Literature Review of Environmental Applications and Impacts of Biotechnology*. Lyngby: Technical University of Denmark, 1986.

Ladrière, J. *L'articulation du sense*. Paris: Aubier-Montaigne, 1970.

———. *I rischi della razionalità*. Turin: SEI, 1978.

Laín Entralgo, P. *Antropología medica para clínicos*. Barcelona: Salvat, 1985.

———. *Historia universal de la medicina*. 8 vols. Barcelona: Salvat, 1972.

———. *Il medico e il paziente*. Milan: Mondadori, 1969.

Lalande, A. *Vocabulaire technique et critique de la philosophie*. Paris: Presses Universitaires de France, 1968.

Lamotte, B. "Le réductionisme: Méthode ou idéologie?" *Lumière et vie* 172 (1985): 5–19.

Largey, G. "Reproductive Technologies: Sex Selection." In *Encyclopedia of Bioethics*, edited by W.T. Reich, 1439–1443. New York: Free Press, 1978.

La Vergata, A. *L'evoluzione biologica: da Linneo a Darwin*. Turin: Loescher, 1979.

Lecaldano, E. "Il contributo di una filosofia 'laica.'" *Biblioteca della Libertà* 99 (1987): 57–66.

———. "Etica e significato: Un bilancio." In *Teorie etiche contemporanee*, edited by C.A. Viano, 58–86. Turin: Bollati Boringhieri, 1990.

———. "'Grande Divisione,' 'legge di Hume' e 'ragionamento in morale.'" *Rivista di Filosofia* 67 (1976): 82.

———. *Hume e la nascita dell'etica contemporanea*. Rome: Laterza, 1991.

———. "Principi e basi razionali di un'etica non religiosa." In *Problemi di etica: Fondazione, norme, orientamenti*, edited by E. Berti, 23–68. Padua: Gregoriana, 1990.

Leibniz, G.W. *Monadology and Other Philosophical Essays*. Translated by P. Schrecker and A.M. Schrecker. New York: Macmillan, 1965.

———. *New Essays on Human Understanding*. New York: Cambridge University Press, 1996.

Leigh and Barron Consulting and Christie Associates for NHSTD on behalf of the UK Department of Health. *Standard for Local Research Ethics Committees: A Framework for Ethical Review*. London: NHS Training Division, 1994.

Leisinger, K.M., and K. Schmitt. *All Our People*. Washington, DC: Island Press, 1994.

Lener, S. "Sul diritto dei malati e dei moribondi: È lecita l'eutanasia?" *La Civiltà Cattolica* 2 (1976): 217–232.

Leone, S. "'Divinum est sedare dolorem': Risorse mediche e implicanze etiche dell'analgesia." *Camillianum* 11ns (2004): 275–312.

784

———. "Sterilizzazione." In *Nuovo dizionario di bioetica*, edited by S. Leone and S. Privitera, 1137–1139. Acireale: EDB-ISB, 2004.

Leopold, A. *A Sand County Almanac and Sketches Here and There*. New York: Oxford University Press, 1949.

Lettellier, P., and G. Gambino, eds. *L'eutanasia*. Vol. 1, *Aspetti etici e umani*. Sapere 2000 Edizioni Multimediali 2004.

Leuzzi, L. "Deontologia medica e fecondazione in vitro." *Medicina e Morale* 33, no. 4 (1983): 365–390.

———. "Il dibattito sulla inseminazione artificiale nella riflessione medico-morale in Italia nell'ultimo decennio." *Medicina e Morale* 32, no. 4 (1982): 343–370.

———. "Riflessione etico-morale sulla fecondazione in vitro." *Ospedale Miulli* 3 (1986): 1–166.

Levi, A. "Intorno ad un corollario del principio di socialità del diritto." In *Scritti minori di filosofia del diritto*, 1:3. Padua: CEDAM, 1957.

Levinas, E. *Ethics and Infinity*. Translated by R.A. Cohen. Pittsburgh: Duquesne University Press, 1985.

———. *Etica e infinito*. Rome: Città Nuova, 1984.

Levine, C., and T.W. Ogletree. "Research Methodology. III. Subjects." In *Encyclopedia of Bioethics*, edited by S.G. Post, 2343–2347. 3rd ed. New York: Macmillan Reference USA, 2004.

Levine, R.J. *Ethics and Regulation of Clinical Research*. 2nd ed. New Haven: Yale University Press, 1988.

———. "Informed Consent: III. Consent Issues in Human Research." In *Encyclopedia of Bioethics*, edited by S.G. Post, 1280–1290. 3rd ed. New York: Macmillan Reference USA, 2004.

———. "Research Ethics Committees." In *Encyclopedia of Bioethics*, edited by W.T. Reich, 4:2266–2270. Rev. ed. New York: Macmillan, 1995.

Lewin, A., A. Simon, R. Rabinowitz, and J.G. Schenker. "Second-Trimester Heterotopic Pregnancy after In Vitro Fertilization and Embryo Transfer: A Case Report and Review of the Literature." *International Journal of Fertility* 36, no. 4 (1991): 227–230.

Li, H.H., U.B. Gyllensten, X.F. Cui, R.K. Saiki, H.A. Erlich, and N. Arnheim. "Amplification and Analysis of DNA Sequences in Single Human Sperm and Diploid Cells." *Nature* 335 (1988): 414–417.

Lichfield, M., and S. Kentish. *Bambini da bruciare*. Catania: Paoline, 1976.

Lichter, P., T. Cremer, J. Borden, L. Manuelidis, and D.C. Ward. "Delineation of Individual Human Chromosomes in Metaphase and Interphase Cells by In Situ Suppression Hybridisation Using Recombinant DNA Libraries." *Human Genetics* 80 (1988): 224–234.

Lifton, R.J. *I medici nazisti: Lo sterminio sotto l'egida della medicina e la psicologia del genocidio*. Milan: Rizzoli, 1988.[*The Nazi Doctors: Medical Killing and the Psychology of Genocide*. New York: Basic Books, 1988.]

Limat, R., C. Josserand, B. Nicod, and M. Ogier. "Du soin à la contrainte: Quelques interrogations éthiques vécues par l'infirmier(e) dans la pratique des soins." *Médecine et Hygiène* 42 (1984): 1177–1182.

Lio, H. *"Humanae vitae" e coscienza: L'insegnamento di K. Wojtyla teologo e Papa*. Vatican City: Libreria Editrice Vaticana, 1980.

List, P. C. *Radical Environmentalism: Philosophy and Tactics*. Belmont, CA: Wadsworth, 1993.

Little, P. F. R. "Structure and Function of the Human Genome." *Genome Research* 15 (2005): 1759–1766.

Liu, H. C., H. W. Jones, and Z. Rosenwaks. "The Efficacy of Human Reproduction after In Vitro Fertilization and Embryo Transfer." *Fertility and Sterility* 49 (1988): 649–653.

Livi, A. "Metafisica." In *Dizionario interdisciplinare di scienza e fede*, edited by G. Tanzella-Nitti and A. Strumia, 939–957. Rome: Urbaniana University Press, 2002.

Llano, A. *La nuova sensibilità*. Milan: Ares, 1995.

Lo, B., and G. Rubenfeld. "Palliative Sedaction in Dying Patients." *Journal of the American Medical Association* 14 (2005): 1810–1816.

Locatelli, F. "Alcune note sulla dimostrazione dell'immortalità dell'anima in S. Tommaso." *Rivista di Filosofia Neoscolastica* 33 (1941): 413–418.

Lock, S., and F. Wells. *Fraud and Misconduct in Medical Research*. London: British Medical Journal Publishing Group, 1996.

Lombardi Ricci, M. *Fabbricare bambini? La questione dell'embrione tra nuova medicina e genetica*. Milan: Vita e Pensiero, 1996.

Lombardi Vallauri, L. "Bioetica, potere, diritto." *Jus* 1–2 (1984): 41–80.

———. "Le culture riduzionistiche nei confronti della vita." In *Il valore della vita: L'uomo di fronte al problema del dolore, della vecchiaia, dell'eutanasia; Atti del 54° corso di aggiornamento culturale dell'Università Cattolica; Roma, 2–7 settembre 1984*, by A. Bausola et al., 41–74. Milan: Vita e Pensiero, 1985.

———. "Manipolazioni genetiche e diritto." *Rivista di Diritto Civile* 1 (January–February 1985): 1–23.

Lonergan, B. *Method in Theology*. London: Darton and Todd, 1972.

Lopéz Guzmán, J. *Objeción de conciencia farmacéutica*. Barcelona: Ediciones Internacionales Universitarias, 1997.

López Trujllo, A., J. Herranz, and E. Sgreccia, eds. *"Evangelium vitae" e diritto*. Vatican City: Libreria Editrice Vaticana, 1997.

Lopéz Trujillo, A., and E. Sgreccia, eds. *Metodi naturali per la regolazione della fertilità: L'alternativa autentica*. Milan: Vita e Pensiero, 1994.

Lovelock, J. *The Ages of Gaia: A Biography of Our Living Earth*. New York: Norton, 1988.

Loverro, G., L. Selvaggi, and F. M. Boscia. "Procedure di indagine prenatale: Significato diagnostico e pericolosità." *Medicina e Morale* 34, no. 4 (1984): 464–487.

Lucan, M., P. Rotariou, D. Neculoiu, and G. Iacob. "Kidney Exchange Program: A Viable Alternative in Countries with Low Rate of Cadaver Harvesting." *Transplantation Proceedings* 35, no. 3 (2003): 933–934.

Lucarelli, G., M. Galimberti, P. Polchi, et al. "Marrow Transplantation in Patients with Advanced Thalassaemia." *New England Journal of Medicine* 316 (1987): 1050–1055.

Lucarelli, G., P. Polchi, M. Galimberti, et al. "Marrow Transplantation for Thalassaemia following Busulphan and Cyclophosphamide." *Lancet* 1 (1985): 1355–1357.

Lucarelli, G., P. Polchi, T. Izzi, et al. "Allogeneic Marrow Transplantation for Thalassaemia." *Experimental Haematology* 12 (1984): 676–681.

Lucas Lucas, R. "Morte encefalica e morte umana." In *Antropologia e problemi bioetici*, 119–158. Cinisello Balsamo: Edizioni San Paolo, 2001.

———. *L'uomo spirito incarnato: Compendio di filosofia dell'uomo*. Milan: Paoline, 1993.

Lucioni, C. "Economia e salute." *Medicina e Morale* 36, no. 4 (1986): 777–786.

Lurie, P., and S. M. Wolfe. "Unethical Trials of Intervention to Reduce Perinatal Transmission of the Human Immunodeficiency Virus in Developing Countries." *New England Journal of Medicine* 337, no. 12 (1997): 853–856.

Luzi, F., ed. *L'ingegneria genetica nella Germania Federale: Raccomandazione del Bundestag, Legge, Regolamenti*. Rome: Senato della Repubblica Italiana, Servizio Studi, Ufficio Ricerche nel Settore Sociale, 1992.

Lynge, E., L. B. Knudsen, and H. Muller. "Vasectomy and Testicular Cancer: Epidemiological Evidence of Association."*European Journal of Cancer* 29A, no. 7 (1993): 1064–1066.

MacDonald, R. M. Wagner, and R. N. Slotnik, "Sensitivity and Specificity of Screening for Down Syndrome with Alpha-Fetoprotein, HCG, Unconjugated Estriol, and Maternal Age." *Obstetrics and Gynecology* 77 (1991): 63–68.

MacIntyre, A. *After Virtue: A Study in Moral Theory*. Notre Dame, IN: University of Notre Dame Press, 1981.

MacIntyre, M. N. "Professional Responsibility in Prenatal Genetic Evaluation." *Birth Defects Original Article Series* 8, no. 4 (1972): 31–35.

MacKinnon, B. "How Important Is Consent for Controlled Clinical Trials?" *Cambridge Quarterly of Healthcare Ethics* 5, no. 2 (1996): 221–227.

Maddox, J. "The Case for the Human Genome." *Nature* 352 (1991): 11–14.

Maffettone, S. "Proposte per uno statuto morale e giuridico dell'embrione." In *Quale statuto per l'embrione umano: Problemi e prospettive; Convegno Internazionale, Milano, gennaio 1991*, edited by M. Mori, 96–107. Milan: Bibliotechne, 1992.

Mahowald, M. B., J. Areen, B. J. Hoffer, et al. "Transplantation of Neural Tissue from Fetuses." *Science* 235 (March 13, 1987): 1307–1308.

Maier, A. W., ed. "Suicide." In *Encyclopedia of Religion and Ethics*, edited by J. Hastings, J. A. Selbie, and L. H. Gray, 12:32. Edinburgh: T. and T. Clark, 1971.

Malherbe, J. F. "L'embryon est-il une personne humaine?" *Lumière et Vie* 172 (1985): 19–31.

———. "Médecine, anthropologie et éthique." *Médecine de l'Homme* 156–157 (1985): 5–12.

———. *Pour une éthique de la médecine*. Paris: Larousse, 1987.

Malthus, T. R. *Essay on the Principal of Population*. London: J. Johnson, 1798. Republished in electronic format in 1998 by Electronic Scholarly Publishing Project and available at http://www.esp.org/books/malthus/population/malthus.pdf.

Mancini, I. *Teologia, ideologia, utopia*. Brescia: Queriniana, 1974.

Mancuso, S., and E. Sgreccia, eds. *Trattamento della sterilità coniugale*. Milan: Vita e Pensiero, 1988.

Mancuso, S., and P. F. van Look. *La regolazione naturale della fertilità oggi*. Milan: Vita e Pensiero, 1992.

Manga, P. "A Commercial Market for Organs? Why Not." *Bioethics* 1, no. 4 (1987): 321–338.

Manghi, S. *Il medico, il paziente e l'altro: Un'indagine sull'interazione comunicativa nelle pratiche mediche*. Milan: F. Angeli, 2005.

Manni, C. "Considerazioni mediche sull'eutanasia: Il pensiero di un medico anestesista-rianimatore." *Federazione Medica* 38, no. 1 (1985): 20–25.

Manni, C., R. Proietti, and F. Della Corte. "La morte cerebrale, aspetti diagnostici." *Medicina e Morale* 43, no. 5 (1993): 903–918.

Mannion, M. T., ed. *Post-Abortion Aftermath*. Kansas City, MO: Sheed and Ward, 1994.

Mantegazza, F. "Tecnologie per la diagnosi prenatale." *Medicina e Morale* 32, no. 1 (1982): 72–75.

Mantegazzini, C. *The Environmental Risks from Biotechnology*. London: Frances Printer, 1986.

Mantovani, F. "La sterilizzazione consensuale irreversibile alla luce del principio personalistico." *Medicina e Morale* 22, no. 2 (1972): 191–198.

Marana R., L. Muzii, M. Rizzi, S. dell'Acqua, and S. Mancuso. "Prognostic Role of Laparoscopic Salpingoscopy of the Only Remaining Tube after Controlateral Ectopic Pregnancy." *Fertility and Sterility* 63, no. 2 (1995): 303–306.

Marana, R., G. F. Catalano, and L. Muzii. "Trattamento chirurgico della sterilità di origine tubarica." *Medicina e Morale* 43, no. 1 (1993): 67–74.

Marcel, G. *L'homme problématique*. Paris: Aubier, 1965.

———. *Homo Viator: Introduction to a Metaphysic of Hope*. Translated by E. Craufurd and P. Seaton. South Bend, IN: St. Augustine's Press, 2010.

———. *Metaphysical Journal*. Translated by B. Wall. Chicago: H. Regnery, 1952.

———. *Problematic Man*. Translated by B. Thompson. New York: Herder and Herder, 1967. Originally published as *L'Homme problèmatique* (Paris: Aubier, 1955).

———. *Du refus à l'invocation*. Paris: Gallimard, 1940.

Marcozzi, V. "Il cristiano di fronte all'eutanasia." *La Civiltà Cattolica* 4 (1975): 322 ff.

———. *Le origini dell'uomo: L'evoluzione oggi*. Milan: Massimo, 1972.

———. *Però l'uomo è diverso*. Milan: Rusconi, 1981.

———. "'Sorella scimmia' e controversie evoluzionistiche." *La Civiltà Cattolica* 1 (1985): 134–145.

———. *L'uomo nello spazio e nel tempo*. Milan: CEA, 1953.

———. *La vita e l'uomo*. Milan: CEA, 1946.

Marcuse, H. *Eros and Civilization: A Philosophical Inquiry into Freud*. Boston: Beacon Press, 1966.

Marino, P. A. "Sviluppi della medicina e progresso umano." In *Trapianto di cuore e trapianto di cervello*, 236–246. Rome: Orizzonte Medico, 1983.

Maritain, J. *Bergsonian Philosophy and Thomism*. Translated by M. L. Andison. New York: Philosophical Library, 1955. Originally published as *De Bergson à St. Thomas d'Aquin* (Neuchâtel: Éditions de la Baconnière 1947).

———. *Integral Humanism: Temporal and Spiritual Problems of a New Christendom*. Translated by J. W. Evans. New York: Charles Scribner's Sons, 1968. Originally published as *Humanisme intégral: Problèmes temporels et spirituels d'une nouvelle chrétienté* (Paris: Aubier, 1936).

———. *Introduction to the Basic Problems of Moral Philosophy*. Albany, NY: Magi Books. 1990. Originally published as *Neuf leçons sur les notions premières de la philosophie morale* (Paris: Téqui, 1951).

———. *Neuf leçons sur les notions premières de la philosophie morale*. Paris: Téqui, 1951.

———. *Nove lezioni sulla legge naturale*. Milan: Jaca Book, 1985.

———. *Quatre essais sur l'esprit dans sa condition charnelle*. Paris: Aubier, 1965.

———. *The Range of Reason*. New York: C. Scribner's Sons, 1952. Originally published as *Raison et raisons: Essais détachés* (Paris: Egloff, 1947).

———. *The Rights of Man and Natural Law*. Translated by D.C. Anson. New York: C. Scribner's Sons, 1943. Originally published as *Les droits de l'homme et la loi naturelle* (New York: Maison, 1942).

———. "Vers une idée thomiste de l'évolution." *Nova et Vetera* 42 (1967): 130–131.

Marmont, A. "Stato attuale del trapianto del midollo osseo allogenico nelle leucemie." In *Atti del Convegno internazionale*: Trapianto di midollo osseo in pediatria; Problematiche etico-giuridiche e scientifiche; Pavia, 15–16 novembre 1986, edited by Direzione Medico-Scientifica della Immuno, 17–22. Pisa: Pacini, 1987.

Marrama, P., C. Cardini, W. Pasini, and I. Baldaro Verde, eds. *L'inseminazione della discordia*. Milan: F. Angeli, 1987.

Martin, R.H. "Comparison of Chromosomal Abnormalities in Hamster Egg and Human Sperm Pronuclei." *Biology of Reproduction* 31, no. 4 (1984): 819–825.

———. "The Risk of Chromosomal Abnormalities following ICSI." *Human Reproduction* 11 (1996): 924–925.

Martin, R.H., M.M. Mahadevan, P.J. Taylor, et al. "Chromosomal Analysis of Unfertilised Human Oocytes." *Journal of Reproduction and Fertility* 78 (1986): 673–678.

Martini, A. *Profili giuridici della procreazione medicalmente assistita*. Naples: Editoriale Scientifica, 2006.

Marziale, F. "Cooperazione europea nell'ambito dei trapianti e problemi etici." In *Trapianto di cuore e trapianto di cervello*, 184–207. Rome: Orizzonte Medico, 1983.

Mascitelli, E. "Per una lettura antropologica della medicina." In *Saggi di medicina e scienze umane*, by E. Mascitelli et al., 4–35. Milan: Istituto Scientifico Ospedale S. Raffaele, 1984.

Mashiach S, D. Bider, O. Moran, M. Goldenberg, and Z. Ben-Rafael. "Adnexal Torsion of Hyperstimulated Ovaries in Pregnancies after Gonadotropin Therapy." *Fertility and Sterility* 53, no. 1 (1990): 76–80.

Masi, G. "Suicidio." In *Enciclopedia Filosofica*, vol. 8, cols. 26–28. 2nd ed. Rome: Epidem, 1979.

Masini, V. *Medicina Narrativa: Comunicazione empatica ed interazione dinamica nella relazione medico-paziente*. Milan: F. Angeli, 2005.

Massari, A., G. Novelli, A. Colosimo, et al. "Non-Invasive Early Prenatal Molecular Diagnosis Using Retrieved Transcervical Trophoblast Cell." *Human Genetics* 97, no. 2 (1996): 150–155.

Massone, A. *Cause e terapie nell'omosessualità*. Brezzo di Bedero: Salcom, 1970.

Masters, W. H., and V. E. Johnson. *Human Sexual Inadequacy*. Boston: Little, Brown, 1970.

Mastroiacovo, P., A. E. Tozzi, S. Agosti, et al. "Transverse Limb Reduction Defects after Chorion Villus Sampling: A Retrospective Cohort Study." *Prenatal Diagnosis* 13 (1993): 1051–1056.

Mastroianni, A. C., R. Faden, and D. Federman, eds. *Women and Health Research: Ethical and Legal Issues of Including Women in Clinical Studies*. 2 vols. Washington, DC: National Academy Press, 1994.

Mastroianni, L. "In Vitro Fertilization." In *Encyclopedia of Bioethics*, edited by W. T. Reich, 1448–1451. New York: Free Press, 1978.

Mastropaolo, F. "Lo statuto dell'embrione." In *A sua immagine e somiglianza?*, edited by A. Mazzoni, 142–179. Rome: Città Nuova, 1997.

Mattei, J. F., M. G. Mattei, S. Ayme, and F. Giraud. "Origin of the Extra Chromosome in Trisomy 21." *Human Genetics* 46 (1979): 107–110.

Mattioli, V. *Laboratorio umano*. Palermo: Augustinus, 1990.

Maturana, H., and F. Varala. *L'albero della conoscenza*. Milan: Garzanti, 1987.

Maximus the Confessor. *De variis difficilibus locis Sanctorum Dyonisii et Gregorii seu ambiguorum liber*. PG vol. XCI, 1335 a.

Maxwell, D. J., P. Johnson, P. Hurley, K. Neales, L. Allan, and P. Knott. "Fetal Blood Sampling and Pregnancy Loss in Relation to Indication." *British Journal of Obstetrics and Gynaecology* 98 (1991): 892–897.

May, W. E. *Catholics Bioethics and the Gift of Human Life*. Huntington, IN: Our Sunday Visitor, 2000.

———— "Maternità surrogata e mercato: Un punto di vista." *Kos* 75 (1991): 34–38.

May, W. F. "Code or Covenant or Philanthropy and Contract." In *Ethics in Medicine: Historical Perspectives and Contemporary Concerns*, ed. S. J. Reiser, A. J. Dyck, W. J. Curran. Cambridge, MA: MIT Press, 1977.

Mayr, E. *Storia del pensiero biologico*. Turin: Bollati Boringhieri, 1990.

Mazza, S. "Aspetti neurologici della morte cerebrale." *Medicina e Morale* 43, no. 5 (1993): 919–932.

Mazzantini, G. *Storia del pensiero antico*. Turin: Bottega d'Erasmo, 1965.

McCormick, R. A. "Notes on Moral Theology." *Theological Studies* 40 (1979): 107 ff.

McGourty, C. "Profiles Bank on the Way." *Nature* 339, no. 6223 (1989): 327.

McKibbon, A., A. Eady, S. Marks. *Guida alla evidence based medecine*. Rome: Il Pensiero Scientifico, 2000.

McLaren, A. "Prelude to Embryogenesis." In *Human Embryo Research: Yes or No?*, edited by G. Bock and M. O'Connor, 5–23. London: Tavistock, 1986.

McLellan, F. "US Surgeons Do First 'Triple-Swap' Kidney Transplantation." *Lancet* 362, no. 9382 (2003): 456.

McNally, E., and A. Cambon-Thomsen. *25 Recommendations on the Ethical, Legal and Social Implications of Genetic Testing*. Luxembourg: Office for Official Publications of the European Communities, 2004. Available online at http://ec.europa.eu/research/conferences/2004/genetic/pdf/recommendations_en.pdf.

McNeill, J. J. *The Church and the Homosexual*. Kansas City: Sheed, Andrews and McMeel, 1976.

McNeill, P. *The Ethics and Politics of Human Experimentation*. Cambridge: Cambridge University Press, 1993.

McSweeney, L. "Preselection of the Sex of Baby in Nigeria Using Billings Method." In *Report to 6th International Institute of the Ovulation Method*. Los Angeles: 1980.

McWhinnie, A. "A Study of Parenting of IVF and DI Children." *Medicine and Law* 14 (1995): 501–508.

Mead, M. *Sex and Temperament in Three Primitive Societies*. New York: Morrow, 1963.

Medical Research Council Working Party on the Evaluation of Chorion Villus Sanpling. "MRC European Trial of Chorion Villus Sampling." *Lancet* 337 (1991): 1491–1499.

Medical Research International Society for Assisted Reproductive Technology and the American Fertility Society. "In Vitro Fertilization-Embryo Transfer (IVF-ET) in the United States: 1990 Results from the IVF-ET Registry." *Fertility and Sterility* 57 (1992): 15–24.

Medical Task Force on Anencephaly. "The Infant with Anencephaly." *New England Journal of Medicine* 322 (1990): 669–673.

Meirow, D., and J. G. Schenker. "Appraisal of Gamete Intrafallopian Transfer." *European Journal of Obstetrics and Gynecology and Reproductive Biology* 58, no. 1 (1995): 59–65.

Meissner, A., and R. Jaenisch. "Generation of Nuclcar Transfer-Derived Pluripotent ES Cells from Cloned Cdx2-Deficient Blastocysts." *Nature* 439 (2006): 212–215.

Melchiorre, V. *Il corpo*. Brescia: La Scuola, 1984.

———, ed. *Amore e matrimonio nel pensiero filosofico e teologico moderno*. Milan: Vita e Pensiero, 1976.

Meldrum, D. R. "Assisted Reproductive Technology: Choice of Patient, Program, and Procedure." In *Infertility: Evaluation and Treatment*, edited by W. R. Keye, R. J. Chang, R. W. Rebar, and M. R. Soules, 879–885. Philadelphia: W. B. Saunders, 1995.

Mele, V. *Bioetica al femminile*. Milan: Vita e Pensiero, 1998.

———. "Diritti e doveri della medicina predittiva." *Medicina e Morale* 41, no. 3 (1991): 71–90.

———. "La geneterapia ed il parere del Comitato Nazionale per la Bioetica." *Vita e Pensiero* 5 (1991): 362–367.

Mele, V., G. Girlando, and E. Sgreccia. "La diagnosi genetica sui lavoratori: Recenti acquisizioni scientifiche, problematiche etiche ed etico-giuridiche." *Medicina e Morale* 40, no. 2 (1990): 301–329.

Mele, V., A. R. Morgani, and C. Giorlandino. "Etica 'pratica' e counseling genetico." In *Trattato di diagnosi prenatale e terapia fetale*, edited by C. Giorlandino, 171–183. Rome: CIC Edizioni Internazionali, 1997.

Melgar Riol, J. "Objeción de conciencia y farmacia." *Cuadernos de Bioética* 2 (1993): 37–47.

Melina, L. *Morale: Tra crisi e rinnovamento*. Milan: Ares, 1993.

———. "Vita." In *Dizionario interdisciplinare di scienza e fede*, edited by G. Tanzella-Nitti and A. Strumia, 1519–1530. Rome: Urbaniana University Press, 2002.

Merkatz, R. B. "Women in Clinical Trials of New Drugs: A Change in Food and Drug Administration Policy." *New England Journal of Medicine* 329, no. 4 (1993): 292–296.

Merleau-Ponty, M. *The Structure of Behavior*. Boston: Beacon Press, 1963.

———. *The Visible and the Invisible: Followed by Working Notes*, edited by C. Lefort. Translated by A. Lingis. Evanston, IL: Northwestern University, 1968.

Merton, R. K. "Priorities in Scientific Discovery." *American Sociological Review* 22 (1966): 235–259.

Metters, J. "Regulations on Bioethics in the United Kingdom." *Funzione e funzionamento dei comitati etici*, edited by G. Gerin, 107–116. Padua: CEDAM, 1991.

Meulembergs, T., J. Vermylen, and P. T. Schotsmans. "The Current State of Clinical Ethics and Healthcare Ethics Committees in Belgium." *Journal of Medical Ethics* 31 (2005): 318–321.

Midgley, M. *Animals and Why They Matter*. Harmondsworth, Middlesex, England: Penguin Books, 1983.

Milani Comparetti, M. "Introduzione alla nuova genetica." In *Ingegneria genetica e biotecnologie nel futuro dell'uomo*, edited by E. Sgreccia and V. Mele, 3–23. Milan: Vita e Pensiero, 1992.

Milano, A. "Freud critico della religione per l'autonomia della morale." *Aspernas* 3, no. 4 (1981): 397–425.

Miles, S. H. *The Hippocratic Oath and the Ethics of Medicine*. New York: Oxford, 2004.

Mill, J. S. *Utilitarianism*. 1781. Reprint, New York: Bobbs-Merrill, 1957.

Mills, M. S., H. A. Eddowes, D. J. Cahill, et al. "A Prospective Controlled Study of In Vitro Fertilization, Gamete Intra-Fallopian Transfer and Intrauterine Insemination Combined with Superovulation." *Human Reproduction* 7, no. 4 (1992): 490–494.

Minacori, R., and A. G. Spagnolo. "Consenso, 'privacy' e ricerca medica." *Medicina e Morale* 47, no. 4 (1997): 811–814.

———. "Il placebo nella pratica clinica e nella ricerca." *Medicina e Morale* 47, no. 2 (1997): 444–447.

Mingone, L. "I trapianti di organi nei dibattiti dell'etica contemporanea." *Medicina e Morale* 44, no. 1 (1994): 11–37.

Minogue, B. "Medical Research Involving Persons." In *Bioethics: A Committee Approach*. Boston: Jones and Bartlett, 1996.

Miranda, G. "'Cultura della morte': Analisi di un concetto e di un dramma." In *Commento interdisciplinare alla "Evangelium vitae,"* edited by E. Sgreccia and R. Lucas Lucas. Vatican City: Libreria Editrice Vaticana, 1997.

———. "I problemi etici dell'eutanasia nell'enciclica *Evangelium vitae*." *Medicina e Morale* 45, no. 4 (1995): 719–738.

———. "Riflessioni etiche intorno alla fine della vita." In *A sua immagine e somiglianza?* edited by A. Mazzoni, 180–202. Rome: Città Nuova, 1997.

Mitcham, C. "Philosophy of Technology." In *Encyclopedia of Bioethics*, edited by W. T. Reich, 5:2477–2484. Rev. ed. New York: Macmillan, 1995.

Mitscherlich, A., and F. Mielke. *Medizine und Menschlichkeit: Dokumenten des Nürnberger Arzteprozesses*. Frankfurt: Fischer, 1960.

Mobili, L., and A. R. Morgani. "Ruolo della cordocentesi in medicina prenatale." *Ultrasonica* 2 (1995): 37–39.

Mohacsi, P. J., C. E. Blumer, S. Quine, and J. F. Thompson. "Aversion to Xenotransplantation." *Nature* 378 (1995): 434.

Moltmann, J. *The Beginnings of Dialectical Theology*, edited by J. M. Robinson. Richmond: John Knox Press, 1968.

Mondin, B. *L'uomo: Chi è? Elementi di antropologia filosofica.* Milan: Massimo, 1982.

Monod, J. *Chance and Necessity: An Essay on the Natural Philosophy of Modern Biology.* Translated by A. Wainhouse. New York: Knopf, 1971. Originally published as *Le hasard et la nécessité* (Paris: Editions du Seuil, 1970).

Montalenti, G. "Storia della biologia e della medicina." In *Storia delle scienze*, edited by N. Abbagnano, 3:1. Turin: UTET, 1962.

Montgomery, R. A., S. E. Gentry, W. H. Marks, et al. "Domino Paired Kidney Donation: A Strategy to Make Best Use of Live Non-Directed Donation." *Lancet* 368, no. 9533 (2006): 419–421.

Moore, G. E. *Principia Ethica.* 1903. Reprint, Cambridge: Cambridge University Press, 1968.

Moore O'Connor, W. M. "Gene Therapy: Clinical Aspects." In *Human Genome: Human Person and the Society of the Future*, edited by J. de D. Vial Correa and E. Sgreccia. Vatican City: Libreria Editrice Vaticana, 1999.

Morchio, R. "La biologia nel XX secolo." In *Storia delle scienze*, edited by E. Agazzi, 11:367–385. Rome: Città Nuova, 1984.

Mordacci, R. "Disponibilità e disposizione: Riflessioni etiche sulla partecipazione di volontari sani alla ricerca biomedica." *Medicina e Morale* 41, no. 4 (1991): 585–611.

Moretti, J.-M. "L'insémination artificielle." *Études* 351, no. 6 (1979): 619–629.

Morgani, A. R. "Diagnosi prenatale invasiva: 'Stato dell'arte' dell'embrioscopia." *Ultrasonica* 3 (1995): 59–61.

Morgani, A. R., and V. Mele. "L'embrioscopia: stato dell'arte sul piano scientifico e valutazione sul piano etico-deontologico." In *Le radici della Bioetica*, edited by E. Sgreccia, V. Mele, and D. Sacchini, 2:71–77. Milan: Vita e Pensiero, 1998.

Mori, M. *Aborto e morale.* Milan: Il Saggiatore, 1996.

———. "Il feto ha diritto alla vita? Un'analisi filosofica dei vari argomenti in materia con particolare riguardo a quello di potenzialità." In *Il meritevole di tutela*, edited by L. Lombardi Vallauri, 735–840. Milan: Giuffrè, 1990.

———. "I limiti dell'etica senza verità." *Biblioteca della Libertà* 99 (1987): 67–76.

———. "Per un'analisi dei problemi morali relativi agli interventi che comportano la morte degli embrioni umani." In *Quale statuto per l'embrione umano: Problemi e prospettive; Convegno Internazionale, Milano, gennaio 1991*, 75–91. Milan: Bibliotechne, 1992.

———. *Utilitarismo e morale razionale.* Milan: Giuffré, 1986.

———, ed. *La bioetica: Questioni morali e politiche per il futuro dell'uomo.* Milan: Bibliotechne, 1991.

———, ed. *Questioni di bioetica.* Rome: Editori Riuniti, 1988.

Morin, E. *L'homme et la mort devant l'histoire.* Paris: Éditions Corrêa, 1951.

———. *Introduzione al pensiero complesso.* 3rd ed. Milan: Sperling and Kupfer, 1995.

———. "La via della complessità." In *La sfida della complessità*, edited by G. Bocchi and M. Ceruti. Milan: Feltrinelli, 1985.

Mork, W. *The Biblical Meaning of Man.* Milwaukee: Bruce, 1967.

Morra, G. F. "¿Por qué la cultura contemporánea no respeta la vida?" *Ecclesia* 1, no. 1 (1987): 53–67.

Morrissey, P. E., C. Dube, R. Gohh, A. Yango, A. Gautam, and A. P. Monaco. "Good Samaritan Kidney Donation." *Transplantation* 80, no. 10 (2005): 1369–1373.

Mosso, S. "Il ruolo della connaturalità affettiva nella conoscenza morale secondo J. Maritain." In *Jacques Maritain oggi: Atti del Convegno, Milano, 20–23 ottobre 1992*, edited by V. Possenti, 525–546. Milan: Vita e Pensiero, 1983.

Mounier, E. *Le personnalisme*. Paris: Seuil, 1950.

Mouroux, J. *Sens chrétien de l'homme*. Paris: Aubier, 1945.

Muñoz Conde, F. "Sterilization of the Mentally Handicapped: Comments on the Ruling of Spain's Constitutional Court (July 14th 1994)." *Law and Human Genome Review* 2 (1995): 175–196.

Munthe, C. *The Moral Roots of Prenatal Diagnosis*. Göteborg: Centre for Research Ethics Graphic Systems, 1996.

Muratore, S. *L'evoluzione cosmologica e il problema di Dio*. Rome: AVE, 1993.

Muret, S. "La réflexion chrétienne et l'avortement." *La Revue Nouvelle* (Jan. 1973).

Musio, A. "La casualità dell'uomo e la rinascita del pensiero aristotelico in margine ad alcune voci della filosofia pratica tedesca." *Rivista di Filosofia Neoscolastica* 1 (2005): 105–130.

Naam, R. "More Than Human." *New York Times*. July 3, 2005. http://www.nytimes.com/2005/07/03/books/chapters/0703-1st-naam.html?pagewanted=print.

Nalesso, A. *L'autoerotismo nell'adolescente*. Turin: Marietti, 1970.

Nancy, J.-L. *The Experience of Freedom*. Stanford, CA: Stanford University Press, 1993.

Nanni, G. "Microallocazione delle risorse: Il punto di vista del chirurgo." In *Etica e allocazione delle risorse in sanità*, edited by E. Sgreccia and A. G. Spagnolo, 135–136. Milan: Vita e Pensiero, 1996.

Nardi, E. *Aborto procurato nel mondo greco-romano*. Milan: Giuffrè, 1971.

———. "L'eredità del mondo antico." In *L'aborto: Riflessioni di studiosi cattolici*, edited by A. Fiori and E. Sgreccia, 23–47. Milan: Vita e Pensiero, 1975.

Nastasi, F. A. *La fecondazione artificiale nella prospettiva antropologica del diritto canonico, del matrimonio e della famiglia*. Rome: Università della Santa Croce, 2005.

National Bioethics Advisory Commission. *Research Involving Human Biological Materials: Ethical Issues and Policy Guidance*. Vol. 1. Rockville, MD: US Government Printing Office, 1999.

Nazari, A., H. A. Askari, J. H. Check, and A. O'Shaughnessy. "Embryo Transfer Technique as a Cause of Ectopic Pregnancy in In Vitro Fertilization." *Fertility and Sterility* 60, no. 5 (1993): 919–921.

Nebuloni, R. "Crisi dell'eros e crisi della civiltà nel pensiero di H. Marcuse." In *Amore e matrimonio nel pensiero filosofico e teologico moderno*, edited by V. Melchiorre, 319–344. Milan: Vita e Pensiero, 1976.

Nelson, J. R. *On the New Frontiers of Genetics and Religion*. Grand Rapids, MI: William B. Eerdmans, 1990.

Nepi, P. *Il valore persona: Linee di un personalismo morale*. Rome: Editrice Universitaria di Roma/La Goliardica, 1993.

Nespor, S., R. Santosuosso, and R. Satolli. *Vita, morte e miracoli*. Milan: Feltrinelli, 1992.

Neuhaus, W., and A. Boltea. "Prognostic Factors for Preoperative Consultation of Women Desiring Sterilization: Findings of a Retrospective Analysis." *Journal of Psychosomatic Obstetrics and Gynaecology* 16, no. 1 (1995): 45–50.

Neumann, P.J., S.D. Gharib, and M.C. Weinstein. "The Cost of a Successful Delivery with In Vitro Fertilization." *New England Journal of Medicine* 331 (1994): 239–243.

Nicholson, R. "*Who Is Vulnerable in Clinical Research?*" *Bulletin of Medical Ethics* 181 (2002): 19–24.

Nicolaides, K.H. "Cordocentesis." *Clinical Obstetrics and Gynecology* 131 (1988): 123–135.

Nicotra Guerra, I. "Il silenzio legale informato nella recente legge sui trapianti dalla regola coercitiva alla norma pedagogica." *Rivista Trimestrale di Diritto Pubblico* 3 (1999): 829–848.

Niekerk, A. van, and L. van Zyl. "Commercial Surrogacy and the Commodification of Children: An Ethical Perspective." *Medicine and Law* 14 (1995): 163–170.

Nietzsche, F. *The Will to Power*. New York: Vintage, 1967.

Nodari, M.V., ed. *Uomo e salute*. Vicenza: Edizioni del Rezzara, 1979.

Noia, G., A. Caruso, and S. Mancuso. *Le terapie fetali invasive*. Rome: SEU, 1998.

Noia, G., L. Masini, M. De Santis, et al. "La cordocentesi: Indicazioni, utilità e rischi." *Medicina e Morale* 41, no. 4 (1991): 625–640.

Nord Italia Transplant. "Dati sull'attività di prelievo e trapianto in Italia e all'estero." In *Servizio Studi Camera dei Deputati: XII Commissione—Indagine conoscitiva sui trapianti, n. 16*, XII Legislatura, edited by D. Chiassi, 1–10. February 1996.

Nørgaard-Pedersen, B., S.O. Larsen, J. Arends, B. Svenstrup, and A. Tabor. "Maternal Serum Markers in Screening for Down Syndrome." *Clinical Genetics* 37 (1990): 35–43.

Noríega, J. *El destino del eros: Perspectivas de moral sexual*. Madrid: Palabra, 2005.

Nozick, R. *Philosophical Explanations*. Cambridge, MA: Belknap Press, 1981.

Nuffield Council on Bioethics. *Report on Genetic Screening: Ethical Issues*. London: Nuffield Council on Bioethics, 1993. Conclusions and recommendations published in *Bulletin of Medical Ethics* (December 1993): 8–11.

Nuyens, F. *L'évolution de la psychologie d'Aristote*. Louvain: Inst. supérieur de philosophie, 1948.

Oddone, A. "L'uccisione pietosa." *La Civiltà Cattolica* 1 (1950): 248 ff.

Olanda. "Introduzione di una disciplina giuridica per la procedura di notifica degli interventi di eutanasia." *Medicina e Morale* 43, no. 2 (1993): 446–448.

Omenn, G.S. "Predictive Identification of Hypersusceptible Individuals." *Journal of Occupational Medicine* 24, no. 5 (1982): 369–374.

"$100 Million Over 5 Years Will Bring Sterilization to 80–100 Million Couples in Developing Lands." *International Family Planning Digest* 2, no. 1 (1976): 3–5.

Oppenheim, F.E. "Non cognitivismo, razionalità e relativismo." *Rivista di Filosofia* 78 (1987): 17–29.

*Ordine morale e ordine giuridico: Rapporto e distinzione tra diritto e morale; Atti del X°
congresso nazionale dei teologi moralisti; Roma, 24–27 aprile 1984.* Bologna:
Dehoniane, 1985.

Organisation for Economic Co-operation and Development. *Good Developmental Practices
for Small Scale Field Research.* Paris: OECD, 1990.

————. *Recombinant DNA Safety Considerations: Safety Considerations for Industrial,
Agricultural and Environmental Applications of Organisms Derived by Recombinant DNA Techniques.* Paris: OECD, 1986.

O'Rourke, K. "The Christian Affirmation of Life." *Hospital Progress* 55 (1974): 65–72.

Osborne, L. W. "Research on Human Subjects: Australian Ethics Committees Take Tentative
Steps." *Journal of Medical Ethics* 9 (1983): 66–68.

"Osservatorio sulla donazione di organi." *Bioetica e Cultura* 9 (1996): 27–61.

Pace, C. A., and E. J. Emanuel. "The Ethics of Research in Developing Countries: Assessing
Voluntariness." *Lancet* 365, no. 9453 (2005): 11–2.

Pachì, A., R. Paesano, and F. Torcia, "Diagnosi prenatale: Clinica e tecniche diagnostiche."
In *La clinica ostetrica e ginecologica*, edited by G. B. Candiani, V. Danesino, and
A. Gastaldi, 1:133–145. Milan: Masson, 1996.

Pagliaro, L. "Trials." In *Nuovo dizionario di bioetica*, edited by S. Leone and S. Privitera,
1211–1214. Palermo: EDB-ISB, 2004.

Palazzani, L. "Bioetica dei principi e bioetica delle virtù: Il dibattito attuale negli Stati Uniti."
Medicina e Morale 42, no. 1 (1992): 59–86.

————. *Il concetto di persona tra bioetica e diritto.* Turin: Giappichelli, 1996.

————. "Dall'etica 'laica' alla bioetica 'laica': Linee per un approfondimento filosofico-
critico del dibattito italiano attuale." *Humanitas* 4 (1991): 413–446.

————. "La natura nel dibattito bioetica." In *La tecnica, la vita, i dilemmi dell'azione*, edited
by V. Possenti, 204–226. Milan: Mondadori, 1998.

Palazzani, L., and E. Sgreccia. "Il dibattito attuale sulla fondazione etica in bioetica." *Medicina e Morale* 42, no. 5 (1992): 847–870.

Palazzini, P. *Dictionarium canonicum et morale.* Rome: Officium Libri Catholici, 1965.

Palermo, V., and E. Ravera. "Note sulla legge 1 aprile 1999, n. 91 'Disposizioni in materia di
prelievi e di trapianti di organi e tessuti'." *Rivista di Diritto delle Professioni Sanitarie* 2, no. 2 (1999): 104–117.

Palmerini, E. "La nuova legge sui trapianti d'organo: Prime notazioni." *Studium Iuris* 12
(1999): 1311–1321.

Palombo, D., A. Ramello, and P. Tappero, eds. *Trapianti e xenotrapianti: Aspetti etici e giuridici; Atti del Convegno; Agliè, 12 ottobre 2002.* Turin: Selcom, 2003.

Palter, S. F. "Ethics of Clinical Trials." *Seminars in Reproductive Endocrinology* 14, no. 2
(1996): 85–92.

Palumbieri, S. *Antropologia filosofica.* 2 vols. Rome: Urbaniana University Press, 1999.

Panarese, F., F. S. Fusco, E. D'Oro, and P. Ricci. "Trapianti, donazione e normativa: Risultati di
una legge controversa; Quattro anni dopo." *Difesa sociale* 82, no. 4–5 (2003): 11–36.

Pangallo, M. "Actus essendi tomistico e spiritualità dell'anima." *Medicina e Morale* 36, no. 2 (1986): 407–414.

Pappalardo, M., and F. Martinelli. "Qualche osservazione sull'assenso alla donazione post mortem di organi e tessuti e sull'art. 14, punto 2 della legge 1 aprile 1999, n. 91." *Sanità Pubblica* 10 (2002): 1215–1224.

Paquin, I. *Morale e Medicina*. Rome: Orizzonte Medico, 1962.

Pareyson, L. *Esistenza e persona*. 7th ed. Genoa: Il Melangolo, 2002.

———. *Ontologia della libertà: Il male e la sofferenza*. Turin: Einaudi, 1995.

Park, K., Moon Jang Il, Kim Soon Il, and Y. S. Kim. "Exchange Donor Programme in Kidney Transplantation." *Transplantation* 67, no. 2 (1999): 336–338.

Parliament of the United Kingdom. *Human Fertilisation and Embryology Act 1990*. London: HMSO, 1990. Available online at http://www.legislation.gov.uk/ukpga/1990/37/pdfs/ukpga_19900037_en.pdf.

Parliament of Victoria (Australia). Infertility (Medical Procedures) Act 1984. Law No. 10163. http://www.austlii.edu.au/au/legis/vic/hist_act/ipa1984311.pdf.

———. Infertility (Medical Procedures) Act 1987. (Amendment) Law No. 86. http://www.austlii.edu.au/au/legis/vic/hist_act/ipa1987391.pdf.

Pascali, V. L., and E. D'Aloja. "Il progetto genoma e le conoscenze sui geni normali e patologici dell'uomo: Problemi etici e deontologici." *Medicina e Morale* 42, no. 2 (1992): 219–232.

Pasetto, N. "Aspetti biologici e clinici della fecondazione in vitro." In *Fecondazione artificiale embryotransfer: Problemi biologici, clinici, giuridici, etici*, edited by G.F. Zuanazzi, 38–48. Verona: Cortina International, 1986.

Pastori, G. *Le leggi dell'ereditarietà biologica*. Brescia: La Scuola, 1958.

Paton, W. *Man and Mouse: Animals in Medical Research*. New York: Oxford University Press, 1984.

Paton, W. "Vivisection, Morals, Medicine: Commentary from a Vivisecting Professor of Pharmacology." *Journal of Medical Ethics* 9 (1983): 102–104.

Paul VI. Address to the Committee of the United Nations on Apartheid. May 22, 1974.

———. Address at the Eleventh National Congress of the Italian Society of Pathology. October 31, 1968. In *Insegnamenti di Paolo VI*, 7:717–720 [Italian]. Vatican City: Tipografia Poliglotta Vaticana, 1969.

———. Address to Participants in the Third World Congress of the International College of Psychosomatic Medicine. September 18, 1975. In *Insegnamenti di Paolo VI*, 13:953–956 [Italian]. Vatican City: Tipografia Poliglotta Vaticana, 1975.

———. Address to Participants in the Twenty-fifth General Assembly of the International Pharmaceutical Federation and the Thirty-fourth International Conference on Pharmaceutical Sciences. September 7, 1974. In *Insegnamenti di Paolo VI*, 12:800 [Italian]. Vatican City: Tipografia Poliglotta Vaticana, 1974.

———. Address to the Participants in the Twenty-third National Conference of the Italian Catholic Jurists Union. December 9, 1972. In *Insegnamenti di Paolo VI*, 10:1260–1264 [Italian]. Vatican City: Tipografia Poliglotta Vaticana, 1972.

———. Encyclical Letter *Humanae vitae*. July 25, 1968.

———. *Insegnamenti di Paolo VI.* 17 vols. Vatican City: Tipografia Poliglotta Vaticana, 1963–1979.

Pavlovich, C. P., and P. N. Schlegel. "Fertility Options after Vasectomy: A Cost-Effectiveness Analysis." *Fertility and Sterility* 67, no. 1 (1997): 133–141.

Peek, J. C., and F. M. Graham. "Ectopic Pregnancy in a Non-Patent Fallopian Tube following Transfer of Embryos to the Contralateral Tube." *Human Reproduction* 7, no. 1 (1992): 136–137.

Pellegrino, E. D. "Altruism vs. Self-Interest: Ethical Models for Medical Professions." *NYU Physicians* 45, no. 1 (1988): 41–43.

———. "The Caring Ethics: The Relationship of Physician to Patient." In *Caring, Curing, Coping: Nurse–Physician–Patient Relationships,* edited by A. H. Bishop and J. R. Scudder, 8–30. Alabama: University of Alabama Press, 1985.

———. "Character, Virtue, and Self-Interest in the Ethics of the Professions." *Journal of Contemporary Health Law and Policy* 5 (1989): 53–73.

———. "Philosophical Groundings for Treating Patient as a Person: A Commentary on Alasdair MacIntyre." In *Changing Values in Medicine,* edited by E. J. Cassell and M. Siegler, 97–104. Frederick, MO: University Publications of America, 1985.

———. "A Philosophical Source of Medical Morality." *Journal of Medicine and Philosophy* 4, no. 1 (1979): 1–7.

———. "Professional Ethics: Moral Decline or Paradigm Shift?" *Religion and Intellectual Life* 4, no. 3 (1987): 21–45.

———. "Trust and Distrust in Professional Ethics." In *Ethics, Trust, and the Professions: Philosophical and Cultural Aspects,* edited by E. D. Pellegrino, R. M. Veatch, and J. Langan , 69–89. Washington, DC: Georgetown University Press, 1991.

———. "The Virtuous Physician, and the Ethics of Medicine." In *Virtue and Medicine: Explorations in the Character of Medicine,* edited by E. E. Shelp, 237–255. Boston: Reidel, 1985.

Pellegrino, E. D., J. C. Harvey, and J. P. Langan, eds. *Gift of Life.* Washington, DC: Georgetown University Press, 1990.

Pellegrino, E. D., and D. C. Thomasma. *For the Patient's Good: The Restoration of Beneficence in Health Care.* New York: Oxford University Press, 1988.

———. *. A Philosophical Basis of Medical Practice: Toward a Philosophy and Ethics of the Healing Professions.* New York: Oxford University Press, 1981.

———. *. Virtues in Medical Practice.* New York: Oxford University Press, 1993.

Pelliccia, G. "L'eutanasia ha una storia?" In *Morire sì, ma quando?,* edited by P. Beretta. Rome: Paoline, 1977.

Penketh, R., and A. McLaren. "Prospects for Prenatal Diagnosis during Preimplantation Human Development." *Baillier's Clinical Obstetrics and Gynaecology* 1 (1987): 747–764.

Pennacchini, M. *Il trapianto eterologo: Storia, problemi etici e impatto psicologico nei pazienti in lista d'attesa.* Rome: Aracne, 2004.

Pennsylvania Catholic Conference. *Living Will and Proxy for Health Care Decisions.* Harrisburg: Pennsylvania Catholic Conference, 2004. Available at http://www.pacatholic.org/wp-content/uploads/livingwill.pdf.

———. "Nutrition and Hydration: Moral Considerations; A Statement of the Catholic Bishops of Pennsylvania." Rev. ed. 1999. http://www.pacatholic.org/bishops-statements/nutrition-and-hydration-moral-considerations.

Percival, T. *Medical Ethics*. Manchester: S. Russell, 1803.

Perico, G. "La cura pastorale delle persone omosessuali in un recente documento della S. Sede." *Aggiornamenti Sociali* 1 (1987): 79–87.

———. *Difendiamo la vita*. 2nd ed. Milan: Centro studi sociali, 1962.

———. *A difesa della vita*. Milan: Centro Studi Sociali, 1965.

———. *Problemi che scottano*. Milan: Ancora, 1976.

———. *Problemi di etica sanitaria*. 2nd ed. Milan: Ancora, 1992.

———. "Testamento biologico e malati terminali." *Aggiornamenti sociali* 43, no. 11 (1992): 677–692.

———. "Testimoni di Geova e trasfusioni di sangue." *Aggiornamenti Sociali* 5 (1986): 323–336.

———. "Trapianti umani." In *Nuovo dizionario di teologia morale*, edited by F. Compagnoni, G. Piana, and S. Privitera, 1383–1391. Cinisello Balsamo: Paoline, 1990.

Perlow, M.J. "Brain Grafting as a Treatment for Parkinson's Disease." *Neurosurgery* 20 (1987): 335–342.

Persidis, A. "The Business of Pharmacogenomics." *Nature Biotechnology* 16 (1998): 209–210.

Persson, J.W., G.B. Peters, and D.M. Saunders. "Genetic Consequences of ICSI." *Human Reproduction* 11 (1996): 921–924.

Pertl, B., A. Davies, P. Soothill, C. Rodeck, and M. Adinolfi. "Detection of Fetal Cells in Endocervical Samples." *Annals of the New York Academy of Sciences* 731 (1994): 186–192.

Pescetto, G., L. De Cecco, D. Pecorari, and A. Ragni. *Manuale di Ginecologia e Ostetricia*. 2 vols. Rome: SEU, 1989.

Pessina, A. "Bioetica e antropologia: Il problema dello statuto ontologico dell'embrione umano." *Vita e Pensiero* 6 (1996): 402–424.

———. *Bioetica: L'uomo sperimentale*. Milan: Bruno Mondadori Editore, 1999.

———. "Il contesto culturale dello sviluppo della biomedicina." In *Etica e giustizia in sanità,* by A.G. Spagnolo, D. Sacchini, A. Pessina, and M. Lenoci, 3–20. Milan: McGraw-Hill, 2004.

———. "Filosofia e scienza al capezzale dell'uomo: La morte cerebrale." In *Bioetica: L'uomo sperimentale*, 159–171. Milan: Bruno Mondadori, 1999.

———. "Linee per una fondazione filosofica del sapere morale." In *Identità e statuto dell'embrione umano*, edited by Pontifical Academy for Life. Vatican City: Libreria Editrice Vaticana, 1998.

———. "Operatori sanitari come agenti morali." In *Etica e giustizia in sanità*, by A.G. Spagnolo, D. Sacchini, A. Pessina, and M. Lenoci, 21–37. Milan: McGraw-Hill, 2004.

———. "Personalismo e ricerca in bioetica." *Medicina e Morale* 47, no. 3 (1997): 443–459.

———. "La relazione fra la ricerca biomedica, l'antropologia e l'etica filosofica." In *Etica della ricerca biomedica: Per una visione cristiana; Atti della nona assemblea generale della Pontificia Accademia per la Vita; Città del Vaticano, 24–26 febbraio*

2003, edited by J. de D. Vial Correa and E. Sgreccia, 144–158. Vatican City: Libreria Editrice Vaticana, 2004.

———, ed. *Scelte di confine in medicina: Sugli orientamenti dei medici rianimatori*. Milan: Vita e Pensiero, 2004.

Petersen, K. "Private Decision and Public Scrutiny: Sterilization and Minors in Australia and England." In *Contemporary Issues in Law, Medicine and Ethics*, edited by S. McLean, 57–77. Brookfield, VT: Dartmouth Publishing Company, 1996.

Petersen, U. H. "The Danish Committee-System." In *Funzione e funzionamento dei comitati etici*, edited by G. Gerin, 131–141. Padua: CEDAM, 1991.

Peterson, H. B., Z. Xia, J. M. Hughes, et al., for the U.S. Collaborative Review of Sterilization Working Group. "The Risk of Ectopic Pregnancy after Tubal Sterilization." *New England Journal of Medicine* 336, no. 11 (1997): 762–767.

Peterson, H. B., and L. S. Wilcox. "Female Sterilization." In *Reproductive Health Care for Women and Babies*, edited by B. P. Sachs, R. Beard, E. Papiernik, and C. Russel, 161–169. New York: Oxford, 1995.

Petrini, M. *Accanto al morente: Prospettive etiche e pastorali*. Milan: Vita e Pensiero, 1990.

———. "L'assistenza al morente: Orientamenti e prospettive." *Medicina e Morale* 35, no. 2 (1985): 365–388.

Petrinovich, L. *Human Evolution, Reproduction, and Morality*. New York: Plenum Press, 1995.

Peyron, R., E. Aubény, V. Targosz, et al. "Early Termination of Pregnancy with Mifepristone (RU486) and the Orally Active Prostaglandin Misoprostol." *New England Journal of Medicine* 328, no. 21 (1993): 1509–1513.

Pezzoli, G., V. Silani, E. Motti, et al. "Human Fetal Adrenal Medulla for Transplantation in Parkinsonian Patients." *Annals of the New York Academy of Sciences* 495 (1987): 771–773.

Phupong, V., S. Taneepanichskul, and T. Rungruxsirivorn. "Bilateral Tubal Pregnancies after Tubal Sterilization in a Human Immunodeficiency Virus Seropositive Woman." *Journal of the Medical Association of Thailand* 85, no. 11 (2002): 1236–1239.

Piana, G. "Libertà." In *Dizionario enciclopedico di teologia morale*, 562–574. Rome: Paoline, 1981.

———. "Orientamenti di etica sessuale." In *Corso di morale*, edited by T. Goffi and G. Piana, 2:271–366. Brescia: Queriniana, 1983.

Pidier, J. M. "La Chiesa e l'aborto." *Il Regno attualità* 2 (1973): 16–17.

Pinckaers, S. "Human Freedom according to St. Thomas Aquinas." In *The Sources of Christian Ethics*, 379–399. Translated by M. T. Noble. Washington, DC: Catholic University of America, 1995.

———. *The Sources of Christian Ethics*. Translated by M. T. Noble. Washington, DC: Catholic University of America, 1995.

Pinkel, D., T. Straume, and J. W. Gray. "Cytogenetic Analysis Using Quantitative, High Sensitivity, Fluorescence Hybridization." *Proceedings of the Natural Academy of Sciences of the USA* 83 (1986): 2934–2938.

Piomelli, S., N. Lerner, A. Cohen, et al. "Bone Marrow Transplantation for Thalassaemia." *New England Journal of Medicine* 317 (1987): 964.

Pitt, B., D. Julian, and S. Pocock. *La sperimentazione clinica.* Rome: Il Pensiero Scientifico Ed., 2000.

Pius XI. Encyclical Letter *Quadragesimo anno.* May 15, 1931.

Pius XII. Address at the Conference of the Family Front and the Federation of Large Family Associations. November 27, 1951. In *Discorsi e Radiomessaggi*, 13:411–418 [Italian]. Vatican City: Tipografia Poliglotta Vaticana, 1961.

———. Address to Conference Participants of the Collegium Internationale Neuro-Psychopharmacologicum. September 9, 1958. In *Discorsi e Radiomessaggi*, 20:321–333 [Italian]. Vatican City: Tipografia Poliglotta Vaticana, 1959.

———. Address to the First International Congress on the Histopathology of the Nervous System. September 14, 1952. http://www.ncbcenter.org/Page.aspx?pid=388.

———. Address to the Italian Medical-Biological Union of St. Luke. November 12, 1944. In *Discorsi e Radiomessaggi*, 6:181–196 [Italian]. Vatican City: Tipografia Poliglotta Vaticana, 1960.

———. Address to Managers and Associates of the Italian Society of Cornea Donors. May 14, 1956. In *Discorsi e Radiomessaggi*, 18:192–201 [Italian]. Vatican City: Tipografia Poliglotta Vaticana, 1969.

———. Address to Participants in the Conference of the Italian Catholic Union of Obstetricians. October 29, 1951. In *Discorsi e Radiomessaggi*, 13:211–221 [Italian]. Vatican City: Tipografia Poliglotta Vaticana, 1961.

———. Address to Participants in the Eighth Assembly of the World Medical Association. September 30, 1954. In *Discorsi e Radiomessaggi*, 17:167–179 [Italian]. Vatican City: Tipografia Poliglotta Vaticana, 1969.

———. Address to the Participants in the First International Symposium on Medical Genetics. September 7, 1953. In *Discorsi e Radiomessaggi*, 15:251–266 [Italian]. Vatican City: Tipografia Poliglotta Vaticana, 1969.

———. Address to Participants in the Fourth International Congress of Catholic Physicians. September 29, 1949. In *Discorsi e Radiomessaggi*, 11:221–225 [Italian]. Vatican City: Tipografia Poliglotta Vaticana, 1960.

———. Address to Participants in the Second World Congress on Fertility and Sterility." May 19, 1956. In *Discorsi e Radiomessaggi*, 18:211–221 [Italian]. Vatican City: Tipografia Poliglotta Vaticana, 1969.

———. Address to Participants in the Seventh International Congress of the Italian Hematology Society. September 12, 1958. In *Discorsi e Radiomessaggi*, 20:341–352 [Italian]. Vatican City: Tipografia Poliglotta Vaticana, 1959.

———. Address to Participants in the Sixth International Congress on Surgery. May 21, 1948. In *Discorsi e Radiomessaggi*, 10:95–100 [Italian]. Vatican City: Tipografia Poliglotta Vaticana, 1960.

———. Address to Participants in the Twenty-sixth Congress of the Italian Urology Association. October 8, 1953. In *Discorsi e Radiomessaggi*, 15:371–379 [Italian]. Vatican City: Tipografia Poliglotta Vaticana, 1969.

———. Address at the Sixteenth Session of the International Office of Documentation on Military Medicine. October 19, 1953. In *Discorsi e Radiomessaggi*, 11:415–428 [Italian]. Vatican City: Tipografia Poliglotta Vaticana, 1969.

———. *Discorsi e Radiomessaggi.* 22 vols. Vatican City: Tipografia Poliglotta Vaticana, 1940–1969.

———. Encyclical Letter *Humani generis.* August 12, 1950.

———. Encyclical Letter *Mystici Corporis.* June 29, 1943.

———. Radio Message to the Seventh International Congress of Catholic Physicians. September 11, 1956. In *Discorsi e Radiomessaggi*, 18:423–435. Vatican City: Tipografia Poliglotta Vaticana, 1969.

———. Response to Questions of the Ninth Congress of the Italian Society of Anesthesiology. February 24, 1957. In *Discorsi e Radiomessaggi*, 18:777–799 [Italian]. Vatican City: Tipografia Poliglotta Vaticana, 1969.

Plachot, M., A.M. Junca, J. Mandelbaum, J. de Grouchy, J. Salat-Baroux, and J. Cohen. "Chromosome Investigation in Early Life: I. Human Oocytes Recovered in an IVF Programme." *Human Reproduction* 1, no. 8 (1986): 547–551.

"A Plea for Beneficent Euthanasia." *Humanist* 34, no. 4 (1974): 4–5. Available at http://users.rcn.com/pknyc/articles/doc68.pdf.

Poincaré, H.J. *Dernières Pensées.* Paris: Flammarion, 1913.

Policar, M. "Fertility Control: Medical Aspects." In *Encyclopedia of Bioethics*, edited by S.G. Post, 891–901. 3rd ed. New York: Macmillan Reference USA, 2004.

Pollack, R. *Signs of Life: The Language and Meanings of DNA.* Boston: Houghton Mifflin, 1994.

Pontifical Academy for Life. "Comunicato finale dell'Assemblea Plenaria su 'Identità e Statuto dell'embrione umano.'" *L'Osservatore Romano.* February 21, 1997.

———. *Declaration on the Production and the Scientific and Therapeutic Use of Stem Cells.* August 25, 2000.

———. "Prospects for Xenotransplantation: Scientific Aspects and Ethical Considerations." September 26, 2001.

———. "Respect for the Dignity of the Dying." December 9, 2000.

Pontifical Academy for Life and World Federation of Catholic Medical Associations. "Joint Statement on the Vegetative State." International Congress on Life Sustaining Treatments and Vegetative State: Scientific Advances and Ethical Dilemmas. March 17–20, 2004. http://www.vatican.va/roman_curia/pontifical_academies/acdlife/documents/rc_pont-acd_life_doc_20040320_joint-statement-veget-state_en.html.

———. "Life-Sustaining Treatments in Vegetative State: Scientific Advances and Ethical Dilemmas." Edited by G.L. Gigli and N.D. Zasler. *NeuroRehabilitation* 19, no. 4 (2004).

Pontifical Academy of Sciences. "Declaration on Prevention of Nuclear War." September 23–24, 1982. http://www.vatican.va/roman_curia/pontifical_academies/acdscien/2010/documenta_3_4_7_11_pas.pdf.

———. *Population and Resources: A Report.* Vatican City: Pontificia Academia Scientiarum, 1994.

Pontifical Council Cor Unum. "Ethical Questions regarding the Seriously Ill and Dying." June 27, 1981. In *Enchiridion Vaticanum*, 7:1133–1173 [Italian]. Bologna: Dehoniane, 1982.

Pontifical Council for Pastoral Assistance to Health Care Workers. *Charter for Health Care Workers.* Boston: Pauline Books and Media, 1995.

Pontifical Council for the Family. *Ethical and Pastoral Dimensions of Population Trends*. Vatican City: Libreria Editrice Vaticana, 1994.

———. *The Truth and Meaning of Human Sexuality*, August 8, 1995.

Pontifical Council for the Interpretation of Legislative Texts. "I diritti del nascituro secondo la legislazione canonica e il Magistero della Chiesa." *Medicina e Morale* 45, no. 3 (1995): 532–542.

Popper, K. *The Logic of Scientific Discovery*. London: Hutchinson, 1968.

———. *Objective Knowledge: An Evolutionary Approach*. Oxford: Oxford University Press, 1972.

———. *The Open Society and Its Enemies*. Vol. 1, *The Spell of Plato*. 5th ed. London: Routledge and Kegan Paul, 1966.

———. *The Open Universe: An Argument for Indeterminism*. Totowa: Rowman and Littlefield, 1982.

Popper, K., and J. Eccles. *The Self and Its Brain*. New York: Springer International, 1977.

"Population: Battle of the Bulge." *Economist*. September 3, 1994. 19–21.

Population Reference Bureau. *World Population Data Sheet 1991*. Washington, DC: PRB, 1991.

Porro, C. "'I cieli narrano la gloria di Dio': Note su cosmologia e teologia." *La Rivista del Clero Italiano* 1 (1996): 453–463.

Porter, J.P. "Informed Consent Issues in International Research Concerns." *Cambridge Quarterly of Healthcare Ethics* 5, no. 2 (1996): 237–243.

Possenti, V. "La bioetica alla ricerca dei principi: La persona." *Medicina e Morale* 42, no. 6 (1992): 1075–1096.

———. "La cultura radicale." In *Filosofia e società: Studi sui progetti etico-politici contemporanei*, 94–134. Milan: Massimo, 1983.

———. *Essere e libertà*. Soveria Mannelli, CZ: Rubbettino, 2004.

———. "Metafisica, problema della verità, pragmatica trascendentale." In *Annuario di filosofia 2000*. Milan: Mondadori, 2000.

———. "Noi che non sappiamo affatto che cosa sia la persona umana." *Filosofia Oggi* 27, no. 1 (2004): 3–28.

———. "Prospettive sull'etica." In *Essere e libertà*, edited by L. Pareyson, 207–246. Soveria Mannelli, CZ: Rubbettino, 2004.

———. *Le società liberali al bivio: Lineamenti di filosofia della società*. Turin: Marietti, 1991.

———. "Vita, natura e teleologia." In *Essere e libertà*. Soveria Mannelli, 115–144. CZ: Rubbettino, 2004.

———. "La vita preconscia dello spirito nella filosofia della persona di J. Maritain." In *Jacques Maritain oggi*, edited by V. Possenti, 228–242. Milan: Vita e Pensiero, 1983.

———, ed. *La questione della verità*. Rome: Armando, 2003.

Potter, V.R. *Bioethics: Bridge to the Future*. Englewood Cliffs, NJ: Prentice Hall, 1971.

———. "Bioethics: The Science of Survival." *Perspectives in Biology and Medicine* 14, no. 1 (1970): 127–153.

―――. *Global Bioethics: Building on the Leopold Legacy*. East Lansing: Michigan University Press, 1988.

Poupard, "Fideismo." In *Dizionario di scienza e fede*, edited by G. Tanzella-Nitti and Strumia. Rome: Città Nuova Editrice, 2002.

Pousset, E. "Être humain déjà." *Études* (November 1970): 512–513.

Premuda, L. *Metodo e conoscenza da Ippocrate ai nostri giorni*. Padua: CEDAM, 1971.

―――. *Storia della medicina*. Padua: CEDAM, 1960.

Preziosi, P. "Valore predittivo della sperimentazione pre-clinica sull'animale." In *Lineamenti di etica della sperimentazione clinica*, edited by A.G. Spagnolo and E. Sgreccia, 71–84. Milan: Vita e Pensiero, 1994.

Privitera, S. "Casistica," "Deontologia/teleologia," "Etica normativa," "Principii morali tradizionali," and "Valori." In *Dizionario di bioetica*, edited by S. Leone and S. Privitera. Bologna: Dehoniane, 1994.

Przewozny, B., O. Todisco, and F. Targonski. *Etica ambientale*. Rome: Miscellanea Francescana, 1991.

Puca, A. "Il caso di Nancy Beth Cruzan." *Medicina e Morale* 42, no. 5 (1992): 911–932.

―――. "I trapianti d'organo." In *Bioetica*, edited by C. Romano and G. Grassani, 487–504. Turin: UTET, 1995.

―――. *Trapianto di cuore e morte cerebrale: Aspetti etici*. Turin: Camilliane, 1993.

Pullman, D., and X. Wang. "*Adaptive Designs, Informed Consent and the Ethics of Research.*" *Controlled Clinical Trials* 22 (2001): 203–210.

Putnam, H. *The Collapse of the Fact/Value Dichotomy and Other Essays*. Cambridge, MA: Harvard University Press, 2002.

Pyrgiotis, E., K. Sultan, G.S. Neal, H.C. Liu, J.A. Grifo, and Z. Rosenwaks. "Ectopic Pregnancies after In Vitro Fertilization and Embryo Transfer." *Journal of Assisted Reproduction and Genetics* 11, no. 2 (1994): 79–84.

"Quale base comune per la riflessione bioetica in Italia? Dibattito sul Manifesto di bioetica laica." *Notizie di Politeia* 12 (1996).

Quarello, E. "Male fisico e male morale nei conflitti di coscienza." *Salesianum* 34 (1972): 295–318.

Quattrocchi, P. "La bioetica: Storia di un progetto." In *Dalla bioetica ai comitati etici*, edited by C.G. Vella, P. Quattrocchi, and A. Bompiani, 57–97. Milan: Ancora, 1988.

―――. *Etica scienza complessità*. Milan: F. Angeli, 1984.

Quelquejeu, B. "La volonté de procréer : Réflexion philosophique." *Lumière et Vie* (August–October 1972): 64.

Quin, P. "Cryopreservation of Embryoys and Oocytes." In *Infertility: Evaluation and Treatment*, edited by W.R. Keye, R.J. Chang, R.W. Rebar, and M.R. Soules, 821–840. Philadelphia: W.B. Saunders, 1995.

Quintavalla, E., and D.E. Raimon, eds. *Aborto, perché?* Milan: Feltrinelli, 1989.

Quintero, R.A., R. Hume, C. Smith, et al. "Percutaneous Fetal Cystoscopy and Endoscopic Fulguration of Posterior Urethral Valves." *American Journal of Obstetrics and Gynecology* 172 (1995): 206–209.

Quintero, R. A., K. S. Puder, and D. B. Cotton. "Embryoscopy and Fetoscopy." *Obstetrics and Gynecology Clinics of North America* 20, no. 3 (1993): 563–581.

Quinzio, S. "Perché la tolleranza non basta." *Biblioteca della Libertà* 99 (1987): 77–81.

Rachels, J. *Created from Animals: The Moral Implications of Darwinism.* New York: Oxford University Press, 1990.

Radestad, A., T. H. Bui, and K. G. Nygren. "Multifetal Pregnancy Reduction in Sweden: Utilization Rate and Pregnancy Outcome (1986–1992)." *Acta Obstetricia et Gynecologica Scandinavica* 73, no. 5 (1996): 403–406.

Raes, J. "A propos de l'avortement." *La Revue Nouvelle* (1971): 90.

Raffard de Brienne, D. *Per finirla con l'evoluzionismo: Dichiarazioni su un mito inconsistente.* Rome: Minotauro, 2003.

Rahner, K. "The Experiment with Man: Theological Observations on Man's Self-Manipulation." In *Theological Investigations,* 9:205–224. Translated by G. Harrison. New York: Seabury Press, 1972.

———. "The Problem of Genetic Manipulation." In *Theological Investigations,* 9:225–252. Translated by G. Harrison. New York: Seabury Press, 1972.

Ramsey, J. T. *Freedom and Immortality.* London: SCM, 1971.

Ramsey, P. *Parenthood and the Future of Man by Artificial Donor Insemination: Fabricated Man.* New Haven: Yale University Press, 1970.

Rapaport, F. T. "The Case for a Living Emotionally Related International Kidney Donor Exchange Registry." *Transplantation Proceedings* 18, suppl. 2 (1986): 5–9.

———. "Exchange Donor Program in Kidney Transplantation." *Transplantation Proceedings* 19, no. 1, part 1 (1987): 169–173.

Ratzinger, J. "Fede e ragione." *L'Osservatore Romano.* November 19, 1998.

Ratzinger, J., and J. Habermas. *Dialectics of Secularization.* San Francisco: Ignatius Press, 2006.

Ravarotto, L., and R. Pegoraro. *Transgenesi, clonazione, xenotrapianto: Analisi scientifica, giuridica ed etica sull'impiego degli animali.* Padua: Piccin Nuova Libraria, 2003.

Ravenholt, R. T. "World Epidemiology and Potential Fertility Impact of Voluntary Sterilization." Presented at the Third International Conference on Voluntary Sterilization. February 1, 1976.

Rawls, J. *A Theory of Justice.* Cambridge, MA: Belknap Press, 1971.

Raymond, J. G., R. Klein, and L. J. Dumble. *RU486: Misconceptions, Myths and Morals.* Cambridge, MA: Institute on Women and Technology, 1991.

Reale, G. *Corpo, anima e salute: Il concetto di uomo da Omero a Platone.* Milan: Vita e Pensiero / Raffaello Cortina Editore, 1999.

Reale, G., and D. Antiseri, *Il pensiero occidentale dalle origini ad oggi.* Vol. 3. Brescia: La Scuola, 1983.

Redmon, R. H. "How Children Can Be Respected as 'Ends' Yet Still Be Used as Subjects in Non-Therapeutic Research." *Journal of Medical Ethics* 12 (1986): 77–82.

Reece, E. A., S. Rotmensch, J. Whetham, M. Cullen, and J. C. Hobbins. "Embryoscopy: A Closer Look at First-Trimester Diagnosis and Treatment." *American Journal of Obstetrics and Gynecology* 166 (1992): 775–780.

Reece, E.A.,C.J. Homko, S. Koch, and L. Chan. "First-Trimester Needle Embryofetoscopy and Prenatal Diagnosis." *Fetal Diagnosis and Therapy* 12 (1997): 136–139.

Regal, P. "The Adaptive Potential of Genetically Engineered Organisms in Nature." *Trends in Biotechnology* 6 (1988): 36–38.

Regan, T. *The Case for Animal Rights*. Berkeley: University of California Press, 1983.

———. "The Nature and Possibility of an Environmental Ethics." *Environmental Ethics* 3 (1981): 19–34.

Reich, W.T. "The Word 'Bioethics': The Struggle over Its Earliest Meanings." *Kennedy Institute of Ethics Journal* 5 (1995): 19–34.

———, ed. *Encyclopedia of Bioethics*. Vol. 5. Rev. ed. New York: Macmillan, 1995.

Reiser, S.J. *Medicine and the Reign of Technology*. New York: Cambridge University Press, 1978.

Rella, W. *Die Wirkungsweise oraler Kontrazeptiva und die Bedeutung ihres noidationshemmenden Effekts*. Vienna: IMABE Studien, 1994.

Remotti, F. "La tolleranza verso i costumi." In *Teorie etiche contemporanee*, edited by C.A. Viano, 165–185. Turin: Bollati Boringhieri, 1990.

Rentchnick, P."Euthanasie: Évolution du concept d''euthanasie' au cours de ces cinquante dernières années." *Médecine et Hygiène* (February 29, 1984): 653–666.

Resnik, D.B. "Research Subjects in Developing Nations and Vulnerability." *American Journal of Bioethics* 4, no. 3 (2004): 63–4.

Rhonheimer, M. *The Ethics of Procreation and the Defense of Human Life: Contraception, Artificial Fertilization, and Abortion*. Washington, DC: Catholic University of America Press, 2010.

———. "Natural Moral Law: Moral Knowledge and Conscience; The Cognitive Structure of the Natural Law and the Truth of Subjectivity." In *The Nature and Dignity of the Human Person as the Foundation of the Right to Life: The Challenges of the Contemporary Cultural Context; Proceedings of the Eighth Assembly of the Pontifical Academy for Life: Vatican City, 25–27 February 2002*, edited by J. de D. Vial Correa and E. Sgreccia, 123–159. Vatican City: Libreria Editrice Vaticana, 2003.

Ribes, B. "Pour une réforme de la législation française relative à l'avortement." *Études* (January 1973): 69.

Ricca, P., ed. *Eutanasia: La legge olandese e commenti*. Turin: Claudiana, 2002.

Richart, R. M., and D.J. Prager, eds. *Human Sterilization*. Springfield, IL:C. C. Thomas, 1972.

Ricoeur, P. *The Conflict of Interpretations: Essay in Hermeneutics*. Evanston, IL: Northwestern University Press, 1974. Originally published as *La conflict des interpretations* (Paris: Seuil, 1969).

———. *Oneself as Another*. Translated by K. Blamey. Chicago: University of Chicago Press, 1992.

———. *Philosophie de la volonté*. Paris: Aubier, 1949.

Rigobello, A.., ed. *La persona e le sue immagini*. Vatican City: Urbaniana University Press, 1999.

Riis, P. "Medical Ethics in the European Community." *Journal of Medical Ethics* 19 (1993): 7–12.

Riquet, I. *La castration*. Paris: Lethielleux, 1948.

Robert, C. J., and C. R. Lowe. "Where Have All the Conceptions Gone?" *Lancet* 305 (1975): 498–499.

Robertson, J. A. "Consent and Privacy in Pharmacogenetic Testing." *Nature Genetics* 28 (2001): 207–209.

Rodeck, C., B. Tutschek, J. Sherlock, and J. Kingdom. "Methods for the Transcervical Collection of Fetal Cells during the First Trimester of Pregnancy." *Prenatal Diagnosis* 15 (1995): 933–942.

Rodeck, C. "Fetoscopia e prelievo di sangue fetale." In *La diagnosi prenatale dei difetti congeniti: Atti del 5º Corso Nazionale di Aggiornamento in Medicina Prenatale*, edited by G. B. Candiani et al., 273–280. Palermo: Codese, 1981.

Rodriguez Luño, A. *Etica*. Florence: Le Monnier, 1992.

———. *"Leggi imperfette e inique."* In *Lexicon: Termini ambigui e discussi su famiglia, vita e questioni etiche*, edited by Pontifical Council for the Family, 523–527. Bologna: EDB, 2003.

———. *"Il parlamentare cattolico di fronte ad una legge gravemente ingiusta: Una riflessione sul n. 73 di Evangelium vitae."* *L'Osservatore Romano*. September 6, 2002.

———. "Rapporti tra il concetto filosofico e il concetto clinico di morte." *Acta Philosophica* 1, no. 1 (1992): 54–68.

———. "La valutazione teologico-morale dell'aborto." In *Commento interdisciplinare alla "Evangelium vitae,"* edited by E. Sgreccia and R. Lucas Lucas, 419–434. Vatican City: Libreria Editrice Vaticana, 1997.

———. *"Veritatis splendor* un anno dopo: Appunti per un bilancio (II)." *Acta Philosophica* 1, no. 5 (1996): 47–75.

Rodriguez Luño, A., and R. López Mondéjar. *La fecondazione in vitro: Aspetti medici e morali*. Rome: Città Nuova, 1986.

Rogers, A., and D. Durand de Bouringen. *Bioethics in Europe*. Strasbourg: Council of Europe Press, 1995.

Rohrmann, S. D., N. Paltoo, E. A. Platz, S. C. Hoffman, G. W. Comstock, and K. J Helzlsouer. "Association of Vasectomy and Prostate Cancer Among Men in a Maryland Cohort." *Cancer Causes Control* 16, no. 10 (2005): 1189–1194.

Rolston, H., ed. *Biology, Ethics and Origins of Life*. Boston: Jones and Bartlett, 1995.

Romano, A. *Sterilizzazione umana e legalità costituzionale*. Naples: Edizioni Scientifiche Italiane, 2000.

Romero Pérez, P., F. J. Merenciano Cortina, W. Rafie Mazketli, M. Amat Cecilia, and M. C. Martínez Hernández. "Vasectomy: Study of 300 Interventions; Review of the National Literature and of Its Complications" [in Spanish]. *Actas Urológicas Españolas* 28, no. 3 (2004): 175–214.

Rose, H. J. "Euthanasia." In *Encyclopedia of Religion and Ethics*, edited by J. Hastings, J. A. Selbie, and L. H. Gray, vol. 5. Edinburgh, T. and T. Clark, 1971.

Roselli, A. "La medicina e le scienze della vita." In *Storia delle scienze*, edited by E. Agazzi, 1:93–113. Rome: Città Nuova, 1984.

Rosenberg, L., J. R. Palmer, A. G. Zauber, et al. "The Relation of Vasectomy to the Risk of Cancer." *American Journal of Epidemiology* 140, no. 5 (1994): 431–438.

Rosenbusch, B., E. Strehler, and K. Sterzik. "Micro-Assisted Fertilization and Sperm Chromosome Abnormalities." *Human Reproduction* 11 (1996): 928–930.

Roses, A. D. "Pharmacogenetics and the Practice of Medicine." *Nature* 40 (2000): 857–865.

Rosner, F. "Hospital Medical Ethics Committees: A Review of their Development." *Journal of the American Medical Association* 253 (1985): 2693–2697.

Ross, D. *Foundations of Ethics*. Oxford: Clarendon Press, 1939.

———. *The Right and the Good*. Oxford: Oxford University Press, 1930.

Ross, G., R. Erickson, D. Knorr, et al. "Gene Therapy in the United States: A Five-Year Status Report." *Human Gene Therapy* 7 (1996): 1781–1790.

Ross, W. D. *The Foundations of Ethics: The Gifford Lectures; Delivered in the University of Aberdeen 1935–6*. Oxford: Clarendon Press, 1968.

Rossano, R., and S. Sibilla, eds. *La tutela giuridica della vita prenatale*. Turin: Giappichelli, 2005.

Rossi, G. "Il rapporto uomo-donna." In *Trattato di etica teologica*, edited by L. Lorenzetti, 2:399–458. Bologna: Dehoniane, 1981.

Rossi, L. "Eutanasia." In *Dizionario enciclopedico di teologia morale*, 380–386. Rome: Paoline, 1981.

———. "Masturbazione." In *Dizionario enciclopedico di teologia morale*, edited by by L. Rossi and A. Valsecchi, 614–625. Rome: Paoline, 1981.

———. "Sterilità e sterilizzazione." In *Dizionario enciclopedico di teologia morale*, 1055–1062. Rome: Paoline, 1981.

Rossi, P. *La nascita della scienza moderna in Europa*. Rome: Laterza, 1997.

Rossi Sciumé, G. "Problemi sociologici emergenti nel merito del dibattito sulla procreazione assistita." *Medicina e Morale* 43, no. 1 (1993): 165–181.

Rothman, K. J., and K. B. Michels. "The Continuing Unethical Use of Placebo Controls." *New England Journal of Medicine* 331, no. 6 (1994): 394–398.

Rothmas, D. J. "Research, Human: Historical Aspects." In *Encyclopedia of Bioethics*, edited by S. G. Post, 2316–2326. 3rd ed. New York: Macmillan Reference USA, 2004.

Rouse, F., S. Johnson, D. W. Brock, L. Emanuel, S. M. Wolf, D. Mason, M. Mezey, R. B. Purtilo, and E. L. McCloskey. "Practicing the PSDA." *Hastings Center Report* 21, no. 5, suppl. (1991): 15–165.

Royal College of Physicians. *Guidelines on the Practice of Ethics Committees in Medical Research Involving Human Subjects*. 2nd ed. London: Royal College of Physicians, 1990.

Rufat, P., F. Olivennes, J. de Mouzon, M. Dehan, and R. Frydman. "Task Force Report on the Outcome of Pregnancies and Children Conceived by In Vitro Fertilization (France: 1987–1989)." *Fertility and Sterility* 61, no. 2 (1994): 324–330.

Ruff, W. "Individualität und Personalität in embryonalen Werden: Die Frage nach dem Zeitpunkt der Geistbeseelung." *Theologie und Philosophie* 45 (1970): 25–49.

Ruiz de la Peña, J. L. "Anthropologie et tentation biologiste." *Communio* 9, no. 6 (1984): 66–80.

Ruof, M. C. "Vulnerability, Vulnerable Populations, and Policy." *Kennedy Institute of Ethics Journal* 14, no. 4 (2004): 411–25.

Russo, M. T. *Corpo, salute, cura: Linee di antropologia biomedica*. Soveria Mannelli, CZ: Rubbettino, 2004.

Sacchini, D. "Sistemi-sanità: Una lettura etica "integrata."" In *Etica e giustizia in sanità*, by A.G. Spagnolo, D. Sacchini, A. Pessina, and M. Lenoci, 79–119. Milan: McGraw-Hill, 2004.

Sailer, R.I. "Our Immigrant Fauna." *ESA Bulletin* 24 (1978): 3.

Salvoni, F. *Sesso e amore nella Bibbia*. Genoa: Lanterna, 1970.

Sandel, M. *Liberalism and the Limits of Justice*. New York: Cambridge University Press, 1982.

Sansone, G. "Recenti progressi in genetica medica." *Medicina e Morale* 32, no. 1 (1982): 14–24.

Santaló, J., J. Badenas, J.M. Calafell, et al. "The Genetic Risks of In Vitro Fertilization Techniques: The Use of an Animal Model." *Journal of Assisted Reproduction and Genetics* 9 (1992): 462–474.

Santosuosso, A., ed. *Il consenso informato: Tra giustificazione per il medico e diritto del paziente*. Milan: Cortina, 1996.

Sardon, I.P. "La stérilisation dans le monde: Aperçus médicaux et legislatifs." *Population* (March–April 1977): 415 ff.

Sass, H. -M. "Fritz Jahr's 1927 Concept of Bioethics." *Kennedy Institute of Ethics Journal* 17, no. 4 (2007): 279-295.

Saunders, D. M., M. Mathews, and P.A. L. Lancaster. "The Australian Register: Current Research and Future Role; A Preliminary Report." In *In Vitro Fertilization and Other Assisted Reproduction*, edited by H.W. Jones and C. Schraedere, 7–21. Annals of the New York Academy of Sciences 541. New York: New York Academy of Sciences, 1988.

Scalabrino Spadea, M. "La tutela del malato di mente nel diritto internazionale dei diritti dell'uomo: Documenti vecchi e nuovi." *Medicina e Morale* 42, no. 6 (1992): 1105–1118.

Scarpelli, U. "La bioetica: Alla ricerca dei principi." *Biblioteca della Libertà* 99 (1987): 7–32.

———. "Bioetica: Prospettive e principi fondamentali." In *La bioetica: Questioni morali e politiche per il futuro dell'uomo*, edited by M. Mori, 20–25. Milan: Bibliotechne, 1991.

———. *Bioetica laica*, edited by M. Mori. Milan: Baldini and Castoldi, 1998.

———. *Etica senza verità*. Bologna: Il Mulino, 1982.

Schaffner, K.F. "Research Methodology: I. Conceptual Issues." In *Encyclopedia of Bioethics*, edited by S.G. Post, 2326–2334. 3rd ed. New York: Macmillan Reference, 2004.

Scheler, M. *The Nature of Sympathy*. Translated by P. Heath. London: Routledge and K. Paul, 1954.

———. *Philosophische Weltanschauung*. Bonn: F. Cohen, 1929.

Schelsky, H. *Soziologie der Sexualität*. Hamburg: Rowohlt, 1955.

Schenker, J. G., and Y. Ezra. "Complications of Assisted Reproductive Techniques." *Fertility and Sterility* 61, no. 3 (1994): 411–422.

Schlegelberger, B. *Rapporto sessuale prima e fuori del matrimonio*. Rome: Paoline, 1974.

Schmidt-Sarosi, C.L. "In Vitro Fertilization with Donor Oocytes." In *Infertility: Evaluation and Treatment*, edited by W.R. Keye, R.J. Chang, R.W. Rebar, and M.R. Soules, 780–787. Philadelphia: W.B. Saunders, 1995.

Schoonenberg, P. "Je crois à la vie éternelle." *Concilium* 5 (1969): 91ff.

Schooyans, M. *Aborto e politica*. Vatican City: Libreria Editrice Vaticana, 1991.

———. *L'avortemen: enjeux politiques*. Longueuil, Quebec: Le Préambule, 1990.

———. *La dérive totalitaire du libéralisme*. Paris: Editions Universitaires, 1991.

———. *L'évangile face au désordre mondial*. Paris: Fayard, 1997.

Scola, A. *Identidad y differencia*. Madrid: Encuentro, 1989.

———. *Il mistero nuziale*. Vol. 1, *Uomo-donna*. Rome: PUL / Mursia, 1998. Vol. 2, *Matrimonio e famiglia*. Rome: PUL / Mursia, 2000.

Secretariat of the French Bishops Conference. "Problème étiques posés aujourd'hui par la mort et le mourir." March 6, 1976. *Bulletin du Secrétariat de la Conférence Episcopale Française* (1976): 6. Published in Italian as "Problemi etici posti oggi dalla morte e dal morire," in *Umanizzare la malattia e la morte: Documenti pastorali dei vescovi francesi e tedeschi*, ed. S. Spinsanti (Rome: Paoline, 1980), 43–44.

Seifert, J. *Essere e persona: Verso una fondazione fenomenologica di una metafisica classica e personalistica*. Milan: Vita e Pensiero, 1989.

———. "Is 'Brain Death' Actually Death?" *Monist* 76 (1993): 175–202.

———. "Is 'Brain Death' Actually Death? A Critique of Redefining Man's Death in Terms of 'Brain Death.'" In *Working Group on the Determination of Brain Death and Its Relationship to Human Death*, ed. R. J. White, H. Angstwurm, and I. Carrasco de Paula, 95–143. Vatican City: Pontifical Academy of Sciences, 1992.

———. *Das Leib-Seele Problem und die gegenwärtige philosophische Diskussion: Eine kritisch-systematische Analyse*. Darmstadt: Wissenschaftliche Buchgesellschaft, 1989.

———. *Leib und Seele: Ein Beitrag zur philosophischen Anthropologie*. Salzburg: Pustet, 1973.

———. *The Philosophical Diseases of Medicine and Their Cure: Philosophy and Ethics of Medicine*. Vol. 1, *Foundations*. Dordrecht: Springer, 2004.

———. "Respect for the Nature and Responsibility of the Person in Acquiring Knowledge about the Human Genome and in the Application of Human Biotechnology." In *Human Genome: Human Person and the Society of the Future*, edited by J. de D. Vial Correa and E. Sgreccia, 351–395. Vatican City: Libreria Editrice Vaticana, 1999.

———. "Substitution of the Conjugal Act or Assistance to It? IVF, GIFT and Some Other Medical Interventions. Philosophical Reflections on the Vatican Declaration *Donum Vitae*." *Anthropotes* 2 (1988): 273–286.

———. *What Is Life? On the Originality, Irreducibility and Value of Life*, edited by H. G. Callaway. Amsterdam: Rodopi, 1997.

Sella, G., and P. Cervella. *L'evoluzione biologica e la formazione della specie*. Turin: SEI, 1987.

Semaine des Intellectuels Catholiques. *Qu'est-ce que la vie*. Paris: P. Horay, 1958.

Seppala, M. "The World Collaborative Report on In Vitro Fertilization and Embryo Replacement: Current State of Art in January 1984." *Annals of New York Academy of Sciences* 442 (1985): 558–563.

Serebrovska, Z., and M. L. Di Pietro. "La sindrome da iperstimolazione ovarica: Tra clinica e etica." *Medicina e Morale* 56, no. 2 (2006): 327–347.

Serebrovska, Z., M.L. Di Pietro, and A. Bompiani. "Fecondazione artificiale e crioconservazione degli embrioni." *Medicina e Morale* 56, no. 1 (2006): 13–39.

Sermonti, F., and R. Fondi. *Dopo Darwin: Critica all'evoluzionismo*. Milan: Rusconi, 1980.

Serra, A. "Le componenti biologiche della sessualità umana." In *Uomo-Donna: Progetto di vita*, edited by Centro Italiano Femminile, 103–136. Rome: CIF-UECI, 1985.

———. "Il concepimento umano in vitro: Aspetti e problemi biologici." In *Fecondazione artificiale embryotransfer: Problemi biologici, clinici, giuridici, etici*, edited by G.F. Zuanazzi, 11–24. Verona: Cortina International, 1986.

———. "La consulenza genetica prima della diagnosi prenatale: Un obbligo deontologico." *Medicina e Morale* 47, no. 5 (1997): 903–921.

———. "Dalle nuove frontiere della biologia e della medicina nuovi interrogativi alla filosofia, al diritto, e alla teologia." In *Nuova genetica e embriopoiesi umana*, edited by A. Serra, E. Sgreccia, and M.L. Di Pietro, 69–70. Milan: Vita e Pensiero, 1990.

———. "Embrione umano, scienza e medicina, In margine al recente documento vaticano." *La Civiltà Cattolica* 2 (1987): 247–261.

———. "Interrogativi etici dell'ingegneria genetica." *Medicina e Morale* 34, no. 3 (1984): 306–321.

———. "La 'nuova genetica': Attualità, prospettive, problemi." In *Medicina e genetica verso il futuro*, edited by S. Cotta, 5–23. L'Aquila: Japadre, 1986.

———. "Pari dignità all'embrione umano nell'enciclica *Evangelium vitae*." In *"Evangelium vitae" e bioetica: Un approccio interdisciplinare*, edited by E. Sgreccia and D. Sacchini, 147–173. Milan: Vita e Pensiero, 1996.

———. "Per uno statuto integrato dell'embrione umano: Alcuni dati della genetica e dell'embriologia." In *Nascita e morte dell'uomo*, edited by S. Biolo, 55–105. Genoa: Marietti, 1993.

———. "Problemi etici della diagnosi prenatale." *Medicina e Morale* 32, no. 1 (1982): 52–61.

———. "Quando comincia un essere umano." In *Il dono della vita*, edited by E. Sgreccia, 99–105. Milan: Vita e Pensiero, 1987.

———. "Quando è iniziata la mia vita?" *La Civiltà Cattolica* 3348 (1989): 582.

———. "La rivoluzione genomica: Conquiste, attese, rischi." *La Civiltà Cattolica* 2, no. 4 (2001): 439–453.

———. "La sperimentazione sull'embrione umano: Una nuova esigenza della scienza e della medicina?" *Medicina e Morale* 43, no. 1 (1993): 97–116.

———. "Lo stato biologico dell'embrione umano: Quando inizia l' 'essere umano'?" In *Commentario interdisciplinare alla "Evangelium vitae,"* edited by E. Sgreccia and R. Lucas Lucas, 573–597. Vatican City: Libreria Editrice Vaticana, 1997.

Serra, A. "La diagnosi prenatale di malattie genetiche." *Medicina e Morale* 34, no. 4 (1984): 433–448.

Serra, A., and G. Bellanova. "Accertamento prenatale di rischio di patologia cromosomica fetale: Aspetti scientifici, etici e deontologici." *Medicina e Morale* 47, no. 1(1997): 15–35.

Serra, A., and G. Neri. eds. *A cento anni dalla morte del fondatore della genetica: Nuove conquiste, nuove responsabilità*, in *Nuova genetica, uomo e società*. Milan: Vita e Pensiero, 1986.

—, eds. *Nuova genetica: Uomo e società*. Milan: Vita e Pensiero, 1986.

Serra, A., E. Sgreccia, and M. L. Di Pietro. *Nuova genetica ed embriopoiesi umana: Prospettive della scienza e riflessioni etiche*. Milan: Vita e Pensiero, 1990.

Serres, M. Preface to *L'oeuf transparent*, by J. Testart. Paris: Flammarion, 1986.

Sgreccia, E. "Autonomia e responsabilità della scienza." In *Lineamenti di etica della sperimentazione clinica*, edited by A. G. Spagnolo and E. Sgreccia, 39–49. Milan: Vita e Pensiero, 1994.

—. "La bioetica, fondamenti e contenuti." *Medicina e Morale* 34, no. 3 (1984): 285–306.

—. "Il comitato etico tra assistenza e ricerca." *Orizzonte Medico* 4 (1987): 2–3.

—. "Corpo e persona." In *Questioni di Bioetica*, edited by S. Rodotà, 113–122. Rome: Laterza, 1993.

—. "Le 'Disposizioni anticipate di trattamento.' " *L'Osservatore Romano*. July 25–26, 2005.

—. "Divieto morale e profezia." In *Il dono della vita*, 205–211. Milan: Vita e Pensiero, 1987.

—, ed. *Il dono della vita*. Milan: Vita e Pensiero, 1987.

—. "Economia e salute: considerazioni etiche." *Medicina e Morale* 36, no. 3 (1986): 31–46.

—. "L'etica: Presupposto di affidabilità dell'ospedale." *Sanare Infirmos* 1 (1987): 12–16.

—. "Etica, ma su quale fondamento?" *Orizzonte Medico* 1 (1987): 1–2.

—. "L'eutanasia in Olanda: Anche per i bambini!" *L'Osservatore Romano*. September 3, 2004.

—. "Fecondazione artificiale: Problemi etici." *Medicina e Morale* 43, no. 1 (1993): 183–204.

—. "L'insegnamento dei Padri della Chiesa." In *L'aborto: Riflessioni di studiosi cattolici*, edited by A. Fiori and E. Sgreccia , 49–68. Milan: Vita e Pensiero, 1975.

—. "Interventi su embrioni e feti umani." In *Commento interdisciplinare alla "Evangelium Vitae,"* edited by E. Sgreccia and R. Lucas Lucas, 617–635. Vatican City: Libreria Editrice Vaticana, 1997.

—. *Manuale di bioetica*. Vol. 2, *Aspetti medico-sociali*. 3rd ed. Milan: Vita e Pensiero, 2002.

—. "La obiezione di coscienza e le implicazioni nella prassi assistenziale e nei consultori familiari." *Anime e Corpi* 77 (1978): 295–315.

—. "Per l'esercizio cristiano della medicina." *Medicina e Morale* 29, no. 2 (1979): 161–190.

—. "La persona e la vita." *Dolentium Hominum* 2 (1986): 38–41.

—. "La persona umana." In *Bioetica*, edited by C. Romano and G. Grassani, 194–195. Turin: UTET, 1995.

—. "La posizione della Chiesa di fronte alla vita e alla salute nell'attuale contesto socio-culturale." *Camillianum* 13 (2005): 9–31.

—. "Potenzialità e limiti del progresso scientifico e tecnologico." *Dolentium Hominum* 37, no. 1 (1998): 137–144.

———. "Problemi dell'insegnamento della bioetica." *Giornale Italiano per la Formazione Permanente del Medico* 15, no. 2 (1987): 104–117.

———. "Procreazione responsabile e metodi di regolazione naturale della fertilità: Aspetti teologici." In *Metodi naturali per la regolazione della fertilità: L'alternativa autentica*, edited by A. Lopéz Trujillo and E. Sgreccia. Milan: Vita e Pensiero, 1994

———. "Il riduzionismo biologico in medicina." *Medicina e Morale* 35, no. 1 (1985): 3–9.

———. "Il rifiuto della verità e la tentazione del suicidio." *Prospettive nel mondo* 75, no. 76 (1982): 35–41.

———. "La risposta nella trascendenza." In *Scienza ed etica: Quali limiti?* edited by J. Jacobelli, 163–173. Rome: Laterza, 1990.

———. "Salute e salvezza cristiana nel contesto dell'educazione sanitaria." *Medicina e Morale* 32, no. 3 (1982): 284–302.

———. "Scienza, medicina, etica." In *Nuova genetica: Uomo e società*, edited by A. Serra and G. Neri, 7–11. Milan: Vita e Pensiero, 1986.

———. "Sterilizzazione volontaria e mentalità contraccettiva." *La Famiglia* 75 (1979): 227–240.

———. "Storia della bioetica e sua giustificazione epistemologica." In *La storia della medicina nella società e nella cultura contemporanea: Atti del convegno internazionale di studio all'Istituto di Studi Politici "S. Pio V"; Frascati, 21–30 giurno 1991*, 69–84. Rome: Apes, 1992.

———, ed. *Storia della medicina e storia dell'etica medica verso il terzo millennio.* Soveria Mannelli, CZ: Rubbettino, 2000.

———. "Uomo e salute." *Anime e Corpi* 91 (1980): 419–444.

———. "Valori morali per la salute dell'uomo." *Rassegna di Teologia* 5 (1979): 390–396.

Sgreccia, E., S. Burgalassi, and G. Fasanella, eds. *Anzianità e valori.* Milan: Vita e Pensiero, 1991.

Sgreccia, E., and M. L. Di Pietro. "L'interruzione volontaria di gravidanza nel pensiero cattolico." *Affari sociali* 4 (1989): 75–92.

———. "Manipolazioni genetiche e procreazione artificiale: Orientamenti giuridici e considerazioni etiche." *Il diritto di famiglia e delle persone* 3, no. 4 (1987): 1351–1447.

———. "Storia del fenomeno dell'eutanasia dall'antichità ai nostri giorni." In *Eutanasia, diritto alla vita*, edited by A. Tarantino and M. L. Tarantino, 13–46. Lecce: Edizioni del Grifo, 1994.

Sgreccia, E., M. L. Di Pietro, and G. Fasanella. "I trapianti d'organo e di tessuti nell'uomo: Aspetti etici." In *Trapianti d'organo*, edited by A. Bompiani and E. Sgreccia, 150–154. Milan: Vita e Pensiero, 1989.

Sgreccia, E., and R. Lucas Lucas, eds. *Commento interdisciplinare alla "Evangelium vitae."* Vatican City: Libreria Editrice Vaticana, 1997.

Sgreccia, E., and V. Mele. "La diagnosi genetica postnatale." In *Ingegneria genetica e biotecnologie nel futuro dell'uomo*, 251–277. Milan: Vita e Pensiero, 1992.

———. "Gli aspetti etici dell'ingegneria genetica." In *Ingegneria genetica e biotecnologie nel futuro dell'uomo*, 131–166. Milan: Vita e Pensiero, 1992.

———, eds. *Ingegneria genetica e biotecnologie nel futuro dell'uomo.* Milan: Vita e Pensiero, 1992.

———, eds. *Rilevanza dei fattori etici e sociali nella prevenzione delle malattie professionali*. Milano: Vita e Pensier, 1994.

Sgreccia, E., and D. Sacchini, eds. *"Evangelium vitae" e bioetica: Un approccio interdisciplinare*. Milan: Vita e Pensiero, 1996.

Sgreccia, E., and A. G. Spagnolo, eds. *Etica e allocazione delle risorse in sanità*. Milan: Vita e Pensiero, 1996.

———. "L'insegnamento di bioetica nel Corso di laurea in Medicina e Chirurgia: L'esperienza nell'Università Cattolica del S. Cuore." *Medicina e Morale* 46, no. 4 (1996): 639–654.

Shaw, W. H. "To Be or Not To Be? That Is the Question." *American Journal of Human Genetics* 36 (1984): 1–9.

Shea, M. C. "Embryonic Life and Human Life." *Journal of Medical Ethics* 11 (1985): 205–209.

Sherman, M., and J. D. van Vleet. "The History of Institutional Review Boards." *Regulatory Affairs* 3 (1991): 615–628.

Sherwin, S. *No Longer Patient: Feminist Ethics in Health Care*. Philadelphia: Temple University Press, 1992.

Shettles, L. B. "Use of the Y Chromosome in Prenatal Sex Determination." *Nature* 230 (1971): 52–53.

Shewmon, D. A. "Anencephaly: Selected Medical Aspects." *Hastings Center Report* 18, no. 5 (1988): 11–18.

———. "'Brain Death': A Valid Theme with Invalid Variations, Blurred by Semantic Ambiguity." In *Working Group on the Determination of Brain Death and Its Relationship to Human Death*, ed. R. J. White, H. Angstwurm, and I. Carrasco de Paula, 23–51. Vatican City: Pontifical Academy of Sciences, 1992.

———. "Recovery from 'Brain Death': A Neurologist's Apologia." *Linacre Quarterly* 64, no. 1 (1997): 30–96.

Shopenhauer, A. "Metafisica dell'amore sessuale." In *Supplementi al "Mondo come volontà e rappresentazione."* Vol. 2. Rome: Laterza, 1986.

Shorter, E. *Bedside Manners: The Troubled History of Doctors and Patients*. New York: Simon and Schuster, 1985.

Shuster, E. "When Genes Determine Motherhood: Problems in Gestational Surrogacy." *Human Reproduction* 7 (1992): 1029–1033.

Simberloff, D. "Community Effects of Introduced Species." In *Bioethics Crises in Ecological and Evolutionary Time*, edited by M. H. Nitecki. Chicago: Academic Press, 1981.

Simili, A. *La fecondazione artificiale umana*. Turin: Minerva Medica, 1961.

Simon, R. "Amore e sessualità, matrimonio e famiglia." In *L'ateismo contemporaneo*, edited by Pontificia Università Salesiana. 3 vols. Turin: SEI, 1969.

Simoni, G., G. Gimelli, C. Cuoco, et al. "First Trimester Fetal Karyotyping: One Thousand Diagnoses." *Human Genetics* 72 (1986): 203–209.

Singer, P. *Animal Liberation: A New Ethics for Our Treatment of Animals*. 2nd ed. New York: Random House, 1990. First edition published in 1975.

———. *Practical Ethics*. Cambridge: Cambridge University Press, 1973.

———. *Practical Ethics*. 2nd ed. New York: Cambridge University Press, 1993. First edition published in 1979.

———. *Rethinking Life and Death: The Collapse of Our Traditional Ethics*. New York: St. Martin's Press, 1994.

Singer, P., and K. Dawson. "Individuals, Humans and Persons: The Issue of Moral Status." In *Embryo Experimentation*, edited by P. Singer et al. London: Cambridge University Press, 1990.

Skinner, R. *One Flesh, Separate Persons*. London: Constable, 1976.

Sloterdijk, P. *Non siamo stati ancora salvati*: *Saggi dopo Heidegger*. Milan: Bompiani, 1994. Originally published as *Nicht gerettet: Versuche nach Heidegger* (Frankfurt: Suhrkamp, 2001).

Smart, J. J., and B. Williams, eds. *Utilitarianism For and Against*. Cambridge: Cambridge University Press, 1973.

Smith, S., L. Scott, and S. Hosid. "Combined Intrauterine Triplet and Ectopic Pregnancy following Pronuclear Embryo Transfer in a Patient with Elevated Serum Progesterone during Ovulation Induction." *Journal of Assisted Reproduction and Genetics* 10, no. 7 (1993): 478–480.

Socci, A., and C. Casini. *In difesa della vita: Legge 40, fecondazione assistita e mass media*. Casale Monferrato, AL: Piemme, 2005.

Società Italiana di Bioetica e dei Comitati Etici. "Statuto." July 1989, http://www.sibce.org/statuto.html.

Società Italiana di Medicina Legale e delle Assicurazioni. "Il Documento di Erice sui rapporti della bioetica e della deontologia medica con la medicina legale." 53rd Course on New Trends in Forensic Haematology and Genetics: Bioethical Problems (February 18–21, 1991); published in *Medicina e Morale* 41, no. 4 (1991): 561–567.

Società Italiana per lo Studio della Fertilità e Sterilità. *Libro bianco sulla riproduzione assistita*. Palermo: SIFES, 1991.

Sokal, D. C., J. Zipper, and T. King. "Transcervical Quinacrine Sterilization: Clinical Experience." *International Journal of Gynecology and Obstetrics* 51, suppl. 1 (1995): S57–S59.

Sokoloff, B. Z. "Alternative Methods of Reproduction." *Clinical Pediatrics* 26, no. 1 (1987): 11–17.

Soloiev, V. *La justification du bien*. Geneva: Ed. Skatkine, 1997.

Soricelli, E. *Il trapianto d'organi*. Genoa: Pantograph, 1994.

Soussis, I., J. C. Harper, A. H. Handyside, and R. M. L. Winston. "Obstetric Outcome of Pregnancies Resulting from Embryos Biopsied for Pre-Implantation Diagnosis of Inherited Disease." *British Journal of Obstetrics and Gynaecology* 103 (1996): 784–788.

Spaemann, R. "Ars longa vita brevis." In *Etica della ricerca biomedica: Per una visione cristiana; Atti della nona assemblea generale della Pontificia Accademia per la Vita; Città del Vaticano, 24–26 febbraio 2003*, edited by J. de D. Vial Correa and E. Sgreccia, 159–174. Vatican City: Libreria Editrice Vaticana, 2004.

———. *Basic Moral Concepts*. Translated by T. J. Armstrong New York: Routledge, 1989. Originally published as *Moralische Grundbegriff* (Munich: C. H. Beck, 1982).

——. "Genetic Manipulation of Human Nature in the Context of Human Personality." In *Human Genome: Human Person and the Society of the Future*, edited by J. de D. Vial Correa and E. Sgreccia. Vatican City: Libreria Editrice Vaticana, 1999.

——. "La morte della persona e la morte dell'essere umano." Dossier "Ai confini della vita." *Lepanto* 21, no. 162 (2002).

——. *Persons: The Difference between "Someone" and "Something."* Translated by O. O'Donovan. New York: Oxford University Press, 2006. Originally published as *Personen: Versuche über den Unterschied zwischen "etwas" und "jemand"* (Stuttgart: Klett-Cotta, 1996).

——. "La responsabilità personale e il suo fondamento." In *Etica teleologica o etica deontologica? Un dibattito al centro della teologia morale odierna*. Rome: 1983.

Spagnolo, A.G. "Aborto e nuova sessualità: La situazione oggi sul piano sociale e politico." *Anime e Corpi* 129 (1987): 33–48.

——. "Ai confini tra atteggiamento eutanasico e terapia palliativa." *Quaderni di cure palliative* 1 (1994): 49–51.

——. "Il bene del paziente e i limiti dei testamenti di vita." *L'Osservatore Romano* 138 (June 17–18, 1996).

——. "Bioetica." In *Dizionario interdisciplinare di scienza e fede*, edited by G. Tanzella-Nitti and A. Strumia, 196–214. Rome: Urbaniana University Press, 2002.

——. "Bioetica (Fondamenti)." In *Dizionario di teologia pastorale sanitaria*. Turin: Camillianum, 1997.

——. *Bioetica nella ricerca e nella prassi medica*. Turin: Camilliane, 1997.

——. "Comitati di bioetica in tema di procreazione artificiale." *Medicina e Morale* 43, no. 1 (1993): 205–230.

——. "Comitati di etica per la genetica." In *Ingegneria genetica e biotecnologie nel futuro dell'uomo*, edited by E. Sgreccia and V. Mele, 411–451. Milan: Vita e Pensiero, 1992.

——. "Comitati di etica per la ricerca: Procedure e qualità della revisione etica." *Etica della ricerca biomedica: Per una visione cristiana; Atti della nona assemblea generale della Pontificia Accademia per la Vita; Città del Vaticano, 24–26 febbraio 2003*, edited by J. de D. Vial Correa and E. Sgreccia, 245–269. Vatican City: Libreria Editrice Vaticana, 2004.

——. "I comitati etici negli ospedali: Sintesi e considerazioni a margine di un recente simposio." *Medicina e Morale* 36, no. 3 (1986): 566–583.

——. "Il dibattito sull'eutanasia in Italia." *Le Scienze* 297 (1993): 9–10.

——. "L'inclusione delle donne nelle sperimentazioni farmacologiche." In *Bioetica nella ricerca e nella prassi medica*, 470–472. Turin: Camilliane, 1997.

——. "L'inevitabile complicità nel trapianto di tessuti fetali da aborti volontari." *L'Osservatore Romano*. January 28, 1995.

——. "Necessità, opportunità, utilità della istituzione di un Comitato Etico presso la Direzione Generale della sanità militare." *Giornale di Medicina Militare* 146, no. 3 (1996): 300–305.

——. "'Norme di buona pratica clinica': Il documento della Comunità Europea sulla sperimentazione di nuovi prodotti farmaceutici." *Medicina e Morale* 41, no. 2 (1991): 201–227.

816

———. "Un nuovo ruolo per il comitato etico istituzionale." In *Etica e giustizia in sanità*, edited by A.G. Spagnolo, D. Sacchini, A. Pessina, and M. Lenoci, 217–223. Milan: McGraw-Hill, 2004.

———. "Predictive and Presymptomatic Genetic Testing: Service or Sentence?" In *Human Genome: Human Person and the Society of the Future*, edited by J. de D. Vial Correa and E. Sgreccia. Vatican City: Libreria Editrice Vaticana, 1999.

———. "I principi della bioetica Nord-Americana, e la critica del 'Principlismo.'" *Camillianum* 20 (1999): 225–246.

———. "Principi etici e metodologie di sperimentazione clinica." In *Lineamenti di etica della sperimentazione clinica*, edited by A.G. Spagnolo and E. Sgreccia, 51–70. Milan: Vita e Pensiero, 1994.

———. "Principios de la bioética norteamericana y critica del principlismo." *Bioética y ciencias de la salud* 3, no. 1 (1998): 102–110.

———. "Procreazione sempre più 'assistita' con l'uso degli spermatidi." *Medicina e Morale* 45, no. 5 (1995): 1107–1111.

———. "Il progetto di 'convenzione' Europea sulla bioetica." *Vita e Pensiero* 4 (1995): 249–268.

———. "La protezione dei soggetti di sperimentazione: Ruolo e procedure operative dei Comitati di Etica." In *Lineamenti di etica della sperimentazione clinica*, edited by A.G. Spagnolo and E. Sgreccia, 113–140. Milan: Vita e Pensiero, 1994.

———. "La relazione medico-paziente nella sanità aziendalizzata." In *Etica e giustizia in sanità*, by A.G. Spagnolo, D. Sacchini, A. Pessina, and M. Lenoci, 155–183. Milan: McGraw-Hill, 2004.

———. "La sperimentazione e i soggetti di sperimentazione." In *Bioetica nella ricerca e nella prassi medica*, 453–482. Turin: Camilliane, 1997.

———. "Testamenti di vita." In *Bioetica in medicina*, edited by A. Bompiani, 340–355. Rome: CIC Edizioni Internazionali, 1996.

———. "L'uso del placebo nelle sperimentazioni farmacologiche." In *Bioetica nella ricerca e nella prassi medica*, 464–468. Turin: Camilliane, 1997.

Spagnolo, A.G., A.A. Bignamini, and A. de Franciscis. "I Comitati di Etica fra linee-guida dell'Unione Europea e decreti ministeriali." *Medicina e Morale* 47, no. 6 (1997): 1059–1098.

Spagnolo, A.G., M. Cicerone, and R. Minacori, "Biobanche: Aspetti clinici della conservazione del materiale biologico umano." *Iustitia* 1 (2006): 63–78.

Spagnolo, A.G., and M.L. Di Pietro. "Bioetica clinica: Quale decisione per l'embrione in una gravidanza tubarica?" *Medicina e Morale* 45, no. 2 (1995): 285–310.

———. "Terapia genica: Il documento 15.2.91 del Comitato nazionale per la Bioetica ed un'analisi comparativa con le esperienze di altri Comitati etici nazionali ed internazionali." *Il Diritto di Famiglia e delle Persone* 21, no. 2 (1992): 323–363.

Spagnolo, A.G., M.L. Di Pietro, L. Palazzani, and E. Sgreccia. "Significato della diagnosi prenatale nella prevenzione delle malattie congenite: aspetti etico-sociali." In *Atti del LXVII Congresso della Società Italiana di Ostetricia e Ginecologia*, 39–40. Brescia: Clas International, 1990.

Spagnolo, A.G., A. Mancini, L. De Marinis, et al. "Valutazione scientifica ed etica di un metodo per il prelievo diagnostico del liquido seminale." *Medicina e Morale* 43, no. 6 (1993): 1189–1202.

Spagnolo, A. G., and R. Minacori. "Farmacogenetica e farmacogenomica: Aspettative e questioni etiche." *Medicina e Morale* 52, no. 5 (2002): 819–866.

Spagnolo, A. G., and D. Sacchini. "Etica della gestione delle risorse per malattie terminali." *Quaderni di cure palliative* 5, no. 3 (1997): 231–236.

Spagnolo, A. G., D. Sacchini, A. Pessina, and M. Lenoci, eds. *Etica e giustizia in sanità.* Milan: McGraw-Hill, 2004.

Spagnolo, A. G., and E. Sgreccia. "I comitati di bioetica: Sviluppo storico, presupposti e tipologie." *Vita e Pensiero* 78 (1989): 500–514.

———. "Comitati e Commissioni di bioetica in Italia e nel mondo." *Vita e Pensiero* 12 (1989): 802–818.

———. "Il feto umano come donatore di tessuti e di organi." *Medicina e Morale* 38, no. 6 (1988): 843–875.

———. "Prelievi di organi e tessuti fetali a scopo di trapianto: Aspetti conoscitivi e istanze etiche." In *Trapianti d'organo*, edited by A. Bompiani and E. Sgreccia, 47–84. Milan: Vita e Pensiero, 1989.

Spagnolo, G. "L'eutanasia: Aspetto etico del problema." *Scienza e Fede* 8 (1983): 1–38.

Spalla, C. *Il progresso delle biotecnologie in Italia.* Milan: Federchimica-Assobiotech, 1990.

Spanish Bishops Conference Committee for the Family and the Defense of Life. *La Eutanasia: 100 cuestiones y respuestas sobre la defensa de la vida y la actitud de los católicos.* Madrid: Paulinas, 1993. Available online at http://www.conferenciaepiscopal .nom.es/archivodoc/jsp/system/win_main.jsp.

Spaziante, E. "L'aborto provocato: Dimensioni planetarie del fenomeno; aspetti epidemiologici, demografici e considerazioni di bioetica sociale." *Medicina e Morale* 46, no. 6 (1996): 1083–1134.

Speidel, J. J., and R. T. Ravenholt, eds. *Atti del workshop sulla sterilizzazione.* San Francisco: December 4–6, 1977.

Spiazzi, R. *Lineamenti di etica della vita.* Bologna: Studio Domenicano, 1990.

Spinsanti, S. *Bioetica e grandi religioni.* Milan: Paoline, 1987.

———. *Il corpo nella cultura contemporanea.* Brescia: Queriniana, 1983.

———, ed. *Documenti di deontologia e etica medica.* Milan: Paoline, 1985.

———. "Incontro con Warren Reich," *L'Arco di Giano* 7 (1995): 215–225.

———. "L'inseminazione artificiale: Scelte deontologiche ed interrogativi etici." *Res Medicae* (March–April 1970): 134–138.

———. "Psicologi incontro ai morenti." *Medicina e Morale* 26, nos. 1–2 (1976): 79–96.

———. "Salute, Malattia, Morte." In *Nuovo dizionario di teologia morale*, edited by F. Compagnoni, G. Piana, and S. Privitera, 1134–1144. Cinisello Balsamo: Paoline, 1990.

———, ed. *Umanizzare la malattia e la morte: Documenti pastorali dei vescovi francesi e tedeschi.* Rome: Paoline, 1980.

———. "Vita fisica." In *Corso di morale*, edited by T. Goffi and G. Piana, 2:127–267. Brescia: Queriniana, 1983.

Spital, A. "Should People Who Donate a Kidney to a Stranger Be Permitted to Choose Their Recipients? Views of the United States." *Transplantation* 76, no. 8 (2003): 1252.

Sprigge, L. S. "Vivisection, Morals, Medicine: Commentary from Anti-Vivisectionist Philosopher." *Journal of Medical Ethics* 9 (1983): 98–101.

Squarise, C. "Corpo." In *Dizionario enciclopedico di teologia morale*, edited by L. Rossi and A. Valsecchi, 149–166. Rome: Paoline, 1981.

Staglianò, A. "Ragione." In *Dizionario interdisciplinare di scienza e fede*, edited by G. Tanzella-Nitti and Strumia, 1167–1180. Rome: Urbaniana University Press, 2002.

Standing Council of the French Bishops Conference. "Respecter l'homme proche de sa mort." *Medicina e Morale* 42, no. 1 (1992): 124–133.

Stanzione, P., ed. *La disciplina giuridica dei trapianti: Legge 1 aprile 1999 n. 91*. Milan: Giuffré, 2000.

Steinbock, B., J. D. Arras, and A. J. London. *Ethical Issues in Modern Medicine*. Boston: McGraw-Hill, 2003.

Stella, F. "L'aborto come illecito penale." *Medicina e Morale* 24, no. 2 (1974): 243–249.

———. "La situazione legislativa in merito alla obiezione sanitaria in Europa." *Medicina e Morale* 35, no. 2 (1985): 281–302.

Sterzik, K. "Assistierte Reproduktion." *Sexualmedizin* 5 (1996): 148–151.

Stevenson, C. L. *Facts and Values*. New Haven, CT: Yale University Press, 1963.

Still, K., T. Kolatat, T. Corbett, and P. Byrne. "Early Third Trimester Selective Feticide of a Compromising Twin." *Fetal Therapy* 4 (1989): 83–87.

Stirrat, G. M. *Aids in Obstetrics and Gynaecology*. London: Churchill Livingstone, 1987.

Stroppiana, L. *Storia della medicina tra arte e scienza*. Rome: Edizioni dell'Ateneo, 1982.

Styczen, T. "Le leggi contro la vita: Analisi etico-culturale." In *"Evangelium vitae" e diritto*, edited by A. López Trujillo, G. Herranz, and E. Sgreccia, 213–227. Vatican City: Libreria Editrice Vaticana, 1997.

Sundstrom, P., O. Nilsson, and P. Liedholm. "Scanning Electron Microscopy of Human Preimplantation Endometrium in Normal and Clomiphene/Human Chorionic Gonadotropin Stimulated Cycles." *Fertility and Sterility* 40 (1983): 642–647.

Surrey, E. S., and J. F. Kerin. "Extended Techniques in Assisted Reproductive Technologies." In *Infertility: Evaluation and Treatment*, edited by W. R. Keye, R. J. Chang, R. W. Rebar, and M. R. Soules, 788–797. Philadelphia: W. B. Saunders, 1995.

Sussmann, M., C. H. Collins, F. A. Skinner, and D. E. Stewart-Tull, eds. *The Release of Genetically Engineered Micro-organisms*. New York: Academic Press, 1988.

Sutton, A. "Ten Years after the Warnock Report: Is the Human Neo-Conceptus a Person?" *Medicina e Morale* 44, no. 3 (1994): 475–490.

Swiss National Advisory Commission on Biomedical Ethics (NEK-CNE). "On the Regulation of Living Donation in the Transplantation Law." Opinion no. 6/2003. November 17, 2003. http://www.bag.admin.ch/nek-cne/04229/04232/index.html?lang=de.

Synod of Bishops. "The Tasks of the Christian Family." October 24, 1980. Propositions 20–25. In *Enchiridion Vaticanum*, 7: 699–707 [Italian]. Bologna: Dehoniane, 1982.

Tabor, A., J. Philip, M. Madsen, J. Bang, E. B. Obel, and B. Nørgaard-Pedersen. "Randomized Controlled Trial of Genetic Amniocentesis in 4606 Low-Risk Women." *Lancet* 1 (1986): 1287–1292.

Tagliapietra, G. "Le banche del seme: Il caso italiano; È necessaria subito una rigida regolamentazione." *Prospettive nel mondo* (1984): 95–98.

Taipale, P., V. Hiilesmaa, R. Salonen, and P. Ylöstalo. "Increased Nuchal Translucency as a Marker for Fetal Chromosomal Defects." *New England Journal of Medicine* 337 (1997): 1654–1658.

Tallo, C.P., B. Vohr, W. Oh, L.P. Rubin, D.B. Seifer, and R.V. Haning Jr. "Maternal and Neonatal Morbidity Associated with In Vitro Fertilization." *Journal of Pediatrics* 127, no. 5 (1995): 794–800.

Tanzella Nitti, G. "Creazione ed evoluzione: Chi ha rimescolato le carte?" *Documentazione Internazionale di Scienza e Fede*. September 2005. http://www.disf.org/Editoriali/ Editoriale0509.asp.

Tarantino, A. "Sul fondamento dei diritti del nascituro: Alcune considerazioni bioetico-giuridiche." *Medicina e Morale* 45, no. 5 (1995): 951–984.

Task Force on Heart Transplantation. *Journal of American College of Cardiology* 22 (1993): 1–64.

Tatarelli, R., E. De Pisa, and P. Girardi. *Curare con il paziente: Metodologia del rapporto medico-paziente*. Milan: F. Angeli, 1998.

Teilhard de Chardin, P. *The Phenomenon of Man*. Translated by B. Wall. New York: Harper and Brothers, 1959. Originally published as *Le phénomène humain* (Paris: Editions du Seuil, 1955).

Teitelbaum, M. S., and J.M. Winter. *The Fear of Population Decline*. Orlando: Academic Press, 1985.

Tertullian. *Quaestiones disputatae: De anima*. C. XIX, PL II, 682.

Testart, J. *Le désir du gène*. Paris: Francois Bourin, 1992.

———. "La fécondation de mieux en mieux assistée." *Les Cahiers du CCNNE* 3 (1995): 24–25.

———. "Procréation médicalement assistée: L'éthique et la loi." *Études* 3816 (1994): 599–610.

Tettamanzi, D. *Alle sorgenti della vita*. Casale Monferrato: Piemme, 1993.

———. *Bambini fabbricati*. Casale Monferrato: Piemme, 1985.

———. *Bioetica: Nuove frontiere per l'uomo*. 2nd ed. Casale Monferrato: Piemme, 1996.

———. *Comunità cristiana e aborto*. Alba: Paoline, 1975.

———. "Diagnosi prenatale e aborto selettivo: Problemi etici." *Anime e Corpi* (1983): 339–360

———. "L'etica sessuale." In *Sessualità da ripensare*, edited by C. Dastoli, 26–33. Milan: Vita e Pensiero, 1990.

———. "Feti umani e sperimentazione biomedica: Problemi etici." *Anime e Corpi* 111 (1984): 37–50.

———. *"Humanae vitae": Commento all'Enciclica sulla regolazione delle nascite*. Milan: Ancora, 1968.

———. "Interventi su embrioni / feti umani: In margine alla Raccomandazione 1046 del Consiglio d'Europa." *La Famiglia* 120 (1986): 31–53.

———. *Nuova bioetica cristiana*. Casale Monferrato: Piemme, 2000.

———. "Nutrizione e idratazione medicalmente assistite nel paziente in stato di incoscienza: Problemi morali." *L'Osservatore Romano*. December 11, 1992.

———. "Problemi etici sulla fertilizzazione in vitro e sull'embryotransfer." *Medicina e Morale* 33, no. 4 (1983): 342–364.

———. "Problemi morali circa alcuni interventi su feti ed embrioni umani." *Medicina e Morale* 35, no. 1 (1985): 23–43.

———. *Il procreare umano: Verità e responsabilità.* Casale Monferrato: Piemme, 1985.

———. "La sessualità umana: Prospettive antropologiche, etiche e pedagogiche." *Medicina e Morale* 34, no. 2 (1984): 129–154.

———. *La sterilizzazione anticoncezionale.* Brezzo di Bedero: Salcom, 1981.

———. "La sterilizzazione antiprocreativa." In *Bioetica: Difendere le frontiere della vita*, 273–285. Casale Monferrato: Piemme, 1996.

———. "La sterilizzazione: Problemi morali oggi." *Rivista del Clero Italiano* 2 (1979): 115–128.

Theoret, S. "The Role of the Bioethics Committee Dealing with HIV Infection." *Canadian Journal of Nursing Research* 84, no. 7 (1988): 41–47.

Thévoz, J.-M. "Research and Hospital Ethics Committees in Switzerland." *HEC Forum* 4, no. 1 (1992): 41–47.

Thiel, G., P. Vogelbach, L. Gürke, et al. "Crossover Renal Transplantation: Hurdles To Be Cleared!" *Transplantation Proceedings* 33, no. 1–2 (2001): 811–816.

Thomas, L.-V. *Anthropologie de la mort.* Paris: Payot, 1975.

Timio, M. *La storia tecnologica del guarire.* Rome: Borla, 1990.

Tiraboschi, P., and A. Spagnoli. "Le indagini sull'uomo sano." *Federazione Medica* 46 (1991): 27–30.

Tonti Filippini, N. "*Donum Vitae* and Gamete Intra Fallopian Transfer." In *"Humanae vitae": 20 anni dopo; Atti del II congresso internazionale di teologia morale; Roma 9–12 novembre 1988*, edited by A. Ansaldo, 791–802. Milan: Ares, 1989.

Tooley, M. "Abortion and Infanticide." *Philosophy and Public Affairs* 2, no. 1 (1972): 37–65.

Torlone, G. "Le sperimentazioni internazionali: Il dibattito sugli studi placebo-controllati." *Medicina e Morale* 54, no. 3 (2004): 555–588;

Torrelli, M. *Le médecin et les droits de l'homme.* Paris: Berger-Levrault, 1983.

Tosti, E., and B. Dale. *Fecondazione in vitro.* Naples: Edizioni Scientifiche Italiane, 1995.

Toth, T. L., S. G. Baka, L. L. Veeck, H. W. Jones Jr., S. Muasher, and S. E. Lanzendorf. "Fertilization and In Vitro Development of Cryopreserved Human Prophase 1 Oocytes." *Fertility and Sterility* 61, no. 5 (1994): 891–894.

Toulmin, S. E. *An Examination of the Place of Reason in Ethics.* Cambridge: Cambridge University Press, 1950.

Traina, V. "Inseminazione artificiale eterologa." In *L'inseminazione artificiale umana: Atti del II Seminario Internazionale; Bari 12–15 maggio 1980*, edited by R. Schoysmann, S. Bettocchi, and F. M. Boscia, 133–150. Palermo: Cofese, 1981.

Traina, V., C. Mancini, and G. Miniello. "L'inseminazione artificiale: Timing e tecnica." In *Atti XIII Congresso Nazionale della Società Italiana per lo Studio della Fertilità e la Sterilità (SIFES).* Salsomaggiore Terme: 1986.

Trask, B., G. Van den Hengh, D. Pinkel, J. Mullikin, F. Waldman, H. Van Dekken, and J. Gray. "Fluorescence In Situ Hybridization to Interphase Cell Nuclei in Suspension Allows

Flow Cytometric Analysis of Chromosome Content and Microscopic Analysis of Nuclear Organisation." *Human Genetics* 78 (1988): 251–254.

Tremblay, E. *L'affaire Rockfeller: L'Europe occidentale en danger*. Malmaison: Reuil, 1979.

Tripaldi, E. *Il cappellano ospedaliero nei Comitati di Bioetica*. Rome: Centro Studi S. Giovanni di Dio, 1995.

Troisfontaines, R. "Faut-il légaliser l'avortement?" *Nouvelle Revue de Théologie* 103 (1971): 500.

Truffino, J.C. "La esterilización en los enfermos mentales: Casos clínicos y consideraciones éticas." *Cuadernos de Bioética* 5, no. 22 (1995): 170–172.

Truog, R.D. "Is it Time to Abandon Brain Death?" *Hastings Center Report* 27, no. 1 (1997): 29–37.

Tuffs, A. "Surgeons Perform Germany's First Crossover Kidney Transplantation." *British Medical Journal* 331, no. 7520 (2005): 798–799.

Tulandi, T. "Tubal Sterilization." *New England Journal of Medicine* 336, no. 11 (1997): 796–797.

Turoldo, F., ed. *Le dichiarazioni anticipate di trattamento: Un testamento per la vita*. Padua: Fondazione Lanza/Gregoriana Libreria Editrice, 2006.

Tutschek, B., J. Sherlock, A. Halder, J. Delhanty, C. Rodeck, and M. Adinolfi. "Isolation of Fetal Cells from Transcervical Samples by Micromanipulation: Molecular Confirmation of Their Fetal Origin and Diagnosis of Fetal Aneuploidy." *Prenatal Diagnosis* 15 (1995): 951–960.

Tuveri, G., ed. *Saper ascoltare, saper comunicare: Come prendersi cura della persona con tumore*. Rome: Il Pensiero Scientifico Editore, 2005.

Udo, S. "*Protecting the Vulnerable: Testing Times for Clinical Research Ethics.*" *Social Science and Medicine* 51 (2000): 969–977.

Ufficio Nazionale per la Pastorale della Sanità. "Le Istituzioni sanitarie cattoliche in Italia: Identità e ruolo." July 7, 2000. http://www.chiesacattolica.it/pls/cci_new_v3/cciv4_doc.edit_documento?id_pagina=9068&p_id=5324.

Umani Ronchi, G., and C. Vecchiotti. *Il laboratorio medico-legale*. Rome: Lombardo, 1994.

"L'UNESCO sul genoma umano: Un segnale di forte significato bioetico." *Medicina e Morale* 48, no. 1 (1998): 9–14.

United Nations. *Report of the Fourth World Conference on Women: Beijing, September 4–15, 1995*. New York: United Nations, 1996. Available at http://www.un.org/womenwatch/daw/beijing/pdf/Beijing%20full%20report%20E.pdf.

———. *World Populations Trends and Policies*. New York: United Nations, 1988.

United Nations Department for Economic and Social Information and Policy Analysis Population Division. *World Contraceptive Use*. New York: United Nations, 1994.

United Nations Educational, Scientific and Cultural Organization. Universal Declaration on the Human Genome and Human Rights. November 11, 1997. http://unesdoc.unesco.org/images/0011/001102/110220e.pdf#page=47.

———. *World Directory of Academic Research Groups in Science Ethics*. Paris: UNESCO, 1993.

United Nations General Assembly. Universal Declaration of Human Rights. December 10, 1948. 217 A (III). Available at http://www.un.org/en/documents/udhr.

United States Advisory Committee on Human Radiation Experiments. *Final Report*. Washington, DC: US Government Printing Office, 1995. Available online at http://hss.energy.gov/healthsafety/ohre/roadmap/achre/report.html.

United States Conference of Catholic Bishops. *Ethical and Religious Directives for Catholic Health Care Services*. 5th ed. Washington, DC: United States Conference of Catholic Bishops, 2009. Available at http://www.usccb.org/issues-and-action/human-life-and-dignity/health-care/upload/Ethical-Religious-Directives-Catholic-Health-Care-Services-fifth-edition-2009.pdf.

United States Congress Office of Technology Assessment. *Commercial Biotechnology: An International Analysis*. Washington, DC: US Government Printing Office, 1984. Available at http://www.fas.org/ota/reports/8407.pdf.

———. *Genetic Monitoring and Screening in the Workplace*. Washington, DC: US Government Printing Office, 1990. Available at http://www.fas.org/ota/reports/9020.pdf.

———. *Genetic Witness: Forensic Use of DNA Tests*. Washington, DC: US Government Printing Office, 1990. Available at http://www.fas.org/ota/reports/9021.pdf.

———. *New Developments in Biotechnology: Field Testing Engineered Organisms; Genetic and Ecological Issues* (Washington, DC: US Government Printing Office, 1988. Available at http://www.fas.org/ota/reports/8816.pdf.

———. *New Developments in Biotechnology: Patenting Life—Special Report*. Washington, DC: US Government Printing Office, 1989. Available at http://www.fas.org/ota/reports/8924.pdf.

United States Department of Agriculture Agricultural Research Service. *Plant Introduction Service*. Washington, DC: Agricultural Research Office, 1986.

United States National Commission for the Protection of Human Subjects of Biomedical and Behavioral Research. *The Belmont Report: Ethical Principles and Guidelines for the Protection of Human Subjects of Research*. Washington, DC: US Government Printing Office, 1979. Available at http://ohsr.od.nih.gov/guidelines/belmont.html.

United States President's Commission for the Study of Ethical Problems in Medicine and Biomedical and Behavioral Research. *Defining Death: A Report on the Medical, Legal and Ethical Issues in the Determination of Death*. Washington, DC: 1981. Available online at http://bioethics.georgetown.edu/pcbe/reports/past_commissions/defining_death.pdf.

———. *Final Report on Studies of the Ethical and Legal Problems in Medicine and Biomedical and Behavioral Research*. Washington, DC: US Government Printing Office, 1983.

———. *Splicing Life: A Report on the Social and Ethical issues of Genetic Engineering with Human Beings*. Washington, DC: President's Commission for the Study of Ethical Problems in Medicine and Biomedical and Behavioral Research, 1982. Available at http://bioethics.georgetown.edu/documents/pcemr/splicinglife.pdf.

———. *Summing Up: Final Report on Studies of Ethical and Legal Problems in Medicine and Biomedical and Behavioral Research*. Washington, DC: US Government Print Office, 1983.

United States President's Council on Bioethics. *Alternative Sources of Pluripotent Stem Cells*. Washington, DC: US President's Council on Bioethics, 2005. Available at https://scholarworks.iupui.edu/bitstream/handle/1805/751/Alternative%20sources%20of%20human%20stem%20cells%202005.pdf.

Vaccaro, C. "I Comitati di Etica in Italia." In *Una Verità in Dialogo: Storia, metodologia e pareri di un comitato di etica*, edited by P. Cattorini. Milan: Istituto Scientifico Ospedale San Raffaele-Europa Scienza Umane, 1994.

Valeriani, A. *Il nostro corpo come comunicazione*. Brescia: La Scuola, 1964.

Valori, P. *Il libero arbitrio: Dio, l'uomo, la libertà*. Milan: Rusconi, 1987.

———. "Può esistere una morale laica?" *La Civiltà Cattolica* 3 (1984): 19–29.

———. "Valore morale." In *Nuovo dizionario di teologia morale*, edited by F. Compagnoni, G. Piana, and S. Privitera, 1416–1427. Cinisello Balsamo: Paoline, 1990.

Vandekerckhove, P., P.A. Donovan, R.J. Lilford, and T.W. Harada. "Infertility Treatment: From Cookery to Science; The Epidemiology of Randomised Controlled Trials." *British Journal of Obstetrics and Gynaecology* 100 (1993): 1005–1036.

Van den Bergh, M., M.K. Hohl, C. De Geyter, A.M. Stalberg, and C. Limoni. "Ten Years of Swiss National IVF Register FIVNAT-CH, Are We Making Progress?" *Reproductive BioMedicine Online* 5, no. 11 (2005): 632–640.

Van Der Maas, P.J., J.J. Van Delden, L. Pijnenborg, and C.W. Looman. "Euthanasia and Other Medical Decisions Concerning the End of Life." *Lancet* 338 (1991): 669–674.

Vanderpool, H.Y. *The Ethics of Research Involving Human Subjects: Facing the 21st Century*. Frederick, MD: University Publishing Group, 1996.

Vanni Rovighi, S. *L'antropologia filosofica di S. Tommaso D'Aquino*. Milan: Vita e Pensiero, 1965.

———. *Elementi di filosofia*. 3 vols. Brescia: La Scuola, 1963.

Van Steirteghen, A.C., ' "Higher Success Rate by Intracytoplasmic Sperm Injection than by Subzonal Insemination: Report of a Second Series of 300 Consecutive Treatment Cycles." *Human Reproduction* 8, no. 7 (1993): 1055 ff.

Vatican Council II. Pastoral Constitution *Gaudium et spes*. December 7, 1965. In *Documents of Vatican II*, edited by A.P. Flannery, 903–1001. Grand Rapids, MI: William B. Eerdmans, 1975.

Veatch, R.M. "Consent, Confidentiality, and Research." *New England Journal of Medicine* 336, no. 12 (1997): 869–870.

———. "Hospital Ethics Committees: Is There a Role?" *Hastings Center Report* 7 (1977): 22–25.

———. *Medical Ethics*. 2nd ed. Boston: Jones and Bartlett, 1997.

———. *Transplantation Ethics*. Washington, DC: Georgetown University Press, 2000.

Vedrinne, J."Éthiques et professions de santé." *Médecine et Hygiène* 11 (1984): 1171–1177.

Vega, M., J. Vega, and P. Martinez Baza. "Comentarios a la legislación española sobre Reproducción Asistida." *Cuadernos de Bioética* 21, no. 5 (1995): 57–64.

Vella, C.G., P. Quattrocchi, and A. Bompiani. *Dalla bioetica ai comitati etici*. Milan: Istituto Scientifico Ospedale San Raffaele, 1988.

Venditti, R. *Le ragioni dell'obiezione di coscienza*. Turin: Gruppo Abele, 1986.

Ventafrida, V. "Cure palliative." In *Dizionario di teologia pastorale sanitaria*, 325–326. Turin: Camilliane, 1997.

Venter, J.C., M.D. Adams, E.W. Myers, et al. "The Sequence of the Human Genome." *Science* 291, no. 5507 (2001): 1304–1351.

Ventura, P. *Problemi attuali della filosofia del diritto*. Turin: Giappichelli, 1991.

Vercellone, P. "Children's Rights and Artificial Procreation." *Medicine and Law* 14 (1995): 13–22.

Verhagen, E., and P.J. J. Sauer. "The Groningen Protocol: Euthanasia in Severely Ill Newborns." *New England Journal of Medicine* 352 (2005): 959–62.

Verspieren, P. "L'aventure de la fécondation 'in vitro.'" *Études* (November 1982): 479–491.

Viafora, C., ed. *Centri di bioetica in Italia: Orientamenti a confronto*. Padua: Gregoriana, 1993.

Viafora, C. *Fondamenti di bioetica*. Milan: Ambrosiana, 1989.

———, ed. *Vent'anni di bioetica: Idee protagonisti istituzioni*. Padua: Gregoriana, 1990.

Vial Correa, J. de D., and E. Sgreccia. "Cellule staminali autologhe e trasferimento di nucleo." *L'Osservatore Romano*. January 5, 2001.

———, eds. *The Dignity of Human Procreation and Reproductive Technologies: Anthropological and Ethical Aspects; Proceedings of the Tenth Assembly of the Pontifical Academy for Life; Vatican City, 20–22 February 2004*. Vatican City: Libreria Editrice Vaticana, 2005.

———, eds. *The Dignity of the Dying Person*. Vatican City: Libreria Editrice Vaticana, 2000.

———, eds. *Etica della ricerca biomedica: Per una visione cristiana; Atti della nona assemblea generale della Pontificia Accademia per la Vita; Città del Vaticano, 24–26 febbraio 2003*. Vatican City: Libreria Editrice Vaticana, 2004.

Vidal, M. *L'atteggiamento morale*. 2 vols. Assisi: Cittadella, 1976.

———. "Ética de la actividad cientifico-técnica." *Moralia* 5 (1983): 419–443.

———. *Morale dell'amore e della sessualità*. Assisi: Cittadella, 1973.

Vigna, C., ed. *Introduzione all'etica*. Milan: Vita e Pensiero, 2001.

———. *La libertà del bene*. Milan: Vita e Pensiero, 1998.

———. "*Sostanza e relazione: una aporetica della persona*." In *L'idea di persona*, edited by V. Melchiorre. Milan: Vita e Pensiero, 1996.

Villa, L. "Etica e deontologia della sperimentazione sull'uomo." *Minerva Medica* 57, no. 89 (1966): 3733–3739.

———. *Medicina oggi: Aspetti di ordine scientifico, filosofico, etico-sociale*. Padua: Piccin, 1980.

Villot, J. Letter to Dr. James Rugia, Secretary General of the International Federation of Catholic Medical Associations. October 3, 1970. Reprinted in *La Civiltà Cattolica* 4 (1970): 275–277.

Viola, F. "La conoscenza della legge naturale nel pensiero di Jacques Maritain." In *Jacques Maritain oggi*, edited by V. Possenti, 560–582. Milan: Vita e Pensiero, 1983.

Viotti, S. "Il problema morale della legge civile." *Studia Moralia* 37 (1999): 321–356.

Visser, J. V. "Pronunciamento ufficiale della S. Sede sull'eutanasia." *Medicina e Morale* 31, no. 3 (1981): 358–372.

Vogel, F., and A.G. Motulsky. *Human Genetics: Problems and Approaches*. Berlin: Springer, 1986.

Volterra, E. "Esposizione dei nati: Diritto greco e romano." In *Novissimo digesto italiano*, edited by A. Azara and E. Eula, 6:878–879. 3rd ed. Turin: UTET, 1960.

Von Balthasar, H. U. *Theo-Drama: Theological Dramatic Theory*. Vol. 2. San Francisco: Ignatius Press, 1990.

Voorzanger, B. "No Norms and No Nature: The Moral Relevance of Evolutionary Biology." *Biology and Philosophy* 2 (1987): 253–270.

Waddington, S.N., M.G. Kramer, R. Hernandez-Alcoceba, et al. "In Utero Gene Therapy: Current Challenges and Perspectives." *Molecular Therapy* 11, no. 5 (2005): 661–676.

Wadlington, W. "Artificial Insemination." In *Encyclopedia of Bioethics*, edited by W.T. Reich, 2216–2221. Rev. ed. New York: Macmillan, 1995.

Wagner, M.G., and P.A. St. Claire. "Are In Vitro Fertilisation and Embryo Transfer of Benefit to All?" *Lancet* 2 (1989): 1027–1030.

Wakitani, S., K. Takaoka, T. Hattori, et al. "Embryonic Stem Cells Injected into the Mouse Knee Joint Form Teratomas and Subsequently Destroy the Joint." *Rheumatology* 42 (2003): 162–165.

Wald, N.J., H.S. Cuckle, J.W. Densem, et al. "Maternal Serum Screening for Down's Syndrome in Early Pregnancy." *British Medical Journal* 297 (1988): 883–887.

Wald, N.J., and A. Kennard. "Screening biochimico prenatale per la sindrome di Down." *Ligand Quarterly: Edizione Italiana* 12, no. 4 (1993): 519–523.

Walters, L. "Bioethics Commissions: International Perspectives." *Journal of Medicine and Philosophy* 14, no. 4 (1989): 363–462.

———. "The Center for Bioethics at the Kennedy Institute." *Georgetown Medical Bulletin* 37, no. 1 (1984): 6–8.

———. "The Ethics of Human Gene Therapy." *Nature* 320 (1986): 225–227.

———. "Religion and the Renaissance of Medical Ethics in USA, 1965–1975." In *Theology and Bioethics*, edited by E.E. Shelp. Dordrecht: Reidel, 1985.

Wang, X.J., G.M. Warnes, R.J. Norman, C.A. Kirby, A.M. Clark, and C.D. Matthews. "Gamete Intra-Falloppian Transfer: Outcome following the Elective or Non Elective Replacement of Two, Three or Four Oocytes." *Human Reproduction* 8 (1993): 1231.

Wang, J.X., A.M. Clark, C.A. Kirby, et al. "The Obstetrics Outcome of Singleton Pregnancies following In Vitro Fertilization/Gamete Intra-Fallopian Transfer." *Human Reproduction* 9, no. 1 (1994): 141–146.

Warnock, M. "A National Ethics Committee." *British Medical Journal* 297 (1988): 1626–1627.

Watt, H. "Human Gene Therapy: Ethical Aspects." In *Human Genome: Human Person and the Society of the Future*, edited by J. de D. Vial Correa and E. Sgreccia, 255–270. Vatican City: Libreria Editrice Vaticana, 1999.

Wattiaux, H. "Insémination artificielle, fécondation in vitro et transplantation embryonnaire—Repères éthiques." *Esprit et Vie* 24 (1983): 354–364.

Wattles, J. *The Golden Rule*. New York: Oxford University Press, 1996.

Weijer, C., B. Dickens, and E.M. Meslin. "Bioethics for Clinicians: 10. Research Ethics." *Canadian Medical Association Journal* 156, no. 8 (1997): 1153–1157.

Weiner, C.P. "Cordocentesis." *Obstetrics and Gynecology Clinics of North America* 15 (1988): 283–301.

Weissman, I.L. "Medicine: Politic Stem Cells." *Nature* 439 (2006): 145–147.

Wernow, J.R. "The Living Will." *Ethics & Medicine* 10 (1994): 27–35.

White, R.J. "Individualità e trapianto cerebrale." In *Trapianto di cuore e trapianto di cervello*, 102–130. Rome: Orizzonte Medico, 1983.

White, R. J., M. S. Albin, J. Verdura, et al. "The Isolation and Transplantation of the Brain: An Historical Perspective Emphasizing the Surgical Solutions to the Design of These Classical Models." *Neurological Research* 18 (1996): 194–203.

White, R. J., H. Angstwurm, and I. Carrasco de Paula, eds. *Working Group on the Determination of Brain Death and Its Relationship to Human Death*. December 10–14, 1989. Vatican City: Pontificia Accademia delle Scienze, 1992.

Whittingham, D. G. "In Vitro Fertilization, Embryotransfer and Storage." *British Medical Bulletin* 35 (1979): 105–111.

Whittle, M. J. "Cordocentesis." *British Journal of Obstetrics and Gynaecology* 96 (1989): 262–264.

Wikler, D. "Bioethics Commissions Abroad." *HEC Forum* 6, no. 4 (1994): 290–304.

Williamson, B. "Gene Therapy." *Nature* 298 (1982): 416–418.

Willison, D. J., K. Keshavjee, K. Nair, C. Goldsmith, A. M. Holbrook. "Patients' Consent Preferences for Research Uses of Information in Electronic Medical Records:Interview and Survey Data." *British Medical Journal* 326, no. 7385 (2003): 373–77.

Wilson, E. O. *Sociobiology: The New Synthesis*. Harvard: Belknap, 1975.

Wilson, E. W. "The Evolution of Methods for Female Sterilization." *International Journal of Gynecology and Obstetrics* 51, suppl. 1 (1995): S3–S13.

Wittgenstein, L. "Conferenza sull'etica." In *Lezioni e conversazioni sull'etica, l'estetica, la psicologia e la credenza religiosa*, edited by M. Ranchetti. Milan: Adelphi, 1967.

———. *Tractatus Logico-Philosophicus*. Translated by D. F. Pears and B. F. McGuinness. 2nd ed. London: Routledge and Kegan Paul, 1963.

Wittle, M. "Ultrasonographic 'Soft Markers' of Fetal Chromosomal Defects." *British Medical Journal* 314 (1997): 918.

Wojtyla, K. *Love and Responsibility*. New York: Farrar, Straus, Giroux, 1981.

———. "La visione antropologica della *Humanae vitae*." *Lateranum* 44 (1978): 129 ff.

Working Party on the Practice of Genetic Manipulation. *Report of the Working Party on the Practice of Genetic Manipulation*. London: Her Majesty's Stationary Office, 1976.

World Health Organization. "Constitution of the World Health Organization." In *Basic Documents*, 1-18. 45th ed. supplement. October 2006. Available at http://www.who.int/governance/eb/who_constitution_en.pdf.

———. *Operational Guidelines for Ethics Committees That Review Biomedical Research*. Geneva: WHO, 2000. Available online at http://www.searo.who.int/LinkFiles/RPC_Operational_Guidlines_Ethics.pdf.

———. *Ottawa Charter for Health Promotion*. November 21, 1986. WHO/HPR/HEP/95.1. http://www.who.int/hpr/NPH/docs/ottawa_charter_hp.pdf.

———. WHA Resolution 50.37. "Cloning in Human Reproduction." May 14, 1997.

World Health Organization Special Programme of Research Development and Research Training in Human Reproduction. 1977.

World Medical Association. Declaration of Helsinki: Ethical Principles for Medical Research Involving Human Subjects. June 1964. Last amended 2008 in Seoul, South Korea. http://www.wma.net/en/30publications/10policies/b3/.

———. Declaration of Sydney on the Determination of Death and the Recovery of Organs. August 1968. Updated October 2006. http://www.wma.net/en/30publications/ 10policies/d2/.

Worsnop, R. L. "Assisted Suicide Controversy." *CQ Researcher* 5, no. 17 (1995): 393–416.

Wramsby, H., K. Fredga, and P. Liedholm. "Chromosome Analysis of Human Oocytes Recovered from Preovulatory Follicles in Stimulated Cycles." *New England Journal of Medicine* 316 (1987): 121–124.

Wright, G. H. von. "Is and Ought." In *Man, Law and Modern Forms of Life*, edited by E. Bulygin, J. L. Gardies, and I. Niiniluoto, 263–282. Dordrecht: Reidel, 1985.

Yaron, Y., J. B. Lessing and M. R. Peyser. "Expectant Management of Ectopic Pregnancy in the Presence of Ovarian Hyperstimulation Syndrome." *Acta Obstetricia et Gynecologica Scandinavica* 74 (1995): 80–81.

Zaggia, C., ed. *Progresso biomedico e diritto matrimoniale canonico*. Padua: Ed. Veneta, 1992.

Zalba, M. "La portata del principio di totalità nella dottrina di Pio XI e Pio XII e la sua applicazione nei casi di violazioni sessuali." *Rassegna di Teologia* 9 (1968): 225–237.

———. "Principia ethica in crisim vocata intra (propter?) crisim morum." *Periodica de re morali, canonica et liturgica* 71 (1982): 319–357.

———. "Totalità (principio di)." In *Dizionario enciclopedico di teologia morale*, 1141–1149. Rome: Paoline, 1981.

Zalel, Y., A. Barash, B. Caspi, and R. Borenstein. "Heterotopic Pregnancy: An Usual Case Report following In Vitro Fertilization and Embryo Transfer." *Journal of Assisted Reproduction and Genetics* 10 (1993): 169–171.

Zanchetti, M. *La legge sull'interruzione della gravidanza: Commentario sistematico alla legge 22 maggio 1978, n. 194*. Padua: CEDAM, 1992.

———. "La responsabilità giuridica del comitato di etica ospedaliero." In *Una verità in dialogo: Storia, metodologia e pareri di un comitato di etica*, edited by P. Cattorini, 78–94. Milan: Istituto Scientifico Ospedale San Raffaele / Europa Scienza Umane, 1994.

Zanjani, E. D., and W. F. Anderson. "Prospects for In Utero Human Gene Therapy." *Science* 283 (1999): 2084–2088.

Zarutskie, P. W., C. H. Muller, M. Magone, and M. R. Soules. "The Clinical Relevance of Sex Selection Techniques." *Fertility and Sterility* 51 (1989): 891–904.

Zatti, M. "Biologia antropica." In *Il principio antropico: Condizioni per l'esistenza dell'uomo nell'universo*, edited by B. Giacomini. Ferrara: Istituto Gramsci, 1991.

Names Index

Subject Index